Organic Reactions

Organic Reactions

VOLUME 66

WILEY

JOHN WILEY & SONS, INC., PUBLICATION

Published by John Wiley & Sons, Inc., Hoboken, New Jersey

Published simultaneously in Canada.

Library of Congress Catalog Card Number 42-20265

ISBN-13 978-0-471-68258-5

ISBN-10 0-471-68258-6

Printed in the United States of America

10 9 8 7 6 5 4 3 2 1

PREFACE TO THE SERIES

In the course of nearly every program of research in organic chemistry, the investigator finds it necessary to use several of the better-known synthetic reactions. To discover the optimum conditions for the application of even the most familiar one to a compound not previously subjected to the reaction often requires an extensive search of the literature; even then a series of experiments may be necessary. When the results of the investigation are published, the synthesis, which may have required months of work, is usually described without comment. The background of knowledge and experience gained in the literature search and experimentation is thus lost to those who subsequently have occasion to apply the general method. The student of preparative organic chemistry faces similar difficulties. The textbooks and laboratory manuals furnish numerous examples of the application of various syntheses, but only rarely do they convey an accurate conception of the scope and usefulness of the processes.

For many years American organic chemists have discussed these problems. The plan of compiling critical discussions of the more important reactions thus was evolved. The volumes of *Organic Reactions* are collections of chapters each devoted to a single reaction, or a definite phase of a reaction, of wide applicability. The authors have had experience with the processes surveyed. The subjects are presented from the preparative viewpoint, and particular attention is given to limitations, interfering influences, effects of structure, and the selection of experimental techniques. Each chapter includes several detailed procedures illustrating the significant modifications of the method. Most of these procedures have been found satisfactory by the author or one of the editors, but unlike those in *Organic Syntheses*, they have not been subjected to careful testing in two or more laboratories.

Each chapter contains tables that include all the examples of the reaction under consideration that the author has been able to find. It is inevitable, however, that in the search of the literature some examples will be missed, especially when the reaction is used as one step in an extended synthesis. Nevertheless, the investigator will be able to use the tables and their accompanying bibliographies in place of most or all of the literature search so often required.

Because of the systematic arrangement of the material in the chapters and the entries in the tables, users of the books will be able to find information desired by reference to the table of contents of the appropriate chapter. In the interest of economy, the entries in the indices have been kept to a minimum, and, in particular, the compounds listed in the tables are not repeated in the indices.

The success of this publication, which will appear periodically, depends upon the cooperation of organic chemists and their willingness to devote time and effort to the preparation of the chapters. They have manifested their interest already by the almost unanimous acceptance of invitations to contribute to the work. The editors will welcome their continued interest and their suggestions for improvements in *Organic Reactions.*

Chemists who are considering the preparation of a manuscript for submission to *Organic Reactions* are urged to write either secretary before they begin work.

BLAINE CHASE MCKUSICK
March 22, 1918–January 5, 2005

Blaine Chase McKusick peacefully passed away in his sleep on January 5, 2005 at the age of 86.

Blaine was born on March 22, 1918 to James Gillespie Blaine McKusick and Marjorie Chase McKusick in Minneapolis, Minnesota. He received his Bachelors degree in Chemical Engineering in 1940 from the University of Minnesota and his Ph.D. in 1944 from the University of Illinois from Professor R. C. Fuson. He did postdoctoral work at Harvard with P. T. Bartlett and the Technical University of Zurich with V. Prelog.

He worked for E. I. DuPont de Nemours and Company for 37 years from 1945 to 1982 in the Central Research Department, the Agricultural Chemicals Department, and in the Haskell Laboratory for Toxicology and Industrial Medicine. He authored 40 papers, 20 patents, and three National Research Council reports. He received a Guggenheim Fellowship in 1950, the American Chemical Society Award of Chemical Health and Safety in 1986, a DuPont Safety and Health Award in 1990, and a Medal of Distinction from the University of Delaware in 1992 for his work in the College of Marine Studies.

He was an editor of *Organic Syntheses* and *Organic Reactions* and later served on the Board of Directors of both organizations. He was chairman of the *Organic Syntheses* Board. He also was an active member of the American Chemical Society, The American Association for the Advancement of Science, and the International Union of Pure and Applied Chemistry.

His work with war gases during World War II led him to a lifelong goal of getting the United States to sign the war-gas treaty entered into by most other countries after World War I. His goal was finally achieved with its ratification by the US Senate in the late 1990's.

Blaine's hobbies included hiking, skiing, and tennis. He was also an avid birder and amateur astronomer.

Blaine was married to Marjorie Jane Kirk in 1952. She was a pediatrician in Wilmington who later was a founder of the Delaware Adolescent Program Incorporated, which became a national model for helping pregnant teenagers get needed child care and complete their high-school education. She died of cancer in 1976.

Blaine and Marjorie had three children: Marshall Kirk McKusick, a computer scientist living in Berkeley, California; James Chase McKusick, an English professor and Honors College director at the University of Maryland Baltimore County; and Kathleen Blaine McKusick, a computing professional in biotechnology living in Redwood City, California. Kathleen is the mother of Blaine's grandchild, Marjorie Rose McKusick.

Blaine remarried to Virginia Walters in 1979. Virginia was an artist and art instructor doing both pottery and painting. Virginia died in 1997. Virginia is survived by her four children and five grandchildren.

Blaine remarried to a lifelong friend Emily Morris in 1999. Emily has four surviving children, seven grandchildren, and one great-grandchild.

Blaine is survived by his wife, Emily, his three children, Kirk, Jim, and Kate, his granddaughter, Marjorie Rose, his brother, Marshall, and his sister, Laura Bell Berthold.

Blaine lived a rich and high quality life. He was rarely sick and seemed barely slowed by advancing age. He was always looking to help others financially and with his upbeat, outgoing, and happy disposition. He will be greatly missed by his friends and colleagues.

CONTENTS

CHAPTER PAGE

1. The Allylic Trihaloacetimidate Rearrangement
   Larry E. Overman and Nancy E. Carpenter 1

2. Asymmetric Dihydroxylation of Alkenes
   Mark C. Noe, Michael A. Letavic, Sheri L. Snow, and Stuart W. McCombie 109

Cumulative Chapter Titles by Volume 627

Author Index, Volumes 1–66 641

Chapter and Topic Index, Volumes 1–66 645

CHAPTER 1

THE ALLYLIC TRIHALOACETIMIDATE REARRANGEMENT

LARRY E. OVERMAN

University of California, Irvine, Irvine, California 92697

NANCY E. CARPENTER

University of Minnesota, Morris, Morris, Minnesota 56267

CONTENTS

	PAGE
ACKNOWLEDGMENTS	3
INTRODUCTION	3
MECHANISM	4
Thermal Rearrangements	4
Metal-Catalyzed Rearrangements	6
SCOPE & LIMITATIONS	9
Preparation and Stability of Allylic Trihaloacetimidates	9
Preparation of Allylic Trichloroacetimidates	9
Preparation of Allylic Trifluoroacetimidates	10
Stability of Allylic Trihaloacetimidates	11
Thermal Rearrangements of Allylic Trihaloacetimidates	12
Reaction Conditions: Temperature, Solvent, and Additives	12
Scope	14
The Halogen Substituents	15
Carbon Skeleton	16
Cyclic Substrates	18
Substituent Effects and Problematic Substituents	20
Regioselectivity	22
Stereochemistry	22
Chiral Secondary Imidates: Chirality Transfer	22
Diastereoselectivity Arising from Stereocenters Outside the Pericyclic Arena	23
Geometry of the New Double Bond	24
Catalyzed Rearrangements of Allylic Trihaloacetimidates	25
General Conditions	26
Scope	27
Stereoselectivity	29

leoverman@uci.edu
carpenne@morris.umn.edu

Organic Reactions, Vol. 66, Edited by Larry E. Overman et al.

Chiral Secondary Imidates: Chirality Transfer . . . 29
Chiral Primary Imidates: Diastereoselectivity . . . 29
Asymmetric Catalysis . . . 31
APPLICATIONS TO SYNTHESIS . . . 33
Overview . . . 33
Preparation of Allylic Amines . . . 33
Other Direct Uses of Allylic Trihaloacetamides . . . 38
Applications in the Total Synthesis of Natural Products . . . 39
(±)-Acivicin . . . 39
(+)-Lactacystin . . . 40
(±)-Pancratistatin . . . 40
COMPARISON WITH OTHER METHODS . . . 41
Other Allylic Rearrangements . . . 41
Other Routes to Allylic Amines . . . 43
Amination of Allylic Electrophiles . . . 43
C-H Activation . . . 44
Hydroamination . . . 44
Addition of Vinyl Nucleophiles to Imine Derivatives . . . 45
EXPERIMENTAL CONDITIONS . . . 45
General Comments . . . 45
Preparation of Allylic Trichloroacetimidates . . . 45
Preparation of Allylic Trifluoroacetimidates . . . 46
Thermal Rearrangements of Allylic Trihaloacetimidates . . . 46
Metal-Catalyzed Rearrangements of Allylic Trihaloacetimidates . . . 47
EXPERIMENTAL PROCEDURES . . . 47
2,2,2-Trichloro-*N*-(3,7-dimethylocta-1,6-dien-3-yl)acetamide (Alkoxide-Catalyzed Procedure for Preparing Allylic Trichloroacetimidates and Thermal Rearrangement of the Crude Imidate Intermediate) . . . 48
3,7-Dimethylocta-1,6-dien-3-amine (Basic Hydrolysis of an Allylic Trichloroacetamide to Form the Allylic Amine) . . . 48
(*Z*)-2-[(6*S*)-6-(2,2-Dimethyl-1,3-dioxolan-4-yl)-4-methylcyclohex-3-enylidene]ethyl 2,2,2-Trichloroacetimidate (Preparation of a Trichloroacetimidate Using DBU) . . . 48
2,2,2-Trichloro-*N*-[(1*R*,6*S*)-6-(2,2-dimethyl-1,3-dioxolan-4-yl)-4-methyl-1-vinylcyclohex-3-enyl]acetamide (Thermal Rearrangement of a Trichloroacetimidate in the Presence of K_2CO_3) . . . 48
(4*S*,5*S*)-4-(*tert*-Butyldimethylsiloxymethyl)-5-[(3*S*,4*S*,1*E*)-4-*tert*-butyldiphenylsiloxy-3-(2,2,2-trifluoroacetimidoyloxy)pentenyl]-2,2-dimethyl-1,3-dioxolane (Preparation of an Allylic Trifluoroacetimidate) . . . 49
(4*S*,5*S*)-4-(*tert*-Butyldimethylsiloxymethyl)-5-[(1*R*,4*S*,2*E*)-4-*tert*-butyldiphenylsiloxy-1-(2,2,2-trifluoroacetylamino)pent-2-enyl]-2,2-dimethyl-1,3-dioxolane (Thermal Rearrangement of an Allylic Trifluoroacetimidate) . . . 50
Cinnamyl 2,2,2-Trifluoroacetimidate ("One-Pot" Procedure for the Preparation of an Allylic Trifluoroacetimidate) . . . 50
(3*R*,4*S*)-4-*tert*-Butoxycarbonylamino-3-(trichloroacetylamino)-1-pentene (Pd(II)-Catalyzed Rearrangement of an Allylic Trichloroacetimidate) . . . 51
(*S*)-2,2,2-Trichloro-*N*-(1-propylallyl)acetamide (Catalytic Asymmetric Rearrangement of an Allylic Trichloroacetimidate) . . . 51
TABULAR SURVEY . . . 51
Catalysts Used in Table 4 . . . 53
Table 1. Rearrangements of Trihaloacetimidates of Primary Allylic Alcohols . . . 54
Table 2A. Rearrangements of Trihaloacetimidates of Acyclic Secondary Allylic Alcohols . 70
Table 2B. Rearrangements of Trihaloacetimidates of Cyclic Secondary Allylic Alcohols . . 87
Table 3. Rearrangements of Trihaloacetimidates of Tertiary Allylic Alcohols . . . 95
Table 4. Transition-Metal-Catalyzed Asymmetric Rearrangements of Allylic Trihaloacetimidates . . . 97
REFERENCES . . . 103

ACKNOWLEDGMENTS

Research at UC Irvine in this area has been supported by grants from the National Science Foundation and the National Institute of Neurological Disorders and Stroke of the National Institutes of Health.

INTRODUCTION

Sigmatropic rearrangements of allylic systems have found wide application in organic synthesis, with carbon-carbon bond forming rearrangements such as the Cope and Claisen rearrangements being particularly well known. The sigmatropic rearrangement of allylic imidates (also known as the "aza-Claisen" or "Claisen-imidate" rearrangement) offers a valuable entry into the preparation of protected allylic amines. Conversion of an imidate to the amide is essentially irreversible, with the transformation of the imidate to the amide being exothermic by about 15 kcal/mol.[1,2] Since the discovery of the thermal allylic imidate rearrangement in 1937,[3] a number of systems have been investigated for the practical preparation of allylic amines by this route, including urethanes, isourethanes, formimidates, benzimidates, isoureas and carbonimidothioates. However, it was the discovery and development of the rearrangement of allylic trichloroacetimidates that overwhelmingly demonstrated the widespread utility of this synthetic method (Eq. 1).[4–6]

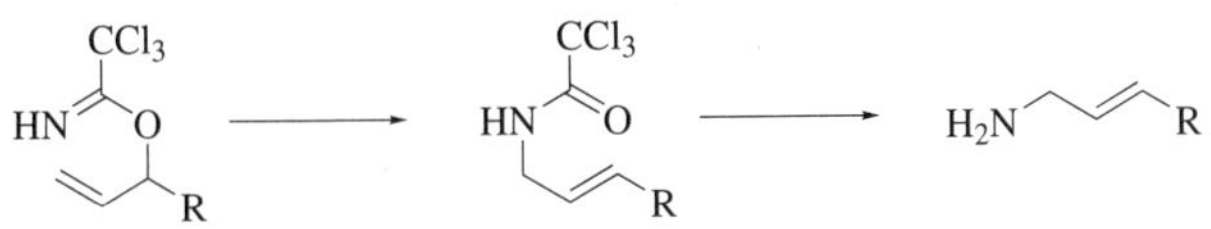

(Eq. 1)

The [3,3]-sigmatropic rearrangement of allylic trichloroacetimidates (now generally referred to as the Overman rearrangement) or trifluoroacetimidates can be conveniently carried out either thermally or with Hg(II) or Pd(II) catalysis. The scope of the rearrangement is such that primary, secondary, and tertiary allylic amides are readily accessible, thus providing entry into a wide variety of nitrogen-containing products including amino sugars, nucleotides, amino acids, peptides, and various nitrogen heterocycles. In addition, the Overman rearrangement has found extensive application in the total synthesis of natural products. The recent development of chiral Pd(II) catalysts to promote asymmetric allylic trichloroacetimidate rearrangements with good enantioselectivity bodes well for the continued broad application of this amine synthesis.[7–9]

This chapter is limited to the discussion of allylic trichloro- and trifluoroacetimidate rearrangements. Several relevant reviews have appeared regarding the allylic imidate rearrangement,[10] the use of allylic imidates in organic synthesis,[11] Hg(II)- and Pd(II)-catalyzed [3,3]-sigmatropic rearrangements,[12] enantioselective rearrangement of allylic imidates,[13,14] and the preparation of allylic amines.[15] Trichloroacetimidates of propargylic alcohols also undergo thermal [3,3]-sigmatropic rearrangement, giving *N*-acylamino-1,3-dienes as a result of tautomerization of the initially formed allenyl trichloroacetamide (Eq. 2).[16] This more limited transformation is not reviewed in this chapter.

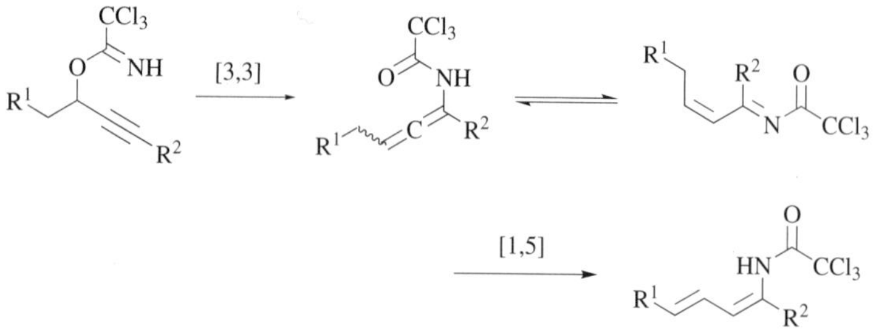

(Eq. 2)

MECHANISM

Thermal Rearrangements

The concerted nature of the allylic trichloroacetimidate rearrangement has been established by examination of thermodynamic parameters, solvent effects, regioselection, and stereoselection. Activation parameters for the thermal rearrangement of the trichloroacetimidate of geraniol (Eq. 3) are consistent with a [3,3]-sigmatropic rearrangement, with the large negative change in entropy being similar to that observed for Cope and Claisen rearrangements.[6] An intermediate is not detected, even in cases where a highly delocalized allylic carbocation would result from the ionization of the trichloroacetimidate, as in the rearrangement of the trichloroacetimidate of cinnamyl alcohol.[6] The observed regiochemical outcome of the allylic trihaloacetimidate rearrangement also lends support to the representation of this transformation as an essentially concerted process. For allylic trichloroacetimidates, the thermal rearrangement occurs with complete allylic oxygen-to-nitrogen transfer; the product arising from ionization and recombination, formal [1,3]-rearrangement, is rarely observed.[17] As discussed in more detail shortly, solvent and substituent effects support the development of some charge separation in the transition state.

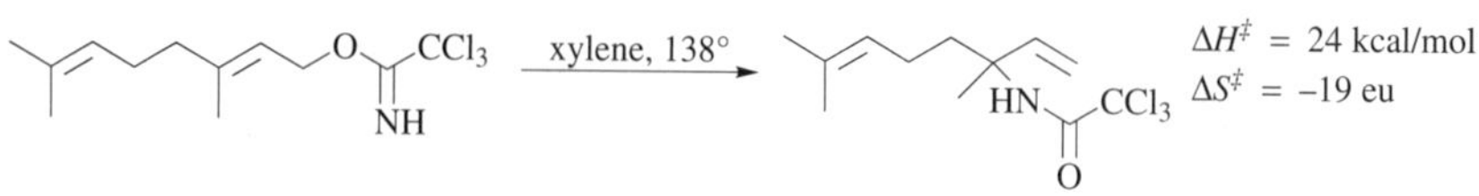

(Eq. 3)

The thermal allylic trihaloacetimidate rearrangement follows first-order kinetics,[6,18] with an assortment of steric and electronic substituent effects influencing the rate of the reaction. Alkenes having E double bonds tend to react more quickly than Z alkenes, a difference embodied in the $\Delta H^{\ddagger}$-term for reaction of the E and Z isomers of imidate **1**.[18] The rearrangement is facile for a wide variety of allylic alcohols, with doubly allylic alcohols reacting at room temperature, tertiary alcohols generally reacting at 80° ($t_{1/2}$ ~ 1 hour), and primary alcohols reacting the least quickly ($t_{1/2}$ ~ 1 hour at 140°).[6] These reactivity trends are attributed to stabilization of positive charge developed on the oxygen-bearing carbon in the transition state.[6] A five-fold rate enhancement is observed in changing the solvent from xylenes to nitrobenzene in the rearrangement of the trichloroacetimidate of geraniol at 132°. This finding is consistent with the postulated charge development in the transition state, that is, partial negative charge on the electronegative HN=C(CCl_3)O fragment and partial positive charge on the all-carbon allyl fragment.[6] One observation is not readily rationalized by this model: the apparent increase in rate seen when para electron-withdrawing substituents are present in the rearrangement of cinnamyl substrates **2** shown in Eq. 4.[19]

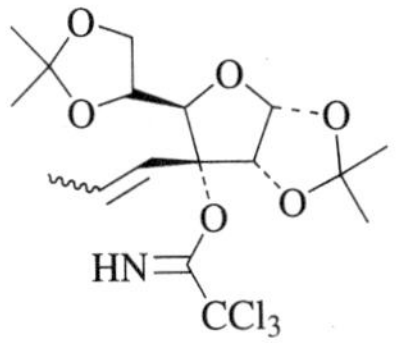

E $\Delta H^{\ddagger}$ = 25.6 kcal/mol
Z $\Delta H^{\ddagger}$ = 28.3 kcal/mol

1

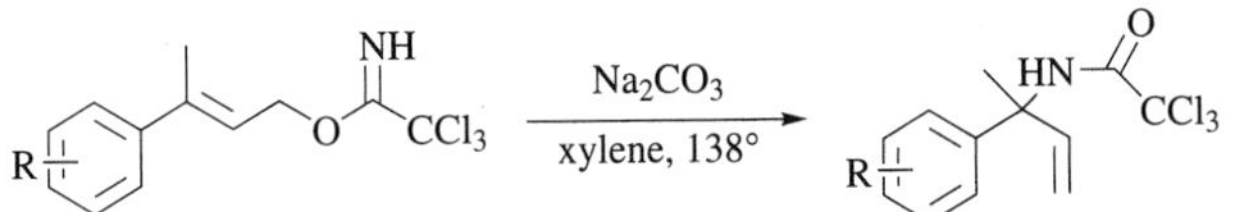

2

(Eq. 4)

The excellent stereoselectivity of the allylic trihaloacetimidate rearrangement is typical of suprafacial [3,3]-sigmatropic processes, as complete transfer of chirality is a hallmark of this reaction. Thus, the trichloroacetimidate of (*R*,*E*)-4-phenyl-3-buten-2-ol rearranges to give the (*R*,*E*)-trichloroacetamide **4** with no loss of enantiomeric purity, an outcome which can be rationalized as arising from the preferred chair-like transition structure **3** (Eq. 5).[20]

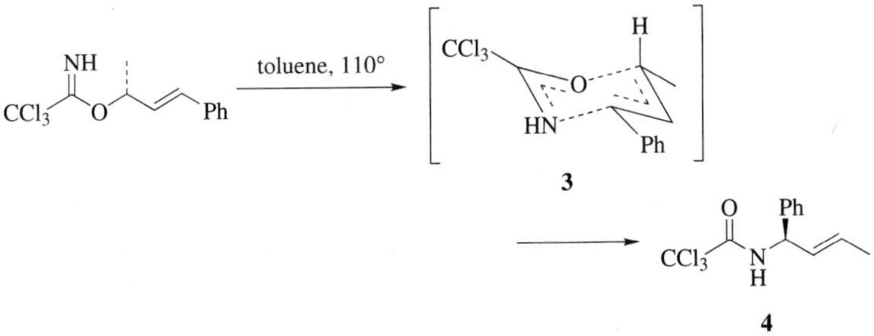

(Eq. 5)

High selectivity for forming the E stereoisomer of the product is seen in rearrangements of virtually all trihaloacetimidates of secondary allylic alcohols, including trisubstituted allylic alcohols such as 4-methyl-3-penten-2-ol,[6,21–23,81] and E and Z disubstituted allylic alcohols.[24–26] For example, rearrangement of the trichloroacetimidate of 1-hepten-3-ol proceeds to give a 92% yield of the E isomer by transition state structure **5** having the *n*-butyl group in the preferred equatorial position (Eq. 6).[6]

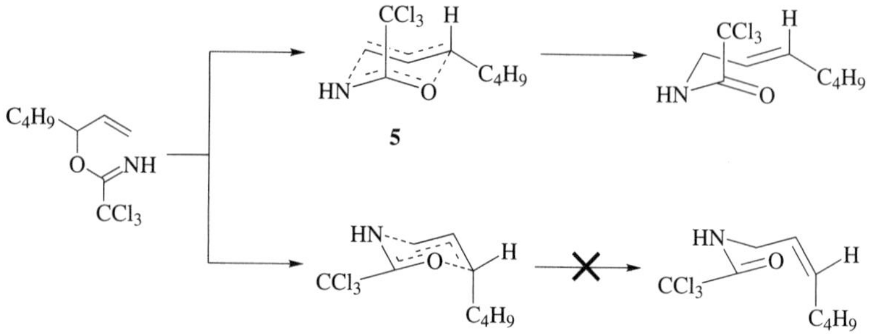

(Eq. 6)

As a result of the suprafacial nature of the rearrangement and the chair-like topography, either enantiomer of an E allylic amine can be prepared from the appropriate enantiomer of the starting allylic alcohol or from a configurationally related pair of alkene stereoisomers. An example of the latter strategy is shown in Eqs. 7 and 8.[24]

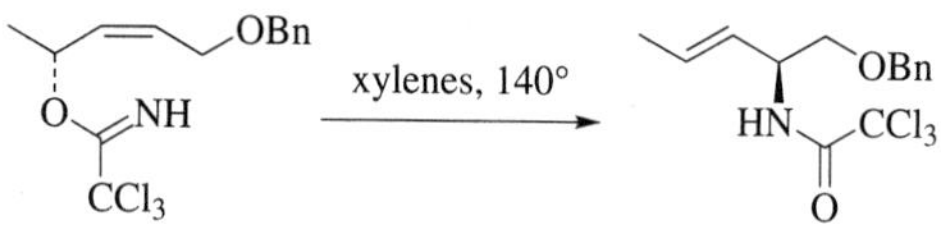

(35% from alcohol) (Eq. 7)

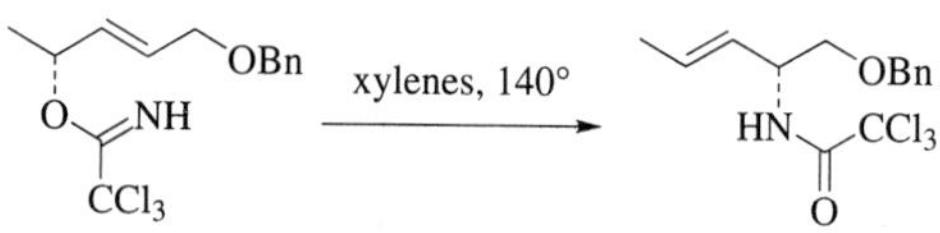

(45% from alcohol) (Eq. 8)

The allylic trichloroacetimidate rearrangement has not been the subject of *ab initio* or DFT theoretical studies. Early MNDO-PM3 semi-empirical molecular orbital calculations of the rearrangement of the trichloroacetimidate of allyl alcohol suggest an ion pair reaction pathway.[18] In this study, an ion pair transition state is calculated to have an enthalpy of formation of 11 kcal/mol, more than 9 kcal/mol lower than the calculated enthalpy of formation for the transition state of the [3,3]-sigmatropic pathway. Nevertheless, experimental evidence is fully consistent with a concerted sigmatropic rearrangement pathway.

Metal-Catalyzed Rearrangements

The rearrangement of allylic trihaloacetimidates can also be induced by using metal catalysts, thus lowering the temperature required for rearrangement and sometimes leading to higher yields, cleaner reactions and/or better stereocontrol. Many allylic trichloroacetimidates, ranging from simple allylic trichloroacetimidates to highly functionalized substrates, rearrange rapidly in the presence of Pd(II) or Hg(II) catalysts. Although the first reports employed $Hg(O_2CCF_3)_2$,[4] soluble complexes of $PdCl_2$ emerged as the most useful metal catalysts.[27–31] Rate accelerations are large (10^{12} is estimated for 1 M $Hg(O_2CCF_3)_2$),[4] allowing many Pd(II)- and Hg(II)-catalyzed trichloroacetimidate rearrangements to be carried out at room temperature.

A cyclization-induced rearrangement mechanism in which the metal coordinates to the allylic double bond to bring about antarafacial intramolecular nucleophilic attack by the imidate nitrogen is believed to be involved in Pd(II)- or Hg(II)-catalyzed rearrangements (see Scheme 1).[4,12,32] This mechanism is closely related to the mechanism originally proposed by Henry[33,34] for Pd(II)-catalyzed allylic acetate rearrangements and subsequently by Overman for $PdCl_2$-catalyzed Cope rearrangements.[35]

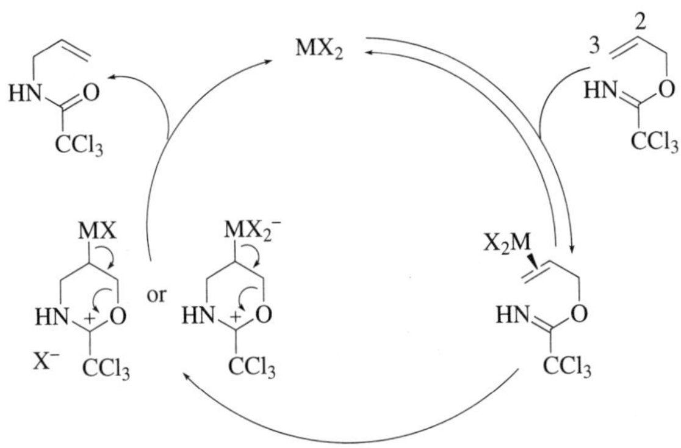

Scheme 1

A cyclization-induced rearrangement mechanism rationalizes the complete 1,3 oxygen-to-nitrogen transfer that is observed in Hg(II)- and Pd(II)-catalyzed rearrangements such as depicted in Eq. 9.[28] In contrast, rearrangements of allylic *N*-phenylformimidates and *N*-phenylbenzimidates catalyzed by Pd(0) complexes provide mixtures of products resulting from formal [1,3]- and [3,3]-sigmatropic

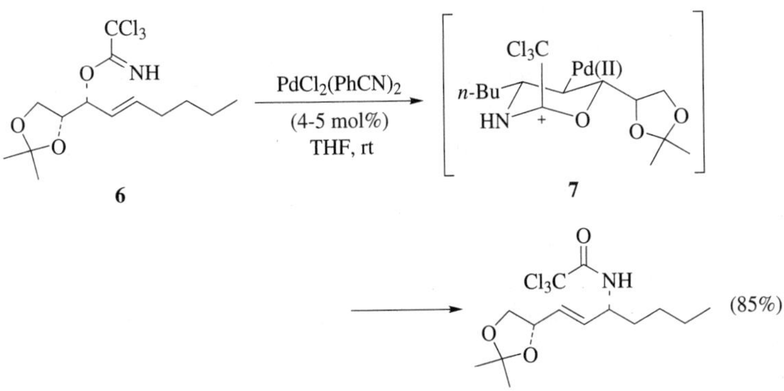

(Eq. 9)

rearrangements that undoubtedly involve the formation of Pd-π-allyl intermediates (Eq. 10).[36]

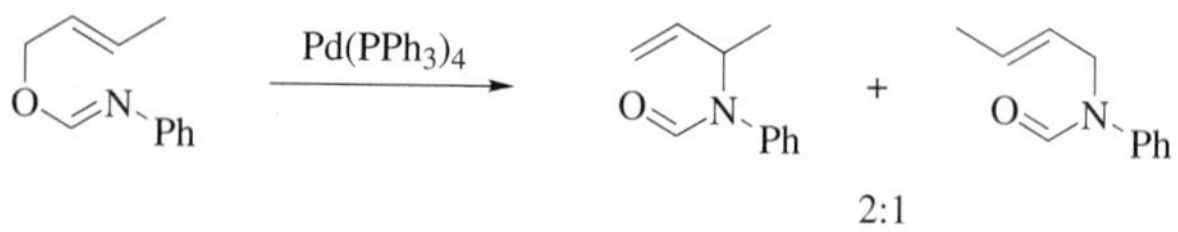

(Eq. 10)

Reactivity trends also support the cyclization-induced rearrangement mechanism for metal-catalyzed allylic trichloroacetimidate rearrangements. For example, substitution at the internal carbon of the allyl double bond (C2) slows the rate of the Pd(II)-catalyzed rearrangement, presumably as a result of the difficulty of generating a tertiary carbon-palladium sigma bond.[32] The limited scope of the Hg(II)-catalyzed process also provides support for a cyclization-induced rearrangement mechanism. Allylic substrates in which alkene substitution does not strongly favor intramolecular nucleophilic attack by the imino nitrogen at the allyl terminus (C3) fail to rearrange or rearrange in low yields under Hg(II) catalysis. Thus, for trichloroacetimidates containing terminal vinyl units, the 2-amino alcohol can be obtained after basic hydrolysis (Eq. 11).[4] This latter limitation is also seen in Pd(II)-catalyzed rearrangements as no reports exist of successful catalyzed rearrangements of allylic imidates containing terminal vinyl units.

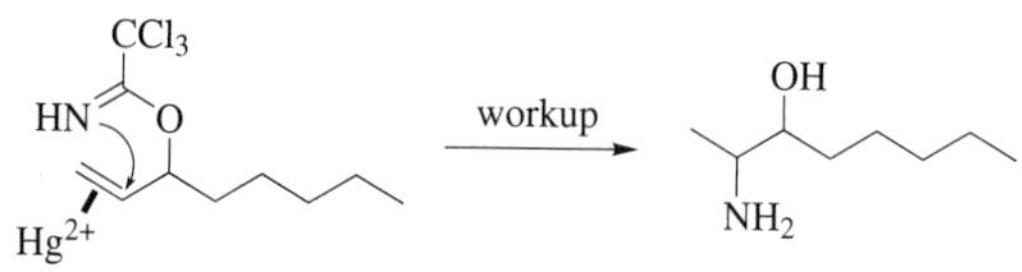

(Eq. 11)

The suprafacial nature and high E stereoselectivity of Pd(II)-catalyzed rearrangements also implicate a cyclization-induced mechanism. Collapse of the most stable chair conformation of intermediate **7** of Eq. 9 predicts the observed stereochemical outcome, just as in corresponding thermal rearrangements. For example, palladium-catalyzed rearrangement of dioxolane **6** gives the trichloroacetamide product with exclusive E geometry and complete transfer of chirality (Eq. 9).[28]

Given the direct involvement of the metal, the potential exists for asymmetric induction by a chiral metal complex in the rearrangement of prochiral allylic imidates. Not surprisingly, the development of suitable chiral metal catalysts has been a focus of intensive recent research in this area. Early results suggest that allylic trichloroacetimidates are generally unsuitable substrates for palladium-catalyzed asymmetric rearrangements, either resulting in poor yields, poor stereoselectivity or both.[37–39] However, recent reports demonstrate that di-μ-chlorobis[(η^5-(*S*)-($_pR$)-2-(2′-(4′-isopropyl)oxazolinylcyclopentadienyl-1-*C*, 3′-*N*))-(η^4-tetraphenylcyclobutadiene)-cobalt]dipalladium (**8**, COP-Cl) and related Pd(II) complexes are excellent catalysts for asymmetric rearrangements of E allylic trichloroacetimidates and *N*-aryl trifluoroacetimidates, e.g., Eq. 12.[7–9]

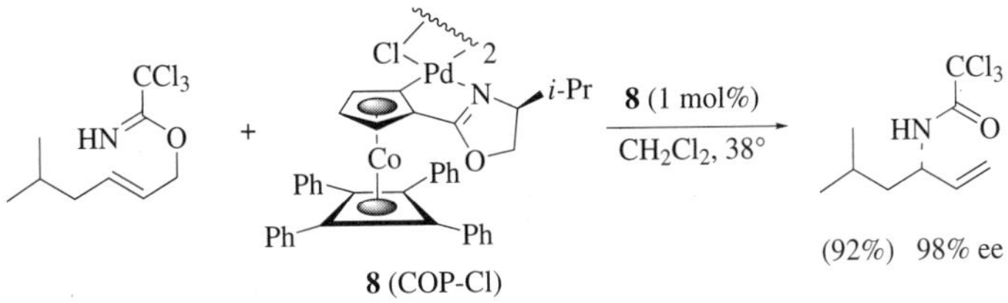

(Eq. 12)

SCOPE AND LIMITATIONS

Preparation and Stability of Allylic Trihaloacetimidates

Preparation of Allylic Trichloroacetimidates. The ease of preparation of trichloroacetimidates is a major reason for the broad synthetic utility of the allylic trichloroacetimidate rearrangement. The Pinner synthesis of imidates, wherein an alcohol is condensed with a nitrile in the presence of one or more equivalents of a strong mineral acid, is not suitable for preparing allylic imidates because ionization to the allylic cation and subsequent Ritter reaction typically occurs.[11] However, allylic trihaloacetimidates can be prepared conveniently by a base-catalyzed method first presented by Cramer.[40,41] Thus, addition of a mixture of the allylic alcohol and 5–20% of its alkoxide to an ether solution of trichloroacetonitrile at low temperature provides the trichloroacetimidate in isolated yields generally greater than 85%.[4,6,42,43] Although many allylic trichloroacetimidates can be purified by silica gel chromatography or vacuum distillation, such purification is frequently bypassed as crude trichloroacetimidates commonly can be used directly in the subsequent rearrangement step.

A wide variety of bases can be used to generate the alkoxide, with alkali metal hydrides often selected.[6,44–46] More recently, the addition of allylic alcohols to trichloroacetonitrile in dichloromethane or other aprotic solvents in the presence of 1,8-diazabicyclo[5.4.0]undec-7-ene (DBU), employed either catalytically[47] or in excess,[48–50] has been the method of choice for preparing allylic trichloroacetimidates. A direct comparison of potassium hydride to DBU in the preparation of aromatic trichloroacetimidates showed that catalytic DBU gave improved yields.[50]

Limitations to the preparation of allylic trichloroacetimidates are rare. Primary allylic alcohols are converted readily to the corresponding trichloroacetimidates at or below room temperature. Although these conditions also succeed with many secondary and tertiary alcohols, some hindered alcohols require more forcing conditions. For example, the reaction of (*R,E*)-4-methylhexa-3,5-dien-2-ol with trichloroacetonitrile requires the addition of 18-crown-6 to the potassium hydride/alcohol mixture and long reaction times.[51] The synthesis of trichloroacetimidate derivatives of cyclic tertiary alcohols, for example 1-vinylcyclopentanol and 1-vinylcyclohexanol, is reported to be problematic,[52] however such imidates have been sucessfully prepared and rearranged.[6,52–54]

An unusual obstruction to the preparation of the trihaloacetimidate is seen when a nucleophilic functional group is proximal to the alcohol. For example, attempted formation of the monotrichloroacetimidate of *cis*-2-butene-1,4-diol by reaction with 1.1 equivalents of trichloroacetonitrile provides instead the dioxepin **9** in 84% yield (Eq. 13). Although heating **9** in *tert*-butylbenzene at 175–180° for 1.5 hours gives the desired rearrangement product in 80% yield,[55] rearrangement of such orthoamides is not always successful. For example, efforts to force the [3,3]-sigmatropic rearrangement of orthoamide **10** (formed in 85% yield from the precursor diol) failed to produce the product of allylic rearrangement.[56]

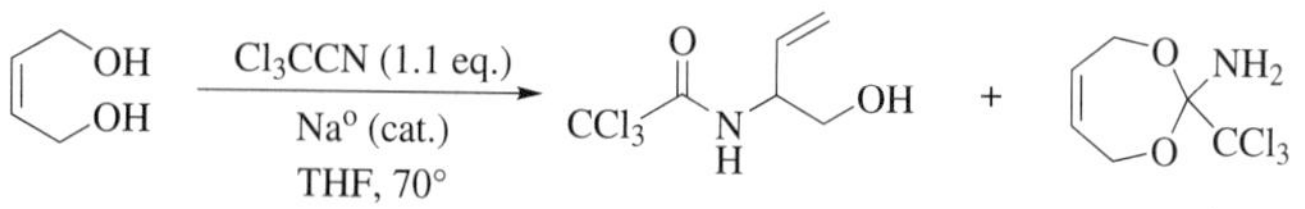

(Eq. 13)

9

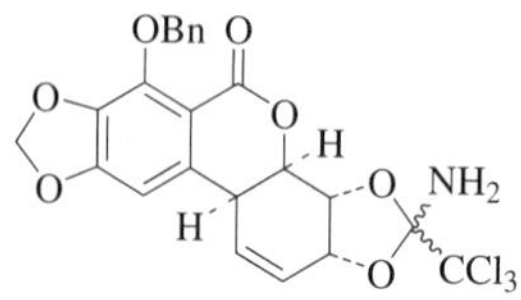

10

The presence of fluoro functionality in the allylic alcohol can prevent successful formation of the trichloroacetimidate, although only in rare situations where the fluoro substituent is suitably positioned. For example, efforts to prepare the trichloroacetimidate of fluoro alcohol **11** led to either Grob fragmentation (Eq. 14) or no reaction.[57] Likewise, 1,1,1-trifluoro-2-phenylbut-3-en-2-ol failed to react with trichloroacetonitrile, presumably because of low nucleophilicity of the alkoxide.[58]

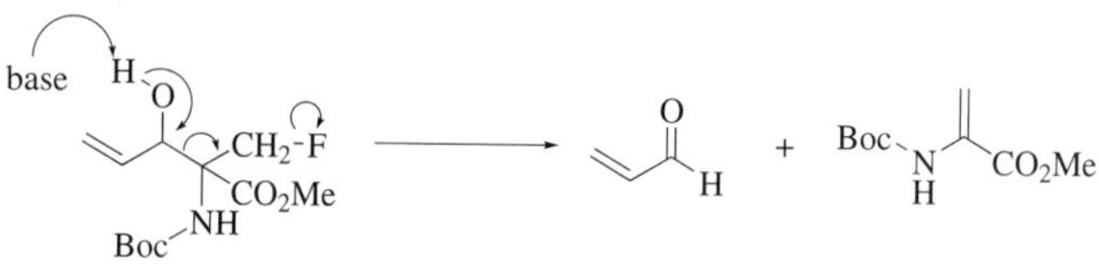

(Eq. 14)

11

Preparation of Allylic Trifluoroacetimidates. The preparation of allylic trifluoroacetimidates is complicated by the fact that trifluoroacetonitrile is a gas at room temperature and pressure.[1] Initial procedures for preparing allylic trifluoroacetimidates involved deprotonation of a THF solution of the allylic alcohol with 20 mol% of *n*-butyllithium followed by addition of an excess of a freshly-prepared THF solution of trifluoroacetonitrile at −78°.[60] More recently, a "one-pot" procedure was developed in which trifluoroacetonitrile is generated in situ by the reaction of trifluoroacetamide with oxalyl chloride, dimethyl sulfoxide, and triethylamine.[61] A mixture of DBU and the allylic alcohol is then added to the crude trifluoroacetonitrile solution, providing the trifluoroacetimidates in good yields (54–92%).[62]

[1] The trifluoroacetonitrile can be generated from trifluoroacetamide by dehydration with phosphorus pentoxide.[59]

Stability of Allylic Trihaloacetimidates. Primary allylic trihaloacetimidates are quite robust. However, if the alcohol is more substituted, trihaloacetimidate derivatives become susceptible to acid-catalyzed ionization at elevated temperatures to form trihaloacetamide and, initially, the corresponding allylic cation. For example, the bis-trichloroacetimidate **12** failed to undergo rearrangement, with imidate cleavage taking place instead (Eq. 15).[52] As will be discussed in more detail subsequently, addition of an acid scavenger such as potassium carbonate can minimize this decomposition pathway, allowing some rearrangements to take place that were previously unsuccessful.[63]

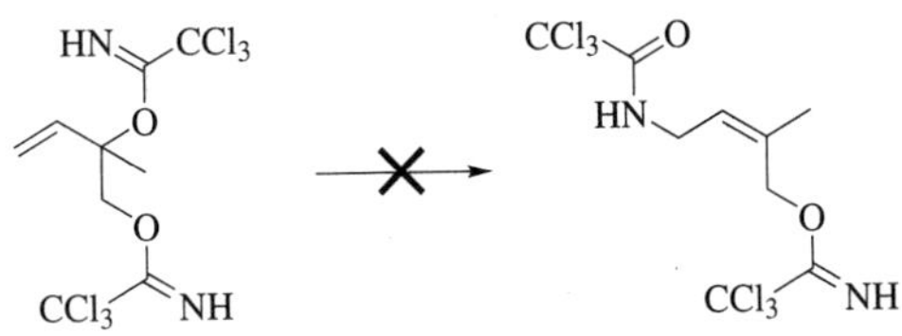

12

(Eq. 15)

The [1,3]-rearrangement product arising from dissociation-recombination is rarely observed, however such products are formed exclusively upon attempted Overman rearrangement of pyranoside **13** (Eq. 16).[64] The failed rearrangements of the trichloroacetimidates of aromatic allylic alcohols **14** and **15** are also attributed to the instability of the trichloroacetimidate,[65] although other cinnamyl substrates have been rearranged in high yields.[20,65,66]

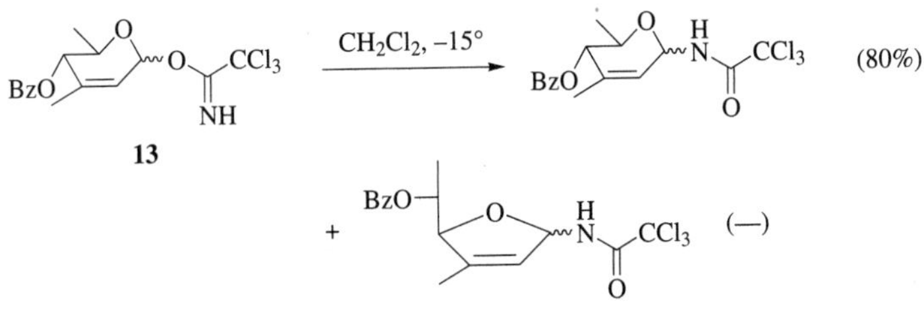

13

(Eq. 16)

14 R = Me, MeO **15**

In some cases, the stability of the imidate can be enhanced by manipulation of the structural features of the allyl substrate. Thus, while trichloroacetimidate **16a**[21] is unstable and fails to rearrange, replacement of the terminal methyl group (R^1) with a sterically more demanding phenyl or isopropyl group apparently enhances the imidate stability to a level that the rearrangement takes place in good yields

(60–98%).[21] The tendency of the trichloroacetimidate of β-hydroxy ester **17** to eliminate to form a dienyl ester is overcome by reduction to diol **18** followed by formation of the bis-trichloroacetimidate and rearrangement to give the desired acetamide (Eq. 17).[67] However, such an alternative route generally appears not to be required. For example, the series of structurally similar alcohols **19** are converted to the corresponding trichloroacetimidates and rearrange without incident.[22] Moreover, the rearrangement of the trichloroacetimidate derived from the unsaturated β-hydroxy ester **20** is reported to proceed in 100% yield.[23]

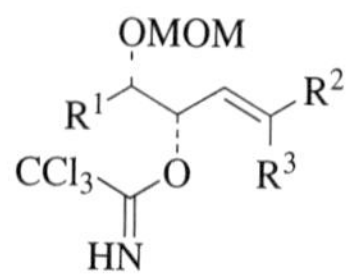

16

R^1 = Me (**a**), Ph, *i*-Pr
R^2, R^3 = *n*-Bu, Me, Ph, Et

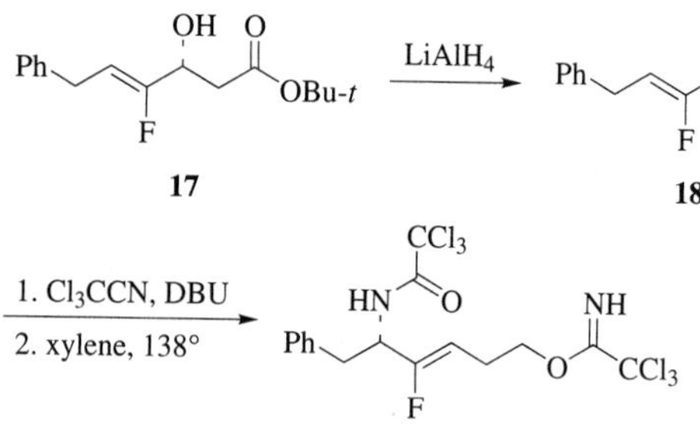

(73% from alcohol **18**) (Eq. 17)

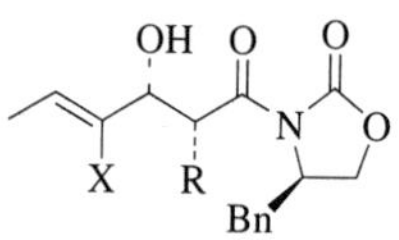

19

X = F, Cl, Me
R = Me, *i*-Pr, Bn

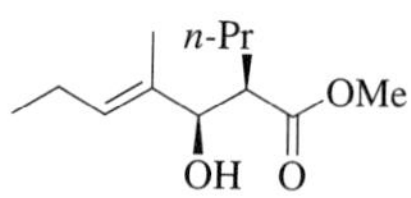

20

Thermal Rearrangements of Allylic Trihaloacetimidates

Reaction Conditions: Temperature, Solvent, and Additives. Thermal rearrangements of allylic trihaloacetimidates are carried out conveniently by dissolving the imidate in an aprotic solvent at ~0.1 M and heating the solution at reflux. Typically, trichloroacetimidates of primary allylic alcohols rearrange at 138–140° (refluxing xylene) in the range of 4–24 hours. Trichloroacetimidates derived from secondary alcohols rearrange at lower temperatures (110°, refluxing toluene) or in shorter reaction times (often in 1–5 hours). Tertiary trichloroacetimidates typically rearrange within a few hours at 80° (refluxing benzene). As noted previously, rearrangements of trichloroacetimidates of doubly allylic alcohols take place at temperatures at or below 0°. For example, the preparation of trichloroacetimidate **21** under normal conditions (addition of the alkoxide/alcohol mixture to trichloroacetonitrile at −5° to 0°) yields the trienylamide rearrangement product directly (Eq. 18).[6] In general, rearrangements of allylic trifluoroacetimidates are carried out under similar conditions (see later discussion).

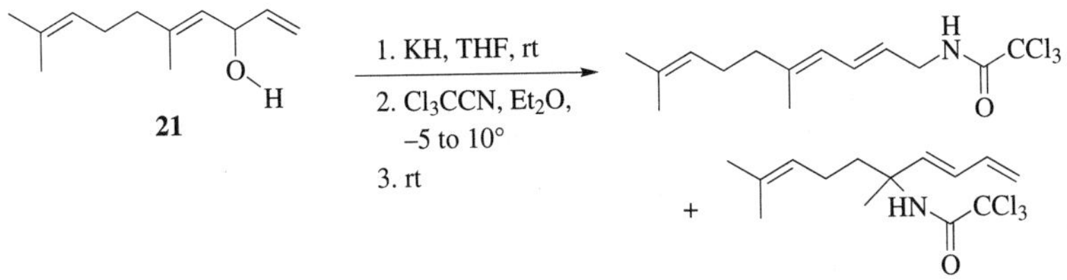

(Eq. 18)

Although allylic trichloroacetimidate rearrangements can be effected under solvent-free conditions by preadsorbing the alcohol onto KF-alumina, reacting this mixture with trichloroacetonitrile, and then allowing the rearrangement to take place at room temperature,[68] the vast majority of thermal rearrangements are run in a solvent at reflux. Frequently the impact of the solvent is related to its reflux temperature. Thus, various cinnamyl trichloroacetimidates do not rearrange at a convenient rate in refluxing chloroform, toluene or THF, whereas refluxing xylenes give the rearrangement products in good to high yields (Eq. 19).[19] Similarly, *tert*-butylbenzene (bp 169°) proved to be a convenient solvent for the rearrangement of imidate **22**; in refluxing xylene the reaction failed to go to completion in a convenient time span (Eq. 20).[69]

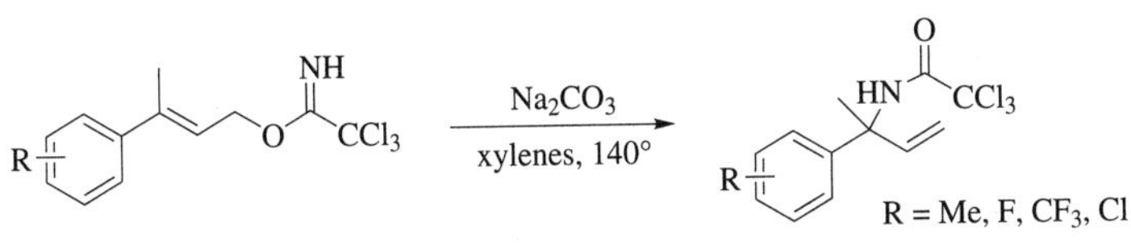

(Eq. 19)

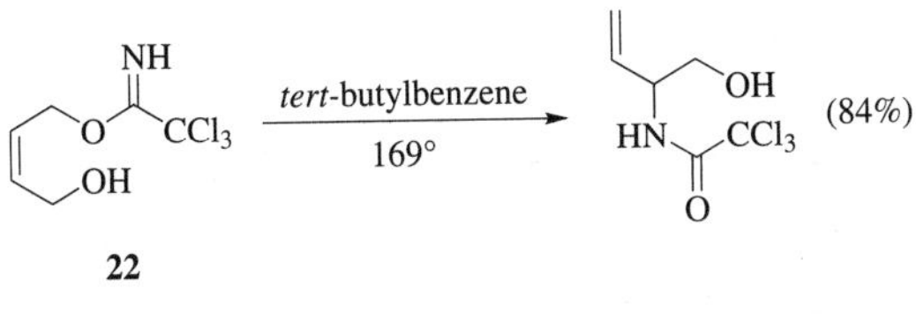

(Eq. 20)

In some cases the polarity of the solvent appears to play an important role. For example, attempted rearrangement of dissaccharide trichloroacetimidate **23** in refluxing xylenes at 140° for 5.5 hours gives the allylic trichloroacetamide product in only a 27% yield, whereas switching to *N,N*-dimethylformamide (with the reaction run at essentially the same temperature) increases the yield to 80% (Eq. 21).[70] The increase in yield in this example likely reflects a faster rate of rearrangement in the

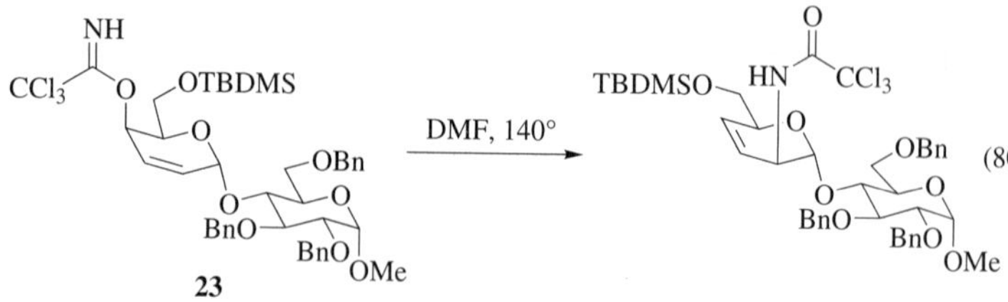

23 (80%) (Eq. 21)

more polar solvent DMF, and perhaps less acid-catalyzed decomposition of the imidate in this Lewis basic solvent.

Acid-catalyzed decomposition of the allylic trichloroacetimidate is typically the problematic reaction pathway that competes with allylic rearrangement. As a result, the addition of sodium or potassium carbonate can dramatically increase the yield of the rearrangement product.[63] For example, rearrangement of trichloroacetimidate **24** in refluxing *para*-xylene in the absence of potassium carbonate yields trichloroacetamide **25** in 74% yield from the starting allylic alcohol, whereas the yield is 90% when the rearrangement step is conducted in the presence of K_2CO_3 (2 mg/mL solvent, Eq. 22).[63] An even more dramatic improvement is seen in the rearrangement of the tetrahydropyridine derivative **26**. Attempts to carry out this rearrangement in several solvents (THF, toluene, or chlorobenzene) give low yields (0–50%); however, upon addition of K_2CO_3 the rearrangement is accomplished in refluxing chlorobenzene in 95% yield (Eq. 23).[71]

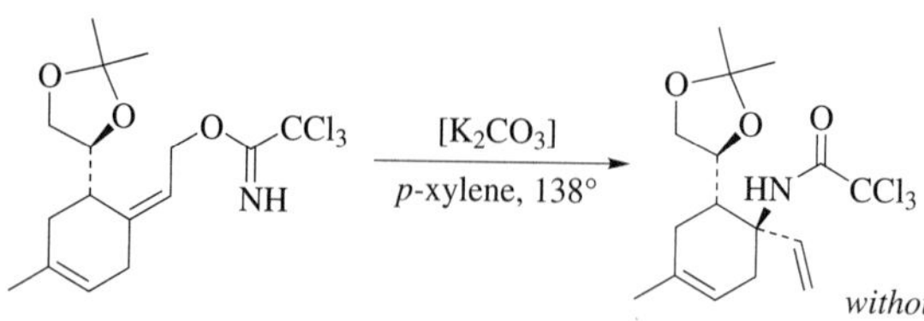

24 **25**

without K_2CO_3 (74% from alcohol)
with K_2CO_3 (90% from alcohol)

(Eq. 22)

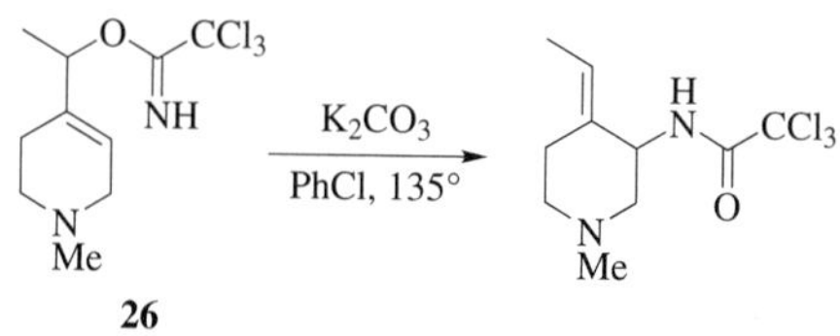

26 (Eq. 23)

Scope. Allylic trihaloacetimidates of primary, secondary, and tertiary allylic alcohols—both cyclic and acyclic—can be prepared and undergo Overman rearrangement in good yields with few limitations. Representative examples are shown in Eqs. 24–28.[23,52,60,63,72–75]

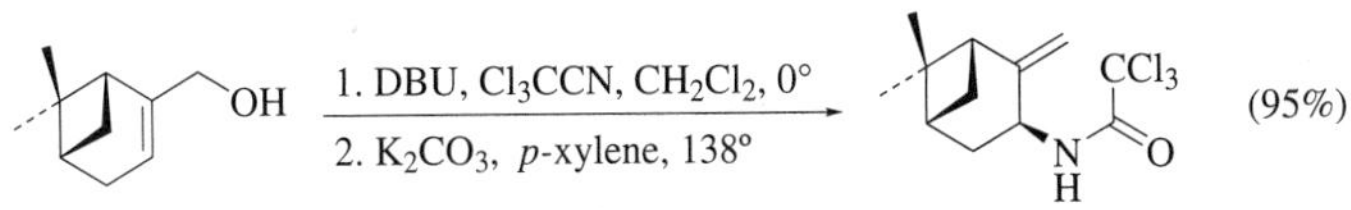

(Eq. 24)

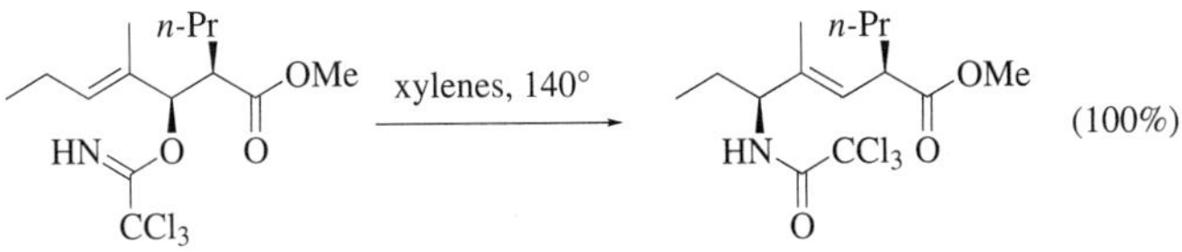

(Eq. 25)

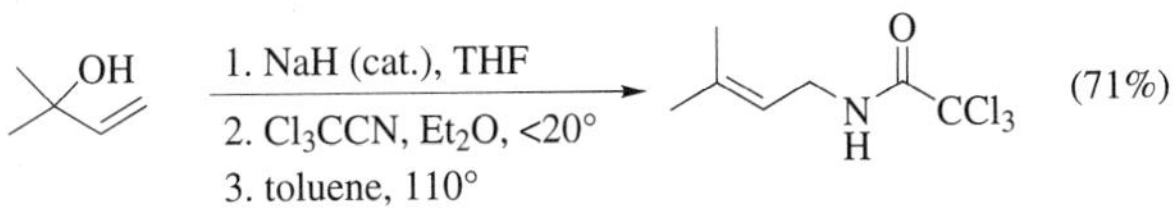

(Eq. 26)

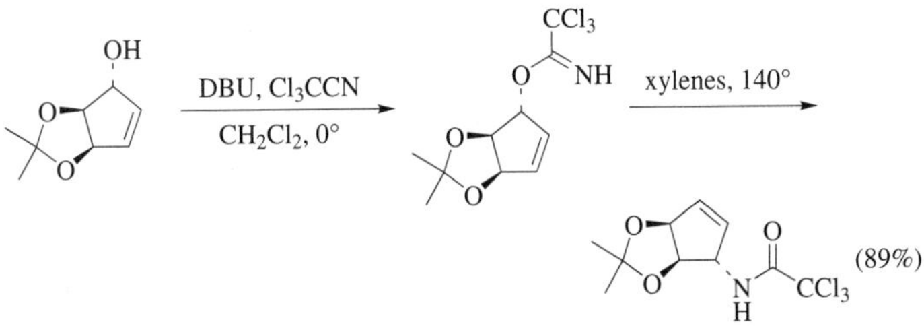

(Eq. 27)

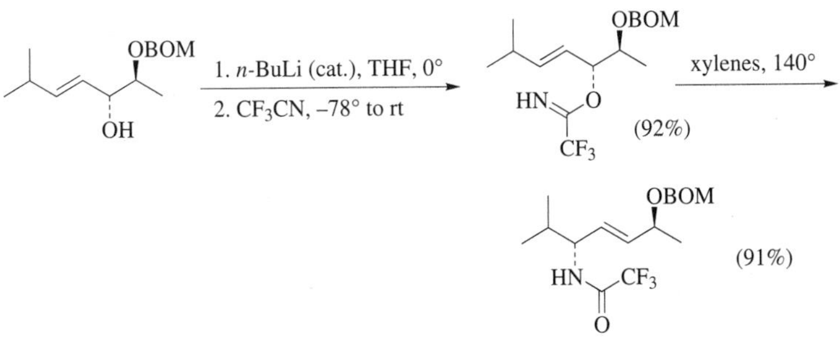

(Eq. 28)

The Halogen Substituents. The halogen substituent on the imidate plays an important role, not only in facilitating the synthesis of the rearrangement precursor, but also in increasing the rate of rearrangement. While thermal allylic rearrangements of several types of imidates are known, the presence of the electron-withdrawing CCl_3 or CF_3 group on the imidate leads to a more facile rearrangement. For example, formimidates and benzimidates require higher temperatures and/or longer reaction times than trihaloacetimidates to effect their allylic transposition.[10]

Trifluoroacetimidates in some cases rearrange slightly faster than the corresponding trichloroacetimidates. This rate enhancement, found to be approximately two-fold in one study,[75] can result in increased product yields. For example, the rearrangement of trichloroacetimidate **27** proceeds slowly in refluxing xylenes, with accompanying decomposition of the imidate under these conditions resulting in a poor yield of allylic trichloroacetamide **28** (Eq. 29). Switching to trifluoroacetimidate **29** results in doubling the yield of the allylic amide product (Eq. 30).[76]

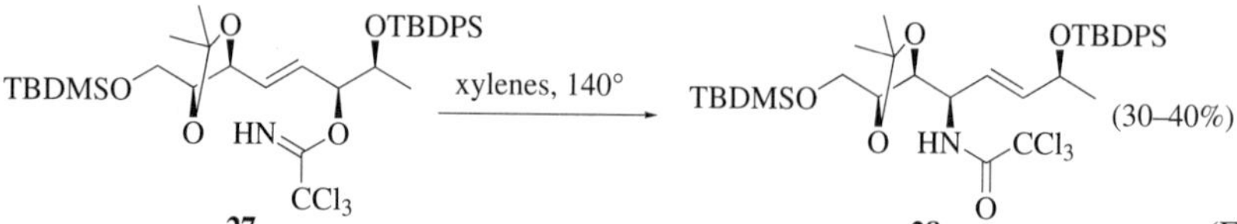

27 **28** (Eq. 29)

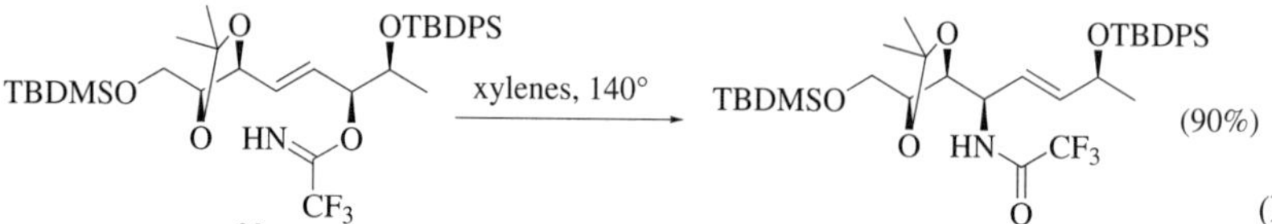

29 (Eq. 30)

However, higher rearrangement yields are not universally observed when trifluoroacetimidates are employed. For example, rearrangement of the trifluoroacetimidate of 2,4-hexadien-1-ol proceeds under thermal conditions to provide the dienyl trifluoroacetamide in low yield (Eq. 31),[60] whereas thermal rearrangement of the trichloro analogue gives the analogous rearrangement product in 63% yield (73% in the presence of K_2CO_3).[63] In another recent study, several trifluoroacetimidates were found to rearrange more slowly and in lower yields than their trichloro congeners. Results obtained for the geranyl to linalyl conversion are shown in Eq. 32.[62]

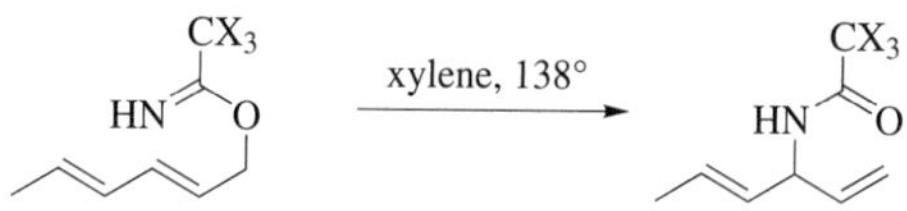

(Eq. 31)

X = F (35%)
X = Cl (63%)

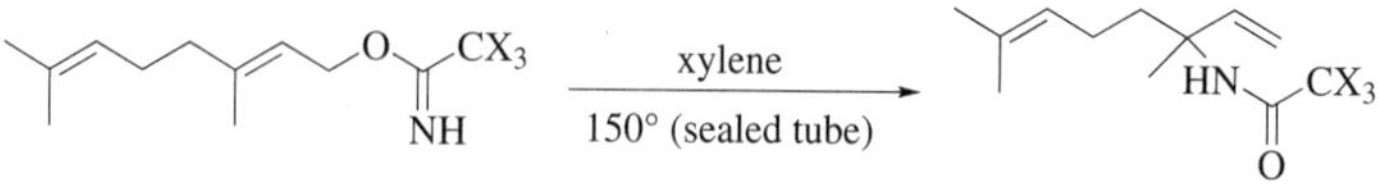

(Eq. 32)

X = F 16 h (69%)
X = Cl 4 h (90%)

Carbon Skeleton. The allylic carbon skeleton, particularly substitution at the α carbon, plays a large role in determining what temperature is required for allylic trihaloacetimidate rearrangements. However, there are only a few reports where structural features present insurmountable difficulties in carrying out the rearrangement.

In most cases, difficulties can be overcome by increasing the reaction temperature. We have already discussed the fact that higher reaction temperatures are required as electron-releasing α substituents are replaced by hydrogen, with 130–160° typically being required for primary allylic imidates and 80° or less for tertiary allylic analogues.

Steric effects at the imidate γ carbon can also play a role. In comparing the rearrangement of the ortho- and para-substituted aromatic imidates **30**, the para-substituted substrate is found to rearrange more quickly, and in higher yield, than the ortho isomer. This outcome is attributed to steric encumbrance to C–N bond formation in the latter case (Eq. 33).[19]

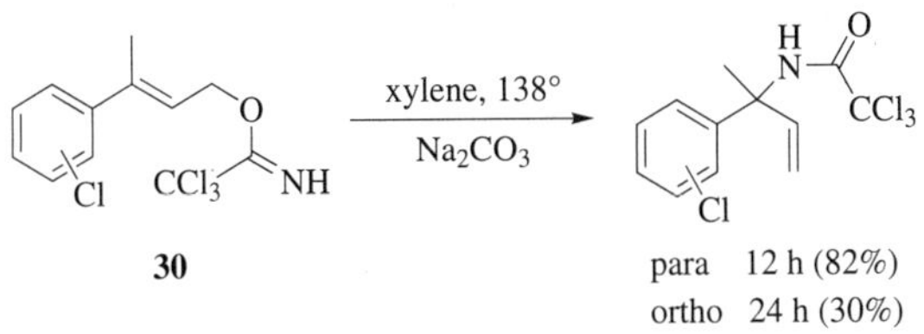

(Eq. 33)

The geometry of the alkene affects both the temperature required for trihaloacetimidate rearrangements and the yields observed. Allylic trihaloacetimidates having a Z 1,2-disubstituted double bond typically require slightly higher temperatures to promote their allylic reorganization than their E counterparts.[24,25] Presumably in part for this reason, many fewer examples of allylic trihaloacetimidate rearrangements are reported in the Z series. For example, the E trichloroacetimidate **31** rearranges in useful yield (Eq. 34), whereas the Z stereoisomer is noted to rearrange with "significantly more byproducts".[77] However, a few Z disubstituted allylic trichloroacetimidates are reported to rearrange in high yield. For example, the rearrangement of the secondary trichloroacetimidates **32** proceeds in good yields in refluxing xylenes (Eq. 35).[78]

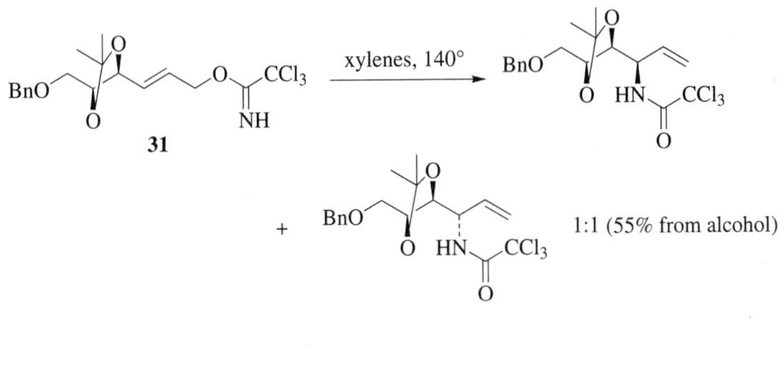

(Eq. 34)

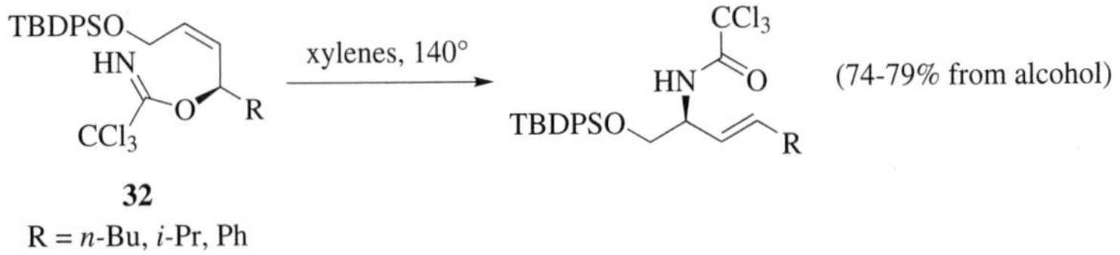

(Eq. 35)

Numerous trisubstituted allylic trihaloacetimidates having a second substituent at either the β or γ carbon undergo thermal rearrangement in useful yields. Three representative examples are shown in Eq. 36–38.[79–81]

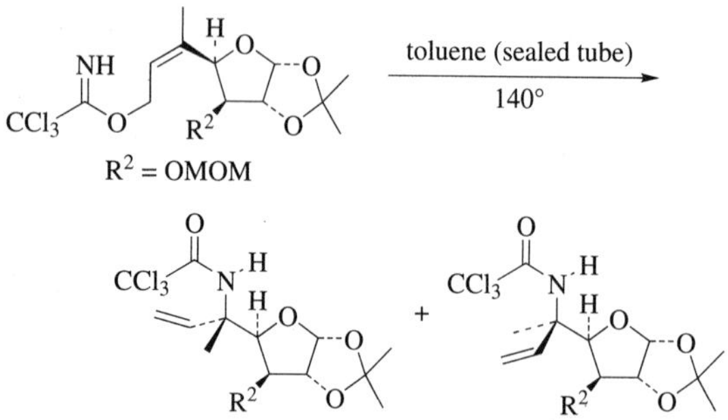

(78% from alcohol)

(Eq. 36)

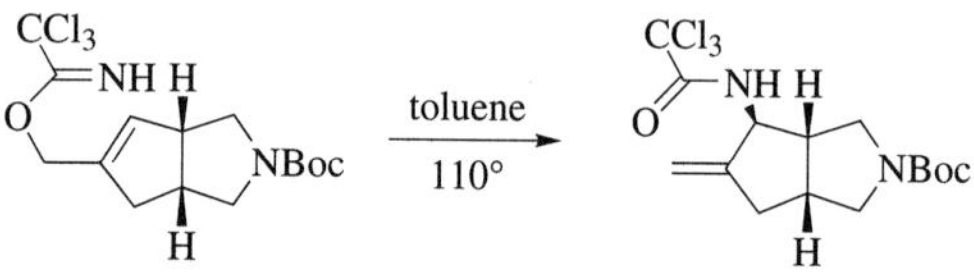

(51% from alcohol) (Eq. 37)

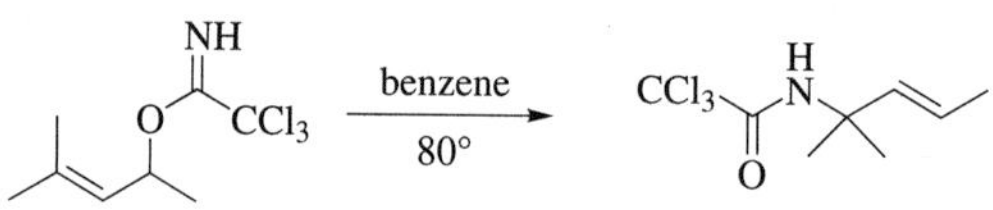

(84%) (Eq. 38)

Cyclic Substrates. Rearrangements of allylic trihaloacetimidates in which the allyl unit is embedded in a ring can be challenging as the transition state conformation required for rearrangement is higher in energy than that of acyclic counterparts. If the ring is not large, the oxygen of the imidate must adopt a quasi-axial orientation to bring the imidate nitrogen and distal alkene carbon within bonding distance. This feature is illustrated in the cyclohexenyl series in Eq. 39. Moreover, in a half-chair transition structure, a destabilizing syn-pentane interaction would exist between the starred atoms in **33**. This interaction would be avoided in a twist-boat transition structure.

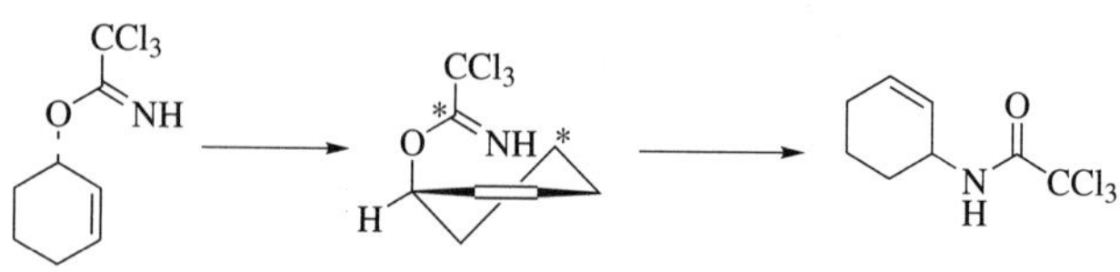

33

(Eq. 39)

Because the trihaloacetimidate must adopt a high-energy conformation to undergo rearrangement, cyclic allylic imidates are particularly prone to decompose at the elevated temperatures required to promote their sigmatropic rearrangement. The addition of K_2CO_3 to thwart acid-catalyzed decomposition of the allylic imidate can be particularly critical in these cases as illustrated in the rearrangement of imidate **34** (Eq. 40).[63]

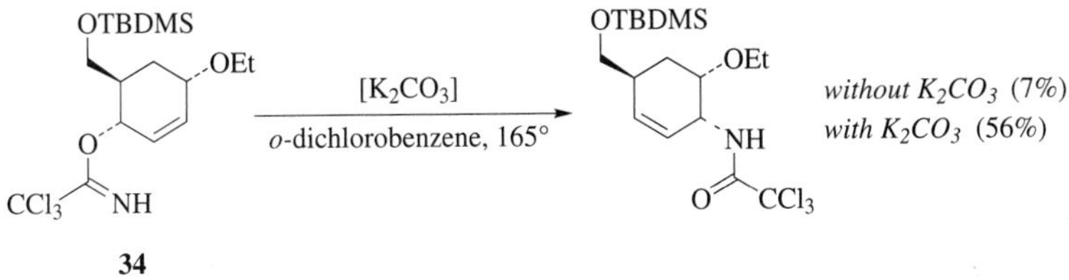

(Eq. 40)

If additional steric impediments exist, rearrangements of six-membered cyclic substrates can be low-yielding. For example, the additional 1,3-diaxial interaction (between O and C) brought about by the gem dimethyl group in trichloroacetimidate **35** is believed to be responsible for the low rearrangement yields realized in this case (Eq. 41);[6] this situation is not improved by adding K_2CO_3.[63]

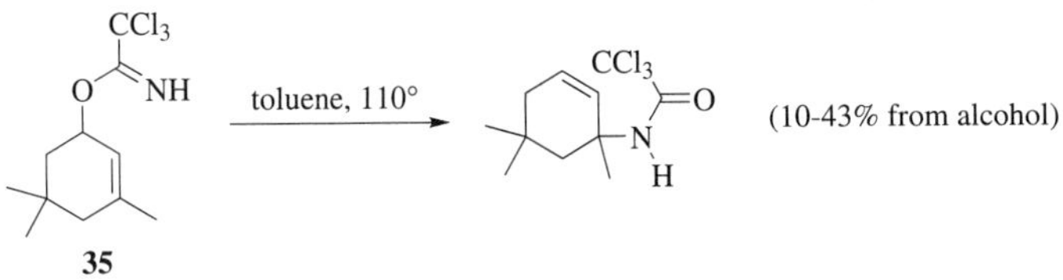

(Eq. 41)

The deleterious effect of adding an additional 1,3-diaxial interaction is also seen in the rearrangements shown in Eq. 42 and Eq. 43. Thus, while cyclohexenyl tri-

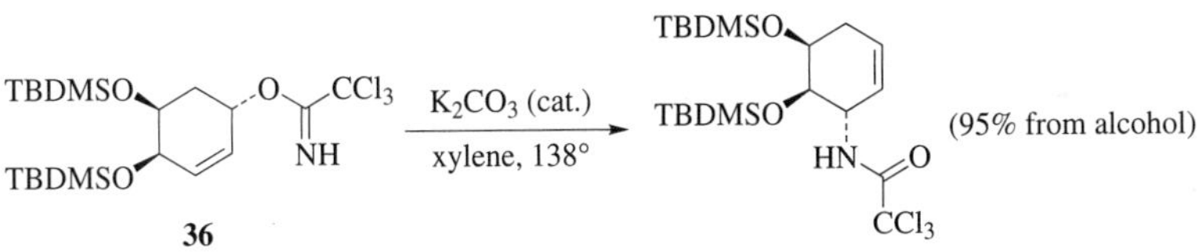

(Eq. 42)

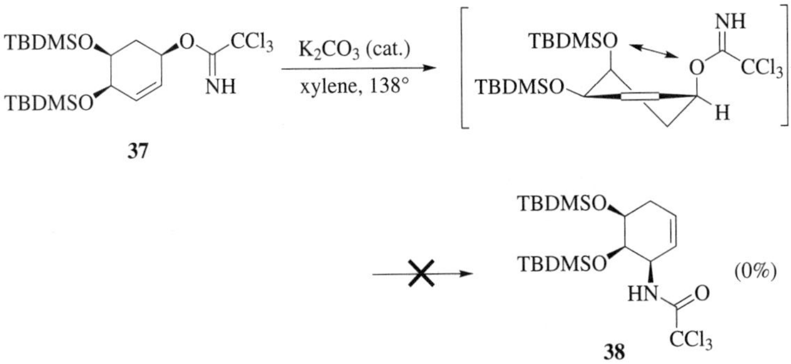

(Eq. 43)

chloroacetimidate **36** rearranges in high yield at 138°, the attempted rearrangement of stereoisomeric imidate **37**, which would suffer a 1,3-diaxial O–O interaction if the imidate is oriented quasi-axially,[96] yields none of the rearrangement product **38**.

Instead, the starting imidate is recovered in 35% yield.[82] Under some circumstances, interactions of this type can completely subvert the rearrangement process (Eq. 44).[83]

(Eq. 44)

Many pyranose[45,64,70,84–86] and furanose[18,53,54,87] substrates have been subjected to the Overman rearrangement with good success, but rearrangements of unsaturated pyranose substrates having the 1α,4α configuration are problematic. For example, the 2β,5α-trichloroacetimidate **39** rearranges stereospecifically in refluxing xylenes in 80% yield, whereas the 2α,5α isomer **40** (R = Me) rearranges much more sluggishly (Eqs. 45, 46).[86] Although these results are ascribed to a more sterically congested transition state in the reaction of isomer **40**, the loss of anomeric stabilization in transition structure **41** is more likely the origin of this difference. However, examples exist where pyranose substrates having the 1α,4α configuration do rearrange in moderate yield. For example, heating **40** (R = Et) at 165° in *ortho*-dichlorobenzene in the presence of K_2CO_3 gives the trichloroacetamide in 56% yield.[63]

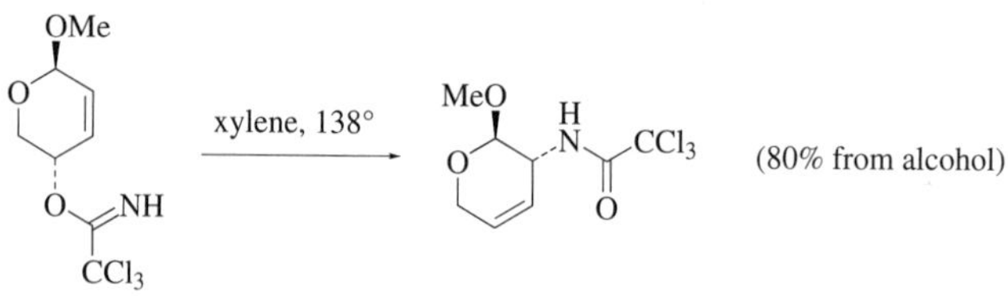

39

(Eq. 45)

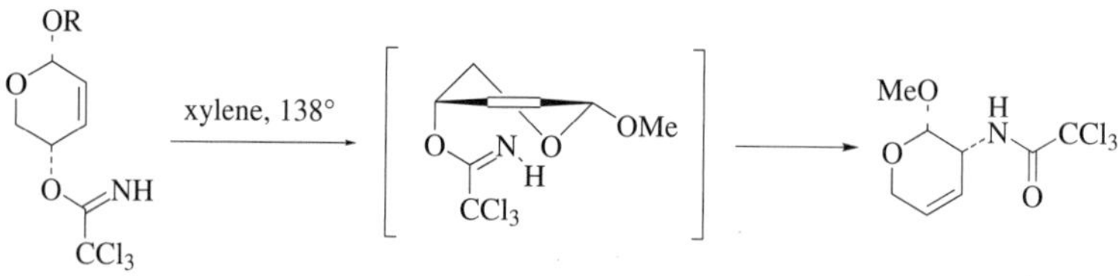

40 (R = Me, Et) **41**

(Eq. 46)

Successful Overman rearrangements of dissaccharides have been reported, particularly in cases where the energetically favored half-chair conformation of the unsaturated pyranose places the trichloroacetimidate in an axial orientation. Such an example is shown in Eq. 21.[70]

Substituent Effects and Problematic Substituents. Allylic trihaloacetimidate rearrangements are impacted by the electronic effects of substituents attached to carbons 2 and 3 of the allylic system. Electron-releasing substituents can favor the rearrangement. For example, reaction of the secondary dihydrofuryl alcohol **42** with trichloroacetonitrile at 0° leads directly to the rearranged trichloroacetamide **43** in 78% yield (Eq. 47); the intermediate imidate derivative is not observed.[88]

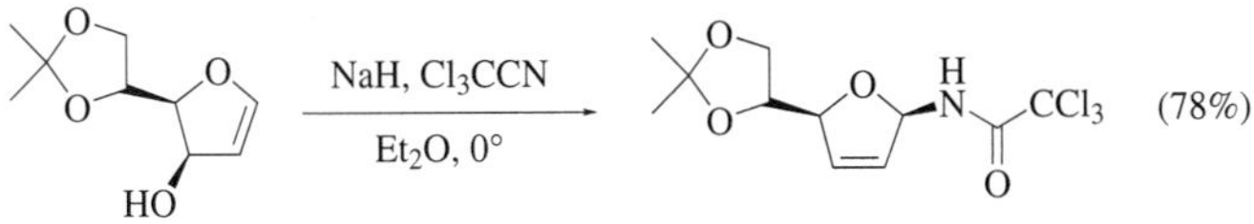

42 43

(Eq. 47)

The presence of electron-withdrawing groups at the distal alkene carbon can be problematic. For example, the trifluoromethylated allylic alcohol **44** undergoes imidate formation without incident, however allylic rearrangement fails (Eq. 48).[58] Some allylic trichloroacetimidates in which the double bond is part of an α,β-unsaturated carbonyl system fail to undergo Overman rearrangement, for example, dienyl ester **45**. The presence of the electron-withdrawing ester is shown to be the problem, as the structurally similar THP-protected dienol **46** rearranges without incident (Eq. 49).[89]

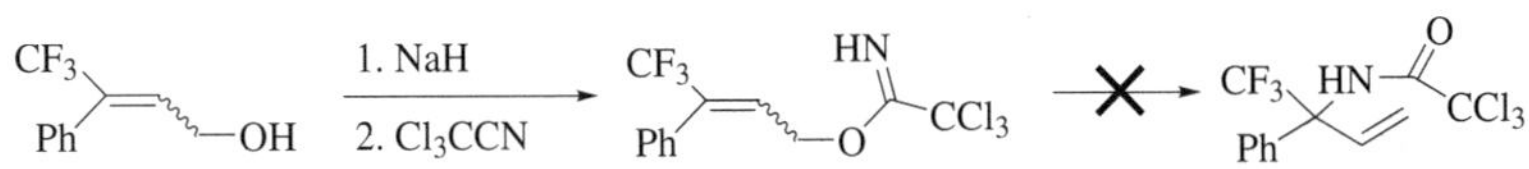

44

(Eq. 48)

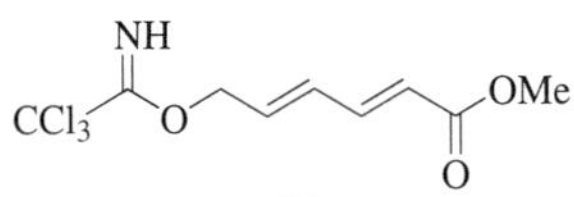

45

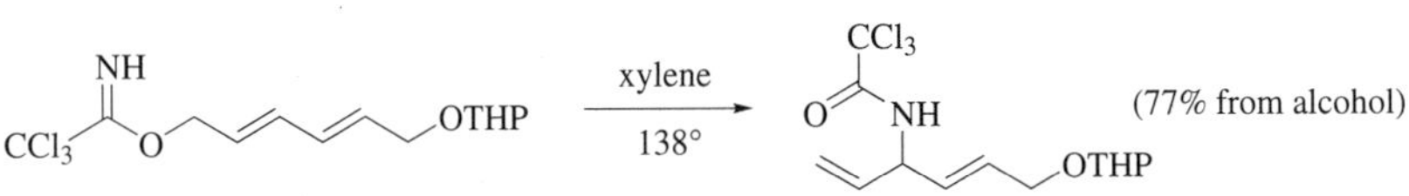

46

(Eq. 49)

The problem in these examples is likely competitive 1,4-addition of the imidate nitrogen to the α,β-unsaturated carbonyl functionality. For example, when unsaturated ester **47** is heated in refluxing toluene, [3,3]-sigmatropic rearrangement is not observed, but instead oxazoline **48** is formed (Eq. 50). As in the previous example, this side reaction is circumvented by reducing the ester to the alcohol, protecting the alcohol, then carrying out the rearrangement with the protected alcohol congener.[29]

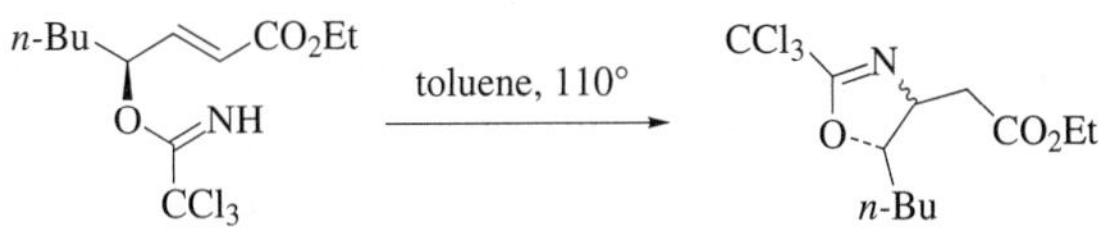

47 48

(Eq. 50)

Regioselectivity

If the trihaloacetimidate is positioned proximal to two different double bonds, two regioisomeric dienyl trichloroacetamides can be formed. In cases of this type, little selectivity is observed. For example, conversion of 5,9-dimethyl-1,4,8-decatriene-3-ol to its trichloroacetimidate derivative results in a 60:40 mixture of regioisomeric trichloroacetamides upon rearrangement (Eq. 51).[6] In the case of the trichloroacetimidate derived from 1,4-hexadien-3-ol, a slight preference for reaction at the more highly substituted alkene is observed (Eq. 52).[90]

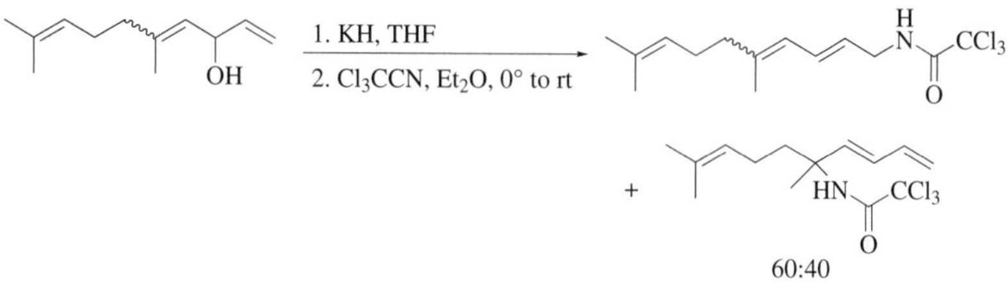

(Eq. 51)

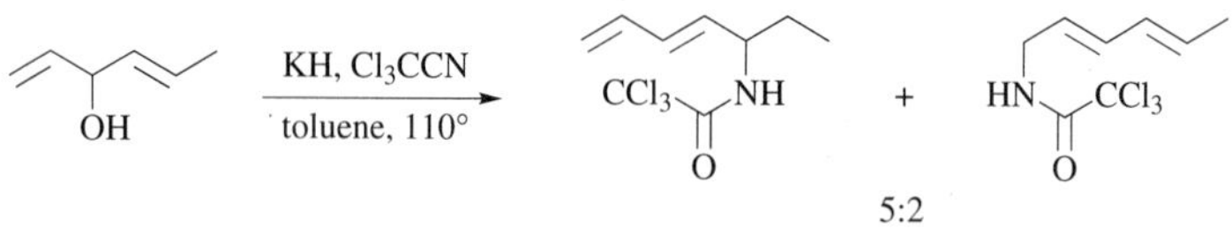

(Eq. 52)

An interesting example in which regioisomers result from participation of the double bond of a heteroaromatic system is known. Thus, attempted rearrangement of the imidazole trichloroacetimidate **49** leads to the desired product **50** in relatively low yield, with the major product arising from competitive rearrangement at the 4,5 double bond of the imidazole ring (Eq. 53).[91] This limitation is not seen with related aromatic substrates and may arise in the imidazole case by an ionization-recombination pathway as the trityl-protected nitrogen is perfectly situated to stabilize the derived allyl cation.

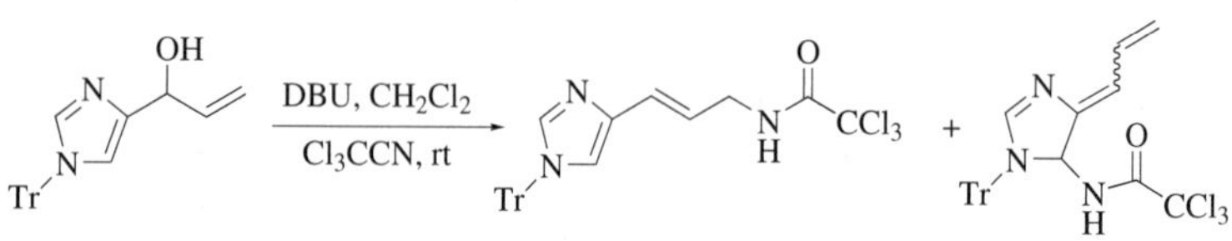

(Eq. 53)

49 **50** (22%) major product

Stereochemistry

Chiral Secondary Imidates: Chirality Transfer. High stereoselection is a hallmark of the trihaloacetimidate rearrangement. Self-immolative[92] transfer of chirality was first demonstrated in the rearrangement of (*R*)-1-methyl-3-phenyl-2*E*-

propenyl trichloroacetimidate, which proceeds smoothly to give the corresponding secondary benzylic trichloroacetamide with complete transfer of chirality (Eq. 5).[20]

There are numerous other examples of this reliable transposition of chirality. Three are shown in Eqs. 54–56.[24,26,93]

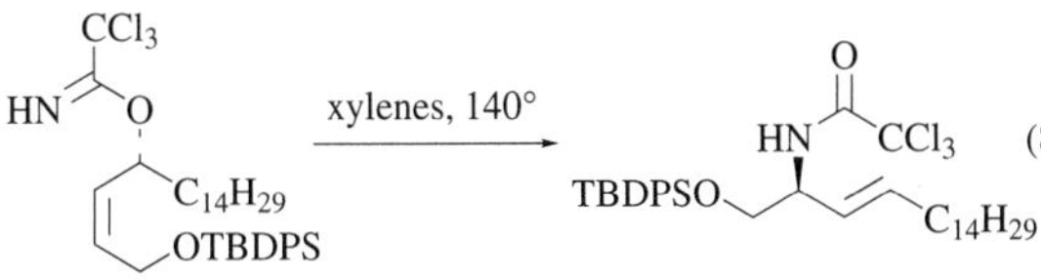

(81% from alcohol) (Eq. 54)

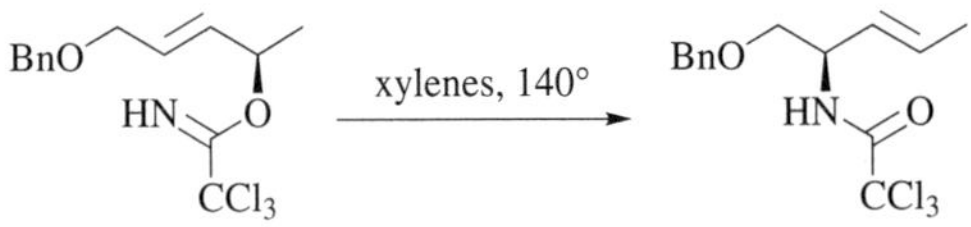

(45% from alcohol) (Eq. 55)

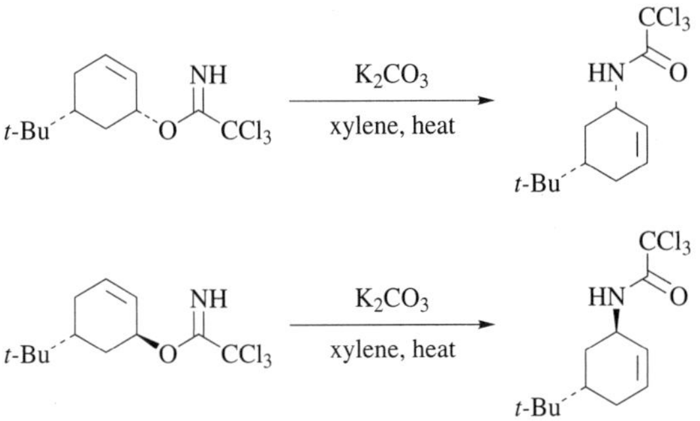

(75% from alcohol)

(71% from alcohol)

(Eq. 56)

Diastereoselectivity Arising from Stereocenters Outside the Pericyclic Arena. The impact of chirality external to the six-centered electrocyclic framework varies widely. The conformational flexibility of the substrate and the temperature for the rearrangement influence the observed degree of diastereoselection. For example, no stereocontrol is achieved in the rearrangement of the chiral allylic dioxolane **31** in refluxing xylenes (Eq. 34).[77] Likewise, rearrangement of trichloroacetimidate **51** at 140° yields a 1.6 : 1 mixture of diastereomeric amides (Eq. 57).[30] As will be discussed later, Pd(II)-catalyzed rearrangements of substrates of this type typically proceed with high stereoselectivity.[30,94]

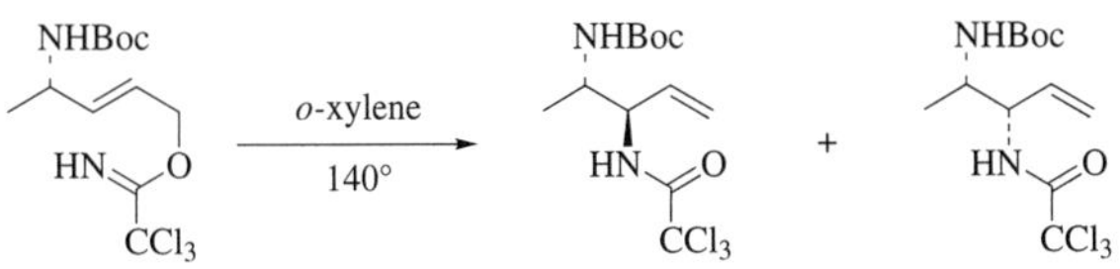

(85%) (Eq. 57)

51 1.6:1

Allylic trihaloacetimidate rearrangements of more conformationally constrained systems often proceed with substantial diastereoselectivity, however. Thus, trifluoroacetimidate **52** rearranges to give an excellent yield of epimeric amides **53** and **54** with 10:1 diastereoselectivity (Eq. 58). The trichloroacetimidate analogue of **52** is reported to rearrange over a period of 30 hours in 47% yield with no diastereoselectivity.[95] No explanation for this surprising difference has been advanced.

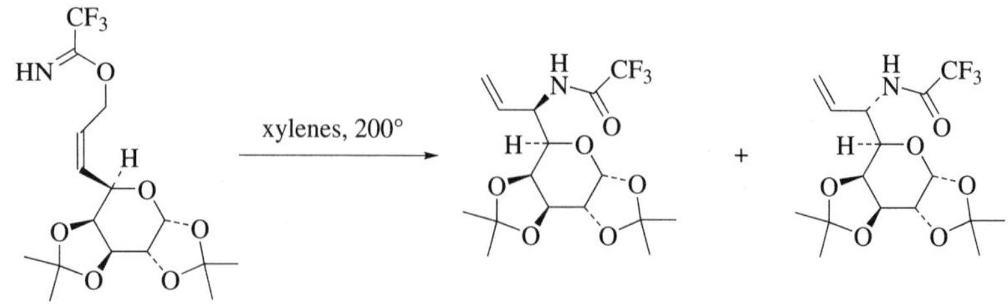

52 **53** (95%) 10:1 **54** (Eq. 58)

Rearrangement of the exo-allylic trichloroacetimidate **24** takes place selectively from the face opposite the dioxolane substituent, which would be oriented quasi-axially[96] to give amide **25** as a single stereoisomer in excellent yield on a 20 gram scale (Eq. 22).[49] This product is a key intermediate in the synthesis of (–)-5,11-dideoxytetrodotoxin. Excellent stereoselectivity also is observed in the rearrangement of the E and Z propenyl trichloroacetimidates **55** and **56** (Eqs. 59 and 60). In both reactions, rearrangement occurs from the same alkene face to provide opposite epimers at the new nitrogen-bearing stereocenter.[53]

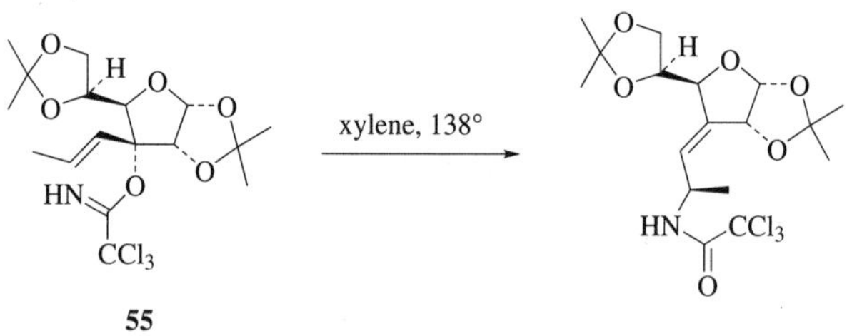

55 (90%) (Eq. 59)

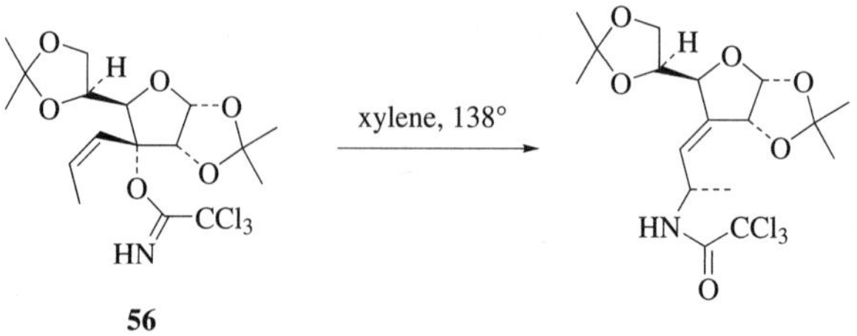

56 (≈100%) (Eq. 60)

Geometry of the New Double Bond. As discussed earlier in the context of favored chair transition structures for allylic trihaloacetimidate rearrangements, a common feature of the [3,3]-sigmatropic rearrangement of secondary allylic trihaloacetimidates is transposition to generate a new E double bond. Although extremely high E stereoselectivity is typically seen, some examples of moderate stereoselection have been reported. For example, in the rearrangement of a series of trichloroacetimidates of alkyl-substituted α-hydroxyphosphonates, the product is obtained as a mixture of E and Z isomers, with the Z isomer typically accounting for ~13% of the product mixture (Eq. 61).[97]

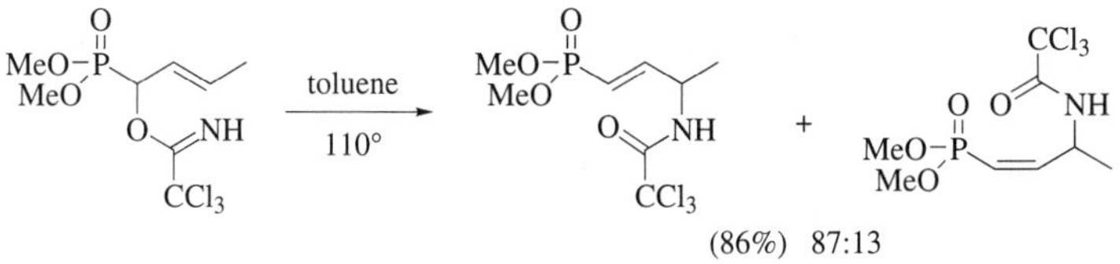

(Eq. 61)

Low stereoselectivity is seen in allylic trichloroacetimidate rearrangements of chiral tertiary allylic alcohols when the two α substituents are of similar size, as the two possible chair transition structures are of similar energy. For example, rearrangement of the trichloroacetimidate of linalool at 80° provides a 60 : 40 mixture of geranyl and neryl trichloroacetamides (Eq. 62).[6]

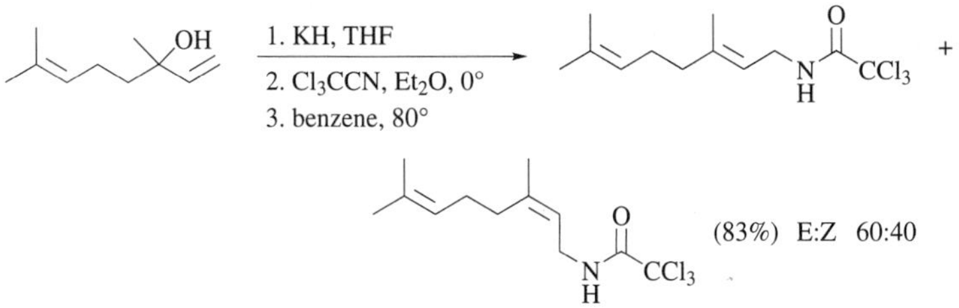

(Eq. 62)

Catalyzed Rearrangements of Allylic Trihaloacetimidates

Catalysis of trichloroacetimidate rearrangements by mercuric trifluoroacetate was disclosed in the inaugural report of this transformation.[4] Whereas the thermal rearrangement of the trichloroacetimidate of (*E*)-2-hexen-1-ol requires refluxing *meta*-xylene (138°) for nine hours to give the allylically rearranged trichloroacetamide in 81% yield, this allylic amide is formed in similar yield within minutes in THF at 0° in the presence of 10 mol% of mercuric trifluoroacetate. This mercury(II) complex is estimated to increase the rate of the rearrangement by a factor greater than 1×10^{12}.[6] Similar rate accelerations are realized in the presence of soluble complexes of $PdCl_2$, which have emerged as the most generally useful catalysts for allylic trihaloacetimidate rearrangements. An early example was reported in 1980,[11] although Pd(II)-catalysis of allylic trihaloacetimidate rearrangements was not studied in detail until several years later.[28]

The utility of the metal-catalyzed rearrangement is limited by the propensity of the catalyst to promote elimination to form dienes and trichloroacetamide, occasioned by competitive coordination of the catalyst to the allylic trichloroacetimidate nitrogen. In general, primary allylic trihaloacetimidates containing trans 1,2-disubstituted double bonds rearrange in good yields in the presence of Hg(II) or Pd(II) catalysts. Fewer examples exist of successful rearrangements of secondary allylic trihaloacetimidates; however, several high-yielding examples are known with Pd(II) catalysts.[25,28,29] Complete suprafacial transfer of chirality is a hallmark of catalyzed versions of the rearrangement, as it is of the thermal variant. Diastereoselection in rearrangements of chiral, δ-substituted, allylic trihaloacetimidates can be significantly enhanced in the presence of $PdCl_2$ catalysts. Of potentially greater significance, useful asymmetric Pd(II) catalysts have been developed quite recently.[7–9]

General Conditions. The rearrangement of allylic trichloroacetimidates with mercuric trifluoroacetate is carried out using 10–30 mol% of this catalyst in THF, e.g., Eq. 63. Hg(II)-catalyzed rearrangements can take place at temperatures as low as −60°, with yields often being high for primary substrates.[6]

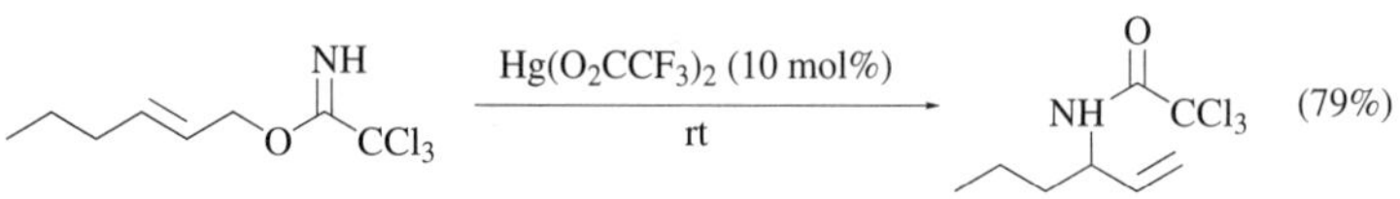

(Eq. 63)

More recently, Pd(II) complexes have been the catalysts of choice, with excellent outcomes being achieved at room temperature using 4–8 mol% of the soluble bis(acetonitrile) or bis(benzonitrile) complexes of $PdCl_2$ in aprotic solvents such as THF or toluene. Palladium acetate and palladium trifluoroacetate have also been employed, although rarely.[25] Pd(II)-catalyzed allylic trichloroacetimidate rearrangements typically take place in a few hours at or below room temperature, as exemplified by the conversion of trichloroacetimidate **57** to amide **58** (Eq. 64).[28]

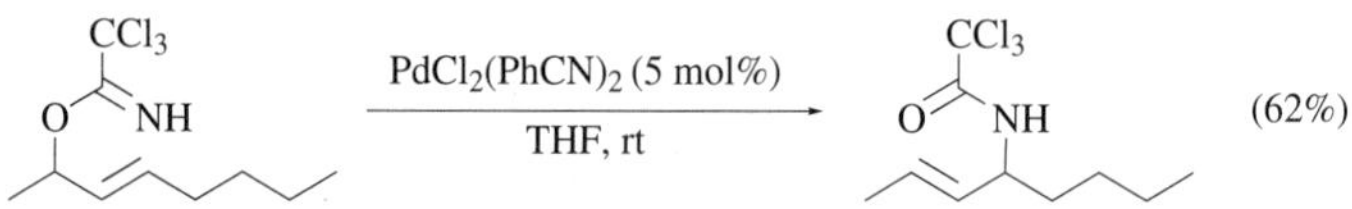

57 **58**

(Eq. 64)

There is a single report of trichloroacetimidate rearrangements being promoted by halogen electrophiles, specifically *N*-bromosuccinimide (NBS) and *N*-iodosuccinimide. For example, the E allylic phosphonate trichloroacetimidate **59** is transformed to the rearrangement product **60** in moderate yield at room temperature in the presence of one equivalent of NBS (Eq. 65). In contrast, the Z isomer reacts slowly under identical conditions to give oxazoline **61** (Eq. 66).[66,2]

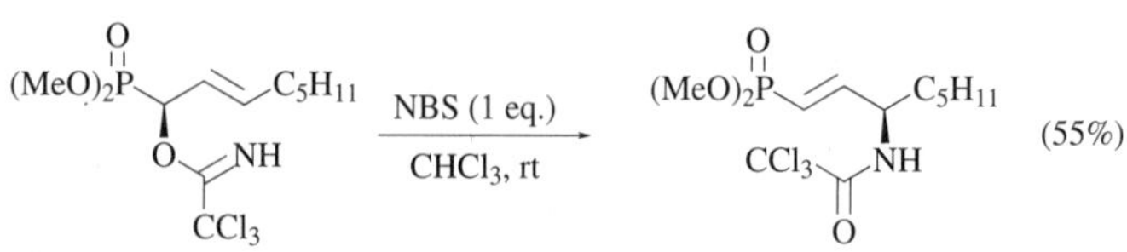

59 **60**

(Eq. 65)

[2] The reported yield in Eq. 66 is after hydrolysis to the hydroxamide.

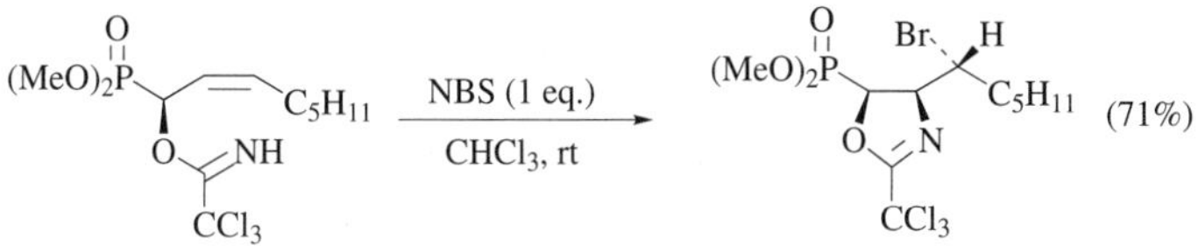

61

(Eq. 66)

Scope. Although metal-catalyzed allylic trihaloacetimidate rearrangements take place under milder conditions than their thermal counterparts, the scope of the catalyzed rearrangement is much more limited. Trichloroacetimidates are typically used and give higher yields than the less nucleophilic trifluoroacetimidates,[62] a result expected for a cyclization-induced rearrangement mechanism. Primary trichloroacetimidates containing trans 1,2-disubstituted double bonds are the best substrates for metal-catalyzed rearrangements, typically undergoing rearrangement in good yields using either $PdCl_2(RCN)_2$ or $Hg(O_2CCF_3)_2$ catalysts.[6,27,62,94] Substitution on the allylic double bond impedes the metal-induced rearrangement. For example, the rearrangement of the trichloroacetimidate of geraniol (**62**, Eq. 67) under thermal conditions provides linalyl trichloroacetamide in 90% yield (Eq. 32), whereas this product is formed in only 66% yield in the presence of $PdCl_2(MeCN)_2$.[62] There have been no reports of successful metal-catalyzed rearrangements of substrates having a substituent on the internal allylic alkene carbon.

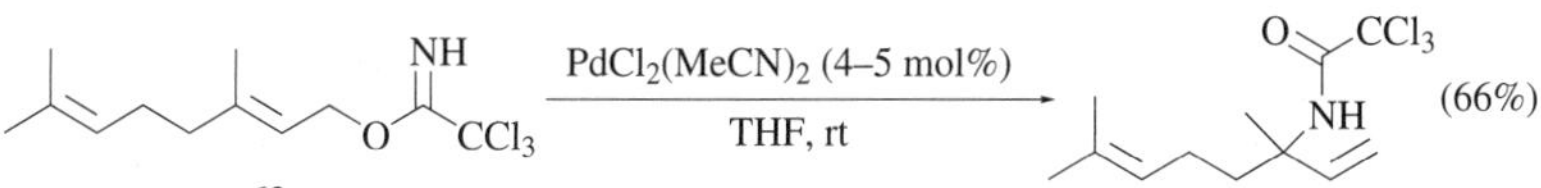

62

(Eq. 67)

More highly functionalized primary allylic trichloroacetimidates often have proven to be resistant to metal catalysis. For example, the ribose-derived imidate **63** fails to yield any rearrangement product in the presence of $PdCl_2(MeCN)_2$ or $Hg(O_2CCF_3)_2$.[98] This result should be contrasted with the successful rearrangement of this imidate under thermal conditions (xylene at 137° in the presence of two percent di(*tert*-butyl)-*para*-cresol; Eq. 68). Similarly, xylose derivative **64** (Eq. 69) fails to rearrange under catalysis by either Pd(II) or Hg(II) species, although again its rearrangement is successfully realized thermally (xylene at 200°).[95]

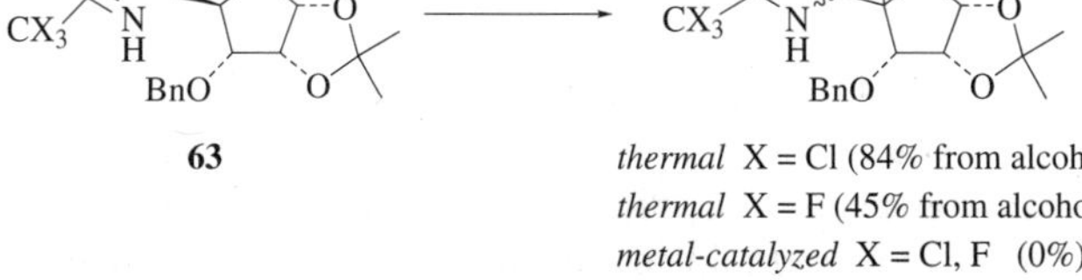

(Eq. 68)

63

thermal X = Cl (84% from alcohol)
thermal X = F (45% from alcohol)
metal-catalyzed X = Cl, F (0%)

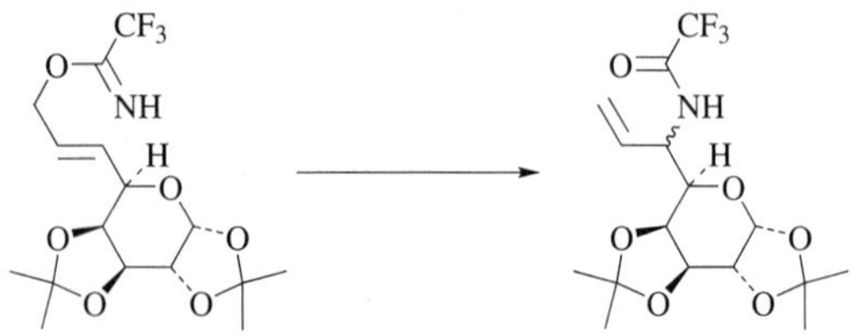

64

thermal (93%)
metal-catalyzed (0%)

(Eq. 69)

An interesting example is shown in Eq. 70, where failure to observe [3,3]-sigmatropic rearrangement arises from an alternate reaction pathway of a Pd(II)-olefin complex. In this case, attempted rearrangement of **65** using $PdCl_2(MeCN)_2$ gives a 58% yield of cyclopropane derivative **67**. This product is postulated to derive from **66**, which would arise if intramolecular attack by the imidate on the nascent Pd(II)–olefin complex took place at the proximal alkene carbon (5-exo, rather than 6-endo cyclization).[98]

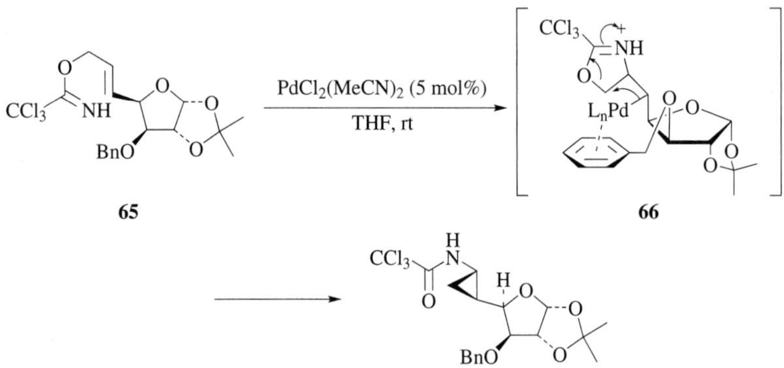

65 66 67

(Eq. 70)

Although there are no examples of high-yielding rearrangements of secondary allylic trichloroacetimidates using $Hg(O_2CCF_3)_2$ as the catalyst, several successful $PdCl_2$-catalyzed rearrangements of secondary allylic trichloroacetimidates containing trans 1,2-disubstituted double bonds are known (see the following section). However, this catalyzed transformation is limited to secondary substrates of this specific type. For example, attempted $PdCl_2$-catalyzed rearrangement of the cis secondary trichloroacetimidates **68** failed due to competing elimination to form the corresponding dienes and trichloroacetamide,[25] an outcome also observed with Z trichloroacetimidates **69**.[78] Low yields (<30%) were also reported for rearrangements of secondary allylic trichloroacetimidates containing trans 1,2-disubstituted double bonds using $Pd(OAc)_2$ as the catalyst; elimination was again the major reaction pathway.[25] Consistent with these observations, there are no reports of successful catalytic rearrangements of cyclic secondary allylic trichloroacetimidates nor secondary allylic trichloroacetimidates containing trisubstituted double bonds.

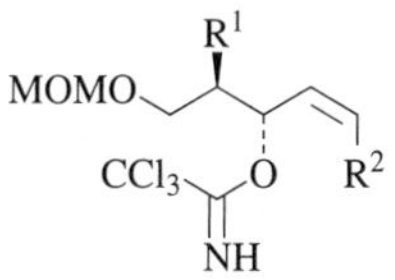

R^1	R^2
Me	*n*-Bu
i-Pr	*n*-Bu
i-Bu	*i*-Bu

68

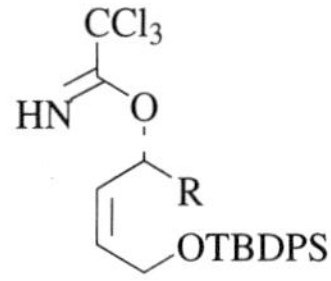

R
n-Bu
i-Pr
Ph

69

Stereoselectivity. *Chiral Secondary Imidates: Chirality Transfer.* Although metal-catalyzed rearrangements of secondary allylic trihaloacetimidates are limited to substrates containing trans 1,2-disubstituted double bonds, such rearrangements of chiral imidates proceed with excellent transfer of chirality. For example, Pd(II)-catalyzed rearrangement of trichloroacetimidate **6** proceeds with complete suprafacial diastereoselection to give the trichloroacetamide in high yield (Eq. 9).[28] In one reaction, the milder conditions associated with Pd(II) catalysis lead to enhanced selectivity when compared to the thermal counterpart. For example, thermal rearrangement of chiral trichloroacetimidate **70** (Eq. 71) at 110° (refluxing toluene) takes place with 7% loss of enantiomeric purity.[29] In contrast, the Pd(II)-catalyzed version of this rearrangement occurs with no loss of enantiomeric purity. As is seen in analogous thermal rearrangements, high E stereoselectivity is observed in forming the new 1,2-disubstituted double bond of the allylic trichloroacetamide products, see, e.g., Eqs. 64 and 71.

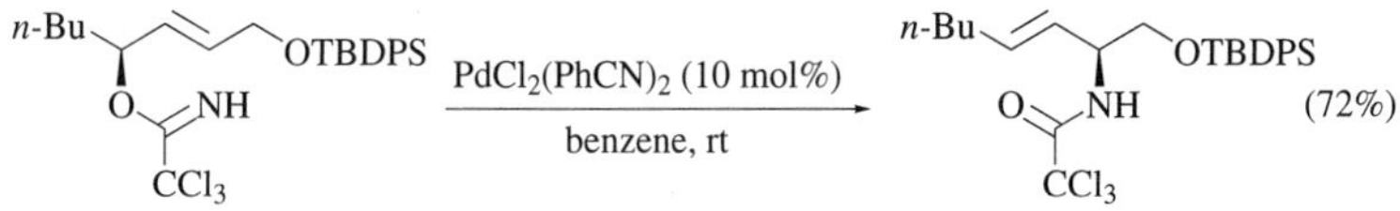

70

(Eq. 71)

Chiral Primary Imidates: Diastereoselectivity. A number of examples of chiral, δ-substituted trichloroacetimidates rearranging with enhanced stereoselectivity in the presence of $PdCl_2$ catalysts have been reported. For example, thermal rearrangement of trichloroacetimidate **71** in refluxing xylene provides the anti and syn stereoisomeric allylic amides in moderate yield and a mediocre 3:2 ratio (Eq. 72).[99] In

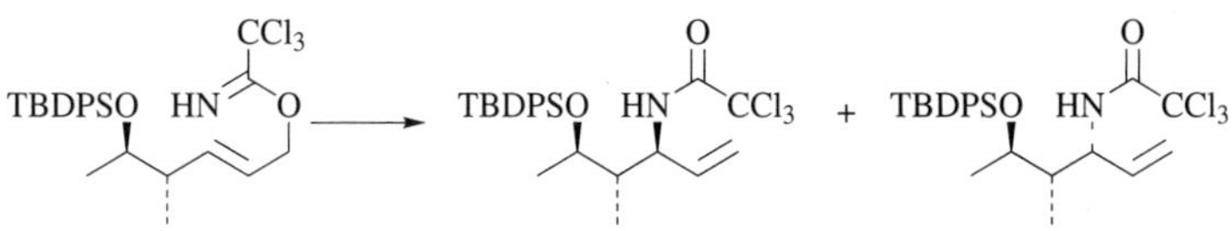

71 **I** **II**

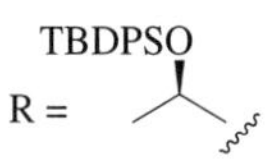

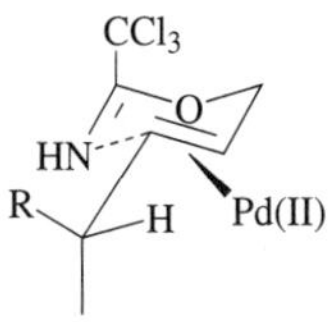

72

I:II	Conditions	
3:2	*p*-xylene, 138°	(54%)
1:7	$PdCl_2(PhCN)_2$ (10 mol%), THF, rt	(71%)

(Eq. 72)

contrast, rearrangement of **71** with either $PdCl_2(MeCN)_2$ or $PdCl_2(PhCN)_2$ takes place with much improved diastereoselectivity, providing the 3,4-anti isomer as the major product. This product results from preferential cyclization of the imidate nitrogen from the si face of the alkene. The authors suggest that this transition structure is favored because it allows coordination of the Pd(II) catalyst to the double bond to occur away from the bulky *tert*-butyldiphenylsilyl protecting group, as depicted in **72**.[99,3]

Several examples show high stereoselectivity in the synthesis of anti vicinal diamines by diastereoselective Pd(II)-catalyzed rearrangements of allylic trichloroacetimidates having a Boc-protected amine substituent at the δ position. For example, rearrangement of **73** in the presence of $PdCl_2(MeCN)_2$ gives exclusive formation of the anti isomer (the syn isomer was undetected; Eq. 73). In contrast, thermal rearrangement of the same substrate gives a 62:38 mixture of the anti:syn products.[30] Coordination of the palladium to the adjacent Boc-protected nitrogen, as depicted in **74**, is invoked to rationalize the stereoselection.

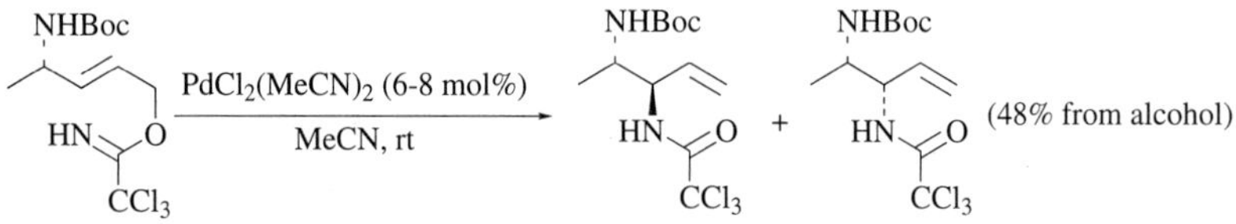

73 *anti:syn* 99:1

(Eq. 73)

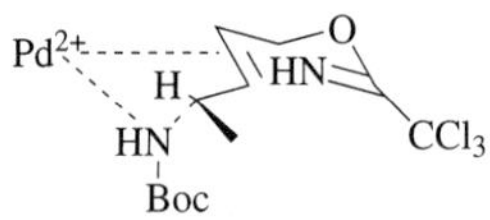

74

In a related example, potential chelation of a Pd(II) catalyst to a δ-alkoxy or δ-siloxy substituent was examined and found to have no impact. The rearrangement results summarized in Eq. 74 lead to the conclusion that diastereoselection, which can be as high as 10:1, results solely from steric effects.[100]

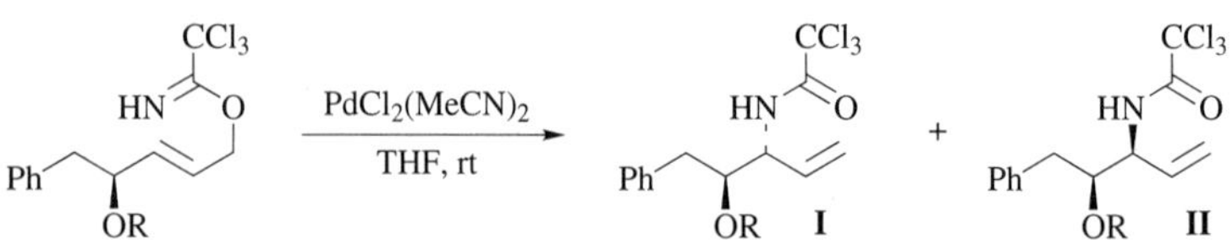

(Eq. 74)

R	I:II	
Me	1.4:1	(63%)
MOM	6.2:1	(58%)
BOM	9.95:1	(57%)
OTBDMS	10:1	(61%)

[3] This intermediate is drawn incorrectly in the original paper; i.e. the configuration shown at C4 is incorrect.

Asymmetric Catalysis. The success of Pd(II) catalysis for allylic trichloroacetimidate rearrangements naturally has led to research on the development of asymmetric Pd(II) catalysts. Early studies employing cationic Pd(II) complexes suggested that allylic trichloroacetimidates were not viable substrates for asymmetric catalysis, as attempts to carry out such transformations were plagued by competing elimination reactions, slow reaction rates, and low enantioselectivities.[101] The first two of these difficulties were ascribed to competitive strong complexation of the small, basic trichloroacetimidate nitrogen to the hard palladium center.[102] Consequently, success in developing asymmetric Pd(II) catalysts for allylic imidate rearrangements was realized initially with less strongly coordinating *N*-arylimidates.[101] Whereas high enantioselectivities can be realized with substrates of this type, for example Eq. 75,[39] transformation of the amide products to the corresponding allylic amines is not high yielding. A wide variety of chiral, enantiopure Pd(II) complexes have been shown to catalyze allylic rearrangements of *N*-arylimidates,[8,37,39,101,103–105] although a survey of these studies is outside the scope of this review. Reviews of early developments in this area have appeared.[14,102]

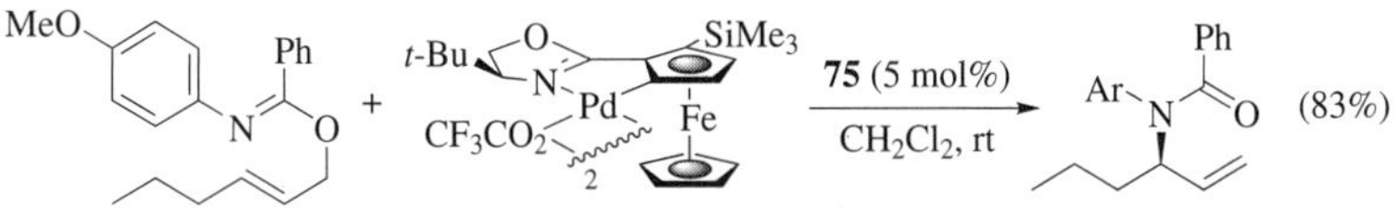

75

91% ee

(Eq. 75)

N-Anisyltrifluoroacetimidates are more attractive substrates for catalytic asymmetric allylic imidate rearrangements as their allylic *N*-anisyltrifluoroacetamide products can be converted to the parent allylic amines in good yield. An initial survey of six asymmetric Pd(II) complexes for catalyzing the rearrangement of **76** to **79** identified the cationic ferrocenyl oxazoline palladacyclic complex **75** (Eq. 75),

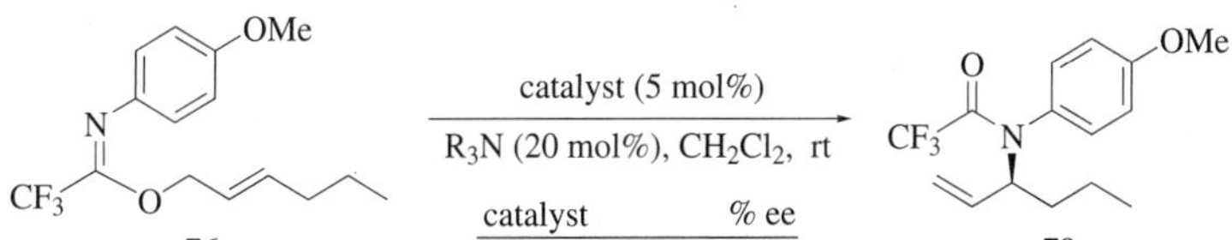

76

catalyst		% ee
77	(79%)	89
8	(92%)	92

79

(Eq. 76)

and the related cationic and neutral palladacyclic catalysts **77**, **78**, and **8** (COP-Cl) containing a chiral oxazoline substituent and a planar chiral cyclopentadienyl(η^4-''' tetraphenylcyclobutadiene)cobalt fragment, as effective catalysts for this transformation (Eq. 76).[8] Subsequent deprotection of the *N*-anisyltrifluoroacetamide **79** to form (*R*)-3-amino-1-hexene is accomplished in 73% yield. Cis allylic *N*-anisyltrifluoroacetimidates are converted also in good yield and high ee to the corresponding secondary *N*-anisyltrifluoroacetamides using the cobalt oxazoline palladacycle (COP) catalysts (Eq. 77). Incorporation of small amounts of tertiary amines, typically i-Pr_2NEt, minimizes acid-catalyzed decomposition of the starting imidates in rearrangements that employ the trifluoroacetate-bridged catalysts **75** and **77**, which are generated in situ by reaction of the corresponding halide-bridged dimers with excess silver trifluoroacetate.

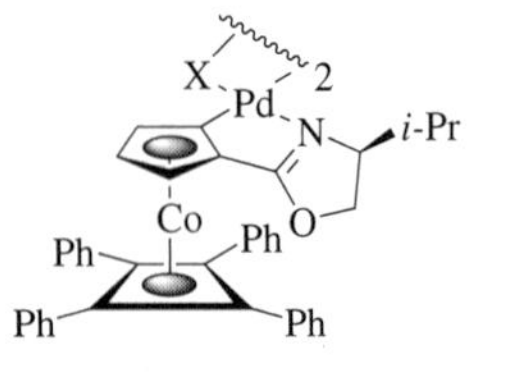

77 X = O_2CCF_3
8 X = Cl (COP-Cl)

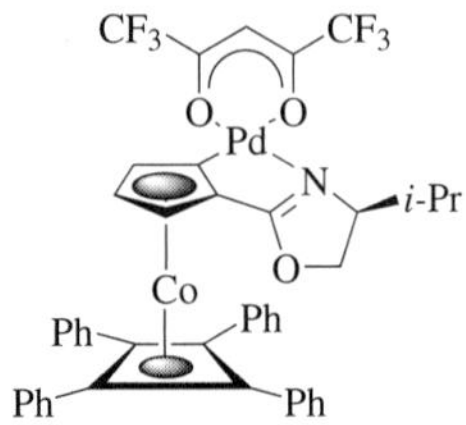

78

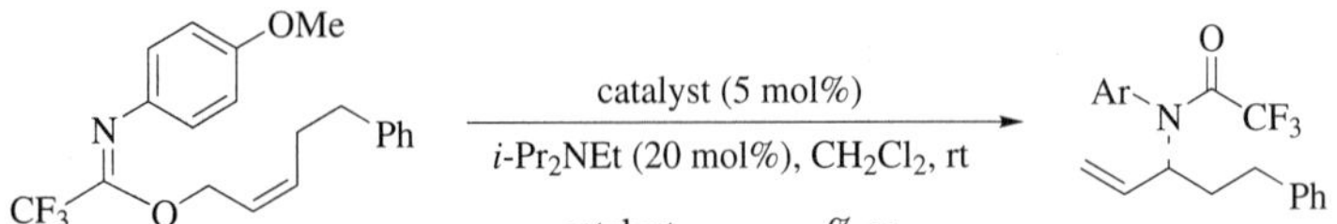

(Eq. 77)

catalyst		% ee
77	(76%)	96
8	(99%)	96

In a recent publication, the neutral chloride-bridged dimer COP-Cl (**8**) was shown to be an excellent catalyst for asymmetric rearrangement of trans 1,2-disubstituted allylic trichloroacetimidates, thus providing the first truly useful catalytic asymmetric method for transforming prochiral allylic alcohols to enantioenriched allylic amines and their analogues, e.g., Eq. 78.[7] Although the scope and limitations of this method are not well explored at this point, a variety of oxygen functionality is well tolerated. Some nitrogen functional groups appear to be also well tolerated, e.g. Eq. 79. However, the allylic rearrangement is prevented (at least at 38°) by tertiary amine functionality at C6, secondary amine functionality at either C6 or C12, or a thio ether substituent at C6 of the (*E*)-2-alkenyl trichloroacetimidate starting material. This method is limited to the rearrangements of allylic trichloroacetimidates containing trans 1,2-disubstituted double bonds as analogous cis substrates rearrange only slowly in the presence of COP-Cl.

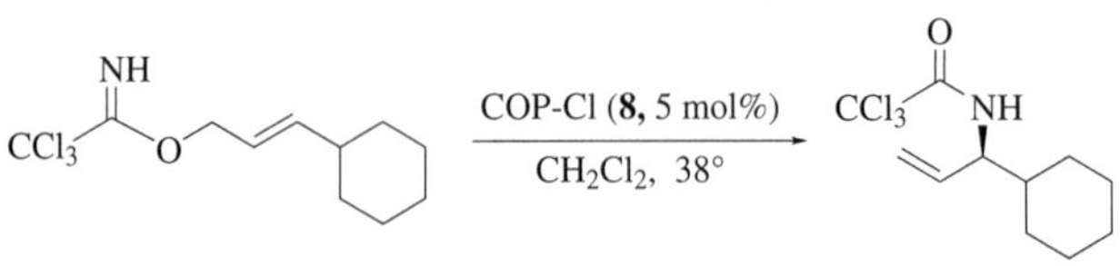

(82%) 96% ee (Eq. 78)

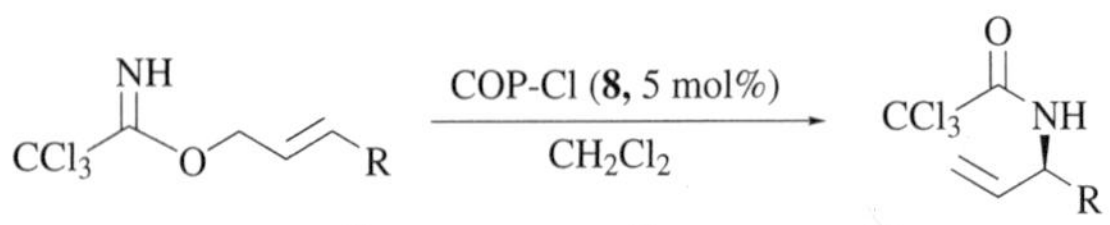

(Eq. 79)

R	Temp.		% ee
$(CH_2)_3NBn(Boc)$	38°	(96%)	95
$(CH_2)_9NBn_2$	rt	(82%)	97

The more soluble monomeric COP-hexafluoroacetylacetonate complex **78** allows a wider variety of solvents to be employed and higher catalyst concentrations, and correspondingly higher catalysis rates, to be achieved.[9] One example is shown in Eq. 80.

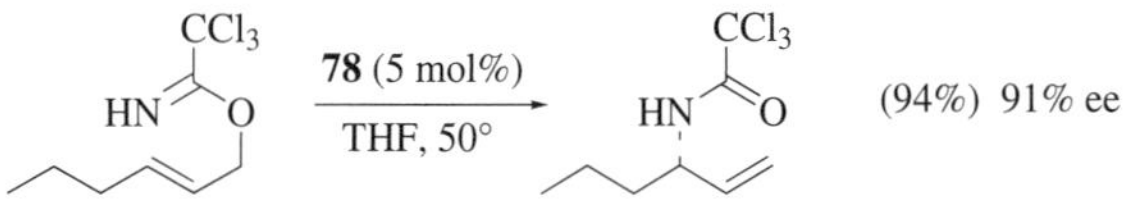

(Eq. 80)

APPLICATIONS TO SYNTHESIS

Overview

The chief significance of the allylic trihaloacetimidate rearrangement lies in the many uses of the allylic trichloroacetamide products. Foremost among these uses is ready access to allylic amines. The trichloroacetyl group is typically removed from allylic trichloroacetamides using either NaOH in mixed organic–aqueous solvent systems, strong mineral acids, or methanolic $NaBH_4$. One potential advantage of employing trifluoroacetimidates in this allylic rearrangement is the ready cleavage of the trifluoroacetyl group under mildly basic conditions.[106]

The trichloroacetyl group has also been exploited to accomplish subsequent elaborations of rearrangement products. For example, the trichloroacetyl group has been used to initiate radical cyclizations; as a precursor of guanidines, carbodiimides, and ureas; and to regulate facial selectivity in the addition of various reagents to the allylic double bond.

Preparation of Allylic Amines

Overman rearrangements have been employed to prepare a broad range of allylic amines for many diverse uses. For instance, a number of selective enzyme inhibitors have been prepared in this way. Examples include a variety of 2-(2-thienyl)allylamines, synthesized for studies of the inhibition of dopamine β-hydroxylase,[107] and dienyl amino acid **81**, an inhibitor of 4-aminobutyrate-2-oxoglutarate aminotransferase.[89] This latter agent is prepared in four steps from dienyl trichloroacetamide **80** (Eq. 81), whose synthesis by allylic trichloroacetimidate rearrangement is summarized in Eq. 49.

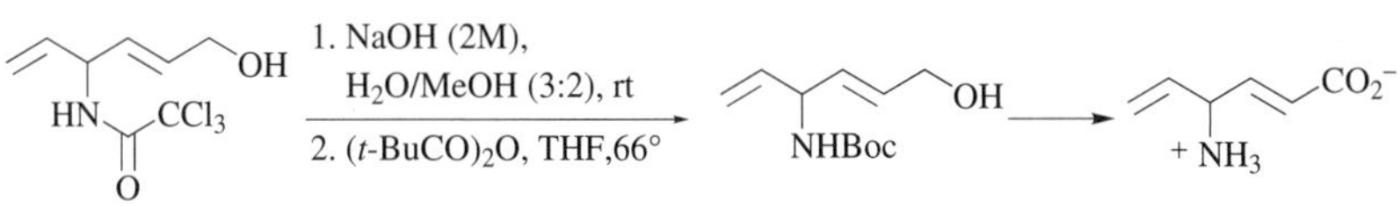

80 (46%) **81**

(Eq. 81)

An example of the application of this rearrangement to the preparation of novel classes of biologically important nitrogen compounds is seen in the synthesis of γ-aminophosphonic acids, for example **83** (Eq. 82).[97] Refluxing 6N HCl was employed to hydrolyze the trichloroacetamide and phosphonic ester functionalities of **82** in good yield.

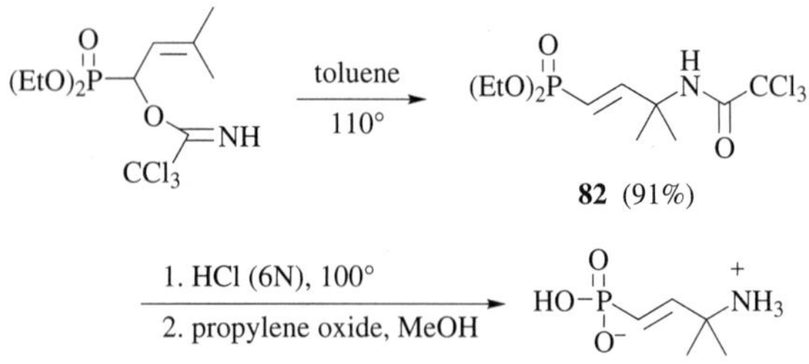

82 (91%)

83 (88%)

(Eq. 82)

A variety of amino sugars, including both monosaccharides and disaccharides, has been prepared using the Overman rearrangement as the key step. In many of these syntheses, the trihaloacetimidate rearrangement is employed to install the amino functionality with high stereocontrol.[45,70,85,108,109] In one approach, the trihaloacetimidate rearrangement is carried out on a carbohydrate framework, as exemplified by the example shown in Eq. 83. This tactic has been employed widely, with several examples of the rearrangement stage of these amino sugar constructions being highlighted earlier in Scope and Limitations (cyclic substrates).

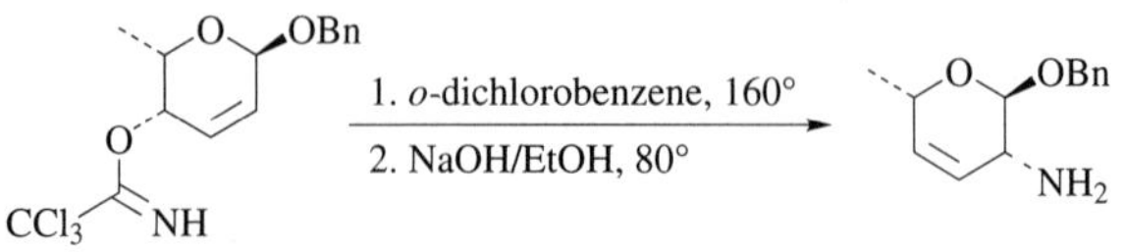

(Eq. 83)

In another construction of amino sugars, the rearrangement step is carried out on an acyclic substrate, with cyclization to form the carbohydrate occurring at a later stage. An early example, the preparation of (±)-*N*-(trichloroacetyl)vancosamine (**84**), is shown in Eq. 84.[51] This strategy has been applied also to the synthesis of several aminocarbasugar derivatives,[27,110] as in preparation of the conduramine analogue **85** (Eq. 85).[27]

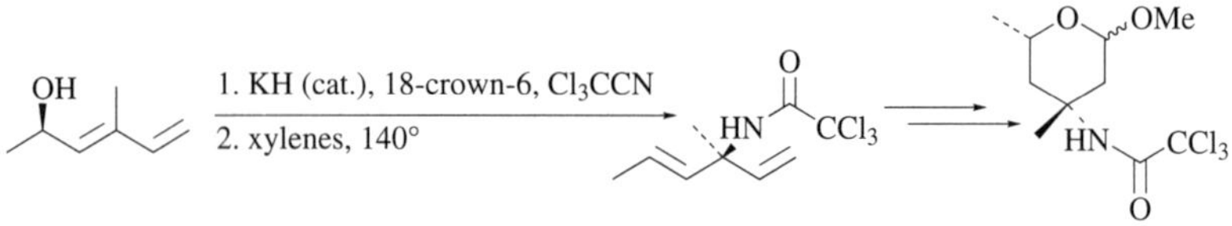

84

(Eq. 84)

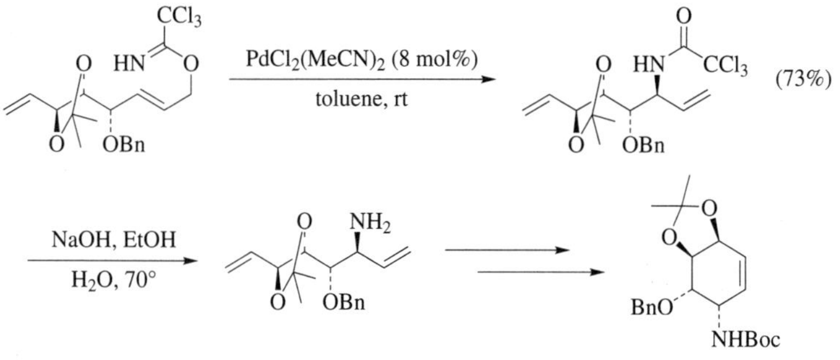

85

(Eq. 85)

Given the promise of nucleoside analogues as therapeutic agents, much effort has been devoted to their synthesis, with trihaloacetimidate rearrangements being central steps in several approaches. For example, the dideoxyribose **43** has been prepared in 78% yield from the allylic alcohol precursor, placing the amino group in the correct position for subsequent construction of the appropriate heterocylic substituent (Eq. 47).[88] Allylic rearrangement of trichloroacetimidate **86** is a central step in the synthesis of a series of 5′-branched 5′-aminothymidines (Eq. 86).[98] This example illustrates the mild removal of the trichloroacetyl group by reaction of allylic trichloroacetamide **87** with ethanolic $NaBH_4$ at room temperature.

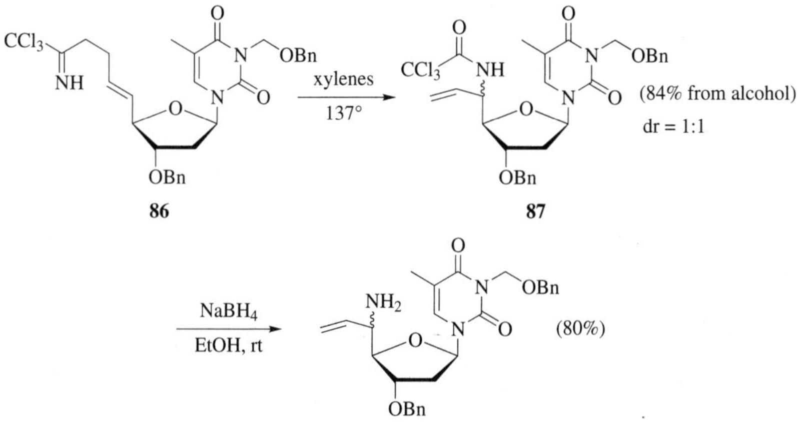

(Eq. 86)

Allylic trichloroacetimidate rearrangements have been central to the synthesis of several peptide analogues. For example, a range of dipeptide olefin isosteres has been synthesized in a study of parathyroid hormone receptor activation by analogues of the *N*-terminal fragment of the natural hormone. These dipeptide isosteres were accessed by the sequence exemplified in Eq. 87.[22] Numerous other dipeptide isosteres have been prepared using allylic trichloroacetimidate rearrangements.[22,23,25,67,111,112] In an additional example, the modified opioid pentapeptide **88** is prepared from an allylic alcohol precursor as summarized in Eq. 88.[196]

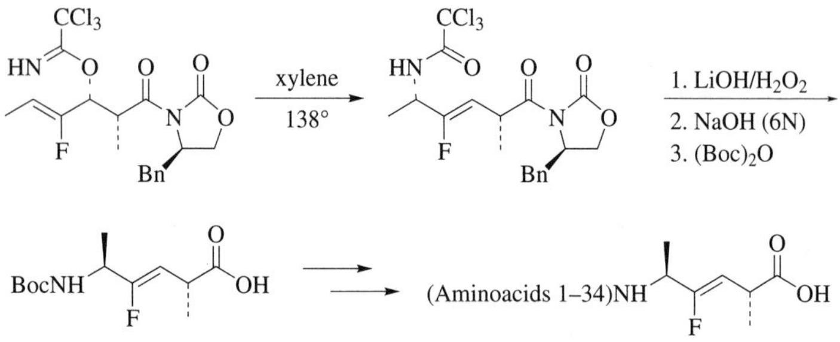

(Eq. 87)

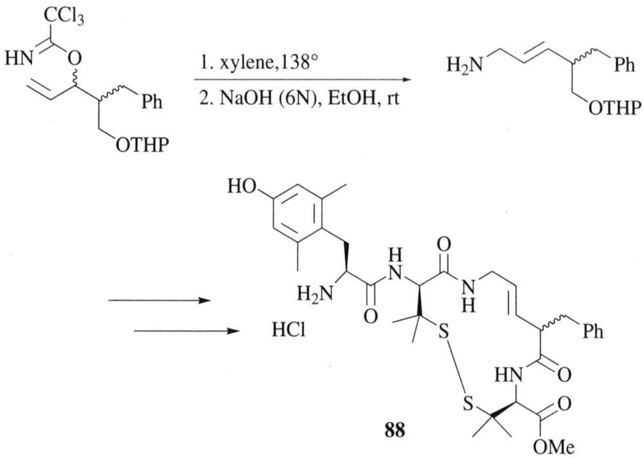

(Eq. 88)

88

Trichloroacetimidate rearrangements have been used in the synthesis of a wide variety of α-amino acids, for example, the Boc-protected β,γ-unsaturated α-amino acid **89** (Eq. 89).[29,86] A more common sequence for assembling α-amino acids introduces the carboxylic acid by oxidative cleavage of the allylic double bond.[113] Use of such a strategy to prepare the biologically important 1-aminocyclopropanecarboxylic acid in high overall yield from allylic alcohol precursor **90** is summarized in Eq. 90.[31]

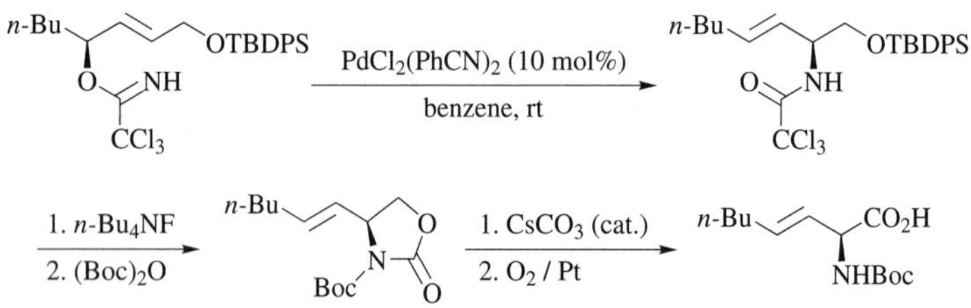

(Eq. 89)

89

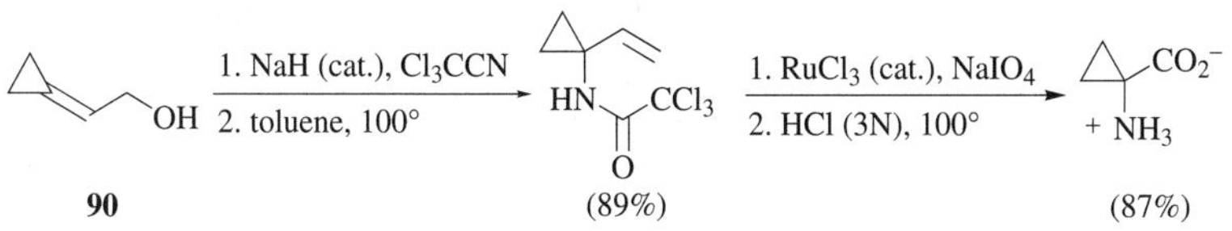

(Eq. 90)

Enantiopure amino acids have been accessed in this way using several strategies. In one method, a diastereoselective allylic imidate rearrangement is orchestrated on a chiral, enantiopure template. An example that employs a carbohydrate scaffold to synthesize (*S*)-(2-^{2}H)glycine is shown in Eq. 91.[53] A second example, where a δ-methoxymethyl substituent regulates face selectivity in the key allylic imidate rearrangement, is summarized in Eq. 92.[21]

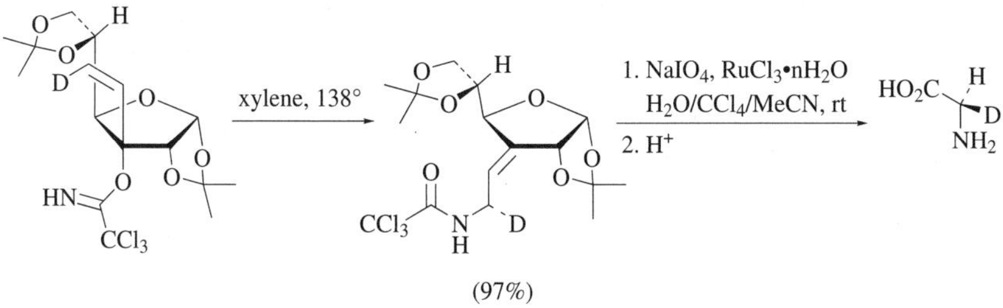

(Eq. 91)

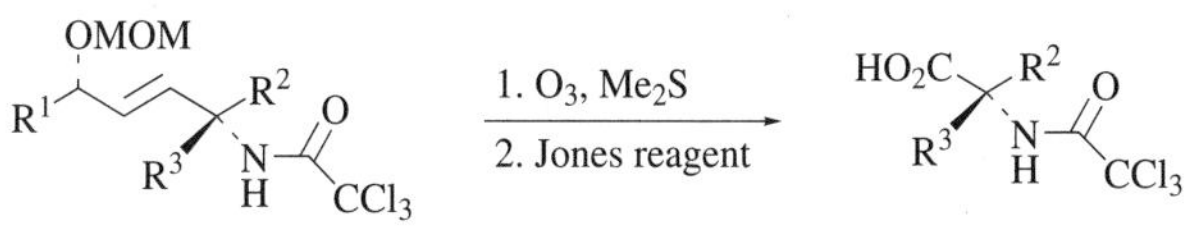

R^1	R^2	R^3	
i-Pr	Me	*n*-Bu	(85%)
i-Pr	Me	Et	(81%)
i-Pr	Ph	Me	(70%)

(Eq. 92)

Another tactic couples direct construction of enantioenriched, chiral secondary allylic alcohols with oxygen-to-nitrogen chirality transfer. In the example shown in Eq. 93, catalytic asymmetric vinylation of benzaldehyde provides **91**, which is transformed in 83% overall yield without loss of enantiomeric purity into trichloroacetyl-protected amino acid **92**.[114]

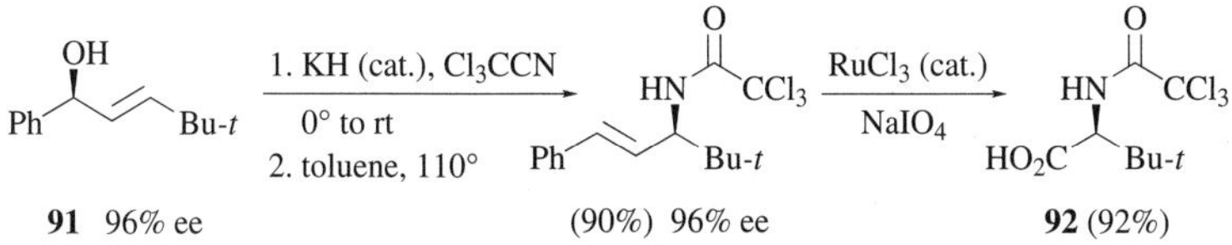

(Eq. 93)

A third appealing strategy exploits asymmetric catalysis to access directly the chiral, enantioenriched allylic trichloroacetamide. For example, the differentially protected S α-amino ester **94** is prepared without loss of enantiopurity from allylic trichloroacetamide **93** (Eq. 94), which in turn is prepared by a COP-Cl catalyzed allylic trichloroacetimidate rearrangement (see Eq. 78).[7]

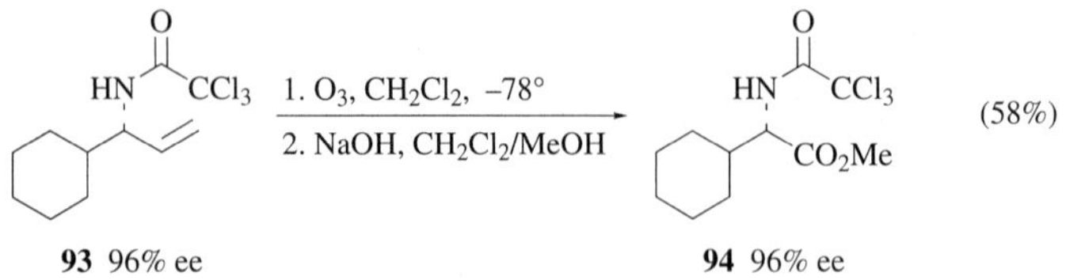

(Eq. 94)

Other Direct Uses of Allylic Trihaloacetamides

Because trichloromethyl is a competent leaving group, allylic trichloroacetamides can be directly converted to congeneric ureas, carbodiimides, and guanidines.[115] One example of this chemistry, which was developed by Isobe and co-workers during their studies of the total synthesis of tetrodotoxin and analogues, is illustrated in Eq. 95; see Eq. 22 for the preparation of the starting allylic trichloroacetamide **25**.

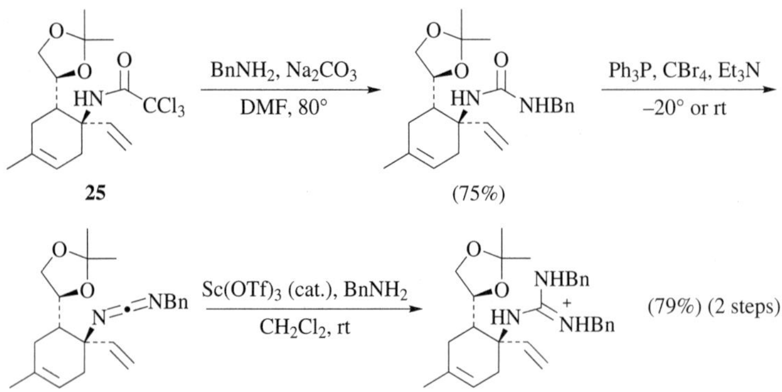

(Eq. 95)

A variety of heterocycles have been assembled from products of allylic trihaloacetimidate rearrangements, employing the double bond as a partner in ring-closing metatheses,[116] or involving the double bond in other C–C bond-forming ring constructions.[90,117–122] A clever strategy of this type which requires no additional manipulation of the allylic trichloroacetamide functionality has been used to prepare a series of γ-lactams.[72] In this case, ring construction is accomplished by an atom transfer radical cyclization (Eq. 96).

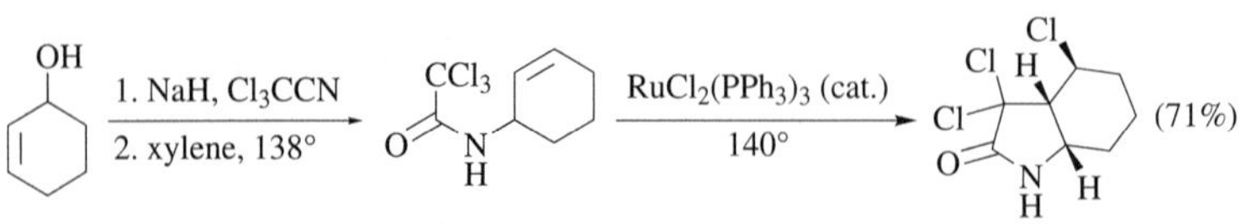

(Eq. 96)

The potential for functionalization of the double bond to be directed by the nearby trichloroacetamide group has also been exploited. The trichloroacetamide group is particularly effective in promoting syn dihydroxylation as a result of the strong propensity of the N-H bond to participate in hydrogen bonding.[93,123,124] This strategy has been employed in syntheses of several amino sugars as exemplified in the preparation of talosamine (Eq. 97).[125,126]

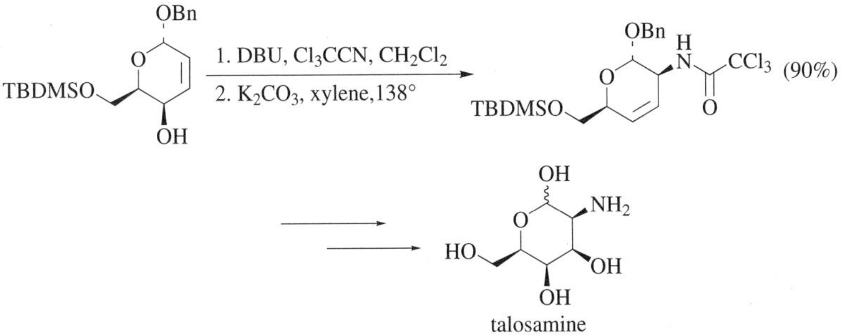

(Eq. 97)

However, the trichloroacetyl group proves to be a hindrance in attempted Sharpless oxyamination of allylic trichloroacetamide **95**, a finding attributed to the "suppressing influence of an electronegative substituent".[45] Conversion of the trichloroacetyl group to the acetate followed by oxyamination with chloramine T and osmium tetroxide yields the desired product **96** in 39% yield from the precursor allylic amide (Eq. 98).[45]

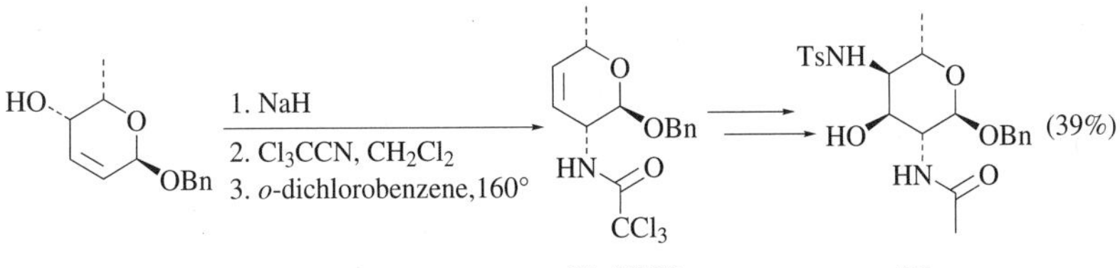

95 (68%) **96**

(Eq. 98)

Applications in the Total Synthesis of Natural Products

(±)-Acivicin. Syntheses of the antitumor, antimetabolites (±)-acivicin and (±)-bromoacivicin take advantage of the trichloroacetimidate rearrangement's residual allylic functionality to elaborate the heterocyclic acivicin framework through a 1,3-dipolar cycloaddition. The crucial building block in this synthesis is vinyl glycine, for which, at the time, no other efficient preparation was available. Thermal rearrangement of the trichloroacetimidate derivative **97** of *cis*-2-butene-1,4-diol is used to prepare the vinyl glycine fragment **98** (Eq. 99).[55,69]

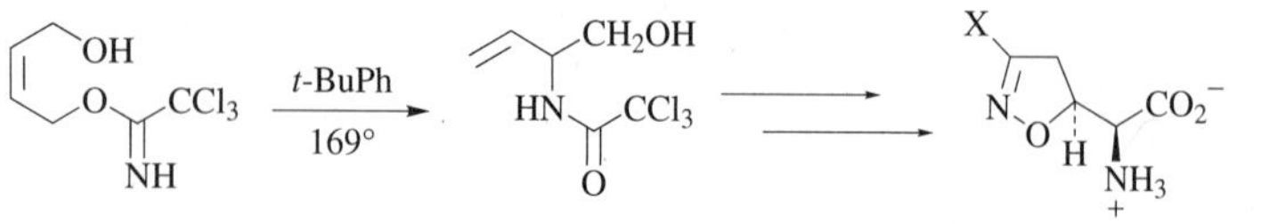

97 98 (84%) X = Cl acivicin X = Br bromoacivicin (Eq. 99)

(+)-Lactacystin. The application of the trichloroacetimidate rearrangement to the synthesis of (+)-lactacystin exemplifies the utility of this reaction for constructing tertiary carbinyl amine stereocenters. The starting material for the synthesis, D-glucose, provides the chiral template for the rearrangement, giving allylic trichloroacetamides **99** as a 4.8 : 1 mixture of diastereomers in 60% yield from the starting allylic alcohol (Eq. 100). After acidic cleavage of the acetonide, these epimers are separated and the major stereoisomer is processed in two steps to pyrrolidine imide **100**. Treatment of intermediate **100** with $NaBH_4$ in methanol at 0° removes both the *N*-trichloroacetyl and *O*-formyl groups to provide late stage lactacystin precursor **101**.[127] A similar strategy is employed to synthesize (–)-sphingofungin E.[47,128]

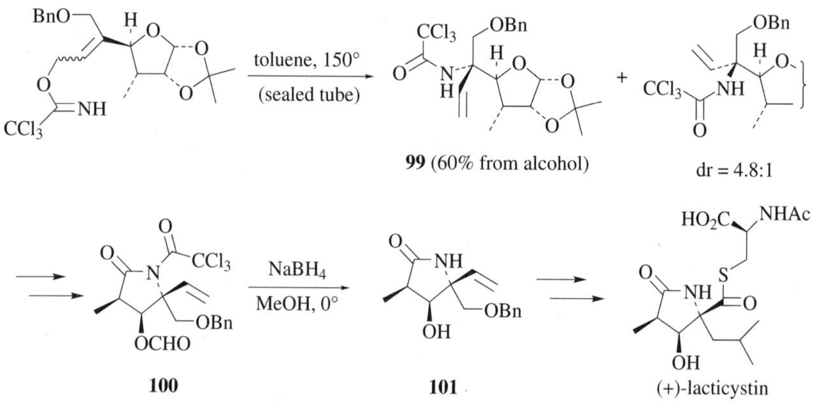

100 101 (Eq. 100)

(±)-Pancratistatin. The reliable suprafacial transfer of chirality of trichloroacetimidate rearrangements was exploited in the synthesis of (±)-pancratistatin. Although it was initially hoped that diol **102** would react with trichloroacetonitrile at the less-hindered allylic oxygen center, orthoamide **10**, formed as a mixture of diastereomers, was the only product produced. Unfortunately, this intermediate stubbornly refused to undergo thermal [3,3]-sigmatropic rearrangement (Eq. 101).[56]

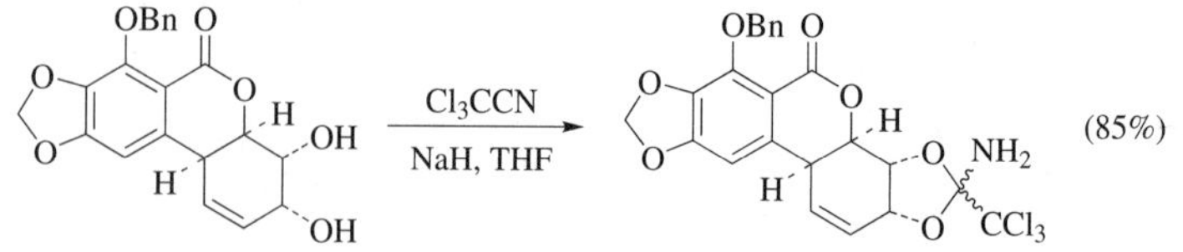

102 10 (Eq. 101)

However, benzyl-protected congener **103** is converted to the late-stage (±)-pancratistatin precursor **104** upon heating at 100–105° under high vacuum (Eq. 102).

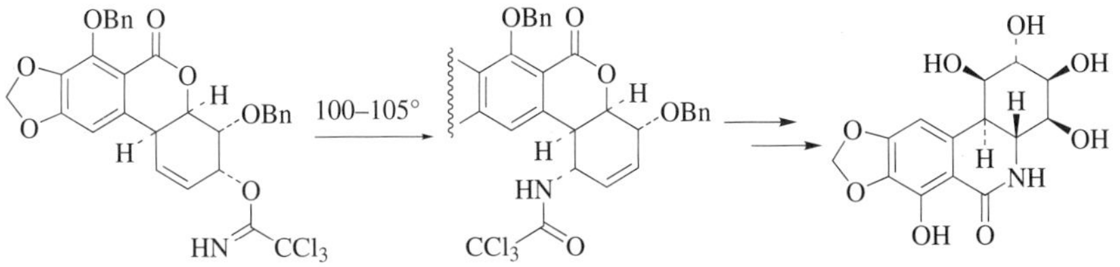

103 **104** (56%) (±)-pancratistatin

(Eq. 102)

COMPARISON WITH OTHER METHODS

The importance of allylic amines in synthesis has led to the development of many methods for their synthesis, several of which take advantage of allylic alcohols as starting materials. Although various attractive methods have been developed, few match the breadth, regiocontrol, and stereocontrol of the trihaloacetimidate rearrangement. The ease by which the trihaloacetyl group can be removed to release the free allylic amine further distinguishes the trihaloacetimidate rearrangement from related methods. A recent review summarizes some of these methods as well as other syntheses of allylic amines.[15] This section covers the more significant developments since that review, including other rearrangement methods, aminolysis of allylic electrophiles, allylic amination of alkenes, hydroamination of alkynes and dienes, and addition of vinyl nucleophiles to imine derivatives.

Other Allylic Rearrangements

In addition to the imidate group, several functional groups are suitable for preparation of allylic amines by [3,3]-sigmatropic rearrangement. Allylic urethanes,[129] isoureas,[130] cyanates,[131] thioimidates,[132,133] and sulfamates[134] were explored early on as potential candidates for the sigmatropic transformation of allylic alcohols into allylic amines.[10] More recently, rearrangements of allylic *N*-benzoyl benzimidates and allylic phosphorimidates have been studied.[101,135] Each of these methods exhibits some limitations, either in the preparation of the rearrangement precursors, conditions required for the rearrangement, scope of the rearrangement, or difficulties encountered in subsequent hydrolysis of the protected amine products. At this point, none of the aforementioned alternatives to the trihaloacetimidate rearrangement has been developed into a broadly useful method for preparing allylic amines.

Thioiminocarbonates have found some application in the synthesis of allylic nitrogen derivatives.[136] However, in a direct comparison, trichloroacetimidate rearrangement of **105** is found to be the preferred method for preparation of the amine **106** (Eq. 103). Rearrangement of thioiminocarbonate **107** gives rearrangement product **108** in low yield (Eq. 104), whereas heating **105** in refluxing xylene provides the desired amine in high yield after mild hydrolysis to remove the trichloroacetyl group.[137]

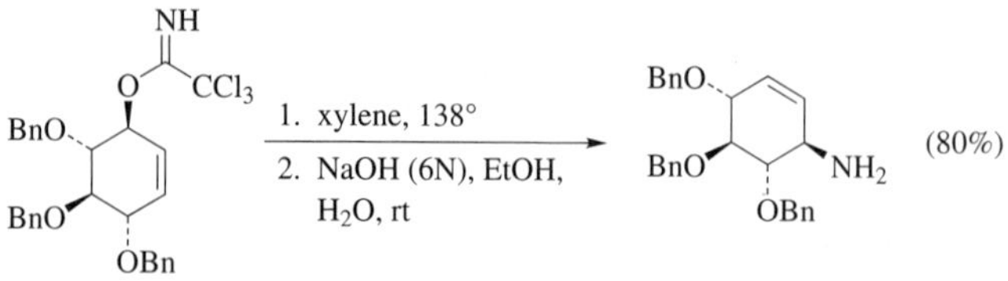

105 **106** (Eq. 103)

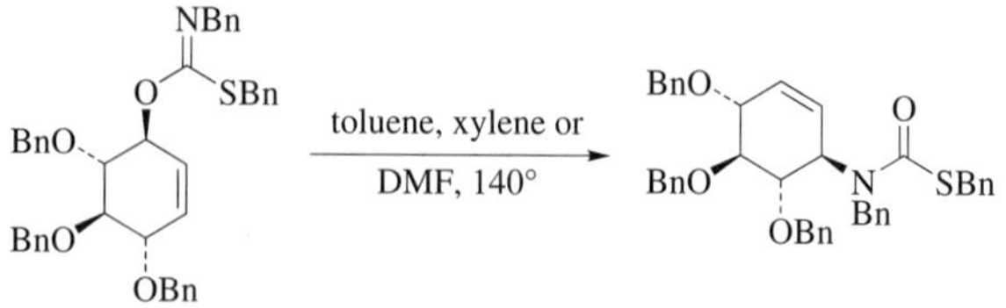

107 **108** (Eq. 104)

A method for converting an allylic alcohol to the corresponding amine with retention of configuration that involves two sequential [3,3]-sigmatropic rearrangement steps is used in the synthesis of an ansamycin.[138] In this study, thionocarbamate **109** is not isolated, but rearranges spontaneously upon its formation at room temperature to give the allylically-transposed thiocarbamate product in 90% yield from the starting allylic alcohol (Eq. 105). This intermediate is then transformed in two

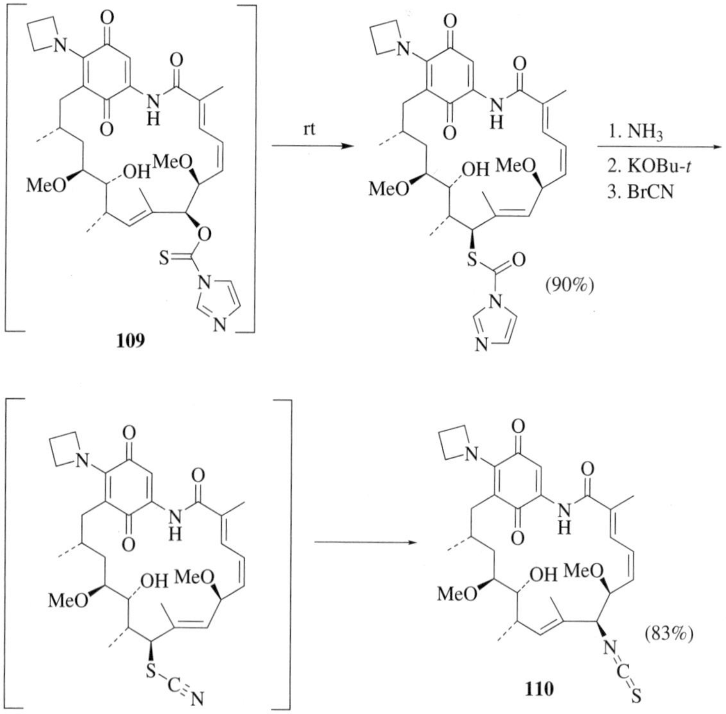

(Eq. 105)

steps to the allylic thiocyanate, which also rearranges spontaneously providing allylic isothiocyanate **110** in 83% yield.

Allylic amine derivatives are also available from allylic alcohols by [2,3]-sigmatropic rearrangements of allylic selenimides[139,140] and allylic selenonium ylides.[141] These promising methods also exhibit the potential for asymmetric induction. For example, rearrangement of chiral, enantioenriched allylic selinimide **111** takes place in situ to yield the allylic carbamate (Eq. 106). Enantiomeric excesses of up to 93% are achieved in some related reactions when the substituent on nitrogen is a *para*-toluenesulfonyl group.[140]

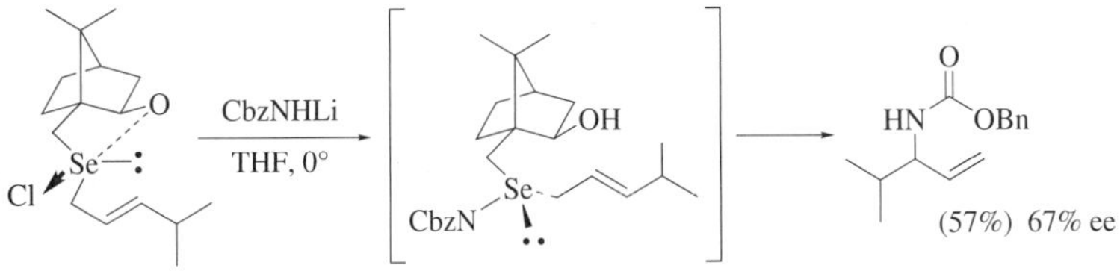

100% ee **111** (Eq. 106)

Allylic sulfoximines undergo either thermal or Pd(II)-catalyzed rearrangement to yield *N*-protected allylic amines. Although the thermal process gives a mixture of isomers, presumably by a dissociation-recombination mechanism,[142] Pd(II)-catalyzed rearrangement of allylic sulfoximine **112** takes place readily with transposition of the double bond to give rearrangement product **113** in 80% overall yield (Eq. 107).[143,144]

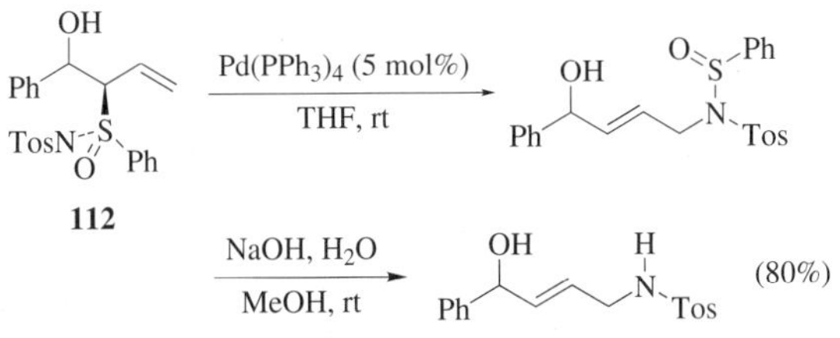

112 **113** (Eq. 107)

Other Routes to Allylic Amines

Amination of Allylic Electrophiles. The most straightforward preparation of allylic amines is unquestionably direct allylation of nitrogen with an allylic electrophile. However, this approach is often compromised by the propensity for di- and triallylation. This problem can be circumvented by delivery of the nitrogen atom in protected form as in the Gabriel synthesis.

An area of intense recent study is the transition-metal-catalyzed reaction of amines with allylic electrophiles.[15] A variety of allyl acetates, carbonates, halides, and even allylic alcohols are converted to allylic amines in this way, with the most widely used catalysts to date being complexes of palladium. When a π-allyl palladium intermediate is unsubstituted at one end, the amine generally attacks at this terminus to give a trans unbranched allylic amine. Recently, complexes of rhodium,[145–147] ruthenium,[148,149] iridium,[150–153] and palladium[154] have been shown to catalyze allylic aminations favoring reaction at the more hindered allyl terminus to give the branched allylic amine product (Eq. 108).[147] However, regioselectivity is typically diminished when there is branching α to the leaving group. These discoveries open up the opportunity to accomplish such aminations in enantioselective fashion,[150,153–156] providing a highly attractive route to enantioenriched allylic amines (Eq. 109).[150,155,156]

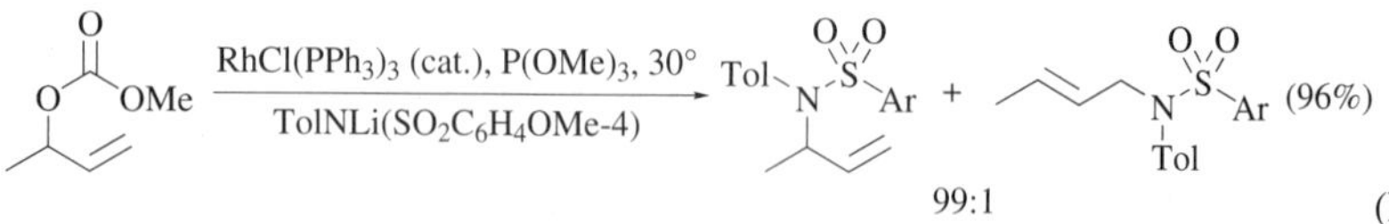

(Eq. 108)

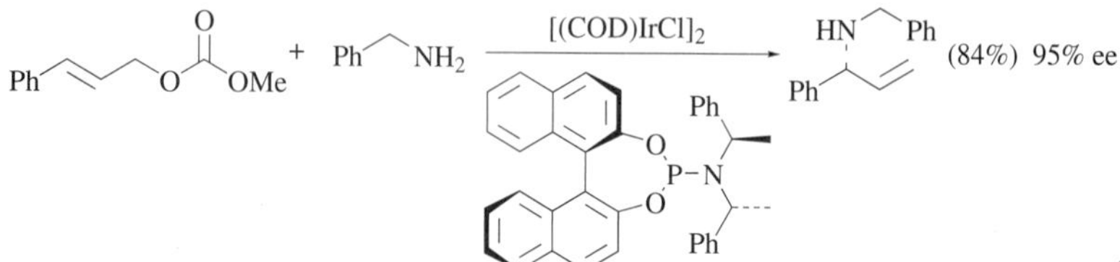

(Eq. 109)

C–H Activation. Direct substitution of an allylic C–H bond of an alkene by nitrogen is a very attractive route to allylic amines. This chemistry is generally associated with ene reactions of N=N species[157] or nitrene addition followed by rearrangement to the allylic amine.[158] Several approaches of these types have met with success, although yields are generally moderate and these methods at present are limited in scope. Thus, the ene reaction of diethyl-azodicarboxylate with alkenes provides good yields (77–95%) of allylic amine adducts (Eq. 110).[157] Efforts to introduce asymmetry into the reaction using di-(–)-menthyl diazodicarboxylate meet with moderate success; however, removal of the menthol chiral auxiliary is difficult.[159] Iron[160,161] and manganese[158] complexes have also been used to successfully introduce nitrogen at the allylic position of alkenes, although yields are modest.

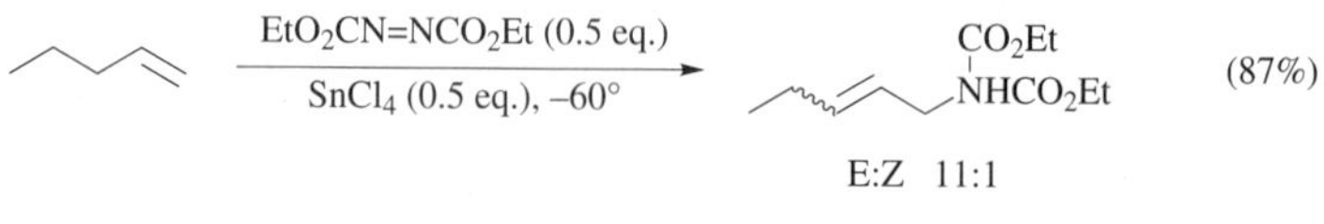

(Eq. 110)

Hydroamination. The application of transition metal catalysis to the synthesis of allylic amines by hydroamination of dienes[162] and alkynes (Eq. 111)[163] holds considerable promise. At this point, the scope of this approach is largely unexplored. Mixtures of regioisomeric products are not uncommon in the hydroamination of dienes, including variable trans/cis ratios of the allylic amine product, depending upon the reaction conditions.

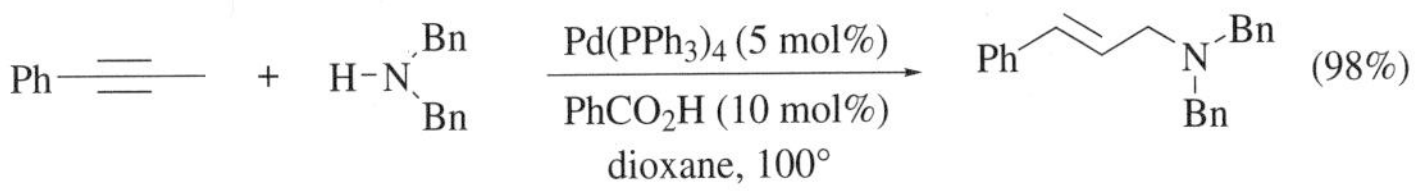

(Eq. 111)

Addition of Vinyl Nucleophiles to Imine Derivatives. Much progress has been recorded in recent years in the synthesis of amines by the addition of organometallic reagents to imines and their derivatives, with several recent reviews summarizing this progress.[164,165] When the nucleophile is vinylic, an allylic amine or a derivative results. An example, illustrating the synthesis of enantioenriched allylic amines, is shown in Eq. 112.[166]

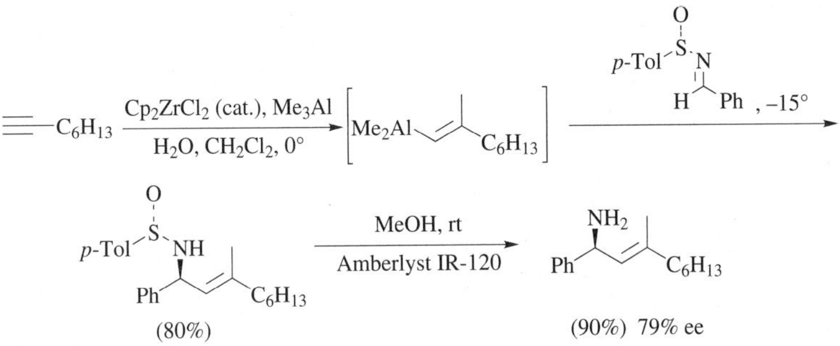

(Eq. 112)

EXPERIMENTAL CONDITIONS

General Comments

An attractive aspect of the Overman rearrangement is its experimental simplicity. Typically the intermediate allylic trichloroacetimidate is not purified, but rearranged directly under either thermal or metal-catalyzed conditions. Despite the progress that has been made in the area of metal catalysis, the thermal rearrangement of allylic trihaloacetimidates is so reliable and convenient that most applications employ these conditions. A wide variety of non-nucleophilic solvents can be used in the imidate preparation and rearrangement steps. Dry solvents (commercial quality) are typically acceptable, and should be employed to minimize hydrolysis of the trichloroacetimidate intermediate.

Preparation of Allylic Trichloracetimidates

Bases promote the addition of alcohols to trichloroacetonitrile. Two general conditions are commonly employed for preparing allylic trichloroacetimidates: catalytic or stoichiometric amounts of 1,8-diazobicyclo[5.4.0]undec-7-ene (DBU),[91,108] or catalytic quantities of a metal alkoxide that is generated in situ by reaction of the allylic alcohol with 5–20 mol% of NaH, KH, sodium metal, or *n*-BuLi. Sodium hydride is employed most commonly in the latter procedure with diethyl ether or

THF being used as the solvent and deprotonation being carried out at temperatures from 0° to −15°. If the alkoxide is generated, it is important to use only catalytic quantities: although the equilibrium constant for generating trichloroacetimidates from the addition of alcohols to trichloroacetonitrile is high, the corresponding equilibrium of an alkoxide and trichloroacetonitrile to form the trichloroacetimidate conjugate base is much less favorable. Dichloromethane has commonly been the solvent when DBU is used as a base, although other dry aprotic solvents could likely be utilized. When DBU is used, it is employed catalytically (10 mol%) or used in excess.

In the alkoxide-catalyzed procedure, the order of addition of the reagents is variable. For secondary and tertiary alcohols, the preferred order of addition is to slowly transfer the alcohol/alkoxide solution by cannula into a solution of trichloroacetonitrile,[6] although the reverse order of addition has been employed without adverse effects.[137,167] The formation of trichloroacetimidates from primary alcohols is best carried out by adding trichloroacetonitrile to the alcohol/alkoxide mixture.

In the DBU-promoted process, the order of addition of the base does not play a critical role: DBU can be pre-mixed with the alcohol,[26,49] added to the alcohol as a mixture with the trichloroacetonitrile,[47,108] or added sequentially with trichloroacetonitrile (first DBU, followed by trichloroacetonitrile). Use of DBU allows unhindered allylic trichloroacetimidates to be formed at low temperatures (down to −78°), although temperatures around −35° to 0° are most common.

Allylic trichloroacetimidates are reasonably stable to standard purification methods, such as either vacuum distillation or silica gel chromatography. Nonetheless, they are almost always used with minimal purification. For example, after extractive workup and quick passage through a plug of silica gel, the solvent is removed and the crude imidate is redissolved in the rearrangement solvent for heating to reflux. Allylic trichloroacetimidates often undergo partial hydrolysis during slow chromatography on silica gel; this is not observed if Davisil grade silica gel (W.R. Grace & Co.) is used.

Preparation of Allylic Trifluoroacetimidates

The preparation of allylic trifluoroacetimidates first requires generation of trifluoroacetonitrile, a colorless gas (bp = −64°). In early studies, trifluoroacetonitrile was generated by the vigorous dehydration of trifluoroacetamide using phosphoric pentoxide.[75] After passing gaseous trifluoroacetonitrile through a sequence of traps and scrubbers, an excess of trifluoroacetonitrile is condensed to prepare an ethereal solution for subsequent addition to a solution of the alcohol/alkoxide at −78°. The alkoxide is typically generated by the action of *n*-BuLi (20 mol%) on the alcohol in THF at 0° or below. A more convenient "one-pot" procedure for preparing allylic trifluoroacetimidates generates trifluoroacetonitrile in situ from trifluoroacetamide by reaction with oxalyl chloride and DMSO.[168]

Thermal Rearrangements of Allylic Trihaloacetimidates

As discussed in the Scope and Limitations section, the thermal rearrangement is somewhat faster in solvents of higher polarity. Nevertheless, high-boiling hydrocarbon solvents such as toluene and xylenes have been employed most widely. The crude allylic trichloroacetimidate typically is dissolved in the solvent of choice and the rearrangement carried out at reflux. Some substrates require higher temperatures and these rearrangements can be carried out in a sealed tube. Alternatively, *ortho*-dichlorobenzene (bp 180°), decalin (bp 190°), *tert*-butylbenzene (bp 169°), or diphenyl ether (bp 259°) can be used to bring about the rearrangement. The concentration at which the thermal rearrangement is conducted appears to be of little importance: most rearrangements are run at about 0.1 M, although some have even been effected at high concentration or neat.

Unwanted decomposition of the allylic trichloroacetimidate by ionization to form trichloroacetamide and the corresponding allylic cation (and eventually diene byproducts) can be problematic when ionization of the allylic C–O σ-bond produces a particularly stable allylic cation, for example with tertiary allylic trichloroacetimidates. This side reaction is likely promoted by acidic impurities. A significant improvement to the rearrangement procedure is the inclusion of powdered anhydrous K_2CO_3, which scavenges acids and prevents the decomposition of the imidate.[63] The amount of K_2CO_3 added varies from a small amount (20 mol% relative to the crude allylic imidate) to a slight excess. Improvements in yield of up to 50% are reported;[63] in some cases, previously nonviable rearrangements are made possible.

Rearrangements of allylic trifluoroacetimidates are carried out under similar conditions. The time required for the rearrangement to proceed to completion may be reduced and/or the yield improved when trifluoroacetimidates, rather than their trichloro congeners, are employed.[75,95]

Metal-Catalyzed Rearrangements of Allylic Trichloroacetimidates

After the initial report that mercuric trifluoroacetate catalyzes the rearrangement of allylic trichloroacetimidates,[4] the catalysts that have been most successfully applied are Pd(II) complexes. The commercially available, or readily prepared,[169] bis(acetonitrile) and bis(benzonitrile) complexes of $PdCl_2$ are employed widely. These complexes are used at 4–10 mol% to effect the rearrangement of primary and secondary allylic trichloroacetimidates; reports of successful Pd(II)-catalyzed rearrangements of secondary imidates are rare. Tetrahydrofuran, toluene, and dichloromethane have been used as solvents, with rearrangements taking place at room temperature in a matter of hours.[27,28,30,62] Catalytic asymmetric rearrangements employing the catalyst COP-Cl (**8**) generally are carried out with catalyst loadings of 1–5 mol% in CH_2Cl_2.[7] The recently introduced COP-hexafluoroacetylacetonate complex **78** is more soluble allowing a wider variety of solvents to be employed and the reactions to be conducted at high substrate concentration (up to 2.6 M).[9] Catalytic asymmetric rearrangements of allylic trichloroacetimidates with the COP catalysts can be conducted at temperatures up to 60° without loss of enantioselectivity.

EXPERIMENTAL PROCEDURES

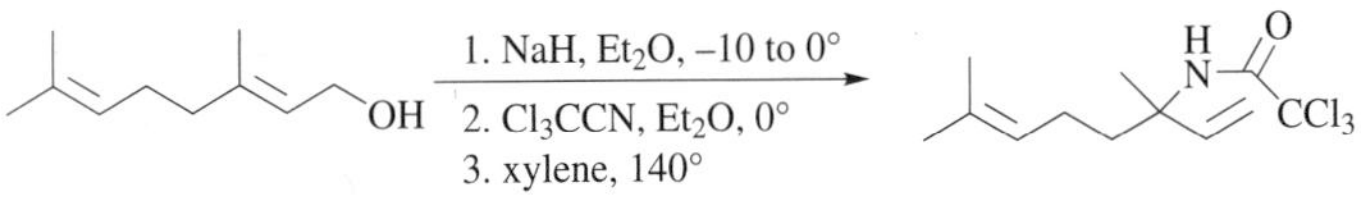

2,2,2-Trichloro-*N*-(3,7-dimethylocta-1,6-dien-3-yl)acetamide (Alkoxide-Catalyzed Procedure for Preparing Allylic Trichloroacetimidates and Thermal Rearrangement of the Crude Imidate Intermediate).[43] This preparation is described in *Organic Synthesis*.[43]

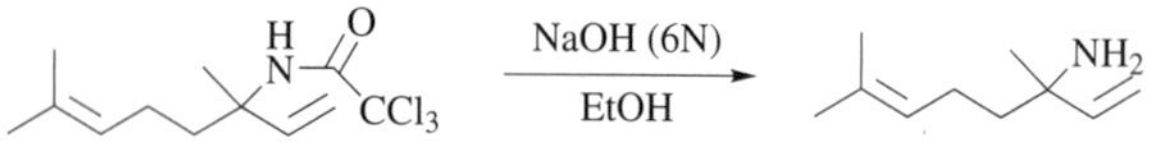

3,7-Dimethylocta-1,6-dien-3-amine (Basic Hydrolysis of an Allylic Trichloroacetamide to Form the Allylic Amine).[43] This preparation is described in *Organic Synthesis*.[43]

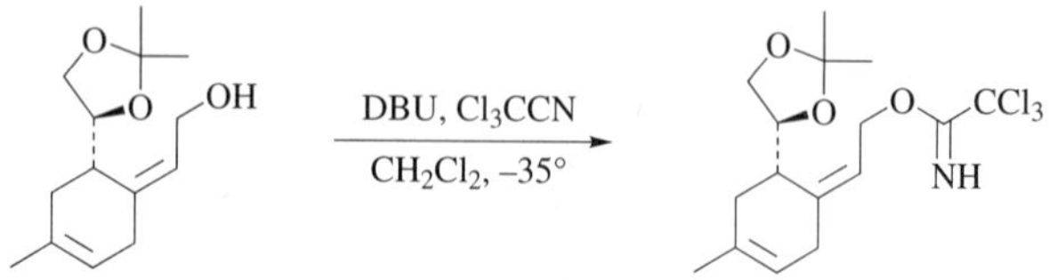

(*Z*)-2-[(6*S*)-6-(2,2-Dimethyl-1,3-dioxolan-4-yl)-4-methylcyclohex-3-enylidene]ethyl 2,2,2-Trichloroacetimidate (Preparation of a Trichloroacetimidate using DBU).[49] To a cooled (−35°) solution of the allylic alcohol (18.5 g, 77.7 mmol) in dry CH_2Cl_2 (370 mL) was added DBU (13.9 mL, 93.2 mmol) followed by the dropwise addition of Cl_3CCN (9.35 mL, 93.2 mmol) over a period of 10 minutes. The reaction mixture was stirred at −35° for 1 hour, then was quenched with NH_4Cl (saturated aqueous, 300 mL) and extracted with CH_2Cl_2. The combined organic extracts were washed with saturated NH_4Cl solution, dried (Na_2SO_4) and concentrated. The residue was dissolved in Et_2O and passed through a short column packed with anhydrous Na_2SO_4 and silica gel 60. The Et_2O was evaporated to yield the imidate as a light yellow oil: IR (KBr) 3345, 2983, 2931, 1661, 1455, 1370, 1289, 1072 cm^{-1}; ^1H NMR (300 MHz, $CDCl_3$) δ 1.33 (s, 3H), 1.41 (s, 3H), 1.64 (br s, 3H), 1.72 (br d, $J = 18$ Hz, 1H), 2.33 (br d, $J = 18$ Hz, 1H), 2.67 (br d, $J = 19$ Hz, 1H), 2.94–3.10 (m, 2H), 3.71 (dd, $J = 8, 6.5$ Hz, 1H), 4.10 (dd, $J = 8, 6$ Hz, 1H), 4.23 (dt, $J = 10, 6.5$ Hz, 1H), 4.77 (ddd, $J = 12, 6, 2$ Hz, 1H), 5.00 (ddd, $J = 12, 8, 1$ Hz, 1H), 5.38 (br s, 1H), 5.71 (ddd, $J = 8, 6, 2$ Hz, 1H), 8.25 (br s, 1H); ^{13}C NMR (100 MHz, $CDCl_3$) δ 23.2, 25.3, 26.7, 32.3, 33.3, 39.8, 65.3, 68.6, 75.7, 91.5, 109.0, 118.2, 120.2, 131.1, 142.5, 162.7.

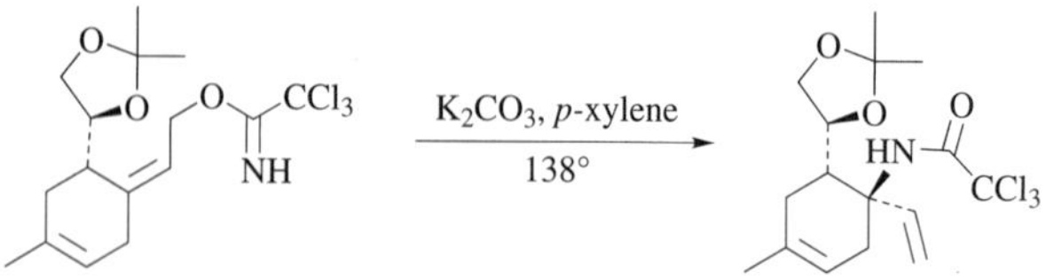

2,2,2-Trichloro-*N*-[(1*R*,6*S*)-6-(2,2-dimethyl-1,3-dioxolan-4-yl)-4-methyl-1-vinylcyclohex-3-enyl]acetamide (Thermal Rearrangement of a Trichloroacetimidate in the Presence of K_2CO_3).[49] Powdered K_2CO_3 (1.2 g) was added to a solution of the crude imidate (prepared as described in the previous procedure) in *para*-xylene (600 mL) and the mixture was heated at reflux with vigorous stirring for 20 hours. After cooling to room temperature, the mixture was filtered through a pad of Super-Cel and the precipitate was washed with toluene. The combined filtrates were concentrated and the residue was purified by column chromatography (SiO_2, 550 g, ether in hexane, 1:10 to 1:5) to yield the amide as colorless crystals (27.4 g, 92% from the alcohol): mp 100–102°; $[\alpha]_D^{27} = +70.2°$ (*c* 0.97, $CHCl_3$); IR (KBr) 3313, 2987, 2924, 1727, 1542, 1261, 1067 cm^{-1}; ^1H NMR (400 MHz, $CDCl_3$) δ 1.39 (s, 3H), 1.42 (s, 3H), 1.64–1.71 (m, 5H), 2.08 (td, $J = 9, 7.5$ Hz, 1H), 2.27 (d quintet, $J = 17.5, 2.5$ Hz, 1H), 3.37 (ddq, $J = 17.5, 6, 1.5$ Hz, 1H), 3.63 (dd, $J = 9, 7.5$ Hz, 1H), 4.03 (td, $J = 9, 5.5$ Hz, 1H), 4.10 (dd, $J = 7.5, 5.5$ Hz, 1H), 5.30 (dd, $J = 17, 1$ Hz, 1H), 5.32 (dd, $J = 11, 1$ Hz, 1H), 5.39 (m, 1H), 5.82 (dd, $J = 17, 11$ Hz, 1H), 9.21 (br s, 1H); ^{13}C NMR (75 MHz, $CDCl_3$) δ 22.5, 26.2, 26.5, 30.0, 35.8, 44.3, 59.9, 68.8, 76.3, 93.8, 109.9, 116.0, 119.0, 130.8, 133.7, 160.4; HRMS (FAB) calcd for $C_{16}H_{22}NO_3Cl_3$ (M+H), 382.0743, found 382.0721.

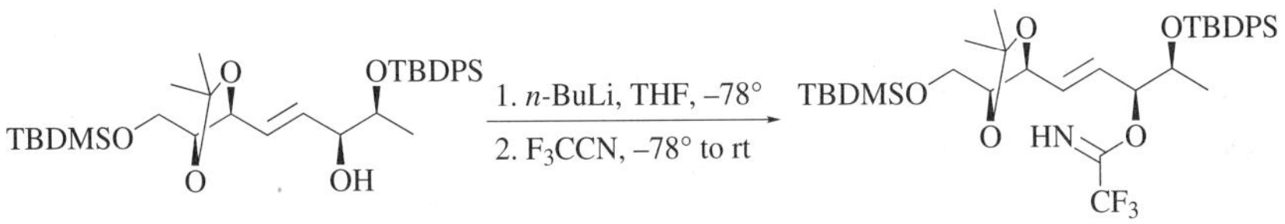

(4*S*,5*S*)-4-(*tert*-Butyldimethylsiloxymethyl)-5-[(3*S*,4*S*,1*E*)-4-*tert*-butyldiphenylsiloxy-3-(2,2,2-trifluoroacetimidoyloxy)pentenyl]-2,2-dimethyl-1,3-dioxolane (Preparation of an Allylic Trifluoroacetimidate).[75] To a solution of the alcohol (0.25 g, 0.43 mmol) in THF (6 mL) at −78° was added *n*-butyllithium (1.6M in hexane; 0.29 mL, 0.46 mmol). The resulting solution was stirred for 1 hour whereupon a stream of trifluoroacetonitrile was allowed to bubble through the reaction mixture for five minutes. [The trifluoroacetonitrile was prepared by mixing powdered trifluoroacetamide (10 g, 88 mmol) with phosphorus pentoxide (24 g, 148 mmol) in a round-bottomed flask fitted with a nitrogen inlet and condenser. A polytetrafluoroethylene tube fitted to the condenser led to a trap cooled in an ice–salt mixture, then to a trap cooled in an ether–N_2 (liquid) mixture, and finally to a bath containing aqueous sodium hydroxide via a tube packed with calcium chloride. The reaction mixture was slowly heated to 150° and held at this temperature for 3 hours under a gentle stream of nitrogen gas. The trifluoroacetonitrile distilled out and was collected as a colorless liquid in the −100° ether/N_2 (liquid) trap.] The reaction was allowed to warm to room temperature over a period of 1 hour, then NH_4Cl (0.2 g, 3.6 mmol) was added and the reaction mixture was concentrated under reduced pressure. The product was purified by column chromatography (silica gel 60; petroleum ether in ether, 9:1): 81%; $[\alpha]_D^{23} = -5.7°$ (*c* 0.5, $CHCl_3$); IR ($CHCl_3$) 3360, 1680, 1470, 1430, 1380, 1200, 1170, 1112, 840 cm^{-1}; ^1H NMR ($CDCl_3$) δ 0.1 (s, 6H), 0.9 (s, 9H), 1.0 (d, $J = 7$ Hz, 3H), 1.1 (s, 9H), 1.45 (s, 6H), 3.74 (m, 3H), 4.2 (quintet, $J = 6$ Hz, 1H), 4.44 (m, 1H), 5.5 (t, $J = 6$ Hz, 1H), 5.85 (dd, $J = 5, 6$ Hz, 1H), 5.9 (dd, $J = 16, 5$ Hz, 1H), 7.3–7.8 (m, 10H), 8.2 (s, 1H).

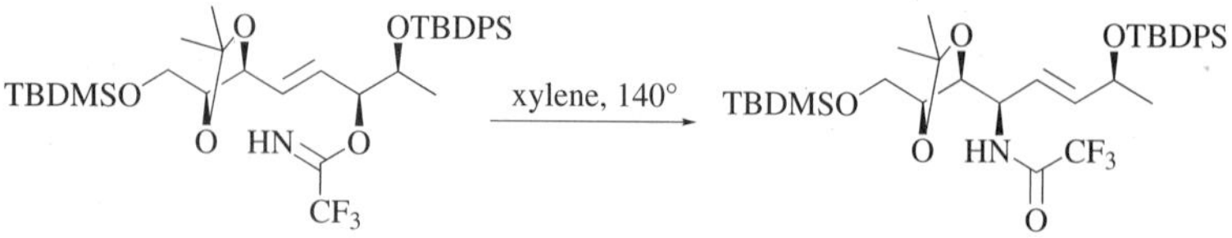

(4*S*,5*S*)-4-(*tert*-Butyldimethylsiloxymethyl)-5-[(1*R*,4*S*,2*E*)-4-*tert*-butyldiphenylsiloxy-1-(2,2,2-trifluoroacetylamino)pent-2-enyl]-2,2-dimethyl-1,3-dioxolane (Thermal Rearrangement of an Allylic Trifluoroacetimidate).[75] A solution of a portion of the trifluoroacetimidate prepared in the previous procedure (90 mg, 0.13 mmol) in xylene (2 mL) was degassed with argon and heated at reflux for 20 hours. The reaction mixture was then concentrated under reduced pressure and the residue was purified by column chromatography (silica gel 60, petroleum ether in ether, 15:1) to give the amide (74 mg, 82%): $[\alpha]_D^{23} = -50.6°$ (*c* 1.1, $CHCl_3$); IR ($CHCl_3$) 3420, 1730, 1530, 1480, 1470, 1430, 1370, 1170, 1110, 1080, 840 cm^{-1}; ^1H NMR ($CDCl_3$) δ 0.1 (s, 6H), 0.9 (s, 9H), 1.1 (s, 9H), 1.2 (d, $J = 6$ Hz, 3H), 1.4 (s, 6H), 3.7 (m, 2H), 3.85 (m, 1H), 4.0 (dd, $J = 8$, 2 Hz, 1H), 4.3 (quintet, $J = 6$ Hz, 1H), 4.67 (t, $J = 7.5$ Hz, 1H), 5.5 (dd, $J = 16$, 7 Hz, 1H), 5.75 (dd, $J = 16$, 6 Hz, 1H), 6.85 (d, $J = 9$ Hz, 1H), 7.3–7.8 (m, 10H). Anal. Calcd for $C_{35}H_{52}F_3NO_5Si_2$: C, 61.8; H, 7.7. Found: C, 61.5; H, 8.05.

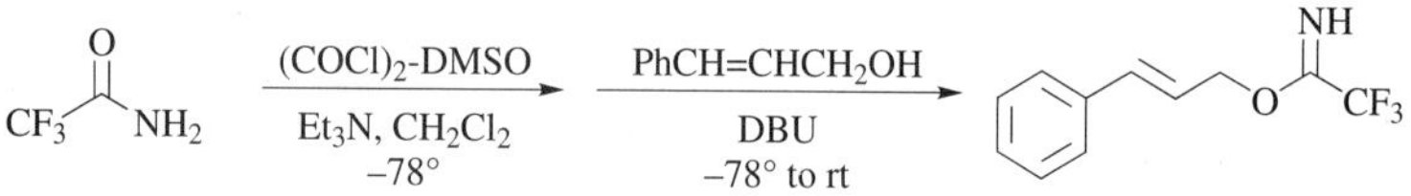

Cinnamyl 2,2,2-Trifluoroacetimidate ("One-Pot" Procedure for the Preparation of an Allylic Trifluoroacetimidate).[62] Into a flame-dried three-necked round-bottomed flask was placed 2,2,2-trifluoroacetamide (734 mg, 3 mmol), DMSO (1.36 mL, 8.7 mmol), and CH_2Cl_2 (30 mL). The mixture was cooled to $-75°$ whereupon oxalyl chloride (0.51 mL, 5.9 mmol) and triethylamine (2.5 mL, 18 mmol) were slowly added. The mixture was stirred at $-78°$ for 30 minutes, then DBU (0.6 mL, 4 mmol) and cinnamyl alcohol (296 mg, 2.2 mmol) were added slowly by syringe. The reaction mixture was allowed to warm to room temperature over 10 hours, then the reaction was quenched by the addition of water. The aqueous layer was extracted (EtOAc), the combined organic layers washed with brine, dried (Na_2SO_4) and filtered. Purification by column chromatography (SiO_2, hexane in ethyl acetate, 10:1) and then Kugelrohr distillation afforded the product as a colorless oil (384 mg, 76%): bp 100–105°/0.9 mm; IR (neat) 3347, 3087, 3063, 3031, 2950, 2887, 1686, 1356, 1202, 1167, 1117, 1076, 967, 847, 747, 735, 693 cm^{-1}; ^1H NMR (400 MHz, $CDCl_3$) δ 4.94 (d, $J = 6.3$ Hz, 2H), 6.37 (td, $J = 6.3$, 16.1 Hz, 1H), 6.73 (d, $J = 16.1$ Hz, 1H), 7.26–7.29 (m, 1H), 7.32–7.36 (m, 2H), 7.40–7.43 (m, 2H), 8.23 (s, 1H); ^{13}C NMR (100 MHz, $CDCl_3$) δ 68.2, 115.6 (q, $J = 280$ Hz), 121.9, 126.7, 128.4, 128.6, 135.0, 136.0, 157.8 (q, $J = 38.0$ Hz); EI-HRMS: [M^+] calcd for $C_{11}H_{10}F_3NO$, 229.0714; found 229.0689.

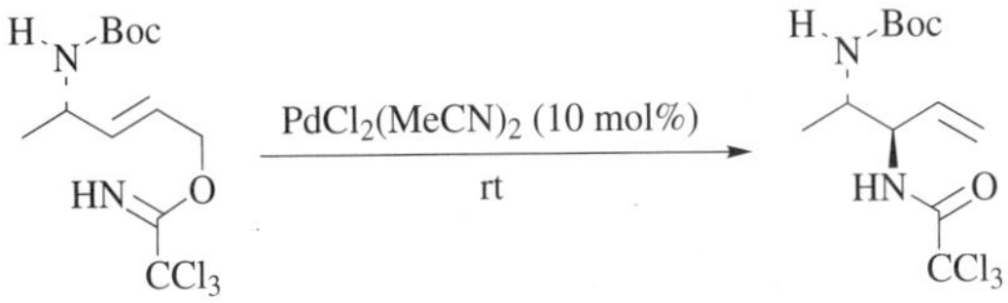

(3*R*,4*S*)-4-*tert*-Butoxycarbonylamino-3-(trichloroacetylamino)-1-pentene (Pd(II)-Catalyzed Rearrangement of an Allylic Trichloroacetimidate).[30] To a solution of the crude allylic trichloroacetimidate, prepared from 4.25 g (21 mmol) of the corresponding allylic alcohol, in THF was added $PdCl_2(MeCN)_2$ (552 mg, 2.13 mmol) under nitrogen. The reaction mixture was stirred at room temperature for 3 hours, whereupon the solvent was removed and the product (3.48 g, 48% from the alcohol) was isolated by column chromatography (silica gel 60, 4 : 1 toluene–ethyl acetate): ^{1}H NMR ($CDCl_3$) δ 1.21 (d, J = 7.0 Hz, 3H), 1.44 (s, 9H), 4.02 (m, 1H), 4.28 (m, 1H), 4.49 (d, J = 7 Hz, 1H), 5.34 (m, 2H), 5.73 (m, 1H), 8.73 (s, 1H); ^{13}C NMR ($CDCl_3$) δ 18.6, 28.7, 49.7, 60.2, 80.9, 93.3, 119.6, 131.7, 157.5, 161.9.

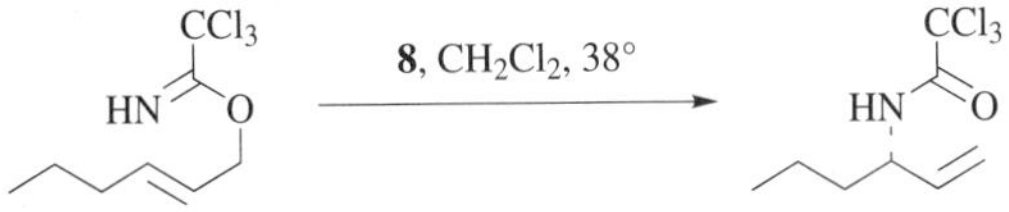

(*S*)-2,2,2-Trichloro-*N*-(1-propylallyl)acetamide (Catalytic Asymmetric Rearrangement of an Allylic Trichloroacetimidate).[170] A round-bottomed flask fitted with a stirring bar was charged with (*E*)-2-hexenyl trichloroacetimidate (6.81 g, 28 mmol, prepared from (*E*)-2-hexenol in 99% yield using the DBU procedure and purified by filtration through a short column of Davisil-grade silica gel using 2% ethyl acetate–hexanes as eluent), (*S*)-COP-Cl (816 mg, 0.56 mmol), and dry methylene chloride (9.3 mL). The flask was sealed with a polyethylene stopper, the stopper secured to the flask with Parafilm, and placed in an oil bath preheated to 38°. After 24 hours, the solution was cooled to room temperature and concentrated using a rotary evaporator to yield a brown oil. This oil was purified by flash chromatography (Davisil-grade silica gel, 0.5% ethyl acetate:hexanes) to give after concentration 6.50 g (95%, 94% ee) of the S allylic trichloroacetamide product as a pale yellow oil: IR (neat) 3329, 2966, 2873, 1699, 1522, 1460, 1251 cm^{-1}; ^{1}H NMR ($CDCl_3$) δ 0.96 (t, J = 7.4 Hz, 3H), 1.35–1.47 (m, 2H), 1.55–1.69 (m, 2H), 4.39–4.47 (m, 1H), 5.19 (d, J = 10.5 Hz, 1H), 5.23 (d, J = 17.2 Hz, 1H), 5.80 (ddd, J = 17.0, 10.4, 5.6 Hz, 1H), 6.50 (broad s, 1H); ^{13}C NMR ($CDCl_3$) δ 13.7, 18.8, 36.6, 53.3, 92.8, 116.0, 136.7, 161.2; HRMS (CI-NH_3): (M–*n*-Pr) calcd for $C_5H_5Cl_3NO$, 199.9437; found 199.9436.

TABULAR SURVEY

The tabular survey in this chapter covers allylic trihaloacetimidate rearrangements reported from 1974 through April, 2004. [3,3]-Sigmatropic rearrangements of other imidates are not included, nor are the [3,3]-sigmatropic rearrangements of propargylic alcohols. The tables are organized by substrate structure (the starting allylic alcohol) and are arranged on the basis of increasing carbon count of the alcohol, exclusive of protecting groups.[4] Secondary alcohols are separated into acyclic (Table 2A) and cyclic (Table 2B) substrates. Both thermal and metal-catalyzed conditions are included in the individual tables with the exception of Table 4, which presents examples of metal-catalyzed asymmetric trihaloacetimidate rearrangements.

For Tables 1–3 the yield presented is the overall yield for the two-step process (preparation of the imidate and subsequent rearrangement) unless otherwise noted. For Table 4 the yield is for the rearrangement step only. The highest yield is generally given in the case of multiple reports for the same rearrangement. A "(—)" entry indicates that no yield was reported.

The following abbreviations are used in the tables:

OAc	Acetate
Bn	Benzyl
Boc	*tert*-Butoxycarbonyl
BOM	Benzyloxymethyl
Bz	Benzoyl
Cbz	Carbobenzyloxy
DBPC	Di(*tert*-butyl)-*para*-cresol
DBU	1,8-Diazabicyclo[5.4.0]undec-7-ene
DMF	Dimethylformamide
DMSO	Dimethylsulfoxide
MOM	Methoxymethyl
MPM	*para*-Methoxybenzyl
NIS	*N*-Iodosuccinimide
NBS	*N*-Bromosuccinimide
TBDMS	*tert*-Butyldimethylsilyl
TBDPS	*tert*-Butyldiphenylsilyl
THF	Tetrahydrofuran
THP	Tetrahydropyran
Tr	Trityl (triphenylmethyl)
Ts	*para*-Toluenesulfonyl

[4] Generally, the methoxy group was not considered as a protecting group, hence anisole, for example, would be categorized as a C7 compound.

CATALYSTS USED IN TABLE 4

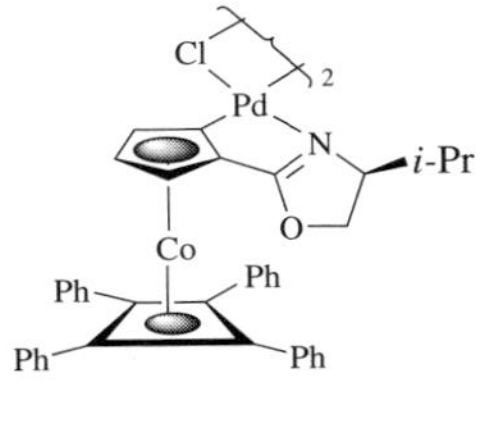

1

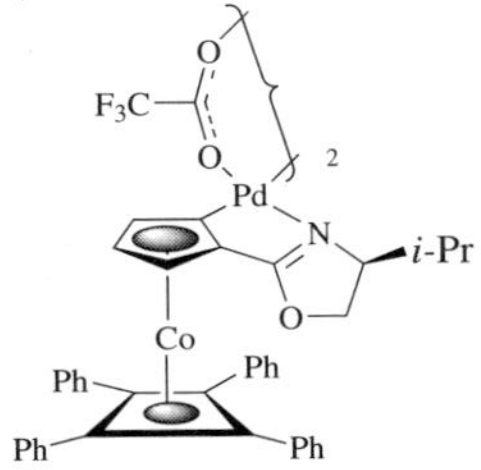

2

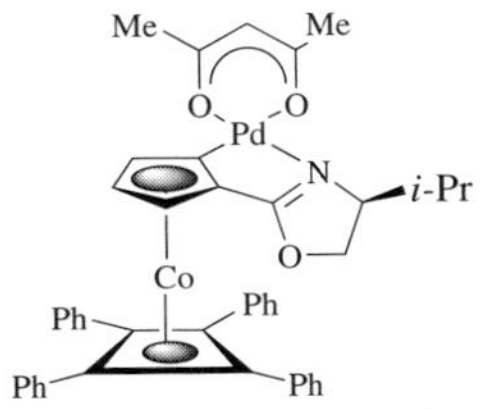

3

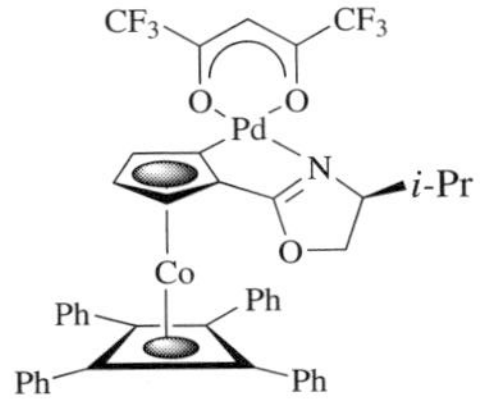

4

$[Pd(MeCN)_4](BF_4)_2$ /

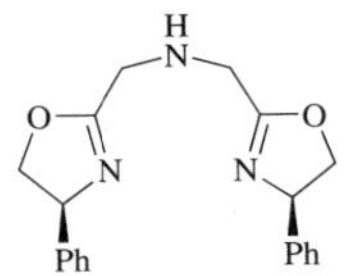

5

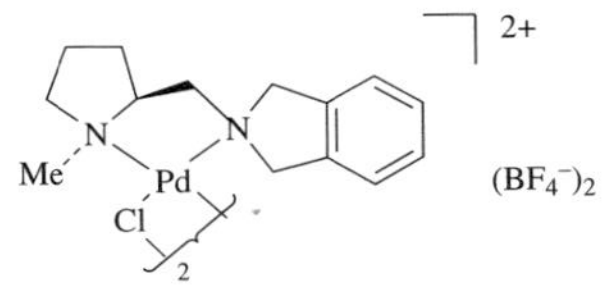

6

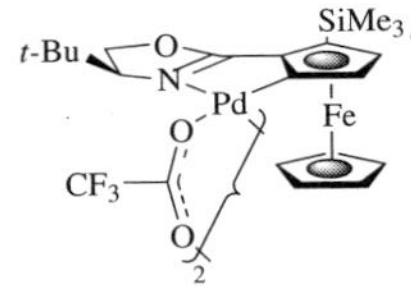

7

TABLE 1. REARRANGEMENTS OF TRIHALOACETIMIDATES OF PRIMARY ALLYLIC ALCOHOLS

Substrate	Conditions	Product(s) and Yield(s) (%)	Refs.
C_3			
	1. NaH (cat.), Et_2O 2. Cl_3CCN, <0° to rt 3. Xylene, 140°, 12 h	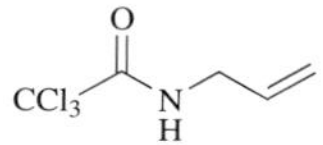 (50)	6, 4, 113
C_4			
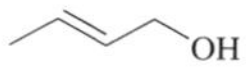	1. NaH (cat.), Et_2O 2. Cl_3CCN, <0° to rt 3. Xylene, 140°, 12 h	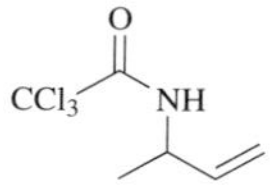 **I** (83)[a]	4, 118, 171, 172, 173
	1. NaH (cat.), THF 2. Cl_3CCN 3. $Hg(O_2CCF_3)_2$ (20 mol%), THF, 0° to rt, 1 h	**I** (69)	6
	1. KF-Al_2O_3 2. Cl_3CCN, rt, 48 h	**I** (85)	68
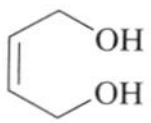	1. Na^0 (cat.), THF 2. Cl_3CCN, –23° 3. *t*-BuC_6H_5, 169°, 1 h	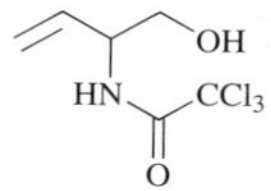 (50)	69, 55
	1. Na^0 (cat.), 0° 2. Cl_3CCN (2.2 eq.), rt 3. 180-185°, 1 h	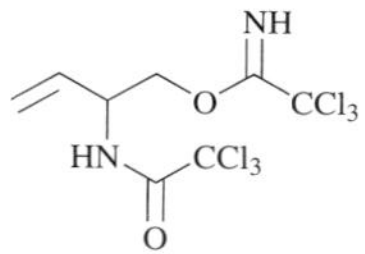 (46)	174, 55
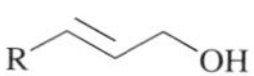	1. NaH (cat.), Cl_3CCN, THF, 0° to rt, 2.5 h 2. $C_{10}H_{18}$, 190°, 12 h	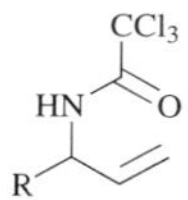 R: $BnOCH_2$ (—)[b]; $HOCH_2$ (—)[b]	117

	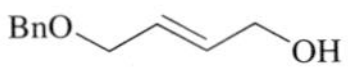	1. NaH (cat.), Et_2O 2. Cl_3CCN, Et_2O, –15 to 0° 3. Xylenes, 150° (sealed tube), 52 h	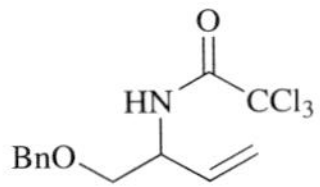 (86)	62, 113
		1. F_3CCN, DBU, CH_2Cl_2, –78° [c] 2. Xylenes, 150° (sealed tube), 68 h	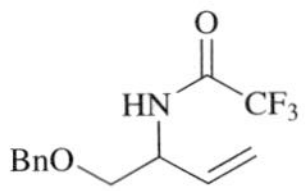 (71)	62
	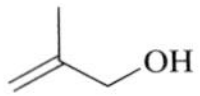	1. NaH (cat.), Et_2O 2. Cl_3CCN, 0° 3. Xylenes, 140°	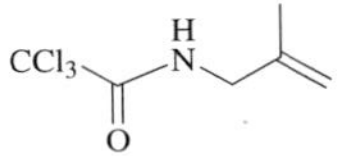 (36)	116
C_5	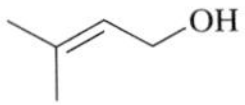	1. KF-Al_2O_3 2. Cl_3CCN, rt, 48 h	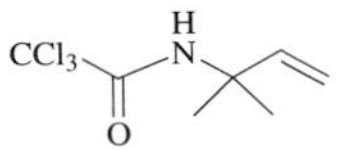 (65)	68
	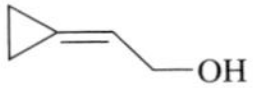	1. NaH (cat.) 2. Cl_3CCN 3. Toluene, 100°, 48 h	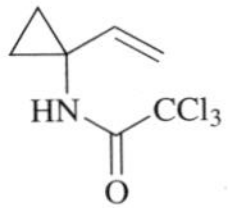 (89)[d]	31
	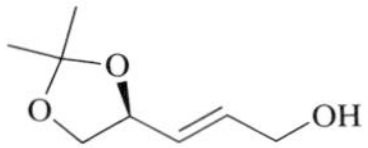	1. NaH (cat.), Et_2O 2. Cl_3CCN, Et_2O, –15 to 0° 3. Xylenes, 150° (sealed tube), 23 h	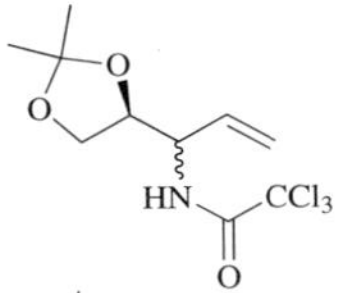 (61)	62
		1. F_3CCN, DBU, CH_2Cl_2, –78° [c] 2. Xylenes, 150° (sealed tube), 96 h	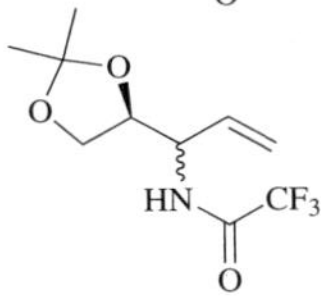 (60)	62

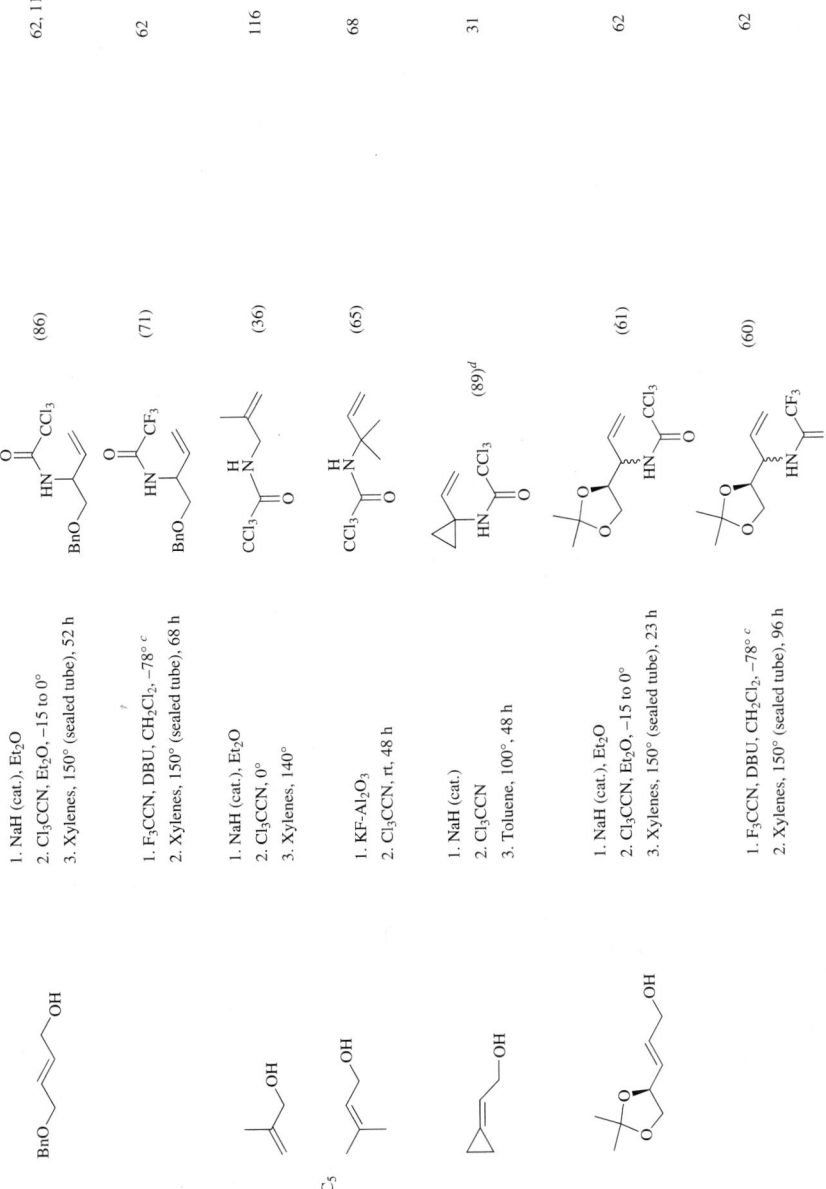

TABLE 1. REARRANGEMENTS OF TRIHALOACETIMIDATES OF PRIMARY ALLYLIC ALCOHOLS (*Continued*)

Substrate	Conditions	Product(s) and Yield(s) (%)	Refs.
C_5			
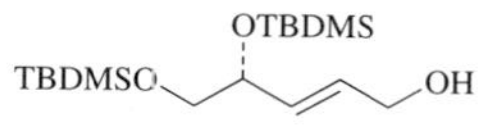	1. DBU, Cl_3CCN, 0° 2. $PdCl_2(MeCN)_2$ (10 mol%), THF, rt, 5 h	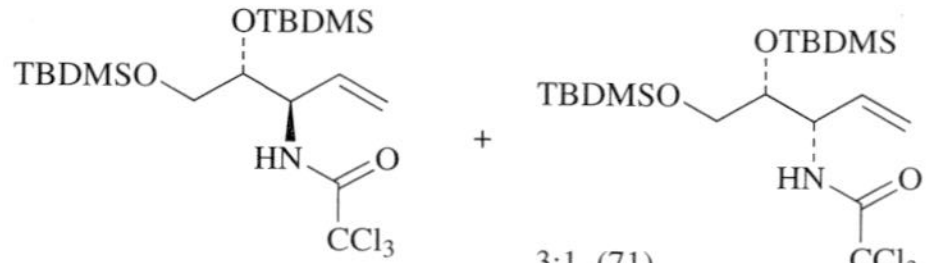 3:1 (71)	100
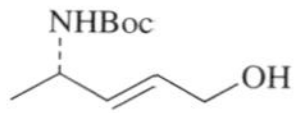	1. NaH (cat.), Et_2O 2. Cl_3CCN, rt 3. Conditions	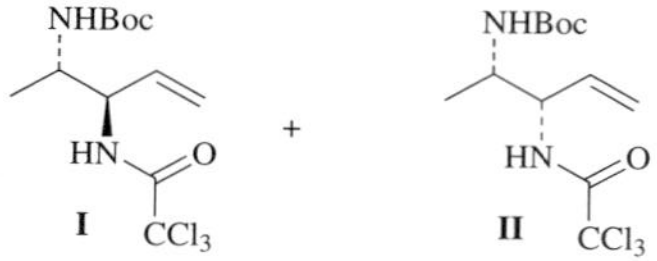 Conditions — I:II *o*-xylene, 140°, 25 h — 62:38 (82) $PdCl_2(MeCN)_2$ (6-8 mol%), MeCN, rt, 3 h — >99:1 (46)	30
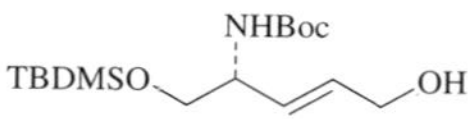	1. NaH (cat.), Et_2O 2. Cl_3CCN, rt 3. $PdCl_2(MeCN)_2$ (6-8 mol%), MeCN, rt, 3 h	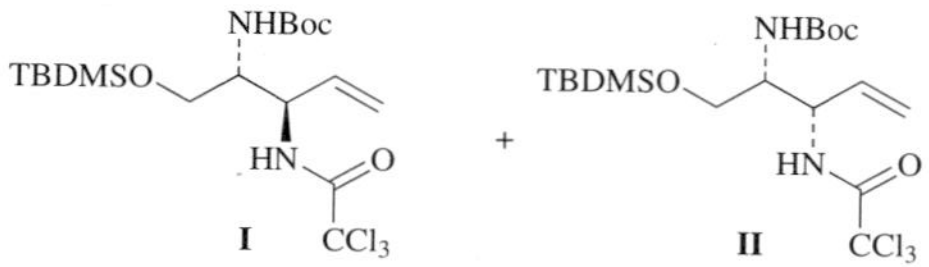 **I**:**II** 99:1 (57)[a]	30
C_6			
	1. NaH (cat.), Et_2O 2. Cl_3CCN 3. Xylenes, 140°, 12 h	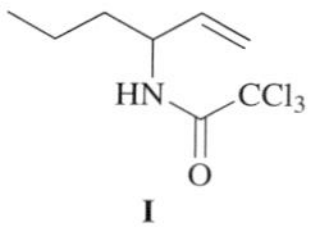 **I** (72)	6, 4, 117, 118, 120, 175, 173
	1. NaH (cat.), Et_2O 2. Cl_3CCN, 0° to rt 3. $Hg(O_2CCF_3)_2$ (10 mol%), THF, 25 min	**I** (62)	6

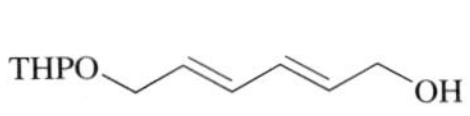	1. NaH (cat.), Et_2O 2. Cl_3CCN, 0° to rt 3. Xylenes, 140°, 20 h	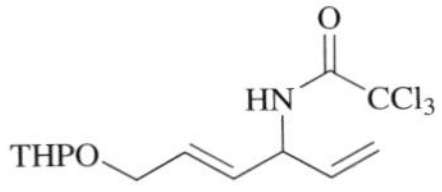 (77)	89
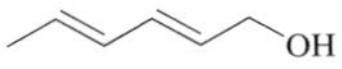	1. NaH (cat.), Et_2O 2. Cl_3CCN, –15° to rt, 4 h 3. Xylenes, 140°, 12 h	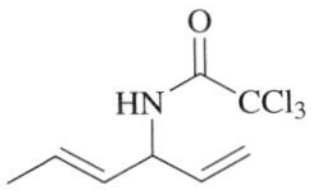 **I** (100)	51
	1. DBU, CH_2Cl_2, 0° 2. Cl_3CCN, CH_2Cl_2, 0°, 1 h 3. K_2CO_3, *p*-xylene, 138°, 13 h	**I** (73)	63
	1. *n*-BuLi (cat.), THF, –78° to 0° 2. F_3CCN (excess), THF, –78° to rt 3. Xylenes, 140°, 9.5 hr	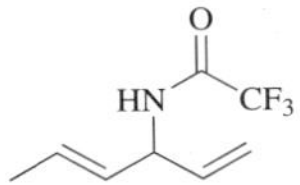 (35)[a]	60, 75
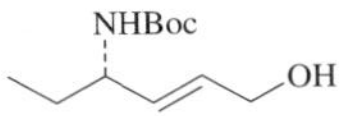	1. NaH (cat.), Et_2O 2. Cl_3CCN, rt 3. $PdCl_2(MeCN)_2$ (6-8 mol%), MeCN, rt, 3 h	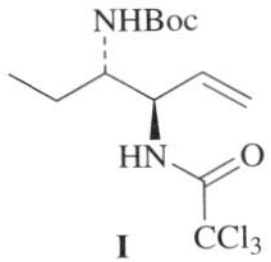 **I** + 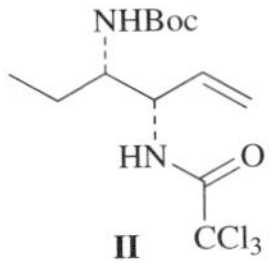 **II** **I**:**II** 99:1 (62)[a]	30
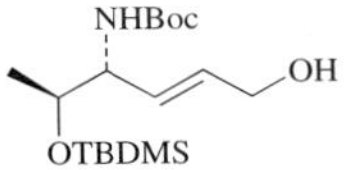	1. NaH (cat.), Et_2O 2. Cl_3CCN, rt 3. $PdCl_2(MeCN)_2$ (6-8 mol%), MeCN, rt, 3 h	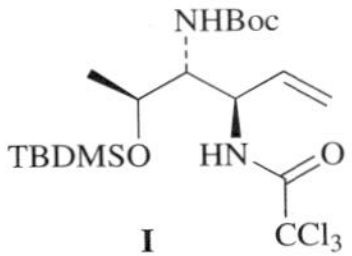 **I** + 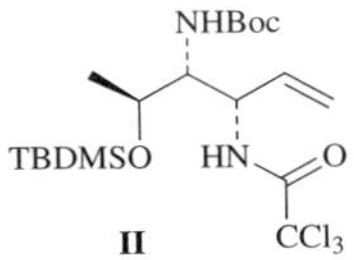 **II** **I**:**II** 99:1 (45)[a]	30

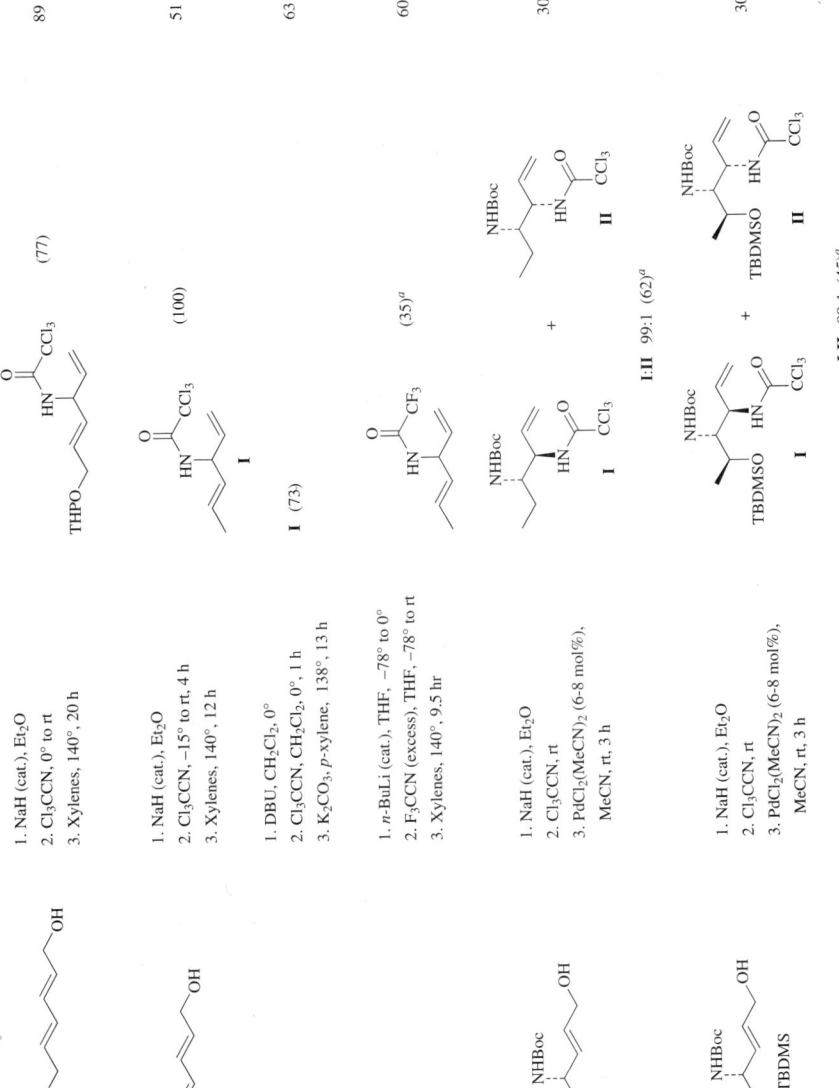

TABLE 1. REARRANGEMENTS OF TRIHALOACETIMIDATES OF PRIMARY ALLYLIC ALCOHOLS (*Continued*)

Substrate	Conditions	Product(s) and Yield(s) (%)	Refs.
C_6			
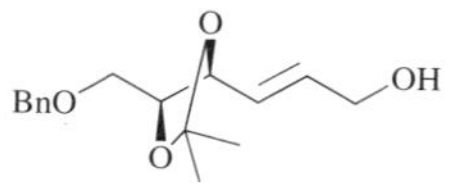	1. NaH (cat.), Et_2O, 15° 2. Cl_3CCN, Et_2O, –15° to rt 3. Xylenes, 140°, 48 h	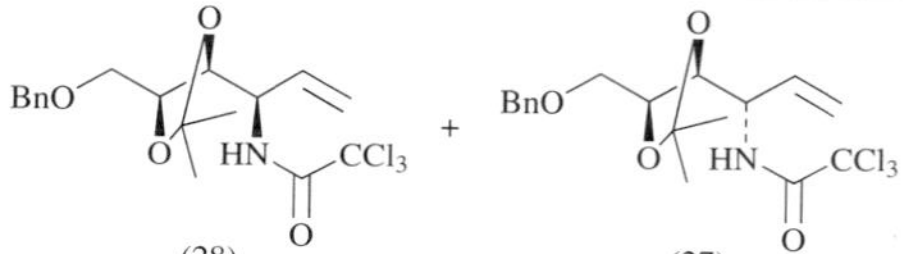 (28) + (27)	77, 176
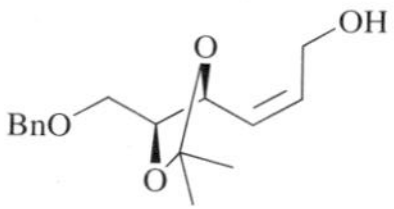	1. NaH (cat.), Et_2O, 15° 2. Cl_3CCN, Et_2O, –15° to rt 3. Xylenes, >140°, 48 h[e]	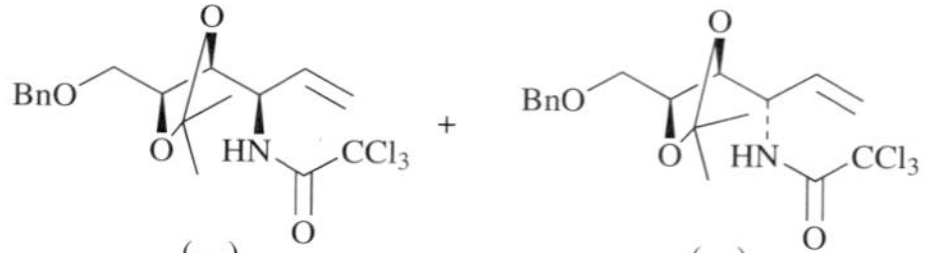 (—) + (—)	77
C_7			
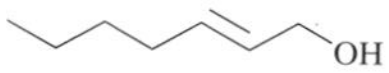	1. NaH (cat.), Et_2O 2. Cl_3CCN 3. Xylenes, 140°	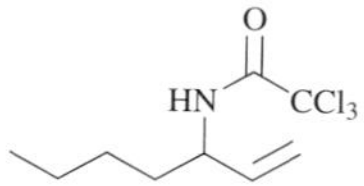 (—)	172
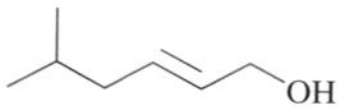	1. NaH (cat.), Et_2O 2. Cl_3CCN 3. Xylenes, 140°	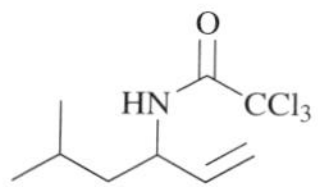 (62)	177
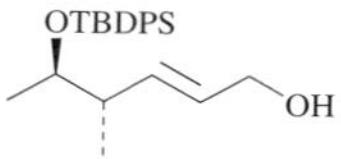	1. NaH (cat.) 2. Cl_3CCN 3. *p*-Xylenes, 140°, 24 h	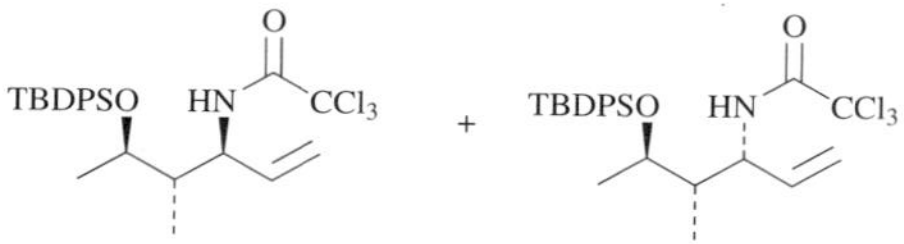 I + II **I:II** 3:2 (45)	99

	1. NaH (cat.) 2. Cl_3CCN 3. $PdCl_2(RCN)_2$ (10 mol%), THF, rt, 3 h	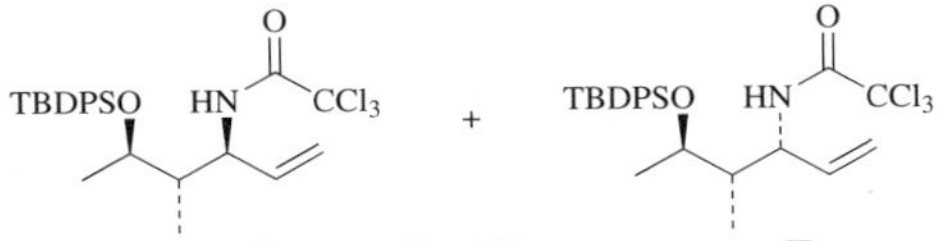 **I** + **II** R **I:II** Me 1:6 (54) Ph 1:7 (59)	99
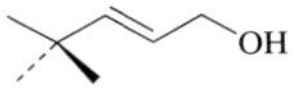	1. NaH (cat.), Et_2O 2. Cl_3CCN 3. Xylenes, 140°	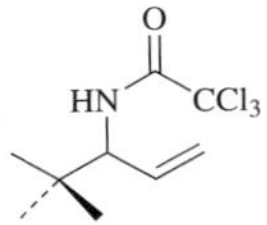 (63)	113
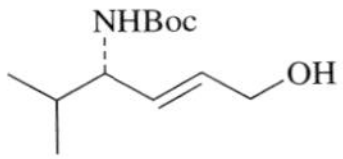	1. NaH (cat.), Et_2O 2. Cl_3CCN, rt 3. $PdCl_2(MeCN)_2$ (6-8 mol%), MeCN, rt, 3 h	 **I** + **II** **I:II** 99:1 (49)[a]	30
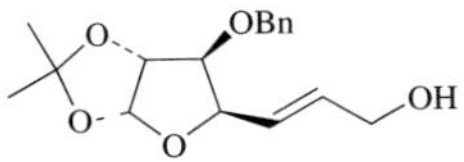	1. NaH (cat.), Et_2O, 0°, 5 min 2. Cl_3CCN, 0° to rt, 45 min 3. DBPC (2%), xylenes, 137°, 13 h	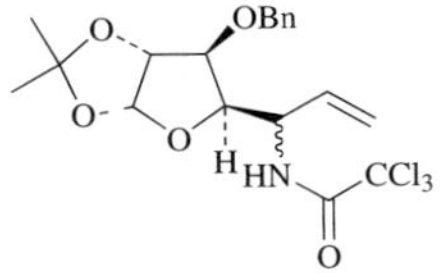 (72) dr = 1.1:1	98
	1. NaH (cat.), THF, 0°, 15 min 2. F_3CCN, –78°, 20 min 3. DBPC (2%), xylenes, 137°, 81 h	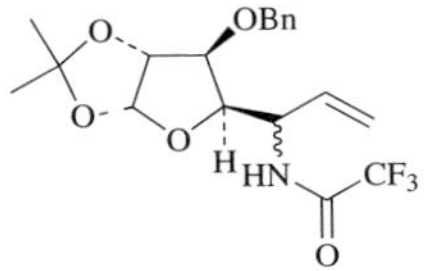 (64) dr = 1.5:1	98

TABLE 1. REARRANGEMENTS OF TRIHALOACETIMIDATES OF PRIMARY ALLYLIC ALCOHOLS *(Continued)*

Substrate	Conditions	Product(s) and Yield(s) (%)	Refs.
C_7			
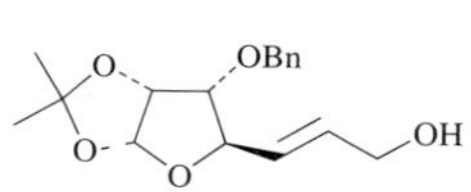	1. NaH (cat.), Et_2O, 0°, 5 min 2. Cl_3CCN, 0° to rt 3. DBPC (%), xylenes, 137°	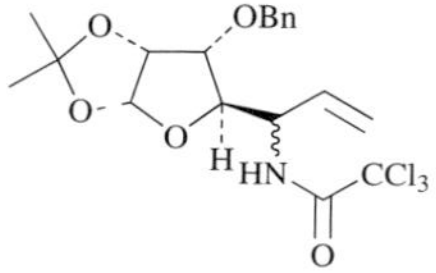 dr = 1:1 % DBPC 0 (23)[a] 2 (84)[a]	98
	1. NaH (cat.), THF, 0°, 15 min 2. F_3CCN, –78°, 20 min 3. DBPC (2%), xylenes, 137°, 81 h	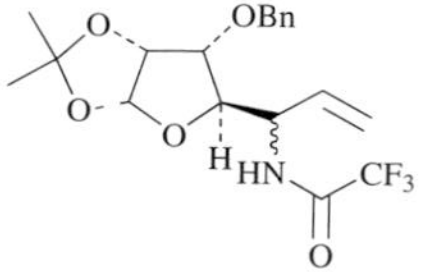 (45) dr = 1.2:1	98
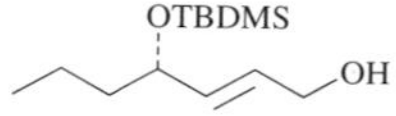	1. DBU, Cl_3CCN, 0° 2. $PdCl_2(MeCN)_2$ (10 mol%), THF, rt, 5 h	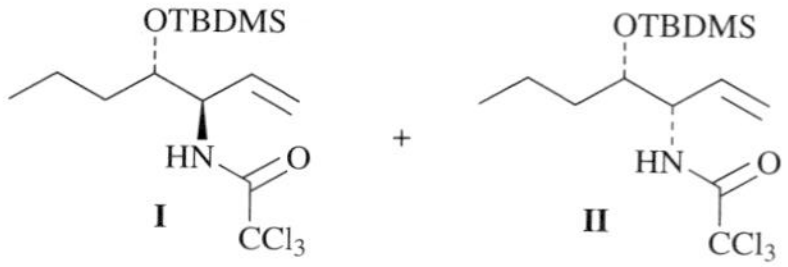 **I** + **II** (52) **I**:**II** 2.2:1	100
C_8			
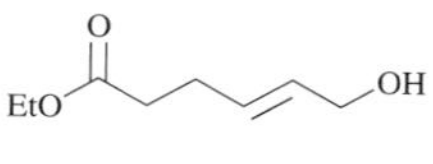	1. NaH (cat.), Et_2O, 0° 2. Cl_3CCN, 0°, 1 hr 3. Xylenes, 140°, 48 h	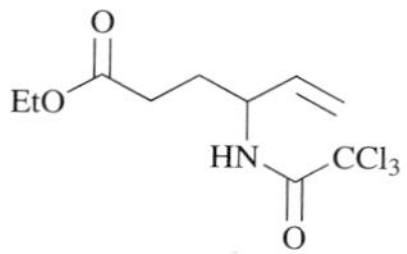 (65)	178
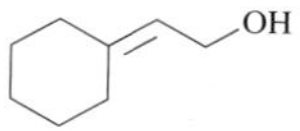	1. NaH (cat.) 2. Cl_3CCN 3. Toluene, 110°	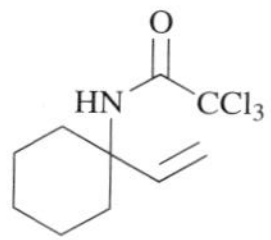 (75)[a]	179
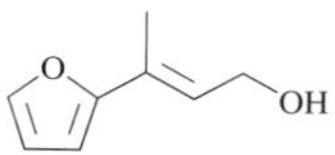	1. DBU, CH_2Cl_2, –10° 2. Cl_3CCN, –10°, 1 h 3. Na_2CO_3, *p*-xylene, 138°	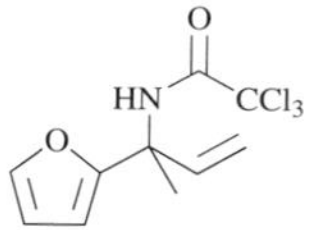 (trace)	19

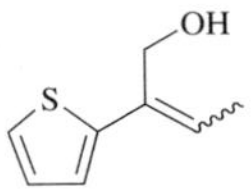

1. DBU (cat.), Cl_3CCN, CH_2Cl_2, rt
2. Xylenes, 140°, 24 h

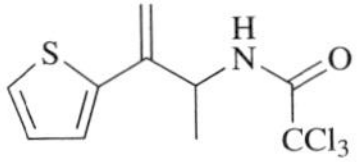

(—) 51

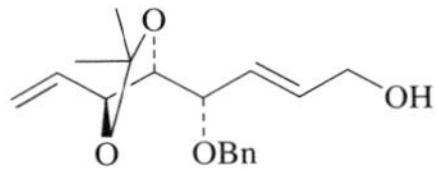

1. DBU (cat.), Cl_3CCN, CH_2Cl_2, 0°
2. $PdCl_2(MeCN)_2$ (8 mol%), toluene, rt, 12 h

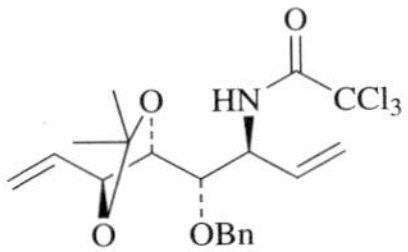

(69) 27

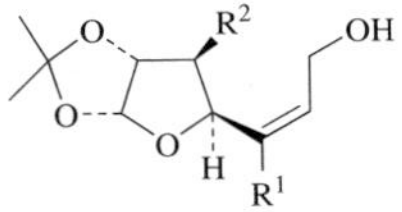

1. DBU (cat.), Cl_3CCN, CH_2Cl_2, 0°
2. Toluene (sealed tube), 140°

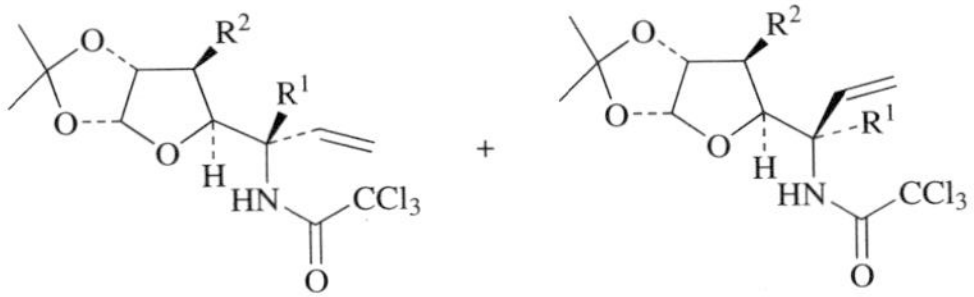

I + II

79

R^1	R^2	Time	I:II	
Me	MOMO	3 d	1.0:8.3	(78)
$BnOCH_2$	MeO	6.5 d	6.3:1.0	(64)
Me	BnO	2 d	1.0:5.4	(71)
$MPMOCH_2$	BnO	2.5 d	5.0:1.0	(69)

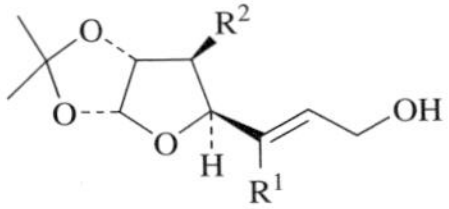

1. DBU (cat.), Cl_3CCN, CH_2Cl_2, 0°
2. Toluene (sealed tube), 140°

I + II

79

R^1	R^2	Time	I:II	
Me	MOMO	3 d	1.3:1.0	(73)
$MPMOCH_2$	BnO	1.5 d	1.0:9.0	(65)

TABLE 1. REARRANGEMENTS OF TRIHALOACETIMIDATES OF PRIMARY ALLYLIC ALCOHOLS (*Continued*)

Substrate	Conditions	Product(s) and Yield(s) (%)	Refs.
C_8 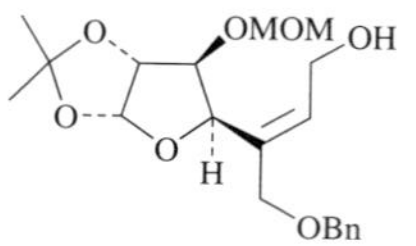	1. DBU (cat.), Cl_3CCN, CH_2Cl_2, 0° 2. K_2CO_3, xylenes, 140°, 140 h	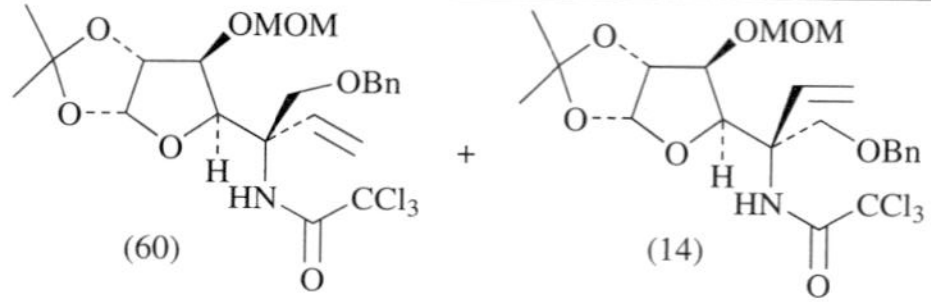 (60) + (14)	128
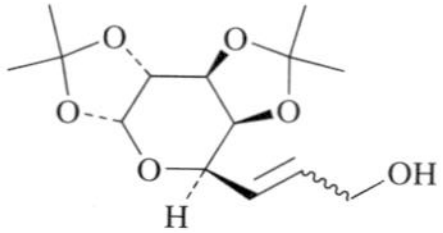	1. NaH (cat.), Et_2O, 0° 2. X_3CCN 3. Xylenes, 200°	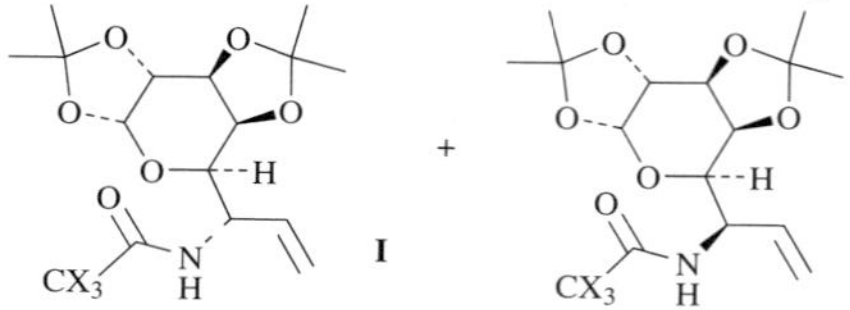 I + II X / E/Z / Time / I:II Cl / E / 30 h / 50:50 (22) F / E / 2 h / 42:58 (93) Cl / Z / 30 h / 50:50 (25) F / Z / 2 h / 9:91 (95)	95
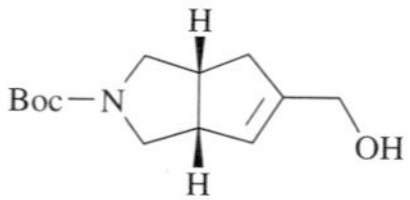	1. NaH (cat.) 2. Cl_3CCN 3. Toluene, 110°	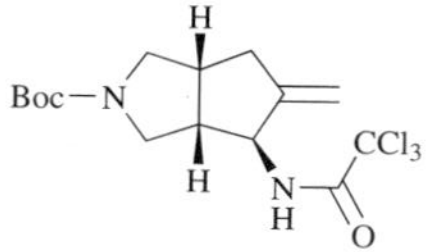 (51)	80
C_9 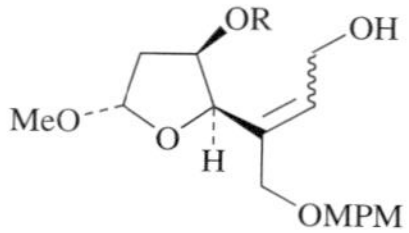	1. DBU (cat.), Cl_3CCN, CH_2Cl_2, 0° 2. K_2CO_3, xylenes, 140°	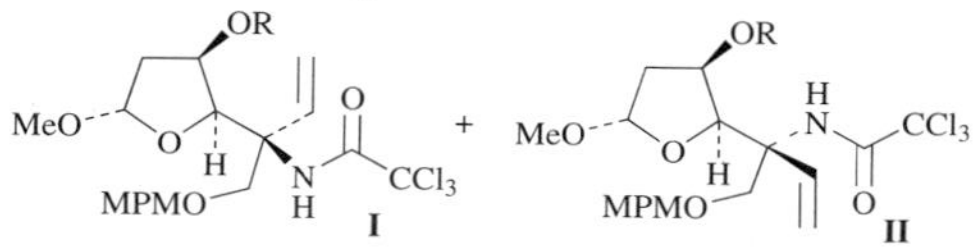 I + II E/Z / R / Time / I:II E / MOM / — / 5:1 (65) E / Bn / 60 h / 7:1 (90) Z / MOM / — / 1:4 (49) Z / Bn / — / 1:5 (70)	47, 180

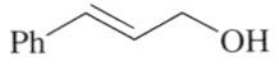	1. NaH (cat.), Et_2O 2. Cl_3CCN 3. Xylenes, 140°, 6 h	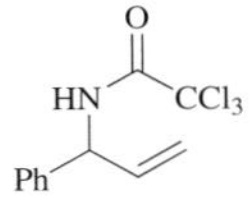 (92) **I**	4, 6, 60, 62, 68, 112, 113, 115, 116, 118, 120, 122, 171, 173,175, 177, 181, 182
	1. KF-Al_2O_3 2. Cl_3CCN, rt, 48 h	**I** (78)	68
	1. NaH (cat.), Et_2O 2. Cl_3CCN 3. $Hg(O_2CCF_3)_2$ (40 mol%), THF, 0° to rt, 1 h	**I** (39)	4, 6
	1. *n*-BuLi (cat.), Et_2O 2. F_3CCN, THF, –78° 3. Xylenes, 140°, 5 h	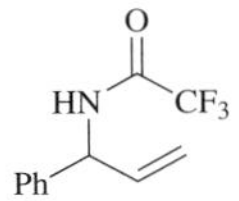 (81)[a]	60, 62
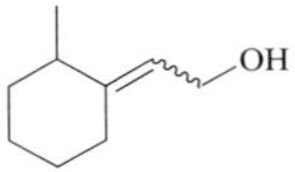	1. NaH (cat.), Et_2O 2. Cl_3CCN 3. Xylenes, 140°	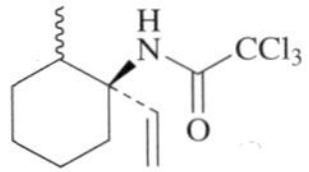 (—)	115
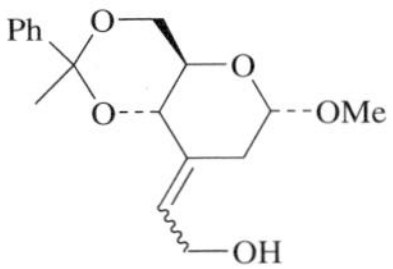	1. NaH, CH_2Cl_2, –6° 2. Cl_3CCN, CH_2Cl_2, –6° 3. *o*-Dichlorobenzene, 165°, 6 h	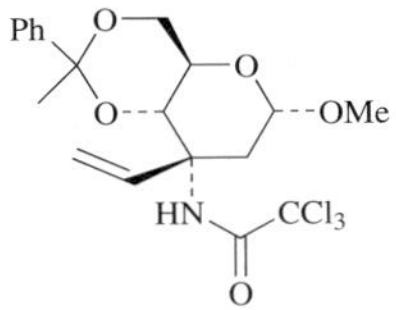 (53)	84, 85

TABLE 1. REARRANGEMENTS OF TRIHALOACETIMIDATES OF PRIMARY ALLYLIC ALCOHOLS (*Continued*)

Substrate	Conditions	Product(s) and Yield(s) (%)	Refs.
C_9 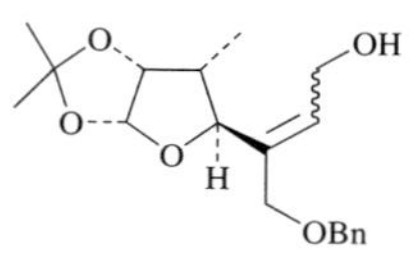	1. NaH (cat.), Et_2O, –15° 2. Cl_3CCN, –15° to rt 3. Toluene, 150° (sealed tube), 89 h	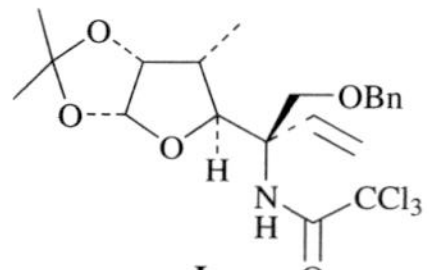 **I** + 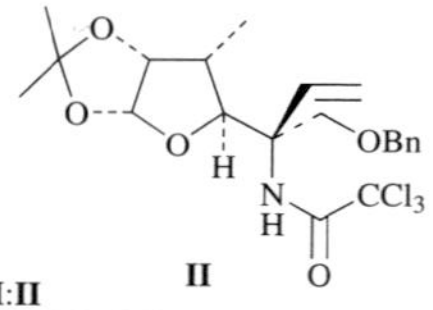 **II** E/Z — I:II E — 2.7:1.0 (—) Z — 5.0:1.0 (—) 1:1 — 4.8:1 (60)	79, 87, 127
C_{10} 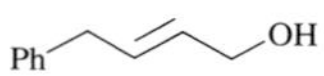	1. NaH (cat.), Et_2O 2. Cl_3CCN, 0° to rt 3. Xylenes, 140°	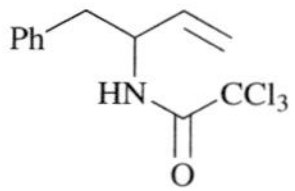 (62)	113
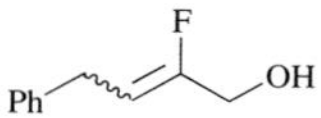	1. NaH (cat.), Et_2O 2. Cl_3CCN 3. Xylenes, 190° (sealed tube), 4 h	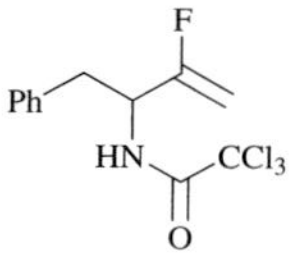 (80)	183
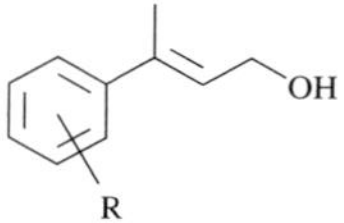	1. DBU, CH_2Cl_2, –10° 2. Cl_3CCN, –10°, 1 h 3. Na_2CO_3, *p*-xylene, 138°, time	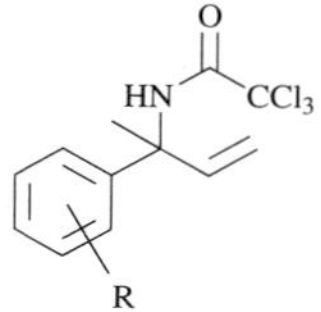 R — Time H — 24 h (89) *p*-F — 12 h (84) *p*-Cl — 12 h (82) *o*-Cl — 24 h (30)	19, 116[f]
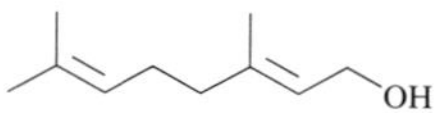	1. NaH (cat.), Et_2O 2. Cl_3CCN, <0° to rt 3. *m*-Xylene, 138°, 6 h	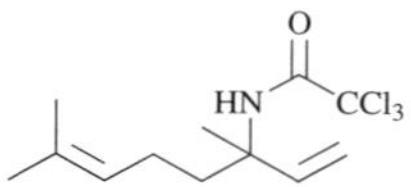 **I** (67–74)	4, 6, 42, 43, 62, 63, 68, 115, 184, 185

	1. DBU, CH_2Cl_2, 0° 2. Cl_3CCN, CH_2Cl_2, 0°, 1 h 3. K_2CO_3, *p*-xylene, 138°, 13 h	**I** (84)	63
	1. KF-Al_2O_3 2. Cl_3CCN, rt, 48 h	**I** (69)	68
	1. NaH (cat.), Et_2O 2. Cl_3CCN 3. $Hg(O_2CCF_3)_2$ (20 mol%), THF, –78° to rt, 1 h	**I** (63–73)	4, 6
	1. NaH (cat.), Et_2O 2. Cl_3CCN 3. $PdCl_2(MeCN)_2$, THF, rt, 4 h	**I** (66)	62
	1. *n*-BuLi (cat.), THF, 0° to –78° 2. F_3CCN, THF, –78° to rt, 1 h 3. Xylenes, 140°, 9.5 h	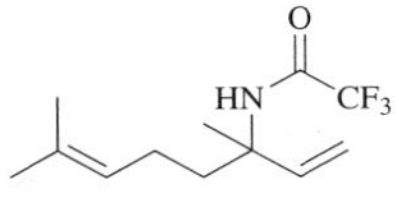 (70) **I**	60, 75
	1. F_3CCN, DBU, CH_2Cl_2, –78° [c] 2. $PdCl_2(MeCN)_2$, THF, rt, 16 h	**I** (19)	62
	1. DBU, CH_2Cl_2, 0° 2. Cl_3CCN, CH_2Cl_2, 0°, 0.5-1 h 3. [K_2CO_3], *p*-xylene, 138°, 13-15 h	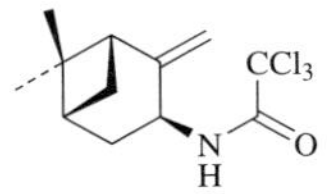 *without K_2CO_3* (72) *with K_2CO_3* (95)	63
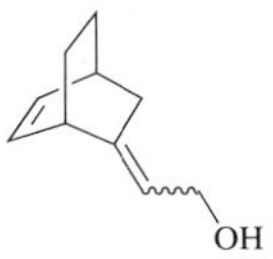	1. NaH (cat.), THF 2. Cl_3CCN 3. Xylenes, 130°, 5 h	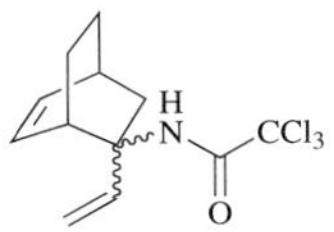 (72) 1:1 exo:endo	186

TABLE 1. REARRANGEMENTS OF TRIHALOACETIMIDATES OF PRIMARY ALLYLIC ALCOHOLS (*Continued*)

Substrate	Conditions	Product(s) and Yield(s) (%)	Refs.
C_{10}			
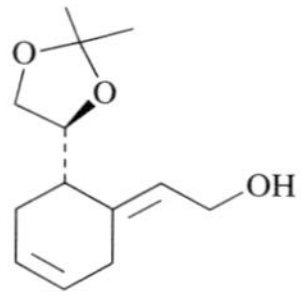	1. NaH (cat.), Et_2O 2. Cl_3CCN, Et_2O 3. Xylene, 140°, 12 h	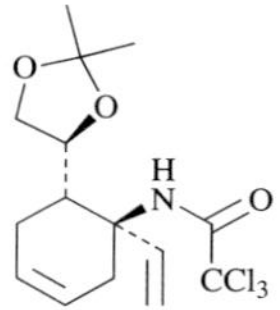 (60)	179
C_{11}			
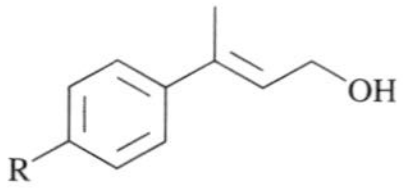	1. DBU, CH_2Cl_2, –10° 2. Cl_3CCN, –10°, 1 h 3. Na_2CO_3, *p*-xylene, 138°, time	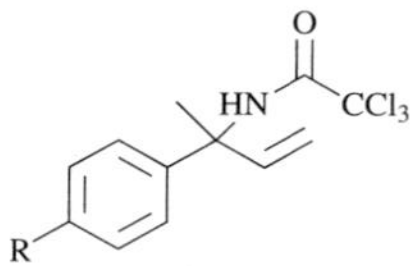 R / Time Me 24 h (72) CF_3 12 h (89) MeO 24 h (trace)	19
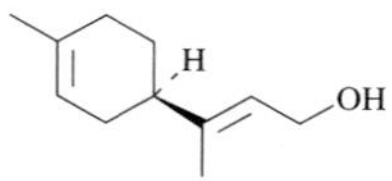	1. NaH (cat.), Et_2O, 0° 2. Cl_3CCN, Et_2O 3. Toluene, 110°, 5 h	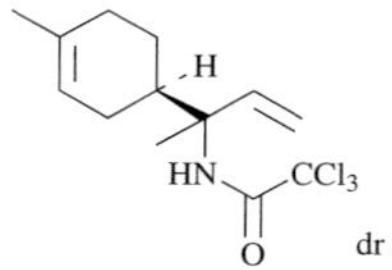 (45) dr = 1:1	187, 188
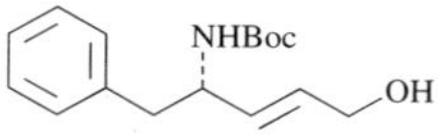	1. NaH (cat.), Et_2O 2. Cl_3CCN, rt 3. $PdCl_2(MeCN)_2$ (6-8 mol%), MeCN, rt, 3 h	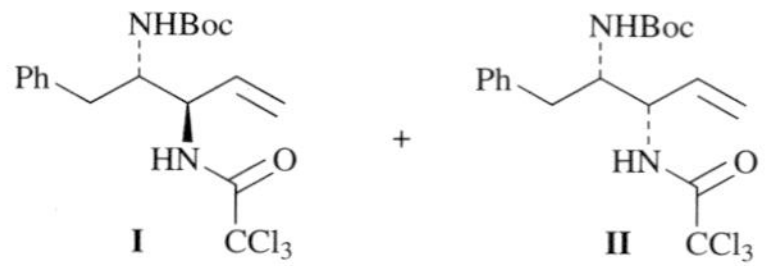 **I** + **II** **I**:**II** 99:1 (43)[a]	30
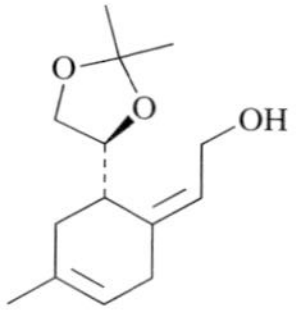	1. NaH (cat.), Et_2O 2. Cl_3CCN 3. Xylenes, 140°	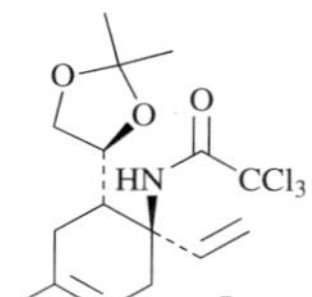 **I** (56)	115

	1. DBU, CH_2Cl_2 2. Cl_3CCN, CH_2Cl_2 3. K_2CO_3, *p*-xylene, 138°	**I** (93)		189, 63
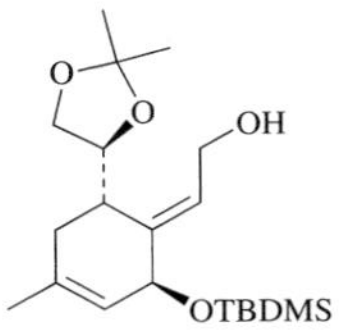	1. DBU, CH_2Cl_2, 0° 2. Cl_3CCN, CH_2Cl_2, 0°, 0.5-1 h 3. [K_2CO_3], Xylenes, 140°, 13-15 h	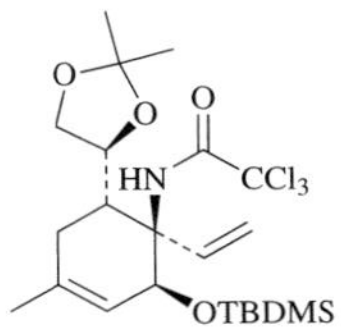	*without K_2CO_3* $(37)^g$ *with K_2CO_3* $(62)^h$	63
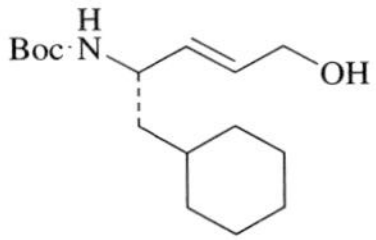	1. NaH (cat.), THF, 0° 2. Cl_3CCN, 0° to rt 3. Xylenes, 140°, 24 h	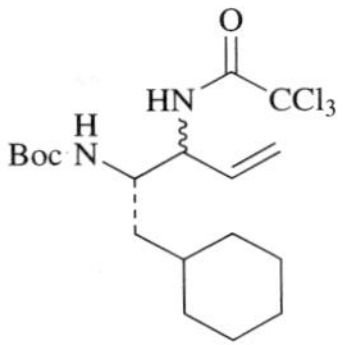	dr = 55:45 (52)	94
	1. NaH (cat.), THF, 0° 2. Cl_3CCN, 0° to rt 3. Conditions	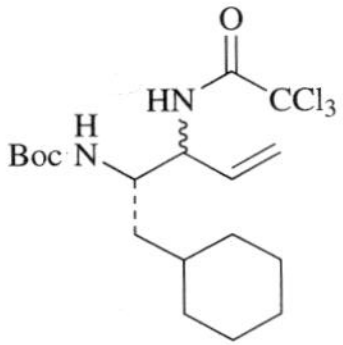		94

Conditions	
I_2, rt, 48 h	$(0)^i$
NBS, THF	$(—)^j$
$Hg(O_2CCF_3)_2$, $NaBH_4$	$(—)^j$

TABLE 1. REARRANGEMENTS OF TRIHALOACETIMIDATES OF PRIMARY ALLYLIC ALCOHOLS *(Continued)*

Substrate	Conditions	Product(s) and Yield(s) (%)	Refs.
C_{11}			
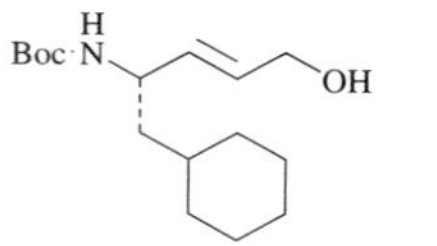	1. NaH (cat.), THF, 0° 2. Cl_3CCN, 0° to rt 3. $Pd(OAc)_2$ (6.4 eq.), THF, rt, 4 h	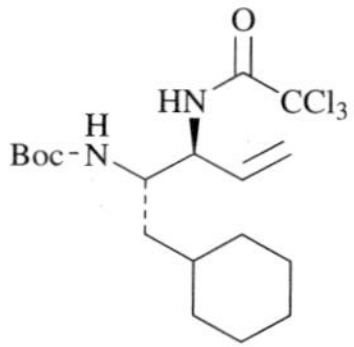 (44)	94
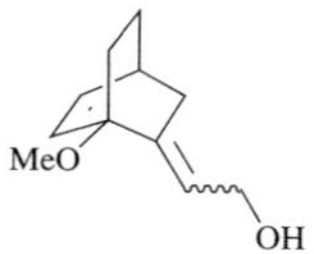	1. NaH (cat.), THF 2. Cl_3CCN 3. Xylenes, 130°, 5 h	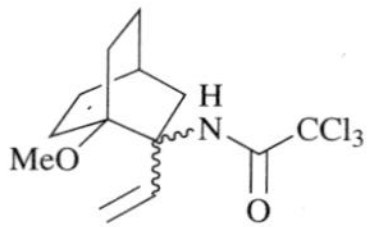 (48) 1:1 exo:endo	186
C_{11-12}			
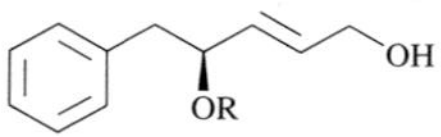	1. DBU, Cl_3CCN, 0° 2. $PdCl_2(MeCN)_2$ (10 mol%), THF rt, 5 h	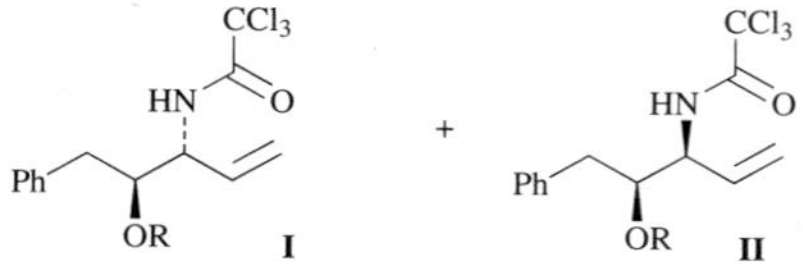	100
C_{14}			
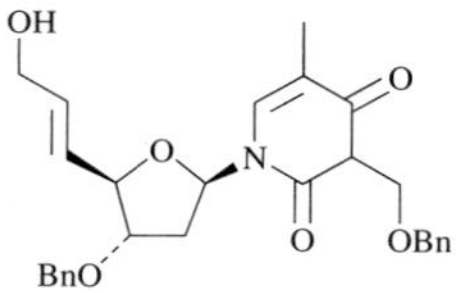	1. NaH (cat.), THF, rt, 15 min 2. Cl_3CCN, 0° to rt, 45 min 3. DBPC (2%), xylene, 137°, 44 h	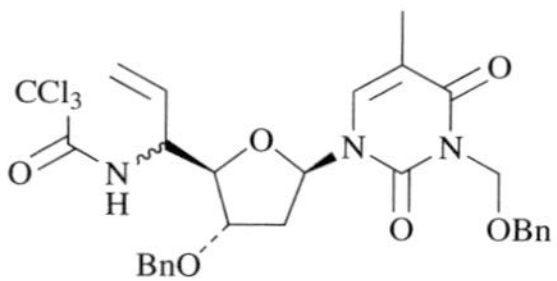 (84) dr = 1:1	98

R	I:II	
Me	1.4:1	(63)
MOM	6.2:1	(58)
BOM	9.95:1	(57)
OTBDMS	10:1	(61)

C_{14} 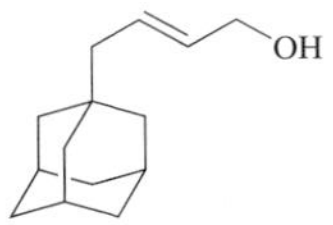	1. NaH (cat.), Et_2O, rt 2. Cl_3CCN, –5° to 0° 3. Xylenes, 140°, 12 h	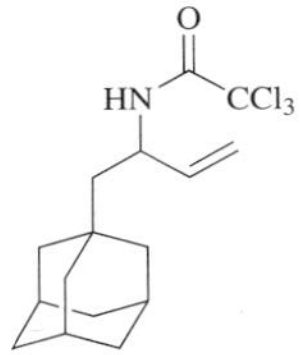	(78)	190
C_{15} 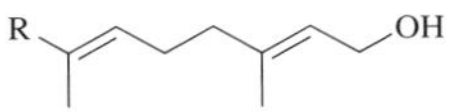 R = 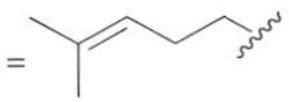	1. KF-Al_2O_3 2. Cl_3CCN, rt, 48 h	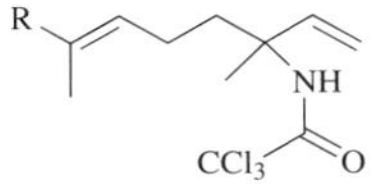	(65)	68

[a] This is the yield for rearrangement only.

[b] The trichloroacetamide products were hydrolyzed and the amine hydrochloride salts were obtained in "very high yields" from the corresponding alcohol.

[c] Trifluoroacetonitrile was generated in situ from trifluoroacetamide by reaction with dimethylsulfoxide/oxalyl chloride.

[d] Heating at 140° (xylene reflux) led to decomposition. Several other variables (solvent, microwave irradiation) were tried with poor success.

[e] This, the Z isomer, required higher temperatures and produced "significantly more byproducts" than its E counterpart.

[f] R = H only.

[g] In addition to the desired product, a 32% yield of 2,2-dimethyl-4-(5-methyl-2-vinylphenyl)-1,3-dioxolane (a product of aromatization) was obtained.

[h] Ten percent of the starting imidate was recovered.

[i] Attempted rearrangement resulted in either no reaction or a mixture of several products.

[j] Several products were obtained.

TABLE 2A. REARRANGEMENTS OF TRIHALOACETIMIDATES OF ACYCLIC SECONDARY ALLYLIC ALCOHOLS

Substrate	Conditions	Product(s) and Yield(s) (%)	Refs.
C_4 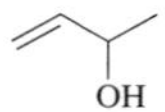	1. NaH (cat.), THF, 5°, 1 h 2. Cl_3CCN, THF, 0°, 2.5 h 3. Toluene, 110°, 2 h	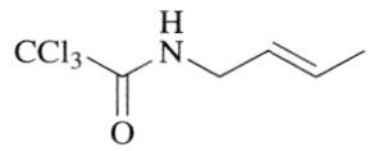 (53)	121
C_5 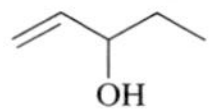	1. KH (cat.) 2. Cl_3CCN 3. Xylenes, 140° 4. NaOH 5. $(Boc)_2O$	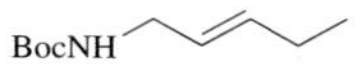 (71)	191
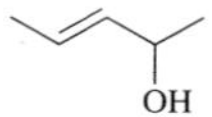	1. KF-Al_2O_3 2. Cl_3CCN, rt, 48 h	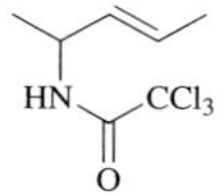 (90)	68
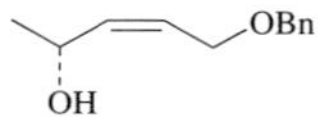	1. KH, 18-crown-6, Et_2O, –15° 2. Cl_3CCN, Et_2O, –15° to rt, 4 h 3. Xylenes, 140°, 12 h	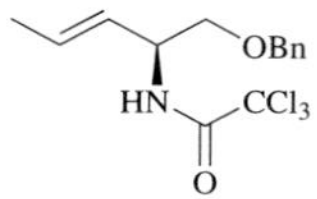 (≈35)	24
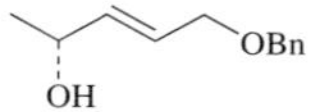	1. KH, Et_2O, –15° 2. Cl_3CCN, Et_2O, –15° to rt, 4 h 3. Xylenes, 140°, 12 h	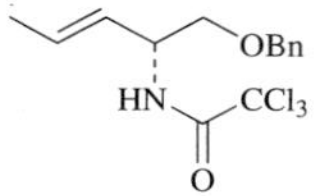 (≈45)	24
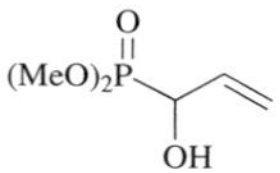	1. DBU, Cl_3CCN, CH_2Cl_2, –35°, 35 min 2. Xylene, 140°, 22 h	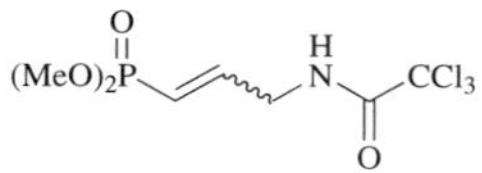 E:Z 86:14 (61)	97
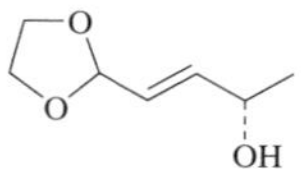	1. *n*-BuLi (20 mol%) 2. F_3CCN, THF, –78° 3. Xylene, 140°	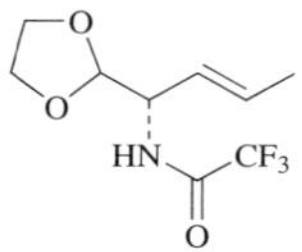 (93)[a]	60

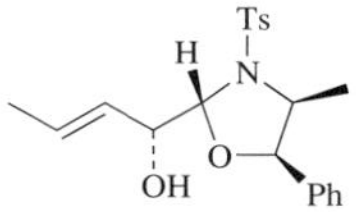	1. KH, THF, Et_2O 2. Cl_3CCN 3. Toluene, 110°	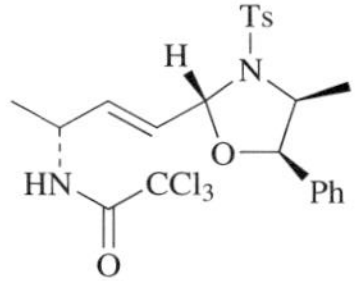 (87)	126
C_6 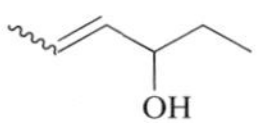	1. NaH, Et_2O 2. Cl_3CCN, Et_2O, <0° 3. Xylene, 140°, 1 h	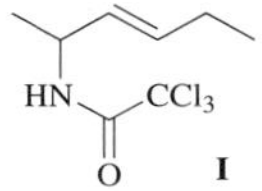 **I** (74)	4, 6
	1. NaH, Et_2O 2. Cl_3CCN, Et_2O, <0° 3. $Hg(O_2CCF_3)_2$ (30 mol%), THF, rt, 4 h	**I** (8)[b]	4, 6
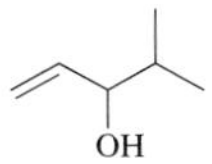	1. KH (cat.) 2. Cl_3CCN 3. Xylenes, 140° 4. NaOH 5. $(Boc)_2O$	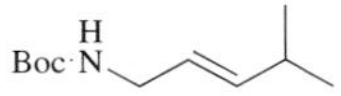 (69)	191
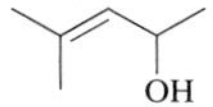	1. NaH (1 eq.) 2. Cl_3CCN 3. Benzene, 80°, 1 h	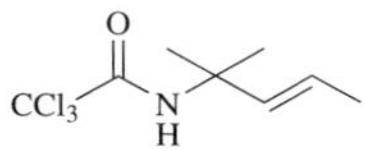 (50)	81
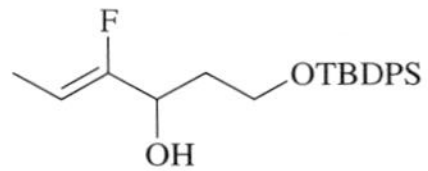	1. NaH (cat.), hexane/Et_2O 2. Cl_3CCN, hexane/Et_2O 3. Xylenes, 140°, 1-3 h	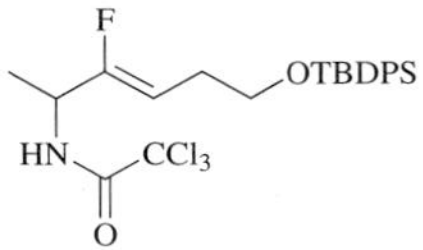 (65)	183
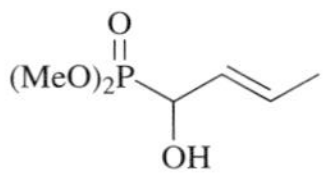	1. Cl_3CCN, DBU, CH_2Cl_2, –35°, 40 min 2. Toluene, 110°, 10 h	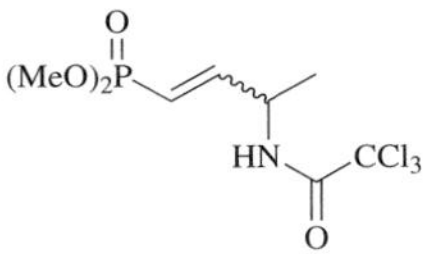 (82) E:Z 87:13	97

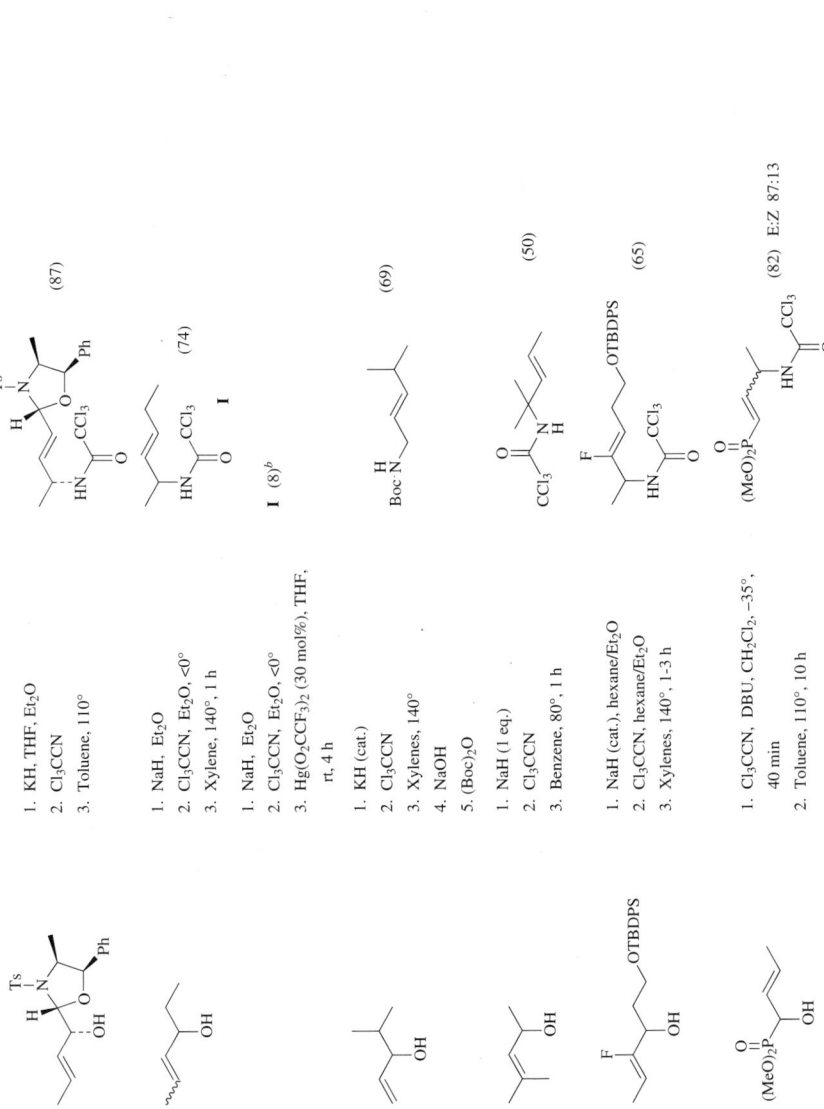

TABLE 2A. REARRANGEMENTS OF TRIHALOACETIMIDATES OF ACYCLIC SECONDARY ALLYLIC ALCOHOLS *(Continued)*

Substrate	Conditions	Product(s) and Yield(s) (%)	Refs.
C_6 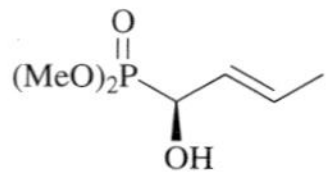	1. DBU, Cl_3CCN 2. NBS (1 eq.), $CHCl_3$, 36 h	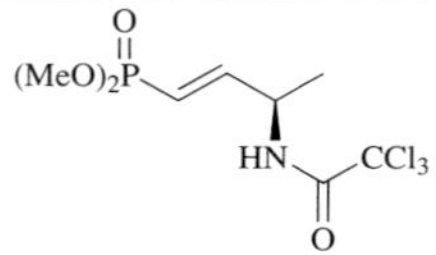 (53)[a, c]	66
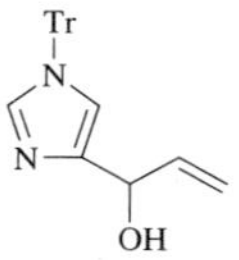	DBU, CH_2Cl_2, Cl_3CCN (2 eq.), 20°, 24 h	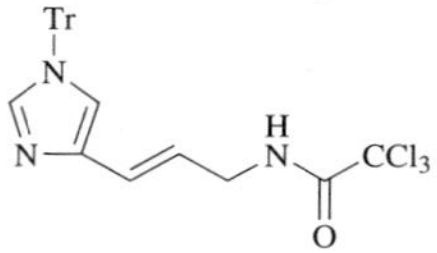 (22)	91
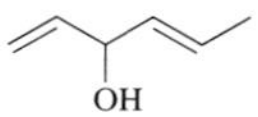	1. KH (cat.), hexane/THF, –10° 2. Cl_3CCN, Et_2O, –10° to rt 3. Toluene, 110°, 3 h	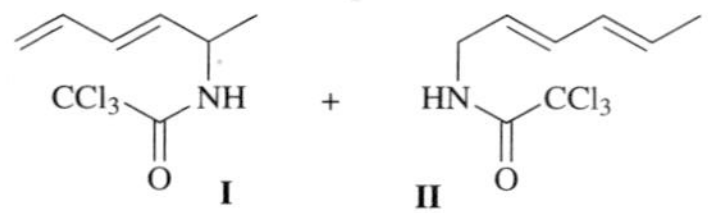 **I** + **II** (84) **I**:**II** 5:2	90
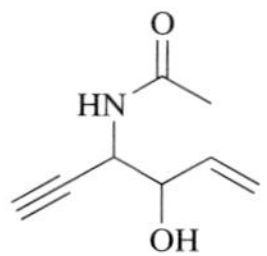	1. NaH (cat.) 2. Cl_3CCN 3. *m*-Xylene, 139°	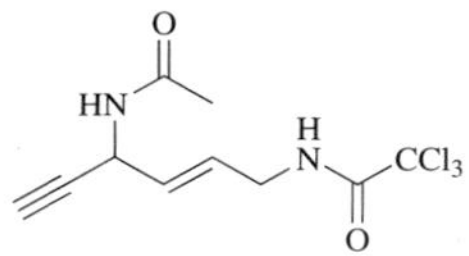 (40)	192
C_7 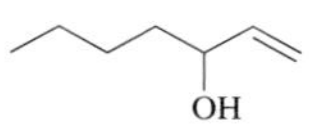	1. NaH (cat.), Et_2O 2. Cl_3CCN, Et_2O, <0° 3. Xylenes, 140°, 2.5 h	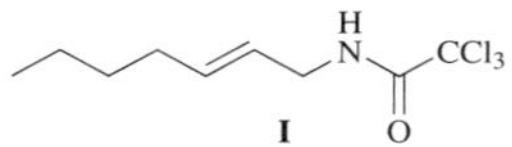 **I** (74)	4, 6
	1. NaH (cat.), Et_2O 2. Cl_3CCN, Et_2O, <0° 3. $Hg(O_2CCF_3)_2$ (30 mol%), THF, rt, 4 h	**I** (<5)[d]	4, 6

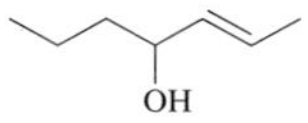	1. NaH (cat.), Et_2O 2. Cl_3CCN, Et_2O, <0° 3. Xylenes, 140°, 2.5 h	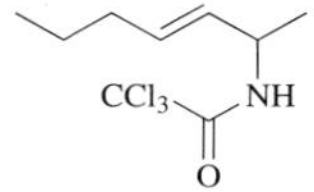	(—)	72, 172
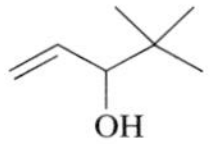	1. KH (cat.) 2. Cl_3CCN 3. Xylenes, 140° 4. NaOH 5. $(Boc)_2O$	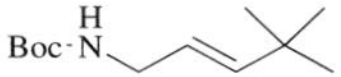	(68)	191
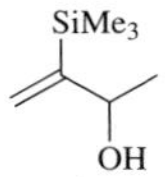	1. KH (cat.) 2. Cl_3CCN 3. Xylenes, 140° 4. NaOH 5. $(Boc)_2O$	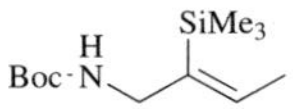	(52)	191
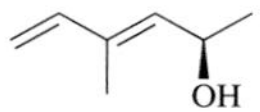	1. KH (cat.), 18-crown-6, Et_2O 2. Cl_3CCN, Et_2O, –15° to rt 3. Xylenes, 140°, 12 h	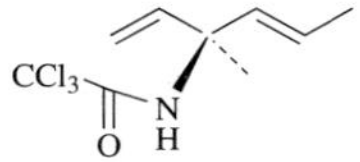	(89)	51
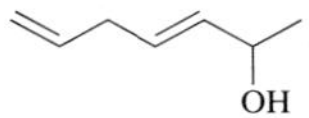	1. KH (cat.), THF 2. Cl_3CCN, Et_2O, 0° to rt 3. Xylenes, 140°, 8 h	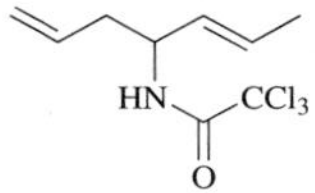	(67)	193
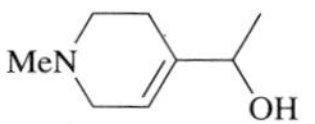	1. NaH (cat.) 2. Cl_3CCN 3. [K_2CO_3], PhCl, 135°, 15 h	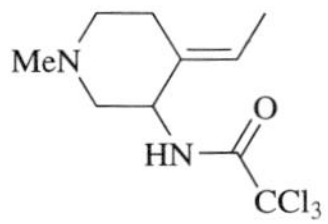	*without K_2CO_3* (28) *with K_2CO_3* (52)	71
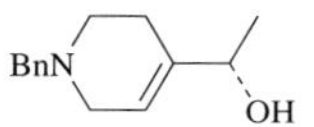	1. DBU, Cl_3CCN 2. K_2CO_3, PhCl, 135°, 5 h	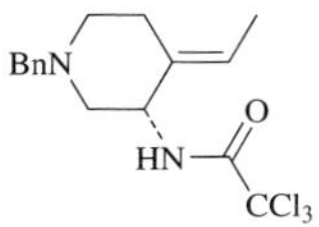	(85)	71

TABLE 2A. REARRANGEMENTS OF TRIHALOACETIMIDATES OF ACYCLIC SECONDARY ALLYLIC ALCOHOLS *(Continued)*

Substrate	Conditions	Product(s) and Yield(s) (%)	Refs.
C_7			
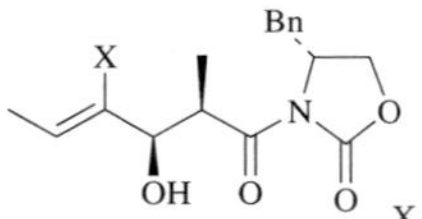 X = F, Cl	1. DBU, Cl_3CCN 2. Xylene, 140°	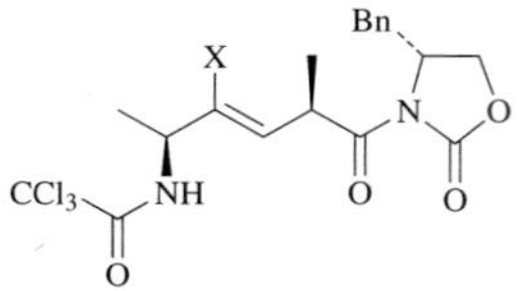 (—)	22
C_8			
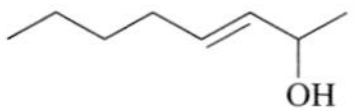	1. NaH (cat.), Et_2O 2. Cl_3CCN, Et_2O, –5° 3. $PdCl_2(PhCN)_2$ (5 mol%), THF, 1.5-3 h	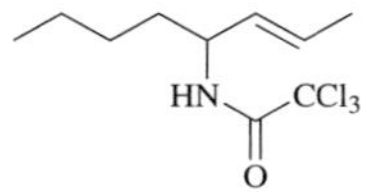 (46)	28
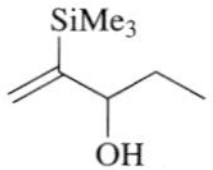	1. KH (cat.) 2. Cl_3CCN 3. Xylenes, 140° 4. NaOH 5. $(Boc)_2O$	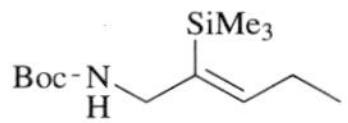 (34)	191
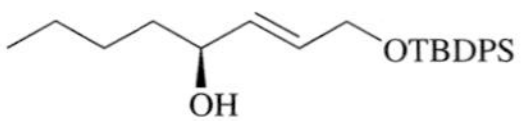 88% ee	1. DBU, Cl_3CCN 2. Toluene, 110°, 20 h	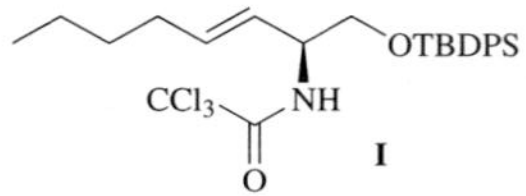 **I** (92)[a] 82% ee	29
	1. DBU, Cl_3CCN 2. $PdCl_2(PhCN)_2$ (10 mol%), benzene, rt	**I** (72)[a] 88% ee	29

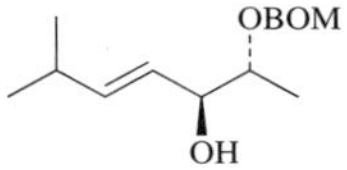	1. n-BuLi (cat.), THF, 0° 2. F_3CCN, –78° to rt 3. Xylenes, 140°, 4 h	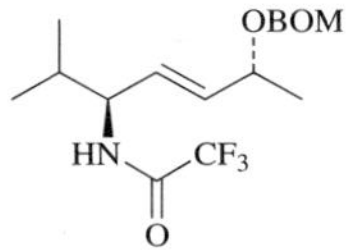	(80)	60, 75
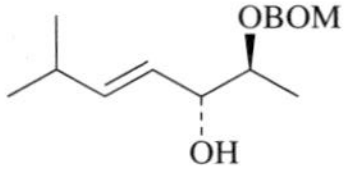	1. n-BuLi (cat.), THF, 0° 2. F_3CCN, –78° to rt 3. Xylenes, 140°, 4 h	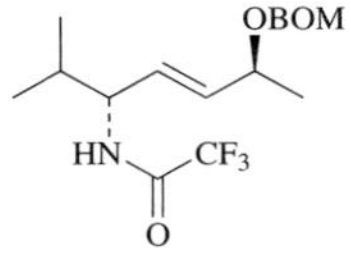	(84)	60, 75
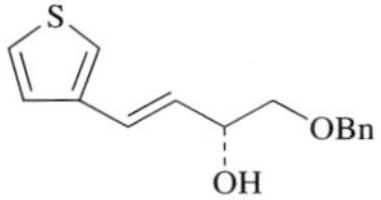	1. KH (cat.), THF 2. Cl_3CCN, THF, <0° 3. Xylenes, 140°	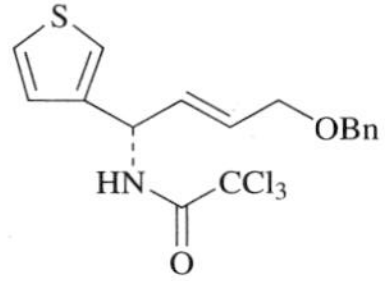	(10)	65
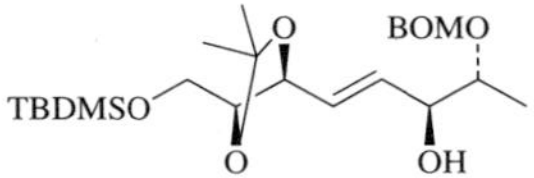	1. n-BuLi (cat.), THF, 0° 2. F_3CCN, –78° to rt 3. Xylenes, 140°, 9.5 h	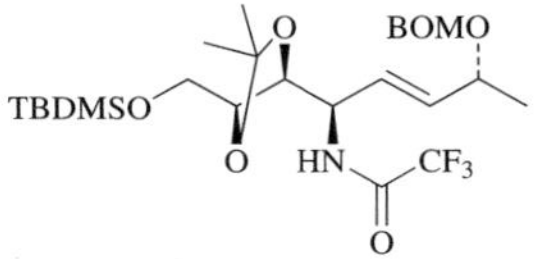	(78)	60, 75
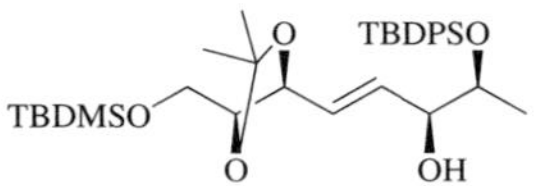	1. n-BuLi, THF, –78° 2. Cl_3CCN, –78° to rt 3. Xylenes, 140°, 48 h	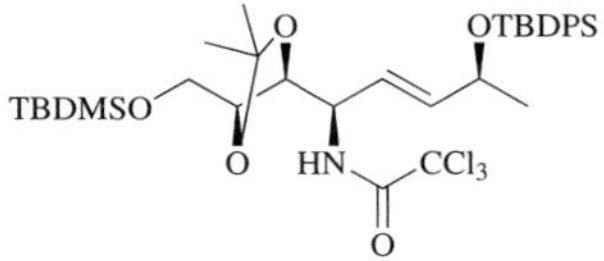	(34)	60, 75, 76
	1. n-BuLi, THF, –78° 2. F_3CCN, –78° to rt 3. Xylenes, 140°, 20 h	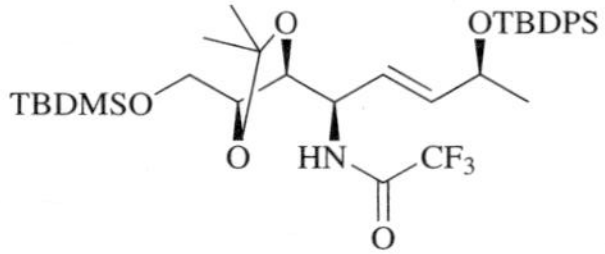	(69)	60, 75, 76

TABLE 2A. REARRANGEMENTS OF TRIHALOACETIMIDATES OF ACYCLIC SECONDARY ALLYLIC ALCOHOLS *(Continued)*

Substrate	Conditions	Product(s) and Yield(s) (%)	Refs.
C_8 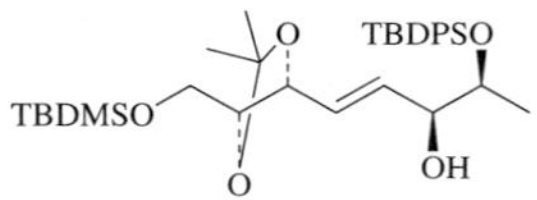	1. *n*-BuLi, THF, –78° 2. F_3CCN, –78° to rt 3. Xylenes, 140°, 20 h	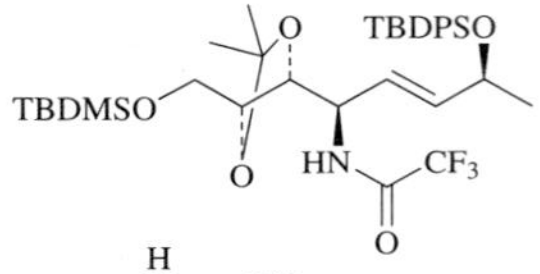 (53)	75, 76
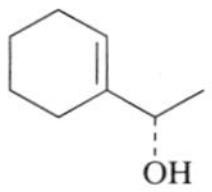	1. NaH (cat.), Et_2O 2. Cl_3CCN, Et_2O, < –5° to rt 3. Toluene, 110°, 3 h	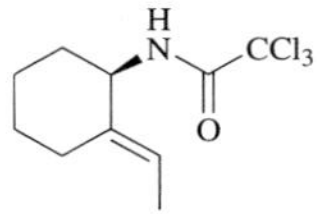 (75)	194
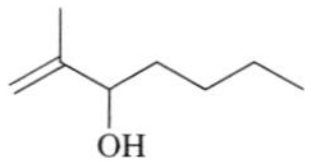	1. NaH (cat.), Et_2O 2. Cl_3CCN, Et_2O, <0° 3. Xylenes, 140°, 2.5 h	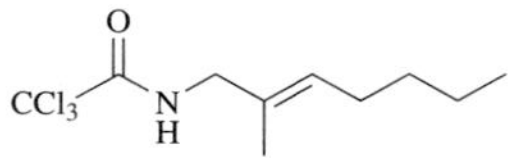 (61)	6
C_9 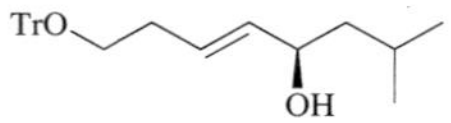	1. DBU (1.2 eq.), Cl_3CCN, CH_2Cl_2, 0° to rt 2. K_2CO_3, *p*-Xylene, 140°	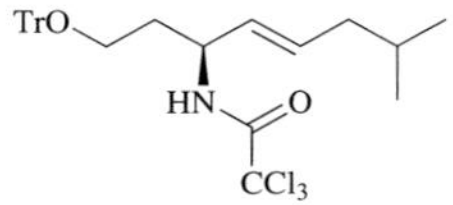 (74)	50
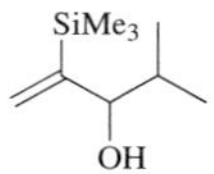	1. KH (cat.) 2. Cl_3CCN 3. Xylenes, 140° 4. NaOH 5. $(Boc)_2O$	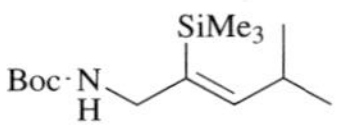 (38)	191
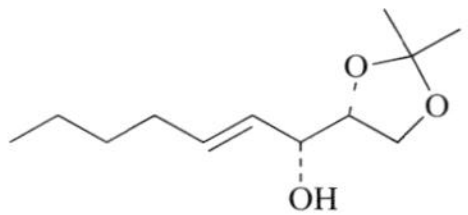	1. NaH (cat.), Et_2O, rt, 15 min 2. Cl_3CCN, –5°, 15 min 3. Toluene, 110°, 8 h	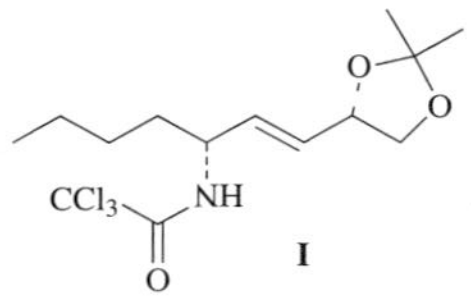 **I** (48)	28
	1. NaH (cat.), Et_2O, rt, 15 min 2. Cl_3CCN, –5°, 15 min 3. $PdCl_2(PhCN)_2$ (4-5 mol%), THF, rt, 1.5-3 h	**I** (67)	28

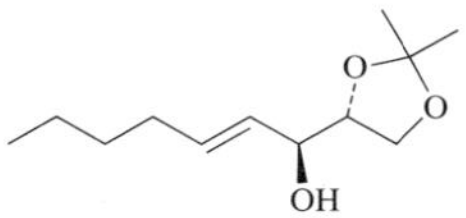	1. NaH (cat.), Et_2O, rt, 15 min 2. Cl_3CCN, –5°, 15 min 3. $PdCl_2(PhCN)_2$ (4-5 mol%), THF, rt, 1.5-3 h	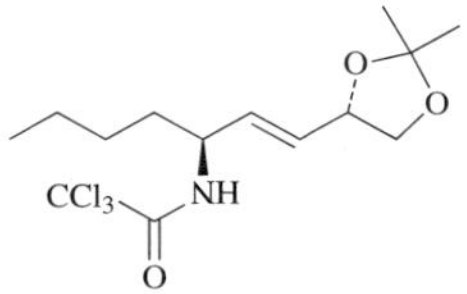 (44)	28
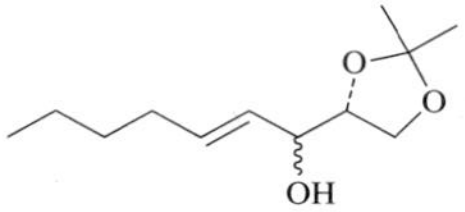	1. NaH (cat.), Et_2O, rt, 15 min 2. Cl_3CCN, –5°, 15 min 3. $PdCl_2(PhCN)_2$ (4-5 mol%), THF, rt, 1.5-3 h	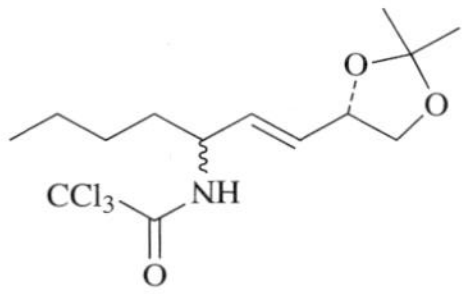 (46)	28
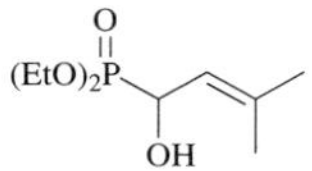	1. DBU, Cl_3CCN, CH_2Cl_2, –35°, 45 min 2. Toluene, 110°, 14 h	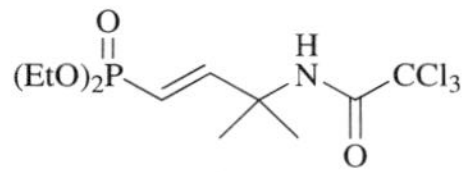 (80)	97
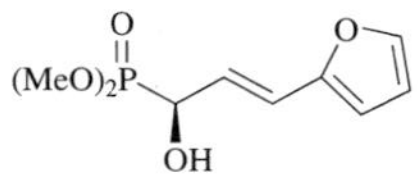	1. DBU, Cl_3CCN, CH_2Cl_2, –35°, 45 min 2. NIS, CH_2Cl_2, rt, 24 h	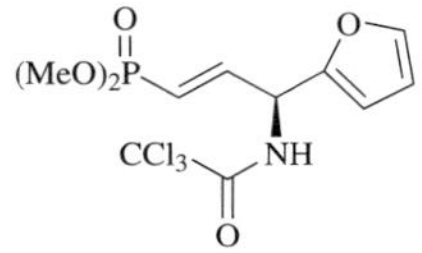 (65)[a]	66
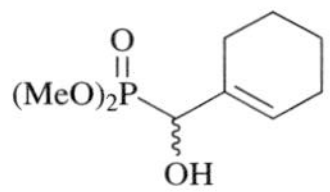	1. DBU, Cl_3CCN, CH_2Cl_2, –35°, 45 min 2. Toluene, 110°, 11 h	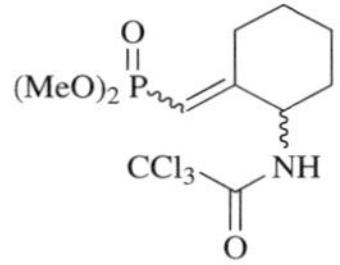 (83) E:Z 87:13	97
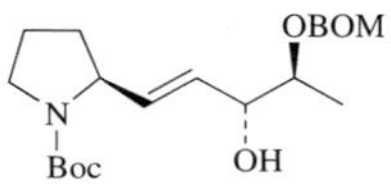	1. *n*-BuLi (cat.), THF, 0° 2. F_3CCN, –78° to rt 3. Xylenes, 140°, 7 h	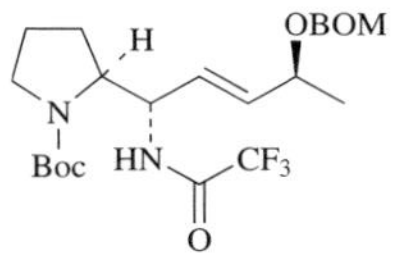 (30)	75

TABLE 2A. REARRANGEMENTS OF TRIHALOACETIMIDATES OF ACYCLIC SECONDARY ALLYLIC ALCOHOLS *(Continued)*

Substrate	Conditions	Product(s) and Yield(s) (%)	Refs.
C_9			
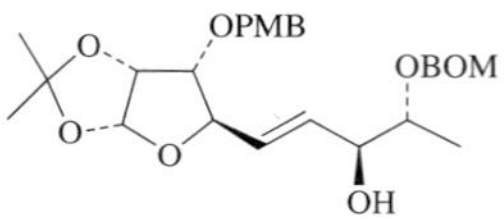	1. *n*-BuLi (20 mol%) 2. F_3CCN, THF, –78° 3. Xylene, 140°	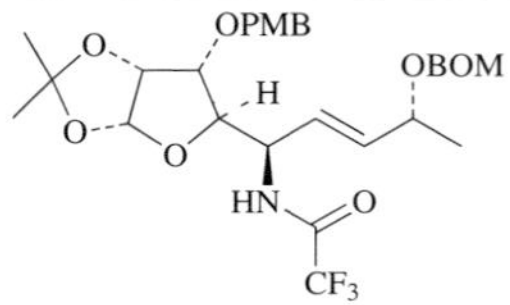 (92)	60
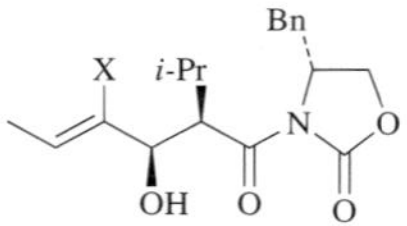 X = F, Cl	1. DBU, Cl_3CCN 2. Xylenes, 140°	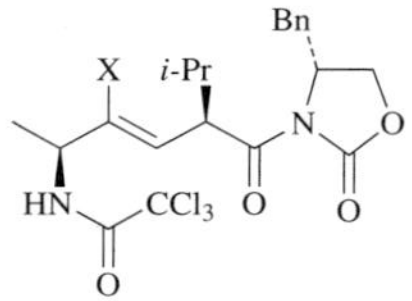 (—)	22
C_{10}			
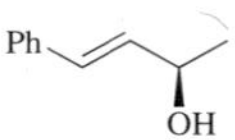	1. NaH (cat.), Et_2O, rt, 15 min 2. Cl_3CCN, –5°, 15 min 3. Toluene, 110°, 8 h	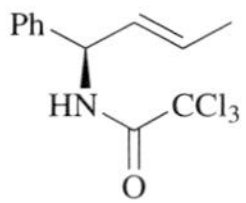 (70)	20
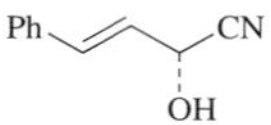	1. DBU, Cl_3CCN 2. NBS (1 eq.), $CHCl_3$, 24 h	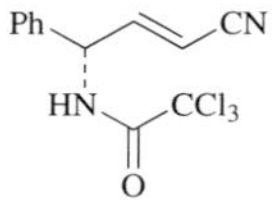 (71)[a]	66
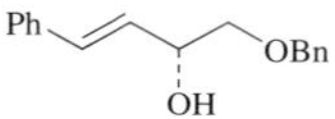	1. KH (cat.), THF 2. Cl_3CCN, THF, <0° 3. Xylenes, 140°	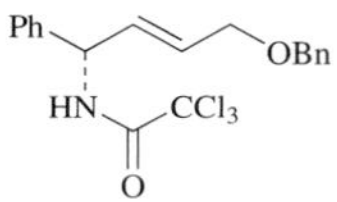 (49.5)	65
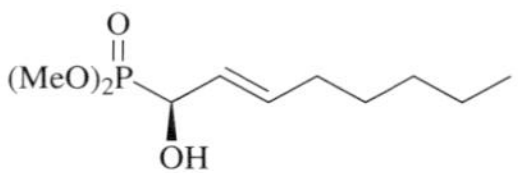	1. DBU, Cl_3CCN 2. NBS (1 eq.), $CHCl_3$, 24 h	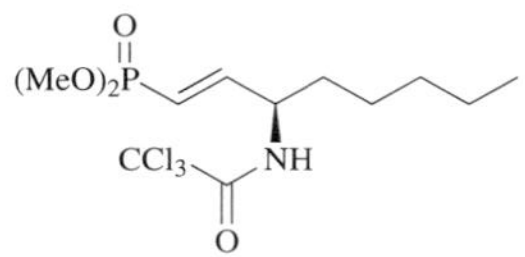 (55)[a, e]	66

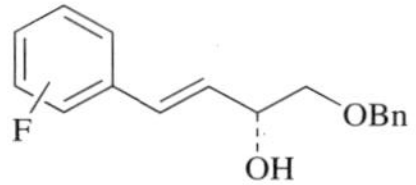	1. KH (cat.), THF 2. Cl_3CCN, THF, <0° 3. Xylenes, 140°	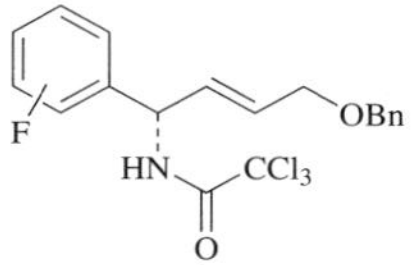	*o* (18.1) *m* (15.5) *p* (31.1)	65
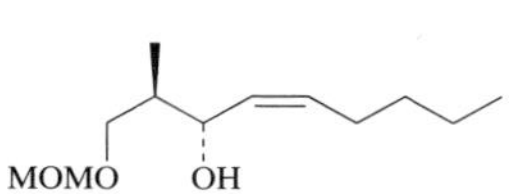	1. DBU, Cl_3CCN, CH_2Cl_2, –60°, 3 h 2. Xylenes, 140°, 12 h	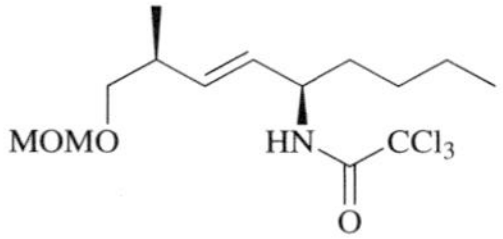	(≈57)[f]	25
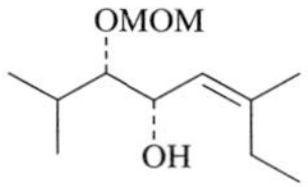	1. DBU, Cl_3CCN, CH_2Cl_2, –78°, 1 h 2. Xylenes, 140°, 4 h	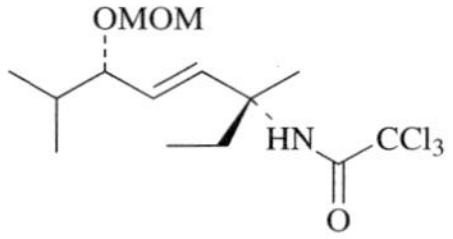	(>57)[g]	21
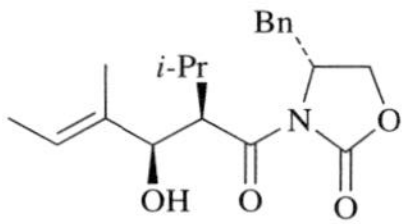	1. DBU, Cl_3CCN 2. Xylenes, 140°	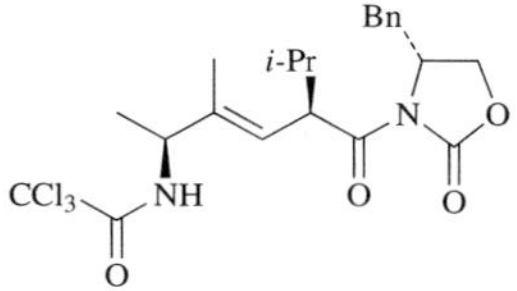	(—)	22
C_{11}				
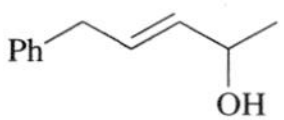	1. KH (cat.), THF 2. Cl_3CCN, Et_2O, 0° 3. Xylenes, 140°, 8 h	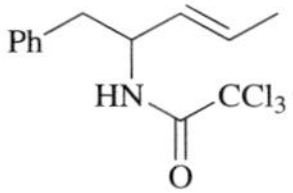	(68)	193
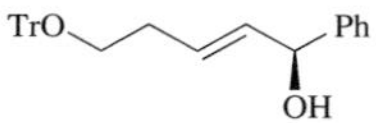	1. DBU (20 mol%), Cl_3CCN, CH_2Cl_2, 0° to rt 2. Toluene, 110°	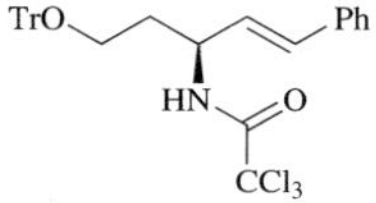	(70)	50

TABLE 2A. REARRANGEMENTS OF TRIHALOACETIMIDATES OF ACYCLIC SECONDARY ALLYLIC ALCOHOLS *(Continued)*

Substrate	Conditions	Product(s) and Yield(s) (%)	Refs.
C_{11}			
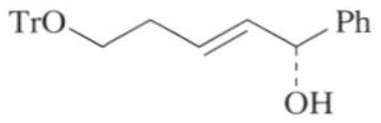	1. DBU (20 mol%), Cl_3CCN, CH_2Cl_2, 0° to rt 2. Toluene, 110°	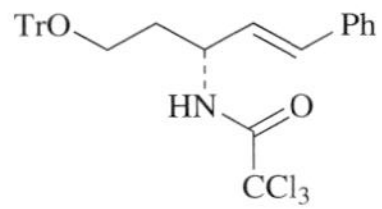 (72)	50
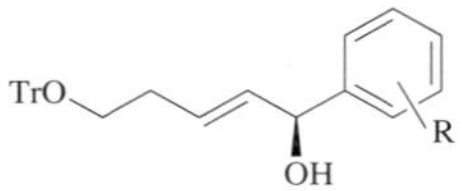	1. DBU (20 mol%), Cl_3CCN, CH_2Cl_2, 0° to rt 2. Toluene, 110°	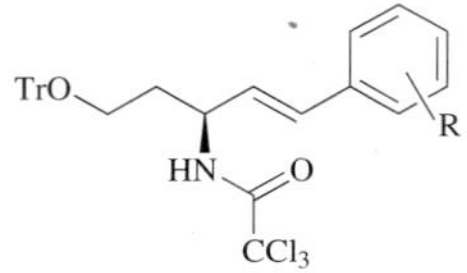 R *o*-Br (71) *p*-Cl (68) *p*-F (81)	50
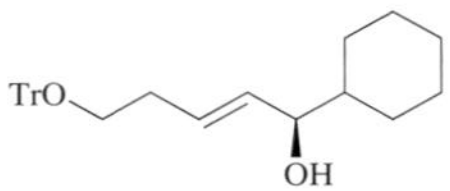	1. DBU (1.2 eq.), Cl_3CCN, CH_2Cl_2, 0° to rt 2. K_2CO_3, *p*-Xylene, 140°	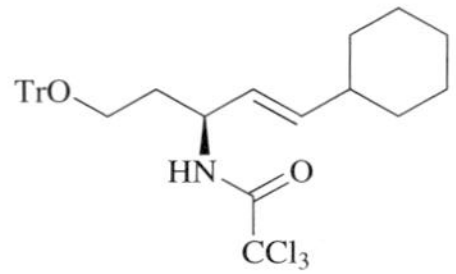 (66)	50
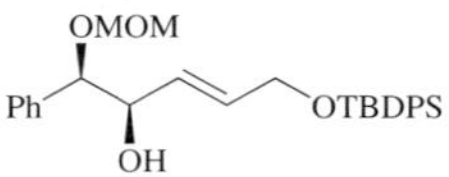	1. NaH (cat.), Et_2O 2. Cl_3CCN, Et_2O, –5° to rt 3. Xylenes, 140°, 6 h	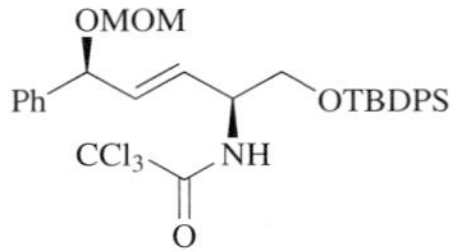 (70)	119
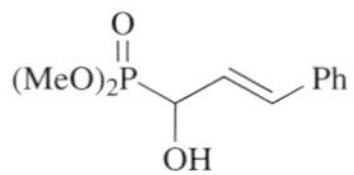	1. DBU, Cl_3CCN, CH_2Cl_2, –35°, 30 min 2. Toluene, 110°, 24 h	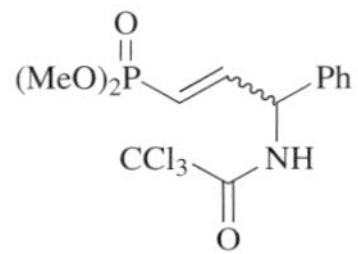 (90) E:Z 90:10	97
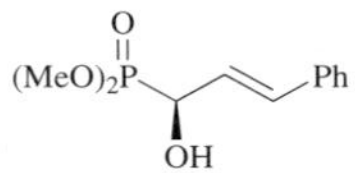	1. DBU, Cl_3CCN 2. NBS (1 eq.), $CHCl_3$, 24 h	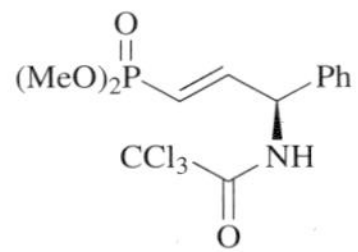 (91)[a]	66

	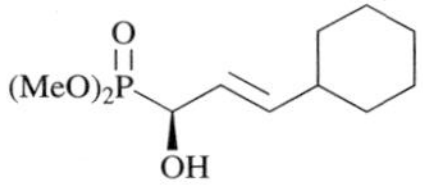	1. DBU, Cl_3CCN 2. NBS (1 eq.), $CHCl_3$, 24 h	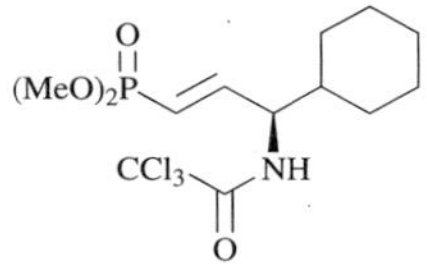 (89)[a]	66
	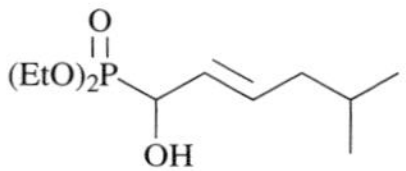	1. DBU, Cl_3CCN, CH_2Cl_2, –35°, 40 min 2. Toluene, 110°, 21 h	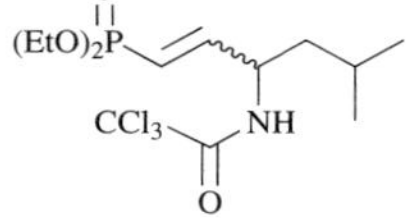 (85) E:Z 87:13	97
	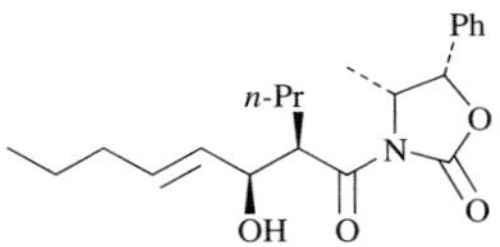	1. DBU (cat.), Cl_3CCN, CH_2Cl_2, 0° 2. Xylenes, 140°	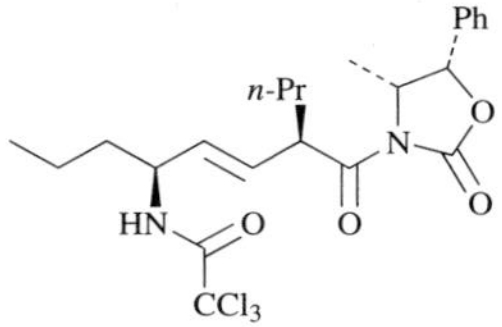 (—)	195
C_{12}	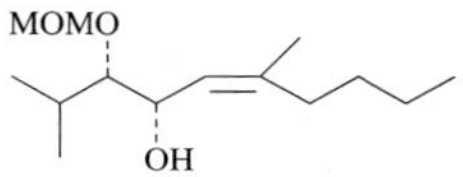	1. DBU, Cl_3CCN, CH_2Cl_2, –78°, 1 h 2. Xylenes, 140°, 4-10 h	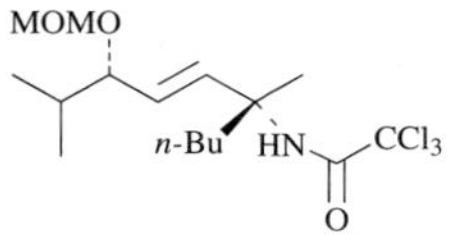 (>59)[g]	21
	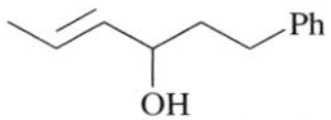	1. NaH (cat.), Et_2O, 0° 2. Cl_3CCN, Et_2O 3. Xylenes, 140°, 14 h 4. NaOH (6 N), EtOH	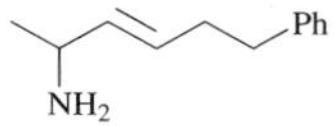 (—)	196

TABLE 2A. REARRANGEMENTS OF TRIHALOACETIMIDATES OF ACYCLIC SECONDARY ALLYLIC ALCOHOLS *(Continued)*

Substrate	Conditions	Product(s) and Yield(s) (%)	Refs.
C_{12} 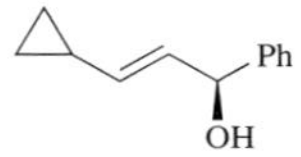	DBU (cat.), Cl_3CCN, Et_2O, 0° to rt, 4 h	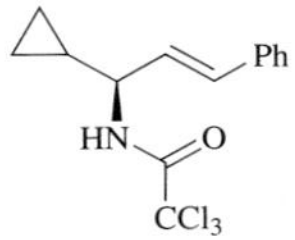 (93)	114
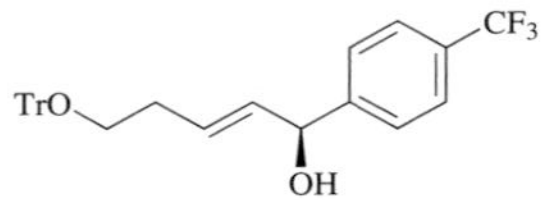	1. DBU (20 mol%), Cl_3CCN, CH_2Cl_2, 0° to rt 2. Toluene, 110°	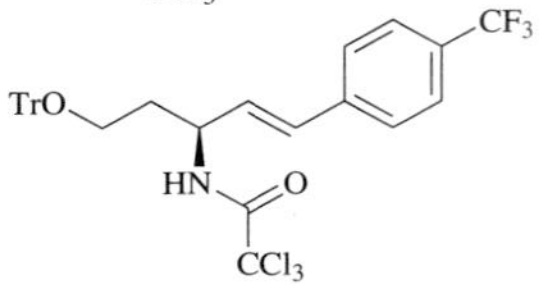 (69)	50
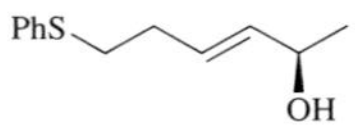	1. KH (cat.) 2. Cl_3CCN 3. Ethyl acetate, 77°, 12 h	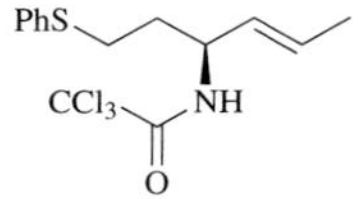 (100)	51
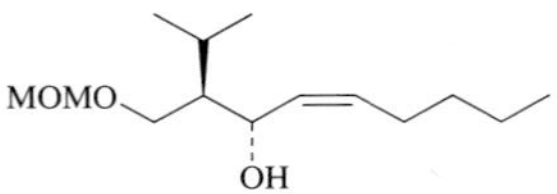	1. DBU, Cl_3CCN, CH_2Cl_2, –60°, 3 h 2. Xylenes, 140°, 12 h	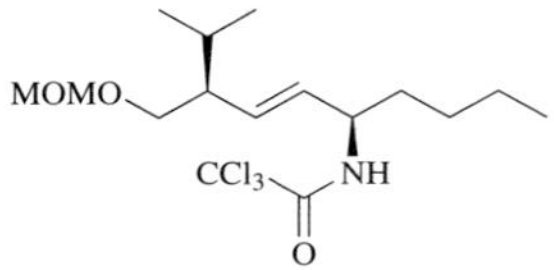 (≈68)[f]	25
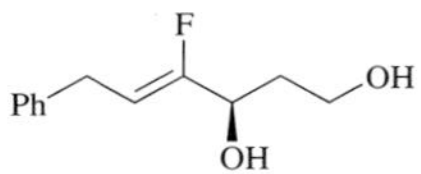	1. NaH (cat.), THF 2. Cl_3CCN (2 eq.), Et_2O 3. Xylenes, 140°	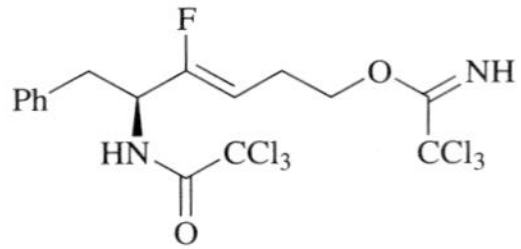 (73)	67, 183
	1. KH (cat.), THF 2. Cl_3CCN, Et_2O, –5° to 0° 3. rt	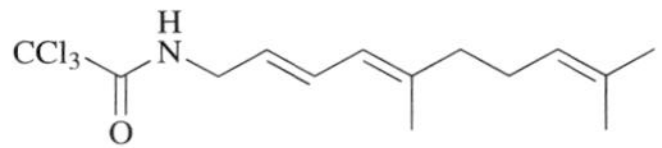 (39) + 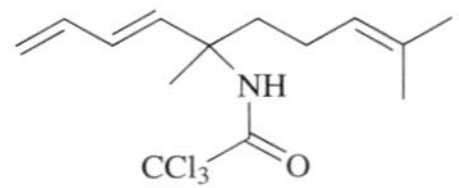 (27)	6

	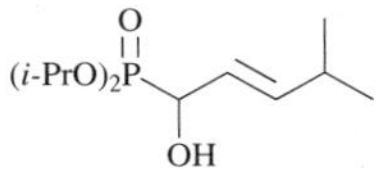	1. DBU, Cl_3CCN, CH_2Cl_2, –35°, 55 min 2. Toluene, 110°, 19 h	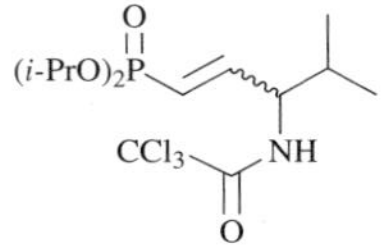	(89) E:Z 86:14	97
	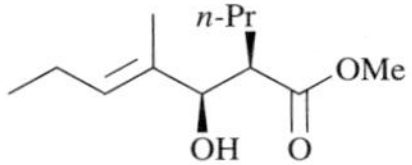	1. DBU, Cl_3CCN, CH_2Cl_2, rt 2. Xylenes, 140°, 18 h	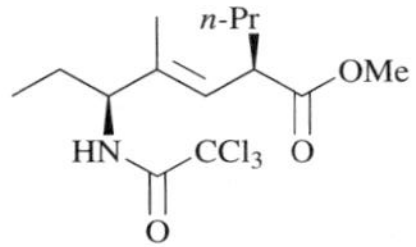	(100)[a]	23
	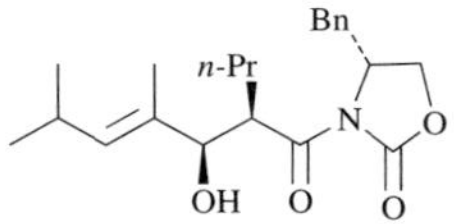	1. DBU, F_3CCN, CH_2Cl_2, –196° to rt 2. Xylenes, 140°, 18 h	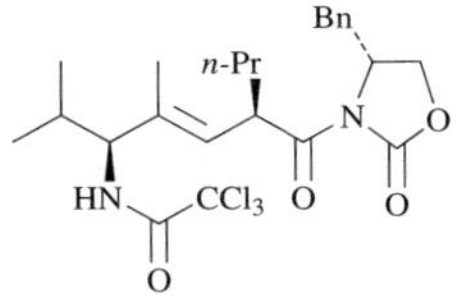	(—)	23
	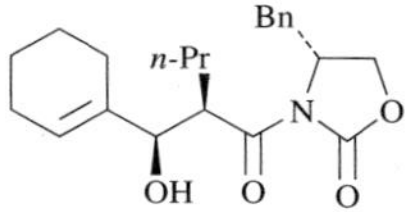	1. DBU, Cl_3CCN, CH_2Cl_2, rt 2. Xylenes, 140°, 18 h	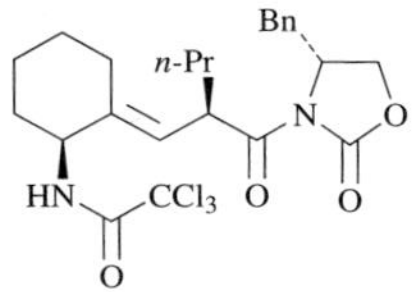	(—)	23
C_{13}	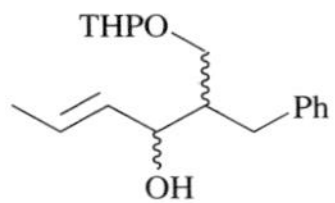	1. NaH (cat.), Et_2O, 0° 2. Cl_3CCN, 0° to rt 3. Xylenes, 140°, 2 h	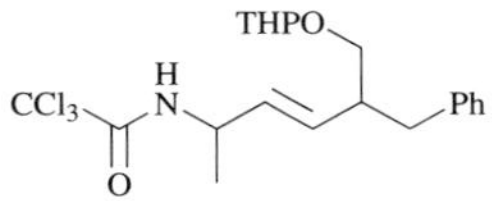	(81)	44
	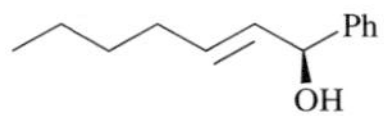	1. KH (cat.), Cl_3CCN, 0° to rt 2. Toluene, 110°	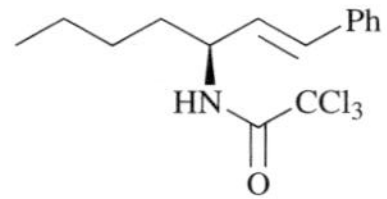	(64)	114

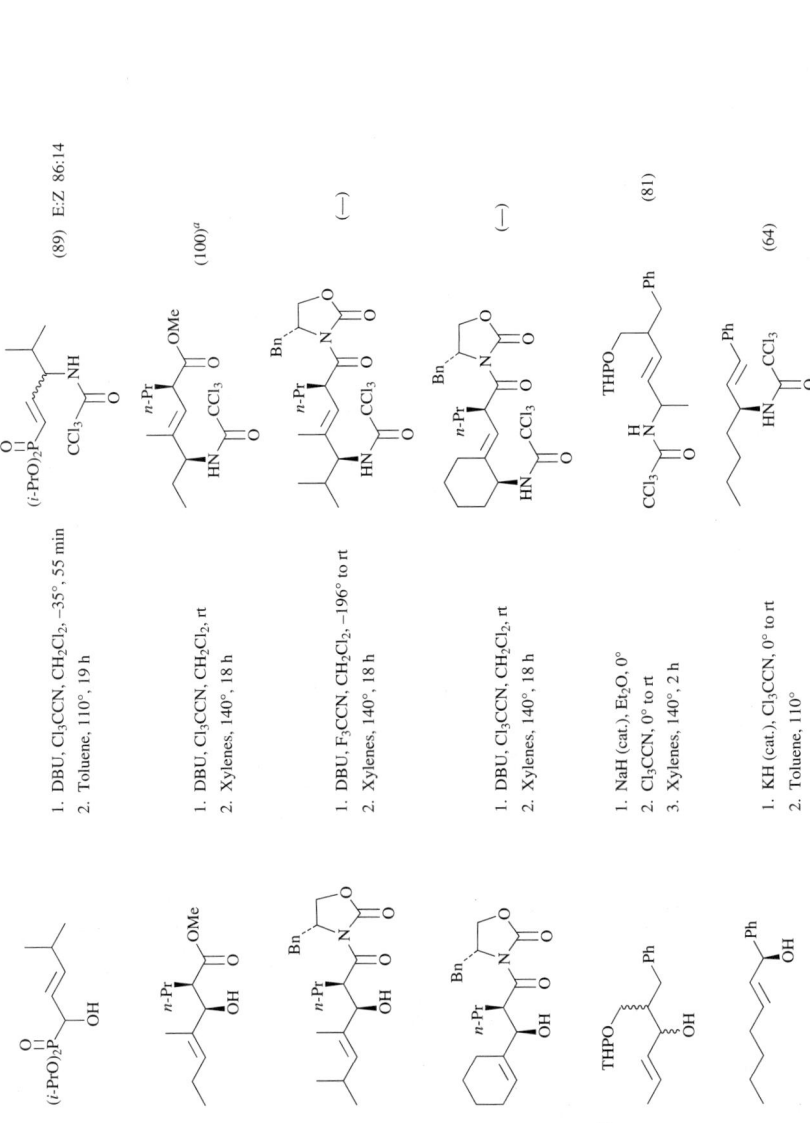

TABLE 2A. REARRANGEMENTS OF TRIHALOACETIMIDATES OF ACYCLIC SECONDARY ALLYLIC ALCOHOLS *(Continued)*

Substrate	Conditions	Product(s) and Yield(s) (%)	Refs.
C_{13} 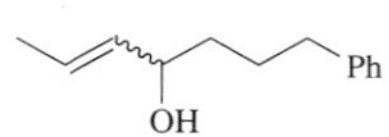	1. NaH (cat.), Et_2O, 0° 2. Cl_3CCN, Et_2O 3. Xylenes, 140°, 14 h 4. NaOH, EtOH	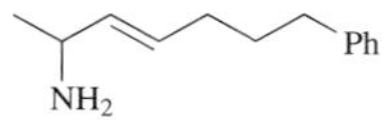 (67)	196
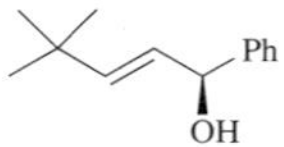	1. KH (cat.), Cl_3CCN, 0° to rt 2. Toluene, 110°	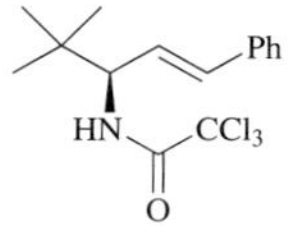 (90)	114
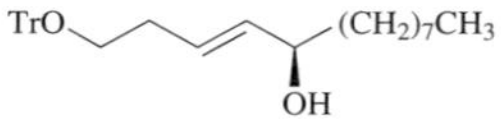	1. DBU (1.2 eq.), Cl_3CCN, CH_2Cl_2, 0° to rt 2. K_2CO_3, *p*-xylene, 140°	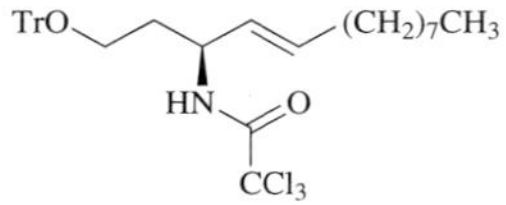 (68)	50
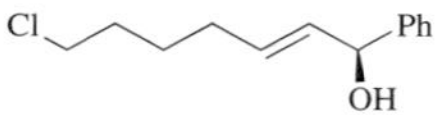	1. KH (cat.), Cl_3CCN, 0° to rt 2. Toluene, 110°	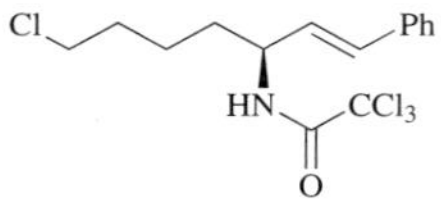 (66)	114
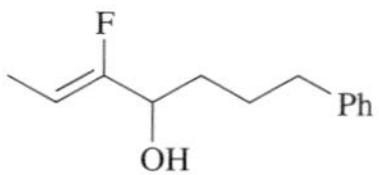	1. NaH (cat.), Et_2O, 0° 2. Cl_3CCN, Et_2O 3. Xylenes, 140°, 14 h 4. NaOH, EtOH	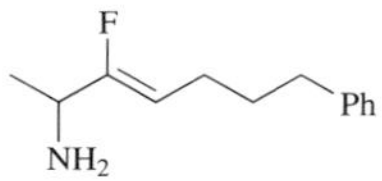 (83)	196
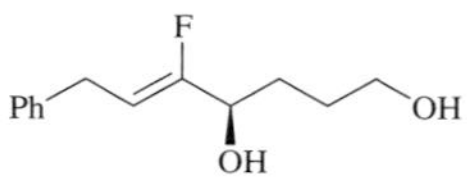	1. NaH 2. Cl_3CCN 3. Xylenes, 140°	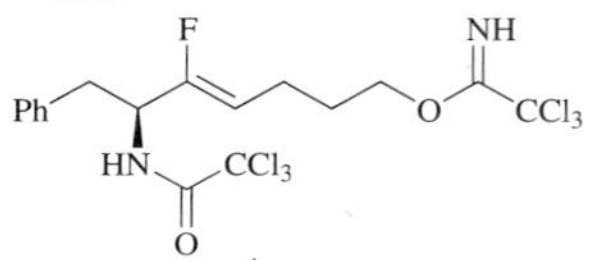 (73)	197
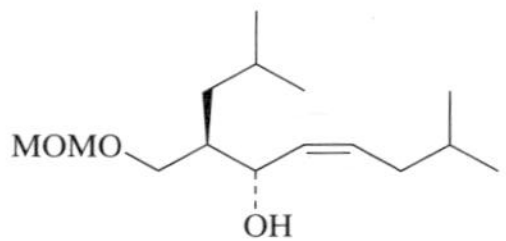	1. DBU, Cl_3CCN, CH_2Cl_2, –60°, 3 h 2. Xylenes, 140°, 12 h	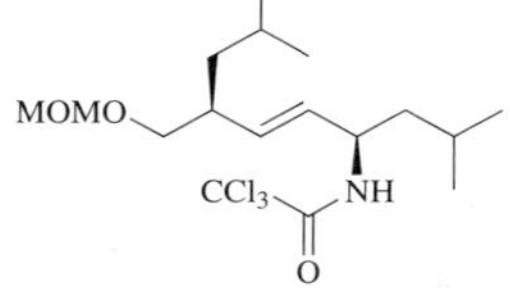 (≈71)[f]	25

	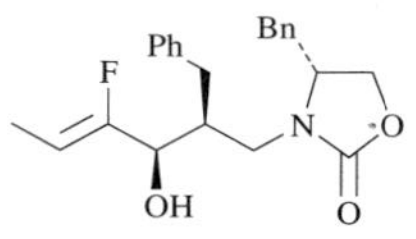	1. DBU, Cl_3CCN 2. Xylenes, 140°	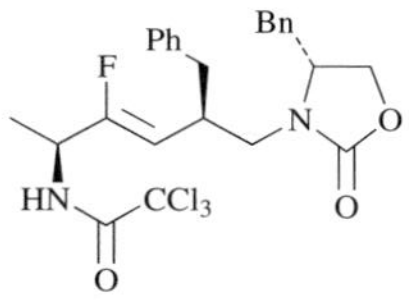 (—)	22
	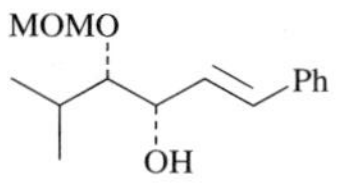	1. DBU, Cl_3CCN, –78°, 1 h 2. Xylenes, 140°, 4 h	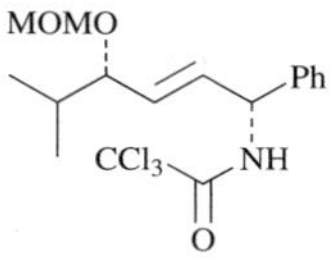 (>66)[g]	21
C_{15}	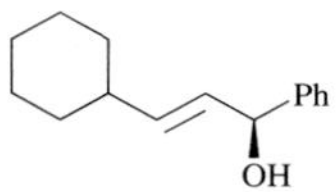	1. KH (cat.), Cl_3CCN, 0° to rt 2. Toluene, 110°	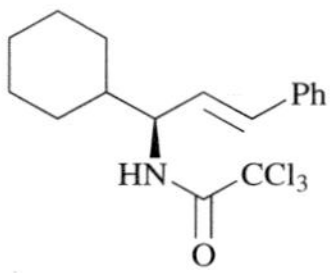 (96)	114
	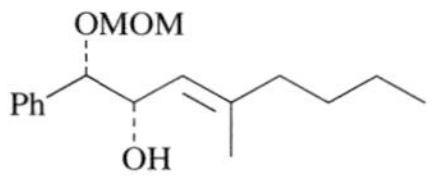	1. DBU, Cl_3CCN, –78°, 1 h 2. Xylenes, 140°, 4 h	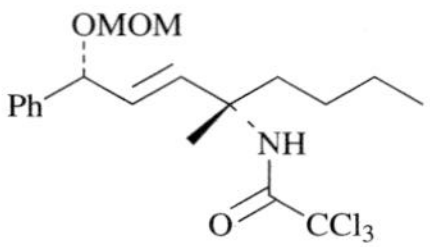 (>93)[f]	21
C_{18}	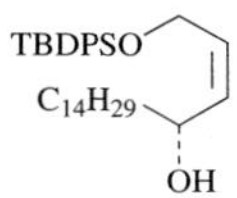	1. DBU, Cl_3CCN, CH_2Cl_2, –5° 2. Xylenes, 140°, 7 h	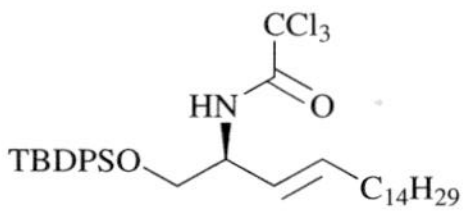 (81)	26
C_{19}	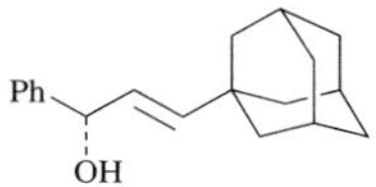	1. KH (cat.), Et_2O, rt, 15 min 2. Cl_3CCN, Et_2O, 0° to rt 3. Toluene, 110°, 1-2 h	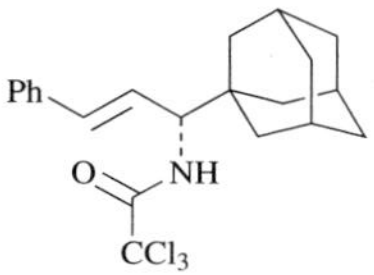 (83)	114

TABLE 2A. REARRANGEMENTS OF TRIHALOACETIMIDATES OF ACYCLIC SECONDARY ALLYLIC ALCOHOLS (*Continued*)

Substrate	Conditions	Product(s) and Yield(s) (%)	Refs.
C_{21} 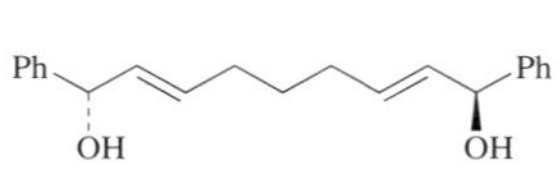	1. KH (cat.), Et_2O, rt, 15 min 2. Cl_3CCN, Et_2O, 0° to rt 3. Toluene, 110°, 1-2 h	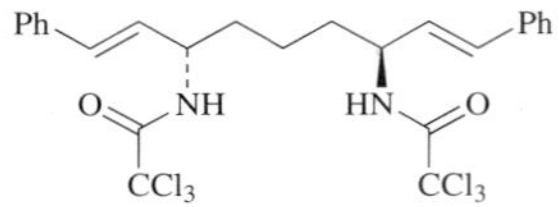 (65)[h]	114
C_{22} 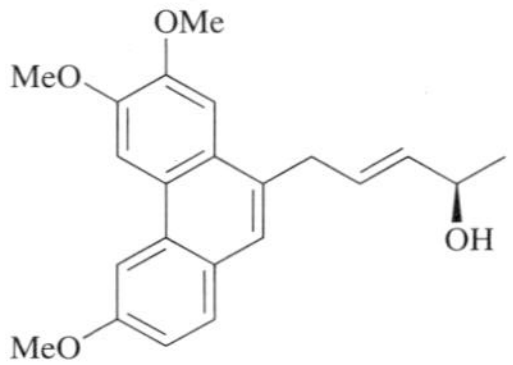	1. DBU (cat.), Cl_3CCN, CH_2Cl_2, 0°, 2 h 2. Toluene, 110°, 12 h	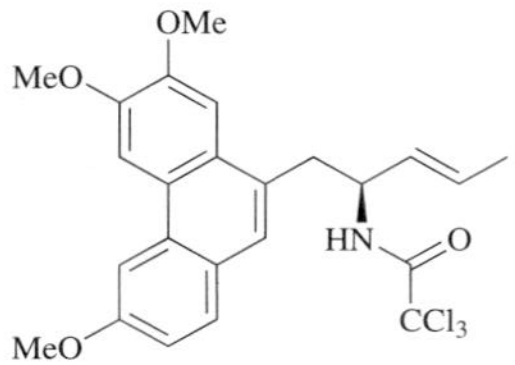 (92)	198

[a] This is the yield for rearrangement only.

[b] This is the NMR yield; elimination was the major reaction.

[c] Product was obtained in 86% crude yield on a 4 gram scale.

[d] This is the NMR yield.

[e] Rearrangement of the Z isomer gave 71% yield of the 5-exo-oxazoline halocyclization product.

[f] Attempted catalysis with Pd(II) or Hg(II) led only to low (≈20-30%) yields of the desired product accompanied by diene formation from elimination of trichloroacetamide.

[g] Attempted catalysis with Pd(II) or Hg(II) led only to diene formation from elimination of trichloroacetamide.

[h] Standard rearrangement conditions led to recovery of 27% of the mono trichloroacetamide, which was recycled to provide a 65% combined yield of the desired bis(allylic trichloroacetamide).

TABLE 2B. REARRANGEMENTS OF TRIHALOACETIMIDATES OF CYCLIC SECONDARY ALLYLIC ALCOHOLS

	Substrate	Conditions	Product(s) and Yield(s) (%)	Refs.
C_5	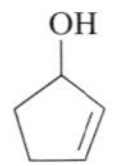	1. NaH (cat.), Et_2O 2. Cl_3CCN 3. Xylene, 140°	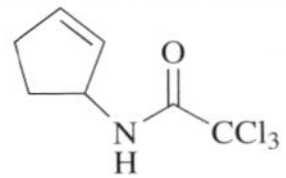 (80)[a]	123
	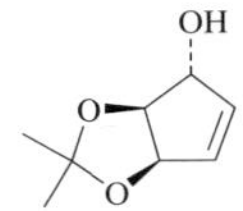	1. DBU, Cl_3CCN, CH_2Cl_2, 0°, 10 min 2. Xylenes, 140°, 4 h	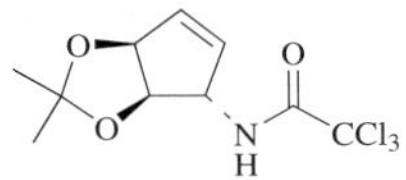 (89)	73, 74
C_6	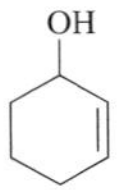	1. NaH (cat.), Et_2O 2. Cl_3CCN 3. Xylene, 140°	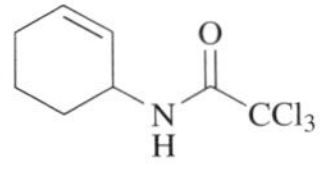 **I** (69)	4, 6, 72, 93, 113, 199
		1. NaH (cat.), Et_2O 2. Cl_3CCN 3. $Hg(O_2CCF_3)_2$ (30 mol%), THF, 0° to rt, 1 h	**I** (<5)[b]	4, 6
	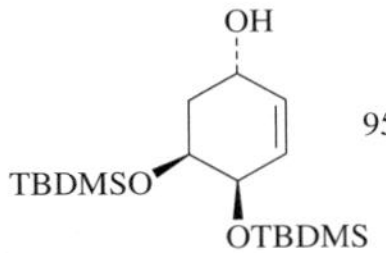 95% ee	1. DBU, Cl_3CCN, THF, rt, 1.5 h 2. K_2CO_3, xylene, 140°, 18 h	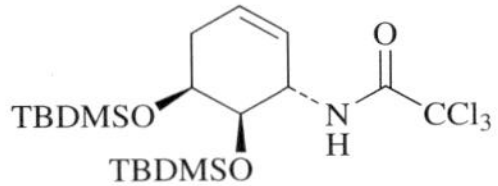 (95)[c] 95% ee	82, 200
	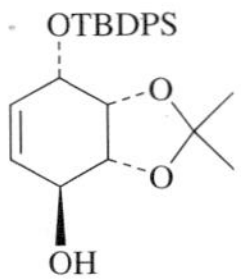	1. KH, Cl_3CCN, THF, rt, 1.5 h 2. Xylene, 140°, 4 h	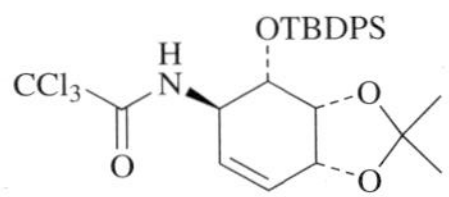 (92)	201
	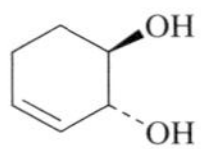	1. NaH (cat.) 2. Cl_3CCN (2 eq.), THF, 0° to rt 3. Xylene, 140°, 6 h	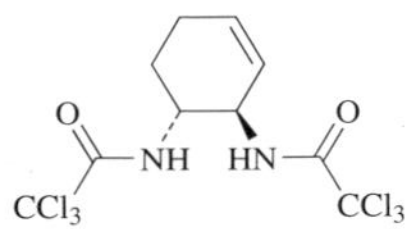 (56)	167

TABLE 2B. REARRANGEMENTS OF TRIHALOACETIMIDATES OF CYCLIC SECONDARY ALLYLIC ALCOHOLS *(Continued)*

	Substrate	Conditions	Product(s) and Yield(s) (%)	Refs.
C_6	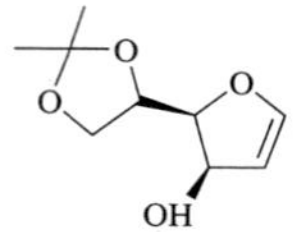	1. NaH (cat.) 2. Cl_3CCN, Et_2O, 0°	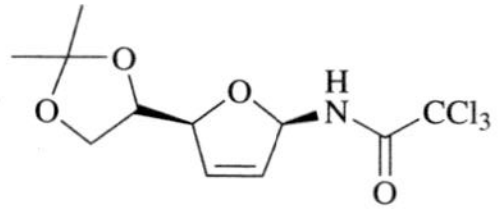 (78)	88
	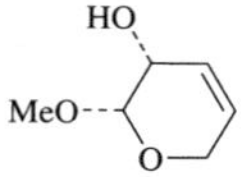	1. KH (cat.) 2. Cl_3CCN, –10° to rt 3. *o*-Dichlorobenzene, 165°, 8 h	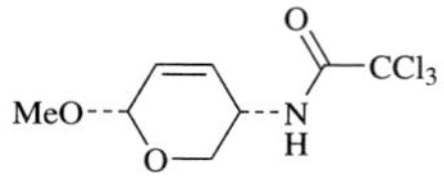 (50)	85
	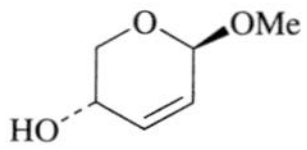	1. NaH 2. Cl_3CCN 3. Xylene, 138°	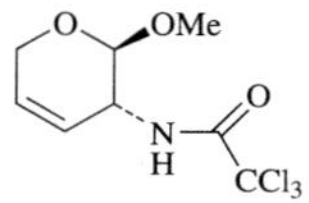 (80)	86
	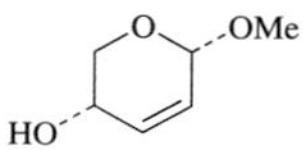	1. NaH 2. Cl_3CCN 3. Xylene	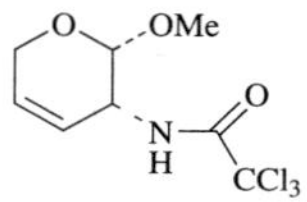 (—)[d]	86
	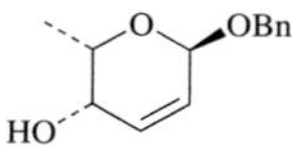	1. NaH, CH_2Cl_2, rt 2. Cl_3CCN, CH_2Cl_2, 0° 3. *o*-Dichlorobenzene, 160°, 5 h	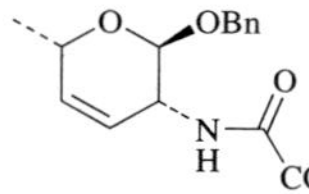 (68)	45
	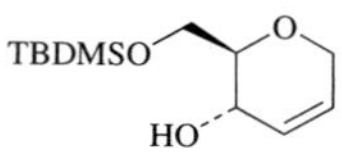	1. DBU, CH_2Cl_2, 0° 2. Cl_3CCN, 0° 3. [K_2CO_3], *p*-xylene, 138°	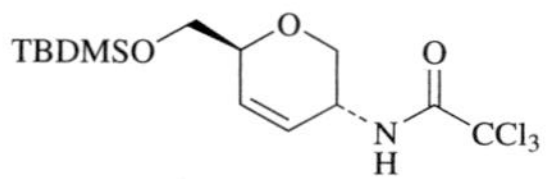 *with or without* *K_2CO_3* (85)	63, 93, 111, 202
	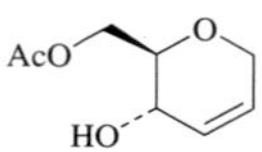	1. DBU, Cl_3CCN, –17° 2. Mol sieves (4Å), Ph_2O, rt to 210°, 35 min	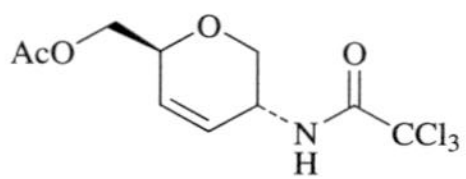 (25)	108

	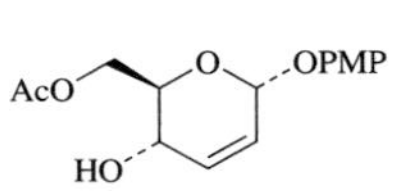	1. DBU, Cl_3CCN, –17° 2. Mol sieves (4Å), Ph_2O, rt to 210°, 35 min	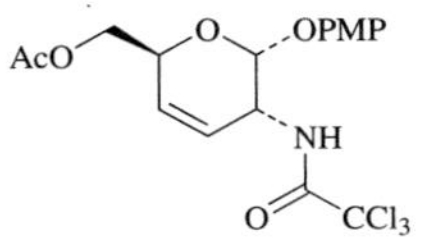	$(30)^e$	108
	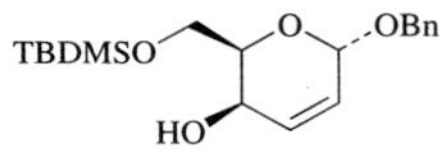	1. DBU, Cl_3CCN, CH_2Cl_2, –20° to rt 2. K_2CO_3 (cat.), xylene, 138°, 18 h	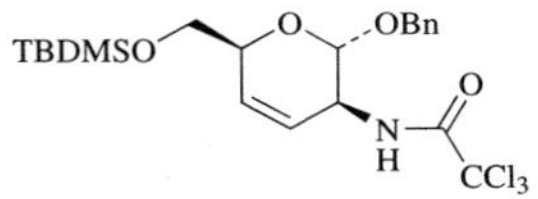	(94)	123, 125, 202
	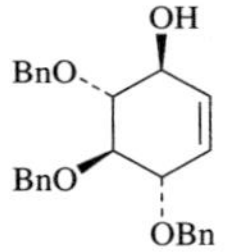	1. NaH, CH_2Cl_2 2. Cl_3CCN 3. Xylene, 140°, 9 h	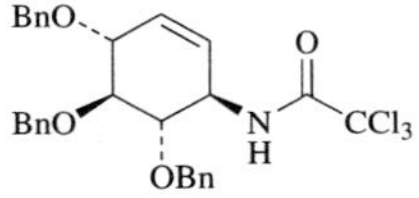	(81)	137
C_7	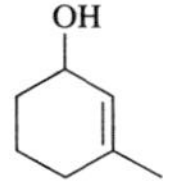	1. DBU, Cl_3CCN, CH_2Cl_2, –20° 2. [K_2CO_3], toluene, 110°	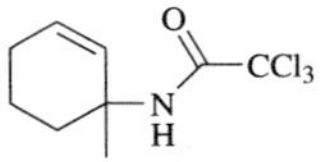	*without K_2CO_3* (72) *with K_2CO_3* (80)	63
	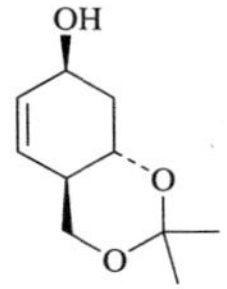	1. DBU, Cl_3CCN, 0°, 1 h 2. K_2CO_3, *p*-xylene, 138°, 12 h	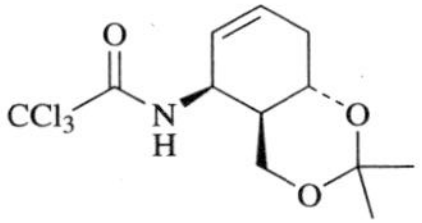	(65)	110
	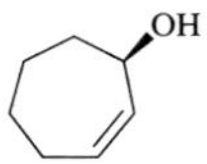	1. NaH 2. Cl_3CCN 3. Xylene, 140°	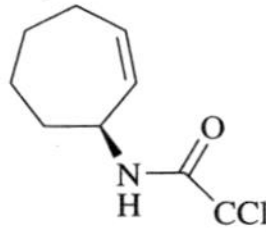	(59)	123, 124
	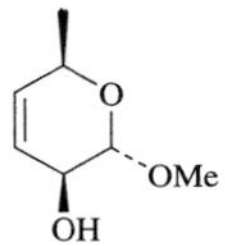	1. KH (cat.), THF 2. Cl_3CCN 3. *o*-Dichlorobenzene, 165°, 11 h	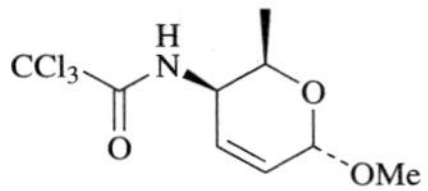	(52)	84

TABLE 2B. REARRANGEMENTS OF TRIHALOACETIMIDATES OF CYCLIC SECONDARY ALLYLIC ALCOHOLS *(Continued)*

	Substrate	Conditions	Product(s) and Yield(s) (%)	Refs.
C_7	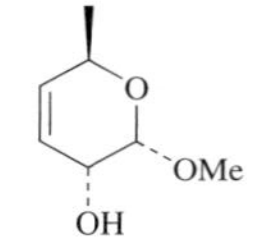	1. KH (cat.), THF 2. Cl_3CCN 3. *o*-Dichlorobenzene, 165°, 6 h	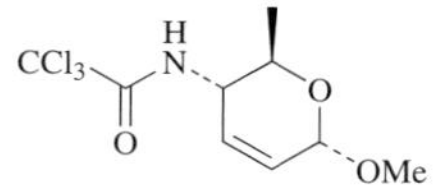 (64)	84
	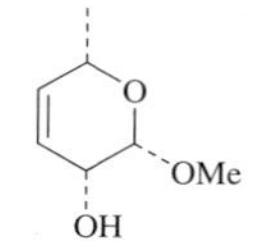	1. KH, CH_2Cl_2 2. Cl_3CCN, –10° 3. *o*-Dichlorobenzene, 165°	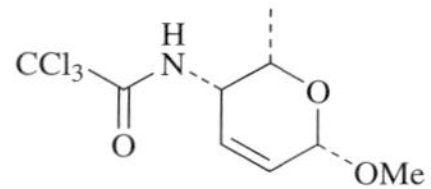 (11)	85
	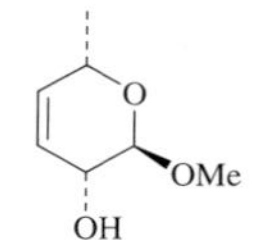	1. KH, CH_2Cl_2 2. Cl_3CCN, –10° 3. *o*-Dichlorobenzene, 165°	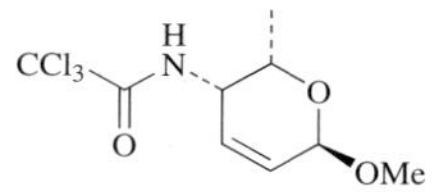 (70)	85
	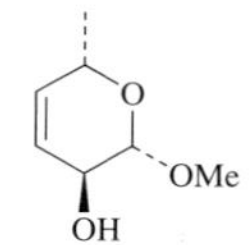	1. KH, CH_2Cl_2 2. Cl_3CCN, –10° 3. *o*-Dichlorobenzene, 165°	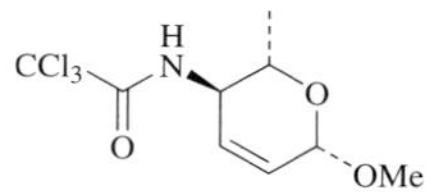 (52)	85
	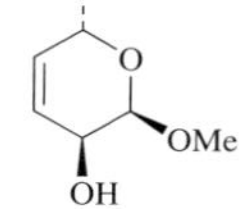	1. KH, CH_2Cl_2 2. Cl_3CCN, –10° 3. *o*-Dichlorobenzene, 165°	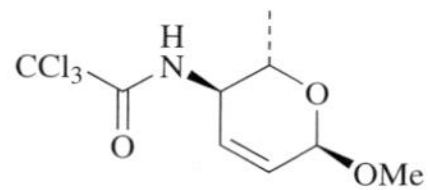 (18)[f]	85
	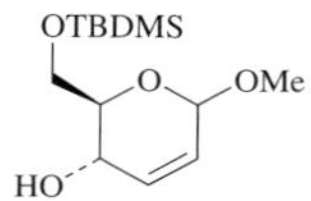	1. DBU (1.2 eq.), Cl_3CCN, CH_2Cl_2, –20° to rt 2. K_2CO_3, Ph_2O, 195°, 3 h	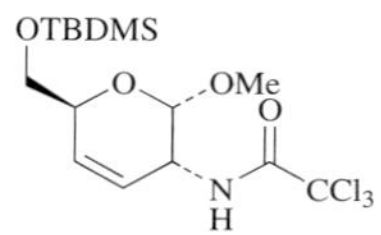 (63)	125

C_8	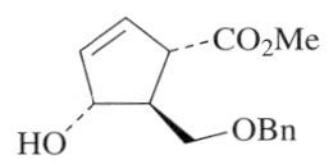	1. NaH (cat.), Et_2O 2. Cl_3CCN, Et_2O, –30° to –10° 3. Xylene, 133°, 6 h	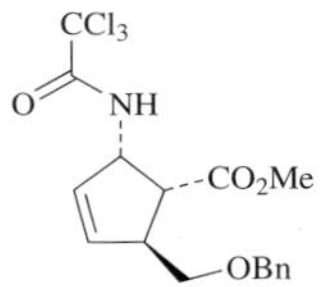 (54)[g]	203
	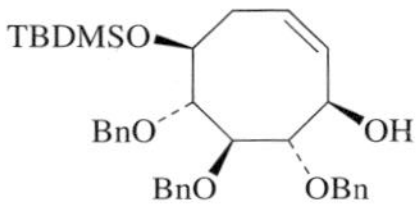	1. NaH, CH_2Cl_2, 0° 2. Cl_3CCN, 0°, 6 h 3. K_2CO_3, *o*-dichlorobenzene, heat, 2 h	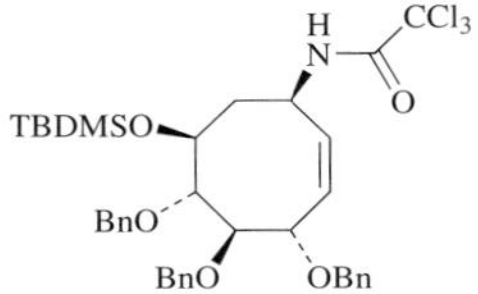 (46)	204
	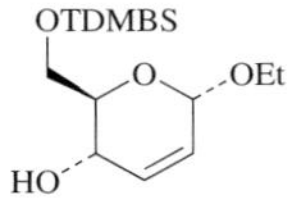	1. DBU, Cl_3CCN, CH_2Cl_2, 0° 2. [K_2CO_3], *o*-dichlorobenzene, 165°	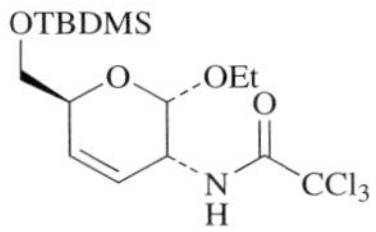 *without K_2CO_3* (7) *with K_2CO_3* (56)	63
C_9	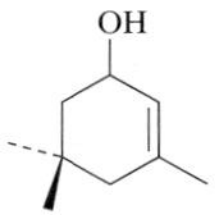	1. DBU, Cl_3CCN, CH_2Cl_2, –20° 2. [K_2CO_3], toluene, 110°	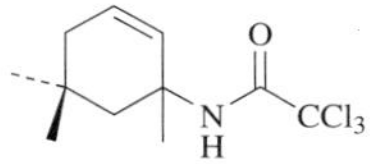 **I** *without K_2CO_3* (10–43) *with K_2CO_3* (19)	4, 6 63, 93
		1. NaH (cat.), Et_2O 2. Cl_3CCN 3. $Hg(O_2CCF_3)_2$ (20 mol%), THF, 0° to rt, 1 h	**I** (20)[h]	4
	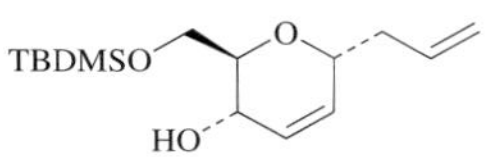	1. DBU (1 eq.), Cl_3CCN, CH_2Cl_2, 0° 2. *o*-Dichlorobenzene, 180°, 5 h	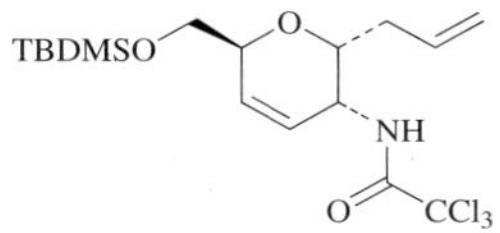 (69)	111

TABLE 2B. REARRANGEMENTS OF TRIHALOACETIMIDATES OF CYCLIC SECONDARY ALLYLIC ALCOHOLS *(Continued)*

Substrate	Conditions	Product(s) and Yield(s) (%)	Refs.
C_{10} 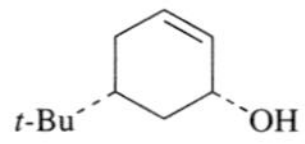	1. DBU, Cl_3CCN 2. K_2CO_3, xylene, heat	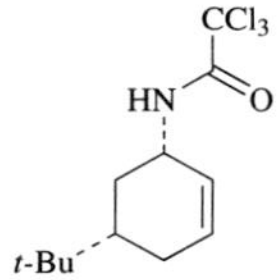 (75)[i]	93
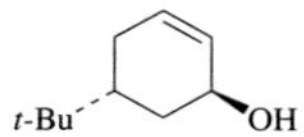	1. DBU, Cl_3CCN 2. K_2CO_3, xylene, heat	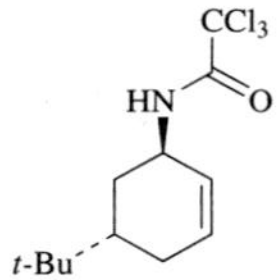 (71)	93, 123, 202
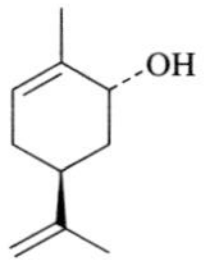	1. KH (cat.), Et_2O 2. Cl_3CCN, Et_2O, –15° to 0°, 1.5 h 3. Toluene, 110°, 10 h	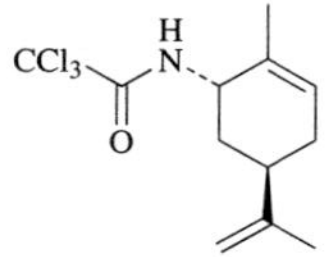 (35)	20
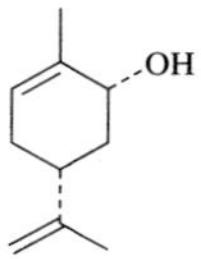	1. KH (cat.), Et_2O 2. Cl_3CCN, Et_2O, –15° to 0°, 1.5 h 3. Xylene, 140°, 8 h	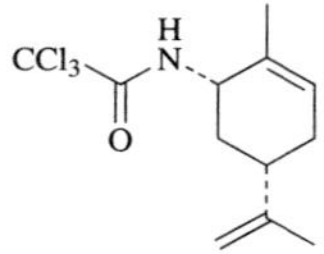 (42)	20
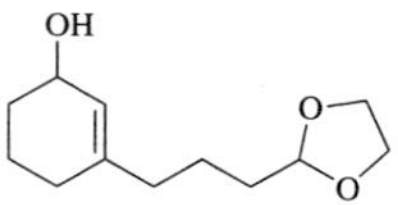	1. NaH (cat.), THF 2. Cl_3CCN, Et_2O, –5° to 0° 3. Hexane, 69°, 5 d	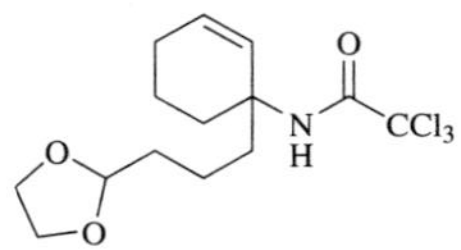 (20)	6, 5

C_{11-13} 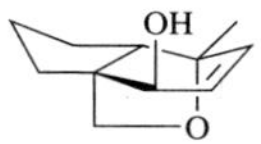	1. NaH, Et_2O 2. Cl_3CCN 3. Toluene, 170°, 20 h	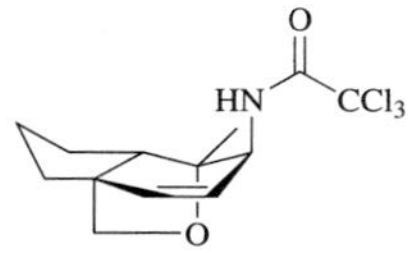 (62)	205
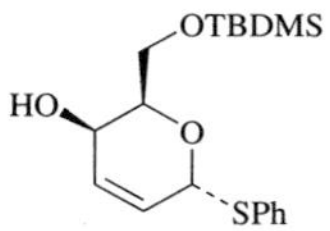	1. NaH, CH_2Cl_2 2. Cl_3CCN, rt, 24 h 3. Xylene, 140°, 24 h	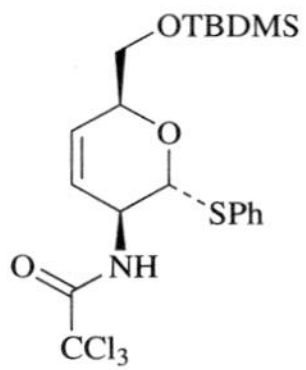 (29)	70
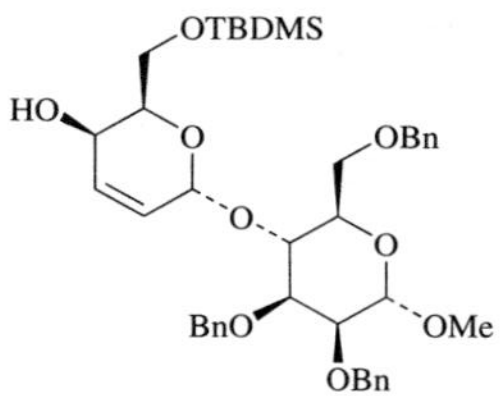	1. NaH, CH_2Cl_2 2. Cl_3CCN, rt, 2.5 h 3. DMF, 140°, 6 h	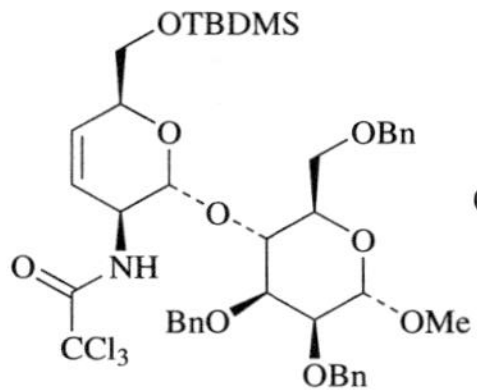 (76)	70
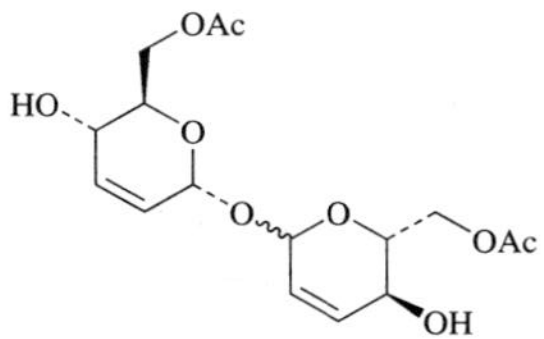	1. DBU, Cl_3CCN, CH_2Cl_2, –17°, 15 min 2. Mol sieves (4Å), Ph_2O, 210°	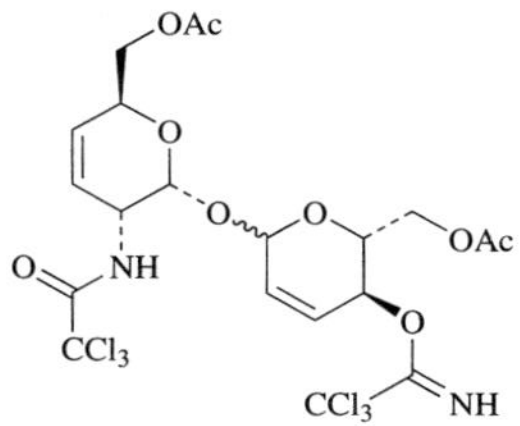 (8)	108

TABLE 2B. REARRANGEMENTS OF TRIHALOACETIMIDATES OF CYCLIC SECONDARY ALLYLIC ALCOHOLS (*Continued*)

	Substrate	Conditions	Product(s) and Yield(s) (%)		Refs.
	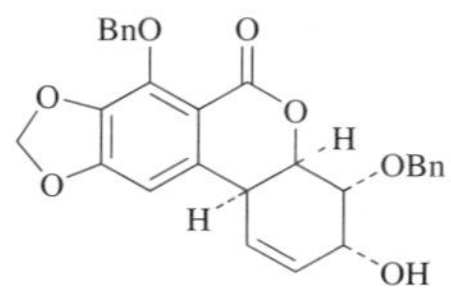	1. NaH (cat.), THF, 0° 2. Cl_3CCN, 0° to rt, 2 h 3. 0.05-0.1 mm Hg, 100-105°, 1.2 h	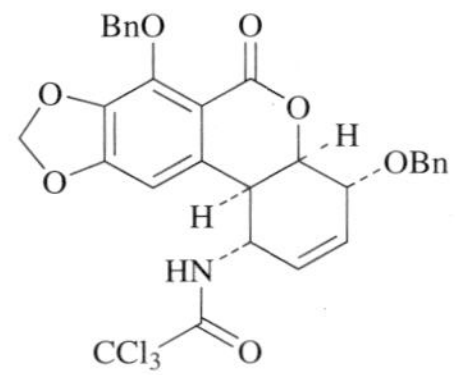	(41)	56
C_{18}	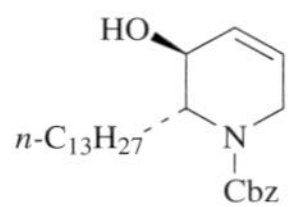	1. NaH (cat.), THF, 66° 2. Cl_3CCN, THF, 0° to rt 3. Xylene, 160-170°, 3 h	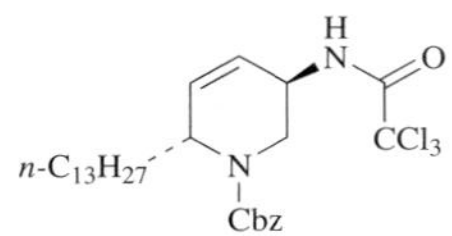	(76)	206

[a] Previous reports[113] noted this substrate failed to rearrange successfully under similar conditions.

[b] NMR yield for rearrangement only.

[c] The (1*S*)-all-cis isomer did not successfully rearrange, giving instead a 34% yield of recovered acetimidate.

[d] This rearrangement was reported to be slow due to steric congestion in the transition state.

[e] The yield is 54% based on consumed imidate.

[f] Fifteen percent of the starting material was recovered.

[g] The β-epimer is incorrectly shown in paper; see *Tetrahedron,* **1981**, *37*, 4391.

[h] Yield for rearrangement only.

[i] Rearrangement of the enantiomer is reported.[123,202]

TABLE 3. REARRANGEMENTS OF TRIHALOACETIMIDATES OF TERTIARY ALLYLIC ALCOHOLS

Substrate	Conditions	Product(s) and Yield(s) (%)	Refs.
C_5 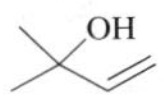	1. NaH (cat.), THF 2. Cl_3CCN, Et_2O, <20°, 1 h 20 min 3. Toluene, 110°, 1 h	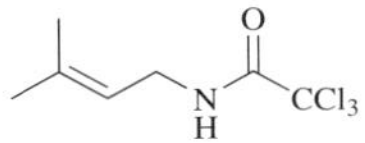 (71–82)	52, 72
C_7 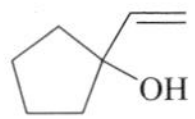	1. NaH (cat.), THF 2. Cl_3CCN, Et_2O, <20° [a]	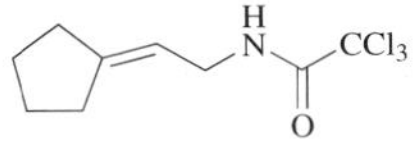 (5)	52
C_8 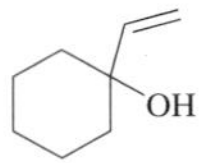	1. KH (cat.), THF, rt 2. Cl_3CCN, Et_2O, –5° to 0°, 1.5 h 3. Benzene, 80°, 2 h	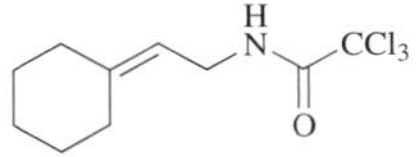 **I** (57)[b]	6
	1. KH (cat.), THF, rt 2. Cl_3CCN, Et_2O, –5° to 0°, 1.5 h 3. $Hg(O_2CCF_3)_2$ (30 mol%), THF, rt, 4 h	**I** (<2)[c]	6
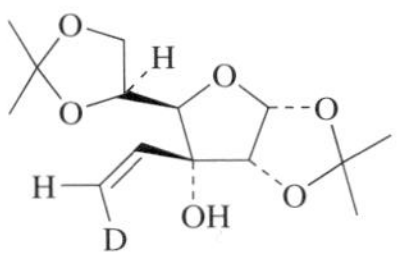	1. NaH 2. Cl_3CCN 3. Xylene, 140°, 36 h	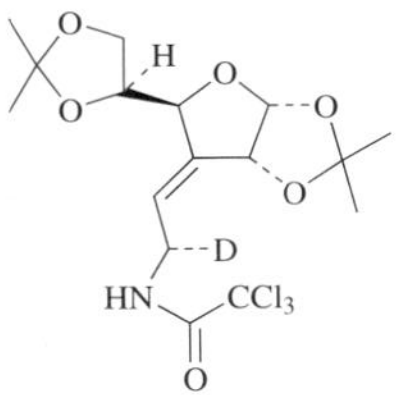 (86)	53, 54
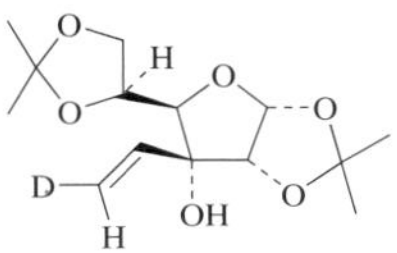	1. NaH 2. Cl_3CCN 3. Xylene, 140°, 36 h	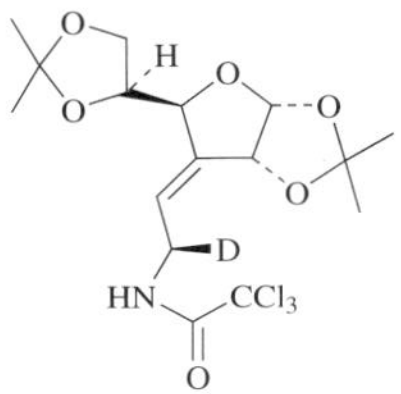 (—)	53, 54

TABLE 3. REARRANGEMENTS OF TRIHALOACETIMIDATES OF TERTIARY ALLYLIC ALCOHOLS (*Continued*)

Substrate	Conditions	Product(s) and Yield(s) (%)	Refs.
C_9 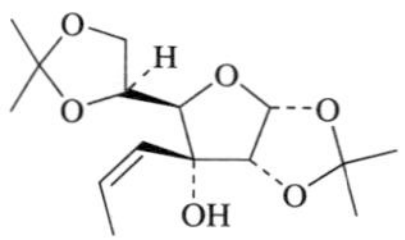	1. NaH 2. Cl_3CCN 3. Xylene, 138°, 36 h	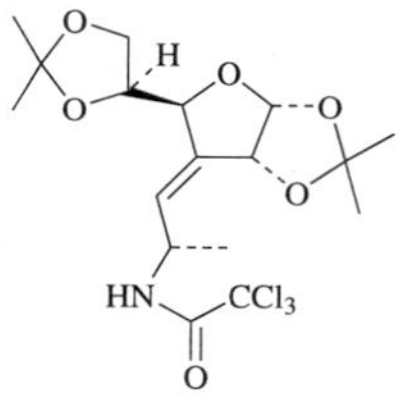 (≈68)	18, 53, 54, 207
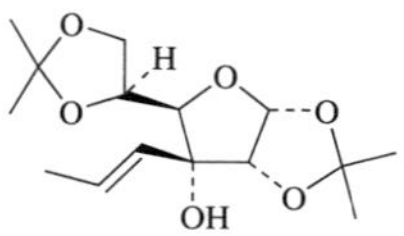	1. NaH 2. Cl_3CCN 3. Xylene, 138°, 36 h	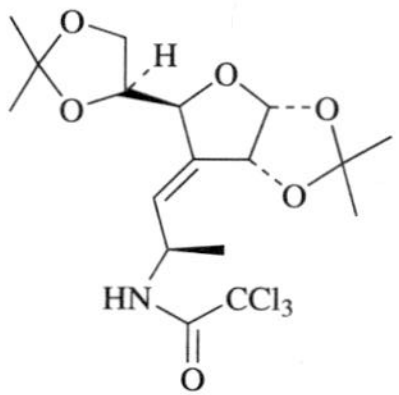 (68)	18, 53, 54, 207
C_{10} 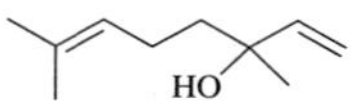	1. KH, THF 2. Cl_3CCN, Et_2O, –5° to 0°, 1.5 h 3. Benzene, 80°, 2 h	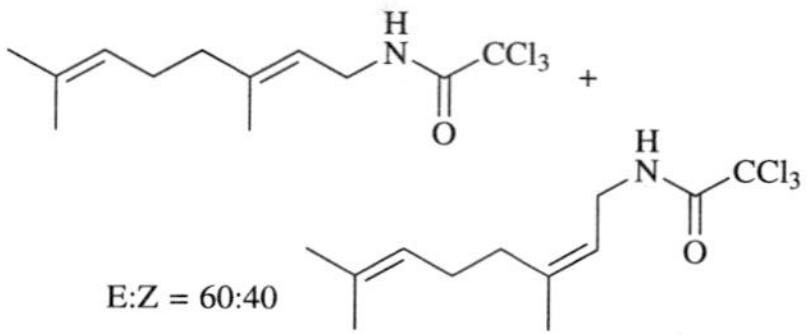 E:Z = 60:40 (83)	6, 72, 208

[a] Imidate rearranged upon workup.

[b] The yield is 67% based on recovered starting material.

[c] This is the NMR yield; elimination was the major reaction.

TABLE 4. TRANSITION-METAL-CATALYZED ASYMMETRIC REARRANGEMENTS OF ALLYLIC TRIHALOACETIMIDATES

Substrate	Conditions	Product(s), Yield(s) (%)[a], and % ee	Refs.
Refer to the chart at the beginning of the Tabular Survey for catalyst (bold numbers) structures.			
C_4			
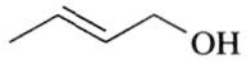	1. NaH, THF, 0° 2. *p*-$MeOC_6H_4N{=}C(Cl)CF_3$ 3. **2** (5 mol%), *i*-Pr_2NEt (20 mol%), CH_2Cl_2, rt, 30 h	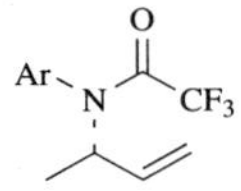 (90), 73	8
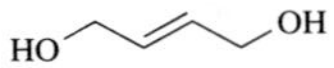	1. DBU (cat.), Cl_3CCN, CH_2Cl_2, rt 2. **1** (5 mol%), CH_2Cl_2, rt, 18 h	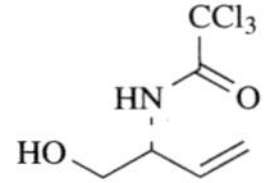 (84), 80	7
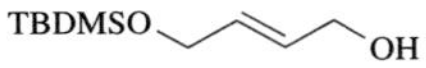	1. DBU (cat.), Cl_3CCN, CH_2Cl_2, rt 2. **1** (5 mol%), CH_2Cl_2, 38°, 18 h	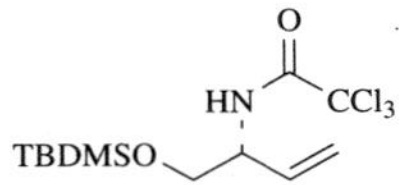 (98), 96	7
	1. DBU (cat.), Cl_3CCN, CH_2Cl_2, rt 2. *ent*-**1** (5 mol%), CH_2Cl_2, rt	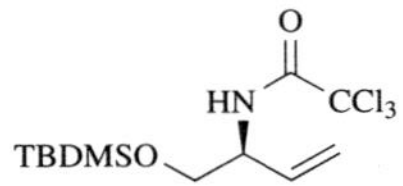 (98), 96	7
C_6			
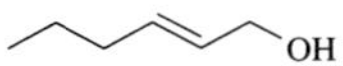	1. DBU, Cl_3CCN 2. **5** (10 mol%), $ClCH_2CH_2Cl$, rt, 7 d	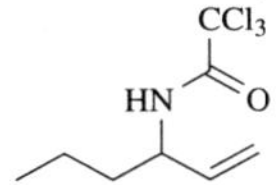 (4), 13 [b]	104
	1. NaH (cat.), Cl_3CCN, Et_2O 2. **6** (5 mol%), CH_2Cl_2, 40°	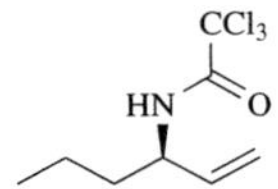 (20), 5	38, 102
	1. DBU (cat.), Cl_3CCN, CH_2Cl_2, rt 2. **6** (5 mol%), CH_2Cl_2, 40°, 6 d	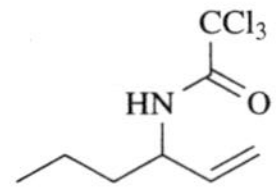 (50), 43 [b]	39

TABLE 4. TRANSITION-METAL-CATALYZED ASYMMETRIC REARRANGEMENTS OF ALLYLIC TRIHALOACETIMIDATES (*Continued*)

Substrate	Conditions	Product(s), Yield(s) (%)[a], and % ee	Refs.
C_6 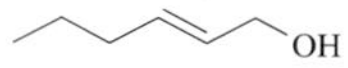	1. DBU (cat.), Cl_3CCN, CH_2Cl_2, rt 2. **1** (5 mol%), CH_2Cl_2, temp, 18 h	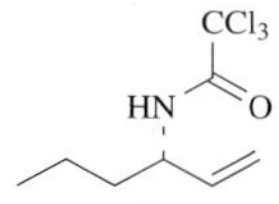 **I** Temp: rt (80), 94; 38° (99), 95	7
	1. DBU (cat.), Cl_3CCN, CH_2Cl_2, rt 2. **4** (5 mol%), THF, 50°, 8 h	**I** (94), 91	9
	1. DBU (cat.), Cl_3CCN, CH_2Cl_2, rt 2. **4** (1 mol%), MeCN, 50°, 22 h	**I** (91), 91	9
	1. DBU (cat.), Cl_3CCN, CH_2Cl_2, rt 2. **3** (1 mol%), THF, 50°, 8 h	**I** (91), 95	9
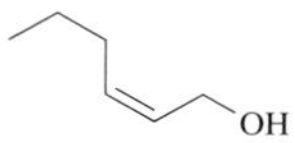	1. DBU (cat.), Cl_3CCN, CH_2Cl_2, rt 2. **1** (5 mol%), CH_2Cl_2, 38°, 18 h	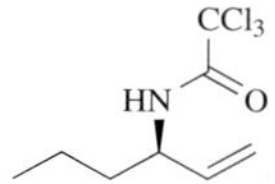 (17), 71	7
	1. NaH, THF, 0° 2. *p*-$MeOC_6H_4N$=C(Cl)CF_3 3. **2** (5 mol%), 1,8-bis(dimethylamino)-naphthalene, CH_2Cl_2, rt, 36 h	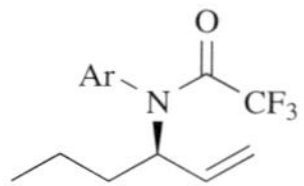 (71), 94	8
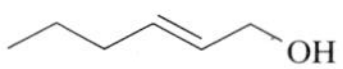	1. NaH, THF, 0° 2. *p*-$MeOC_6H_4N$=C(Cl)CF_3 3. **2** (5 mol%), 1,8-bis(dimethylamino)-naphthalene, CH_2Cl_2, 23°, 36 h	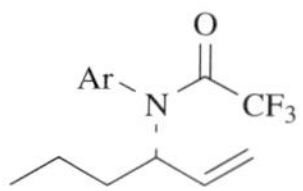 (84), 84	8
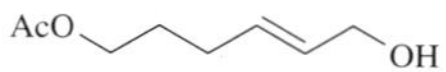	1. DBU (cat.), Cl_3CCN, CH_2Cl_2, rt 2. **1** (5 mol%), CH_2Cl_2, temp, 18 h	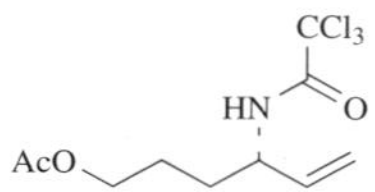 **I** Temp: rt (74), 92; 38° (97), 92	7

	1. DBU (cat.), Cl_3CCN, CH_2Cl_2, rt 2. **4** (5 mol%), THF, 50°, 8 h	**I** (95), 92		9
	1. DBU (cat.), Cl_3CCN, CH_2Cl_2, rt 2. **4** (1 mol%), MeCN, 50°, 24 h	**I** (93), 93		9
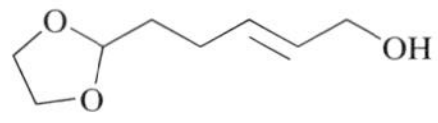	1. DBU (cat.), Cl_3CCN, CH_2Cl_2, rt 2. **1** (5 mol%), CH_2Cl_2, rt, 18 h	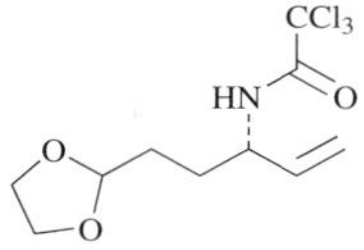 **I**	(85), 95	7
	1. DBU (cat.), Cl_3CCN, CH_2Cl_2, rt 2. **4** (5 mol%), THF, 50°, 9 h	**I** (80), 91		9
	1. DBU (cat.), Cl_3CCN, CH_2Cl_2, rt 2. **4** (1 mol%), MeCN, 50°, 29 h	**I** (86), 93		9
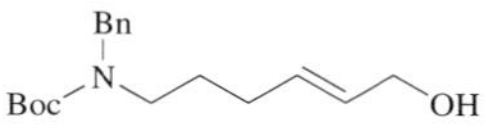	1. DBU (cat.), Cl_3CCN, CH_2Cl_2, rt 2. **1** (5 mol%), CH_2Cl_2, temp, 18 h	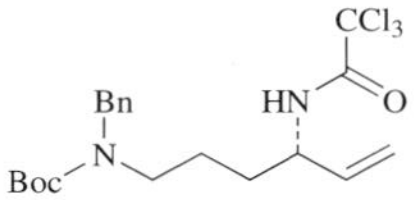	Temp rt (87), 95 38° (96), 95	7
C_7 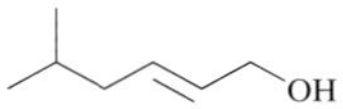	1. DBU (cat.), Cl_3CCN, CH_2Cl_2, rt 2. **1** (1 mol%), CH_2Cl_2, 38°, 18 h	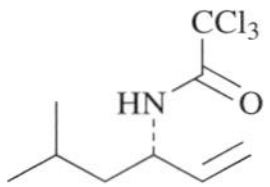 **I**	(92), 98	7
	1. DBU (cat.), Cl_3CCN, CH_2Cl_2, rt 2. **2** (5 mol%), *i*-Pr_2NEt (20 mol%), CH_2Cl_2, rt, 30 h	**I** (80), 92		8

TABLE 4. TRANSITION-METAL-CATALYZED ASYMMETRIC REARRANGEMENTS OF ALLYLIC TRIHALOACETIMIDATES (*Continued*)

Substrate	Conditions	Product(s), Yield(s) (%)[a], and % ee	Refs.
C_7 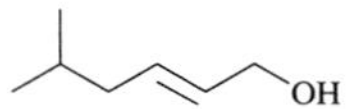	1. DBU (cat.), Cl_3CCN, CH_2Cl_2, rt 2. **4** (5 mol%), THF, 50°, 8 h	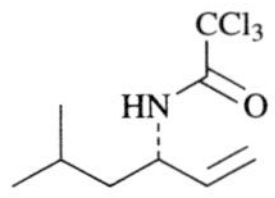 **I**	9
	1. DBU (cat.), Cl_3CCN, CH_2Cl_2, rt 2. **4** (1 mol%), MeCN, 50°, 29 h	**I** (95), 95	9
	1. NaH, THF, 0° 2. *p*-$MeOC_6H_4N{=}C(Cl)CF_3$ 3. **1** (5 mol%), CH_2Cl_2, rt, 60 h	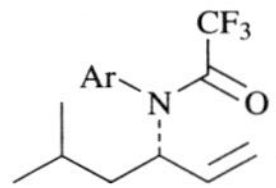 (88), 94	8
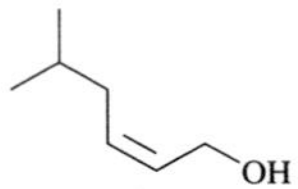	1. DBU (cat.), Cl_3CCN, CH_2Cl_2, rt 2. **1** (5 mol%), CH_2Cl_2, 38°, 18 h	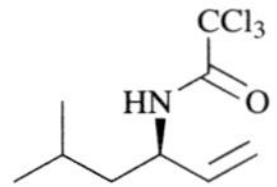 (8), 73	7
	1. NaH, THF, 0° 2. *p*-$MeOC_6H_4N{=}C(Cl)CF_3$ 3. **1** (5 mol%), CH_2Cl_2, rt, 60 h	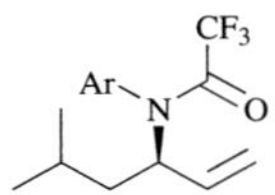 (58), 90 **I**	8
	1. NaH, THF, 0° 2. *p*-$MeOC_6H_4N{=}C(Cl)CF_3$ 3. **2** (5 mol%), 20 mol % *i*-Pr_2NEt, CH_2Cl_2, rt, 30 h	**I** (67), 97	8
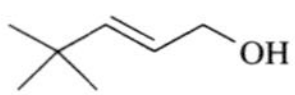	1. DBU (cat.), Cl_3CCN, CH_2Cl_2, rt 2. **1** (5 mol%), CH_2Cl_2, 38°, 18 h	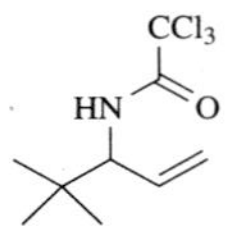 (7)[c]	7

	Substrate	Conditions	Product(s) and Yield(s) (%), ee (%)		Refs.
		1. DBU (cat.), Cl_3CCN, CH_2Cl_2, rt 2. **1** (5 mol%), CH_2Cl_2, rt, 18 h	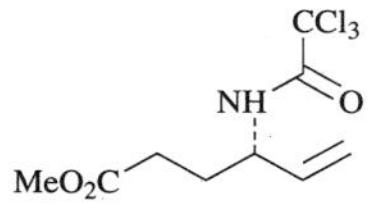	(73), 95	7
	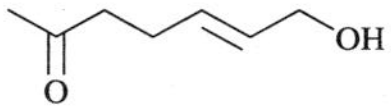	1. DBU (cat.), Cl_3CCN, CH_2Cl_2, rt 2. **1** (5 mol%), CH_2Cl_2, temp, 18 h	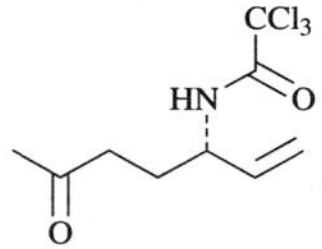	Temp rt (80), 94 38° (98), 95	7
C_9	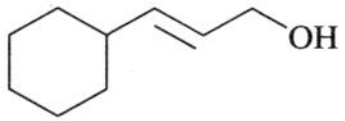	1. DBU (cat.), Cl_3CCN, CH_2Cl_2, rt 2. **1** (5 mol%), CH_2Cl_2, 38°, 18 h	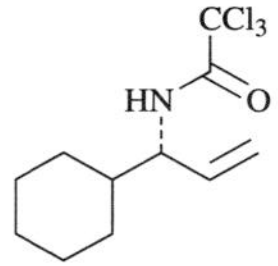	(82), 96	7
C_{11}	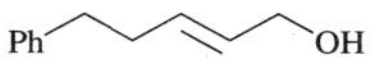	1. DBU (cat.), Cl_3CCN, CH_2Cl_2, rt 2. **1** (5 mol%), CH_2Cl_2, temp, 18 h	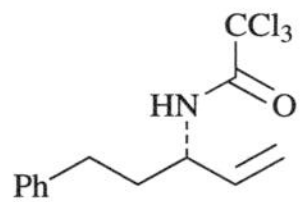	Temp rt (83), 96 38° (93), 93	7
		1. NaH, THF, 0° 2. *p*-$MeOC_6H_4N{=}C(Cl)CF_3$ 3. **1** (5 mol%), CH_2Cl_2, rt, 60 h	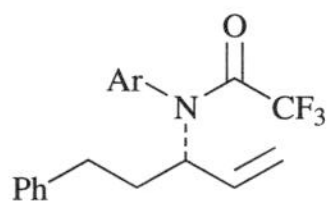	(86), 93	8

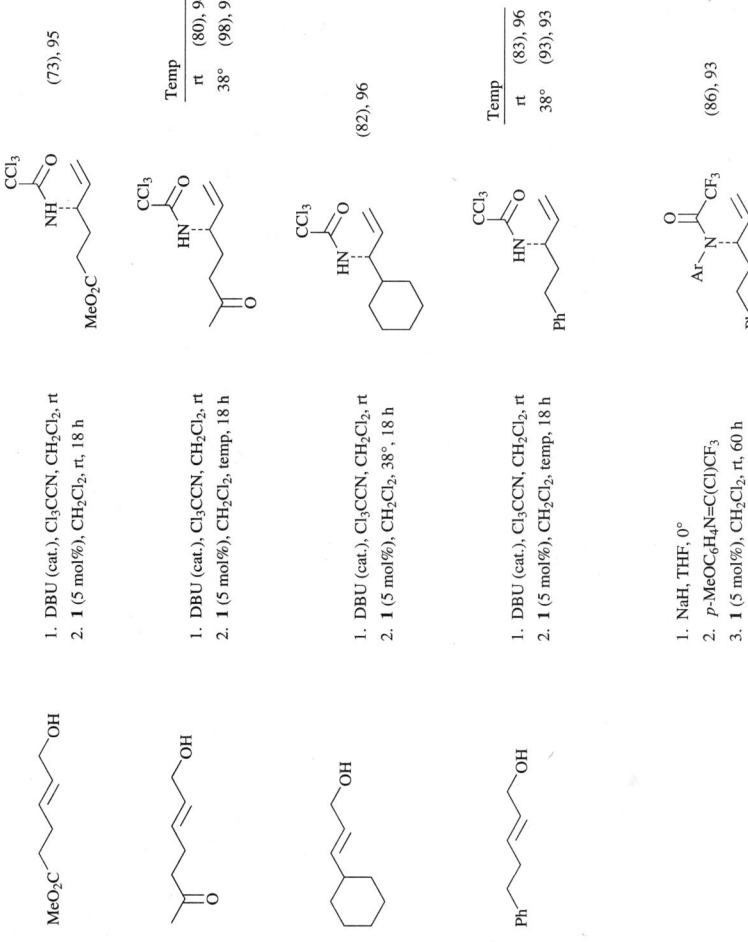

TABLE 4. TRANSITION-METAL-CATALYZED ASYMMETRIC REARRANGEMENTS OF ALLYLIC TRIHALOACETIMIDATES (*Continued*)

Substrate	Conditions	Product(s), Yield(s) (%)[a], and % ee	Refs.
C_{11}			
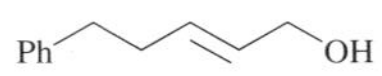	1. NaH, THF, 0° 2. p-MeOC$_6$H$_4$N=C(Cl)CF$_3$ 3. **2** (5 mol%), i-Pr$_2$NEt (20 mol%), CH$_2$Cl$_2$, rt, 30 h	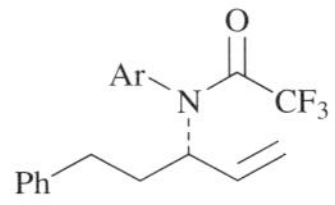 (80), 88	8
	1. DBU (cat.), Cl$_3$CCN, CH$_2$Cl$_2$, rt 2. **4** (5 mol%), solvent, 50°, time	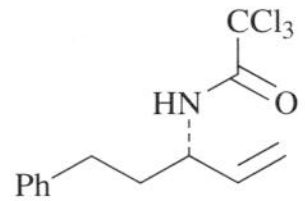 Solvent / Time / % ee: THF, 6 h, (93) 96 cyclohexane, 12 h, (88) 95 toluene, 15 h, (85) 97 acetone, 9 h, (95) 96 acetonitrile, 9 h, (93) 92	9
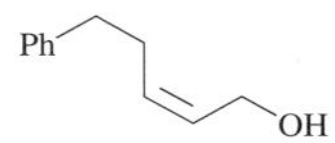	1. NaH, THF, 0° 2. p-MeOC$_6$H$_4$N=C(Cl)CF$_3$ 3. **1** (5 mol%), CH$_2$Cl$_2$, temp, 60 h	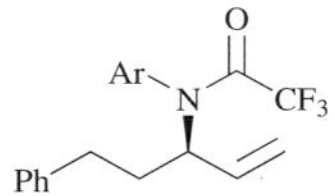 **I** Temp: rt (77), 97 38° (99), 96	8
	1. NaH, THF, 0° 2. p-MeOC$_6$H$_4$N=C(Cl)CF$_3$ 3. **2** (5 mol%), i-Pr$_2$NEt (20 mol%), CH$_2$Cl$_2$, rt, 60 h	**I** (76), 96	8
C_{12}			
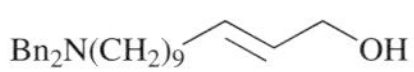	1. DBU (cat.), Cl$_3$CCN, CH$_2$Cl$_2$, rt 2. **4** (5 mol%), THF, 50°, 9 h	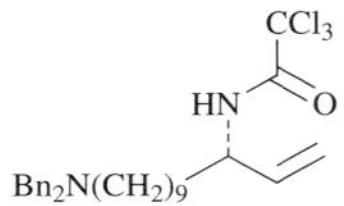 (98), 92	9

[a] Yields are for the rearrangement step only.

[b] Absolute configuration is not reported.

[c] Neither enantiomer excess nor configuration was determined.

REFERENCES

[1]Beak, P.; Bonham, J.; Lee, J. T. *J. Am. Chem. Soc.* **1968**, *90*, 1569.

[2]Beak, P.; Mueller, D. S.; Lee, J. *J. Am. Chem. Soc.* **1974**, *96*, 3867.

[3]Mumm, O.; Möller, F. *Ber. Dtsch. Chem. Ges.* **1937**, *70*, 2214.

[4]Overman, L. E. *J. Am. Chem. Soc.* **1974**, *96*, 597.

[5]Overman, L. E. *Tetrahedron Lett.* **1975**, *13*, 1149.

[6]Overman, L. E. *J. Am. Chem. Soc.* **1976**, *98*, 2901.

[7]Anderson, C. E.; Overman, L. E. *J. Am. Chem. Soc.* **2003**, *125*, 12412.

[8]Overman, L. E.; Owen, C. E.; Pavan, M. M.; Richards, C. J. *Org. Lett.* **2003**, *5*, 1809.

[9]Kirsch, S. F.; Overman, L. E.; Watson, M. P. *J. Org. Chem.* **2004**, *69*, 8101.

[10]Ritter, K. In *Methoden der organischen Chemie/Organic Chemistry (Houben-Weyl) Stereoselective Synthesis*; 21 ed.; Helmchen, G., Hoffmann, R. W., Mulzer, J., Schaumann, E., Eds.; G. Thieme: Stuttgart, New York, 1996; Vol. 9, pp. 5677–5699.

[11]Overman, L. E. *Acc. Chem. Res.* **1980**, *13*, 218.

[12]Overman, L. E. *Agnew. Chem., Int. Ed. Engl.* **1984**, *23*, 579.

[13]Calter, M. A.; Hollis, T. K.; Overman, L. E.; Ziller, J.; Zipp, G. G. *J. Org. Chem.* **1999**, *64*, 1428.

[14]Nubbemeyer, U. *Synthesis* **2003**, 961.

[15]Johannsen, M.; Jørgensen, K. A. *Chem. Rev.* **1998**, *98*, 1689.

[16]Overman, L. E.; Clizbe, L. A.; Freerks, R. L.; Marlowe, C. K. *J. Am. Chem. Soc.* **1981**, *103*, 2807.

[17]Dyong, I.; Merten, H.; Thiem, J. *Tetrahedron Lett.* **1984**, *25*, 277.

[18]Eguchi, T.; Koudate, T.; Kakinuma, K. *Tetrahedron* **1993**, *49*, 4527.

[19]Cho, C.-G.; Lim, Y.-K.; Lee, K.-S.; Jung, I.-H.; Yoon, M.-Y. *Synth. Commun.* **2000**, *30*, 1643.

[20]Yamamoto, Y.; Shimoda, H.; Oda, J.; Inouye, Y. *Bull. Chem. Soc. Jpn.* **1976**, *49*, 3247.

[21]Imogai, H.; Petit, Y.; Larcheveque, M. *Tetrahedron Lett.* **1996**, *37*, 2573.

[22]Waelchili, R.; Gamse, R.; Bauer, W.; Meigel, H.; Lier, E.; Feyen, J. H. M. *Bioorg. Med. Chem. Lett.* **1996**, *6*, 1151.

[23]Wai, J. S.; Fisher, T. E.; Embrey, M. W. *Tetrahedron Lett.* **1995**, *36*, 3461.

[24]Tanner, D.; He, H. M. *Acta Chem. Scand.* **1993**, *47*, 592.

[25]Imogai, H.; Petit, Y.; Larcheveque, M. *Synlett* **1997**, 615.

[26]Martin, C.; Prunck, W.; Bortolussi, M.; Bloch, R. *Tetrahedron: Asymmetry* **2000**, *11*, 1585.

[27]Ovaa, H.; Codee, J. D. C.; Lastdrager, B.; Overkleeft, H. S.; van der Marel, G. A.; van Boom, J. H. *Tetrahedron Lett.* **1999**, *40*, 5063.

[28]Metz, P.; Mues, C.; Schoop, A. *Tetrahedron* **1992**, *48*, 1071.

[29]Mehmandoust, M.; Petit, Y.; Larcheveque, M. *Tetrahedron Lett.* **1992**, *33*, 4313.

[30]Gonda, J.; Helland, A. C.; Ernst, B.; Bellus, D. *Synthesis* **1993**, *7*, 729.

[31]Estieu, K.; Ollivier, J.; Salaun, J. *Tetrahedron Lett.* **1995**, *36*, 2975.

[32]Bosnich, B.; Schenck, T. G. *J. Am. Chem. Soc.* **1985**, *107*, 2058.

[33]Henry, P. M. *Acc. Chem. Res.* **1973**, *6*, 16.

[34]Henry, P. M. *Adv. Organomet. Chem.* **1975**, *13*, 363.

[35]Overman, L. E.; Jacobsen, E. J. *J. Am. Chem. Soc.* **1982**, *104*, 7225.

[36]Ikariya, T.; Ishikawa, Y.; Hirai, K.; Yoshikawa, S. *Chem. Lett.* **1982**, *11*, 1815.

[37]Cohen, F.; Overman, L. E. *Tetrahedron: Asymmetry* **1998**, *9*, 3213.

[38]Calter, M.; Hollis, T. K.; Overman, L. E.; Ziller, J.; Zipp, G. G. *J. Org. Chem.* **1997**, *62*, 1449.

[39]Donde, Y.; Overman, L. E. *J. Am. Chem. Soc.* **1999**, *121*, 2933.

[40]Cramer, F.; Pawelzik, K.; Baldauf, H. J. *Chem. Ber.* **1958**, *91*, 1049.

[41]Cramer, F.; Baldauf, H. *Chem. Ber.* **1959**, *92*, 370.

[42]Clizbe, L. A.; Overman, L. E. *Org. Synth.* **1978**, *58*, 4.

[43]Clizbe, L. A.; Overman, L. E. *Org. Synth.Coll. Vol. 6* **1988**, 507.

[44]Chandrakumar, N. S.; Yonan, P. K.; Stapelfeld, A.; Savage, M.; Rorbacher, E.; Contreras, P. C.; Hammond, D. *J. Med. Chem.* **1992**, *35*, 223.

[45]Banaszek, A.; Pakulski, Z.; Zamojski, A. *Carbohydr. Res.* **1995**, *279*, 173.

[46]Patil, V. J. *Tetrahedron Lett.* **1996**, *37*, 1481.

[47]Oishi, T.; Ando, K.; Inomiya, K.; Sato, H.; Iida, M.; Chida, N. *Bull. Chem. Soc. Jpn.* **2002**, *75*, 1927.

[48]Kumata, M.; Sugimoto, M.; Koike, K.; Ogawa, T. *Carbohydr. Res.* **1987**, *163*, 209.
[49]Nishikawa, T.; Asai, M.; Ohyabu, N.; Yamamoto, N.; Fukuda, Y.; Isobe, M. *Tetrahedron* **2001**, *57*, 3875.
[50]Lurain, A. E.; Walsh, P. J. *J. Am. Chem. Soc.* **2003**, *125*, 10677.
[51]Hauser, F. M.; Ellenberger, S. R.; Glusker, J. P.; Smart, C. J.; Carrell, H. L. *J. Org. Chem.* **1986**, *51*, 50.
[52]Jordis, U.; Grohmann, F.; Küenburg, B. *Org. Prep. Proced. Int.* **1997**, *29*, 549.
[53]Kakinuma, K.; Koudate, T.; Li, H.-Y.; Eguchi, T. *Tetrahedron Lett.* **1991**, *32*, 5801.
[54]Eguchi, T.; Kakinuma, K. *Yuki Gosei Kagaku Kyokaishi* **1997**, 814.
[55]Vyas, D. M.; Chiang, Y.; Doyle, T. W. *J. Org. Chem.* **1984**, *49*, 2037.
[56]Danishefsky, S.; Lee, J. Y. *J. Am. Chem. Soc.* **1989**, *111*, 4829.
[57]Hijfte, L. V.; Heydt, V.; Kolb, M. *Tetrahedron Lett.* **1993**, *34*, 4793.
[58]Felix, C.; Laurent, A.; Lebideau, F.; Mison, P. *J. Chem. Res. (S)* **1993**, 389.
[59]Gilman, H.; Jones, R. G. *J. Am. Chem. Soc.* **1943**, *65*, 1458.
[60]Chen, A.; Savage, I.; Thomas, E. J.; Wilson, P. D. *Tetrahedron Lett.* **1993**, *34*, 6769.
[61]Nakajima, N.; Ubukata, M. *Tetrahedron Lett.* **1998**, *39*, 2099.
[62]Nakajima, N.; Saito, M.; Kudo, M.; Ubukata, M. *Tetrahedron* **2002**, *58*, 3579.
[63]Nishikawa, T.; Asai, M.; Isobe, M. *J. Org. Chem.* **1998**, *63*, 188.
[64]Dyong, I.; Merten, H.; Thiem, J. *Liebigs Ann. Chem.* **1986**, 600.
[65]Takano, S.; Akiyama, M.; Sugihara, T.; Ogasawara, K. *Heterocycles* **1992**, *33*, 831.
[66]Shabany, H.; Spilling, C. D. *Tetrahedron Lett.* **1998**, *39*, 1465.
[67]Allmendinger, T.; Felder, E.; Hungerbuehler, E. *Tetrahedron Lett.* **1990**, *31*, 7301.
[68]Villemin, D.; Hachemi, M. *Synth. Commun.* **1996**, *26*, 1329.
[69]Vyas, D. M.; Chiang, Y.; Doyle, T. W. *Tetrahedron Lett.* **1984**, *25*, 487.
[70]Takeda, K.; Kaji, E.; Nakamura, H.; Akiyama, A.; Konda, Y.; Mizuno, Y.; Takayanagi, H.; Harigaya, Y. *Synthesis* **1996**, *3*, 341.
[71]Reilly, M.; Anthony, D. R.; Gallagher, C. *Tetrahedron Lett.* **2003**, *44*, 2927.
[72]Nagashima, H.; Wakamatsu, H.; Ozaki, N.; Ishii, T.; Watanabe, M.; Tajima, T.; Itoh, K. *J. Org. Chem.* **1992**, *57*, 1682.
[73]Ovaa, H.; Codee, J. D. C.; Lastdrager, B.; Overkleeft, H. S.; van der Marel, G. A.; van Boom, J. H. *Tetrahedron Lett.* **1998**, *39*, 7987.
[74]Ovaa, H.; Lastdrager, B.; Codee, J. D. C.; van der Marel, G. A.; Overkleeft, H. S.; van Boom, J. H. *J. Chem. Soc., Perkin Trans. 1* **2002**, 2370.
[75]Savage, I.; Thomas, E. J.; Wilson, P. D. *J. Chem. Soc., Perkin Trans. 1* **1999**, *22*, 3291.
[76]Savage, I.; Thomas, E. J. *J. Chem. Soc., Chem. Commun.* **1989**, 717.
[77]Saksena, A. K.; Lovey, R. G.; Girijavallabhan, V. M.; Ganguly, A. K.; McPhail, A. T. *J. Org. Chem.* **1986**, *51*, 5024.
[78]Martin, C.; Bortolussi, M.; Bloch, R. *Tetrahedron Lett.* **1999**, *40*, 3735.
[79]Chida, N.; Takeoka, J.; Ando, K.; Tsutsumi, N.; Ogawa, S. *Tennen Yuki Kagobutsu Toronkai Koen Yoshishu* **1996**, 259.
[80]Flynn, D. L.; Becker, D. P.; Nosal, R.; Zabrowski, D. L. *Tetrahedron Lett.* **1992**, *33*, 7283.
[81]Stavinoka, J. L.; Mariano, P. S.; Leone-Bay, A.; Swanson, R.; Brachen, C. *J. Am. Chem. Soc.* **1981**, *103*, 3148.
[82]de Sousa, S.; O'Brien, P.; Pilgram, C. D. *Tetrahedron Lett.* **2001**, *42*, 8081.
[83]Pelyvas, I. S.; Toth, Z. G.; Vereb, G.; Balla, A.; Kovacs, E.; Gorzsas, A.; Sztaricskai, F.; Gergely, P. *J. Med. Chem.* **2001**, *44*, 627.
[84]Dyong, I.; Weigand, J.; Merten, H. *Tetrahedron Lett.* **1981**, *22*, 2965.
[85]Dyong, I.; Weigand, J.; Thiem, J. *Liebigs Ann. Chem.* **1986**, 577.
[86]Campbell, M. M.; Floyd, A. J.; Lewis, T.; Mahon, M. F.; Ogilvie, R. J. *Tetrahedron Lett.* **1989**, *30*, 1993.
[87]Chida, N.; Takeoka, J.; Ando, K.; Tsutsumi, N.; Ogawa, S. *Tetrahedron* **1997**, *53*, 16287.
[88]Armstrong, P. L.; Coull, I. C.; Hewson, A. T.; Slater, M. J. *Tetrahedron Lett.* **1995**, *36*, 4311.
[89]Bey, P.; Gerhart, F.; Jung, M. *J. Org. Chem.* **1986**, *51*, 2835.
[90]Birtwistle, D. H.; Brown, J. N.; Foxton, M. W. *Tetrahedron* **1988**, *44*, 7309.
[91]Commerçon, A.; Ponsinet, G. *Tetrahedron Lett.* **1990**, *31*, 3871.

[92]Mislow, K. *Introduction to Stereochemistry*; W.A. Benjamin: New York, 1965.
[93]Donohoe, T. J.; Blades, K.; Helliwell, M.; Moore, P. R.; Winter, J. J. G. *J. Org. Chem.* **1999**, *64*, 2980.
[94]Doherty, A. M.; Kornberg, B. E.; Reily, M. D. *J. Org. Chem.* **1993**, *58*, 795.
[95]Gonda, J.; Zavacka, E.; Budesinsky, M.; Cisarova, I.; Podlaha, J. *Tetrahedron Lett.* **2000**, *41*, 525.
[96]Hoffmann, R. W. *Chem. Rev.* **1989**, *89*, 1841.
[97]Oehler, E.; Kotzinger, S. *Synthesis* **1993**, *5*, 497.
[98]Ammenn, J.; Altmann, K. H.; Bellus, D. *Helv. Chim. Acta* **1997**, *80*, 1589.
[99]Jamieson, A. G.; Sutherland, A.; Willis, C. L. *Org. Biomol. Chem.* **2004**, *2*, 808.
[100]Yoon, Y.-J.; Chun, M.-H.; Joo, J.-E.; Kim, Y.-H.; Oh, C.-Y.; Lee, K.-Y.; Lee, Y.-S.; Ham, W.-H. *Arch. Pharmacal. Res.* **2004**, *27*, 136.
[101]Overman, L. E.; Zipp, G. G. *J. Org. Chem.* **1997**, *62*, 2288.
[102]Hollis, T. K.; Overman, L. E. *J. Organomet. Chem.* **1999**, *576*, 290.
[103]Uozumi, Y.; Kato, K.; Hayashi, T. *Tetrahedron: Asymmetry* **1998**, *9*, 1065.
[104]Jiang, Y.; Longmire, J. M.; Zhang, X. *Tetrahedron Lett.* **1999**, *40*, 1449.
[105]Leung, P.-H.; Ng, K.-H.; Li, Y.; White, A. J. P.; Williams, D. J. *Chem. Commun.* **1999**, 2435.
[106]Greene, T. W.; Wuts, P. G. M. *Protective Groups in Organic Synthesis*; 3rd ed.; Wiley: New York, 1999.
[107]Bargar, T. M.; Broersma, R. J.; Creemer, L. C.; McCarthy, J. R.; Hornsperger, J.-M.; Palfreyman, M. G.; Wagner, J.; Jung, M. J. *J. Med. Chem.* **1986**, *29*, 315.
[108]Sugai, T.; Okazaki, H.; Kuboki, A.; Ohta, H. *Bull. Chem. Soc. Jpn.* **1997**, *70*, 2535.
[109]Takeda, K.; Kaji, E.; Konda, Y.; Sato, N.; Nakamura, H.; Miya, N.; Morizane, A.; Yanagisawa, Y.; Akiyama, A.; Zen, S.; Harigaya, Y. *Tetrahedron Lett.* **1992**, *47*, 7145.
[110]Mehta, G.; Lakshminath, S.; Talukdar, P. *Tetrahedron Lett.* **2002**, *43*, 335.
[111]Kriek, N. M. A. M.; van der Hout, E.; Kelly, P.; van Meijgaarden, K. E.; Geluk, A.; Ottenhoff, T. H. M.; van der Marel, G. A.; Overhand, M.; van Boom, J. H.; Valentijn, A. R. P. M.; Overkleeft, H. S. *Eur. J. Chem.* **2003**, 2418.
[112]Boucard, V.; Sauriat-Dorizon, H.; Guibe, F. *Tetrahedron* **2002**, *58*, 7275.
[113]Takano, S.; Akiyama, M.; Ogasawara, K. *J. Chem. Soc., Chem. Commun.* **1984**, 770.
[114]Chen, Y. K.; Lurain, A. E.; Walsh, P. J. *J. Am. Chem. Soc.* **2002**, *124*, 12225.
[115]Yamamoto, N.; Isobe, M. *Chem. Lett.* **1994**, *12*, 2299.
[116]Bujard, M.; Briot, A.; Gouverneur, V.; Mioskowski, C. *Tetrahedron Lett.* **1999**, *40*, 8785.
[117]Cardillo, G.; Orena, M.; Sandri, S. *J. Org. Chem.* **1986**, *51*, 713.
[118]Cancho, Y.; Martin, J. M.; Martinez, M.; Liebaria, A.; Moreto, J. M.; Delgado, A. *Tetrahedron* **1998**, *54*, 1221.
[119]Haddad, M.; Imogaie, H.; Larcheveque, M. *J. Org. Chem.* **1998**, *63*, 5680.
[120]Guy, A.; Barbetti, J.-F. *J. Chem. Res. (S)* **1994**, 278.
[121]Jacobson, M. A.; Williard, P. G. *J. Org. Chem.* **2002**, *67*, 32.
[122]Bertozzi, S.; Salvadori, P. *Synth. Commun.* **1996**, *26*, 2959.
[123]Donohoe, T. J.; Blades, K.; Moore, P. R.; Waring, M. J.; Winter, J. J. G.; Helliwell, M.; Newcombe, N. J.; Stemp, G. *J. Org. Chem.* **2002**, *67*, 7946.
[124]Donohoe, T. J. *Synlett* **2002**, *8*, 1223.
[125]Donohoe, T. J.; Blades, K.; Helliwell, M. *Chem. Commun.* **1999**, 1733.
[126]Poli, G.; Pagni, I.; Maffioli, S. I.; Scolastico, C.; Zanda, M. *Gazz. Chim. Ital.* **1995**, *125*, 505.
[127]Chida, N.; Takeoka, J.; Tsutsumi, N.; Ogawa, S. *J. Chem. Soc., Chem. Commun.* **1995**, *7*, 793.
[128]Oishi, T.; Ando, K.; Inomiya, K.; Sato, H.; Iida, M.; Chida, N. *Org. Lett.* **2002**, *4*, 151.
[129]Synerholm, M. E.; Gilman, N. W.; Morgan, J. W.; Hill, R. K. *J. Org. Chem.* **1968**, *33*, 1111.
[130]Tsuboi, S.; Stromquist, P.; Overman, L. E. *Tetrahedron Lett.* **1976**, *17*, 1144.
[131]Overman, L. E.; Kakimoto, M.-A. *J. Org. Chem.* **1978**, *43*, 4564.
[132]Tamura, Y.; Kagotani, M.; Yoshida, Z. *J. Org. Chem.* **1980**, *45*, 5221.
[133]Tamura, Y.; Kagotani, M.; Yoshida, Z. *Tetrahedron Lett.* **1981**, *22*, 4245.
[134]White, E. H.; Elliger, C. A. *J. Am. Chem. Soc.* **1965**, *87*, 5261.
[135]Chen, B.; Mapp, A. K. *J. Am. Chem. Soc.* **2004**, *126*, 5364.
[136]Knapp, S.; Naughton, A. B. J.; Murali Dhar, T. G. *Tetrahedron Lett.* **1992**, *33*, 1025.
[137]McDonough, M. J.; Stick, R. V.; Tilbrook, D. M. G. *Aust. J. Chem.* **1999**, *52*, 143.
[138]Schnur, R. C.; Corman, M. L. *J. Org. Chem.* **1994**, *59*, 2581.

[139]Shea, R. G.; Fitzner, J. N.; Fankhauser, J. E.; Spaltenstein, A.; Carpino, P. A.; Peevey, R. M.; Pratt, D. V.; Tenge, B. J.; Hopkins, P. B. *J. Org. Chem.* **1986**, *51*, 5243.
[140]Kurose, N.; Takahashi, T.; Koizumi, T. *J. Org. Chem.* **1996**, *61*, 2932.
[141]Kurose, N.; Takahashi, T.; Koizumi, T. *J. Org. Chem.* **1997**, *62*, 4562.
[142]Gais, H.-J.; Scommoda, M.; Lenz, D. *Tetrahedron Lett.* **1994**, *35*, 7361.
[143]Pyne, S. T.; Dong, Z. *Tetrahedron Lett.* **1995**, *36*, 3029.
[144]Pyne, S. G.; Dong, Z. *J. Org. Chem.* **1996**, *61*, 5517.
[145]Evans, P. A.; Robinson, J. E.; Nelson, J. D. *J. Am. Chem. Soc.* **1999**, *121*, 6761.
[146]Evans, P. A.; Robinson, J. E. *Org. Lett.* **1999**, *1999*, 1929.
[147]Evans, P. A.; Robinson, J. E.; Moffett, K. K. *Org. Lett.* **2001**, *3*, 3269.
[148]Kondo, T.; Ono, H.; Satake, N.; Mitsudo, T.; Watanabe, Y. *Organometallics* **1995**, *14*, 1945.
[149]Trost, B. M.; Fraisse, P. L.; Ball, Z. T. *Angew. Chem., Int. Ed. Engl.* **2002**, *41*, 1059.
[150]Ohmura, T.; Hartwig, J. F. *J. Am. Chem. Soc.* **2002**, *124*, 15164.
[151]Takeuchi, R.; Shiga, N. *Org. Lett.* **1999**, *1*, 265.
[152]Takeuchi, R.; Ue, K.; Tanabe, K.; Yamashita, K.; Shiga, N. *J. Am. Chem. Soc.* **2001**, *123*, 9525.
[153]Croset, K.; Polet, D.; Alexakis, A. *Agnew. Chem., Int. Ed. Engl.* **2004**, *43*, 2426.
[154]Lou, Y.-M.; Hou, X.-L.; Dai, L.-X. *J. Am. Chem. Soc.* **2001**, *123*, 7471.
[155]Guiry, P. J.; Malone, Y. M. *J. Organomet. Chem.* **2000**, *603*, 110.
[156]Ding, K.; Wang, Y. *J. Org. Chem.* **2001**, *66*, 3238.
[157]Brimble, M. A.; Heathcock, C. H. *J. Org. Chem.* **1993**, *58*, 5261.
[158]Katsuki, T.; Khomura, Y. *Tetrahedron Lett.* **2001**, *42*, 3339.
[159]Brimble, M. A.; Heathcock, C. H.; Nobin, G. N. *Tetrahedron: Asymmetry* **1996**, *7*, 2007.
[160]Hogan, G. A.; Gallo, A. A.; Nicholas, K. M.; Srivastava, R. S. *Tetrahedron Lett.* **2002**, *43*.
[161]Srivastava, R. S. *Tetrahedron Lett.* **2003**, *44*, 3271.
[162]Müller, T. E.; Beller, M. *Chem. Rev.* **1998**, *98*, 675.
[163]Kadota, I.; Shibuya, A.; Lutete, L. M.; Yamamoto, Y. *J. Org. Chem.* **1999**, *64*, 4570.
[164]Reggelin, M.; Zur, D. *Synthesis* **2000**, 1.
[165]Enders, D.; Reinhold, U. *Tetrahedron: Asymmetry* **1997**, *8*, 1895.
[166]Wipf, P.; Nunes, R. L.; Ribe, S. *Helv. Chim. Acta* **2002**, *85*, 3478.
[167]Demay, S.; Kotschy, A.; Knochel, P. *Synthesis* **2001**, 863.
[168]Nakajima, N.; Ubukata, M. *Tetrahedron Lett.* **1997**, *38*, 2099.
[169]Anderson, G. K.; Lin, M. In *Reagents for Transition Metal Comlexes and Organometallic Syntheses*; Wiley: New York, 1990; Vol. 28, pp. 60–63.
[170]Anderson, C. E.; Kirsch, S. F.; Overman, L. E.; Richards, C. J.; Watson, M. P. *Org. Synth.* **2004** (submitted).
[171]Ghelfi, F.; Bellesia, F.; Forti, L.; Ghirardini, G.; Grandi, R.; Libertini, E.; Montemaggi, M. C.; Pagnoni, U. M.; Pinetti, A.; DeBuyck, L.; Parsons, A. F. *Tetrahedron* **1999**, *55*, 5839.
[172]Nagashima, H.; Ozaki, N.; Ishii, M.; Seki, K.; Washiyama, M.; Itoh, K. *J. Org. Chem.* **1993**, *58*, 464.
[173]Grison, C.; Thomas, A.; Coutrot, F.; Coutrot, P. *Tetrahedron* **2003**, *59*, 2101.
[174]Vyas, D. M.; Doyle, T. W.; Chiang, Y.; U.S. Patent 4,692,538 (1987);
[175]Barta, N. S.; Cook, G. R.; Landis, M. S.; Stille, J. R. *J. Org. Chem.* **1992**, *57*, 7188.
[176]Kochetkov, K. A.; Sviridov, A. F. *Bioorg. Khim.* **1991**, *17*, 149.
[177]Halling, K.; Torssell, K. B. G.; Hazell, R. G. *Acta Chem. Scand.* **1991**, *45*, 736.
[178]Casara, P. *Tetrahedron Lett.* **1994**, *35*, 3049.
[179]Isobe, M.; Fukuda, Y.; Nishikawa, T.; Chabert, P.; Kawai, T.; Goto, T. *Tetrahedron Lett.* **1990**, *31*, 3327.
[180]Oishi, T.; Ando, K.; Chida, N. *Chem. Commun.* **2001**, 1932.
[181]Pallavicini, M.; Valoti, E.; Villa, L.; Piccolo, O. *Tetrahedron: Asymmetry* **2000**, *11*, 4017.
[182]Oh, J. S.; Park, D. Y.; Song, B. S.; Bae, J. G.; Yoon, S. W.; Kim, Y. G. *Tetrahedron Lett.* **2002**, *43*, 7209.
[183]Allmendinger, T.; Angst, C.; Karfunkel, H. *J. Fluorine Chem.* **1995**, *72*, 247.
[184]Nishikawa, T.; Ohyabu, N.; Yamamoto, N.; Isobe, M. *Tetrahedron* **1999**, *55*, 4325.
[185]Overman, L. E.; Campbell, C. B.; Knoll, F. M. *J. Am. Chem. Soc.* **1978**, *100*, 4822.
[186]Sprules, T. J.; Galpin, J. D.; Macdonald, D. *Tetrahedron Lett.* **1993**, *34*, 247.
[187]Ichikawa, Y. *Chem. Lett.* **1990**, *8*, 1347.
[188]Ichikawa, Y. *J. Synth. Org. Chem. Jpn* **1997**, *55*, 281.

[189]Nishikawa, T.; Ohyabu, N.; Yamamoto, N. In *41st Symposium on the Chemistry of Natural Products* Nagoya, Japan, 1999, p. 7–12.
[190]Luly, J. R.; Dellaria, J. F.; Plattner, J. J.; Soderquist, J. L.; Yi, N. *J. Org. Chem.* **1987**, *52*, 1487.
[191]Anderson, J. C.; Siddons, D. C.; Smith, S. C.; Swarbrick, M. E. *J. Org. Chem.* **1996**, *61*, 4820.
[192]Metcalf, B. W.; Bey, P.; Danzin, C.; Jung, M. J.; Casara, P.; Vevert, J. P. *J. Am. Chem. Soc.* **1978**, *100*, 2551.
[193]Roush, W. R.; Straub, J. A.; Brown, R. J. *J. Org. Chem.* **1987**, *52*, 5127.
[194]Grigg, R.; Santhakumar, V.; Sridharan, V.; Thornton-Pett, M.; Bridge, A. W. *Tetrahedron* **1993**, *49*, 5177.
[195]Adams, A. D.; Jones, A. B.; U. S. Patent 5,840,835 (1998).
[196]Chandrakumar, N. S.; Stapelfeld, A.; Beardsley, P. M.; Lopez, O. T.; Drury, B.; Anthony, E.; Savage, M. A.; Williamson, L. N.; Reichman, M. *J. Med. Chem.* **1992**, *35*, 2928.
[197]Ibuka, T. *J. Synth. Org. Chem. Jpn.* **1992**, 953.
[198]Kim, S.; Lee, T.; Lee, E.; Lee, J.; Fan, G.-J.; Lee, S. K.; Kim, D. *J. Org. Chem.* **2004**, *69*, 3144.
[199]O'Brien, P.; Childs, A. C.; Ensor, G. J.; Hill, C. L.; Kirby, J. P.; Dearden, M. J.; Oxenford, S. J.; Rosser, C. M. *Org. Lett.* **2003**, *5*, 4955.
[200]O'Brien, P.; Pilgram, C. D. *Org. Biomol. Chem.* **2003**, *1*, 523.
[201]Pingli, L.; Vandewalle, M. *Tetrahedron* **1994**, *50*, 7061.
[202]Blades, K.; Donohoe, T. J.; Winter, J. J. G.; Stemp, G. *Tetrahedron Lett.* **2000**, *41*, 4701.
[203]Guzman, A.; Martinez, E.; Velarde, E.; Maddox, M. L.; Muchowski, J. M. *Can. J. Chem.* **1987**, *65*, 2164.
[204]van Hooft, P. A. V.; Litjens, R. E. J. N.; van der Marel, G. A.; van Boeckel, C. A. A.; van Boom, J. H. *Org. Lett.* **2001**, *3*, 731.
[205]Ellis, D. A.; Hart, D. J.; Zhao, L. *Tetrahedron Lett.* **2000**, *41*, 9357.
[206]Utsunomiya, I.; Ogawa, M.; Natsume, M. *Heterocycles* **1992**, *33*, 349.
[207]Eguchi, T.; Koudate, T.; Takagi, I.; Kakinuma, K. *Tennen Yuki Kagobutsu Toronkai Koen Yoshishu* **1992**, 39.
[208]Chalk, A. J.; Wertheimer, V.; Magennis, S. A. *J. Mol. Catal.* **1983**, *19*, 189.

CHAPTER 2

ASYMMETRIC DIHYDROXYLATION OF ALKENES

Mark C. Noe, Michael A. Letavic, and Sheri L. Snow

Pfizer Global Research and Development, Groton Laboratories
Groton, Connecticut 06340

Stuart W. McCombie

CONTENTS

	Page
Acknowledgments	111
Introduction	111
Mechanism and Stereochemistry	113
Scope and Limitations	120
Terminal Alkenes	120
Disubstituted Alkenes	127
1,1-Disubstituted Alkenes	127
E-1,2-Disubstituted Alkenes	132
Z-1,2-Disubstituted Alkenes	142
Trisubstituted Alkenes	144
Acyclic Trisubstituted Alkenes	146
Exocyclic Trisubstituted Alkenes	147
Endocyclic Trisubstituted Alkenes	148
Tetrasubstituted Alkenes	151
Polyalkene Substrates	152
Kinetic Resolutions	160
Functional Group Compatibility	162
Experimental Conditions	164
Stoichiometric Enantioselective Dihydroxylation Using Chiral Diamine Ligands	165
Selection of Ligand	165
Solvent, Temperature, and Concentration	165
Recovery of Ligand and Osmium	165
Catalytic Enantioselective Dihydroxylation Using Cinchona Alkaloid Ligands	166
Selection of Ligand	166
Solid-Supported Cinchona Alkaloid Catalysts	171
Osmium Sources	176

mark.c.noe@pfizer.com

Organic Reactions, Vol. 66, Edited by Larry E. Overman et al.

Secondary Oxidants . . . 177
Solvent and Concentration . . . 182
Reaction Temperature and Catalyst Loading . . . 183
Premixed Reagents for Catalytic Enantioselective Dihydroxylation . . . 183
EXPERIMENTAL PROCEDURES . . . 184
Procedures for the Synthesis of Ligands for Enantioselective Dihydroxylation . . . 184
(1*R*,2*R*)-*N*,*N*′-Bis(3,3-dimethylbutyl)cyclohexane-1,2-diamine . . . 184
1,4-Bis(9-*O*-dihydroquinidyl)-6,7-diphenylphthalazine [$(DHQD)_2$DP-PHAL] . . . 184
1,4-Bis(9-*O*-dihydroquinyl)-6,7-diphenylphthalazine [$(DHQ)_2$DP-PHAL] . . . 185
5,8-Bis(9-*O*-dihydroquinidyl)-2,3-diphenylpyrazino[2,3-*d*]pyridazine [$(DHQD)_2$DPP] . . . 185
5,8-Bis(9-*O*-dihydroquinyl)-2,3-diphenylpyrazino[2,3-*d*]pyridazine [$(DHQ)_2$DPP] . . . 186
1,4-Bis[*O*-6′-(4-heptyl)hydrocupreidyl]naphthopyridazine . . . 187
3,6-Bis(hydroquinidyl)pyridazine-mono-9-anthracenylmethyl Chloride . . . 187
6-Dihydroquinidyl-3-[1(*S*)-anthracen-1-yl-2,2-dimethylpropoxy]pyridazine [DHQD-PYDZ-(*S*)-Anthryl Ligand] . . . 188
2,5-Diphenyl-bis(9-*O*-dihydroquinidyl)pyrimidine [$(DHQD)_2$PYR] . . . 189
2,5-Diphenyl-4,6-bis(dihydroquinyl)pyrimidine [$(DHQ)_2$PYR] . . . 189
1,4-Bis(dihydroquinidyl)benzo[*g*]phthalazine-5,10-dione [$(DHQD)_2$AQN] . . . 190
1,4-Bis(dihydroquinyl)benzo[*g*]phthalazine-5,10-dione [$(DHQ)_2$AQN] . . . 190
$(DHQD)_2$PHAL—EGDMA—HEMA Block Copolymer Ligand . . . 191
Procedures for the Enantioselective Dihydroxylation of Alkenes . . . 191
S,*S*-Diethyl Tartrate [Stoichiometric Enantioselective Dihydroxylation Using a Chiral 1,2-Diamine Ligand] . . . 191
2-(2′-Isopropoxy-3′-methoxyphenyl)-2-hydroxyethanol [Catalytic Asymmetric Dihydroxylation Using NMO as the Secondary Oxidant] . . . 192
(*R*,*R*)-(+)-1,2-Diphenyl-l,2-ethanediol [Solid to Solid Asymmetric Dihydroxylation with NMO] . . . 192
Buffered Asymmetric Dihydroxylation Protocol . . . 193
General Procedure for the Asymmetric Dihydroxylation of Allylic 4-Methoxybenzoates . 193
(*R*)-(–)-1-Phenyl-2-propen-1-yl 4-Methoxybenzoate [Kinetic Resolution] . . . 194
(10*R*)-10,11-Dihydroxy-l0,11-dihydrofarnesyl Acetate [Asymmetric Dihydroxylation of Non-Conjugated Polyalkenes] . . . 194
2-Phenyl-1,2-propanediol [Asymmetric Dihydroxylation under Atmospheric Oxygen Pressure] . . . 195
(4*S*)-4-Ethyl-4-hydroxy-8-methoxy-1,4-dihydropyrano[3,4-c]pyridin-3-one [Asymmetric Dihydroxylation of Enol Ethers] . . . 195
(1*R*,2*R*)-1,2-Diphenyl-1,2-ethanediol [Asymmetric Dihydroxylation with Iodine as Secondary Oxidant] . . . 196
(l*R*,2*R*)-1,2-Diphenyl-1,2-ethanediol [Electrochemical Asymmetric Dihydroxylation] . 196
Asymmetric Dihydroxylation Using ABS-MC OsO_4 . . . 197
Preparation of ABS-MC OsO_4 . . . 197
(*R*)-1-Phenyl-1,2-ethanediol [Asymmetric Dihydroxylation Using a Polymer Bound Ligand] . . . 197
(2*R*,3a*R*,4*R*,5*R*,7a*R*)-2-Phenyl-3a,4,5,7a-tetrahydrobenzo[1,3]dioxole 4,5-Diacetate [Asymmetric Dihydroxylation Using AD-mix] . . . 198
TABULAR SURVEY . . . 198
Chart 1. Ligands Used in Tables 1–9 . . . 200
Chart 2. Ligands and Additives Used in Table 10 . . . 209
Table 1. Reactions of Terminal Alkenes . . . 214
Table 2. Reactions of 1,1-Disubstituted Alkenes . . . 285
Table 3. Reactions of Trans 1,2-Disubstituted Alkenes . . . 310
Table 4. Reactions of Cis 1,2-Disubstituted Alkenes . . . 424
Table 5. Reactions of Trisubstituted Alkenes . . . 444
Table 6. Reactions of Tetrasubstituted Alkenes . . . 483

Table 7. Reactions of Conjugated Polyalkenes 492
Table 8. Reactions of Unconjugated Polyalkenes 509
Table 9. Kinetic Resolutions 540
Table 10. Supplemental Table Entries: 2001–2004 544
Table 10A. Reactions of Terminal Alkenes 557
Table 10B. Reactions of 1,1-Disubstituted Alkenes 563
Table 10C. Reactions of Trans 1,2-Disubstituted Alkenes 564
Table 10D. Reactions of Cis 1,2-Disubstituted Alkenes 589
Table 10E. Reactions of Trisubstituted Alkenes 592
Table 10F. Reactions of Conjugated Polyalkenes 600
Table 10G. Reactions of Unconjugated Polyalkenes 602
Table 10H. Reactions of Allenes 607
Table 10I. Kinetic Resolutions 608
References 609

ACKNOWLEDGMENTS

The authors thank Ms. Darra Waller for helpful assistance in collecting references, Ms. Crystal Jordan for assistance with table entries, and the Pfizer Global Research and Development Information Sciences Group for supplying copies of the primary literature used in this work. We are also greatly indebted to Dr. Linda Press for invaluable editorial assistance, to Professor James Rigby for helpful guidance, and to the Organic Reactions reviewers for their insightful comments.

INTRODUCTION

The oxidation of alkenes to vicinal diols using osmium tetroxide (OsO_4) is one of the most selective and reliable transformations in organic synthesis. The reaction stereospecifically produces a cis-1,2-glycol and is tolerant of a wide array of functional groups.[1] Methods have been developed to oxidize alkenes stoichiometrically, as well as in the presence of catalytic amounts of OsO_4 when a suitable secondary oxidant is present. The latter process is particularly useful considering the expense and toxicity of OsO_4. The utility of dihydroxylation in organic synthesis is enhanced by the availability of facile transformations of the cis-1,2-diol products into other synthetically useful intermediates. Among the most versatile intermediates are the corresponding cyclic sulfates, which serve as reactive epoxide equivalents that can be singly or doubly displaced with amine-, oxygen-, sulfur-, or carbon-based nucleophiles.[2–5]

The reaction of OsO_4 with alkenes is accelerated by several orders of magnitude in the presence of coordinating amine ligands such as triethylamine, quinuclidine, or diazabicyclooctane (DABCO).[6,7] These ligands form complexes (**1**) with OsO_4 that subsequently react with the alkene to produce Os(VI) ester intermediate **2** (Eq. 1). When ligand binding to the Os(VI) ester is tight, a reductive quench and workup affords the 1,2-diol product **3** along with reduced osmium species and recoverable ligand. In certain cases, the Os(VI) ester can be oxidatively cleaved, allowing catalytic turnover of the ligand and osmium.

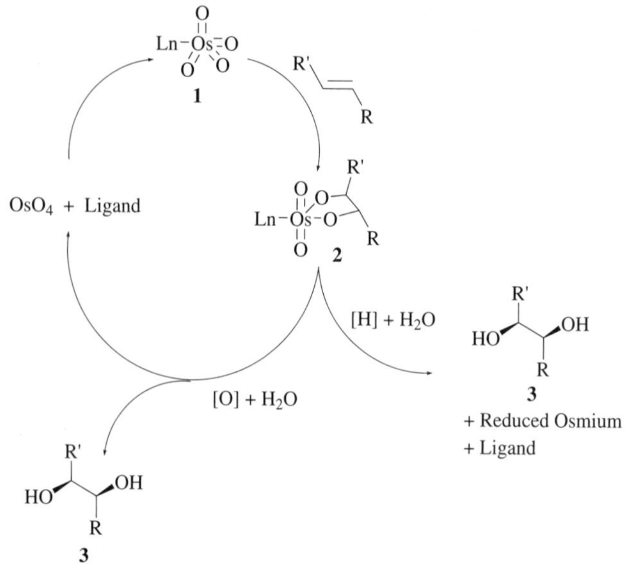

(Eq. 1)

The logical extension to asymmetric osmylation of alkenes in the presence of chiral amine bases spurred the study of asymmetric dihydroxylation. In 1980 the first asymmetric dihydroxylation of alkenes was reported using a cinchona alkaloid as the ligand coordinated to OsO_4.[8] This initial report detailed the stoichiometric dihydroxylation of various alkene substrates using dihydroquinidyl acetate resulting in modest to good levels of enantioselection. The availability of the pseudo-enantiomeric alkaloids quinine and quinidine allows for convenient preparation of either enantiomer of the product glycol using the appropriate ligand. Chiral 1,2-diamine ligands have also been used successfully, and ligands derived from 1,2-diaminocyclohexane,[9,10] 1,2-diphenylethylenediamine,[11] *N,N′*-dineohexyl-2,2′-bipyrrolidine,[12,13] and 1,2-*bis*-(pyrrolidinyl)ethane[14−17] provide high levels of enantioselectivity. The tight binding of these diamine ligands to the Os(VI) ester intermediate precludes oxidative catalytic turnover, requiring the use of stoichiometric amounts of the chiral ligand and OsO_4.

The observation of catalytic turnover in the cinchona alkaloid-OsO_4 system was a breakthrough discovery that revolutionized the field of asymmetric dihydroxylation. The first highly enantioselective catalytic asymmetric dihydroxylation using *N*-methylmorpholine *N*-oxide (NMO) as the secondary oxidant was reported in 1988.[18] Subsequent process improvements and changes in ligand design dramatically extended the scope and utility of the reaction. The use of $K_3Fe(CN)_6$ as the secondary oxidant in the presence of aqueous potassium carbonate further enhanced the generality of the process and obviated the requirement of gradual addition of the alkene substrate in cases where hydrolysis of the osmate ester intermediate was slow. By 1991, the *p*-chlorobenzoate (DHQD-*p*-chlorobenzoate) (**4**), the methylquinoline ether (DHQD-MEQ) (**5**), and the 9-phenanthryl ether (DHQD-PHN) (**6**) of dihydroquinidine and dihydroquinine were the most commonly used ligands for catalytic asymmetric dihydroxylation.

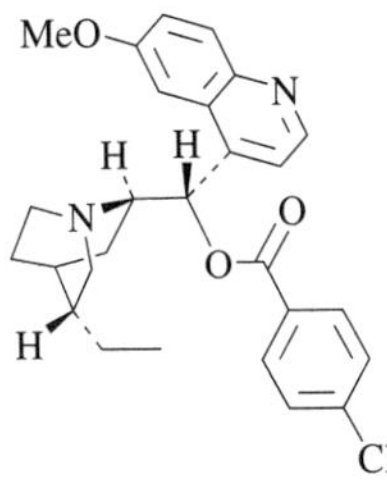

DHQD-*p*-chlorobenzoate (**4**)

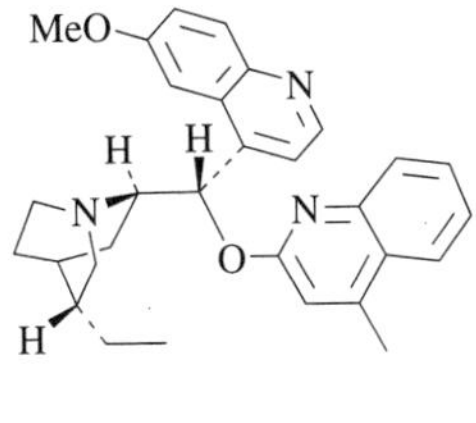

DHQD-MEQ (**5**)

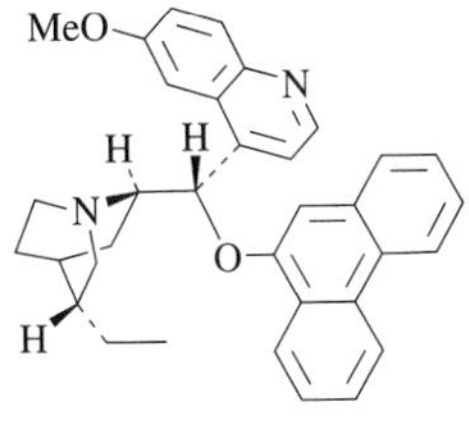

DHQD-PHN (**6**)

The discovery of bis-cinchona alkaloid ligands that afford higher enantioselectivity and increased generality with regard to alkene substitution was reported in 1992.[19] Since then, specialized ligands have been developed that provide position selectivity in the dihydroxylation of polyenes, efficient kinetic resolution of racemic substrates, and high levels of enantioselectivity for each of the six alkene classes (terminal, 1,1-disubstituted, cis-1,2-disubstituted, trans-1,2-disubstituted, trisubstituted, and tetrasubstituted). The most commonly used ligands for catalytic asymmetric dihydroxylation are $(DHQD)_2PHAL$ (**7**) and $(DHQ)_2PHAL$ (**8**). These ligands are commercially available and can be purchased pre-mixed with potassium osmate, potassium ferricyanide, and potassium carbonate as AD-mix α [containing $(DHQ)_2PHAL$ (**8**)] or AD-mix β [containing $(DHQD)_2PHAL$ (**7**)] for added convenience.

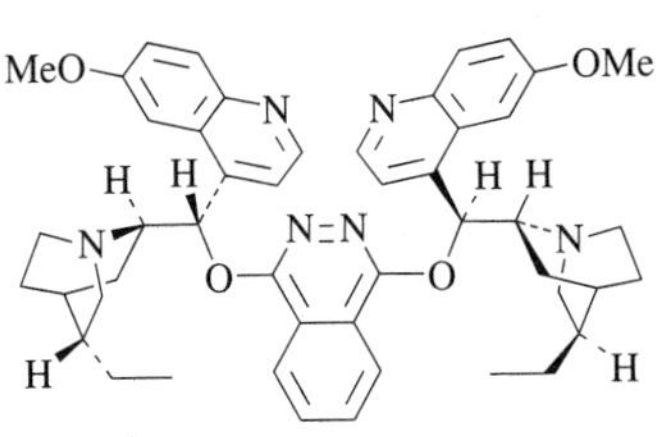

$(DHQD)_2PHAL$ (**7**)

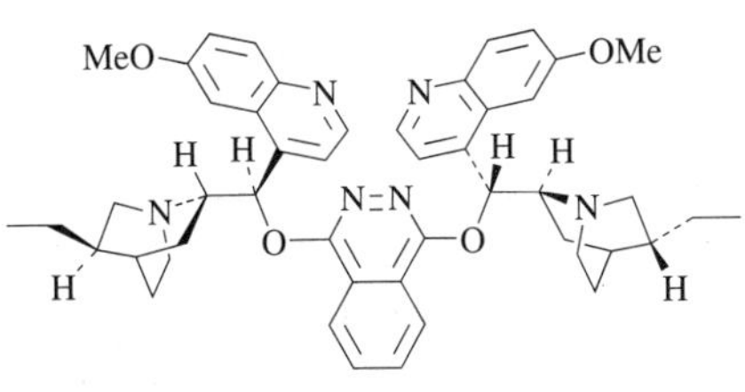

$(DHQ)_2PHAL$ (**8**)

Several reviews have appeared covering various aspects of asymmetric dihydroxylation, the most extensive being that of Kolb, VanNieuwenhze, and Sharpless in 1994.[20–29] This review provides a comprehensive treatment of enantioselective dihydroxylation in the presence of a chiral ligand coordinated to OsO_4. The subject of diastereoselective dihydroxylation under substrate control has been covered extensively elsewhere and is not treated here.[30]

MECHANISM AND STEREOCHEMISTRY

The mechanism of the reaction of OsO_4 with alkenes has been a matter of debate for several decades. At the heart of the controversy is the mechanism of cycloaddition of OsO_4 (or one of its complexes) to the substrate double bond (Eq. 2). Criegee proposed a concerted [3+2] cycloaddition that directly produces the observed

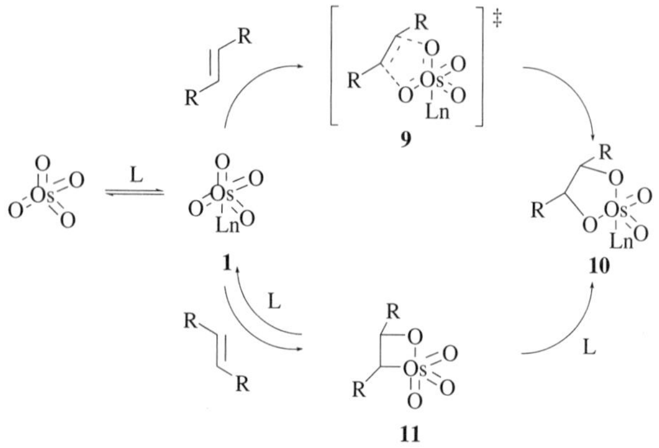

(Eq. 2)

Os(VI) intermediate **10** via transition state **9**.[31–33] Many of these ligand-bound Os(VI) esters are stable, colored solids that have been characterized by NMR spectroscopy. In some cases, crystal structures have been determined by X-ray methods.[34–38] The presence of a chiral ligand on osmium affects the stereochemical course of the reaction, and the enantioselectivity of the dihydroxylation is determined by the relative energies of the diastereomeric transition states for the cycloaddition. An understanding of the substrate and catalyst features that are important in determining the relative energies of these diastereomeric transition states is essential for designing more efficient catalysts and for predicting the direction and amount of enantioselectivity that can be expected in the dihydroxylation of a given substrate. An alternative model has been proposed involving an initial [2+2] cycloaddition of one of the Os=O bonds across the substrate double bond to produce an unstable intermediate (**11**) that subsequently rearranges to afford the observed Os(VI) ester.[39] In this model the chiral ligand can be involved either in the initial [2+2] cycloaddition or in the subsequent rearrangement to produce the Os(VI) ester.[8] The proposed [2+2] cycloadduct has never been observed experimentally, and while the [2+2] cycloaddition reaction is consistent with frontier molecular orbital theory, the latest computational studies favor the Criegee [3+2] cycloaddition pathway.[40–44]

The origin of enantioselectivity in the reaction of 1,2-diamine-OsO_4 complexes with alkenes has been postulated to be governed by steric interactions between the OsO_4 complex and the substituents of the double bond of the substrate. The chelated 20-electron species **12** resulting from coordination of the 1,2-diamine ligand to OsO_4 has been proposed as the active species in these reactions.[11] Both kinetic and structural studies provide support for this hypothesis. Kinetic studies show that the dihydroxylation of alkenes in the presence of one equivalent of a 1,2-diamine ligand is several orders of magnitude faster than the corresponding reaction using two equivalents of a monoamine ligand. Moreover, a crystal structure of 1,2-bis(pyrrolidinyl)cyclohexane coordinated to OsO_4 (**13**) has been reported, providing evidence of the existence of these 20-electron species.[45] A mechanism involving concerted

[3+2] cycloaddition of the alkene across one axial and one equatorial oxygen atom of the bidentate complex **12** predicts the correct stereochemical outcome of the reaction for each of the known reaction systems (Eq. 3).

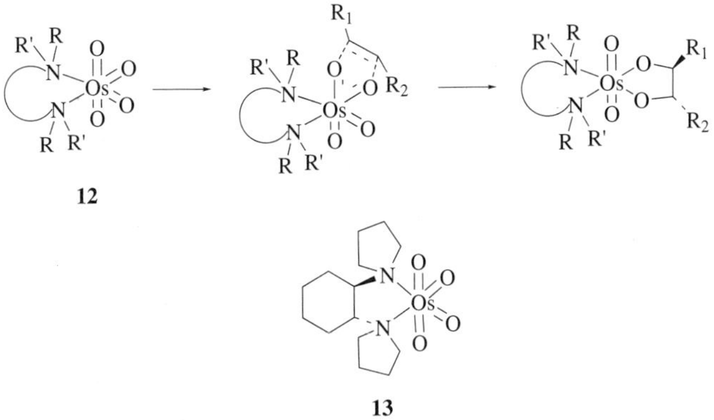

12

(Eq. 3)

13

The origin of enantioselectivity in catalytic asymmetric dihydroxylation has been a topic of intense study and debate throughout much of the last decade.[28,29] Critical to the understanding of the reaction mechanism are kinetic studies that show that the reaction proceeds through a monomeric ligand-osmium complex (**14**; Q = cinchona alkaloid) and point to the existence of a rapid pre-equilibrium that precedes the rate-determining formation of the Os(VI) ester (**15**) (Eq. 4).[46–49]

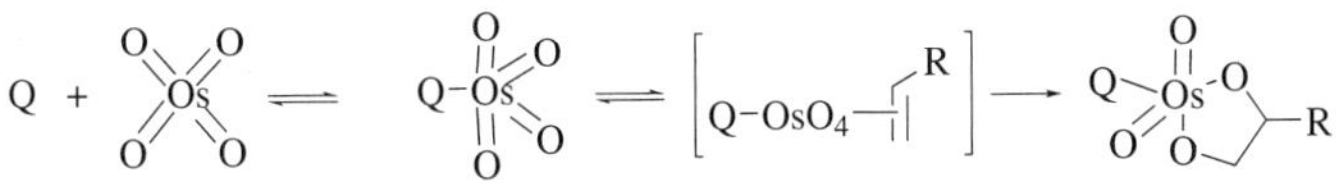

(Eq. 4)

14 15

Because of the sterically demanding environment around the quinuclidine nitrogen atom, the equilibrium constant for formation of the cinchona alkaloid-OsO_4 complex **14** is relatively small. Both enantioselectivity and reaction rate are dependent on cinchona alkaloid ligand concentration, and saturation behavior is observed for both parameters at high ligand concentration.[46] The reaction is first order in both ligand and OsO_4 and exhibits a dramatic ligand acceleration effect relative to the corresponding reaction using quinuclidine as ligand.[50] Moreover, the extent of ligand acceleration directly parallels the level of enantioselectivity generally observed with each of the alkene classes. The observation of Michaelis-Menten kinetics in the catalytic asymmetric dihydroxylation suggests the existence of a rapid pre-equilibrium step prior to the rate-determining formation of the Os(VI) ester **15**.[49] The observed inversion phenomenon in Eyring plots of enantioselectivity as a function of temperature is consistent with this observation and suggests the existence of more than one enantioselectivity-determining process in the reaction.[51,52] Recently reported kinetic isotope effects on the rate of dihydroxylation are more consistent with theoretical calculations for the [3+2] cycloaddition process than for current proposals of a stepwise [2+2] cycloaddition and subsequent rearrangement of a metallaoxetane intermediate to the observed Os(VI) ester **15**.[53,54]

Structural and kinetic studies suggest the formation of a monomeric complex of OsO_4 with the quinuclidine nitrogen of the cinchona alkaloid ligand as the catalytically active species.[55,56] The observation that reactions using the mono-quaternary ammonium salt **16** as ligand behave identically with respect to enantioselectivity and rate to those employing the corresponding free base **7** establishes that a single ligand-OsO_4 complex is involved in the reaction.[47,57]

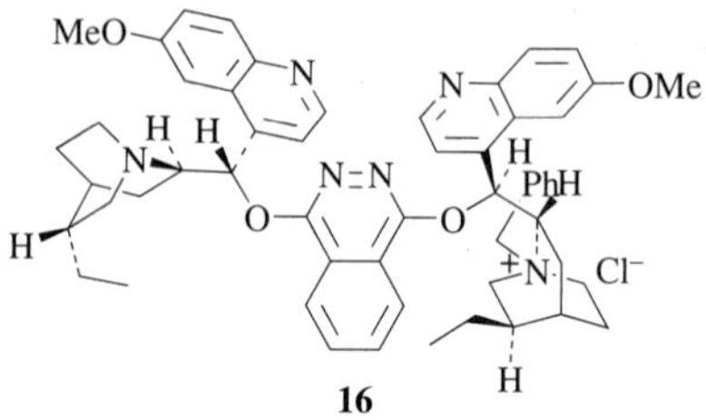

16

The observation that the tethered bis-cinchona alkaloid ligand **17** behaves identically with regard to rate and enantioselectivity to the corresponding non-tethered ligand **18** rigorously establishes the active conformation of the bis-cinchona alkaloid catalyst.[58,59] X-ray crystallographic and computational modeling studies suggest that the nitrogen atoms of the phthalazine and pyridazine linker groups are critical to enforcing this active conformation through stereoelectronic effects.[60]

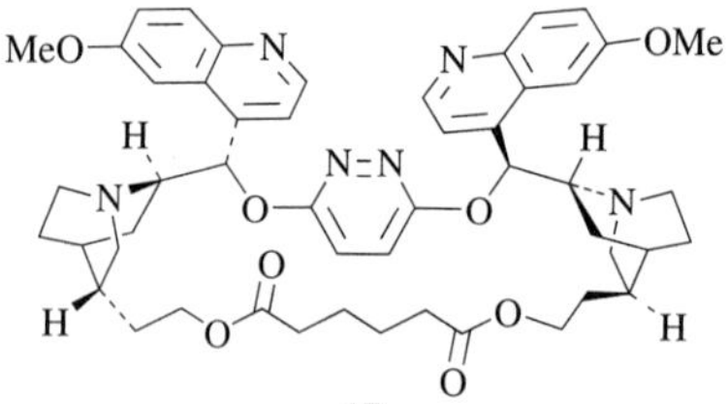

17

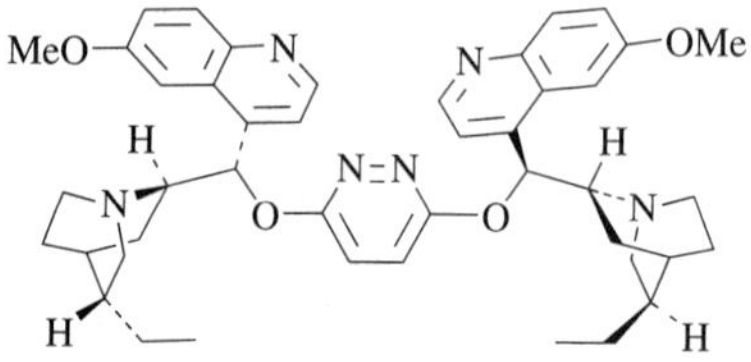

$(DHQD)_2PYDZ$ (**18**)

Attractive interactions between the cinchona alkaloid ligand and the alkene substrate influence enantioselectivity in the catalytic asymmetric dihydroxylation. It has been proposed that the high levels of enantioselectivity and rate accelerations observed in the cinchona alkaloid catalyzed asymmetric dihydroxylation are the result of attractive interactions between the alkene substrate and a U-shaped binding pocket established by the methoxyquinoline "walls" and the pyridazine or phthalazine linker "floor" of the catalyst.[58] The proposed transition state for the dihydroxylation of styrene is depicted in complex **19**.

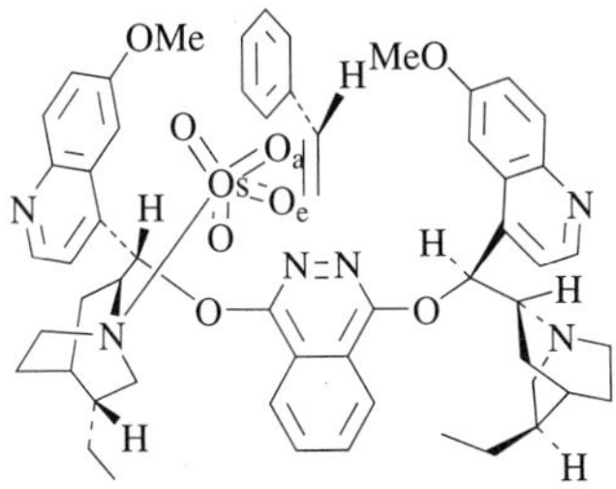

19

Binding of the substrate in this pocket positions the double bond of the substrate in a perfect orientation for the [3+2] cycloaddition across the axial and one of the equatorial oxygen atoms of the coordinated OsO_4 molecule. According to this model, the rate acceleration for the observed enantioselective pathway relative to other modes is due to the favorable free energy of activation for the reaction from the complex **19** in a manner that is ideal for the formation of the thermodynamically more stable osmate ester. Dihydroxylation of the opposite alkene face to that shown in **19** is unfavorable, as there is no three-dimensional arrangement for π-facial approach of the alkene to the oxygens labeled O_a and O_e while maintaining favorable Van der Waals interactions with the binding pocket. The observation of Michaelis-Menten kinetics and inversion phenomena in Eyring plots of enantioselectivity vs. temperature can be understood in terms of rapid, reversible formation of the complex **19**, followed by irreversible formation of the Os(VI) ester. The sense and magnitude of enantioselectivity for a given dihydroxylation reaction can be anticipated as follows:

(1) Orient the substrate and catalyst such that the alkene substituent with the greatest potential for binding (usually an aromatic or aliphatic substituent with little steric demand) is positioned within the U-shaped pocket.
(2) Allow the carbon atoms of the reacting double bond to overlap with the oxygen atoms of the coordinated OsO_4 molecule as illustrated in **19**.
(3) Assess the degree of unfavorable steric interactions that remain between the catalyst and the substrate.

This model for the asymmetric dihydroxylation reaction has been successfully used to predict the highly enantioselective dihydroxylation of allylic 4-methoxybenzoates,[61] to design a ligand for enantioselective and regioselective dihydroxylation of terpenes,[62] to optimize catalyst and substrate pairs in kinetic resolutions,[63] and to incorporate appropriate hydroxy protecting groups for the regioselective dihydroxylation of dienyl alcohols.[64]

An alternative mechanistic model was proposed that invokes a different mode of cycloaddition of the substrate double bond across one of the Os=O bonds as well as a different catalytic binding pocket that governs enantioselectivity. According to this model, ligand-accelerated catalysis occurs through reduction of the activation energy required for either the formation of the putative metallaoxetane intermediate **20** or its subsequent rearrangement to the Os(VI) ester. Enantioselectivity in the asymmetric dihydroxylation is determined by the difference in transition-state energies for either the formation or rearrangement of the diastereomeric metallaoxetanes. The differential stabilization of these diastereomeric intermediates is thought to occur through favorable Van der Waals interactions between one of the substrate substituents and an L-shaped domain composed of the phthalazine linker group and one of the methoxyquinoline rings of the catalyst.[48,65] Computational models were developed to describe quantitatively the interactions between the ligand and substrate within the context of this model that lead to enantioselectivity in the reaction.[48]

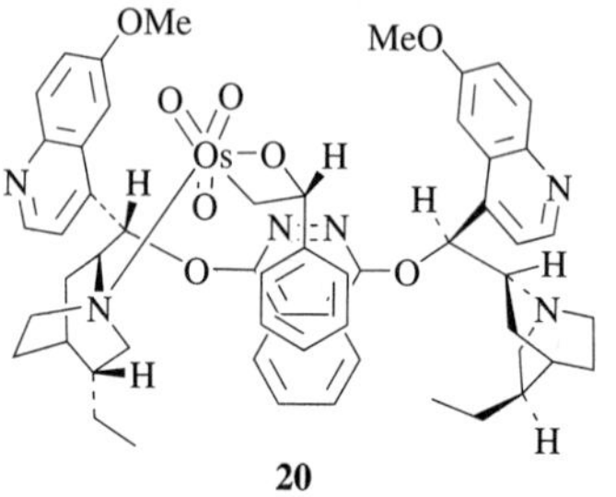

20

A comparison of the two models for dihydroxylation describes experimental observations of enantioselectivity, position selectivity, and efficiency of kinetic resolution that are more easily understood in the context of the U-shaped binding pocket model.[66] Computational studies using hybrid quantum mechanics/molecular mechanics (QM/MM) descriptors provide further support for the [3+2] model in the cinchona alkaloid catalyzed asymmetric dihydroxylation.[67,68] Subsequent studies of kinetic isotope effects[53] were inconsistent with the intermediacy of a metallaoxetane intermediate that has been shown by low level computational studies to have higher transition state energies for its formation and rearrangement than the transition state energy for the [3+2] pathway.[54]

Elucidation of the mechanism of catalytic turnover has been more straightforward and has resulted in several improvements in both process and scope for the cinchona alkaloid catalyzed asymmetric dihydroxylation. The Upjohn NMO-based secondary oxidant system was the first to be successfully applied to the reaction.[18] After formation of the Os(VI) ester **2**, oxidation to the Os(VIII) ester **21** precedes the hydrolysis step that produces the product glycol and recycles OsO_4 (Eq. 5). In cases where this hydrolysis step is slow, the intermediate Os(VIII) ester **21** can undergo a second reaction with the alkene substrate, which produces a bis-glycolate ester **22**. Hydrolysis of the osmium bis-glycolate produces the diol product and regenerates the osmate ester **2**. This "second cycle" results in deterioration of both enantioselectivity and rate, as the second alkene addition occurs with poor facial selectivity, and the hydrolysis of the bis-glycolate ester **22** can be extremely slow.[69] The addition of tetraalkylammonium acetate salts can accelerate the hydrolysis of the Os(VIII) ester

21, but slow addition of substrate is often required to circumvent the second cycle for substrates in which each carbon atom of the double bond is substituted.

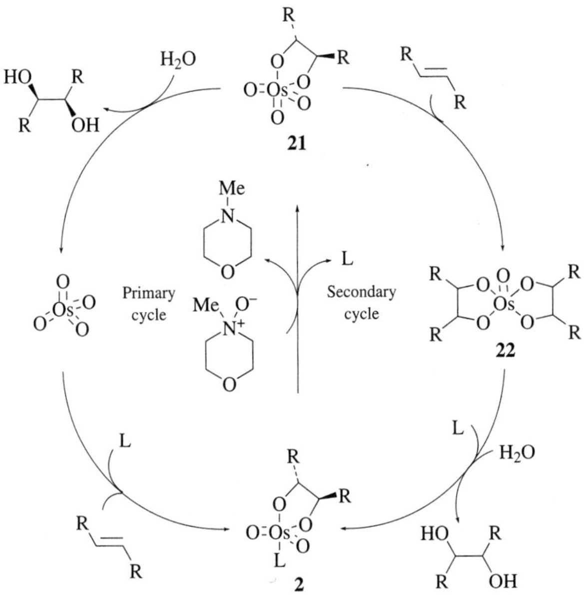

(Eq. 5)

The Tsuji $K_3Fe(CN)_6$–K_2CO_3 secondary oxidant system has also been used successfully for catalytic asymmetric dihydroxylation and is currently the preferred system (Eq. 6). Mechanistic studies indicate that hydrolysis of Os(VI) ester **2** occurs

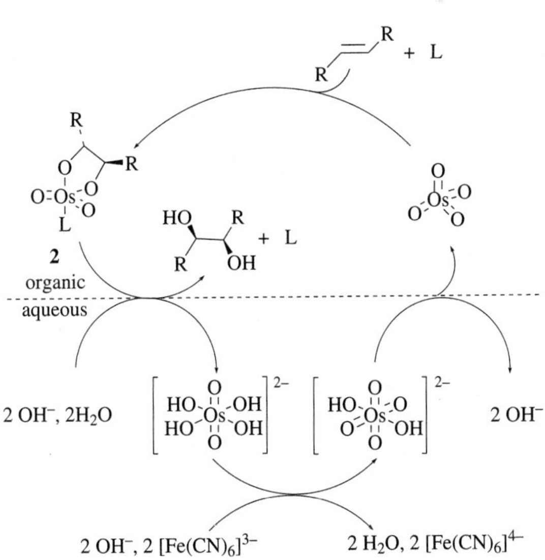

(Eq. 6)

prior to outer-sphere oxidation of Os(VI) to Os(VIII) by $[Fe(CN)^{-}_{6}]^{3-}$, thereby precluding the possibility of a second catalytic cycle.[70] Where each carbon atom of the substrate double bond is substituted, hydrolysis of the corresponding Os(VI) esters can be accelerated by addition of methanesulfonamide, which presumably functions as a nucleophile at the moderately basic pH of the aqueous mixture, or by running the reaction under slightly more basic conditions (pH 12).[19,71] In situations where Os(VI) salts are used as the osmium source in the reaction, oxidation to Os(VIII) precedes entry into the catalytic cycle.

SCOPE AND LIMITATIONS

Asymmetric dihydroxylation of alkenes using either 1,2-diamine ligands or the cinchona alkaloid system is perhaps one of the most general asymmetric processes. The reaction is tolerant of a diverse array of functional groups that may be present on the substrate, and a number of specially developed ligands can be employed for special substrate classes and processes, such as: regioselective and enantioselective dihydroxylation of substrates containing more than one double bond, kinetic resolution of racemic substrates, and dihydroxylation of sterically hindered alkenes. Perhaps the main factor governing the success of an asymmetric dihydroxylation is the substitution pattern of the double bond undergoing transformation. It has long been recognized that the substitution pattern of a carbon-carbon double bond undergoing dihydroxylation by OsO_4 influences the rate of the initial addition of OsO_4 as well as the hydrolysis of the resultant osmate ester intermediate. Steric and electronic effects of substituents can influence both the rate and enantioselectivity of asymmetric dihydroxylation. In order to simplify the analysis, this section will discuss the scope of the reaction as defined for different alkene classes. The six basic alkene classes are: terminal, 1,1-disubstituted, cis-1,2-disubstitiuted, trans-1,2-disubstituted, trisubstituted, and tetrasubstituted. Within each section, both stoichiometric and catalytic asymmetric dihydroxylation will be addressed with the main focus on the latter process, as this is most general and useful to the synthetic chemist.

Terminal Alkenes

The enantioselectivity and rate of dihydroxylation of terminal alkenes is strongly influenced by the nature of the single substituent of the substrate double bond. There are few examples of highly enantioselective dihydroxylations of these substrates using 1,2-diamine-OsO_4 complexes. The most commonly examined substrates include styrene, 1-hexene, and 1-heptene. The dihydroxylation of styrene proceeds with generally high enantioselectivity (>90% ee), and the best reported case employed *N,N′*-bis(3,3-dimethylbutyl)-1,2-cyclohexanediamine (23) as the chiral ligand to afford styrene glycol with 99% ee and 70% yield (Eq. 7).[9] Substrates bearing simple aliphatic substituents react with significantly lower enantioselectivity. The best reported case for the dihydroxylation of 1-heptene used a related ligand (*N,N′*-dimethyl-1,2-cyclohexanediamine) (**24**) and afforded the product glycol in 86% ee (Eq. 7).[10] In general, these reactions are run in aprotic solvents (methylene chloride, tetrahydrofuran, or toluene) at low temperatures (−78°) and require stoichiometric amounts of ligand and OsO_4.

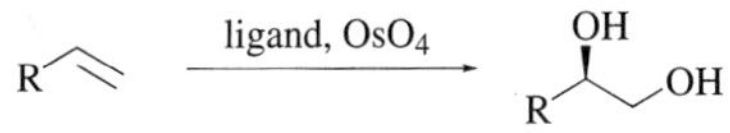

R	Ligand	Conditions		% ee
Ph	**23**	toluene, −90°	(70%)	99
n-C_5H_{11}	**24**	CH_2Cl_2, rt	(75%)	86

(Eq. 7)

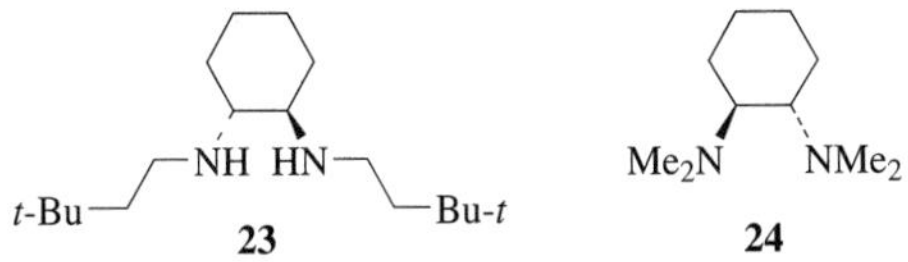

23 **24**

The asymmetric dihydroxylation of terminal alkenes using the catalytic system developed by Sharpless has the greatest synthetic utility due to its breadth of scope and the fact that high levels of enantioselectivity can be routinely obtained. Dimeric ligands, such as the commercially available $(DHQD)_2PHAL$ (**7**) and $(DHQ)_2PHAL$ (**8**), afford glycol products of very high enantiomeric purity in cases where the substrate bears an aromatic substituent (styrene-like alkenes) (Eq. 8).[19] This enantioselectivity is attributed to the ability of the substrate to participate in π-stacking and hydrophobic interactions with the binding pocket of the catalyst. These dimeric ligands give substantially higher enantioselectivities than earlier monomeric ligands, such as DQHD-PHN (**6**).[72] Enantioselectivities greater than 90% can be routinely obtained, and selectivity is largely unaffected by electronic effects imparted by substituents on the aromatic ring. Only substrates possessing unusually large substituents (e.g. 3,5-di-*tert*-butylstyrene)[60] are oxidized with modest enantioselectivity (Eq. 9), and this is attributed to their inability to interact effectively with the catalyst's binding pocket.

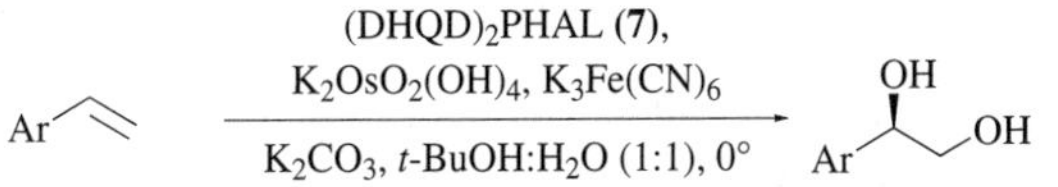

(67-99%), >95% ee (Eq. 8)

Ar = simple substituted phenyl

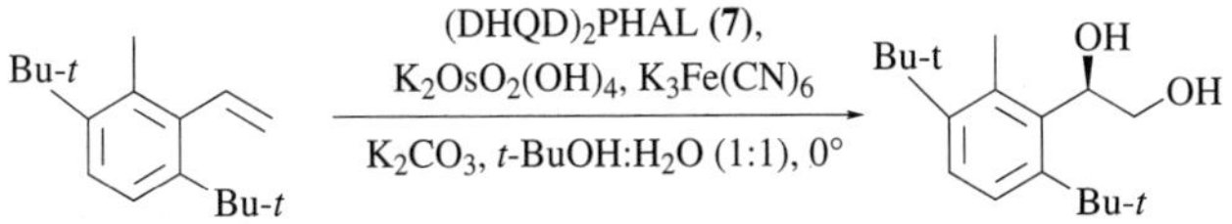

49% ee (Eq. 9)

In addition to simple substituted styrenes, the reaction affords very high levels of enantioselectivity in the dihydroxylation of substrates bearing fused aromatic and heteroaromatic rings. Thus, vinylnaphthalene and 9-vinylanthracene are exceptionally good substrates (Eq. 10).[66,73] Dihydroxylation of vinylheterocycles generally occurs with high enantioselectivity and produces synthetically useful products. The dihydroxylation of 2-vinylfuran was used to establish a non-carbohydrate route to levoglucosenone, an important chiral building block,[74] and provides convenient access to carbohydrates (Eq. 11).[75] Like their carbocyclic counterparts, substrates possessing a fused heteroaromatic substituent are also oxidized with high enantioselectivity (Eq. 12), and the asymmetric dihydroxylation of these substrates has wide scope.[76]

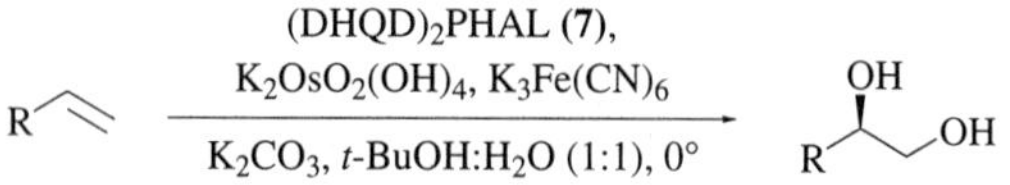

(Eq. 10)

R		% ee
2-naphthyl	(98%)	99
9-anthryl	(75%)	>98

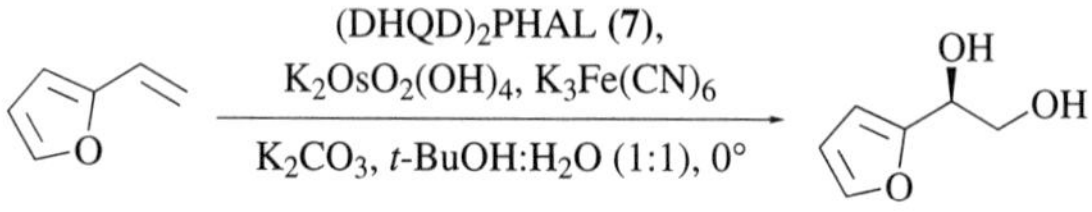

(89%), 93% ee (Eq. 11)

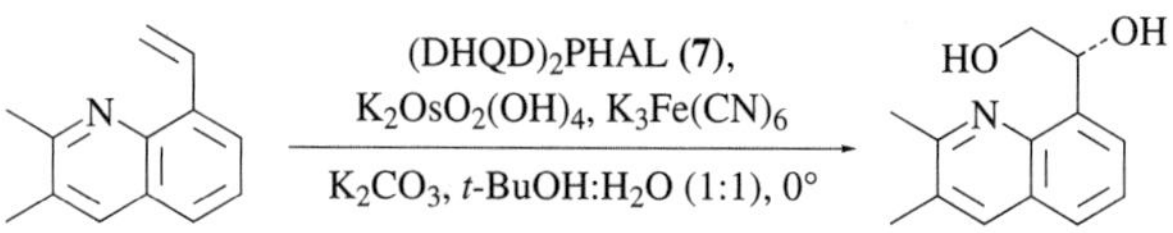

(96%), 95.2% ee (Eq. 12)

The asymmetric dihydroxylation of terminal alkenes bearing simple aliphatic substituents using (DHQD)$_2$PHAL (**7**) or (DHQ)$_2$PHAL (**8**) occurs with somewhat lower enantioselectivity. Some of the best results are obtained for substrates with a straight chain substituent, such as 1-hexene. For these substrates, enantioselectivity is a direct function of substituent chain length as shown in Figure 1.[20] This observation can be understood in terms of the increasing hydrophobic interactions between the substrate and the catalyst, which become saturated with substituent chains longer than C_5.

For substrates in which the double bond substituent is cyclic or branched, asymmetric dihydroxylation using the 2,5-diphenylpyrimidine-based ligands [(DHQD)$_2$PYR (**25**) or (DHQ)$_2$PYR (**26**)] affords better enantioselectivity than the corresponding phthalazine-linked ligands (Eq. 13).[77] These two sets of ligands are complementary in that the phthalazine-linked ligands are far superior for the dihydroxylation of terminal alkenes that bear an aromatic substituent such as styrene.

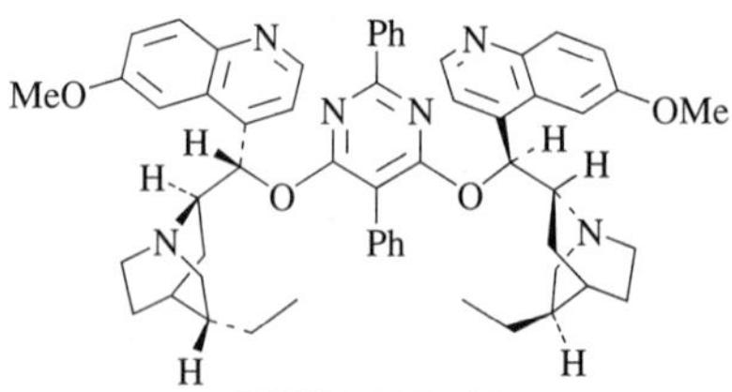

(DHQD)$_2$PYR (**25**)

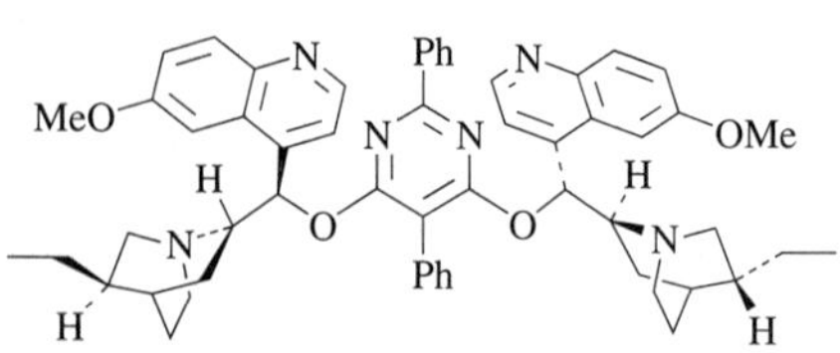

(DHQ)$_2$PYR (**26**)

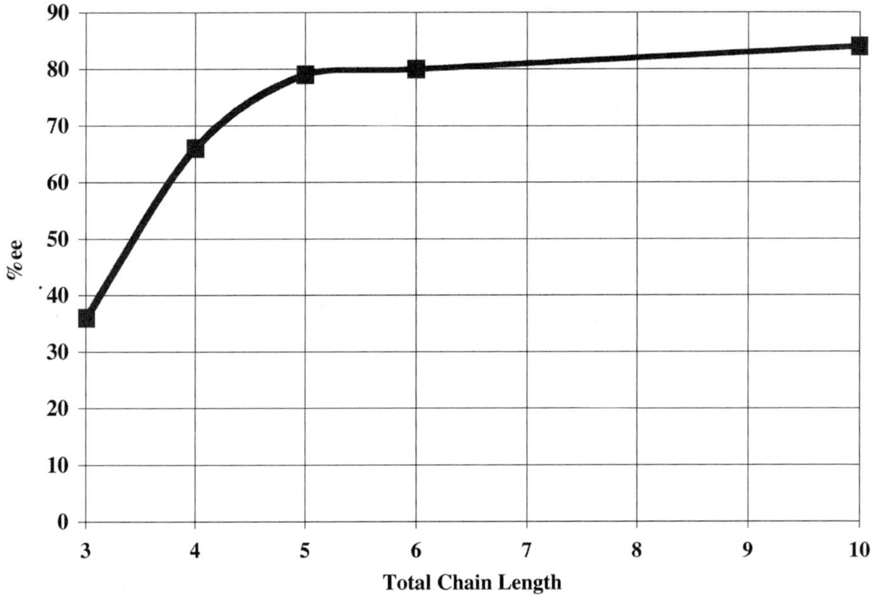

Figure 1. Enantioselectivity as a Function of Total Chain Length for the Asymmetric Dihydroxylation of *n*-Alkyl-substituted Terminal Olefins Catalyzed by $(DHQD)_2PHAL$.

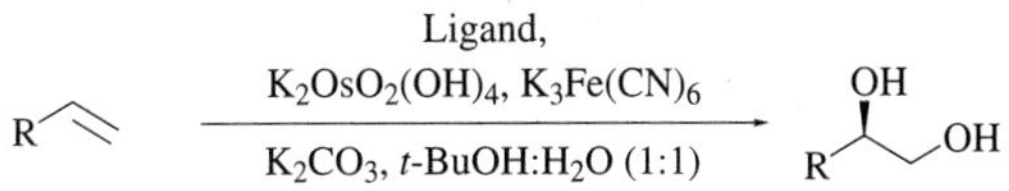

R	Ligand		% ee
Ph	$(DHQD)_2PHAL$ (**7**)	(—%)	97
	$(DHQD)_2PYR$ (**25**)	(—%)	80
n-C_8H_{17}	$(DHQD)_2PHAL$ (**7**)	(—%)	84
	$(DHQD)_2PYR$ (**25**)	(—%)	89
t-Bu	$(DHQD)_2PHAL$ (**7**)	(80%)	64
	$(DHQD)_2PYR$ (**25**)	(80%)	92
c-C_5H_9	$(DHQD)_2PHAL$ (**7**)	(—%)	80
	$(DHQD)_2PYR$ (**25**)	(—%)	93
c-C_6H_{11}	$(DHQD)_2PHAL$ (**7**)	(—%)	88
	$(DHQD)_2PYR$ (**25**)	(—%)	96

(Eq. 13)

The asymmetric dihydroxylation of monosubstituted alkenes is largely tolerant of protected nitrogen or oxygen functionality on the substituent chain. Substrates containing azides, thioethers, amides, ethers, ketones, and imines can be oxidized without destruction of the additional functionality. Steric effects of alkene substitution on enantioselectivity are consistent with the trends outlined above. Dihydroxylation of hydroxyl-containing substrates proceeds with somewhat lower enantioselectivity, although several protecting groups have been discovered that allow highly enantioselective dihydroxylation, and these are discussed below. Reactions where the substrate hydroxy group is protected as an ether or ester have been extensively evaluated.[61,78,79]

Suitably protected allyl alcohols can be oxidized with very high levels of enantioselectivity, providing synthetically useful chiral glycerol equivalents. The dihydroxylation of allyl aryl ethers proceeds with good enantioselectivity and is largely unaffected by the presence of functional groups present at the 4-position of the aromatic ring.[79] The most synthetically useful example is the dihydroxylation of allyl 4-methoxyphenyl ether using $(DHQD)_2PHAL$ (**7**), which affords the product glycol in 95% yield and 90% ee (Eq. 14).[79,80] Significantly better enantioselectivities can be obtained in the dihydroxylation of allyl 4-methoxybenzoate [98% ee with $(DHQD)_2PYDZ$ (**18**)], affording a chiral glycerol equivalent with a readily removed protecting group (Eq. 15).[61] Under these conditions, the reaction time must be minimized, and the isolation of the product from the reaction must be carried out with

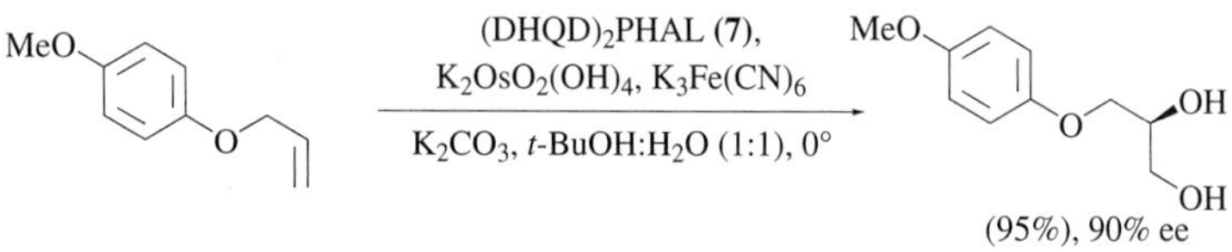

(Eq. 14)

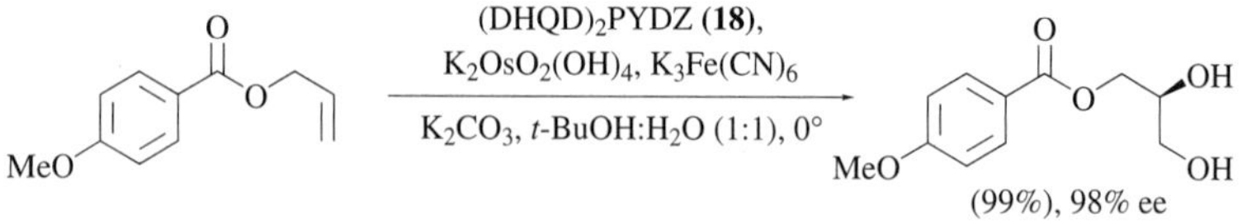

(Eq. 15)

care (by washing the organic phase several times with brine) to remove traces of base that can cause racemization through intramolecular transesterification upon concentration in vacuo. The use of the 4-methoxybenzoyl protecting group as opposed to benzoyl retards transesterification by increasing the kinetic barrier to this side reaction. The enantioselective dihydroxylation of allyl *N*-phenyl carbamate also proceeds with very high enantioselectivity, and the reduced tendency of the carbamoyl group to undergo oxygen to oxygen acyl migration obviates the need to minimize reaction time (Eq. 16).[81] Protection of the allylic alcohol as a bulky silyl or aliphatic ether results in dramatically reduced enantioselectivity, presumably resulting from the inability of the bulky silyl group to fit adequately in the catalyst binding pocket. The 4-methoxybenzoyl group is also the optimum protecting group for 4-pentenol in the asymmetric dihydroxylation, affording the chiral diol **27** with 82% ee (Eq. 17).[61,82]

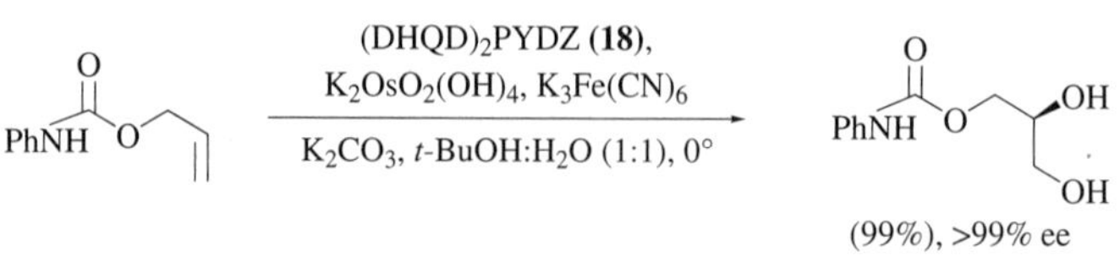

(Eq. 16)

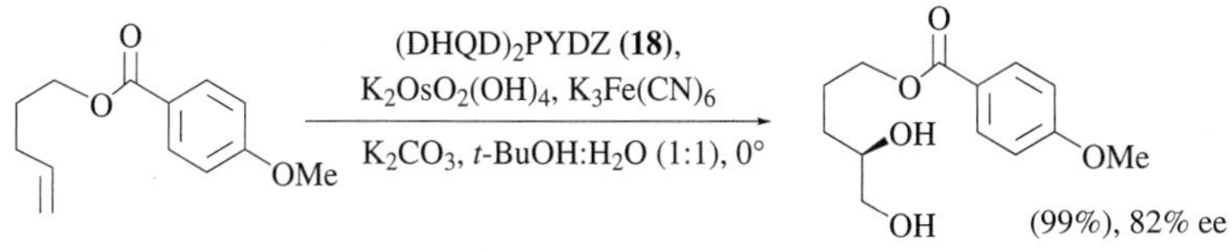

27

(Eq. 17)

Selectivity in the dihydroxylation of protected homoallylic alcohol derivatives is highly sensitive to the nature of the protecting group.[61,83] Unlike allylic and bis-homoallylic alcohols that afford products of high enantiomeric purity when protected as the corresponding 4-methoxybenzoates, homoallylic esters are poor substrates for (DHQD)$_2$PYDZ (**18**). For reactions using this ligand, protection of the substrate hydroxy group as its 4-methoxyphenyl ether is necessary for high enantioselectivity (Eq. 18).[61] The 4-methoxyphenyl group is easily removed by oxidation with ceric ammonium nitrate or DDQ, allowing further elaboration of this useful synthetic piece. The poor enantioselectivity obtained in the reaction of homoallylic 4-methoxybenzoates is attributed to repulsive electronic interactions between the ester carbonyl group and the linker atom nitrogens of the catalyst, as these groups come in close proximity during the [3+2] cycloaddition transition state for the reaction. This inference was experimentally verified by the poor enantioselectivity obtained in the dihydroxylation of homoallyl 2-pyrimidyl ether (50% ee) (Eq. 19),[61,83] as opposed to that observed with the corresponding 4-methoxyphenyl ether (91% ee) (Eq. 18).

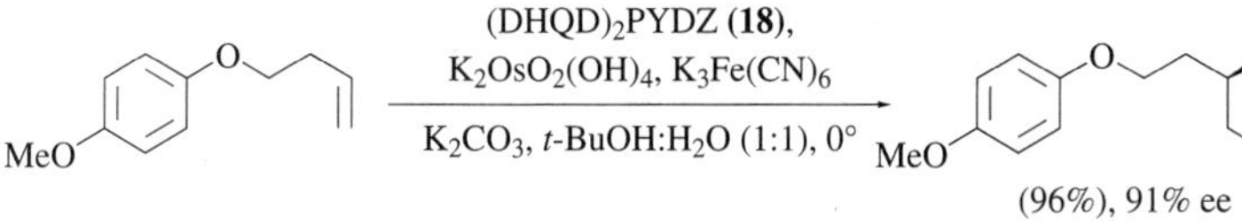

(Eq. 18)

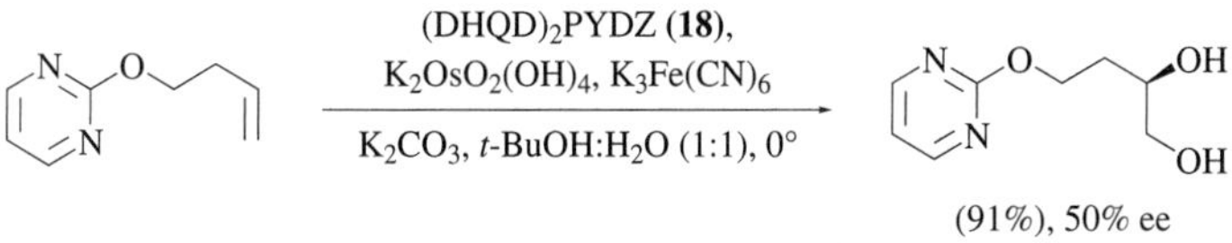

(Eq. 19)

The asymmetric dihydroxylation of acrolein acetals and acrylic esters has also been investigated. Modest levels of enantioselectivity are observed for reaction of acrylic esters catalyzed by (DHQD)$_2$PHAL (**7**) (Eq. 20); significantly lower enantioselectivity was obtained using (DHQD)$_2$PYR (**25**) as the chiral ligand.[19] The asymmetric dihydroxylation of acrolein acetals affords synthetically useful glyceraldehyde acetals. The optimum substrate for this reaction is acrolein acetal **28**, which is oxidized to **29** in 86% ee (Eq. 21). The enantiomeric purity of this product can be enhanced by recrystallization, affording glyceraldehyde acetal **29** of 97% ee in 60% overall yield.[84]

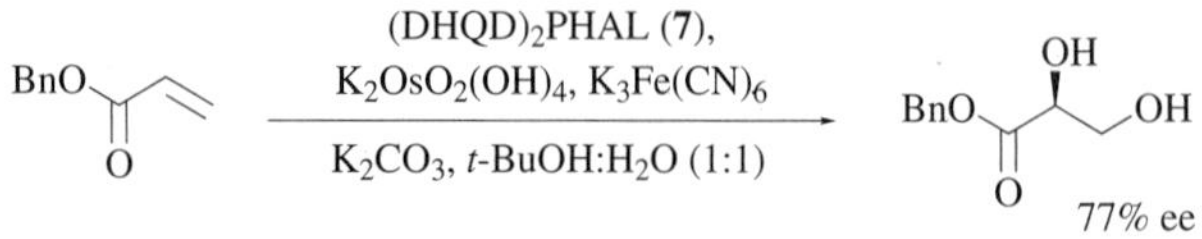

(Eq. 20)

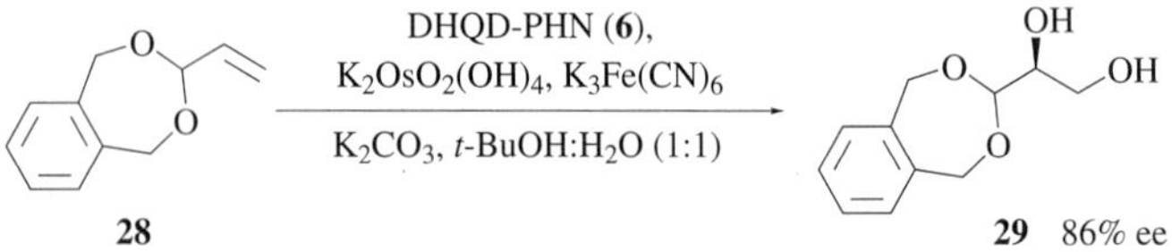

(Eq. 21)

28 **29** 86% ee

The asymmetric dihydroxylation of nitrogen-containing substrates has been studied with substrates possessing amide, imine, and azide groups.[61,85] These non-basic substrates react in high yield, and enantioselectivity is strongly dependent on the nature of the nitrogen substituent(s). As observed for allylic alcohol derivatives, protection with flat, aromatic systems results in higher enantioselectivity than protection with bulkier or conformationally flexible groups. For allylic amines, the planar fluorenone imine (Eq. 22a) is superior to the non-planar, benzophenone imine (Eq. 22b) for high enantioselectivity and can be easily removed by treatment with aqueous acid or by hydrogenolysis. The product of this dihydroxylation is a useful synthetic building block for the preparation of cardiovascular β-blocker drugs and other medicinally active substances.

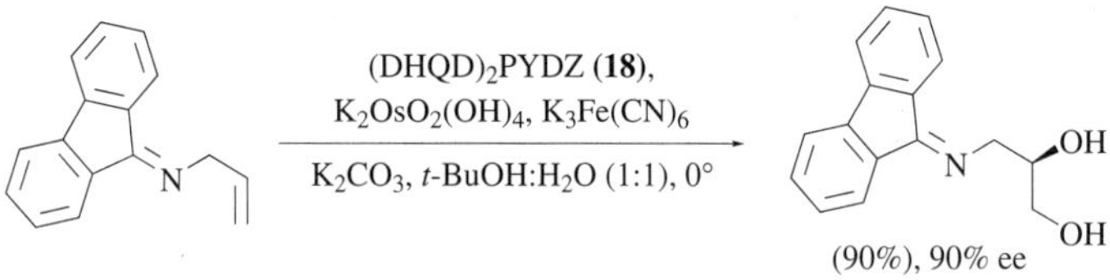

(Eq. 22a)

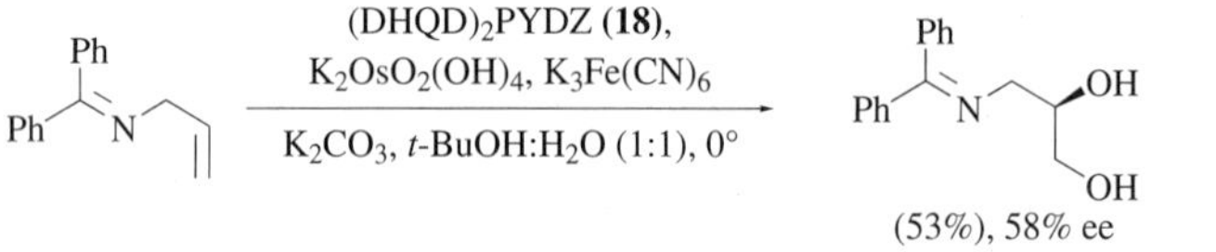

(Eq. 22b)

Whereas the majority of reported catalytic enantioselective dihydroxylations of terminal alkenes use the conventional PHAL and PYR based ligands, the anthraquinone ligands $(DHQD)_2AQN$ (**30**) and $(DHQ)_2AQN$ (**31**) afford higher enantioselectivities in reactions of many allylically substituted terminal alkenes (Eq. 23).[86] The diols that are derived from these substrates are enantiomerically enriched, functionalized chiral glycerol derivatives that are important building blocks for asymmetric synthesis. While the difference in enantioselectivity obtained in the dihydroxylations with the AQN ligands as compared with the PHAL ligands is typically on the order of 10%, it can be as high as 40% for the dihydroxylation of certain substrates such as allyl tosylate. The AQN ligands are not as effective as the

PHAL ligands in the dihydroxylation of substituted styrenes, but they do provide higher levels of enantioselectivity in the oxidation of allylbenzene (78% ee). The AQN ligands are thus complementary to the PHAL and PYR ligands for the enantioselective dihydroxylation of certain terminal alkenes.

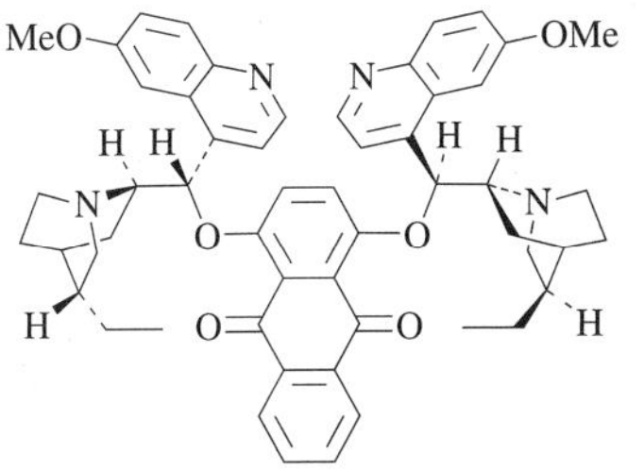

$(DHQD)_2AQN$ (**30**)

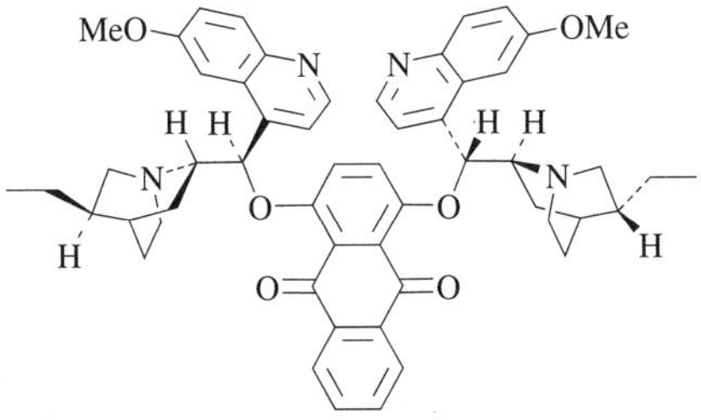

$(DHQ)_2AQN$ (**31**)

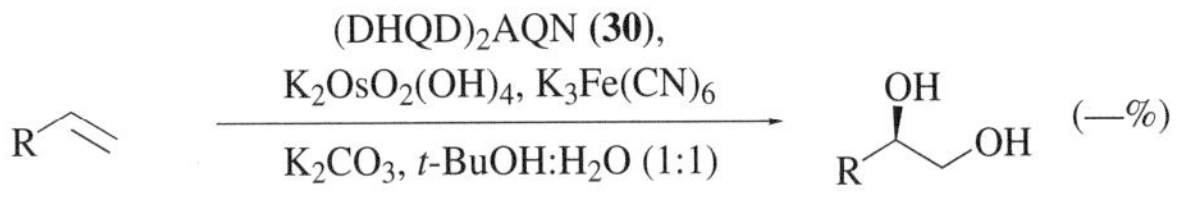

R	% ee
CH_2Cl	90
CH_2Br	89
CH_2I	83
CH_2OTs	83
CH_2OMs	84
$CHCl_2$	72
CF_3	81
CO_2Bn	88
Bn	78

(Eq. 23)

Disubstituted Alkenes

1,1-Disubstituted Alkenes. Enantiofacial discrimination in the enantioselective dihydroxylation of 1,1-disubstituted alkenes is dependent not only on the efficiency of the catalyst, but also on its ability to differentiate the two alkene substituents. When the alkene substituents possess similar steric and electronic properties, the energetic differences for oxidation at either enantiomeric face of the alkene are diminished, resulting in reduced enantioselectivity. While the asymmetric dihydroxylation of these substrates with simple chiral 1,2-diamine ligands has not been reported, the catalytic dihydroxylation of 1,1-disubstituted alkenes with cinchona alkaloids has been extensively studied. For these substrates, enantioselectivities are generally lower than those observed for terminal alkene dihydroxylation due to competition of each alkene substituent for interaction with the catalyst's binding pocket. This effect is dramatically evident when comparing the enantioselectivities for the dihydroxylation of a variety of α-substituted styrenes. The highest enantioselectivities in the dihydroxylation with $(DHQD)_2PHAL$ (**7**) are realized with bulky, short, or hydrophilic α-substituents (Eqs. 24a-c), whereas substrates such as 2-

phenyl-1-octene, which possesses two groups that can interact effectively with the catalyst binding pocket, react with poor facial selectivity (Eq. 25).[87,88]

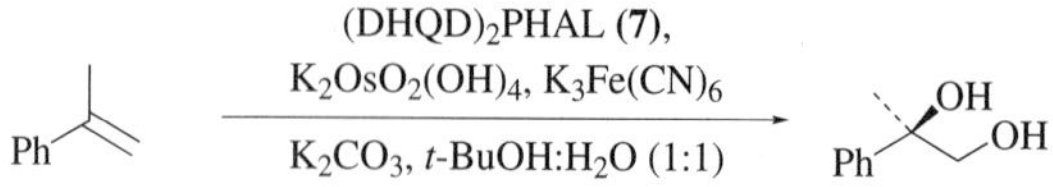

(80-98%), 97% ee (Eq. 24a)

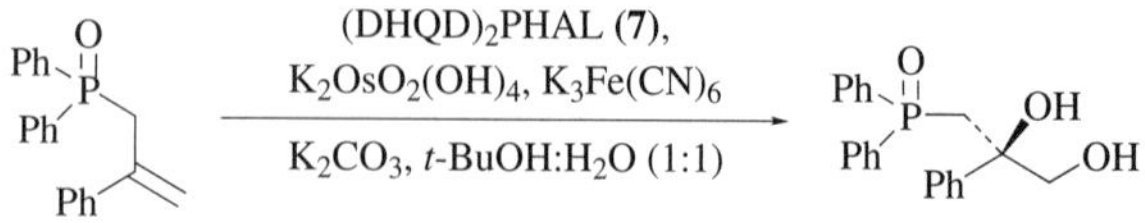

(75%), 86% ee (Eq. 24b)

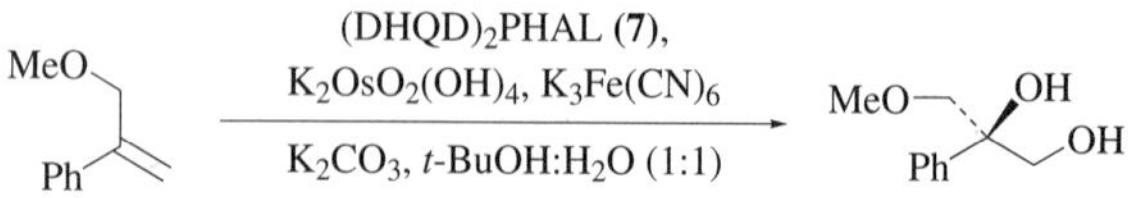

(79%), 97% ee (Eq. 24c)

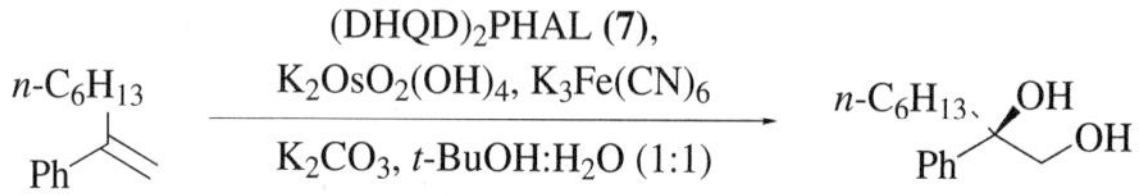

(85-95%), 37% ee (Eq. 25)

The highly enantioselective dihydroxylation of α-substituted styrenes affords a short route to a Mosher's acid precursor in which the dihydroxylation of α-trifluoromethylstyrene (**32**) catalyzed by $(DHQD)_2DPP$ (**35**) gives the product diol **33** in 94% yield and 91% ee (Eq. 26).[89] Oxidation of the diol with oxygen in the presence of platinum affords carboxylic acid derivative **34**, which crystallizes as a conglomerate, allowing the preparation of enantiomerically pure material after a single recrystallization.

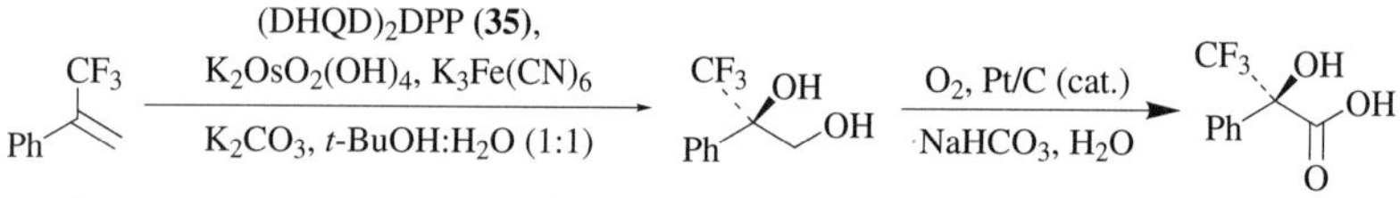

32   **33** (94%), 91% ee   **34** (95%)

(Eq. 26)

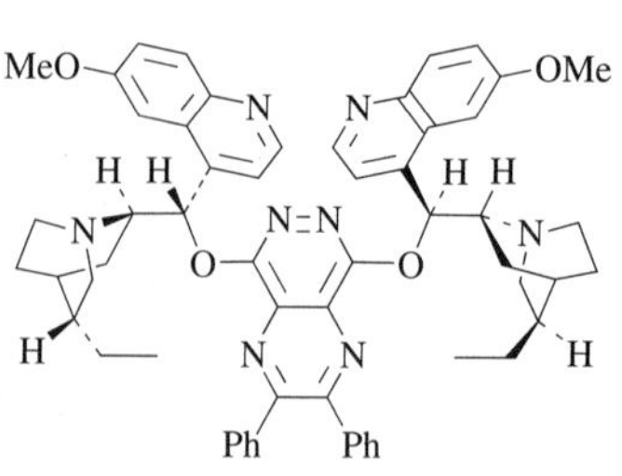

$(DHQD)_2DPP$ (**35**)

Simple 2-substituted propenes are oxidized with moderate to high enantioselectivity that is strongly dependent on the properties of the alkene substituent. In general, substrates possessing aromatic or heteroaromatic substituents react with higher enantioselectivity than those containing aliphatic groups (Eqs. 27a and 27b).[19,77,90] This trend is similar to that observed with enantioselective terminal alkene dihydroxylation.

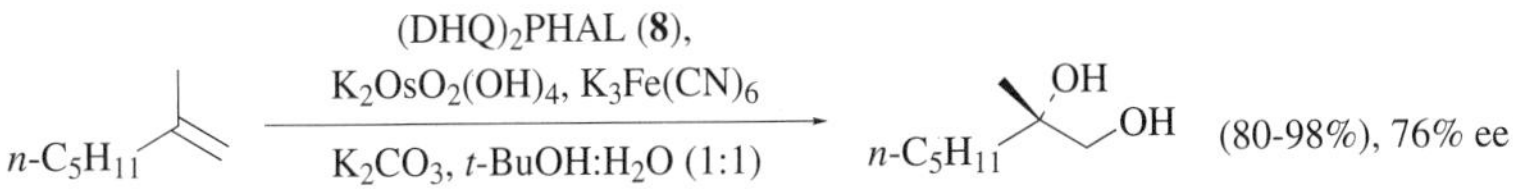

(Eq. 27a)

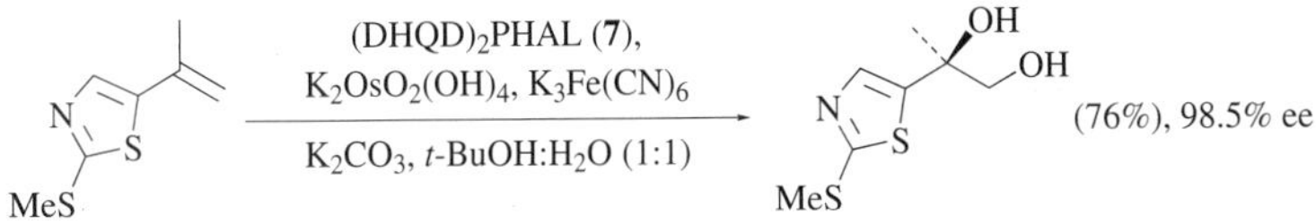

(Eq. 27b)

The diol products derived from the asymmetric dihydroxylation of these substrates are important chiral building blocks for the synthesis of medicinal and natural products. For example, the enantioselective dihydroxylation of 2-(6-methoxynaphthyl)propene (**36**) using $(DHQ)_2$PHAL (**8**) produces diol **37**, which, after deoxygenation and oxidation, affords the important non-steroidal antiinflammatory agent Naproxen (**38**) (Eq. 28).[91,92]

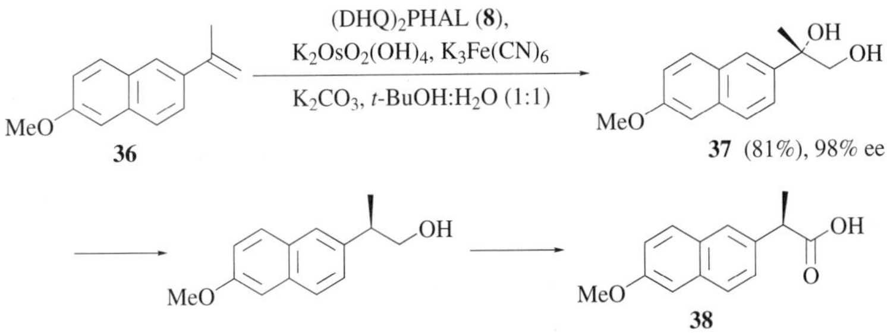

(Eq. 28)

While the direction of enantioselectivity for the vast majority of cinchona alkaloid catalyzed asymmetric dihydroxylations is dictated by the configuration of the cinchona alkaloid subunit (DHQD or DHQ), the preference of the PHAL linked ligands for aromatic substituents vs. the preference of PYR linked ligands for branched aliphatic substituents results in an interesting ligand-dependent reversal in enantioselectivity for 1,1-disubstituted alkenes bearing each substituent type (Eq. 29).[88] Thus, dihydroxylation of α-substituted styrenes bearing simple hydrocarbon chains with $(DHQD)_2$PHAL (**7**) produces the *R* diol preferentially, whereas the use of $(DHQD)_2$PYR (**25**) affords the *S* diol, even though both ligands bear the same cinchona alkaloid subunit. This phenomenon is a dramatic illustration of the effect of alkene substituent competition for the catalyst's binding pocket.

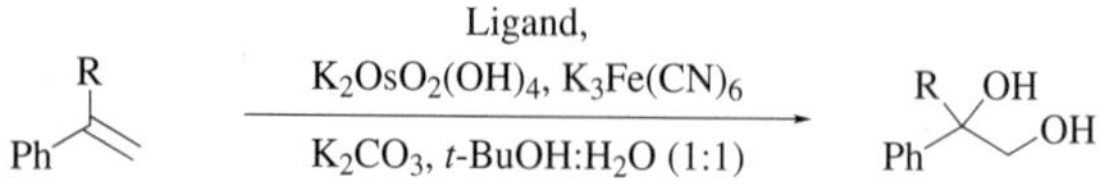

R	$(DHQD)_2$PHAL (**7**) (yield) % ee, config.	$(DHQD)_2$PYR (**25**) (yield) % ee, config.
n-Pr	(85-95%) 60, R	(85-95%) 16, S
n-Bu	(85-95%) 56, R	(85-95%) 28, S
n-C_5H_{11}	(85-95%) 48, R	(85-95%) 30, S
n-C_6H_{13}	(85-95%) 37, R	(85-95%) 35, S
i-Pr	(85-95%) 82, R	(85-95%) 8, S
c-C_3H_5	(85-95%) 70, R	(85-95%) 24, S
c-C_4H_7	(85-95%) 58, R	(85-95%) 59, S
c-C_5H_9	(85-95%) 55, R	(85-95%) 66, S
c-C_6H_{11}	(85-95%) 57, R	(85-95%) 68, S
t-Bu	(30-40%) 8, R	(30-40%) 33, S

(Eq. 29)

The reversal in enantiofacial selectivity is especially pronounced for exocyclic 1,1-disubstituted alkenes, such as 2,2-dimethyl-1-methylenetetrahydronaphthalene (**39**), a rigid analog of *α-tert*-butylstyrene (Eq. 30).[88] The poor enantioselectivity observed in the dihydroxylation of the latter substrate (Eq. 29) is attributed to poor presentation of the phenyl group to the catalyst's binding pocket as a result of conformational restrictions imposed by the *tert*-butyl group. Another important example in this series is the asymmetric dihydroxylation of **40** to produce diol **41**, which was used as an intermediate in the synthesis of a pharmaceutical compound (Eq. 31).[93]

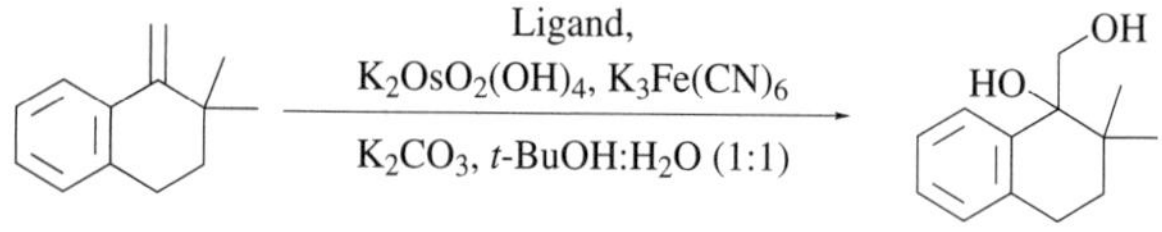

39 (20-40%) (Eq. 30)

Ligand	% ee	Config.
$(DHQD)_2$PHAL (**7**)	82	R
$(DHQD)_2$PYR (**25**)	59	S

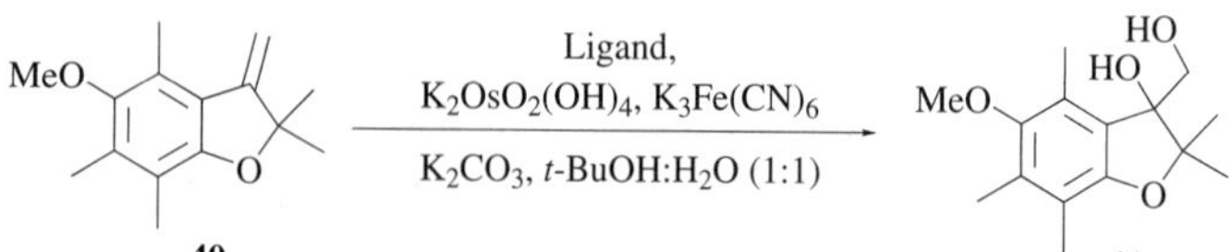

40 **41** (Eq. 31)

Ligand	% ee	Config.
$(DHQD)_2$PHAL (**7**)	36	R
$(DHQD)_2$PYR (**25**)	93	S

The enantioselective dihydroxylation of 2-methallyl alcohol derivatives and homologs proceeds with high levels of enantioselectivity, provided that the appropriate ligand and protecting group are used (Eq. 32).[61,94] In general, enantioselectivity trends for this class of substrates follow those observed for allyl alcohol derivatives. Protection of the hydroxy group as an aromatic ester that can interact favorably with the catalyst binding pocket results in high enantioselectivity in the asymmetric dihydroxylation, whereas substrates possessing bulky silyl ether or pivaloyl protecting groups react with much lower selectivity.[61,94] Moreover, the sense of enantioselectivity changes depending on the preference of the catalyst's binding pocket to accommodate the methyl substituent vs. the other olefin substituent. Much more subtle changes in the protecting group can profoundly influence enantioselectivity, as illustrated by a study of the asymmetric dihydroxylation of methoxy-substituted aryl methallyl ethers (Eq. 33).[95] Here enantioselectivity falls as the methoxy group is moved from the para to the meta and finally to the ortho position, with the latter substrate affording the diol product in only 24% ee.

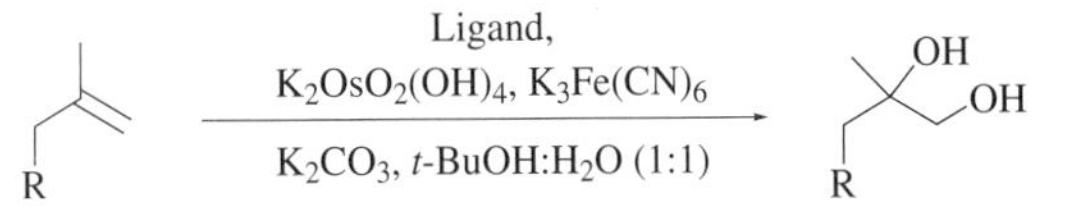

R	Ligand		% ee	Config.
OTBDMS	$(DHQ)_2PHAL$ (**8**)	(92%)	43	S
OBn	$(DHQ)_2PHAL$ (**8**)	(93%)	45	R
OPiv	$(DHQ)_2PHAL$ (**8**)	(63%)	15	S
OTBDPS	$(DHQ)_2PHAL$ (**8**)	(94%)	47	S
$OCOC_6H_4OMe$-4	$(DHQD)_2PYDZ$ (**18**)	(99%)	97	S

(Eq. 32)

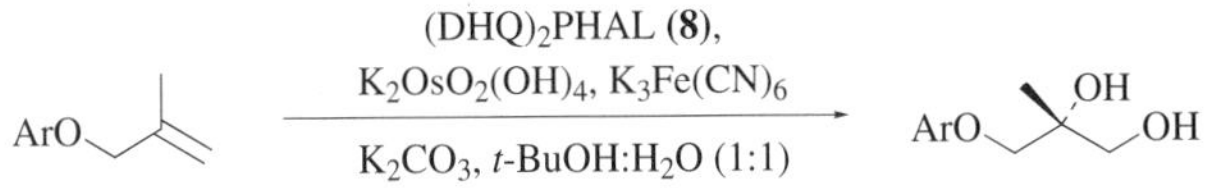

Ar		% ee
4-$MeOC_6H_4$	(95%)	90
3-$MeOC_6H_4$	(91%)	85
2-$MeOC_6H_4$	(91%)	24

(Eq. 33)

The preferential binding of aromatic substituents is exemplified by the highly enantioselective dihydroxylation of ester **42** by $(DHQD)_2PYDZ$ (**18**) to afford product **43**, a differentially protected derivative of tris(hydroxymethyl)methanol, which is a synthetically useful chiral building block (Eq. 34).[61]

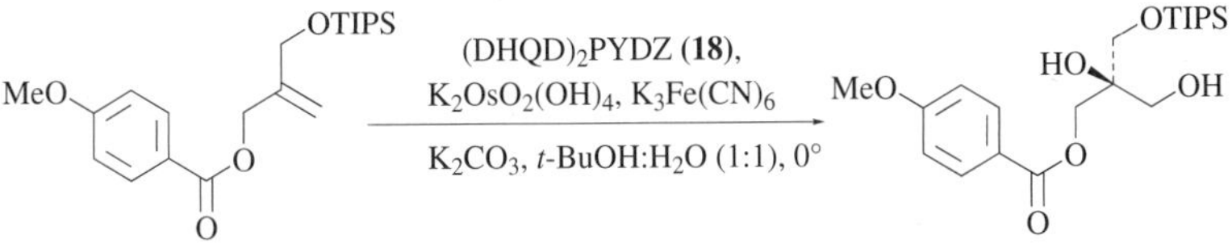

42

43 (99%), 97% ee

(Eq. 34)

As was observed for terminal alkenes, the selection of an appropriate protecting group is critical for high enantioselectivity in the dihydroxylation of homoallylic and bis(homoallylic) alcohol derivatives of this substrate class. In contrast to what is observed for substituted methallyl aryl ethers, subtle changes in substitution of 2-methylbutenyl aryl ethers modestly affect enantioselectivity (Eq. 35),[95] and the preferred protecting group for homoallylic alcohols is the 4-methoxyphenyl ether (Eq. 36). Bis(homoallylic) alcohols should be protected as the corresponding 4-methoxybenzoate esters (Eq. 37).[61,82,83]

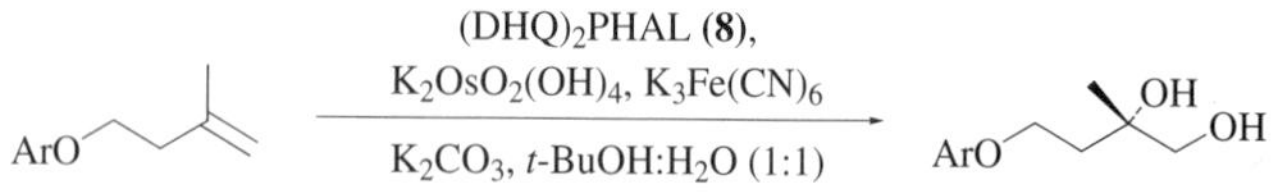

Ar		% ee	Config.
4-MeOC$_6$H$_4$	(93%)	96	S
3-MeOC$_6$H$_4$	(94%)	86	S
2-MeOC$_6$H$_4$	(93%)	80	S

(Eq. 35)

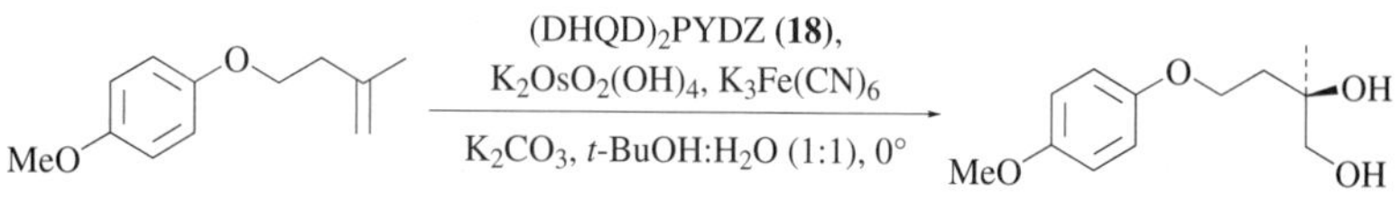

(99%), 96% ee

(Eq. 36)

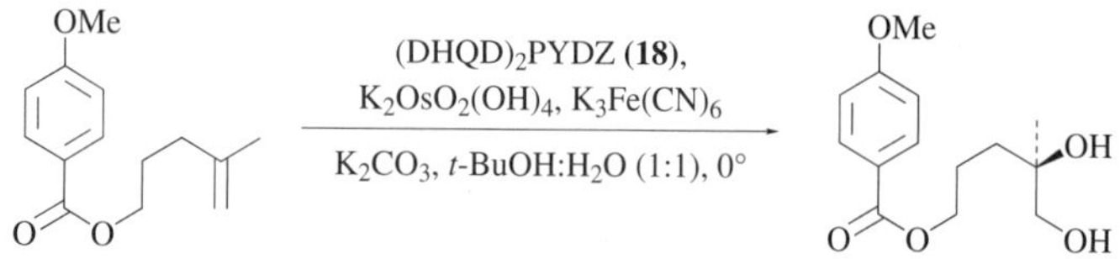

(Eq. 37)

(95%), 79% ee

E-1,2-Disubstituted Alkenes. The dihydroxylation of E-1,2-disubstituted alkenes has been extensively investigated with chiral 1,2-diamine ligands. Enantioselectivity is generally highest with this substrate class, and this can in part be attributed to the C_2 symmetry of substitution about the alkene carbon atoms of these substrates, which is complementary to a C_2-symmetrical chiral environment presented by the ligand. The most extensively studied substrate for these reactions is *E*-stilbene, which is oxidized with high enantioselectivity using ligands derived from tartaric acid, diphenyldiaminoethane, *N,N′*-dialkyl-2,2′-bipyrrolidine, and 1,2-bis(3,4-diphenylpyrrolidino)ethane (Eq. 38).[10–12,14,96] Catalytic turnover is not observed in any of these systems, and stoichiometric amounts of ligand and OsO_4 are required in these oxidations. In addition to that of *E*-stilbene, highly enantioselective dihydroxylation (ee > 90%) of other trans alkenes such as *E*-β-methylstyrene, dimethyl fumarate, and *E*-3-hexene occurs with each of the aforementioned ligands.

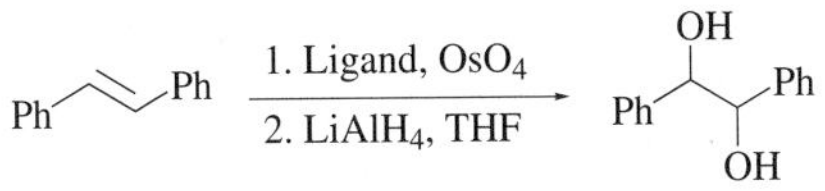

Ligand		% ee	Config.
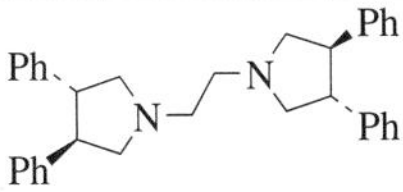	(85%)	97	S,S
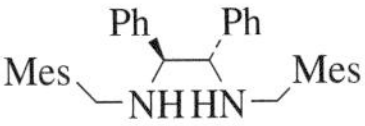	(95%)	92	S,S
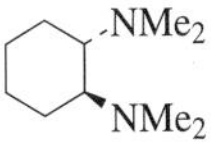	(69%)	34	R,R
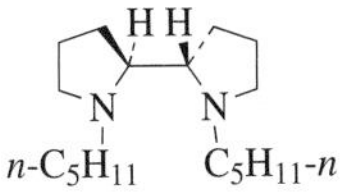	(67%)	89	S,S
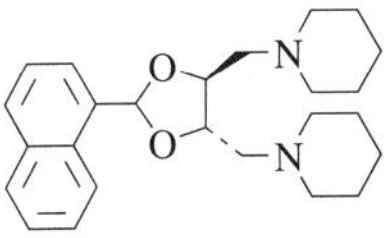	(71%)	90	R,R

(Eq. 38)

Enantioselective dihydroxylation of trans alkenes using *N,N'*-bis(2,4,6-trimethylbenzyl)-1,2-diphenyl-1,2-diaminoethane (**44**) occurs with generally high levels of enantioselectivity to produce synthetically useful products.[11] This ligand, which is also useful for asymmetric Diels-Alder, aldol, and carbonyl allylation processes, is particularly effective for the enantioselective osmylation of α,β-unsaturated esters (Eqs. 39a and 39b).[11] A simple extraction procedure followed by chromatography allows efficient recovery of the diamine ligand and recycling of osmium.

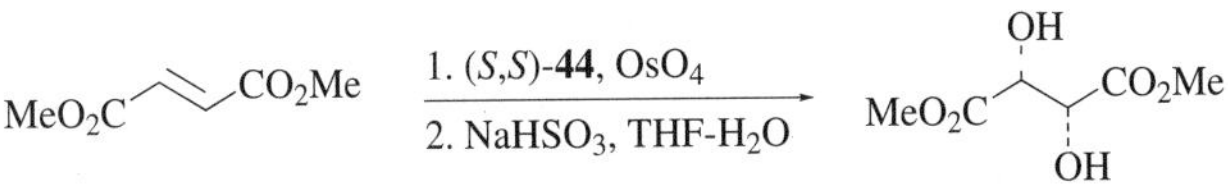

(75%), 92% ee

(Eq. 39a)

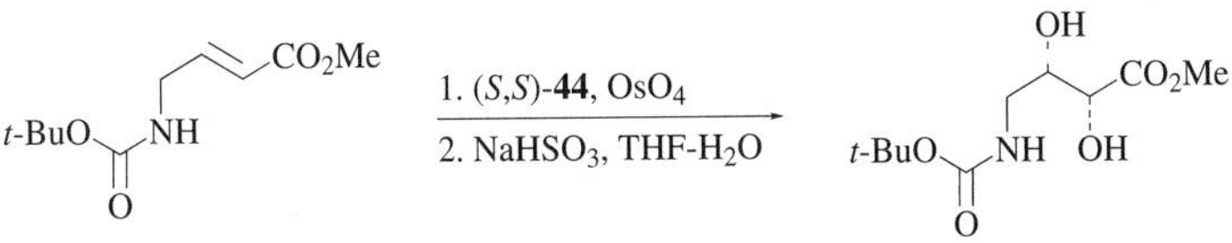

(91%), 97% ee

(Eq. 39b)

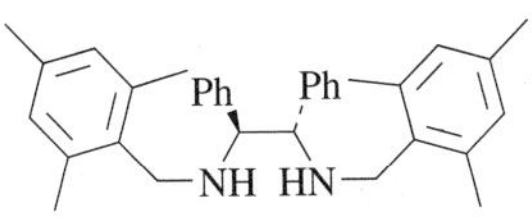

(*S,S*)-**44**

The catalytic asymmetric dihydroxylation of E-1,2-disubstituted alkenes with cinchona alkaloid catalysts generally proceeds with high levels of asymmetric induction, and the level of enantioselectivity is much less sensitive to the nature of the alkene substituents than is seen for oxidations of terminal or 1,1-disubstituted alkenes. While the rate of osmylation of these substrates is higher than the corresponding rate of reaction of terminal alkenes, the overall catalytic reaction can be slower compared to monosubstituted alkenes. The presence of a substituent at each olefinic carbon atom of the substrate results in a dramatic reduction in the rate of hydrolysis of the Os(VI) ester intermediate. The reaction is accelerated by addition of a stoichiometric equivalent of methanesulfonamide to the standard AD-mix where potassium ferricyanide is used as the terminal oxidant, or tetraalkylammonium acetate when NMO is used as the stoichiometric oxidant.[19,69,72]

The breadth of scope for the highly enantioselective dihydroxylation of E-1,2-disubstituted alkenes can be understood on examination of the chiral template presented by both monomeric and dimeric cinchona alkaloid catalysts represented by Figure 2.[20] Positioning the two alkene substituents into the open regions of the catalyst template results in the observed face-selective dihydroxylation. The dihydroxylation of the opposite alkene face is highly disfavored due to severe steric interactions between the double bond substituents and a sterically congested region of the catalyst. In the U-shaped binding pocket model, correct positioning of the substrate results in binding of one of the substituents in the U-shaped pocket, whereas the other alkene substituent resides in the open space in front of the linker group. Dihydroxylation of the opposite alkene face is highly disfavored due to severe steric interactions between one of the alkene substituents and the linker group. Thus, hydrophobic binding interactions favoring dihydroxylation of the observed alkene face and prohibitive steric interactions disfavoring osmylation of the opposite alkene face act in concert to produce the observed high enantioselectivity. This effect is best illustrated by comparison of the enantioselectivity observed for the dihydroxylation of terminal alkenes compared with that of the corresponding C_2 symmetrically substituted trans alkenes (Eq. 40).[20,77,86]

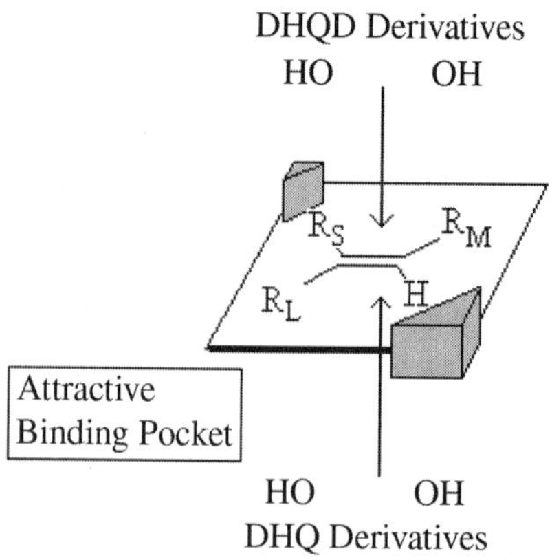

Figure 2. Representation of the Chiral Template of Both Monomeric and Dimeric Cinchona Alkaloid Catalysts.

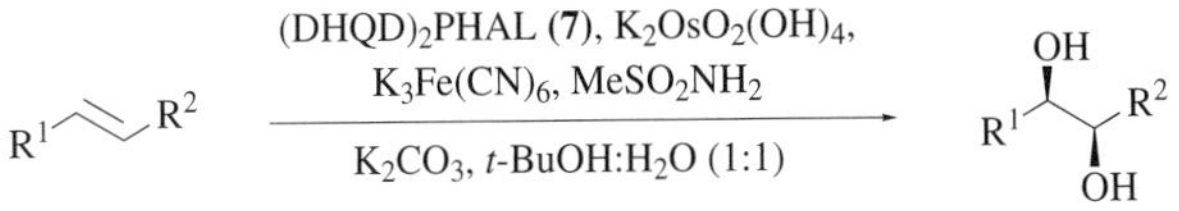

R^1	R^2		% ee	R^1	R^2		% ee
Me	H	(47%)	36	Me	Me	(—%)	72
n-Bu	H	(98%)	80	*n*-Bu	*n*-Bu	(80-98%)	97
Ph	H	(80-98%)	97	Ph	Ph	(80-98%)	99.8
$ClCH_2$	H	(—%)	63	$ClCH_2$	$ClCH_2$	(88%)	94
t-Bu	H	(80%)	64	*t*-Bu	*t*-Bu	(—%)	95

(Eq. 40)

A wide variety of substrates of this alkene class are oxidized by the dimeric PHAL-linked ligands with high levels of enantioselectivity provided that one of the alkene substituents can interact favorably with the catalyst binding pocket. Ligands such as $(DHQD)_2PHAL$ (**7**) prefer aromatic substituents, and trans β-styrenes are oxidized to diols of high enantiomeric purity regardless of the steric or electronic properties of the β-substituent (Eq. 41).[97–101]

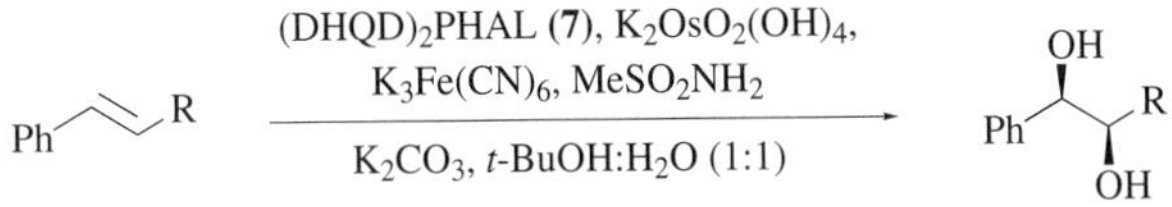

R		% ee
Me	(80%)	>99
CH_2OH	(—%)	97
COMe	(69%)	92
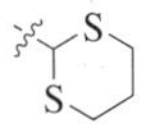	(78%)	97
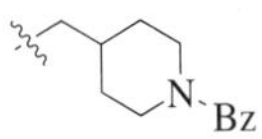	(72%)	>95

(Eq. 41)

The asymmetric dihydroxylation of trans β-substituted acrylic esters using dimeric PHAL-based ligands is very general and proceeds with high enantioselectivity (Eq. 42).[102–105] A diverse array of protected β-substituent functionality is tolerated, including ethers, acetals, phthalimides, and arenes. This breadth of scope, combined with tolerance of a variety of ester groups, allows convenient protecting group manipulation of the highly functionalized products. The α-hydroxy acid products are synthetically versatile intermediates, allowing for convenient preparation of enantiomerically pure α-amino acid derivatives through a four-step sequence (Eq. 43).[105,106] The product diols **45** can be converted into cyclic sulfate intermediates **47**, which are versatile synthetic epoxide equivalents, allowing replacement of one or both of the hydroxy groups with a nucleophile. Cyclic sulfate formation is typically accomplished by treatment of the appropriate diol with thionyl chloride and triethylamine to form the cyclic sulfite **46**, followed by oxidation to the cyclic sulfate **47** using sodium periodate and catalytic ruthenium (III) chloride. Thus, asymmetric osmylation of methyl 6-phthalimido-2-hexenoate (**48**), followed by bromohydrin formation and nucleophilic displacement of the bromide by sodium azide

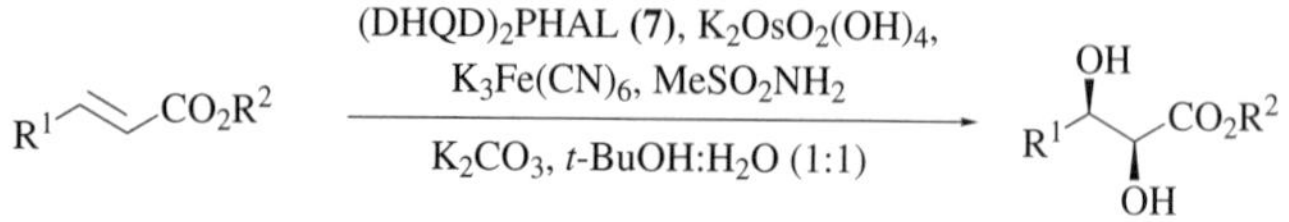

R^1	R^2		% ee
$BnOCH_2$	Et	(82-98%)	98[a]
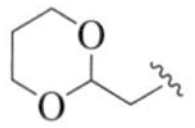	Bn	(80%)	>99
	Et	(90%)	>95[a]
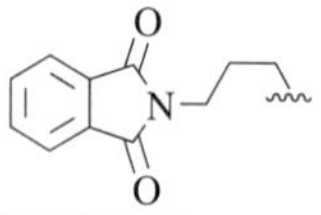	Me	(90%)	>99

[a] The reaction was conducted using $(DHQ)_2PHAL$ (**8**), giving the opposite enantiomer to that depicted.

(Eq. 42)

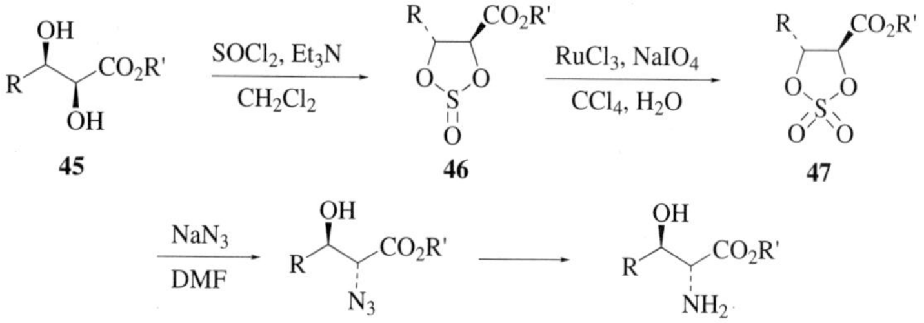

45 **46** **47**

(Eq. 43)

establishes both stereocenters of 3-hydroxylysine (**49**), a naturally occurring amino acid and putative intermediate in the biosynthesis of balanol (Eq. 44).[105] A variety of synthetically important β-hydroxy amino acids can be similarly prepared, such as the allothreonine derivative **50** (Eq. 45) and hydroxyvaline.[106] The enantioselective dihydroxylation of methyl *trans* β-phenylacrylate (**51**) using $(DHQ)_2PHAL$ (**7**) is the key step in a short, industrially scaleable synthesis of the Taxol® side chain **52** (Eq. 46).[107]

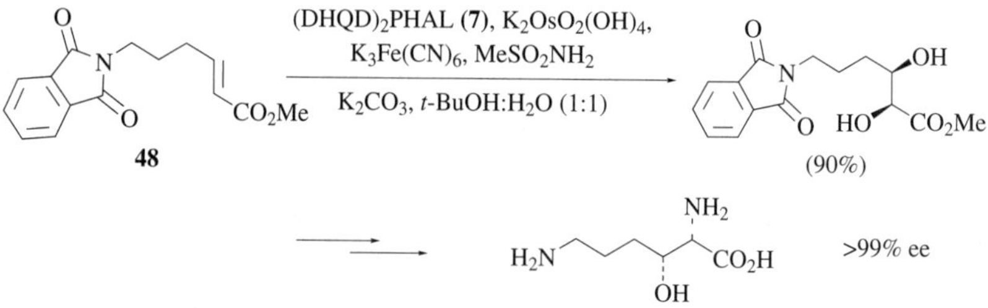

48 **49**

(Eq. 44)

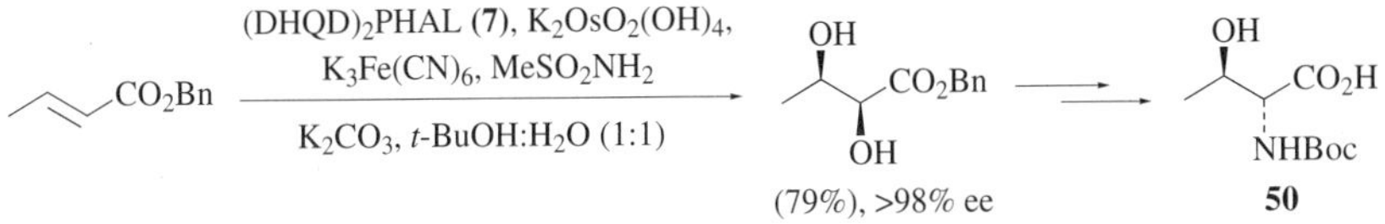

(Eq. 45)

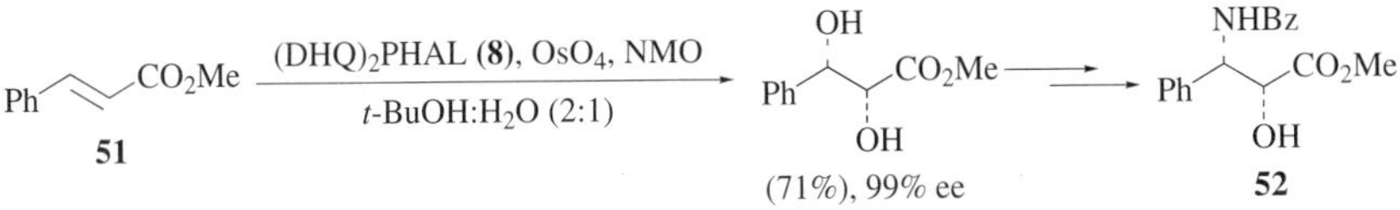

(Eq. 46)

The asymmetric dihydroxylation of β,γ or γ,δ-unsaturated esters proceeds with high enantioselectivity, and subsequent cyclization of the ester group and one of the newly installed hydroxy groups provides an elegant solution to the problem of functional differentiation of the diol (Eq. 47). Treatment of the hydroxylactone intermediate with triethylamine and methanesulfonyl chloride in methylene chloride at 0° results in elimination of the remaining hydroxy group and provides convenient access to synthetically useful chiral butenolides in high yield.[108] Under the basic reaction conditions of the $K_3Fe(CN)_6$-K_2CO_3 secondary oxidant system, the diol esters derived from γ,δ-unsaturated esters spontaneously cyclize to produce functionalized γ-lactones, which are important intermediates for the synthesis of a number of natural products and medicinally active substances, including precursors for HIV-protease inhibitors.[108,109] Although the presence of two hydroxy groups offers the possibility of formation of γ and δ lactones, the formation of the five-membered ring is kinetically favored and occurs selectively (Eq. 48). The cyclization of the diol ester products can be prevented by utilization of the corresponding γ,δ-unsaturated *tert*-butyl esters, which are sufficiently hindered to prevent lactone formation under the slightly basic reaction conditions.[110]

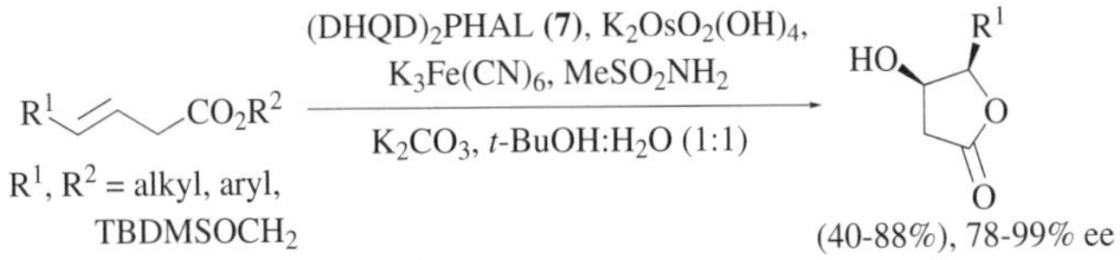

(Eq. 47)

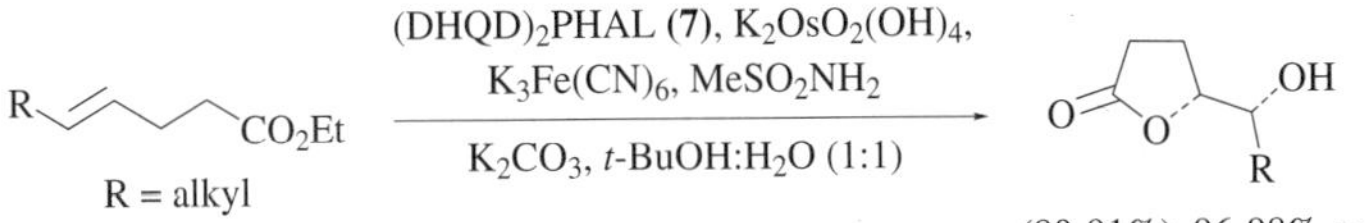

(Eq. 48)

The dihydroxylation of α,β- and β,γ-unsaturated amides also proceeds with high enantioselectivity using $(DHQD)_2$PHAL (**7**) as the chiral ligand (Eq. 49).[111] These alkenes react more sluggishly than the corresponding esters, presumably due to slower hydrolysis of the intermediate osmate ester. This problem is easily solved by increasing the catalyst loading to 5 mol % of chiral ligand and 1 mol % of OsO_4 in addition to the use of stoichiometric methanesulfonamide, which is recommended for the reaction of all trans alkenes using the $K_3Fe(CN)_6$-K_2CO_3 counteroxidant system. Because *N*-methoxy-*N*-methyl amides (Weinreb amides) can be readily converted into aldehydes and ketones by reduction or nucleophilic substitution, the dihydroxylation of unsaturated Weinreb amides can be used to produce masked dihydroxy aldehydes and dihydroxy ketones.

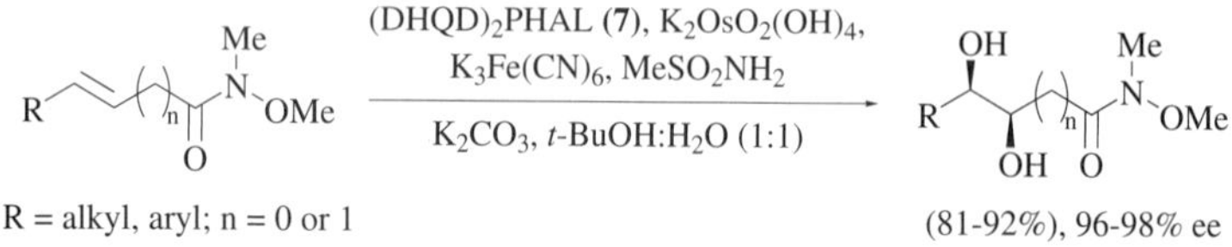

(Eq. 49)

The enantioselective dihydroxylation of α,β-unsaturated ketals and acetals provides an alternative access to protected dihydroxy aldehydes and ketones. The only reported general study of the cinchona alkaloid catalyzed asymmetric dihydroxylation of these substrates utilizes the DHQD-*p*-chlorobenzoate (**4**) ligand (Eq. 50),

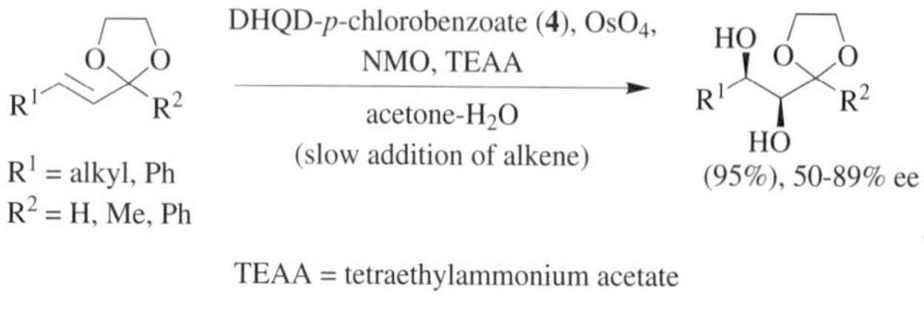

(Eq. 50)

which is known to be inferior to the dimeric ligands such as $(DHQD)_2$PHAL (**7**), for the enantioselective dihydroxylation of many substrates.[112] These reactions were also conducted using NMO as the stoichiometric oxidant, requiring the addition of tetraethylammonium acetate and slow addition of the alkene to preclude the second cycle that is known to cause a reduction in enantioselectivity and reaction rate. Thus, although the reported enantioselectivities are in the 50–89% ee range, the use of one of the more recently discovered ligands and the ferricyanide secondary oxidant system may produce higher enantioselectivity for the oxidation of these substrates. Indeed, the dihydroxylation of α,β-unsaturated acetals derived from 1,2-phenylenedimethanol using the dimeric PHAL ligands with potassium ferricyanide as the stoichiometric oxidant affords the diol products in high yield and enantioselectivity (Eq. 51).[84]

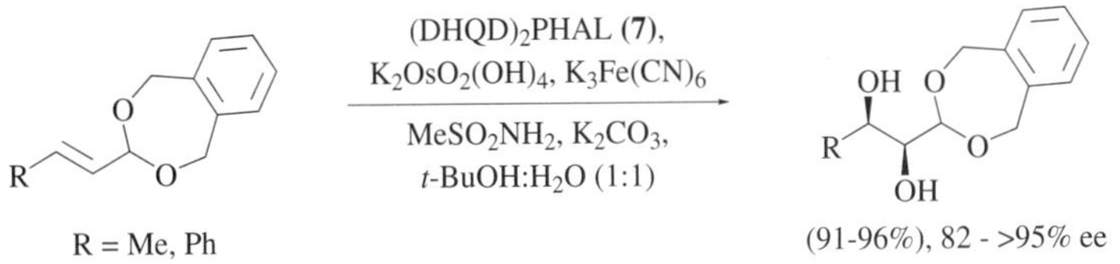

(Eq. 51)

The reactions of E-1,2-disubstituted *N,N*-di(*tert*-butoxycarbonyl)allylic or homoallylic amines also proceed with high levels of enantioselectivity using $(DHQD)_2PHAL$ (**7**) as ligand (Eq. 52).[85] The diols derived from these substrates also undergo spontaneous cyclization with one of the *tert*-butoxycarbonyl groups to provide cyclic *N*-Boc carbamates. This in situ protection of one of the hydroxy groups allows convenient differentiation of the newly installed hydroxy groups in subsequent synthetic transformations. The cyclic carbamate can be selectively hydrolyzed and decarboxylated under mildly basic conditions, such as treatment with K_2CO_3 in methanol, producing the corresponding Boc-protected amino diols. In a particularly interesting application of this methodology, the base-catalyzed hydrolysis of the diol **53** occurs with concomitant reaction of the chlorohydrin to form the highly functionalized chiral epoxide **54** in high yield (Eq. 53).[85]

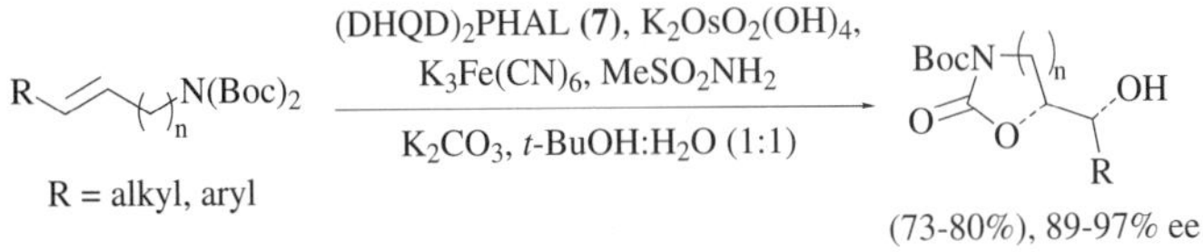

(Eq. 52)

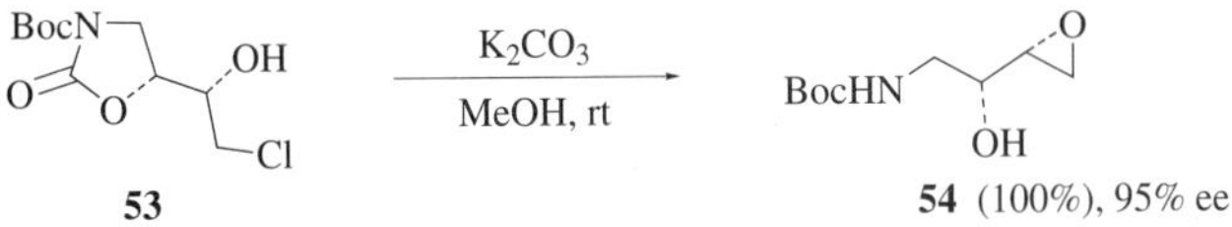

(Eq. 53)

The asymmetric dihydroxylation of E-1,2-disubstituted vinyl- or allyltrimethylsilanes generally proceeds with high enantioselectivity using either PHAL- or PHN-linked ligands, although it has been reported that oxidation of allylsilanes occurs with higher enantioselectivity using the PHN class of ligands.[113–115] Treatment of the diol product with potassium hydride in ether or THF forces Peterson elimination of trimethylsilanol to produce synthetically useful chiral allylic alcohols in high yield (Eq. 54).[115] Because bulky trialkylsilanes do not interact favorably with the catalyst's binding pocket, the presence of a second alkene substituent that can serve as a modest to good binding group is required to attain synthetically useful levels of enantioselectivity when using the $(DHQ)_2PHAL$ (**8**) ligand. Thus, whereas 3-butyl- and 3-isopropylallyltrimethylsilane react with only modest enantioselectivity using $(DHQ)_2PHAL$ (**8**), 3-phenylallyltrimethylsilane is oxidized in 86% yield with 95% ee (Eq. 55).[113] Vinylsilanes of this alkene class are also good substrates for the asymmetric dihydroxylation catalyzed by $(DHQD)_2PHAL$ (**7**).[113] Attempts to convert these diol products into the corresponding cyclic sulfates, thereby producing synthetically useful epoxysilane equivalents, were unsuccessful.

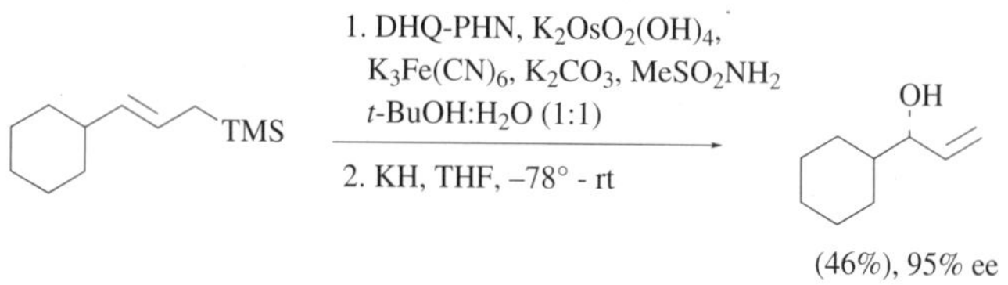

(Eq. 54)

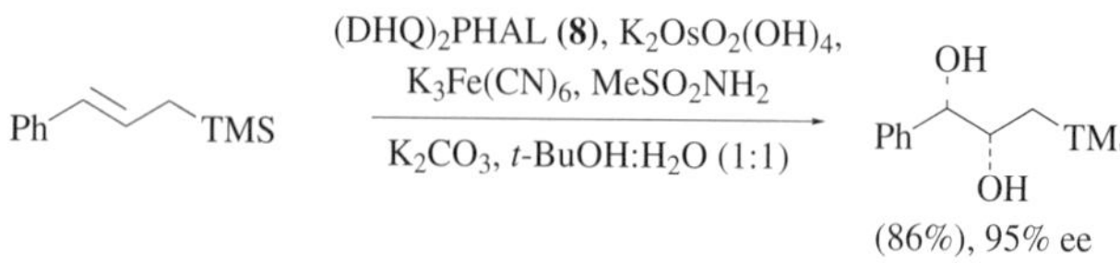

(Eq. 55)

The cinchona alkaloid catalyzed asymmetric dihydroxylation of trans allylic alcohols affords chiral triols of moderate to high enantiomeric purity.[98,116] The presence of the allylic hydroxy group reduces enantioselectivity compared with the reaction of the corresponding unsubstituted trans alkenes.[98] Thus, the enantioselective dihydroxylation of 4,4-dimethylpent-2-en-1-ol (**56**) proceeds with 74% ee using the $(DHQD)_2PHAL$ (**7**) ligand, whereas the parent 4,4-dimethylpent-2-ene (**55**) is oxidized with 95% ee under similar conditions (Eq. 56).[98] The observation of stereo-directing effects of the hydroxy group for cyclic allylic alcohols favoring syn hydroxylation suggests that alkene substituents that can function as hydrogen bond donors may interact with OsO_4, and for these substrates, reduce enantioselectivity.[98] Protection of the allylic hydroxy group as its benzoate derivative easily solves this problem, and very high levels of enantioselectivity are obtained with $(DHQD)_2PHAL$ (**7**) (e.g., compare **57** with **56** and **59** with **58**). The beneficial effect of benzoate protection may result in part from result the ability of the allylic benzoate to interact favorably with the catalyst's binding pocket, as was found with terminal allylic 4-methoxybenzoates.

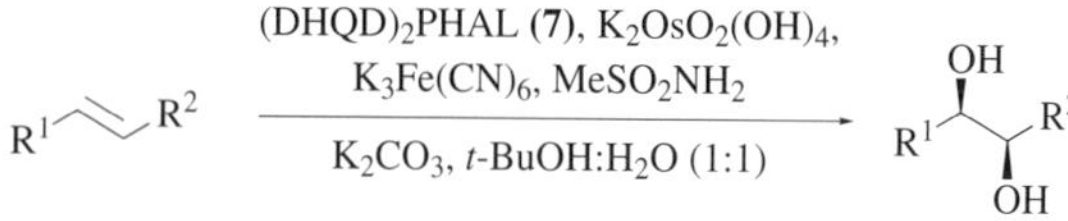

	R^1	R^2		% ee
55	t-Bu	Me	(—%)	95
56	t-Bu	CH_2OH	(—%)	74
57	t-Bu	CH_2OBz	(—%)	95
58	n-C_7H_{15}	CH_2OH	(—%)	93
59	n-C_7H_{15}	CH_2OBz	(—%)	99
	$C(CH_3)_2OH$	$C(CH_3)_2OH$	(70%)	90
	n-C_5H_{11}	$C(CH_3)_2OH$	(87%)	90
	Ph	$C(CH_3)_2OH$	(83%)	77
	$PhCH_2CH_2$	$C(CH_3)_2OH$	(83%)	91

(Eq. 56)

The highly enantioselective dihydroxylation of ferrocenyl-substituted trans alkenes illustrates the diverse functionality tolerated by the dimeric cinchona alkaloid catalysts (Eq. 57).[117] These products have potential utility in the synthesis of chiral ligands for other asymmetric transformations. The proper choice of cinchona alkaloid ligand is critical for achieving high levels of enantioselectivity with these substrates. Unlike the enantioselective dihydroxylation of other substrates, careful optimization of ligand and osmium stoichiometry is also important for obtaining good results. Generally, the use of 10 mol% of the alkaloid ligand and 4–10 mol% of potassium osmate gives the highest enantioselectivities, and 1 : 1 acetonitrile-water is a better solvent system than 1 : 1 *t*-BuOH-water. The $(DHQD)_2PYR$ (**25**) ligand generally provides higher enantioselectivity than the corresponding PHAL-linked ligand. This observation is consistent with the known preference of the PYR ligands for bulky alkene substituents as compared to the PHAL ligands.[77]

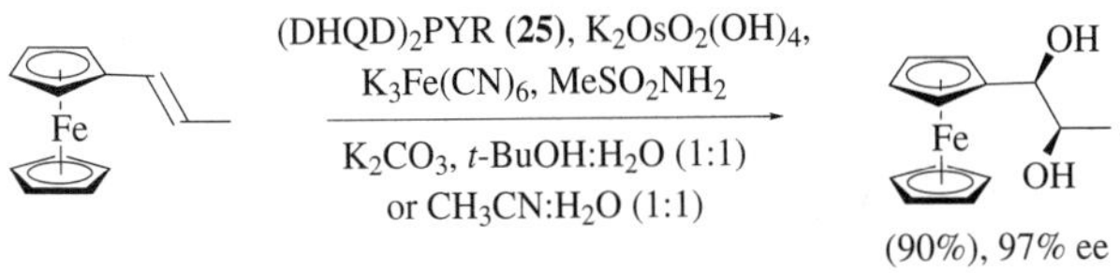

(Eq. 57)

The enantioselectivity of the dihydroxylation of trans α,β-unsaturated alkyl phosphonates catalyzed by $(DHQD)_2PHAL$ (**7**) is highly dependent on the properties of the other alkene substituent. Substrates possessing aromatic substituents are oxidized with high enantioselectivity (Eq. 58), whereas alkenes possessing small or bulky aliphatic substituents react with modest selectivity.[118] The preference of the PHAL-linked ligands for aromatic substituents is consistent with the observed trends in enantioselectivity. The asymmetric dihydroxylation of α,β-unsaturated diphenylphosphinates proceeds with modest enantioselectivity for substrates possessing small aliphatic substituents (Eq. 59).[119] Interestingly, the DHQD-*p*-chlorobenzoate (**4**) and DHQ-*p*-chlorobenzoate ligands were found to afford higher enantioselectivity than the dimeric cinchona alkaloids for the reactions of these substrates.[120]

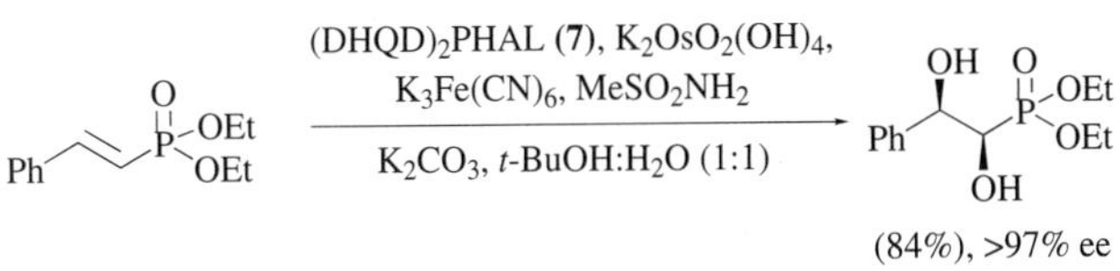

(Eq. 58)

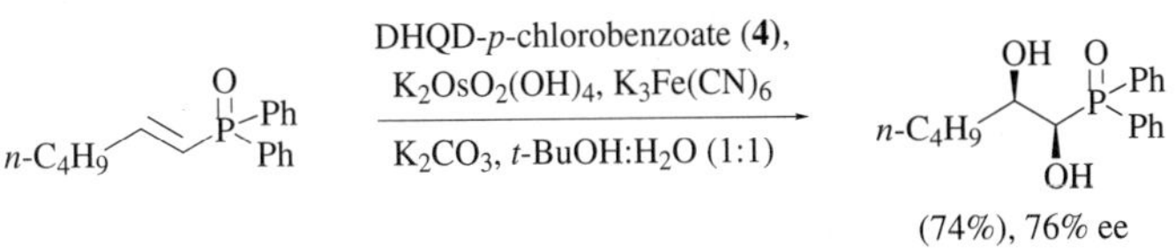

(Eq. 59)

Z-1,2-Disubstituted Alkenes. Z-1,2-Disubstituted alkenes are generally the poorest substrates for enantioselective dihydroxylation. The substrate scope for which dihydroxylation occurs with preparatively useful enantioselectivity is somewhat limited, and specialized ligands have been developed to enhance reactivity and enantioselectivity. The narrow scope observed for this alkene class can be understood in terms of the inability of the ligand-OsO_4 complex to discriminate between the two alkene substituents, particularly when they have similar steric and electronic properties. This phenomenon has been referred to as the meso effect, as Z-1,2-disubstituted alkenes that possess identical substituents produce meso diols upon cis-dihydroxylation.

There are several examples of enantioselective dihydroxylation of Z-1,2-disubstituted alkenes using chiral 1,2-diamine ligands. The most useful of these ligands is **23**, which is derived from (*R,R*)-*trans*-1,2-diaminocyclohexane. This ligand has been used to oxidize a variety of alkenes in moderate to high enantiomeric excess, including acyclic (Eq. 60) and endocyclic (Eq. 61) cis alkenes.[9] Unfortunately, simple aliphatic cis alkenes do not give good selectivity under these conditions. Dihydroxylations using 1,2-diamine ligands such as *N,N′*-bis(3,3-dimethylbutyl)-1,2-cyclohexanediamine (**23**) are typically performed at low temperature and require stoichiometric OsO_4 and ligand, thereby limiting their utility for large-scale reactions.

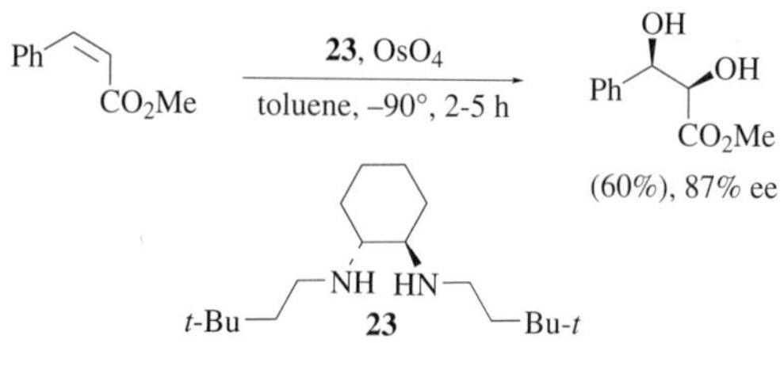

(Eq. 60)

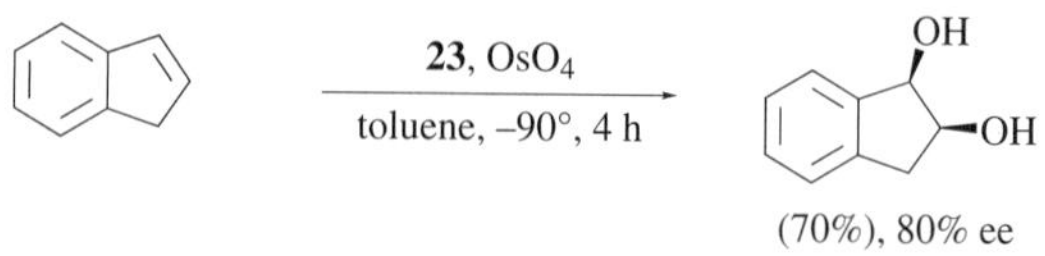

(Eq. 61)

The optimum cinchona alkaloid ligands for the enantioselective dihydroxylation of Z-1,2-disubstituted alkenes are DHQD-IND (**60**) and DHQ-IND (**61**). The catalytic enantioselective dihydroxylation of Z-1,2-disubstituted alkenes typically proceeds with low to moderate enantioselection, and there are only a few substrate-catalyst combinations that produce diols of greater than 80% ee (Eq. 62).[121] As observed for other cinchona alkaloid ligands, reactions using DHQ-IND (**61**) consistently result in lower enantioselectivity than the pseudoenantiomeric DHQD-IND (**60**).[121]

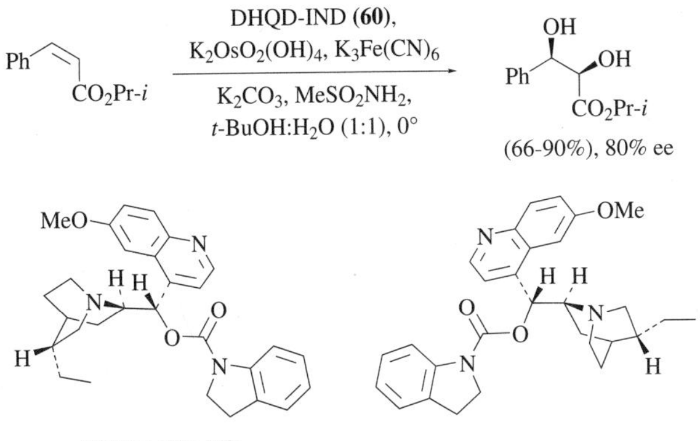

DHQD-IND (**60**) DHQ-IND (**61**)

(Eq. 62)

Although the asymmetric dihydroxylation of Z-1,2-disubstituted alkenes can generally be expected to proceed with moderate enantioselectivity, there are isolated examples where high enantioselectivity is observed. Thus, the catalytic asymmetric dihydroxylation of alkene **62** using the $(DHQ)_2$PYR (**26**) ligand under standard conditions produces diol **63** in 86% ee (Eq. 63).[122] This intermediate was used to prepare 3′,4′-di-*O*-(–)-camphanoyl-(+)-*cis*-khellactone, a potent anti-HIV agent.[122]

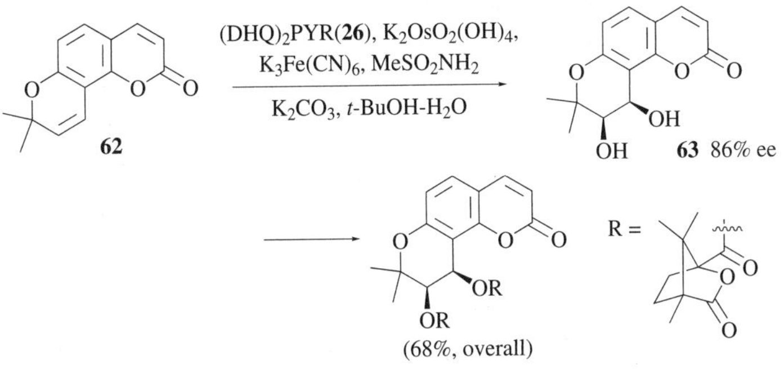

(Eq. 63)

The dihydroxylation of Z-allylic and homoallylic alcohols has been reported to proceed with moderate enantioselectivity using the PHAL class of ligands. The higher enantioselectivity observed with these substrates than that generally seen with cis alkenes presumably results from hydrogen-bonding interaction between the catalyst and the hydroxy group of the substrate. This hypothesis is supported by the fact that *cis*-3-hexen-1-ol (**64**) reacts to give the triol with 54% ee in spite of the fact that the two alkene substituents are of very similar size (Eq. 64).[123] Conversely, the corresponding methyl ether **65** reacts under the same conditions to give racemic triol (Eq. 65).[123] Certain endocyclic dienes are also effective substrates for the catalytic enantioselective dihydroxylation. These examples are treated in the section describing polyalkene substrates, as both regioselectivity and facial selectivity are important parameters for these reactions.

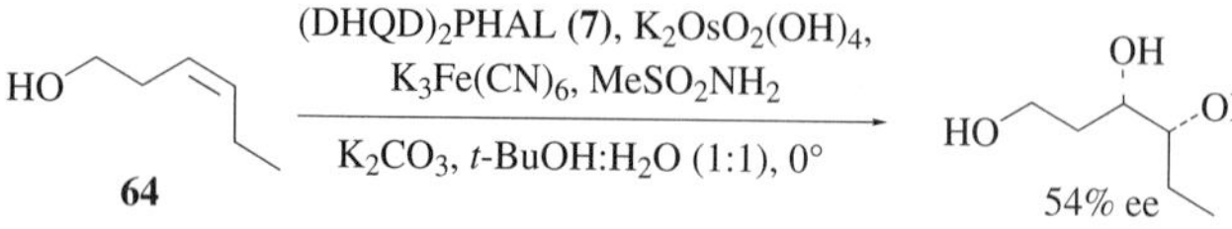

(Eq. 64)

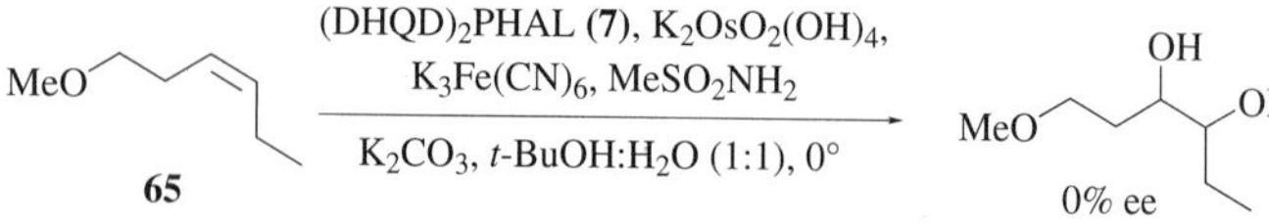

(Eq. 65)

In order to overcome the difficulties associated with enantioselective dihydroxylation of Z-1,2-disubstituted alkenes, a strategy has been developed that utilizes the highly enantioselective dihydroxylation of E-1,2-disubstituted alkenes followed by stereoinversion of one of the resulting hydroxy groups. The three-step process involves (1) asymmetric dihydroxylation of an E-1,2-disubstituted alkene, (2) formation of a cyclic sulfate[2] and, (3) inversion of one of the alcohols via a Payne-type rearrangement followed by nucleophilic opening of the resulting epoxide (Eq. 66). The sequence is illustrated by the conversion of alkene **66** into product **67**.[124,125] The entire process, post isolation of the diol, can be completed in a single reaction vessel and generally results in the formation of the formal products of dihydroxylation of a cis alkene in high enantioselectivity and good overall chemical yield. Nucleophiles that have been used to open the epoxide formed from the Payne-type rearrangement include sulfide, azide, acetate, cyanide, iodide, and organometallic reagents.[124,125]

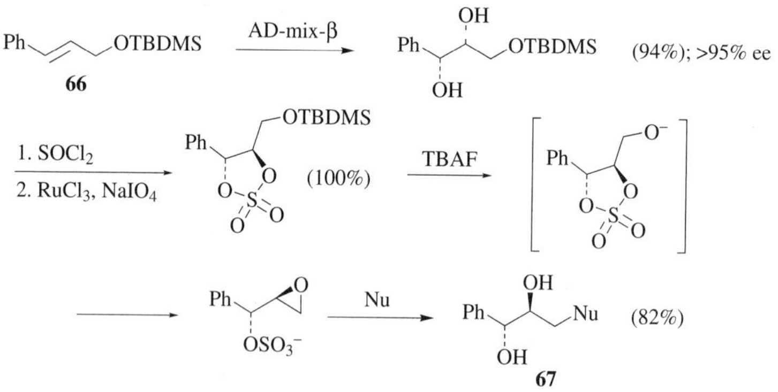

(Eq. 66)

Trisubstituted Alkenes

The enantioselective osmylation of trisubstituted alkenes has been examined using a variety of 1,2-diamine ligands. Both 1-methylcyclohexene and 1-phenylcyclohexene are commonly used substrates for the study of this reaction (Eqs. 67 and 68).[9,10,12] The *N,N′*-di(*tert*-butylethyl)-1,2-cyclohexanediamine ligand (**23**) is the most effective chiral controller for this reaction, affording 90% ee in the osmylation of phenylcyclohexene (Eq. 68).[9] The dihydroxylation of other trisubstituted alkenes using this ligand has not been reported.

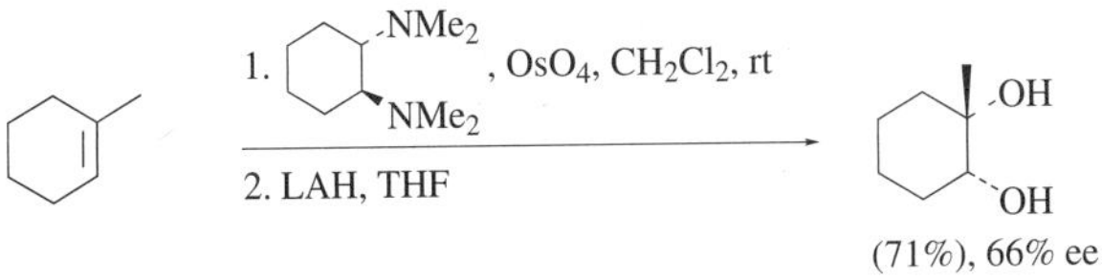

(Eq. 67)

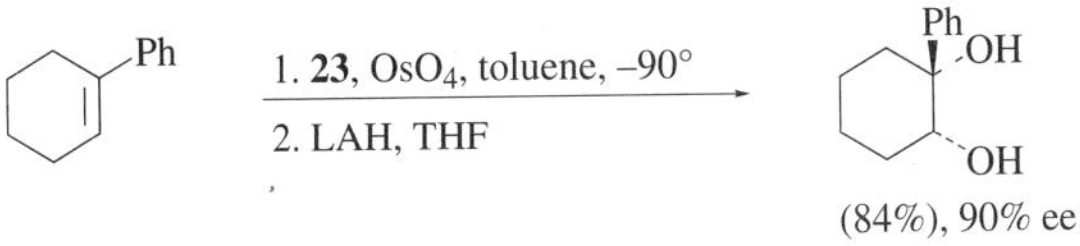

(Eq. 68)

Chiral diamine ligands derived from tartaric acid are moderately effective for the asymmetric dihydroxylation of silyl ketene acetals, affording synthetically useful α-hydroxy esters (Eq. 69).[96] These reactions are conducted in methylene chloride at $-78°$ to $-100°$, and the intermediate osmate ester is converted into the α-hydroxy ester using H_2S in methanol due to the sensitivity of the ester groups to $LiAlH_4$. The enantioselectivity of the reaction appears to be independent of the properties of the ketene acetal group, but is strongly influenced by the nature of the remaining alkene substituent.

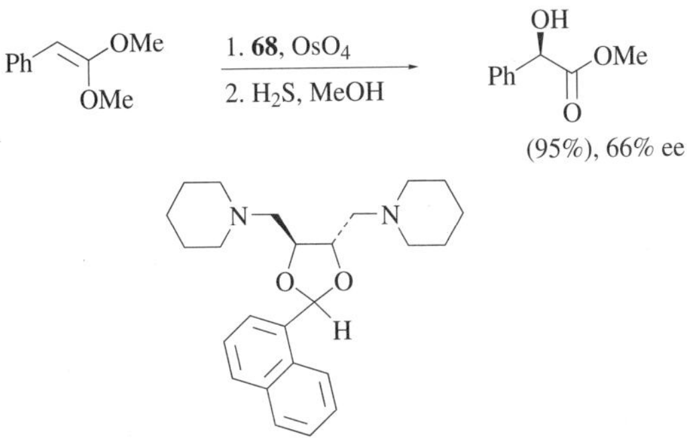

68

(Eq. 69)

The catalytic enantioselective dihydroxylation of trisubstituted alkenes using cinchona alkaloid ligands is very general; endocyclic, exocyclic, and acyclic alkenes all react with high enantioselectivity depending on the nature of the alkene substituents and the alkene configuration. Often the best results are obtained with substrates wherein the group being presented to the catalyst's binding pocket is cis to the vinylic proton of the substrate, although there are several exceptions to this trend. In general, trisubstituted alkenes are electronically activated relative to less substituted alkenes toward the osmylation by cinchona alkaloid-OsO_4 complexes; however, the hydrolysis of the trisubstituted Os(VI) ester is often slow as a result of steric hindrance. A stoichiometric equivalent of methanesulfonamide is typically added to ferricyanide-supported oxidations of trisubstituted alkenes in order to accelerate the hydrolysis of the intermediate Os(VI) ester. The use of NMO as the stoichiometric oxidant typically requires the addition of tetraethylammonium acetate and slow addition of the alkene to preclude the second cycle that degrades both the rate and enantioselectivity of the reaction. Since the dihydroxylation of trisubstituted alkenes produces chiral diols with dense and often diverse functionality, the reaction of these substrates has often been used to establish the first chiral centers in the synthesis of complex natural products. This review of the catalytic enantioselective dihydroxylation of trisubstituted alkenes is divided into three sections: acyclic trisubstituted alkenes, exocyclic trisubstituted alkenes, and endocyclic trisubstituted alkenes.

Acyclic Trisubstituted Alkenes. The catalytic enantioselective dihydroxylation of acyclic trisubstituted alkenes has been extensively studied within the context of oxidation of prenyl groups present in polyalkene substrates, such as terpenes, and additional material can be found in the section describing the reaction of polyunsaturated substrates. The reactions of simple prenyl groups proceed with high enantioselectivity using the PHAL and PYR class ligands and are representative of the high selectivities typically observed for the reactions of higher polyunsaturated substrates. Thus, the reaction of 2-methyl-2-heptene (**69**) using $(DHQD)_2PHAL$ (**7**) proceeds with 98% ee to give the 1,2-diol in 98% yield (Eq. 70).[19] A variety of aliphatic and aromatic substituents are tolerated, and the oxidation of these substrates provides synthetically useful products with high levels of enantioselectivity. For example, the diol derived from the oxidation of benzyl 3,3-dimethylacrylate (**70**) using $(DHQD)_2PHAL$ (**7**) is a useful intermediate for the synthesis of enantiomerically pure β-hydroxyvaline (Eq. 71).[106]

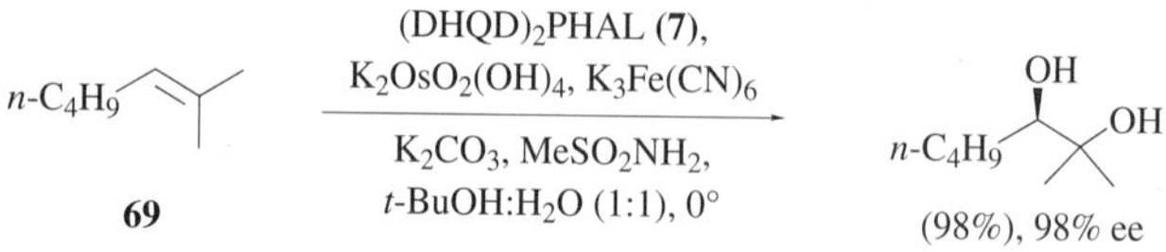

(Eq. 70)

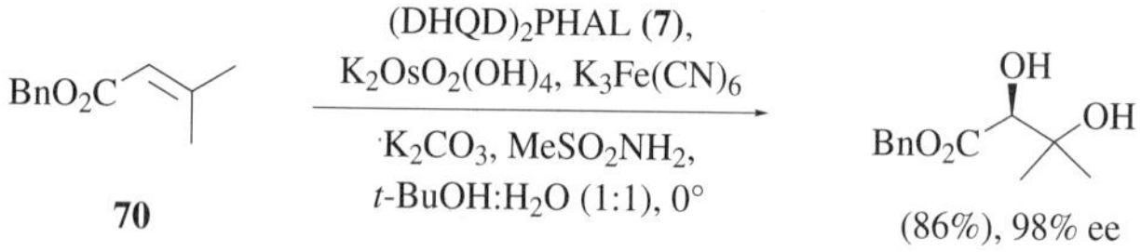

(Eq. 71)

Trisubstituted enol ethers and silyl enol ethers are generally good substrates for catalytic asymmetric dihydroxylation using $(DHQD)_2PHAL$ (**7**) (Eq. 72). The diol products spontaneously hydrolyze, affording α-hydroxy ketones of high enantiomeric purity.[126] There is a slight dependence of enantioselectivity on alkene geometry, and the Z-isomer is generally preferred for high selectivity. Highly Z-rich enol ethers derived from aliphatic ketones can be prepared by treatment of the parent ketone with trimethyl orthoformate and catalytic *p*-toluenesulfonic acid.[127] The treatment of aryl ketones with trimethyl orthoformate affords mixtures of enol ethers containing substantial amounts of the E-isomers. In these cases, the corresponding silyl enol ethers may be prepared as mainly the Z-isomers by treatment of the corresponding ketone with LDA in THF-HMPA and trapping the resultant enolate with *tert*-butyldimethylsilyl chloride.[128] These silyl enol ethers afford the same α-hydroxy ketone products after asymmetric dihydroxylation as the corresponding methyl enol ethers, but with greater enantiomeric purity.

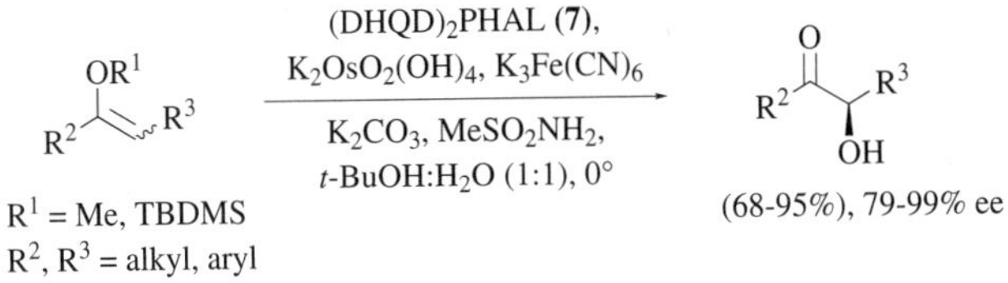

(Eq. 72)

Exocyclic Trisubstituted Alkenes. The enantioselective dihydroxylation of substituted methylenecyclopropanes has been extensively studied, and the diol products can be easily converted into enantiomerically enriched substituted cyclobutanones by pinacol rearrangement catalyzed by $BF_3{\bullet}OEt_2$ in THF at ambient temperature.[129] It is critical that the pinacol rearrangement be conducted under carefully selected conditions in order to avoid racemization. The conversion of such diols to enantiomerically enriched cyclobutanones using thionyl chloride and triethylamine in methylene chloride is believed to occur with substantial racemization.[130,131] For these substrates, enantioselectivity is highly dependent on the characteristics of the cyclopropylidene substituent, with aromatic substituents providing the highest selectivity (Eq. 73).[131] The reaction of aliphatic methylenecyclopropanes typically affords diol products with moderate enantioselectivity (50–70% ee). The reduction in enantioselectivity observed on switching from the $(DHQD)_2PHAL$ (**7**) ligand to the pseudoenantiomeric $(DHQ)_2PHAL$ (**8**) ligand is more pronounced with these substrates (10–20%) than is typically observed for other trisubstituted alkenes (5–10%).

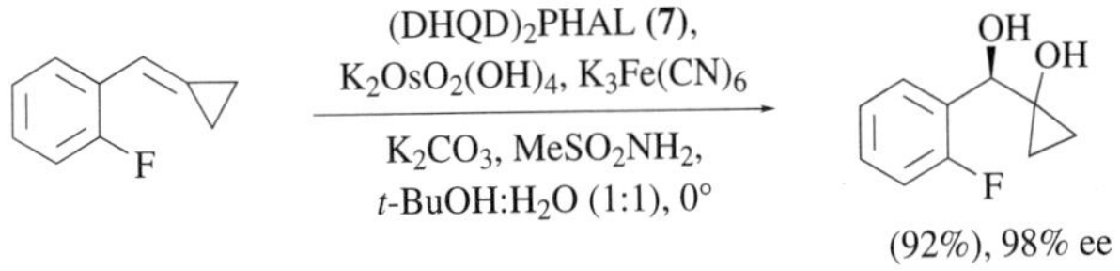

(Eq. 73)

The reactions of other exocyclic trisubstituted alkenes have not been as thoroughly studied. There are several examples where high enantioselectivities have been observed using the $(DHQD)_2PHAL$ (**7**) or $(DHQD)_2PYDZ$ (**18**) ligands, and facial selectivity can be strongly dependent on alkene geometry. One of the most striking demonstrations of this effect is the difference in enantioselectivity observed for the dihydroxylation of the E and Z exocyclic alkenes **71** and **74,** which were model substrates used in synthetic studies of the DE ring fragment of the antineoplastic agent camptothecin™ (Eq. 74).[132] Although the results observed for the reaction of the E isomer **71** were disappointing, the corresponding Z isomer **74** affords the diol product in 68% yield and 99% ee using $(DHQD)_2PHAL$ (**7**)

(Eq. 75).[132] Moreover, the reduced enantioselectivity noted in the reaction of **74** with $(DHQ)_2$PHAL (**7**) is less significant than that noted for the reaction of substrate **71**. These results can be rationalized in terms of the hydrophobicity of the group that is cis relative to the vinylic proton of the substrate and is presented to the catalyst's binding pocket. In the case of the E substrate **71**, that group is the carbonyl moiety of the lactone ring. In the case of substrate **74**, the phenyl ring of the substrate is cis to the vinyl proton and is presented to the catalyst's binding pocket. The preference of the PHAL ligands to bind aromatic substituents makes this an especially favorable interaction, resulting in the observed higher levels of enantioselectivity. A similar orientation of an aromatic group relative to the vinylic proton of **77** results in high enantioselectivity in the asymmetric dihydroxylation using the related catalyst $(DHQD)_2$PYDZ (**18**) (Eq. 76).[66]

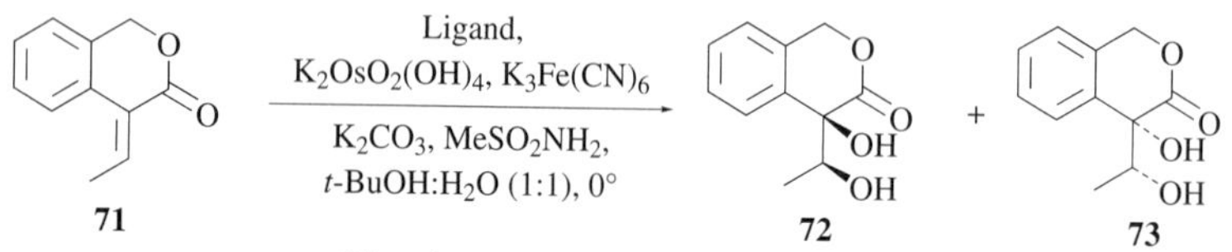

(Eq. 74)

Ligand	Major Enantiomer	
$(DHQD)_2$PHAL (**7**)	**73**	(78%), 50% ee
$(DHQ)_2$PHAL (**8**)	**72**	(70%), 28% ee

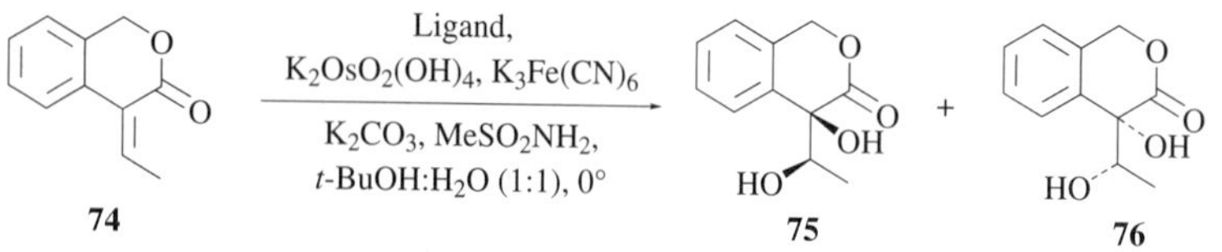

(Eq. 75)

Ligand	Major Enantiomer	
$(DHQD)_2$PHAL (**7**)	**75**	(68%), 99% ee
$(DHQ)_2$PHAL (**8**)	**76**	(73%), 97% ee

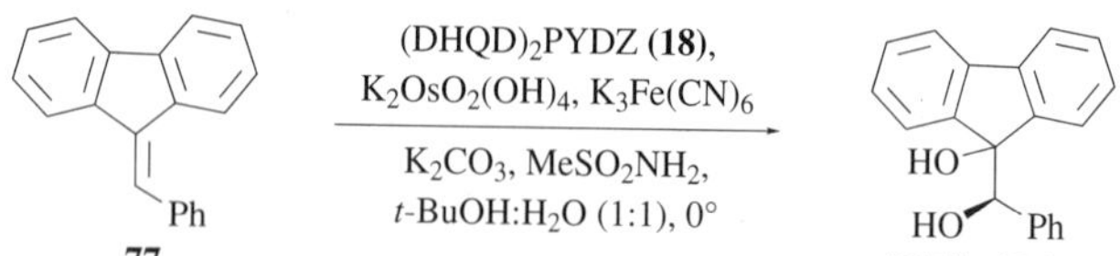

(61%), 97% ee

(Eq. 76)

Endocyclic Trisubstituted Alkenes. The catalytic enantioselective dihydroxylation of endocyclic trisubstituted alkenes has been thoroughly studied, and the diol products derived from these reactions have been widely used in the total synthesis of complex natural products. The reaction of simple, hydrocarbon-substituted substrates of this class generally proceeds with moderate to high enantioselectivity and is dependent on both the nature of the exocyclic substituent as well as the ring size (Eq. 77).[19,20] The effect of ring size on enantioselectivity is illustrated by comparing the results obtained from simple phenylcycloalkenes using $(DHQD)_2$PHAL (**7**).

Enantioselectivity is optimum for the six-membered ring, with a slight reduction noted as ring size increases, especially for the reaction of 1-phenylcyclooctene.

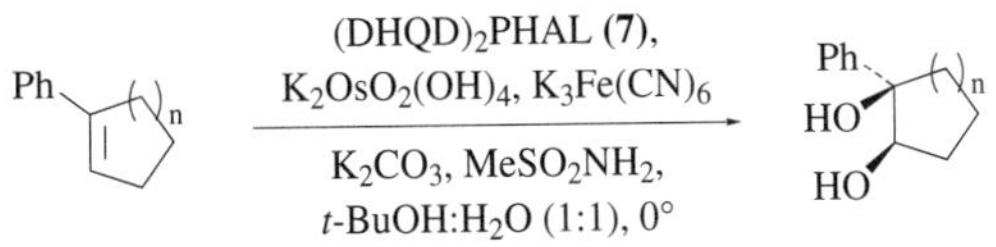

n	% ee
1	97
2	99
3	95
4	83

(Eq. 77)

While most substrates belonging to this alkene class yield good results in the asymmetric dihydroxylation catalyzed by PHAL or PYR ligands, enantioselectivity can be affected by the nature of the exocyclic substituent (Eq. 78).[19,20,61,99,115] Thus, for a series of 1-substituted cyclohexenes, enantioselectivity is highly dependent on the ability of the pendant substituent to interact effectively with the catalyst's binding pocket. For the PHAL ligand class, substituent preferences mirror those observed for other alkene classes, and non-bulky hydrophobic or aromatic groups are preferred to more hydrophilic or sterically demanding groups. Thus, phenylcyclohexene reacts with high enantioselectivity using $(DHQD)_2PHAL$ (**7**) as catalyst, but the corresponding reaction of 1-naphthyl- and 9-phenanthryl-substituted cyclohexenes proceeds with lower enantioselectivity. This reduction in selectivity is presumably due to the steric demand of the twisted and extended aromatic substituent.[20] Similarly, whereas methylcyclohexene reacts with only 52% ee, protected allylic and homoallylic alcohol derivatives are oxidized with very high enantioselectivity, allowing convenient access to cyclic chiral glycerol derivatives. This effect was utilized for the enantioselective synthesis of the angiogenesis inhibitor ovalicin (**79**), where the initial chirality was established using an asymmetric dihydroxylation of the allylic 4-methoxybenzoate **78** (Eq. 79).[133]

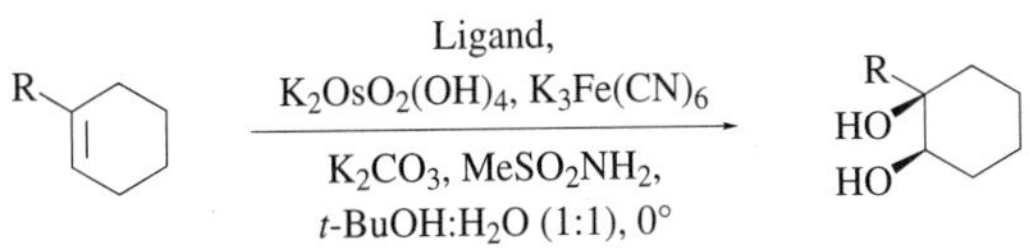

R	Ligand		% ee
Me	$(DHQD)_2PHAL$ (**7**)	(—%)	52
CH_2TMS	$(DHQD)_2PHAL$ (**7**)	(55%)	15
COMe	$(DHQD)_2PHAL$ (**7**)	(73%)	98
$CH_2O_2CC_6H_4OMe$-4	$(DHQD)_2PYDZ$ (**18**)	(99%)	98
$(CH_2)_2OC_6H_4OMe$-4	$(DHQD)_2PYDZ$ (**18**)	(99%)	95
Ph	$(DHQD)_2PHAL$ (**7**)	(80-98%)	99
1-naphthyl	$(DHQD)_2PHAL$ (**7**)	(—%)	86
9-phenanthryl	$(DHQD)_2PHAL$ (**7**)	(—%)	74

(Eq. 78)

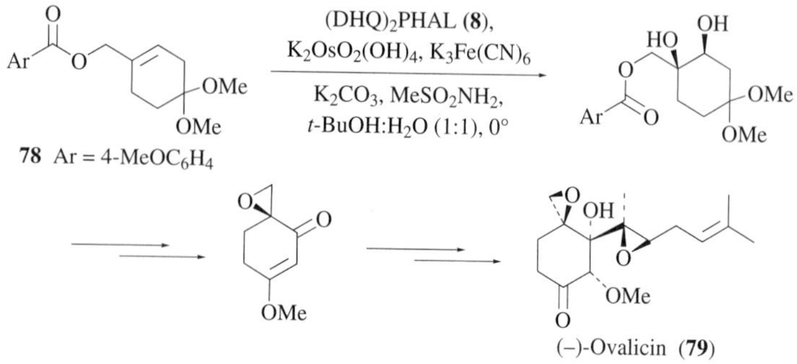

(Eq. 79)

Cyclic enol ethers are oxidized to lactol derivatives with moderate to high enantioselectivity depending on alkene substitution, ring size, and choice of ligand (Eq. 80).[132] Moderate enantioselectivity is typically obtained with the exception of substrate **80**, which is converted into the corresponding lactol in 91% ee using $(DHQD)_2$PYR (**25**) (Eq. 81). The enantioselectivity observed in the reaction of this substrate strongly depends on the type of cinchona alkaloid ligand used, and the corresponding reaction catalyzed by $(DHQD)_2$PHAL (**7**) produces the lactol product with only 74% ee. The lactol products can be easily oxidized to the corresponding lactones using iodine in ether in the presence of calcium carbonate.

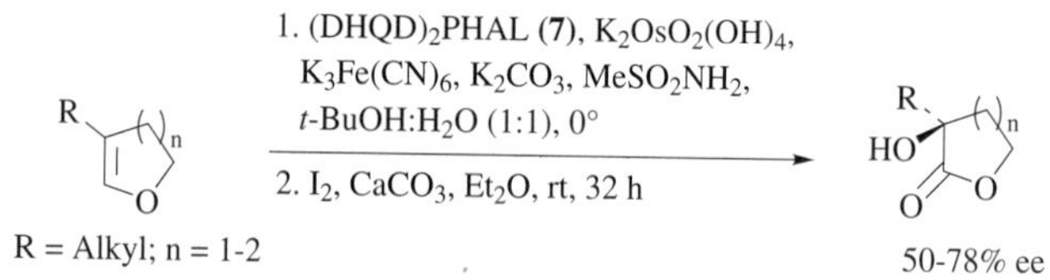

(Eq. 80)

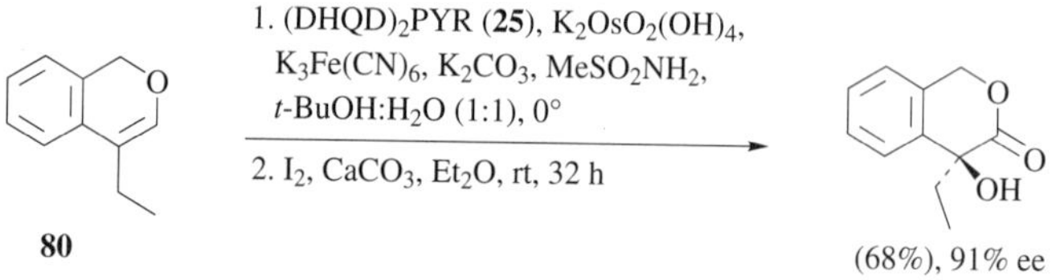

(Eq. 81)

The chiral α-hydroxy lactones that are produced are useful intermediates for natural product synthesis. The catalytic enantioselective dihydroxylation of substrate **81** was studied as part of an investigation toward an industrially scaleable route to the DE ring of camptothecin (**82**) and its analogs.[134,135] Whereas the reaction of **81** using $(DHQD)_2$PHAL (**7**) affords the lactone product with only 26% ee after oxidation with I_2 and $CaCO_3$, the corresponding reaction using $(DHQD)_2$PYR (**25**) affords the same product with 94% ee (Eq. 82).[134,135] Comparing these results with those obtained for the non-heterocyclic analog underscores the importance of a systematic investigation of ligand effects on enantioselectivity where the first choice affords poor results and further illustrates the complementary nature of the PHAL and PYR ligand classes with respect to substrate preferences.

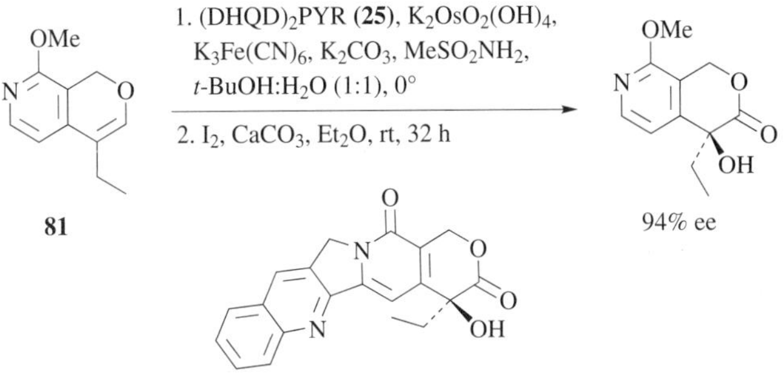

81

(20*S*)-Camptothecin (**82**)

(Eq. 82)

Trisubstituted ketene *O,O*-acetals have been shown to be excellent substrates for the catalytic asymmetric dihydroxylation utilizing the $(DHQD)_2PHAL$ (**7**) and $(DHQ)_2PHAL$ (**8**) ligands (Eq. 83). The ketene *O,O*-acetals are prepared from the corresponding aldehydes by the Horner-Wittig reaction with dialkoxymethyldiphenylphosphine oxides. After the dihydroxylation sequence, the intermediate diol hydrolyzes to the corresponding α-hydroxy ester.[136] This sequence has been successfully used to produce α-hydroxy esters starting with *O,O*-ketene acetals derived from aromatic, α,β-unsaturated, and aliphatic aldehydes, and is an attractive alternative to the conventional alkylation of aldehydes with nucleophilic carbanions, which can be challenging due to difficulties controlling the configuration of the newly created stereocenter.

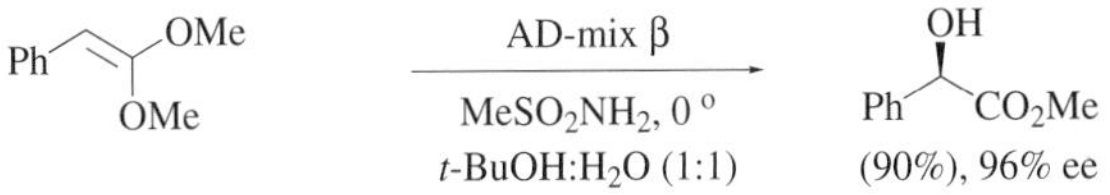

(Eq. 83)

Tetrasubstituted Alkenes

Very few asymmetric dihydroxylations of tetrasubstituted alkenes have been reported, and all utilize cinchona alkaloid ligands under catalytic conditions. This limited scope is due to the sluggish rate of hydrolysis of the osmate ester formed during the catalytic cycle as well as to the crowded asymmetric dihydroxylation transition state would not effectively accommodate tetrasubstituted alkenes. In the limited number of examples that have been reported, the use of at least one equivalent of methanesulfonamide and 1 mol% osmium catalyst has helped to overcome the turnover problem.[137] The reactions of tetrasubstituted enol ethers proceed at 0° with one equivalent of methanesulfonamide;[132] however, the asymmetric dihydroxylation of all-carbon tetrasubstituted alkenes requires three equivalents of methanesulfonamide, and reactions of these substrates are typically performed at room temperature. The reported reactions of all-carbon tetrasubstituted alkenes proceed with moderate enantioselectivity using either the PHAL or PYR class of ligands, and endocyclic alkenes generally give the best results (Eqs. 84 and 85).[137] However, there are examples where the chemical yields are low due to incomplete reaction. Due to the lack of known compounds for comparison, the absolute configuration of the diol products was tentatively assigned using the face-selection mnemonic.

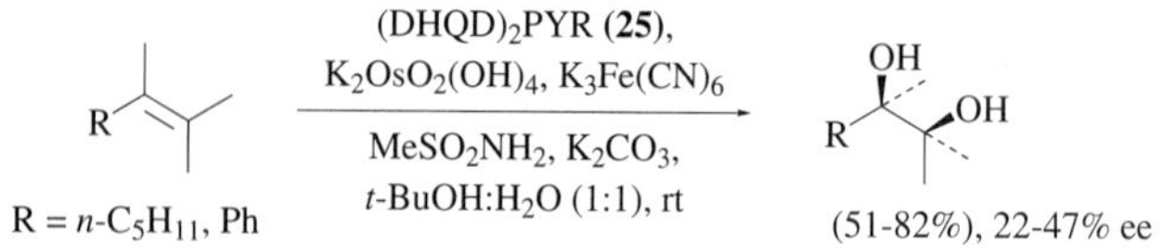

(Eq. 84)

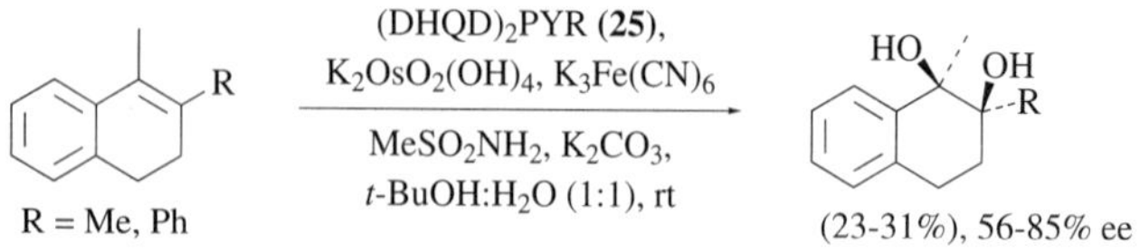

(Eq. 85)

Tetrasubstituted enol ethers react to give α-hydroxy ketones in good yield and good to excellent enantioselectivity using either the PHAL or PYR chiral ligands (Eq. 86).[137]

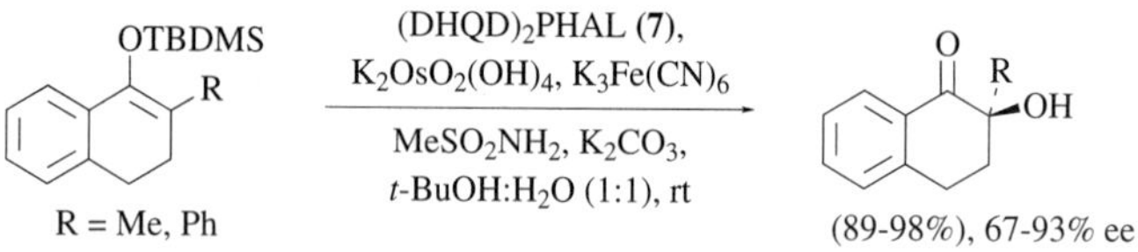

(Eq. 86)

A limited number of examples of the asymmetric dihydroxylation of tetrasubstituted ketene acetals and enol esters such as **83** have been reported. These reactions also require higher catalyst loading and proceed with moderate enantioselectivity and chemical yield to form the α-hydroxy esters **84** (Eq. 87).[132]

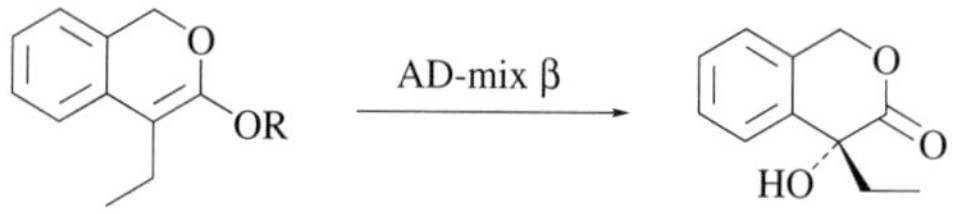

83 **84**

R = TBDMS, CO_2Bu-*t* (70-100%), 40-78% ee

(Eq. 87)

Polyalkene Substrates

Catalytic enantioselective dihydroxylation has been successfully used to control both the facial selectivity and regioselectivity of the oxidation of polyalkene substrates. The position selectivity of these oxidations is controlled by both the steric and the electronic properties of the individual double bond units. This section covers position and face selectivity observed in the catalytic asymmetric dihydroxylation of conjugated polyalkenes, non-conjugated polyalkenes, and endocyclic polyalkenes. Recent reviews provide additional detail regarding the regioselective asymmetric oxidation of these substrates.[138]

There are numerous examples of the regioselective asymmetric dihydroxylation of simple aliphatic polyenes. These reactions typically proceed with high enantioselectivity at the more electron-rich alkene using the $(DHQD)_2PHAL$ (**7**) ligand. Both conjugated and non-conjugated alkenes are good substrates for the reaction, and the order of reactivity can be predicted based on kinetic trends noted for the osmylation of simple alkenes by cinchona alkaloid-OsO_4 complexes.[50] Thus, E-alkenes are dihydroxylated in preference to Z-alkenes, and 1,1-disubstituted double bonds are oxidized in preference to terminal alkenes (Eqs. 88a-88c).[139]

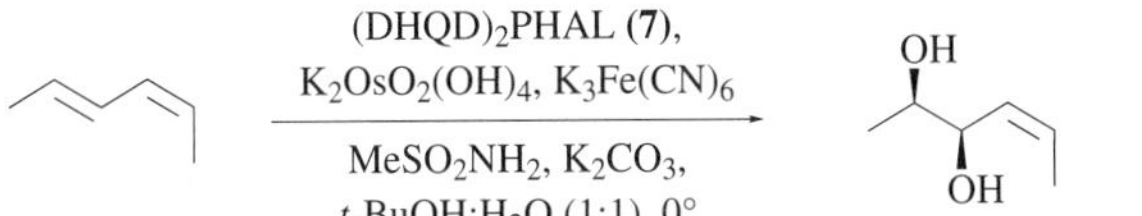

major diol (15:1)
(88%), 98% ee

(Eq. 88a)

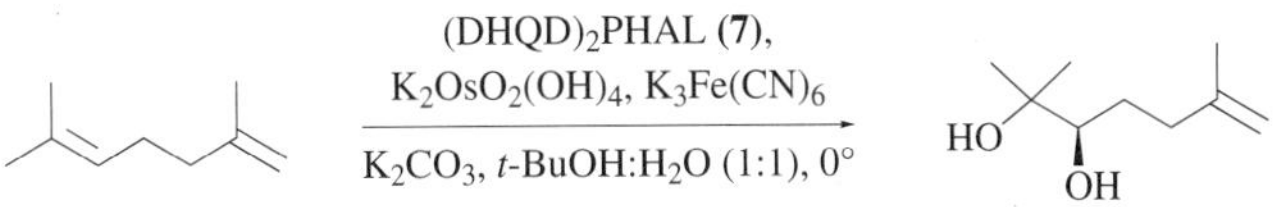

(70%), 98% ee

(Eq. 88b)

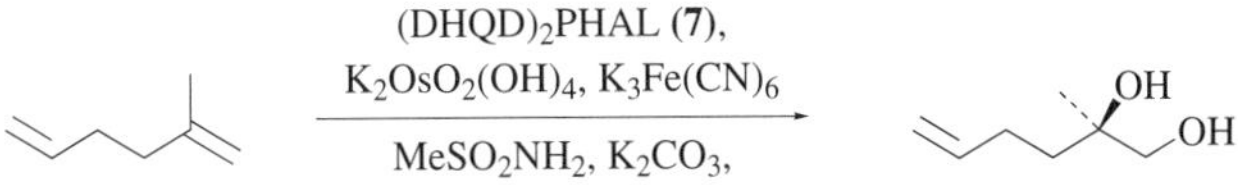

major diol (5:1)
(42%), 74% ee

(Eq. 88c)

Conjugated substrates are preferentially oxidized such that a minimal disruption of conjugation occurs. For example, the reaction of triene **85** gives primarily diol **86** along with a small amount of the diol resulting from oxidation at the terminal alkene (Eq. 89).[138] This regioisomer is expected on the basis of a reaction at the most electron-rich trans alkene with minimal disruption of conjugation. In contrast, when triene **87**

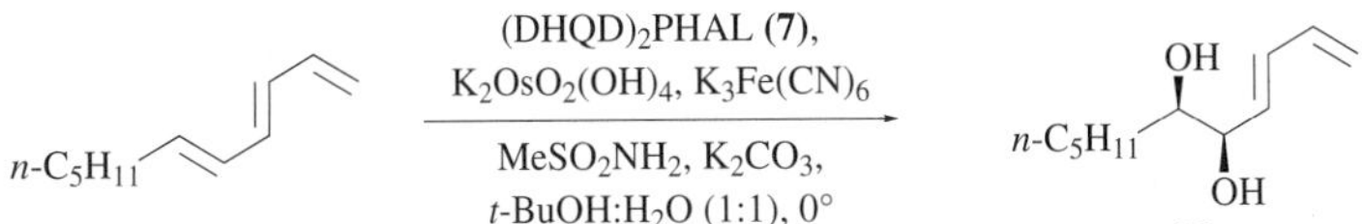

85 **86**
(60%), 6:1 position selectivity

(Eq. 89)

is subjected to the same conditions reaction occurs exclusively at the terminal alkene (Eq. 90).[138] This regioselectivity is most likely due to a preference not to disrupt conjugation via reaction at the internal trans alkene as well as a strong kinetic preference of the cinchona alkaloid-OsO_4 catalyst for oxidation of terminal alkenes compared to cis alkenes. Strong electron-withdrawing groups can also influence the regioselectivity of these reactions. In substrate **88** the benzoyl group directs the dihydroxylation to the more distal alkene with 14:1 regioselectivity, affording the diol product in high enantiomeric excess (Eq. 91).[138] Similarly, conjugated esters and aldehydes react preferentially at the most electron-rich alkene forming the corresponding diol in high chemical yield with excellent enantioselectivity (Eq. 92).[139,140] This regiochemical preference is also consistent with minimal disruption of conjugation.

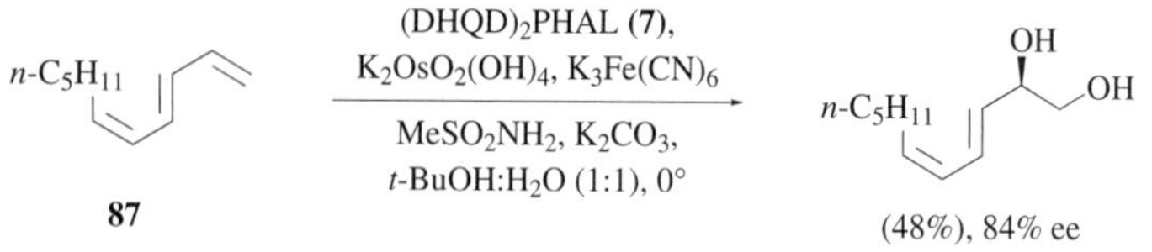

(Eq. 90)

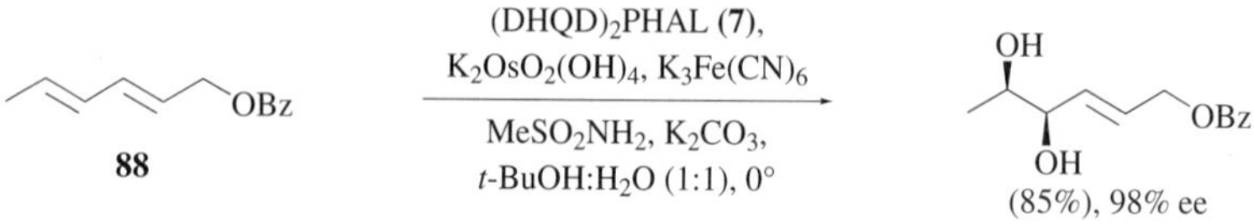

(Eq. 91)

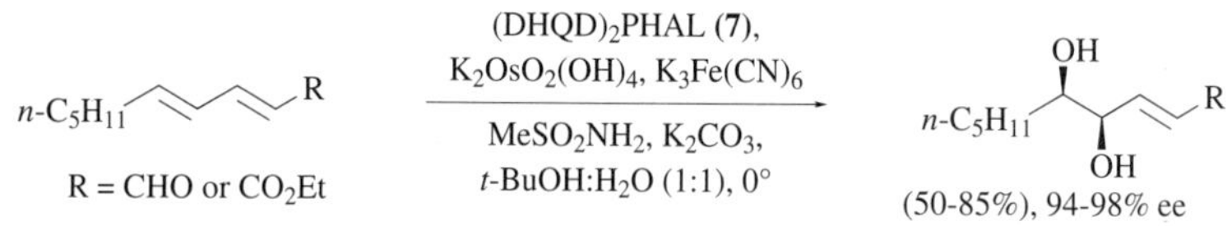

(Eq. 92)

The catalytic asymmetric dihydroxylation reactions of dienes containing an aryl substituent generally proceed with high enantioselectivity due to favorable interactions between the catalyst's binding pocket and the aromatic ring of the substrate. The positional selectivity of these reactions is partially governed by the nature of the substituents on the aromatic ring (e.g. substrate **90**) (Eq. 93). Generally, reactions occur at the double bond that is distal to the aromatic ring in order to preserve conjugation, and regioselectivity is further improved when bulky groups are present on the aromatic ring. Interestingly, replacement of phenyl (**89**) with *β*-naphthyl (**91**) causes a reversal in regioselectivity that favors oxidation of the internal double bond (Eq. 93). This reversal of position-selectivity can be understood in terms of a combination of highly favorable interactions between the catalyst and the naphthyl ring of **91** as well as differences in the conjugation energies of the phenyl- (**89**) versus the naphthyl- (**91**) substituted dienes.[138]

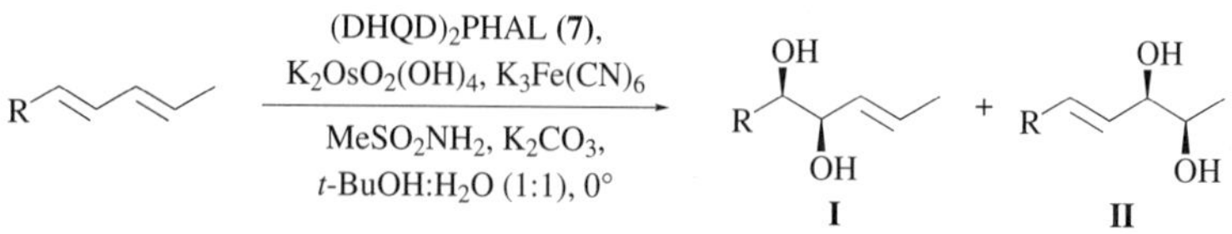

(Eq. 93)

R			I:II	% ee (major isomer)
Ph	**89**	(83%)	1:4	92
3,5-(t-Bu)$_2$C$_6$H$_3$	**90**	(76%)	1:15	92
2-naphthyl	**91**	(71%)	4:1	98

There are numerous examples of the position-selective dihydroxylation of geranyl, farnesyl, and other oligoprenyl substrates. Use of the $(DHQD)_2PHAL$ (**7**) ligand gives good positional selectivity in the dihydroxylation of monoterpenes, such as geraniol, geranyl acetate, and neryl acetate due to the electronic differentiation of the two substrate double bonds that favors dihydroxylation of the terminal prenyl unit (Eq. 94).[141] With higher terpenes, such as 2,6-*E*,*E*-farnesyl acetate, the electronic similarity of the terminal vs. internal prenyl units requires the use of modified bis(cinchona) alkaloid ligands in order to obtain high positional selectivity.[62] Thus, whereas enantioselective dihydroxylation of farnesyl acetate using $(DHQD)_2PYDZ$ (**18**) affords a 2 : 1 mixture of diols derived from oxidation of the terminal and internal prenyl units, the use of ligand **92** affords the product of terminal oxidation with 80% yield, 120 : 1 positional selectivity and 96% ee (Eq. 95).[62] This highly selective terminal oxidation of oligoterpenes has been used in the synthesis of complex terpenoid natural products, such as dammaranediol II (**93**) (Eq. 96).[142]

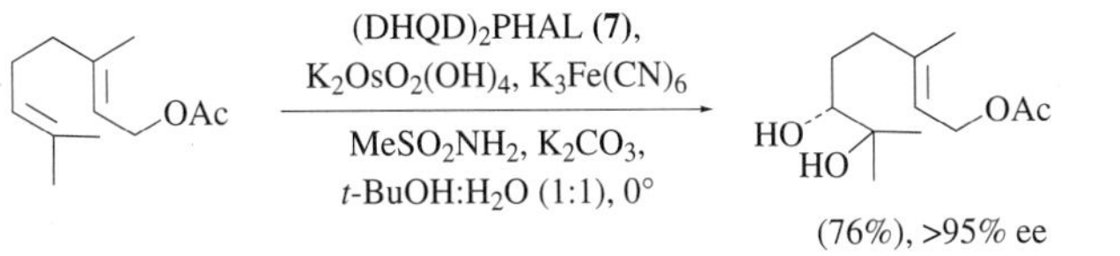

(Eq. 94)

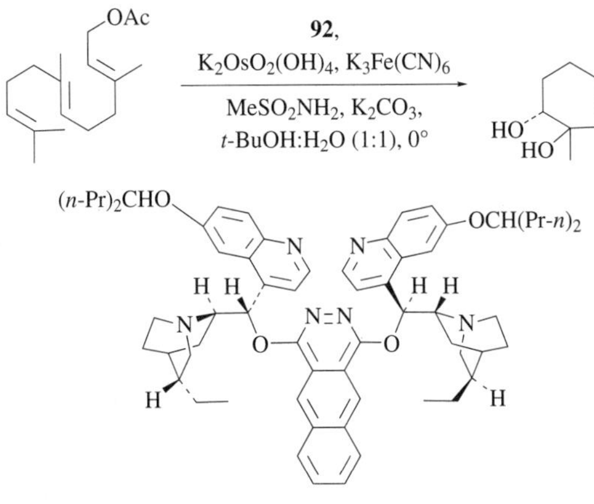

92

(Eq. 95)

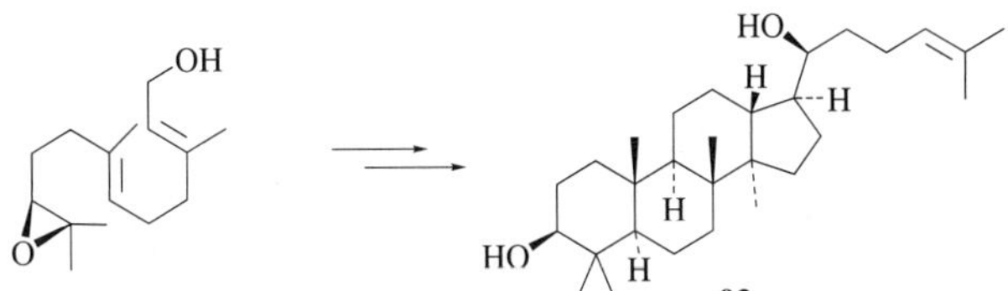

93

(Eq. 96)

The discovery that *p*-methoxybenzoate esters of allylic alcohols are excellent substrates for the catalytic asymmetric dihydroxylation reaction led to the investigation of regioselective dihydroxylation of suitably protected bis unsaturated alcohols.[64] The observation that allylic and bis(homoallylic) 4-methoxybenzoates are dihydroxylated with high enantioselectivity, as opposed to the poor selectivity noted for homoallylic benzoates, prompted an investigation of the utility of this protecting group to control position-selectivity in the oxidation of conjugated dienyl alcohols. For example, the reaction of diene **94** catalyzed by the $(DHQD)_2PYDZ$ (**18**) ligand favors oxidation at the double bond proximal to the 4-methoxybenzoyl group (Eq. 97).[64] This observation can be explained in terms of unfavorable interactions between the catalyst's pyridazine ring nitrogen atoms and the ester carbonyl group of the substrate that prevent reaction at the distal alkene. Oxidation of the other double bond of the substrate is favored by attractive interactions between the allylic 4-methoxybenzoyl group of the substrate and the catalyst's U-shaped pocket. Using the corresponding *p*-methoxyphenyl ether **95** as the substrate reverses this regioselectivity. Similarly, the reaction of substrate **96** favors the product of oxidation at the bis(homoallylic) position,[82] whereas position-selectivity is reversed in the reaction of the corresponding *p*-methoxyphenyl ether **97**.

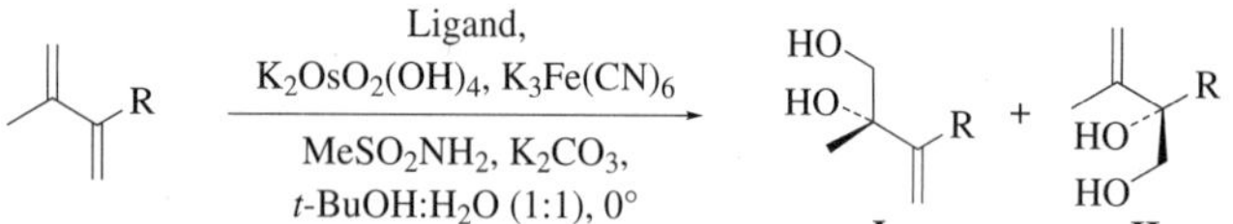

I **II**

	R	Ligand		I:II	% ee
94	$CH_2O_2CC_6H_4OMe$-4	$(DHQD)_2PYDZ$ (**18**)	(80%)	1:33	84
95	$CH_2OC_6H_4OMe$-4	$(DHQD)_2PYDZ$ (**18**)	(70%)	8:1	93
96	$(CH_2)_2O_2CC_6H_4OMe$-4	**98**	(74%)	12:1	97
97	$(CH_2)_2OC_6H_4OMe$-4	$(DHQD)_2PYDZ$ (**18**)	(70%)	1:8	74

(Eq. 97)

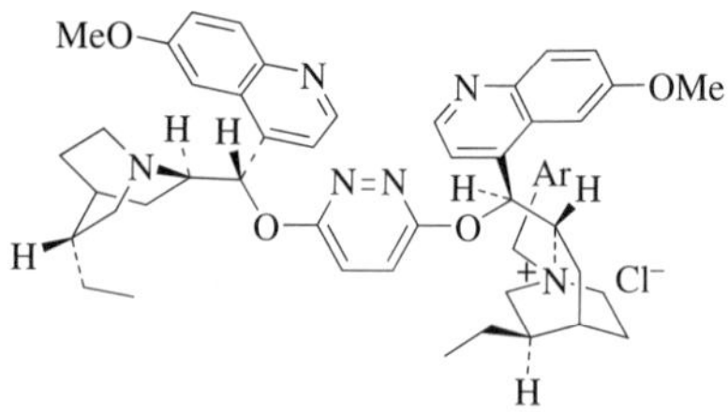

98 Ar = 9-anthryl

The utility of *p*-methoxybenzoate esters and *p*-methoxyphenyl ethers in directing position-selective dihydroxylation of dienes is demonstrated by the synthesis of hexose and 6-deoxyhexose derivatives.[143] Thus, the asymmetric dihydroxylations of dienes **99** and **101** using the $(DHQD)_2PYDZ$ (**18**) ligand proceed in high yield and with high regio- and enantioselectivity affording the diols **100** and **102**, which can be subsequently protected and converted into the corresponding hexoses via a reagent-controlled, selective dihydroxylation (Eqs. 98a and 98b).

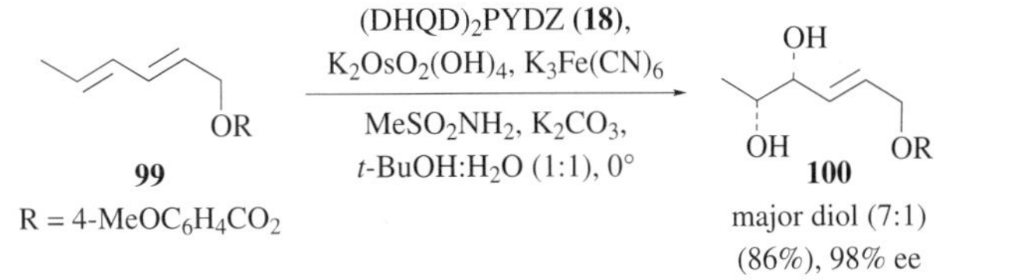

(Eq. 98a)

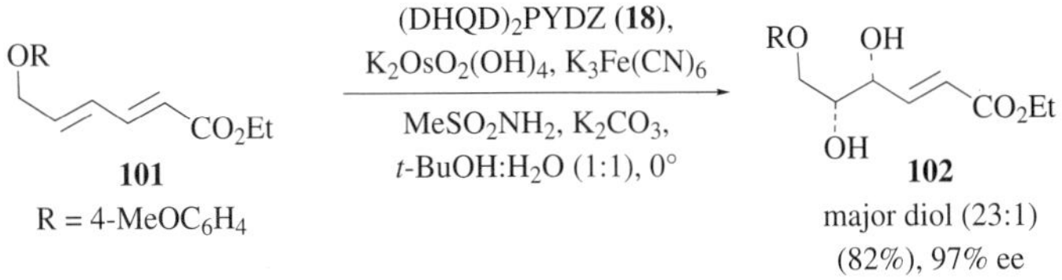

(Eq. 98b)

The asymmetric dihydroxylation of polyunsaturated substrates has been successfully used to access synthetically useful enantiomerically enriched polyalcohols. In these cases, overall enantiomeric purity is enhanced due to minimal influence of the diol that is installed first on the facial selectivity of the second dihydroxylation. If either one of the first or the second dihydroxylation reactions proceeds from the wrong face of the double bond, the resulting product is the meso diastereomer, which can be separated chromatographically. Thus, the asymmetric dihydroxylation of diene **103** using AD-mix β produces tetraol **104** in 88% yield. This material can be subsequently converted into piperidine **105**, which is obtained in 93% ee. Related piperidine derivatives are potentially useful C2-symmetric chiral directors (Eq. 99).[144]

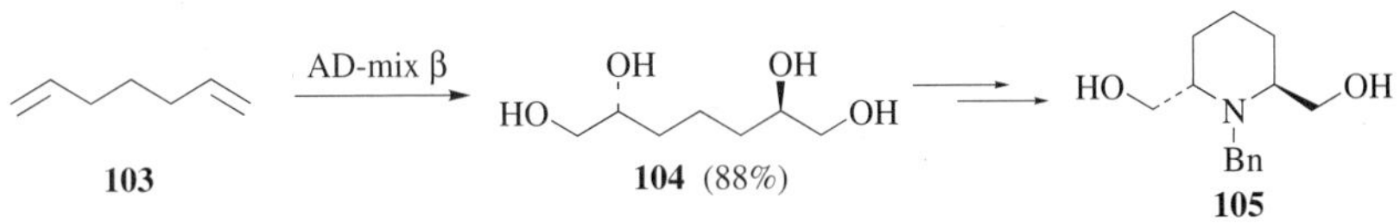

(Eq. 99)

In another example triene **106** is efficiently converted into tetraol **107** in 89% yield using AD-mix β ; the bis-acetonide of **107** was subsequently found to be 83% enantiopure. Dihydroxylation of the remaining terminal alkene of **107** gave an intermediate useful in the synthesis of the lichen macrolide (+)-aspicilin (**108**) (Eq. 100).[145]

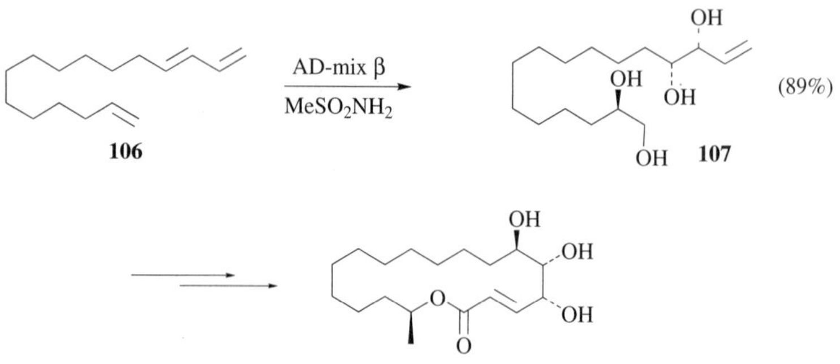

(Eq. 100)

(+)-Aspicilin (**108**)

The double asymmetric dihydroxylation of two terminal alkenes tethered by a linker group, followed by removal of that linker group, is a strategy to efficiently access enantiomerically pure triols. Typically, the process gives triols of higher enantiomeric excess than the corresponding dihydroxylation of the monomer subunit alone (Eqs. 101a and 101b).[146] This enhancement of enantiomeric purity results from the fact that chromatography or crystallization can remove the diastereomeric product obtained from improper facial approach in the second dihydroxylation. The use of potassium ferricyanide as the stoichiometric oxidant generally gives superior results to iodine, and no single ligand is preferred in all cases.[146]

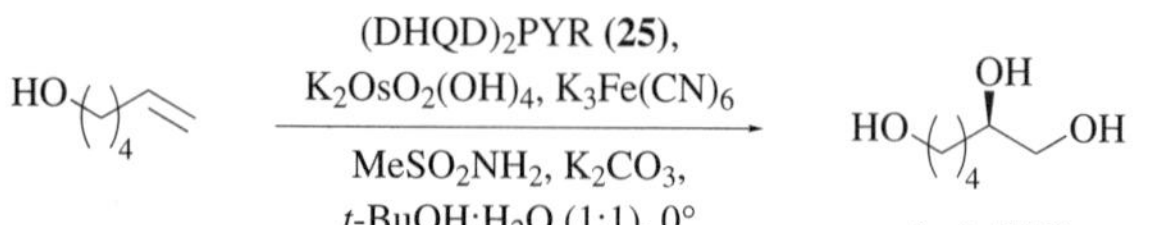

(—), 74% ee

(Eq. 101a)

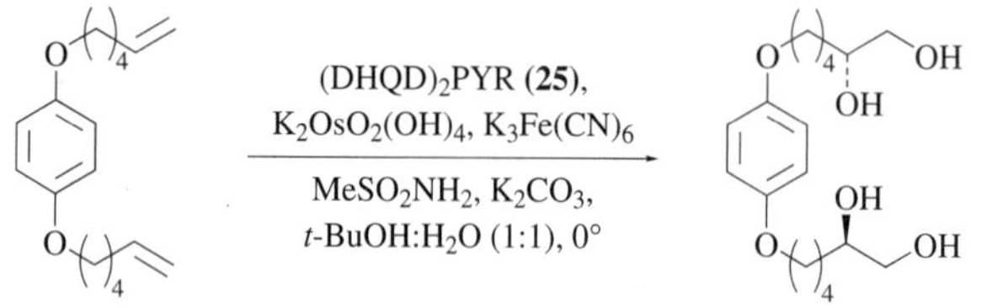

(99%), 90% ee

(Eq. 101b)

Small polyalkene ring systems generally give poor enantioselectivity in the asymmetric dihydroxylation reaction due primarily to the fact that Z-alkenes are among the poorest classes of alkenes for this reaction. In a few cases good to excellent enantioselectivity has been observed using the $(DHQD)_2PHAL$ (**7**) and $(DHQD)_2PYR$ (**25**) ligands.[147] The selectivity is enhanced when one of the alkenes is sterically hindered. Two of the best substrates are 1-phenylcyclopentadiene and 1-phenylcyclohexadiene, which react in the presence $(DHQD)_2PHAL$ (**7**) at the more electron-rich double bond to give the corresponding diols in 97 and 91% ee, respectively (Eq. 102).

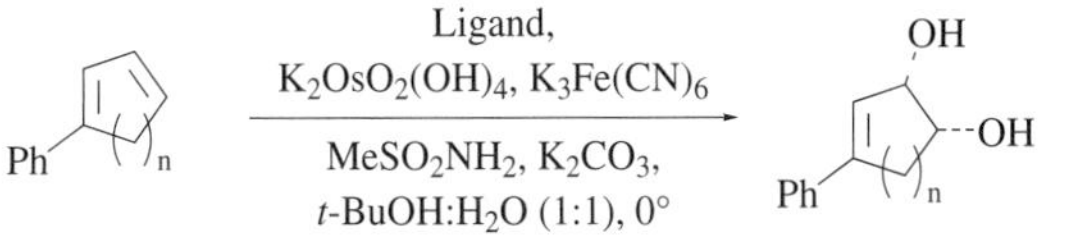

(Eq. 102)

n	Ligand	
1	$(DHQD)_2PYR$ (**25**)	(34-84%), 97% ee
2	$(DHQD)_2PHAL$ (**7**)	(34-84%), 91% ee

This dihydroxylation strategy has been used to complete an enantioselective synthesis of conduritol E.[148] Thus, enantioselective dihydroxylation of prochiral precursor **109** affords **110** with 85% yield and 85% ee. Removal of the benzylidene protecting group by hydrolysis affords conduritol E (**111**), an important intermediate for the preparation of cyclitols (Eq. 103).[147,148]

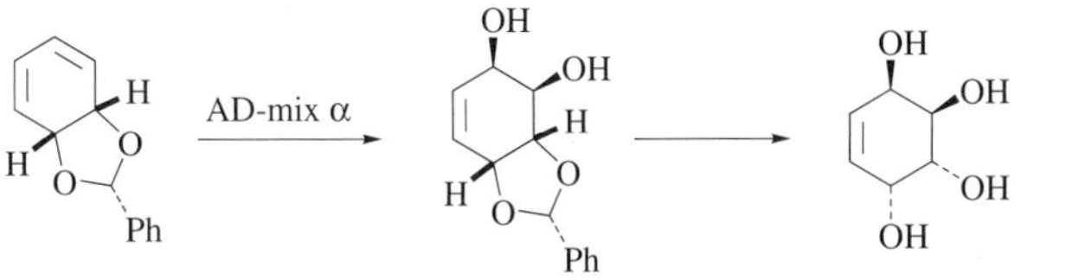

(Eq. 103)

109 **110** (85%), 85% ee (+)-Conduritol E (**111**)

Medium and large polyalkene rings can be good substrates for the asymmetric dihydroxylation reaction, provided that at least one of the double bonds is trans. The dihydroxylation of large rings with multiple E-double bonds proceeds with high enantioselectivity; however, because the reaction must be stopped at low conversion in order to suppress bis-dihydroxylation, the reactions of some of these substrates are not synthetically useful. Typically the pyrimidine ligands are superior to the phthalazine ligands (Eqs 104a-104c).[138]

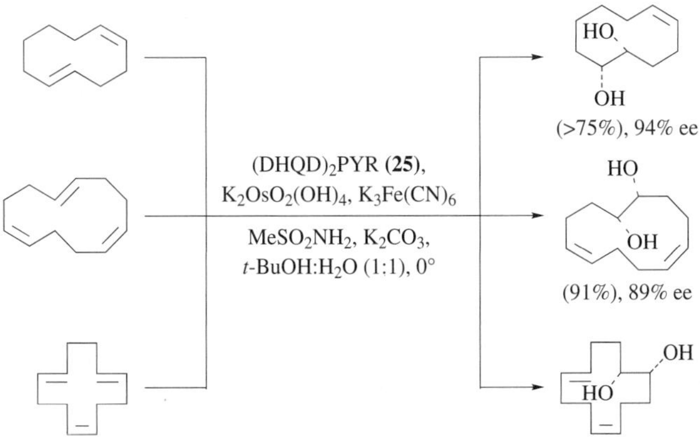

(>75%), 94% ee (Eq. 104a)

(91%), 89% ee (Eq. 104b)

(13%), 88% ee (Eq. 104c)

Kinetic Resolutions

Because the catalytic asymmetric dihydroxylation of alkenes is a highly efficient and selective process for a wide variety of substrates, it seems logical to apply the reaction to the kinetic resolution of racemic alkenes. This approach can be used for the preparation of enantiopure diols and for the recovery of enantiopure alkenes when installation of the chiral center is difficult by other means or when separation of the alkene enantiomers is difficult. Kinetic resolutions are inherently inefficient processes, as the undesired enantiomer cannot be directly converted into the desired stereoisomer and is usually discarded. Despite its generality for enantioselective dihydroxylation of achiral alkenes, few examples exist of successful kinetic resolutions using cinchona alkaloid-based catalytic systems. Only some of the examples reported are synthetically useful, as the existing cinchona alkaloid catalysts poorly discriminate between the alkene enantiomers. The k_{rel} parameter (the ratio of the rate constant for the fast vs. slow reacting enantiomer) is a good indicator of the synthetic utility of a kinetic resolution. For example, in order to obtain greater than 40% recovery (80% of the theoretical) of alkene that is of high enantiomeric excess, the k_{rel} for the reaction must be greater than 25.[149] The generally low observed k_{rel} for catalytic asymmetric dihydroxylations requires reactions to be run to much greater than 50% conversion to obtain recovered substrate with high enantiomeric purity.

The kinetic resolution of the axially dissymmetric racemic methylenecyclohexane derivatives **112** and **113** was performed using $(DHQD)_2PHAL$ (**7**) and $(DHQ)_2PHAL$ (**8**) (Eq. 105).[149] The k_{rel} value (k_{fast}/k_{slow}) for these reactions ranges from 5.0 to 32.0, depending on the catalyst used and the exocyclic alkene substituent. Dihydroxylation using cinchona alkaloid catalysts produces the diol diastereomer resulting from axial attack on the alkene. In contrast, when no chiral ligand is used, the product of equatorial attack is the major dihydroxylation product observed in this reaction.[149]

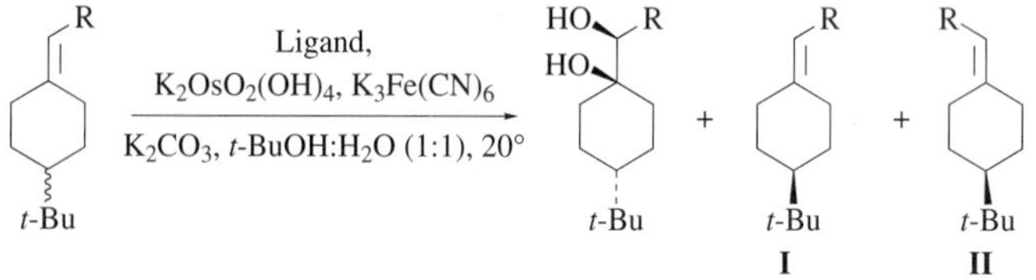

(Eq. 105)

	R	Ligand	k_{rel}	Recovered Olefin
112	Ph	$(DHQD)_2PHAL$ (**7**)	9.7	I
112	Ph	$(DHQ)_2PHAL$ (**8**)	5.0	II
113	CO_2Et	$(DHQD)_2PHAL$ (**7**)	32.0	I
113	CO_2Et	$(DHQ)_2PHAL$ (**8**)	26.5	II

The asymmetric dihydroxylation of a series of racemic alkenylphosphonates **114** has also been examined using the AD-mix reagents. The k_{rel} values are generally moderate (4–15) allowing the alkenes to be recovered in high enantiomeric excess only if the reactions are run to high conversion (Eq. 106). Therefore, the yields of recovered alkene with high enantiomeric excess are typically less than 40%.[150]

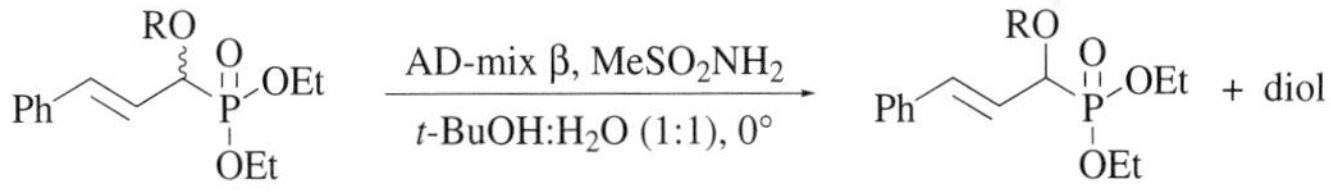

114
R = acetyl, benzoyl

unreacted **114**
k_{rel} = 4.4-15.6

(Eq. 106)

A number of kinetic resolutions of allylic acetates have been performed. The reactions typically proceed with moderate k_{rel} values between 3 and 10; therefore these reactions have limited synthetic utility. For example, in the kinetic resolution of acetic acid 1-cyclohexyl-3-phenylallyl ester (**115**) using ligand **116**, the starting material is recovered in only 88% ee at 60% conversion, and the reaction needs to proceed to 70% conversion in order to recover the alkene at >98% ee (Eq. 107).[151,152] In this example the k_{rel} value is ~25, which was the best observed and significantly higher than that of a typical allylic acetate substrate.[151,152]

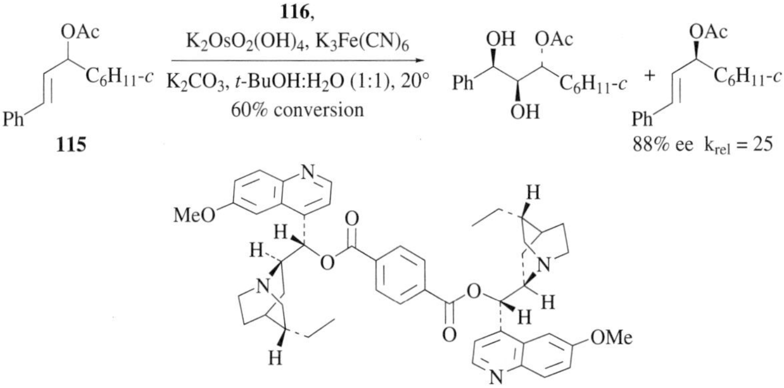

115

116

(Eq. 107)

Significant improvements in efficiency are observed in the reactions of allylic 4-methoxybenzoates, particularly when ligand **117** is used (Eq. 108).[63] Use of this catalyst with the methyl-substituted 4-methoxybenzoate **118** gives a k_{rel} value of 20. Upon replacing the methyl substituent with the more bulky phenyl group (substrate **119**), the k_{rel} value for this reaction improves to 79. The proper choice of catalyst is critical for efficient kinetic resolutions of these substrates. Note that when the conventional C2 symmetric $(DHQD)_2$PYDZ (**18**) ligand is used, the k_{rel} value drops below five for each of these substrates. Thus, with proper choice of ligands and substrates, synthetically useful kinetic resolutions are quite feasible.[63]

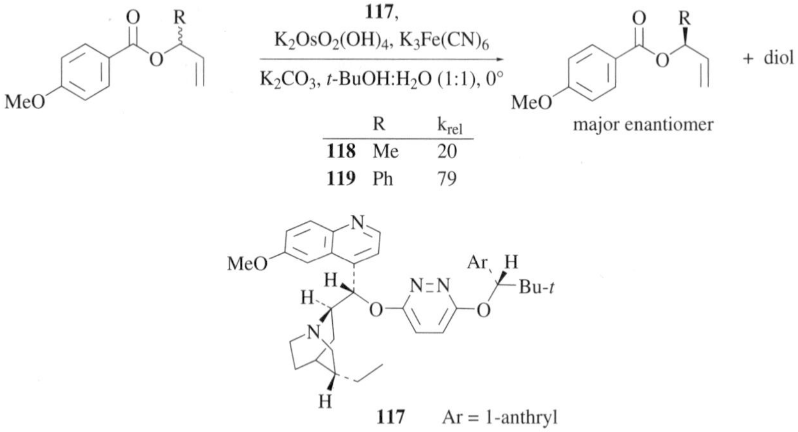

(Eq. 108)

Functional Group Compatibility

The dihydroxylation of alkenes by OsO_4 is one of the most general transformations in organic chemistry and is tolerant of a wide variety of functionality on the alkene substrate. The standard conditions for catalytic asymmetric dihydroxylation using cinchona alkaloid ligands are compatible with a wide variety of substrate functional groups, including: aliphatic and aromatic substituents, alkynes, alcohols, ethers, esters, amides, carbamates, nitriles, and sulfonamides. In addition to the high levels of enantioselectivity that are observed for a broad range of substrates, the cinchona alkaloid catalyzed asymmetric dihydroxylation offers improved chemoselectivity for the dihydroxylation of alkenes over other functional groups that are normally oxidized by OsO_4 alone. One of the most striking examples of this chemoselectivity is the asymmetric dihydroxylation of alkenes containing sulfur substituents. Osmium tetroxide is known to oxidize aromatic and aliphatic sulfides to sulfones preferentially to the dihydroxylation of alkenes present on the same substrate. In contrast, sulfides are oxidized slowly under the standard cinchona alkaloid catalyzed asymmetric dihydroxylation conditions, whereas the rate of alkene dihydroxylation is accelerated. Thus, chemoselective dihydroxylation of carbon-carbon double bonds is possible with substrates containing sulfur functionality.[100] High yields and enantioselectivity are obtained for substrates containing aromatic and aliphatic sulfides, dithianes, or disulfides (Eq. 109).[100]

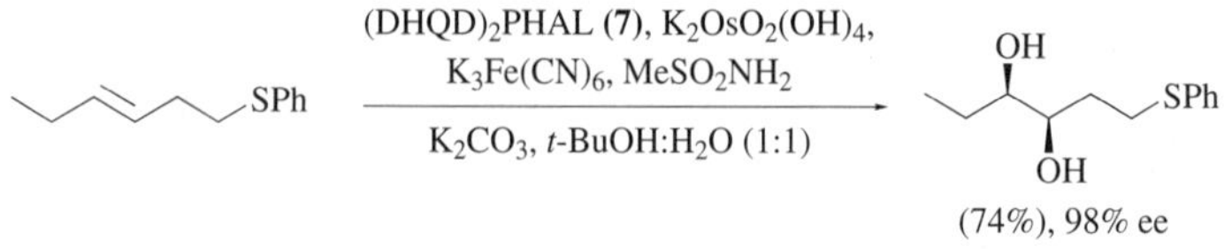

(Eq. 109)

In contrast to the chemoselective dihydroxylation of allylic sulfides and dithianes, the catalytic enantioselective dihydroxylation of allylic selenides affords products derived from the reaction of both the substrate double bond and the selenide functional group.[153] Oxidation at selenium is followed by elimination of the selenoxide, affording chiral allylic alcohols with low enantioselectivity. Since the diol product resulting from chemoselective alkene oxidation is formed with high enantioselectivity, it is unclear whether the low enantiomeric purity of the allylic alcohol products results from racemization of the newly formed chiral center or from poor facial discrimination in the dihydroxylation of the allylic selenoxides. The rate of oxidation at selenium can be controlled by the electronic properties of the selenide, and allylic 2-nitrophenylselenides are selectively oxidized at the carbon-carbon double bond (Eq. 110).[153]

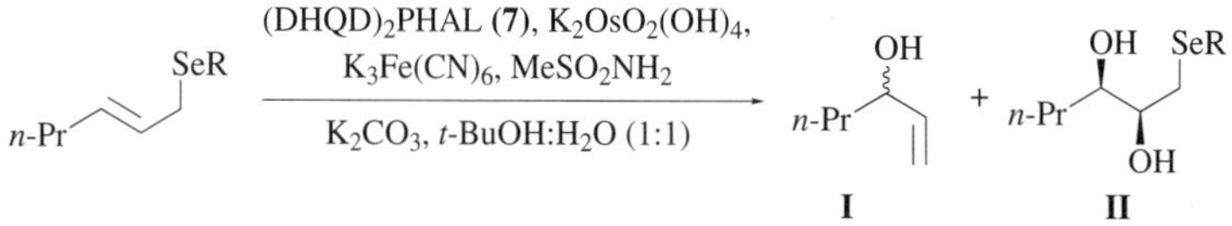

(Eq. 110)

R	I	% ee	II	% ee
Ph	(81%)	0	(7%)	94
Me	(100%)	0	(0%)	—
o-NO_2Ph	(0%)	—	(78%)	94

Whereas the enantioselective dihydroxylation of many nitrogen-containing substrates has been reported using cinchona alkaloid catalysts, the reaction of allylic and homoallylic amines does not proceed with useful yields or enantioselectivity. For example, the enantioselective dihydroxylation of *N,N*-dimethylcinnamylamine using modified AD-mix β, wherein the loading of ligand and potassium osmate is increased to 5 mol% and 1 mol%, respectively, affords only trace amounts of the diol product after 36 hours at room temperature.[111] Similarly, 4-phenyl-1-allylpiperazine gives the corresponding diol product in 70% conversion and only 55% ee after 7 days at room temperature.[111] In sharp contrast to the above results, several successful catalytic asymmetric dihydroxylations using cinchona alkaloid ligands have been reported for substrates where the allylic or homoallylic nitrogen functionality is protected as an imine, carbamate, or amide (Eq. 111).[61,111] Thus, it appears that turnover in the cinchona alkaloid catalyzed enantioselective dihydroxylation is dependent on the basicity of the nitrogen functionality that is proximal to the reactive double bond.

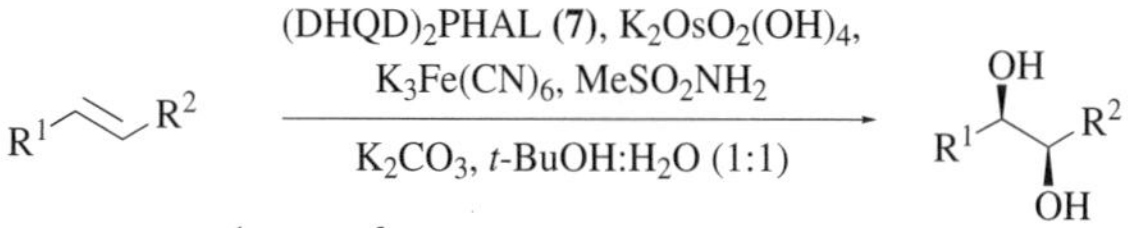

(Eq. 111)

R^1	R^2		% ee
Ph	CH_2NMe_2	(<5%)	—
H		(70%)	55
Ph	$CONEt_2$	(96%)	96
H	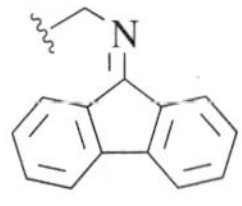	(90%)	90

Certain substrates possessing base-sensitive functionality decompose under the moderately basic conditions of the catalytic enantioselective dihydroxylation using potassium ferricyanide as the secondary oxidant. The basicity of the reaction can be reduced somewhat by using a mixed sodium bicarbonate-potassium carbonate buffer system. With these modified conditions, the asymmetric dihydroxylation of allylic halides proceeds with good yields and enantioselectivities, especially when the AQN ligands are used (Eq. 112).[86,154] Unfortunately, total replacement of potassium carbonate with sodium bicarbonate results in a loss of catalytic turnover for the cinchona alkaloid catalyzed dihydroxylation. The reaction of unsaturated ketones proceeds well using this mixed buffer system, which, in most cases, circumvents the problems of epimerization at the α-carbon or retro-aldol fragmentation of the product, a side reaction that may occur under more basic conditions (Eq. 113).[155]

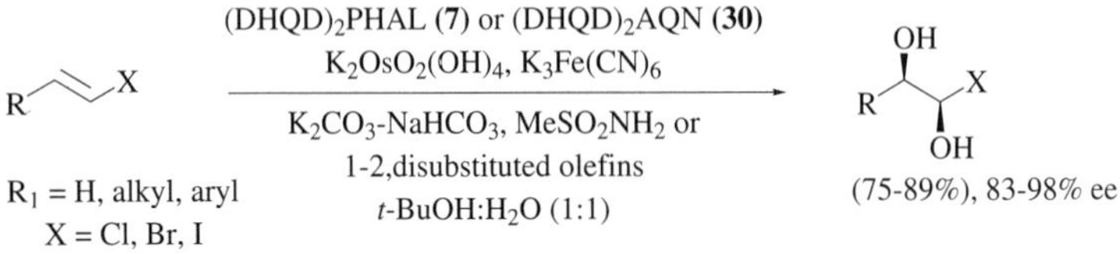

(Eq. 112)

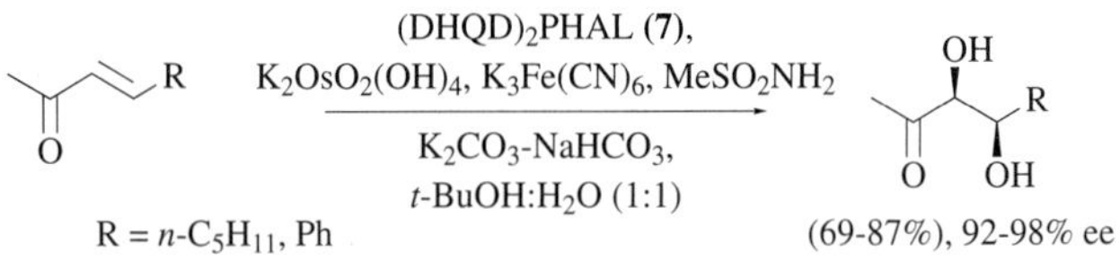

(Eq. 113)

Whereas the enantioselective dihydroxylation of α,β-unsaturated ketones, amides, and esters affords products of high enantiomeric purity in good yield, the corresponding reactions of α,β-unsaturated aldehydes are generally not successful. However, α,β-unsaturated *N*-methyl-*N*-methoxy amides (Weinreb amides) are good substrates for the cinchona alkaloid catalyzed asymmetric dihydroxylation and can serve as useful synthons for enantiomerically enriched 2,3-dihydroxyaldehydes (Eq. 114).[111]

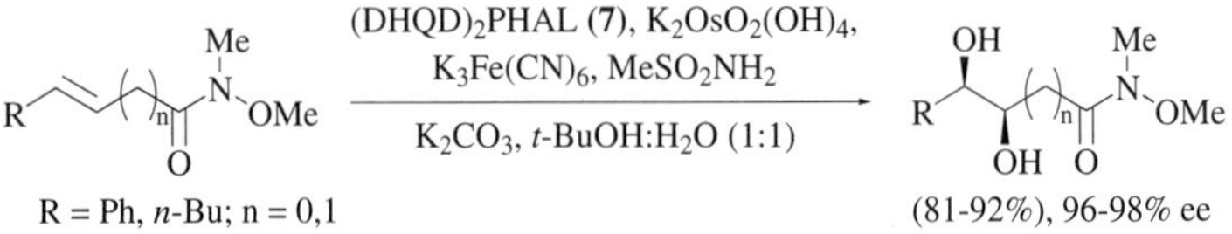

(Eq. 114)

EXPERIMENTAL CONDITIONS

Note: Osmium and its salts are highly toxic and must be handled using appropriate personal protective equipment. OsO_4 *is very toxic by inhalation, in contact with skin, and if swallowed. Osmium tetroxide is a highly volatile (bp 130°) low melting solid (mp 40°) that should only be handled in a fume cupboard by qualified individ-*

uals using chemical-resistant gloves. Osmium salts should be properly disposed of in specially designated containers. Further information can be obtained from the Material Safety Data Sheet (MSDS) available from the supplier.

Stoichiometric Enantioselective Dihydroxylation Using Chiral Diamine Ligands

Selection of Ligand. Significant research has been devoted to the study of enantioselective dihydroxylation using stoichiometric amounts of chiral 1,2-diamine ligands and OsO_4. Although catalytic turnover is precluded by the high affinity of the diamine ligand for osmium, very effective chiral ligands have been developed for the enantioselective dihydroxylation of a variety of alkene substrates. The 1,2-diamine ligands that have been surveyed thus far belong to three ligand classes: (1) acyclic 1,2-diamines, as represented by ligand **44** derived from chiral 1,2-diphenyl-1,2-diaminoethane; (2) heterocyclic diamines, exemplified by the bis(pyrrolidino)ethane ligand **118**; and (3) cyclic diamines, such as ligand **23**, derived from chiral 1,2-diaminocyclohexane.

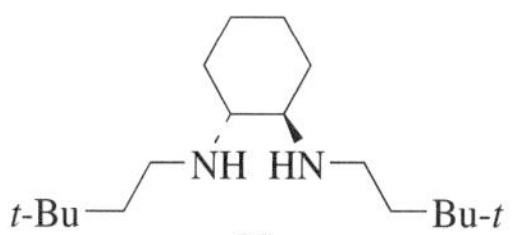

23

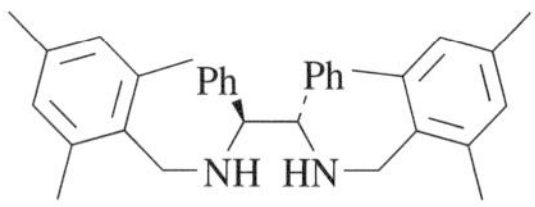

(*S*,*S*)-**44**

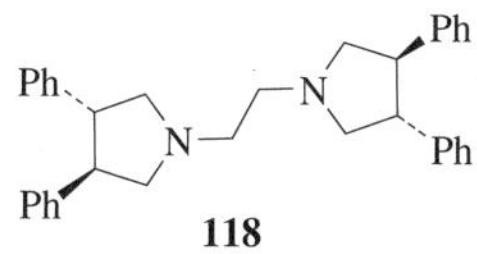

118

These three ligands represent the best members of each ligand class based on substrate scope and average enantioselectivity in the asymmetric dihydroxylation. While subtle differences in enantioselectivity exist for each of the ligands in the reaction of common substrates, such as styrene or dimethyl fumarate, ligand **23** has the widest reported scope, with enantioselectivities for many substrates rivaling the cinchona alkaloid catalyzed asymmetric dihydroxylations. The enantioselectivities observed for the asymmetric dihydroxylation of cis alkenes are among the highest reported, making **23** the ligand of choice for the oxidation of these substrates.

Solvent, Temperature, and Concentration. The choice of solvent for these reactions has a significant effect on enantioselectivity and is best guided by empirical observations for the ligand under consideration. The reaction is typically conducted in an aprotic organic solvent, such as THF, toluene, or methylene chloride, under anhydrous conditions. Complexation of the 1,2-diamine ligand with OsO_4 produces a bright red or orange complex that is soluble and highly reactive at temperatures as low as $-110°$. These reactions are generally conducted at these very low temperatures to ensure low background reaction rates and optimum enantioselectivity. The effect of concentration on rate and enantioselectivity has not been explicitly studied for these reactions. Laboratory-scale reactions are generally conducted at relatively low concentrations (0.04 to 0.2 M).

Recovery of Ligand and Osmium. Because these reactions require the use of a stoichiometric amount of OsO_4 and the chiral ligand, procedures for the recovery of each of these components are critical for economic reasons. The tightly associated ligand-bound osmate ester complex is typically broken up by a reductive quench with either lithium aluminum hydride or aqueous sodium bisulfite. The chiral ligand can typically be recovered through either extraction of the mixture with aqueous acid, basification of the aqueous layer, and extraction of the ligand into an organic solvent, or by chromatographic separation of the crude reaction mixture.[11] The recovery of OsO_4 is somewhat more complicated. Isolation of spent osmium is accomplished by adsorption of the crude reaction mixture onto silica gel, followed by separation of the dark-colored silica after elution of the diol product and the diamine ligand. Oxidation of the osmium-impregnated silica with a mixture of 30% hydrogen peroxide and methylene chloride, separation of the organic phase, and drying with $MgSO_4$ produces a concentrated solution of OsO_4 in methylene chloride that contains >80% of the originally used osmium.[11]

Catalytic Enantioselective Dihydroxylation Using Cinchona Alkaloid Ligands

Selection of Ligand. Extensive studies on the effects of ligand structure on enantioselectivity have resulted in the discovery of specialized ligands for the highly enantioselective oxidation of specific substrate classes. The choice of cinchona alkaloid ligand is critical for highly enantioselective dihydroxylation and to a certain extent can be empirical when the first selection does not yield satisfactory results. Indeed, many extensions to the scope of the asymmetric dihydroxylation have been discovered by surveying the panel of commercially available ligands in the oxidation of a new substrate. Many cinchona alkaloid derivatives have been reported for the asymmetric dihydroxylation of various substrates, and only those that are commercially available or offer optimum selectivity for special substrate classes will be reported here. The proper choice of cinchona alkaloid ligand can be guided by a few simple observations of ligand-substrate preferences. Many of these observations were detailed earlier in the description of the dihydroxylation of various substrate classes and will only be briefly summarized here. The determination of which cinchona alkaloid derivative to be used (DHQ or DHQD) is dictated by the desired direction of enantiofacial selectivity. The sense of enantioselectivity is readily predicted using the mechanistic models detailed earlier or can be simply estimated using the mnemonic device depicted in Figure 2.

Recovery of the ligands upon workup is typically achieved by extraction of the organic phase with aqueous acid, followed by basification of the aqueous phase and extraction into organic solvent. In cases where the diol product is very hydrophilic or easily extracted into aqueous acid, the cinchona alkaloid ligand may be recovered upon chromatography of the product and typically elutes from silica gel with 5–10% methanol in methylene chloride with 1% aqueous ammonium hydroxide added as a modifier.

The bis(cinchona) alkaloids are optimum ligands for the dihydroxylation of the vast majority of substrates and have replaced the first-generation mono-cinchona alkaloids for most applications. In general, the PHAL-linked ligands $(DHQD)_2PHAL$ (**7**) or $(DHQ)_2PHAL$ (**8**) are often good first choices to explore the enantioselective dihydroxylation of a new substrate.[19]

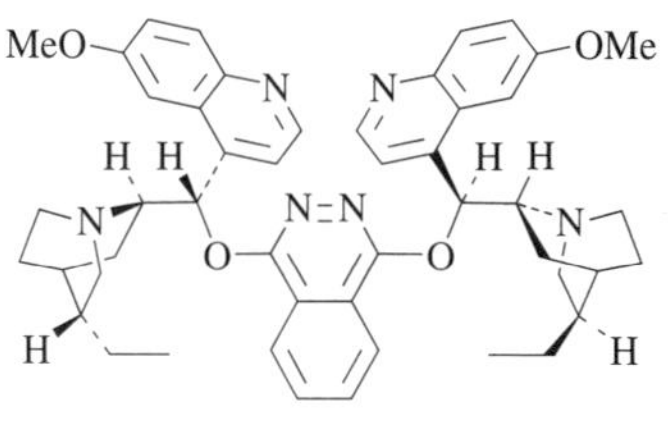

$(DHQD)_2PHAL$ (**7**)

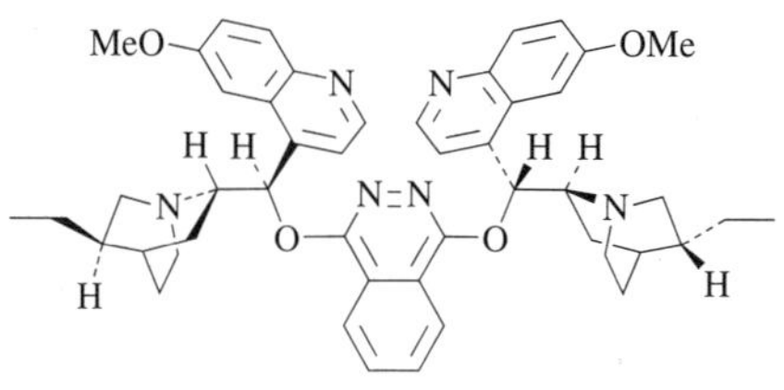

$(DHQ)_2PHAL$ (**8**)

These ligands afford high enantioselectivities in the asymmetric dihydroxylation of terminal, 1,1-disubstituted, E-1,2-disubstituted, trisubstituted, and tetrasubstituted alkenes, and as such, offer the broadest substrate scope of the cinchona alkaloid derivatives studied to date. The highest enantioselectivities are obtained for alkenes that possess an aromatic or non-bulky aliphatic substituent that can favorably interact with the catalyst's binding pocket. Terminal alkenes possessing bulky substituents and Z-1,2-disubstituted alkenes react with modest to poor enantioselectivity. Several variants of the PHAL-linked ligands, such as $(DHQD)_2PYDZ$ (**18**), $(DHQD)_2DPP$ (**35**), and $(DHQD)_2DP$-PHAL (**119**) have been evaluated in the enantioselective dihydroxylation of simple hydrocarbon substrates.[156,157] Although the $(DHQD)_2DPP$ (**35**) and $(DHQD)_2DP$-PHAL (**119**) ligands afford somewhat higher enantioselectivities for the oxidation of certain cis-alkenes, they generally offer comparable performance to the commercially available PHAL ligands.

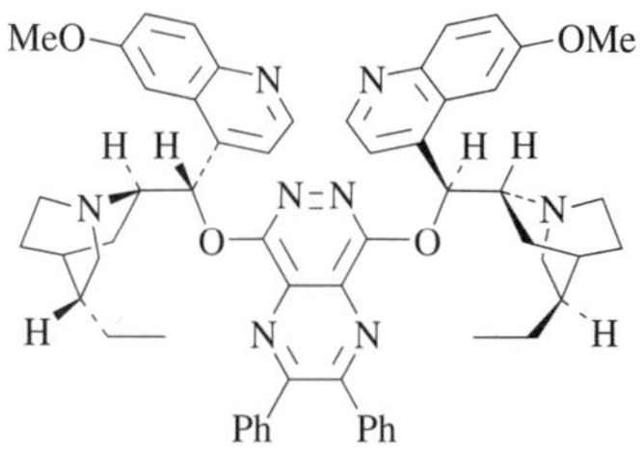

$(DHQD)_2DPP$ (**35**)

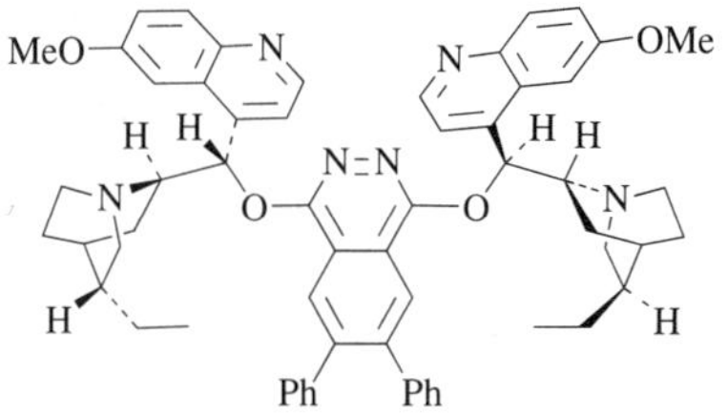

$(DHQD)_2DP$-PHAL (**119**)

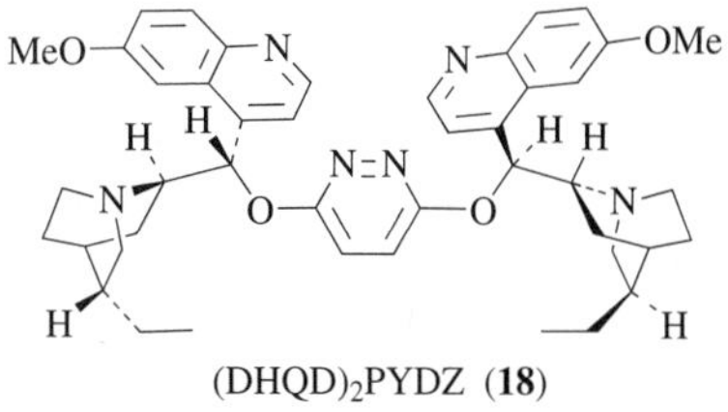

$(DHQD)_2PYDZ$ (**18**)

The pyrimidine-linked ligands $(DHQD)_2PYR$ (**25**) and $(DHQ)_2PYR$ (**26**) address some of the limitations of the PHAL-linked ligands for the dihydroxylation of terminal alkenes possessing bulky aliphatic substituents.[77] However, the PYR ligands afford significantly lower enantioselectivity than the PHAL ligands in the dihydroxylation of substrates possessing aromatic substituents. The PYR ligands are truly complementary to the PHAL ligands in that alkenes that work well for one ligand class are generally worse substrates for the other.

The PYR-linked alkaloids are the ligands of choice for the dihydroxylation of terminal alkenes possessing branched aliphatic substituents, such as *tert*-butylethylene and vinylcyclohexane, where enantioselectivities are significantly higher with this class of alkaloids as compared to the PHAL-linked ligands. In some cases the mono cincona alkaloid ligands, such as DHQD-PHN (**6**), offer comparable performance for the enantioselective dihydroxylation of terminal alkenes possessing a bulky aliphatic substituent. Each of these ligands is commercially available.

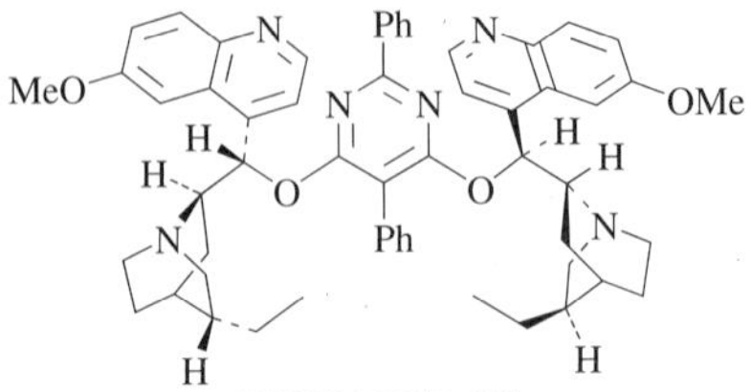

$(DHQD)_2PYR$ (**25**)

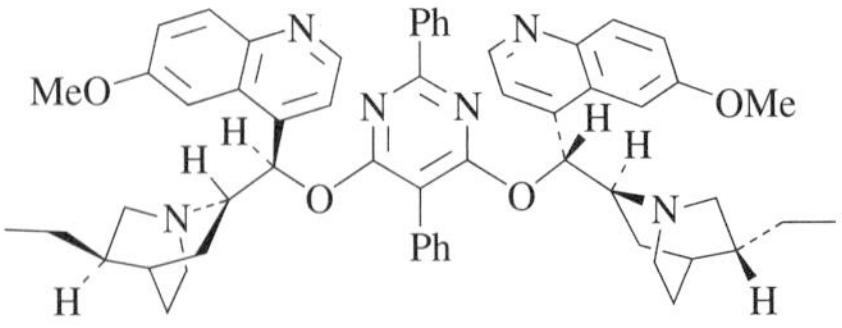

$(DHQ)_2PYR$ (**26**)

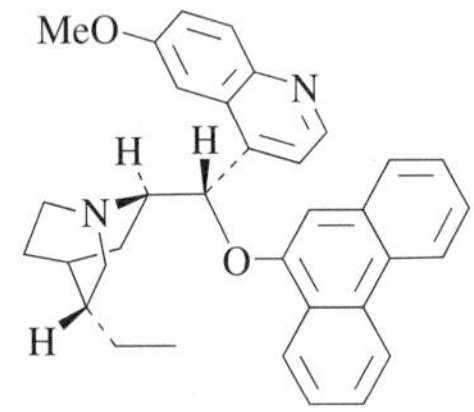

DHQD-PHN (**6**)

The anthraquinone-linked ligands $(DHQD)_2AQN$ (**30**) and $(DHQ)_2AQN$ (**31**) offer superior performance in the enantioselective dihydroxylation of allylically functionalized terminal alkenes.[86] Allylic halides and sulfonates are oxidized with 83–90% ee, affording functionalized chiral glycerol derivatives. This important substrate class affords a variety of small chiral non-racemic building blocks for asymmetric synthesis. The AQN-linked ligands provide the highest enantioselectivities for the catalytic asymmetric dihydroxylation of indene and allylbenzene, two difficult substrates for the PHAL-linked ligands, and afford enantioselectivities comparable to the PYR-linked ligands for the oxidation of *n*-alkyl-substituted terminal alkenes. Like the PYR-linked ligands, the AQN ligands exhibit worse performance than the PHAL-linked ligands in the enantioselective oxidation of substituted styrenes and other alkenes possessing aromatic substituents.

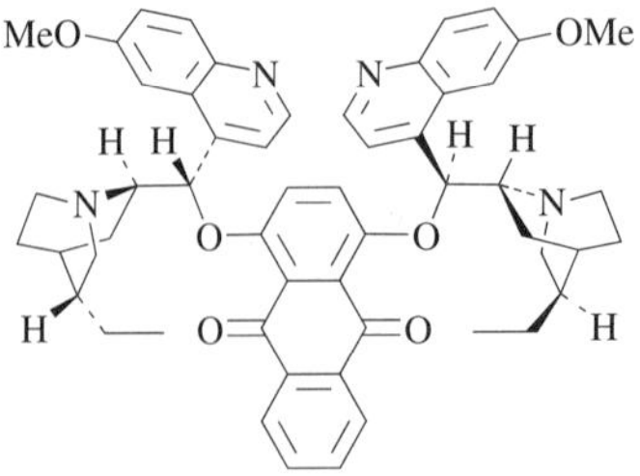

$(DHQD)_2AQN$ (**30**)

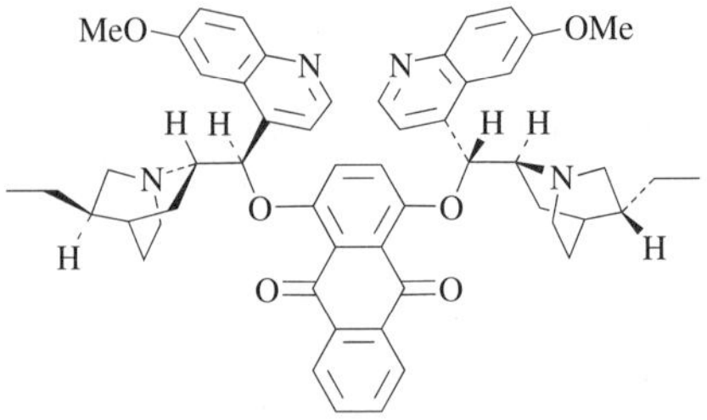

$(DHQ)_2AQN$ (**31**)

While the commercially available $(DHQD)_2PHAL$ (**7**) and $(DHQ)_2PHAL$ (**8**) ligands and the related PYDZ-linked ligands provide excellent enantioselectivity in the dihydroxylation of allylic 4-methoxybenzoates and homoallylic 4-methoxyphenyl ethers, the reactions of bis(homoallylic) 4-methoxybenzoates proceed with substantially lower facial selection using either of these cinchona alkaloid ligands. The reactions of these substrates can be significantly improved by using ligand **98**, which is a mono-9-anthracenylmethyl quaternary ammonium salt of the $(DHQD)_2PYDZ$ (**18**) ligand (Eq. 115).[82] Although this ligand is not commercially available, it is readily prepared by reaction of $(DHQD)_2PYDZ$ (**18**) with 9-chloromethylanthracene in acetonitrile at 40°, followed by chromatographic purification. The improved facial selectivity in the dihydroxylation of these substrates using the 9-anthracenylmethyl quaternary ammonium salt derivative **98** is presumably derived from this ligand's deeper binding pocket, which can interact with functional groups of the substrate that are remotely positioned relative to the double bond that is being oxidized.

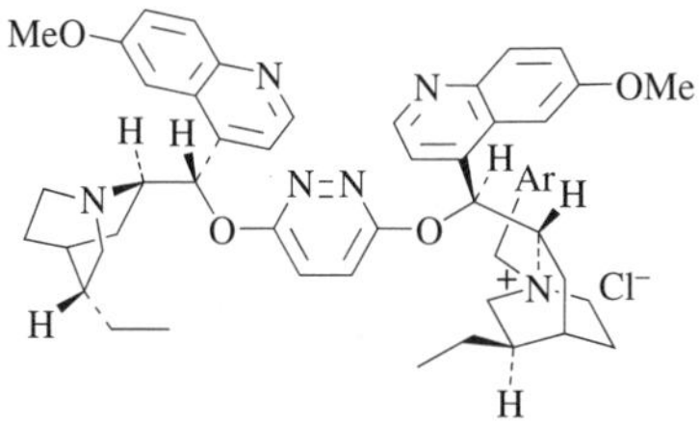

98
Ar = 9-anthryl

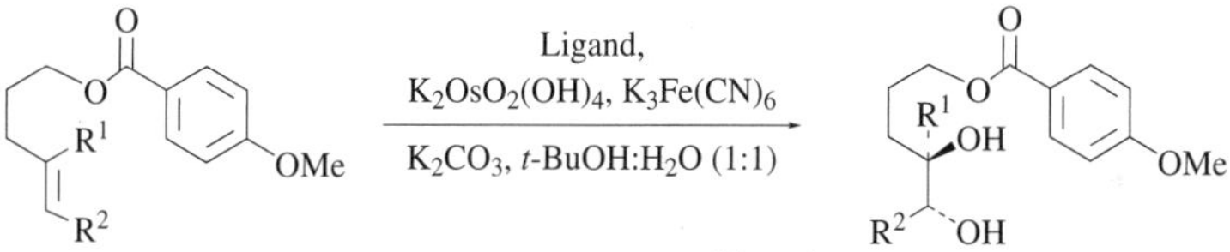

(Eq. 115)

		Ligand	
R^1	R^2	$(DHQD)_2PYDZ$ (**18**)	**98**
H	H	(99%), 82% ee	(91%), 86% ee
Me	H	(95%), 79% ee	(99%), 90% ee
n-Bu	H	(99%), 33% ee	(99%), 62% ee
t-Bu	H	(88%), 92% ee	(91%), 92% ee
H	Me	(97%), 89% ee	(98%), 90% ee

The low enantioselectivity observed for the reaction of Z-alkenes remains a major hurdle for the cinchona alkaloid catalyzed asymmetric dihydroxylation. Enantioselectivities above 80% ee are rare, and the mono cinchona alkaloid ligands DHQD-IND (**60**) or DHQ-IND (**61**) generally afford the highest enantioselectivities in the dihydroxylation of these substrates.[121] In general, Z-β-substituted styrenes react with higher facial selectivity as compared to non-aromatic substrates using this ligand. Interestingly, dihydronaphthalene reacts with substantially lower enantiofacial selectivity as compared to Z-β-methylstyrene, suggesting that conformational flexibility in the substrate, which allows the aromatic ring to twist out of the plane of the reactive double bond, may be important for achieving high enantioselectivity. The reduction in facial selectivity that is normally observed with the DHQ-derived ligands as compared to the corresponding DHQD-derived ligands is most striking for the IND ligand class. Thus, while the enantioselective dihydroxylation of Z-β-methylstyrene using DHQD-IND (**60**) affords the 1*R*,2*S* enantiomer with 72% ee, the corresponding reaction with DHQ-IND (**61**) gives the 1*S*,2*R* enantiomer with only 59% ee (Eq. 116).[121]

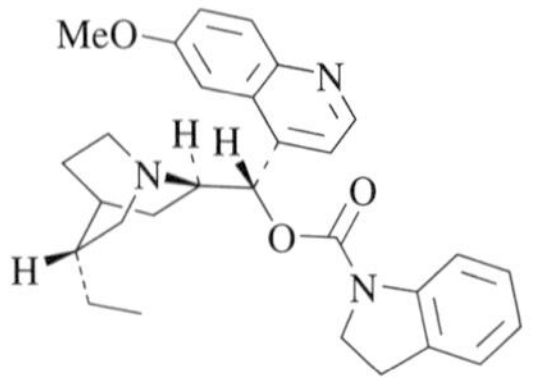

DHQD-IND (**60**)

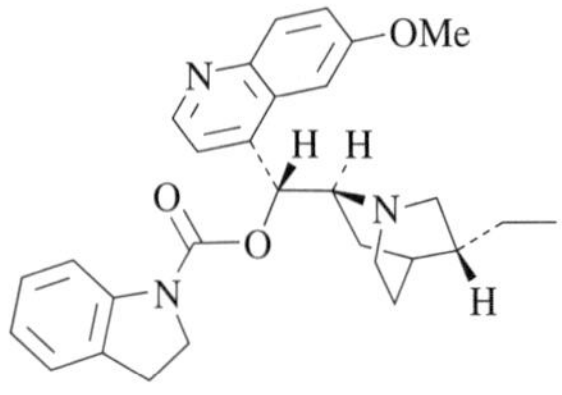

DHQ-IND (**61**)

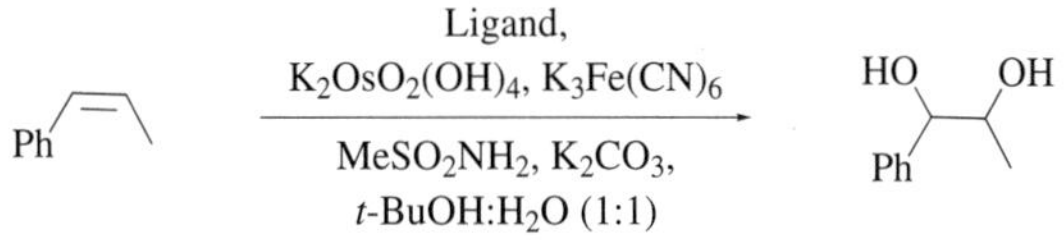

(Eq. 116)

Ligand		Config.
DHQD-IND (**60**)	(66-90%), 72% ee	1R,2S
DHQ-IND (**61**)	(66-90%), 59% ee	1S,2R

Position-selective dihydroxylation of substrates possessing multiple unconjugated double bonds with similar substitution patterns is often problematic for the PHAL-linked cinchona alkaloid ligands. The modified cinchona alkaloid ligands **92** and **120** offer superior position-selective asymmetric dihydroxylation of terpene substrates such as farnesol and geranylgeraniol (Eq. 117).[62] Position selectivities on

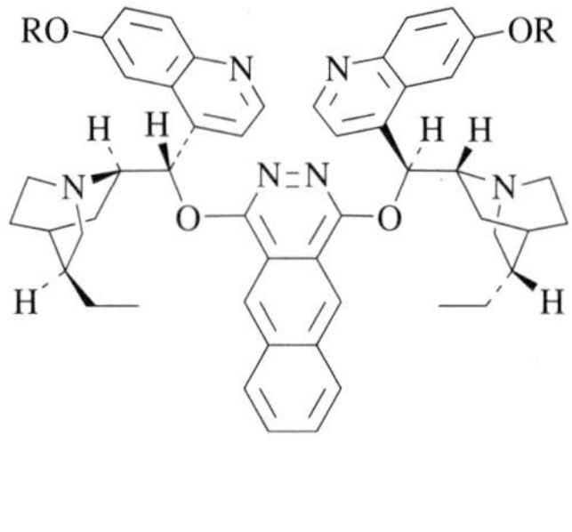

	R
92	CH(Pr-*n*)$_2$
120	CH(Bu-*n*)$_2$

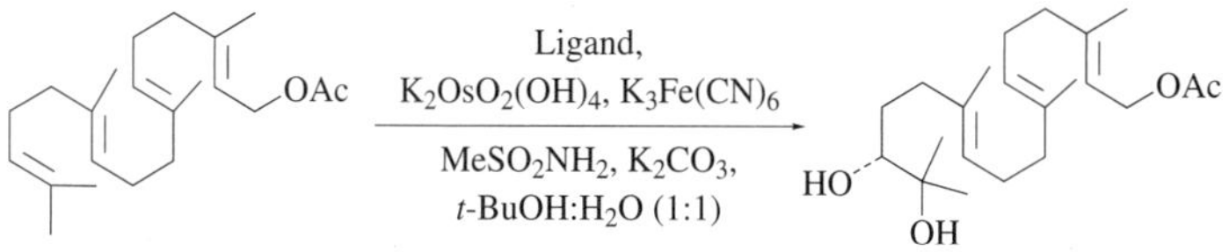

(Eq. 117)

Ligand		Position Selectivity	% ee
$(DHQD)_2$PYDZ (**18**)	(10%)	1:1.1	—
92	(54%)	50:1	95

the order of 50:1 to 100:1 in favor of the oxidation of the terminal prenyl unit are observed using these ligands. The corresponding oxidations using $(DHQD)_2$PHAL (**7**) are not position-selective and afford mixtures of diols and polyols in the reaction of these substrates. Although these ligands are not commercially available, they are easily prepared and offer convenient access to functionalized terpenes that are intermediates for cation-π-cyclization reactions that afford polycyclic steroid precursors.

Solid-Supported Cinchona Alkaloid Catalysts

While the cinchona alkaloid-catalyzed asymmetric dihydroxylation is both very efficient and highly enantioselective for a wide variety of substrates, the recovery of the expensive chiral ligand from large-scale reactions can be difficult. Due to the high cost of both the cinchona alkaloid ligands and OsO_4, the development of facile methods for efficient catalyst recovery has been the focus of much research. The use of polymer-bound cinchona alkaloids as efficient and selective ligands for the enantioselective dihydroxylation represents a major advance. Polymer-bound catalysts for the asymmetric dihydroxylation can be divided into three classes: (1) insoluble polymer-bound ligands; (2) soluble polymer-bound ligands; and (3) silica-anchored ligands. Each of these classes has unique characteristics affecting reaction rate, enantioselectivity, and ease of recovery of the chiral ligand and/or osmium catalyst. The subject of solid-supported cinchona alkaloids and their use in catalytic asymmetric dihydroxylation reactions has been reviewed elsewhere, and only state of the art methods representing each of the classes of solid-supported ligands are presented herein.[158]

The use of insoluble polymer-supported ligands for the catalytic enantioselective dihydroxylation reactions offers the advantage of facile ligand recovery and the potential for osmium recovery based on complexation with the polymer-supported ligand. The effectiveness of polymer-supported cinchona alkaloid ligands is dependent on the point of attachment of the cinchona alkaloid to the polymer support, the characteristics of the polymer support, and the extent of ligand incorporation. In general, the most effective polymer-bound cinchona alkaloid ligands are those in which the polymer support is attached to the pendant vinyl group of the parent cinchona alkaloid. This attachment leaves open the possibility of incorporating alkaloid ligands with different O(9) ether groups, allowing optimization of enantioselectivity and substrate specificity by modification of this critical site. The characteristics of the polymer support affect the ability of the polymer to swell in various media. More extensive swelling generally leads to higher reaction rates by increasing the accessibility of the cinchona alkaloid OsO_4 complex to the substrate in solution. Polymer supports incorporating polar functionality are more effective than those with hydrophobic groups for catalytic asymmetric dihdyroxylation.[159] The extent of ligand incorporation is another important variable, and polymer-bound systems with *ca.* 10–15 mol% ligand incorporation are optimum.

Among the most effective insoluble polymer-bound cinchona alkaloid ligands is the copolymer **121**.[160] The hydrophilic functionality present on the polymer backbone allows efficient swelling in either the acetone-water or aqueous *tert*-butyl alcohol solvent systems.[161] Crosslinking of the polymer is essential to prevent gelling of the insoluble ligand that complicates recovery, but the incorporation of crosslinking agent should be limited to approximately 20% of the total polymer to prevent deterioration of enantioselectivity.[162] Since some of the osmium catalyst is lost in the mother liquor and the methanol that is used to wash the catalyst, an additional 0.2 mol% of OsO_4 must be added to the recovered catalyst for subsequent reactions. The catalyst can thus be recovered and reused for at least five recycles without deterioration of either rate or enantioselectivity.

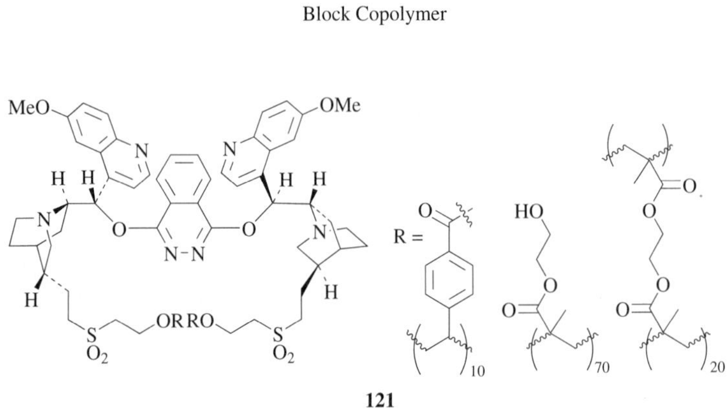

121

The rate and enantioselectivity obtained for the dihydroxylation of a variety of alkenes parallels that observed for homogeneous reactions using the corresponding soluble cinchona alkaloid ligands. As observed for the conventional asymmetric dihydroxylation, the PHAL-based ligand **121** provides superior enantioselectivity for the dihydroxylation of the alkenes studied to date compared to polymer-supported ligands with other linker groups, and reflects the catalyst preference observed for the homogeneous reaction (Eq. 118).[163]

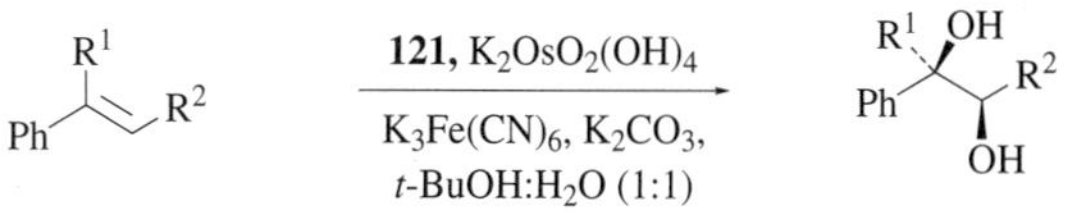

R^1	R^2		% ee	% ee (DHQD)$_2$PHAL, homogeneous)
H	H	(86%)	91	97
Me	H	(88%)	94	94
H	Ph	(90%)	>99	>99
—$(CH_2)_4$—		(85%)	97	99

(Eq. 118)

The immobilized PYR ligand **122** is also an efficient catalyst for the enantioselective dihydroxylation reaction of branched terminal alkenes (Eq. 119).[164] Unlike the other polymer-supported ligands described above, a slight deterioration of enantioselectivity is observed relative to the corresponding homogeneous reactions.

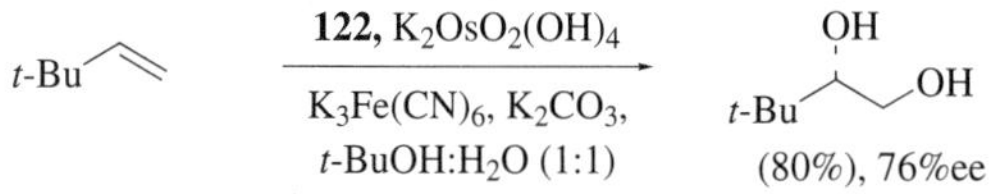

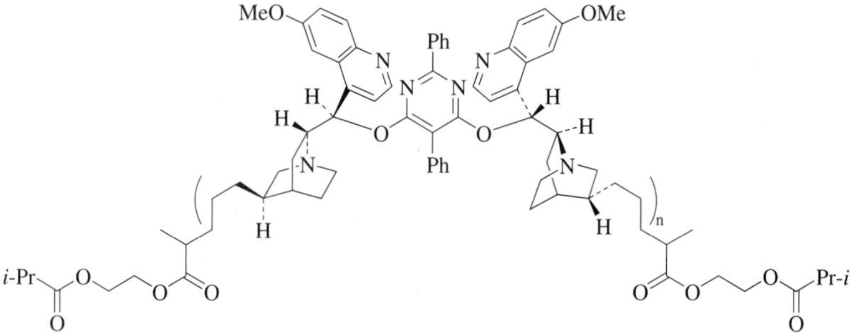

122

(Eq. 119)

Silica gel supported bis(cinchona) alkaloid ligands are also effective for the catalytic enantioselective dihydroxylation reaction. Because the silica-supported ligands reside on the surface of the silica, substrates can access the catalytic sites easily, and reaction rates are comparable to those observed in the corresponding homogeneous reactions. These solid-supported ligands are easily prepared by the reaction of functionalized silica with the chiral monomer. Careful selection of comonomers and control of crosslinking is obviated by the use of silica as the solid support. Thus, treatment of silica gel with (3-mercaptopropyl)trimethoxysilane in 1:1 pyridine-toluene affords the functionalized silica support **123**, which, when treated with the chiral monomer 1,4-bis(9-*O*-quinyl)phthalazine and AIBN, affords the immobilized ligand **124** with approximately 16 wt% incorporation of the alkaloid (Eq. 120).[165] This solid-supported alkaloid is an excellent ligand for the catalytic enantioselective dihydroxylation of aromatic alkenes under the standard conditions, as, for example, in Eq. 121.[165] The ligand can be recovered partially complexed with osmium by simple filtration and reused with a modest reduction in reaction rate, but no change in enantioselectivity.

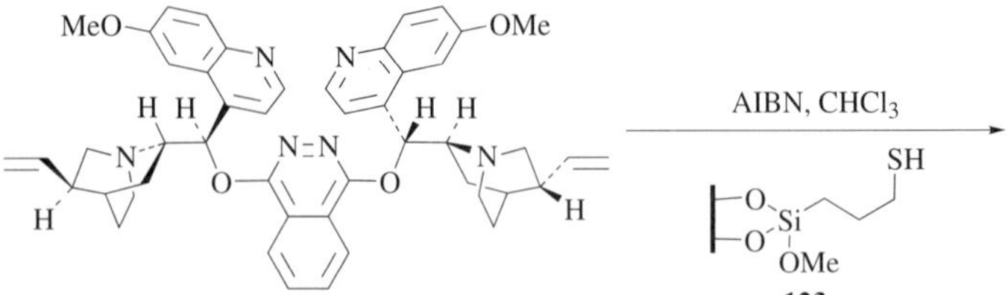

123

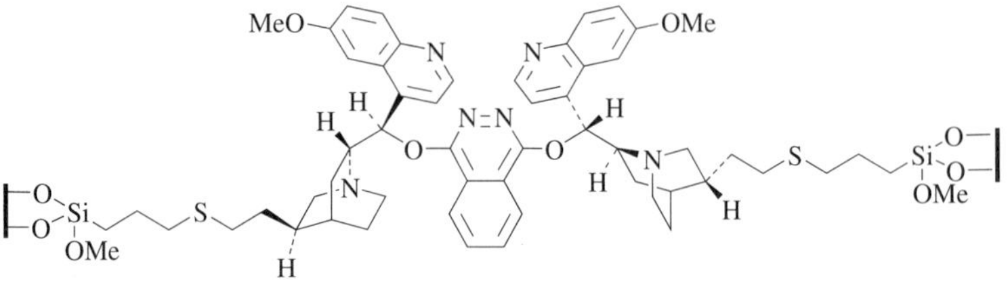

124

(Eq. 120)

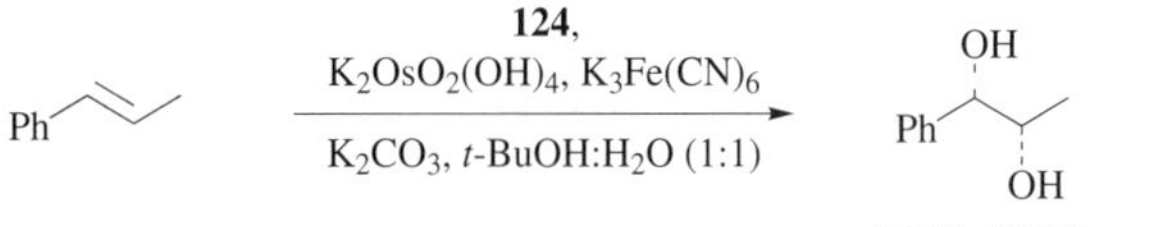

(92%), 96.5% ee

(Eq. 121)

Attachment of the bis(cinchona) alkaloid ligand to the silica support by the heteroaromatic linker group has also been accomplished, and effective solid-supported catalysts based on the PHAL and PYR scaffolds have been successfully used for the catalytic enantioselective dihydroxylation of aryl- and alkylsubstituted alkenes.[166] Several functionalized silica supports can be used to immobilize the ligands, allowing either ester or ether functionality at the point of attachment, although attachment via an ether group is preferred owing to its improved stability under the basic reaction conditions used in the dihydroxylation. Recovery of the immobilized ligand results in a loss of the osmium catalyst due to the weak association constant

(K_{eq} = 15–30) of the ligand-OsO_4 complex. Like the polymer-bound PYR-ligand **122**, the silica-supported PYR ligand **125** is an effective catalyst for the enantioselective dihydroxylation of aliphatic alkenes (Eq. 122),[166] but enantioselectivity is sometimes substantially lower than in the corresponding homogeneous reactions.

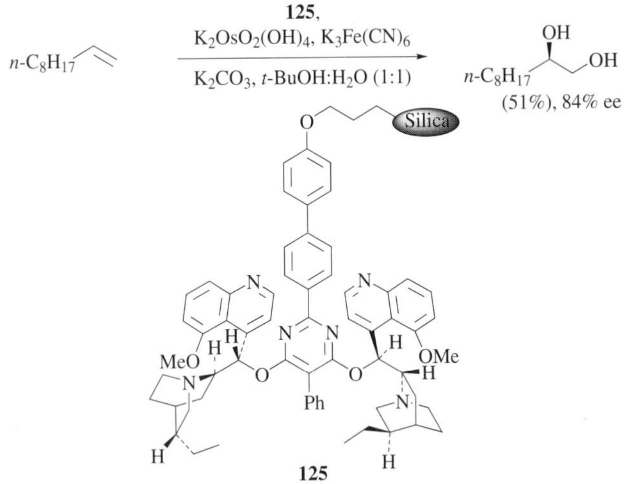

(Eq. 122)

Several soluble polymer-supported cinchona alkaloid ligands have been developed for the catalytic enantioselective dihydroxylation of alkenes that circumvent the prolonged reaction times and lower enantioselectivities associated with some of the insoluble polymer-supported ligands. Soluble polymer-supported cinchona alkaloid ligands are all based on a poly(ethylene glycol) monomethyl ether backbone, which is completely soluble in either the aqueous *tert*-butyl alcohol or acetone solvent systems used for the vast majority of catalytic enantioselective dihydroxylation reactions. This solvent compatibility allows the use of either the NMO or ferricyanide counteroxidants in the reaction. The solubility of the polymer requires an additional precipitation step to allow recovery of the ligand by filtration. Typically, *tert*-butyl methyl ether or diethyl ether is added to the reaction mixture to precipitate the catalyst, which is then recovered by filtration. An important advantage of using soluble polymer-bound ligands is the ability to use them in conjunction with polymer-bound substrates, enabling high-throughput, automated synthesis with recovery of both the ligand and the diol products by separate filtration and precipitation steps.[167] The bis(cinchona)alkaloids **126** and **127**, which are attached to the soluble polymer backbone via either the quinuclidine sidechain or the heteroaryl linker group, are superior ligands for the enantioselective dihydroxylation and offer selectivities and reaction rates that are very similar to those of homogeneous reactions (Eq. 123).[168,169]

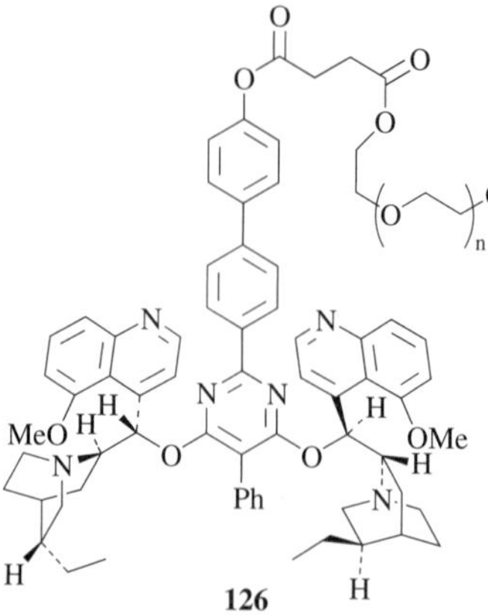

126

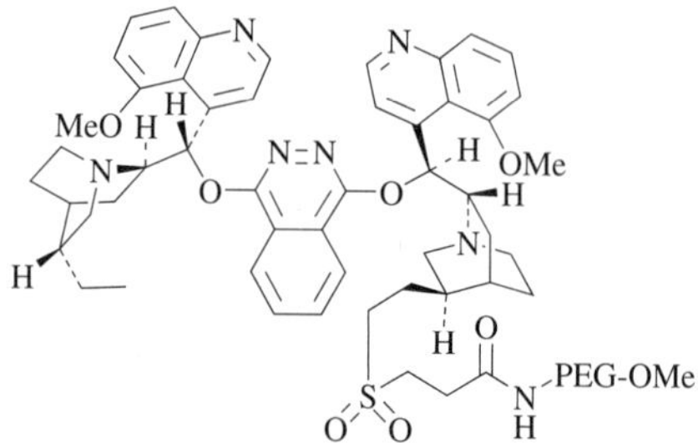

127

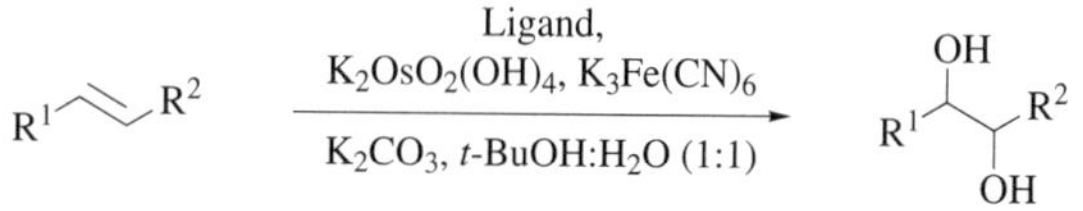

R^1	R^2	Ligand		% ee	Config.
n-C_8H_{17}	H	**126**	(86%)	87	R
t-Bu	H	**126**	(84%)	90	R
Ph	H	**127**	(88%)	98	R
n-Bu	n-Bu	**127**	(80%)	97	R,R
Ph	Ph	**127**	(95%)	99	R,R

(Eq. 123)

Osmium Sources. Osmium is the only known transition metal for the enantioselective dihydroxylation of alkenes. For many of the stoichiometric enantioselective dihydroxylations using chiral 1,2-diamine complexes of OsO_4, OsO_4 is used directly as the osmium source. Osmium tetroxide is a light yellow, volatile, crystalline solid. It is typically stored in sealed ampoules at 0° to prevent loss to sublimation. It is available as both the free solid as well as a solution either in toluene, *tert*-butyl alcohol, or water. Because solid OsO_4 typically arrives as a single large crystal that must be broken up prior to use, many prefer to use solutions of OsO_4 to minimize exposure during measurement and addition of the reagent. Osmium tetroxide is both expensive ($110.00 per gram) and toxic. Since many of the enantioselective dihydroxylations using OsO_4-1,2-diamine complexes are carried out under anhydrous conditions, either solid OsO_4 or OsO_4 in toluene is typically used. Osmium tetroxide is also the preferred osmium source for catalytic reactions where NMO is used as the secondary oxidant. Catalytic asymmetric dihydroxylation using potassium ferricyanide as the secondary oxidant allows the use of potassium osmate (VI) dihydrate as the osmium source. This easily handled, free flowing, purple crystalline solid is non-volatile and easily measured, thereby minimizing risk of exposure.

While procedures exist for the recovery and reuse of OsO_4, they can be somewhat cumbersome and result in poor recovery due to the volatility of the reagent. The development of polymer-supported osmium catalysts is one approach to allowing convenient recovery and reuse of osmium. Microencapsulation of OsO_4 onto an acrylonitrile-butadiene-polystyrene (ABS) copolymer has been successfully used in both achiral and enantioselective catalytic dihydroxylations.[170] The stability of some of the microencapsulated osmium sources is somewhat dependent on the nature of the substrate. For example, polystyrene microencapsulated OsO_4 (PS-MC OsO_4) dissolves in the presence of styrene. The ABS-MC OsO_4 is stable in the presence of a variety of substrates and is the preferred polymer-supported osmium source for enantioselective dihydroxylation (Eq. 124).[170] This microencapsulated polymer-supported osmium source works well for both the NMO and potassium ferricyanide supported dihydroxylations. Recovery of the osmium source is conveniently accomplished by simple filtration, and no deterioration in either rate or enantioselectivity is noted even after five recyclings.

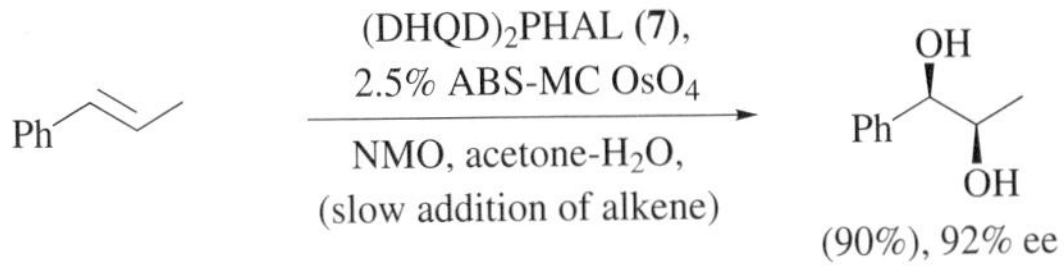

(Eq. 124)

Secondary Oxidants. Several secondary oxidant systems have been developed for the cinchona alkaloid catalyzed asymmetric dihydroxylation reaction. The first catalytic process that was developed utilized NMO as the stoichiometric oxidant. High conversion and enantioselectivity can be obtained using this oxygen source provided that experimental conditions are carefully selected. Two catalytic cycles exist that result in the conversion of the alkene substrate to the diol product with differing rates and enantioselectivities. The primary catalytic cycle results in highly selective and rapid dihydroxylation, whereas a secondary cycle resulting from alkene oxidation by the Os(VIII) ester intermediate results in a deterioration of the overall reaction rate and enantioselectivity. As discussed previously, the secondary cycle can be circumvented by minimizing both the concentration of alkene and Os(VIII) ester intermediate. Optimum conditions typically utilize a stoichiometric amount of tetraalkylammonium acetate to accelerate hydrolysis of the Os(VIII) ester. In many instances, slow addition of the substrate alkene is also necessary to circumvent the secondary cycle. The use of NMO as the secondary oxidant has significant process advantages for large-scale reactions: (1) high concentrations can be used; (2) there are no large amounts of salts required in the reaction and needing handling in the subsequent workup; and (3) the *N*-methyl morpheline (NMM) byproduct from the reaction is easily removed and can be recycled. Under careful control of substrate addition rate, the reaction can produce diols with very high enantiomeric purity when run in aqueous *tert*-butyl alcohol (Eq. 125).[171] A solid to solid process for the highly enantioselective oxidation of stilbene to hydrobenzoin has recently been developed using this oxidant system (Eq. 126).[172]

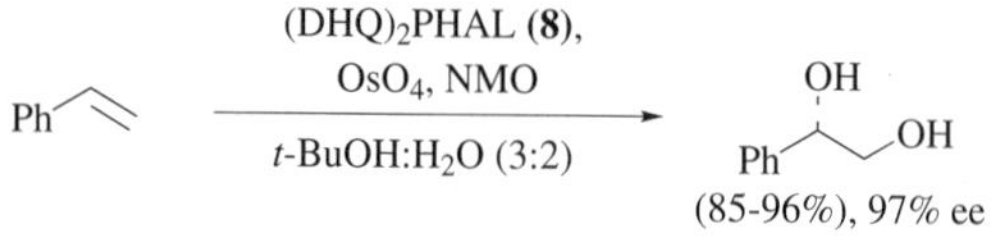

(Eq. 125)

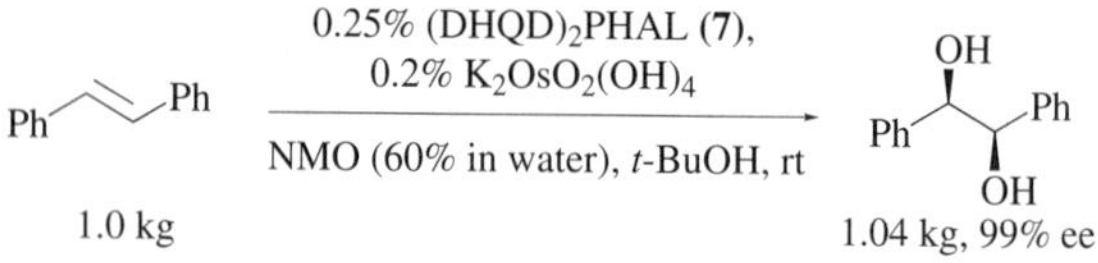

(Eq. 126)

Catalytic amounts of NMO may be used for the asymmetric dihydroxylation in a new coupled catalytic system that utilizes hydrogen peroxide as the terminal oxidant.[173,174] Oxygen or hydrogen peroxide are attractive terminal oxidants for industrial applications, as they are both inexpensive and environmentally friendly. A flavin catalyst is necessary for the reoxidation of NMM to NMO. The proposed catalytic cycle detailing the transfer of oxygen from H_2O_2 to the flavin catalyst and subsequently to NMM to regenerate NMO is shown in Eq. 127.[174] Initial oxidation of the flavin catalyst **128** by air produces the flavin peroxide **129,** which rapidly recycles NMM to NMO. The system leads to a mild, kinetically-controlled electron transfer from the substrate alkene to hydrogen peroxide at ambient temperature. This oxidant system can be applied to the enantioselective dihydroxylation using cinchona alkaloid ligands. The reaction of both styrene and (*E*)-stilbene was reported using $(DHQD)_2PHAL$ (**7**) as the chiral ligand, affording the diol products in 84–87% yield and 88% ee.[174] Slow addition of the alkene substrate is still required with this system in order to achieve optimum reaction rate and enantioselectivity.

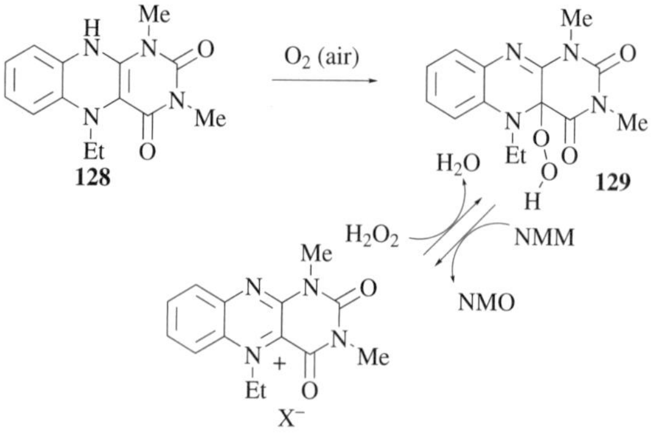

(Eq. 127)

Potassium ferricyanide has been successfully used as the stoichiometric oxidant for a wide variety of alkene substrates and is the secondary oxidant of choice for small-scale reactions. Because the second cycle is precluded by obligatory hydrolysis of the Os(VI) ester intermediate prior to oxidation to Os(VIII), there is no need for slow addition of the substrate alkene during the course of the reaction. Enantioselectivities are generally higher with the use of this counteroxidant as compared to

NMO. The reaction medium must be kept basic in order for catalytic turnover to be achieved, and this can result in undesirable side reactions with base-sensitive substrates. Potassium carbonate is typically used to maintain a buffered alkaline medium during the course of the reaction. With sensitive substrates, the pH can be slightly lowered with the use of a mixed bicarbonate-carbonate buffer. No turnover is observed when bicarbonate is used as the sole buffer. In general, there is no need to add a catalyst to accelerate the hydrolysis of the Os(VI) ester intermediate when the substrate has one unsubstituted alkene carbon atom. However, when both carbon atoms of the substrate double bond bear a substituent, it is typically necessary to add a stoichiometric equivalent of methanesulfonamide to accelerate the hydrolysis step. In cases where the hydrolysis step is exceptionally slow (e.g. with tetrasubstituted alkenes with all-carbon substituents), it is necessary to add three equivalents of methanesulfonamide to achieve reasonable substrate conversion. Presumably methanesulfonamide functions as a nucleophilic catalyst under the basic conditions of the reaction. Alternatively, the rate of hydrolysis may be accelerated for these substrates by careful control of the pH. Thus, under conventional conditions (3 equivalents of $K_3Fe(CN)_6$, 3 equivalents of K_2CO_3), the pH of the reaction mixture slowly decreases from 12.2 at the start of the reaction to a final value of 9.9. However, with the use of an automatic titration apparatus to maintain a pH of 12.0, the overall reaction rate can be dramatically increased without the need for methanesulfonamide. Thus, for the catalytic enantioselective dihydroxylation of α-methylstilbene, complete conversion of the alkene is observed after only 1.5 hours of reaction at pH 12.0, whereas 21 hours are required under the conventional reaction conditions. There is a slight reduction in enantioselectivity at the higher pH, and this lower selectivity presumably results from competition of hydroxide ion with the chiral ligand for binding to OsO_4. These results are summarized in Eq. 128.[71] This decrease in enantioselectivity can be overcome by increasing the loading of chiral ligand from 1 mol% to 4 mol%. Over-oxidation of diols derived from substituted styrenes and stilbenes to benzoins is sometimes problematic under these more basic reaction conditions.[71]

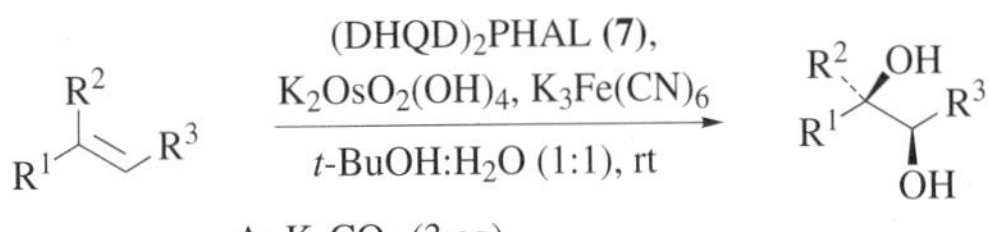

A: K_2CO_3 (3 eq)
B: Automatic titration to pH 12.0

(Eq. 128)

R^1	R^2	R^3		Time		% ee
n-Bu	H	*n*-Bu	A	3 h	(95%)	93
n-Bu	H	*n*-Bu	B	1.8 h	(95%)	90
Ph	Me	Ph	A	21 h	(82%)	99
Ph	Me	Ph	B	1.5 h	(62%)	99

While the use of potassium ferricyanide as the counteroxidant has advantages of convenience, disposal issues associated with large quantities of iron salts and cyanide have prompted several investigations into the use of coupled oxygen sources. In these systems, a catalytic amount of potassium ferricyanide is used as the secondary oxidant, and the oxidative regeneration of Fe(III) is mediated either electrochemically or by a tertiary oxidant. The direct electrochemical oxidation of Os(VI) to Os(VIII) occurs in two stages with potentials of $E_{pa}(O_1) = -0.115$V (Eq. 129) and $E_{pa}(O_2) = +0.225$V (Eq. 130), respectively.[175] The first oxidation produces a species that is adsorbed at the electrode surface, and therefore the rate of the second electron transfer is not diffusion-controlled. The species resulting from the first electron transfer is also reduced at $E_{pc}(R_1) = -0.225$ V (Eq. 131).[175]

$$Os(VI)O_2(OH)_4^{2-} \xrightarrow{-e} Os(VII)O_3(OH)_3^{2-} + H^+ \quad \text{(Eq. 129)}$$

$$Os(VII)O_3(OH)_3^{2-} \xrightarrow{H^+,\ e} Os(VI)O_2(OH)_4^{2-} \quad \text{(Eq. 130)}$$

$$Os(VII)O_3(OH)_3^{2-} \xrightarrow{-e} Os(VIII)O_4(OH)_2^{2-} + H^+ \quad \text{(Eq. 131)}$$

It is not necessary to use a divided cell when the electrochemical oxidation is conducted in the presence of a mediator, such as $K_3Fe(CN)_6$, as the direct electrochemical oxidation of $K_2OsO_2(OH)_4$, which results in electrodeposition of Os, is slow relative to its oxidation by the mediator.[175] A detailed investigation of the electrochemical dihydroxylation of stilbene revealed that an anodic potential of 400 mV or larger is necessary to regenerate either ferricyanide or perosmate ion.[176] Direct electrochemical oxidation of hydrobenzoin to benzaldehyde can occur at these potentials, but does not significantly contribute to the observed current inefficiency. The perosmate-catalyzed oxidation of hydrobenzoin to benzil is a more significant side reaction. Preparative oxidation of stilbene was reported using 1.6 mol% OsO_4 and 40 mol% $K_3Fe(CN)_6$ with the DHQD-*p*-chlorobenzoate (**4**) ligand in aqueous *tert*-butyl alcohol-water. A broad maximum in yield is noted when 85–90% of the theoretical charge is passed, corresponding to 80–95% alkene conversion and 17 turnovers of ferricyanide ion. The yield of hydrobenzoin decreases beyond this point, presumably as a result of competitive perosmate-catalyzed oxidation of hydrobenzoin. Under optimum conditions, hydrobenzoin is produced in 94% yield with 90–95% ee using this system.

The enantioselectivity and scope of the electrochemical osmium-catalyzed enantioselective dihydroxylation is dramatically enhanced by the use of the PHAL-class ligands.[177] The reaction is conducted in an undivided cell using platinum electrodes, and the diol product may be obtained in high yield and enantiomeric purity using as little as 10 mol% of potassium ferricyanide. Preparative runs are conducted at a constant current of 2 mA/cm^2 (applied voltage 1–3 V) until 2.33 F/mol (1.17 equivalents) of electricity are passed. Yields and enantioselectivities for the dihydroxylation of several alkenes of the terminal, E-1,2-disubstituted and trisubstituted alkene classes are comparable to those obtained under the conventional reaction conditions with potassium ferricyanide as the counteroxidant. Thus, the electrochemical oxidation of styrene affords the corresponding glycol in 95% yield and 97% ee (Eq. 132).[177]

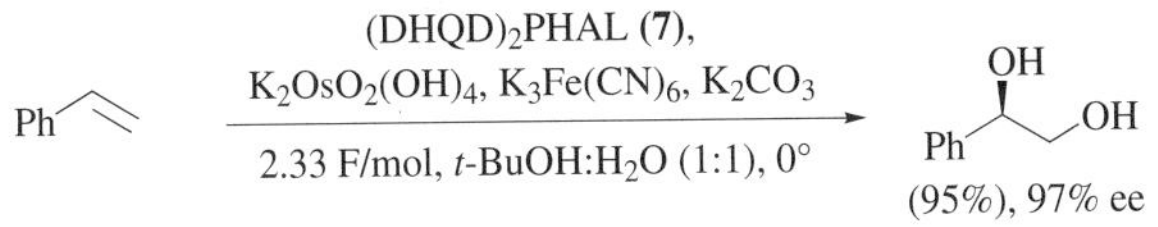

(Eq. 132)

Iodine has been successfully used in both chemical and electrochemical dihydroxylations of alkenes using cinchona alkaloid catalysts and potassium osmate.[175] The reaction must be conducted under suitably basic conditions using either potassium carbonate or potassium phosphate as the buffer salt, since hydrolysis of the osmate ester precedes oxidation of Os(VI) to Os(VIII) and requires an alkaline pH (>11.0). Optimum conversion for the chemical oxidation using iodine is observed when 1.5 equivalents of the counteroxidant are used, and yields and enantioselectivities for the dihydroxylation of substrates belonging to a variety of alkene classes are generally very high (Eq. 133).[176,177] The yield for the oxidation of several α,β-unsaturated esters under these conditions is somewhat lower than that observed for reactions using $K_3Fe(CN)_6$ and can be optimized by using a more weakly basic mixed buffer system (K_2CO_3-$KHCO_3$).

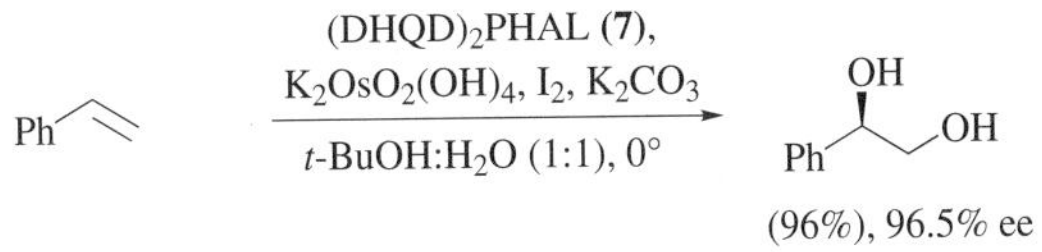

(Eq. 133)

Electrochemical regeneration of I_2 from I^- is also possible, and as little as 0.5 equivalents of I_2 based on substrate can be used.[175] The use of iodide rather than iodine as the initial mediator results in lower conversions and isolated yields. This is because the lower redox potential of the Os(VI)-Os(VIII) couple relative to the I^--I_2 couple precludes oxidation of I^- to I_2 in the presence of Os(VI) . As indicated earlier, this direct electrochemical oxidation of Os(VI) to Os(VIII) can result in electrodeposition of the osmium catalyst, resulting in lower turnover; however, when iodine is used as the initial mediator, it rapidly oxidizes Os(VI) to Os(VIII), resulting in a low concentration of Os(VI) in the diffusion layer that is replenished with I^- ions that can be oxidized at the electrode. Thus, under optimum electrolysis conditions, $K_2Os(VI)O_2(OH)_4$ undergoes a two-electron oxidation mediated by the electrochemically produced active iodine oxidizing species. The resultant $K_2Os(VIII)O_6(OH)_2$ is in equilibrium with OsO_4, which is transferred to the organic phase and oxidizes the substrate.

The most attractive reagents for the reoxidation of Os(VI) species are air or oxygen, since they are the least expensive and most environmentally friendly oxidants. Oxygen can be used as a secondary oxidant for the catalytic enantioselective dihydroxylation without the use of an intermediate oxygen transfer reagent. Alkene substrates in five of the six alkene classes have been successfully oxidized without the use of methanesulfonamide or other additives to accelerate the hydrolysis of the intermediate osmium ester species. Substrates possessing hydrogen atoms alpha to an aromatic substituent are prone to osmium-catalyzed over-oxidation of the diol product, giving benzaldehyde, benzoic acid, and other oxygenated products.[178] There is a strong dependence of reaction rate, chemoselectivity, and enantioselectivity on the pH of the aqueous phase. Under optimum conditions, the reaction is conducted with a 3:1 ratio of ligand to potassium osmate, and the pH is maintained between 10.4 and 11.2. More alkaline conditions result in sharply lower enantioselectivity, presumably due to competition of hydroxide with the cinchona alkaloid for coordination to osmium. *The reaction rate can be enhanced by increasing the oxygen pressure, but reactions run under a pure oxygen atmosphere should never be conducted under >10 bar of pressure due to the risk of explosion.* For each of the substrate classes that were evaluated, the enantioselectivity of the reaction is generally lower than that obtained using the conventional $K_3Fe(CN)_6$-K_2CO_3 system. The observed substrate functional group compatibility, direction of enantioselectivity, optimum ligand, and substrate structure-enantioselectivity relationships are similar to those observed under conventional conditions. The proposed catalytic cycle is shown in Eq. 134 and is similar to that of the potassium ferricyanide based oxidant system.[178]

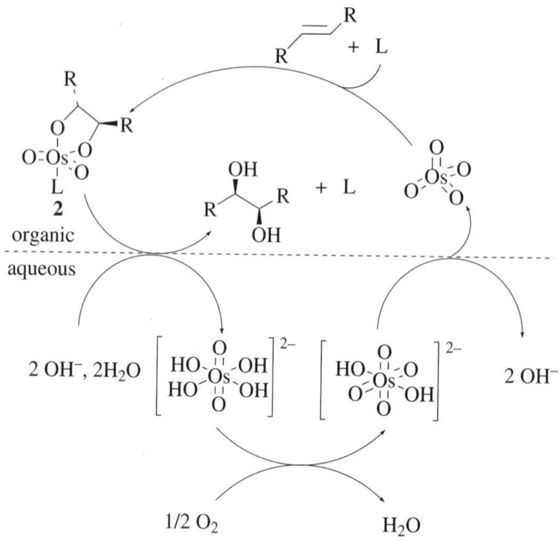

(Eq. 134)

Solvents and Concentration

Catalytic oxidations of alkenes with OsO_4 require an aqueous medium to effect hydrolysis of the intermediate osmate ester. Asymmetric dihydroxylations using potassium ferricyanide as the counteroxidant are typically triphasic at 0° (solid, aqueous, and organic phases), with the hydrolysis and reoxidation steps occurring in the aqueous phase, while the dihydroxylation step occurs in the organic phase. Oxidations using NMO as the counteroxidant are typically homogeneous due to the absence of salts. Dramatic effects of organic solvent on enantioselectivity have been shown, and *tert*-butyl alcohol is by far the preferred solvent for highly enantioselective dihydroxylations. A survey of other solvents, such as toluene, methylene chloride, acetone, acetonitrile, and chloroform in the enantioselective dihydroxylation of simple substituted styrenes indicated that each of these solvents was inferior to *tert*-butyl alcohol with regard to the enantiomeric purity of the diol product.[58] Oxidations with NMO as the secondary oxidant are typically performed in mixtures of acetone and water, although the enantioselectivity of the reaction is dramatically improved when conducted in aqueous *tert*-butyl alcohol.[171] The concentration of alkene in the enantioselective dihydroxylation using potassium ferricyanide as the counteroxidant is typically 0.1 M. For industrial scale applications, higher concentrations are desired, and reactions using NMO as the stoichiometric oxidant can be carried out at concentrations as high as 2.5 M.[172]

Reaction Temperature and Catalyst Loading

The enantioselectivity of the catalytic enantioselective dihydroxylation can be sensitive to temperature, and reactions are typically conducted at 0 to 4°, except in cases where catalyst turnover is slow, in which case the reaction is conducted at room temperature (e.g. tetrasubstituted alkenes with all carbon substituents). The aqueous *tert*-butyl alcohol mixture typically used for the potassium ferricyanide secondary oxidant system freezes at temperatures lower than 0°.

Typical loadings for laboratory scale reactions are 1 mol% of ligand and 0.2 mol% of osmium. Interestingly, enantioselectivity is markedly insensitive to the relative amounts of ligand and osmium. Successful dihydroxylations have been conducted with as little as 0.01 mol % of ligand without a catastrophic reduction in enantioselectivity. Thus for the dihydroxylation of *trans*-stilbene using $(DHQD)_2PHAL$ (**7**), use of 0.01 mol % of $(DHQD)_2PHAL$ (**7**) affords the diol product with 96% ee as opposed to 99.8% ee under normal conditions.[19] Alternatively, it is possible to increase the amount of osmium in cases where catalytic turnover is slow. For sluggish substrates, such as unactivated tetrasubstituted alkenes and α,β-unsaturated amides, the catalyst loading is typically increased to 5 mol% of ligand and 1 mol% of osmium. The enantioselective dihydroxylation of *cis*-1,2-disubstituted alkenes requires 2 mol% of the IND ligand and 0.2 mol% of osmium.

Premixed Reagents for Catalytic Enantioselective Dihydroxylation

Premixed reagents are available for the catalytic enantioselective dihydroxylation reaction and are sold as AD-mix α and AD-mix β corresponding to dihydroxylation of the α or β alkene faces according to the mnemonic for prediction of enantioselectivity (Figure 2). These commercially available mixtures are adjusted to provide 0.4 mol% of osmium and 1 mol% of ligand when used in the recommended ratio of 1.4 g of AD-mix for each millimole of alkene substrate.[20] For particularly sluggish reactions, additional ligand (ca. 5 mol%) and osmium (ca. 1 mol%) are used, and pre-mixed solid reagents at these ratios are sometimes referred to as Super AD-mix. These AD-mixes are pre-mixed PHAL-based ligand, potassium osmate, potassium ferricyanide, and potassium carbonate and are sold as AD-mix α, containing $(DHQ)_2PHAL$ (**8**) or AD-mix β, containing $(DHQD)_2PHAL$ (**7**). The recommended contents of 1 kg of AD-mix are as follows: $K_3Fe(CN)_6$: 699.6 g; K_2CO_3: 293.9 g;

(DHQD)$_2$PHAL (**7**) or (DHQ)$_2$PHAL (**8**): 5.52 g; and $K_2OsO_2(OH)_4$: 1.04 g.[20] This mixture should be thoroughly mixed in a blender and can be stored indefinitely at ambient temperature in a dry environment. There are no pre-mixed reagents available at present containing other cinchona alkaloid ligands.

EXPERIMENTAL PROCEDURES

Procedures for the Synthesis of Ligands for Enantioselective Dihydroxylation

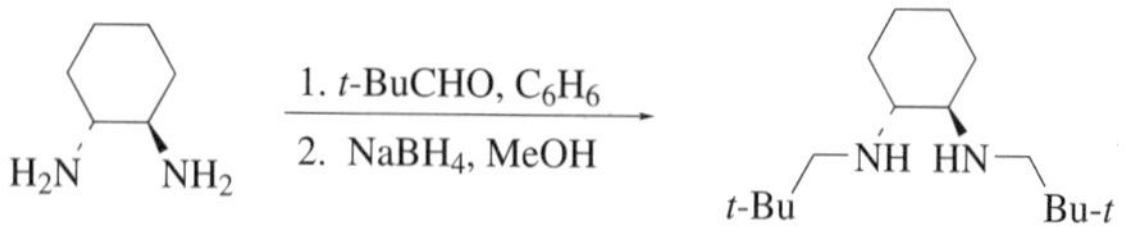

(1*R*,2*R*)-*N*,*N*′-Bis(3,3-dimethylpropyl)cyclohexane-1,2-diamine.[9] To a solution of (1*R*,2*R*)-cyclohexane-1,2-diamine (1.45 g, 12.7 mmol) in benzene (50 mL) was added freshly distilled 3,3-dimethylbutyraldehyde (2.8 g, 28 mmol). The flask was fitted with a Dean-Stark apparatus, and the mixture was refluxed for 1 hour. The solution was then evaporated to dryness, and the residue was dissolved in methanol (50 mL). Sodium borohydride (1.92 g, 50.8 mmol) was added portionwise at 0°. The mixture was stirred overnight, the solution was acidified with 10% HCl to pH 2, concentrated, and extracted with CH_2Cl_2. The aqueous solution was basified to pH 12 with 10% NaOH and extracted three times with CH_2Cl_2. The organic layers were dried over $MgSO_4$, filtered, and concentrated in vacuo. The residue was distilled using a Kugelrohr apparatus to obtain 2.66 g (75%) of (1*R*,2*R*)-*N*,*N*′-bis(3,3-dimethylpropyl)cyclohexane-1,2-diamine as a colorless solid: mp 67-68°; $[\alpha]^{20}_{D}$ −91.7° (*c* 2.1, $CHCl_3$); ^{1}H NMR (300 MHz, $CDCl_3$) δ 2.73 (td, *J* = 5.7 Hz, 10.5 Hz, 2H), 2.40 (td, *J* = 5.7 Hz, 10.5 Hz, 2H), 2.12-2.03 (m, 4H), 1.75-1.65 (m, 2H), 1.45-1.17 (m, 8H), 1.05-0.90 (m, 2H), 0.89 (s, 18H).

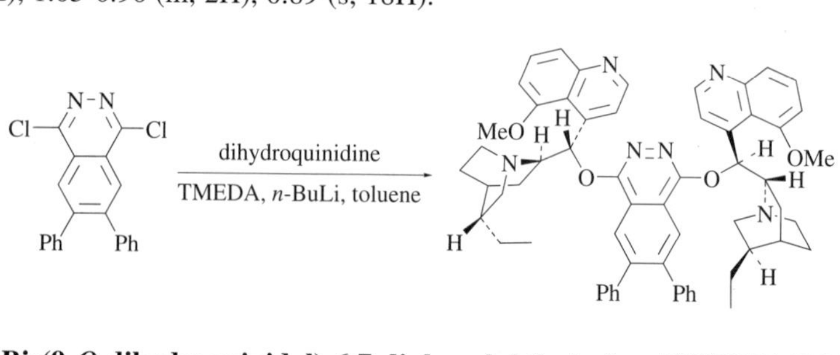

1,4-Bis(9-*O*-dihydroquinidyl)-6,7-diphenylphthalazine [(DHQD)$_2$DP-PHAL (119)].[157] In an oven-dried 50-mL round-bottom flask, a 2.5 M solution of *n*-BuLi in hexanes (1.1 mL, 2.75 mmol) was added dropwise under N_2 to a suspension of dihydroquinidine (0.815 g, 2.5 mmol) and TMEDA (0.58 g, 0.75 mL, 5.0 mmol) in toluene (25 mL) at 0°. After 40 minutes, 6,7-diphenyl-1,4-dichlorophthalazine[157] (0.35 g, 1.0 mmol) was added, and the mixture was heated to reflux. After 6 hours, the mixture was cooled to room temperature, water (15 mL) and ethyl acetate

(10 mL) were added, and the layers were separated. The organic phase was washed with water (10 mL), and the combined aqueous layers were extracted with ethyl acetate (10 mL). The combined organic layers were dried over $MgSO_4$, filtered, and concentrated to give a solid which was purified by flash chromatography (CH_2Cl_2 + 5% CH_3OH + 1% NH_4OH) to give the title compound (0.45 g, 48%) as a colorless solid: mp 154–157°; R_f 0.32 (CH_2Cl_2, 5% CH_3OH, 1% NH_4OH); $[\alpha]_D^{23}$ −273.3° (*c* 0.99, $CHCl_3$); ^1H NMR (400 MHz, $CDCl_3$) δ 8.63 (d, J = 4.5 Hz, 2H), 8.34 (s, 2H), 7.98 (d, J = 9.2 Hz, 2H), 7.56 (d, J = 2.6 Hz, 2H), 7.41 (d, J = 4.6 Hz, 2H), 7.37-7.31 (m, 8H), 7.26-7.23 (m, 4H), 7.06 (d, J = 5.3 Hz, 2H), 3.86 (s, 6H), 3.40 (q, J = 8.5 Hz, 2H), 2.83–2.71 (m, 8H), 2.04 (br s, 2H), 1.71 (br s, 2H), 1.58-1.39 (m, 12H), 0.73 (t, J = 7.0 Hz, 6H).

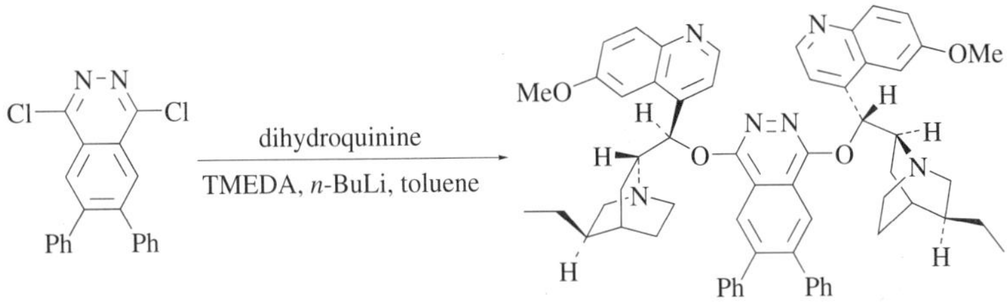

1,4-Bis(9-*O*-dihydroquinyl)-6,7-diphenylphtalazine [(DHQ)$_2$DP-PHAL].[157] Using dihydroquinine (0.815g, 2.5 mmol) and the identical procedure used to prepare 1,4-bis-(9-*O*-dihydroquinidyl)-6,7-diphenylphthalazine, a crude solid was produced which was purified by flash chromatography (CH_2Cl_2 + 5% CH_3OH + 1% NH_4OH) to give the title compound (0.64 g, 68%) as a white solid. This solid could be further purified by crystallization from methylene chloride:hexane (1:5): mp 231-234°; R_f 0.32 (CH_2Cl_2 + 5% CH_3OH + 1% NH_4OH); $[\alpha]_D^{23}$ +320.1°(*c* 0.95, $CHCl_3$); ^1H NMR (400 MHz, $CDCl_3$) δ 8.64 (d, J = 4.5 Hz, 2H), 8.29 (s, 2H), 7.99 (d, J = 9.2 Hz, 2H), 7.62 (d, J = 2.5 Hz, 2H), 7.45 (d, J = 4.5 Hz, 2H), 7.37-7.30 (m, 8H), 7.25-7.20 (m, 4H), 7.07 (d, J = 5.9 Hz, 2H), 3.87 (s, 6H), 3.49 (q, J = 6.3 Hz, 2H), 3.18 (m, 2H), 3.02 (dd, J = 13.6, 9.7 Hz, 2H), 2.59 (m, 2H), 2.33 (d, J = 15.4 Hz, 2H), 1.78 (br s, 6H), 1.44-1.22 (m, 10H), 0.81 (t, J = 7.3 Hz, 6H).

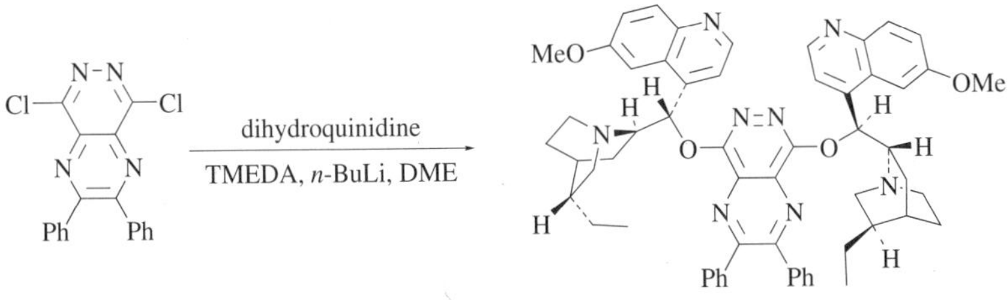

5,8-Bis(9-*O*-dihydroquinidyl)-2,3-diphenylpyrazino[2,3-*d*]pyridazine [(DHQD)$_2$DPP (35)].[157] A solution of dihydroquinidine (13.9 g, 42.7 mmol) and

N,N,N′,N′-tetramethylethylendiamine (TMEDA) (17.1 mL, 95 mmol) in dimethoxyethane (300 mL, distilled from sodium) was cooled to −50°, and *n*-butyllithium (2.5M in hexanes, 17.1 mL, 42.7 mmol, 2.25 equiv) was added slowly. The mixture was stirred for 15 minutes, warmed to room temperature, and 5,8-dichloro-2,3-diphenylpyrazino[2,3-*d*]pyridazine (6.70 g, 18.95 mmol) was added. After heating at reflux for 4 hours, the mixture was cooled to room temperature, and H_2O (20 mL) was added together with ethyl acetate (500 mL). The organic phase was washed with saturated aqueous $NaHCO_3$ solution (100 mL). The aqueous layers were extracted with ethyl acetate (3 × 100 mL), and the combined organic phases were dried over $MgSO_4$ and concentrated in vacuo. The residue was purified by flash chromatography (5 cm × 15 cm pad of silica, $CHCl_3$ + 0.5% MeOH + 0.5% NH_4OH) and was dried overnight in vacuo at 50° to afford 16.8 g (18.0 mmol, 95%) of the title product as a pale yellow solid: mp 173-177°; R_f 0.15 ($CHCl_3$+5% MeOH+0.5% NH_4OH); $[\alpha]_D^{23}$ 345.6° (*c* 1.022, $CHCl_3$); ^1H NMR (400 MHz, $CDCl_3$) δ 8.64 (d, *J* = 4.5 Hz, 2H), 7.97 (d, *J* = 9.2 Hz, 2H), 7.63-7.58 (m, 6H), 7.50-7.44 (m, 4H), 7.42-7.38 (m, 4H), 7.31 (d, *J* = 9.2, 2.6 Hz, 2H), 7.00 (d, *J* = 4.0 Hz, 2H), 3.74 (s, 6H), 3.43 (m, 2H), 2.82-2.78 (m, 6H), 2.70-2.65 (m, 2H), 2.15 (t, *J* = 10.6 Hz, 2H), 1.71 (br s, 2H), 1.56-1.33 (m, 12H), 0.67 (t, *J* = 7.2 Hz, 6H).

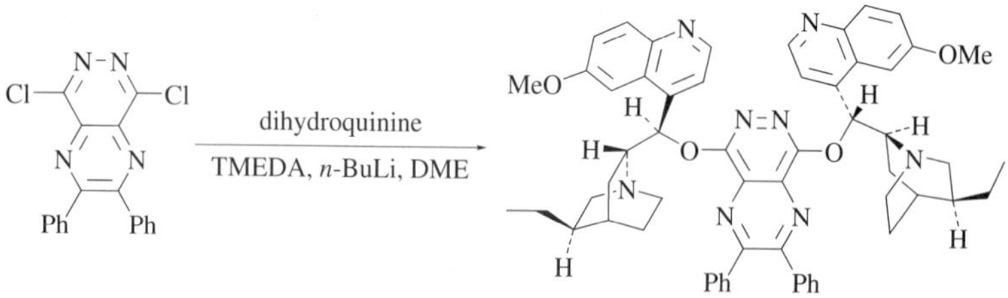

5,8-Bis(9-*O*-dihydroquinyl)-2,3-diphenylpyrazino[2,3-*d*]pyridazine [(DHQ)$_2$DPP].[157] The title compound was prepared analogously to (DHQD)$_2$DPP (**35**) (procedure above) using 5,8-dichloro-2,3-diphenylpyrazino[2,3-d]pyridazine (1.035 g, 2.93 mmol), dihydroquinine (2.39 g, 7.32 mmol, 2.5 equiv), *n*-butyllithium (2.5 M in hexanes, 2.93 mL, 7.32 mmol, 2.5 equiv) and TMEDA (2.21 mL, 14.7 mmol, 5 equiv) in dimethoxyethane (100 mL, distilled from sodium). After a reaction time of 3 hours, analogous workup, flash chromatography (2.5 cm × 15 cm pad of silica, $CHCl_3$ + 5% MeOH + 0.5% NH_4OH), and drying overnight under high vacuum at 50°, 2.554 g (2.74 mmol, 95%) of the title compound was obtained as a pale yellow solid: R_f 0.16 ($CHCl_3$ + 5% MeOH + 0.5% NH_4OH); mp 175-180°; $[\alpha]_D^{23}$ +441.7° (*c* 1.037, $CHCl_3$); ^1H NMR (400 MHz, $CDCl_3$) δ 8.64 (d, *J* = 4.5 Hz, 2H), 7.97 (d, *J* = 9.2 Hz, 2H), 7.65 (br s, 2H), 7.61-7.58 (m, 4H), 7.51–7.45 (m, 4H), 7.43–7.38 (m, 4H), 7.29 (dd, *J* = 9.2, 2.7 Hz, 2H), 6.93 (d, *J* = 5.5 Hz, 2H), 3.73 (s, 6H), 3.56-3.52 (m, 2H), 3.29-3.19 (m, 2H), 3.04 (dd, *J* = 13.5, 9.8 Hz, 2H),

2.61-2.52 (m, 2H), 2.33-2.27 (m, 2H), 1.96-1.71 (m, 8H), 1.49-1.23 (m, 8H), 0.83 (t, $J = 7.3$ Hz, 6H).

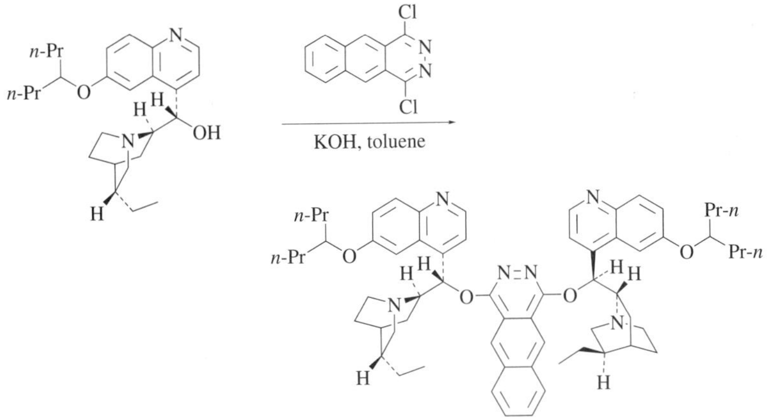

1,4-Bis[*O*-6′-(4-heptyl)hydrocupreidyl]naphthopyridazine (92).[62] To a solution of *O*-6′-(4-heptyl)hydrocupreidine (1.7g, 4.2 mmol) in 140 mL of toluene was added 1,4-dichloronaphthopyridazine[179] (0.51 g, 2.1 mmol) and powdered KOH (1.7g, 30 mmol). The mixture was heated to reflux with azeotropic removal of water using a Dean-Stark trap. After 3 hours, the mixture was cooled to room temperature, diluted with 100 mL of H_2O, and extracted with ethyl acetate (3 × 100 mL). The combined organic extracts were dried over Na_2SO_4, filtered, and concentrated in vacuo. The residue was purified by radial chromatography (8 mm silica plate, 10 mL/minute, $CHCl_3$, followed by 97:3:0.3 $CHCl_3$+MeOH+NH_4OH) to afford 1.55 g (75%) of **(92)** a colorless syrup: R_f 0.37 ($CHCl_3$+MeOH+NH_4OH 9:1:0.1); $[\alpha]_D^{23}$ −275° (*c* 0.51, $CHCl_3$).

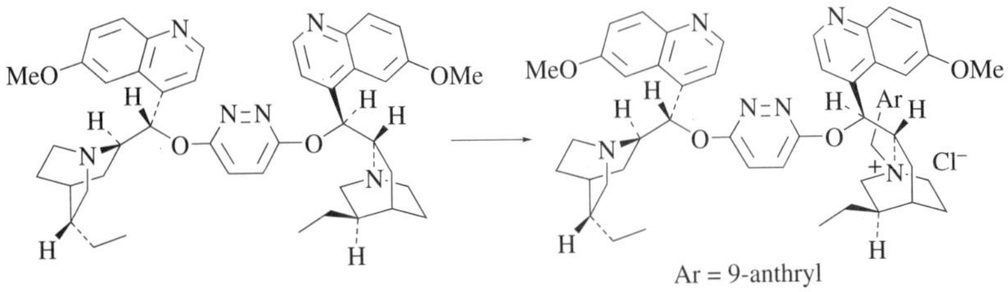

3,6-Bis(hydroquinidyl)pyridazine-mono-9-anthracenylmethyl Chloride.[82] To a solution of 3,6-bis(hydroquinidyl)pyridazine (1.6 g, 2.2 mmol) in 4.4 mL of acetonitrile was added 9-chloromethylanthracene (0.50 g, 2.2 mmol), and the resulting mixture was stirred for 12 hours at 40°. After concentration in vacuo, the residue was purified by flash column chromatography ($CHCl_3$+MeOH+NH_4OH, 90:10:1), followed by radial chromatography (4 mm plate, $CHCl_3$+MeOH+NH_4OH, 98:2:0.2 to 90:10:1), giving 0.85 g (40%) of a light yellow solid: R_f 0.28 ($CHC1_3$+MeOH+NH_4OH, 85:15:1.5); $[\alpha]_D^{23}$ +48° (*c* 0.18, $CHCl_3$); ^1H NMR (400 MHz, CD_3OD) δ 8.73 (d, *J* = 4.6 Hz, 1H), 8.67 (s, 1H), 8.39 (d, *J* = 9.0 Hz, 1H), 8.34 (d, *J* = 4.6 Hz, 1H), 8.27 (s, 1H), 8.10 (m, 3H), 7.88-7.82 (m, 2H), 7.72-7.66 (m, 2H), 7.64 (d, *J* = 9.3 Hz, 1H), 7.61 (d, *J* = 2.2 Hz, 1H), 7.58-7.52 (m, 3H), 7.51 (d, *J* = 4.6 Hz, 1H), 7.44 (t, *J* = 8.2 Hz, 1H), 7.34 (d, *J* = 1.9 Hz, 1H), 7.17 (dd, *J* = 2.5, 9.2 Hz, 1H,), 7.00 (t, *J* = 7.4 Hz, 1H), 6.91 (s, 1H), 5.96 (d, *J* = 14.0 Hz, 1H), 5.66 (d, *J* = 14.0 Hz, 1H), 4.53 (m, 1H), 4.09 (s, 3H), 4.05 (m, 1H), 3.75 (s, 3H), 3.59 (m, 1H), 3.18 (m, 1H), 2.92-2.67 (m, 6H), 2.28 (m, 1H), 1.94 (s, 1H), 1.79 (m, 2H), 1.67-1.53 (m, 11H), 1.37 (m, 1H), 0.94 (t, *J* = 7.1 Hz, 3H), 0.79 (t, *J* = 7.3 Hz, 3H).

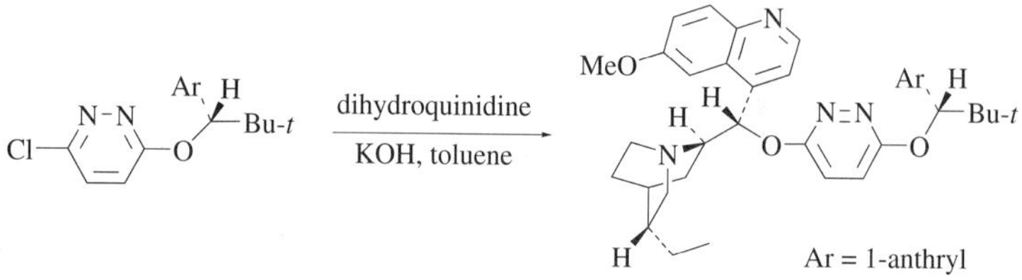

6-Dihydroquinidyl-3-[1(*S*)-anthracen-1-yl-2,2-dimethylpropoxy]pyridazine [DHQD-PYDZ-(*S*)-Anthryl Ligand].[73] To a suspension of dihydroquinidine (0.114 g, 0.350 mmol) in toluene (3 mL) was added KOH (0.070 g, 1.3 mmol, pulverized prior to use) and 3-[(*S*)-l-anthracen-1-yl-2,2-dimethylpropoxy]-6-chloropyridazine[73] (0.12 g. 0.32 mmol). The resulting mixture was heated at reflux (140° bath temperature) for 45 minutes (larger scale reactions require azeotropic removal of water using a Dean-Stark trap). After the mixture was cooled to room temperature, 15 mL of H_2O was added, and the mixture was extracted three times with ethyl acetate (30 mL). The combined organic extracts were dried over Na_2SO_4, filtered, and concentrated in vacuo. The residue was purified by flash chromatography (MeOH+$CHC1_3$+NH_4OH, 3:97:0.3) to give 0.19 (88%) of a light yellow syrup: $[\alpha]_D^{23}$ +188.5° (*c* 0.20, MeOH); ^1H NMR ($CDCl_3$, 500 MHz) δ 8.85 (s, 1H), 8.53 (dd, *J* = 1.5, 4.4 Hz, 1H), 8.37 (s, 1H), 8.06 (d, *J* = 5.7 Hz, 1H), 7.95 (d, *J* = 7.4 Hz, 1H), 7.84 (m, 2H), 7.56 (d, *J* = 4.9 Hz, 1H), 7.45 (m, 2H), 7.36 (m, 2H), 7.27 (d, *J* = 7.2 Hz, 1H), 7.18 (d, *J* = 8.5 Hz, 1H), 7.08 (s, 1H), 7.00 (d, *J* = 8.9 Hz, 1H), 6.91 (d, *J* = 9.4 Hz, 1H), 6.88 (s, 1H), 3.50 (s, 3H), 3.35 (m, 1H), 2.81 (m, 2H), 2.63 (m, 2H), 1.90 (m, 1H), 1.71 (s, 1H), 1.47 (m, 6H), 1.08 (s, 9H), 0.89 (t, *J* = 7.2 Hz, 3H).

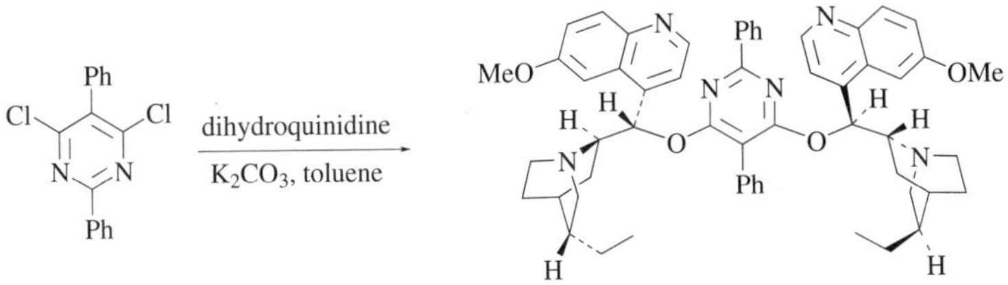

2,5-Diphenyl-bis(9-*O*-dihydroquinidyl)pyrimidine [$(DHQD)_2$PYR (25)].[77] Hydroquinidine hydrochloride (75 g) was added to concentrated aqueous NH_4OH (500 mL). The resulting suspension was extracted with CH_2Cl_2 (3 × 200 mL). The combined organic extracts were washed with aqueous NH_4OH (200 mL) and H_2O (200 mL), dried over $MgSO_4$, filtered, and concentrated in vacuo, affording dihydroquinidine (54 g) as a colorless solid.

A 50-mL flame-dried one-neck round-bottom flask was charged with dihydroquinidine (2.58 g, 7.66 mmol), 2,5-diphenyl-4,6-dichloropyrimidine (1.15 g, 3.83 mmol)[77], K_2CO_3 (1.6 g, 11.5 mmol), and anhydrous toluene (30 mL). The flask was flushed with nitrogen and equipped with a Dean-Stark condenser. After the mixture was stirred at reflux (oil bath temperature 135°) for 2 hours, KOH pellets (729 mg, 13.0 mmol) were added, and the mixture was refluxed with azeotropic removal of water for an additional 12 hours. The light orange solution was cooled to room temperature, diluted with H_2O (100 mL), the layers were separated, and the aqueous layer was extracted with CH_2Cl_2 (3 × 50 mL). The combined organic layers were dried over $MgSO_4$, filtered, and concentrated in vacuo. The crude, pale yellow solid was crystallized from acetonitrile (100 mL), affording the title compound as fluffy white crystals (2.6 g. 77%): mp 253–254°; ^{1}H NMR (400 MHz, $CDCl_3$) δ 8.72 (d, J = 4.4 Hz, 2H), 8.01 (d, J = 9.2 Hz, 2H), 7.59–7.44 (m, 9H), 7.40–7.37 (m, 4H), 7.20 (t, J = 7.3 Hz, 1H), 7.03–6.97 (m, 4H), 3.82 (s, 6H), 3.13–3.11 (m, 2H), 2.81–2.55 (m, 8H), 1.81–1.75 (m, 2H), 1.59 (br s, 2H), 1.47–1.25 (m, 8H), 0.97 (m, 4H), 0.69 (t, J = 7.2 Hz, 6H).

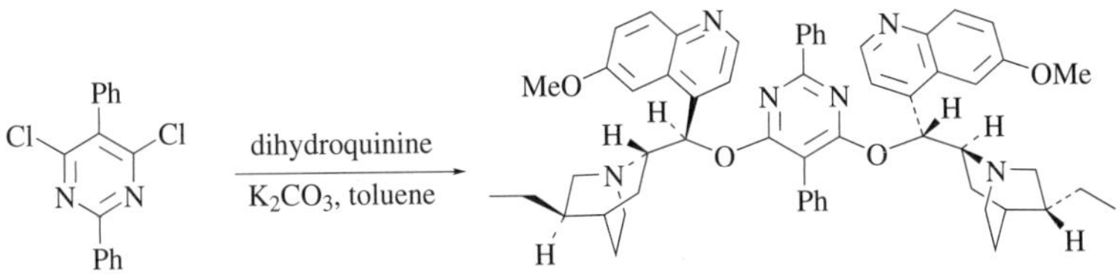

2,5-Diphenyl-4,6-bis(dihydroquinyl)pyrimidine [$(DHQ)_2$PYR (26)].[77] This ligand was prepared following the same procedure described in the section above. Starting from dihydroquinine (1.0 g), the title compound (0.98 g, 72%) was obtained as white crystals after crystallization from ethyl acetate (10 mL): ^{1}H NMR (400 MHz, $CDCl_3$) δ 8.69 (d, J = 4.5 Hz, 2H), 8.01 (d, J = 9.2 Hz, 2H), 7.59–7.55

(m, 4H), 7.50–7.46 (m, 5H), 7.39 (dd, J = 2.6, 9.2 Hz, 2H), 7.33 (d, J = 4.5 Hz, 2H), 7.18 (t, J = 7.3 Hz, 1H), 6.97 (t, J = 7.6 Hz, 2H), 6.90 (br s, 2H), 3.83 (s, 6H), 3.15–3.09 (m, 4H), 2.96 (dd, J = 10.0, 13.4 Hz, 2H), 2.56–2.49 (m, 2H), 2.29 (d, J = 12.7 Hz, 2H), 1.63–1.54 (m, 6H), 1.30 (br s, 2H), 1.24–1.15 (m, 8H), 0.75 (t, J = 7.3 Hz, 6H).

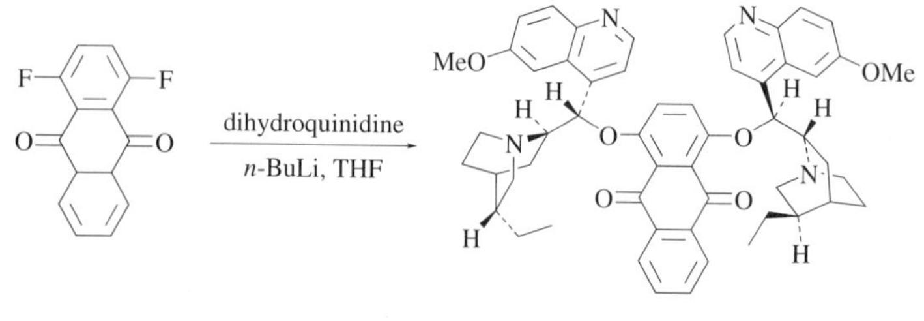

1,4-Bis(dihydroquinidyl)benzo[*g*]phthalazine-5,10-dione [(DHQD)$_2$AQN (30)].[86] A solution of dihydroquinidine (3.33g, 10.3 mmol) in dry THF was cooled to −50°, and a solution of *n*-butyllithium in hexane (1.6 M, 6.4 mL, 10.25 mmol) was added slowly over 10 minutes. The pale red solution was stirred for 15 minutes, warmed to 0°, and difluoroanthraquinone[86] (1.00g. 4.10 mmol) was added as a solid. The mixture was warmed to room temperature and was stirred for 15 hours. After warming to 40°, the mixture was stirred for another 2 hours. For reactions conducted on a larger scale, the solvent was removed in vacuo at this point. Ethyl acetate (200 mL) was added along with saturated aqueous $NaHCO_3$ solution (100 mL). The aqueous phase was extracted twice with 100 mL of ethyl acetate, and the combined organic layers were dried over $MgSO_4$, filtered, and concentrated in vacuo. Purification of the crude product by flash chromatography on silica gel (4 × 15 cm, $CHCl_3$+MeOH+NH_4OH, 95:5:0.5) afforded (DHQD)$_2$AQN (**30**) as yellow solid (3.107 g, 3.625 mmol, 88%): mp 152–157°; $[\alpha]^{25}_{D}$ −487° (*c* 1.1, $CHCl_3$); ^{1}H NMR ($CDCl_3$, 400 MHz) δ 8.65 (d, J = 4.5 Hz, 2H), 8.22 (dd, J = 5.8, 3.3 Hz, 2H), 8.02 (d, J = 9.4 Hz, 2H), 7.79 (dd, J = 5.7, 3.3 Hz, 2H), 7.49 (d, J = 4.2 Hz, 2H), 7.36 (dd, J = 9.4, 2.2 Hz, 4H), 6.65 (s, 2H), 5.87 (br s, 2H), 3.92 (s, 6H), 3.24 (br s, 2H), 2.81–2.41 (m, 8H), 2.41 (br s, 2H), 1.77–1.25 (m, 14H), 0.84 (t, J = 7.3, 6H).

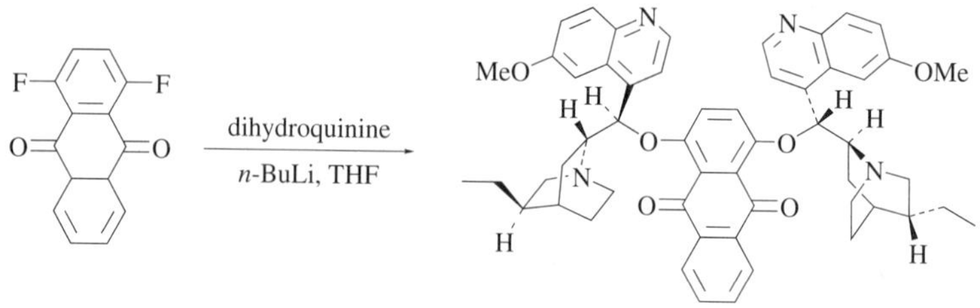

1,4-Bis(dihydroquinyl)benzo[*g*]phthalazine-5,10-dione [(DHQ)$_2$AQN (31)].[86] The synthesis is analogous to that of (DHQD)$_2$AQN (**30**) described above. Dihydroquinine (3.35 g, 10.25 mmol), *n*-butyllithium in hexane (1.6 M, 6.4 mL, 10.25 mmol), and difluoroanthraquinone (1.00 g, 4.10 mmol) afforded (DHQ)$_2$AQN (**31**) (2.967 g, 3.462 mmol, 84%) as a yellow solid: mp 177–180°; $[\alpha]^{25}_{D}$ +579° (*c* 1.18, $CHCl_3$); ^{1}H

NMR ($CDCl_3$, 400 MHz) δ 8.64 (d, J = 4.5 Hz, 2H), 8.28 (dd, J = 5.7, 3.3 Hz, 2H), 8.03 (d, J = 9.2 Hz, 2H), 7.79 (dd, J = 5.7, 3.3 Hz, 2H), 7.43 (d, J = 4.5 Hz, 2H), 7.37 (dd, J = 9.2, 2.5 Hz, 2H), 7.32 (br s, 2H), 6.61 (br s, 2H), 5.95 (br s, 2H), 3.92 (s, 6H), 3.23 (br s, 4H), 3.05 (dd, J = 13.4, 10.0 Hz, 2H), 2.35–2.61 (m, 6H), 2.05 (br s, 2H), 1.92 (br s, 2H), 1.59 (br s, 2H), 1.14–1.44 (m, 8H), 0.80 (t, J = 7.3 Hz, 6H).

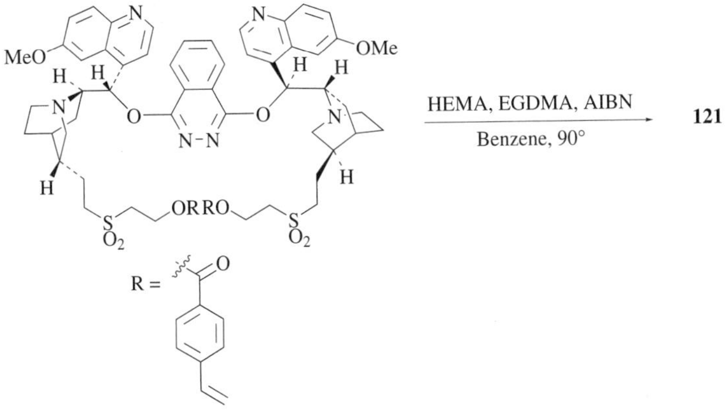

(DHQD)$_2$PHAL—EGDMA—HEMA Block Copolymer Ligand (121).[162] A solution of 1,4-bis-[12-(4-vinylbenzoyloxyethanesulfonyl)-9-*O*-dihydroquinidyl]-phthalazine (0.32 g, 0.25 mmol), hydroxyethyl methacrylate (HEMA, 0.22 ml, 1.7 mmol), and ethylene glycol dimethacrylate (EGDMA, 0.1 mL, 0.5 mmol) in benzene (10 mL) was added to benzene (90 ml) at 80°. The polymerization was initiated by the addition of AIBN (0.063 g, 0.038 mmol) and the mixture was stirred for 24 hours. The precipitated polymer was filtered, extracted with methanol and acetone using a Soxhlet extractor, and concentrated in vacuo, affording a 76% yield of the title compound. Nitrogen analysis indicated a 9.51 mol% loading of chiral alkaloid.

Procedures for the Enantioselective Dihydroxylation of Alkenes

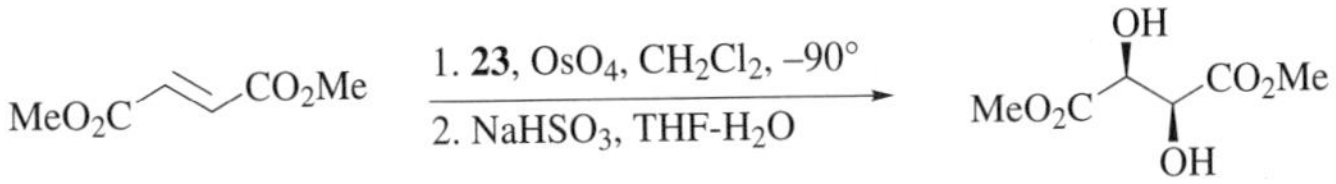

***S,S*-Diethyl Tartrate [Stoichiometric Enantioselective Dihydroxylation Using a Chiral 1,2-Diamine Ligand].**[9] To a solution of (1*R*,2*R*)-*N,N*′-bis(3,3-dimethylbutyl)cyclohexane-1,2-diamine (117 mg, 0.416 mmol) in CH_2Cl_2 (4 mL) at −90° was added OsO_4 (4.1 mL of a 25 mg/mL CH_2Cl_2 solution, 0.403 mmol). After the resulting mixture was stirred for 30 minutes at −90°, dimethyl fumarate (50 mg, 0.347 mmol) in CH_2Cl_2 (5 mL) was added dropwise over 30 minutes. The resulting mixture was stirred at −90° for 5 hours. Sodium bisulfite (1.15 g) was then added, and the mixture was concentrated in vacuo. The residue was dissolved in THF

(8 mL) and H_2O (0.5 mL), and the solution was refluxed for 2 hours. Evaporation of the solvents afforded a black residue that was dissolved in CH_2Cl_2 (20 mL), and the solution was dried with Na_2SO_4. Flash column chromatography (ethyl acetate:hexanes, 55:45) gave (*S,S*) diethyl tartrate (41 mg, 67%): $[\alpha]_D^{23}$ −16.4° (*c* 1.28, H_2O) of 96% ee based on both the ^1H NMR and ^{19}F NMR of its Mosher's ester derivative.

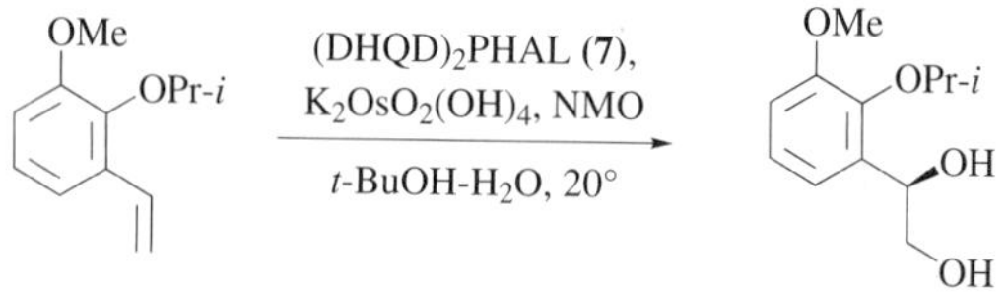

2-(2′-Isopropoxy-3′-methoxyphenyl)-2-hydroxyethanol [Catalytic Asymmetric Dihydroxylation Using NMO as the Secondary Oxidant].[171] To a 50-L three-neck flask was added $K_2OsO_2(OH)_4$ (34.5 g, 0.09 mol), $(DHQD)_2PHAL$ (**7**) (81.6 g, 0.10 mol), NMO (60 wt% in H_2O, 3.0 L, 17.4 mol), *tert*-butyl alcohol (14 L), and H_2O (10 L). The flask was fitted with a mechanical stirrer, and the reaction mixture was stirred until the solution cleared. 2-Isopropoxy-3-methoxystyrene (2.5 kg, 2.2 L, 13.0 mol) was then added at a rate of 5.6 mL/minute using a peristaltic pump, such that the tip of the tubing was immersed in the solution. The temperature of the solution was kept at 20 ± 5° by an external temperature control. Samples were taken at 1-hour intervals and checked to make sure that the alkene content did not exceed 3% and the enantiomeric purity did not fall below 92% by GC and HPLC (chiral column), respectively. At this rate of addition, the alkene content never exceeded 0.7%, and the enantiomeric purity was never below 95%. After addition of the alkene, the resulting orange solution was stirred until the alkene content was less than 0.5%, after which time were added toluene (12 L) and a solution of Na_2SO_3 (1.9 kg) in H_2O (4.7 L). After stirring overnight, the phases were separated and the organic phase was washed with an aqueous solution of Na_2SO_4 (0.8 kg in 5 L of H_2O). The chiral ligand was extracted from the organic phase using H_2SO_4 (0.38 L) in aqueous Na_2SO_4 (1.6 kg in 10 L of H_2O), and the resulting acidic solution was made basic with NaOH and then extracted with toluene (0.70 L). The ligand (61 g) was recovered pure (98% according to HPLC) as a white powder after drying of the solution and evaporation of the solvent. The organic phase remaining after acid extraction was dried with K_2CO_3 (1.0 kg), and the solvent was evaporated under vacuum at 60° to yield 2.5 kg of a light brown oil (94% pure according to GC and an enantiomeric purity of 95%), which crystallized upon standing. ^1H NMR (500 MHz, $CDC1_3$) δ 7.05-6.80 (m, 3H), 5.15 (m, 1H), 4.65 (m, 1H), 3.85 (s, 3H), 3.75 (m, 1H), 3.65 (m, 1H), 3.10 (br s, 1H), 2.50 (br s, 1H), 1.30 (dd, 6H).

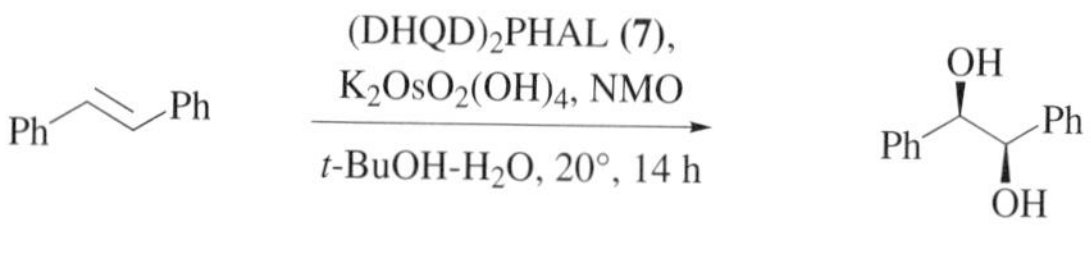

(*R,R*)-(+)-1,2-Diphenyl-l,2-ethanediol [Solid to Solid Asymmetric Dihydroxylation with NMO].[172] A 5-L round-bottomed flask, equipped with a large magnetic stir bar, was charged with $(DHQD)_2PHAL$ (**7**) (10.89 g, 0.25 mol%), *trans*-stilbene (1 kg, 5.6 mol), NMO (1.4 L of 60% aqueous, NMO, Aldrich), and *tert*-butyl alcohol (2.24 L). The flask was placed in a water bath (initially at about 20°, and no further efforts were made to control the temperature). Potassium osmate(VI) dihydrate (4.12 g, 0.2 mol%) was added under stirring. (*Caution: substitution of OsO_4 results in a substantial exotherm which reduces the yield and enantiomeric purity of the product.*) The reaction mixture was then stirred until all the stilbene was consumed as monitored by TLC (about 14 hours). The reaction was quenched by the addition of 4,5-dihydroxy-1,3-benzenedisulfonic acid disodium salt monohydrate (Tiron, Aldrich, 10 g) followed by stirring at room temperature for 3 hours. The resulting mixture was poured into H_2O (3 L) and stirred for another 3 hours. The crude product was collected by filtration, washed with H_2O until colorless, and dried under vacuum, yielding (*R*,*R*)-1,2-diphenyl-1,2-ethanediol as a colorless powder (910 g, 76%, 99% ee). To recover the ligand, the largely aqueous filtrate was stirred with ethyl acetate (2 L) for 1 hour. The resulting organic phase was separated and then extracted with 0.5 M H_2SO_4 (2 × 250 mL). To these combined acidic extracts was added CH_2Cl_2 (500 mL), and this mixture was stirred while sodium carbonate was added until the pH of the aqueous phase was *ca.* 10 or 11. The organic phase was washed with H_2O, dried over $MgSO_4$, and concentrated to give $(DHQD)_2PHAL$ (**7**) (10 g, 95% recovery). An additional 130 g of (*R*,*R*)-1,2-diphenyl-1,2-ethanediol was obtained from the ethyl acetate phase after evaporation of the solvent (11% yield, 99% ee), for an overall yield of 87% (1.04 kg). The enantiomeric excess was determined by HPLC analysis of the bis(MTPA) ester of the diol using a Chiralcel® OD column: mp 148-150°; $[\alpha]_D^{25}$ +95° (*c* 0.87, EtOH); ^1H NMR (400 MHz, $CDCl_3$) δ 7.22 (m, 6H), 7.11 (m, 4H), 4.67 (s, 2H), 3.00 (br s, 2H).

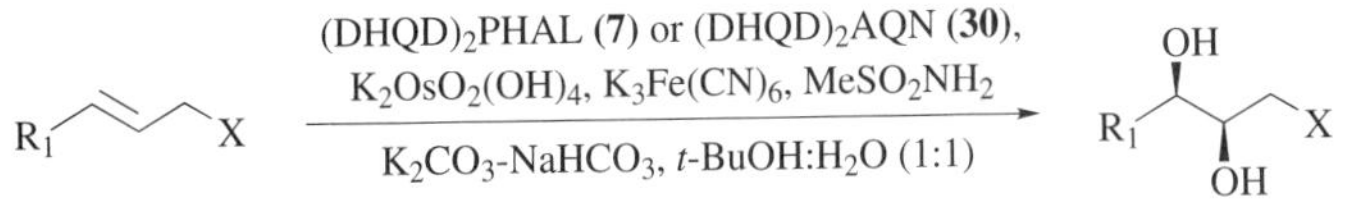

Buffered Asymmetric Dihydroxylation Protocol.[154] To a well-stirred solution of $(DHQD)_2PHAL$ (**7**) (8 mg, 1 mol%), $K_2OsO_2(OH)_4$ (1.8 mg, 0.5 mol%), $K_3Fe(CN)_6$ (988 mg, 3 mmol), K_2CO_3 (415 mg, 3 mmol), $NaHCO_3$ (252 mg, 3 mmol), and $CH_3SO_2NH_2$ (95 mg, 1 mmol) in *tert*-butyl alcohol:water (1 : 1, 10 mL) at 0° was added the appropriate allylic halide (1 mmol). After the reaction was finished (TLC), 1.0 g of $Na_2S_2O_4$ was added and stirring was continued for 30 minutes. The layers were separated, and the aqueous layer was extracted with ethyl acetate (30 mL). The combined organic layers were washed with 1 N KOH (5% aqueous), HCl, and brine, and then dried over $MgSO_4$ and concentrated. The crude halo diol was purified by flash chromatography on silica gel.

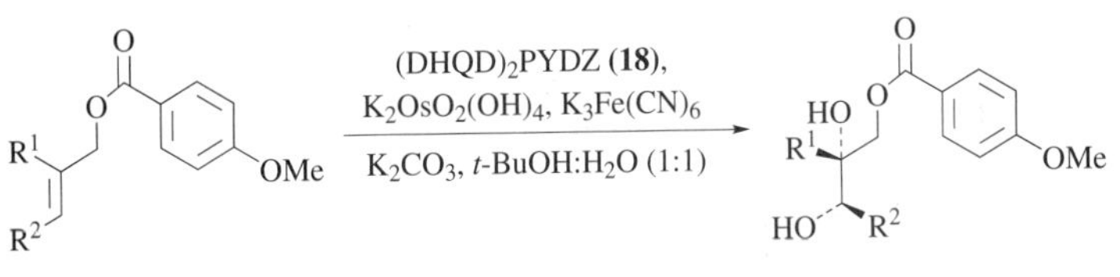

General Procedure for the Asymmetric Dihydroxylation of Allylic 4-Methoxybenzoates.[61] A mixture of K_2CO_3 (3.00 equiv), $K_3Fe(CN)_6$ (3.00 equiv),

$K_2OsO_2(OH)_4$ (0.005 equiv), and $(DHQD)_2PYDZ$ (0.01 equiv) in *tert*-butyl alcohol:water (1 : 1) was cooled to 0°. The resulting suspension was treated with the corresponding alkene (1 equiv, 0.08 M alkene concentration with respect to total reaction volume), stirred until the reaction was complete (TLC), and was quenched by addition of Na_2SO_3 (12 equiv). The resulting mixture was stirred for 5 minutes, warmed to room temperature over 5 minutes, and was partitioned between ethyl acetate and minimal H_2O. The organic extract was washed twice with brine, dried with Na_2SO_4, and concentrated in vacuo. The residue was filtered through a silica gel plug eluting with ethyl acetate, and the filtrate was concentrated in vacuo to afford the product.

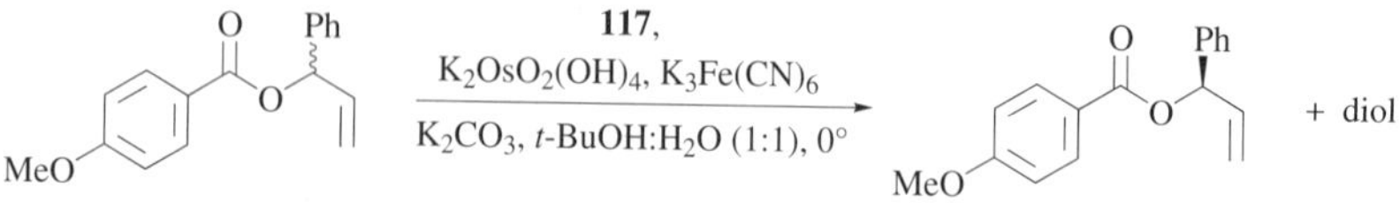

(*R*)-(−)-1-Phenyl-2-propen-1-yl 4-Methoxybenzoate [Kinetic Resolution].[63] A mixture of $K_3Fe(CN)_6$ (0.24 g, 0.73 mmol), K_2CO_3 (0.10 g, 0.73 mmol), DHQD-PYDZ-(*S*)-anthryl ligand (1.6 mg, 0.0024 mmol), (+/−)-1-phenyl-2-propen-1-yl 4-methoxybenzoate (0.065 g, 0.24 mmol), and dibutyl phthalate (0.05 mL, added as an internal standard) in *tert*-butyl alcohol:water (1 : 1, 3 mL) was stirred for 20 minutes at 0°. Approximately 0.025 mL of this mixture was quenched with saturated aqueous Na_2SO_3 (0.05 mL) and extracted with ethyl acetate (0.1 mL). The organic layer was concentrated (reduced pressure, 23° bath temperature) and analyzed by HPLC (Regis Whelk O1 column at 23°, 5% 2-propanol-hexane, 1 mL/minute flow rate; λ235 nm: retention times (*R*) 9.9 minutes, (*S*) 16.6 minutes, dibutyl phthalate 11.9 minutes). The reaction was initiated by addition of $K_2OsO_2(OH)_7$ (0.45 mg, 0.0012 mmol) to the reaction mixture, and aliquots were taken and analyzed using the above procedure every 5-10 minutes. After all of one enantiomer had reacted, the mixture was quenched with 2 mL of saturated aqueous Na_2SO_3, and was extracted with ethyl acetate (3 x). The combined organic layers were washed with brine (2 × 10 mL), dried over Na_2SO_4, filtered, and concentrated in vacuo. Flash chromatography [ethyl acetate:hexane (1 : 1) to elute alkene, followed by ethyl acetate to elute diols] gave recovered (*R*)-(–)-1-phenyl-2-propen-1-yl 4-methoxybenzoate (0.026) g: $[\alpha]_D^{23}$ −21.7° (*c* 0.35, EtOH), and diol (0.036 g, 50%).

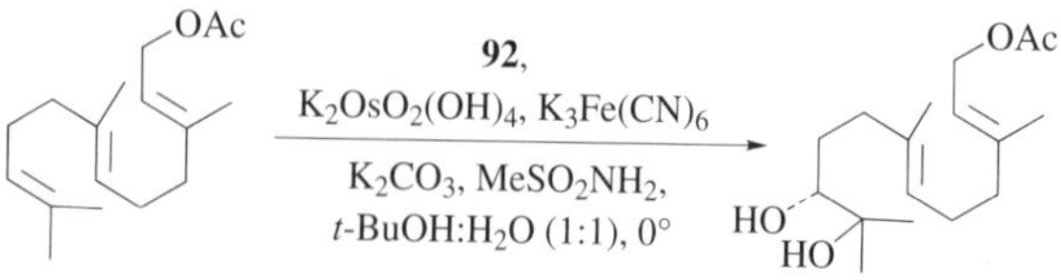

(10*R*)-10,11-Dihydroxy-l0,11-dihydrofarnesyl Acetate [Asymmetric Dihydroxylation of Non-Conjugated Polyalkenes].[62,142] A mixture of 1,4-bis[*O*-6′-(4-heptyl)hydrocupreidyl]naphthopyridazine (**92**) (46 mg, 0.040 mmol), $K_2OsO_2(OH)_4$ (7.4 mg, 0.0020 mmol), $K_3Fe(CN)_6$ (1.98 g, 12 mmol), K_2CO_3 (1.07 g, 12 mmol), $CH_3SO_2NH_2$ (0.38 g, 4.0 mmol), 2,6-*E,E*-farnesyl acetate (1.06 g, 4.0 mmol), and *tert*-butyl alcohol:water (1 : 1, 40 mL) was stirred at 0° for 19 hours. The mixture was treated with saturated Na_2SO_3 solution (15 mL) and then aqueous $Na_2S_2O_3$ solution (15 mL) at 0°, and was warmed to room temperature over 45 minutes. The resulting mixture was extracted with ethyl acetate (4x20 mL), and the combined extracts were washed with 3 M aqueous KOH, brine, dried over Na_2SO_4, and concentrated in vacuo. The residue was separated by silica gel chromatography (1:1 ethyl acetate:hexane) to give unreacted farnesyl acetate (195 mg, 20%), 6,7-dihydroxy-6,7-dihydrofarnesyl acetate (6 mg, 0.5%), and the title compound (754 mg, 80%; 64% uncorrected for recovered farnesyl acetate; 96% ee): $[\alpha]_D^{23}$ +20° (*c* 0.72, MeOH). Elution with ethyl acetate:ethanol (5:1) afforded 6,7,10,11-tetrahydroxy-6,7,10,11-tetrahydrofamesyl acetate (184 mg, 14%). The ligand (35 mg, 76%) was recovered by eluting the column with $CHCl_3$+MeOH+NH_4OH, 20:1:0.1.

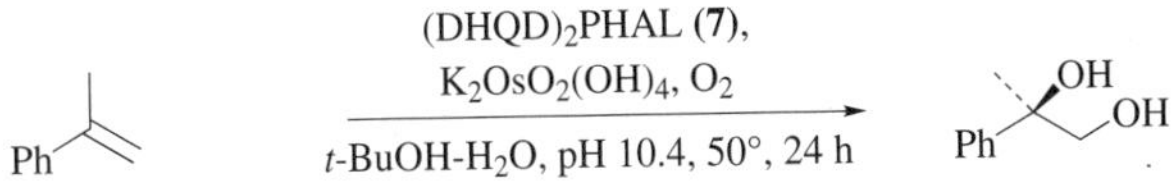

2-Phenyl-1,2-propanediol [Asymmetric Dihydroxylation Under Atmospheric Oxygen Pressure].[180] In a 100-mL Schlenk tube, $K_2OsO_2(OH)_4$ (3.7 mg, 0.01 mmol) and $(DHQD)_2PHAL$ (**7**) (23.4 mg, 0.03 mmol) were dissolved in a mixture of *tert*-butyl alcohol (10 mL) and an aqueous buffer solution (25 mL, pH 10.4). The Schlenk tube was then purged with O_2, and the biphasic mixture was warmed to 50° in an oil bath. Then α-methylstyrene (260 μL, 2 mmol) was added in one portion by a syringe and the tube was connected to a graduated gas buret filled with O_2. The reaction mixture was stirred vigorously with a magnetic stirring bar, and the O_2 uptake was observed to follow the reaction. After 24 hours, 22 mL (*ca.* 1 mmol) of O_2 had been consumed. A small amount of Na_2SO_3 was added, and the mixture was cooled to room temperature while stirring. The mixture was then extracted with ethyl acetate (2 × 20 mL). The combined organic layers were dried over $MgSO_4$, concentrated under reduced pressure, and the crude diol was purified by column chromatography (hexane:ethyl acetate 2:1) to give 2-phenyl-1,2-propanediol (257 mg, 93%, 88% ee) as a colorless solid: ^1H NMR ($CDCl_3$) δ 7.23-7.41 (m, 5H), 3.74 (d, J = 11.1 Hz, 1H), 3.58 (d, J = 11.1 Hz, 1H), 2.39 (br s, 2H), 1.50 (s, 3H); HPLC (diol): (R,R)-Whelk-O1, 2% EtOH in hexane, flow rate 1.0 mL/minute, t_R = 14.4 minutes (*S*), 16.7 minutes (*R*).

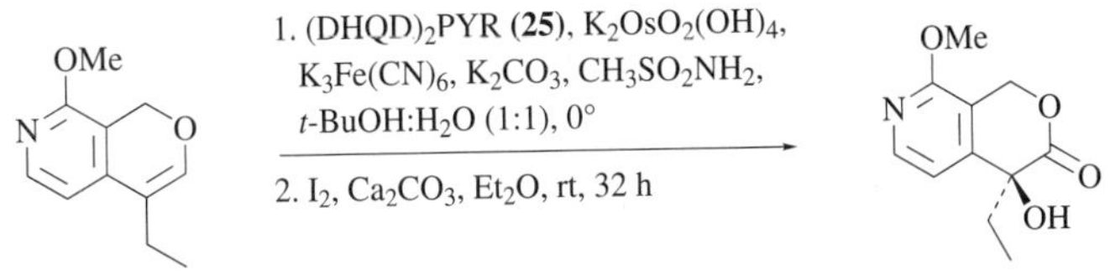

81

(4*S*)-4-Ethyl-4-hydroxy-8-methoxy-1,4-dihydropyrano[3,4-*c*]pyridin-3-one [Asymmetric Dihydroxylation of Enol Ethers].[134,135] A 500-mL flask was charged with $(DHQD)_2PYR$ (**25**) (230 mg, 0.261 mmol) as a solution in *tert*-butyl alcohol (126 mL), followed by successive addition of deionized H_2O (131 mL), $K_3Fe(CN)_6$ (25.8 g, 78.4 mmol), anhydrous K_2CO_3 (10.8g, 78.4 mmol), potassium osmate (VI) dihydrate (19.2 mg, 0.0522 mmol), and $CH_3SO_2NH_2$ (2.49 g, 26.1 mmol). After stirring for 3 minutes at room temperature, the nearly homogeneous mixture was cooled to 0°, at which temperature it turned into a red slurry. A solution of **81** (5.00 g, 26.1 mmol) in *tert*-butyl alcohol (5 mL) was added, and the thick slurry was vigorously stirred for 48 hours at 0°. The mixture was treated with iodine (33.2 g, 131 mmol) and calcium carbonate (13.1 g, 131 mmol). After being warmed to room temperature, the brown mixture was stirred for 48 hours. The mixture was subsequently cooled to 0° and sodium sulfite (25 g) was added in three portions over 5 minutes. The stirring was continued for 30 minutes. The now green mixture was filtered by suction through a pad of Celite® 545 and the pad washed with ethyl acetate:methanol (90:10, 250 mL). The filtrate was washed with brine (2 x), dried over anhydrous Na_2SO_4, and concentrated in vacuo to give crude product (5.79 g) as a yellow oil. The oil had a higher than 90% chemical purity by HPLC: Chiral HPLC (Chiralcel® OD, ethanol:hexane 2:99, λ 276 nm, 1.0 mL/minute) showed an S/R ratio of 34:1 (t_R for S-enantiomer 14.92 minutes, R-enantiomer 16.26 minutes). An analytical sample was obtained by flash chromatography on silica gel eluting with 3% MeOH-CH_2Cl_2; ^1H NMR (400 MHz, $CDCl_3$) δ 8.21 (d, $J = 5$ Hz, 1H), 7.16 (d, $J = 5$ Hz, 1H), 5.58 (d, $J = 16$ Hz, 1H), 5.27 (d, $J = 16$ Hz, 1H), 3:99 (s, 3H), 3.62 (s, 1H), 1.80 (m, 2H), 0.96 (t, $J = 7$ Hz, 3H).

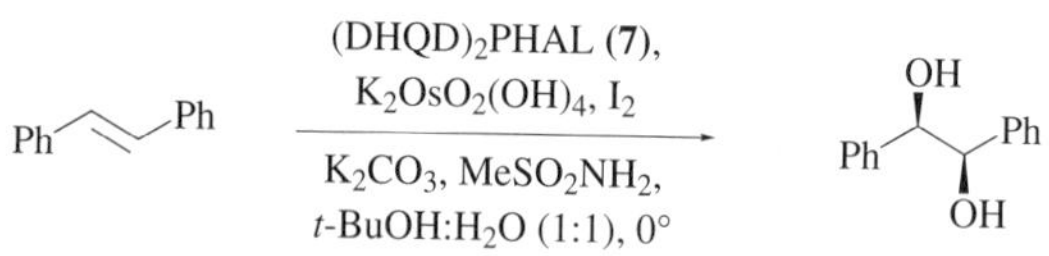

(1*R*,2*R*)-1,2-Diphenyl-1,2-ethanediol (Asymmetric Dihydroxylation with Iodine as Secondary Oxidant).[175] To a stirred solution of *tert*-butyl alcohol:H_2O (1:1, 100 mL), K_2CO_3 (30 mmol), iodine (15 mmol), $CH_3SO_2NH_2$ (10 mmol), $K_2OsO_2(OH)_4$ (0.02 mmol), and $(DHQD)_2PHAL$ (**7**) (0.1 mmol) was added *trans*-stilbene (10 mmol) in one portion at 0°. The mixture was stirred vigorously for more than 30 hours (monitored by TLC), and was then quenched with solid sodium sulfite (10 g). The two phases were separated, and the aqueous phase was extracted with ethyl acetate (3 × 50 mL). The combined organic phases were washed with 2 M NaOH and dried over $MgSO_4$. Concentration and flash chromatography afforded the title compound in 98.4% yield: $[\alpha]_D^{25}$ +91° ($CDCl_3$). The enantiomeric excess of the R,R-diol was determined by HPLC analysis [Chiralcel® OB-H column (Daicel), 10% *i*-PrOH in hexane, 0.5 mL/min] to be 99.5%.

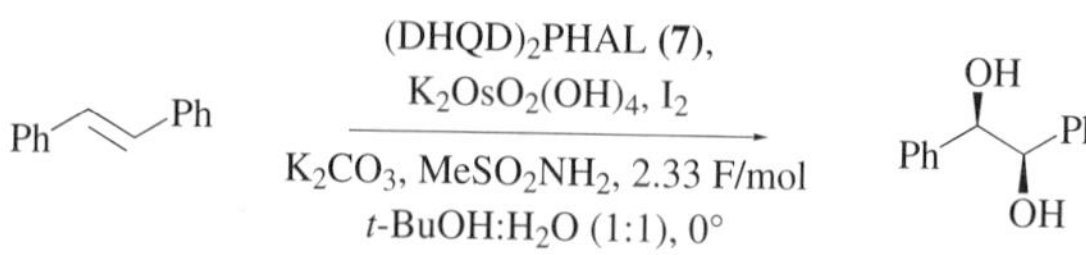

(1*R*,2*R*)-1,2-Diphenyl-1,2-ethanediol [Electrochemical Asymmetric Dihydroxylation].[175] A mixture of *tert*-butyl alcohol:H_2O (1:1, 50 mL), $CH_3SO_2NH_2$ (5 mmol), K_2CO_3 (15 mmol), iodine (2.0 mmol), $K_2OsO_2(OH)_4$ (0.01 mmol), $(DHQD)_2PHAL$ (**7**) (0.05 mmol), and *trans*-stilbene (5 mmol) was electrolyzed in an undivided flow cell (Micro-Flow Cell) equipped with two platinum electrodes

(3 × 4 cm). After passage of 2.33 F/mol of electricity at room temperature (progress was monitored by TLC), solid sodium sulfite (5 g) was added, and the solution was stirred for 40 minutes. The two phases were separated, and the aqueous phase was extracted with ethyl acetate (3 × 25 mL). The combined organic phases were washed with 2 M NaOH and dried over $MgSO_4$. Concentration and flash chromatography afforded the title compound in 82.4% yield, $[\alpha]_D^{25}$ +91° ($CDCl_3$). The enantiomeric excess was determined by HPLC analysis [Chiralcel® OB-H column (Daicel), 10% *i*-PrOH in hexane, 0.5 mL/min] to be 99.5%.

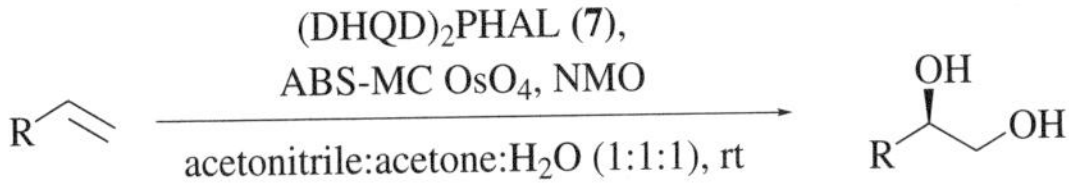

Asymmetric Dihydroxylation using ABS-MC OsO_4.[170] ABS-MC OsO_4 (76.7 mg, 5 mol%), $(DHQD)_2PHAL$ **(7)** (21.5 mg, 5 mol%), and NMO (0.72 mmol) were combined in water:acetone:acetonitrile (1 : 1 : 1, 3.5 mL) at room temperature. To this mixture was added the alkene (0.55 mmol) slowly over a period of about 24 hours. Methanol (10 mL) was added, and the mixture was stirred for 10 minutes. ABS-MC OsO_4 was separated by filtration. After washing with methanol, the combined filtrates were concentrated under reduced pressure. The chiral ligand was recovered from the aqueous layer after acidification with 1N HCl. The concentrated organic layer was purified by chromatography on silica gel to afford the cis-diol product.

Preparation of ABS-MC OsO_4.[170] ABS polymer [Stylac© 200 (Asahi Chemical), 1.00 g] was dissolved in THF (20 mL) at 70-80°, and to this solution was added OsO_4 (0.200 g). The mixture was stirred for 1 hour at this temperature and then slowly cooled to 0°. Coacervates (phase separation) were found to envelop the core dispersed in the medium, and methanol (30 mL) was added to harden the capsule walls. After 8 hours, the capsules were washed with methanol several times and dried at room temperature for 24 hours to afford ABS-MC OsO_4 (1.18 g). Based on the recovered weight, 0.180 g of OsO_4 was microencapsulated according to the above procedure. Unencapsulated OsO_4 was recovered from the washings.

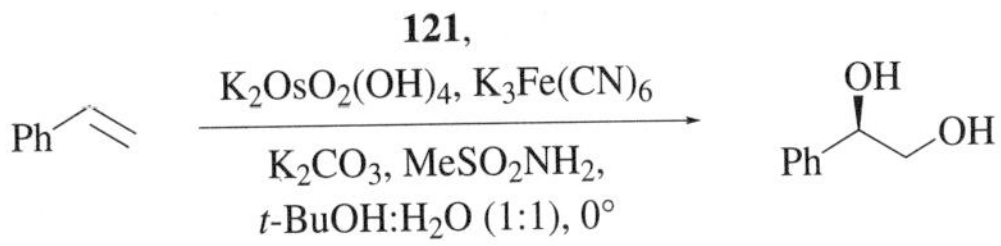

(*R*)-1-Phenyl-1,2-ethanediol [Asymmetric Dihydroxylation Using a Polymer-Bound Ligand].[162] To a solution of $K_3Fe(CN)_6$ (3 equiv) and K_2CO_3 (3 equiv) in *tert*-BuOH:H_2O (1 : 1, 6 mL), was added OsO_4 (0.0125 equiv) and polymeric ligand **121** (0.25 equiv). After 30 min, styrene (2 mmol) was added, and the heterogeneous mixture was stirred at 0° for 24 hours. After addition of H_2O (3.0 mL), the mixture was centrifuged, and the centrifugate was extracted with CH_2Cl_2. The solvent was removed in vacuo, and the residue was purified by silica gel chromatography, affording the title compound in 86% yield and 91% ee.

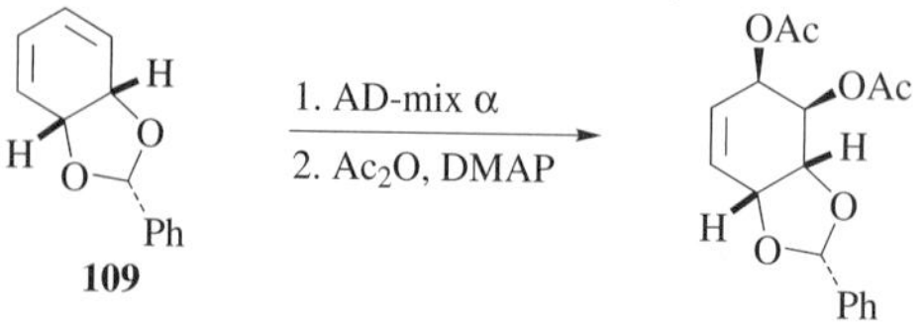

109

(2*R*,3a*R*,4*R*,5*R*,7a*R*)-2-Phenyl-3a,4,5,7a-tetrahydrobenzo[1,3]dioxole 4,5-Diacetate [Asymmetric Dihydroxylation Using AD-mix].[147,148] A mixture of **109** (261 mg, 1.31 mmol), AD-mix α (1.8 g), and $CH_3SO_2NH_2$ (125 mg, 1.31 mmol) in *tert*-butyl alcohol:H_2O (1 : 1, 13.2 mL) was stirred at 0° for 38 hours. To the mixture were added Na_2SO_3 (1.97 g) and KOH (720 mg), and stirring was continued for 1 hour at room temperature. The mixture was diluted with ethyl acetate, and the organic layer was washed with brine, dried over $MgSO_4$, and evaporated under reduced pressure to leave the crude diol (383 mg) as a colorless oil. This material was then stirred with acetic anhydride (0.37 mL, 3.93 mmol), triethylamine (0.64 mL, 4.59 mmol), and 4-(*N,N*-dimethylamino)pyridine (DMAP) (16 mg, 0.13 mmol) in CH_2Cl_2 (10 mL) at room temperature for 10 minutes. The mixture was washed with brine, dried over $MgSO_4$, evaporated under reduced pressure, and chromatographed on a silica gel column (20 g, elution with 1 : 2 v/v ether-hexane) to give unreacted **109** (25 mg, 10%) and the title compound (336 mg, 81%) as a colorless oil: $[\alpha]_D^{23}$ −256° (*c* 1.32, $CHCl_3$); 87% ee by chiral HPLC: (Chiralcel® OD, elution with 1 : 9 v/v *i*-PrOH:hexane); FTIR (neat) 1745 cm^{-1}; ^1H NMR (500 MHz, $CDCl_3$) δ 7.47-7.45 (m, 2H), 7.39-7.36 (m, 3H), 6.17 (dd, J = 9.76, 3.66 Hz, 1H), 6.03 (ddd, J = 9.77, 5.49, 1.22 Hz, 1H), 5.90 (s, 1H), 5.55 (dd, J = 5.49, 3.66 Hz, 1H), 5.27 (dd, J = 8.54, 3.66 Hz, 1H), 4.81 (ddd, J = 9.16, 3.67, 1.22 Hz, 1H), 4.57 (dd, J = 8.55, 6.71 Hz, 1H), 2.09 (s, 3H), 2.08 (s, 3H).

TABULAR SURVEY

Tables 1–9 include examples of stoichiometric and catalytic asymmetric dihydroxylation reactions that have appeared in the literature up to the end of 2000. Supplementary Table 10 is a survey of the literature from 2001 through 2004. The tables are arranged in the same order as the text discussion. Entries in the tables are in the order of increasing number of carbon atoms, although some exceptions occur when a single structure covers a series with different R groups. Polymeric substrates are listed at the end of a section under the category C_n. The symbol (—) indicates that no yield was reported.

There are a number of entries where an absolute configuration of the diol is given but for which the authors did not report ee or de values. These entries are faithful to the original literature. For some products, absolute configurations were assigned in the original literature using the mnemonic of Figure 2. Such examples are marked with an asterisk by the product. OsO_4 was used in stoichiometric amounts unless otherwise indicated, or unless a secondary oxidant is listed in the conditions. There are some polymeric ligands (see Ligand Charts) for which no values are given for the number of repeating units; for conditions including these ligands, additional information (mol% of one or more components of these polymers) may be found in the original literature.

Abbreviations used in the charts and tables are as follows:

ABS MC	acrylonitrile-butadiene-polystyrene microencapsulated copolymer
Ac	acetyl
Bn	benzyl
Boc	*tert*-butoxycarbonyl
Bz	benzoyl
Cbz	benzyloxycarbonyl
DHQ	dihydroquininyl
DHQD	dihydroquinidinyl
DMAP	4-dimethylaminopyridine
DMF	dimethylformamide
DME	1,2-dimethoxyethane
DMPM	3,4-dimethoxyphenylmethyl
Ether	diethyl ether
LAH	lithium aluminum hydride
LDH	layered double hydroxide supported (for catalyst)
MOM	methoxymethyl
PMB	*p*-methoxybenzyl
NMM	*N*-methylmorpholine
NMO	*N*-methylmorpholine *N*-oxide
Os EnCat	polyurea microencapsulated osmium tetroxide
PEG	polyethylene glycol
PEM-MC	poly(phenoxyethoxymethylstyrene)co-styrene
Piv	pivaloyl
PMP	*para*-methoxyphenyl
Phth	phthaloyl
SEM	trimethylsilylethoxymethyl
TBDMS	*tert*-butyldimethylsilyl
TBDPS	*tert*-butyldiphenylsilyl
TEAA	tetraethylammonium acetate
TEMPO	2,2,6,6-tetramethyl-1-piperidinyloxy, free radical
T_f	trifluoromethanesulfonyl
THF	tetrahydrofuran
THP	tetrahydropyranyl
TIPS	tri-iso-propylsilyl
TMS	trimethylsilyl
TON	turnover number
Tr	triphenylmethyl (trityl)
Troc	trichloroethoxycarbonyl
Ts	*p*-toluenesulfonyl
XAD	prefix for a series of Amberlite™ resins

CHART 1. LIGANDS USED IN TABLES 1-9

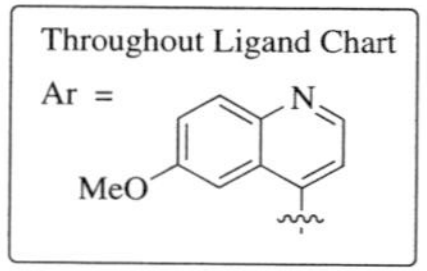

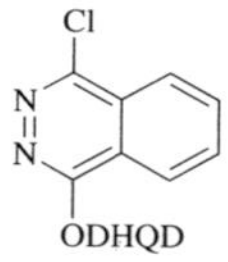

1

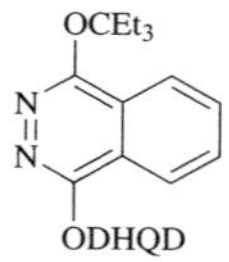

2

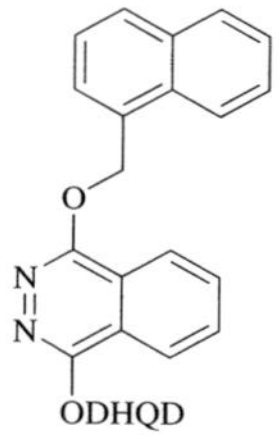

3

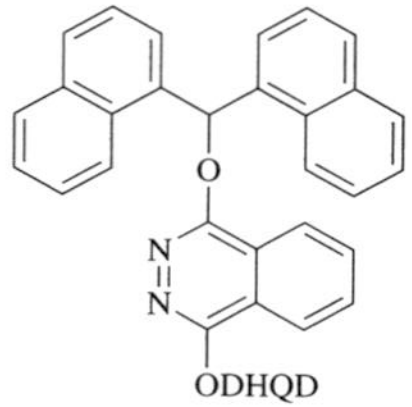

4

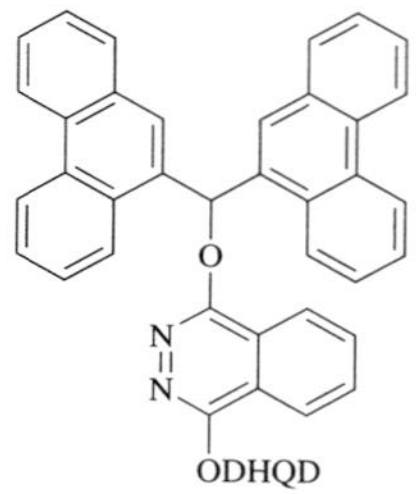

5

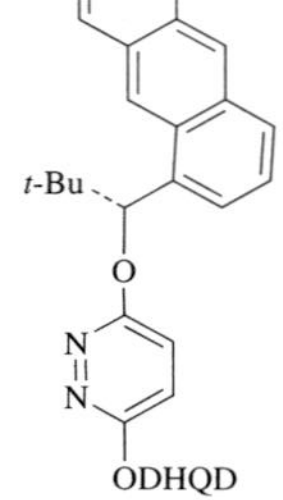

6

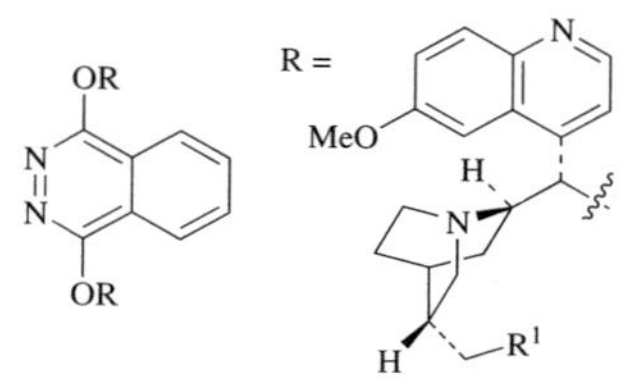

	R^1
7a	Et
7b	Bn
7c	OAc
7d	OCOPr-*i*

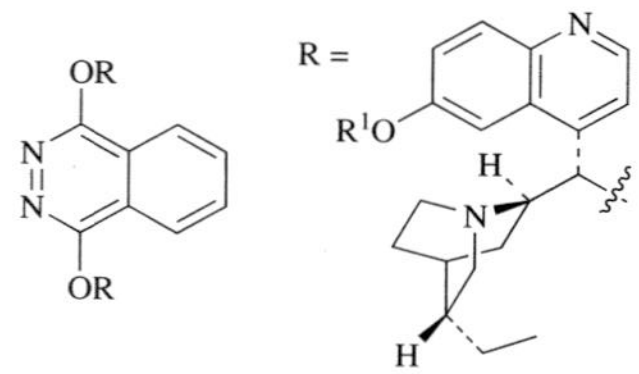

	R^1
8a	Et
8b	*n*-C_5H_{11}
8c	*i*-Pr
8d	*i*-Bu
8e	*i*-C_5H_{11}

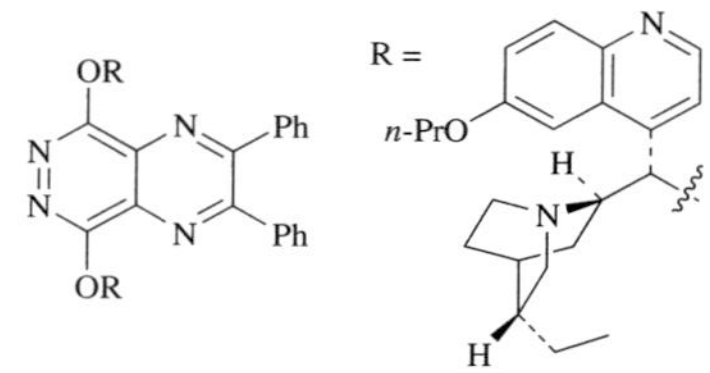

9

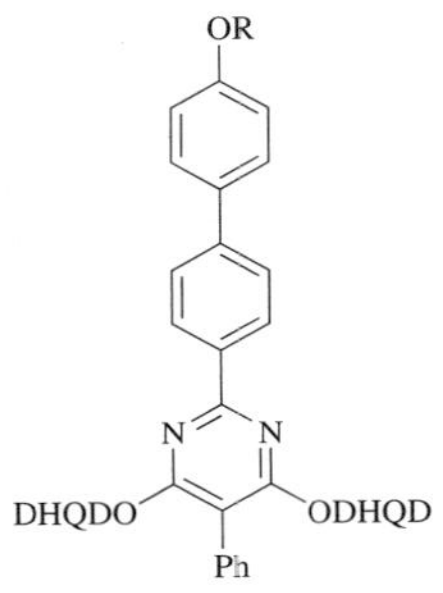

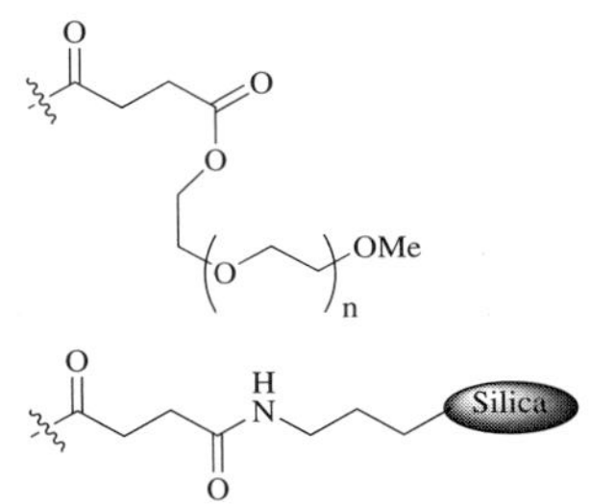

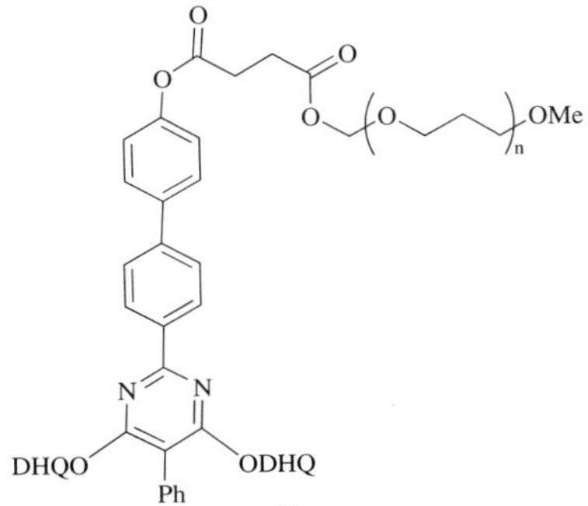

11

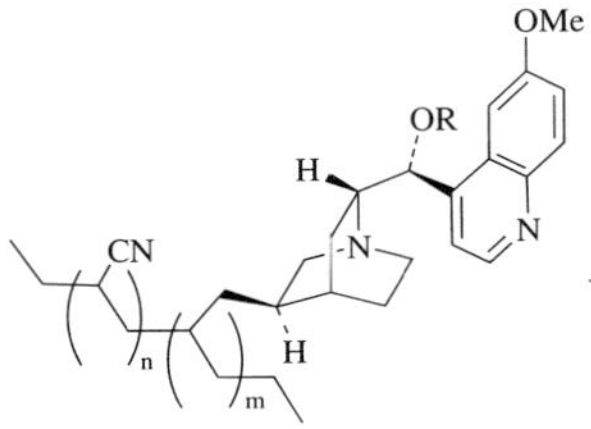

	R
12a	Ac
12b	$C(O)C_6H_3(MeO)_2$-3,5
12c	$C(O)C_6H_4Cl$-*p*

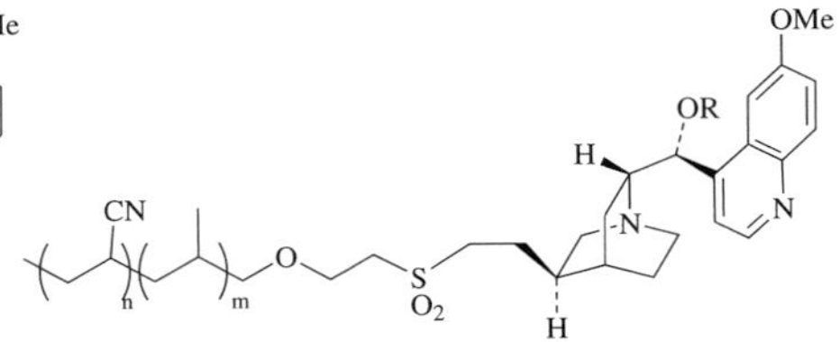

	R
13a	$CONMe_2$
13b	$C(O)C_6H_4Cl$-*p*

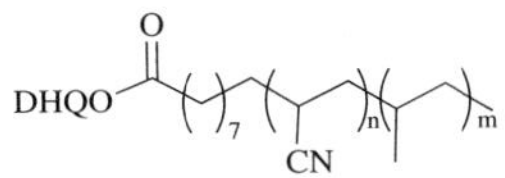

14

CHART 1. LIGANDS USED IN TABLES 1-9 *(Continued)*

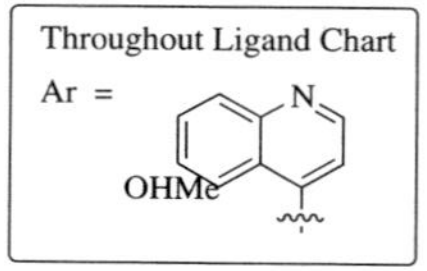

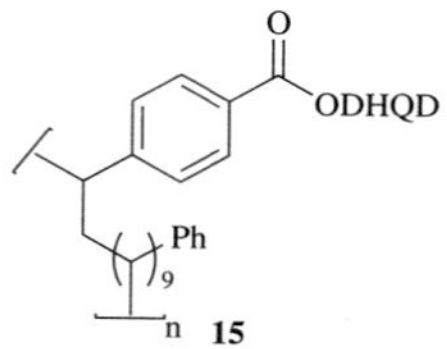

15

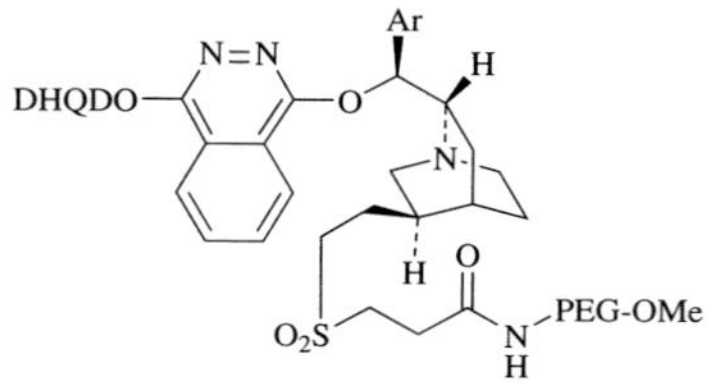

16

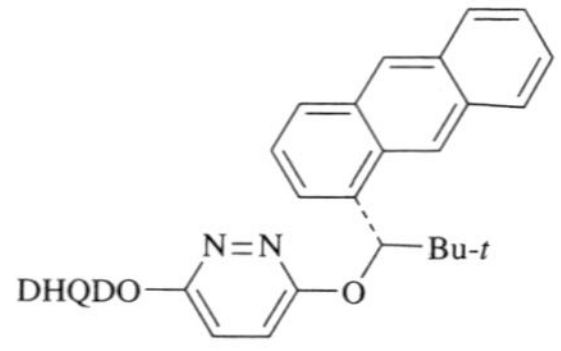

17

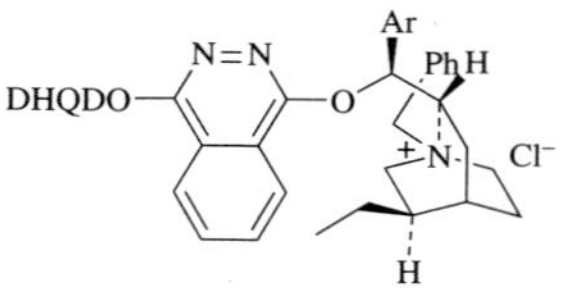

18

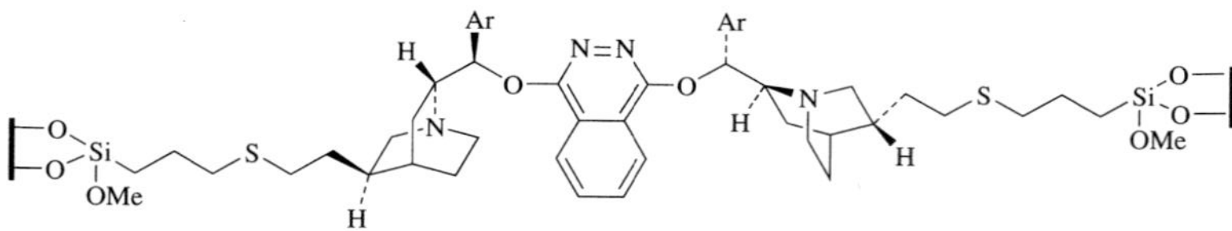

19

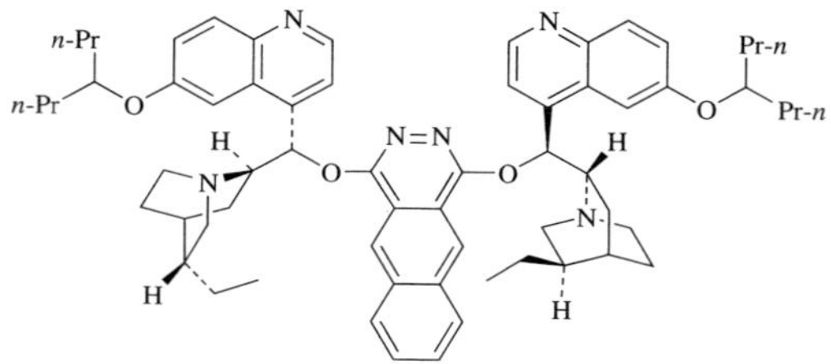

20

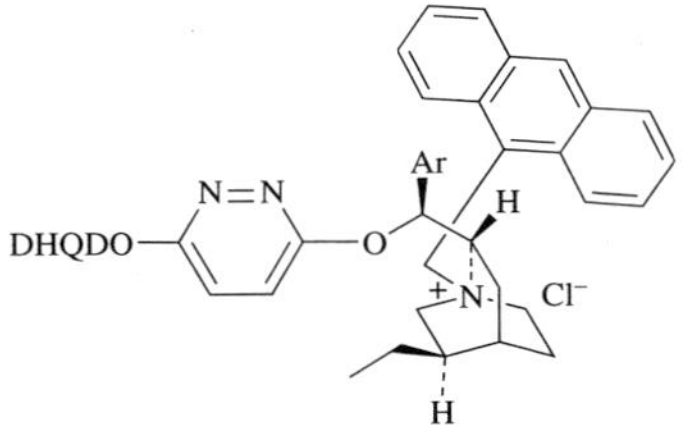

21

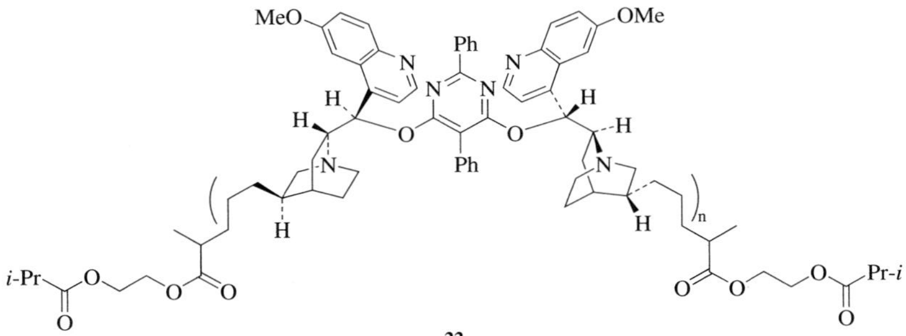

22

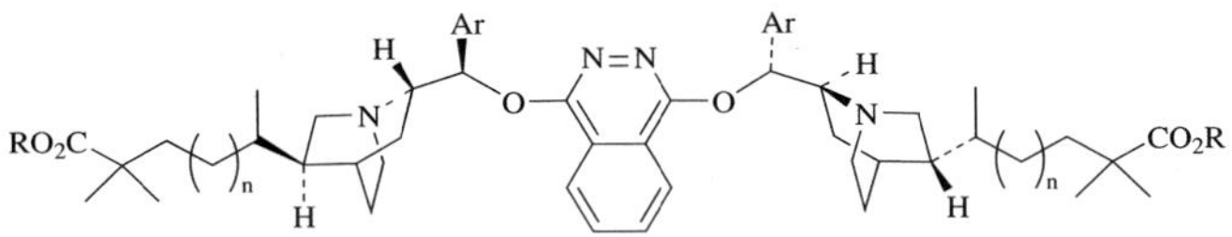

	R
23a	Me
23b	CH_2CH_2OH

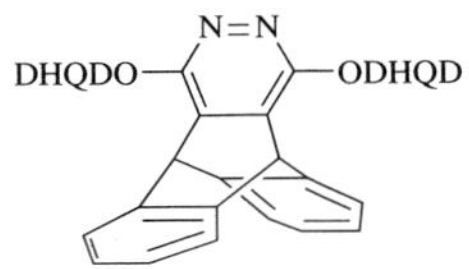

24

CHART 1. LIGANDS USED IN TABLES 1-9 *(Continued)*

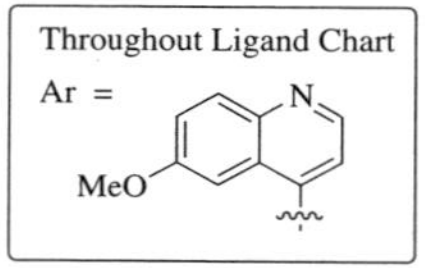

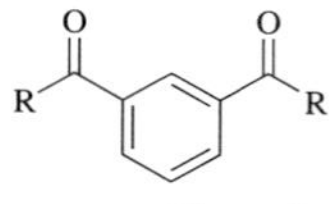

25 R = *O*-9-quininyl

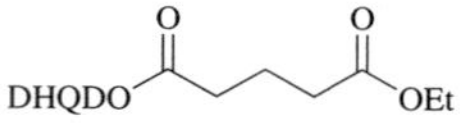

26

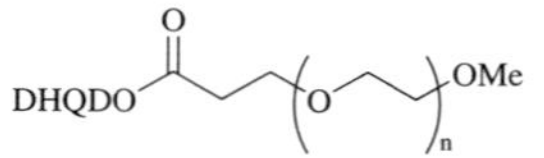

27

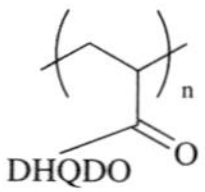

28

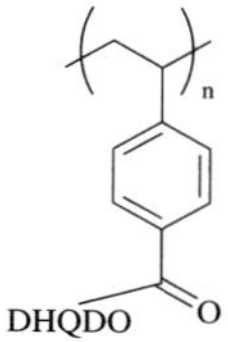

29

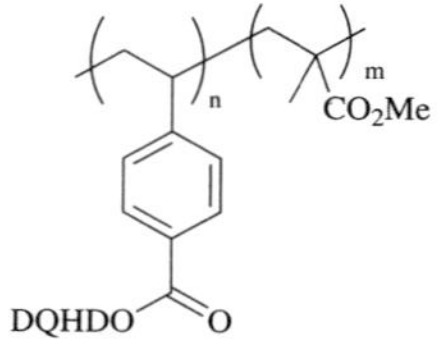

	R
30a	Me
30b	$(CH_2)_2OH$

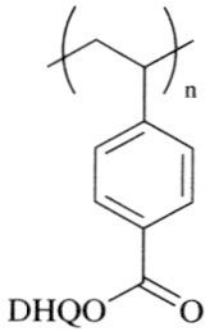

31

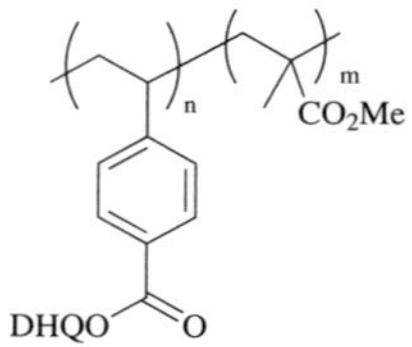

32

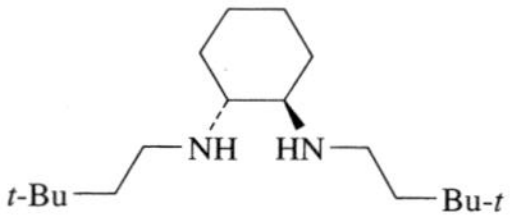

33

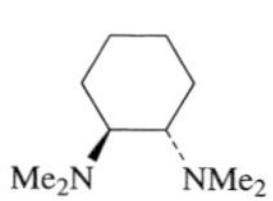

34

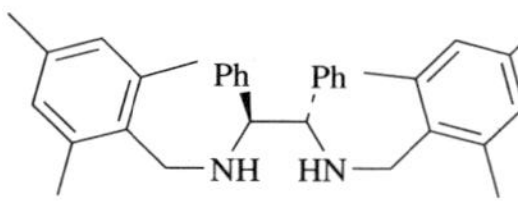

35

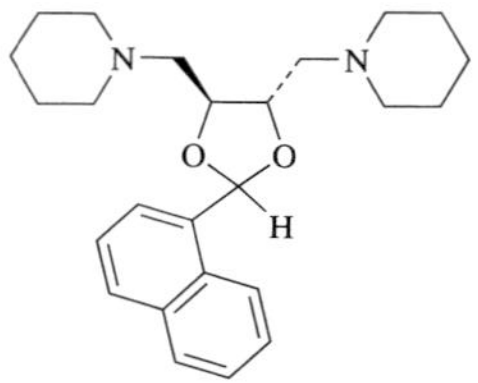

36

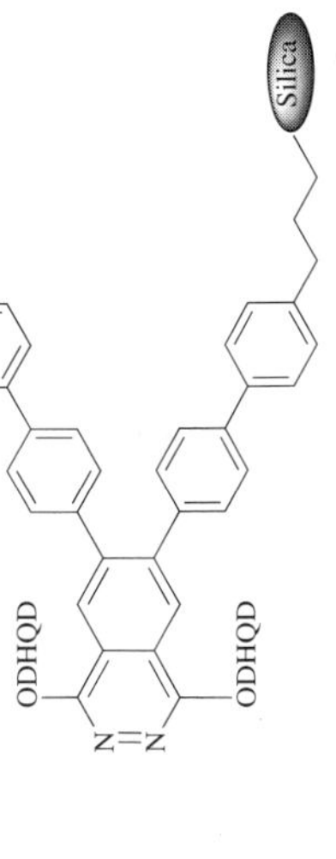

37

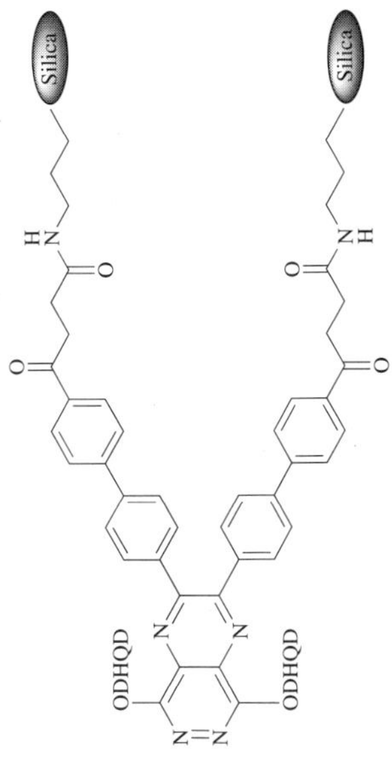

38

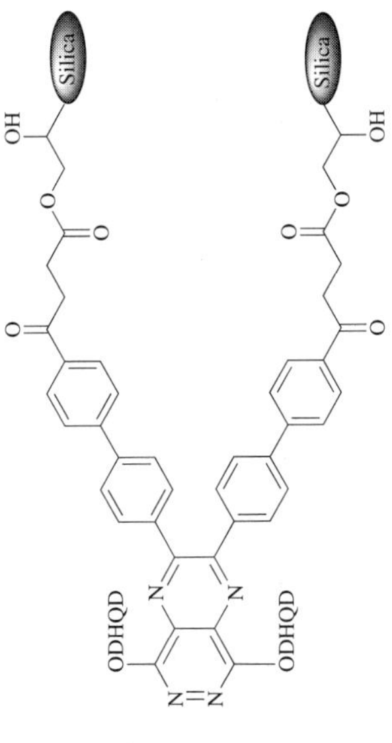

39

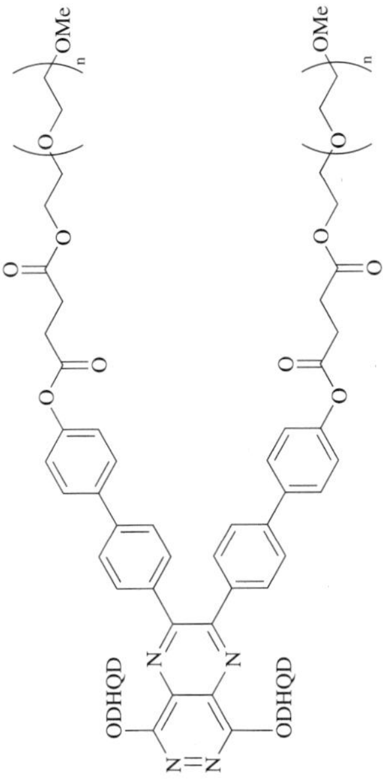

40

CHART 1. LIGANDS USED IN TABLES 1-9 *(Continued)*

Block Co-Polymers

Throughout Ligand Chart

Ar =

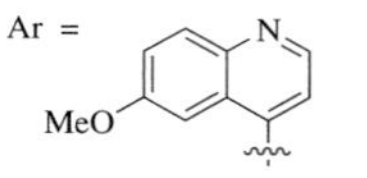

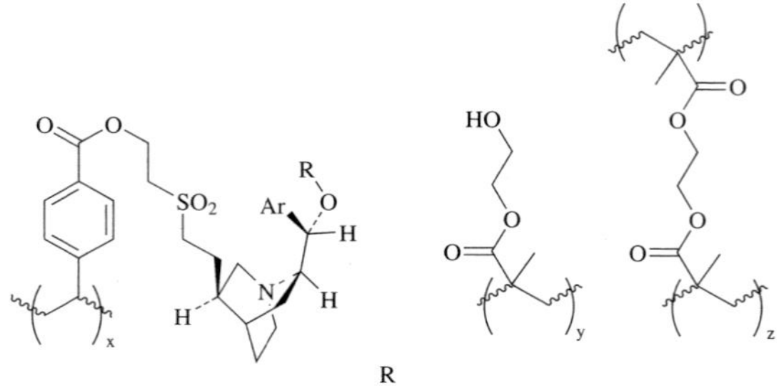

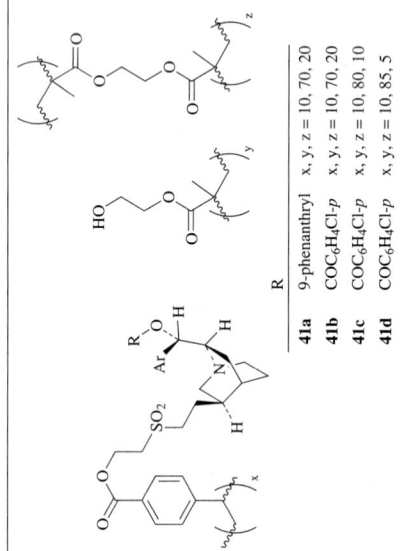

	R	
41a	9-phenanthryl	x, y, z = 10, 70, 20
41b	COC_6H_4Cl-*p*	x, y, z = 10, 70, 20
41c	COC_6H_4Cl-*p*	x, y, z = 10, 80, 10
41d	COC_6H_4Cl-*p*	x, y, z = 10, 85, 5

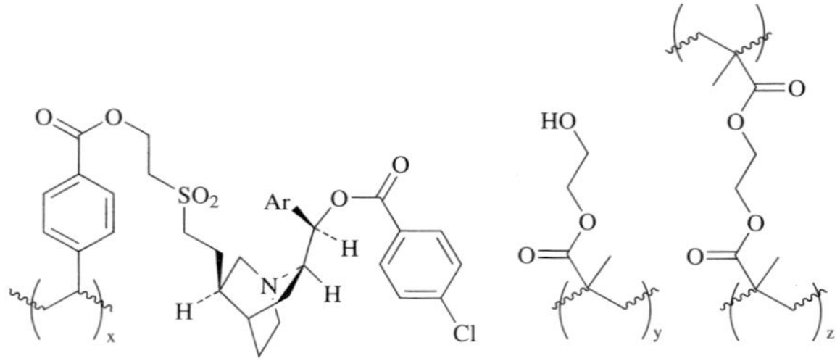

42a x, y, z = not defined

42b x, y, z, = 10, 70, 20

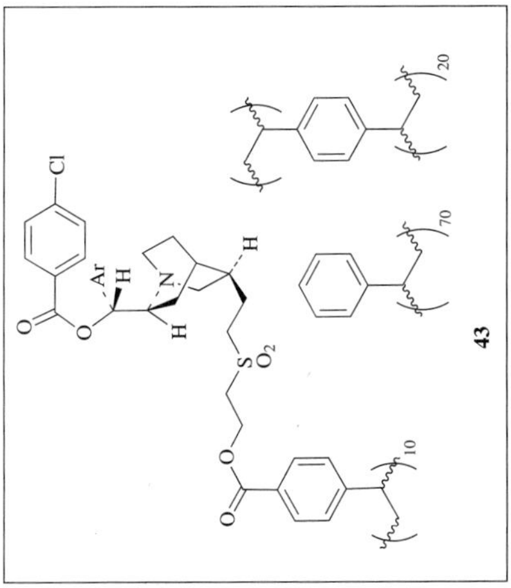

43

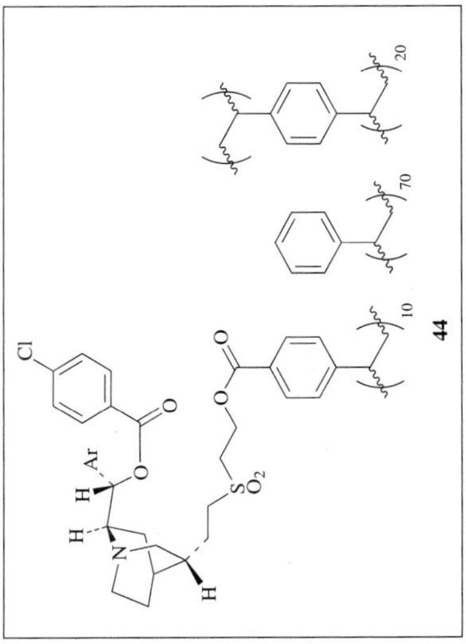

44

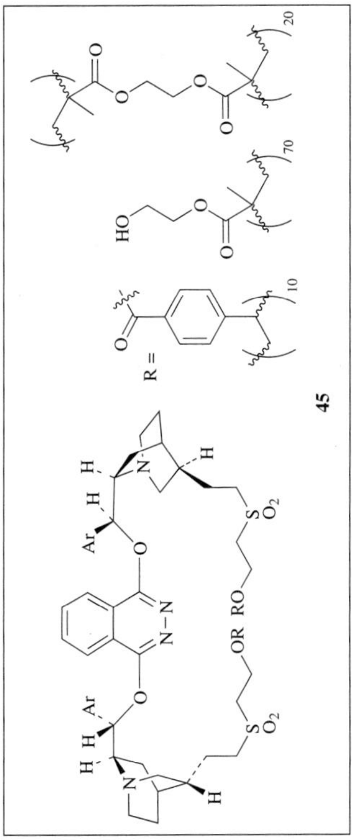

45

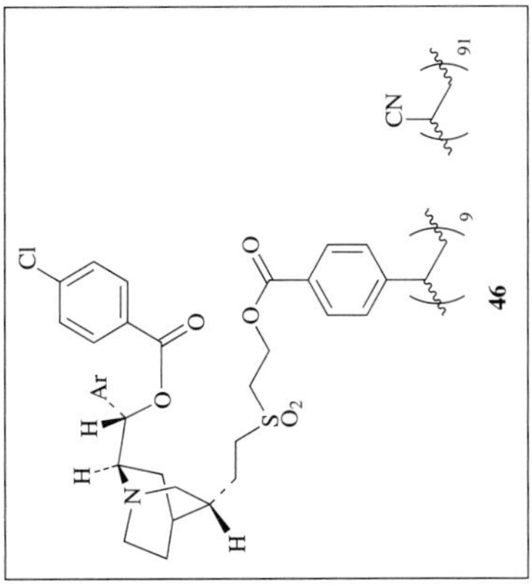

46

CHART 1. LIGANDS USED IN TABLES 1-9 *(Continued)*

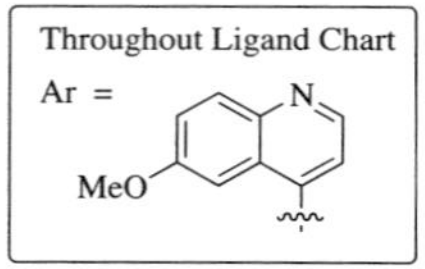

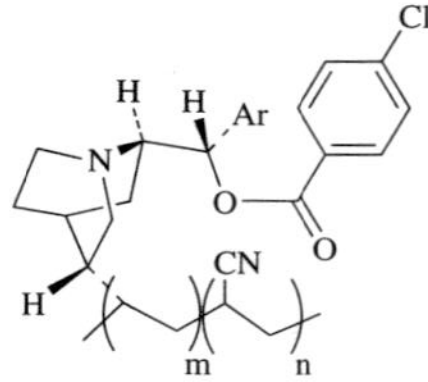

47

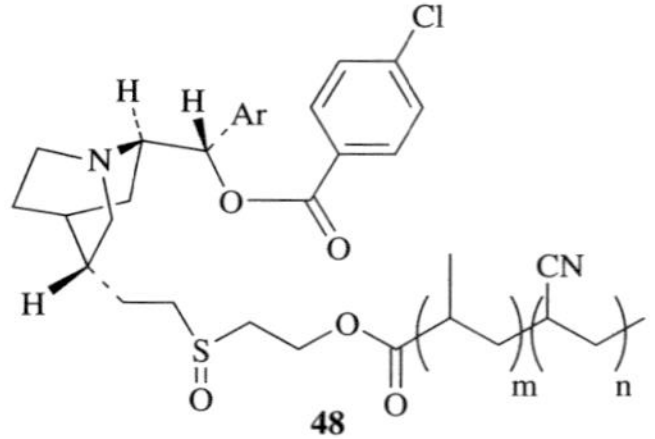

48

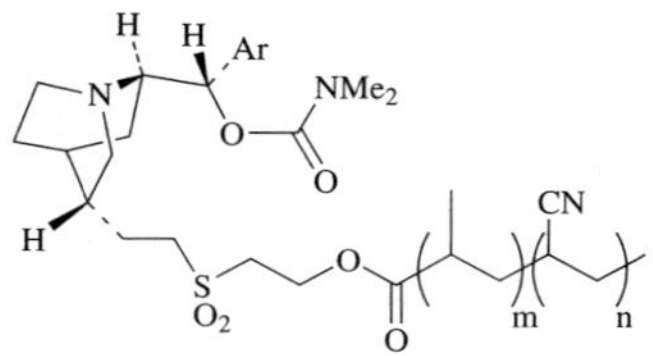

49

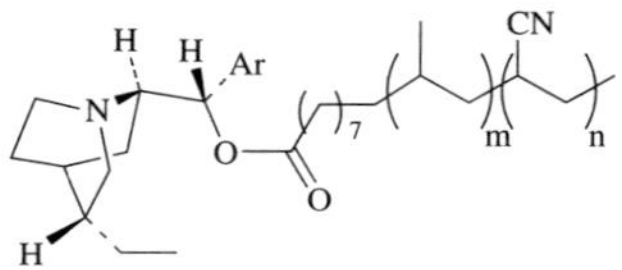

50

CHART 2. LIGANDS AND ADDITIVES USED IN TABLE 10

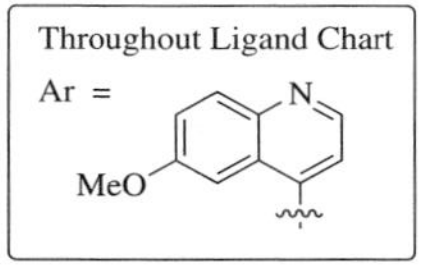

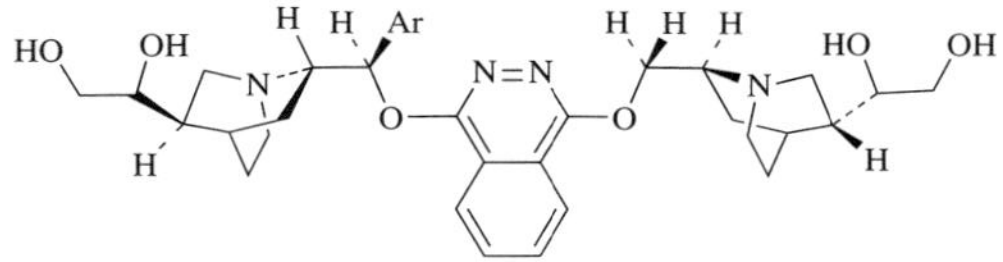

51

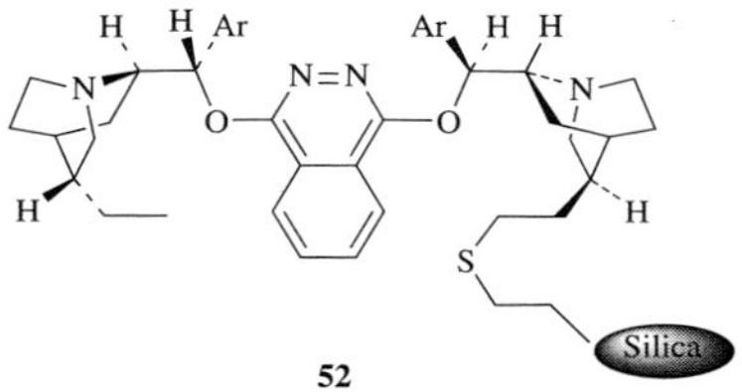

52

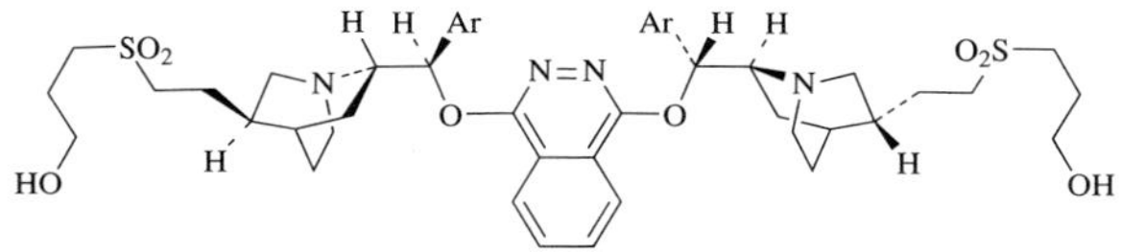

53

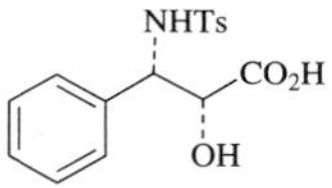

54

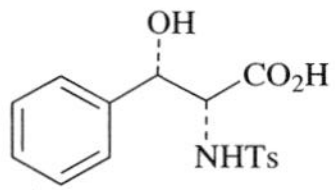

55

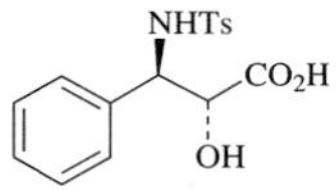

56

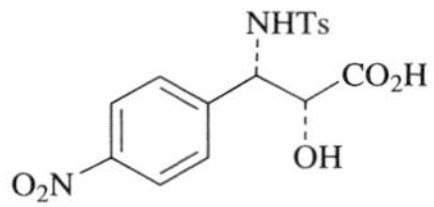

57

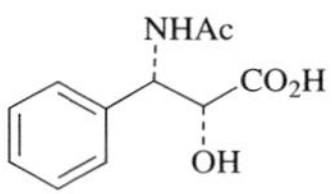

58

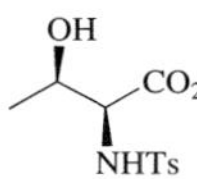

59

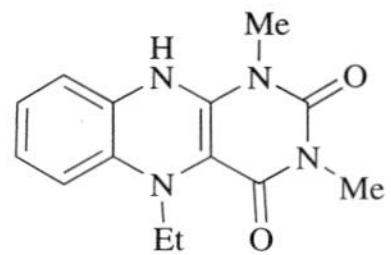

60

CHART 2. LIGANDS AND ADDITIVES USED IN TABLE 10 *(Continued)*

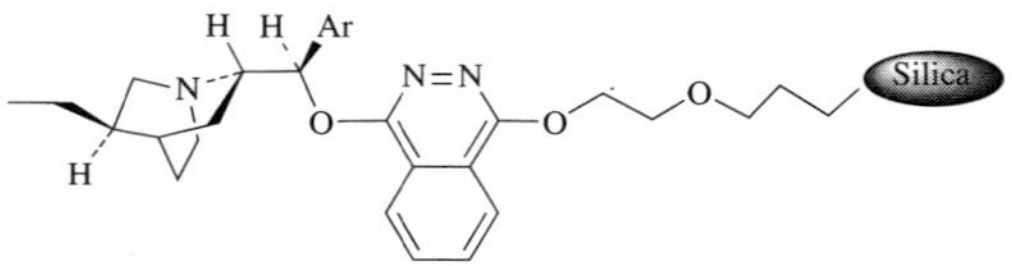

61

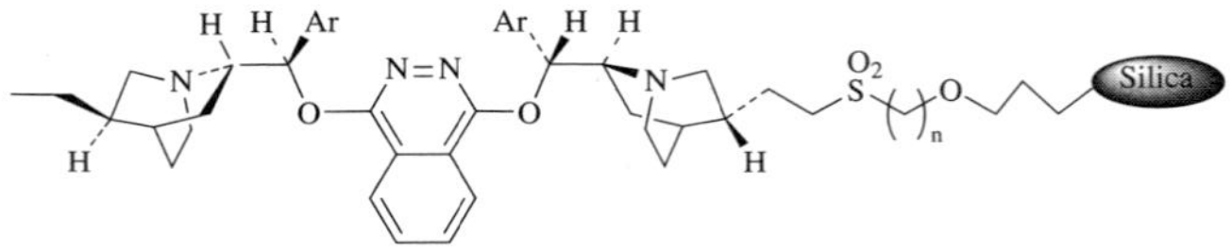

62 n = 2

63 n = 6

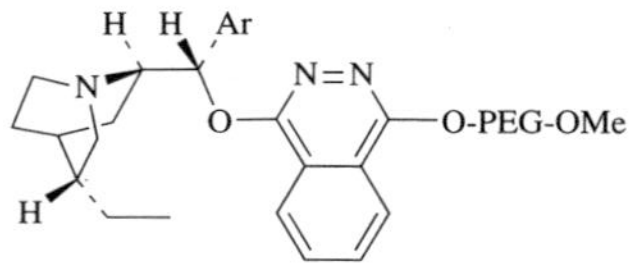

64

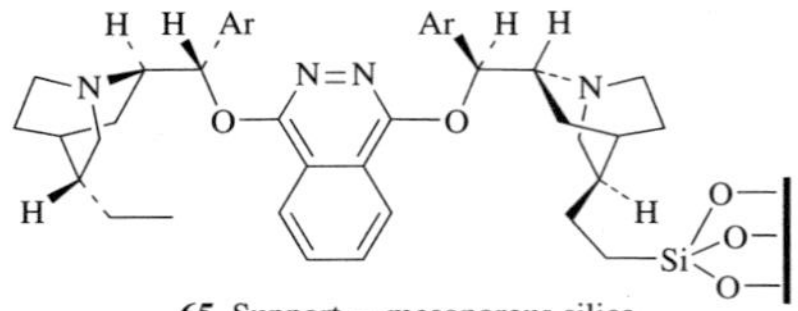

65 Support = mesoporous silica

66 Support = silica gel

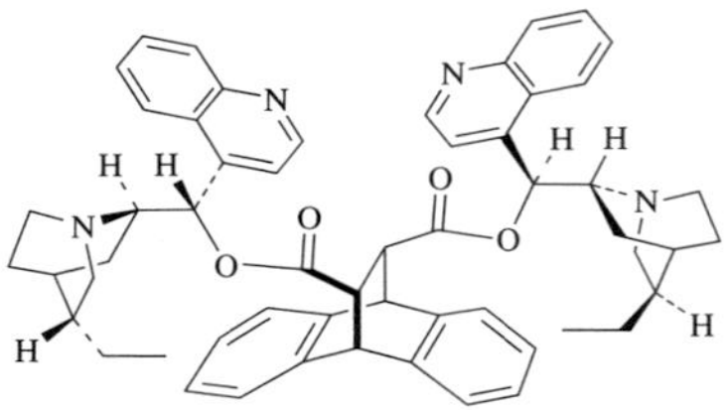

67

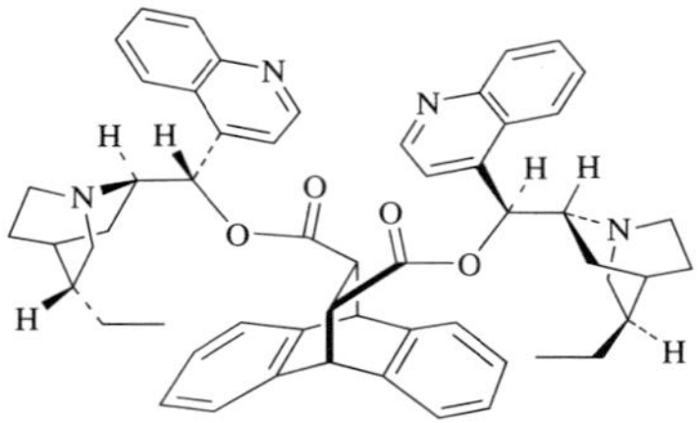

68

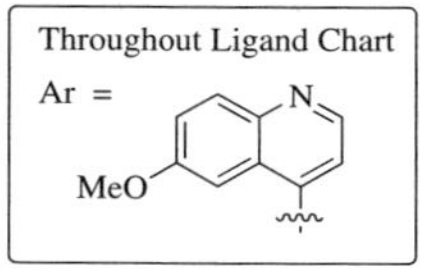

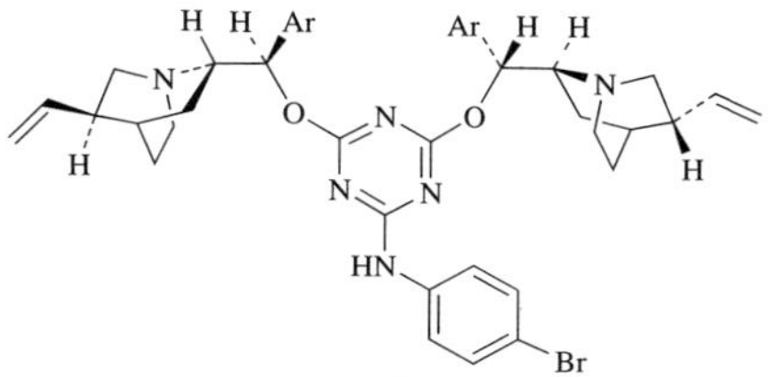

69

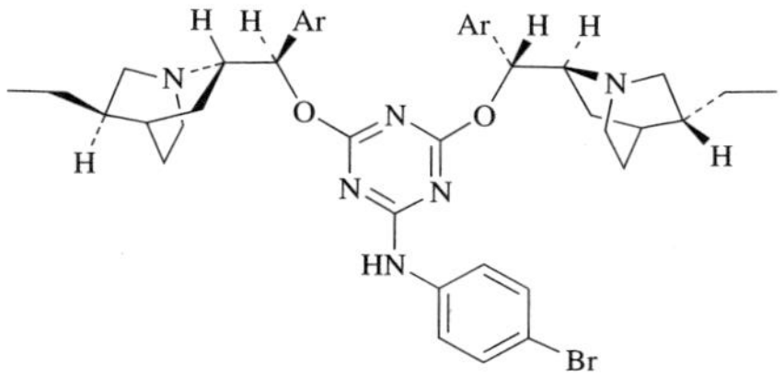

70

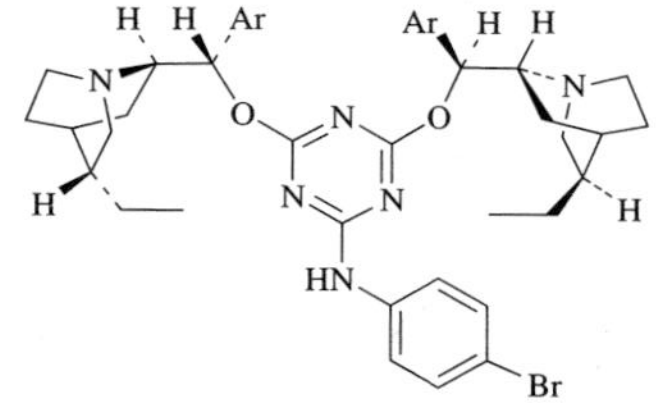

71

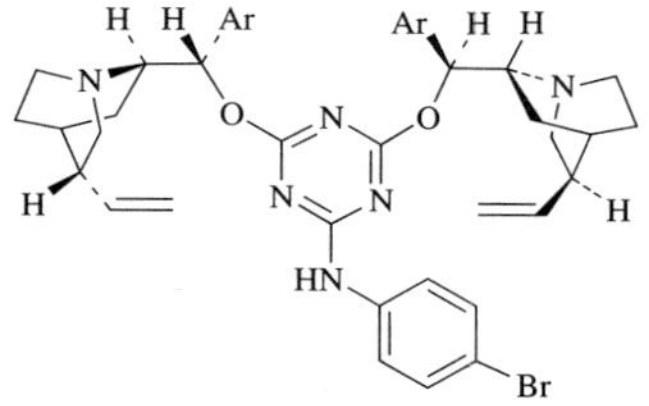

72

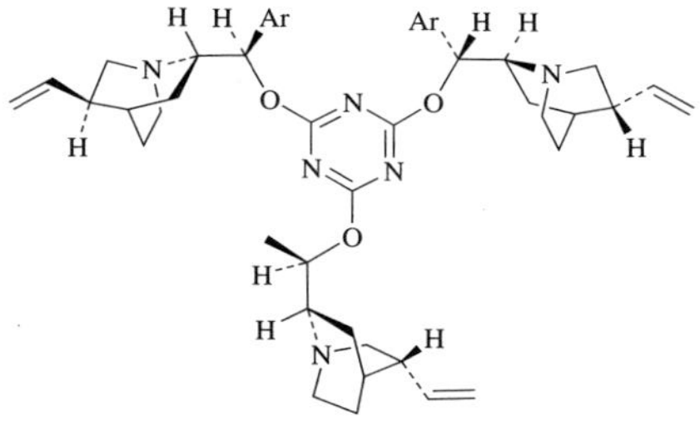

73

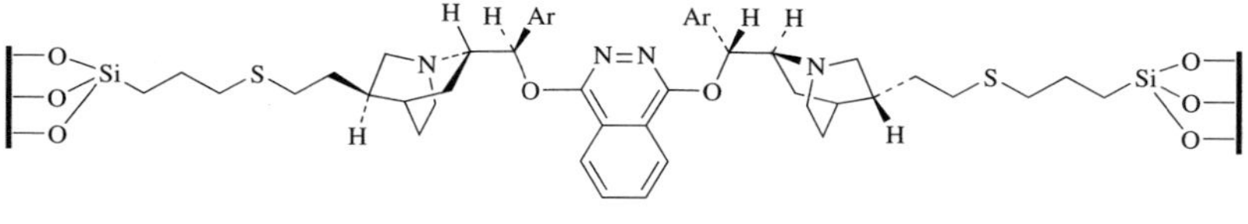

74 (support: MCM-41 silica)

CHART 2. LIGANDS AND ADDITIVES USED IN TABLE 10 *(Continued)*

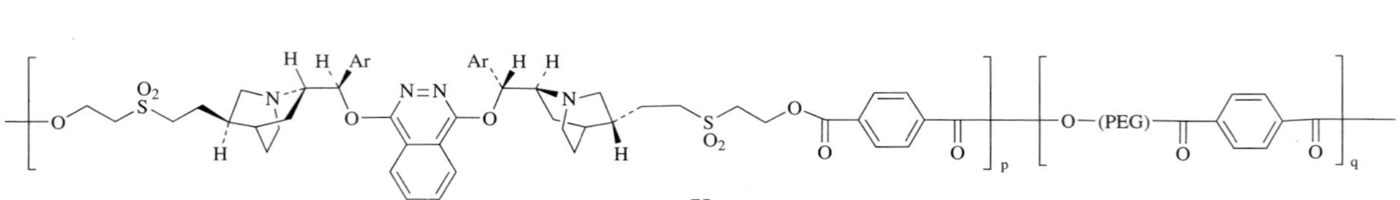

75

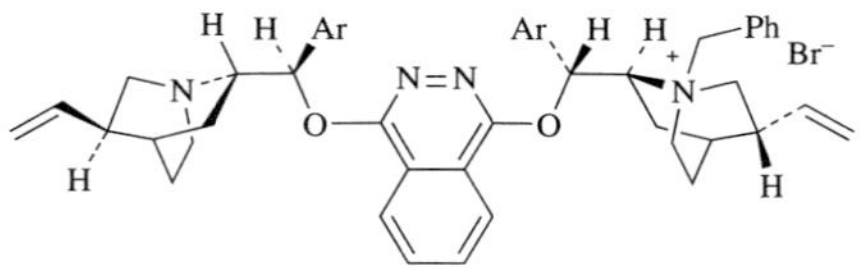

76

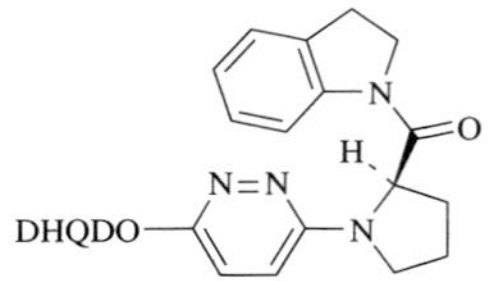

77

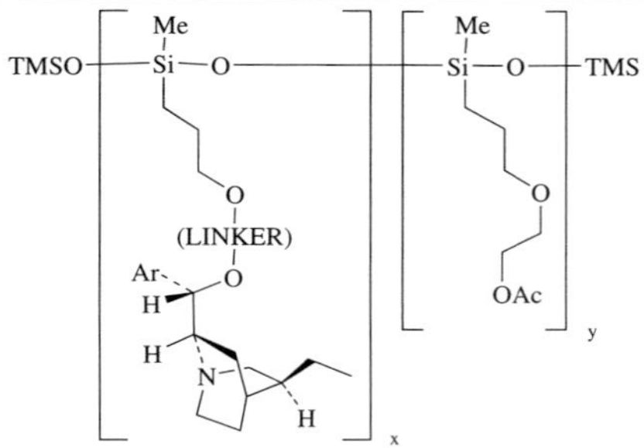

78-82

	LINKER	x:y
78	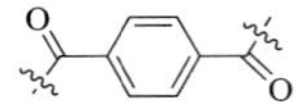	1:0
79		5:7
80	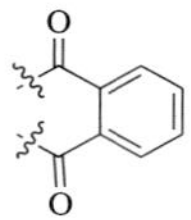	1:0
81		3:2
82	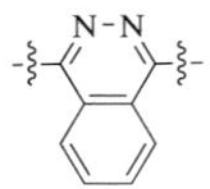	1:0

Throughout Ligand Chart

Ar =

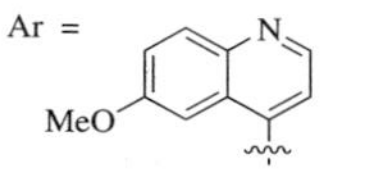

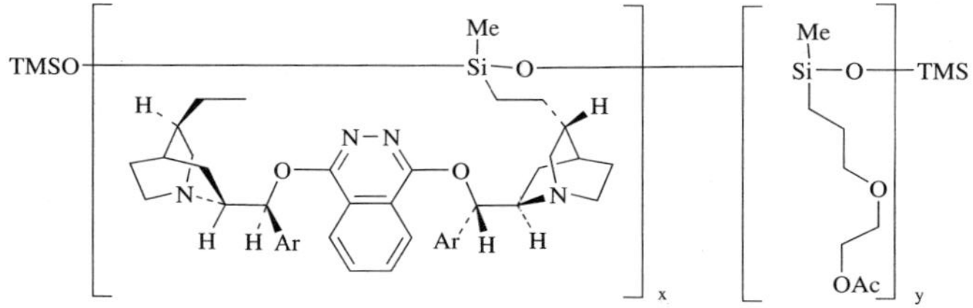

83 x:y = 1:0

84 x:y = 5:7

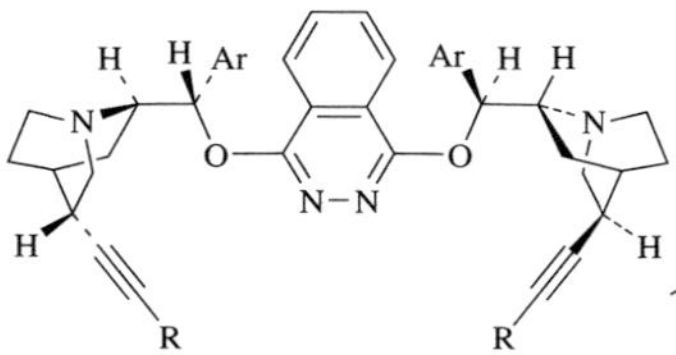

85 R = Ph

86 R = I

87 R = quinolyl

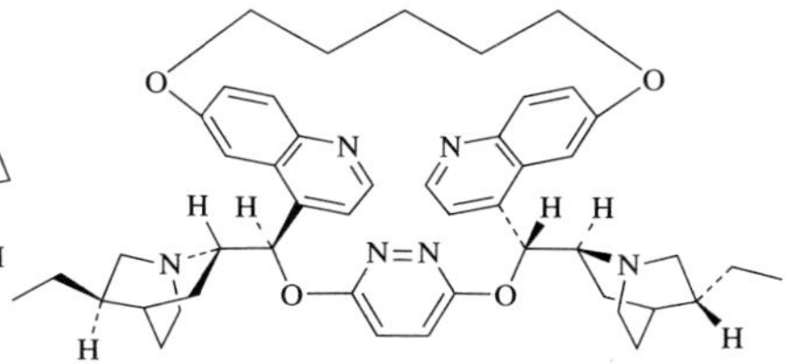

88

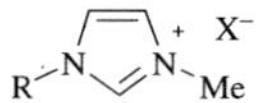

89 R = n-Bu, X = PF_6

90 R = n-Bu, X = BF_4

91 R = n-Bu, X = $N(Tf)_2$

92 R = n-Bu, X = SbF_6

93 R = n-Bu, X = $OCOCF_3$

94 R = n-Oct, X = PF_6

95 R = n-Oct, X = BF_4

96 R = n-$C_{10}H_{21}$, X = PF_6

97 R = n-$C_{10}H_{21}$, X = BF_4

98 R = $MeO(CH_2)_2O(CH_2)_2$, X = PF_6

99 R = $MeO(CH_2)_2O(CH_2)_2$, X = BF_4

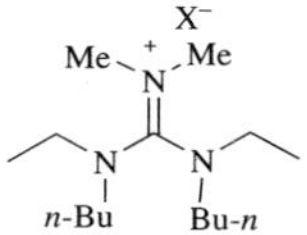

100 X = PF_6

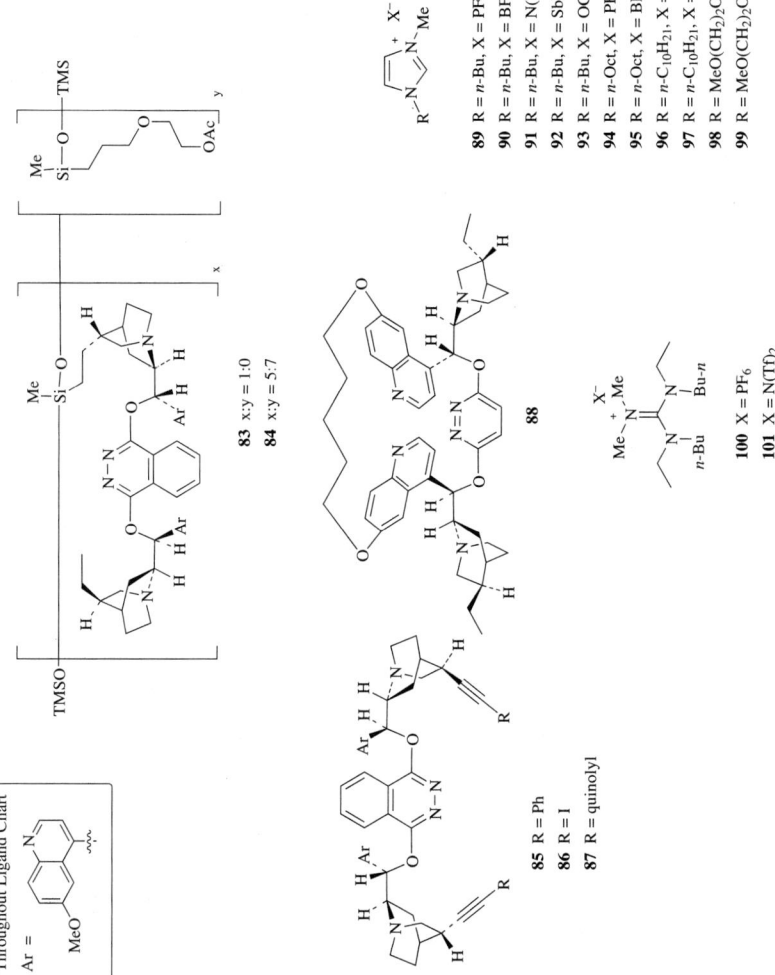

101 X = $N(Tf)_2$

TABLE 1. REACTIONS OF TERMINAL ALKENES

Substrate	Conditions	Product(s) and Yield(s) (%), % ee	Refs.
*See Chart 1 at the beginning of the Tabular Survey for ligand structures that are indicated by **bold** numbers.*			
C_3			
	Ligand, $K_2OsO_2(OH)_4$, $K_3Fe(CN)_6$, K_2CO_3, $NaHCO_3$, t-BuOH:H_2O (1:1)	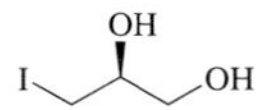	
	Ligand / Temp / Time		
	$(DHQD)_2PHAL$ / 0° / —	(70), 70	154, 157
	$(DHQD)_2AQN$ / 0° / 12 h	(65-90), 83	86
	$(DHQD)_2DPPHAL$ / — / —	(—), 77	68, 157
	$(DHQD)_2PYR$ / 0° / 6-24 h	(—) , 70	157
	$(DHQD)_2DPP$ / 0° / 6-24 h	(—), 68	157
	$(DHQ)_2AQN$, $K_2OsO_2(OH)_4$, $K_3Fe(CN)_6$, K_2CO_3, $NaHCO_3$, t-BuOH:H_2O (1:1), 0°, 12 h	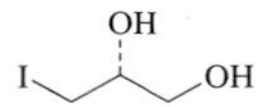 (65-90), 82	86
	Ligand, $K_2OsO_2(OH)_4$, $K_3Fe(CN)_6$, K_2CO_3, $NaHCO_3$, t-BuOH:H_2O (1:1)	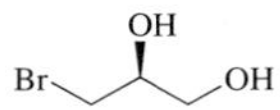	
	Ligand / Temp / Time		
	$(DHQD)_2AQN$ / 0° / 12 h	(65-90), 89	86
	$(DHQD)_2DPP$ / — / —	(—) , 72	86
	$(DHQD)_2DPPHAL$ / — / —	(74), 72	181
	$(DHQD)_2PHAL$ / — / —	(40), 66	181
	$(DHQD)_2PHAL$ / — / —	(74), 66	181
	(DHQD)PHN / — / —	(74), 60	181
	$(DHQ)_2AQN$, $K_2OsO_2(OH)_4$, $K_3Fe(CN)_6$, K_2CO_3, $NaHCO_3$, t-BuOH:H_2O (1:1), 0°, 12 h	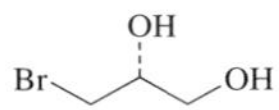 (65-90), 85	86

	22, $K_2OsO_2(OH)_4$, $K_3Fe(CN)_6$, K_2CO_3, *t*-BuOH:H_2O (1:1), 20°, 20 h	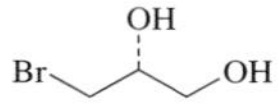	(62), 61	164
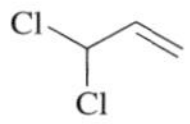	Ligand, $K_2OsO_2(OH)_4$, $K_3Fe(CN)_6$, K_2CO_3, $NaHCO_3$, *t*-BuOH:H_2O (1:1)	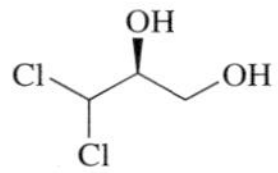		86
	Ligand / Temp / Time			
	$(DHQD)_2AQN$ / 0° / 12 h	(65-90), 72		
	$(DHQD)_2PHAL$ / — / —	(—), 63		
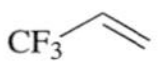	Ligand, $K_2OsO_2(OH)_4$, $K_3Fe(CN)_6$, K_2CO_3, $NaHCO_3$, *t*-BuOH:H_2O (1:1)	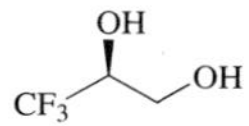		
	Ligand / Temp / Time			
	$(DHQD)_2AQN$ / 0° / 12 h	(65-90), 81		86
	$(DHQD)_2PYR$ / — / —	(75), 64		86, 182
	$(DHQD)_2PHAL$, $K_2OsO_2(OH)_4$, $K_3Fe(CN)_6$, K_2CO_3, *t*-BuOH-H_2O, O_2, 12 h	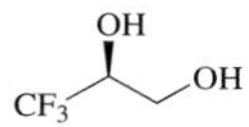	(75), 63	182
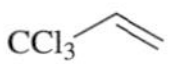	Ligand, $K_2OsO_2(OH)_4$, $K_3Fe(CN)_6$, K_2CO_3, $NaHCO_3$, *t*-BuOH:H_2O (1:1)	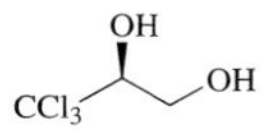		
	Ligand / Temp / Time			
	$(DHQD)_2AQN$ / 0° / 12 h	(65-90), 77		86
	$(DHQD)_2PYR$ / — / —	(86), 86		86, 182
	$(DHQD)_2PHAL$, $K_2OsO_2(OH)_4$, $K_3Fe(CN)_6$, K_2CO_3, *t*-BuOH-H_2O, O_2, 12 h	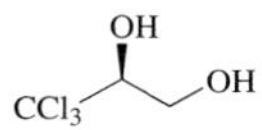	(86), 70	182

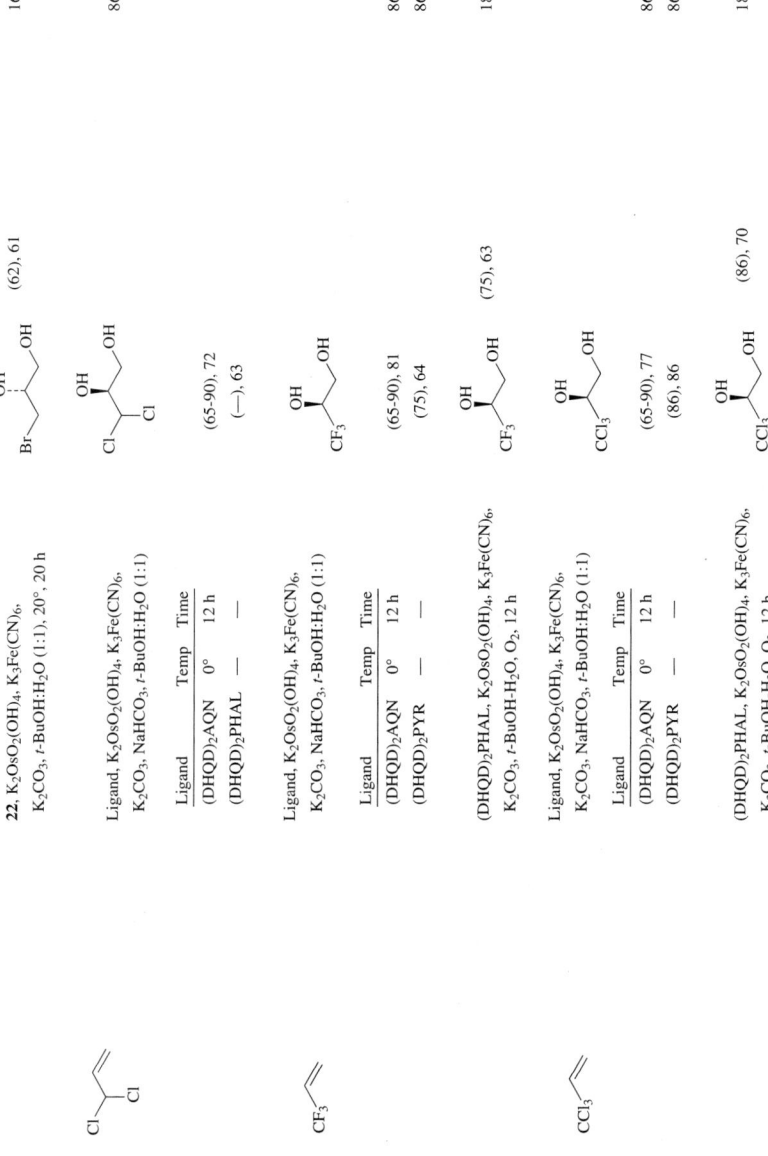

TABLE 1. REACTIONS OF TERMINAL ALKENES (*Continued*)

Substrate	Conditions	Product(s) and Yield(s) (%), % ee	Refs.
C_3			
	Ligand, $K_2OsO_2(OH)_4$, $K_3Fe(CN)_6$, K_2CO_3, $NaHCO_3$, t-BuOH:H_2O (1:1)	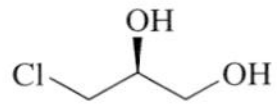	86
	Ligand / Temp / Time		
	$(DHQD)_2AQN$ / 0° / 12 h	(65-90), 90	
	$(DHQD)_2PHAL$ / — / —	(—), 63	
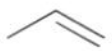	Ligand, $K_2OsO_2(OH)_4$, $K_3Fe(CN)_6$, K_2CO_3, t-BuOH-H_2O, O_2, 12 h	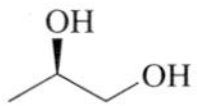	
	Ligand		
	$(DHQD)_2PHAL$	(47), 35	182, 20
	$(DHQD)_2PYR$	(47), 49	182
C_4			
	$(DHQD)_2AQN$, $K_2OsO_2(OH)_4$, $K_3Fe(CN)_6$, K_2CO_3, $NaHCO_3$, t-BuOH:H_2O (1:1), 0°, 12 h	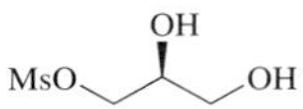 (65-90), 84	86
	Ligand, $K_2OsO_2(OH)_4$, $K_3Fe(CN)_6$, K_2CO_3, t-BuOH-H_2O, O_2, 12 h	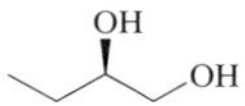	20
	Ligand		
	$(DHQD)_2PHAL$	(—), 65	
	$(DHQD)_2PYR$	(—), 70	
C_5			
	$(DHQ)_2PHAL$, $K_2OsO_2(OH)_4$, $K_3Fe(CN)_6$, K_2CO_3, t-BuOH:H_2O (1:1), 0°, 12-14 h	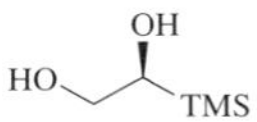 (74), 44	114, 113

	Ligand, $K_2OsO_2(OH)_4$, $K_3Fe(CN)_6$, K_2CO_3, *t*-BuOH-H_2O, 0°	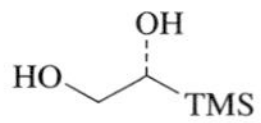	
	Ligand / Time		
	$(DHQD)_2PHAL$ 14 h	(80), 34	113
	$(DHQD)_2PHAL$ 12 h	(83), 46	182
	$(DHQD)_2PYR$ 12 h	(83), 89	182
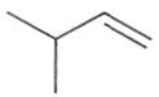	1. 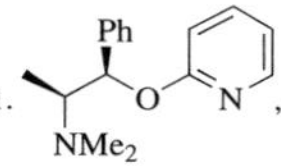 , OsO_4, toluene, –78°, 10 h 2. LAH, ether, –78° to rt, 8 h	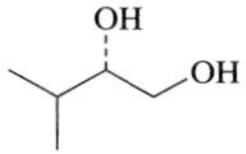 (27), 25	183
	Ligand, $K_2OsO_2(OH)_4$, $K_3Fe(CN)_6$, K_2CO_3, *t*-BuOH:H_2O (1:1), 0°, 6-24 h	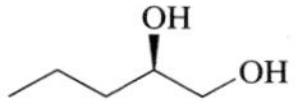	
	Ligand		
	$(DHQD)_2PHAL$	(—), 79	157, 20
	$(DHQD)_2PYR$	(—), 88	157, 20
	$(DHQD)_2DPP$	(—), 78	157
	$(DHQD)_2DPPHAL$	(—), 80	157
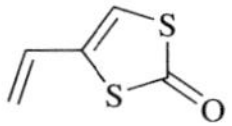	AD-Mix α, *t*-BuOH:H_2O (1:1), 4°, 17 h	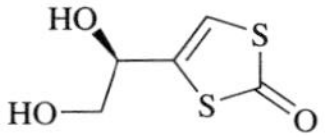 (96), 94	184
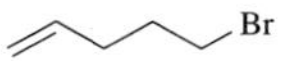	AD-Mix α, *t*-BuOH:H_2O (1:1)°	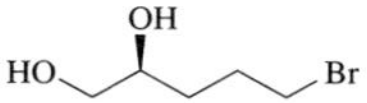	
	Temp		
	4°	(94), 65	185
	0°	(71), 78-97	186
	$(DHQD)_2PYR$, $K_2OsO_2(OH)_4$, $K_3Fe(CN)_6$, K_2CO_3, *t*-BuOH:H_2O (1:1), 0°, 4 d	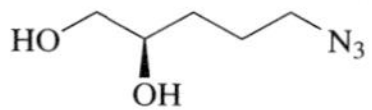 (5-20), —	187

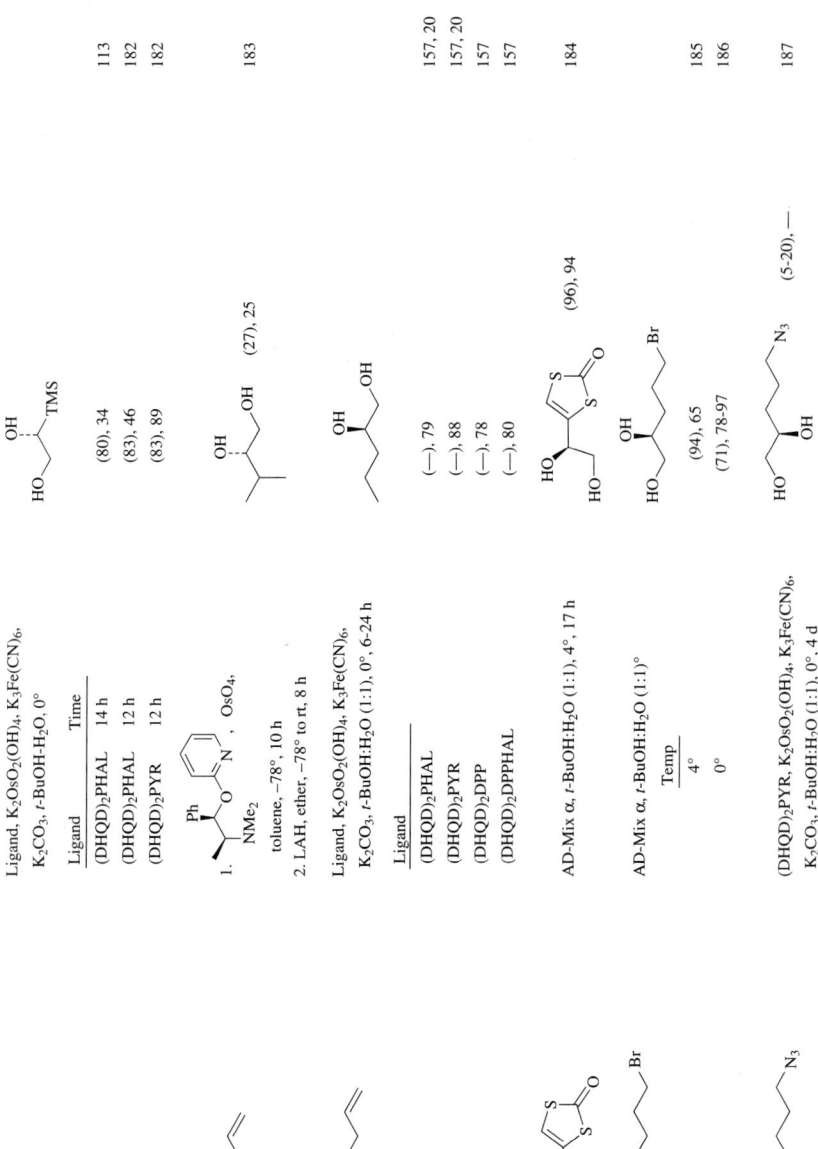

Substrate	Conditions	Product(s) and Yield(s) (%), % ee	Refs.
C_6 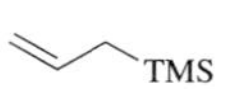	Ligand, $K_2OsO_2(OH)_4$, $K_3Fe(CN)_6$, K_2CO_3, *t*-BuOH:H_2O (1:1), 0°, 14 h	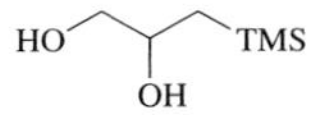	113
	Ligand		
	$(DHQD)_2PHAL$	(82), 34	
	$(DHQ)_2PHAL$	(81), 34	
	$(DHQ)_2PHAL$, $K_2OsO_2(OH)_4$, $K_3Fe(CN)_6$, K_2CO_3, *t*-BuOH:H_2O (1:1), 0°, 24 h	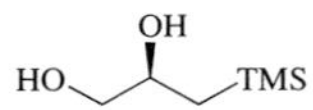 (60), 28	114
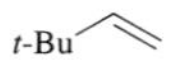	$(DHQ)_2PHAL$, $K_2OsO_2(OH)_4$, $K_3Fe(CN)_6$, K_2CO_3, *t*-BuOH:H_2O (1:1), 0°, 6-24 h	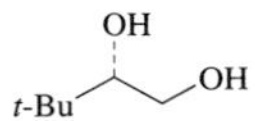 **I** (—), 66	171, 157
	$(DHQ)_2PHAL$, $K_2OsO_2(OH)_4$, NMO, *t*-BuOH:H_2O (1.4:1), rt, 6 h (slow addition of olefin)	**I** (85-96), 46	171
	$(DHQ)_2PYR$, $K_2OsO_2(OH)_4$, $K_3Fe(CN)_6$, K_2CO_3, *t*-BuOH:H_2O (1:1), 0°, 3 h	**I** (—), 87	77, 157
	22, $K_2OsO_2(OH)_4$, $K_3Fe(CN)_6$, K_2CO_3, *t*-BuOH:H_2O (1:1), 20°, 20 h	**I** (80), 76	164
	1. DHQ-OAc , OsO_4, toluene, rt, 12 h 2. LAH, ether	**I** (87), 26.2	8
	$(DHQ)_2DPP$, $K_2OsO_2(OH)_4$, $K_3Fe(CN)_6$, K_2CO_3, *t*-BuOH:H_2O (1:1), 0°, 6-24 h	**I** (—), 65	157
	$(DHQ)_2DPPHAL$, $K_2OsO_2(OH)_4$, $K_3Fe(CN)_6$, K_2CO_3, *t*-BuOH:H_2O (1:1), 0°, 6-24 h	**I** (—), 73	157
	DHQ-MEQ, $K_2OsO_2(OH)_4$, $K_3Fe(CN)_6$, K_2CO_3, *t*-BuOH:H_2O (1:1), 0°	**I** (38), 89	97

	Ligand, $K_2OsO_2(OH)_4$, $K_3Fe(CN)_6$, K_2CO_3, t-BuOH:H_2O (1:1)	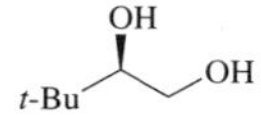		

Ligand	Temp	Time	Product(s) and Yield(s)	Refs.
$(DHQD)_2PYR$	0°	3 h	(—), 92	77, 157
$(DHQD)_2PHAL$	0°	3 h	(80), 64	77, 157
10c	—	—	(62), 70	166
DHQD-PHN	0°	18-24 h	(75-95), 79	72
DHQD-MEQ	0°	18-24 h	(75-95), 79	72
DHQD-CLB	rt	18-24 h	(75-95), 44	72
$(DHQD)_2DPP$	0°	6-24 h	(—), 59	157
$(DHQD)_2DPPHAL$	0°	6-24 h	(—), 67	157
10b	—	<5 h	(84), 90	169
$(DHQD)_2AQN$	0°	12 h	(65-90), 87	86
$(DHQD)_2PYR$	—	—	(—), 87	86

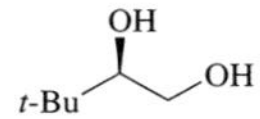

I

Substrate	Conditions	Product(s) and Yield(s)		Refs.
	$(DHQD)_2PYR$, $OsCl_3$, $K_2Fe(CN)_6$, K_2CO_3, t-BuOH:H_2O (1:1), 0°, 3 d	**I** (98), 84		188
	33, OsO_4, CH_2Cl_2, –90°, 5 h	**I** (70), 65		9
	$(DHQD)_2PHAL$, $K_2OsO_2(OH)_4$, $K_3Fe(CN)_6$, K_2CO_3, t-BuOH:H_2O (1:1), O_2, 12 h	**I** (—), 82		182
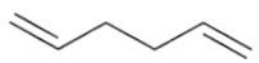	$(DHQD)_2PYR$, $K_2OsO_2(OH)_4$, $K_3Fe(CN)_6$, K_2CO_3, t-BuOH:H_2O (1:1), 0°, 24 h	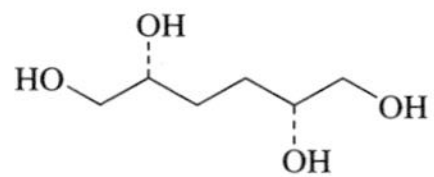	(—), 88	144
	$(DHQ)_2PYR$, $K_2OsO_2(OH)_4$, $K_3Fe(CN)_6$, K_2CO_3, t-BuOH:H_2O (1:1), 0°, 24 h	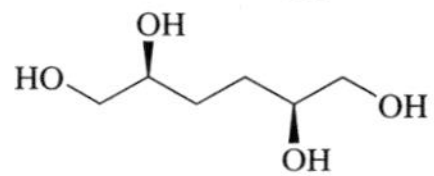	(—), 82	144

TABLE 1. REACTIONS OF TERMINAL ALKENES (*Continued*)

Substrate	Conditions	Product(s) and Yield(s) (%), % ee	Refs.
C_6			
	Ligand, $K_2OsO_2(OH)_4$, $K_3Fe(CN)_6$, K_2CO_3, *t*-BuOH:H_2O (1:1), 0°, 24 h	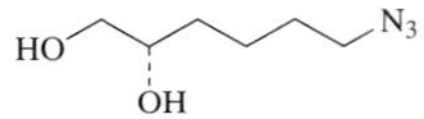	189
	Ligand		
	$(DHQ)_2PYR$	(84), 76	
	$(DHQ)_2PHAL$	(85), 74	
	$(DHQ)_2AQN$	(96), 76	
	Ligand, $K_2OsO_2(OH)_4$, $K_3Fe(CN)_6$, K_2CO_3, *t*-BuOH:H_2O (1:1), 0°, 24 h	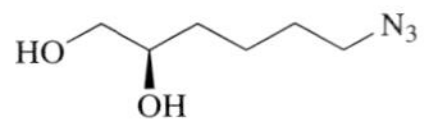	189
	Ligand		
	$(DHQD)_2PYR$	(88), 88	
	$(DHQD)_2PHAL$	(80), 79	
	$(DHQD)_2AQN$	(94), 87	
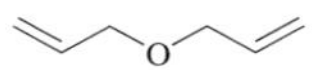	$(DHQ)_2PYR$, $K_2OsO_2(OH)_4$, $K_3Fe(CN)_6$, K_2CO_3, *t*-BuOH:H_2O (1:1), 0°, 24 h	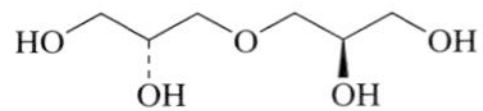 (—), 53	144
	$(DHQD)_2PYR$, $K_2OsO_2(OH)_4$, $K_3Fe(CN)_6$, K_2CO_3, *t*-BuOH:H_2O (1:1), 0°, 24 h	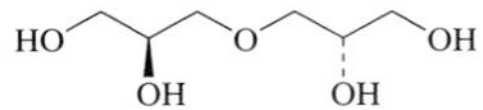 (—), 59	144
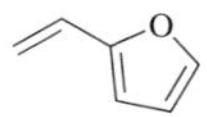	$(DHQD)_2PHAL$, $K_2OsO_2(OH)_4$, $K_3Fe(CN)_6$, K_2CO_3, $MeSO_2NH_2$, *t*-BuOH:H_2O (1:1), 0°, 96 h	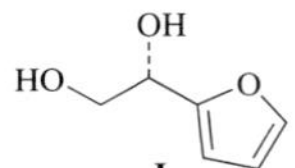 **I** (89), 93	74
	$(DHQD)_2PHAL$, $K_2OsO_2(OH)_4$, $K_3Fe(CN)_6$, K_2CO_3, *t*-BuOH:H_2O (1:1), 0°, 10 h	**I** (92), 92	190, 75
	AD-Mix β, *t*-BuOH:H_2O (1:1), 0°, 12 h	**I** (85), 90	191

	(DHQ)$_2$PHAL, $K_2OsO_2(OH)_4$, $K_3Fe(CN)_6$, K_2CO_3, $MeSO_2NH_2$, *t*-BuOH:H_2O (1:1), 0°, 96 h	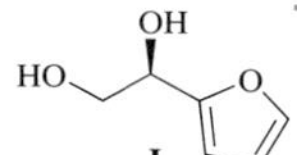 **I**	(81), 90	74
	(DHQ)$_2$PHAL, $K_2OsO_2(OH)_4$, $K_3Fe(CN)_6$, K_2CO_3, *t*-BuOH:H_2O (1:1), 0°, 10 h	**I** (85), 90		75
	AD-Mix α, *t*-BuOH:H_2O (1:1), 0°, 12 h	**I** (85), 92		191
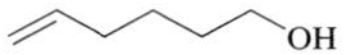	Ligand, $K_2OsO_2(OH)_4$, $K_3Fe(CN)_6$, K_2CO_3, *t*-BuOH:H_2O (1:1), 0°, 12 h	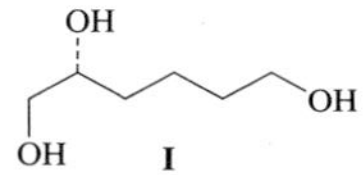 **I**		146
	Ligand	**I**		
	(DHQD)$_2$PHAL	(—), 56		
	(DHQD)$_2$PYR	(—), 74		
	(DHQD)$_2$AQN	(—), 56		
	Ligand, $K_2OsO_2(OH)_4$, I_2, K_2CO_3, *t*-BuOH:H_2O (1:1), 0°, 18 h			146
	Ligand	**I**		
	(DHQD)$_2$PHAL	(—), 53		
	(DHQD)$_2$PYR	(—), 62		
	(DHQD)$_2$AQN	(—), 29		
C_7 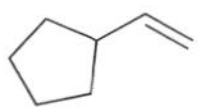	Ligand, $K_2OsO_2(OH)_4$, $K_3Fe(CN)_6$, K_2CO_3, *t*-BuOH:H_2O (1:1), 0°, 3 h	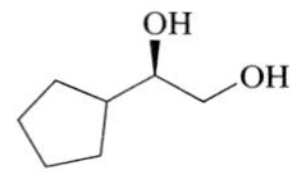		77
	Ligand			
	(DHQD)$_2$PYR	(—), 93		
	(DHQD)$_2$PHAL	(—), 80		

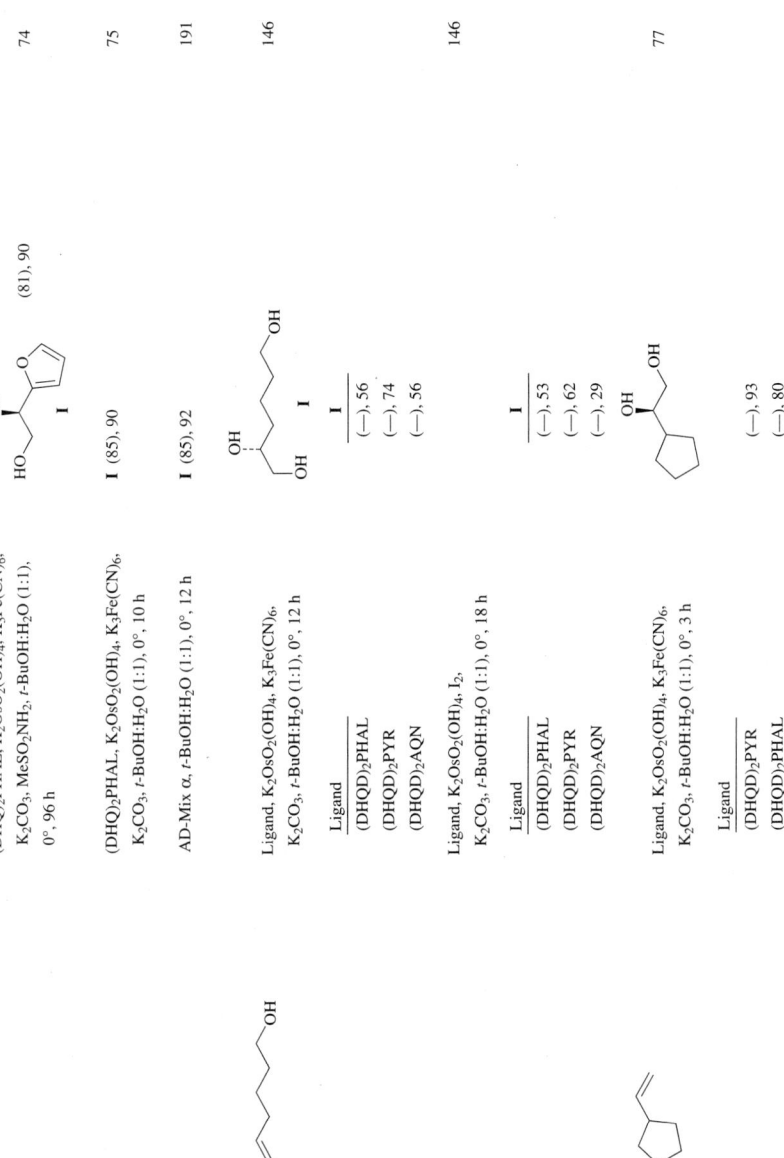

TABLE 1. REACTIONS OF TERMINAL ALKENES (*Continued*)

Substrate	Conditions	Product(s) and Yield(s) (%), % ee	Refs.
C_7			
	1. 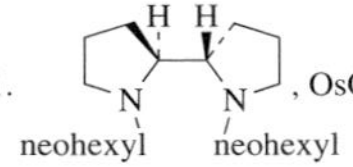, OsO_4, CH_2Cl_2, −78°, 18 h 2. $NaHSO_3$	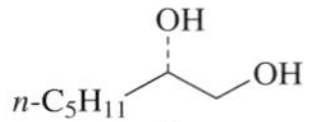 **I** (90), 91	13
	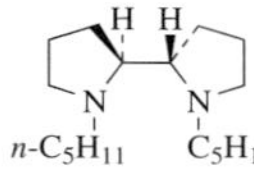, OsO_4, CH_2Cl_2, −78°	**I** (90), 68	12
	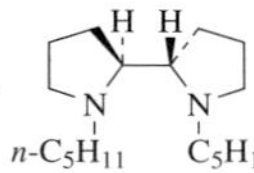, OsO_4, toluene, −78°	**I** (95), 40	12
	34, OsO_4, CH_2Cl_2, rt	**I** (75), 86	10
	12a, NMO, OsO_4, acetone:H_2O (10:1), 40°, 16 h	**I** (10), 6	192
	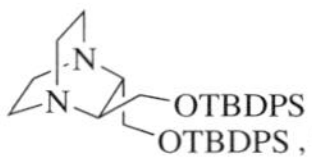, OsO_4, $K_3Fe(CN)_6$, K_2CO_3, *t*-BuOH:H_2O (1:1), rt, 8-24 h	**I** (92), 27	193
	$(DHQ)_2PYR$, OsO_4, $K_3Fe(CN)_6$, K_2CO_3, *t*-BuOH:H_2O (1:1), rt, 12 h	**I** (80), 85	194
	AD-Mix β, *t*-BuOH:H_2O (1:1), 0°	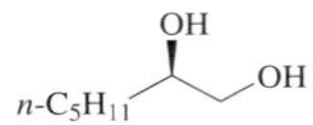 (94), 84	195. 196
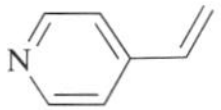	AD-Mix β, *t*-BuOH:H_2O (1:1), 0°, 5 h	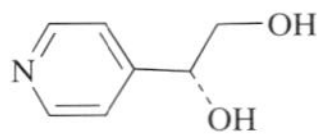 (75-85), >95	197
	AD-Mix α, *t*-BuOH:H_2O (1:1), 0°, 24 h	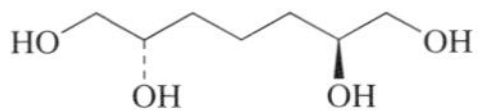 (98), 93	144

C_8 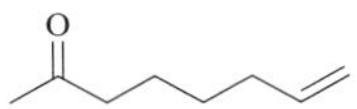	1. $(DHQ)_2PHAL$, $K_2OsO_2(OH)_4$, $K_3Fe(CN)_6$, K_2CO_3, t-BuOH:H_2O (1:1), 0°, 18 h 2. HCl, rt	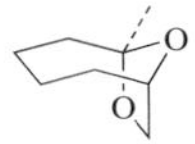 (71-68), 60-70	198
	1. $(DHQD)_2PHAL$, $K_2OsO_2(OH)_4$, $K_3Fe(CN)_6$, K_2CO_3, t-BuOH:H_2O (1:1), 0°, 18 h 2. HCl, rt	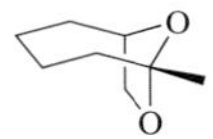 (—), ~60	198
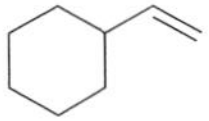	Ligand, $K_2OsO_2(OH)_4$, $K_3Fe(CN)_6$, K_2CO_3, t-BuOH:H_2O (1:1), 0°	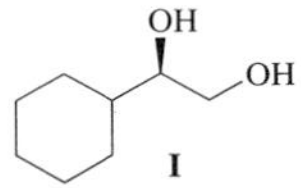 **I**	
	Ligand / Time	**I**	
	$(DHQD)_2PYR$ / 3 h	(—), 96	77, 86, 157
	$(DHQD)_2PHAL$ / 3 h	(—), 88	77, 157
	$(DHQD)_2DPP$ / 6-24 h	(—), 89	157
	$(DHQD)_2DPPHAL$ / 6-24 h	(—), 91	157
	DHQD-p-chlorobenzoate, $K_2OsO_2(OH)_4$, NMO, acetone: H_2O, 0°, 1 h	**I** (80-95), 46	18
	$(DHQD)_2AQN$, $K_2OsO_2(OH)_4$, $K_3Fe(CN)_6$, K_2CO_3, $NaHCO_3$, t-BuOH:H_2O (1:1), 0°, 12 h	**I** (65-90), 86	86
	$(DHQD)_2PYR$, $OsCl_3$, $K_2Fe(CN)_6$, K_2CO_3, t-BuOH:H_2O (1:1), 0°, 3 d	**I** (72), 89	188
	22, $K_2OsO_2(OH)_4$, $K_3Fe(CN)_6$, K_2CO_3, t-BuOH:H_2O (1:1), 20°, 20 h	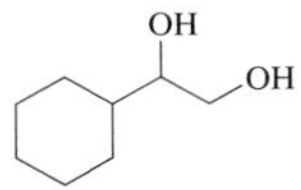 (78), 67	164

TABLE 1. REACTIONS OF TERMINAL ALKENES (*Continued*)

Substrate	Conditions	Product(s) and Yield(s) (%), % ee	Refs.
C_8 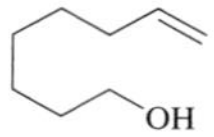	1. AD-Mix β, *t*-BuOH:H_2O (1:1) 2. Recrystallization	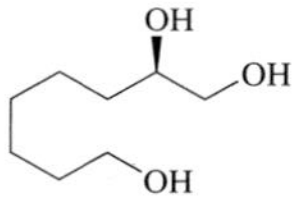 (64), >95	199
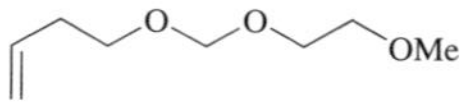	$(DHQD)_2PYDZ$, $K_2OsO_2(OH)_4$, $K_3Fe(CN)_6$, K_2CO_3, *t*-BuOH:H_2O (1:1), 0°, 4-24 h	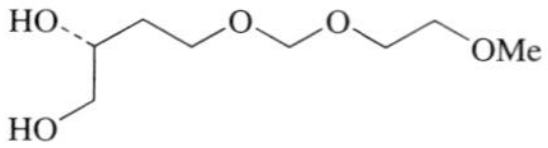 (99), 38	83, 61
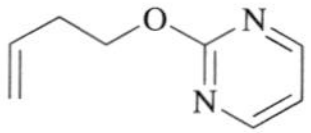	$(DHQD)_2PYDZ$, $K_2OsO_2(OH)_4$, $K_3Fe(CN)_6$, K_2CO_3, *t*-BuOH:H_2O (1:1), 0°, 4-24 h	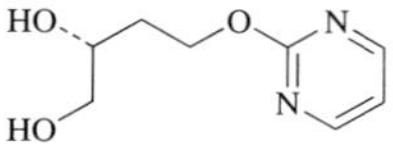 (91), 50	83, 61
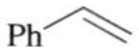	1. DHQ-OAc (1.1 eq), OsO_4 (1.1 eq), toluene, rt, 12 h 2. LAH, ether	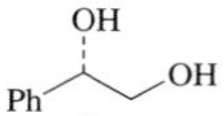 **I** (90), 64.5	8
	35, OsO_4, CH_2Cl_2, –90°, 2 h	**I** (81), 92	11
	1. 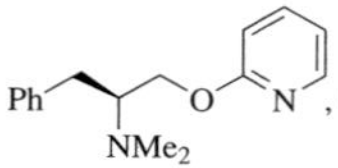, OsO_4, toluene, 21° 2. LAH, ether, –78° to rt, 8 h	**I** (81), 6	183
	1. 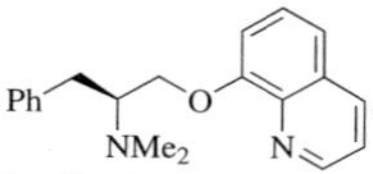, OsO_4, toluene, 21° 2. LAH, ether, –78° to rt, 8 h	**I** (88), 3	183

1. 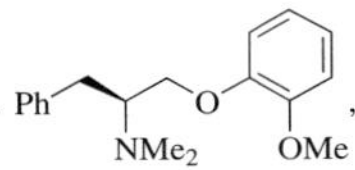, OsO_4, toluene, 21° 2. LAH, ether, –78° to rt, 8 h	**I** (74), 4	183
1. 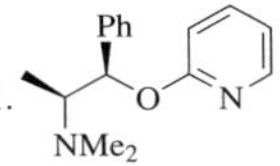, OsO_4, toluene, 21° 2. LAH, ether, –78° to rt, 8 h	**I** (83), 34	183
1. 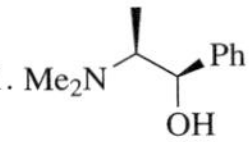, OsO_4, toluene, 21° 2. LAH, ether, –78° to rt, 8 h	**I** (75), <1	183
1. 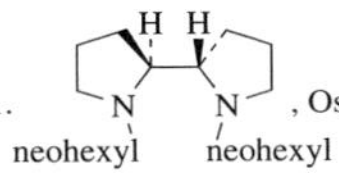, OsO_4, toluene, –78°, 18 h 2. $NaHSO_3$	**I** (90), 88	13
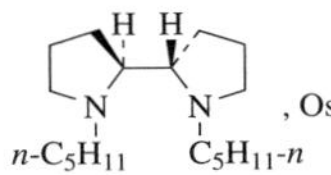, OsO_4, CH_2Cl_2, –78°	**I** (76), 27	12
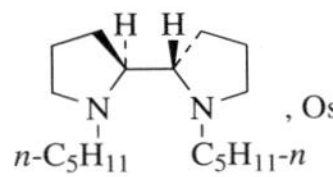, OsO_4, toluene, –78°	**I** (49), 58	12

TABLE 1. REACTIONS OF TERMINAL ALKENES (*Continued*)

Substrate	Conditions	Product(s) and Yield(s) (%), % ee	Refs.
C_8 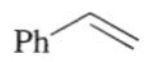	DHQ-*p*-chlorobenzoate, OsO_4, NMO, acetone-H_2O, 0°, 7 h	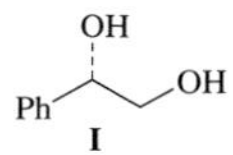 **I** (80-95), 54	18, 171
	DHQ-*p*-chlorobenzoate, OsO_4, NMO, acetone:H_2O (10:1), 0°, (slow addition of olefin 5 h), 1 h	**I** (85-95), 50-55	69
	$(DHQ)_2PHAL$, $K_2OsO_2(OH)_4$, $K_3Fe(CN)_6$, K_2CO_3, *t*-BuOH:H_2O (1:1), 0°, 6-24 h	**I** (83), 97-99	77, 200, 157, 152, 97, 171
	$(DHQ)_2PHAL$, $K_2OsO_2(OH)_4$, NMO, *t*-BuOH:H_2O (1.4:1), rt, 6 h, (slow addition of olefin)	**I** (85-96), 97	171
	Ligand, $K_2OsO_2(OH)_4$, $K_3Fe(CN)_6$, K_2CO_3, *t*-BuOH:H_2O (1:1), 0°, 6-24 h	**I**	
	Ligand	**I**	
	$(DHQ)_2DPPHAL$	(—), 96	157
	$(DHQ)_2DPP$	(—), 97	157
	$(DHQ)_2PYDZ$	(—), 93	152
	$(DHQ)_2TP$	(—), 79	152
	25, $K_2OsO_2(OH)_4$, $K_3Fe(CN)_6$, K_2CO_3, 0°, 24 h	**I** (84), 84.3	201
	42a, $K_3Fe(CN)_6$, K_2CO_3, OsO_4 *t*-BuOH:H_2O (1:1), 0°, 48 h	**I** (75), 65	161
	Ligand, NMO, OsO_4, acetone:H_2O (10:1)	**I**	
	Ligand / Temp / Time	**I**	
	42a / 0° / 7 h	(65), 40	161
	12a / 40° / 16 h	(65), 22	192
	43 / 0° / 5 h	(72), 54	202

Conditions	Product(s) and Yield(s) (%), % ee	Refs.
Ligand, $K_2OsO_2(OH)_4$, $K_3Fe(CN)_6$, K_2CO_3, *t*-BuOH:H_2O (1:1)	**I**	

Ligand	Temp	Time	**I**	Refs.
$(DHQ)_2AQN$	0°	12 h	(65-90), 85	86
$(DHQ)_2DPP$	—	—	(—), 97	86
22	20°	12 h	(92), 67	164

Conditions	Product(s) and Yield(s) (%), % ee	Refs.
1. 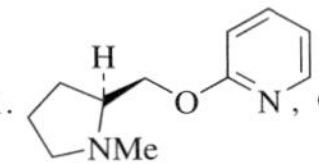 , OsO_4, toluene, 21°, 10 h 2. LAH, ether, –78° to rt, 8 h	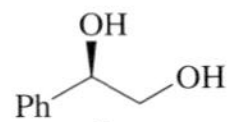 **I** (71), 4	183
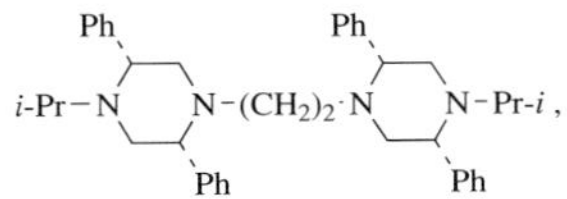 , OsO_4 (1.1 eq), toluene, –78°	**I** (91), 89	215
33, OsO_4, toluene, –90°, 4 h	**I** (70), 99	9
1. DHQD-OAc (1.1 eq), OsO_4 (1.1 eq), toluene, rt, 12 h 2. LAH, ether	**I** (62), 61	8
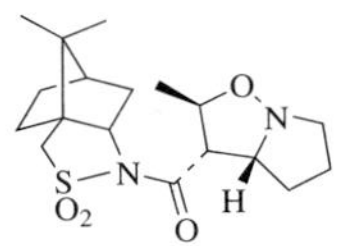 (30%), OsO_4 (1%), $K_3Fe(CN)_6$, K_2CO_3, *t*-BuOH:H_2O (1:1), rt, 24 h	**I** (99), 22	203

TABLE 1. REACTIONS OF TERMINAL ALKENES (*Continued*)

Substrate	Conditions	Product(s) and Yield(s) (%), % ee	Refs.
C_8 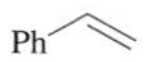	Ligand, $K_2OsO_4(OH)_7$, $K_3Fe(CN)_6$, K_2CO_3, *t*-BuOH:H_2O (1:1)	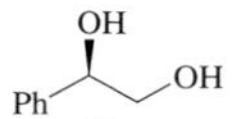 **I**	
	Ligand / Temp / Time	**I**	
	$(DHQD)_2$PYDZ / 0° / —	(95), 96	152, 66
			63
	$(DHQD)_2$DPP / 0° / 6-24 h	(—), 99	157
	$(DHQD)_2$DPPHAL / 0° / 6-24 h	(—), 98	157
	DHQD-PHN / 0° / 18-24 h	(75-95), 78	72
	DHQD-MEQ / 0° / 18-24 h	(75-95), 87	72
	DHQD-*p*-chlorobenzoate / rt / 18-24 h	(75-95), 74	72, 70
	$(DHQD)_2$TP / 0° / 6-24 h	(—), 80	152
	AD-Mix β, *t*-BuOH:H_2O (1:1), 0°, 5 h	**I** (75-85), >95	197
	$(DHQD)_2$PHAL, $K_2OsO_2(OH)_4$, pH 10.4, *t*-BuOH-H_2O, O_2, 50°, 14-18 h	**I** (52), 90	178
	$(DHQD)_2$PHAL, $K_2OsO_2(OH)_4$, $K_3Fe(CN)_6$, K_2CO_3, *t*-BuOH:H_2O (1:1), 0°, 6-24 h	**I** (100), 97	77, 47, 200, 152, 66, 157, 204, 97
	DHQD-*p*-chlorobenzoate (13), OsO_4, NMO, acetone-H_2O, 0°, 3 h	**I** (80-95), 56-62	18, 69, 112, 70
	Ligand, $K_2OsO_2(OH)_4$, $K_3Fe(CN)_6$, K_2CO_3, *t*-BuOH:H_2O (1:1)	**I**	

Conditions	Product(s) and Yield(s) (%)	Refs.

Ligand	Temp	Time	I	Refs.
$(DHQD)_2PYR$	0°	6-24 h	(—), 80	77, 157
6	0°	—	(76), 91	63, 73
37	—	—	(93), 98	166
38	—	—	(93), 90	166
39	—	—	(92), 97	166

40, $K_2OsO_2(OH)_4$, $K_3Fe(CN)_6$, K_2CO_3, *t*-BuOH:H_2O (1:1), <5 h — **I** (92), 98 — 169

Ligand, NMO, OsO_4, TEAA, acetone:H_2O (10:1), 10°, 5 h (slow addition of olefin) — **I** — 205

Ligand	I
26	(80), 60
27	(80), 60

16, NMO, OsO_4, TEAA, acetone:H_2O (10:1), 10°, 5 h, (slow addition of olefin) — **I** (87), 72 — 168

Ligand, $K_3Fe(CN)_6$, K_2CO_3, OsO_4, *t*-BuOH:H_2O (1:1) — **I**

Ligand	Temp	Time	I	Refs.
16	rt	18 h	(88), 98	168
41a	0°	24 h	(68-80), 68	206
45	0°	20 h	(86), 91	160
41b	0°	20 h	(75), 65	160
41a	0°	20 h	(84), 68	160

TABLE 1. REACTIONS OF TERMINAL ALKENES (*Continued*)

Substrate	Conditions	Product(s) and Yield(s) (%), % ee	Refs.
C_8 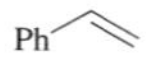	Ligand, NMO, OsO_4, acetone:H_2O (10:1), 0°, 24 h	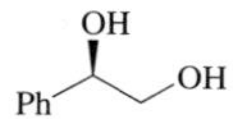 **I**	162
	Ligand	**I**	
	46	(82), 33	
	44	(76), 57	
	41b	(67), 41	
	Ligand, $K_3Fe(CN)_6$, K_2CO_3, OsO_4, *t*-BuOH:H_2O (1:1), 0°, 24 h	**I**	162
	Ligand	**I**	
	46	(73), 48	
	41c	(50), 68	
	41d	(68), 80	
	44	(—), —	
	$(DHQD)_2PYDZ$, $K_3Fe(CN)_6$, K_2CO_3, $K_2OsO_2(OH)_4$, *t*-BuOH:H_2O (1:1), 0°	**I** (95), 94-96	156, 73
	15, NMO, OsO_4, CH_3CN:H_2O (8:2), 5°, 12 h	**I** (87), 48	207
	24, $K_2OsO_2(OH)_4$, $K_3Fe(CN)_6$, K_2CO_3, *t*-BuOH:H_2O (1:1), 0°, 6-24 h	**I** (—), 87	60
	18, $K_2OsO_2(OH)_4$, $K_3Fe(CN)_6$, K_2CO_3, *t*-BuOH:H_2O (1:1)	**I** (—), 98	47

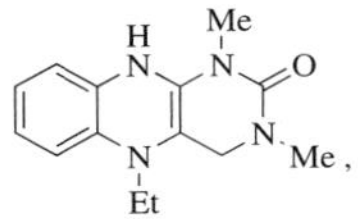

(DHQD)$_2$PHAL, NMM, OsO_4, Et_4NH_4Ac, H_2O_2, t-BuOH:H_2O (3:1), 20°, 7-15 h (slow addition of olefin and peroxide)

I (80-84), 88-95 — 174, 208

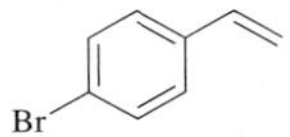

(DHQ)$_2$PHAL, $K_2OsO_2(OH)_4$, $K_3Fe(CN)_6$, K_2CO_3, t-BuOH:H_2O (1:1), 0°, 16 h

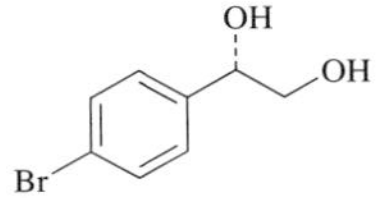

(74), >95 — 209

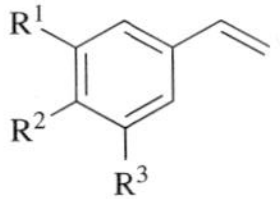

Ligand, $K_2OsO_2(OH)_4$, $K_3Fe(CN)_6$, K_2CO_3, t-BuOH:H_2O (1:1), 0°

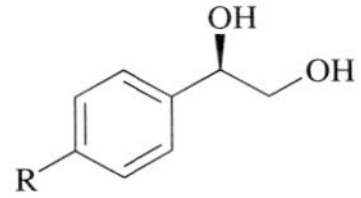

R^1	R^2	R^3	Ligand	Temp	Time		Refs.
H	Br	H	(DHQD)$_2$PHAL	0°	16 h	(73), >95	209
H	Br	H	**17**	0°	—	(82), 80	63
H	NO_2	H	(DHQD)$_2$PYDZ	0°	6-12 h	(78), 98	73
H	NO_2	H	**6**	0°	6-12 h	(79), 99	73
NO_2	H	H	(DHQD)$_2$PYDZ	0°	6-12 h	(89), 96	73
NO_2	H	H	**6**	0°	6-12 h	(99), 95	73
NO_2	H	NO_2	(DHQD)$_2$PYDZ	0°	6-12 h	(83), 94	73
NO_2	H	NO_2	**6**	0°	6-12 h	(90), 96	73
H	Cl	H	(DHQD)$_2$PHAL	—	—	(—), 98	210

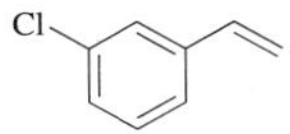

AD-Mix β, t-BuOH:H_2O (1:1), 0°

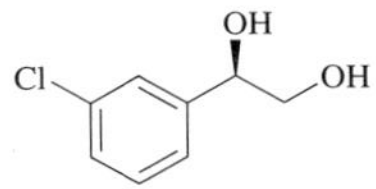

(95-98), 98-99 — 211

TABLE 1. REACTIONS OF TERMINAL ALKENES (*Continued*)

Substrate	Conditions	Product(s) and Yield(s) (%), % ee	Refs.
C_8			
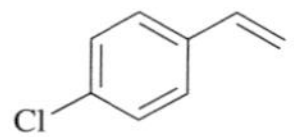	AD-Mix β, *t*-BuOH:H_2O (1:1), 0°, 6 h	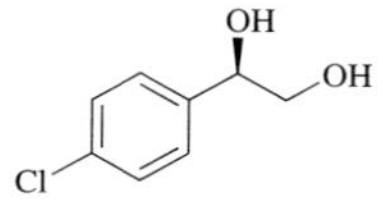 (>95), >95	212
	AD-Mix α, *t*-BuOH:H_2O (1:1), 0°, 6 h	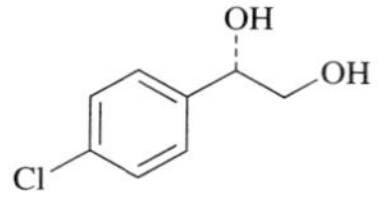 (>95), >95	212
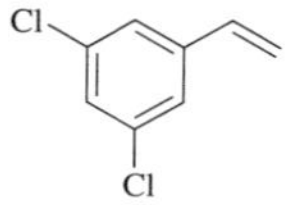	AD-Mix β, *t*-BuOH:H_2O (1:1), 0°	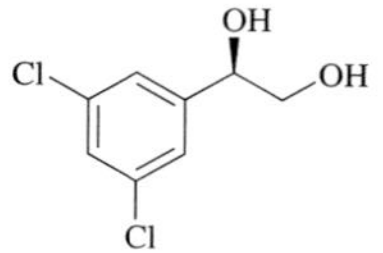 (95-98), 98-99	211
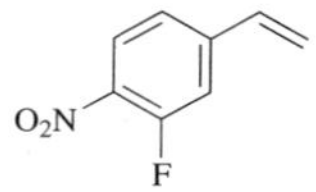	$(DHQ)_2PHAL$, OsO_4, K_2CO_3, $K_3Fe(CN)_6$, *t*-BuOH:H_2O (1:1), 0°, 8 h	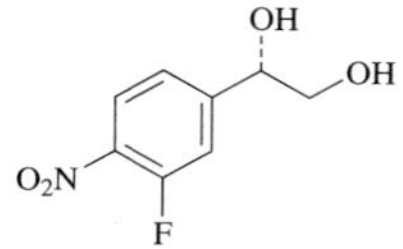 (90), >98	213
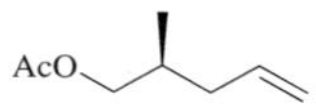	AD-Mix β, *t*-BuOH:H_2O (1:1), 0°, 6 h	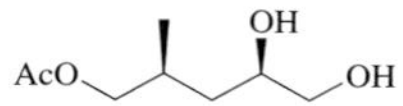 (88), 50 de	214
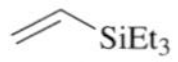	$(DHQ)_2PHAL$, $K_2OsO_2(OH)_4$, $K_3Fe(CN)_6$, K_2CO_3, *t*-BuOH:H_2O (1:1), 0°, 20 h	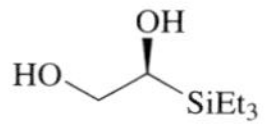 (72), 35	114
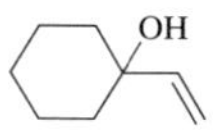	$(DHQD)_2PYR$, $K_2OsO_2(OH)_4$, $K_3Fe(CN)_6$, K_2CO_3, *t*-BuOH:H_2O (1:1), 0°, 6-36 h	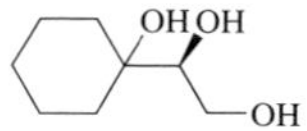 (88), 90	116

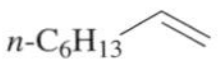	$(DHQD)_2AQN$, $K_2OsO_2(OH)_4$, $K_3Fe(CN)_6$, K_2CO_3, t-BuOH:H_2O (1:1), 0°, 12 h	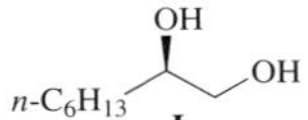 **I** (65-90), 92	86
	$(DHQD)_2PYR$, $K_2OsO_2(OH)_4$, $K_3Fe(CN)_6$, K_2CO_3, t-BuOH:H_2O (1:1)	**I** (—), 87	86
	$(DHQD)_2PHAL$, $K_2OsO_2(OH)_4$, pH 10.4, t-BuOH-H_2O, O_2, 50°, 14-18 h	**I** (98), 65	178
	$(DHQ)_2PHAL$, $K_2OsO_2(OH)_4$, pH 10.4, t-BuOH-H_2O, O_2, 50°, 14-18 h	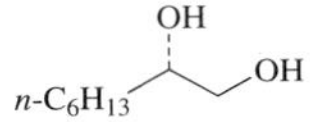 (97), 54	178
C_9 	1. 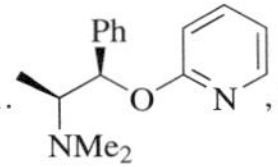, OsO_4, toluene, 21° 2. LAH, ether, –78° to rt, 8 h	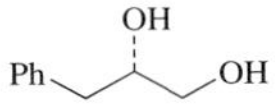 (30), 78	183
	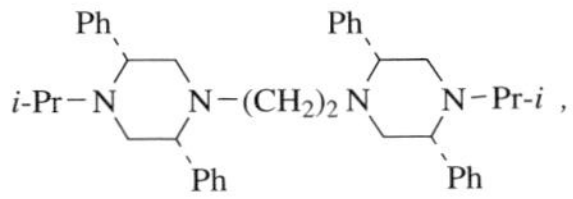 OsO_4, toluene, –78°	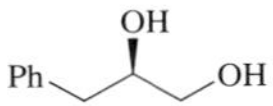 **I** (63), 18	215
	$(DHQD)_2AQN$, $K_2OsO_2(OH)_4$, $K_3Fe(CN)_6$, K_2CO_3, t-BuOH:H_2O (1:1), 0°, 12 h	**I** (65-90), 78	86
	$(DHQD)_2PHAL$, $K_2OsO_2(OH)_4$, $K_3Fe(CN)_6$, K_2CO_3, t-BuOH:H_2O (1:1)	**I** (—), 44	86
	DHQD-p-chlorobenzoate, $K_2OsO_2(OH)_4$, $K_3Fe(CN)_6$, K_2CO_3, electrolysis, t-BuOH:H_2O (4:3), rt	**I** (90), 34	176

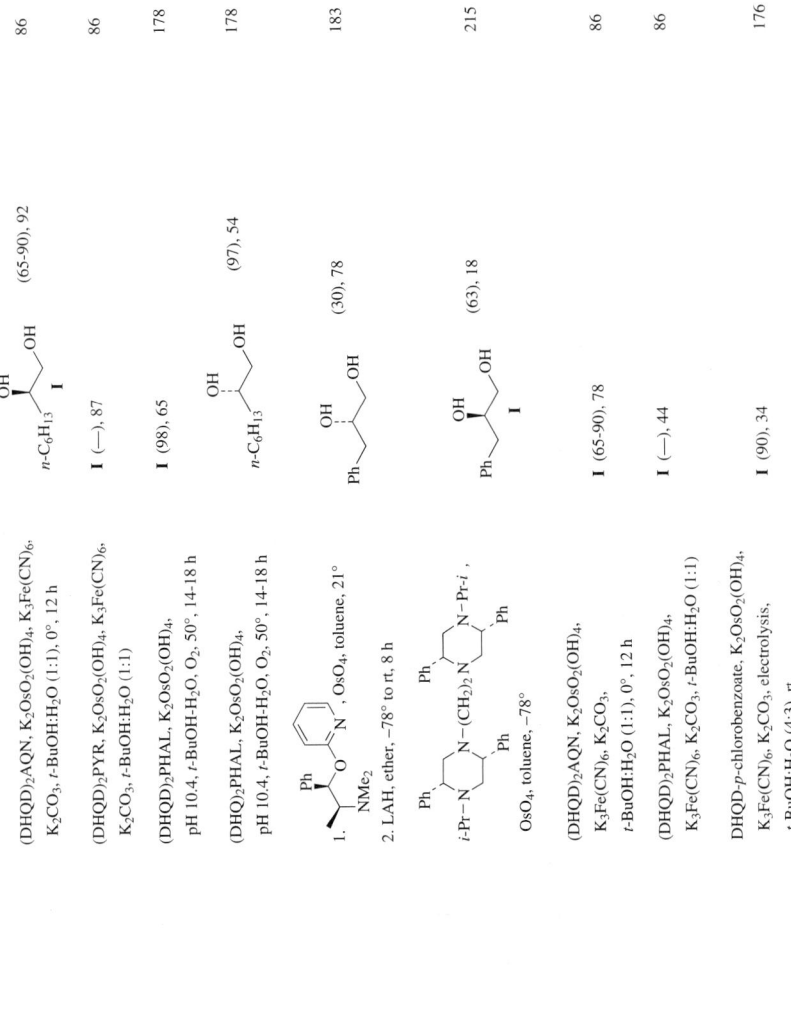

TABLE 1. REACTIONS OF TERMINAL ALKENES (*Continued*)

Substrate	Conditions	Product(s) and Yield(s) (%), % ee	Refs.
C_9			
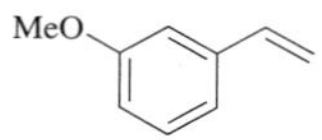	AD-Mix α, *t*-BuOH:H_2O (1:1), 0°	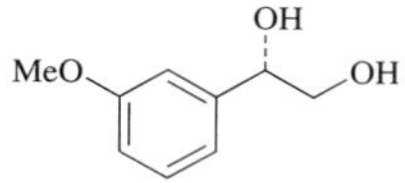 (99), >99	216
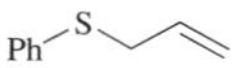	$(DHQ)_2PHAL$, $K_2OsO_2(OH)_4$, $K_3Fe(CN)_6$, K_2CO_3, *t*-BuOH:H_2O (1:1), 0°	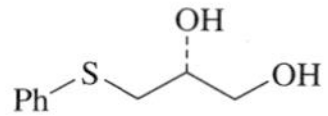 (87), 61	100
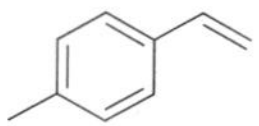	$(DHQD)_2PHAL$, $K_2OsO_2(OH)_4$, $K_3Fe(CN)_6$, K_2CO_3, *t*-BuOH:H_2O (1:1), 0°, 16 h	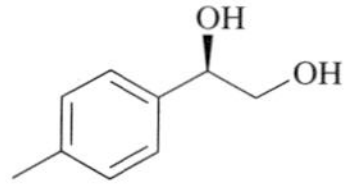 (67), >95	209
	$(DHQ)_2PHAL$, $K_2OsO_2(OH)_4$, $K_3Fe(CN)_6$, K_2CO_3, *t*-BuOH:H_2O (1:1), 0°, 16 h	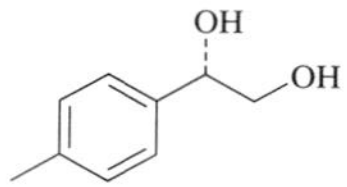 (74), >95	209
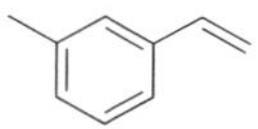	**12a**, OsO_4, NMO, acetone:H_2O (10:1), 40°, 16 h	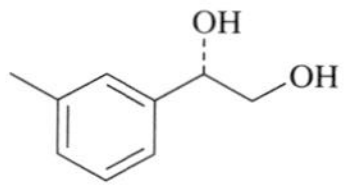 (70), 28	192
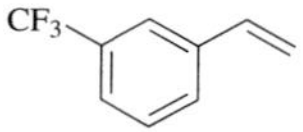	AD-Mix β, *t*-BuOH:H_2O (1:1), 0°	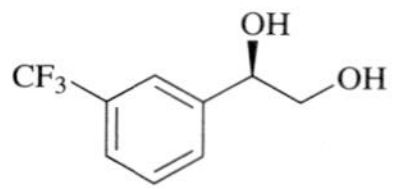 (95-98), 98-99	211
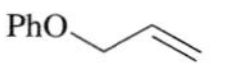	$(DHQD)_2PHAL$, $K_2OsO_2(OH)_4$, $K_3Fe(CN)_6$, K_2CO_3, *t*-BuOH:H_2O (1:1)	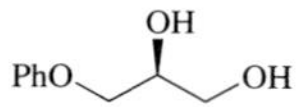 **I**	
	Temp / Time / pH		
	0° / 24 h / —	(85-95), 88	79
	rt / 1.5 h / —	(95), 82	71
	rt / 2 h / 10	(94), 84	71
	33, OsO_4, CH_2Cl_2, –90°, 4 h	**I** (56), 64	9

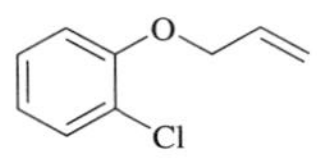	(DHQD)$_2$PHAL, $K_2OsO_2(OH)_4$, $K_3Fe(CN)_6$, K_2CO_3, t-BuOH:H_2O (1:1), 0°, 24 h	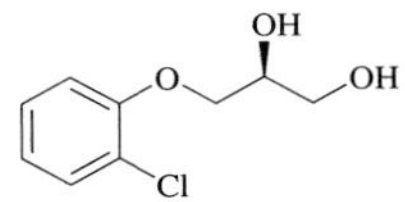	(85-95), 29	79
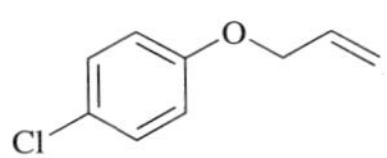	(DHQD)$_2$PHAL, $K_2OsO_2(OH)_4$, $K_3Fe(CN)_6$, K_2CO_3, t-BuOH:H_2O (1:1), 0°, 24 h	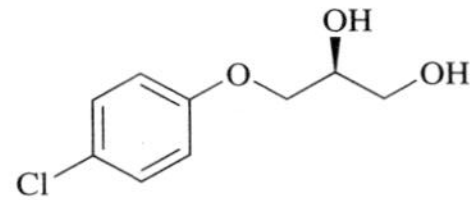	(85-95), 92	79
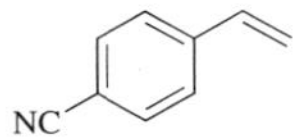	(DHQD)$_2$PYDZ, $K_2OsO_2(OH)_4$, $K_3Fe(CN)_6$, K_2CO_3, t-BuOH:H_2O (1:1), 0°, 18 h	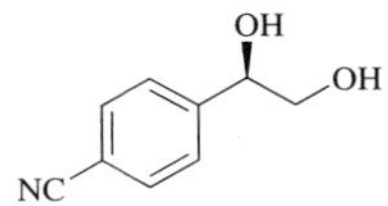	(92), 97	49
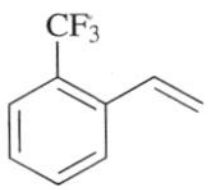	(DHQD)$_2$PYDZ, $K_2OsO_2(OH)_4$, $K_3Fe(CN)_6$, K_2CO_3, t-BuOH:H_2O (1:1), 0°	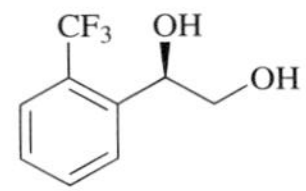 **I**	(—), 97	156
	DHQD-PYDZ-OMe, $K_2OsO_2(OH)_4$, $K_3Fe(CN)_6$, K_2CO_3, t-BuOH:H_2O (1:1), 0°	**I** (—), 39		156
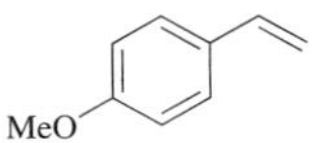	Ligand, $K_2OsO_2(OH)_4$, $K_3Fe(CN)_6$, K_2CO_3, t-BuOH:H_2O (1:1), 0°	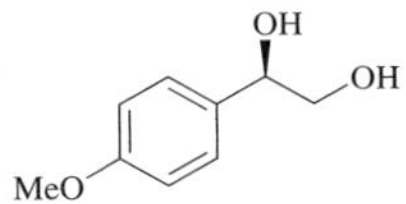		

Ligand	Temp	Time		
(DHQD)$_2$PYDZ	0°	—	(97), 97	49, 73
6	0°	6-12 h	(88), 97	73
(DHQD)$_2$PHAL	—	—	(—), 98	210

TABLE 1. REACTIONS OF TERMINAL ALKENES (*Continued*)

Substrate	Conditions	Product(s) and Yield(s) (%), % ee	Refs.
C_9			
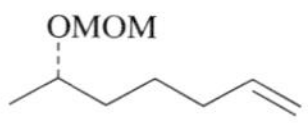	(DHQD)$_2$PYR, $K_2OsO_2(OH)_4$, $K_3Fe(CN)_6$, K_2CO_3, t-BuOH:H_2O (1:1), 0°	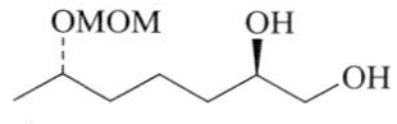 (85), —	217
	(DHQ)$_2$PYR, $K_2OsO_2(OH)_4$, $K_3Fe(CN)_6$, K_2CO_3, t-BuOH:H_2O (1:1), 0°	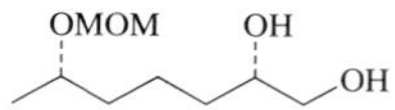 (79), —	217
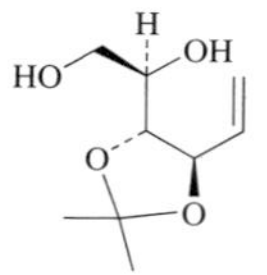	DHQ-MHQ, OsO_4, NMO, acetone-H_2O, rt, 7 min	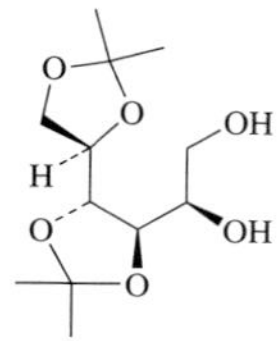 (80-85), 86 de	218
C_{10}			
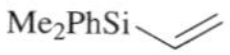	Ligand, $K_2OsO_2(OH)_4$, $K_3Fe(CN)_6$, K_2CO_3, t-BuOH:H_2O (1:1), 0°, 48 h	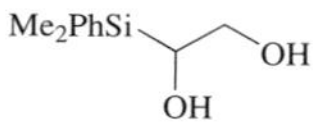	113
	Ligand		
	(DHQ)$_2$PHAL	(84), 25	
	(DHQD)$_2$PHAL	(83), 27	
	Ligand, $K_2OsO_2(OH)_4$, $K_3Fe(CN)_6$, K_2CO_3, t-BuOH:H_2O (1:1), 0°, 15-30 h	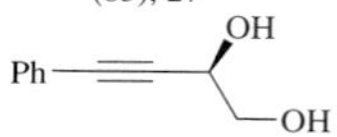	219
	Ligand		
	DHQD-PHN	(91), 53	
	(DHQD)$_2$PHAL	(91), 73	
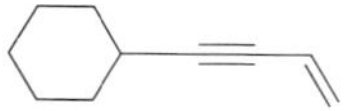	Ligand, $K_2OsO_2(OH)_4$, $K_3Fe(CN)_6$, K_2CO_3, t-BuOH:H_2O (1:1), 0°, 15-30 h		219
	Ligand		
	DHQD-PHN	(67), 44	
	(DHQD)$_2$PHAL	(—), 72	

Substrate	Conditions	Product(s) and Yield(s) (%)	Refs.
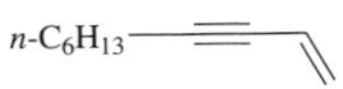	Ligand, $K_2OsO_2(OH)_4$, $K_3Fe(CN)_6$, K_2CO_3, t-BuOH:H_2O (1:1), 0°, 15-30 h		219
	Ligand		
	DHQD-PHN	(76), 38	
	$(DHQD)_2$PHAL	(—), 54	
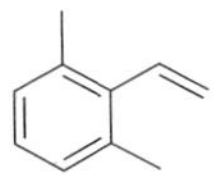	DHQD-p-chlorobenzoate, $K_2OsO_2(OH)_4$, NMO, acetone-H_2O, 0°, 3 h	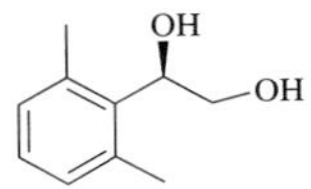 (80-95), 65	18
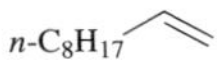	Ligand, $K_2OsO_2(OH)_4$, $K_3Fe(CN)_6$, K_2CO_3, t-BuOH:H_2O (1:1)	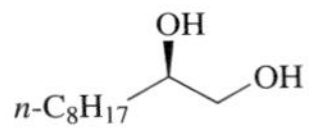	

Ligand	Temp	Time	Product(s) and Yield(s) (%)	Refs.
$(DHQD)_2$PYR	0°	3 h	(—), 89	77, 157
$(DHQD)_2$PHAL	0°	3 h	(—), 84	77, 200, 152, 66, 157
DHQD-PHN	0°	18 h	(75-95), 74	72, 220
DHQD-PHN	rt	24 h	(75-95), 66	220
DHQD-MEQ	0°	18 h	(75-95), 65	72
DHQD-p-chlorobenzoate	rt	18 h	(75-95), 45	72, 220
$(DHQD)_2$PYDZ	0°	6-24 h	(—), 83	152, 66
$(DHQD)_2$TP	0°	6-24 h	(—), 39	152
DHQD-9-phenyl ether	rt	24 h	(75-95), 53	220
DHQD-9-(1-naphthyl) ether	rt	24 h	(75-95), 66	220
$(DHQD)_2$AQN	0°	12 h	(65-90), 92	86
$(DHQD)_2$DPP	—	—	(—), 89	86, 157
$(DHQD)_2$DPPHAL	0°	6-24	(—), 67	157

TABLE 1. REACTIONS OF TERMINAL ALKENES (Continued)

Substrate	Conditions			Product(s) and Yield(s) (%), % ee	Refs.
C_{10} 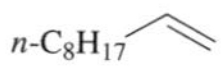	Ligand, $K_2OsO_2(OH)_4$, $K_3Fe(CN)_6$, K_2CO_3, t-BuOH:H_2O (1:1)			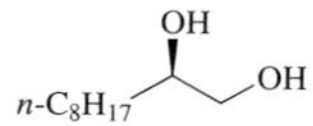	
	Ligand	Temp	Time		
	(DHQD)$_2$PHAL	—	—	(—), 84	47, 221
	18	—	—	(—), 85	47
	1	0°	12 h	(—), 40	47
	2	0°	12 h	(—), 66	47
	3	0°	12 h	(—), 57	47
	4	0°	12 h	(—), 78	47
	5	0°	12 h	(—), 84	47
	(DHQD)$_2$PYDZ	0°	6-12 h	(95), 79	73
	6	0°	6-12 h	(86), 78	73
	17	0°	6-12 h	(—), 44	73
	7a	0°	12 h	(—), 88	221
	7b	0°	12 h	(—), 86	221
	7c	0°	12 h	(—), 88	221
	7d	0°	12 h	(—), 86	221
	8a	0°	12 h	(—), 88	221
	8b	0°	12 h	(—), 91	221
	8c	0°	12 h	(—), 82	221
	8d	0°	12 h	(—), 90	221
	8e	0°	12 h	(—), 92	221
	(DHQD)$_2$DPP	0°	12 h	(—), 89	221
	9	0°	12 h	(—), 92	221
	10a	—	—	(51), 84	166
	10c	—	—	(53), 61	166
	10b	—	<5 h	(86), 87	169

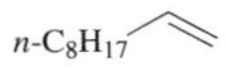

Reagents and Conditions	Product(s) and Yield(s) (%), % ee	Refs.
Ligand, OsO_4, $K_3Fe(CN)_6$, K_2CO_3, t-BuOH:H_2O (1:1), 0°, 48 h	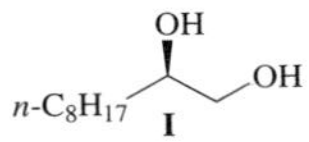 **I**	

Ligand	**I**	Refs.
41a	(68-80), 60	163, 206,
44	(68-80), 70	163

Reagents and Conditions	Product(s) and Yield(s) (%), % ee	Refs.
15, OsO_4, NMO, CH_3CN:H_2O (8:2), 5°, 19 h	**I** (84), 22	207
Ligand, $K_2OsO_2(OH)_4$, $K_3Fe(CN)_6$, K_2CO_3, t-BuOH:H_2O (1:1)	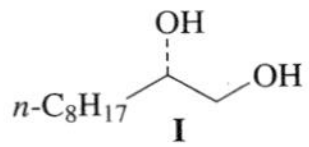 **I**	

Ligand	Temp	Time	**I**	Refs.
$(DHQ)_2$DPP	0°	6-24 h	(—), 81-85	157, 86
$(DHQ)_2$DPPHAL	0°	6-24 h	(—), 73	157
$(DHQ)_2$PHAL	—	—	(—), 80	200, 157
DHQ-PHN	0°	24 h	(75-95), 69	220
DHQ-9-phenyl ether	rt	24 h	(75-95), 44	220
DHQ-9-(1-naphthyl) ether	rt	24 h	(75-95), 56	220
$(DHQ)_2$AQN	0°	12 h	(65-90), 87	86
$(DHQ)_2$PYR	0°	3 h	(—), 76	77, 157
22	20°	20 h	(80), 64	164

Reagents and Conditions	Product(s) and Yield(s) (%), % ee	Refs.
11, $K_2OsO_2(OH)_4$, $K_3Fe(CN)_6$, K_2CO_3, t-BuOH:H_2O (1:1), <5 h	**I** (88), 74	169
42a, OsO_4, NMO, acetone: H_2O (10:1), 0°, 7 h	**I** (89), 14	161
42a, OsO_4, $K_3Fe(CN)_6$, K_2CO_3, t-BuOH:H_2O (1:1), 0°, 48 h	**I** (65), 50	161

TABLE 1. REACTIONS OF TERMINAL ALKENES (*Continued*)

Substrate	Conditions	Product(s) and Yield(s) (%), % ee	Refs.
C_{10} 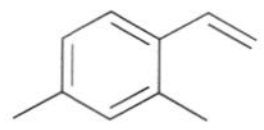	AD-Mix β, *t*-BuOH:H_2O (1:1), 0°	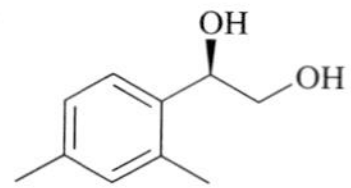 (85), >90	222
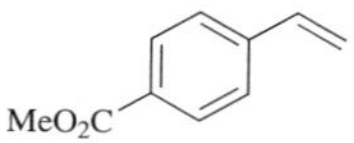	$(DHQD)_2PHAL$, $K_2OsO_2(OH)_4$, $K_3Fe(CN)_6$, K_2CO_3, *t*-BuOH:H_2O (1:1), 0°	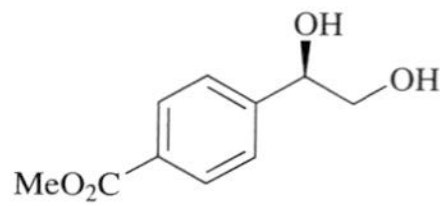 (77), >99	97
	$(DHQ)_2PHAL$, $K_2OsO_2(OH)_4$, $K_3Fe(CN)_6$, K_2CO_3, *t*-BuOH:H_2O (1:1), 0°	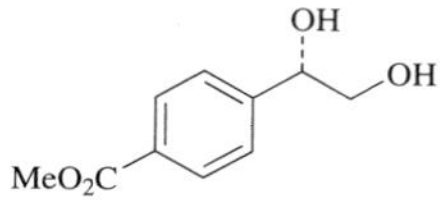 (83), >99	97
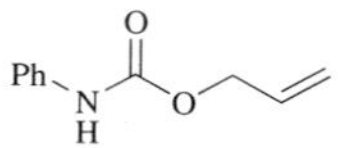	$(DHQ)_2PHAL$, $K_2OsO_2(OH)_4$, $K_3Fe(CN)_6$, K_2CO_3, *t*-BuOH:H_2O (1:1), 0°, 18 h	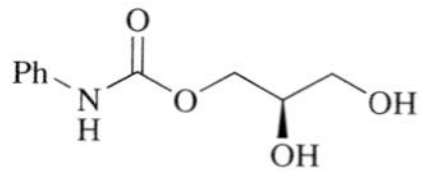 (99), >99	81
	$(DHQD)_2PHAL$, $K_2OsO_2(OH)_4$, $K_3Fe(CN)_6$, K_2CO_3, *t*-BuOH:H_2O (1:1), 0°, 18 h	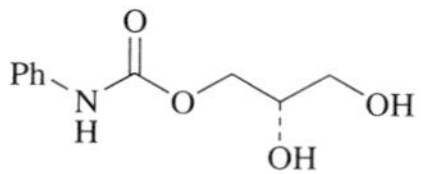 (99), >99	81
	Ligand, $K_2OsO_2(OH)_4$, $K_3Fe(CN)_6$, K_2CO_3, $NaHCO_3$, *t*-BuOH:H_2O (1:1), 0°	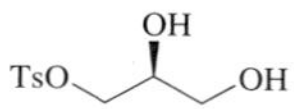	
	Ligand: $(DHQD)_2PHAL$; Time: 6-24 h	(—), 40	19, 86
	Ligand: $(DHQD)_2AQN$; Time: 12 h	(65-90), 83	86

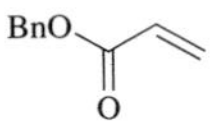	(DHQ)$_2$PHAL, $K_2OsO_2(OH)_4$, $K_3Fe(CN)_6$, K_2CO_3, t-BuOH:H_2O (1:1), 0°, 2-24 h	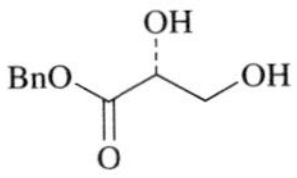	(—), 70	19, 223
	(DHQD)$_2$PHAL, $K_2OsO_2(OH)_4$, $K_3Fe(CN)_6$, K_2CO_3, t-BuOH:H_2O (1:1), 0°, 2-24 h	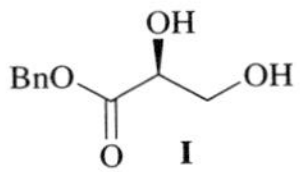 I	(—), 77	19, 86, 223
	(DHQD)$_2$AQN, $K_2OsO_2(OH)_4$, $K_3Fe(CN)_6$, K_2CO_3, $NaHCO_3$, t-BuOH:H_2O (1:1) , 0°, 12 h	**I** (65-90), 88		86
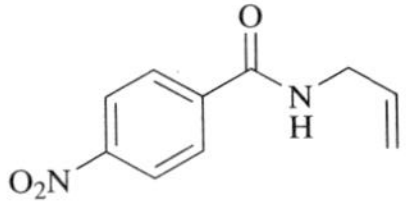	(DHQD)$_2$PYDZ, $K_2OsO_2(OH)_4$, $K_3Fe(CN)_6$, K_2CO_3, t-BuOH:H_2O (1:1), 0°, 1.5-3.5 h	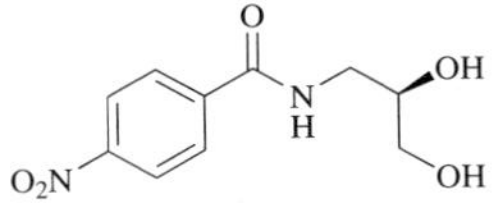	(96), 81	61
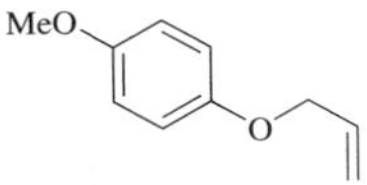	AD-Mix β, t-BuOH:H_2O (1:1), 0°, 24 h	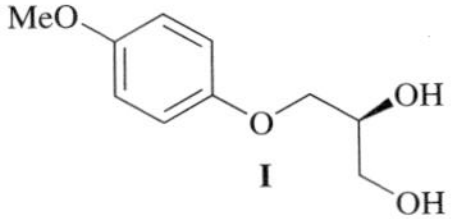 **I**	(85), 92	80
	(DHQD)$_2$PHAL, $K_2OsO_2(OH)_4$, $K_3Fe(CN)_6$, $K_2S_2O_8$, K_2CO_3, t-BuOH:H_2O (1:1), 0°, 24 h	**I** (85), —		224
	(DHQD)$_2$PHAL, $K_2OsO_2(OH)_4$, $K_3Fe(CN)_6$, K_2CO_3, t-BuOH:H_2O (1:1), 0°, 24 h	**I** (85-95), 90		79
	AD-Mix α, t-BuOH:H_2O (1:1), 0°, 24 h	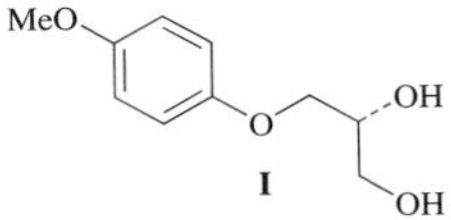 **I**	(85-90), 90-95	80, 225
	(DHQ)$_2$PHAL, $K_2OsO_2(OH)_4$, $K_3Fe(CN)_6$, $K_2S_2O_8$, K_2CO_3, t-BuOH:H_2O (1:1), 0°, 24 h	**I** (84), —		224

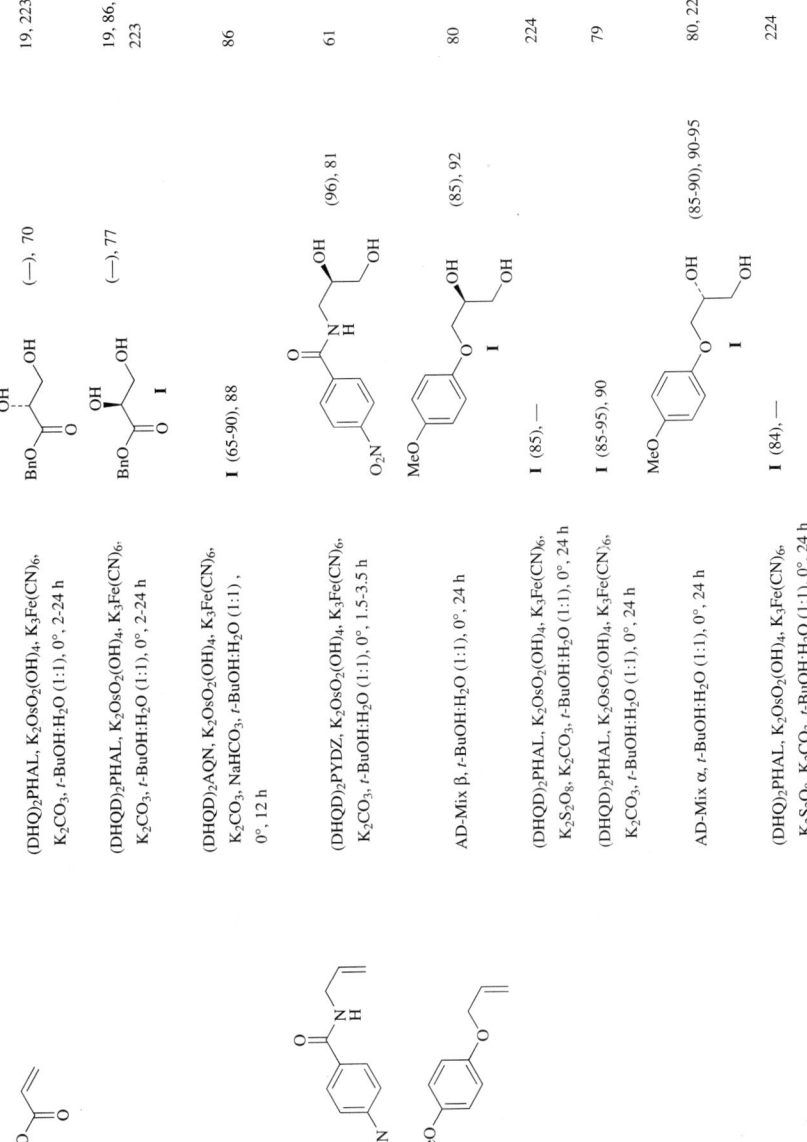

TABLE 1. REACTIONS OF TERMINAL ALKENES (*Continued*)

Substrate	Conditions	Product(s) and Yield(s) (%), % ee	Refs.
C_{10}			
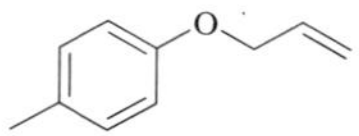	$(DHQD)_2PHAL$, $K_2OsO_2(OH)_4$, $K_3Fe(CN)_6$, K_2CO_3, *t*-BuOH:H_2O (1:1), 0°, 24 h	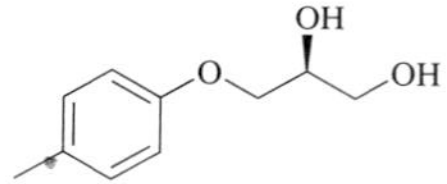 (85-95), 89	79
	$(DHQD)_2PHAL$, $K_2OsO_2(OH)_4$, $K_3Fe(CN)_6$, K_2CO_3, *t*-BuOH:H_2O (1:1), 0°, 24 h	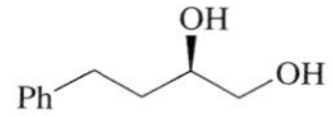 (85-95), 84	79
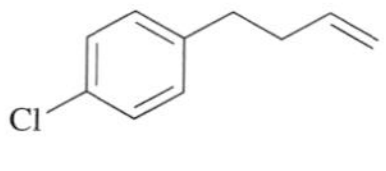	$(DHQD)_2PHAL$, $K_2OsO_2(OH)_4$, $K_3Fe(CN)_6$, K_2CO_3, *t*-BuOH:H_2O (1:1), 0°, 24 h	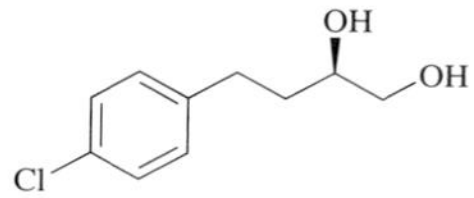 (85-95), 84	79
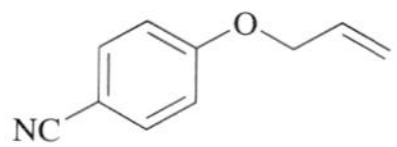	$(DHQD)_2PHAL$, $K_2OsO_2(OH)_4$, $K_3Fe(CN)_6$, K_2CO_3, *t*-BuOH:H_2O (1:1), 0°, 24 h	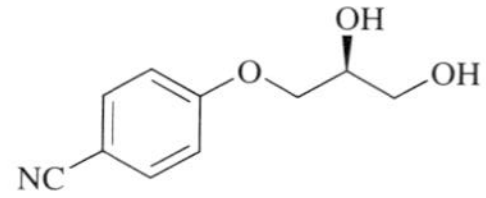 (85-95), 95	79
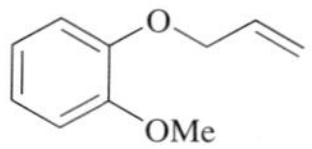	$(DHQD)_2PHAL$, $K_2OsO_2(OH)_4$, $K_3Fe(CN)_6$, K_2CO_3, *t*-BuOH:H_2O (1:1), 0°, 24 h	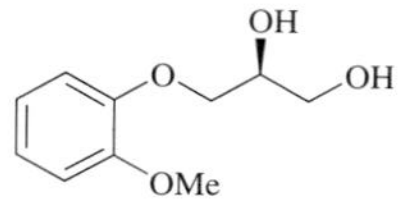 (85-95), 63	79
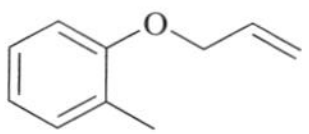	$(DHQD)_2PHAL$, $K_2OsO_2(OH)_4$, $K_3Fe(CN)_6$, K_2CO_3, *t*-BuOH:H_2O (1:1), 0°, 24 h	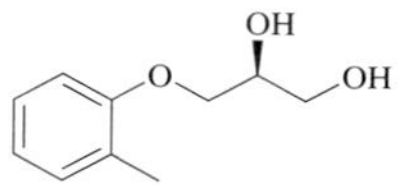 (43-48), 94	79
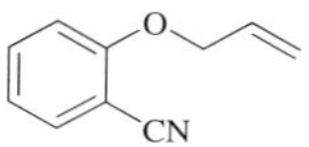	$(DHQD)_2PHAL$, $K_2OsO_2(OH)_4$, $K_3Fe(CN)_6$, K_2CO_3, *t*-BuOH:H_2O (1:1), 0°, 24 h	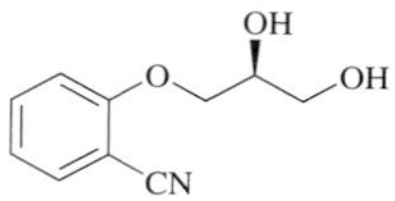 (85-95), 28	79

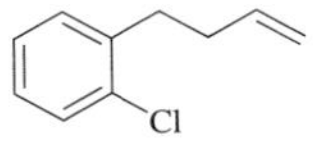

$(DHQD)_2$PHAL, $K_2OsO_2(OH)_4$, $K_3Fe(CN)_6$, K_2CO_3, *t*-BuOH:H_2O (1:1), 0°, 24 h

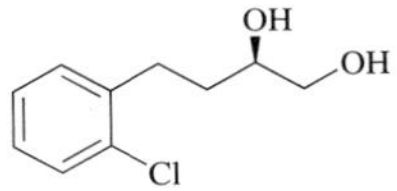

(85-95), 77 | 79

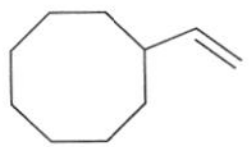

Ligand, $K_2OsO_2(OH)_4$, $K_3Fe(CN)_6$, K_2CO_3, *t*-BuOH:H_2O (1:1)

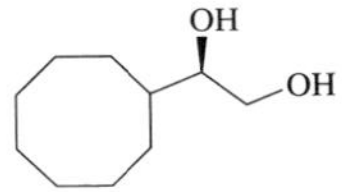

Ligand	Temp	Time		
$(DHQD)_2$PYR	0°	3 h	(75-95), 93-94	77, 72
DHQD-MEQ	0°	18-24 h	(75-95), 85	72
DHQD-*p*-chlorobenzoate	rt	18-24 h	(75-95), 64	72, 220
$(DHQD)_2$PHAL	0°	3 h	(—), 87	77
DHQD-9-phenyl ether	rt	24 h	(75-95), 70	220
DHQD-9-(1-naphthyl) ether	rt	24 h	(75-95), 84	220
DHQD-PHN	rt	24 h	(75-95), 88	220
DHQD-PHN	0°	24 h	(75-95), 93	220

Ligand, $K_2OsO_2(OH)_4$, $K_3Fe(CN)_6$, K_2CO_3, *t*-BuOH:H_2O (1:1)

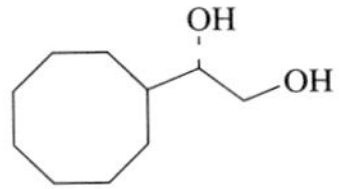

Ligand	Temp	Time		
DHQ-9-(1-naphthyl) ether	rt	24 h	(75-95), 73	220
DHQ-PHN	rt	24 h	(75-95), 83	220
DHQ-PHN	0°	24 h	(75-95), 88	220
DHQ-*p*-chlorobenzoate	rt	18-24 h	(75-95), 54	220
$(DHQ)_2$PYR	0°	3 h	(75-95), 87-88	77, 72
DHQ-9-phenyl ether	rt	24 h	(75-95), 58	220

TABLE 1. REACTIONS OF TERMINAL ALKENES (*Continued*)

Substrate	Conditions	Product(s) and Yield(s) (%), % ee	Refs.
C_{10}			
	$(DHQD)_2PHAL$, $K_2OsO_2(OH)_4$, $K_3Fe(CN)_6$, K_2CO_3, t-BuOH:H_2O (1:1), 0°, 4 h	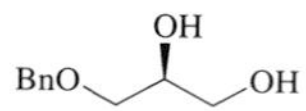 **I** (—), 17	78
	AD-Mix β, t-BuOH:H_2O (1:1), 0°, 4 h	**I** (—), 17	78
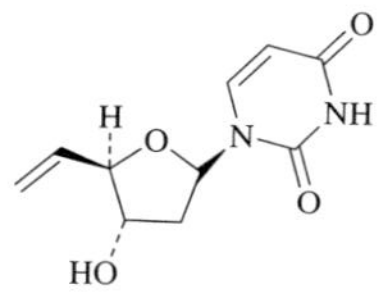	AD-Mix α, t-BuOH:H_2O (1:1)	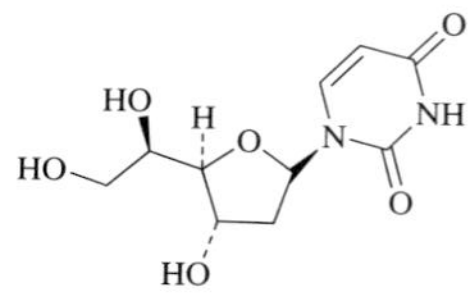 (72), 67 de	226
	AD-Mix β, t-BuOH:H_2O (1:1)	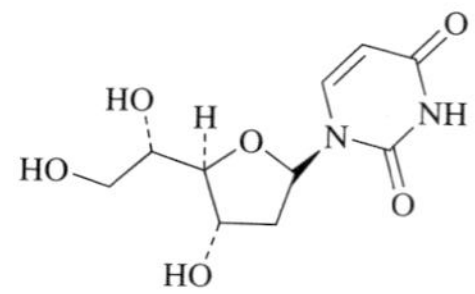 (91), 60 de	226
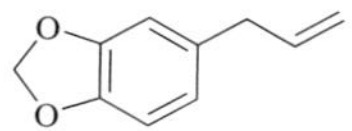	AD-Mix β, $MeSO_2NH_2$, t-BuOH:H_2O (1:1), 20°, 24 h	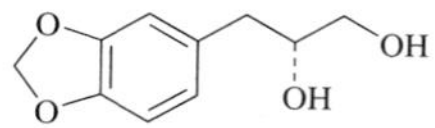 (87), 88	227
C_{11}			
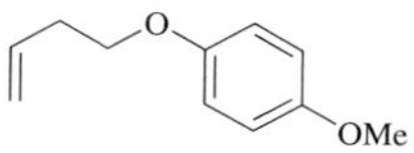	$(DHQD)_2PYDZ$, $K_2OsO_2(OH)_4$, $K_3Fe(CN)_6$, K_2CO_3, t-BuOH:H_2O (1:1), 0°, 1.5-4 h	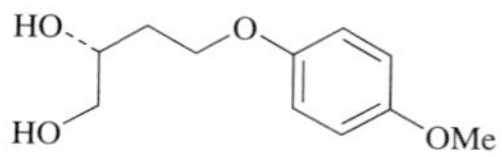 (96), 91	83, 61, 66
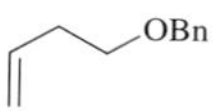	$(DHQD)_2PYDZ$, $K_2OsO_2(OH)_4$, $K_3Fe(CN)_6$, K_2CO_3, t-BuOH:H_2O (1:1), 0°, 1.5-4 h	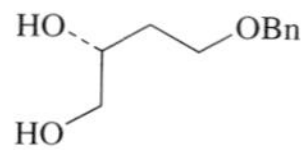 (97), 52	83, 61

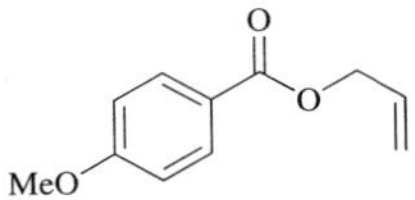	Ligand, $K_2OsO_2(OH)_4$, $K_3Fe(CN)_6$, K_2CO_3, *t*-BuOH:H_2O (1:1), 0°	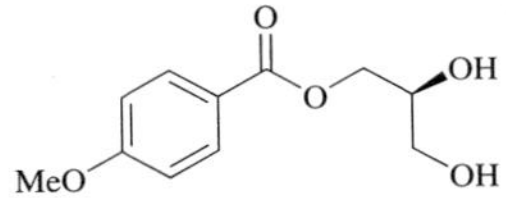	
	Ligand / Time		
	$(DHQD)_2PYDZ$ / 1.5-3.5 h	(99), 98	61, 63, 133, 66
	6 / 18 h	(99), 95	63
	17 / 18 h	(98), 90	63
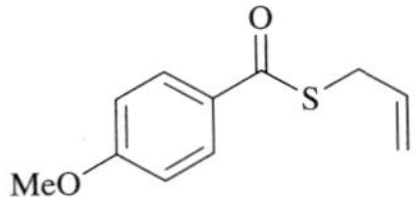	$(DHQD)_2PYDZ$, $K_2OsO_2(OH)_4$, $K_3Fe(CN)_6$, K_2CO_3, *t*-BuOH:H_2O (1:1), 0°, 1.5-3.5 h	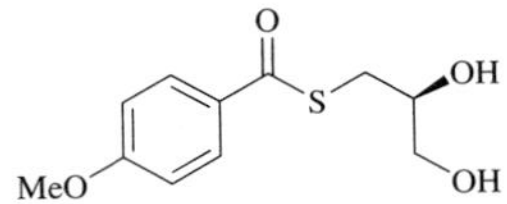 (46), 77	61
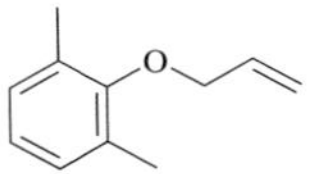	$(DHQD)_2PHAL$, $K_2OsO_2(OH)_4$, $K_3Fe(CN)_6$, K_2CO_3, *t*-BuOH:H_2O (1:1), 0°, 24 h	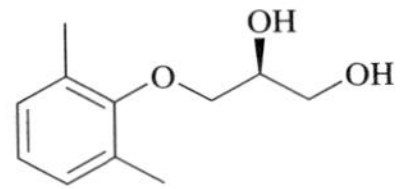 (85-95), 36	79
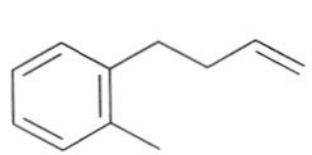	$(DHQD)_2PHAL$, $K_2OsO_2(OH)_4$, $K_3Fe(CN)_6$, K_2CO_3, *t*-BuOH:H_2O (1:1), 0°, 24 h	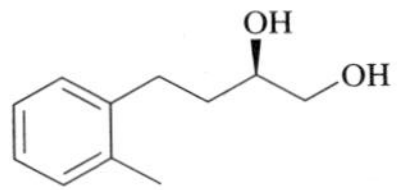 (85-95), 81	79
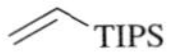	$(DHQ)_2PHAL$, $K_2OsO_2(OH)_4$, $K_3Fe(CN)_6$, K_2CO_3, *t*-BuOH:H_2O (1:1), 0°, 48 h	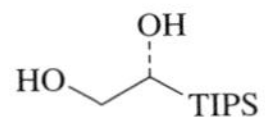 (74), 6	114
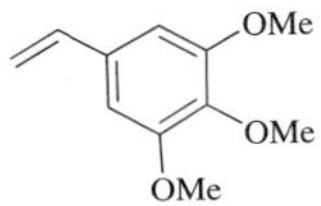	AD-Mix α, *t*-BuOH:H_2O (1:1), 0°, 18 h	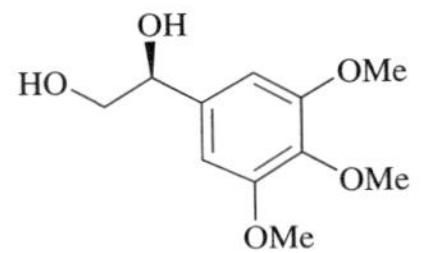 (96), 97	228

TABLE 1. REACTIONS OF TERMINAL ALKENES (*Continued*)

Substrate	Conditions	Product(s) and Yield(s) (%), % ee	Refs.
C_{11} 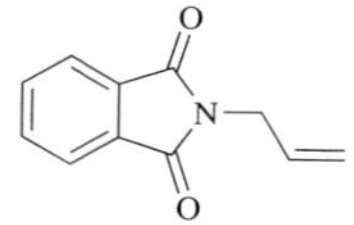	Ligand, OsO_4, $K_3Fe(CN)_6$, K_2CO_3, *t*-BuOH:H_2O (1:1), 20°	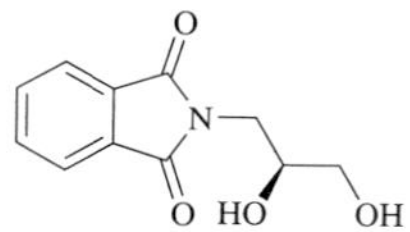	229
	Ligand		
	$(DHQD)_2PYR$	(72-80), 37	
	$(DHQD)_2PHAL$	(72-80), 25-37	
	$(DHQD)_2DPP$	(72-80), 37	
	$(DHQD)_2TP$	(72-80), 25-37	
	$(DHQD)_2ITP$	(72-80), 37	
	$(DHQD)_2PY$	(72-80), 25-37	
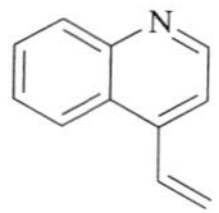	AD-Mix β, *t*-BuOH:H_2O (1:1), 0°, 5 h	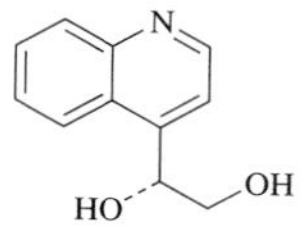 (75-85), >95	197
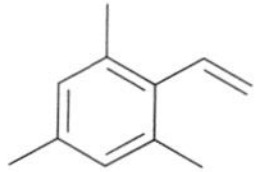	AD-Mix β, *t*-BuOH:H_2O (1:1), 0°	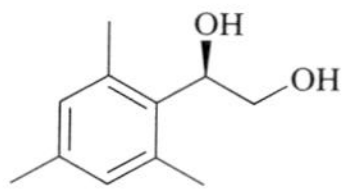 (80), —	222
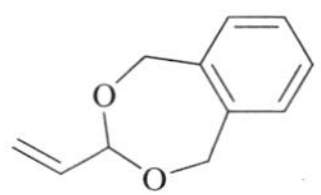	$(DHQ)_2PHAL$, $K_2OsO_2(OH)_4$, $K_3Fe(CN)_6$, K_2CO_3, *t*-BuOH:H_2O (1:1), rt, 48 h	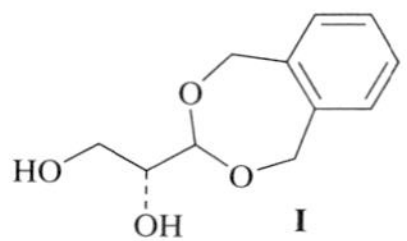 **I** (93), >95	84
	AD-Mix α, *t*-BuOH:H_2O (1:1), 4°	**I** (—), 97	185

Ligand, $K_2OsO_2(OH)_4$, $K_3Fe(CN)_6$, K_2CO_3, t-BuOH:H_2O (1:1), rt

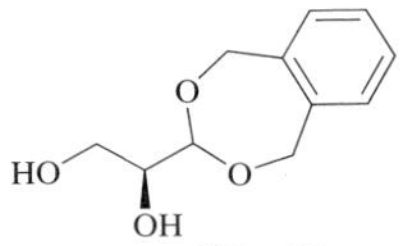

Ligand	Temp	Time	Product	Ref.
(DHQD)$_2$PHAL	rt	48 h	(90), >95	84
DHQD-PHN	0°	24 h	(90), 86	230

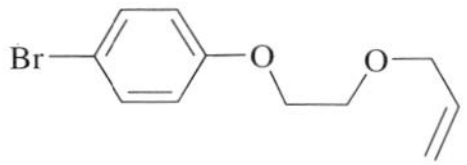

AD-Mix β, t-BuOH:H_2O (1:1), 0°, 24 h

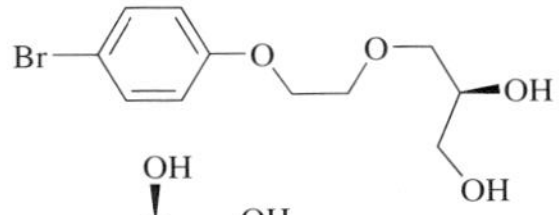

(—), 64 | 231

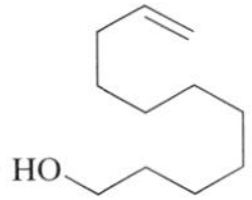

(DHQD)$_2$AQN, $K_2OsO_2(OH)_4$, $K_3Fe(CN)_6$, K_2CO_3, t-BuOH:H_2O (1:1), rt, 18 h

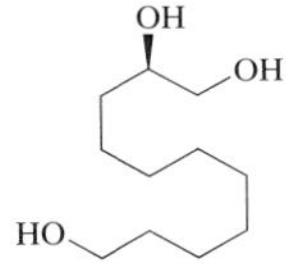

(78), 88 | 232

C_{12}

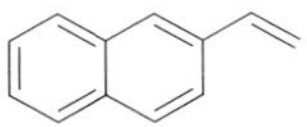

Ligand, $K_2OsO_2(OH)_4$, $K_3Fe(CN)_6$, K_2CO_3, t-BuOH:H_2O (1:1)

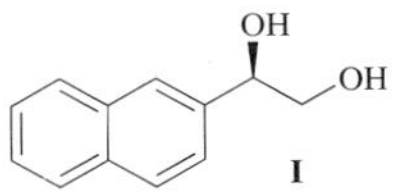

I

Ligand	Temp	Time	**I**	Ref.
DHQD-PHN	0°	18-24 h	(75-95), 83	72
DHQD-MEQ	0°	18-24 h	(75-95), 93	72
DHQD-*p*-chlorobenzoate	rt	18-24 h	(75-95), 88	72
17	0°	18 h	(99), 91	63, 73
6	0°	18 h	(93), 98	63, 73
(DHQD)$_2$PYDZ	0°	18 h	(98), 99	63, 73
(DHQD)$_2$PHAL	0°	6-24 h	(98), 99	157, 210
(DHQD)$_2$DPP	0°	6-24 h	(—), >99.5	157
(DHQD)$_2$DPPHAL	0°	6-24 h	(—), 97	157

AD-Mix β, t-BuOH:H_2O (1:1), 0° — **I** (54-90), >95 — 222, 197

(DHQD)$_2$PHAL, $K_2OsO_2(OH)_4$, pH 10.4, t-BuOH:H_2O, O_2, 50°, 14-18 h — **I** (55), 96 — 178

TABLE 1. REACTIONS OF TERMINAL ALKENES (*Continued*)

Substrate	Conditions	Product(s) and Yield(s) (%), % ee	Refs.
C_{12} 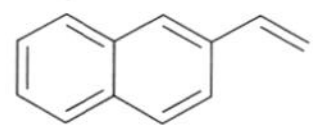	(DHQ)$_2$PHAL, $K_2OsO_2(OH)_4$, $K_3Fe(CN)_6$, K_2CO_3, *t*-BuOH:H_2O (1:1), 3°, 4 h	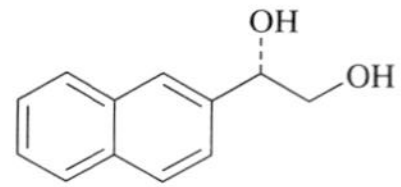 (91), 92.4	76
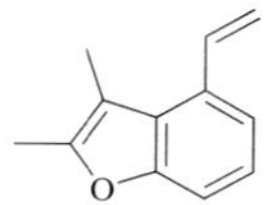	(DHQ)$_2$PHAL, $K_2OsO_2(OH)_4$, $K_3Fe(CN)_6$, K_2CO_3, *t*-BuOH:H_2O (1:1), 3°, 4 h	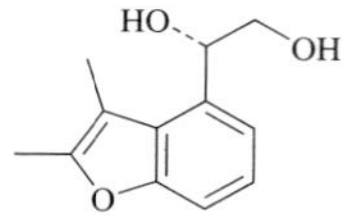 (95), 98.8	76
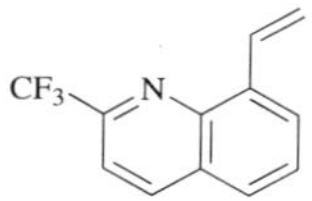	(DHQ)$_2$PHAL, $K_2OsO_2(OH)_4$, $K_3Fe(CN)_6$, K_2CO_3, *t*-BuOH:H_2O (1:1), 3°, 4 h	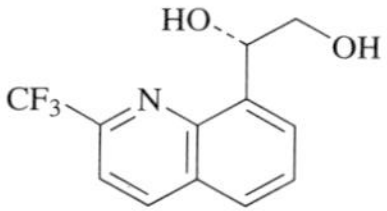 (62), 96	76
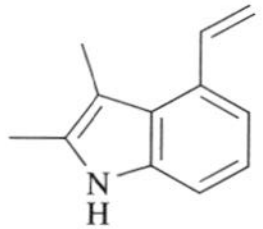	(DHQ)$_2$PHAL, $K_2OsO_2(OH)_4$, $K_3Fe(CN)_6$, K_2CO_3, *t*-BuOH:H_2O (1:1), 3°, 4 h	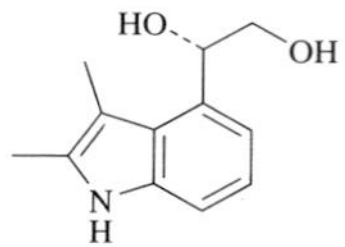 (87), 99.6	76
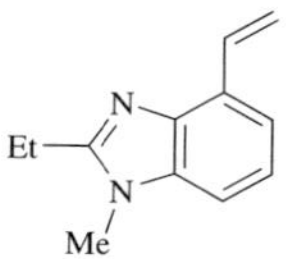	(DHQ)$_2$PHAL, $K_2OsO_2(OH)_4$, $K_3Fe(CN)_6$, K_2CO_3, *t*-BuOH:H_2O (1:1), 3°, 4 h	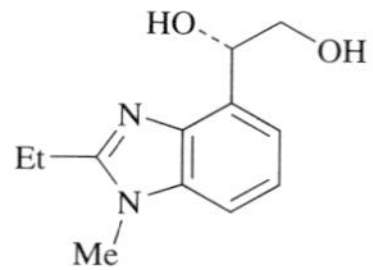 (80), 99.0	76
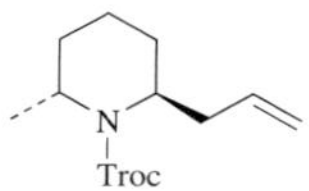	(DHQD)$_2$PYR, $K_2OsO_2(OH)_4$, $K_3Fe(CN)_6$, K_2CO_3, *t*-BuOH:H_2O (1:1), 0°	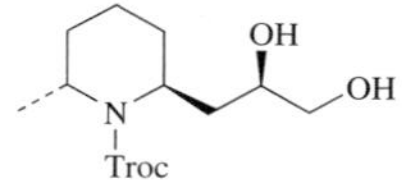 (90), 53 de	217
	(DHQ)$_2$PYR, $K_2OsO_2(OH)_4$, $K_3Fe(CN)_6$, K_2CO_3, *t*-BuOH:H_2O (1:1), 0°	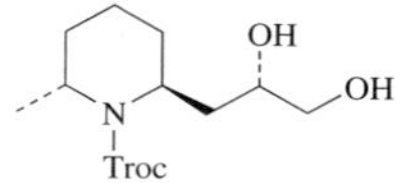 (74), 46 de	217

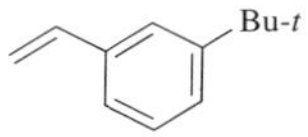	(DHQD)$_2$PHAL, $K_2OsO_2(OH)_4$, $K_3Fe(CN)_6$, K_2CO_3, *t*-BuOH:H_2O (1:1), 0°, 6-24 h	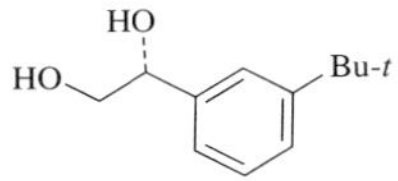	(—), 95	60
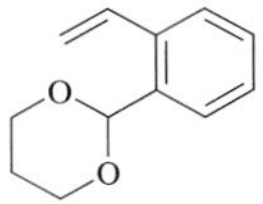	AD-Mix α, *t*-BuOH:H_2O (1:1)	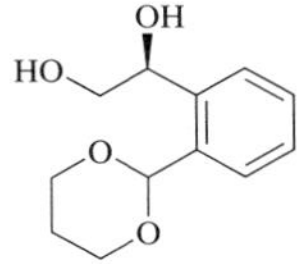	(—), 100	233
	AD-Mix β, *t*-BuOH:H_2O (1:1)	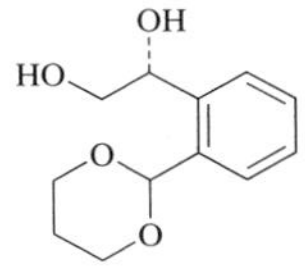	(—), 100	223
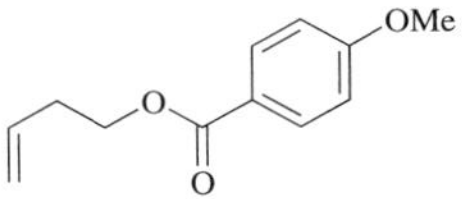	(DHQD)$_2$PYDZ, $K_2OsO_2(OH)_4$, $K_3Fe(CN)_6$, K_2CO_3, *t*-BuOH:H_2O (1:1), 0°, 2-24 h	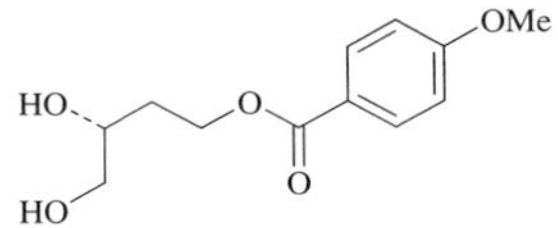	(99), 40	83, 61
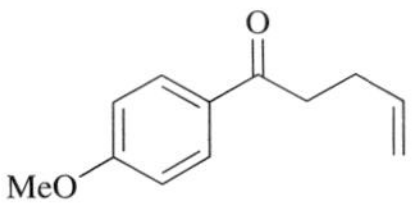	(DHQD)$_2$PYDZ, $K_2OsO_2(OH)_4$, $K_3Fe(CN)_6$, K_2CO_3, *t*-BuOH:H_2O (1:1), 0°, 2-24 h	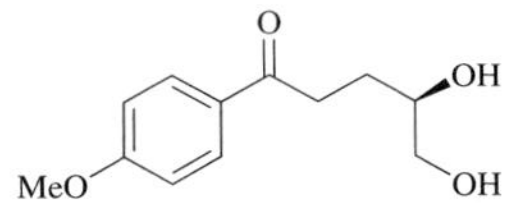	(96), 98	61, 66
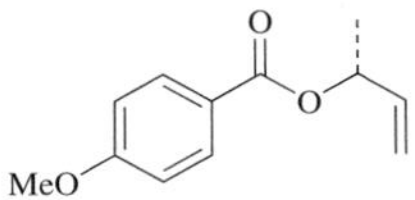	**17**, $K_2OsO_2(OH)_4$, $K_3Fe(CN)_6$, K_2CO_3, *t*-BuOH:H_2O (1:1), 0°, 2 h	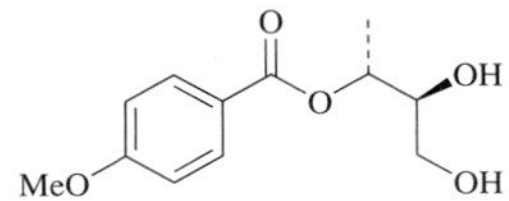	(99), 98 de	63
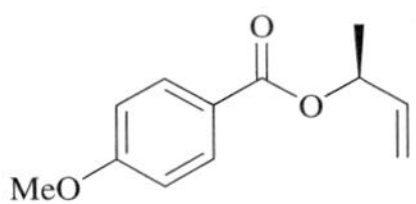	**17**, $K_2OsO_2(OH)_4$, $K_3Fe(CN)_6$, K_2CO_3, *t*-BuOH:H_2O (1:1), 0°	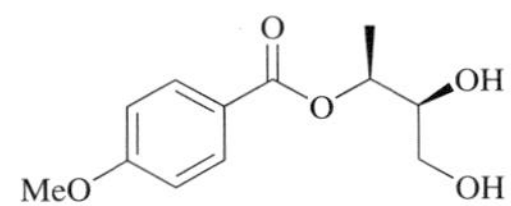	(—), 26 de	63

TABLE 1. REACTIONS OF TERMINAL ALKENES (*Continued*)

Substrate	Conditions	Product(s) and Yield(s) (%), % ee	Refs.
C_{12} 	(DHQ)$_2$PHAL, $K_2OsO_2(OH)_4$, $K_3Fe(CN)_6$, K_2CO_3, *t*-BuOH:H_2O (1:1), 0°, 57 h	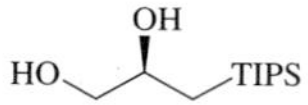 (98), 16	114
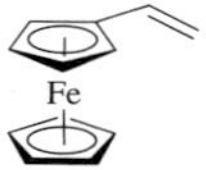	(DHQD)$_2$PHAL, $K_2OsO_2(OH)_4$, $K_3Fe(CN)_6$, K_2CO_3, *t*-BuOH:H_2O (1:1), 0°, 24 h	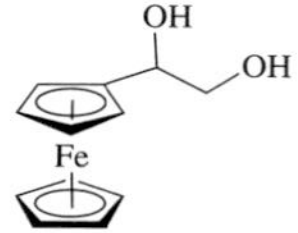 **I** (12), O	117
	(DHQD)$_2$PHAL, 30 mol % $K_2OsO_2(OH)_4$, $K_3Fe(CN)_6$, K_2CO_3, $MeSO_2NH_2$, acetone:*t*-BuOH:H_2O (2:1:1), 0°, 24 h	**I** (25), 5	117
	(DHQD)$_2$PHAL, $K_2OsO_2(OH)_4$, $K_3Fe(CN)_6$, K_2CO_3, acetone:H_2O (1:1), 20°, 48 h	**I** (70), 72	117
	(DHQD)$_2$PYR, $K_2OsO_2(OH)_4$, $K_3Fe(CN)_6$, K_2CO_3, CH_3CN:H_2O (1:1), 20°, 23 h	**I** (79), 96	117
	(DHQ)$_2$PHAL, $K_2OsO_2(OH)_4$, $K_3Fe(CN)_6$, K_2CO_3, $MeSO_2NH_2$, acetone:H_2O (1:1), 20°, 24 h	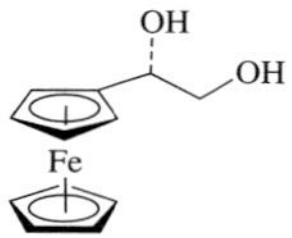 **I** (54), 24	117
	(DHQ)$_2$PYR, $K_2OsO_2(OH)_4$, $K_3Fe(CN)_6$, K_2CO_3, CH_3CN:H_2O (1:1), 20°, 3 h	**I** (87), 84	117
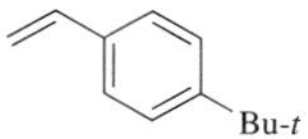	(DHQD)$_2$PHAL, $K_2OsO_2(OH)_4$, $K_3Fe(CN)_6$, K_2CO_3, *t*-BuOH:H_2O (1:1)	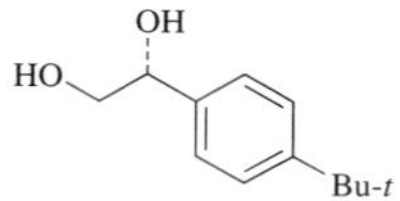 (—), 98	210

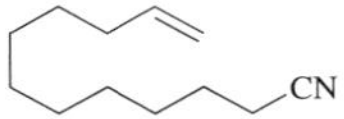	AD-Mix β, t-BuOH:H_2O (1:1), 0°	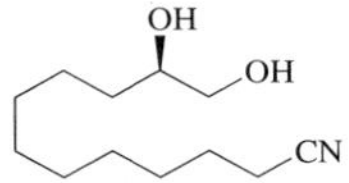	(94), 87	234
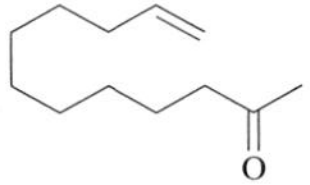	AD-Mix β, t-BuOH:H_2O (1:1), 0°	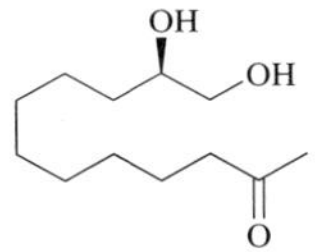	(99), 86	234
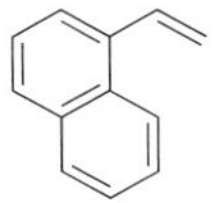	$(DHQD)_2PHAL$, $K_2OsO_2(OH)_4$, $K_3Fe(CN)_6$, K_2CO_3, t-BuOH:H_2O (1:1), 0°	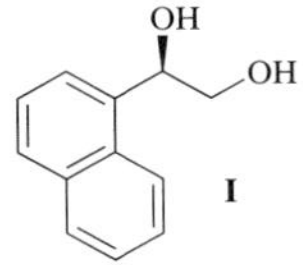 **I**	(100), 96	204
	AD-Mix β, t-BuOH:H_2O (1:1), 0°	**I**	(70-90), 98-99	222, 197, 235
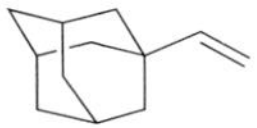	$(DHQD)_2PHAL$, $K_2OsO_2(OH)_4$, $K_3Fe(CN)_6$, K_2CO_3, t-BuOH-H_2O, 0°, 12 h	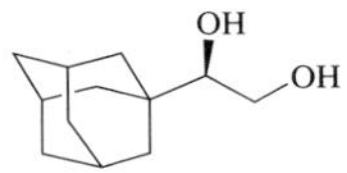 **I**	(85), 87	182
	$(DHQD)_2PYR$, $K_2OsO_2(OH)_4$, $K_3Fe(CN)_6$, K_2CO_3, t-BuOH-H_2O, 0°, 12 h	**I**	(85), 97	182
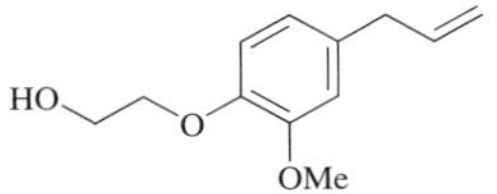	$(DHQD)_2AQN$, $K_2OsO_2(OH)_4$, $K_3Fe(CN)_6$, K_2CO_3, THF:H_2O (1:1), rt, 18 h	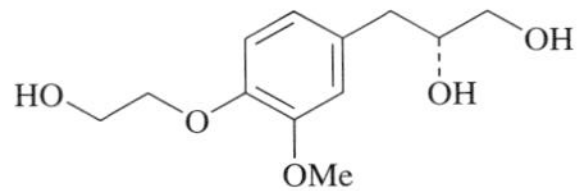	(75), 32	232

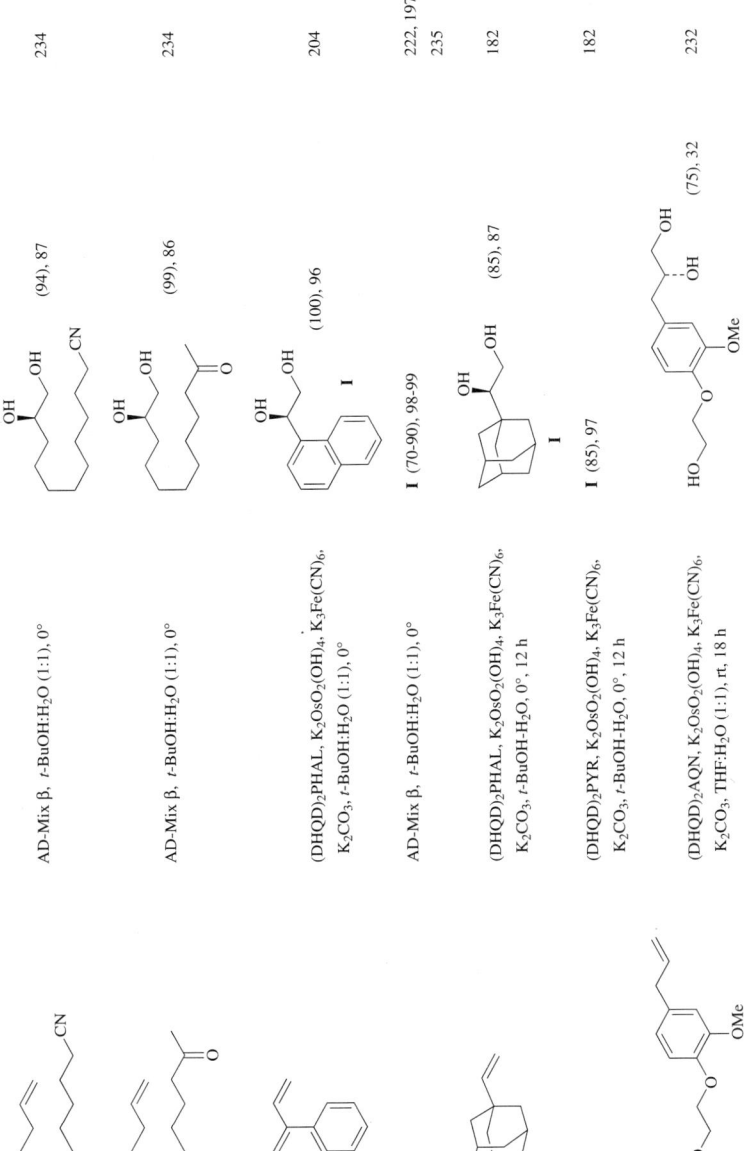

TABLE 1. REACTIONS OF TERMINAL ALKENES (*Continued*)

Substrate	Conditions	Product(s) and Yield(s) (%), % ee	Refs.
C_{12}			
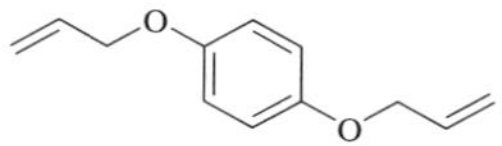	Ligand, $K_2OsO_2(OH)_4$, $K_3Fe(CN)_6$, K_2CO_3, t-BuOH:H_2O (1:1), 0°, 12 h	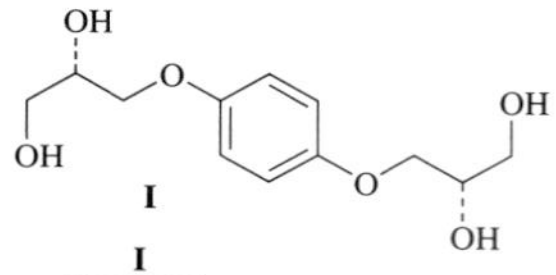 **I**	146
	Ligand	**I**	
	$(DHQD)_2PHAL$	(—), 88	
	$(DHQD)_2PYR$	(—), 56	
	$(DHQD)_2AQN$	(—), 96	
	Ligand, $K_2OsO_2(OH)_4$, I_2, K_2CO_3, t-BuOH:H_2O (1:1), 0°, 18 h	**I**	146
	Ligand	**I**	
	$(DHQD)_2PHAL$	(—), 72	
	$(DHQD)_2PYR$	(—), 44	
	$(DHQD)_2AQN$	(—), 94	
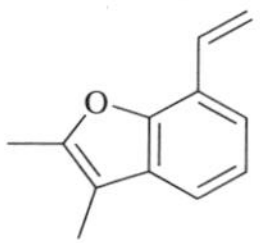	AD-Mix α, t-BuOH:H_2O (1:1), 3°, 4 h	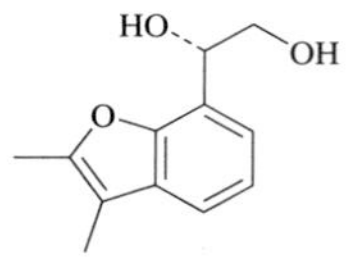 (95), 98.3	76
	AD-Mix β, t-BuOH:H_2O (1:1), 3°, 4 h	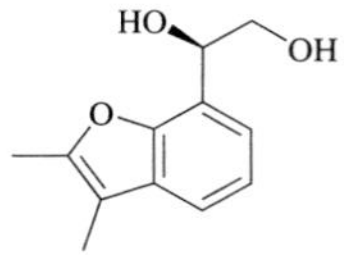 (99), —	76
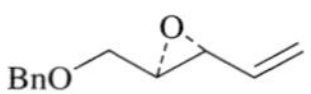	1. $(DHQ)_2PYR$, OsO_4, $K_3Fe(CN)_6$, K_2CO_3, t-BuOH:H_2O (1:1) 2. $(MeO)_2CMe_2$, TsOH, CH_2Cl_2	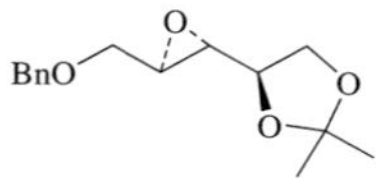 (73), 83 de	236

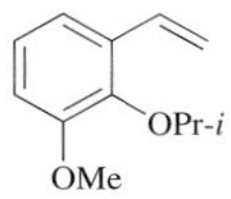	$(DHQ)_2PHAL$, $K_2OsO_2(OH)_4$, $K_3Fe(CN)_6$, K_2CO_3, t-BuOH:H_2O (1:1), 0°, 6-24 h	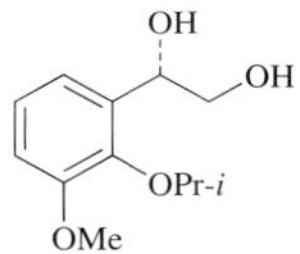 (—), 90	171
	$(DHQ)_2PHAL$, $K_2OsO_2(OH)_4$, NMO, t-BuOH:H_2O (1.4:1), rt, 6.5 h (slow addition of olefin)	**I** (95), 94	171
	AD-Mix α, t-BuOH:H_2O (1:1), 0°, 3 h	**I** (95), >90	237
	AD-Mix β, t-BuOH:H_2O (1:1), 0°, 3 h	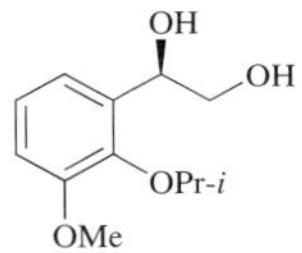 (95), >90	237
C_{12} 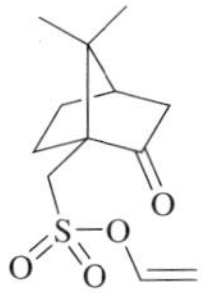	$(DHQD)_2PHAL$, $K_2OsO_2(OH)_4$, $K_3Fe(CN)_6$, K_2CO_3, t-BuOH:H_2O (1:1), 0°, 20 h	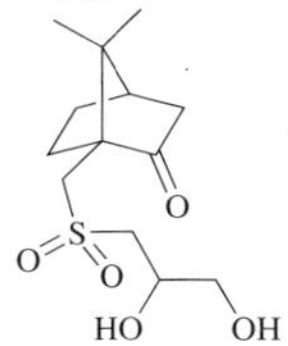 **I** (64), 30 de	81
	DHQ-PHN, $K_2OsO_2(OH)_4$, $K_3Fe(CN)_6$, K_2CO_3, t-BuOH:H_2O (1:1), 0°, 24 h	**I** (13), 40 de	81
C_{13} 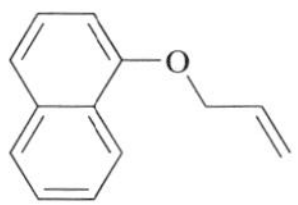	$(DHQ)_2PHAL$, $K_2OsO_2(OH)_4$, $K_3Fe(CN)_6$, K_2CO_3, t-BuOH:H_2O (1:1), 3°, 6-24 h	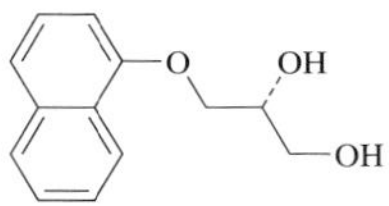 (80-98), 88	19
	$(DHQD)_2PHAL$, $K_2OsO_2(OH)_4$, $K_3Fe(CN)_6$, K_2CO_3, t-BuOH:H_2O (1:1), 0°, 6-24 h	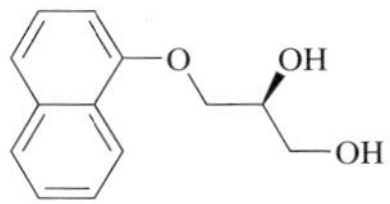 (80-98), 91	19, 79

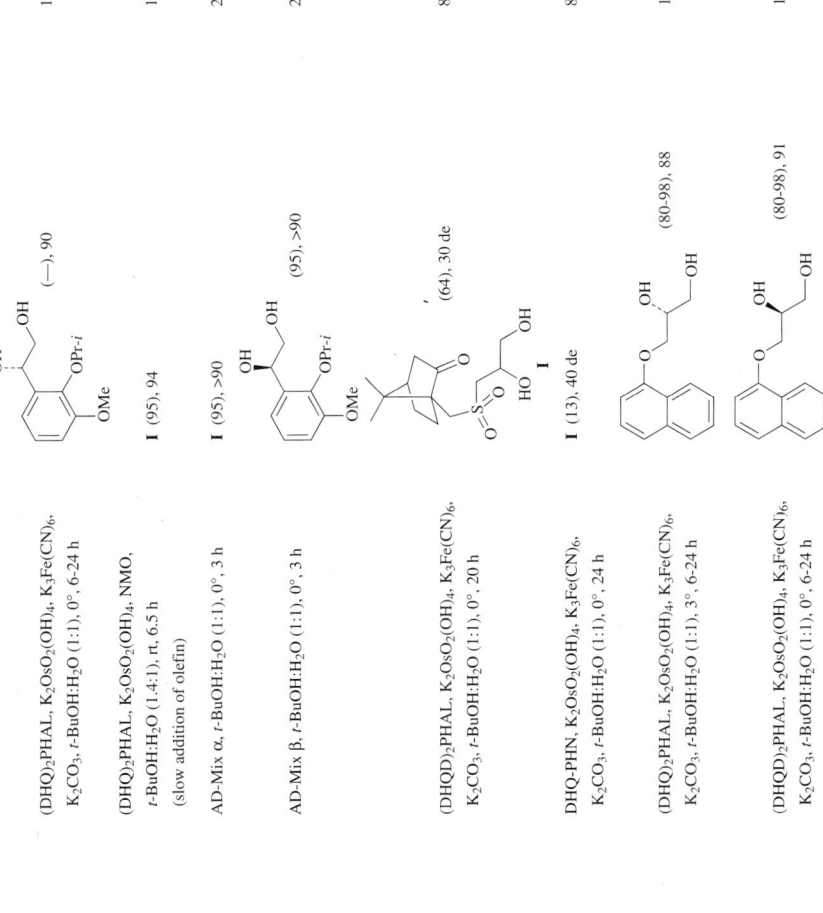

TABLE 1. REACTIONS OF TERMINAL ALKENES (*Continued*)

Substrate	Conditions	Product(s) and Yield(s) (%), % ee	Refs.
C_{13}			
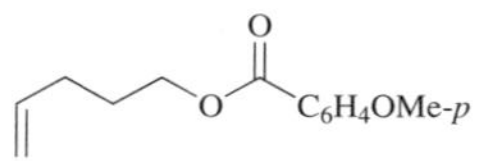	$(DHQD)_2PYDZ$, $K_2OsO_2(OH)_4$, $K_3Fe(CN)_6$, K_2CO_3, *t*-BuOH:H_2O (1:1), 0°, 4-22 h	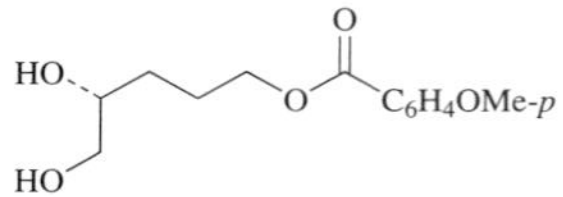 (99), 82	61, 66, 82
	21, $K_2OsO_2(OH)_4$, $K_3Fe(CN)_6$, K_2CO_3, *t*-BuOH:H_2O (1:1), 0°, 4-22 h	**I** (91), 86	82
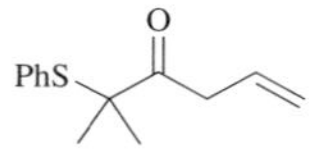	$(DHQD)_2PYR$, $K_2OsO_2(OH)_4$, $K_3Fe(CN)_6$, K_2CO_3, $MeSO_2NH_2$, *t*-BuOH:H_2O (1:1), 0°	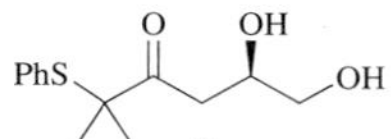 **I** (—), 73	238
	AD-Mix β, $MeSO_2NH_2$, 1:1 *t*-BuOH:H_2O, rt	**I** (77), 27	238
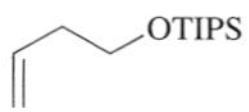	$(DHQD)_2PYDZ$, $K_2OsO_2(OH)_4$, $K_3Fe(CN)_6$, K_2CO_3, *t*-BuOH:H_2O (1:1), 0°, 4-22 h	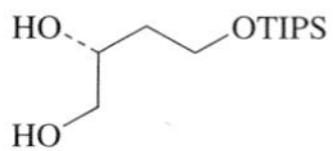 (97), 16	83, 61
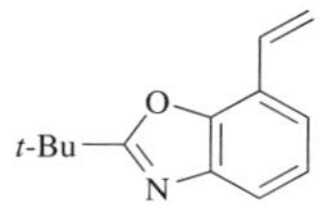	$(DHQ)_2PHAL$, $K_2OsO_2(OH)_4$, $K_3Fe(CN)_6$, K_2CO_3, *t*-BuOH:H_2O (1:1), 3°, 4 h	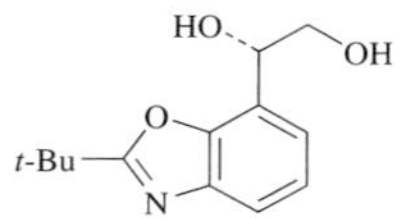 (90), 91.6	76
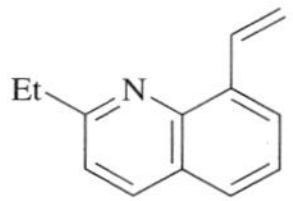	$(DHQ)_2PHAL$, $K_2OsO_2(OH)_4$, $K_3Fe(CN)_6$, K_2CO_3, *t*-BuOH:H_2O (1:1), 3°, 4 h	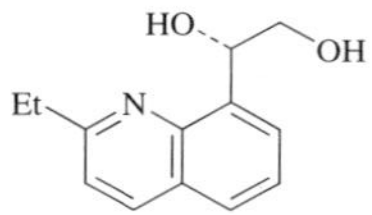 (88), 98.2	76

	$(DHQD)_2PHAL$, $K_2OsO_2(OH)_4$, $K_3Fe(CN)_6$, K_2CO_3, t-BuOH:H_2O (1:1), 3°, 4 h	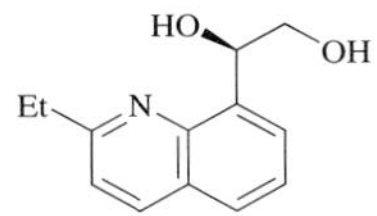	(—), —	76
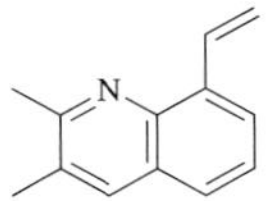	$(DHQ)_2PHAL$, $K_2OsO_2(OH)_4$, $K_3Fe(CN)_6$, K_2CO_3, t-BuOH:H_2O (1:1), 3°, 4 h	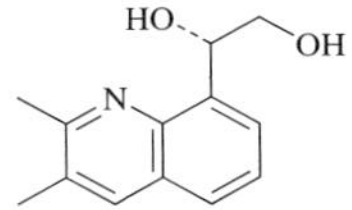	(96), 95.2	76
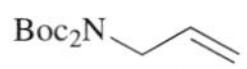	$(DHQD)_2PHAL$, $K_2OsO_2(OH)_4$, $K_3Fe(CN)_6$, K_2CO_3, t-BuOH:H_2O (1:1), 0°	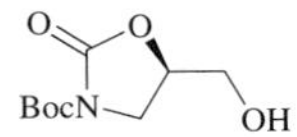	(93), 34	85
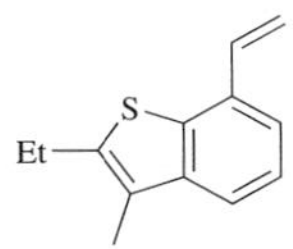	$(DHQ)_2PHAL$, $K_2OsO_2(OH)_4$, $K_3Fe(CN)_6$, K_2CO_3, t-BuOH:H_2O (1:1), 3°, 4 h	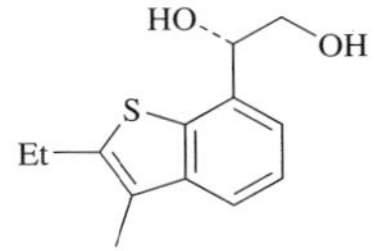	(95), 97.3	76
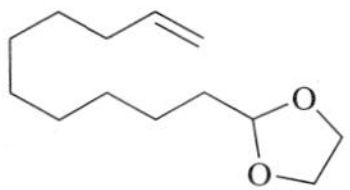	AD-Mix β, t-BuOH:H_2O (1:1), 0°	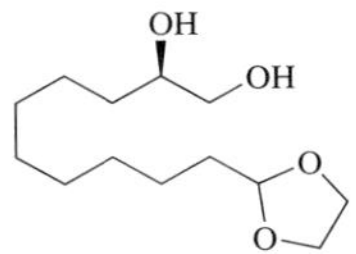	(93), 89	234
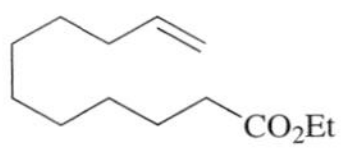	AD-Mix β, t-BuOH:H_2O (1:1), 0°	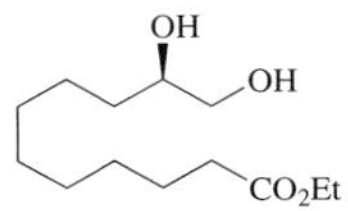	(96), 90	234

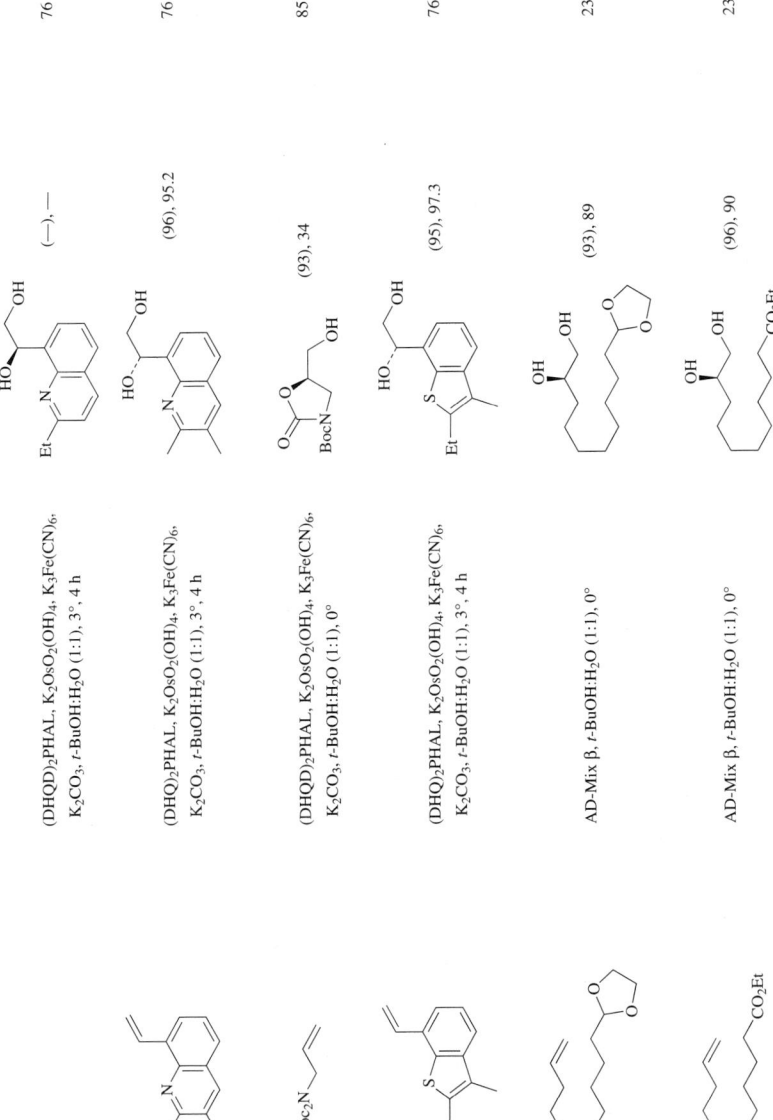

TABLE 1. REACTIONS OF TERMINAL ALKENES (*Continued*)

Substrate	Conditions	Product(s) and Yield(s) (%), % ee	Refs.
C_{13} 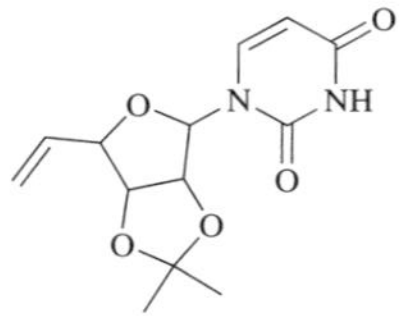	AD-Mix α, *t*-BuOH:H_2O (1:1)	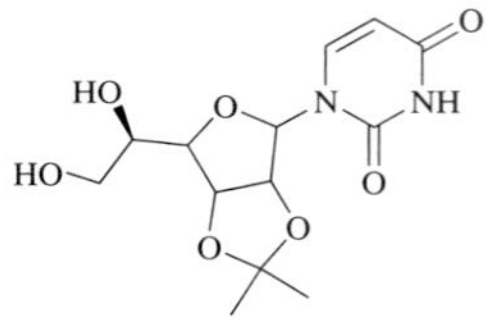 (89), >95 de	239
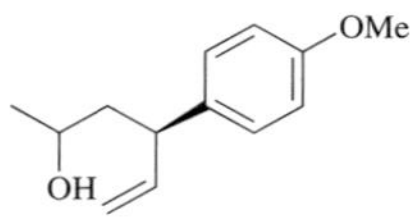	DHQ-*p*-chlorobenzoate, $K_2OsO_2(OH)_4$, $K_3Fe(CN)_6$, K_2CO_3, *t*-BuOH:H_2O (1:1)	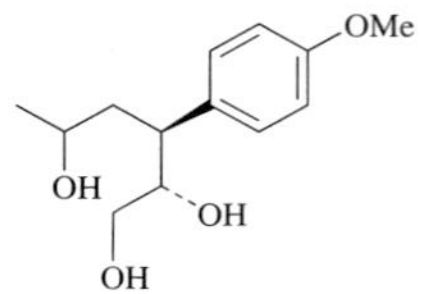 (76), —	240
	AD-Mix α, *t*-BuOH:H_2O (1:1), rt, 24 h	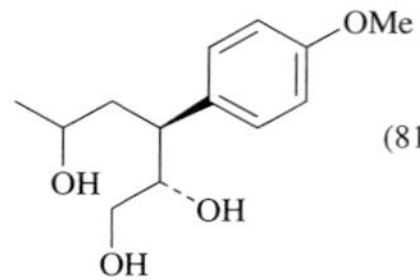 (81), >90 de	240
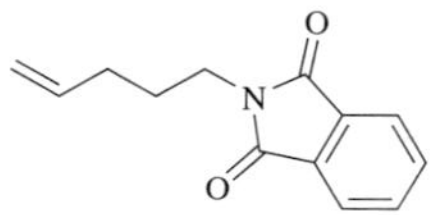	$(DHQD)_2PYR$, $K_2OsO_2(OH)_4$, $K_3Fe(CN)_6$, K_2CO_3, *t*-BuOH:H_2O (1:1), 0°, 42 h	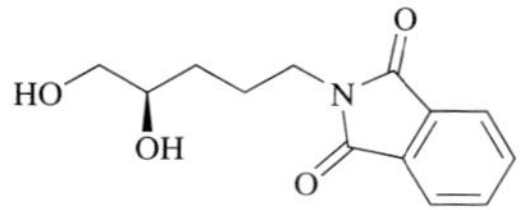 (93), 94	187
	$(DHQ)_2PYR$, $K_2OsO_2(OH)_4$, $K_3Fe(CN)_6$, K_2CO_3, *t*-BuOH:H_2O (1:1), 0°, 42 h	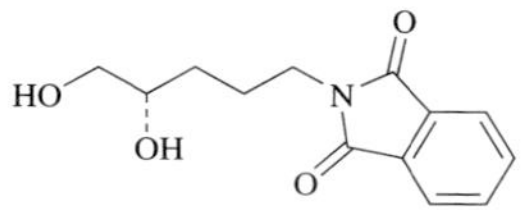 (91), —	187

C$_{14}$

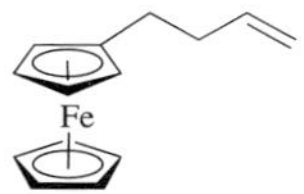

Ligand, $K_2OsO_2(OH)_4$, $K_3Fe(CN)_6$, K_2CO_3, solvent

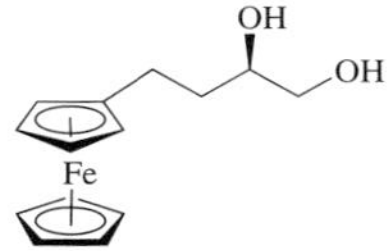

117

Ligand	Solvent	Temp	Time	
$(DHQD)_2PHAL$	t-BuOH:H_2O (1:1)	0°	12 h	(33), 50
$(DHQD)_2PHAL$	acetone:t-BuOH:H_2O (2:1:1)	0°	24 h	(50), 53
$(DHQD)_2PHAL$	acetone:H_2O (1:1)	20°	24 h	(99), 59
$(DHQD)_2PYR$	CH_3CN:H_2O (1:1)	20°	12 h	(61), 49
$(DHQD)_2PHAL$	CH_3CN:H_2O (1:1)	50°	5 h	(75), 56

Ligand, $K_2OsO_2(OH)_4$, $K_3Fe(CN)_6$, K_2CO_3, $MeSO_2NH_2$, solvent, 0°, 24 h

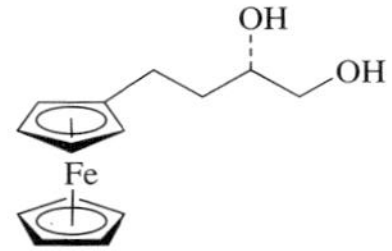

I

117

Ligand	Solvent	
$(DHQ)_2PHAL$	acetone:t-BuOH:H_2O (2:1:1)	(24), 57
$(DHQ)_2PHAL$	acetone:H_2O (1:1)	(54), 46

$(DHQ)_2PYR$, $K_2OsO_2(OH)_4$, $K_3Fe(CN)_6$, K_2CO_3, CH_3CN:H_2O (1:1), 20°, 12 h — **I** (71), 13 — 117

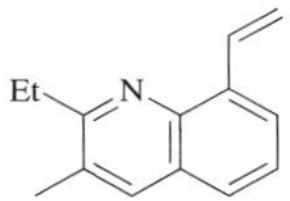

$(DHQ)_2PHAL$, $K_2OsO_2(OH)_4$, $K_3Fe(CN)_6$, K_2CO_3, t-BuOH:H_2O (1:1), 3°, 4 h

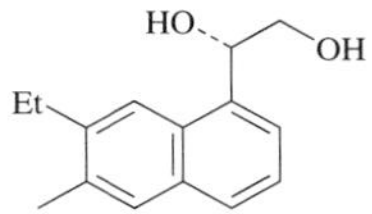

(87), 96 — 76

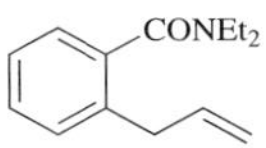

$(DHQD)_2PHAL$, $K_2OsO_2(OH)_4$, $K_3Fe(CN)_6$, K_2CO_3, t-BuOH:H_2O (1:1), 0°, 6-24 h

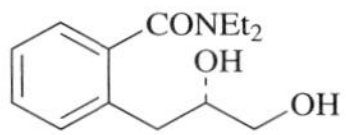

(>60-71), 16 — 241

TABLE 1. REACTIONS OF TERMINAL ALKENES (*Continued*)

Substrate	Conditions	Product(s) and Yield(s) (%), % ee	Refs.
C_{14} 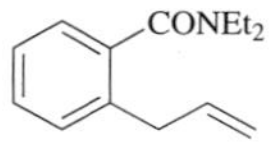	Ligand, $K_2OsO_2(OH)_4$, $K_3Fe(CN)_6$, K_2CO_3, *t*-BuOH:H_2O (1:1), 0°, 18 h	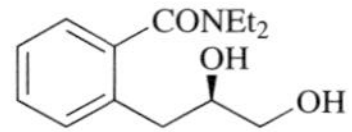	241
	Ligand		
	DHQD-PHN	(>60-71), 44	
	DHQD-*p*-chlorobenzoate	(>60-71), 15	
	$(DHQD)_2AQN$, $K_2OsO_2(OH)_4$, $K_3Fe(CN)_6$, K_2CO_3, *t*-BuOH:H_2O (1:1), 0°, 48 h	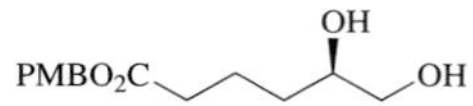 (79), 88-92	242
	$(DHQ)_2AQN$, OsO_4, $K_3Fe(CN)_6$, K_2CO_3, $NaHCO_3$, *t*-BuOH:H_2O (1:1), 0°, 12 h	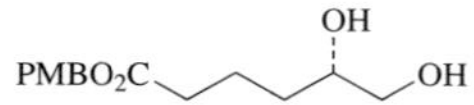 (83), 91	243
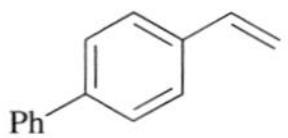	AD-Mix β, *t*-BuOH:H_2O (1:1), 0°	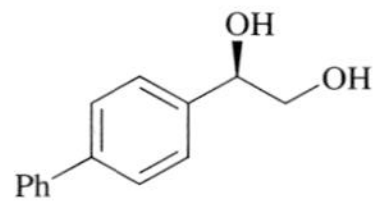 (57), >90	222
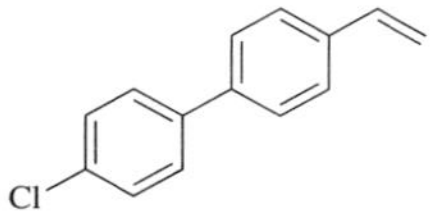	$(DHQD)_2PHAL$, $K_2OsO_2(OH)_4$, $K_3Fe(CN)_6$, K_2CO_3, *t*-BuOH:H_2O (1:1), 0°	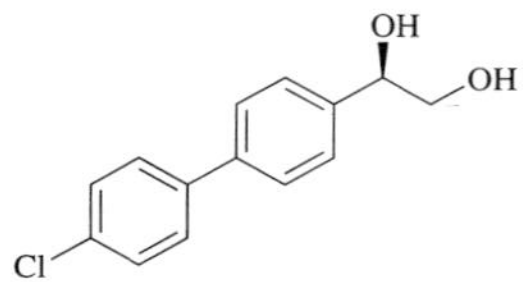 (72), >99	97
	$(DHQ)_2PHAL$, $K_2OsO_2(OH)_4$, $K_3Fe(CN)_6$, K_2CO_3, *t*-BuOH:H_2O (1:1), 0°	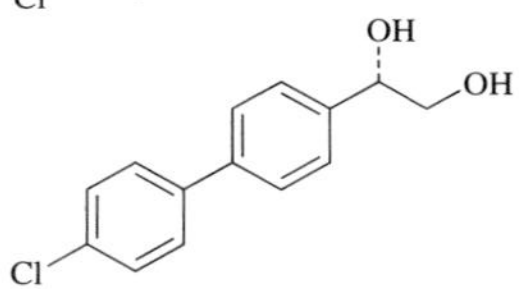 (71), >99	97

	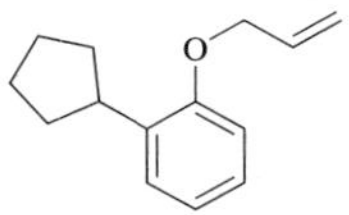	(DHQ)$_2$PHAL, OsO_4, $K_3Fe(CN)_6$, K_2CO_3, *t*-BuOH:H_2O (1:1), rt, 24 h	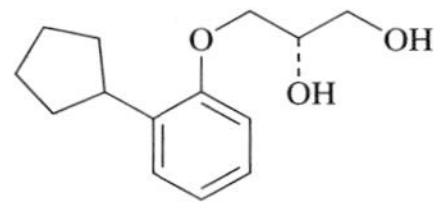 (90), 95	244
	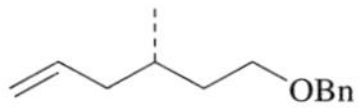	AD-Mix β, *t*-BuOH:H_2O (1:1), 0°, 5 h	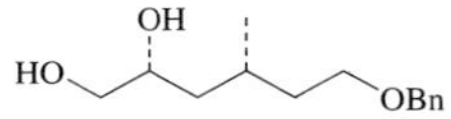 (70), 50 de	245
C_{15}	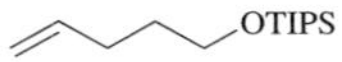	(DHQD)$_2$PYR, $K_2OsO_2(OH)_4$, $K_3Fe(CN)_6$, K_2CO_3, *t*-BuOH:H_2O (1:1), 0°	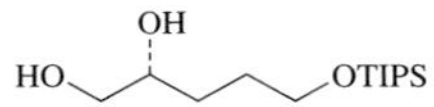 (70), 70	246
	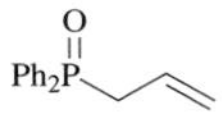	(DHQ)$_2$PHAL, $K_2OsO_2(OH)_4$, $K_3Fe(CN)_6$, K_2CO_3, *t*-BuOH:H_2O (1:1), 0° to rt, 6 days	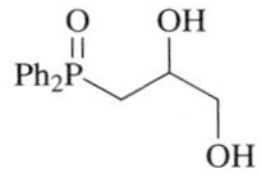 (53), 5	119
	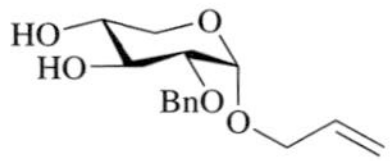	AD-Mix β, *t*-BuOH:H_2O (1:1), 0°	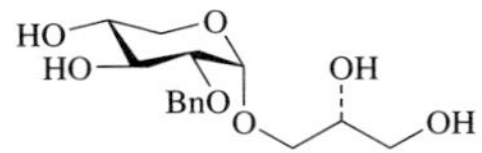 (—), 10 de	247, 248
	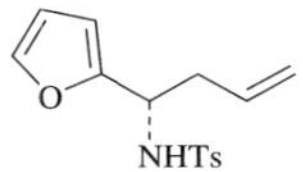	(DHQ)$_2$PYR, $K_2OsO_2(OH)_4$, $K_3Fe(CN)_6$, K_2CO_3, *t*-BuOH:H_2O (1:1), rt, 24 h	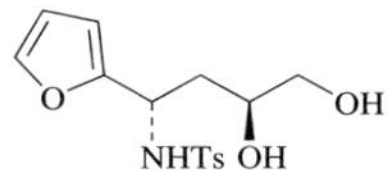 **I** (95), —	249
		(DHQ)$_2$PYR, $K_2OsO_2(OH)_4$, $K_3Fe(CN)_6$, K_2CO_3, *t*-BuOH:H_2O (1:1), 0°	**I** (95), — de	258
		(DHQD)$_2$PYR, $K_2OsO_2(OH)_4$, $K_3Fe(CN)_6$, K_2CO_3, *t*-BuOH:H_2O (1:1), rt, 24 h	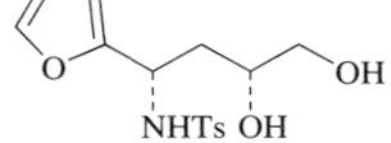 **I** (93), —	249
		(DHQD)$_2$PYR, $K_2OsO_2(OH)_4$, $K_3Fe(CN)_6$, K_2CO_3, *t*-BuOH:H_2O (1:1), 0°	**I** (93), — de	258

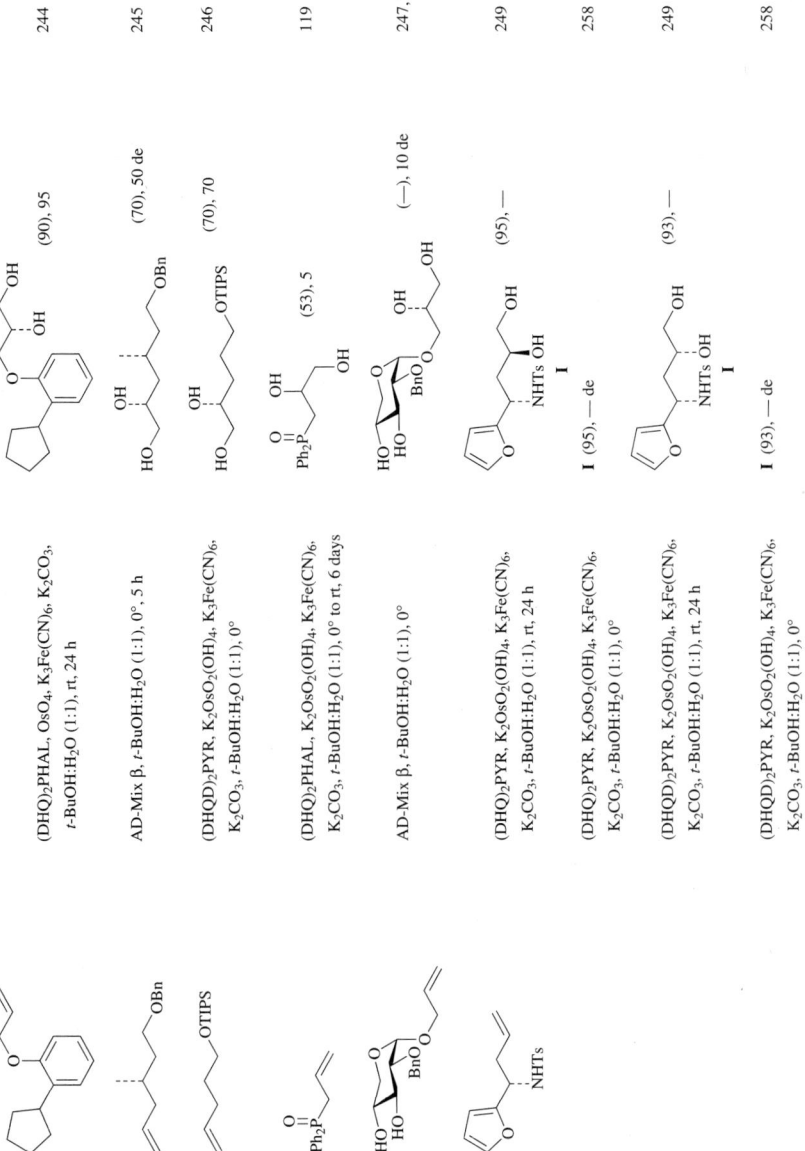

TABLE 1. REACTIONS OF TERMINAL ALKENES (*Continued*)

Substrate	Conditions	Product(s) and Yield(s) (%), % ee	Refs.
C_{15} 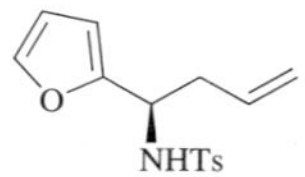	(DHQ)$_2$PYR, $K_2OsO_2(OH)_4$, $K_3Fe(CN)_6$, K_2CO_3, *t*-BuOH:H_2O (1:1), rt, 24 h	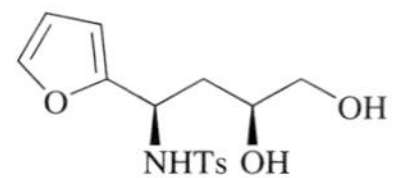 **I** (93), —	249
	(DHQ)$_2$PYR, $K_2OsO_2(OH)_4$, $K_3Fe(CN)_6$, K_2CO_3, *t*-BuOH:H_2O (1:1), 0°	**I** (94), — de	258
	(DHQD)$_2$PYR, $K_2OsO_2(OH)_4$, $K_3Fe(CN)_6$, K_2CO_3, *t*-BuOH:H_2O (1:1), rt, 24 h	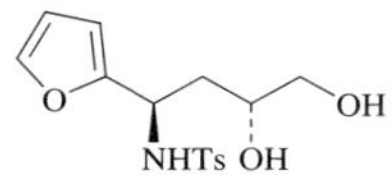 **I** (94), —	249
	(DHQD)$_2$PYR, $K_2OsO_2(OH)_4$, $K_3Fe(CN)_6$, K_2CO_3, *t*-BuOH:H_2O (1:1), 0°	**I** (93), — de	258
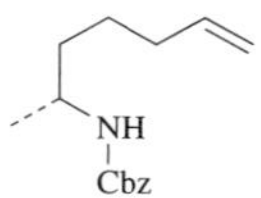	(DHQD)$_2$PHAL, $K_2OsO_2(OH)_4$, $K_3Fe(CN)_6$, K_2CO_3, *t*-BuOH:H_2O (1:1), 0°, 24 h	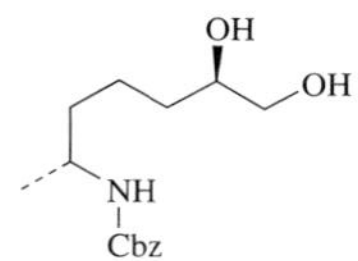 (95), 50 de	250
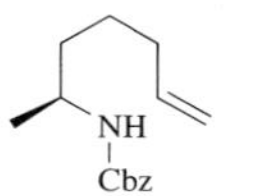	(DHQ)$_2$PHAL, $K_2OsO_2(OH)_4$, $K_3Fe(CN)_6$, K_2CO_3, *t*-BuOH:H_2O (1:1), 0°, 24 h	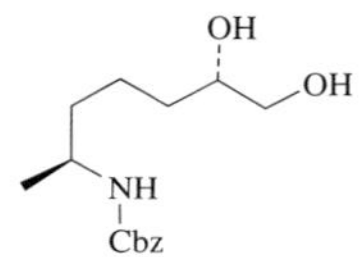 (99), 60 de	250
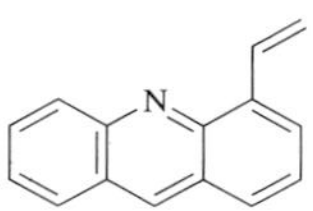	(DHQ)$_2$PHAL, $K_2OsO_2(OH)_4$, $K_3Fe(CN)_6$, K_2CO_3, *t*-BuOH:H_2O (1:1), 3°, 4 h	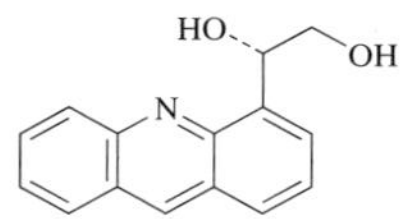 (61), 98	76

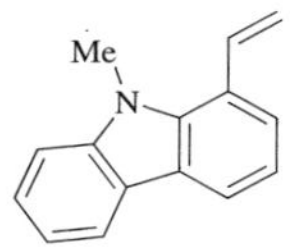	(DHQ)$_2$PHAL, $K_2OsO_2(OH)_4$, $K_3Fe(CN)_6$, K_2CO_3, *t*-BuOH:H_2O (1:1), 3°, 4 h	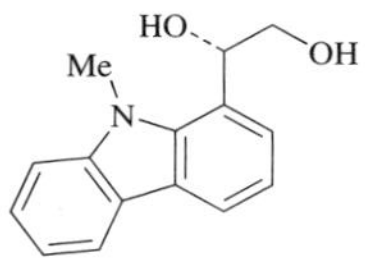	(76), 99.4	76
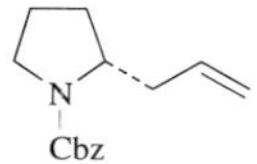	(DHQ)$_2$PYR, $K_2OsO_2(OH)_4$, $K_3Fe(CN)_6$, K_2CO_3, *t*-BuOH:H_2O (1:1), 0°, 24 h	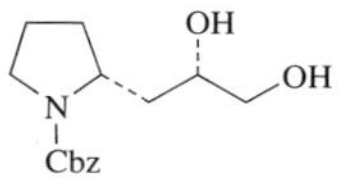	(64), 20 de, >98 ee	251
	(DHQD)$_2$PYR, $K_2OsO_2(OH)_4$, $K_3Fe(CN)_6$, K_2CO_3, *t*-BuOH:H_2O (1:1), 0°, 24 h	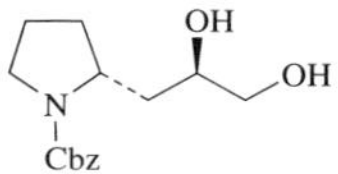	(75), 54 de, >98 ee	251
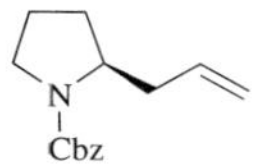	(DHQ)$_2$PYR, $K_2OsO_2(OH)_4$, $K_3Fe(CN)_6$, K_2CO_3, *t*-BuOH:H_2O (1:1), 0°, 24 h	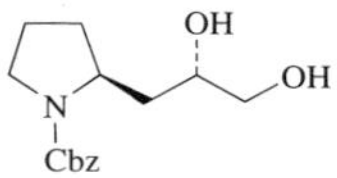	(78), 66 de, >98 ee	251
	(DHQD)$_2$PYR, $K_2OsO_2(OH)_4$, $K_3Fe(CN)_6$, K_2CO_3, *t*-BuOH:H_2O (1:1), 0°, 24 h	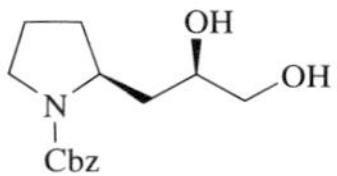	(75), 64 de, >98 ee	251
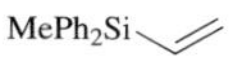	(DHQD)$_2$PHAL, $K_2OsO_2(OH)_4$, $K_3Fe(CN)_6$, K_2CO_3, *t*-BuOH:H_2O (1:1), 0°, 62 h	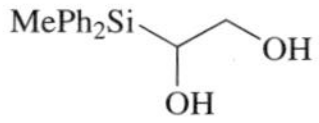 **I**	(30), 0	113
	(DHQ)$_2$PHAL, $K_2OsO_2(OH)_4$, $K_3Fe(CN)_6$, K_2CO_3, *t*-BuOH:H_2O (1:1), 0°, 62 h	**I** (31), 0		113

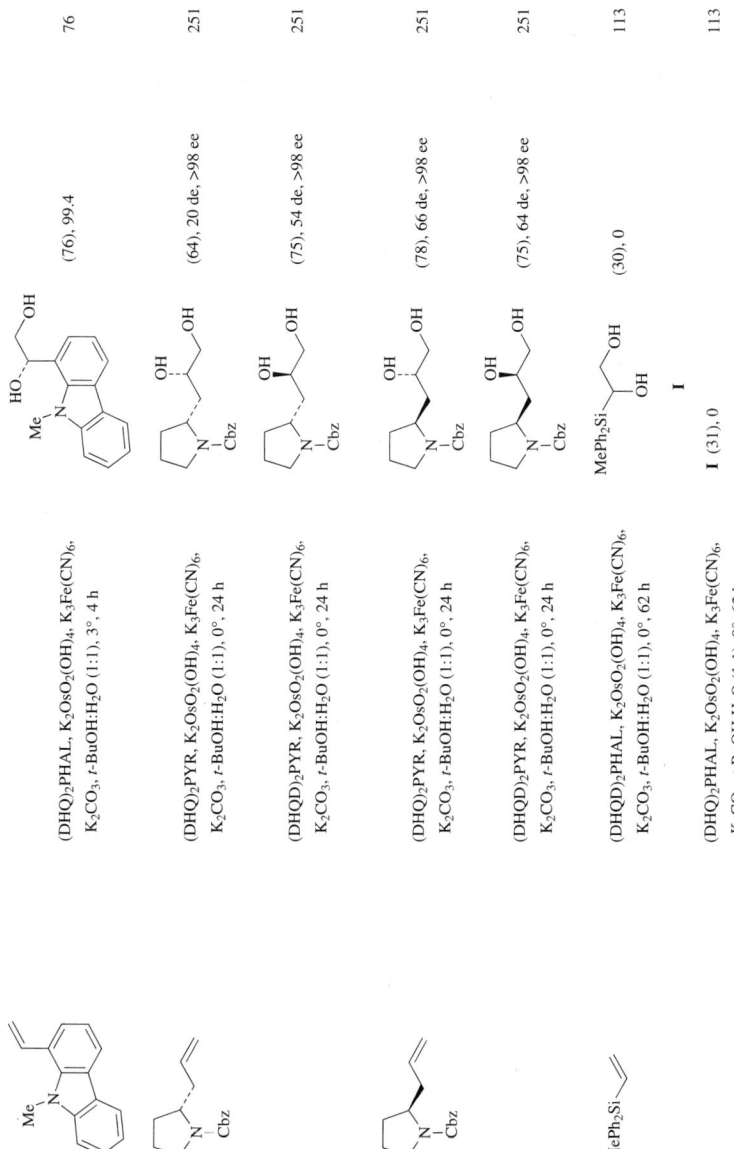

TABLE 1. REACTIONS OF TERMINAL ALKENES (*Continued*)

Substrate	Conditions	Product(s) and Yield(s) (%), % ee	Refs.
C_{15}			
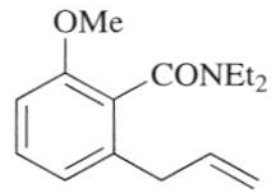	(DHQD)$_2$PHAL, $K_2OsO_2(OH)_4$, $K_3Fe(CN)_6$, K_2CO_3, *t*-BuOH:H_2O (1:1), 0°, 6-24 h	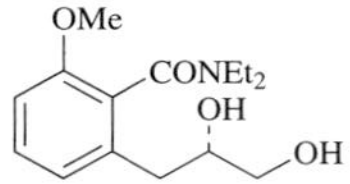 (>60-71), 16	241
	DHQD-PHN, $K_2OsO_2(OH)_4$, $K_3Fe(CN)_6$, K_2CO_3, *t*-BuOH:H_2O (1:1), 0°, 18 h	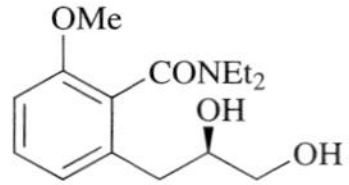 (>60-71), 40	241
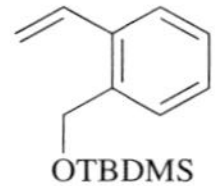	(DHQD)$_2$PHAL, $K_2OsO_2(OH)_4$, $K_3Fe(CN)_6$, K_2CO_3, *t*-BuOH:H_2O (1:1)	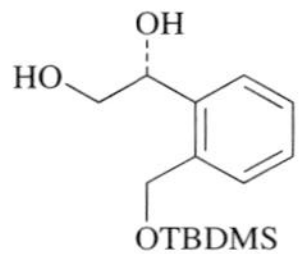 (82), 82	252, 254
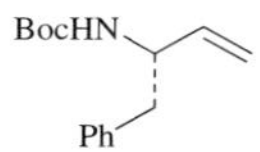	AD-Mix β, *t*-BuOH:H_2O (1:1)	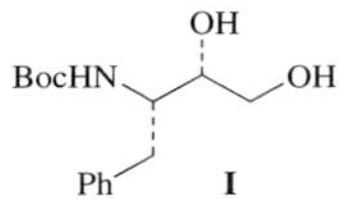 **I** (95), 20 de	253
	AD-Mix α, *t*-BuOH:H_2O (1:1)	**I** (44), 20 de	253
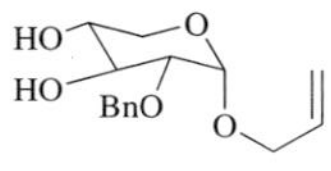	AD-Mix β, *t*-BuOH:H_2O (1:1), 0°, 24 h	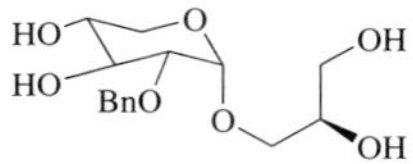 (89), 9 de	248

	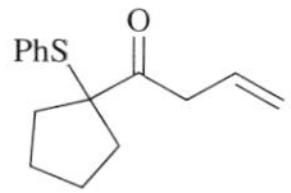	AD-Mix β, $MeSO_2NH_2$, t-BuOH:H_2O (1:1), rt	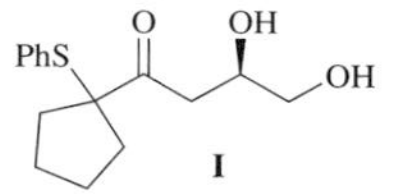 **I** (91), 29	255
		$(DHQD)_2PYR$, $K_2OsO_2(OH)_4$, $K_3Fe(CN)_6$, K_2CO_3, $MeSO_2NH_2$, t-BuOH:H_2O (1:1), 0°	**I** (—), 84	255
	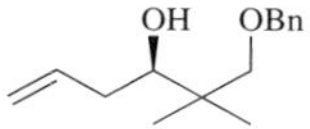	Ligand, $K_2OsO_2(OH)_4$, $K_3Fe(CN)_6$, K_2CO_3, t-BuOH:H_2O (1:1), 0°, 24 h Ligand $(DHQ)_2PHAL$ $(DHQ)_2PYR$	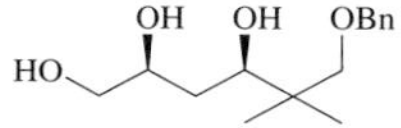 (82), 12 de (90), 42 de	256
	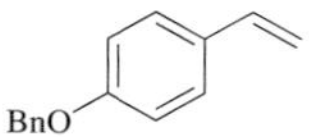	AD-Mix α, t-BuOH:H_2O (1:1)	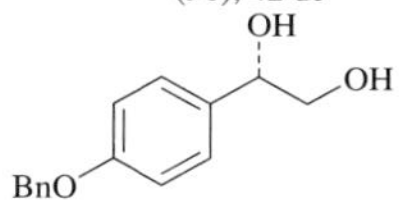 (96), 97	257
C_{16}	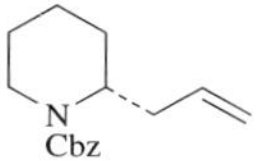	Ligand, $K_2OsO_2(OH)_4$, $K_3Fe(CN)_6$, K_2CO_3, t-BuOH:H_2O (1:1), 0°, 24 h Ligand $(DHQ)_2PYR$ $(DHQD)_2PYR$	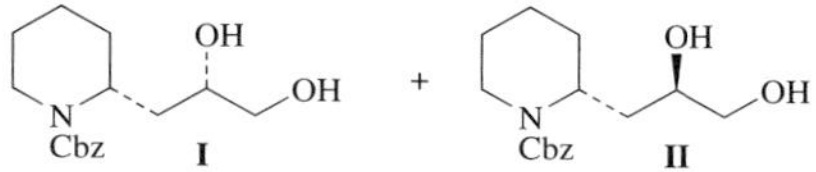 **I** / **II** (70), >98 / (21), 54 (22), 54 / (69), >98	259, 260
	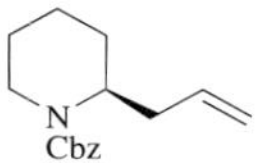	Ligand, $K_2OsO_2(OH)_4$, $K_3Fe(CN)_6$, K_2CO_3, t-BuOH:H_2O (1:1), 0°, 24 h Ligand $(DHQ)_2PYR$ $(DHQD)_2PYR$	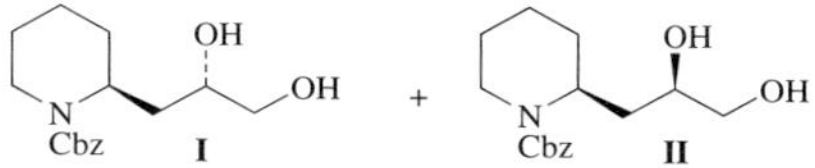 **I** / **II** (70), >98 / (22), 72 (14), 47 / (74), >98	259, 260

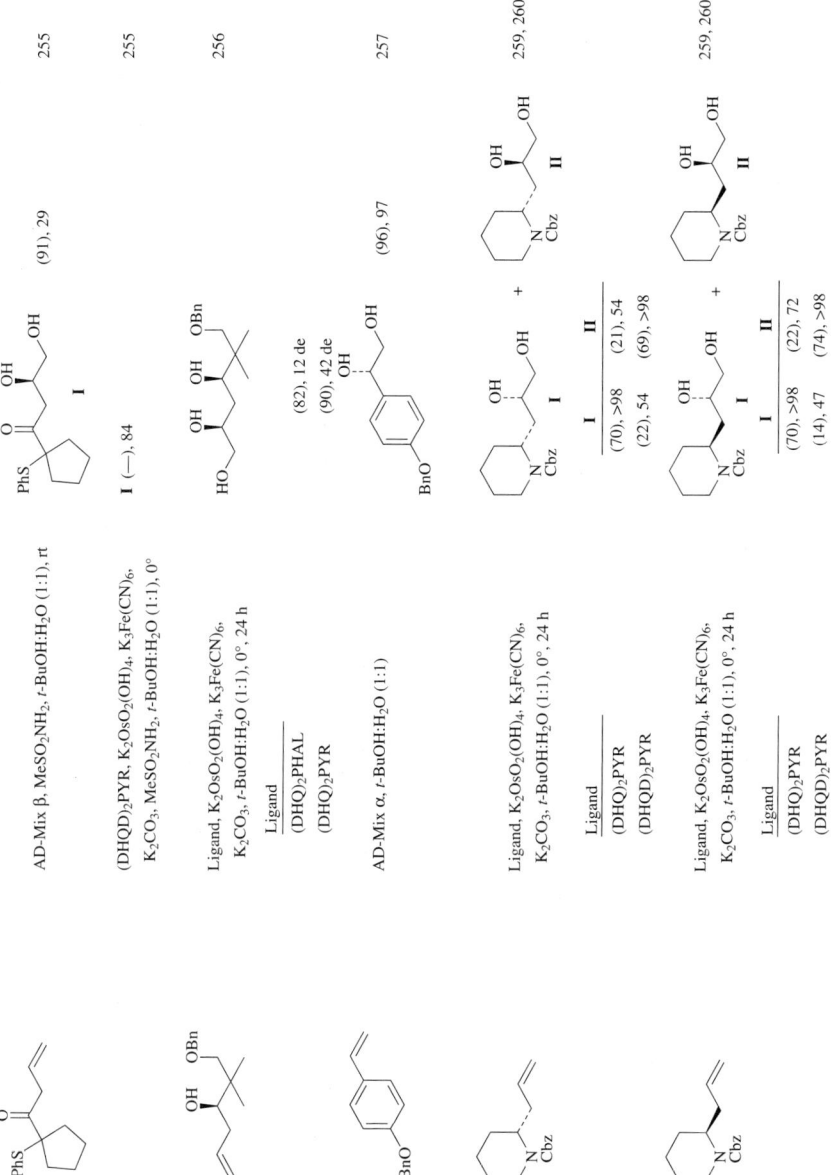

TABLE 1. REACTIONS OF TERMINAL ALKENES (*Continued*)

Substrate	Conditions	Product(s) and Yield(s) (%), % ee	Refs.
C_{16} 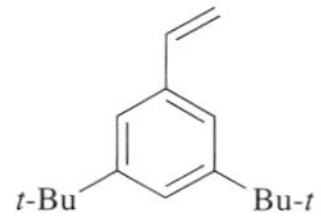	$(DHQD)_2PHAL$, $K_2OsO_2(OH)_4$, $K_3Fe(CN)_6$, K_2CO_3, *t*-BuOH:H_2O (1:1), 0°, 6-24 h	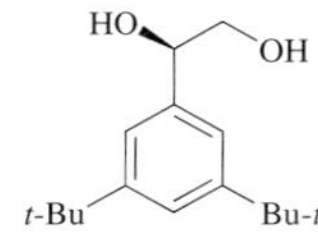 (—), 49	60
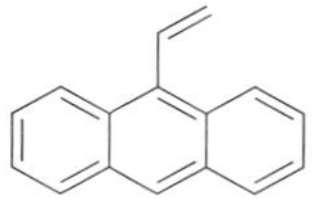	$(DHQD)_2PYDZ$, $K_2OsO_2(OH)_4$, $K_3Fe(CN)_6$, K_2CO_3, *t*-BuOH:H_2O (1:1), 0-4°, 4 days	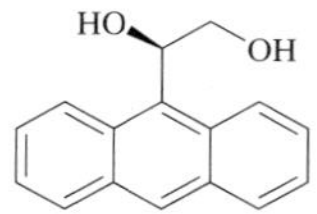 (75), >98	66
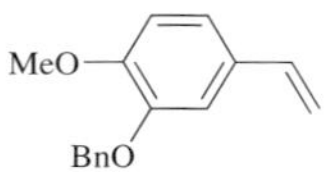	$(DHQ)_2PHAL$, $K_2OsO_2(OH)_4$, $K_3Fe(CN)_6$, K_2CO_3, *t*-BuOH:H_2O (1:1), 0°, 6 h	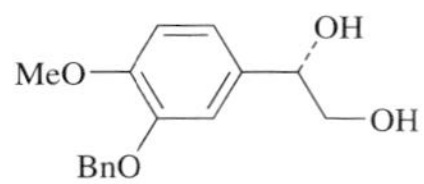 (92), >99	216
	$(DHQD)_2PHAL$, $K_2OsO_2(OH)_4$, $K_3Fe(CN)_6$, K_2CO_3, *t*-BuOH:H_2O (1:1), 0°, 6 h	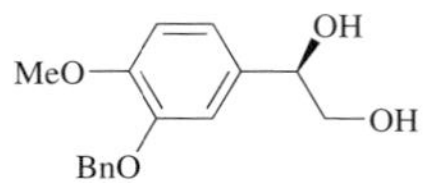 (95), >99	216
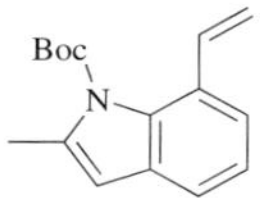	$(DHQ)_2PHAL$, $K_2OsO_2(OH)_4$, $K_3Fe(CN)_6$, K_2CO_3, *t*-BuOH:H_2O (1:1), 3°, 4 h	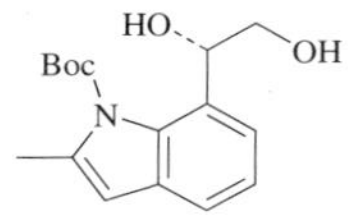 (0), —	76
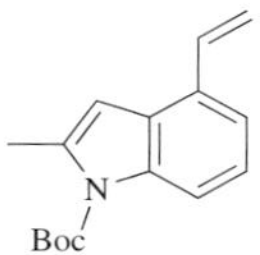	$(DHQ)_2PHAL$, $K_2OsO_2(OH)_4$, $K_3Fe(CN)_6$, K_2CO_3, *t*-BuOH:H_2O (1:1), 3°, 4 h	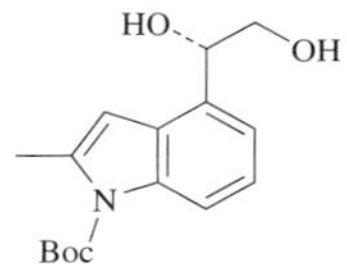 (88), 41.4	76

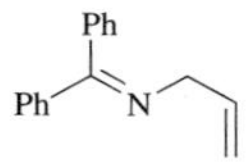	(DHQD)$_2$PYDZ, $K_2OsO_2(OH)_4$, $K_3Fe(CN)_6$, K_2CO_3, *t*-BuOH:H_2O (1:1), 0°	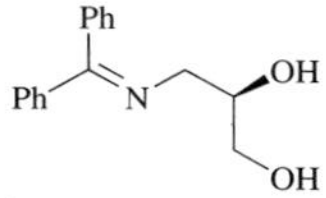	(53), 58	61
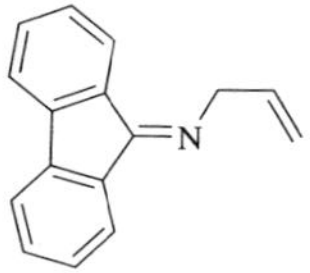	(DHQD)$_2$PYDZ, $K_2OsO_2(OH)_4$, $K_3Fe(CN)_6$, K_2CO_3, *t*-BuOH:H_2O (1:1), 0°, 1.5-3.5 h	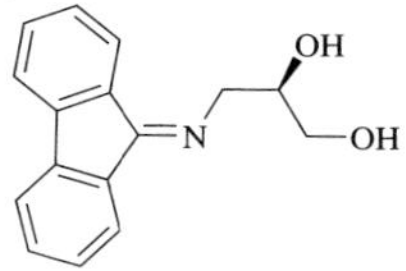	(90), 90	61, 66
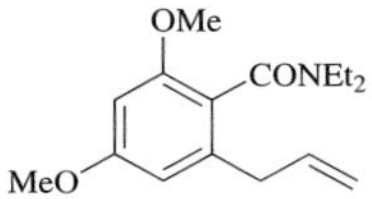	Ligand, $K_2OsO_2(OH)_4$, $K_3Fe(CN)_6$, K_2CO_3, *t*-BuOH:H_2O (1:1), 0°	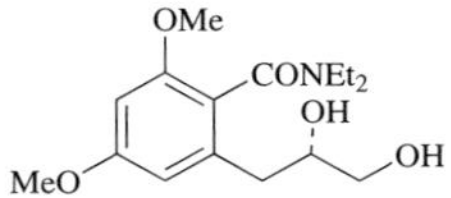		241
	Ligand — Time			
	(DHQD)$_2$PHAL — 6-24 h	(>60-71), 10		
	DHQ-PHN — 18 h	(>60-71), 32		
	DHQD-*p*-chlorobenzoate — 18 h	(>60-71), 11		
	Ligand, $K_2OsO_2(OH)_4$, $K_3Fe(CN)_6$, K_2CO_3, *t*-BuOH:H_2O (1:1), 0°	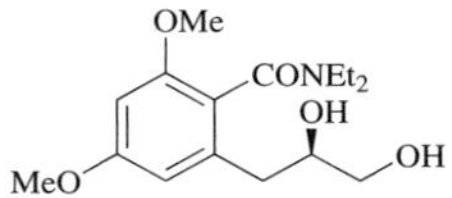		241
	Ligand — Time			
	DHQD-PHN — 18 h	(>60-71), 40		
	(DHQ)$_2$PHAL — 6-24 h	(>60-71), 7		
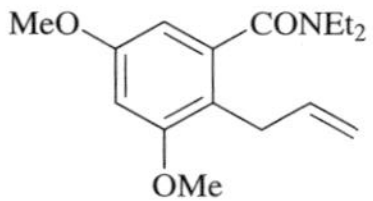	Ligand, $K_2OsO_2(OH)_4$, $K_3Fe(CN)_6$, K_2CO_3, *t*-BuOH:H_2O (1:1), 0°	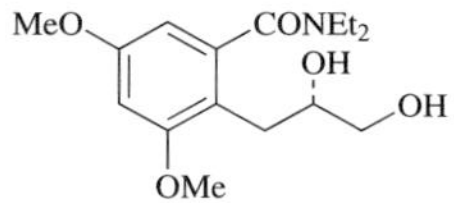		241
	Ligand — Time			
	(DHQD)$_2$PHAL — 6-24 h	(>60-71), 64		
	DHQD-PHN — 18 h	(>60-71), 30		

TABLE 1. REACTIONS OF TERMINAL ALKENES (*Continued*)

Substrate	Conditions	Product(s) and Yield(s) (%), % ee	Refs.
C_{16}			
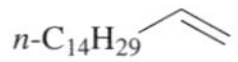	$(DHQD)_2PHAL$, $K_2OsO_2(OH)_4$, $K_3Fe(CN)_6$, K_2CO_3, t-BuOH:H_2O (1:1), 0°	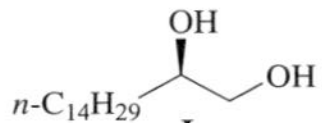 **I** (95), >95	261
	AD-Mix β, t-BuOH:H_2O (1:1), 4°, 40 h	**I** (76-95), 80-85	261a, 261
	$(DHQ)_2PHAL$, $K_2OsO_2(OH)_4$, $K_3Fe(CN)_6$, K_2CO_3, t-BuOH:H_2O (1:1), 0°	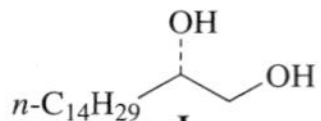 **I** (97), >95	261
	AD-Mix α, t-BuOH:H_2O (1:1), 4°, 40 h	**I** (97), 79	261
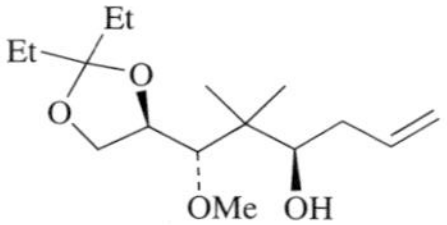	AD-Mix α, t-BuOH:H_2O (1:1)	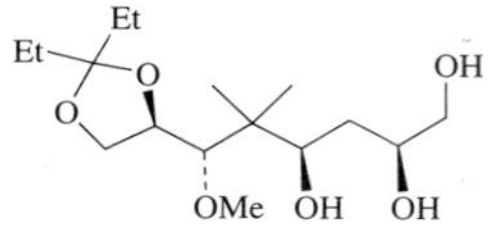 (—), <60	262
C_{17}			
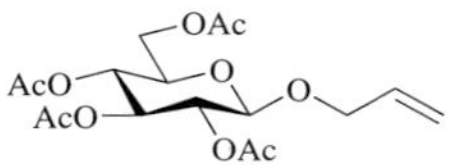	Ligand, $K_2OsO_2(OH)_4$, $K_3Fe(CN)_6$, K_2CO_3, t-BuOH:H_2O (1:1), 0°	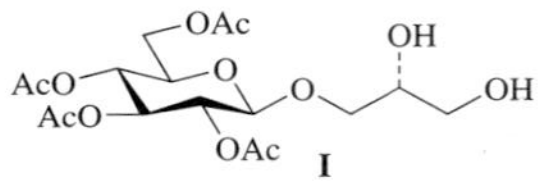 **I**	263
	Ligand		
	Quinuclidin-3-ol	(74), 20 de	
	$(DHQ)_2PHAL$	(50), 0 de	
	$(DHQD)_2PHAL$	(51), 90 de	
	Ligand, OsO_4, NMO, acetone-H_2O, 0°	**I**	264
	Ligand		
	DHQD-p-chlorobenzoate	(—), 50 de	
	DHQ-p-chlorobenzoate	(—), 12 de	

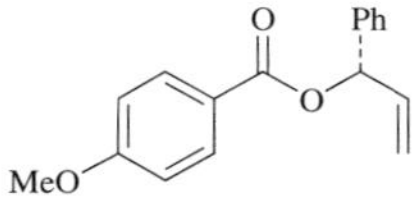	**17**, $K_2OsO_2(OH)_4$, $K_3Fe(CN)_6$, K_2CO_3, *t*-BuOH:H_2O (1:1), 0°, 2.5 h	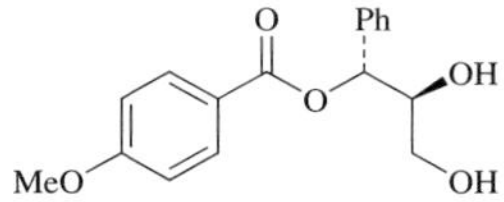	(98), 98 de	63
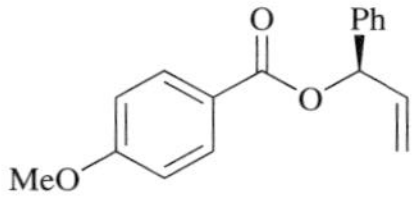	**17**, $K_2OsO_2(OH)_4$, $K_3Fe(CN)_6$, K_2CO_3, *t*-BuOH:H_2O (1:1), 0°	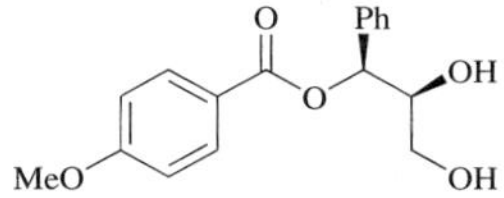	(—), 31 de	63
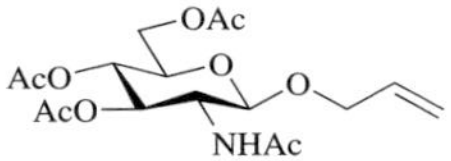	Ligand, $K_2OsO_2(OH)_4$, $K_3Fe(CN)_6$, K_2CO_3, *t*-BuOH:H_2O (1:1), 0°, 4 h; rt, 18 h	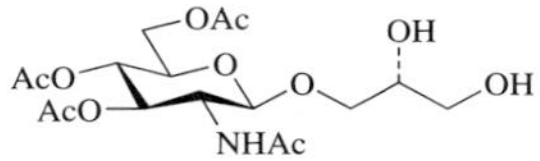		265
	Ligand			
	$(DHQ)_2PHAL$	(—), 26 de		
	$(DHQD)_2PHAL$	(—), 40 de		
	$(DHQ)_2PYR$	(—), 38 de		
	$(DHQD)_2PYR$	(—), 24 de		
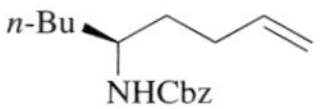	$(DHQ)_2PHAL$, $K_2OsO_2(OH)_4$, $K_3Fe(CN)_6$, K_2CO_3, *t*-BuOH:H_2O (1:1), 0°, 24 h	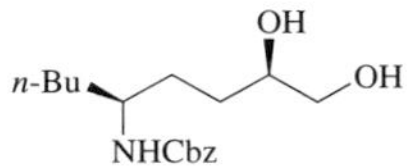	(98), 60 de	250
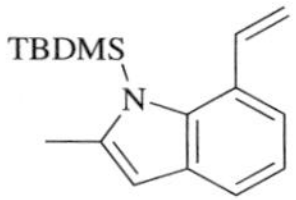	$(DHQ)_2PHAL$, $K_2OsO_2(OH)_4$, $K_3Fe(CN)_6$, K_2CO_3, *t*-BuOH:H_2O (1:1), rt, 18 h	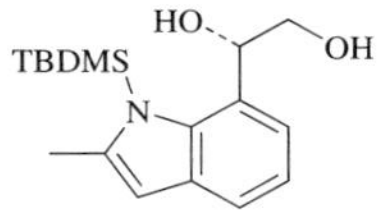	(21), 54.2	76
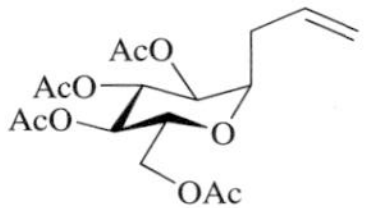	DHQD-PHN, $K_2OsO_2(OH)_4$, $K_3Fe(CN)_6$, K_2CO_3, *t*-BuOH:H_2O (1:1)	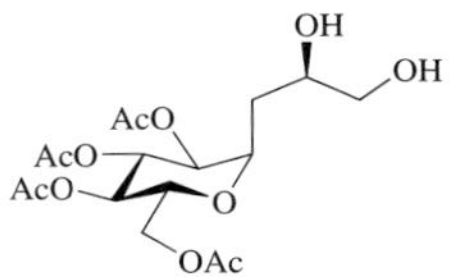	(70), 40	266

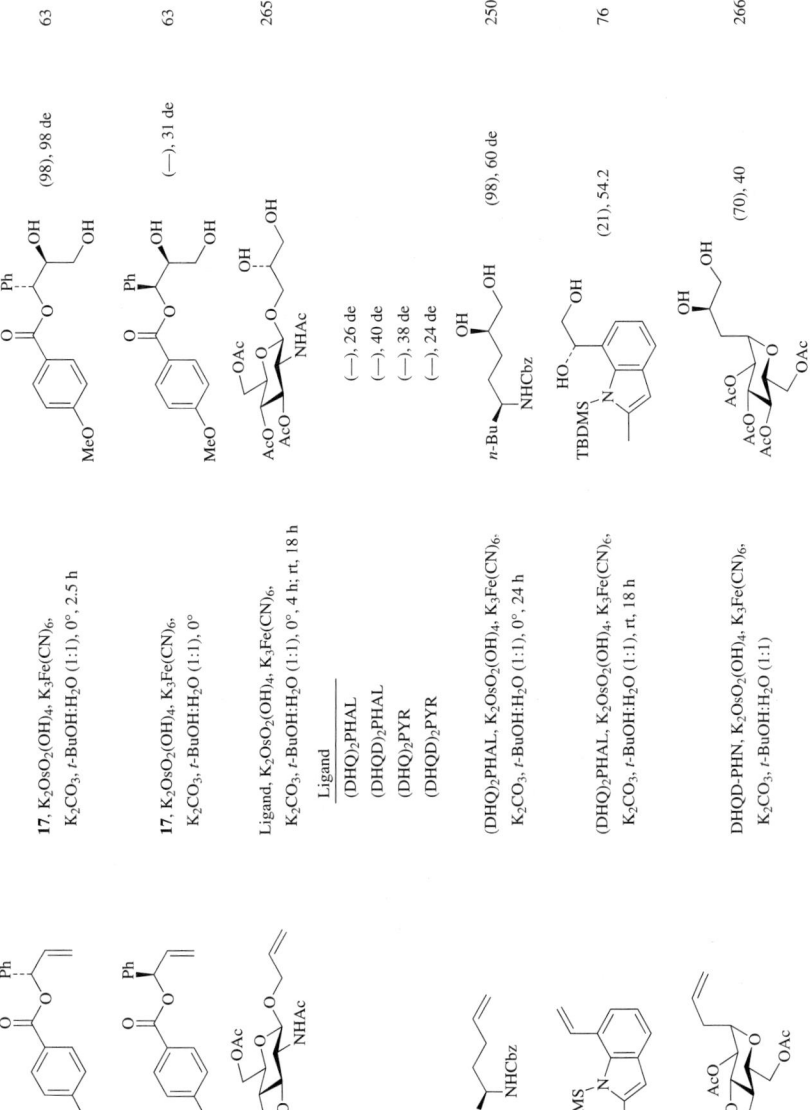

TABLE 1. REACTIONS OF TERMINAL ALKENES (*Continued*)

Substrate	Conditions	Product(s) and Yield(s) (%), % ee		Refs.
C_{17}				
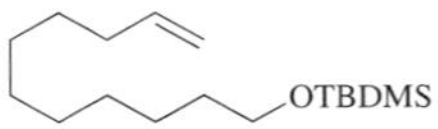	AD-Mix β, *t*-BuOH:H_2O (1:1), 0°	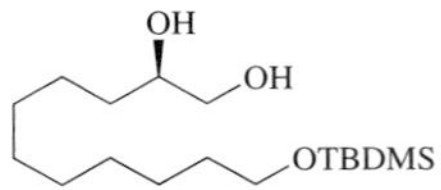	(96), 90	234
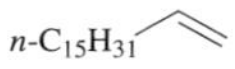	Ligand, $K_2OsO_2(OH)_4$, $K_3Fe(CN)_6$, K_2CO_3, *t*-BuOH:H_2O (1:1)	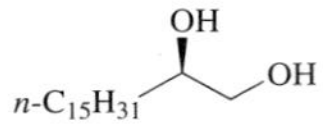		267
	Ligand / Temp			
	$(DHQD)_2PYR$ / 0°	(—), 78		
	$(DHQD)_2PYR$ / rt	(—), 74		
	$(DHQD)_2AQN$ / rt	(—), 73		
	Ligand, $K_2OsO_2(OH)_4$, $K_3Fe(CN)_6$, K_2CO_3, *t*-BuOH:H_2O (1:1)	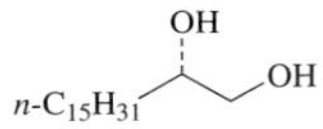		267
	Ligand / Temp			
	$(DHQ)_2PYR$ / 0°	(—), 63		
	$(DHQ)_2AQN$ / rt	(—), 69		
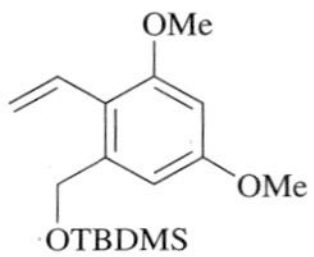	$(DHQD)_2PHAL$, $K_2OsO_2(OH)_4$, $K_3Fe(CN)_6$, K_2CO_3, *t*-BuOH:H_2O (1:1), 0°, 9 h	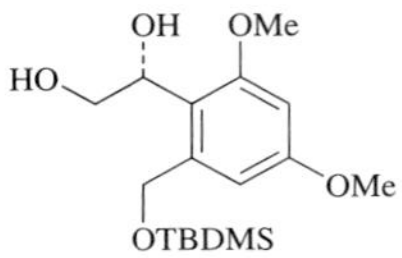	(100), 87	254
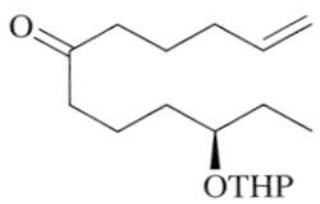	1. AD-Mix α, *t*-BuOH:H_2O (1:1), 0°, 6 days 2. H^+	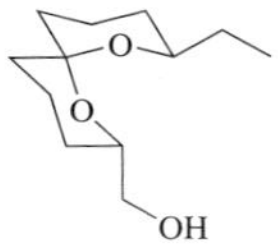	(30), >95 de	268

	1. AD-Mix β, t-BuOH:H_2O (1:1), 0°, 6 days 2. H^+	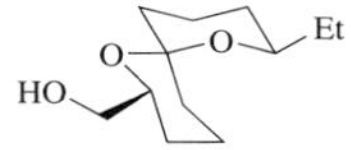 (14), — +  (6), —	268
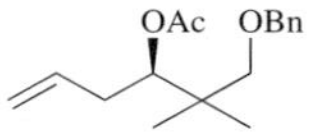	$(DHQ)_2PHAL$, $K_2OsO_2(OH)_4$, $K_3Fe(CN)_6$, K_2CO_3, t-BuOH:H_2O (1:1), 0°, 24 h	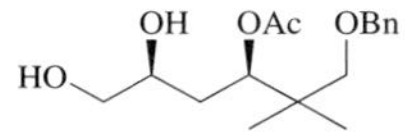 (59), 42 de	256
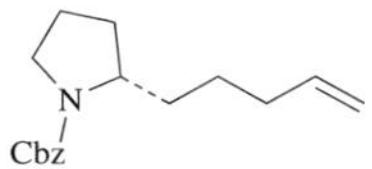	$(DHQ)_2PYR$, $K_2OsO_2(OH)_4$, $K_3Fe(CN)_6$, K_2CO_3, t-BuOH:H_2O (1:1), 0°	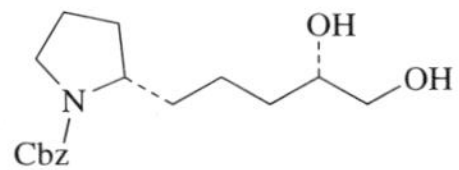 (85), 68 de	269
	$(DHQD)_2PYR$, $K_2OsO_2(OH)_4$, $K_3Fe(CN)_6$, K_2CO_3, t-BuOH:H_2O (1:1), 0°	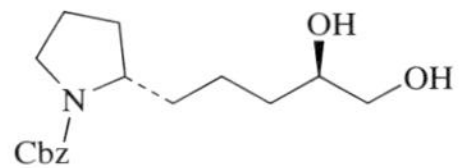 (85), 68 de	269
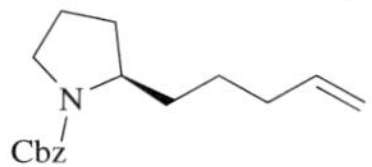	$(DHQ)_2PYR$, $K_2OsO_2(OH)_4$, $K_3Fe(CN)_6$, K_2CO_3, t-BuOH:H_2O (1:1), 0°	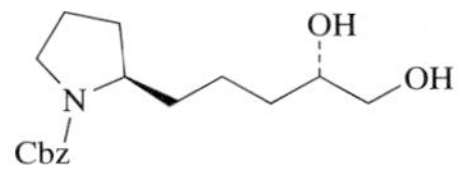 (95), 68 de	269
	$(DHQD)_2PYR$, $K_2OsO_2(OH)_4$, $K_3Fe(CN)_6$, K_2CO_3, t-BuOH:H_2O (1:1), 0°	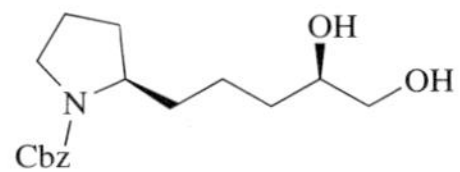 (90), 68	269, 270
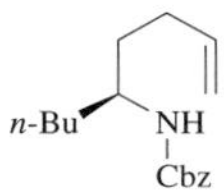	AD-Mix α, t-BuOH:H_2O (1:1)	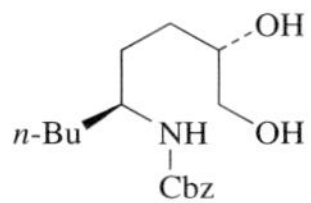 (84), 60 de	271

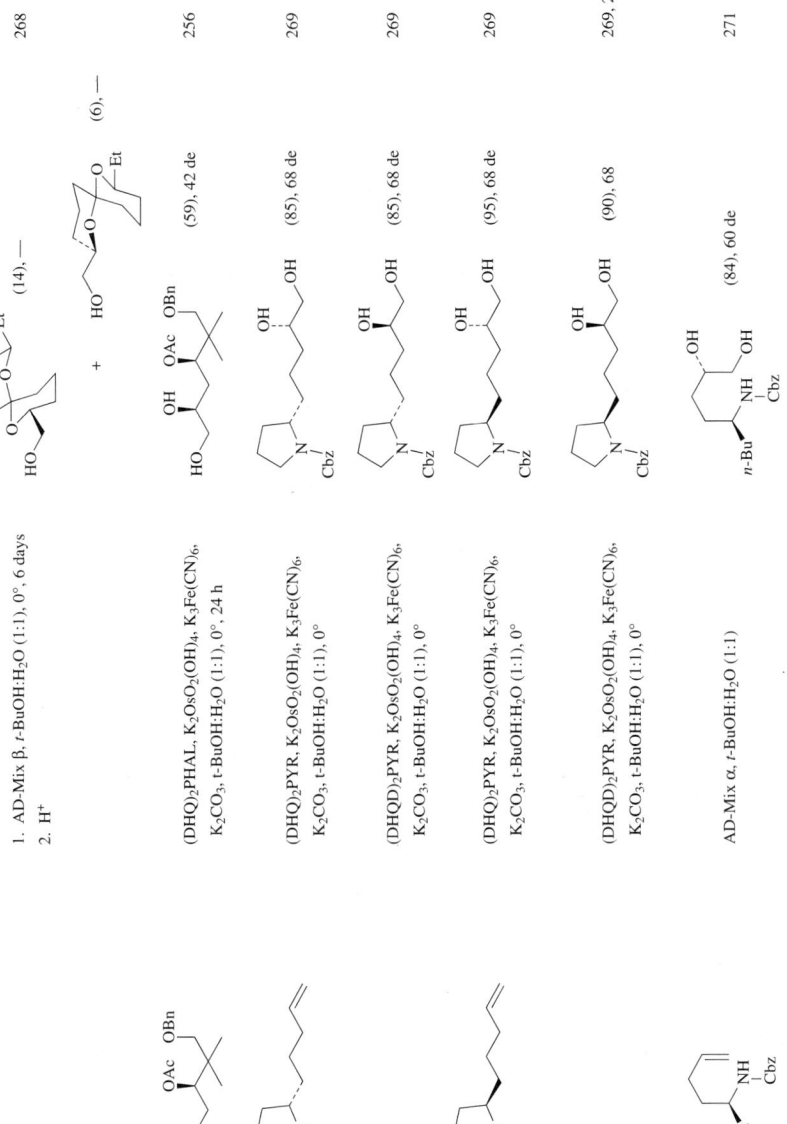

TABLE 1. REACTIONS OF TERMINAL ALKENES (*Continued*)

Substrate	Conditions	Product(s) and Yield(s) (%), % ee	Refs.
C_{18} 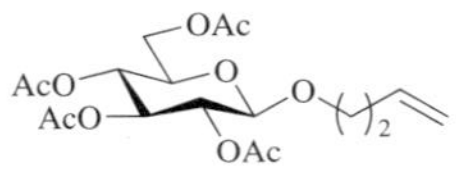	Ligand, $K_2OsO_2(OH)_4$, $K_3Fe(CN)_6$, K_2CO_3, *t*-BuOH:H_2O (1:1), 0°	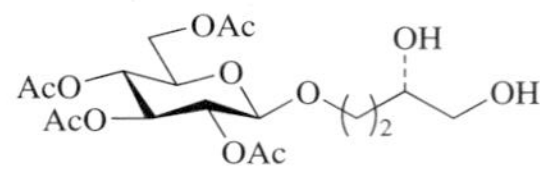	263
	Ligand		
	Quinuclidin-3-ol	(62), 0 de	
	$(DHQ)_2PHAL$	(68), 20 de	
	$(DHQD)_2PHAL$, $K_2OsO_2(OH)_4$, $K_3Fe(CN)_6$, K_2CO_3, *t*-BuOH:H_2O (1:1), 0°	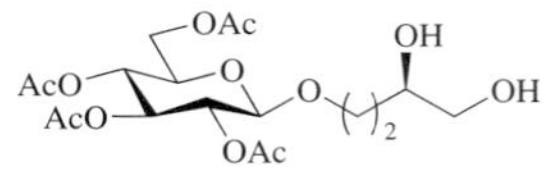 (68), 60 de	263
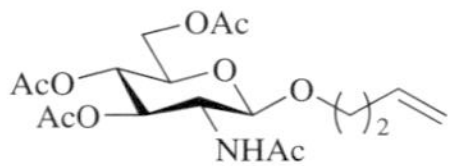	$(DHQ)_2PHAL$, $K_2OsO_2(OH)_4$, $K_3Fe(CN)_6$, K_2CO_3, *t*-BuOH:H_2O (1:1), 0°, 4 h; rt, 18 h	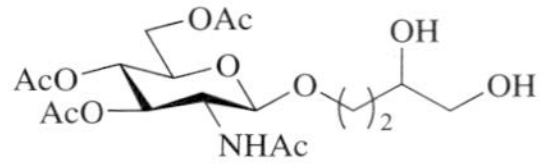 (70), 0 de	265
	$(DHQD)_2PHAL$, $K_2OsO_2(OH)_4$, $K_3Fe(CN)_6$, K_2CO_3, *t*-BuOH:H_2O (1:1), 0°, 4 h; rt, 18 h	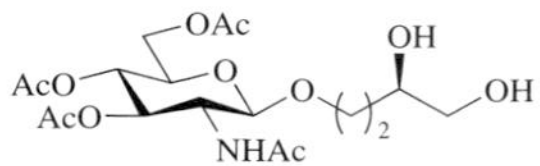 (70), 26 de	265
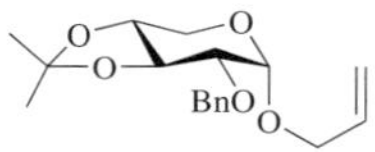	AD-Mix β, *t*-BuOH:H_2O (1:1), 0°, 36 h	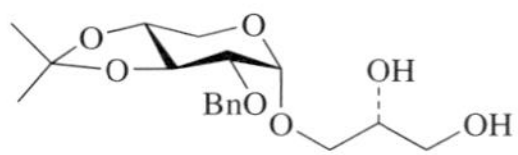 (86), 35 de	248, 270
	AD-Mix β, *t*-BuOH:H_2O (1:1), 0°, 36 h	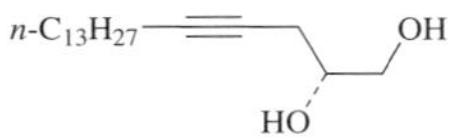 (85), 64	272
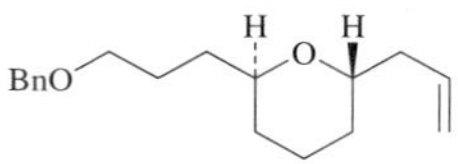	AD-Mix α, $MeSO_2NH_2$, *t*-BuOH:H_2O (1:1), 0°	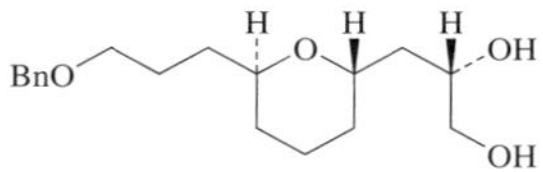 (35), 20 de	273

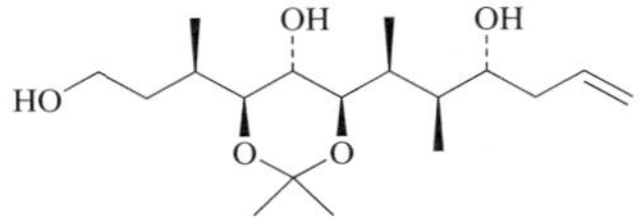

AD-Mix α, t-BuOH:H_2O (1:1), 0-20°, 24 h

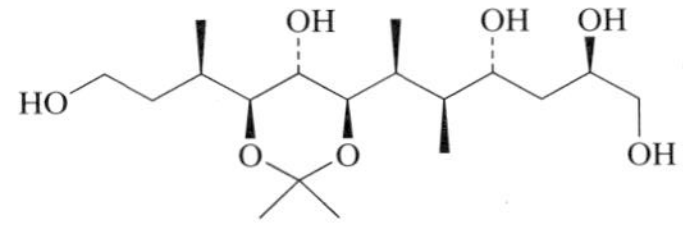

(73), 26 de

274

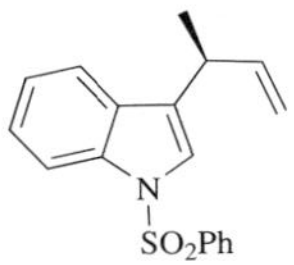

AD-Mix α, t-BuOH:H_2O (1:1), 0°

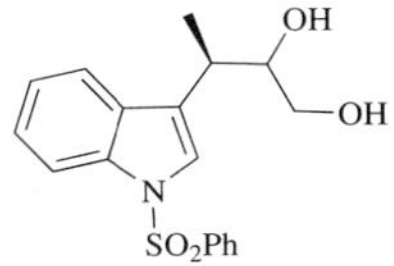

(100), 0 de

275

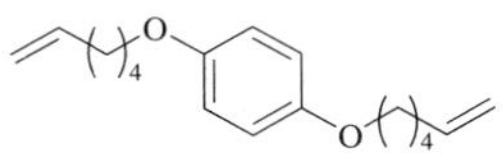

Ligand, $K_2OsO_2(OH)_4$, $K_3Fe(CN)_6$, K_2CO_3, t-BuOH:H_2O (1:1), 0°

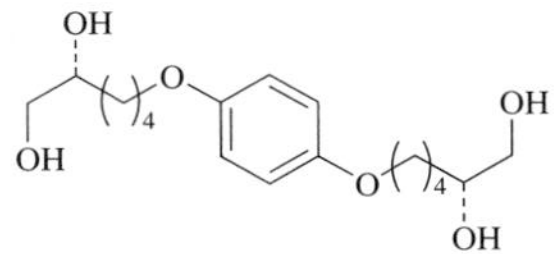

146

Ligand	Time	Product(s) and Yield(s) (%)
$(DHQD)_2PHAL$	12 h	(—), 86
$(DHQD)_2PHAL$	18 h	(—), 78
$(DHQD)_2PYR$	12 h	(—), 90
$(DHQD)_2PYR$	18 h	(—), 41
$(DHQD)_2AQN$	12 h	(—), 82
$(DHQD)_2AQN$	18 h	(—), 80

C_{19}

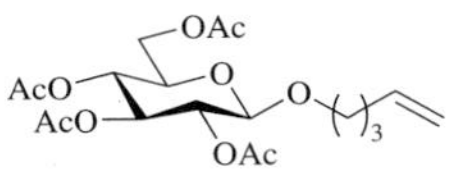

Ligand, $K_2OsO_2(OH)_4$, $K_3Fe(CN)_6$, K_2CO_3, t-BuOH:H_2O (1:1)

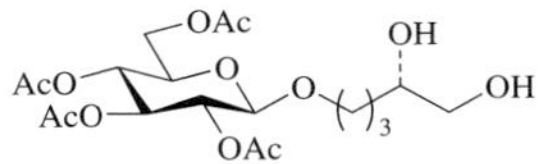

263

Ligand	Temp/Time	Product(s) and Yield(s) (%)
Quinuclidin-3-ol	0°/—	(55), 0 de
$(DHQ)_2PHAL$	0°/—	(60), 50 de
$(DHQ)_2PHAL$	0°/4 h; rt/18 h	(—), 44 de

TABLE 1. REACTIONS OF TERMINAL ALKENES (*Continued*)

Substrate	Conditions	Product(s) and Yield(s) (%), % ee	Refs.
C_{19} 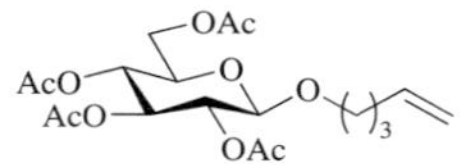	Ligand, $K_2OsO_2(OH)_4$, $K_3Fe(CN)_6$, K_2CO_3, t-BuOH:H_2O (1:1)	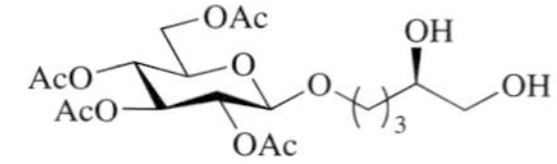	
	Ligand — Temp/Time		
	$(DHQ)_2PHAL$ — 0°/—	(61), 40 de	263
	$(DHQ)_2PHAL$ — 0°/4 h; rt/18 h	(—), 52 de	265
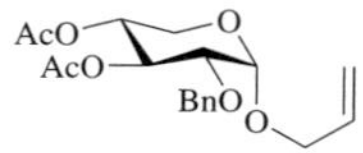	AD-Mix β, t-BuOH:H_2O (1:1), 0°	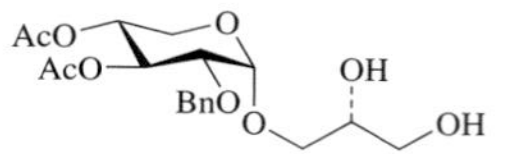 (90), 16 de	247, 248
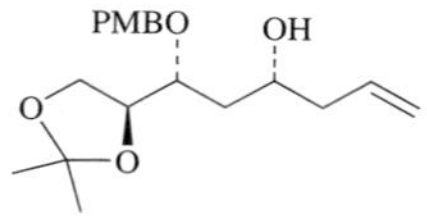	DHQ-PHN, $K_2OsO_2(OH)_4$, $K_3Fe(CN)_6$, K_2CO_3, t-BuOH:H_2O (1:1), rt, 3 h	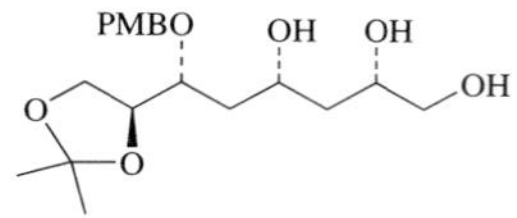 (95), 60 de	276
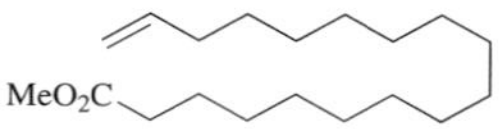	AD-Mix β, $MeSO_2NH_2$, t-BuOH:H_2O (1:1), 0°, 24 h	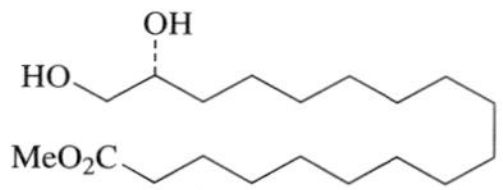 (90), 84	277
	AD-Mix α, $MeSO_2NH_2$, t-BuOH:H_2O (1:1), 0°, 24 h	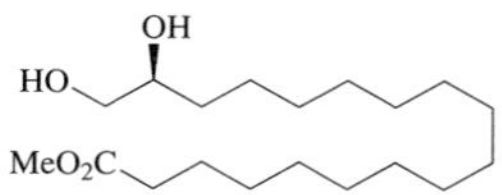 (90), 92	277
C_{20} 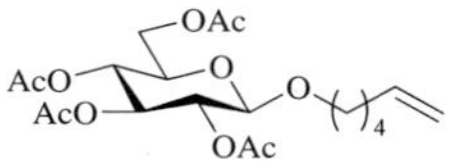	Ligand, $K_2OsO_2(OH)_4$, $K_3Fe(CN)_6$, K_2CO_3, t-BuOH:H_2O (1:1), 0°	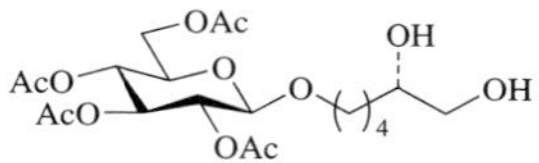	263
	Ligand		
	Quinuclidin-3-ol	(62), 0 de	
	$(DHQ)_2PHAL$	(68), 50 de	
	$(DHQD)_2PHAL$	(68), 70 de	

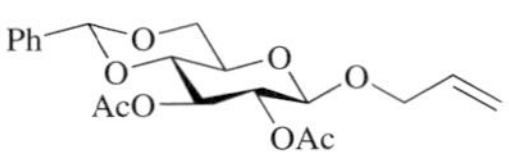	DHQD-*p*-chlorobenzoate, $K_2OsO_2(OH)_4$, $K_3Fe(CN)_6$, K_2CO_3, *t*-BuOH:H_2O (1:1), 0°	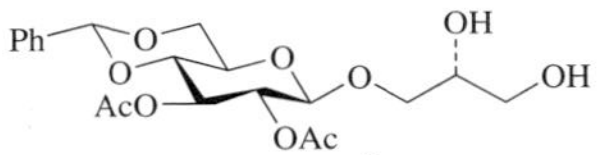 **I**	(—), 26 de	264
	DHQD-*p*-chlorobenzoate, $K_2OsO_2(OH)_4$, NMO, acetone-H_2O, 0°	**I** (—), 8 de		265
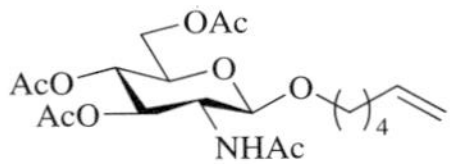	$(DHQ)_2PHAL$, $K_2OsO_2(OH)_4$, $K_3Fe(CN)_6$, K_2CO_3, *t*-BuOH:H_2O (1:1), 0°, 4 h; rt, 18 h	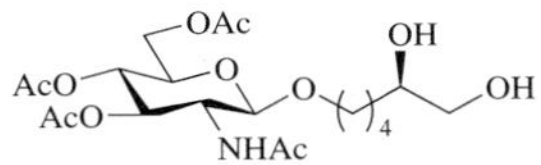	(—), 56 de	265
	$(DHQD)_2PHAL$, $K_2OsO_2(OH)_4$, $K_3Fe(CN)_6$, K_2CO_3, *t*-BuOH:H_2O (1:1), 0°, 4 h; rt, 18 h	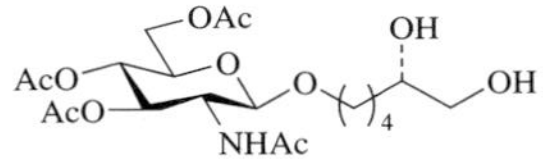	(—), 60 de	265
	Ligand, $K_2OsO_2(OH)_4$, $K_3Fe(CN)_6$, K_2CO_3, *t*-BuOH:H_2O (1:1), 0°, 96 h	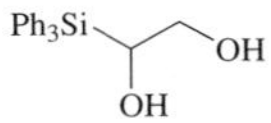		113
	Ligand			
	$(DHQD)_2PHAL$	(25), 0		
	$(DHQ)_2PHAL$	(0), 0		
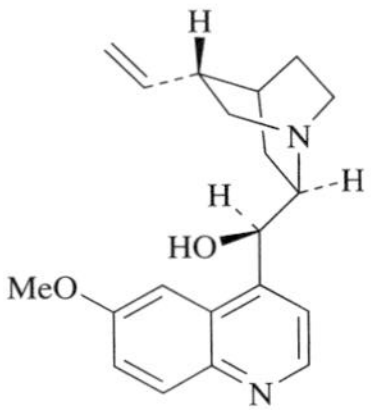	Ligand, $K_2OsO_2(OH)_4$, $K_3Fe(CN)_6$, K_2CO_3, *t*-BuOH:H_2O (1:1), 0°, 4 days	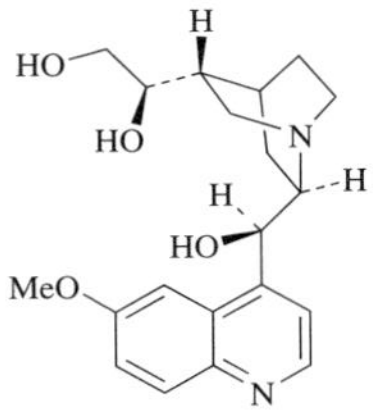		278
	Ligand			
	$(DHQD)_2PHAL$	(91), —		
	$(DHQD)_2PYR$	(91), —		

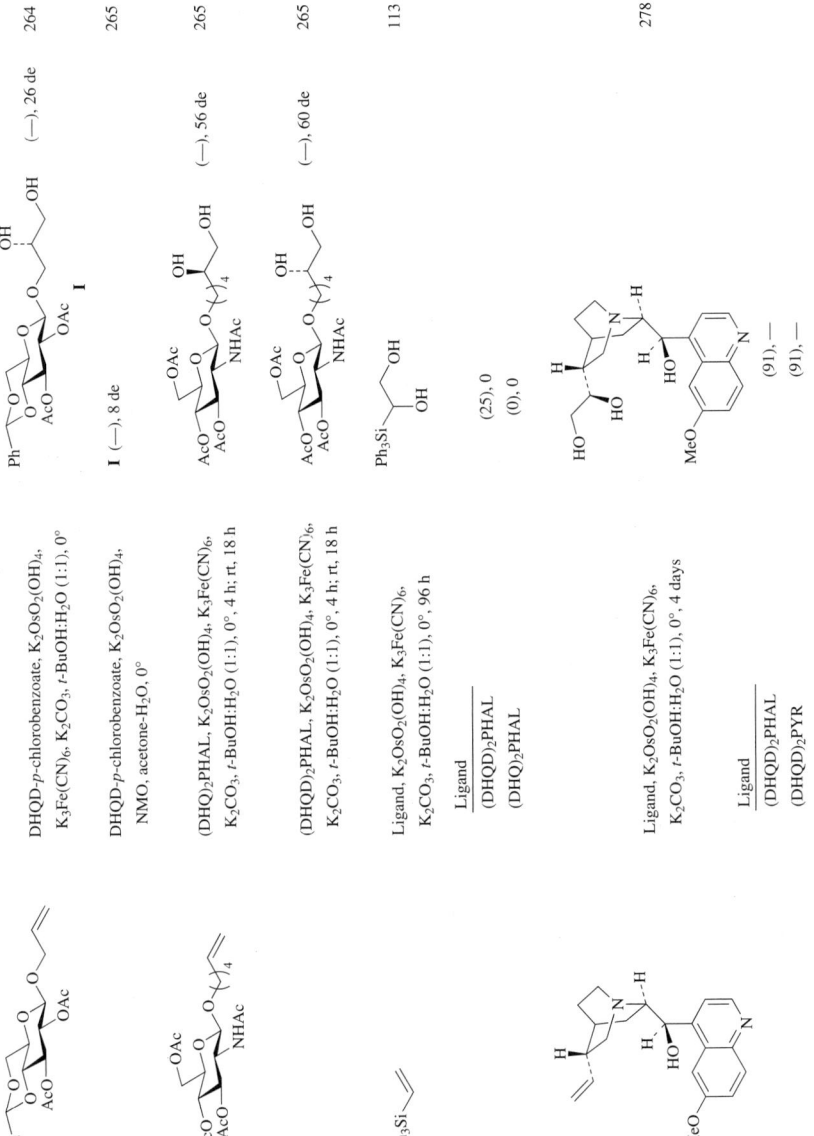

TABLE 1. REACTIONS OF TERMINAL ALKENES (*Continued*)

Substrate	Conditions	Product(s) and Yield(s) (%), % ee	Refs.
C_{20} 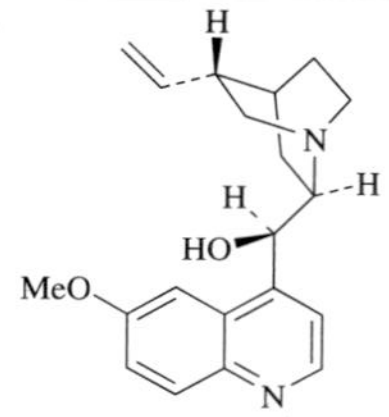	Ligand, $K_2OsO_2(OH)_4$, $K_3Fe(CN)_6$, K_2CO_3, t-BuOH:H_2O (1:1), 0°, 4 days Ligand $(DHQ)_2PYR$ $(DHQD)_2PYR$	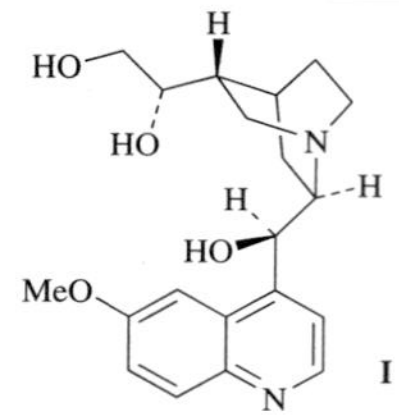 **I** or 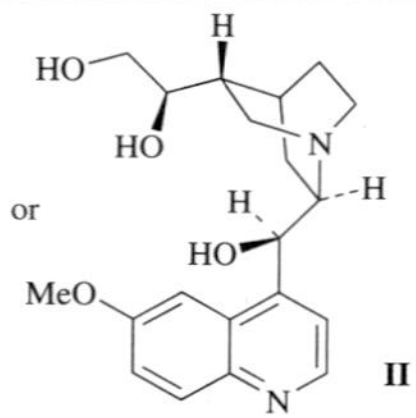 **II** **I** (75), 40 de **II** (91), >95 de	278
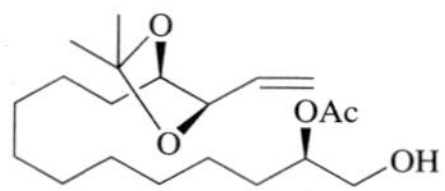	AD-Mix α, $MeSO_2NH_2$, t-BuOH:H_2O (1:1), 0°, 72 h	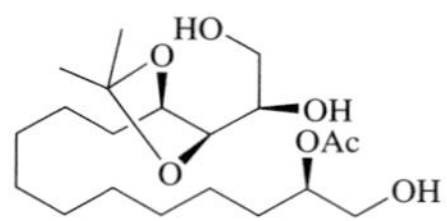 (74), 78 de	279
C_{21} 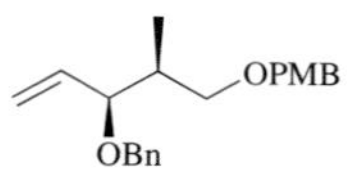	AD-Mix β, t-BuOH:H_2O (1:1)	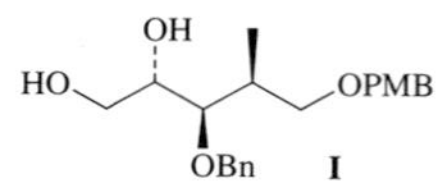 **I** (91), 91	280
	AD-Mix α, t-BuOH:H_2O (1:1)	**I** (90), 67	280
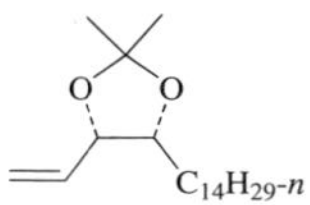	$(DHQ)_2PHAL$, $K_2OsO_2(OH)_4$, $K_3Fe(CN)_6$, K_2CO_3, t-BuOH:H_2O (1:1)	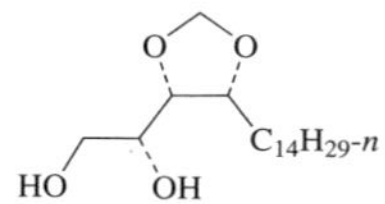 (92), 60 de	281
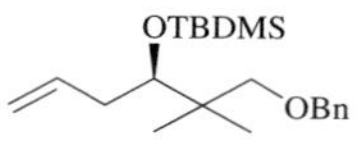	$(DHQ)_2PHAL$, $K_2OsO_2(OH)_4$, $K_3Fe(CN)_6$, K_2CO_3, t-BuOH:H_2O (1:1), 0°, 24 h	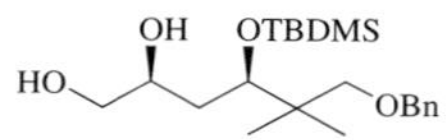 (50), 42 de	256

Substrate	Conditions	Product(s) and Yield(s) (%), ee (%)	Refs.
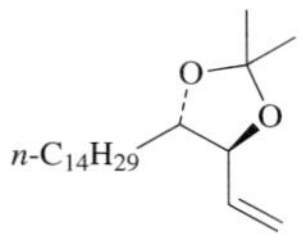	$(DHQD)_2PHAL$, OsO_4, $K_3Fe(CN)_6$, K_2CO_3, $MeSO_2NH_2$, t-BuOH:H_2O (1:1), 0°, 24 h	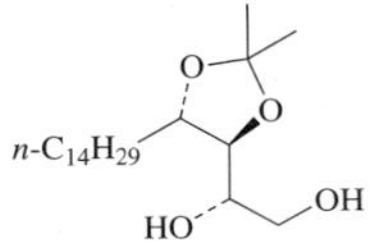 (92), 67 de	282
	$(DHQ)_2PHAL$, OsO_4, $K_3Fe(CN)_6$, K_2CO_3, $MeSO_2NH_2$, t-BuOH:H_2O (1:1), 0°, 24 h	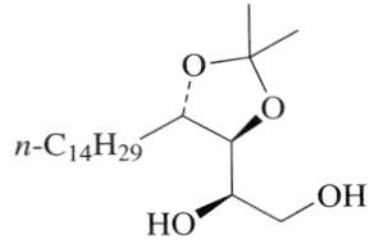 (89), 33 de	282
C_{22}			
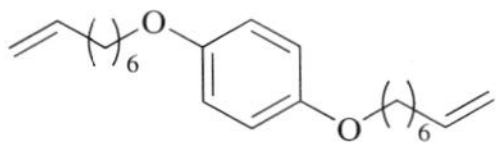	Ligand, $K_2OsO_2(OH)_4$, $K_3Fe(CN)_6$, K_2CO_3, t-BuOH:H_2O (1:1), 0°, 12 h	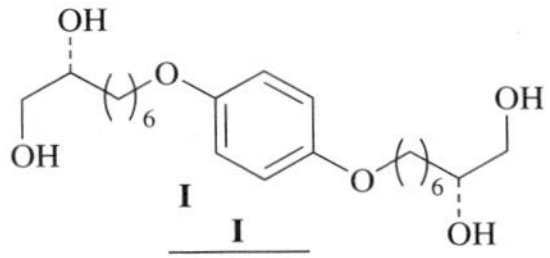 **I**	146
	Ligand	**I**	
	$(DHQD)_2PHAL$	(—), 80	
	$(DHQD)_2PYR$	(—), 74	
	$(DHQD)_2AQN$	(—), 86	
	Ligand, $K_2OsO_2(OH)_4$, I_2, K_2CO_3, t-BuOH:H_2O (1:1), 0°, 18 h	**I**	146
	Ligand	**I**	
	$(DHQD)_2PHAL$	(—), 74	
	$(DHQD)_2PYR$	(—), 62	
	$(DHQD)_2AQN$	(—), 76	

TABLE 1. REACTIONS OF TERMINAL ALKENES (*Continued*)

Substrate	Conditions	Product(s) and Yield(s) (%), % ee	Refs.
C_{22}			
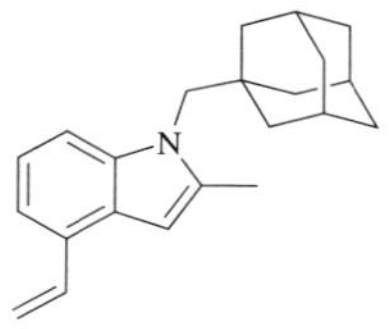	$(DHQ)_2PHAL$, $K_2OsO_2(OH)_4$, $K_3Fe(CN)_6$, K_2CO_3, MeCN:*t*-BuOH:H_2O (1:1:1), rt, 4 h	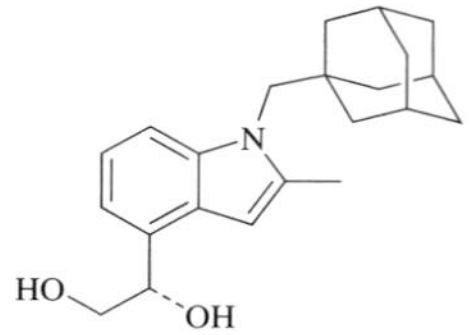 (80), 70.7	76
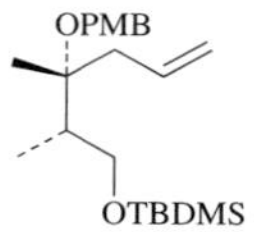	$(DHQ)_2PYR$, OsO_4, $K_3Fe(CN)_6$, K_2CO_3, *t*-BuOH:H_2O (1:1), 0°, 24 h	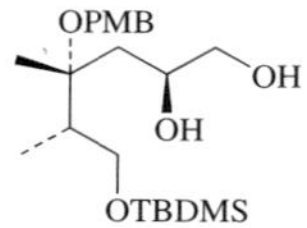 (100), 50 de	283
C_{23}			
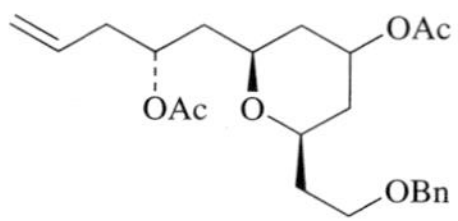	$(DHQD)_2PYR$, $K_2OsO_2(OH)_4$, $K_3Fe(CN)_6$, K_2CO_3, *t*-BuOH:H_2O (1:1)	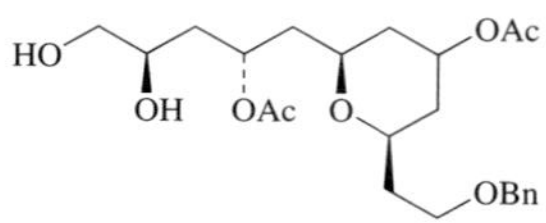 (—), 78	284
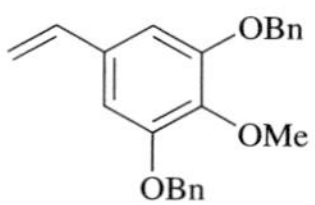	AD-Mix α, *t*-BuOH:H_2O (1:1), rt, 20 h	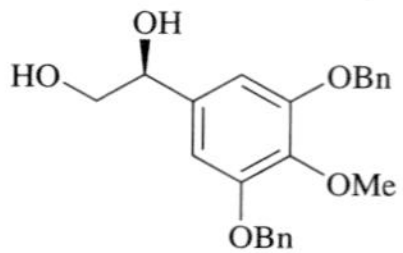 (97), 87	285
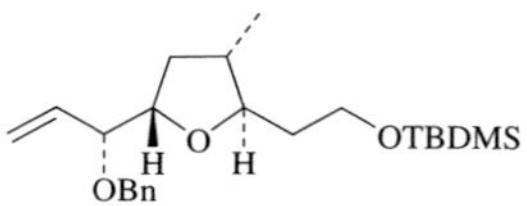	1. AD-Mix α, $NaHCO_3$, *t*-BuOH:H_2O (1:1), 4°, 24 h 2. PhNCO, DMAP, pyridine	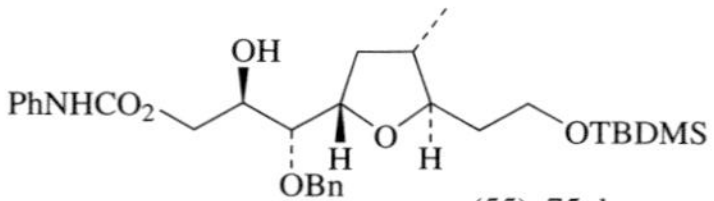 (55), 75 de	286

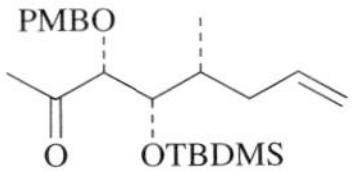	AD-Mix β, t-BuOH:H_2O (1:1), 0°, 24 h	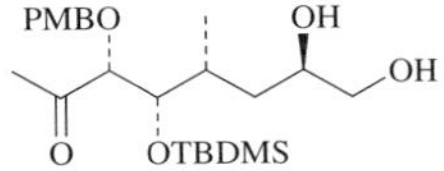	(>65), 50 de	287
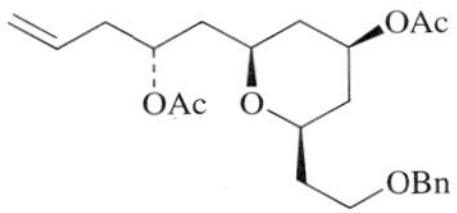	1. $(DHQD)_2PYR$, $K_2OsO_2(OH)_4$, $K_3Fe(CN)_6$, K_2CO_3, t-BuOH-H_2O 2. Moffat's Reagent	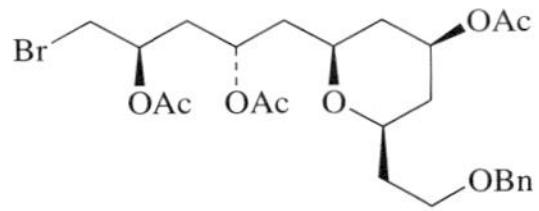	(67), 78 de	288
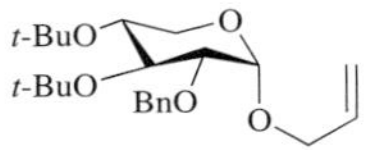	AD-Mix β, t-BuOH:H_2O (1:1), 0°, 24 h	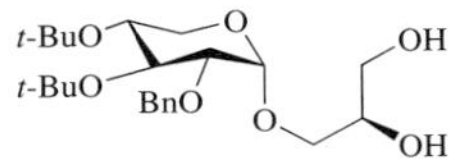	(86), 35 de	248
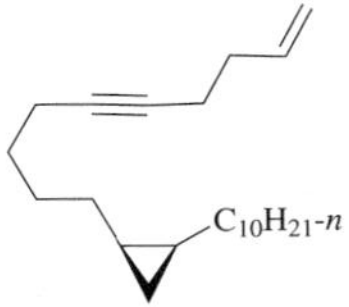	$(DHQD)_2AQN$, $K_2OsO_2(OH)_4$, $K_3Fe(CN)_6$, K_2CO_3, t-BuOH:H_2O (1:1), 0°, 8 h	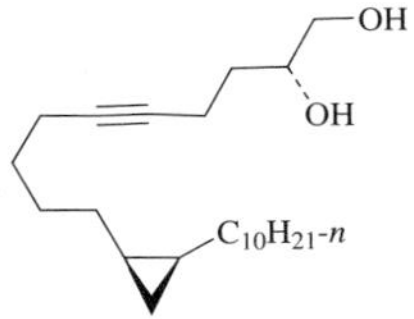	(91), — de	289
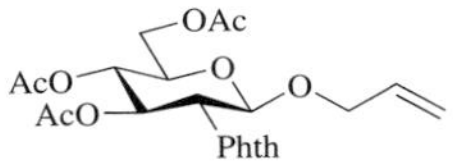	$(DHQ)_2PHAL$, $K_2OsO_2(OH)_4$, $K_3Fe(CN)_6$, K_2CO_3, t-BuOH:H_2O (1:1), 0°, 4 h; rt, 18 h	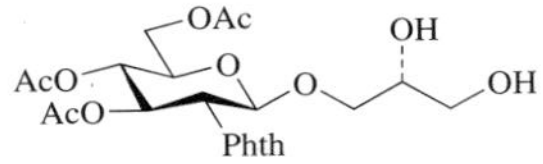	(—), 20 de	265
	$(DHQD)_2PHAL$, $K_2OsO_2(OH)_4$, $K_3Fe(CN)_6$, K_2CO_3, t-BuOH:H_2O (1:1), 0°, 4 h; rt, 18 h	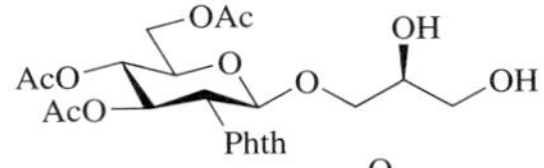	(—), 20 de	265
C_{24} 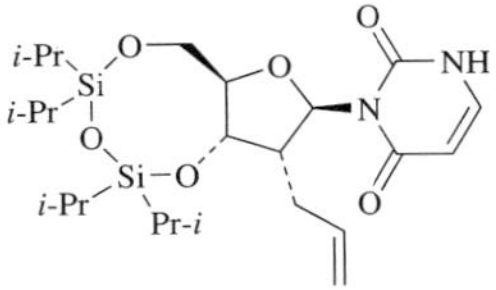	Ligand, $K_2OsO_2(OH)_4$, $K_3Fe(CN)_6$, K_2CO_3, t-BuOH:H_2O (1:1), 0° Ligand $(DHQD)_2PYR$ $(DHQ)_2PYR$	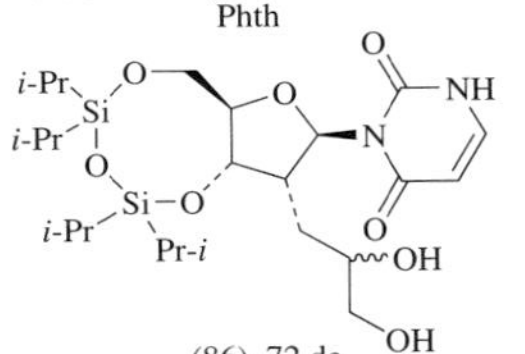 (86), 72 de (72), 52 de		290

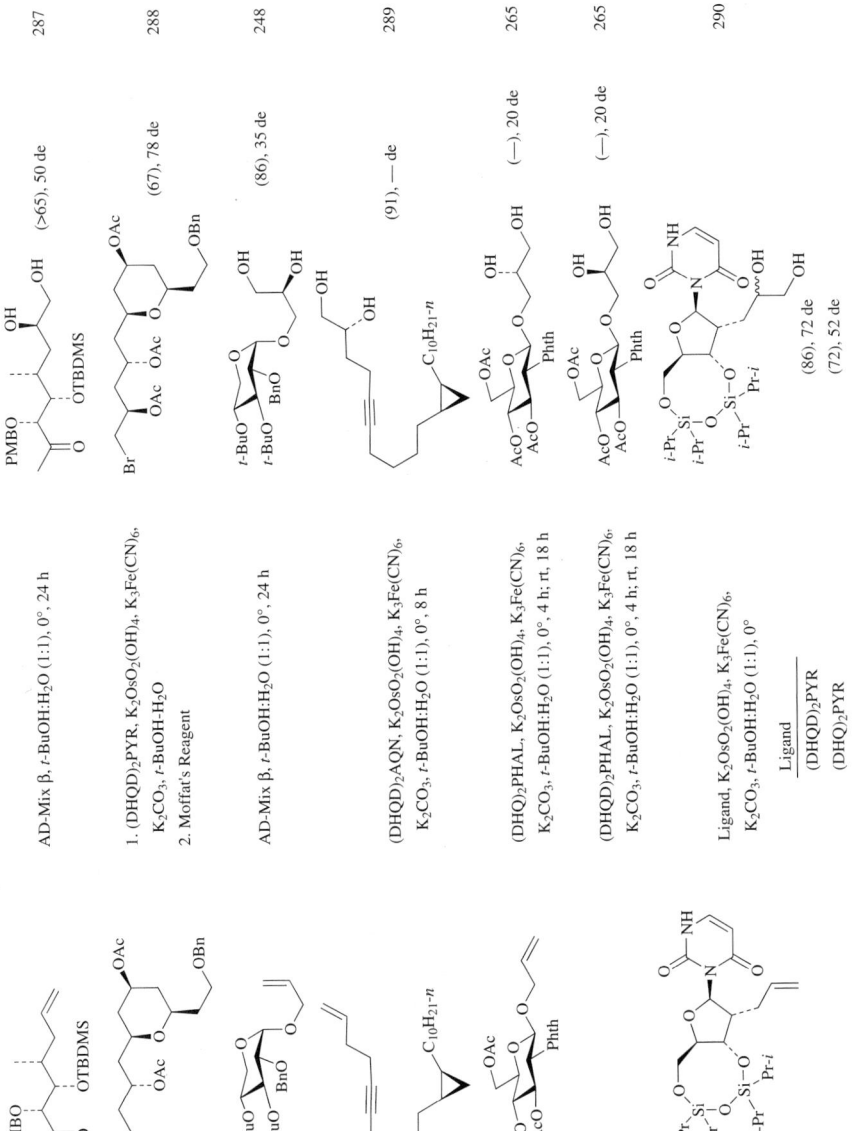

TABLE 1. REACTIONS OF TERMINAL ALKENES (*Continued*)

Substrate	Conditions	Product(s) and Yield(s) (%), % ee	Refs.
C_{24} 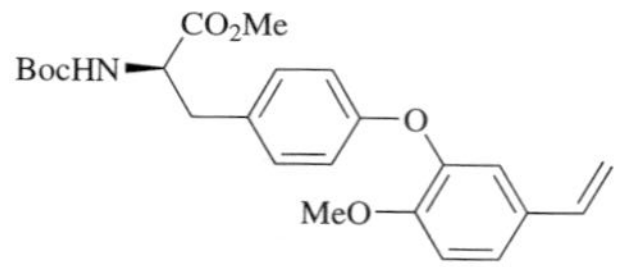	DHQD-*p*-chlorobenzoate, OsO_4, $K_3Fe(CN)_6$, K_2CO_3, *t*-BuOH:H_2O (1:1), rt, 3 h	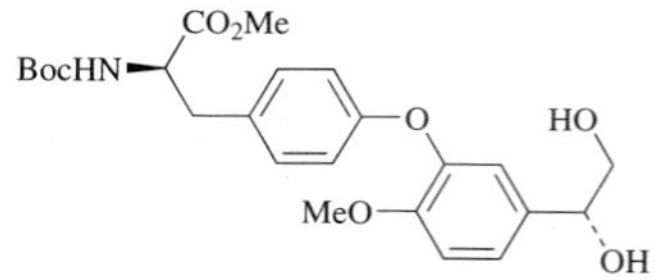 (94), 62 de	291
C_{25} 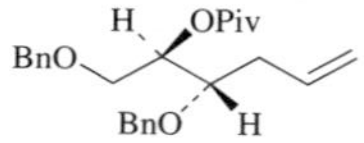	AD-Mix β, *t*-BuOH:H_2O (1:1), 0°-rt, 24 h	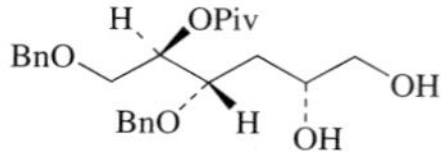 (91), 0-39 de	292, 293
	AD-Mix α, *t*-BuOH:H_2O (1:1), 0°	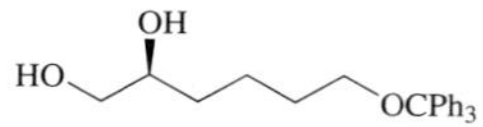 (90), 62	294
C_{26} 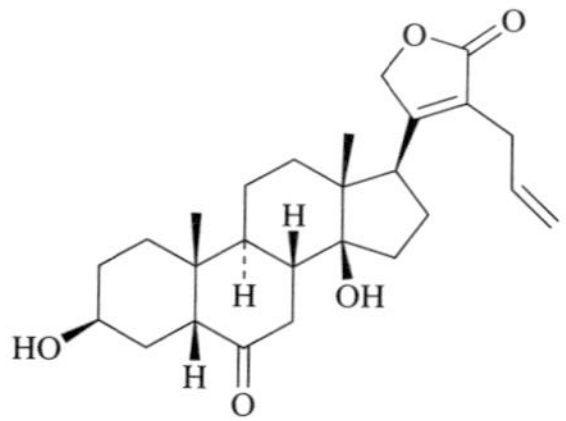	AD-Mix β, *t*-BuOH:H_2O (1:1), 0°, 24 h	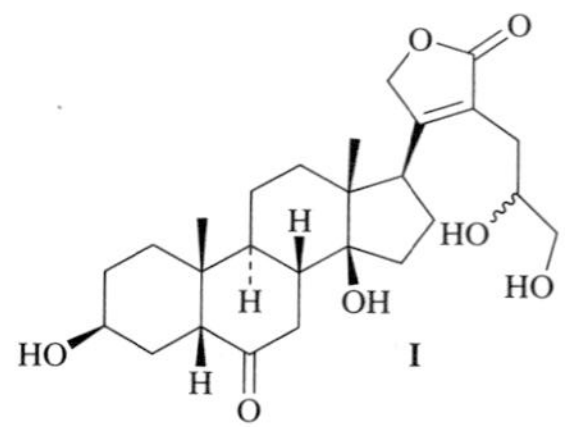 **I** (80), 10 de	295
	AD-Mix α, *t*-BuOH:H_2O (1:1), 0°, 24 h	**I** (80), 0 de	295
C_{27} 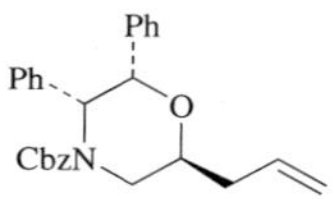	AD-Mix α, *t*-BuOH:H_2O (1:1), 0°, 72 h	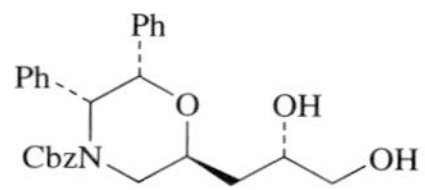 (78), 48 de	296
	AD-Mix β, *t*-BuOH:H_2O (1:1), 0°, 8-72 h	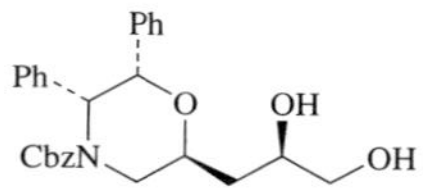 (42-55), 26-30 de	296

	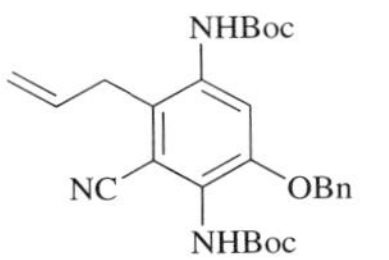	$(DHQ)_2PHAL$, $K_2OsO_2(OH)_4$, $K_3Fe(CN)_6$, K_2CO_3, t-BuOH:H_2O (2:1), 0°, 24 h	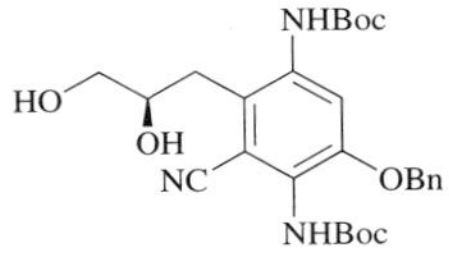 (93), 56	297
	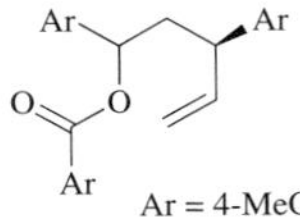 Ar = 4-$MeOC_6H_4$	AD-Mix α, t-BuOH:H_2O (1:1), rt, 16 h	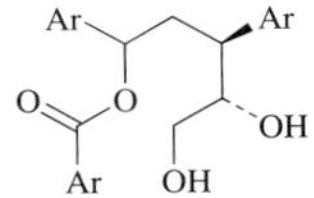 (94), >95de	240
C_{28}	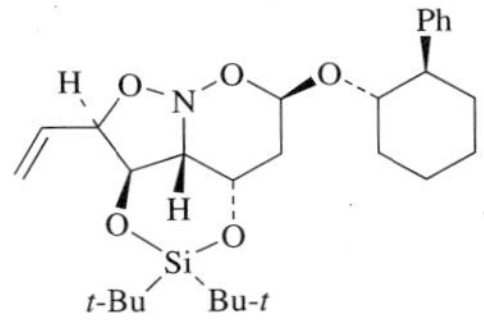	Ligand, OsO_4, $K_3Fe(CN)_6$, K_2CO_3, $MeSO_2NH_2$, t-BuOH:H_2O (1:1) Ligand DHQ-PHN $(DHQ)_2PHAL$ DHQ-p-chlorobenzoate $(DHQ)_2PYR$	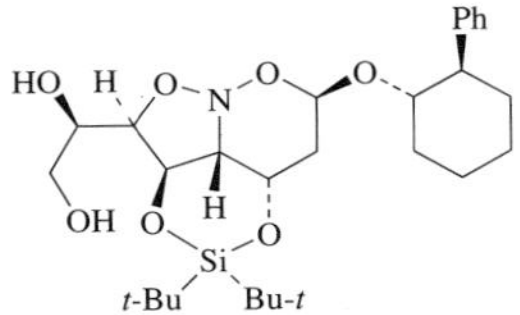 (—), 99 (—), 0 (—), 71 (—), 92	298
		Ligand, OsO_4, $K_3Fe(CN)_6$, K_2CO_3, $MeSO_2NH_2$, t-BuOH:H_2O (1:1) Ligand DHQD-IND $(DHQD)_2PHAL$ $(DHQD)_2AQN$	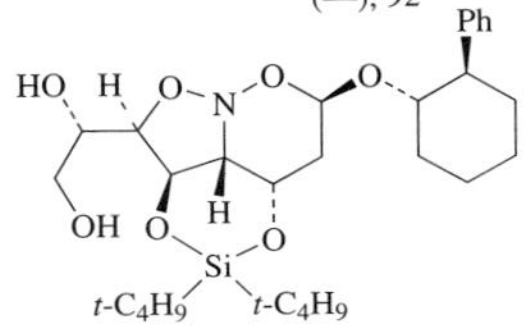 (—), 9 (—), 29 (—), 85	298

TABLE 1. REACTIONS OF TERMINAL ALKENES (*Continued*)

Substrate	Conditions	Product(s) and Yield(s) (%), % ee	Refs.
C_{29} 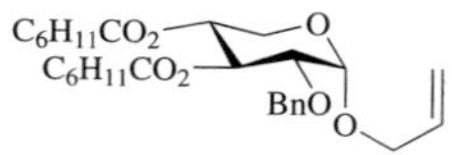	AD-Mix β, t-BuOH:H_2O (1:1), 0°	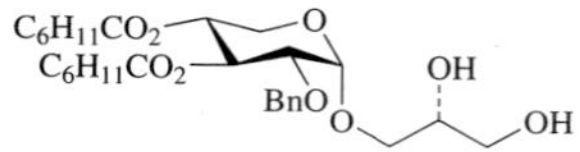 (95), 70 de	247, 248
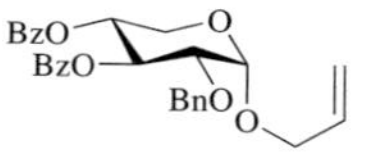	AD-Mix β, t-BuOH:H_2O (1:1), 0°	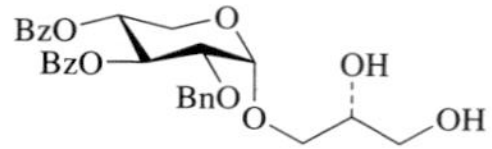 (90), 62 de	247, 248
	AD-Mix β, $(DHQD)_2PYR$, t-BuOH:H_2O (1:1), 0°, 24 h	**I** (—), 78 de	247, 248
	AD-Mix α, t-BuOH:H_2O (1:1), 0°	**I** (90), 5 de	247, 248
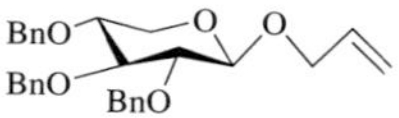	AD-Mix α, t-BuOH:H_2O (1:1), 0°	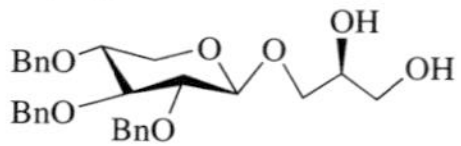 (99), 13 de	247, 248
	AD-Mix β, t-BuOH:H_2O (1:1), 0°	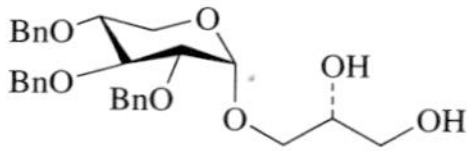 (99), 9 de	247, 248
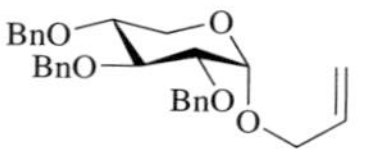	AD-Mix β, t-BuOH:H_2O (1:1), 0°	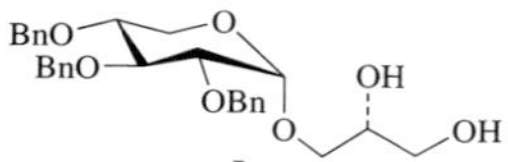 **I** (74-99), 64 de	247, 248
	AD-Mix β, $(DHQD)_2PYR$, t-BuOH:H_2O (1:1), 0°, 24 h	**I** (95), 68 de	247, 248
	$(DHQD)_2PYR$ (1%), $K_2OsO_2(OH)_4$, $K_3Fe(CN)_6$, K_2CO_3, t-BuOH:H_2O (1:1), 0°	**I** (—), 46 de	247, 248
	$(DHQD)_2PYR$ (4%), $K_2OsO_2(OH)_4$, $K_3Fe(CN)_6$, K_2CO_3, t-BuOH:H_2O (1:1), 0°	**I** (95), 68 de	247, 248

C_{31}				
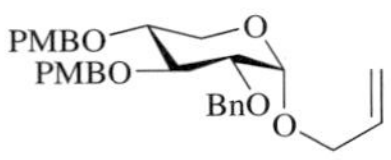	AD-Mix β, t-BuOH:H_2O (1:1), 0°	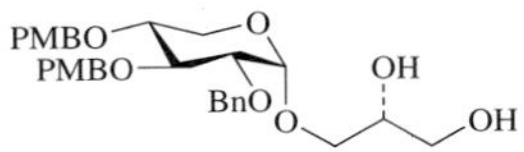	(95), 39 de	247, 248
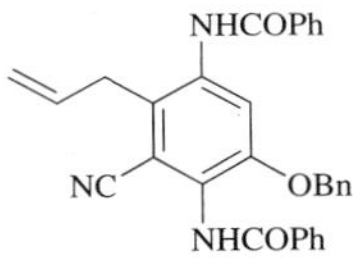	(DHQD)$_2$PHAL, $K_2OsO_2(OH)_4$, $K_3Fe(CN)_6$, K_2CO_3, t-BuOH:H_2O (1:1)	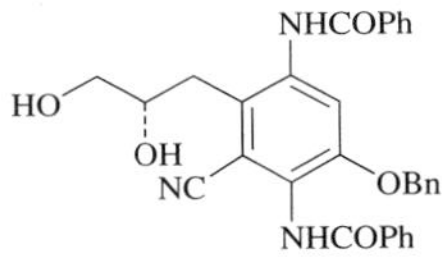	(92), 78	298a, 297
	(DHQ)$_2$PHAL, $K_2OsO_2(OH)_4$, $K_3Fe(CN)_6$, K_2CO_3, t-BuOH:H_2O (1:1)	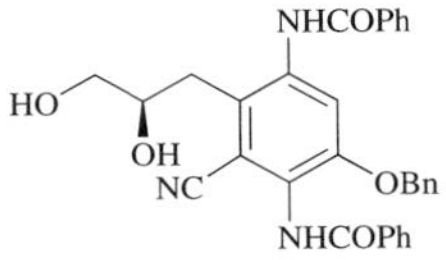	(59-86), 49-78	298a, 297
C_{33}				
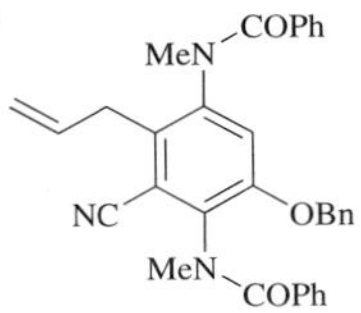	(DHQ)$_2$PHAL, $K_2OsO_2(OH)_4$, $K_3Fe(CN)_6$, K_2CO_3, t-BuOH:H_2O (2:1), 0°, 24 h	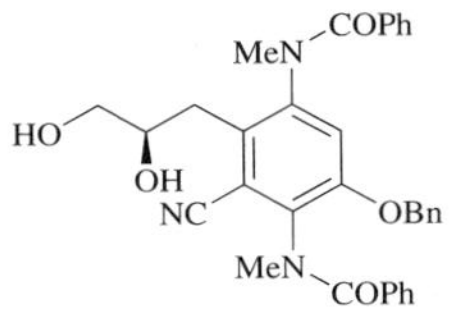	(82), 56	297

TABLE 1. REACTIONS OF TERMINAL ALKENES (*Continued*)

Substrate	Conditions	Product(s) and Yield(s) (%), % ee	Refs.
C_{34}			
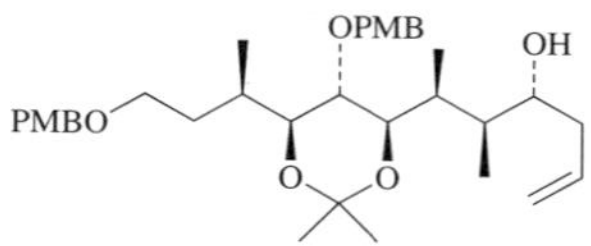	1. AD-Mix α, *t*-BuOH:H_2O (1:1), 0-rt, 48 h 2. HCl (conc.), acetone, rt, 15 min	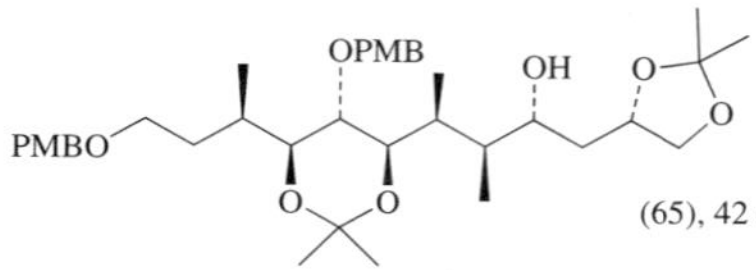 (65), 42 de	299
	DHQD-*p*-chlorobenzoate, OsO_4, $K_3Fe(CN)_6$, K_2CO_3, *t*-BuOH:H_2O (1:1), rt	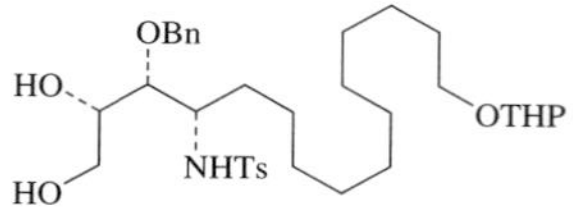 (84), >95 de	300
C_{37}			
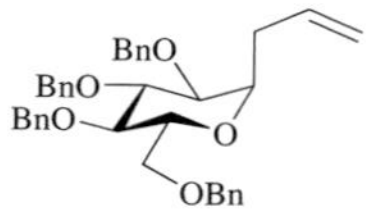	AD-Mix β, *t*-BuOH:H_2O (1:1), 0°-rt, 18 h	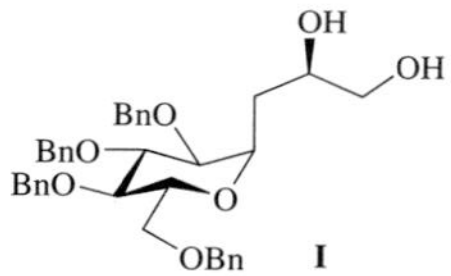 **I** (73), 22 de	266
	DHQD-PHN, $K_2OsO_2(OH)_4$, $K_3Fe(CN)_6$, K_2CO_3, *t*-BuOH:H_2O (1:1)	**I** (83), 56 de	266
	DHQD-*p*-chlorobenzoate, $K_2OsO_2(OH)_4$, $K_3Fe(CN)_6$, K_2CO_3, *t*-BuOH:H_2O (1:1)	**I** (84), 28 de	266
	DHQ-PHN, $K_2OsO_2(OH)_4$, $K_3Fe(CN)_6$, K_2CO_3, *t*-BuOH:H_2O (1:1)	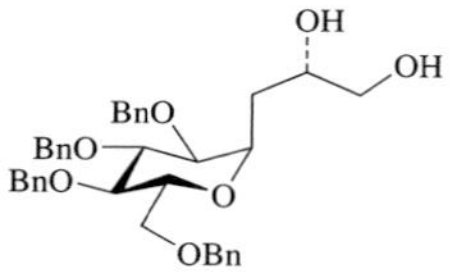 (75), 18	266

C_{39}	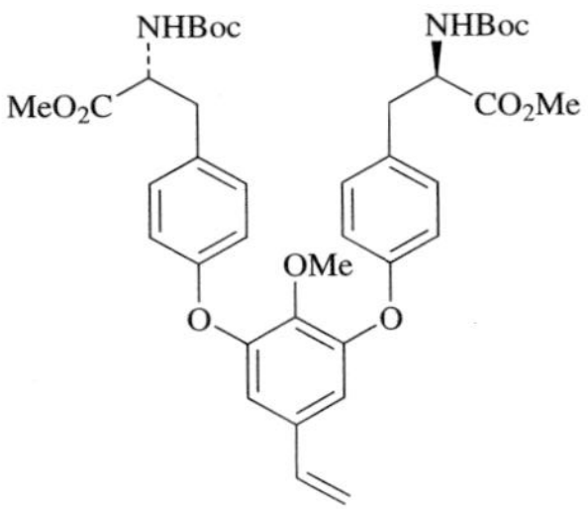	DHQD-*p*-chlorobenzoate, OsO_4, $K_3Fe(CN)_6$, K_2CO_3, *t*-BuOH:H_2O (1:1), rt, 24 h	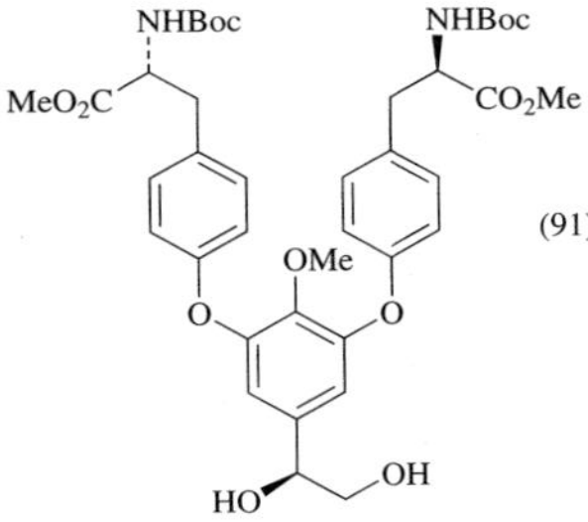 (91), 50 de	301
C_{65}	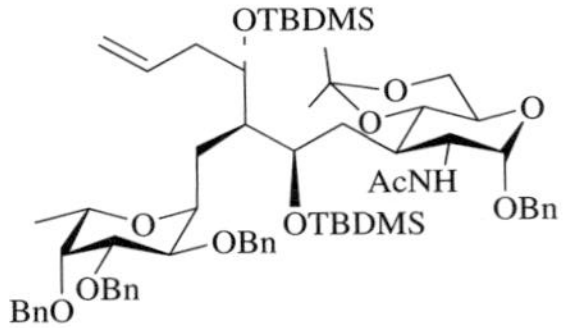	$(DHQD)_2$-PYR, $K_2OsO_2(OH)_4$, $K_3Fe(CN)_6$, K_2CO_3, *t*-BuOH:H_2O (1:1), 0°	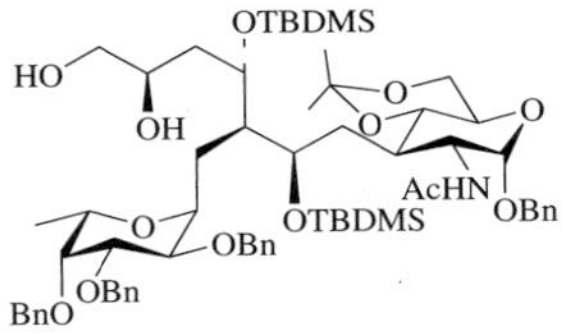 (77), 67 de	302, 303
C_n	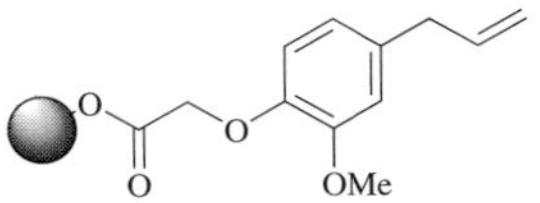 Wang Resin	Ligand, $K_2OsO_2(OH)_4$, $K_3Fe(CN)_6$, $MeSO_2NH_2$, K_2CO_3, THF:H_2O (1:1), rt	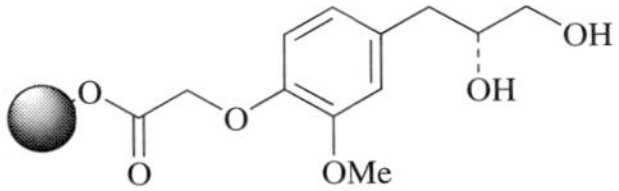	232

Ligand	Time	
$(DHQD)_2$AQN	18 h	(83), 0
$(DHQD)_2$PYR	18 h	(62), 0
$(DHQD)_2$PHAL	12 h	(73), 3

TABLE 1. REACTIONS OF TERMINAL ALKENES (*Continued*)

Substrate	Conditions	Product(s) and Yield(s) (%), % ee	Refs.
C_n 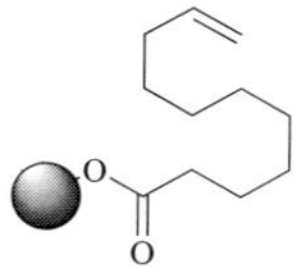 Wang Resin	Ligand, $K_2OsO_2(OH)_4$, $K_3Fe(CN)_6$, $MeSO_2NH_2$, K_2CO_3, THF:H_2O (1:1), rt Ligand / Time $(DHQD)_2AQN$ 18 h $(DHQD)_2PYR$ 18 h $(DHQD)_2PHAL$ 12 h	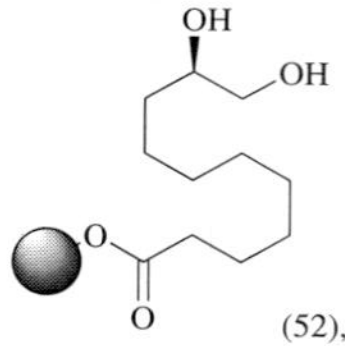 (52), 32 (44), 34 (96), 41	232
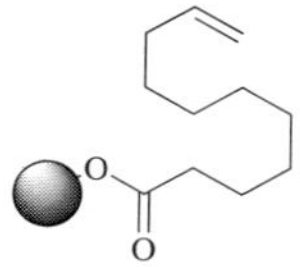 TentaGel S-OH	$(DHQD)_2PHAL$, $K_2OsO_2(OH)_4$, $K_3Fe(CN)_6$, $MeSO_2NH_2$, K_2CO_3, THF:H_2O (1:1), rt, 24 h	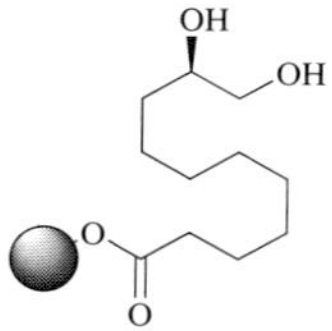 (44), 45	232

TABLE 2. REACTIONS OF 1,1-DISUBSTITUTED ALKENES

Substrate	Conditions	Product(s) and Yield(s) (%), % ee	Refs.
*See Chart 1 at the beginning of the Tabular Survey for ligand structures that are indicated by **bold** numbers.*			
C_4			
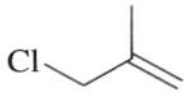	(DHQD)$_2$PHAL, $K_2OsO_2(OH)_4$, $K_3Fe(CN)_6$, K_2CO_3, $NaHCO_3$, t-BuOH:H_2O (1:1), 0°	 (70), 40	154
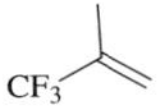	Ligand, $K_2OsO_2(OH)_4$, $K_3Fe(CN)_6$, K_2CO_3, t-BuOH:H_2O (1:1), 0°		20
	Ligand		
	(DHQD)$_2$PHAL	(—), 13	
	(DHQD)$_2$PYR	(—), 25	
C_6			
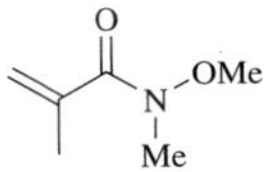	(DHQD)$_2$PHAL, $K_2OsO_2(OH)_4$, $K_3Fe(CN)_6$, K_2CO_3, t-BuOH:H_2O (1:1), 0°	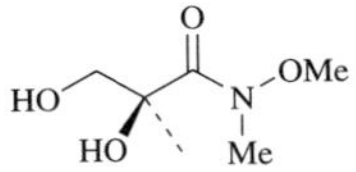 (81), 93	111
C_7			
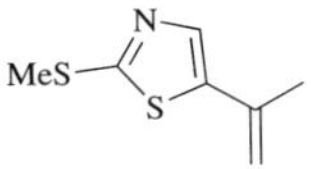	(DHQD)$_2$PHAL, $K_2OsO_2(OH)_4$, $K_3Fe(CN)_6$, K_2CO_3, t-BuOH:H_2O (1:1), 0°, 18 h	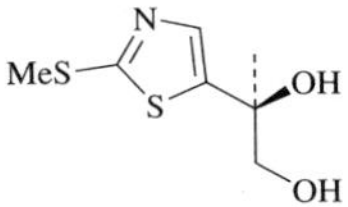 (76), 98.5	90
	(DHQ)$_2$PHAL, $K_2OsO_2(OH)_4$, $K_3Fe(CN)_6$, K_2CO_3, t-BuOH:H_2O (1:1), 0°, 18 h	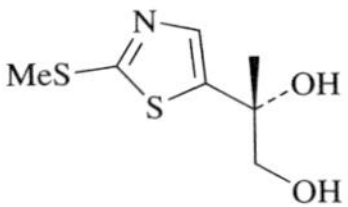 (95), 98.5	90
C_8			
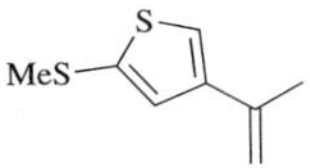	(DHQD)$_2$PHAL, $K_2OsO_2(OH)_4$, $K_3Fe(CN)_6$, K_2CO_3, t-BuOH:H_2O (1:1), 0°, 18 h	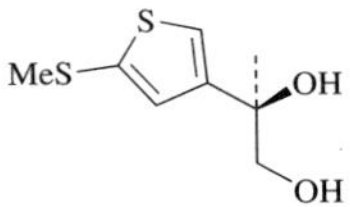 (92), 98	90
	(DHQ)$_2$PHAL, $K_2OsO_2(OH)_4$, $K_3Fe(CN)_6$, K_2CO_3, t-BuOH:H_2O (1:1), 0°, 18 h	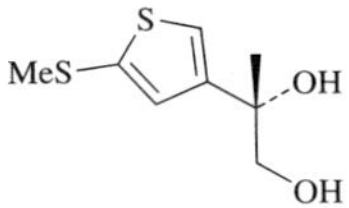 (93), 98	90

TABLE 2. REACTIONS OF 1,1-DISUBSTITUTED ALKENES (*Continued*)

Substrate	Conditions	Product(s) and Yield(s) (%), % ee	Refs.
C_8 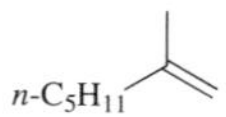	Ligand, $K_2OsO_2(OH)_4$, $K_3Fe(CN)_6$, K_2CO_3, *t*-BuOH:H_2O (1:1), 0°	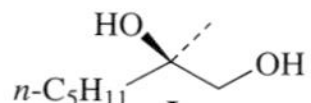 **I**	
	Ligand / Time	**I**	
	$(DHQD)_2PYR$ / 6-24 h	(80-98), 78	19, 157
	$(DHQD)_2PYDZ$ / 12 h	(—), 65	66
	$(DHQD)_2PHAL$ / 6-24 h	(—), 78	157
	$(DHQD)_2DPP$ / 6-24 h	(—), 78	157
	Ligand, $K_2OsO_2(OH)_4$, $K_3Fe(CN)_6$, K_2CO_3, $NaHCO_3$, *t*-BuOH:H_2O (1:1), 0°	**I**	
	Ligand / Time	**I**	
	$(DHQD)_2AQN$ / 12 h	(65-90), 85	86
	$(DHQD)_2DPPHAL$ / —	(—), 81	86, 157
	$(DHQD)_2PYDZ$, $K_2OsO_2(OH)_4$, $K_3Fe(CN)_6$, K_2CO_3, $MeSO_2NH_2$, *t*-BuOH:H_2O (1:1), 0°	**I** (—), 62	156
	$(DHQD)_2PHAL$, ABC MC OsO_4, NMO, acetone:H_2O:CH_3CN (1:1:1), rt, 24 h	**I** (90), 60	170
	$(DHQ)_2PHAL$, $K_2OsO_2(OH)_4$, $K_3Fe(CN)_6$, K_2CO_3, *t*-BuOH:H_2O (1:1), 0°, 6-24 h	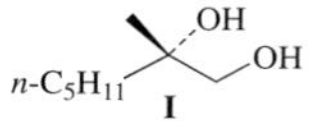 **I** (80-98), 76	19, 66
	AD-Mix α, *t*-BuOH:H_2O (1:1), 0°	**I** (95), 71	304
	1. $(DHQ)_2PHAL$, $K_2OsO_2(OH)_4$, $K_3Fe(CN)_6$, K_2CO_3, *t*-BuOH:H_2O (1:1), rt, 18 h 2. TEMPO, NaOCl, $NaOCl_2$, CH_3CN:phosphate buffer (5:4, pH 5.6), 55°, 4 d	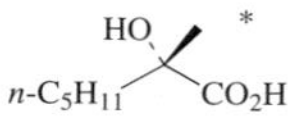 (73), 74	305

Reactant	Conditions	Product(s) and Yield(s) (%)	Refs.
	1. (DHQD)$_2$PHAL, $K_2OsO_2(OH)_4$, $K_3Fe(CN)_6$, K_2CO_3, t-BuOH:H_2O (1:1), rt, 18 h 2. TEMPO, NaOCl, $NaOCl_2$, CH_3CN:phosphate buffer (5:4, pH 5.6), 55°, 4 d	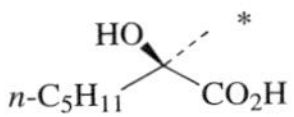 (80), 79	305
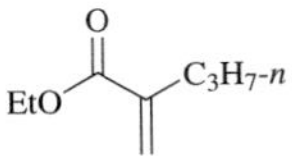	AD-Mix β, t-BuOH:H_2O (1:1), 0°, 8 h	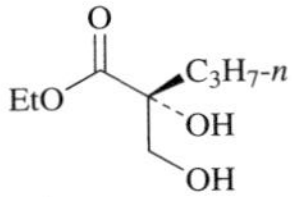 (98), 44	306
C_9 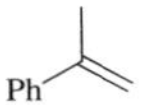	Ligand, $K_2OsO_2(OH)_4$, $K_3Fe(CN)_6$, K_2CO_3, t-BuOH:H_2O (1:1)	 **I**	

Ligand	pH	Temp	Time	I	Refs.
(DHQD)$_2$PHAL	—	0°	3 h	(80-98), 97	77, 200, 157, 152, 66, 307
(DHQD)$_2$PYR	—	0°	3 h	(—), 69	77, 157
(DHQD)$_2$PYDZ	—	0°	6-24 h	(—), 93	152, 66
(DHQD)$_2$TP	—	0°	6-24 h	(—), 72	152
(DHQD)$_2$AQN	—	0°	6-24 h	(65-90), 82	86
(DHQD)$_2$DPP	—	0°	—	(—), 96	86, 157
(DHQD)$_2$DPPHAL	—	0°	6-24 h	(—), 94	157
40	—	0°	<5 h	(89), 95	169
(DHQD)$_2$PHAL	—	rt	0.6 h	(95), 89	71
(DHQD)$_2$PHAL	10	rt	1.6 h	(95), 93	71, 307

Conditions	Product(s) and Yield(s) (%)	Refs.
DHQD-p-chlorobenzoate, OsO_4, NMO, acetone-H_2O, 0°, 1.5 h	**I** (80-95), 33	18
(DHQD)$_2$PHAL, ABC MC OsO_4, NMO, acetone:H_2O:CH_3CN (1:1:1), rt, 24 h	**I** (98), 78	170
(DHQD)$_2$PHAL, $K_2OsO_2(OH)_4$, O_2, $PhSeCH_2Ph$, t-BuOH:H_2O (1:1), 12°	**I** (89), 96	307

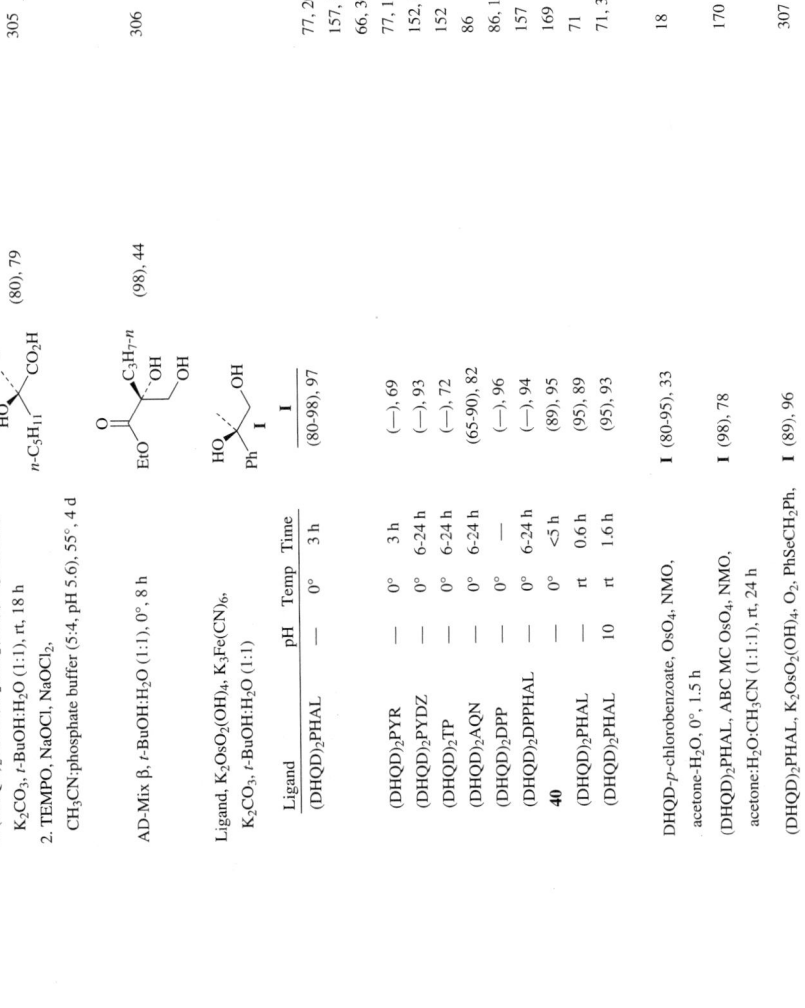

TABLE 2. REACTIONS OF 1,1-DISUBSTITUTED ALKENES (*Continued*)

Substrate	Conditions	Product(s) and Yield(s) (%), % ee	Refs.
C_9 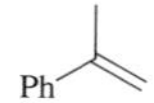	Ligand, OsO_4, $K_3Fe(CN)_6$, K_2CO_3, *t*-BuOH-H_2O (1:1), 0°, 24 h	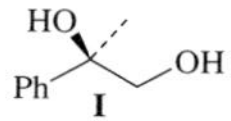 **I**	
	Ligand / pH / Temp / Time		
	41a / — / 0° / 24 h	(68-80), 58	206
	45 / — / 0° / 24 h	(88), 94	160
	$(DHQD)_2PHAL$ / 10.4 / 50° / 14-18 h	(96), 80	178
	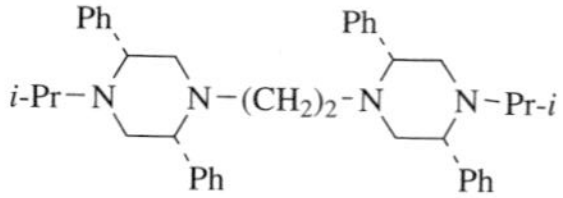 , OsO_4, toluene, –78°	**I** (77), 82	215
	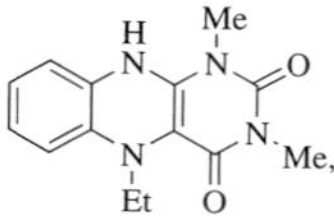 Me, NMM, TEAA, H_2O_2, $(DHQD)_2PHAL$, OsO_4, *t*-BuOH:H_2O (3:1), 0°, 9 h	**I** (88), 99	208
	Ligand, $K_2OsO_2(OH)_4$, $K_3Fe(CN)_6$, K_2CO_3, *t*-BuOH:H_2O (1:1), 0°, 6 h		
	Ligand / Temp / Time		
	$(DHQ)_2PHAL$ / 0° / 6 h	(100), 92	91, 200, 157
	$(DHQ)_2DPP$ / 0° / 6 h	(—), 92	157
	$(DHQ)_2DPPHAL$ / 0° / 6-24 h	(—), 94	157
	Poly$(DHQ)_2PYR$ / 20° / 20 h	(84), 52	164

	42b, OsO_4, $K_3Fe(CN)_6$, K_2CO_3, t-BuOH:H_2O (1:1), 0°, 24 h	**I** (68-80), 56	206
	1. 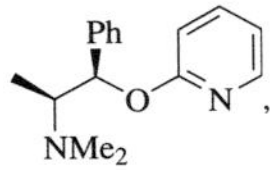 , OsO_4, toluene, –78°, 10 h 2. LAH, ether, –78° to rt, 8 h	**I** (78), 30	183
	1. $(DHQD)_2PHAL$, $K_2OsO_2(OH)_4$, $K_3Fe(CN)_6$, K_2CO_3, t-BuOH:H_2O (1:1), rt, 18 h 2. TEMPO, NaOCl, $NaOCl_2$, CH_3CN:phosphate buffer (5:4, pH 5.6), 55°, 4 d	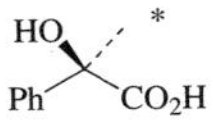 (65), 79	305
	1. $(DHQ)_2PHAL$, $K_2OsO_2(OH)_4$, $K_3Fe(CN)_6$, K_2CO_3, t-BuOH:H_2O (1:1), rt, 18 h 2. TEMPO, NaOCl, $NaOCl_2$, CH_3CN:phosphate buffer (5:4, pH 5.6), 55°, 4 d	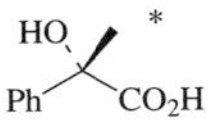 (60), 98	305
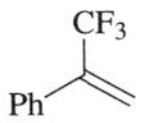	Ligand, OsO_4, $K_3Fe(CN)_6$, K_2CO_3, t-BuOH:H_2O (1:1), 0°, 72 h	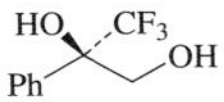	89
	Ligand		
	$(DHQD)_2PHAL$	(94), 83	
	$(DHQD)_2DPP$	(94), 83-91	
	1. $(DHQD)_2PHAL$, $K_2OsO_2(OH)_4$, $K_3Fe(CN)_6$, K_2CO_3, t-BuOH:H_2O (1:1), rt, 18 h 2. TEMPO, NaOCl, $NaOCl_2$, CH_3CN:phosphate buffer (5:4, pH 5.6), 55°, 4 d	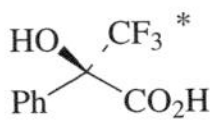 (52), 50	305
	1. $(DHQ)_2PHAL$, $K_2OsO_2(OH)_4$, $K_3Fe(CN)_6$, K_2CO_3, t-BuOH:H_2O (1:1), rt, 18 h 2. TEMPO, NaOCl, $NaOCl_2$, CH_3CN:phosphate buffer (5:4, pH 5.6), 55°, 4 d	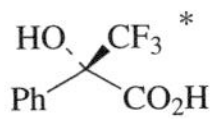 (62), 96	305

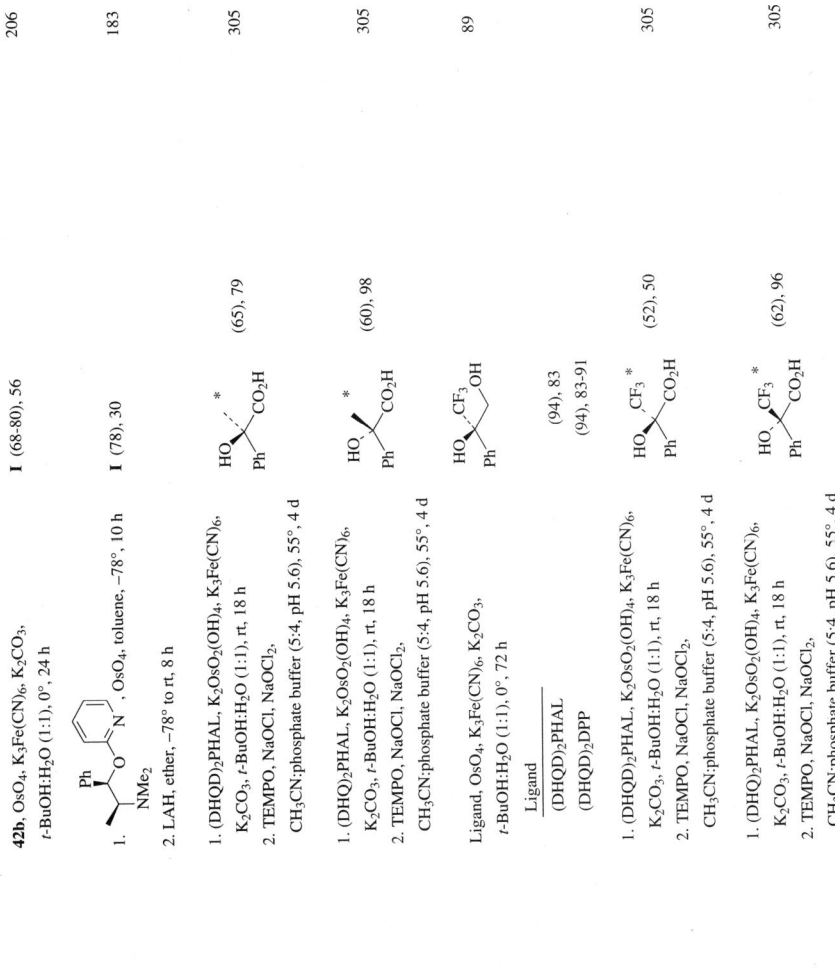

TABLE 2. REACTIONS OF 1,1-DISUBSTITUTED ALKENES (*Continued*)

Substrate	Conditions	Product(s) and Yield(s) (%), % ee	Refs.
C_9 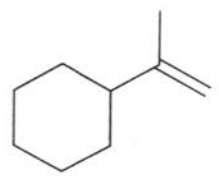	Ligand, $K_2OsO_2(OH)_4$, $K_3Fe(CN)_6$, K_2CO_3, *t*-BuOH:H_2O (1:1), 0°, 18-24 h	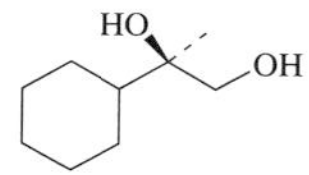	72
	Ligand		
	DHQD-PHN	(75-95), 82	
	DHQD-MEQ	(75-95), 73	
	DHQD-*p*-chlorobenzoate	(75-95), 37	
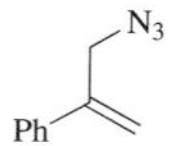	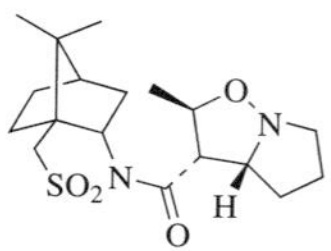, OsO_4, $K_3Fe(CN)_6$, K_2CO_3, *t*-BuOH:H_2O (1:1), rt, 24 h	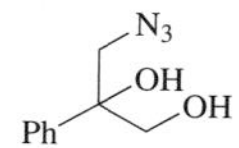 (99), 33	203
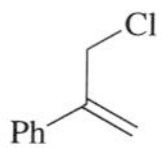	$(DHQD)_2PHAL$, $K_2OsO_2(OH)_4$, $K_3Fe(CN)_6$, K_2CO_3, $NaHCO_3$, *t*-BuOH:H_2O (1:1), 0°	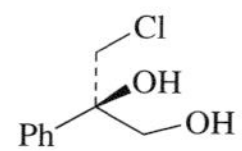 (60), 88	87
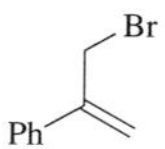	$(DHQD)_2PHAL$, $K_2OsO_2(OH)_4$, $K_3Fe(CN)_6$, K_2CO_3, $NaHCO_3$, *t*-BuOH:H_2O (1:1), 0°	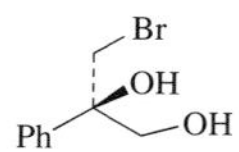 (70), 87	87
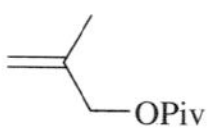	$(DHQ)_2PHAL$, $K_2OsO_2(OH)_4$, $K_3Fe(CN)_6$, K_2CO_3, *t*-BuOH:H_2O (1:1), 0°	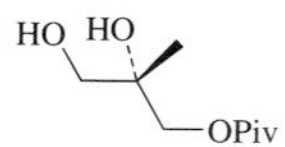 (63), 15	94
C_{10} 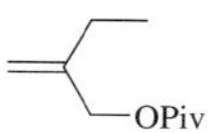	$(DHQD)_2PHAL$, $K_2OsO_2(OH)_4$, $K_3Fe(CN)_6$, K_2CO_3, *t*-BuOH:H_2O (1:1), 0°	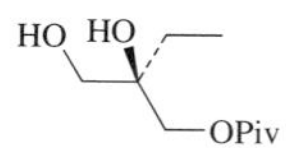 (63), 11	94
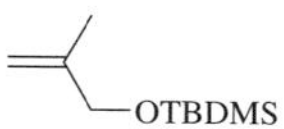	$(DHQ)_2PHAL$, $K_2OsO_2(OH)_4$, $K_3Fe(CN)_6$, K_2CO_3, *t*-BuOH:H_2O (1:1)	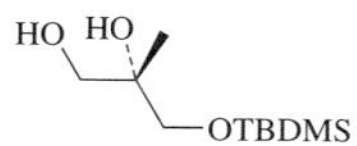 (92), 43	94

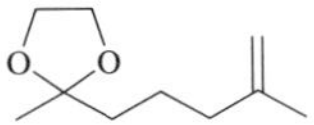	$(DHQ)_2PHAL$, $K_2OsO_2(OH)_4$, $K_3Fe(CN)_6$, K_2CO_3, t-BuOH:H_2O (1:1), 0°, 19 h	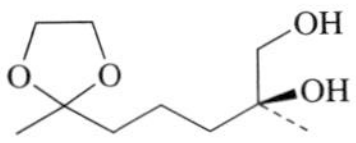	(72), 35	308
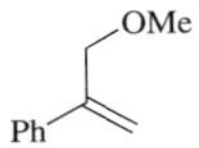	$(DHQD)_2PHAL$, $K_2OsO_2(OH)_4$, $K_3Fe(CN)_6$, K_2CO_3, t-BuOH:H_2O (1:1), 0°	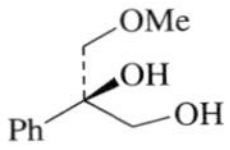	(79), 97	87
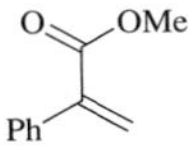	$(DHQD)_2PHAL$, $K_2OsO_2(OH)_4$, $K_3Fe(CN)_6$, K_2CO_3, t-BuOH:H_2O (1:1), 0°, 6.5 h	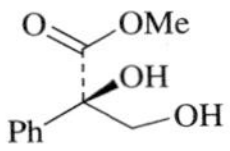	(86), 88	87
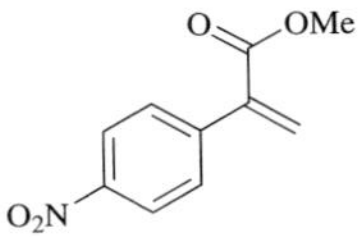	$(DHQD)_2PHAL$, $K_2OsO_2(OH)_4$, $K_3Fe(CN)_6$, K_2CO_3, t-BuOH:H_2O (1:1), 0°, 48 h	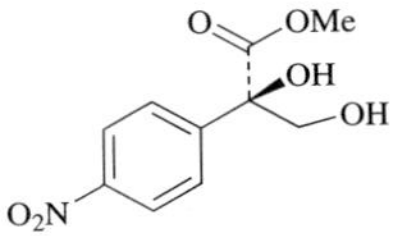	(91), 93	87
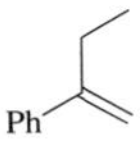	Ligand, $K_2OsO_2(OH)_4$, $K_3Fe(CN)_6$, K_2CO_3, t-BuOH:H_2O (1:1), 0°, 12 h	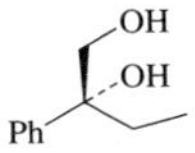		88
	Ligand			
	$(DHQD)_2PHAL$		(85-95), 78	
	$(DHQD)_2PYR$		(85-95), 20	
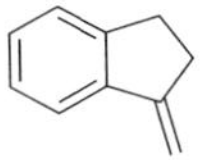	Ligand, $K_2OsO_2(OH)_4$, $K_3Fe(CN)_6$, K_2CO_3, t-BuOH:H_2O (1:1), 0°, 12 h	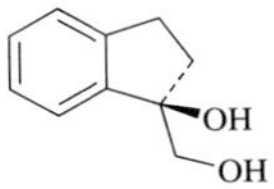		88
	Ligand			
	$(DHQD)_2PHAL$		(85-95), 92	
	$(DHQD)_2PYR$		(85-95), 78	

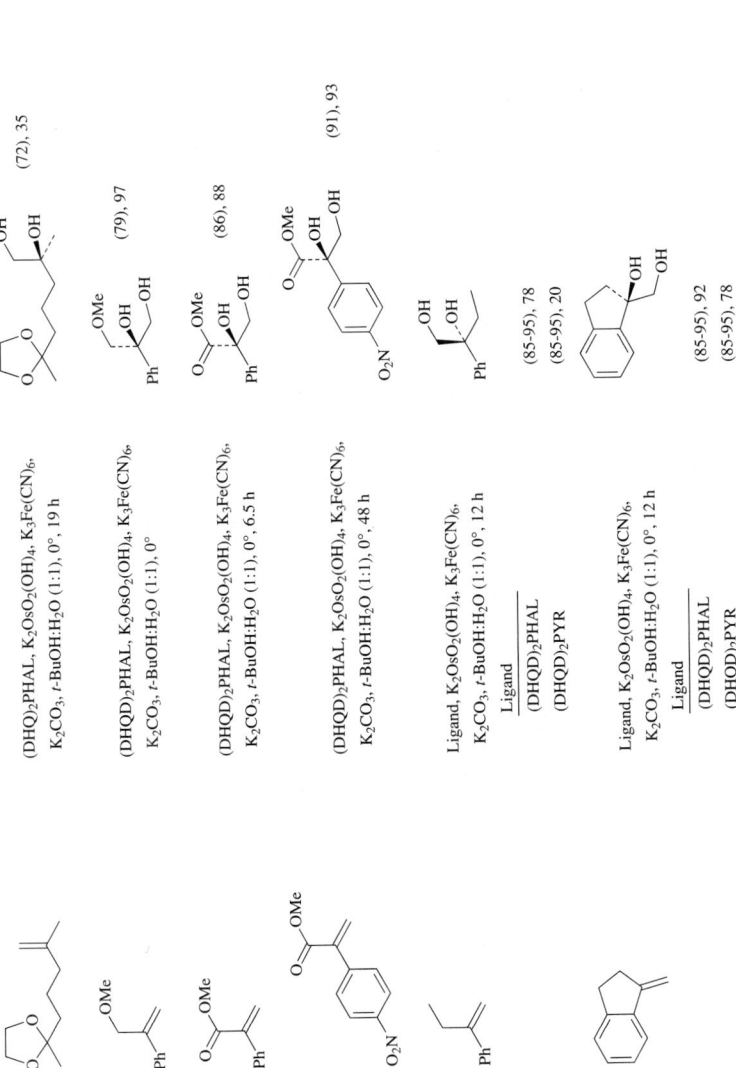

TABLE 2. REACTIONS OF 1,1-DISUBSTITUTED ALKENES (*Continued*)

Substrate	Conditions	Product(s) and Yield(s) (%), % ee	Refs.
C_{10}			
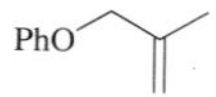	(DHQ)$_2$PHAL, $K_2OsO_2(OH)_4$, $K_3Fe(CN)_6$, K_2CO_3, t-BuOH:H_2O (1:1), 4°, 24 h	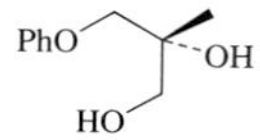 (84), 77	95
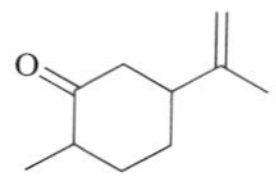	1. (DHQ)$_2$PHAL, $K_2OsO_2(OH)_4$, $K_3Fe(CN)_6$, K_2CO_3, t-BuOH:H_2O (1:1), rt, 18 h 2. TEMPO, NaOCl, $NaOCl_2$, CH_3CN:phosphate buffer (5:4, pH 5.6), 55°, 4 d	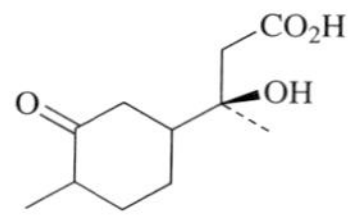 * (55), —	305
	1. (DHQD)$_2$PHAL, $K_2OsO_2(OH)_4$, $K_3Fe(CN)_6$, K_2CO_3, t-BuOH:H_2O (1:1), rt, 18 h 2. TEMPO, NaOCl, $NaOCl_2$, CH_3CN:phosphate buffer (5:4, pH 5.6), 55°, 4 d	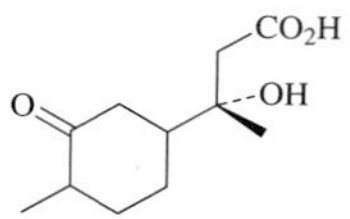 * (75), —	305
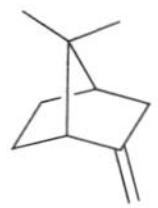	1. (DHQ)$_2$PHAL, $K_2OsO_2(OH)_4$, $K_3Fe(CN)_6$, K_2CO_3, t-BuOH:H_2O (1:1), rt, 18 h 2. TEMPO, NaOCl, $NaOCl_2$, CH_3CN:phosphate buffer (5:4, pH 5.6), 55°, 4 d	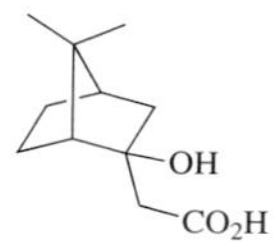 (67), 40 de	305
	1. (DHQD)$_2$PHAL, $K_2OsO_2(OH)_4$, $K_3Fe(CN)_6$, K_2CO_3, t-BuOH:H_2O (1:1), rt, 18 h 2. TEMPO, NaOCl, $NaOCl_2$, CH_3CN:phosphate buffer (5:4, pH 5.6), 55°, 4 d	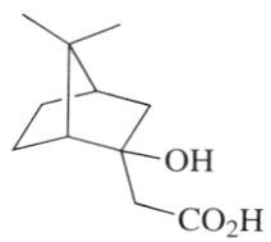 (61), 78 de	305

C_{11}

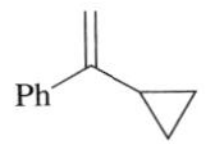	$(DHQD)_2PHAL$, $K_2OsO_2(OH)_4$, $K_3Fe(CN)_6$, K_2CO_3, t-BuOH:H_2O (1:1), 0°, 12 h	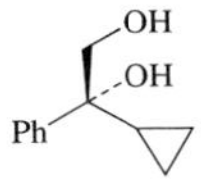	(85-95), 70	88
	$(DHQD)_2PYR$, $K_2OsO_2(OH)_4$, $K_3Fe(CN)_6$, K_2CO_3, t-BuOH:H_2O (1:1), 0°, 12 h	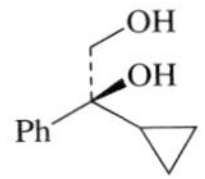	(85-95), 24	88
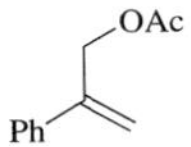	$(DHQD)_2PHAL$, $K_2OsO_2(OH)_4$, $K_3Fe(CN)_6$, K_2CO_3, $NaHCO_3$, t-BuOH:H_2O (1:1), 0°	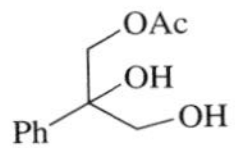	(87), 0	87
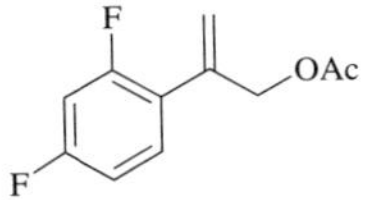	$(DHQD)_2PHAL$, $K_2OsO_2(OH)_4$, $K_3Fe(CN)_6$, K_2CO_3, $MeSO_2NH_2$, t-BuOH:H_2O (1:1), rt, 120 h	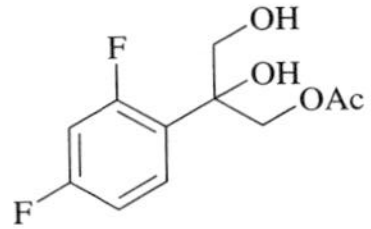	(79), 0	309
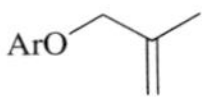	$(DHQ)_2PHAL$, $K_2OsO_2(OH)_4$, $K_3Fe(CN)_6$, K_2CO_3, t-BuOH:H_2O (1:1), 4°, 24 h	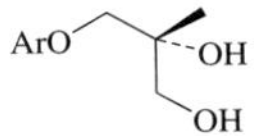		95

Ar	
p-$MeOC_6H_4$	(95), 90
m-$MeOC_6H_4$	(91), 85
o-$MeOC_6H_4$	(91), 24

TABLE 2. REACTIONS OF 1,1-DISUBSTITUTED ALKENES (*Continued*)

Substrate	Conditions	Product(s) and Yield(s) (%), % ee	Refs.
C_{11}			
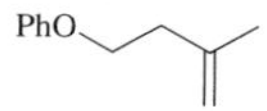	(DHQ)$_2$PHAL, $K_2OsO_2(OH)_4$, $K_3Fe(CN)_6$, K_2CO_3, *t*-BuOH:H_2O (1:1), 4°, 24 h	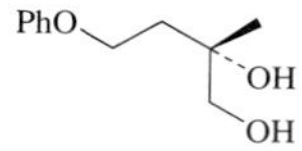 (86), 96	95
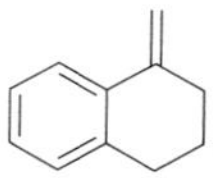	Ligand, $K_2OsO_2(OH)_4$, $K_3Fe(CN)_6$, K_2CO_3, *t*-BuOH:H_2O (1:1), 0°, 12 h	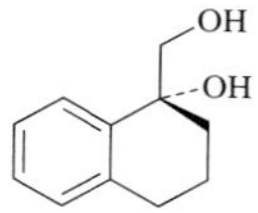	
	Ligand		
	(DHQD)$_2$PYDZ	(90), 94	66
	(DHQD)$_2$PHAL	(85-95), 95	88
	(DHQD)$_2$PYR	(85-95), 60	88
	Ligand, $K_2OsO_2(OH)_4$, $K_3Fe(CN)_6$, K_2CO_3, *t*-BuOH:H_2O (1:1), 0°, 15-30 h	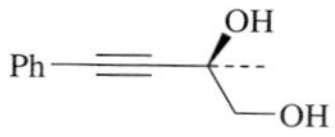	219
	Ligand		
	DHQD-PHN	(91), 48	
	(DHQD)$_2$PHAL	(—), 79	
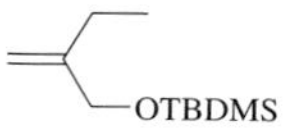	(DHQ)$_2$PHAL, $K_2OsO_2(OH)_4$, $K_3Fe(CN)_6$, K_2CO_3, *t*-BuOH:H_2O (1:1), 0°	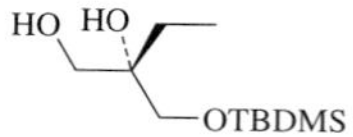 (67), 79	94
	(DHQD)$_2$PHAL, $K_2OsO_2(OH)_4$, $K_3Fe(CN)_6$, K_2CO_3, *t*-BuOH:H_2O (1:1), 0°	 **I** (82), 80-83	94
	AD-Mix β, *t*-BuOH:H_2O (1:1), 0-4°	**I** (82), 80	99

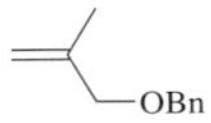	AD-Mix α, t-BuOH:H_2O (1:1), 0°	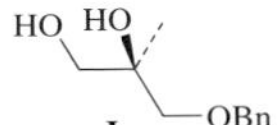 **I**	(93), 45	94
	AD-Mix α, t-BuOH:H_2O (1:1), 4°, 24 h	**I** (92), 53		
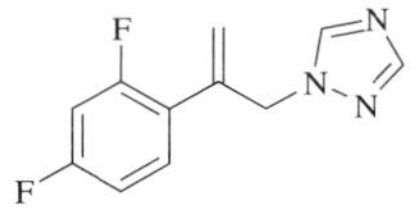	$(DHQ)_2$PHAL, $K_2OsO_2(OH)_4$, $K_3Fe(CN)_6$, K_2CO_3, $MeSO_2NH_2$, t-BuOH:H_2O (1:1), rt, 144 h	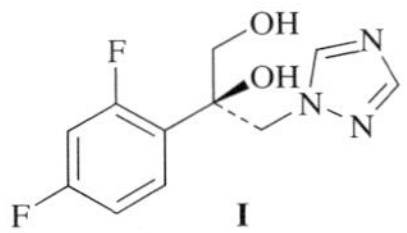 **I**	(72), 85	309
	$(DHQ)_2$PHAL, $K_2OsO_2(OH)_4$, $K_3Fe(CN)_6$, K_2CO_3, $MeSO_2NH_2$, t-BuOH:H_2O (1:1), rt, 12 h	**I** (92), 97		309
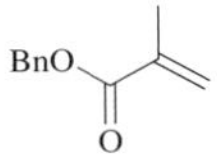	$(DHQ)_2$PHAL, $K_2OsO_2(OH)_4$, $K_3Fe(CN)_6$, K_2CO_3, t-BuOH:H_2O (1:1), 0°, 3 h	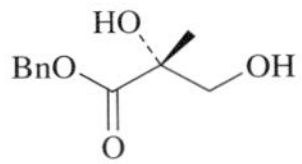	(—), 43-54	223
	$(DHQD)_2$PHAL, $K_2OsO_2(OH)_4$, $K_3Fe(CN)_6$, K_2CO_3, t-BuOH:H_2O (1:1), 0°, 3 h	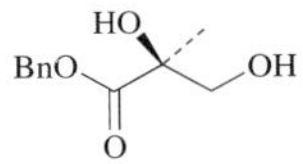	(—), 57-67	223
	1. $(DHQ)_2$PHAL, $K_2OsO_2(OH)_4$, $K_3Fe(CN)_6$, K_2CO_3, t-BuOH:H_2O (1:1), rt, 18 h 2. TEMPO, NaOCl, $NaOCl_2$, CH_3CN:phosphate buffer (5:4, pH 5.6), 55°, 4 d	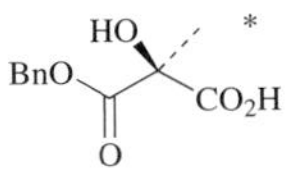	(72), —[a]	305
	1. $(DHQD)_2$PHAL, $K_2OsO_2(OH)_4$, $K_3Fe(CN)_6$, K_2CO_3, t-BuOH:H_2O (1:1), rt, 18 h 2. TEMPO, NaOCl, $NaOCl_2$, CH_3CN:phosphate buffer (5:4, pH 5.6), 55°, 4 d	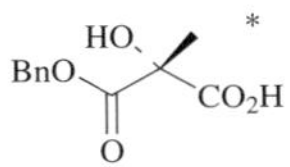	(44), —[a]	305

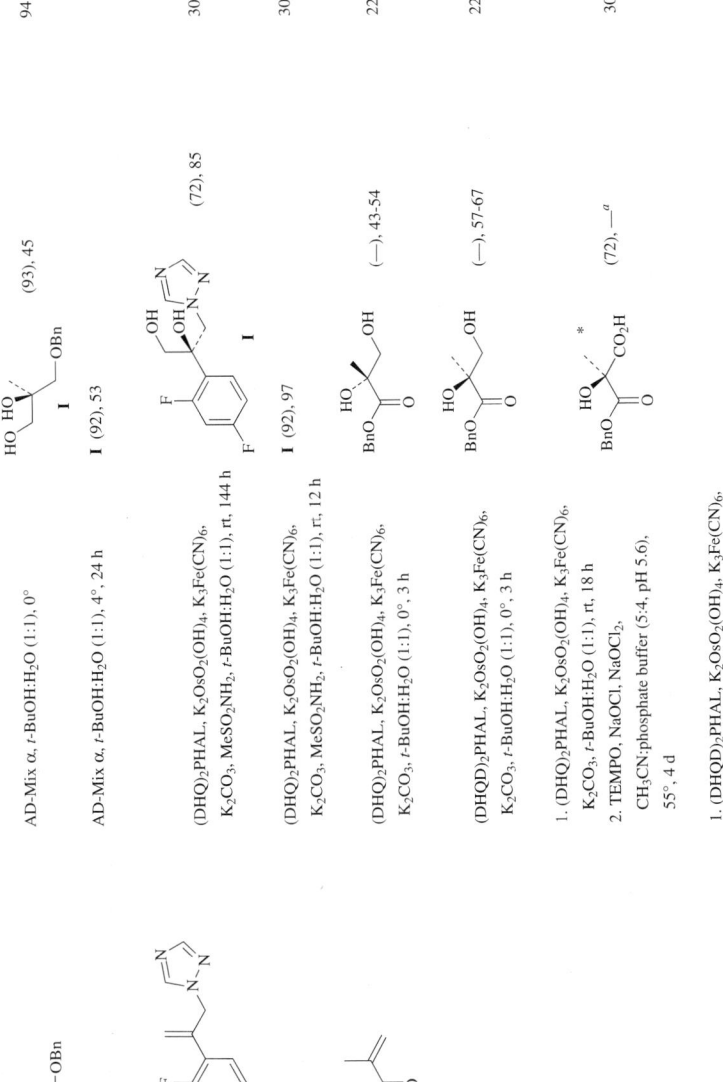

TABLE 2. REACTIONS OF 1,1-DISUBSTITUTED ALKENES (*Continued*)

Substrate	Conditions	Product(s) and Yield(s) (%), % ee	Refs.
C_{11} 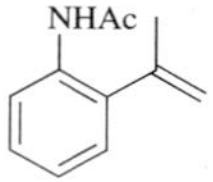	1. $(DHQ)_2PHAL$, $K_2OsO_2(OH)_4$, $K_3Fe(CN)_6$, K_2CO_3, *t*-BuOH:H_2O (1:1), rt, 18 h 2. TEMPO, NaOCl, $NaOCl_2$, CH_3CN:phosphate buffer (5:4, pH 5.6), 55°, 4 d	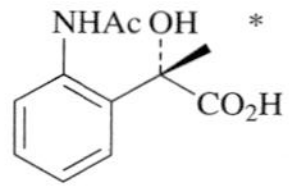 * (54), —[a]	305
	1. $(DHQD)_2PHAL$, $K_2OsO_2(OH)_4$, $K_3Fe(CN)_6$, K_2CO_3, *t*-BuOH:H_2O (1:1), rt, 18 h 2. TEMPO, NaOCl, $NaOCl_2$, CH_3CN:phosphate buffer (5:4, pH 5.6), 55°, 4 d	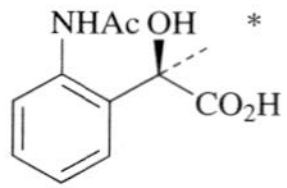 * (40), —[a]	305
$C_{11\text{-}14}$ 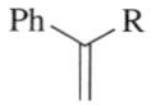	$(DHQD)_2PHAL$, $K_2OsO_2(OH)_4$, $K_3Fe(CN)_6$, K_2CO_3, *t*-BuOH:H_2O (1:1), 0°, 12 h	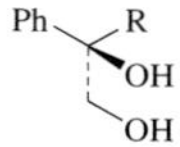 R *n*-Pr (85-95), 60 *i*-Pr (85-95), 82 *n*-Bu (85-95), 56 *t*-Bu (20-40), 8 *c*-C_4H_7 (85-95), 58 *n*-C_5H_{11} (85-95), 48 *c*-C_5H_9 (85-95), 55 *n*-C_6H_{13} (85-95), 37 *c*-C_6H_{11} (85-95), 57	88
	$(DHQD)_2PYR$, $K_2OsO_2(OH)_4$, $K_3Fe(CN)_6$, K_2CO_3, *t*-BuOH:H_2O (1:1), 0°, 12 h	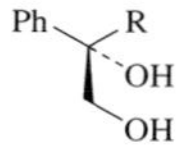 R *n*-Pr (85-95), 16 *i*-Pr (85-95), 8 *n*-Bu (85-95), 28 *t*-Bu (20-40), 37 *c*-C_4H_7 (85-95), 59 *n*-C_5H_{11} (85-95), 30 *c*-C_5H_9 (85-95), 66 *n*-C_6H_{13} (85-95), 35 *c*-C_6H_{11} (85-95), 68	88

C_{12}			
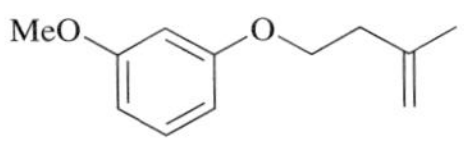	$(DHQ)_2PHAL$, $K_2OsO_2(OH)_4$, $K_3Fe(CN)_6$, K_2CO_3, t-BuOH:H_2O (1:1), 4°, 24 h	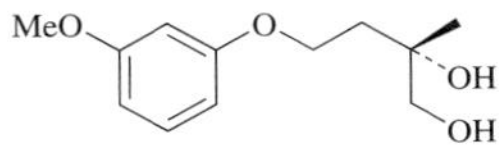 (94), 86	95
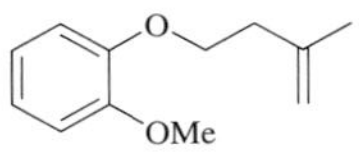	$(DHQ)_2PHAL$, $K_2OsO_2(OH)_4$, $K_3Fe(CN)_6$, K_2CO_3, t-BuOH:H_2O (1:1), 4°, 24 h	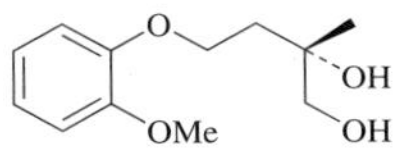 (93), 80	95
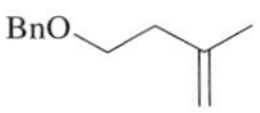	$(DHQ)_2PHAL$, $K_2OsO_2(OH)_4$, $K_3Fe(CN)_6$, K_2CO_3, t-BuOH:H_2O (1:1), 4°, 24 h	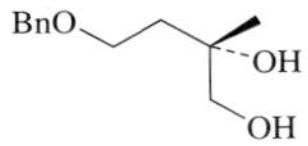 (92), 53	95
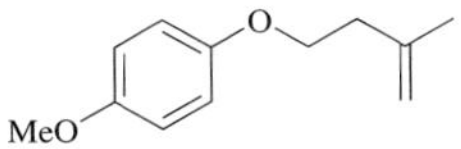	$(DHQD)_2PYDZ$, $K_2OsO_2(OH)_4$, $K_3Fe(CN)_6$, K_2CO_3, t-BuOH:H_2O (1:1), 0°, 4-22 h	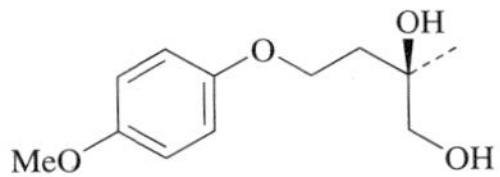 (99), 96	83, 66, 61
	$(DHQ)_2PHAL$, $K_2OsO_2(OH)_4$, $K_3Fe(CN)_6$, K_2CO_3, t-BuOH:H_2O (1:1), 4°, 24 h	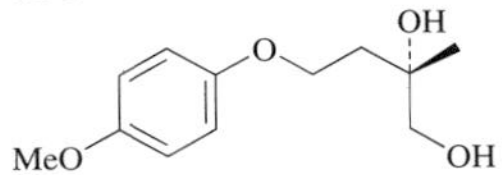 (93), 96	95
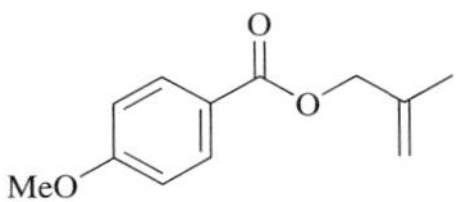	$(DHQD)_2PYDZ$, $K_2OsO_2(OH)_4$, $K_3Fe(CN)_6$, K_2CO_3, t-BuOH:H_2O (1:1), 0°, 1.5-4 h	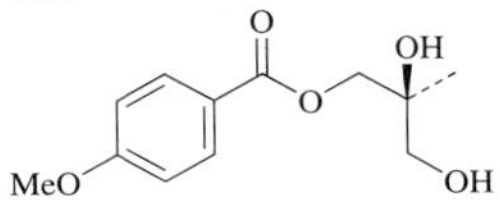 (98-99), 97	61, 66, 133
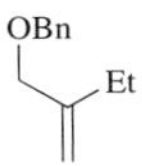	$(DHQD)_2PHAL$, $K_2OsO_2(OH)_4$, $K_3Fe(CN)_6$, K_2CO_3, t-BuOH:H_2O (1:1), 0°	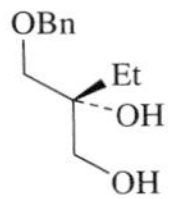 (63), 31	94
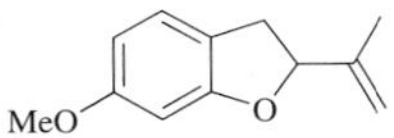	AD-Mix α, t-BuOH:H_2O (1:1), rt, 24 h	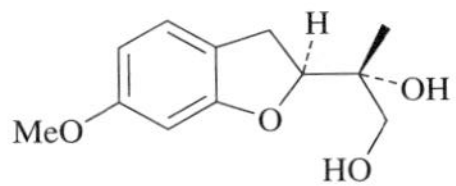 (85), 20 de	310
C_{13}			
	AD-Mix β, t-BuOH:H_2O (1:1), 0°, 8 h	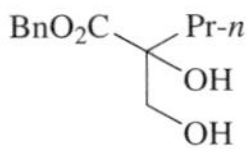 (—), 0	306

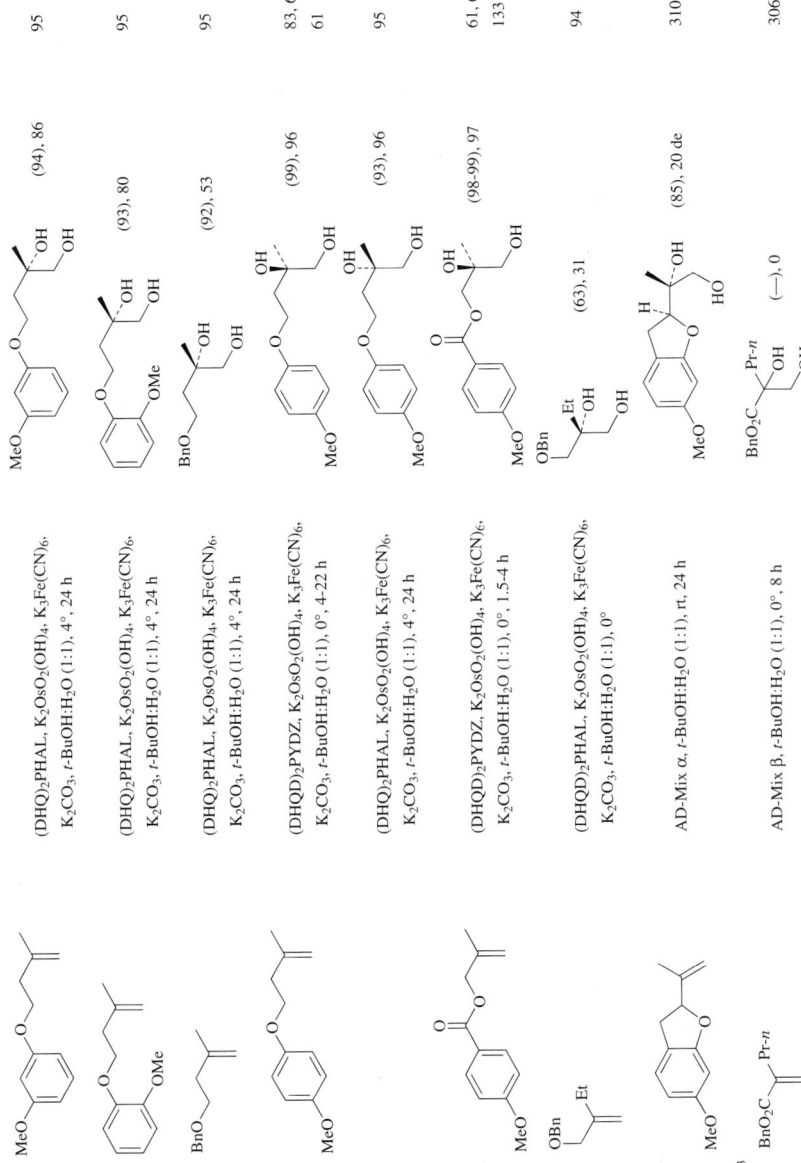

TABLE 2. REACTIONS OF 1,1-DISUBSTITUTED ALKENES (Continued)

Substrate	Conditions	Product(s) and Yield(s) (%), % ee	Refs.
C_{12} 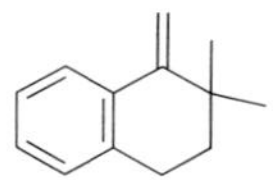	(DHQD)$_2$PHAL, $K_2OsO_2(OH)_4$, $K_3Fe(CN)_6$, K_2CO_3, *t*-BuOH:H_2O (1:1), 0°, 12 h	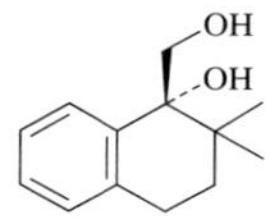 (20-40), 82	88
	(DHQD)$_2$PYR, $K_2OsO_2(OH)_4$, $K_3Fe(CN)_6$, K_2CO_3, *t*-BuOH:H_2O (1:1), 0°, 12 h	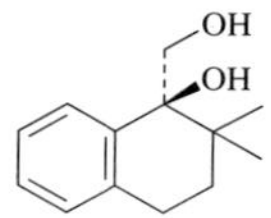 (20-40), 59	88
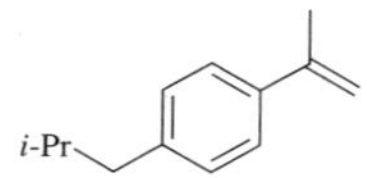	AD-mix α, *t*-BuOH:H_2O (1:1), 0°, 6 h	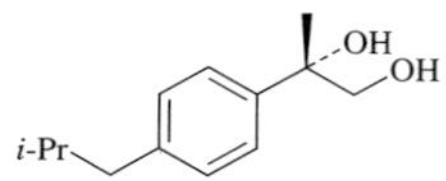 (99), 94	91
C_{13} 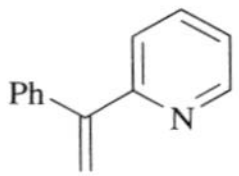	Ligand, $K_2OsO_2(OH)_4$, $K_3Fe(CN)_6$, K_2CO_3, *t*-BuOH:H_2O (1:1), rt, 10 d	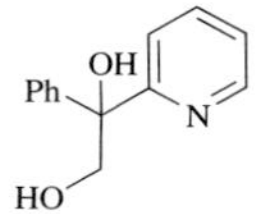	311
	Ligand		
	(DHQD)$_2$PHAL	(60), 20	
	(DHQD)$_2$PYR	(<60), 35	
	(DHQD)$_2$AQN	(<60), 23	
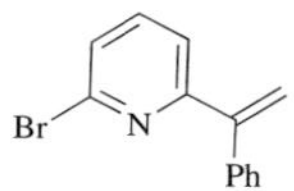	Ligand, $K_2OsO_2(OH)_4$, $K_3Fe(CN)_6$, K_2CO_3, *t*-BuOH:H_2O (1:1), rt, 24-72 h	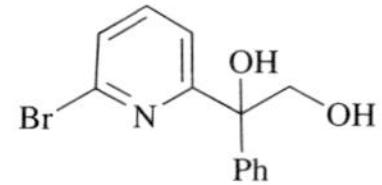	311
	Ligand		
	(DHQD)$_2$PHAL	(69), 31	
	(DHQD)$_2$PYR	(73), 62	
	(DHQD)$_2$AQN	(81), 95	
	(DHQ)$_2$AQN	(77), 92	

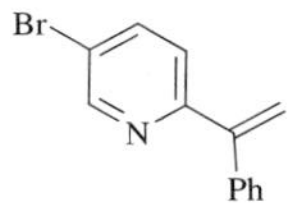

Ligand, $K_2OsO_2(OH)_4$, $K_3Fe(CN)_6$, K_2CO_3, t-BuOH:H_2O (1:1), 24-72 h

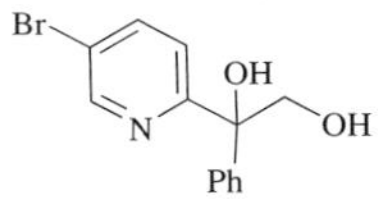

311

Ligand	Temp	
$(DHQD)_2$PHAL	rt	(75), 16
$(DHQD)_2$PYR	rt	(73), 33
$(DHQD)_2$AQN	rt	(79), 75
$(DHQD)_2$AQN	0°	(70), 80

C_{14}

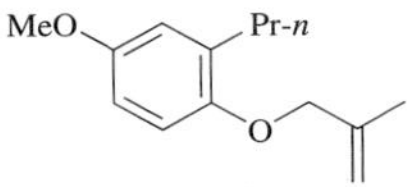

AD-mix α, t-BuOH:H_2O (1:1), 4°, 24 h

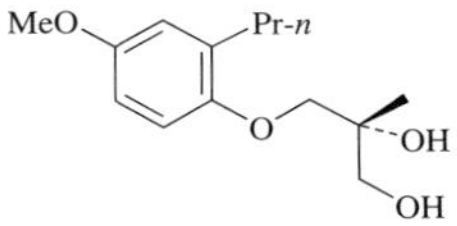

(94), 14 — 95

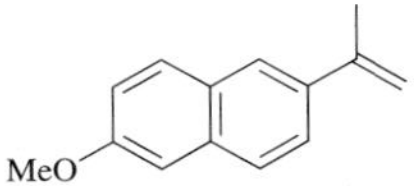

AD-mix α, t-BuOH:H_2O (1:1), 0°, 6 h

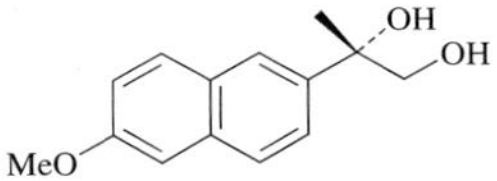

(81), 98 — 91

Ligand, $K_2OsO_2(OH)_4$, $K_3Fe(CN)_6$, K_2CO_3, t-BuOH:H_2O (1:1), 18-24 h

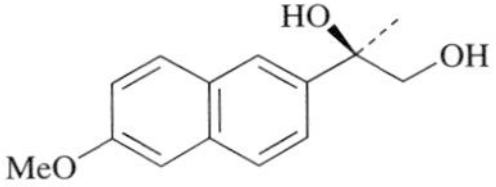

72

Ligand	Temp	
DHQD-PHN	0°	(75-95), 69
DHQD-MEQ	0°	(75-95), 88
DHQD-*p*-chlorobenzoate	rt	(75-95), 74

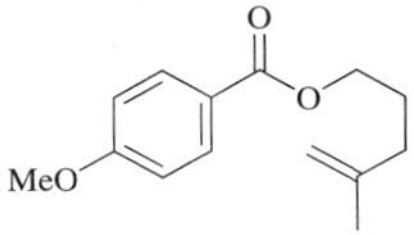

Ligand, $K_2OsO_2(OH)_4$, $K_3Fe(CN)_6$, K_2CO_3, t-BuOH:H_2O (1:1), 0°, 4-22 h

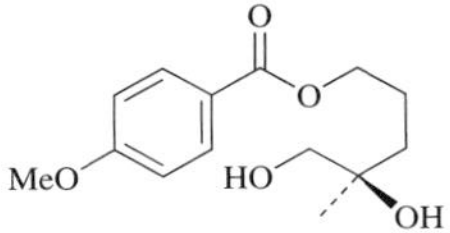

82

Ligand	
$(DHQD)_2$PYDZ	(95), 79
21	(99), 90

TABLE 2. REACTIONS OF 1,1-DISUBSTITUTED ALKENES (*Continued*)

Substrate	Conditions	Product(s) and Yield(s) (%), % ee	Refs.
C_{14}			
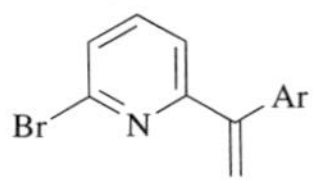	Ligand, $K_2OsO_2(OH)_4$, $K_3Fe(CN)_6$, K_2CO_3, *t*-BuOH:H_2O (1:1), 24-72 h	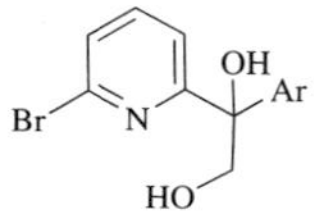	311

Ar	Ligand	Temp	Yield, % ee
p-$MeOC_6H_4$	$(DHQD)_2PYR$	rt	(87), 70
p-$MeOC_6H_4$	$(DHQD)_2AQN$	rt	(87), 96
p-$MeOC_6H_4$	$(DHQ)_2AQN$	rt	(85), 92
p-$MeOC_6H_4$	$(DHQD)_2PHAL$	0°	(83), 54
o-$MeOC_6H_4$	$(DHQD)_2PHAL$	rt	(61), 5
o-$MeOC_6H_4$	$(DHQD)_2PYR$	rt	(75), 31
o-$MeOC_6H_4$	$(DHQD)_2AQN$	rt	(99), 96->99

Substrate	Conditions	Product(s) and Yield(s) (%), % ee	Refs.
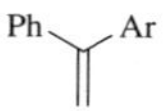	Ligand, $K_2OsO_2(OH)_4$, $K_3Fe(CN)_6$, K_2CO_3, *t*-BuOH:H_2O (1:1), rt	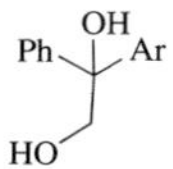	311

Ar	Ligand	Time	Yield, % ee
o-BrC_6H_4	$(DHQD)_2PHAL$	10 d	(40), 22
o-BrC_6H_4	$(DHQD)_2PYR$	10 d	(<40), 15
o-BrC_6H_4	$(DHQD)_2AQN$	10 d	(<40), 1
m-BrC_6H_4	$(DHQD)_2PHAL$	24 h	(93), 2
m-BrC_6H_4	$(DHQD)_2PYR$	24 h	(84), 9
m-BrC_6H_4	$(DHQD)_2AQN$	24 h	(82), 38

Substrate	Conditions	Product(s) and Yield(s) (%), % ee	Refs.
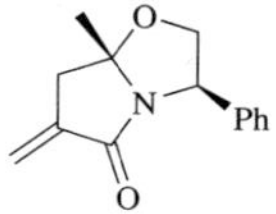	AD-Mix β, *t*-BuOH:H_2O (1:1)	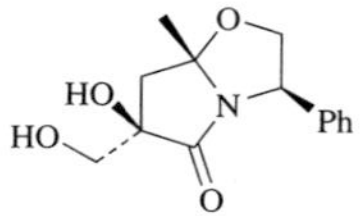 (98), 88	312

	AD-Mix α, t-BuOH:H_2O (1:1)	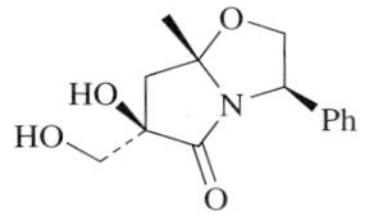	(100), 88	312
C_{15}				
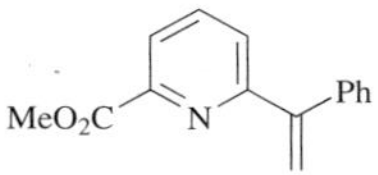	Ligand, $K_2OsO_2(OH)_4$, $K_3Fe(CN)_6$, K_2CO_3, t-BuOH:H_2O (1:1), rt, 24-72 h	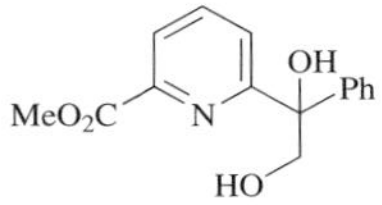		311
	Ligand			
	$(DHQD)_2PHAL$	(76), 9		
	$(DHQD)_2PYR$	(76), 76		
	$(DHQD)_2AQN$	(74-94), 94-98		
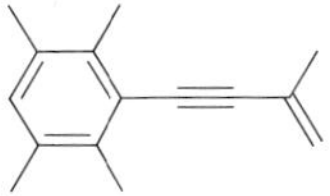	AD-Mix α, t-BuOH:H_2O (1:1), 4°, 48 h	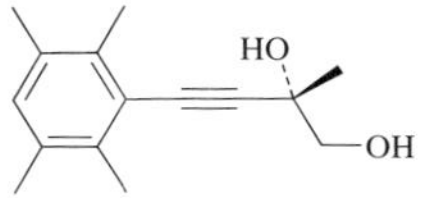	(95), 85	313
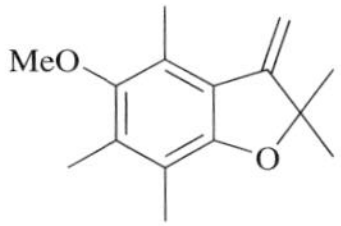	Ligand, $K_2OsO_2(OH)_4$, $K_3Fe(CN)_6$, K_2CO_3, $TsNH_2$, t-BuOH:H_2O (1:1), rt	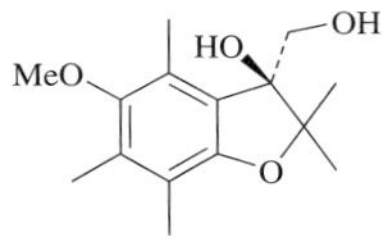		93
	Ligand			
	$(DHQ)_2PHAL$	(85-90), 32		
	$(DHQD)_2PYR$	(85-90), 93		
	Ligand, $K_2OsO_2(OH)_4$, $K_3Fe(CN)_6$, K_2CO_3, $TsNH_2$, t-BuOH:H_2O (1:1), rt	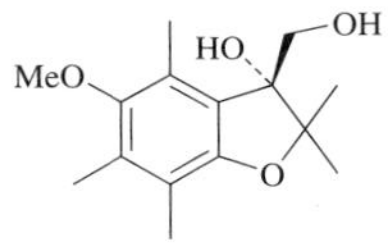		93
	Ligand			
	$(DHQD)_2PHAL$	(85-90), 36		
	$(DHQ)_2PYR$	(85-90), 86		

Substrate	Conditions	Product(s) and Yield(s) (%), % ee	Refs.
C_{15} 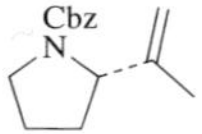	Ligand, $K_2OsO_2(OH)_4$, $K_3Fe(CN)_6$, K_2CO_3, $MeSO_2NH_2$, t-BuOH:H_2O (1:1), 0°	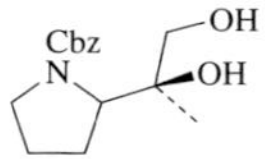	314
	Ligand		
	$(DHQ)_2PHAL$	(—), 33 de	
	$(DHQ)_2PYR$	(—), 28 de	
	$(DHQD)_2PHAL$	(—), 56 de	
	$(DHQD)_2PYR$, $K_2OsO_2(OH)_4$, $K_3Fe(CN)_6$, K_2CO_3, $MeSO_2NH_2$, t-BuOH:H_2O (1:1), 0°	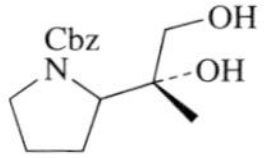 (—), 42 de	314
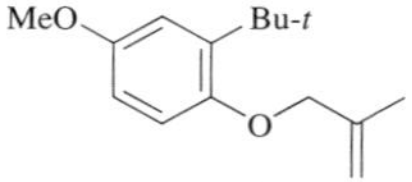	$(DHQ)_2PHAL$, $K_2OsO_2(OH)_4$, $K_3Fe(CN)_6$, K_2CO_3, t-BuOH:H_2O (1:1), 4°, 24 h	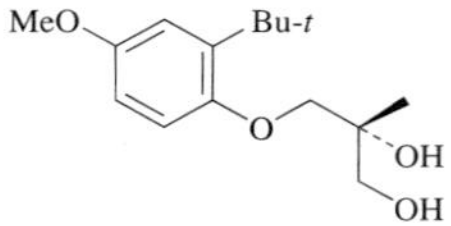 (79), 8	95
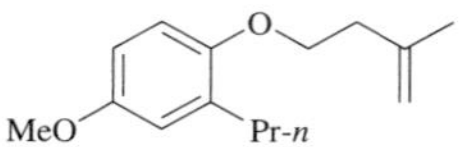	$(DHQ)_2PHAL$, $K_2OsO_2(OH)_4$, $K_3Fe(CN)_6$, K_2CO_3, t-BuOH:H_2O (1:1), 4°, 24 h	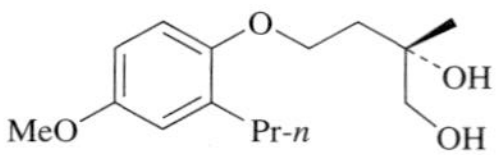 (92), 92	95
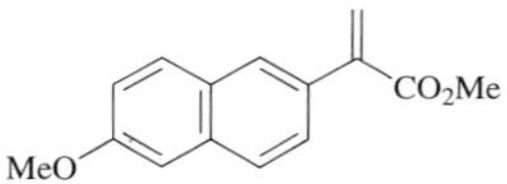	$(DHQD)_2PYDZ$, $K_2OsO_2(OH)_4$, $K_3Fe(CN)_6$, K_2CO_3, $MeSO_2NH_2$, t-BuOH:H_2O (1:1), 0°	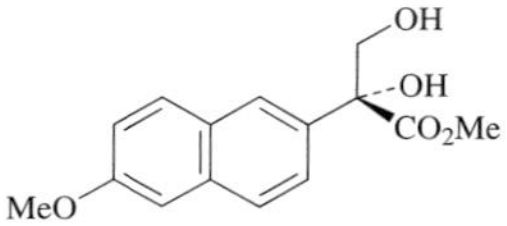 (66), 97	49
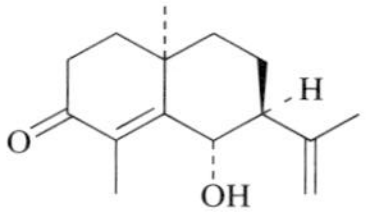	1. $(DHQ)_2PHAL$, $K_2OsO_2(OH)_4$, $K_3Fe(CN)_6$, K_2CO_3, t-BuOH:H_2O (1:1), rt, 18 h 2. TEMPO, NaOCl, $NaClO_2$, CH_3CN:phosphate buffer (5:4, pH 5.6), 55°, 4 d	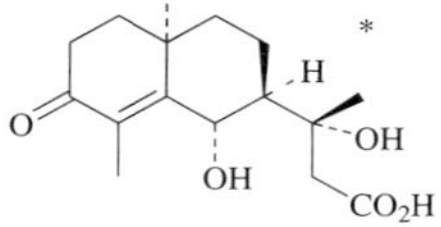 (47), >98 de	305

	1. $(DHQD)_2PHAL$, $K_2OsO_2(OH)_4$, $K_3Fe(CN)_6$, K_2CO_3, *t*-BuOH:H_2O (1:1), rt, 18 h 2. TEMPO, NaOCl, $NaClO_2$, CH_3CN:phosphate buffer (5:4, pH 5.6), 55°, 4 d	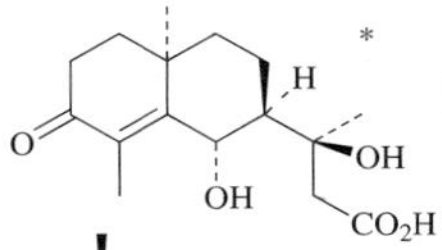 * (48), >98 de	305
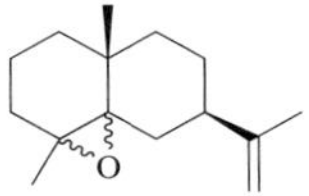	AD-Mix α, *t*-BuOH:H_2O (1:1), 0°, 24 h	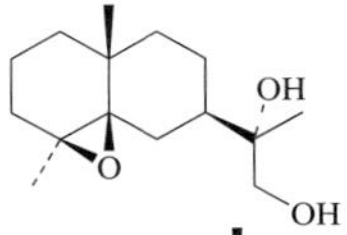 (34), 42-45 de + 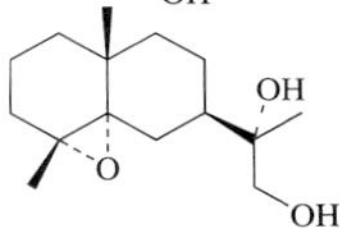 (64), 39 de	315, 316
	AD-Mix β, *t*-BuOH:H_2O (1:1), 0°, 24 h	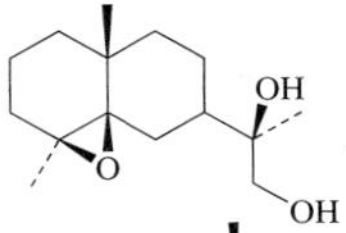 (32), 30 de + 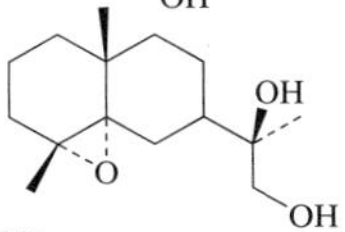 (63), 33 de	316
C_{16} 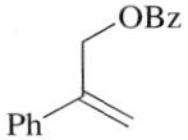	$(DHQD)_2PHAL$, $K_2OsO_2(OH)_4$, $K_3Fe(CN)_6$, K_2CO_3, *t*-BuOH:H_2O (1:1), 0°	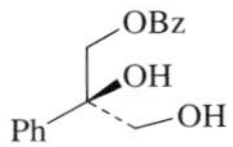 (82), 10	87
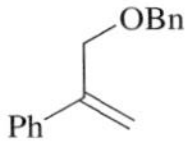	$(DHQD)_2PHAL$, $K_2OsO_2(OH)_4$, $K_3Fe(CN)_6$, K_2CO_3, *t*-BuOH:H_2O (1:1), 0°	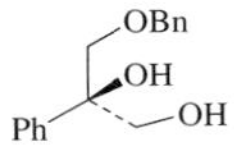 (92), 78	87, 20
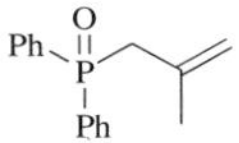	$(DHQD)_2PHAL$, $K_2OsO_2(OH)_4$, $K_3Fe(CN)_6$, K_2CO_3, *t*-BuOH:H_2O (1:1), 0°, 60 h	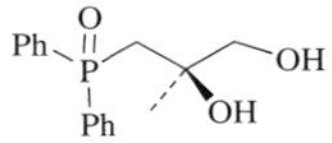 (74), 55	120, 317, 318, 319

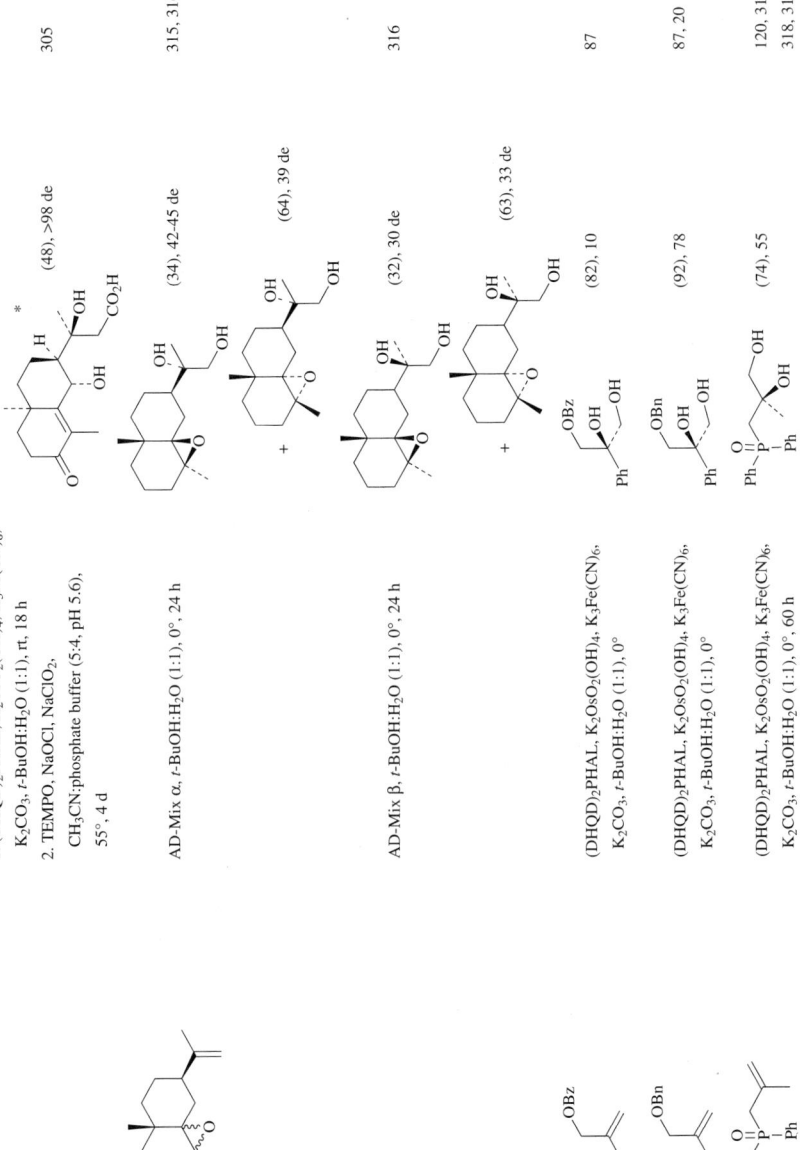

TABLE 2. REACTIONS OF 1,1-DISUBSTITUTED ALKENES (*Continued*)

Substrate			Conditions	Product(s) and Yield(s) (%), % ee		Refs.
C_{16} 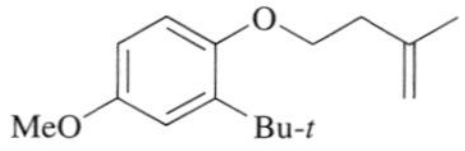			$(DHQ)_2PHAL$, $K_2OsO_2(OH)_4$, $K_3Fe(CN)_6$, K_2CO_3, t-BuOH:H_2O (1:1), 4°, 24 h	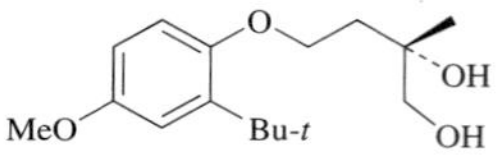	(94), 77	95
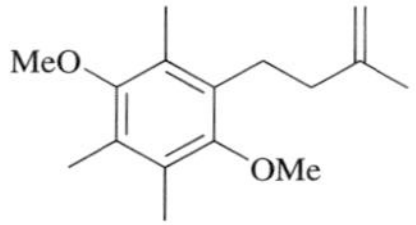			$(DHQ)_2PHAL$, $K_2OsO_2(OH)_4$, $K_3Fe(CN)_6$, K_2CO_3, t-BuOH:H_2O (1:1), 4°, 8-24 h	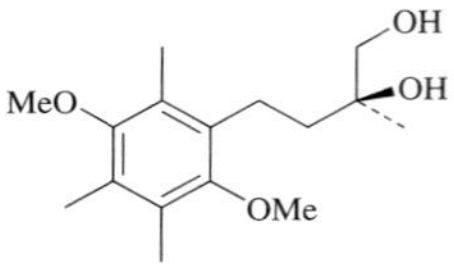	(82), 28	320
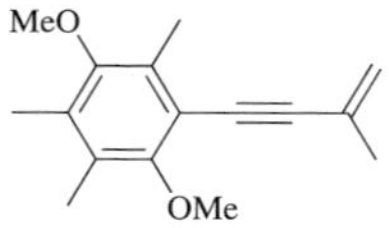			$(DHQ)_2PHAL$, $K_2OsO_2(OH)_4$, $K_3Fe(CN)_6$, K_2CO_3, t-BuOH:H_2O (1:1), 4°, 8-24 h	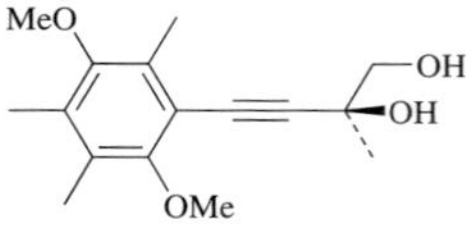	(—), 84	320
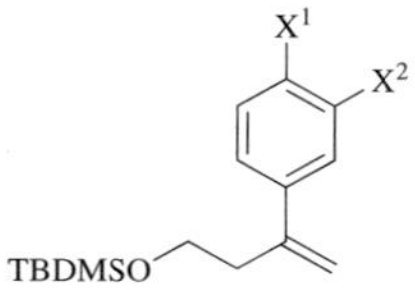	X^1	X^2	AD-Mix β, t-BuOH:H_2O (1:1), 0°	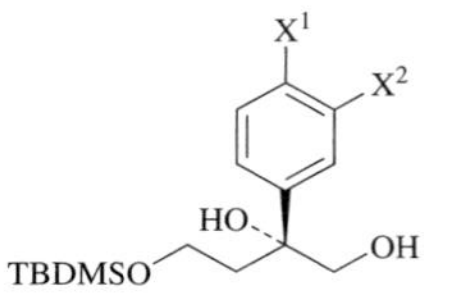		
	Cl	Cl		(94), 98		321, 322
	Cl	H		(95), 98		321
	F	F		(98), 99		321, 322
	F	H		(96), 97		321
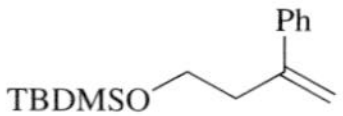			AD-Mix β, t-BuOH:H_2O (1:1)	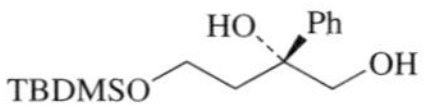	(—), >97	322

	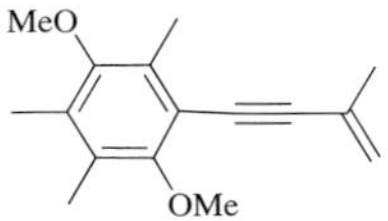	AD-Mix α, *t*-BuOH:H_2O (1:1), 4°, 48 h	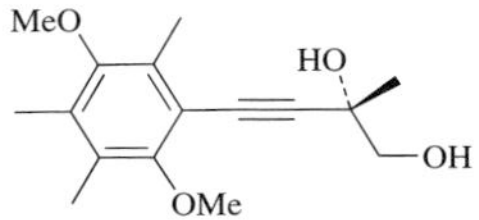 (93), 84	313
C_{17}	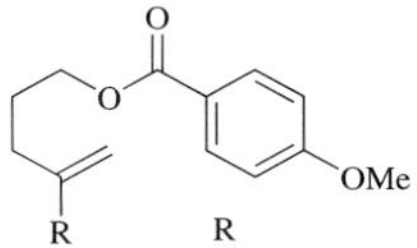	Ligand, $K_2OsO_2(OH)_4$, $K_3Fe(CN)_6$, K_2CO_3, *t*-BuOH:H_2O (1:1), 0°, 4-22 h	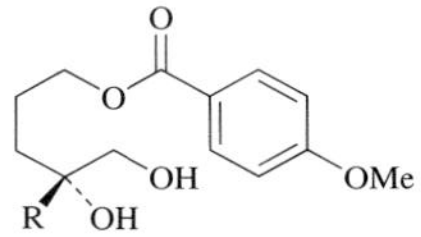	82
	R	Ligand		
	n-Bu	$(DHQD)_2PYDZ$	(99), 33	
	n-Bu	**21**	(99), 62	
	t-Bu	$(DHQD)_2PYDZ$	(88), 92	
	t-Bu	**21**	(91), 92	
	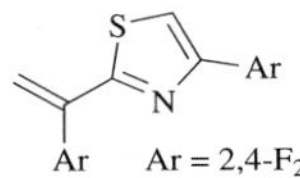 Ar = 2,4-$F_2C_6H_4$	AD-Mix α, *t*-BuOH:H_2O (1:1), rt, 22 h	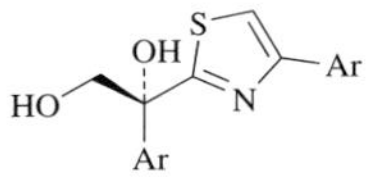 (57), 81	323
	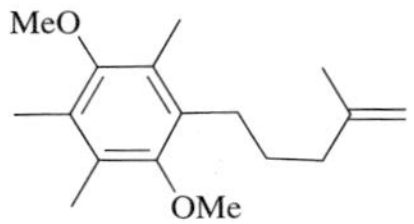	AD-Mix α, *t*-BuOH:H_2O (1:1), 4°, 48 h	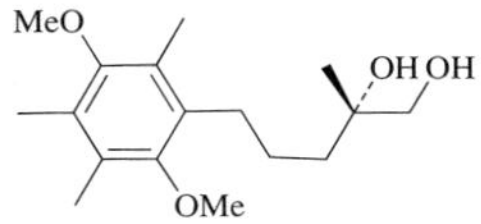 (82), 28	313
	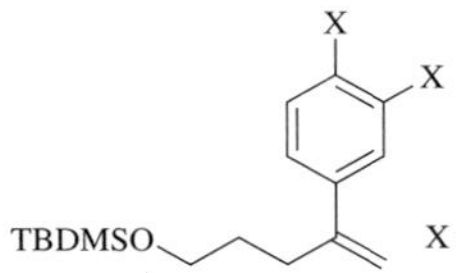 X = H, F, or Cl	AD-Mix β, *t*-BuOH:H_2O (1:1)	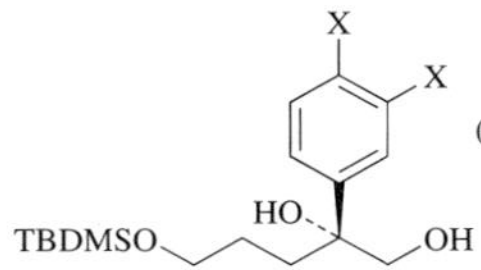 (—), >97	322
C_{18}	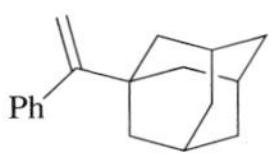	$(DHQD)_2PHAL$, $K_2OsO_2(OH)_4$, $K_3Fe(CN)_6$, K_2CO_3, *t*-BuOH:H_2O (1:1), 0°, 12 h	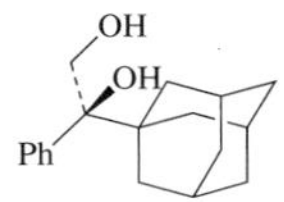 (85-95), 53	88

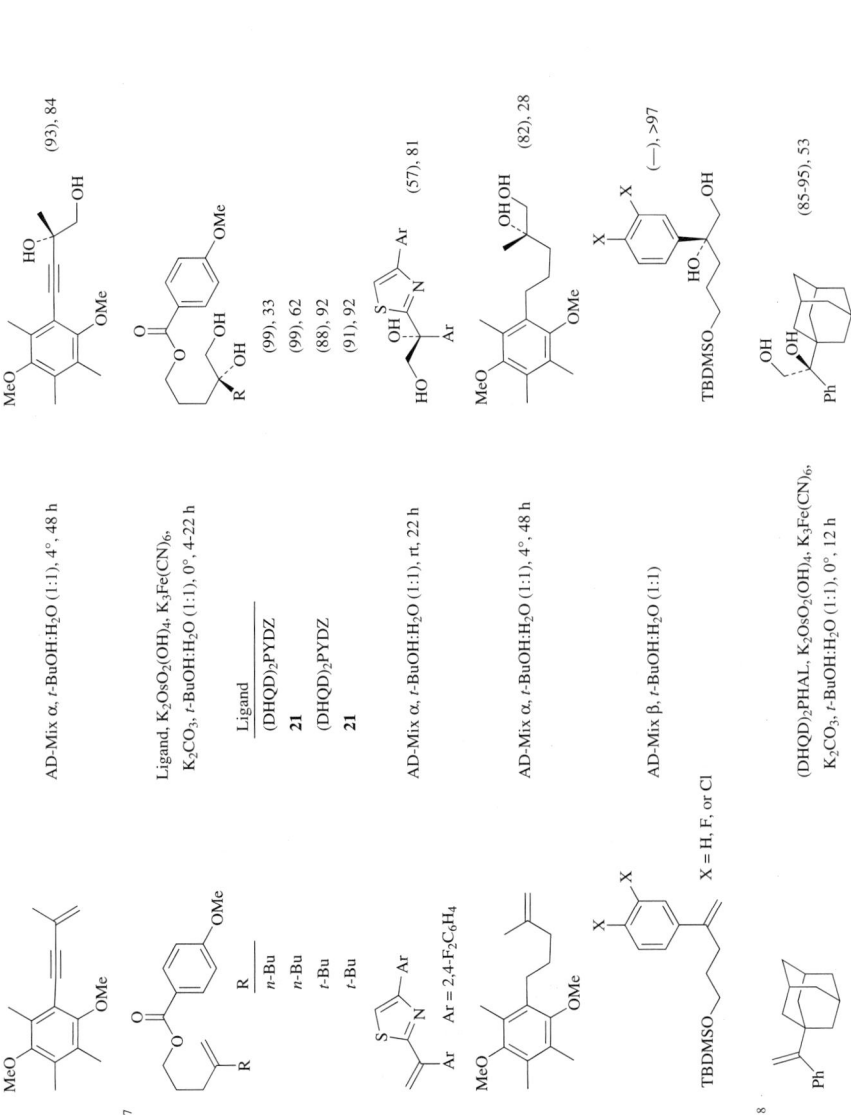

TABLE 2. REACTIONS OF 1,1-DISUBSTITUTED ALKENES (*Continued*)

Substrate	Conditions	Product(s) and Yield(s) (%), % ee	Refs.
C_{18} 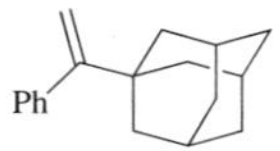	$(DHQD)_2PYR$, $K_2OsO_2(OH)_4$, $K_3Fe(CN)_6$, K_2CO_3, t-BuOH:H_2O (1:1), 0°, 12 h	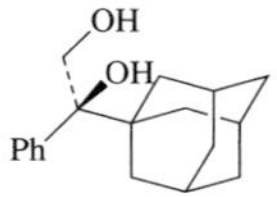 (85-95), 77	88
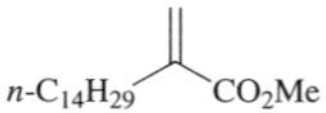	$(DHQD)_2PHAL$, $K_2OsO_2(OH)_4$, $K_3Fe(CN)_6$, K_2CO_3, t-BuOH:H_2O (1:1), 0°, 15-23 h	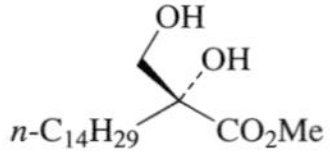 (62), 95	324
	$(DHQ)_2PHAL$, $K_2OsO_2(OH)_4$, $K_3Fe(CN)_6$, K_2CO_3, t-BuOH:H_2O (1:1), 0°, 15-23 h	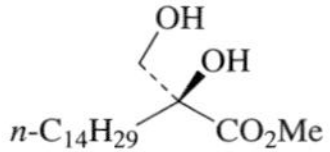 (72), 91	324
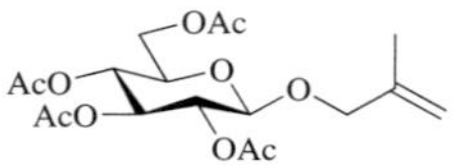	DHQD-*p*-chlorobenzoate, $K_2OsO_2(OH)_4$, $K_3Fe(CN)_6$, K_2CO_3, t-BuOH:H_2O (1:1), 0°	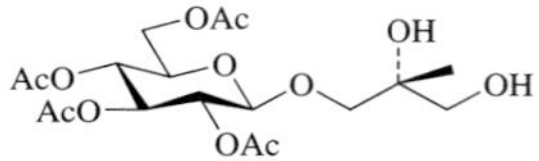 (—), 64 de	264
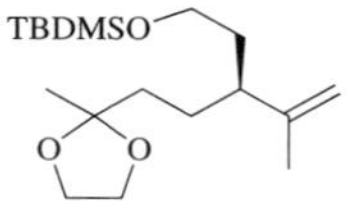	AD-Mix β, t-BuOH:H_2O (1:1), 14 d	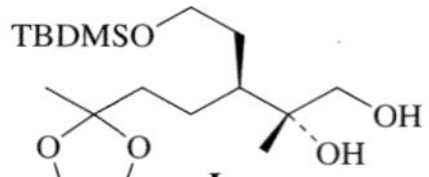 **I** (70), 2 de	325
	AD-Mix α, t-BuOH:H_2O (1:1), 14 d	**I** (70), 16 de	325
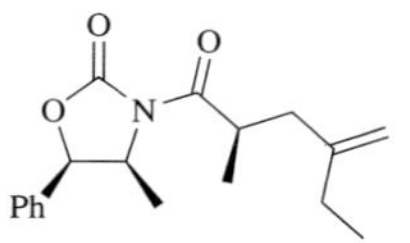	$(DHQ)_2PHAL$, $K_2OsO_2(OH)_4$, $K_3Fe(CN)_6$, K_2CO_3, t-BuOH:H_2O (1:1), 0°	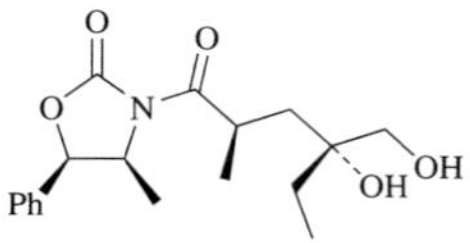 (92), — de	326
C_{19} 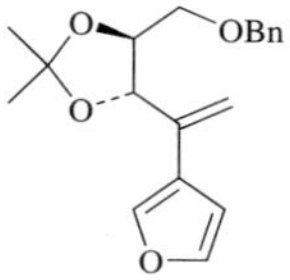	AD-Mix β, t-BuOH:H_2O (1:1), 0°, 10 h	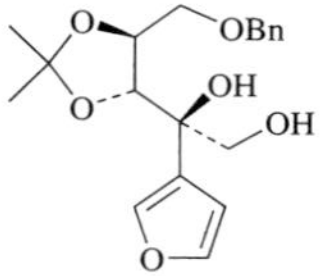 (95), 91	327

C_{20}				
	(DHQ)$_2$PHAL, $K_2OsO_2(OH)_4$, $K_3Fe(CN)_6$, K_2CO_3, *t*-BuOH:H_2O (1:1), 0°	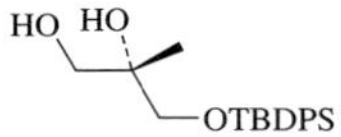	(94), 47	94
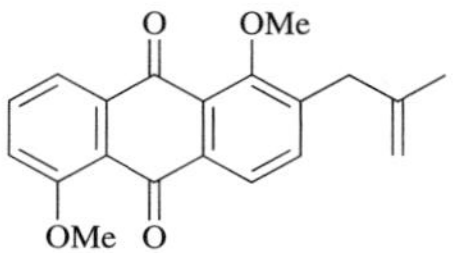	AD-Mix β, *t*-BuOH:H_2O (1:1), 0°, 50 h	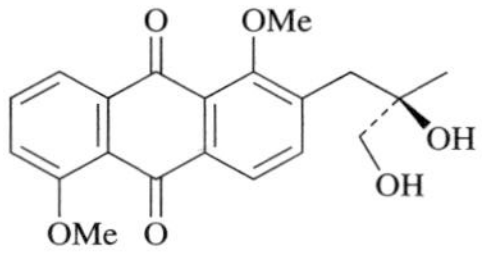	(88), 67	328
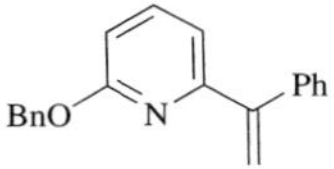	Ligand, $K_2OsO_2(OH)_4$, $K_3Fe(CN)_6$, K_2CO_3, *t*-BuOH:H_2O (1:1), 24-72 h	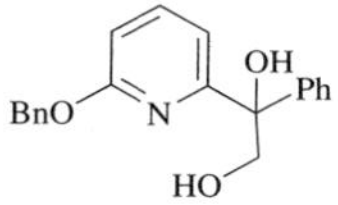		311
	Ligand / Temp			
	(DHQD)$_2$PHAL / rt	(89), 64		
	(DHQD)$_2$PYR / rt	(94), 91		
	(DHQD)$_2$PYR / 0°	(85), 94		
	(DHQ)$_2$PYR / 0°	(93), 87		
	(DHQD)$_2$AQN / rt	(87), 19		
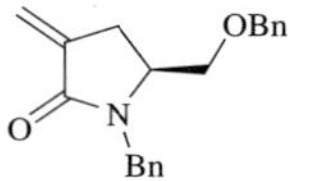	AD-Mix α, *t*-BuOH:H_2O (1:1)	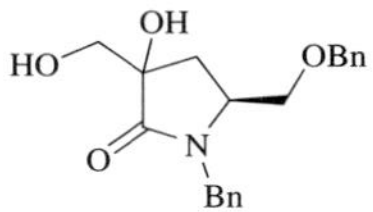	(—), 0	312
	AD-Mix β, *t*-BuOH:H_2O (1:1)	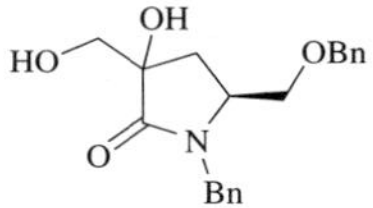	(—), 0	312
C_{21}				
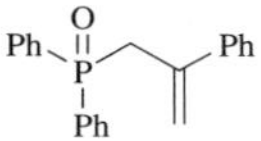	AD-Mix β, *t*-BuOH:H_2O (1:1), 0°-rt, 60-96 h	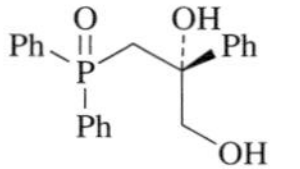	(75), 86	120, 317, 318, 319

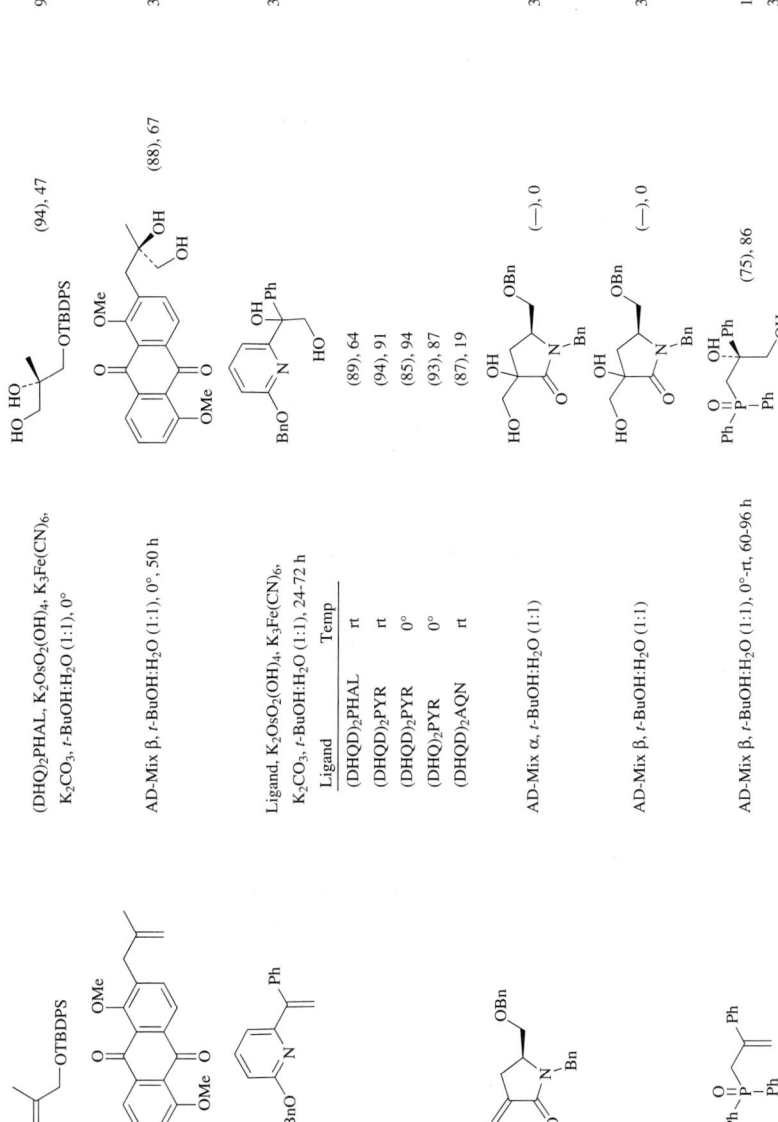

TABLE 2. REACTIONS OF 1,1-DISUBSTITUTED ALKENES (*Continued*)

Substrate	Conditions	Product(s) and Yield(s) (%), % ee		Refs.
C_{21} 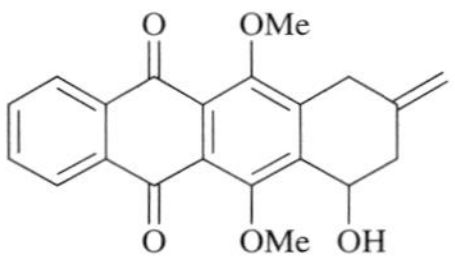	AD-Mix α, $MeSO_2NH_2$, *t*-BuOH:H_2O (1:1), 0°, 24 h	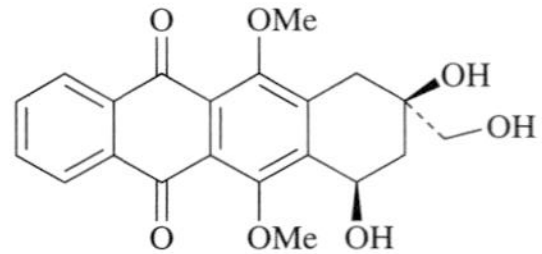	(21), —	329
		+ 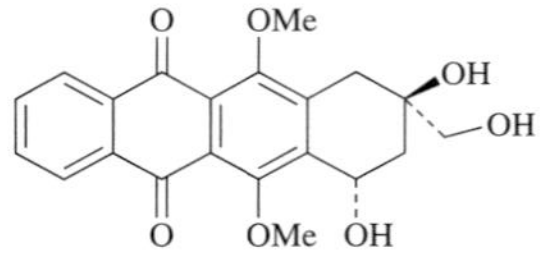	(46), —	
	AD-Mix β, $MeSO_2NH_2$, *t*-BuOH:H_2O (1:1), 0°, 24 h	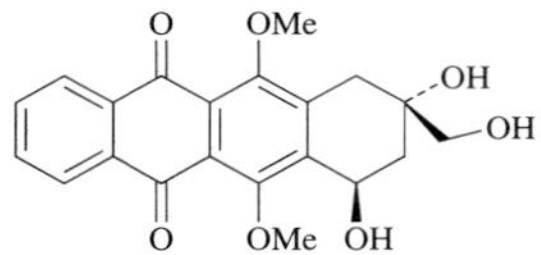	(21), —	329
		+ 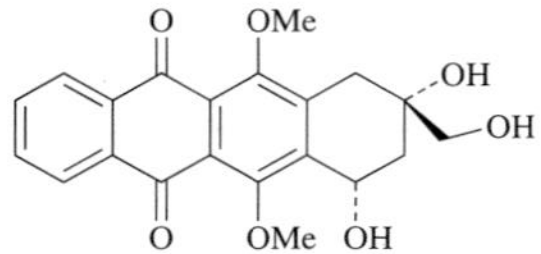	(49), —	
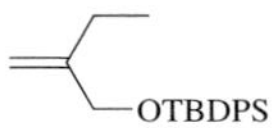	$(DHQD)_2PHAL$, $K_2OsO_2(OH)_4$, $K_3Fe(CN)_6$, K_2CO_3, *t*-BuOH:H_2O (1:1), 0°	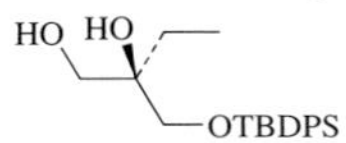	(70), 91	94
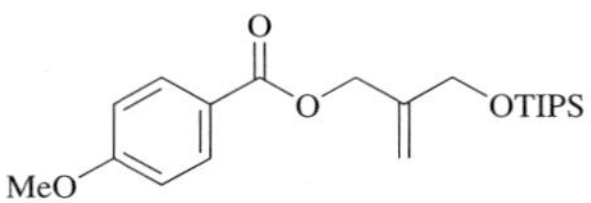	$(DHQD)_2PYDZ$, $K_2OsO_2(OH)_4$, $K_3Fe(CN)_6$, K_2CO_3, *t*-BuOH:H_2O (1:1), 0°, 1.5-3.5 h	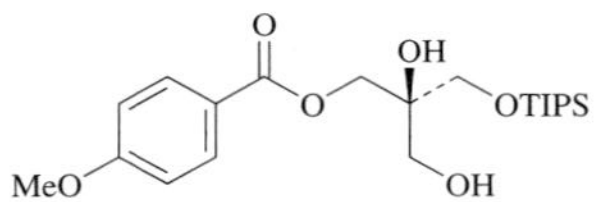	(99), 97	61, 66

C_{22}	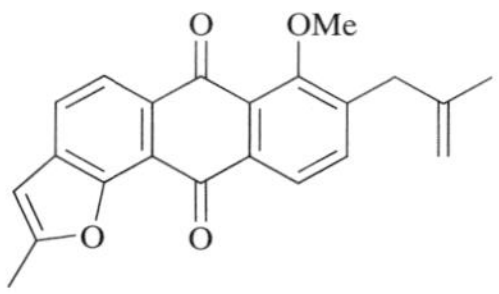	$(DHQD)_2PHAL$, OsO_4, $K_3Fe(CN)_6$, K_2CO_3, *t*-BuOH:H_2O (1:1), 20°, 16 h	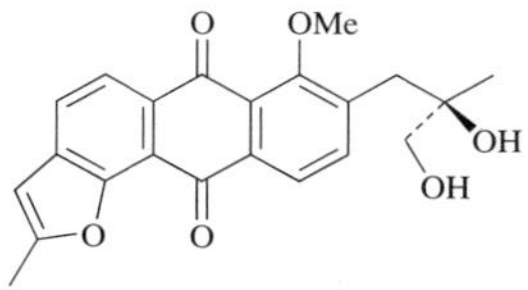 (64), —	328
C_{23}	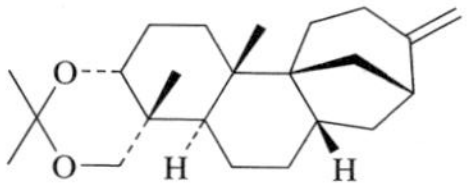	DHQD-*p*-chlorobenzoate, OsO_4, NMO, TEAA, acetone:H_2O (1:1), 4°, 16 h	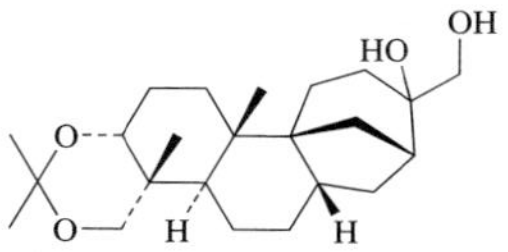 **I** (85), 0 de	330
		DHQ-*p*-chlorobenzoate, OsO_4, NMO, TEAA, acetone:H_2O (1:1), 4°	**I** (85), 0 de	330
C_{28}	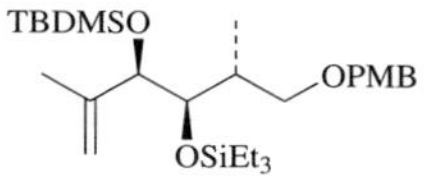	Super AD-Mix β, *t*-BuOH:H_2O (1:1), 0°	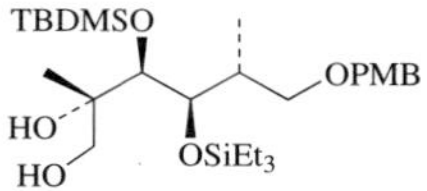 (88), 78 de	331
C_{30}	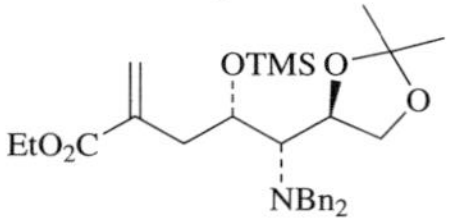	AD-Mix α, *t*-BuOH:H_2O (1:1), 18°, 22 h	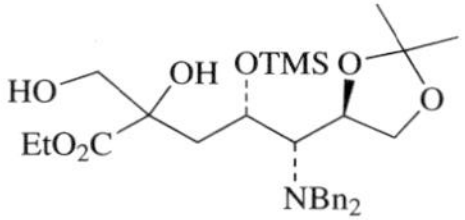 (83), 0 de	332
C_{35}	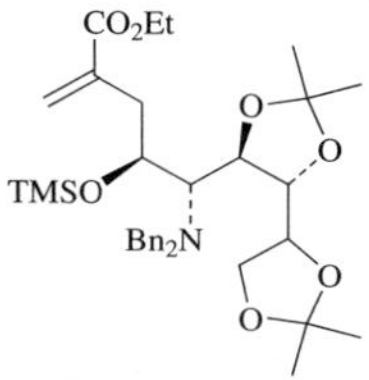	1. AD-Mix α, $MeSO_2NH_2$, *t*-BuOH:H_2O (1:1), 4°, 4 h 2. $Pb(OAc)_4$, $CaCO_3$, CH_2Cl_2, 18°, 20 min	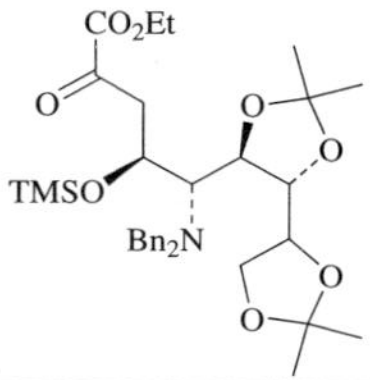 (83), —	333

[a] The authors assigned the absolute configuration using a mechanistic model. The authors were unable to determine % ee due to decomposition of the product upon attenpted derivatization.

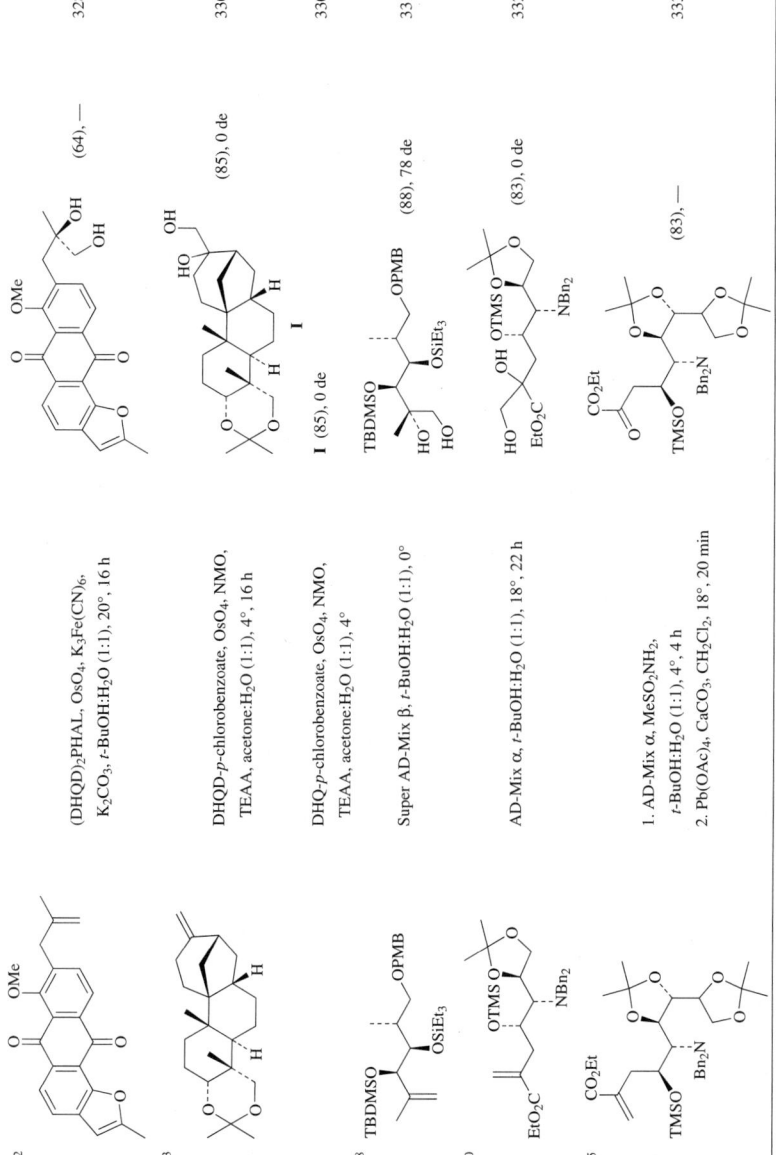

TABLE 3. REACTIONS OF TRANS 1,2-DISUBSTITUTED ALKENES

See Chart 1 at the beginning of the Tabular Survey for ligand structures that are indicated by **bold** numbers.

Substrate	Conditions	Product(s) and Yield(s) (%), % ee	Refs.
C_4 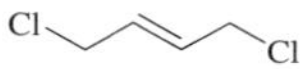	(DHQD)$_2$PHAL, $K_2OsO_2(OH)_4$, $K_3Fe(CN)_6$, K_2CO_3, $MeSO_2NH_2$, $NaHCO_3$, t-BuOH:H_2O (1:1), 0°	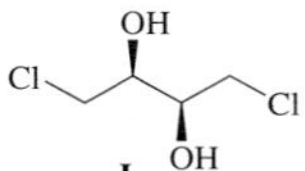 **I** (88), 94	154, 86
	(DHQD)$_2$AQN, $K_2OsO_2(OH)_4$, $K_3Fe(CN)_6$, K_2CO_3, $NaHCO_3$, t-BuOH:H_2O (1:1), 0°, 12 h	**I** (65-90), 96	86
	(DHQ)$_2$PHAL, $K_2OsO_2(OH)_4$, $K_3Fe(CN)_6$, K_2CO_3, $NaHCO_3$, t-BuOH:H_2O (1:1), 0°	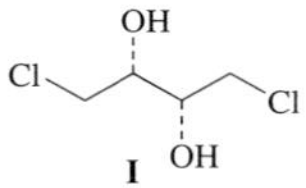 **I** (84), >94	334
	(DHQ)$_2$PHAL, $K_2OsO_2(OH)_4$, $K_3Fe(CN)_6$, K_2CO_3, $MeSO_2NH_2$, $NaHCO_3$, t-BuOH:H_2O (1:1), 0°	**I** (84), >94	335, 334a
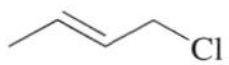	(DHQD)$_2$PHAL, $K_2OsO_2(OH)_4$, $K_3Fe(CN)_6$, K_2CO_3, $MeSO_2NH_2$, $NaHCO_3$, t-BuOH:H_2O (1:1), 0°	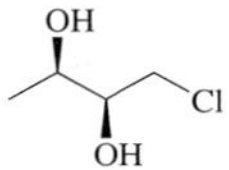 (75), 95	154
	AD-Mix α, $NaHCO_3$, t-BuOH:H_2O (1:1), 0°, 3 d	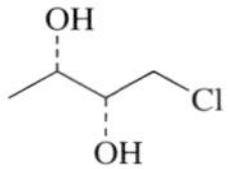 (86), 81	336
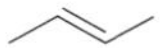	(DHQD)$_2$PHAL, $K_2OsO_2(OH)_4$, $K_3Fe(CN)_6$, K_2CO_3, $MeSO_2NH_2$, t-BuOH:H_2O (1:1), 0°	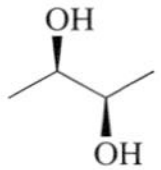 (—), 72	20
C_5			
	**35**, OsO_4, CH_2Cl_2, –90°, 2 h	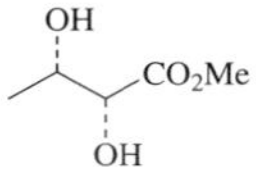 (83), 97	11

C6

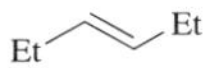

	DHQD-*p*-chlorobenzoate, NMO, OsO_4, acetone:H_2O (10:1), 0°, 17 h	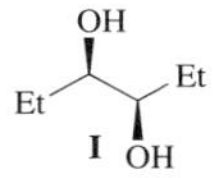 **I** (80-95), 20	18, 69
	DHQD-*p*-chlorobenzoate, OsO_4, NMO, acetone:H_2O (10:1), 0°, 11 h (slow addition of olefin)	**I** (85-95), 70	69, 112
	DHQD-*p*-chlorobenzoate, OsO_4 (stoichiometric) acetone:H_2O (10:1, 0.15M), 0°	**I** (—), 69	69, 112
	DHQD-*p*-chlorobenzoate, OsO_4, NMO, $Et_4NOAc{\bullet}4\ H_2O$, acetone:H_2O (10:1), 0°	**I** (—), 64	69
	DHQD-9-phenyl ether, OsO_4, NMO, acetone:H_2O (10:1), 0°, 121 h (slow addition of olefin)	**I** (85-95), 85	337
	DHQD-9-phenyl ether, OsO_4, NMO, TEAA, acetone:H_2O (10:1), 0°, 16 h	**I** (85-95), 82	337
	DHQD-9-phenyl ether, $K_2OsO_2(OH)_4$, $K_3Fe(CN)_6$, K_2CO_3, *t*-BuOH:H_2O (1:1), rt, 20 h	**I** (85-90), 83	337, 220
	DHQD-9-*o*-methoxyphenyl ether, $K_2OsO_2(OH)_4$, $K_3Fe(CN)_6$, K_2CO_3, *t*-BuOH:H_2O (1:1), rt, 20 h	**I** (85-90), 89	337
	1. Ligand, OsO_4 (stoichiometric) toluene 2. LAH	**I**	337

Ligand	Temp (1)	**I**
DHQD-9-phenyl ether	–78°	(—), 95
DHQD-9-phenyl ether	0°	(—), 85
DHQD-9-methyl ether	0°	(—), 37
DHQD-9-*o*-methoxyphenyl ether	0°	(—), 88
DHQD-9-*o*-trifluoromethyl phenyl ether	0°	(—), 81
DHQD-9-*p*-methoxyphenyl ether	0°	(—), 76
DHQD-9-*p*-nitrophenyl ether	0°	(—), 75
DHQD-*p*-chlorobenzoate	0°	(—), 71

TABLE 3. REACTIONS OF TRANS 1,2-DISUBSTITUTED ALKENES (*Continued*)

Substrate	Conditions	Product(s) and Yield(s) (%), % ee	Refs.
C_6 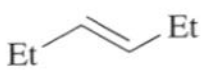	Ligand, $K_2OsO_2(OH)_4$, $K_3Fe(CN)_6$, K_2CO_3, t-BuOH:H_2O (1:1), rt, 24 h	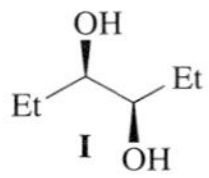 **I**	220
	Ligand		
	DHQD-*p*-chlorobenzoate	(75-95), 74	
	DHQD-9-(1-naphthyl) ether	(75-95), 92	
	DHQD-PHN	(75-95), 92	
	Ligand, $K_2OsO_2(OH)_4$, $K_3Fe(CN)_6$, K_2CO_3, $MeSO_2NH_2$, t-BuOH:H_2O (1:1), 0°	**I**	
	Ligand — Time	**I**	
	$(DHQD)_2$PYDZ — 12 h	(—), 93	66
	$(DHQD)_2$PYDZ — —	(58), 93-94	156
	$(DHQD)_2$PHAL — 12 h	(—), 93	66
	6 — 6-12 h	(74), 96	73
	33, OsO_4, CH_2Cl_2, –90°, 5 h	**I** (78), 90	9
	17, $K_3Fe(CN)_6$, K_2CO_3, $K_2OsO_2(OH)_4$, $MeSO_2NH_2$, t-BuOH:H_2O (1:1), 0°, 6-12 h	**I** (—), 78	73
	DHQ-*p*-chlorobenzoate, OsO_4, NMO, acetone:H_2O (10:1), 0°, 11 h (slow addition of olefin)	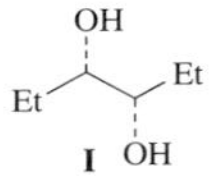 **I** (85-95), 55-60	69

Ligand, $K_2OsO_2(OH)_4$, $K_3Fe(CN)_6$, K_2CO_3, *t*-BuOH:H_2O (1:1), rt, 24 h	**I**	220
Ligand	**I**	
DHQ-9-phenyl ether	(85-90), 75	
DHQ-*p*-chlorobenzoate	(75-95), 67	
DHQ-9-(1-naphthyl) ether	(75-95), 82	
DHQ-PHN	(75-95), 85	
1. DHQ-OAc, OsO_4 (stoichiometric), toluene, rt, 12 h 2. LAH, ether	**I** (69), 50.2	8
1. 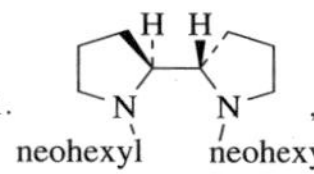 , OsO_4 (stoichiometric), CH_2Cl_2, –78°, 18 h 2. $NaHSO_3$	**I** (82), 96	13
35, OsO_4, CH_2Cl_2, –90°, 2 h	**I** (90), 98	11
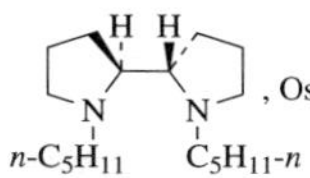 , OsO_4, CH_2Cl_2, –78°	**I** (86), 92	12
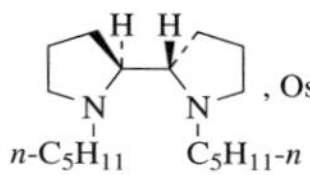 , OsO_4, toluene, –78°	**I** (88), 57	12

Substrate	Conditions	Product(s) and Yield(s) (%), % ee	Refs.
C_6 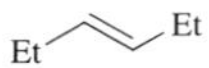	 , OsO_4, $K_3Fe(CN)_6$, K_2CO_3, t-BuOH:H_2O (1:1), rt, 8-24 h	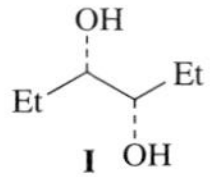 **I** (83), 41	193
	1. 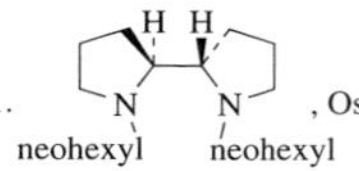 , OsO_4 (stoichiometric), CH_2Cl_2, –78°, 18 h 2. $NaHSO_3$	**I** (82), 96	13
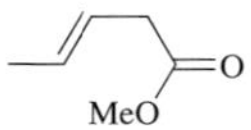	$(DHQ)_2PHAL$, $K_2OsO_2(OH)_4$, $K_3Fe(CN)_6$, K_2CO_3, $MeSO_2NH_2$, t-BuOH:H_2O (1:1), 0°, 5-10 d	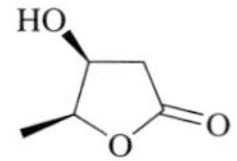 (40-47), 78	108, 338 339
	$(DHQD)_2PHAL$, $K_2OsO_2(OH)_4$, $K_3Fe(CN)_6$, K_2CO_3, $MeSO_2NH_2$, t-BuOH:H_2O (1:1), 0°, 2 d	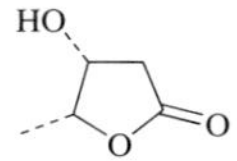 (48), 80	339
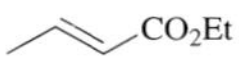	1. 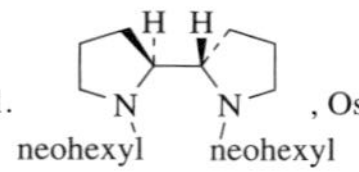 , OsO_4 (stoichiometric), CH_2Cl_2, –78°, 18 h 2. $NaHSO_3$	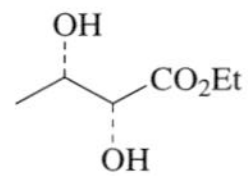 (90), 98	13
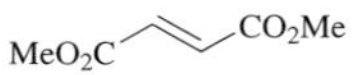	**33**, OsO_4, CH_2Cl_2, –90°, 4 h	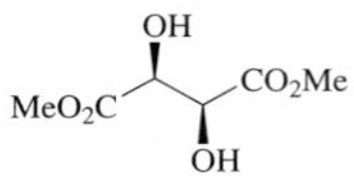 (67), 96	9
	35, OsO_4, CH_2Cl_2, –90°, 2 h	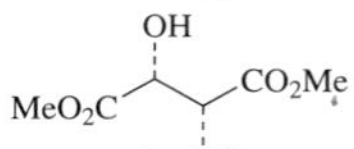 **I** (75), 92	11
	34, OsO_4, CH_2Cl_2, rt	**I** (59), 48	10

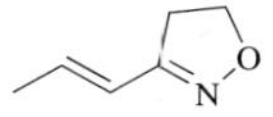	(DHQ)$_2$PHAL, $K_2OsO_2(OH)_4$, $K_3Fe(CN)_6$, K_2CO_3, $PhSO_2NH_2$, *t*-BuOH:H_2O (1:1), rt, 26 h	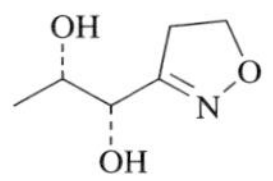 (76), 92	340
	DHQD-*p*-chlorobenzoate, OsO_4, NMO, acetone:H_2O, 0° (slow addition of olefin)	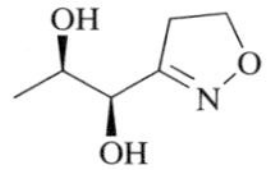 (—), 76	341
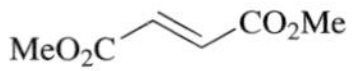	1. 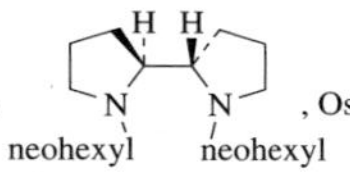, OsO_4, CH_2Cl_2, –78°, 18 h 2. $NaHSO_3$	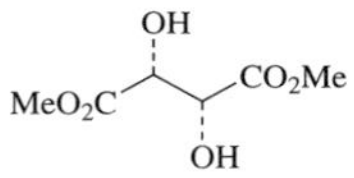 **I** (79), 98	13
	12a, NMO, OsO_4, acetone:H_2O (10:1), 40°, 16 h	**I** (76), 30	192
	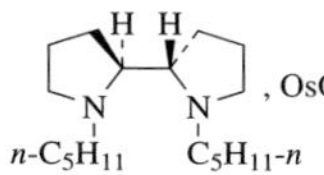, OsO_4, CH_2Cl_2, –78°	**I** (77), 91	12
	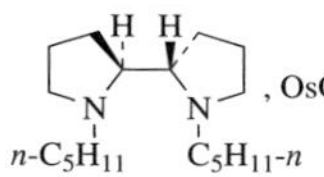, OsO_4, toluene, –78°	**I** (75), 63	12
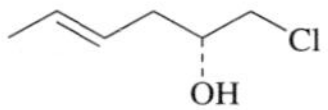	AD-Mix α, *t*-BuOH:H_2O (1:1)	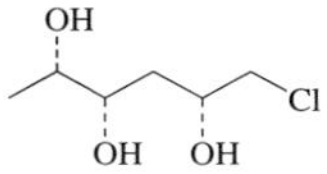 (90), 20 de	342
	AD-Mix β, *t*-BuOH:H_2O (1:1)	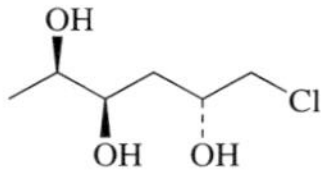 (73), 82 de	342

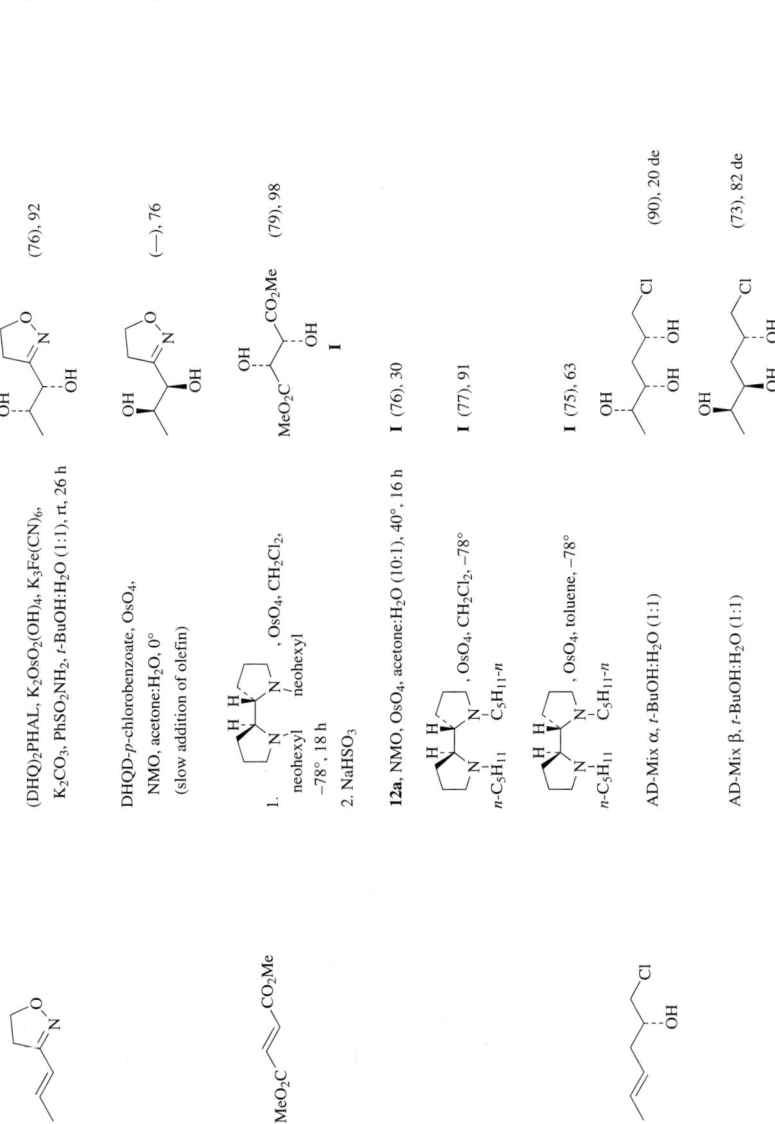

TABLE 3. REACTIONS OF TRANS 1,2-DISUBSTITUTED ALKENES (*Continued*)

Substrate	Conditions	Product(s) and Yield(s) (%), % ee	Refs.
C_7			
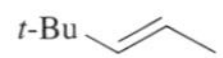	$(DHQD)_2PHAL$, $K_2OsO_2(OH)_4$, $K_3Fe(CN)_6$, K_2CO_3, $MeSO_2NH_2$, t-BuOH:H_2O (1:1), 0°	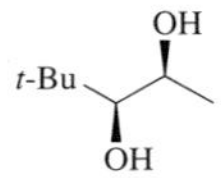 (—), 95	98
	1. DHQD-OAc, OsO_4 (stoichiometric), toluene, rt, 12 h 2. LAH, ether	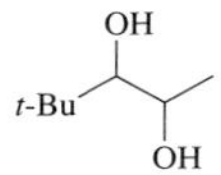 (78), 37	8
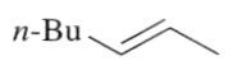	1. 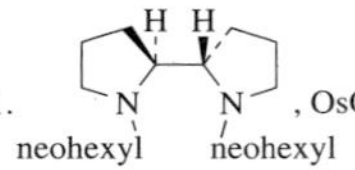 , OsO_4, CH_2Cl_2, –78°, 18 h 2. $NaHSO_3$	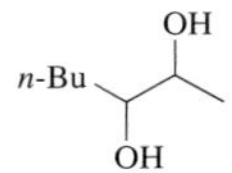 (93), 98	13
	33, OsO_4, toluene, –90°, 2 h	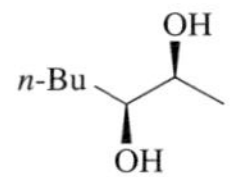 (61), 90	9
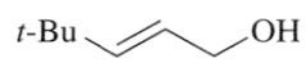	$(DHQD)_2PHAL$, $K_2OsO_2(OH)_4$, $K_3Fe(CN)_6$, K_2CO_3, $MeSO_2NH_2$, t-BuOH:H_2O (1:1), 0°	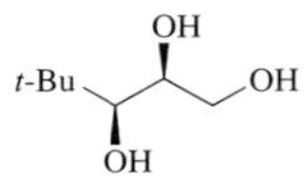 (—), 74	98
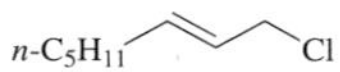	$(DHQ)_2PHAL$, $K_2OsO_2(OH)_4$, $K_3Fe(CN)_6$, K_2CO_3, $MeSO_2NH_2$, t-BuOH:H_2O (1:1), 0°, 17 h	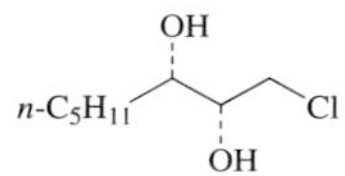 (83), 90	343

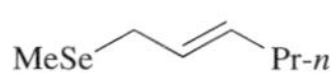	(DHQD)$_2$PHAL, $K_2OsO_2(OH)_4$, $K_3Fe(CN)_6$, K_2CO_3, $MeSO_2NH_2$, *t*-BuOH:H_2O (1:1), 0°, 24 h	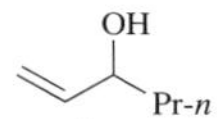 **I**	(100), 0	153
	AD-Mix β, $MeSO_2NH_2$, *t*-BuOH:H_2O (1:1), 0°, 24 h	**I** (100), 0		153
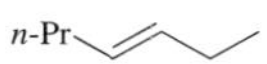	1. 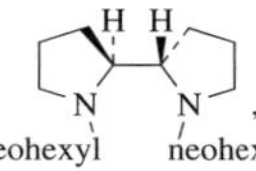 , OsO_4, CH_2Cl_2, –78°, 18 h 2. $NaHSO_3$	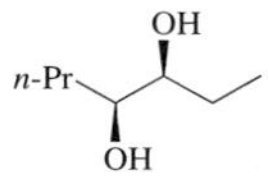	(90), 93	13
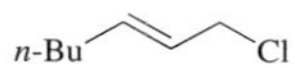	(DHQD)$_2$PHAL, $K_2OsO_2(OH)_4$, $K_3Fe(CN)_6$, K_2CO_3, $MeSO_2NH_2$, $NaHCO_3$, *t*-BuOH:H_2O (1:1), 0°	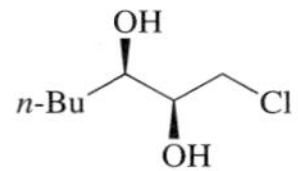	(89), 94	154
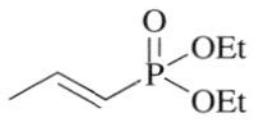	Ligand, OsO_4, $K_3Fe(CN)_6$, K_2CO_3, *t*-BuOH:H_2O (1:1), 20°	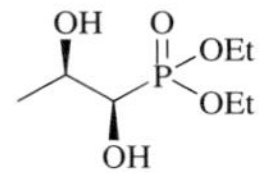		118
	Ligand			
	(DHQD)$_2$PHAL	(50), 40		
	DHQD-MEQ	(62), 80		
	(DHQ)$_2$PHAL, OsO_4, $K_3Fe(CN)_6$, K_2CO_3, *t*-BuOH:H_2O (1:1), 20°	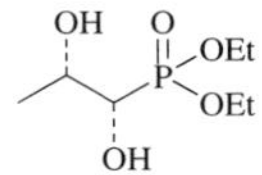	(45), 32	118, 344, 345
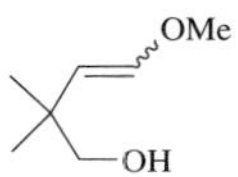	1. (DHQD)$_2$PHAL, $K_2OsO_2(OH)_4$, $K_3Fe(CN)_6$, K_2CO_3, *t*-BuOH:H_2O (1:1), rt 2. H^+	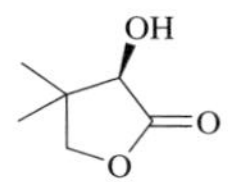	(<10), 45	346

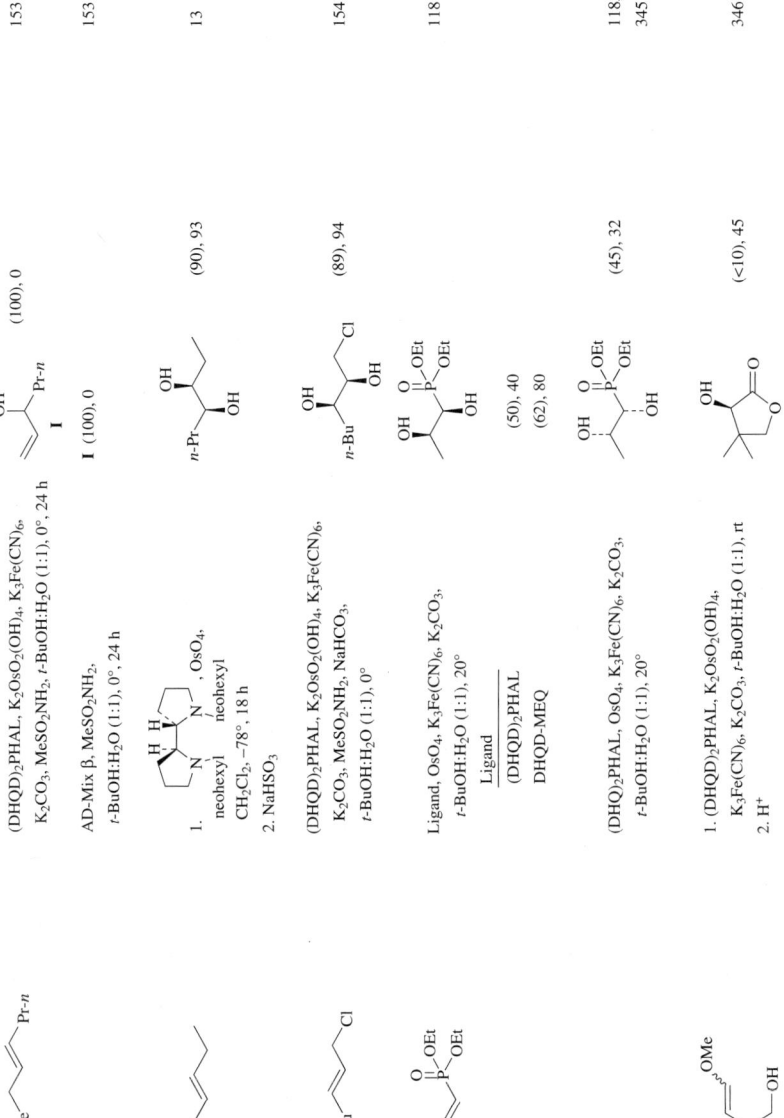

TABLE 3. REACTIONS OF TRANS 1,2-DISUBSTITUTED ALKENES (*Continued*)

Substrate	Conditions	Product(s) and Yield(s) (%), % ee	Refs.
C_8			
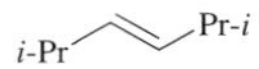	DHQD-*p*-chlorobenzoate, $K_2OsO_2(OH)_4$, NMO, acetone-H_2O, 0°, 5 days	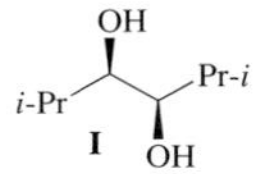 **I** (80-95), 12	69
	1. DHQD-9-phenyl ether, OsO_4, toluene, –78° 2. LAH	**I** (60-95), 94	337
	DHQD-*p*-chlorobenzoate, OsO_4, NMO, $Et_4NOAc{\bullet}4H_2O$, acetone:H_2O (10:1), 0°, 25 h (slow addition of olefin)	**I** (85-95), 76	69, 112
	DHQD-*p*-chlorobenzoate, OsO_4, NMO, acetone:H_2O (10:1), 0°, 25 h (slow addition of olefin)	**I** (85-95), 59	69
	DHQD-*p*-chlorobenzoate, OsO_4, acetone:H_2O (10:1, 0.15 M), 0°	**I** (—), 80	69, 112
	DHQD-*p*-chlorobenzoate, OsO_4, NMO, $Et_4NOAc{\bullet}4H_2O$, acetone-H_2O, 0°	**I** (—), 61	69
	$(DHQD)_2PHAL$, $K_2OsO_2(OH)_4$, $K_3Fe(CN)_6$, $MeSO_2NH_2$, K_2CO_3, *t*-BuOH:H_2O (1:1)	**I** (95), 98	347
	DHQ-*p*-chlorobenzoate, OsO_4, NMO, $Et_4NOAc{\bullet}4H_2O$, acetone:H_2O (10:1), 0°, 25 h (slow addition of olefin)	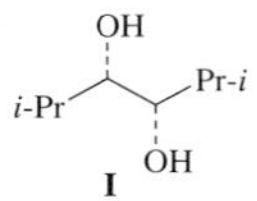 **I** (85-95), 66-71	69
	DHQ-*p*-chlorobenzoate, OsO_4, NMO, acetone:H_2O (10:1), 0°, 11 h (slow addition of olefin)	**I** (85-95), 49-54	69

	(DHQD)$_2$PHAL, $K_2OsO_2(OH)_4$, $K_3Fe(CN)_6$, K_2CO_3, $MeSO_2NH_2$, *t*-BuOH:H_2O (1:1), 0°, 24-36 h	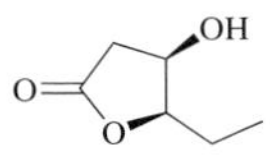 (60), 96	109
	(DHQ)$_2$PHAL, $K_2OsO_2(OH)_4$, $K_3Fe(CN)_6$, K_2CO_3, $MeSO_2NH_2$, *t*-BuOH:H_2O (1:1), 0°, 24-36 h	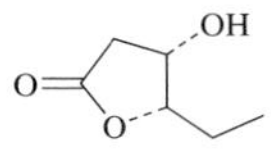 (65), 92	109
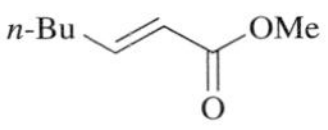	(DHQD)$_2$PHAL, $K_2OsO_2(OH)_4$, $K_3Fe(CN)_6$, K_2CO_3, $MeSO_2NH_2$, *t*-BuOH:H_2O (1:1), rt, 4 h	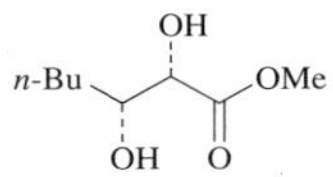 (79), 95	348
	(DHQ)$_2$PHAL, $K_2OsO_2(OH)_4$, $K_3Fe(CN)_6$, K_2CO_3, $MeSO_2NH_2$, *t*-BuOH:H_2O (1:1), 0°, 4 h	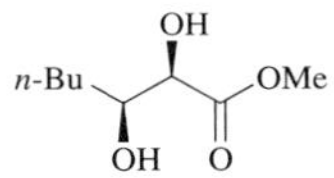 (94), 90	348
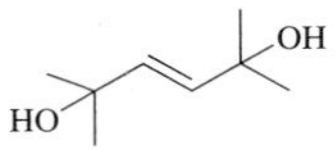	(DHQD)$_2$PHAL, $K_2OsO_2(OH)_4$, $K_3Fe(CN)_6$, K_2CO_3, $MeSO_2NH_2$, *t*-BuOH:H_2O (1:1), 0°, 6-36 h	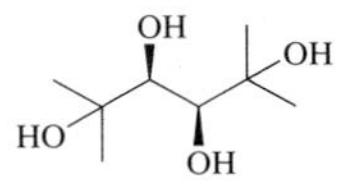 (70), 90	116
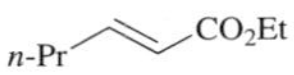	Ligand, $K_2OsO_2(OH)_4$, $K_3Fe(CN)_6$, K_2CO_3, *t*-BuOH:H_2O (1:1), rt, 24 h	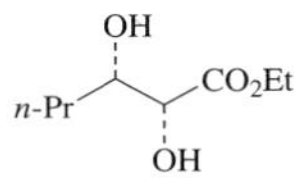	220
	Ligand		
	DHQ-*p*-chlorobenzoate	(75-95), 64	
	DHQ-9-phenyl ether	(75-95), 70	
	DHQ-9-(1-naphthyl) ether	(75-95), 83	
	DHQ-PHN	(75-95), 91	
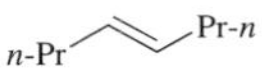	**15**, OsO_4, NMO, CH_3CN:H_2O (4:1), 5°, 24 h (slow addition of olefin)	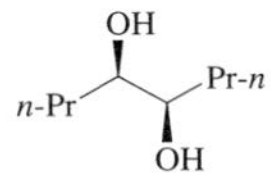 (79), 60	207

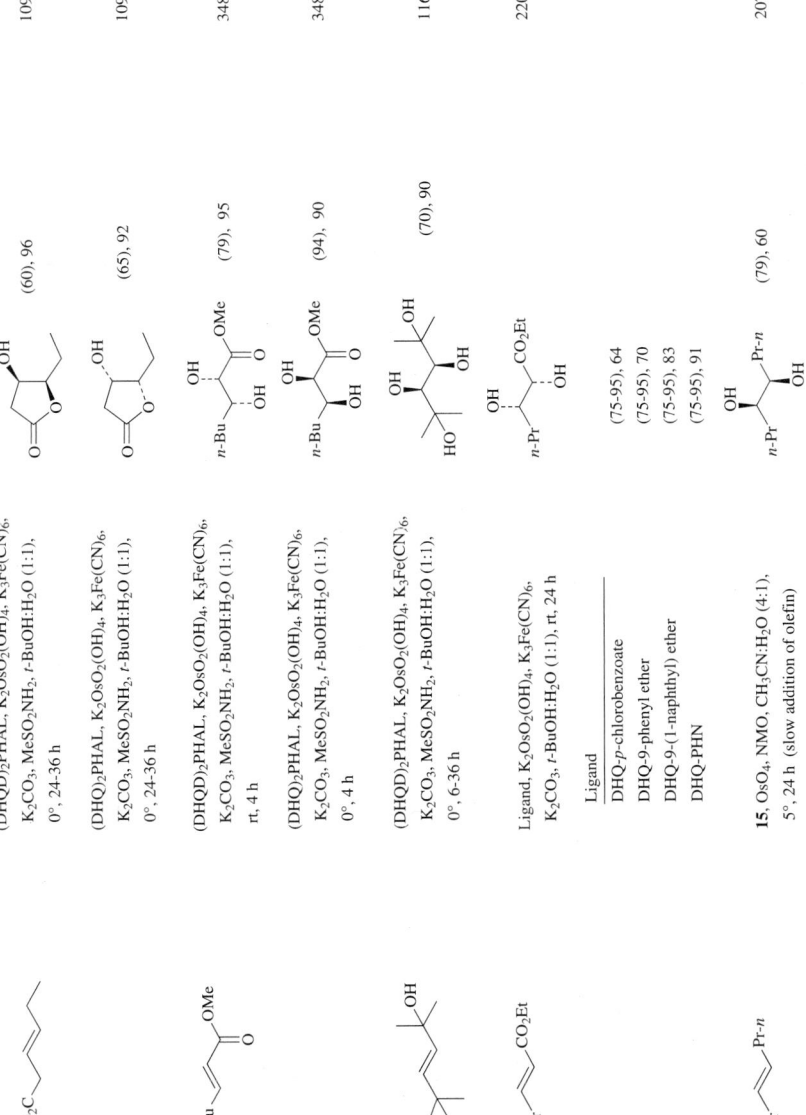

TABLE 3. REACTIONS OF TRANS 1,2-DISUBSTITUTED ALKENES (*Continued*)

Substrate	Conditions	Product(s) and Yield(s) (%), % ee	Refs.
C_8			
	**15**, NMO, OsO_4, $CH_3CN:H_2O$ (4:1), 5°, 36 h (slow addition of olefin)	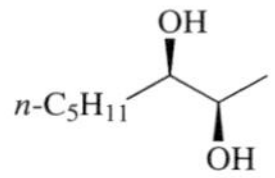 (86), 69	207
	$(DHQD)_2PHAL$, $K_2OsO_2(OH)_4$, $K_3Fe(CN)_6$, K_2CO_3, $MeSO_2NH_2$, t-BuOH:H_2O (1:1), 0°	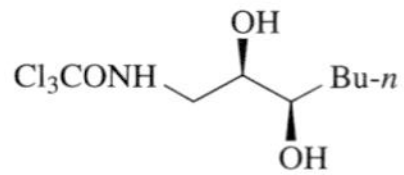 (81), 75	85
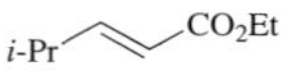	AD-Mix α, $MeSO_2NH_2$, t-BuOH:H_2O (1:1), 0°, 96 h	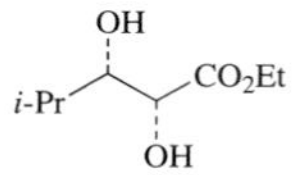 (95), 92	349
	AD-Mix β, $MeSO_2NH_2$, t-BuOH:H_2O (1:1), 0°, 96 h	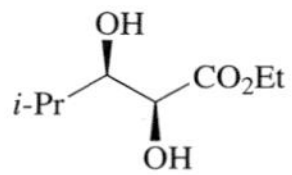 (95), 92	349
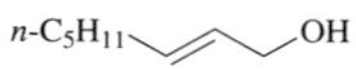	$(DHQ)_2PHAL$, $K_2OsO_2(OH)_4$, $K_3Fe(CN)_6$, K_2CO_3, $MeSO_2NH_2$, t-BuOH:H_2O (1:1), 0°	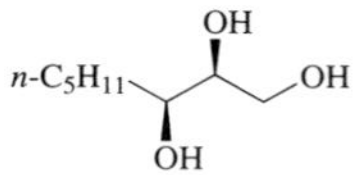 (84-88), 96	350
C_9			
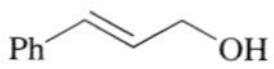	$(DHQD)_2PHAL$, $K_2OsO_2(OH)_4$, $K_3Fe(CN)_6$, K_2CO_3, $MeSO_2NH_2$, t-BuOH:H_2O (1:1), 0°	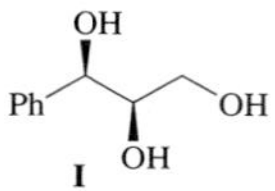 **I** (—), 97	98
	DHQD-*p*-chlorobenzoate, OsO_4, NMO, acetone:H_2O (10:1, 0.15 M), 0°	**I** (—), 66	112
	DHQD-*p*-chlorobenzoate, OsO_4, NMO, acetone:H_2O (10:1), 0°, 17 h (slow addition of olefin)	**I** (85-95), 66	112

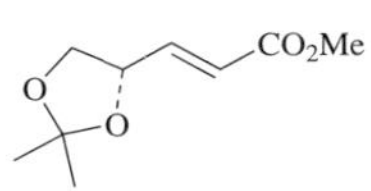	1. DHQD-9-phenyl ether, OsO_4 toluene, –78° 2. $NaHSO_3$, THF:H_2O	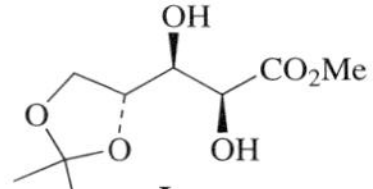 **I** (60-95), 97 de	337
	DHQD-*p*-chlorobenzoate, OsO_4, NMO, acetone:H_2O (10:1), 0°, 43 h (slow addition of olefin)	**I** (80-95), 76 de	112
	DHQD-*p*-chlorobenzoate, OsO_4, NMO, acetone:H_2O (10:1, 0.15 M), 0°	**I** (—), 77 de	112
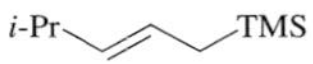	$(DHQ)_2PHAL$, $K_2OsO_2(OH)_4$, $K_3Fe(CN)_6$, K_2CO_3, $MeSO_2NH_2$, *t*-BuOH:H_2O (1:1), 12 h	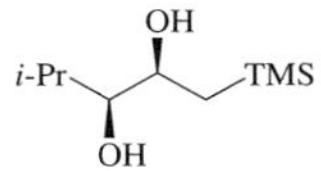 (84), 68	114
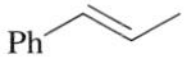	**35**, OsO_4, CH_2Cl_2, –90°, 2 h	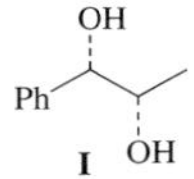 **I** (95), 93	11
	1. 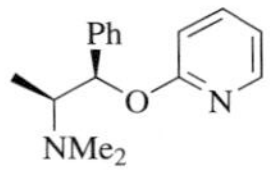, OsO_4, toluene, –78°, 10 h 2. LAH, ether, –78° to rt, 8 h	**I** (91), 8	183
	DHQ-*p*-chlorobenzoate, OsO_4, NMO, acetone:H_2O, 0°, 12 h	**I** (80-95), 55	18, 171
	DHQ-*p*-chlorobenzoate, OsO_4, NMO, acetone:H_2O (10:1), 0°, 6 h (slow addition of olefin)	**I** (85-95), 76-81	69
	1. 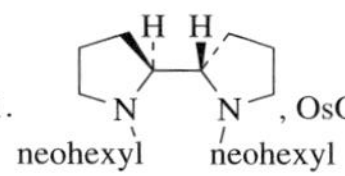, OsO_4, toluene, –78°, 18 h 2. $NaHSO_3$	**I** (92), 95	13

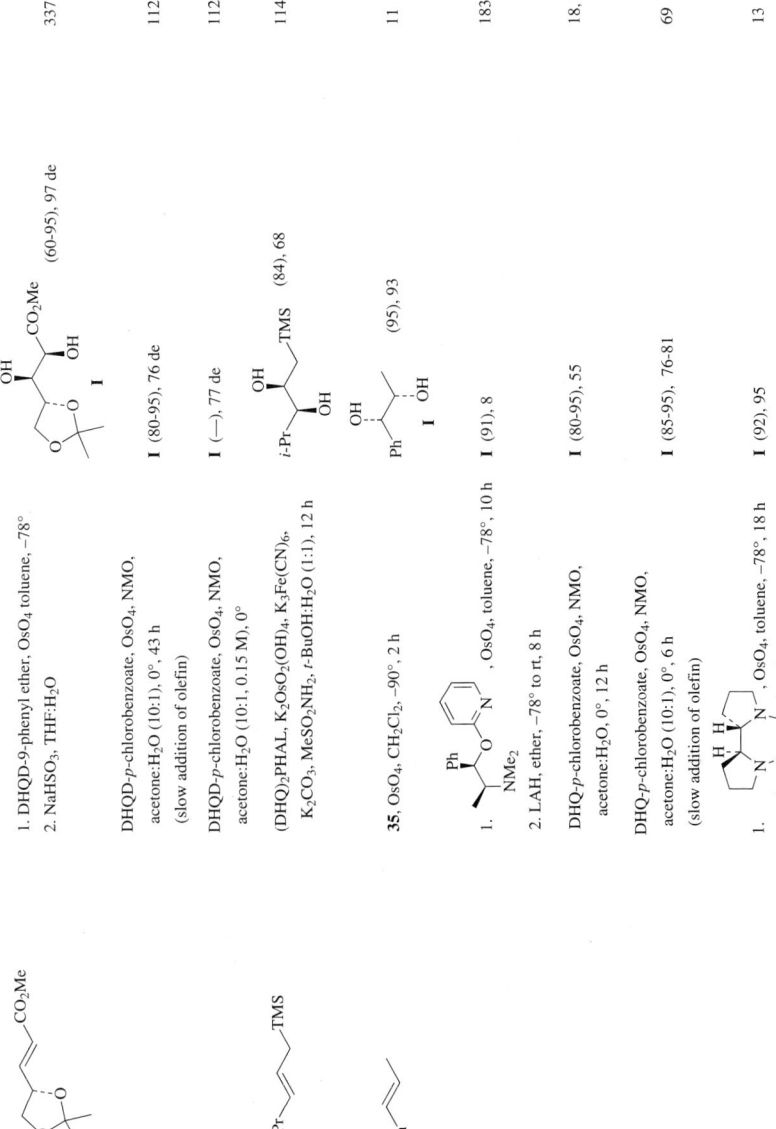

TABLE 3. REACTIONS OF TRANS 1,2-DISUBSTITUTED ALKENES (*Continued*)

Substrate	Conditions	Product(s) and Yield(s) (%), % ee	Refs.
C_9			
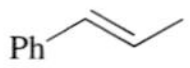	(DHQ)$_2$PHAL, $K_2OsO_2(OH)_4$, $K_3Fe(CN)_6$, K_2CO_3, *t*-BuOH:H_2O (1:1), 0°, 6-24 h	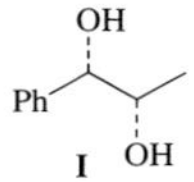 **I** (—), 97	171
	(DHQ)$_2$PHAL, $K_2OsO_2(OH)_4$, NMO, *t*-BuOH:H_2O (1.4:1), 0°, 6 h (slow addition of olefin)	**I** (85-96), 98	171
	1. DHQ-OAc, OsO_4 toluene, rt, 12 h 2. LAH, ether	**I** (90), 45.5	8
	19, OsO_4, $K_3Fe(CN)_6$, K_2CO_3, $MeSO_2NH_2$, *t*-BuOH:H_2O (1:1), 10°, 15 h	**I** (92), 96.5	165
	43, OsO_4, NMO, acetone:H_2O (10:1), 0°, 6 h	**I** (69), 60	202
	42b, OsO_4, NMO, acetone:H_2O (10:1), 0°, 7 h	**I** (69), 60	161
	42b, OsO_4, $K_3Fe(CN)_6$, K_2CO_3, *t*-BuOH:H_2O (1:1), 0°, 48 h	**I** (71), 81	161
	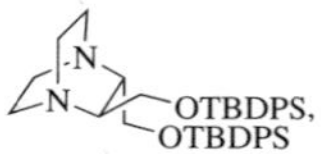 OsO_4, $K_3Fe(CN)_6$, K_2CO_3, *t*-BuOH:H_2O (1:1), rt, 8-24 h	**I** (95), 19	193
	Ligand, $K_2OsO_2(OH)_4$, $K_3Fe(CN)_6$, K_2CO_3, *t*-BuOH:H_2O (1:1)	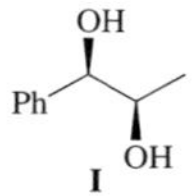 **I**	

Conditions			Product(s) and Yield(s) (%), % ee	Refs.
Ligand	Temp	Time		
DHQD-9-(1-naphthyl) ether	rt	24 h	(75-95), 91	220
DHQD-PHN	rt	24 h	(75-95), 93	220
$(DHQD)_2AQ$	0°	12 h	(65-90), 92	86
$(DHQD)_2DPP$	—	—	(—), 98	86
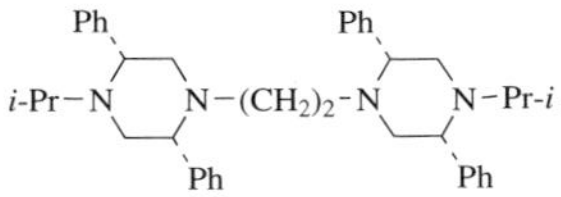 , OsO_4, toluene, –78°			**I** (81), 95	215
33, OsO_4, toluene, –90°, 6 h			**I** (82), 95	9
15, OsO_4, NMO, $CH_3CN:H_2O$ (4:1), 5°, 17 h (slow addition of olefin)			**I** (82), 50	207
1. DHQD-OAc, OsO_4, toluene, rt, 12 h 2. LAH, ether			**I** (66), 48.6	8
$(DHQD)_2PHAL$, $K_2OsO_2(OH)_4$, $K_3Fe(CN)_6$, K_2CO_3, $MeSO_2NH_2$, t-BuOH:H_2O (1:1), 0°			**I** (80), >99	97
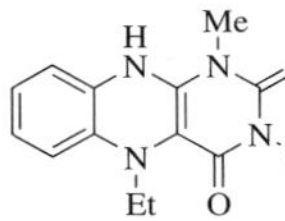 Me, $(DHQD)_2PHAL$, OsO_4, H_2O_2, NMM, TEAA, t-BuOH:H_2O (3:1), 0°, 7 h (slow addition of H_2O_2 and olefin)			**I** (67), 96	208
$(DHQD)_2PHAL$, ABC MC OsO_4, NMO, acetone:H_2O:CH_3CN (1:1:1), rt, 24 h (slow addition of olefin)			**I** (97), 94	170

TABLE 3. REACTIONS OF TRANS 1,2-DISUBSTITUTED ALKENES (*Continued*)

Substrate	Conditions	Product(s) and Yield(s) (%), % ee	Refs.
C_9 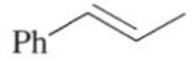	$(DHQD)_2PHAL$, OsO_4, NMO, acetone:H_2O:CH_3CN (1:1:1), rt, 24 h (slow addition of olefin)	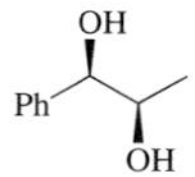 I (73), 95	170
	Ligand, OsO_4,NMO, $Me_4NOAc{\bullet}4H_2O$, acetone:H_2O (10:1), 10°, 5 h (slow addition of olefin)	I	
	Ligand: **27**	I: (80), 84	205
	26	(80), 85	205
	16	(87), 91	168
	16, OsO_4, $K_3Fe(CN)_6$, K_2CO_3, *t*-BuOH:H_2O (1:1), rt, 18 h	**I** (83), 99	168
	DHQD-*p*-chlorobenzoate, OsO_4, NMO, acetone-H_2O, 0°, 5 h	**I** (80-95), 65	18, 70, 351, 69
	DHQD-*p*-chlorobenzoate, OsO_4, NMO, acetone:H_2O (10:1), 0°, 6 h (slow addition of olefin)	**I** (85-95), 86	69, 112, 70
	DHQD-*p*-chlorobenzoate, OsO_4, NMO, acetone:H_2O (10:1, 0.15 M), 0°	**I** (—), 87	69, 112
	DHQD-*p*-chlorobenzoate, OsO_4, NMO, $Et_4NOAc{\bullet}4H_2O$, acetone:H_2O, 0°	**I** (—), 73	69
	DHQD-*p*-chlorobenzoate, $K_2OsO_2(OH)_4$, $K_3Fe(CN)_6$, K_2CO_3, *t*-BuOH:H_2O (1:1), rt, 24 h	**I** (75-95), 91	220, 70
	DHQD-9-phenyl ether, $K_2OsO_2(OH)_4$, $K_3Fe(CN)_6$, K_2CO_3, *t*-BuOH:H_2O (1:1), rt, 24 h	**I** (75-95), 89	220

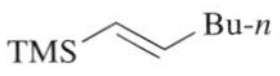	(DHQD)$_2$PHAL, $K_2OsO_2(OH)_4$, $K_3Fe(CN)_6$, K_2CO_3, $MeSO_2NH_2$, *t*-BuOH:H_2O (1:1), 0°, 28 h	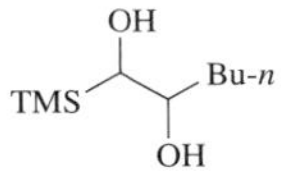 (76), 96	113
	(DHQ)$_2$PHAL, $K_2OsO_2(OH)_4$, $K_3Fe(CN)_6$, K_2CO_3, $MeSO_2NH_2$, *t*-BuOH:H_2O (1:1), 12 h	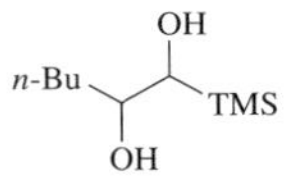 (79-89), 88-95	113, 114
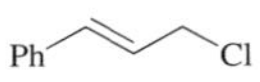	DHQD-*p*-chlorobenzoate, OsO_4, NMO, acetone:H_2O (10:1), 0°, 11 h (slow addition of olefin)	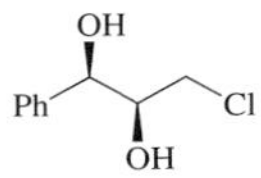 **I** (80-95), 78	112
	DHQD-*p*-chlorobenzoate, OsO_4, NMO, acetone:H_2O (10:1, 0.15 M), 0°	**I** (80-95), 79	112
	(DHQD)$_2$PHAL, $K_2OsO_2(OH)_4$, $K_3Fe(CN)_6$, K_2CO_3, $MeSO_2NH_2$, *t*-BuOH:H_2O (1:1), 0°	**I** (80), 98	154
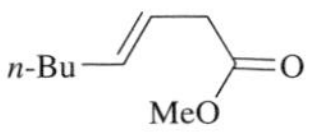	(DHQD)$_2$PHAL, $K_2OsO_2(OH)_4$, $K_3Fe(CN)_6$, K_2CO_3, $MeSO_2NH_2$, *t*-BuOH:H_2O (1:1), 0°, 36 h	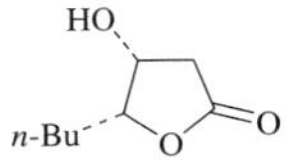 (92), 97	108
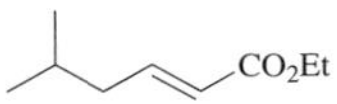	(DHQ)$_2$PHAL, $K_2OsO_2(OH)_4$, $K_3Fe(CN)_6$, K_2CO_3, $MeSO_2NH_2$, *t*-BuOH:H_2O (1:1), 0°, 72 h	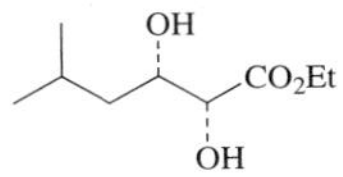 (82), 80	352
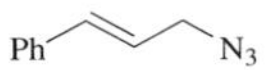	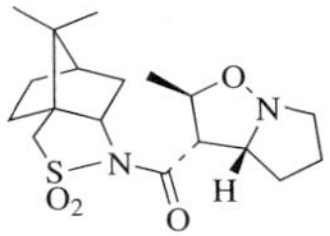 , OsO_4, $K_3Fe(CN)_6$, K_2CO_3, *t*-BuOH:H_2O (1:1), rt, 24 h	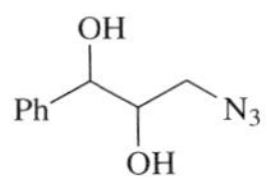 (99), 32	203

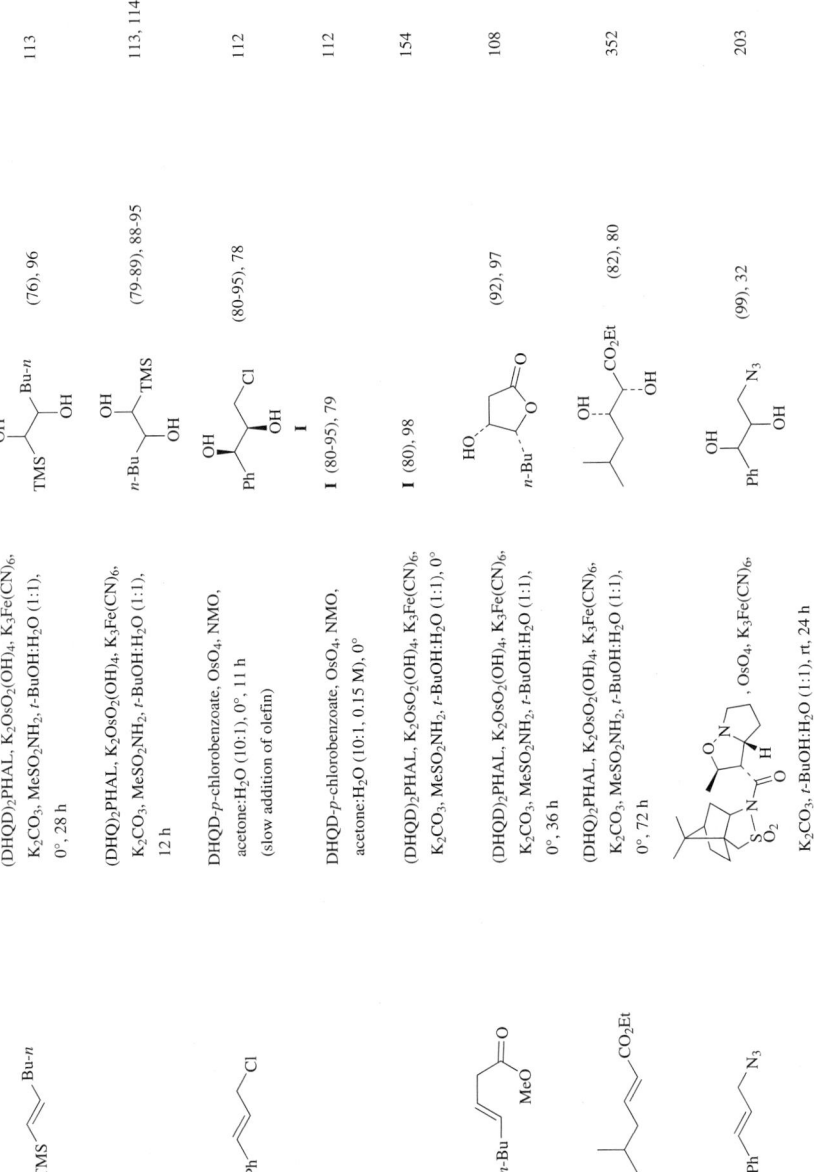

TABLE 3. REACTIONS OF TRANS 1,2-DISUBSTITUTED ALKENES (*Continued*)

Substrate	Conditions	Product(s) and Yield(s) (%), % ee	Refs.
C_9			
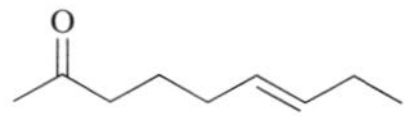	(DHQ)$_2$PHAL, $K_2OsO_2(OH)_4$, $K_3Fe(CN)_6$, K_2CO_3, $MeSO_2NH_2$, *t*-BuOH:H_2O (1:1), 0°, 24 h	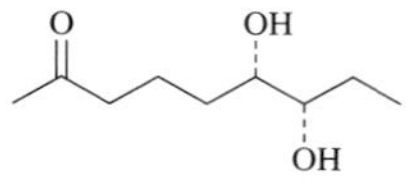 (—), 74	353
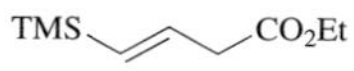	AD-Mix β, $MeSO_2NH_2$, *t*-BuOH:H_2O (1:1)	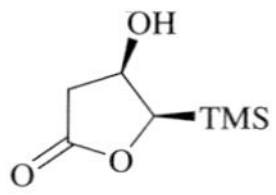 (75), 86	354
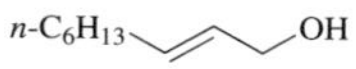	(DHQ)$_2$PHAL, $K_2OsO_2(OH)_4$, $K_3Fe(CN)_6$, K_2CO_3, $MeSO_2NH_2$, *t*-BuOH:H_2O (1:1), 0°	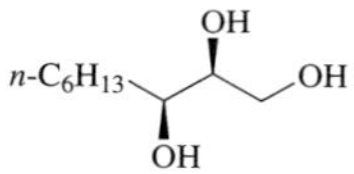 (84-88), 96	350
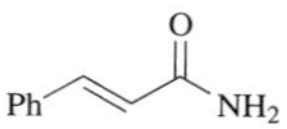	(DHQ)$_2$PHAL, OsO_4, $K_3Fe(CN)_6$, K_2CO_3, $MeSO_2NH_2$, *t*-BuOH:H_2O (1:1), 0°, 4 d	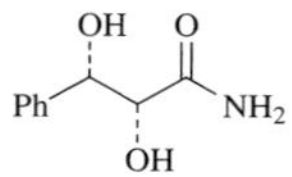 (48), 95	355
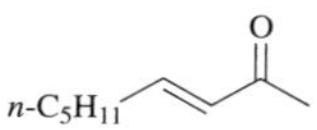	(DHQD)$_2$PHAL, $K_2OsO_2(OH)_4$, $K_3Fe(CN)_6$, $NaHCO_3$-K_2CO_3, $MeSO_2NH_2$, *t*-BuOH:H_2O (1:1), 0°, 24 h	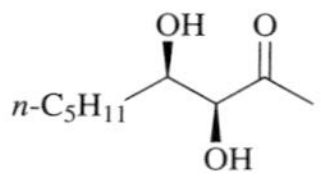 (87), 98	155
C_{10}			
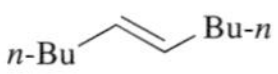	Ligand, $K_2OsO_2(OH)_4$, $K_3Fe(CN)_6$, K_2CO_3, $MeSO_2NH_2$, *t*-BuOH:H_2O (1:1)	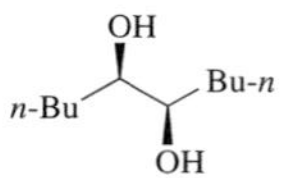 **I**	

Ligand	Temp	Time		Refs.
$(DHQD)_2$PHAL	0°	3-10 h	(80-98), 97-98	77, 19, 86, 157, 152, 66
$(DHQD)_2$PYR	0°	3 h	(—), 88	77, 157
DHQD-PHN	rt	24 h	(78), 95	72
DHQD-MEQ	rt	24 h	(75-95), 90	72, 220
DHQD-*p*-chlorobenzoate	rt	24 h	(75-95), 79	72, 220
$(DHQD)_2$PYDZ	0°	6-24 h	(—), 96	152, 66
$(DHQD)_2$TP	0°	6-24 h	(—), 82	152

DHQD-*p*-chlorobenzoate, $K_2OsO_2(OH)_4$, NMO, $Et_4NOAc\cdot 4H_2O$, acetone:H_2O (10:1), 0°, 41 h (slow addition of olefin)	**I** (85-95), 69	112, 70
DHQD-*p*-chlorobenzoate, OsO_4, NMO, acetone:H_2O (10:1, 0.15 M), 0°	**I** (—), 73	112, 337
1. DHQD-9-phenyl ether, OsO_4, toluene, –78° 2. LAH, ether	**I** (60-95), 93	337
Ligand, $K_2OsO_2(OH)_4$, $K_3Fe(CN)_6$, K_2CO_3, *t*-BuOH:H_2O (1:1), rt	**I**	

Ligand	Time	**I**	
DHQD-9-(1-naphthyl) ether	24 h	(75-95), 94	220
DHQD-PHN	24 h	(75-95), 95	220
DHQD-9-phenyl ether	24 h	(75-95), 88	220
DHQD-9-(4-quinolinyl) ether	24 h	(75-95), 87	220
DHQD-9-(4-methylnaphthyl) ether	24 h	(75-95), 95	220
$(DHQD)_2$PHAL	34 h	(94), 93	71
$(DHQD)_2$PHAL	1.8 h	(95), 90	71

TABLE 3. REACTIONS OF TRANS 1,2-DISUBSTITUTED ALKENES (*Continued*)

Substrate	Conditions	Product(s) and Yield(s) (%), % ee	Refs.
C_{10} 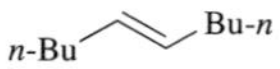	Ligand, OsO_4, $K_3Fe(CN)_6$, K_2CO_3, t-BuOH:H_2O (1:1)	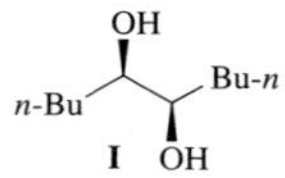 I	
	Ligand / Temp / Time		
	41a 0° 48 h	(68-80), 88	163
	41a 0° 24 h	(68-80), 88	206
	44 0° 48 h	(68-80), 86	163
	16 rt 18 h	(80), 97	168
	Ligand, OsO_4, NMO, $Me_4NOAc•4H_2O$, acetone:H_2O (10:1), 10° (slow addition of olefin)	**I**	
	Ligand / Time	**I**	
	27 10 h	(62), 42	205
	26 10 h	(65), 43	205
	16 5 h	(84), 80	168
	15, OsO_4, NMO, CH_3CN:H_2O (8:2), 5°, 24 h (slow addition of olefin)	**I** (81), 59	207
	AD-Mix β, $MeSO_2NH_2$, t-BuOH:H_2O (1:1), 0°, 24 h	**I** (93), >95	356
	Ligand, $K_2OsO_2(OH)_4$, $K_3Fe(CN)_6$, K_2CO_3, $MeSO_2NH_2$, t-BuOH:H_2O (1:1)	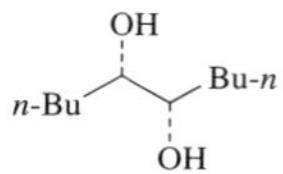	

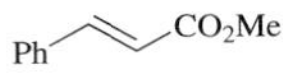

Ligand	Temp	Time		
$(DHQ)_2$PHAL	0°	10 h	(80-98), 93	19, 157, 86, 152
$(DHQ)_2$PYDZ	0°	6-24 h	(—), 83	152
$(DHQ)_2$TP	0°	6-24 h	(—), 75	152
DHQ-*p*-chlorobenzoate	rt	24 h	(75-95), 70	220
DHQ-9-(1-naphthyl) ether	rt	24 h	(75-95), 86	220
DHQ-PHN	rt	24 h	(75-95), 91	220
DHQ-9-phenyl ether	rt	24 h	(75-95), 75	220

Ligand, $K_2OsO_2(OH)_4$, $K_3Fe(CN)_6$, K_2CO_3, $MeSO_2NH_2$, *t*-BuOH-H_2O (1:1), 24 h

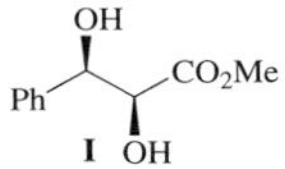

I

Ligand	Temp		
DHQD-PHN	rt	(75-95), 98	220, 72
DHQD-MEQ	rt	(75-95), 98	72
DHQD-*p*-chlorobenzoate	rt	(75-95), 91-95	72, 220, 70

DHQD-*p*-chlorobenzoate, $K_2OsO_2(OH)_4$, NMO, acetone:H_2O (10:1), 0°, 19 h — **I** (80-95), 86 — 112, 70

DHQD-*p*-chlorobenzoate, $K_2OsO_2(OH)_4$, NMO, acetone:H_2O (10:1, 0.15 M), 0° — **I** (—), 89 — 112

DHQD-*p*-chlorobenzoate, OsO_4, NMO, acetone:H_2O (10:1), 0° — **I** (80-95), 60 — 70

Ligand, $K_2OsO_2(OH)_4$, $K_3Fe(CN)_6$, K_2CO_3, *t*-BuOH:H_2O (1:1), rt, 24 h — **I** — 220

Ligand	I
DHQD-9-phenyl ether	(75-95), 91
DHQD-9-(1-naphthyl) ether	(75-95), 96

TABLE 3. REACTIONS OF TRANS 1,2-DISUBSTITUTED ALKENES (*Continued*)

Substrate	Conditions	Product(s) and Yield(s) (%), % ee	Refs.
C_{10}			
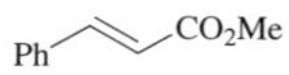	AD-Mix β, $MeSO_2NH_2$, t-BuOH:H_2O (1:1)	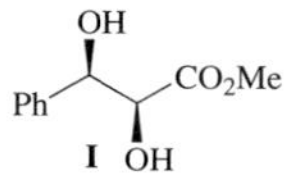 **I** (68), >98	357
	15, OsO_4, NMO, CH_3CN:H_2O (4:1), 5°, 24 h	**I** (78), 45	207
	Ligand, OsO_4, NMO, TEAA, acetone:H_2O (10:1), 10°	**I**	159
	Ligand / Time	**I**	
	28 / 30 h	(74), 30	
	29 / 60 h	(73), 45	
	30a / 18 h	(80), 89	
	30b / 28 h	(87), 85	
	Ligand, OsO_4, $K_3Fe(CN)_6$, K_2CO_3, $MeSO_2NH_2$, t-BuOH:H_2O (1:1), 10°	**I**	159
	Ligand / Time	**I**	
	28 / 24 h	(80), 41	
	29 / 18 h	(85), 91	
	30a / 26 h	(84), 79	
	30b / 20 h	(86), 90	
	33, OsO_4, toluene, –90°, 6 h	**I** (80), 99	9
	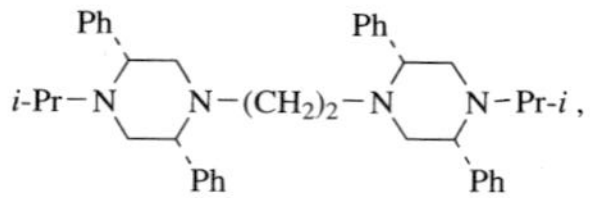 OsO_4, toluene, –78°	**I** (88), 97	215

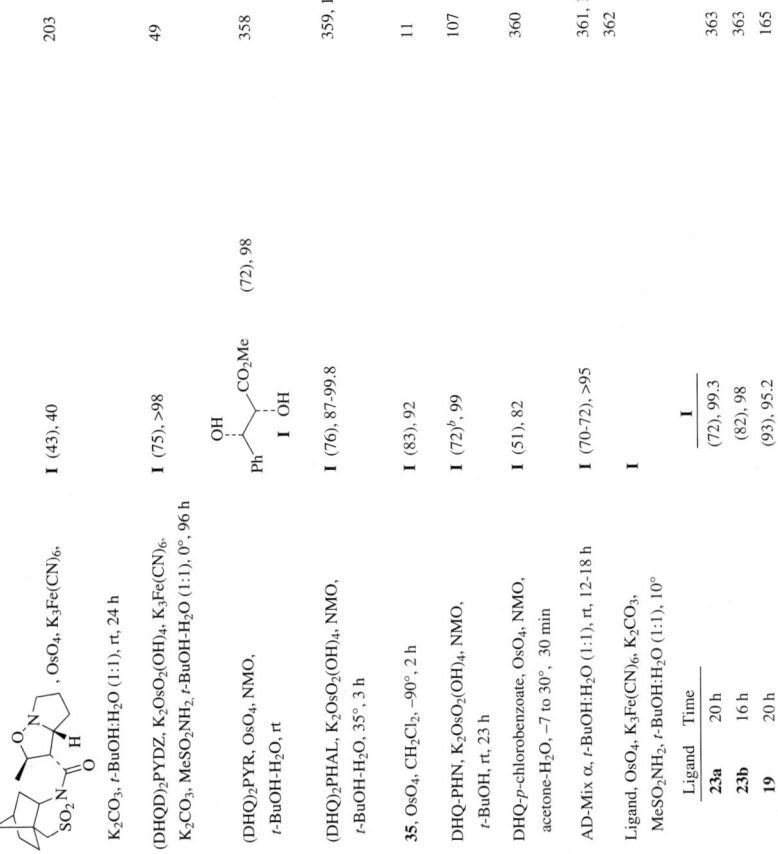

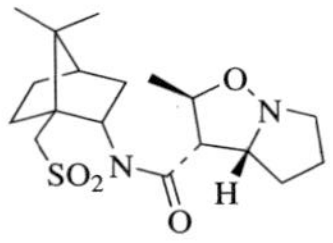, OsO_4, $K_3Fe(CN)_6$, K_2CO_3, t-BuOH:H_2O (1:1), rt, 24 h	**I** (43), 40		203
$(DHQD)_2$PYDZ, $K_2OsO_2(OH)_4$, $K_3Fe(CN)_6$, K_2CO_3, $MeSO_2NH_2$, t-BuOH-H_2O (1:1), 0°, 96 h	**I** (75), >98		49
$(DHQ)_2$PYR, OsO_4, NMO, t-BuOH-H_2O, rt	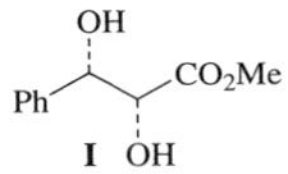 **I**	(72), 98	358
$(DHQ)_2$PHAL, $K_2OsO_2(OH)_4$, NMO, t-BuOH-H_2O, 35°, 3 h	**I** (76), 87-99.8		359, 107
35, OsO_4, CH_2Cl_2, –90°, 2 h	**I** (83), 92		11
DHQ-PHN, $K_2OsO_2(OH)_4$, NMO, t-BuOH, rt, 23 h	**I** (72)[b], 99		107
DHQ-p-chlorobenzoate, OsO_4, NMO, acetone-H_2O, –7 to 30°, 30 min	**I** (51), 82		360
AD-Mix α, t-BuOH:H_2O (1:1), rt, 12-18 h	**I** (70-72), >95		361, 104, 362
Ligand, OsO_4, $K_3Fe(CN)_6$, K_2CO_3, $MeSO_2NH_2$, t-BuOH:H_2O (1:1), 10°	**I**		

Ligand	Time	**I**	
23a	20 h	(72), 99.3	363
23b	16 h	(82), 98	363
19	20 h	(93), 95.2	165

TABLE 3. REACTIONS OF TRANS 1,2-DISUBSTITUTED ALKENES (*Continued*)

Substrate	Conditions	Product(s) and Yield(s) (%), % ee	Refs.
C_{10}			
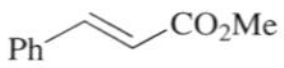	**42b**, OsO_4, $K_3Fe(CN)_6$, K_2CO_3, *t*-BuOH:H_2O (1:1), 0°, 24 h; rt, 24 h	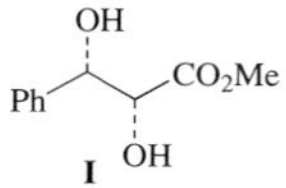 **I** (40), 80	161
	42b, OsO_4, NMO, acetone:H_2O (10:1), 0°, 7 h	**I** (81), 60	161
	Ligand, OsO_4, NMO, TEAA, acetone:H_2O (10:1), 10°	**I**	159
	Ligand / TIme	**I**	
	31 / 60 h	(80), 51	
	32 / 18 h	(83), 89	
	Ligand, OsO_4, $K_3Fe(CN)_6$, K_2CO_3, $MeSO_2NH_2$, *t*-BuOH:H_2O (1:1), 10°	**I**	159
	Ligand / Time	**I**	
	31 / 18 h	(84), 87	
	32 / 26 h	(80), 83	
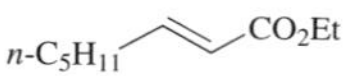	Ligand, $K_2OsO_2(OH)_4$, $K_3Fe(CN)_6$, K_2CO_3, $MeSO_2NH_2$, *t*-BuOH:H_2O (1:1)	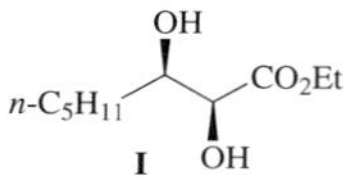 **I**	
	Ligand / Temp / Time		
	$(DHQD)_2AQN$ / 0° / 12 h	(65-90), 99	86
	DHQD-*p*-chlorobenzoate / rt / 24 h	(75-95), 67	220, 72
	DHQD-9-phenyl ether / rt / 24 h	(75-95), 88	220
	DHQD-9-(1-naphthyl) ether / rt / 24 h	(75-95), 94	220, 72

	Ligand, $K_2OsO_2(OH)_4$, $K_3Fe(CN)_6$, K_2CO_3, $MeSO_2NH_2$, t-BuOH:H_2O (1:1)	**I**	
	Ligand / Temp / Time	**I**	
	DHQD-PHN / rt / 24 h	(75-95), 94	220
	$(DHQD)_2PHAL$ / 0° / 6-24 h	(80-98), 99	200, 86, 19
	DHQD-MEQ / rt / 24 h	(75-95), 85	72
	DHQD-*p*-chlorobenzoate, OsO_4, NMO, acetone:H_2O (10:1), 0°, 32 h (slow addition of olefin)	**I** (80-95), 67	112
	DHQD-*p*-chlorobenzoate, OsO_4, NMO, acetone:H_2O (10:1, 0.15 M), 0°	**I** (—), 67	112, 337
	1. DHQD-9-phenyl ether, OsO_4, toluene, –78° 2. LAH	**I** (60-95), 90	337
	AD-Mix β, t-BuOH:H_2O (1:1), 0-4°, 24 h	**I** (89), 95	140
	$(DHQ)_2PHAL$, $K_2OsO_2(OH)_4$, $K_3Fe(CN)_6$, K_2CO_3, $MeSO_2NH_2$, t-BuOH:H_2O (1:1) 0°, 6-24 h	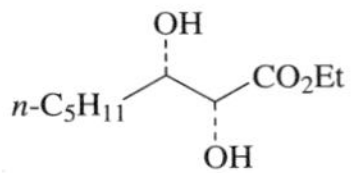 (80-98), 96	200, 19
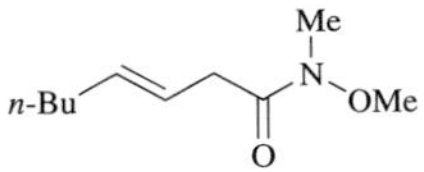	$(DHQD)_2PHAL$, $K_2OsO_2(OH)_4$, $K_3Fe(CN)_6$, K_2CO_3, $MeSO_2NH_2$, t-BuOH:H_2O (1:1), 0°	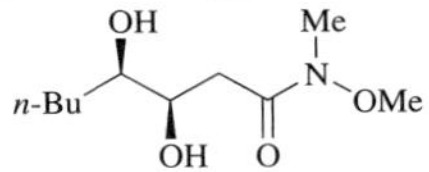 (84), 96	111
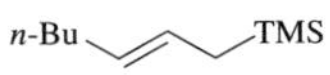	$(DHQ)_2PHAL$, $K_2OsO_2(OH)_4$, $K_3Fe(CN)_6$, K_2CO_3, $MeSO_2NH_2$, t-BuOH:H_2O (1:1), 4 h	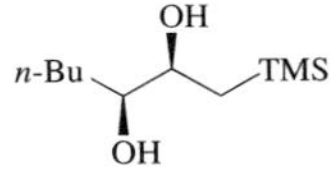 (91), 70	114
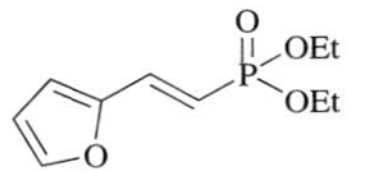	$(DHQ)_2PHAL$, $K_2OsO_2(OH)_4$, $K_3Fe(CN)_6$, K_2CO_3, $MeSO_2NH_2$, t-BuOH:H_2O (1:1), 0-24°, 50 h	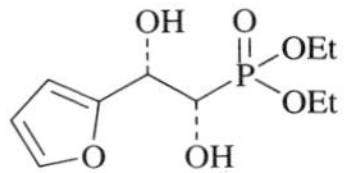 (17), 88	345

TABLE 3. REACTIONS OF TRANS 1,2-DISUBSTITUTED ALKENES (*Continued*)

Substrate	Conditions	Product(s) and Yield(s) (%), % ee	Refs.
C_{10}			
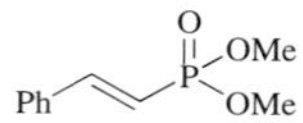	$(DHQD)_2PHAL$, $K_2OsO_2(OH)_4$, $K_3Fe(CN)_6$, K_2CO_3, $MeSO_2NH_2$, t-BuOH:H_2O (1:1), 0-24°, 50 h	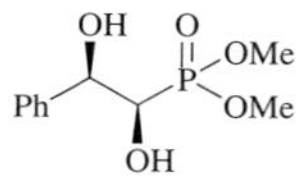 (52), 97	345
	$(DHQD)_2PHAL$, $K_2OsO_2(OH)_4$, $K_3Fe(CN)_6$, K_2CO_3, $MeSO_2NH_2$, t-BuOH:H_2O (1:1), 0-24°, 24-36 h	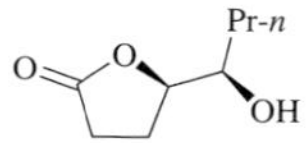 (80), 98	109
	$(DHQ)_2PHAL$, $K_2OsO_2(OH)_4$, $K_3Fe(CN)_6$, K_2CO_3, $MeSO_2NH_2$, t-BuOH:H_2O (1:1), 0-24°, 24-36 h	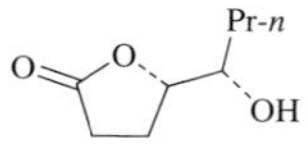 (78), 96	109
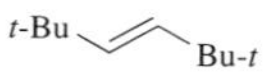	1. DHQD-OAc, OsO_4, toluene, rt, 12 h 2. LAH, ether	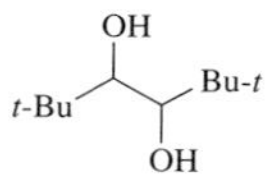 (86), 62	8
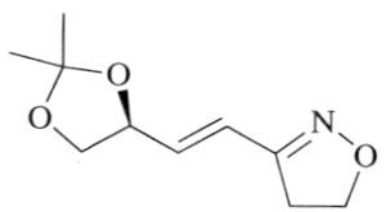	DHQD-MEQ, OsO_4, NMO, THF:H_2O (9:1), 20°	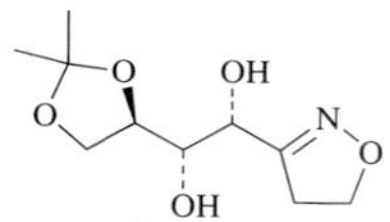 **I** (52), 76 de	364
	DHQD-MEQ, OsO_4, toluene, 20°	**I** (48), 56 de	364
	$(DHQD)_2PHAL$, $K_2OsO_2(OH)_4$, $K_3Fe(CN)_6$, K_2CO_3, $MeSO_2NH_2$, t-BuOH:H_2O (1:1), 0°, 24 h	**I** (53), 92 de	364
	$(DHQ)_2PHAL$, $K_2OsO_2(OH)_4$, $K_3Fe(CN)_6$, K_2CO_3, $MeSO_2NH_2$, t-BuOH:H_2O (1:1), 0°, 24 h	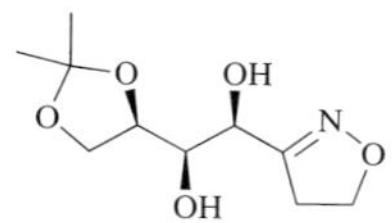 (62), 76 de	364

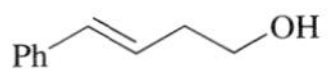	(DHQD)$_2$PHAL, $K_2OsO_2(OH)_4$, $K_3Fe(CN)_6$, K_2CO_3, $MeSO_2NH_2$, t-BuOH:H_2O (1:1), 0°	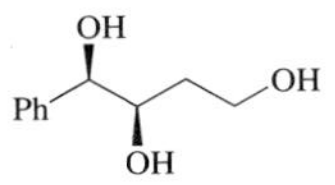	(—), 98	98
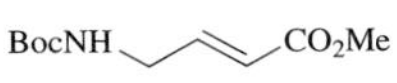	**35**, OsO_4, CH_2Cl_2, –90°, 2 h	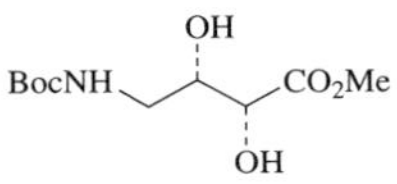	(91), 97	11
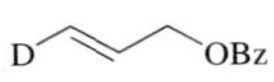	AD-Mix β, t-BuOH:H_2O (1:1), 0°, 4 h	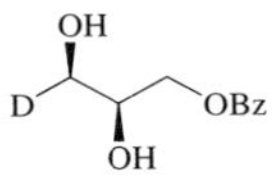	(93), 74	78
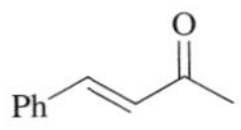	(DHQD)$_2$PHAL, $K_2OsO_2(OH)_4$, $K_3Fe(CN)_6$, $NaHCO_3$-K_2CO_3, $MeSO_2NH_2$, t-BuOH:H_2O (1:1), 0°, 24 h	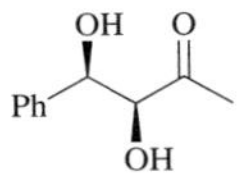	(69), 92	155
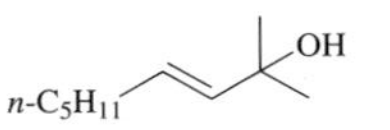	(DHQ)$_2$PHAL, $K_2OsO_2(OH)_4$, $K_3Fe(CN)_6$, K_2CO_3, $MeSO_2NH_2$, t-BuOH:H_2O (1:1), 0°, 6-36 h	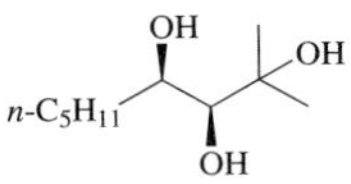	(87), 90	116
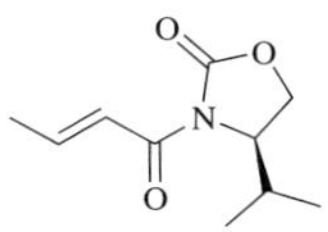	(DHQD)$_2$PHAL, $K_2OsO_2(OH)_4$, $K_3Fe(CN)_6$, K_2CO_3, $MeSO_2NH_2$, t-BuOH:H_2O (1:1), 0°	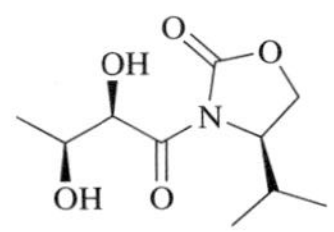	(86), 90 de	365
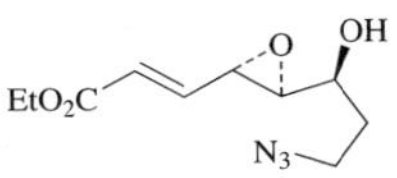	(DHQ)$_2$PHAL, $K_2OsO_2(OH)_4$, $K_3Fe(CN)_6$, K_2CO_3, $MeSO_2NH_2$, t-BuOH:H_2O (1:1), rt, 12 h	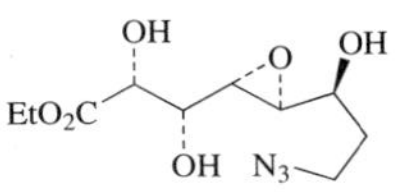	(—), 82 de	366

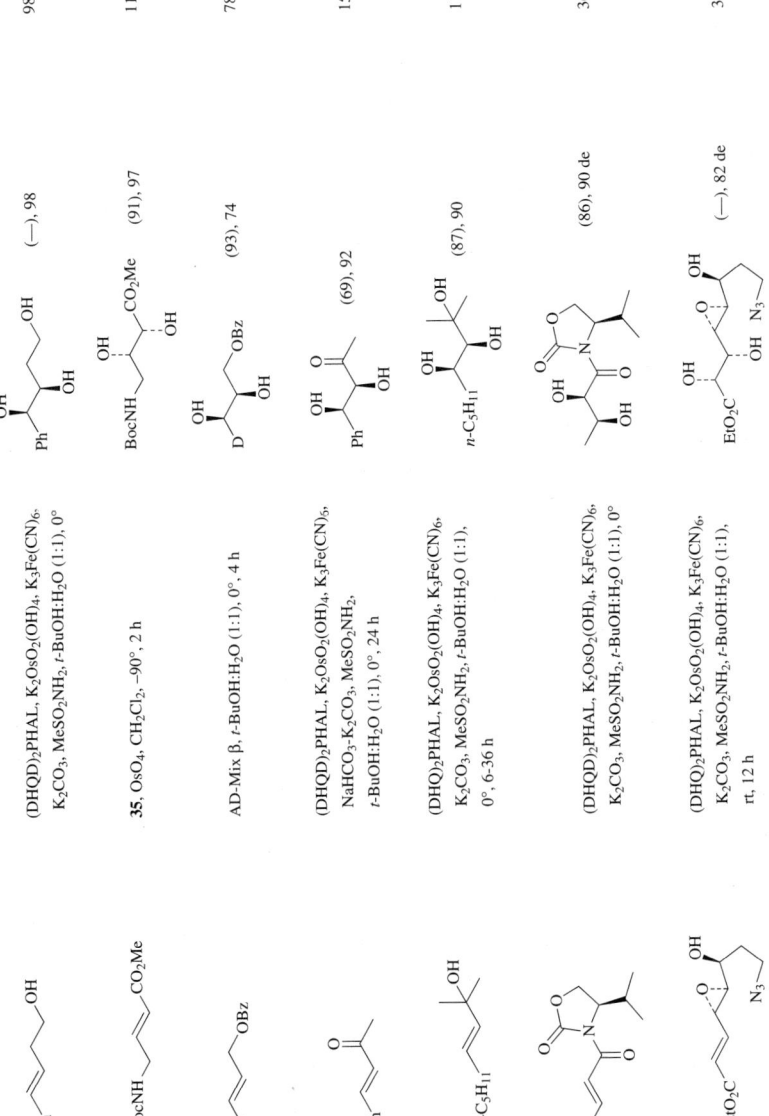

Substrate	Conditions	Product(s) and Yield(s) (%), % ee	Refs.
C_{10}			
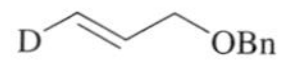	AD-Mix β, t-BuOH-H_2O, 0°, 12 h	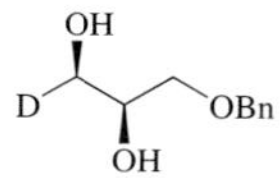 (94), 56	367
	AD-Mix α, t-BuOH-H_2O, 0°, 12 h	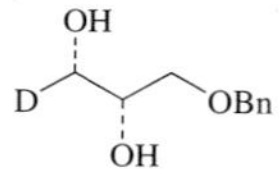 (55), 60	367
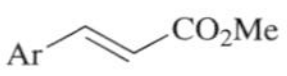	AD-Mix α, t-BuOH:H_2O (1:1), rt, 18 h	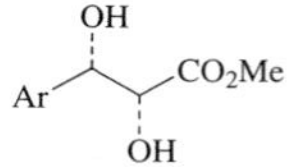	104
Ar			
4-FC_6H_4		(72), >95	
3-FC_6H_4		(67), >95	
2,4-$F_2C_6H_3$		(71), >95	
	AD-Mix α, $MeSO_2NH_2$, t-BuOH:H_2O (1:1), 0°, 20 h	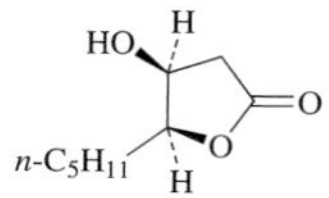 (82), 96	368
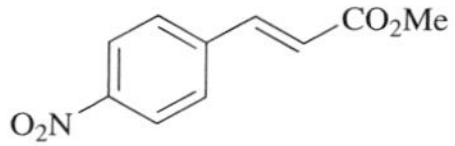	DHQD-p-chlorobenzoate, OsO_4, $K_3Fe(CN)_6$, K_2CO_3, t-BuOH:H_2O (1:1)	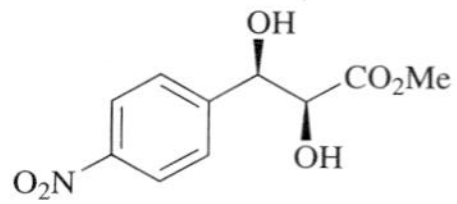 (89), 96	369
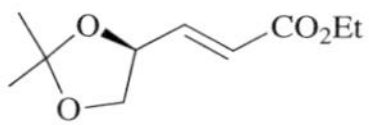	AD-Mix α, $MeSO_2NH_2$, t-BuOH:H_2O (1:1)	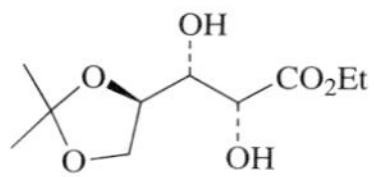 (82-90), >94 de	370
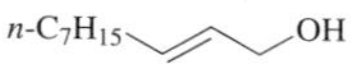	$(DHQ)_2PHAL$, $K_2OsO_2(OH)_4$, $K_3Fe(CN)_6$, $MeSO_2NH_2$, K_2CO_3, t-BuOH:H_2O (1:1), 0°	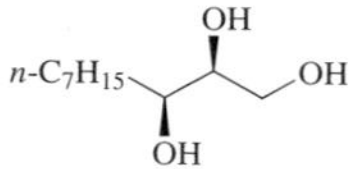 (84-88), 96	350

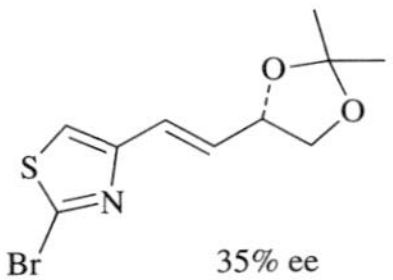 35% ee	AD-Mix β, $(DHQD)_2PHAL$, *t*-BuOH:H_2O (1:1), 0°, 5 days	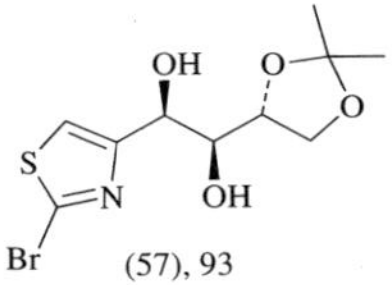 (57), 93 + 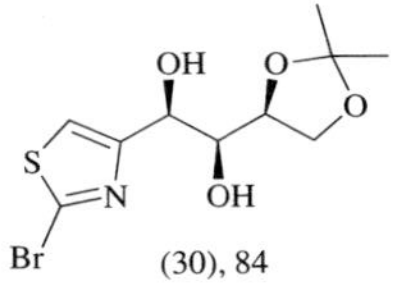 (30), 84	371
C_{11}			
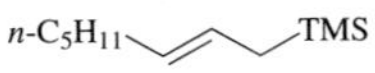	$(DHQ)_2PHAL$, $K_2OsO_2(OH)_4$, $K_3Fe(CN)_6$, K_2CO_3, $MeSO_2NH_2$, *t*-BuOH:H_2O (1:1), rt, 24 h	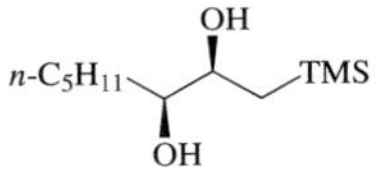 (77), 66	115
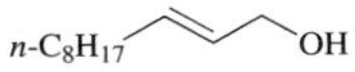	$(DHQD)_2PHAL$, $K_2OsO_2(OH)_4$, $K_3Fe(CN)_6$, K_2CO_3, $MeSO_2NH_2$, *t*-BuOH:H_2O (1:1), 0°	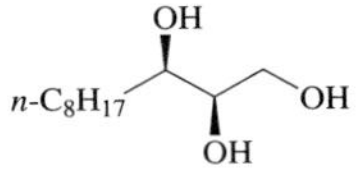 (—), 93	98
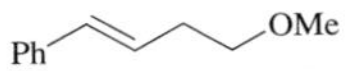	$(DHQD)_2PHAL$, $K_2OsO_2(OH)_4$, $K_3Fe(CN)_6$, K_2CO_3, $MeSO_2NH_2$, *t*-BuOH:H_2O (1:1), 0°	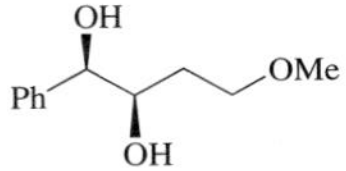 (—), 99	98
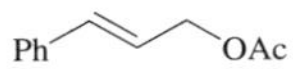	DHQD-*p*-chlorobenzoate, NMO, OsO_4, acetone-H_2O, 0°, 7 h	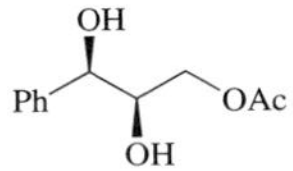 **I** (80-95), 76	18, 70
	DHQD-*p*-chlorobenzoate, OsO_4, NMO, acetone:H_2O (10:1), 0°, 17 h (slow addition of olefin)	**I** (80-95), 79	112, 70
	DHQD-*p*-chlorobenzoate, OsO_4, NMO, acetone:H_2O (10:1, 0.15 M), 0°	**I** (—), 82	112
	DHQD-*p*-chlorobenzoate, $K_2OsO_2(OH)_4$, $K_3Fe(CN)_6$, K_2CO_3, *t*-BuOH:H_2O (1:1), rt, 24 h	**I** (85-95), 91	70
	12a, NMO, OsO_4, acetone:H_2O (10:1), 40°, 16 h	**I** (70), 30	192

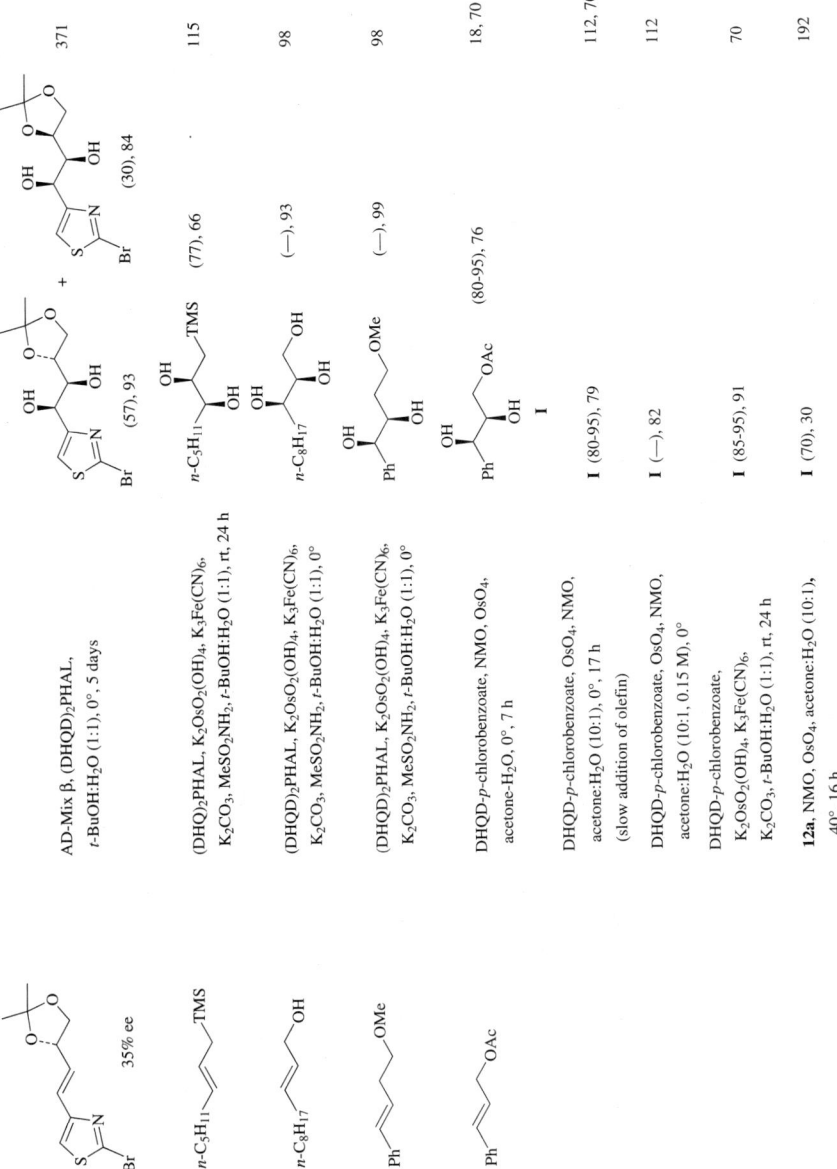

TABLE 3. REACTIONS OF TRANS 1,2-DISUBSTITUTED ALKENES (*Continued*)

Substrate	Conditions	Product(s) and Yield(s) (%), % ee	Refs.
C_{11}			
	DHQD-PHN, $K_2OsO_2(OH)_4$, $K_3Fe(CN)_6$, K_2CO_3, *t*-BuOH:H_2O (1:1), 0°, 15-30 h	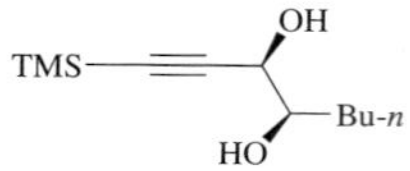 (56), 93	219
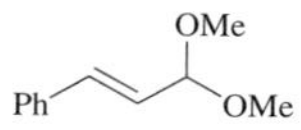	DHQD-*p*-chlorobenzoate, OsO_4, NMO, acetone:H_2O (10:1), 0°, 19 h (slow addition of olefin)	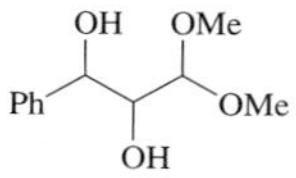 **I** (80-95), 75	112
	DHQD-*p*-chlorobenzoate, OsO_4, NMO, $Et_4NOAc{\bullet}4H_2O$, acetone:H_2O (10:1), 0°, 19 h (slow addition of olefin)	**I** (80-95), 83	112
	DHQD-*p*-chlorobenzoate, OsO_4, NMO, acetone:H_2O (10:1, 0.15 M), 0°	**I** (—), 82	112
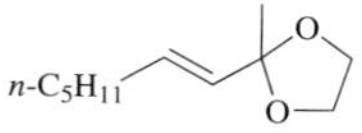	DHQD-*p*-chlorobenzoate, OsO_4, NMO, acetone:H_2O (10:1), 0°, 19 h (slow addition of olefin)	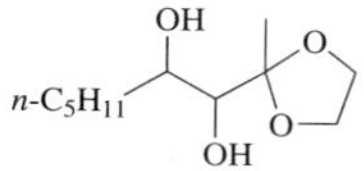 **I** (80-95), 46	112
	DHQD-*p*-chlorobenzoate, OsO_4, NMO, $Et_4NOAc{\bullet}4H_2O$, acetone:H_2O (10:1), 0°, 19 h (slow addition of olefin)	**I** (80-95), 50	112
	DHQD-*p*-chlorobenzoate, NMO, OsO_4, toluene (0.15 M), 0°	**I** (—), 50	112
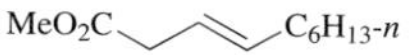	$(DHQD)_2PHAL$, $K_2OsO_2(OH)_4$, $K_3Fe(CN)_6$, K_2CO_3, $MeSO_2NH_2$, *t*-BuOH:H_2O (1:1), 0°, 24-36 h	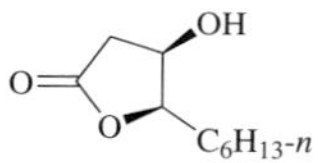 (80), 97	109
	$(DHQ)_2PHAL$, $K_2OsO_2(OH)_4$, $K_3Fe(CN)_6$, K_2CO_3, $MeSO_2NH_2$, *t*-BuOH:H_2O (1:1), 0°, 24-36 h	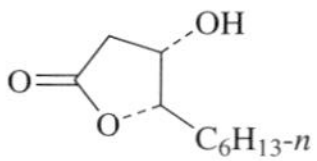 (85), 95	109

	(DHQD)$_2$PHAL, $K_2OsO_2(OH)_4$, $K_3Fe(CN)_6$, K_2CO_3, $MeSO_2NH_2$, t-BuOH:H_2O (1:1), 0°, 24-36 h	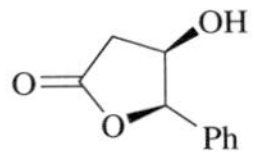	(84), 99	109
	(DHQ)$_2$PHAL, $K_2OsO_2(OH)_4$, $K_3Fe(CN)_6$, K_2CO_3, $MeSO_2NH_2$, t-BuOH:H_2O (1:1), 0°, 24-36 h	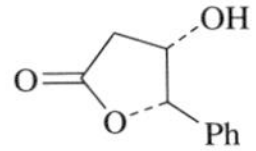	(85), >99	109
	42b, OsO_4, NMO, acetone:H_2O (10:1), 0°, 7 h	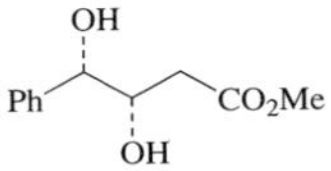 **I**	(81), 43	161
	42b, OsO_4, $K_3Fe(CN)_6$, K_2CO_3, t-BuOH:H_2O (1:1), 0°, 24 h; rt, 24 h	**I** (40), 75		161
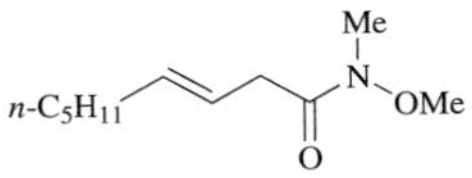	(DHQD)$_2$PHAL, $K_2OsO_2(OH)_4$, $K_3Fe(CN)_6$, K_2CO_3, $MeSO_2NH_2$, t-BuOH:H_2O (1:1), 0°	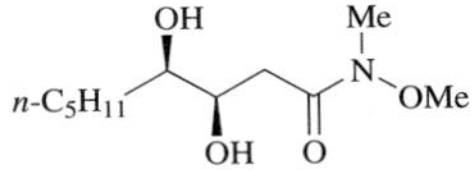	(83), 96	111
	(DHQ)$_2$PHAL, $K_2OsO_2(OH)_4$, $K_3Fe(CN)_6$, K_2CO_3, $MeSO_2NH_2$, t-BuOH:H_2O (1:1), 0°	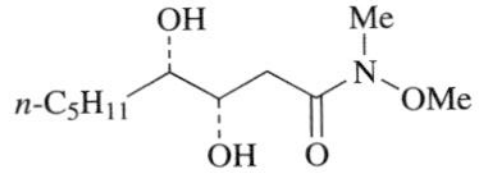	(83), 94	372, 111
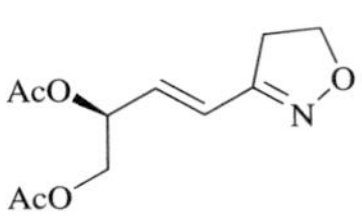	1. (DHQ)$_2$PHAL, $K_2OsO_2(OH)_4$, $K_3Fe(CN)_6$, K_2CO_3, $MeSO_2NH_2$, t-BuOH:H_2O (1:1), rt, 26 h 2. Ac_2O, pyridine, DMAP, CH_2Cl_2, 0-5°, 2 h	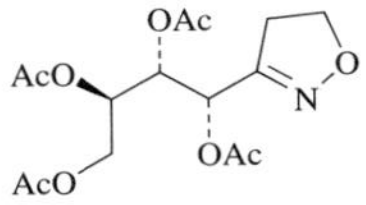	(85), 90	340
	1. (DHQD)$_2$PHAL, $K_2OsO_2(OH)_4$, $K_3Fe(CN)_6$, K_2CO_3, $MeSO_2NH_2$, t-BuOH:H_2O (1:1), rt, 26 h 2. Ac_2O, pyridine, DMAP, CH_2Cl_2, 0-5°, 2 h	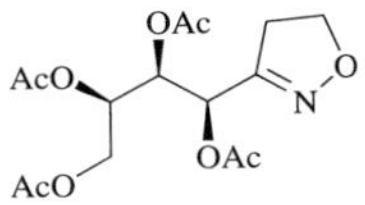	(82), 96	340

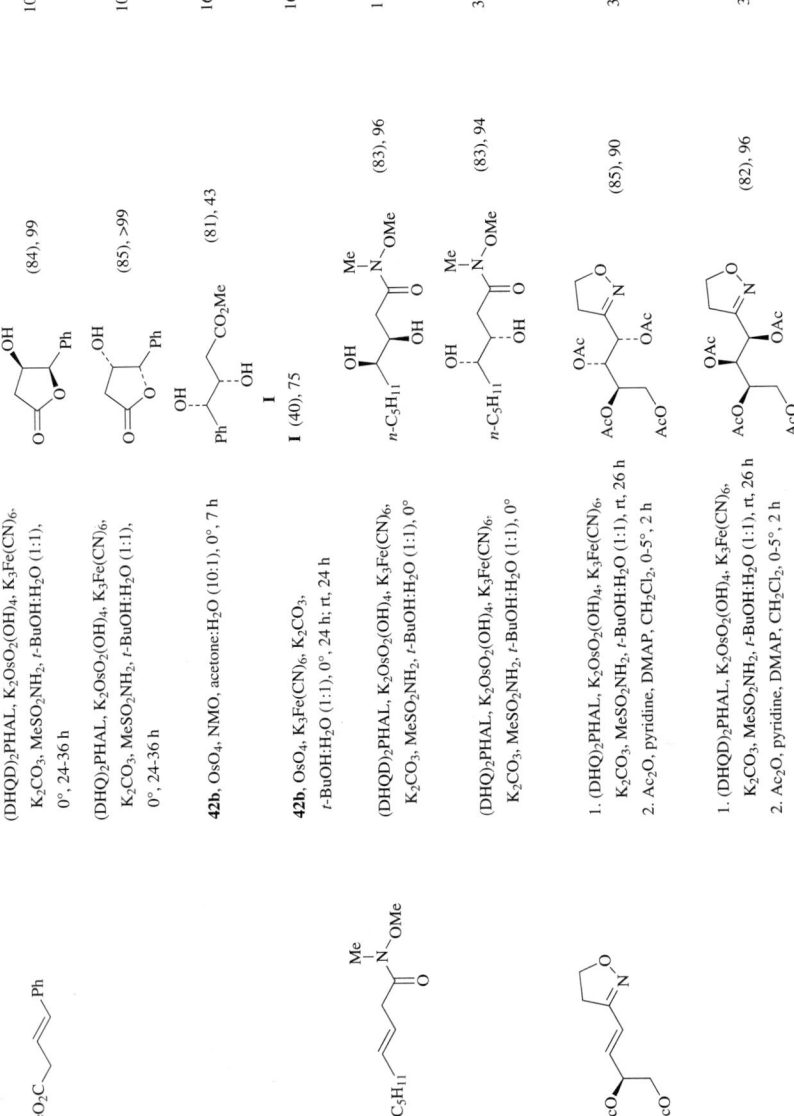

TABLE 3. REACTIONS OF TRANS 1,2-DISUBSTITUTED ALKENES (*Continued*)

Substrate	Conditions	Product(s) and Yield(s) (%), % ee	Refs.
C_{11}			
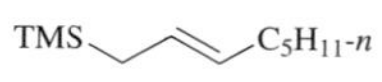	Ligand, $K_2OsO_2(OH)_4$, $K_3Fe(CN)_6$, K_2CO_3, $MeSO_2NH_2$, t-BuOH:H_2O (1:1), rt, 24 h	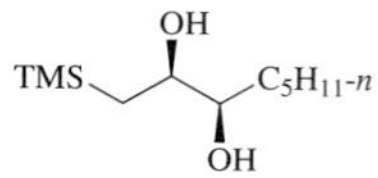	115
	Ligand		
	DHQD-p-chlorobenzoate	(93), 71	
	DHQD-PHN	(63), 91	
	$(DHQD)_2$PHAL	(71), 72	
	DHQ-PHN, $K_2OsO_2(OH)_4$, $K_3Fe(CN)_6$, K_2CO_3, $MeSO_2NH_2$, t-BuOH:H_2O (1:1), rt, 24 h	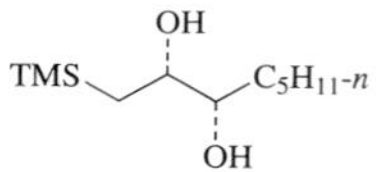 (75), 87	115
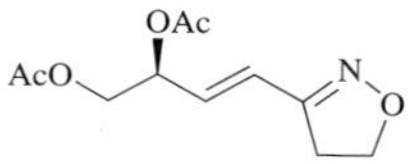	Ligand, $K_2OsO_2(OH)_4$, $K_3Fe(CN)_6$, K_2CO_3, $MeSO_2NH_2$, t-BuOH:H_2O (1:1), 0°, 24 h	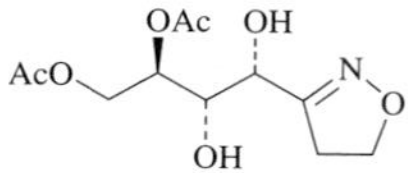	364
	Ligand		
	$(DHQD)_2$PHAL	(82), 96 de	
	DHQ-MEQ	(66), 4 de	
	$(DHQ)_2$PHAL, $K_2OsO_2(OH)_4$, $K_3Fe(CN)_6$, K_2CO_3, $MeSO_2NH_2$, t-BuOH:H_2O (1:1), 0°, 24 h	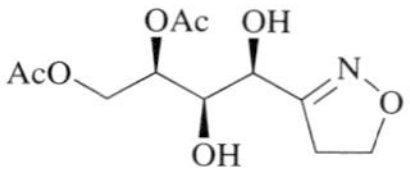 (85), 90 de	364
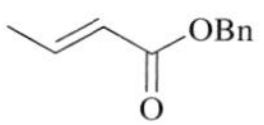	$(DHQD)_2$PHAL, $K_2OsO_2(OH)_4$, $K_3Fe(CN)_6$, K_2CO_3, $MeSO_2NH_2$, t-BuOH:H_2O (1:1), 0°	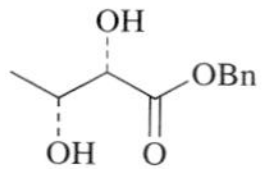 (79), 98	365, 106
	$(DHQ)_2$PHAL, $K_2OsO_2(OH)_4$, $K_3Fe(CN)_6$, K_2CO_3, $MeSO_2NH_2$, t-BuOH:H_2O (1:1), 0°	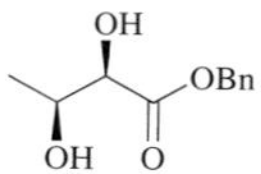 (—), 98	365, 106

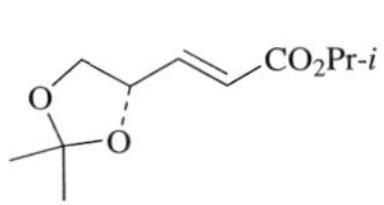	Ligand, OsO_4, $K_3Fe(CN)_6$, K_2CO_3, $MeSO_2NH_2$, t-BuOH:H_2O (1:1), 0°, 24 h	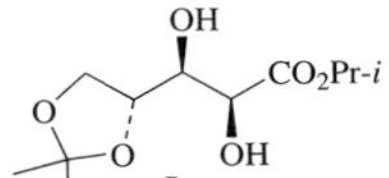 I	373
	Ligand: DHQD-*p*-chlorobenzoate	(87), 82 de	
	Ligand: $(DHQD)_2PYR(OMe)_3$	(89), 84 de	
	Ligand, $K_2OsO_2(OH)_4$, $K_3Fe(CN)_6$, K_2CO_3, $MeSO_2NH_2$, t-BuOH:H_2O (1:1), 0°, 24 h	I	373
	Ligand: $(DHQD)_2PHAL$	I: (84), 95 de	
	Ligand: $(DHQD)_2PYR$	I: (90), 75 de	
	Ligand, $K_2OsO_2(OH)_4$, $K_3Fe(CN)_6$, K_2CO_3, $MeSO_2NH_2$, t-BuOH:H_2O (1:1), 0°, 24 h	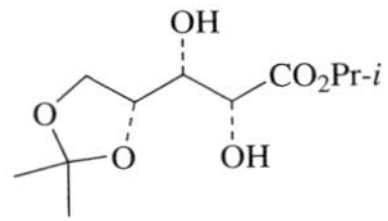	373
	Ligand: $(DHQ)_2PHAL$	(95), 40 de	
	Ligand: DHQ-*p*-chlorobenzoate	(85), 0 de	
	Ligand: $(DHQ)_2PYR$	(86), 61 de	
	Ligand: $(DHQ)_2PYR(OMe)_3$	(90), 75 de	
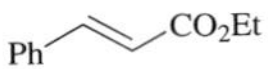	$(DHQD)_2PHAL$, $K_2OsO_2(OH)_4$, $K_3Fe(CN)_6$, K_2CO_3, $MeSO_2NH_2$, t-BuOH:H_2O (1:1), rt, 6-24 h	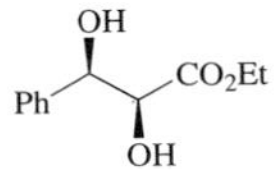 (80-98), 97	19, 223, 374
	$(DHQ)_2PHAL$, $K_2OsO_2(OH)_4$, $K_3Fe(CN)_6$, K_2CO_3, $MeSO_2NH_2$, t-BuOH:H_2O (1:1), rt, 6-24 h	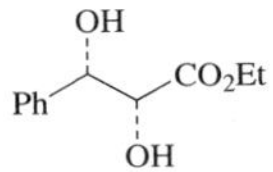 (80-98), 95	19, 223, 374

TABLE 3. REACTIONS OF TRANS 1,2-DISUBSTITUTED ALKENES (*Continued*)

Substrate	Conditions	Product(s) and Yield(s) (%), % ee	Refs.
C_{11}			
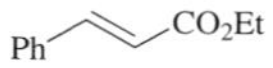	AD-Mix α, $MeSO_2NH_2$, t-BuOH:H_2O (1:1)	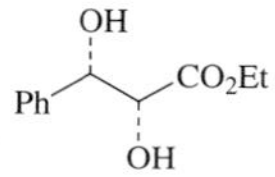 (82-90), 99	370, 375
	1. 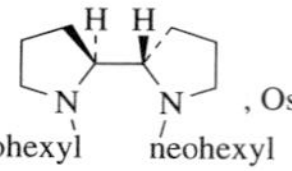 , OsO_4, toluene, –78°, 18 h 2. $NaHSO_3$	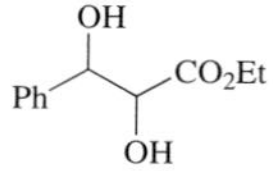 (99), 97	13
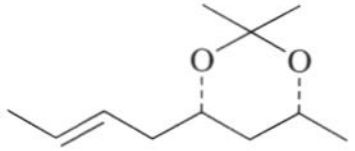	AD-Mix α, t-BuOH:H_2O (1:1), 0°, 3 days	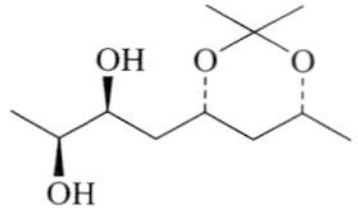 (58), 36 de	336
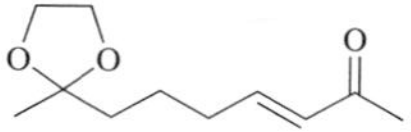	AD-Mix β, $MeSO_2NH_2$, $NaHCO_3$, t-BuOH:H_2O (1:1), 0°, 36 h	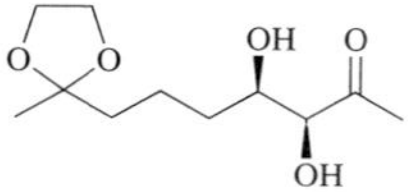 (82), 97	376
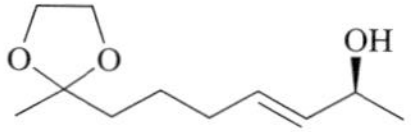	AD-Mix β, $MeSO_2NH_2$, t-BuOH:H_2O (1:1), 0°, 55 h	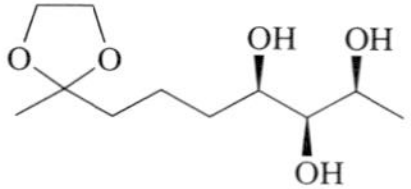 (100), —	376
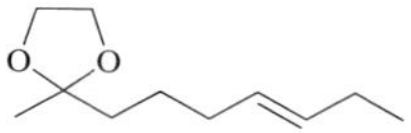	$(DHQD)_2PHAL$, $K_2OsO_2(OH)_4$, $K_3Fe(CN)_6$, K_2CO_3, $MeSO_2NH_2$, t-BuOH:H_2O (1:1), 0°, 24 h	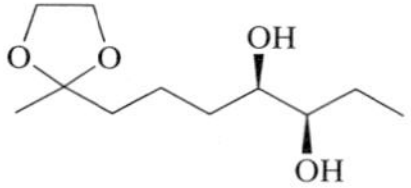 (96), 95	353
	$(DHQ)_2PHAL$, $K_2OsO_2(OH)_4$, $K_3Fe(CN)_6$, K_2CO_3, $MeSO_2NH_2$, t-BuOH:H_2O (1:1), 0°, 24 h	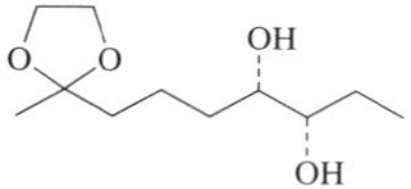 (97), 95	353

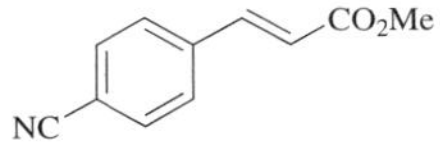	(DHQD)$_2$PYDZ, $K_2OsO_2(OH)_4$, $K_3Fe(CN)_6$, K_2CO_3, $MeSO_2NH_2$, *t*-BuOH:H_2O (1:1), 0°, 96 h	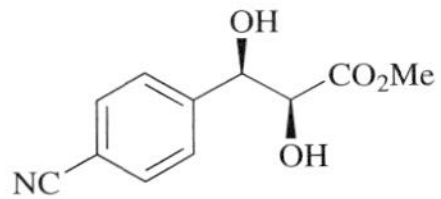 (83), >98	49
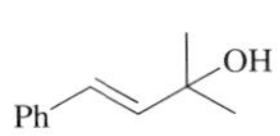	(DHQD)$_2$PHAL, $K_2OsO_2(OH)_4$, $K_3Fe(CN)_6$, K_2CO_3, $MeSO_2NH_2$, *t*-BuOH:H_2O (1:1), 0°, 6-36 h	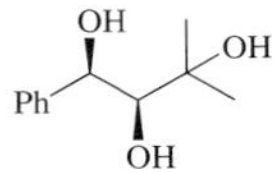 (83), 77	116
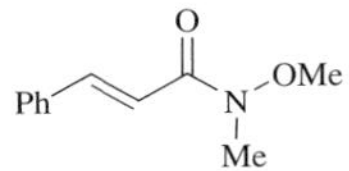	(DHQD)$_2$PHAL, $K_2OsO_2(OH)_4$, $K_3Fe(CN)_6$, K_2CO_3, $MeSO_2NH_2$, *t*-BuOH:H_2O (1:1), rt	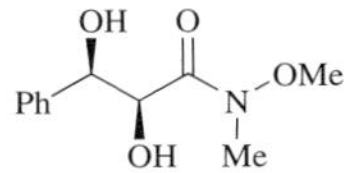 (92), 96	111
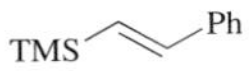	(DHQD)$_2$PHAL, $K_2OsO_2(OH)_4$, $K_3Fe(CN)_5$, K_2CO_3, $MeSO_2NH_2$, *t*-BuOH:H_2O (1:1), 0°, 28 h	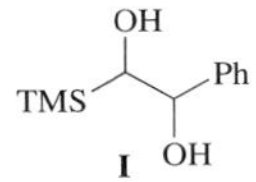 **I** (83), 97	113
	(DHQ)$_2$PHAL, $K_2OsO_2(OH)_4$, $K_3Fe(CN)_6$, K_2CO_3, $MeSO_2NH_2$, *t*-BuOH:H_2O (1:1), 0°, 28 h	**I** (82), 97	113
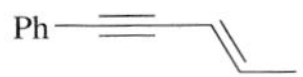	DHQD-PHN, $K_2OsO_2(OH)_4$, $K_3Fe(CN)_6$, K_2CO_3, *t*-BuOH:H_2O (1:1), 0°, 15-30 h	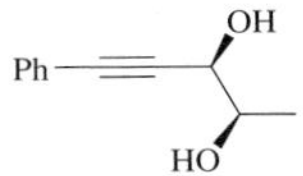 (98), 73	219
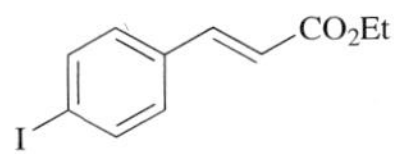	AD-Mix α, $MeSO_2NH_2$, *t*-BuOH:H_2O (1:1), rt, 20 h	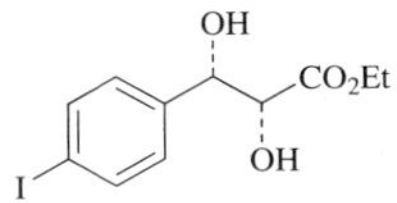 (90), >99	377
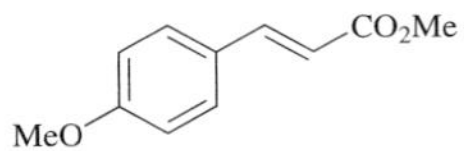	DHQ-OAc, OsO_4, NMO, acetone-H_2O	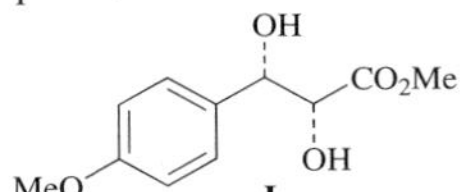 **I** (95), >88	378
	AD-Mix α, $MeSO_2NH_2$, *t*-BuOH:H_2O (1:1), rt, 15 h	**I** (85), >99	362

TABLE 3. REACTIONS OF TRANS 1,2-DISUBSTITUTED ALKENES (*Continued*)

Substrate	Conditions	Product(s) and Yield(s) (%), % ee	Refs.
C_{11}			
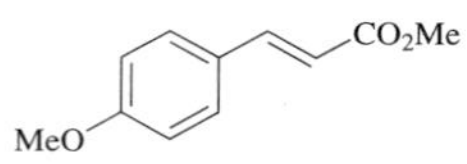	$(DHQD)_2PHAL$, $K_2OsO_2(OH)_4$, $K_3Fe(CN)_6$, K_2CO_3, t-BuOH:H_2O (1:1), 0°, 18 h	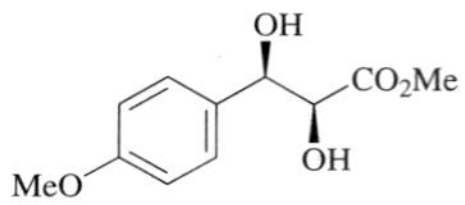 (72), >98	379
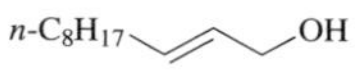	$(DHQ)_2PHAL$, $K_2OsO_2(OH)_4$, $K_3Fe(CN)_6$, K_2CO_3, $MeSO_2NH_2$, t-BuOH:H_2O (1:1), 0°	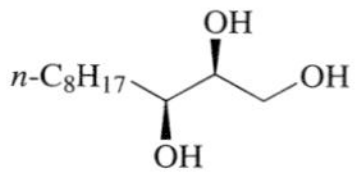 (84-88), 96	350
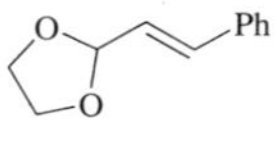	DHQD-OAc, OsO_4, toluene, –20°	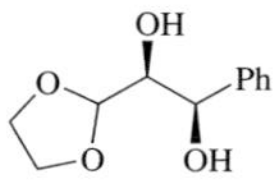 (60), >90	380
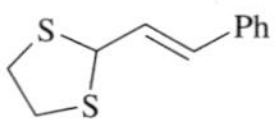	DHQD-OAc, OsO_4, toluene, –20°	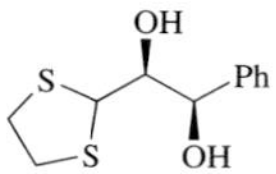 (63), >90	380
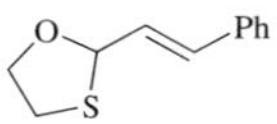	DHQD-OAc, OsO_4, toluene, –20°	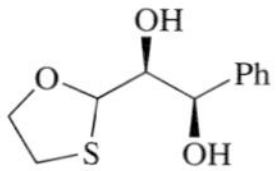 (56), >90, — de	380
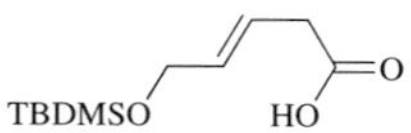	$(DHQD)_2PHAL$, $K_2OsO_2(OH)_4$, $K_3Fe(CN)_6$, K_2CO_3, $MeSO_2NH_2$, t-BuOH:H_2O (1:1), 0°, 36 h	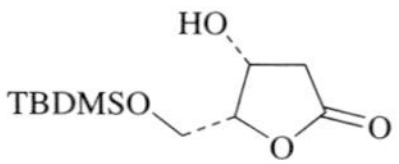 (88), 92	108
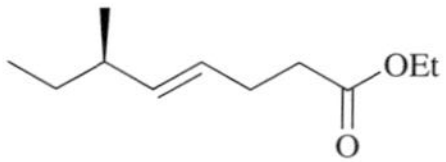	AD-Mix α, t-BuOH:H_2O (1:1)	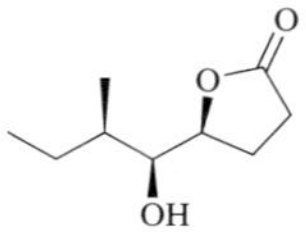 (75), 92 de	381

	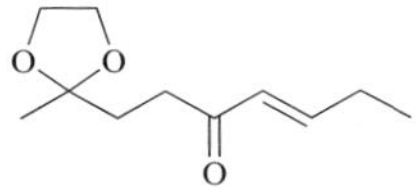	AD-Mix β, $MeSO_2NH_2$, $NaHCO_3$, *t*-BuOH:H_2O (1:1), 0-4°, 16 h	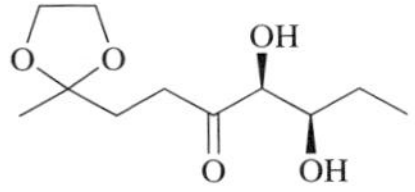	(75), 95	382
	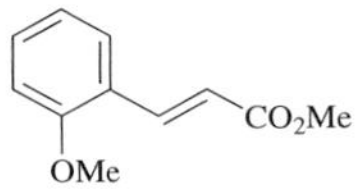	$(DHQD)_2$PHAL, $K_2OsO_2(OH)_4$, $K_3Fe(CN)_6$, K_2CO_3, *t*-BuOH:H_2O (1:1), 0°, 18 h	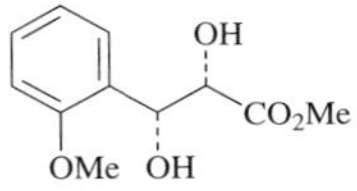	(63), >98	379
	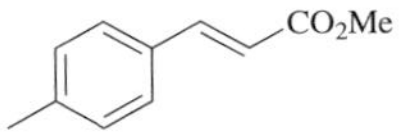	AD-Mix α, $MeSO_2NH_2$, *t*-BuOH:H_2O (1:1), rt, 15 h	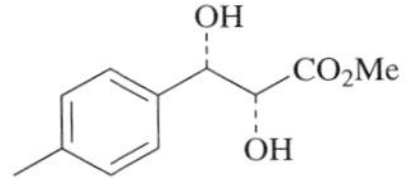	(—), >99	362
	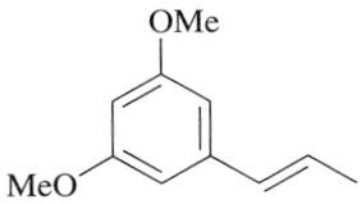	AD-Mix α, $MeSO_2NH_2$, *t*-BuOH-H_2O	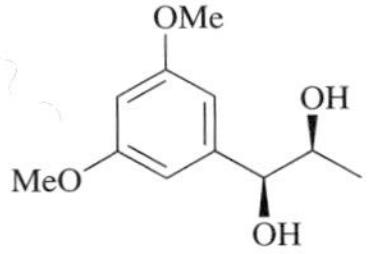	(99), >98	383
C_{12}	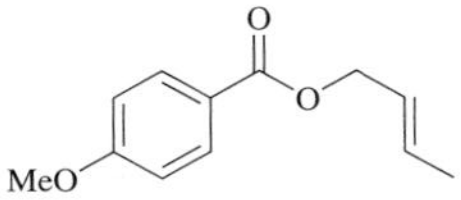	$(DHQD)_2$PYDZ, $K_2OsO_2(OH)_4$, $K_3Fe(CN)_6$, K_2CO_3, $MeSO_2NH_2$, *t*-BuOH:H_2O (1:1), 0°, 1.5-3.5 h	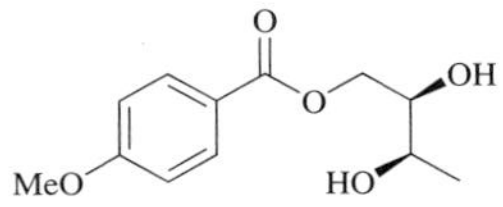	(96), >99	61
	 E:Z = 75:25	DHQD-PHN, $K_2OsO_2(OH)_4$, $K_3Fe(CN)_6$, K_2CO_3, *t*-BuOH:H_2O (1:1), 0°, 15-30 h	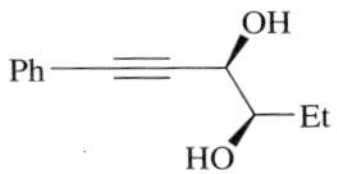	(94), 90	219
	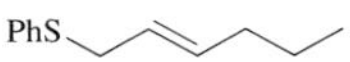	$(DHQD)_2$PHAL, $K_2OsO_2(OH)_4$, $K_3Fe(CN)_6$, K_2CO_3, *t*-BuOH:H_2O (1:1), 0°	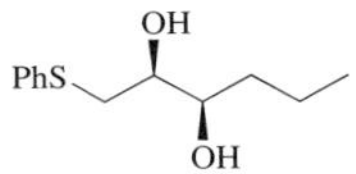	(68), 84	100
	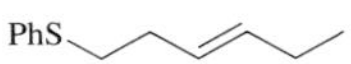	$(DHQD)_2$PHAL, $K_2OsO_2(OH)_4$, $K_3Fe(CN)_6$, K_2CO_3, *t*-BuOH:H_2O (1:1), 0°	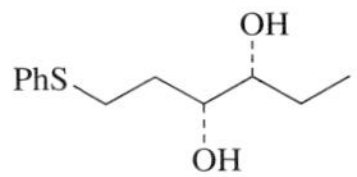	(74), 98	100

TABLE 3. REACTIONS OF TRANS 1,2-DISUBSTITUTED ALKENES (*Continued*)

Substrate	Conditions	Product(s) and Yield(s) (%), % ee	Refs.
C_{12}			
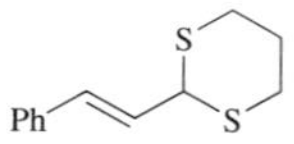	(DHQD)$_2$PHAL, $K_2OsO_2(OH)_4$, $K_3Fe(CN)_6$, K_2CO_3, *t*-BuOH:H_2O (1:1), 0°	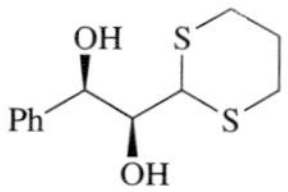 (78), 97	100
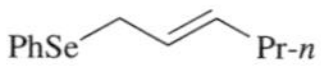	(DHQD)$_2$PHAL, $K_2OsO_2(OH)_4$, $K_3Fe(CN)_6$, K_2CO_3, $MeSO_2NH_2$, *t*-BuOH:H_2O (1:1), 0°, 24 h	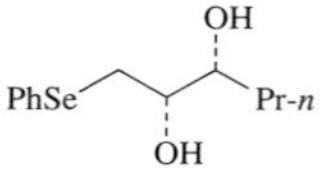 (7), 94 + 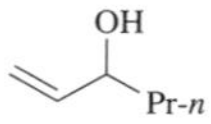 (80), 0	153
	(DHQ)$_2$PHAL, $K_2OsO_2(OH)_4$, $K_3Fe(CN)_6$, K_2CO_3, $MeSO_2NH_2$, *t*-BuOH:H_2O (1:1), 0°, 24 h	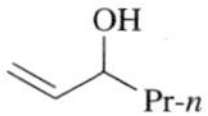 (83), 0	153
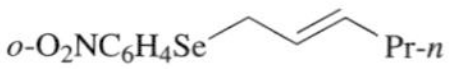	(DHQD)$_2$PHAL, $K_2OsO_2(OH)_4$, $K_3Fe(CN)_6$, K_2CO_3, $MeSO_2NH_2$, *t*-BuOH:H_2O (1:1), 0°, 24 h	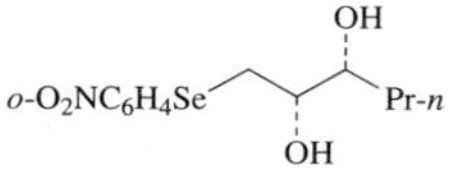 (78), 94	153
	(DHQ)$_2$PHAL, $K_2OsO_2(OH)_4$, $K_3Fe(CN)_6$, K_2CO_3, $MeSO_2NH_2$, *t*-BuOH:H_2O (1:1), 0°, 24 h	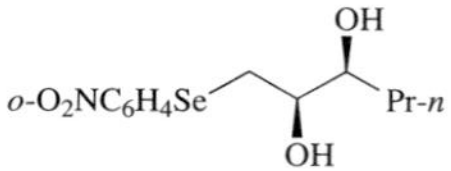 (53), 94	153
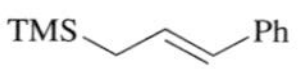	(DHQD)$_2$PHAL, $K_2OsO_2(OH)_4$, $K_3Fe(CN)_6$, K_2CO_3, $MeSO_2NH_2$, *t*-BuOH:H_2O (1:1), 0°, 28 h	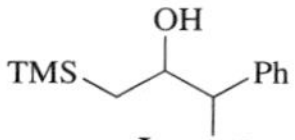 **I** (86), 95	113
	(DHQ)$_2$PHAL, $K_2OsO_2(OH)_4$, $K_3Fe(CN)_6$, K_2CO_3, $MeSO_2NH_2$, *t*-BuOH:H_2O (1:1), 0°, 28 h	**I** (86), 95[c]	113

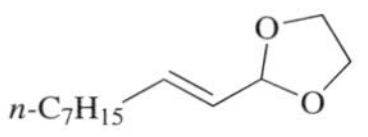	DHQD-*p*-chlorobenzoate, $K_2OsO_2(OH)_4$, NMO, acetone:H_2O (10:1), 0°, 21 h (slow addition of olefin)	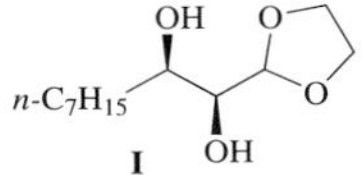 **I**	(80-95), 56[a]	112
	DHQD-*p*-chlorobenzoate, $K_2OsO_2(OH)_4$, NMO, $Et_4NOAc{\bullet}4H_2O$, acetone:H_2O (10:1), 0°, 21 h (slow addition of olefin)	**I** (80-95), 66		112
	DHQD-*p*-chlorobenzoate, $K_2OsO_2(OH)_4$, NMO, toluene, 0°	**I** (—), 83		112
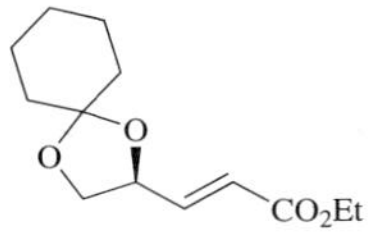	DHQD-OAc, OsO_4, toluene, –20 °, 18 h	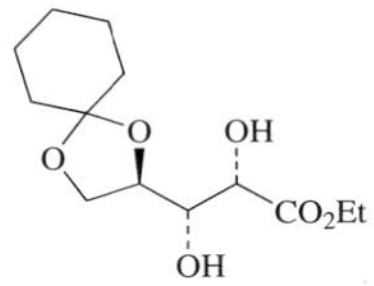	(80), 90 de	384
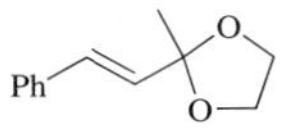	DHQD-*p*-chlorobenzoate, $K_2OsO_2(OH)_4$, NMO, acetone:H_2O (10:1), 0°, 11 h (slow addition of olefin)	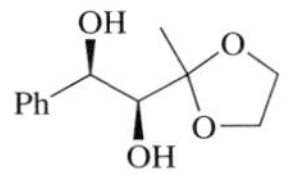	(80-95), 60[a]	112
	DHQD-*p*-chlorobenzoate, $K_2OsO_2(OH)_4$, NMO, $Et_4NOAc{\bullet}4H_2O$, acetone:H_2O (10:1), 0°, 11 h (slow addition of olefin)	**I** (80-95), 89		112
	DHQD-*p*-chlorobenzoate, $K_2OsO_2(OH)_4$, NMO, acetone:H_2O (10:1), 0°	**I** (—), 85		112
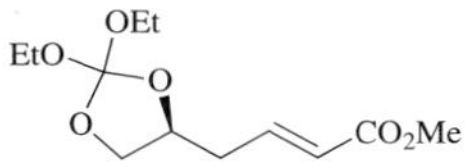	DHQD-*p*-chlorobenzoate, $K_2OsO_2(OH)_4$, NMO, acetone:H_2O (10:1), 0°, 49 h (slow addition of olefin)	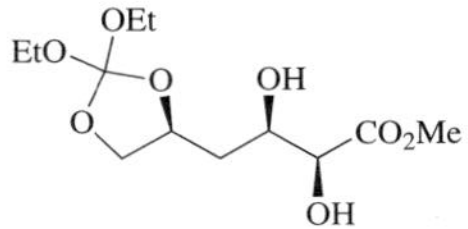	(80-95), 52 de[a]	112
	DHQD-*p*-chlorobenzoate, $K_2OsO_2(OH)_4$, NMO, acetone:H_2O (10:1), 0°	**I** (—), 54 de		112

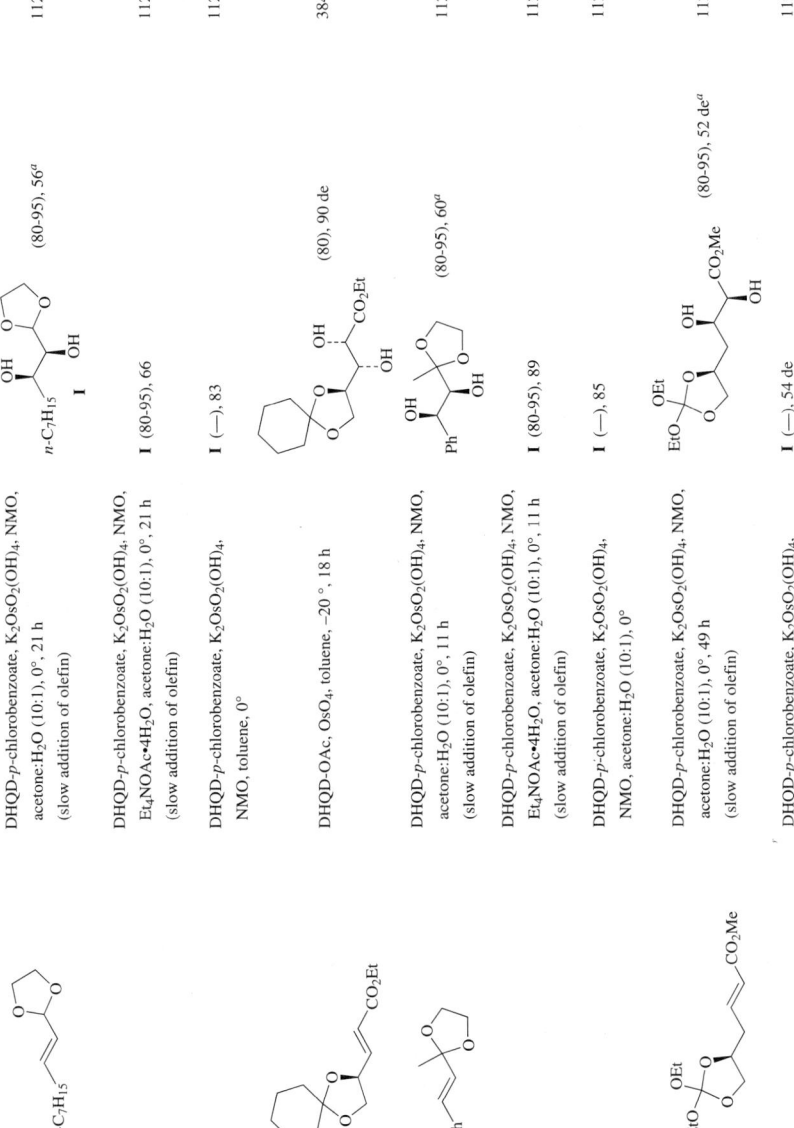

TABLE 3. REACTIONS OF TRANS 1,2-DISUBSTITUTED ALKENES (*Continued*)

Substrate	Conditions	Product(s) and Yield(s) (%), % ee	Refs.
C_{12}			
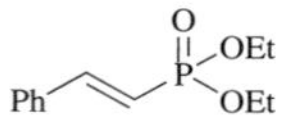	(DHQD)$_2$PHAL, $K_2OsO_2(OH)_4$, $K_3Fe(CN)_6$, K_2CO_3, *t*-BuOH:H_2O (1:1), ~20°	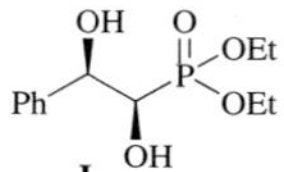 **I** (84), >97	118
	DHQD-*p*-chlorobenzoate, $K_2OsO_2(OH)_4$, $K_3Fe(CN)_6$, K_2CO_3, *t*-BuOH:H_2O (1:1), ~20°	**I** (82), >95	118
	(DHQD)$_2$PHAL, $K_2OsO_2(OH)_4$, $K_3Fe(CN)_6$, K_2CO_3, $MeSO_2NH_2$, *t*-BuOH:H_2O (1:1), 0-24°, 50 h	**I** (45), 92	345
	(DHQ)$_2$PHAL, $K_2OsO_2(OH)_4$, $K_3Fe(CN)_6$, K_2CO_3, $MeSO_2NH_2$, *t*-BuOH:H_2O (1:1), 0-24°, 50 h	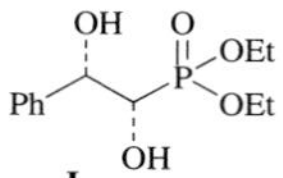 **I** (42), 91	345, 344
	(DHQ)$_2$PHAL, $K_2OsO_2(OH)_4$, $K_3Fe(CN)_6$, K_2CO_3, *t*-BuOH:H_2O (1:1), ~20°	**I** (82), >95	118
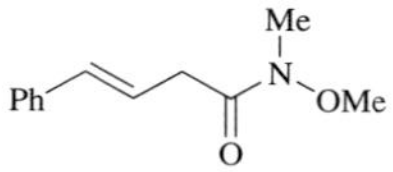	(DHQD)$_2$PHAL, $K_2OsO_2(OH)_4$, $K_3Fe(CN)_6$, K_2CO_3, $MeSO_2NH_2$, *t*-BuOH:H_2O (1:1), rt	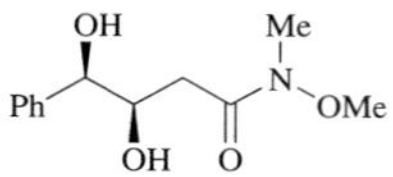 (81), 98	111
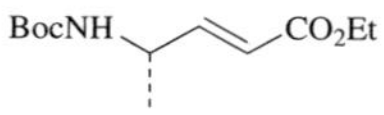	(DHQ)$_2$PHAL, $K_2OsO_2(OH)_4$, $K_3Fe(CN)_6$, K_2CO_3, *t*-BuOH:H_2O (1:1), rt, 1.5 days	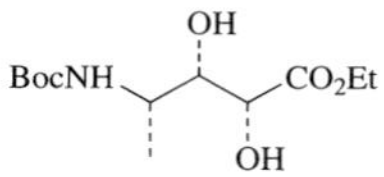 (95), 94 de	385
	(DHQD)$_2$PHAL, $K_2OsO_2(OH)_4$, $K_3Fe(CN)_6$, K_2CO_3, *t*-BuOH:H_2O (1:1), rt, 1.5 days	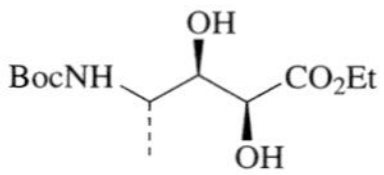 (90), 82 de	385

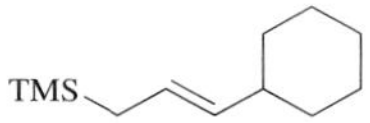	Ligand, $K_2OsO_2(OH)_4$, $K_3Fe(CN)_6$, K_2CO_3, $MeSO_2NH_2$, *t*-BuOH:H_2O (1:1), rt, 24 h	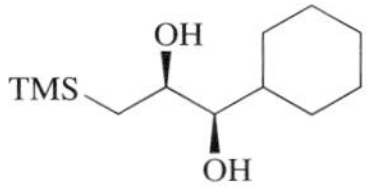		115
	Ligand			
	DHQD-*p*-chlorobenzoate	(77), 82		
	DHQD-PHN	(46), 95		
	$(DHQD)_2PHAL$	(—), 62		
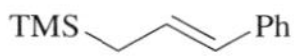	Ligand, $K_2OsO_2(OH)_4$, $K_3Fe(CN)_6$, K_2CO_3, $MeSO_2NH_2$, *t*-BuOH:H_2O (1:1), rt, 24 h	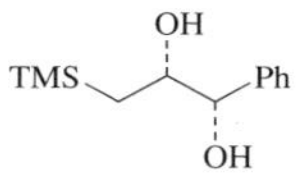		115
	Ligand			
	DHQD-*p*-chlorobenzoate	(95), 94		
	DHQD-PHN	(87), 94		
	$(DHQD)_2PHAL$	(98), 96		
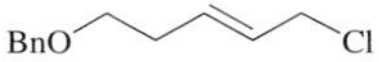	$(DHQD)_2PHAL$, $K_2OsO_2(OH)_4$, $K_3Fe(CN)_6$, K_2CO_3, $MeSO_2NH_2$, *t*-BuOH:H_2O (1:1), 0°, 24 h	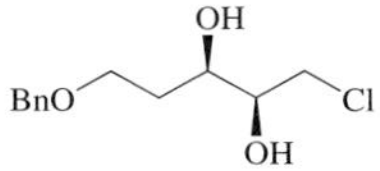	(95), 96	386
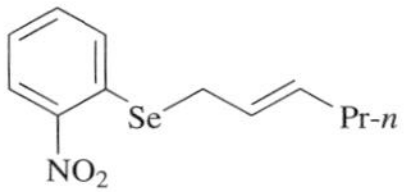	AD-Mix β, $MeSO_2NH_2$, *t*-BuOH:H_2O (1:1), 0°, 24 h	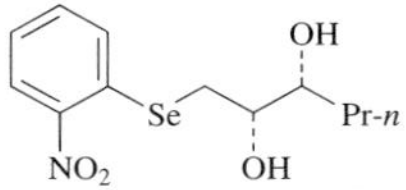	(78), 94	153
	AD-Mix α, $MeSO_2NH_2$, *t*-BuOH:H_2O (1:1), 0°, 24 h	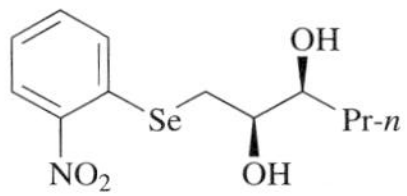	(53), 86	153

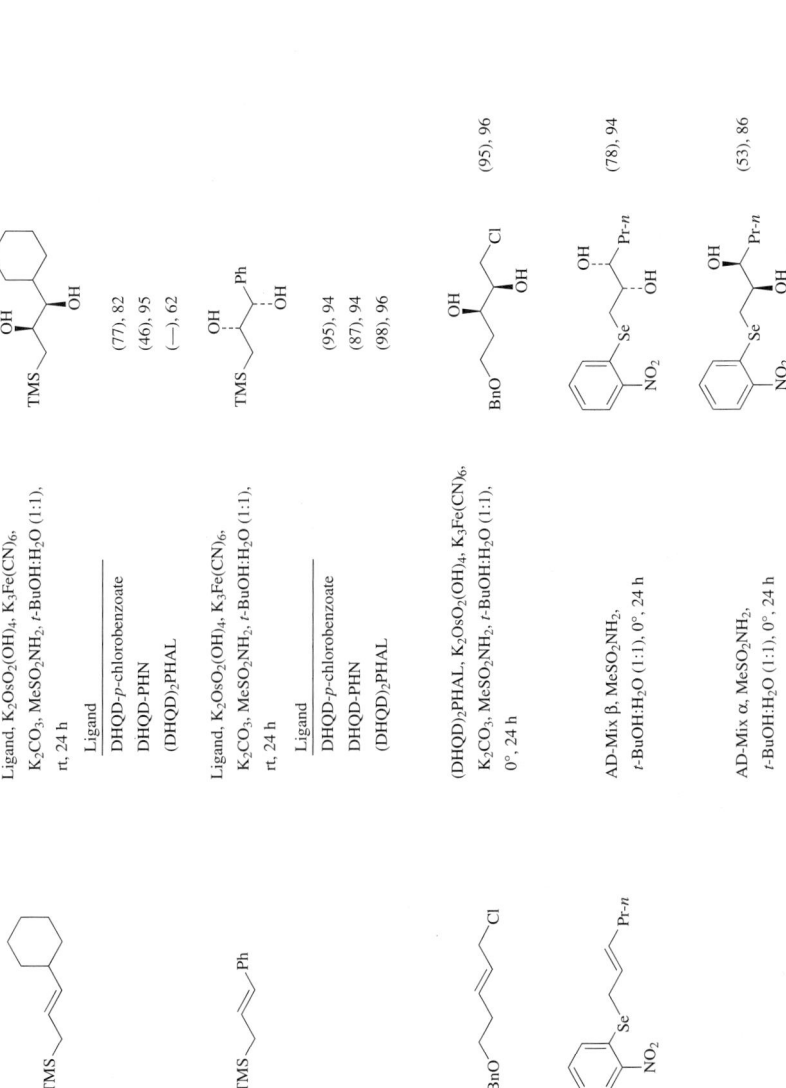

TABLE 3. REACTIONS OF TRANS 1,2-DISUBSTITUTED ALKENES (*Continued*)

Substrate	Conditions	Product(s) and Yield(s) (%), % ee	Refs.
C_{12}			
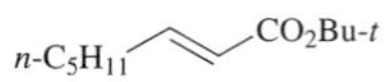	AD-Mix β, $MeSO_2NH_2$, t-BuOH:H_2O (1:1), 0-4°, 24 h	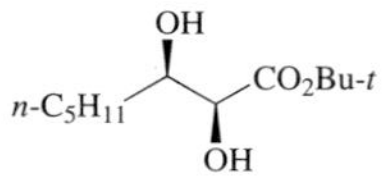 (89), 85	387
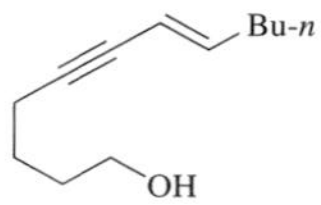	AD-Mix α, $MeSO_2NH_2$, t-BuOH:H_2O (1:1), 0°	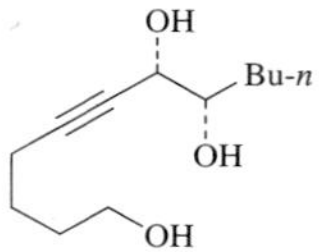 (96), 94	388
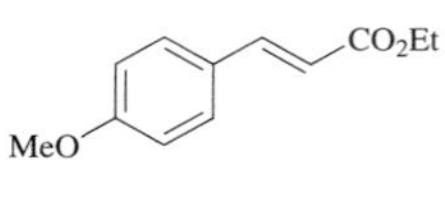	AD-Mix α, $MeSO_2NH_2$, t-BuOH:H_2O (1:1), rt	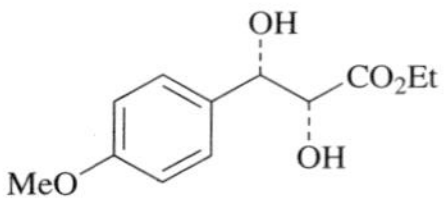 (93), 99	389
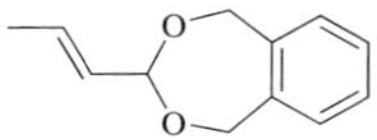	$(DHQD)_2PHAL$, $K_2OsO_2(OH)_4$, $K_3Fe(CN)_6$, K_2CO_3, t-BuOH:H_2O (1:1), rt, 48 h	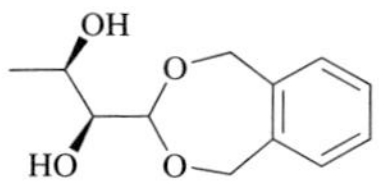 (96), 82	84
	$(DHQ)_2PHAL$, $K_2OsO_2(OH)_4$, $K_3Fe(CN)_6$, K_2CO_3, t-BuOH:H_2O (1:1), rt, 48 h	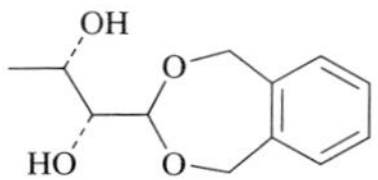 (100), 79	84
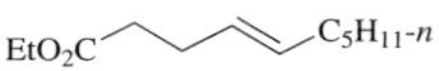	$(DHQD)_2PHAL$, $K_2OsO_2(OH)_4$, $K_3Fe(CN)_6$, K_2CO_3, $MeSO_2NH_2$, t-BuOH:H_2O (1:1), 0°, 24-36 h	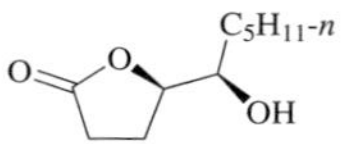 (91), 98	109
	$(DHQ)_2PHAL$, $K_2OsO_2(OH)_4$, $K_3Fe(CN)_6$, K_2CO_3, $MeSO_2NH_2$, t-BuOH:H_2O (1:1), 0°, 24-36 h	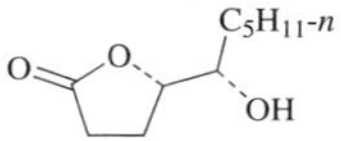 (91), 96	109

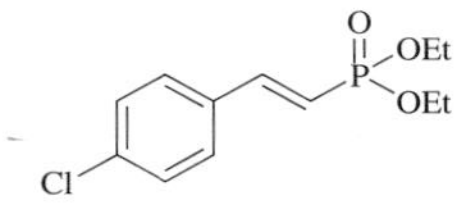	$(DHQ)_2PHAL$, $K_2OsO_2(OH)_4$, $K_3Fe(CN)_6$, K_2CO_3, $MeSO_2NH_2$, *t*-BuOH:H_2O (1:1), 0-24°, 50 h	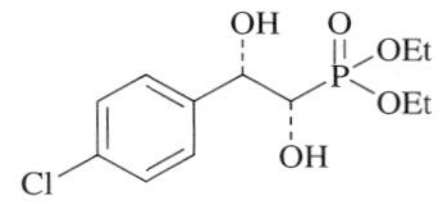 (65), 98	345
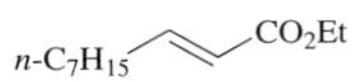	AD-Mix β, $MeSO_2NH_2$, *t*-BuOH:H_2O (1:1), 0-4°, 24 h	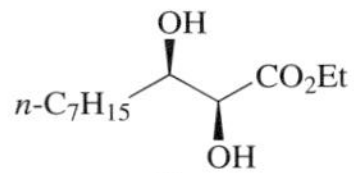 **I** (94), 98	390
	$(DHQD)_2PHAL$, $K_2OsO_2(OH)_4$, $K_3Fe(CN)_6$, K_2CO_3, $MeSO_2NH_2$, *t*-BuOH:H_2O (1:1), 0°, 5 h	**I** (70), —	391
	AD-Mix α, $MeSO_2NH_2$, *t*-BuOH:H_2O (1:1), 0-4°, 24 h	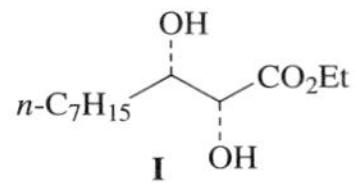 **I** (96), 98	390
	$(DHQ)_2PHAL$, $K_2OsO_2(OH)_4$, $K_3Fe(CN)_6$, K_2CO_3, $MeSO_2NH_2$, *t*-BuOH:H_2O (1:1), 0°, 5 h	**I** (70), —	391
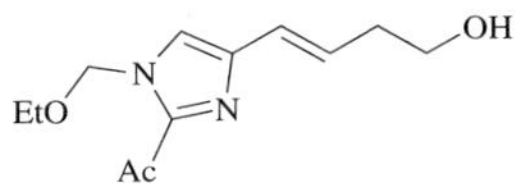	AD-Mix β, $MeSO_2NH_2$, *t*-BuOH:H_2O (1:1), 0°, 3 d	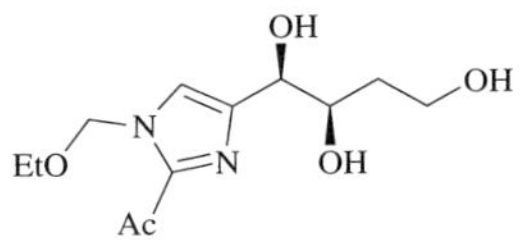 (79), 95	392
	AD-Mix α, $MeSO_2NH_2$, *t*-BuOH:H_2O (1:1), 0°, 3 d	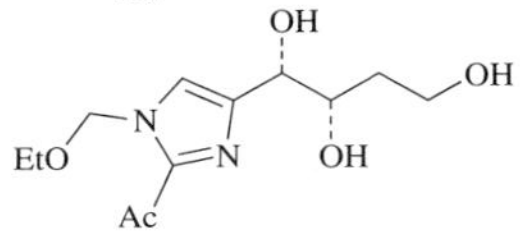 (76), 90	392

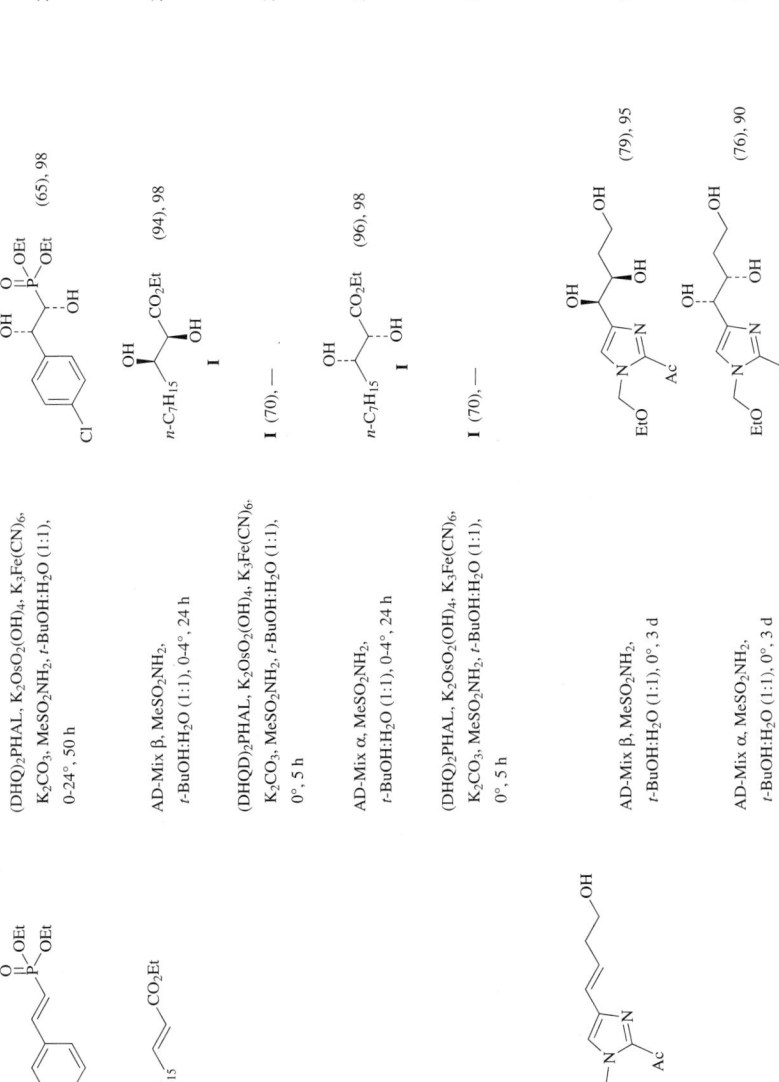

Substrate	Conditions	Product(s) and Yield(s) (%), % ee	Refs.
C_{12}			
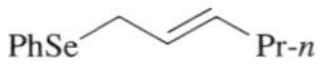	AD-Mix β, $MeSO_2NH_2$, t-BuOH:H_2O (1:1), 0°, 24 h	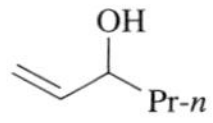 (81), 0 + 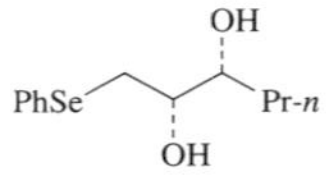 (7), 94	153
	AD-Mix α, $MeSO_2NH_2$, t-BuOH:H_2O (1:1), 0°, 24 h	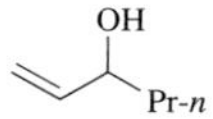 (83), 0	153
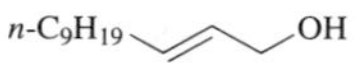	$(DHQ)_2PHAL$, $K_2OsO_2(OH)_4$, $K_3Fe(CN)_6$, K_2CO_3, $MeSO_2NH_2$, t-BuOH:H_2O (1:1), 0°	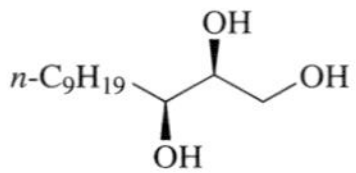 (84-88), 96	350
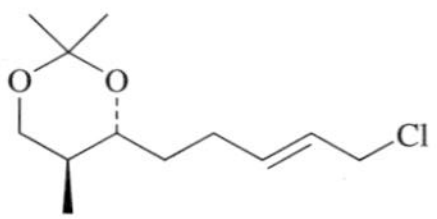	AD-Mix β, $MeSO_2NH_2$, t-BuOH:H_2O (1:1), rt	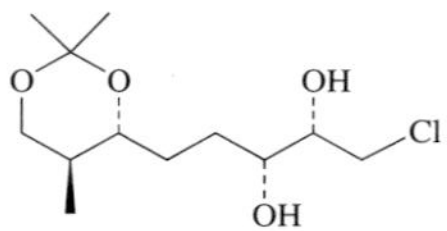 (65), — de	394a
C_{13}			
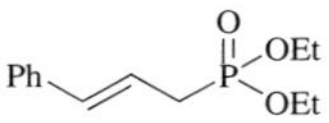	DHQ-9-(4-quinolinyl) ether, $K_2OsO_2(OH)_4$, $K_3Fe(CN)_6$, K_2CO_3, t-BuOH:H_2O (1:1), 20°	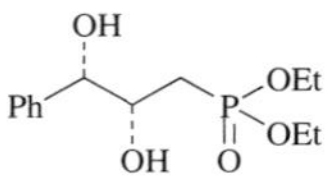 **I** (88), 70	118
	$(DHQ)_2PHAL$, $K_2OsO_2(OH)_4$, $K_3Fe(CN)_6$, K_2CO_3, $MeSO_2NH_2$, t-BuOH:H_2O (1:1), rt, 48 h	**I** (72), 98	150

	(DHQD)$_2$PHAL, $K_2OsO_2(OH)_4$, $K_3Fe(CN)_6$, K_2CO_3, $MeSO_2NH_2$, *t*-BuOH:H_2O (1:1), rt, 48 h	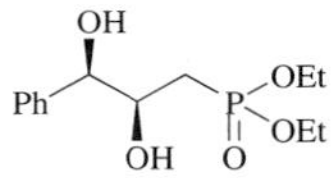 (69-80), 95	150, 118
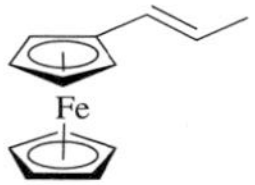	Ligand, $K_2OsO_2(OH)_4$, $K_3Fe(CN)_6$, K_2CO_3, *t*-BuOH:H_2O (1:1), 20°	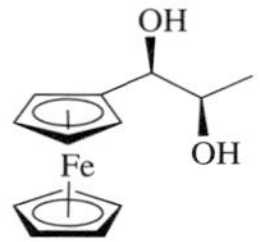	117
	Ligand / Time: (DHQD)$_2$PHAL 12 h	(46), 44	
	(DHQD)$_2$PHAL 3 h	(78), 36	
	(DHQD)$_2$PYR 3 h	(99), 97	
	(DHQ)$_2$PYR, $K_2OsO_2(OH)_4$, $K_3Fe(CN)_6$, K_2CO_3, CH_3CN:H_2O (1:1), 20°, 3 h	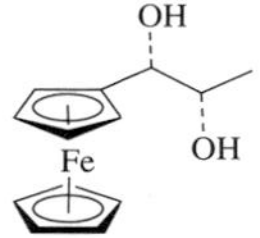 (99), 97	117
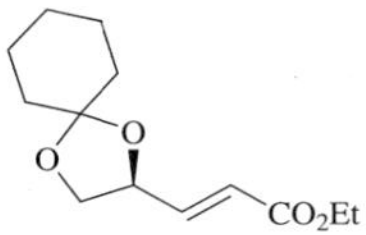	DHQD-OAc, OsO_4, toluene, –20°, 18 h	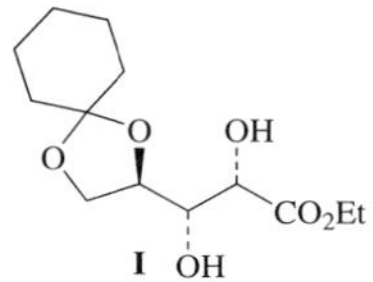 **I** (80), 90 de	384
	DHQD-*p*-chlorobenzoate, OsO_4, toluene, –20°, 18 h	**I** (80), 96 de	384
	DHQD-*p*-chlorobenzoate, OsO_4, Me_3NO, acetone:H_2O (5:1), 0°, 5-18 h	**I** (75), 36 de	384
	DHQ-OAc, OsO_4, toluene, –20°, 18 h	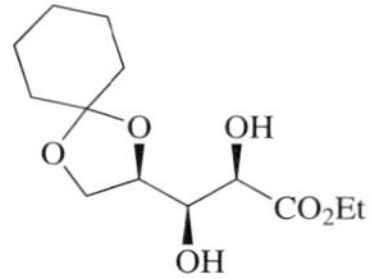 (44), 10 de	384

TABLE 3. REACTIONS OF TRANS 1,2-DISUBSTITUTED ALKENES (*Continued*)

Substrate	Conditions	Product(s) and Yield(s) (%), % ee	Refs.
C_{13} 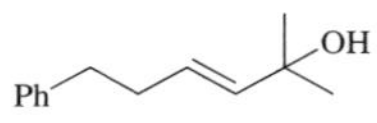	(DHQD)$_2$PHAL, $K_2OsO_2(OH)_4$, $K_3Fe(CN)_6$, K_2CO_3, $MeSO_2NH_2$, t-BuOH:H_2O (1:1), 0°, 6-36 h	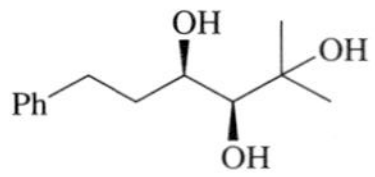 (83), 91	116
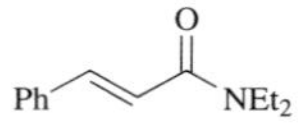	(DHQD)$_2$PHAL, $K_2OsO_2(OH)_4$, $K_3Fe(CN)_6$, K_2CO_3, $MeSO_2NH_2$, t-BuOH:H_2O (1:1), rt	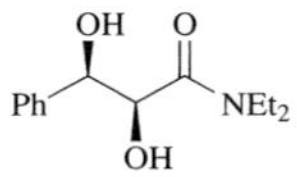 (96), 96	111
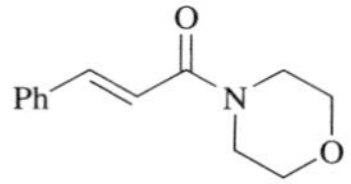	(DHQD)$_2$PHAL, $K_2OsO_2(OH)_4$, $K_3Fe(CN)_6$, K_2CO_3, $MeSO_2NH_2$, t-BuOH:H_2O (1:1), rt	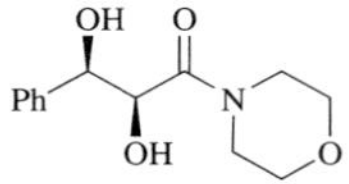 (97), 97	111
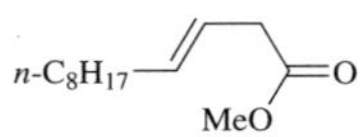	(DHQD)$_2$PHAL, $K_2OsO_2(OH)_4$, $K_3Fe(CN)_6$, K_2CO_3, $MeSO_2NH_2$, t-BuOH:H_2O (1:1), 0°, 48 h	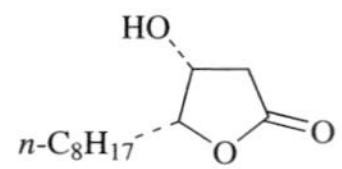 (81), 95	108
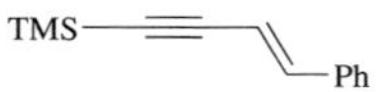	DHQD-PHN, $K_2OsO_2(OH)_4$, $K_3Fe(CN)_6$, K_2CO_3, t-BuOH:H_2O (1:1), 0°, 15-30 h	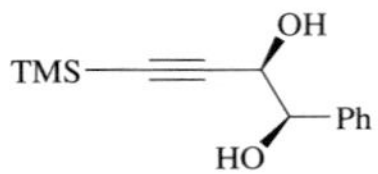 (66), 94	219
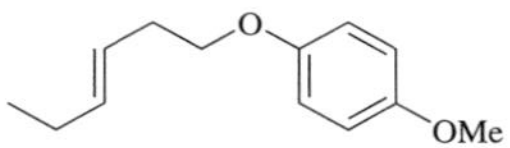	(DHQD)$_2$PYDZ, $K_2OsO_2(OH)_4$, $K_3Fe(CN)_6$, K_2CO_3, $MeSO_2NH_2$, t-BuOH:H_2O (1:1), 0°, 4-22 h	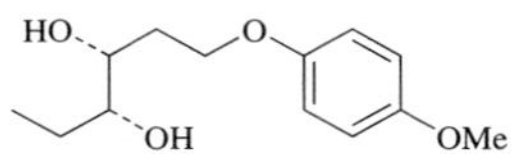 (>99), >95	61

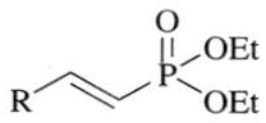	(DHQ)$_2$PHAL, $K_2OsO_2(OH)_4$, $K_3Fe(CN)_6$, K_2CO_3, $MeSO_2NH_2$, *t*-BuOH:H_2O (1:1)	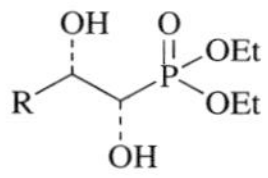	
R	Temp / Time		
3-$MeOC_6H_4$	0°-rt / 50 h	(67), 96	345
4-$MeOC_6H_4$	0°-rt / 48 h	(71), >95	344, 345
$TBDMSOCH_2$	rt / 48 h	(65), 38	344
n-C_7H_{15}	0°-rt / 50 h	(46), 84	345
	(DHQD)$_2$PHAL, $K_2OsO_2(OH)_4$, $K_3Fe(CN)_6$, K_2CO_3, $MeSO_2NH_2$, *t*-BuOH:H_2O (1:1)	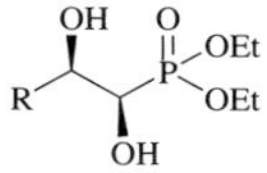	
R	Temp / Time		
4-$MeOC_6H_4$	rt / 48 h	(69), >98	344, 345
$TBDMSOCH_2$	0°-rt / 50 h	(65), 38	345
n-C_7H_{15}	0°-rt / 50 h	(63), 84	345
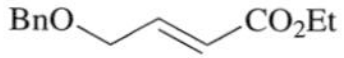	AD-Mix α, $MeSO_2NH_2$, *t*-BuOH:H_2O (1:1), 4°	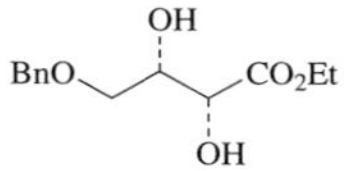 (82-98), 94-98	102, 370
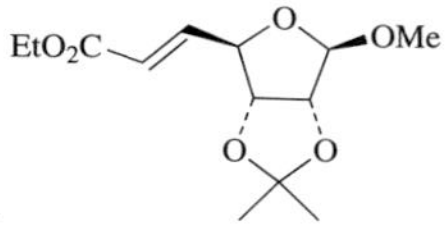	DHQD-*p*-chlorobenzoate, OsO_4, NMO, acetone:H_2O	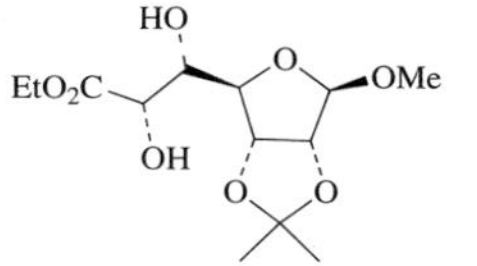 (91), 42 de	395
	AD-Mix α, $MeSO_2NH_2$, *t*-BuOH:H_2O (1:1), 0°, 12 h	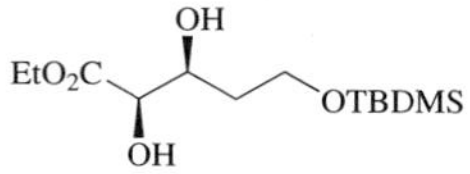 (94), 91	396
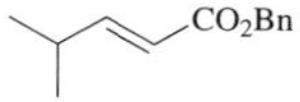	AD-Mix β, $MeSO_2NH_2$, *t*-BuOH:H_2O (1:1)	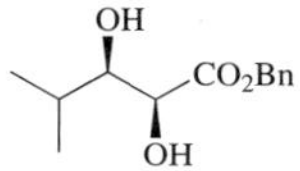 (90), 95	397

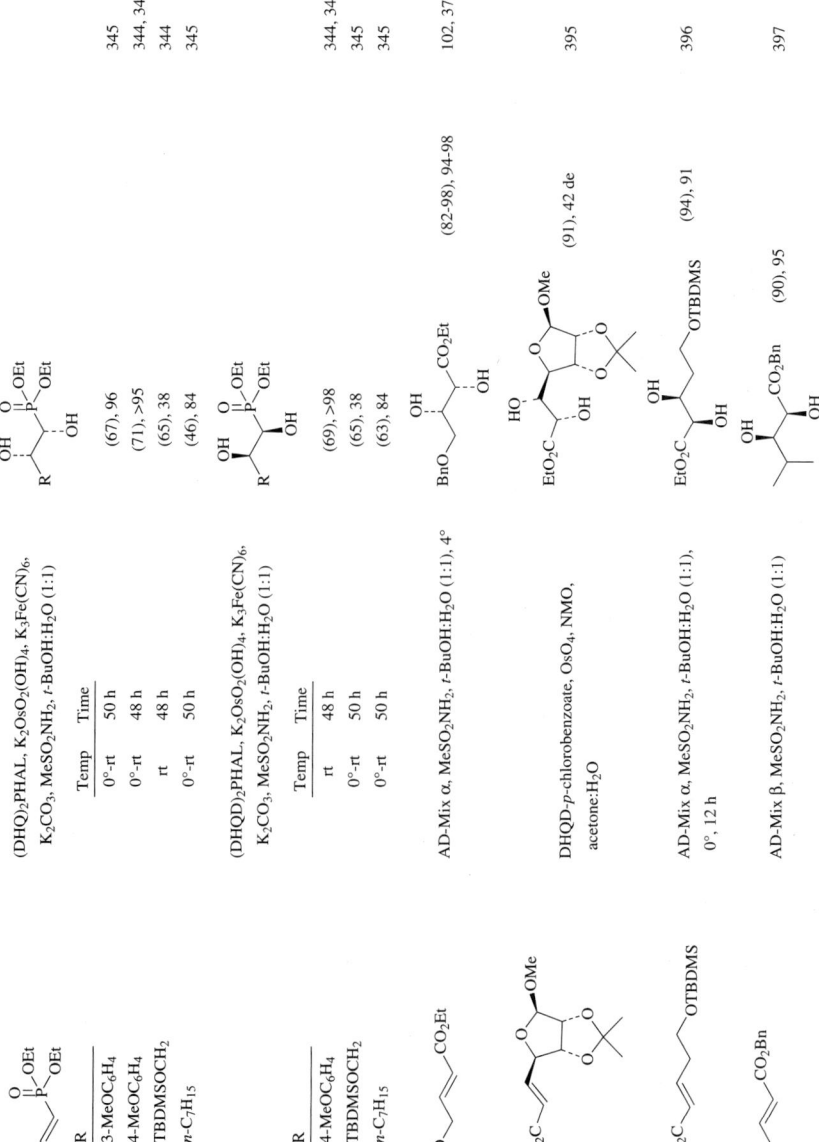

TABLE 3. REACTIONS OF TRANS 1,2-DISUBSTITUTED ALKENES (*Continued*)

Substrate	Conditions	Product(s) and Yield(s) (%), % ee		Refs.
C_{13} 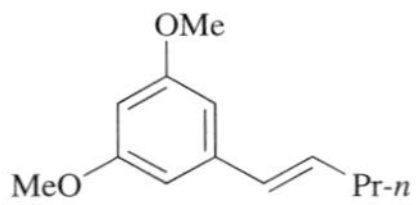	AD-Mix β, $MeSO_2NH_2$, t-BuOH:H_2O (1:1), 4°, overnight	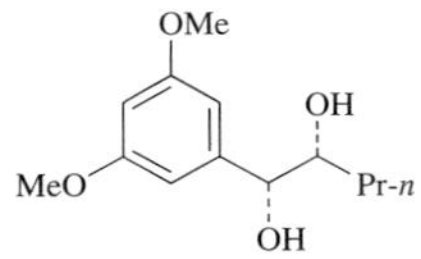	(99), 97	398
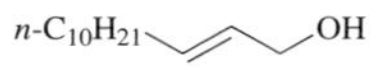	$(DHQ)_2PHAL$, $K_2OsO_2(OH)_4$, $K_3Fe(CN)_6$, K_2CO_3, $MeSO_2NH_2$, t-BuOH:H_2O (1:1), 0°	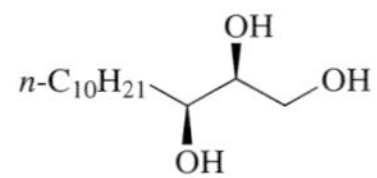	(84-88), 96	350
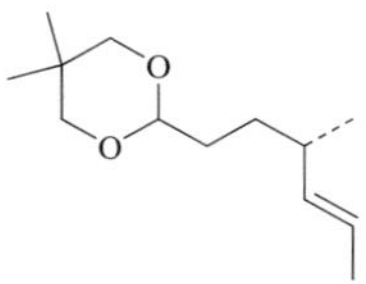	AD-Mix α, $MeSO_2NH_2$, t-BuOH:H_2O (2:3), 0°, 18 h	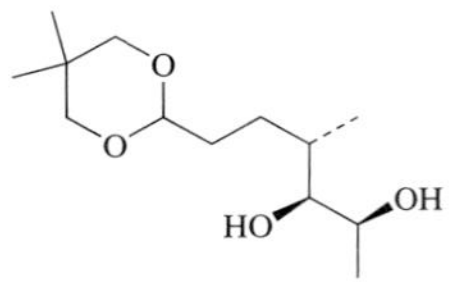	(83), 86 de	399
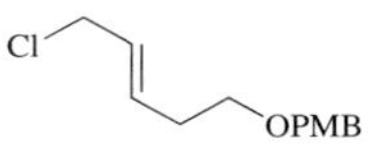	AD-Mix α, $MeSO_2NH_2$, t-BuOH:H_2O (1:1), 0°, 48 h	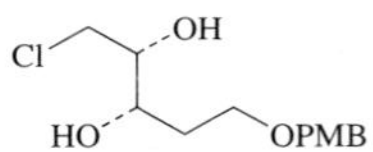	(100), 88	400
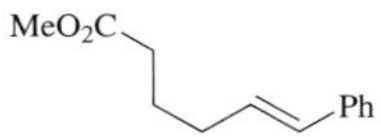	$(DHQ)_2PHAL$, $K_2OsO_2(OH)_4$, $K_3Fe(CN)_6$, K_2CO_3, t-BuOH:H_2O (1:1), 0°, 18 h	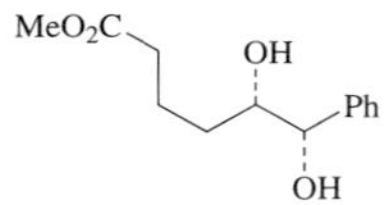	(80), 90	401
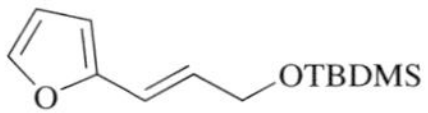	AD-Mix β, $MeSO_2NH_2$, t-BuOH:H_2O (1:1), 0°, 95 h	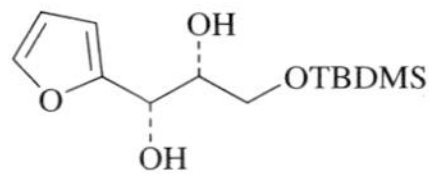	(94), >99	402
	AD-Mix α, $MeSO_2NH_2$, t-BuOH:H_2O (1:1), 0°, 95 h	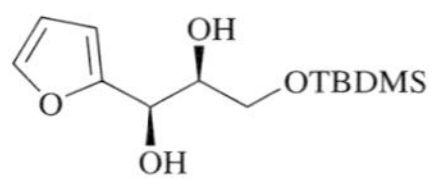	(94), >99	402

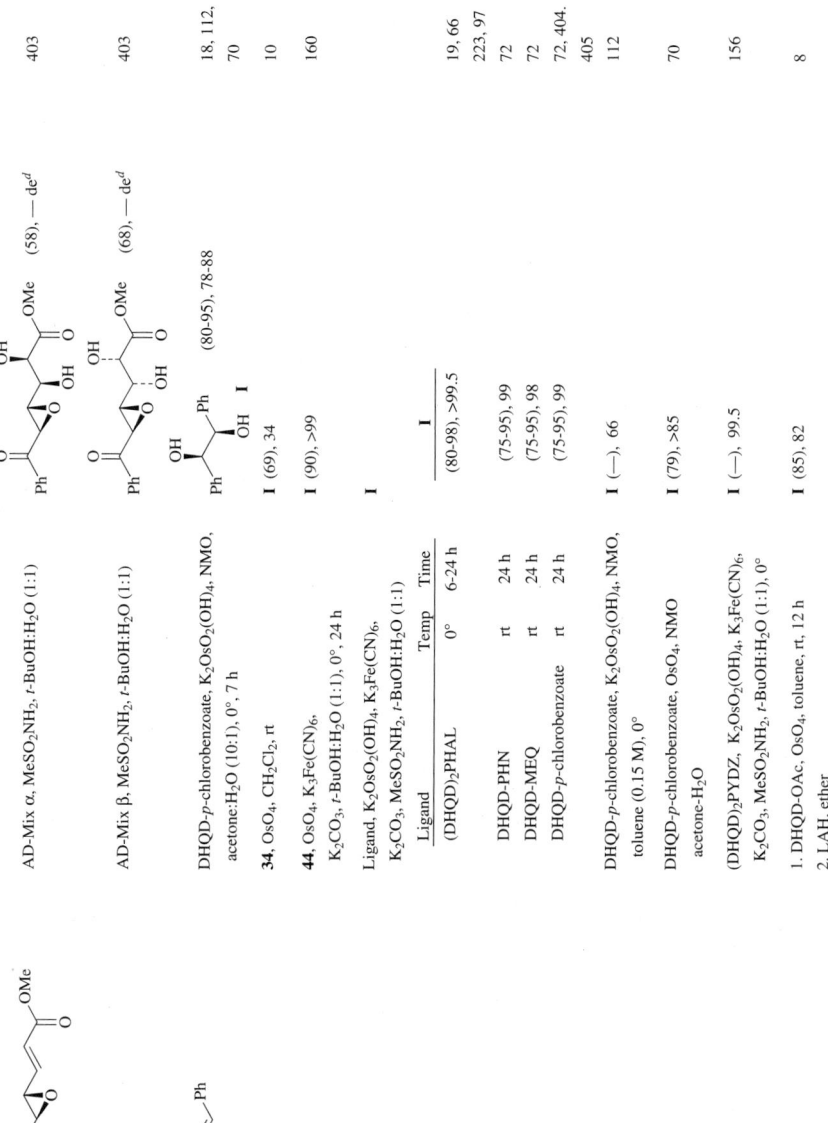

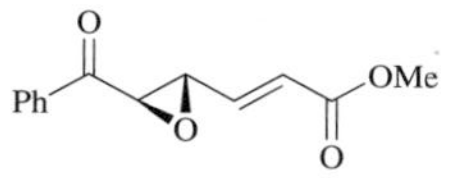	AD-Mix α, $MeSO_2NH_2$, t-BuOH:H_2O (1:1)	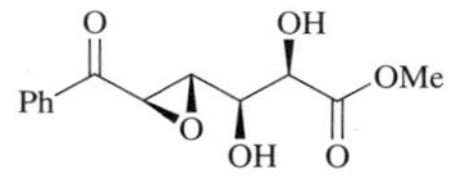 (58), — de[d]	403
	AD-Mix β, $MeSO_2NH_2$, t-BuOH:H_2O (1:1)	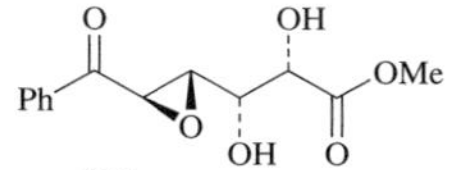 (68), — de[d]	403
C_{14} 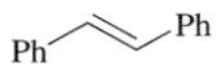	DHQD-p-chlorobenzoate, $K_2OsO_2(OH)_4$, NMO, acetone:H_2O (10:1), 0°, 7 h	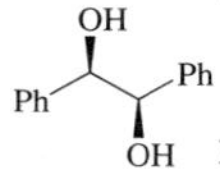 **I** (80-95), 78-88	18, 112, 70
	34, OsO_4, CH_2Cl_2, rt	**I** (69), 34	10
	44, OsO_4, $K_3Fe(CN)_6$, K_2CO_3, t-BuOH:H_2O (1:1), 0°, 24 h	**I** (90), >99	160
	Ligand, $K_2OsO_2(OH)_4$, $K_3Fe(CN)_6$, K_2CO_3, $MeSO_2NH_2$, t-BuOH:H_2O (1:1)	**I**	
	Ligand / Temp / Time	**I**	
	$(DHQD)_2PHAL$ / 0° / 6-24 h	(80-98), >99.5	19, 66, 223, 97
	DHQD-PHN / rt / 24 h	(75-95), 99	72
	DHQD-MEQ / rt / 24 h	(75-95), 98	72
	DHQD-p-chlorobenzoate / rt / 24 h	(75-95), 99	72, 404. 405
	DHQD-p-chlorobenzoate, $K_2OsO_2(OH)_4$, NMO, toluene (0.15 M), 0°	**I** (—), 66	112
	DHQD-p-chlorobenzoate, OsO_4, NMO acetone-H_2O	**I** (79), >85	70
	$(DHQD)_2PYDZ$, $K_2OsO_2(OH)_4$, $K_3Fe(CN)_6$, K_2CO_3, $MeSO_2NH_2$, t-BuOH:H_2O (1:1), 0°	**I** (—), 99.5	156
	1. DHQD-OAc, OsO_4, toluene, rt, 12 h 2. LAH, ether	**I** (85), 82	8

TABLE 3. REACTIONS OF TRANS 1,2-DISUBSTITUTED ALKENES (*Continued*)

Substrate	Conditions	Product(s) and Yield(s) (%), % ee	Refs.
C_{14} 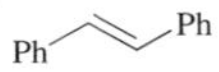	Ligand, $K_2OsO_2(OH)_4$, $K_3Fe(CN)_6$, K_2CO_3, t-BuOH:H_2O (1:1), rt, 24 h	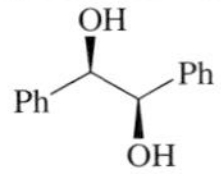 **I**	
	Ligand	**I**	
	DHQD-*p*-chlorobenzoate	(75-95), 99	220, 70
	DHQD-9-(1-naphthyl) ether	(75-95), 99	220
	DHQD-PHN	(75-95), 99	220
	DHQD-9-phenyl ether	(75-95), 94	220
	DHQD-9-(2-pyridyl) ether	(75-95), 96	220
	DHQD-9-(2-tolyl) ether	(75-95), 95	220
	DHQD-9-(2-naphthyl) ether	(75-95), 97	220
	DHQD-9-(3-quinolinyl) ether	(75-95), 94	220
	DHQD-9-(4-quinolinyl) ether	(75-95), 96	220
	DHQD-MEQ	(75-95), 96	220
	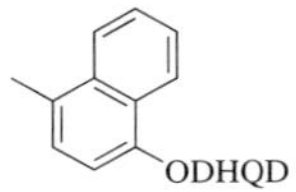	(75-95), 98	220
	DHQD-*p*-chlorobenzoate, OsO_4, NMO, acetone:H_2O (10:1), 0°, 7 h (slow addition of olefin)	**I** (85-95), 95	70
	$(DHQD)_2$PYDZ, K_2OsO_4, $K_3Fe(CN)_6$, K_2CO_3, $MeSO_2NH_2$, t-BuOH:H_2O (1:1), 0°, 12 h	**I** (—), >98	66, 73
	AD-Mix β, t-BuOH:H_2O (1:1), 0°	**I** (80), 90-99	408, 409
	17, $K_2OsO_2(OH)_4$, $K_3Fe(CN)_6$, K_2CO_3, $MeSO_2NH_2$, $NaHCO_3$, t-BuOH:H_2O (1:1), 0°, 6-12 h	**I** (—), >98	73
	DHQD-*p*-chlorobenzoate, $K_2OsO_2(OH)_4$, $K_3Fe(CN)_6$, K_2CO_3, electrolysis, t-BuOH:H_2O (4:3), rt	**I** (94), 90-95	176

1. 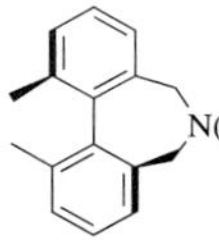 $N(CH_2)_2NMe_2$, OsO_4, THF, –80°, 12 h 2. LAH, THF, rt, 18 h	**I** (94), 96	406
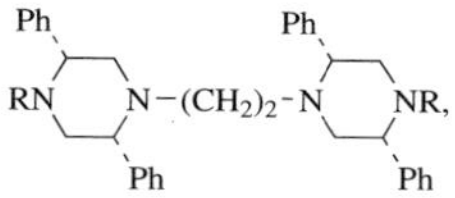 OsO_4, toluene, –78°	**I**	215

R	I
H	(47), 0
Me	(84), 94
Et	(87), 96
i-Pr	(81), 98
t-$BuCH_2$	(87), 91
t-$BuCH_2CH_2$	(94), 90
Bn	(85), 79
2,4,6-$Me_3C_6H_2CH_2$	(93), 92
CO_2Me	(44), 0
CO_2Bu-*t*	(48), 0

36, OsO_4, CH_2Cl_2, –100°	**I** (71), 90	96
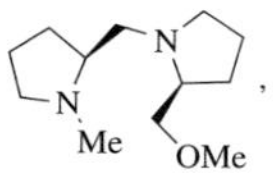, OsO_4, toluene, –78°	**I** (56), 35	96

TABLE 3. REACTIONS OF TRANS 1,2-DISUBSTITUTED ALKENES (*Continued*)

Substrate	Conditions	Product(s) and Yield(s) (%), % ee	Refs.
C_{14} 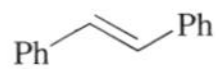	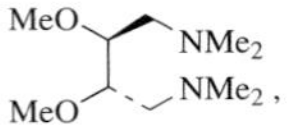 OsO_4, toluene, –78°	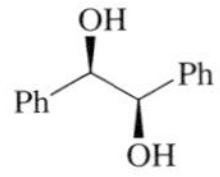 **I** (100), 34	96
	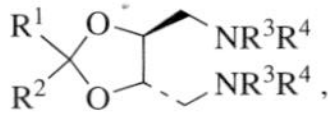 OsO_4, toluene, –78°	**I**	96

R^1	R^2	R^3	R^4	**I**
Me	Me	Me	Me	(84), 58
Me	Me	—$(CH_2)_4$—		(87), 65
Me	Me	—$(CH_2)_5$—		(76), 70
Ph	H	—$(CH_2)_5$—		(76), 78
1-naphthyl	H	—$(CH_2)_5$—		(71), 90

Substrate	Conditions	Product(s) and Yield(s) (%), % ee	Refs.
	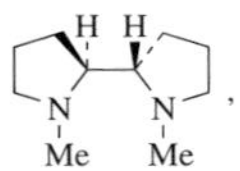 OsO_4, toluene, –78°	**I** (55), 7	12
	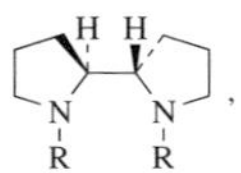 OsO_4, CH_2Cl_2, –78°	**I**	12

R	**I**
Me	(49), 40
Et	(63), 43
n-Pr	(54), 52

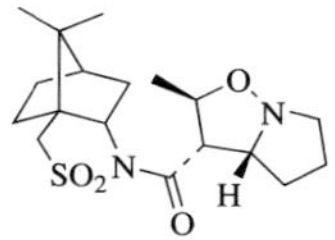 , OsO_4, $K_3Fe(CN)_6$, K_2CO_3, *t*-BuOH:H_2O (1:1), rt, 24 h	**I** (100), 73	203
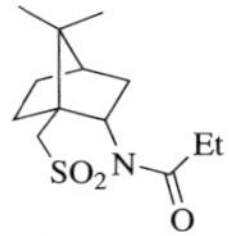 , OsO_4, $K_3Fe(CN)_6$, K_2CO_3, *t*-BuOH:H_2O (1:1), rt, 24 h	**I** (73), 2	203
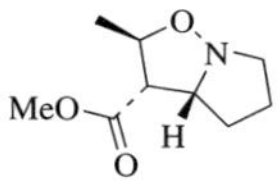 , OsO_4, $K_3Fe(CN)_6$, K_2CO_3, *t*-BuOH:H_2O (1:1), rt, 24 h	**I** (93), 10	203
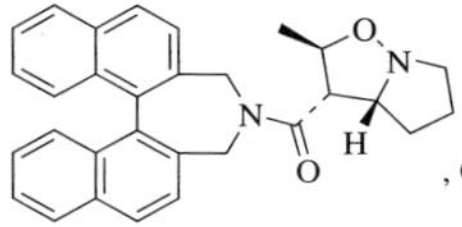 , OsO_4, $K_3Fe(CN)_6$, K_2CO_3, *t*-BuOH:H_2O (1:1), rt, 24 h	**I** (59), 62	203

Substrate	Conditions	Product(s) and Yield(s) (%), % ee	Refs.
C_{14} 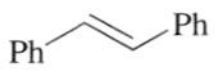	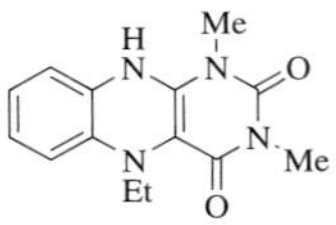 , $(DHQD)_2PHAL$, OsO_4, NMM, TEAA, H_2O_2, *t*-BuOH:H_2O (3:1), 0°, 9 h; slow addition of H_2O_2 and olefin, 2-7 h	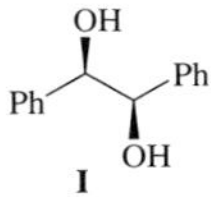 **I** (94), 91	208
	Ligand, $K_2OsO_2(OH)_4$, $K_3Fe(CN)_6$, K_2CO_3, $MeSO_2NH_2$, *t*-BuOH:H_2O (1:1)	**I**	
	Ligand / Temp / Time	**I**	
	6 / 0° / 6-12 h	(92), >98	73
	37 / — / —	(77), 99	166
	40 / — / <5 h	(91), 99	169
	Ligand, OsO_4, NMO, TEAA, acetone:H_2O (10:1)	**I**	
	Ligand / Temp / Time	**I**	
	47 / rt / 7 days	(68), —	407
	48 / 10° / 2-3 days	(81-87), 85-93	407
	49 / 0° / 48 h	(85), 80	407
	50 / 10° / 48 h	(87), 82	407
	26 / 10° / 5 h	(89), 88	205
	27 / 10° / 5 h	(89), 88	205
	16 / 10° / 5 h	(95), 99	168
	Ligand, OsO_4, $K_3Fe(CN)_6$, K_2CO_3, *t*-BuOH:H_2O (1:1)	**I**	
	Ligand / Temp / Time	**I**	
	49 / rt / 18 h	(96), 87	407
	50 / rt / 48 h	(91), 86	407
	16 / rt / 18 h	(98), 94	168
	41a / 0° / 24 h	(68-80), 93	206

15, OsO_4, NMO, CH_3CN:H_2O (4:1), 0°, 18 h; slow addition of olefin, 36 h	**I** (82), 85	207
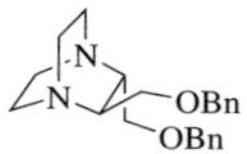 OBn, OBn, OsO_4, $K_3Fe(CN)_6$, K_2CO_3, *t*-BuOH:H_2O (1:1), rt, 8-24 h	**I** (—), 24	193
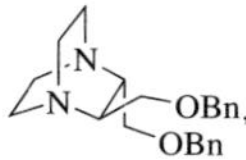 OBn, OBn, OsO_4, NMO, acetone-H_2O, 0°	**I** (—), 7	193
15, OsO_4, NMO, TEAA, CH_3CN:H_2O (4:1), 0°, 18 h; slow addition of olefin, 24 h	**I** (32), 31	207
15, OsO_4, $K_3Fe(CN)_6$, K_2CO_3, *t*-BuOH:H_2O (1:1), 20°, 24 h	**I** (32), 31	207
29, OsO_4, NMO, TEAA, acetone:H_2O (10:1), 10°, 20 h	**I** (81), 70	159
29, OsO_4, $K_3Fe(CN)_6$, K_2CO_3, $MeSO_2NH_2$, *t*-BuOH:H_2O (1:1), 10°, 20 h	**I** (80), 89	159

TABLE 3. REACTIONS OF TRANS 1,2-DISUBSTITUTED ALKENES (*Continued*)

Substrate	Conditions	Product(s) and Yield(s) (%), % ee	Refs.
C_{14} 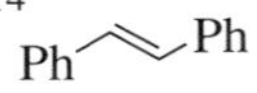	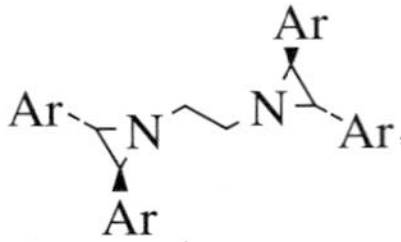, OsO_4, toluene, –78°	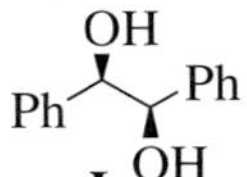 **I**	
	Ar: Ph	(90), 95	410, 411
	Ar: 4-FC_6H_4	(75), 83	410
	Ar: Bn	(78), 33	410
	Ar: $BnOCH_2$	(61), 6	410
	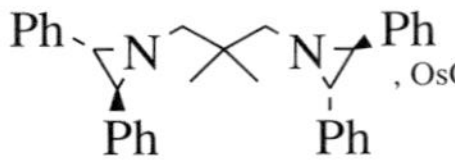, OsO_4, toluene, –78°	**I** (51), <5	410
	DHQ-*p*-chlorobenzoate, $K_2OsO_2(OH)_4$, NMO, acetone-H_2O, 0°, 17 h	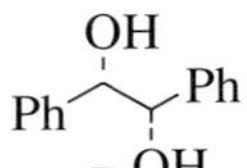 **I** (80-95), 78	18
	DHQ-*p*-chlorobenzoate, $K_2OsO_2(OH)_4$, $K_3Fe(CN)_6$, K_2CO_3, *t*-BuOH:H_2O (1:1), rt, 24 h	**I** (75-95), 97	220, 405
	$(DHQ)_2PHAL$, $K_2OsO_2(OH)_4$, $K_3Fe(CN)_6$, K_2CO_3, $MeSO_2NH_2$, *t*-BuOH:H_2O (1:1), 0°, 6-24 h	**I** (80-98), >99.5	19, 223, 97
	Ligand, $K_2OsO_2(OH)_4$, $K_3Fe(CN)_6$, K_2CO_3, *t*-BuOH:H_2O (1:1), rt, 24 h	**I**	220
	Ligand: DHQ-PHN	**I**: (75-95), 96	
	Ligand: DHQ-9-(1-naphthyl) ether	**I**: (75-95), 94	
	Ligand: DHQ-9-phenyl ether	**I**: (75-95), 93	

Conditions	Product(s) and Yield(s) (%), % ee	Refs.
35, OsO_4, CH_2Cl_2, –90°, 2 h	**I** (95), 92	11
1. 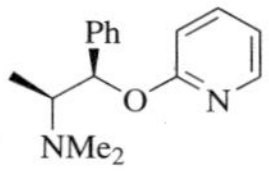, OsO_4, toluene, –78°, 10 h 2. LAH-ether, –78° to rt, 8 h	**I** (82), 70	183
1. 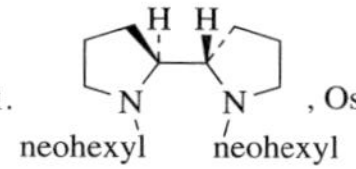, OsO_4, toluene, –78°, 18 h 2. $NaHSO_3$	**I** (100), 96	13
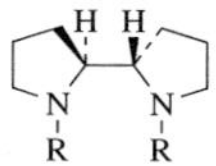, OsO_4, –78°	**I**	12

R	Solvent	I
Et	toluene	(71), 28
n-Pr	toluene	(65), 54
n-Bu	toluene	(67), 85
n-Bu	CH_2Cl_2	(43), 26
n-C_5H_{11}	toluene	(67), 89
n-C_5H_{11}	CH_2Cl_2	(68), 28

Conditions	Product(s) and Yield(s) (%), % ee	Refs.
1. DHQ-OAc, OsO_4, toluene, rt, 12 h 2. LAH, ether	**I** (90), 83.2	8
25, $K_2OsO_2(OH)_4$, $K_3Fe(CN)_6$, K_2CO_3, 20°, 24 h	**I** (93), 99	201
23b, OsO_4, $K_3Fe(CN)_6$, K_2CO_3, $MeSO_2NH_2$, *t*-BuOH:H_2O (1:1), 10°, 20 h	(87), >99.9	363

TABLE 3. REACTIONS OF TRANS 1,2-DISUBSTITUTED ALKENES (*Continued*)

Substrate	Conditions	Product(s) and Yield(s) (%), % ee	Refs.
C_{14} 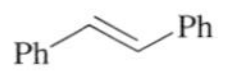	**19**, OsO_4, $K_3Fe(CN)_6$, K_2CO_3, $MeSO_2NH_2$, *t*-BuOH:H_2O (1:1), 10°, 25 h	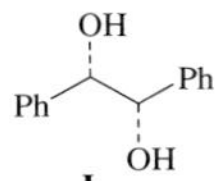 **I** (88), >99	165
	Ligand, OsO_4, NMO, acetone:H_2O (10:1), 0°	**I**	
	Ligand / Time	**I**	
	43 / 7 h	(85), 87	202
	42b / 7 h	(60), 90	161
	12a / 16 h	(73), 38	192
	12c / 16 h	(71), 27	192
	12b / 16 h	(86), 45	192
	31, OsO_4, NMO, TEAA, acetone:H_2O (10:1), 10°, 20 h	**I** (80), 59	159
	Ligand, OsO_4, $K_3Fe(CN)_6$, K_2CO_3, *t*-BuOH:H_2O (1:1), rt, 25 h	**I**	
	Ligand / Temp / Time	**I**	
	43 / rt / 25 h	(—), —	202
	42b / 0° / 48 h	(70), 95	161
	31, OsO_4, $K_3Fe(CN)_6$, K_2CO_3, $MeSO_2NH_2$, *t*-BuOH:H_2O (1:1), 10°, 20 h	**I** (82), 93	159
	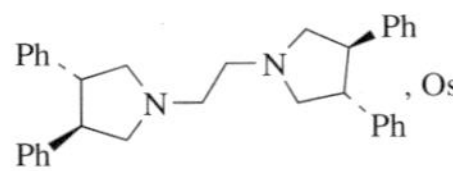 , OsO_4, THF, –110°, 6 h	**I** (85), 97	15, 16, 17, 14

	23a, OsO_4, $K_3Fe(CN)_6$, K_2CO_3, $MeSO_2NH_2$, *t*-BuOH:H_2O (1:1), 10°, 15 h	**I** (93), >99.9		363
	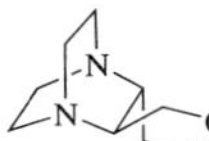 , OsO_4, $K_3Fe(CN)_6$, K_2CO_3, *t*-BuOH-H_2O (1:1), rt, 8-24 h	**I** (85), 40		193
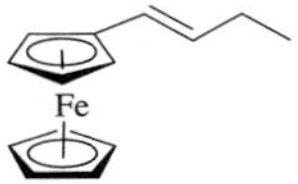	$(DHQD)_2PHAL$, $K_2OsO_2(OH)_4$, $K_3Fe(CN)_6$, K_2CO_3, *t*-BuOH:H_2O (1:1), 20°, 12 h	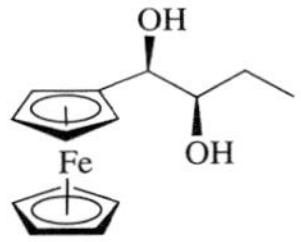 **I**	(51), 50	117
	$(DHQD)_2PYR$, $K_2OsO_2(OH)_4$, $K_3Fe(CN)_6$, K_2CO_3, CH_3CN:H_2O (1:1), 20°, 3 h	**I** (76), 93		117
	$(DHQ)_2PYR$, $K_2OsO_2(OH)_4$, $K_3Fe(CN)_6$, K_2CO_3, CH_3CN:H_2O (1:1), 20°, 3 h	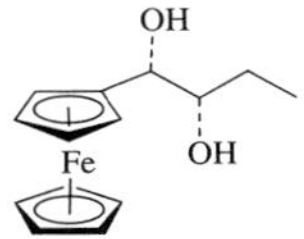	(75), 93	117
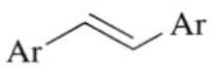	$(DHQD)_2PHAL$, $K_2OsO_2(OH)_4$, $K_3Fe(CN)_6$, K_2CO_3, $MeSO_2NH_2$, *t*-BuOH:H_2O (1:1), 0°	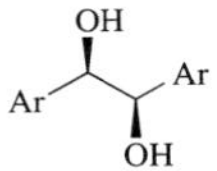		
Ar	**Time**			
4-ClC_6H_4	—	(85), 93		412
4-BrC_6H_4	—	(84), 92		412
2-BrC_6H_4	20 h	(—), 98		413
4-$NO_2C_6H_4$	—	(77), 83		412

TABLE 3. REACTIONS OF TRANS 1,2-DISUBSTITUTED ALKENES (*Continued*)

Substrate	Conditions	Product(s) and Yield(s) (%), % ee	Refs.
C_{14} 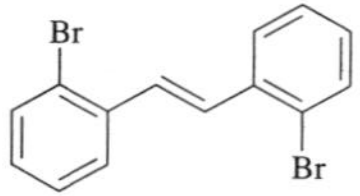	Quinidine-*p*-chlorobenzoate, $K_2OsO_2(OH)_4$, $K_3Fe(CN)_6$, K_2CO_3, $MeSO_2NH_2$, *t*-BuOH:H_2O (1:1), 0°, 4-6 days	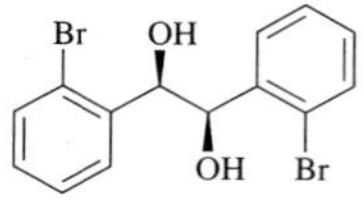 **I** (60), >90	414
	DHQD-*p*-chlorobenzoate, OsO_4, NMO, acetone-H_2O	**I** (94), 72	415
	DHQD-*p*-chlorobenzoate, OsO_4, toluene	**I** (83), 86	415
	AD-Mix β, *t*-BuOH:H_2O (1:1), 0°	**I** (94), >99	416
	AD-Mix β, *t*-BuOH:H_2O (1:1), rt, 96 h	**I** (94), >99	319
	Ligand, $K_2OsO_2(OH)_4$, $K_3Fe(CN)_6$, K_2CO_3, $MeSO_2NH_2$, *t*-BuOH:H_2O (1:1), 0°	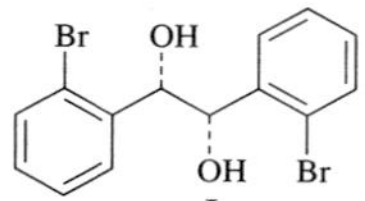 **I**	
	Ligand / Time	**I**	
	$(DHQ)_2PHAL$ / 20 h	(87), 98	413
	DHQ-*p*-chlorobenzoate / 24 h	(62), >95	417
	DHQ-*p*-chlorobenzoate, $K_2OsO_2(OH)_4$, NMO, acetone:H_2O (1:1), 0°, 20 h	**I** (—), 91	413
	DHQ-*p*-chlorobenzoate, OsO_4, toluene	**I** (80), 95	415
	DHQ-*p*-chlorobenzoate, OsO_4, NMO, acetone-H_2O	**I** (94), 79	415
	Quinine-*p*-chlorobenzoate, $K_2OsO_2(OH)_4$, $K_3Fe(CN)_6$, K_2CO_3, $MeSO_2NH_2$, *t*-BuOH:H_2O (1:1), 0°, 4-6 days	**I** (40), >90	414

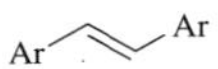	AD-Mix β, t-BuOH:H_2O (1:1)	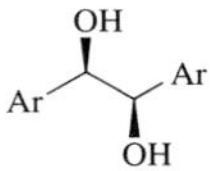	
Ar	**Temp** / **Time**		
4-ClC_6H_4	0° / —	(—), >90	408
3-FC_6H_4	0° / —	(89), >99	416
3-FC_6H_4	rt / 96 h	(89), >99	319
3-FC_6H_4	0° / 72 h	(97), >99	417
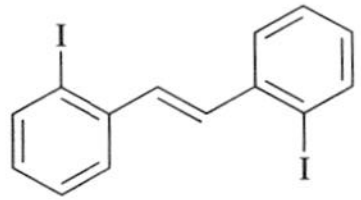	DHQD-p-chlorobenzoate, OsO_4, toluene	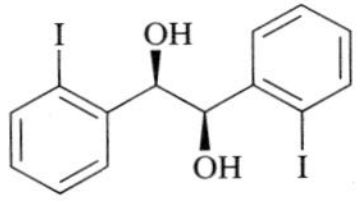 (82), 77	415
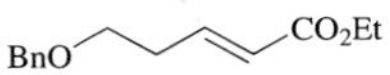	AD-Mix α, $MeSO_2NH_2$, t-BuOH:H_2O (1:1), 4°	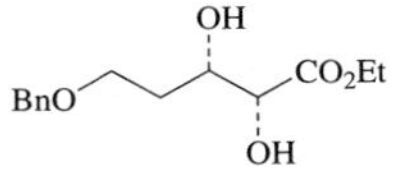 (98), 93	102
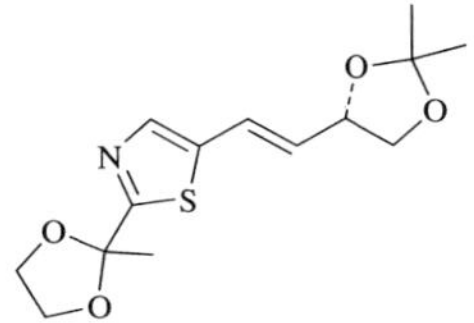	$(DHQD)_2PHAL$, $K_2OsO_2(OH)_4$, $K_3Fe(CN)_6$, K_2CO_3, $MeSO_2NH_2$, t-BuOH:H_2O (1:1), 0°, 5 d	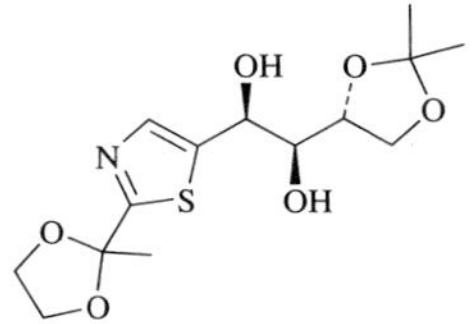 (84), 74 de	418
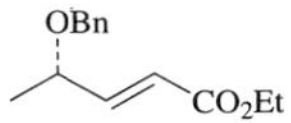	Ligand, OsO_4, toluene, –20°, 18 h	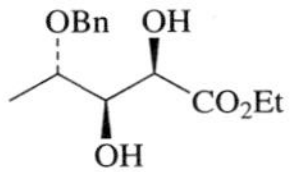	384
	Ligand		
	DHQ-OAc	(50), 73 de	
	DHQ-p-chlorobenzoate	(88), 94 de	

TABLE 3. REACTIONS OF TRANS 1,2-DISUBSTITUTED ALKENES (*Continued*)

Substrate	Conditions	Product(s) and Yield(s) (%), % ee	Refs.
C_{14}			
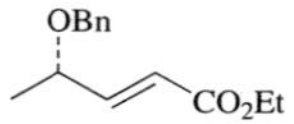	DHQ-*p*-chlorobenzoate, OsO_4, acetone:H_2O (5:1), 0°, 5-18 h	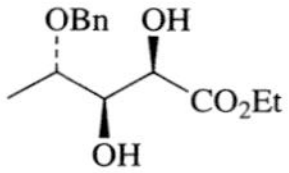 (83), 48 de	384
	AD-Mix β, $MeSO_2NH_2$, *t*-BuOH-H_2O, 0°	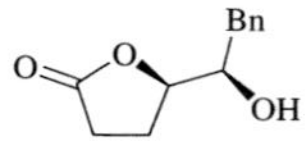 **I** (92), 97	109
	1. AD-Mix β, $MeSO_2NH_2$, *t*-BuOH-H_2O, 0° 2. AcOH (catalytic), toluene 115°, 6 h	**I** (87), —	419
	AD-Mix α, $MeSO_2NH_2$, *t*-BuOH-H_2O, 0°	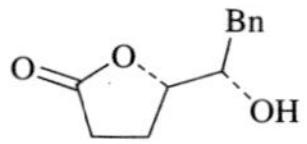 (90), 96	109
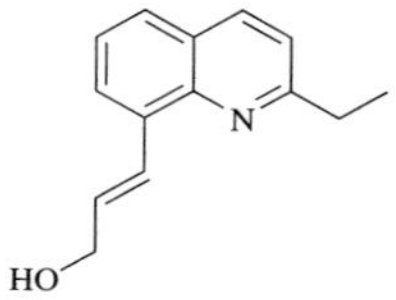	$(DHQ)_2PHAL$, $K_2OsO_2(OH)_4$, $K_3Fe(CN)_6$, K_2CO_3, $MeSO_2NH_2$, *t*-BuOH:H_2O (1:1), rt, 18 h	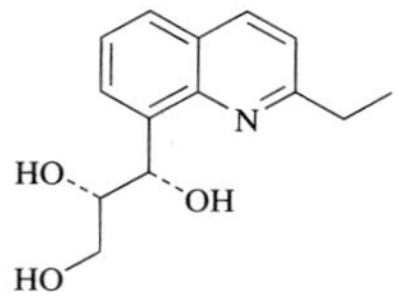 (67), 88.0	76
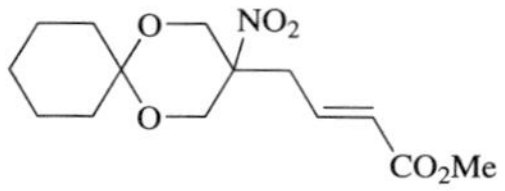	$(DHQD)_2PHAL$, $K_2OsO_2(OH)_4$, $K_3Fe(CN)_6$, AcOH, K_2CO_3, *t*-BuOH:H_2O (1:1), rt, 28 h	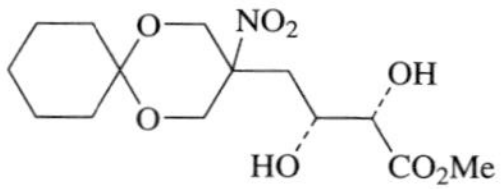 (73), 82	420

	$(DHQ)_2PHAL$, $K_2OsO_2(OH)_4$, $K_3Fe(CN)_6$, K_2CO_3, AcOH, t-BuOH:H_2O (1:1), rt, 24 h	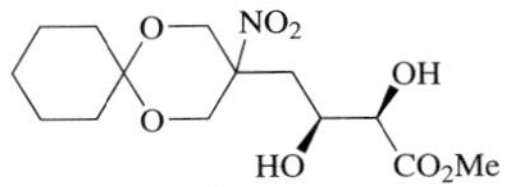 **I** (90), 72	420
	$(DHQ)_2PHAL$, $K_2OsO_2(OH)_4$, $K_3Fe(CN)_6$, $NaHCO_3$, K_2CO_3, $MeSO_2NH_2$, t-BuOH:H_2O (1:1), 0°, 48 h	**I** (66), 79	420
	$(DHQ)_2PHAL$, $K_2OsO_2(OH)_4$, $K_3Fe(CN)_6$, K_2CO_3, $MeSO_2NH_2$, t-BuOH:H_2O (1:1), rt, 48 h	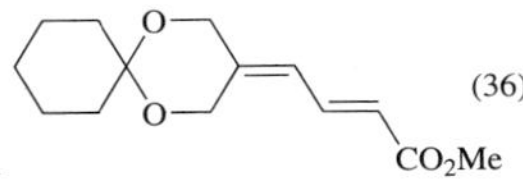 (36), —	420
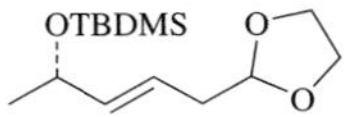	AD-Mix α, $MeSO_2NH_2$, t-BuOH:H_2O (1:1), 0°, 37 h	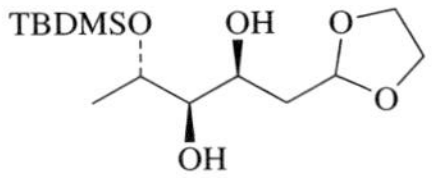 (>81), 87 de	421
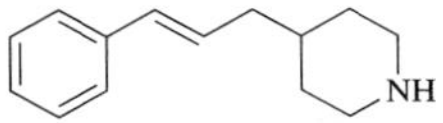	AD-Mix α, $MeSO_2NH_2$, t-BuOH:H_2O (1:1), 0°	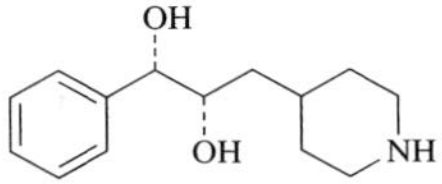 (72), >95	422
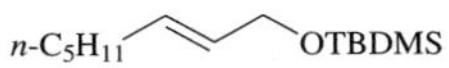	AD-Mix β, t-BuOH:H_2O (1:1), 0-4°, 24 h	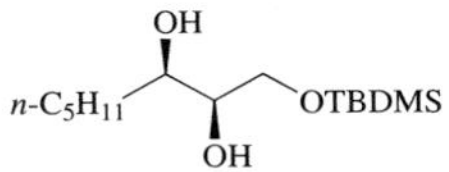 (92), 96	140
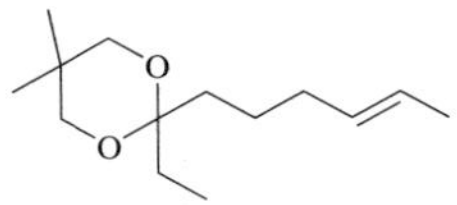	AD-Mix α, $MeSO_2NH_2$, t-BuOH:H_2O (1:1), 4°, 9 h	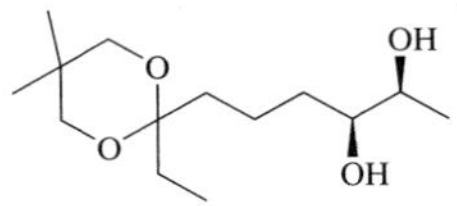 (94), 75	423
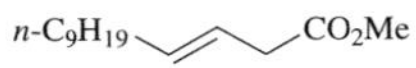	AD-Mix α, $MeSO_2NH_2$, t-BuOH:H_2O (1:1), 0°, 39 h	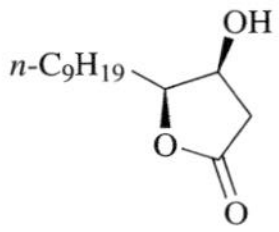 (88), 96	424

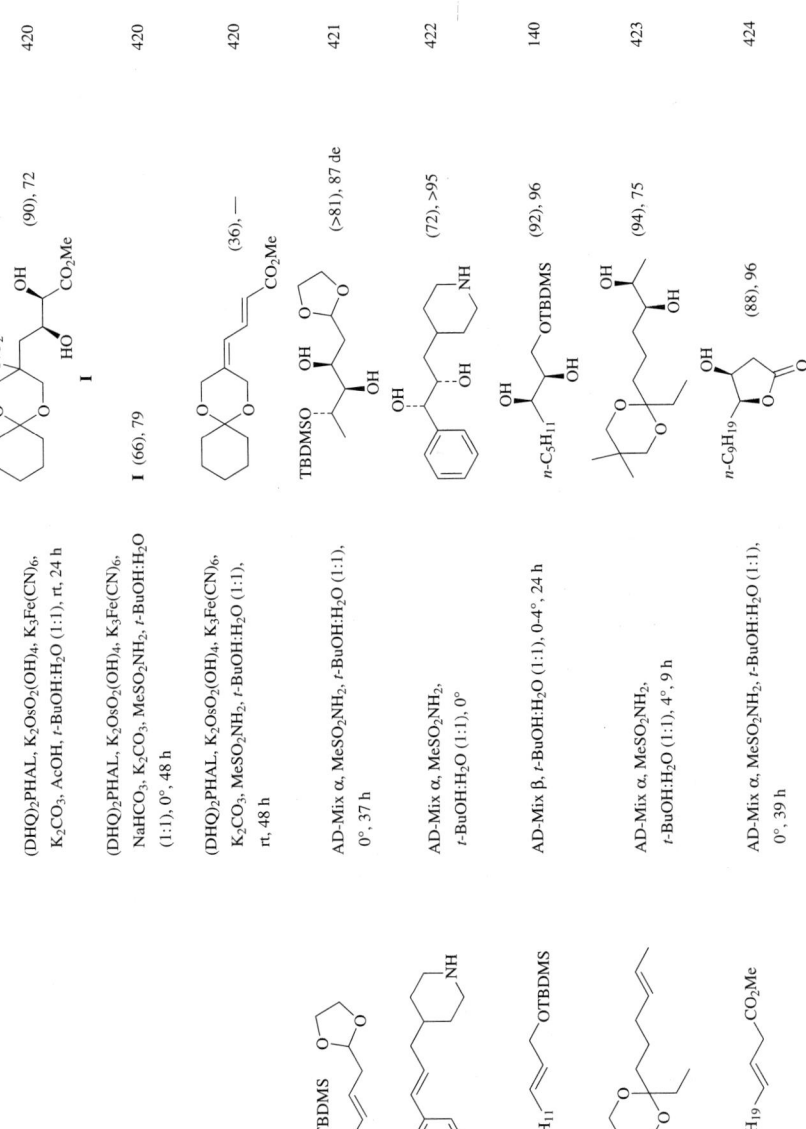

TABLE 3. REACTIONS OF TRANS 1,2-DISUBSTITUTED ALKENES (*Continued*)

Substrate	Conditions	Product(s) and Yield(s) (%), % ee	Refs.
C_{14}			
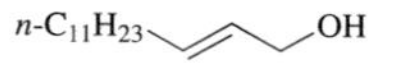	(DHQ)$_2$PHAL, $K_2OsO_2(OH)_4$, $K_3Fe(CN)_6$, K_2CO_3, $MeSO_2NH_2$, *t*-BuOH:H_2O (1:1), 0°	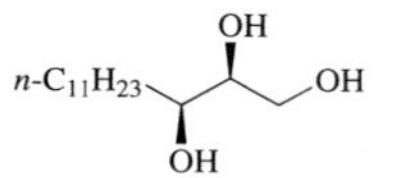 (84-88), 96	350
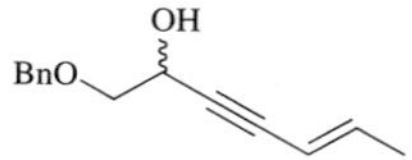	1. AD-Mix α, $MeSO_2NH_2$, *t*-BuOH:H_2O (1:1), 0°, 24 h 2. 2,2-Dimethoxypropane	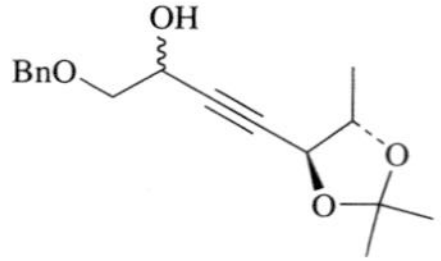 (71), 73	425
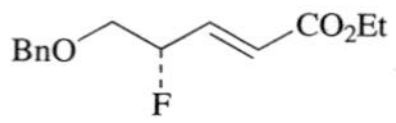	AD-Mix β, $MeSO_2NH_2$, *t*-BuOH:H_2O (1:1), rt, 2 d	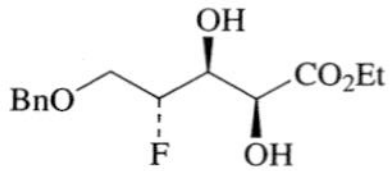 (85-96), 94 de	426
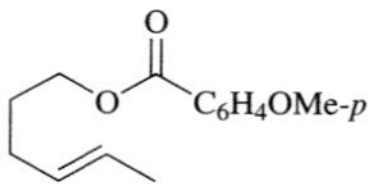	Ligand, $K_2OsO_2(OH)_4$, $K_3Fe(CN)_6$, K_2CO_3, $MeSO_2NH_2$, *t*-BuOH:H_2O (1:1), 0°, 4-22 h	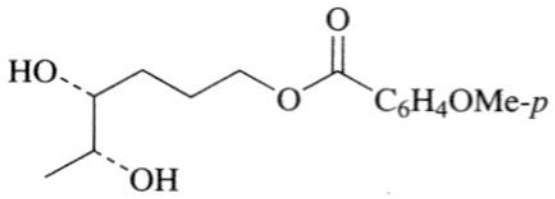	82
	Ligand		
	(DHQD)$_2$PYDZ	(97), 89	
	21	(98), 90	
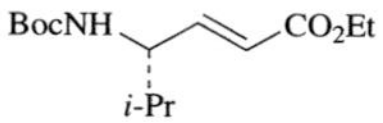	(DHQ)$_2$PHAL, $K_2OsO_2(OH)_4$, $K_3Fe(CN)_6$, K_2CO_3, *t*-BuOH:H_2O (1:1), rt, 16 d	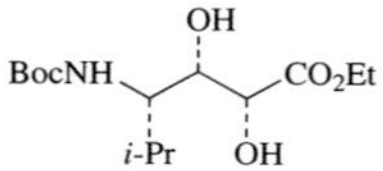 (80), 90 de	385
	(DHQD)$_2$PHAL, $K_2OsO_2(OH)_4$, $K_3Fe(CN)_6$, K_2CO_3, *t*-BuOH:H_2O (1:1), rt, 16 d	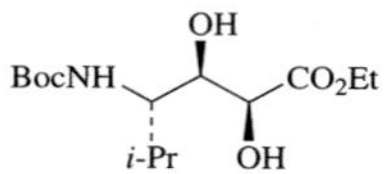 (73), 26 de	385

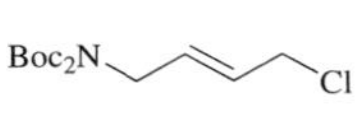	(DHQD)$_2$PHAL, $K_2OsO_2(OH)_4$, $K_3Fe(CN)_6$, K_2CO_3, $MeSO_2NH_2$, t-BuOH:H_2O (1:1), 0°	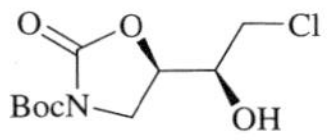	(80), 95	85
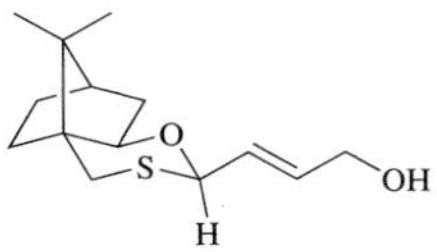	DHQD-OAc, OsO_4, toluene, –20°	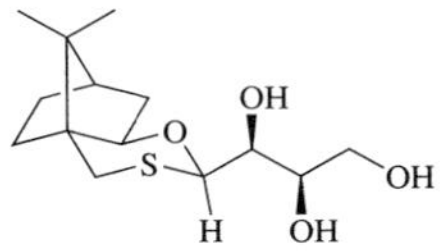	(62), 87	380
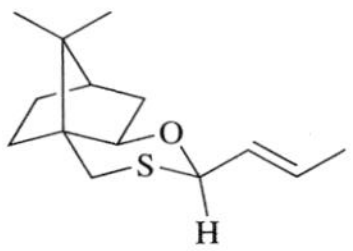	DHQD-OAc, OsO_4, toluene, –20°	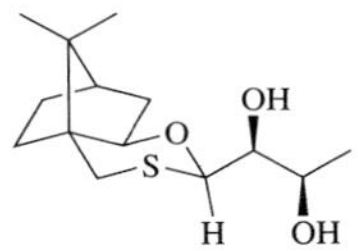	(60), 56	380
	DHQ-OAc, OsO_4, toluene, –20°	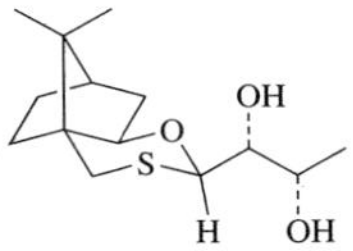	(58), 33	380
C_{15}				
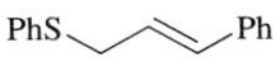	(DHQD)$_2$PHAL, $K_2OsO_2(OH)_4$, $K_3Fe(CN)_6$, K_2CO_3, t-BuOH:H_2O (1:1), 0°	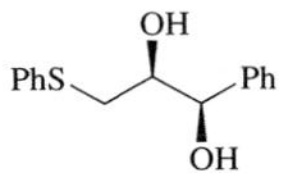	(75), 98	100
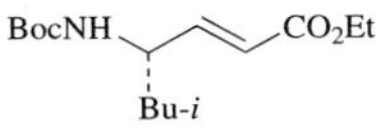	(DHQ)$_2$PHAL, $K_2OsO_2(OH)_4$, $K_3Fe(CN)_6$, K_2CO_3, t-BuOH:H_2O (1:1), rt, 5 d	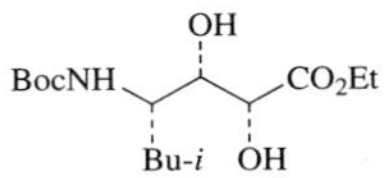	(99), 90 de	385
	(DHQD)$_2$PHAL, $K_2OsO_2(OH)_4$, $K_3Fe(CN)_6$, K_2CO_3, t-BuOH:H_2O (1:1), rt, 5 d	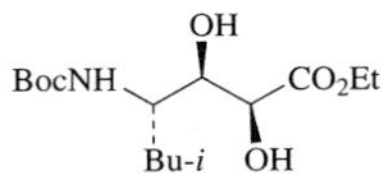	(99), 72 de	385

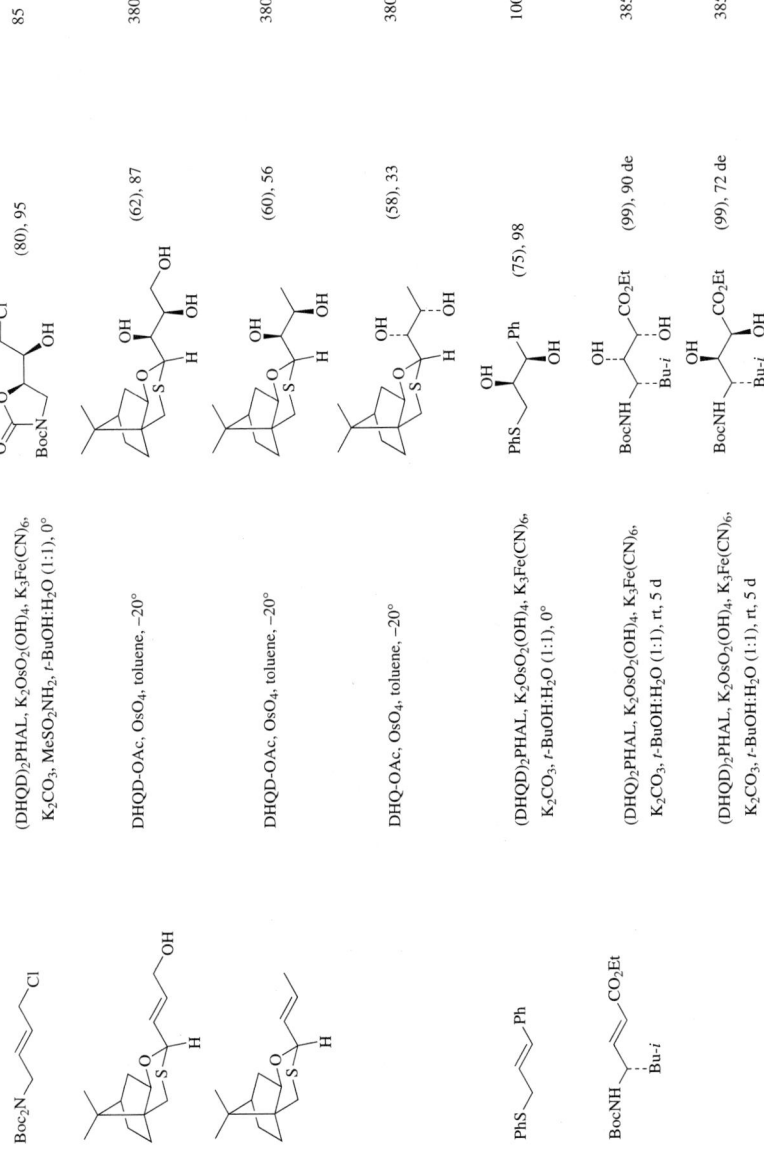

TABLE 3. REACTIONS OF TRANS 1,2-DISUBSTITUTED ALKENES (*Continued*)

Substrate	Conditions	Product(s) and Yield(s) (%), % ee	Refs.
C_{15}			
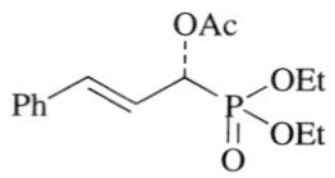	1. $(DHQD)_2PHAL$, $K_2OsO_2(OH)_4$, $K_3Fe(CN)_6$, K_2CO_3, $MeSO_2NH_2$, t-BuOH:H_2O (1:1), rt, 48 h 2. Ac_2O, pyridine, DMAP, CH_2Cl_2	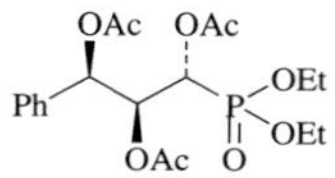 (49), 98	150
	1. $(DHQ)_2PHAL$, $K_2OsO_2(OH)_4$, $K_3Fe(CN)_6$, K_2CO_3, $MeSO_2NH_2$, t-BuOH:H_2O (1:1), rt, 48 h 2. Ac_2O, pyridine, DMAP, CH_2Cl_2	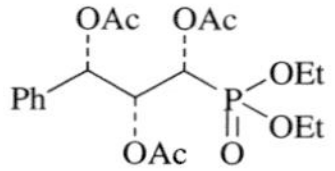 (28), 78	150
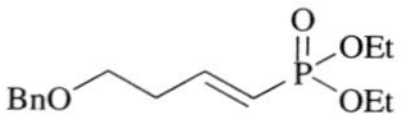	$(DHQ)_2PHAL$, $K_2OsO_2(OH)_4$, $K_3Fe(CN)_6$, K_2CO_3, $MeSO_2NH_2$, t-BuOH:H_2O (1:1), rt, 48 h	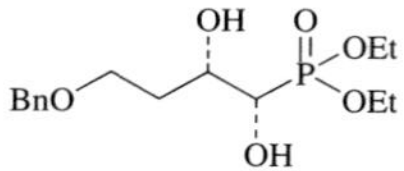 (30), 44	344
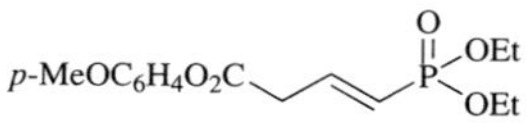	$(DHQ)_2PHAL$, $K_2OsO_2(OH)_4$, $K_3Fe(CN)_6$, K_2CO_3, $MeSO_2NH_2$, t-BuOH:H_2O (1:1), 0°-rt, 50 h	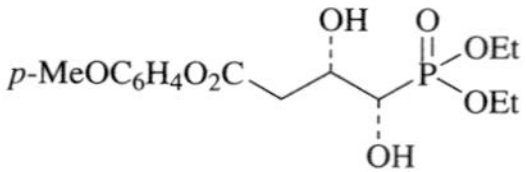 (10), 97	345
	AD-Mix β, t-BuOH:H_2O (1:1)	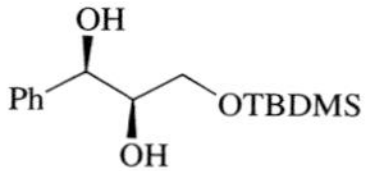 (94), >95	427
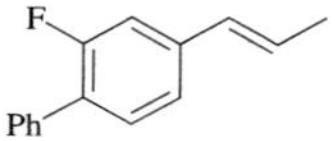	AD-Mix β, $MeSO_2NH_2$, t-BuOH:H_2O (1:1), 0°, 24 h	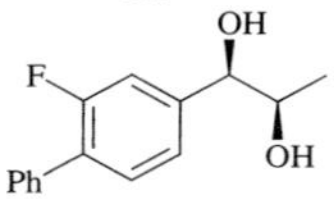 (99), 99	428, 429
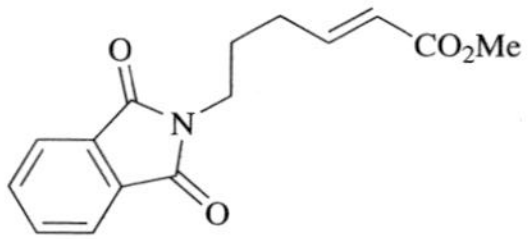	AD-Mix β, t-BuOH:H_2O (1:1), 0°, 4 h	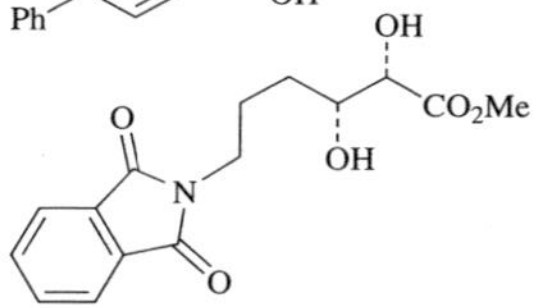 (90), >99	105

Reactant	Conditions	Product(s) and Yield(s) (%)	Refs.
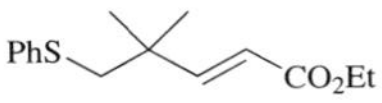	Ligand, $K_2OsO_2(OH)_4$, $K_3Fe(CN)_6$, K_2CO_3, $MeSO_2NH_2$, t-BuOH:H_2O (1:1)	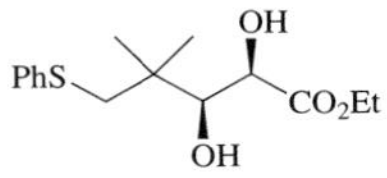	430
	Ligand		
	$(DHQ)_2$PHAL	(—), 79	
	$(DHQ)_2$PYR	(—), 83	
	DHQ-*p*-chlorobenzoate	(—), 64	
	DHQ-MEQ	(—), 85	
	DHQ-PHN	(94), 90	
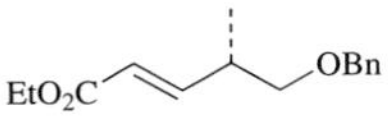	AD-Mix α, $MeSO_2NH_2$, t-BuOH-H_2O, 0°, 8 h	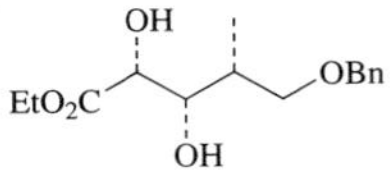 (85), 71 de	431
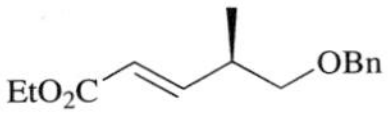	$(DHQD)_2$PHAL, OsO_4, $K_3Fe(CN)_6$, K_2CO_3, $MeSO_2NH_2$, t-BuOH:H_2O (1:1), rt, 24 h	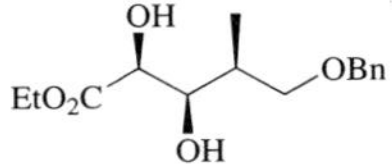 **I** (100), 56 de	432
	AD-Mix β, t-BuOH:H_2O (1:1), 0°	**I** (85), 60 de	433
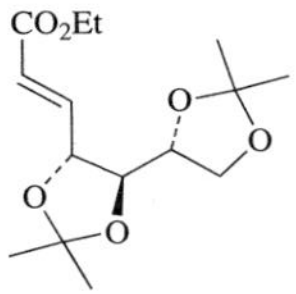	$(DHQ)_2$PHAL, $K_2OsO_2(OH)_4$, $K_3Fe(CN)_6$, K_2CO_3, $MeSO_2NH_2$, t-BuOH:H_2O (1:1), 0°, 25 h	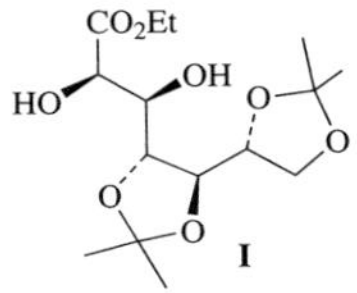 **I** (93), >95 de	434
	AD-Mix α, $MeSO_2NH_2$, t-BuOH:H_2O (1:1)	**I** (82-90), >94 de	370
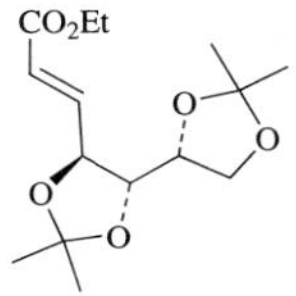	$(DHQ)_2$PHAL, $K_2OsO_2(OH)_4$, $K_3Fe(CN)_6$, K_2CO_3, $MeSO_2NH_2$, t-BuOH:H_2O (1:1), rt, 18 h	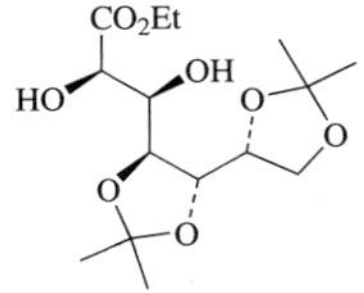 (94), 98 de	374

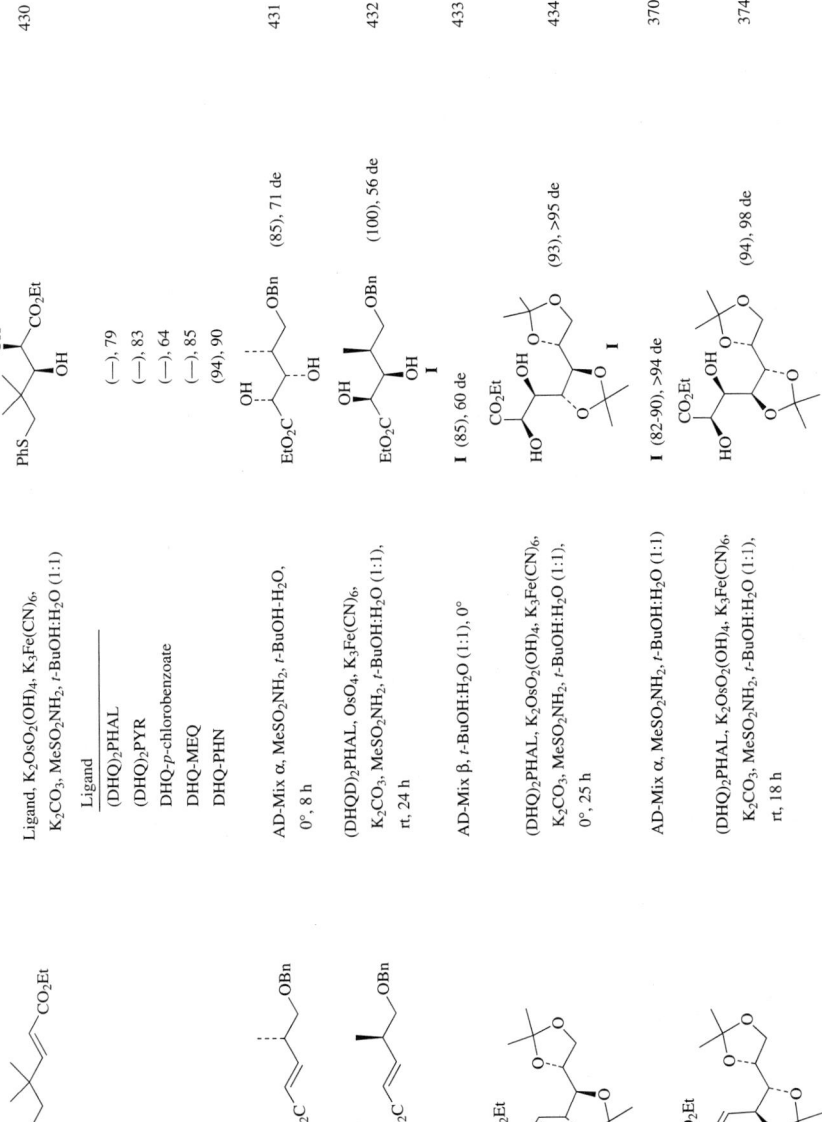

TABLE 3. REACTIONS OF TRANS 1,2-DISUBSTITUTED ALKENES (*Continued*)

Substrate	Conditions	Product(s) and Yield(s) (%), % ee	Refs.
C_{15}			
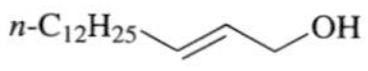	(DHQ)$_2$PHAL, $K_2OsO_2(OH)_4$, $K_3Fe(CN)_6$, K_2CO_3, $MeSO_2NH_2$, t-BuOH:H_2O (1:1), 0°	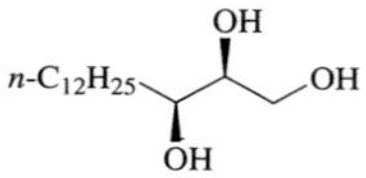 (84-88), 96	350
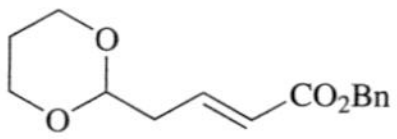	AD-Mix α, $MeSO_2NH_2$, t-BuOH:H_2O (1:1), 4°, 24 h	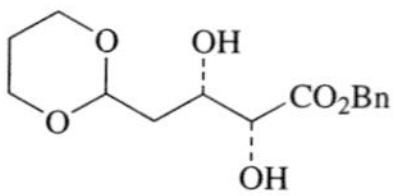 (80), >99	103
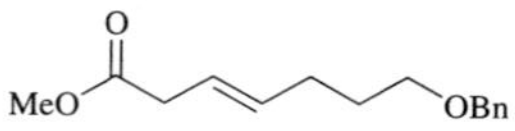	AD-Mix β, t-BuOH:H_2O (1:1)	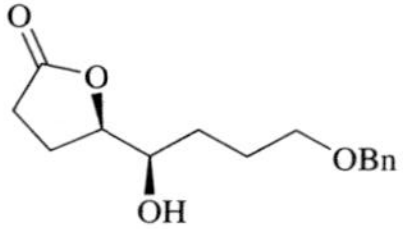 (95), >95	435
C_{16}			
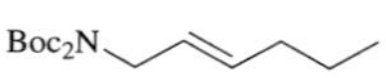	(DHQD)$_2$PHAL, $K_2OsO_2(OH)_4$, $K_3Fe(CN)_6$, K_2CO_3, $MeSO_2NH_2$, t-BuOH:H_2O (1:1), 0°	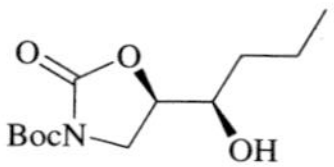 (80), 90	85
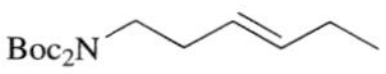	(DHQD)$_2$PHAL, $K_2OsO_2(OH)_4$, $K_3Fe(CN)_6$, K_2CO_3, $MeSO_2NH_2$, t-BuOH:H_2O (1:1), 0°	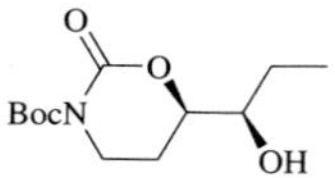 (73), 89	85
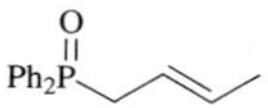	(DHQ)$_2$PHAL, $K_2OsO_2(OH)_4$, $K_3Fe(CN)_6$, K_2CO_3, $MeSO_2NH_2$, t-BuOH:H_2O (1:1), 0°, 6-24 h	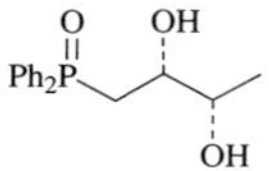 (84), 10	120, 119
	DHQD-p-chlorobenzoate, $K_2OsO_2(OH)_4$, $K_3Fe(CN)_6$, K_2CO_3, $MeSO_2NH_2$, t-BuOH:H_2O (1:1), rt, 1-5 d	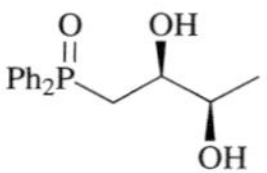 (94), 46	120, 119

	DHQD-PHN, $K_2OsO_2(OH)_4$, $K_3Fe(CN)_6$, K_2CO_3, t-BuOH:H_2O (1:1), 0°, 15-30 h	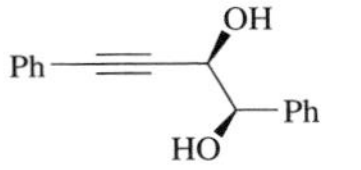 (67), 97	219
	$(DHQD)_2$PHAL, $K_2OsO_2(OH)_4$, $K_3Fe(CN)_6$, K_2CO_3, t-BuOH:H_2O (1:1), 0°	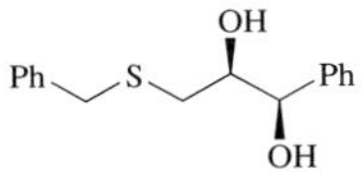 (72), 98	100
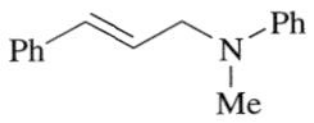	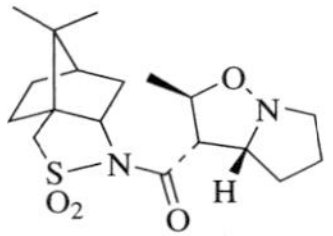 , OsO_4, $K_3Fe(CN)_6$, K_2CO_3, t-BuOH:H_2O (1:1), rt, 24 h	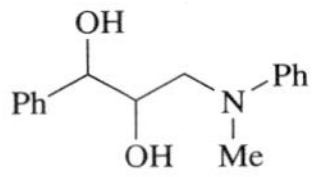 (99), 30	203
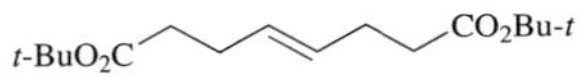	$(DHQD)_2$PHAL, $K_2OsO_2(OH)_4$, $K_3Fe(CN)_6$, K_2CO_3, $MeSO_2NH_2$, t-BuOH:H_2O (1:1), 0°, 24 h	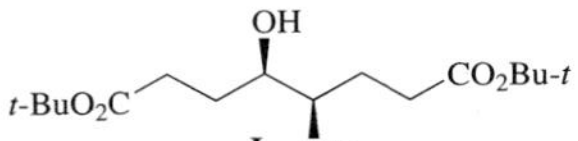 **I** (95), 98	110
	AD-Mix β, t-BuOH:H_2O (1:1), 0°-rt, 24 h	**I** (89), >98	436
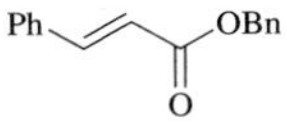	$(DHQD)_2$PHAL, $K_2OsO_2(OH)_4$, $K_3Fe(CN)_6$, K_2CO_3, $MeSO_2NH_2$, t-BuOH:H_2O (1:1), 0°	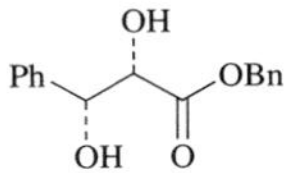 (62), 94	365
	$(DHQ)_2$PHAL, $K_2OsO_2(OH)_4$, $K_3Fe(CN)_6$, K_2CO_3, $MeSO_2NH_2$, t-BuOH:H_2O (1:1), 0°	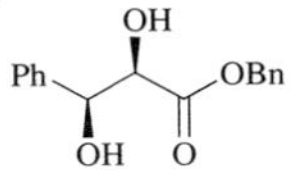 (71), 90	365
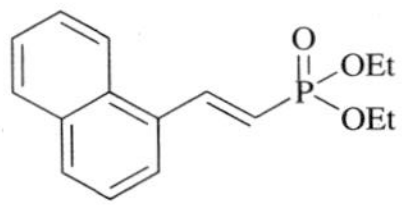	$(DHQ)_2$PHAL, $K_2OsO_2(OH)_4$, $K_3Fe(CN)_6$, K_2CO_3, $MeSO_2NH_2$, t-BuOH:H_2O (1:1), 0-24°, 50 h	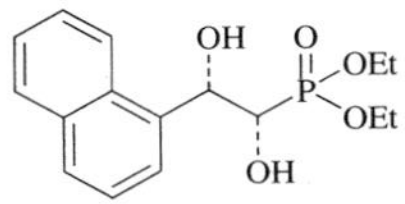 (80), 93	345

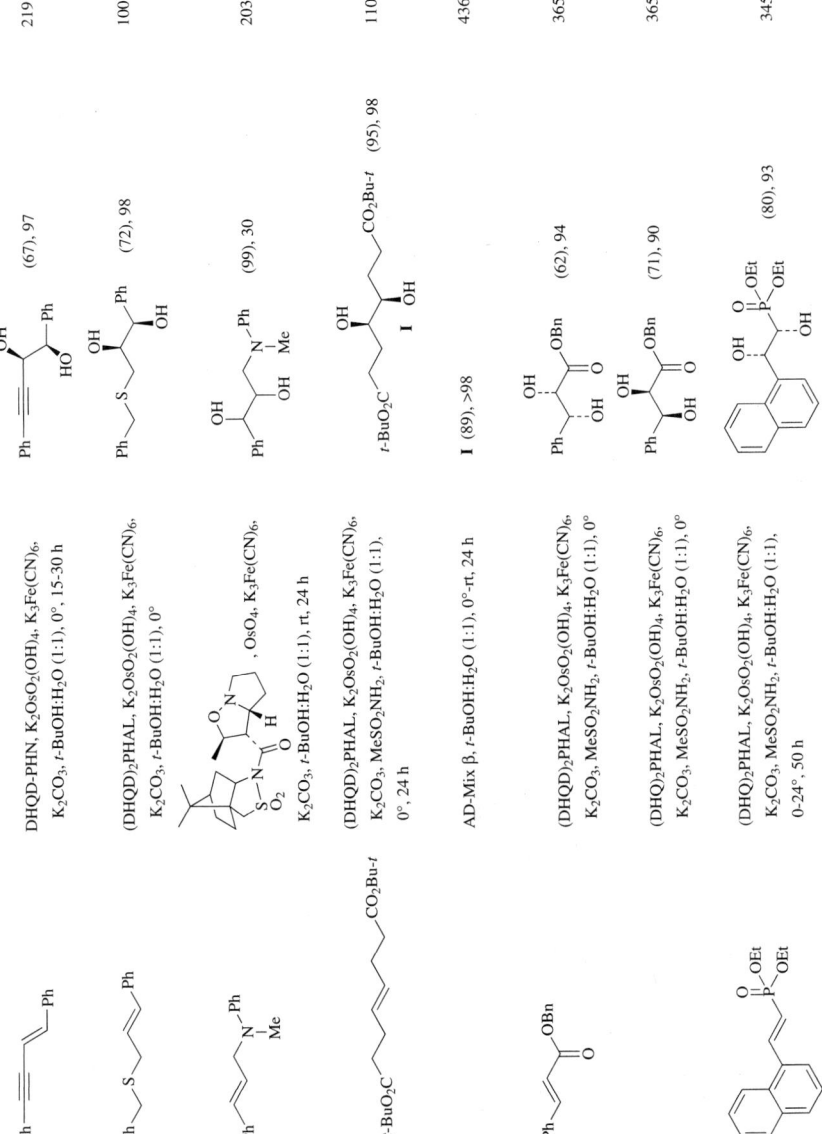

TABLE 3. REACTIONS OF TRANS 1,2-DISUBSTITUTED ALKENES (*Continued*)

Substrate	Conditions	Product(s) and Yield(s) (%), % ee	Refs.
C_{16}			
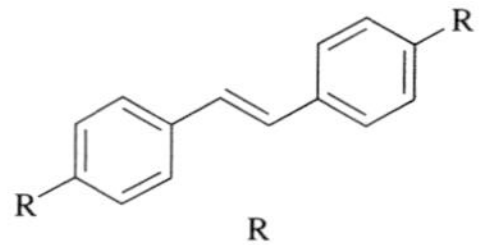	$(DHQD)_2PHAL$, $K_2OsO_2(OH)_4$, $K_3Fe(CN)_6$, K_2CO_3, $MeSO_2NH_2$, t-BuOH:H_2O (1:1), 0°	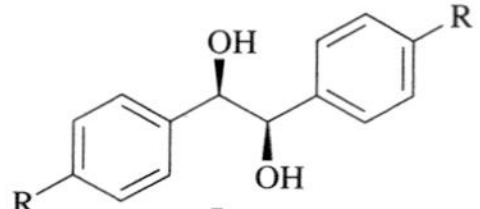 **I**	412
R: CF_3		(79), 82	
R: Me		(87), 91	
R = CF_3, Me, MeO	AD-Mix β, t-BuOH:H_2O (1:1), 0°	**I** (—), >90	408
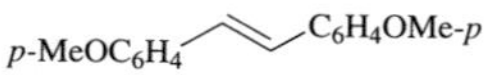	**35**, OsO_4, CH_2Cl_2, –90°, 2 h	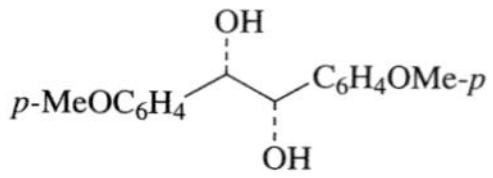 (90), 82	11
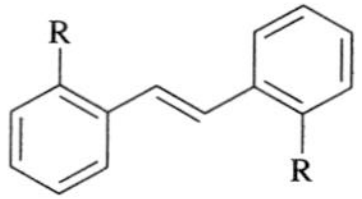	Ligand, $K_2OsO_2(OH)_4$, $K_3Fe(CN)_6$, K_2CO_3, $MeSO_2NH_2$, t-BuOH:H_2O (1:1), 0°, 24 h	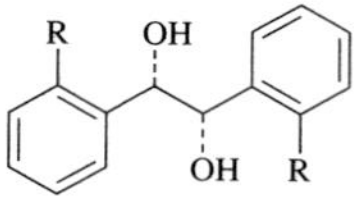	417
R	Ligand		
Me	DHQ-p-chlorobenzoate	(92), >98	
Me	$(DHQ)_2PHAL$	(88), >98	
MeO	$(DHQ)_2PHAL$	(41), >95	
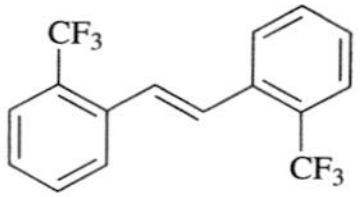	$(DHQD)_2PHAL$, $K_2OsO_2(OH)_4$, $K_3Fe(CN)_6$, K_2CO_3, $MeSO_2NH_2$, t-BuOH:H_2O (1:1), 0°, 24 h	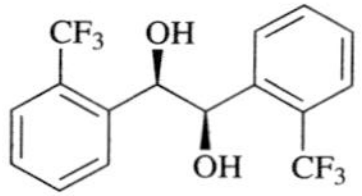 (4), —	417
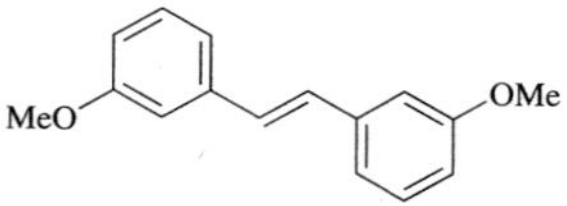	AD-Mix β, t-BuOH:H_2O (1:1), rt, 96 h	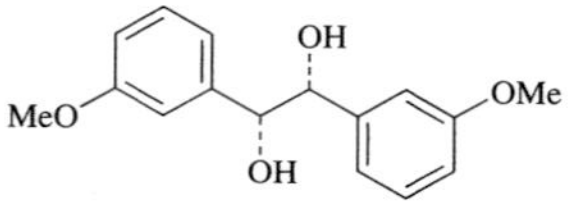 (96), >99	319, 416

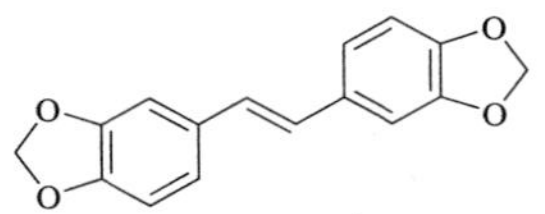	AD-Mix β, t-BuOH:H_2O (1:1), 0°	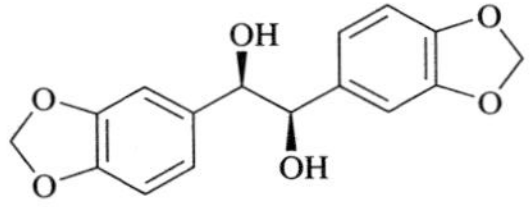 (84), >99	416
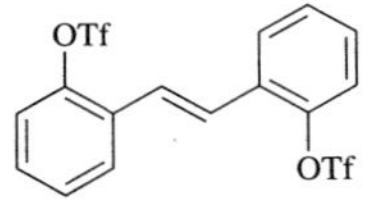	DHQ-p-chlorobenzoate, OsO_4, toluene	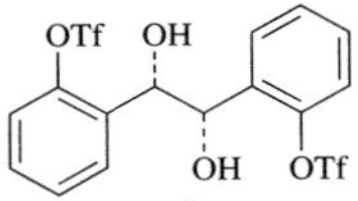 **I** (70), 82	415
	DHQ-p-chlorobenzoate, OsO_4, NMO, acetone-H_2O	**I** (80), 45	415
	DHQD-p-chlorobenzoate, OsO_4, toluene	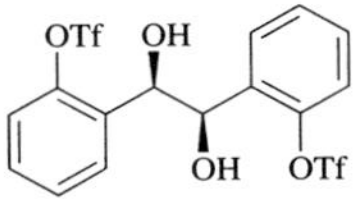 (70), 70	415
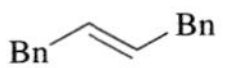	$(DHQD)_2PHAL$, $K_2OsO_2(OH)_4$, $K_3Fe(CN)_6$, K_2CO_3, $MeSO_2NH_2$, t-BuOH:H_2O (1:1), 0°, 24 h	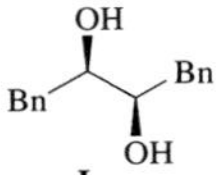 **I** (84), —	417
	AD-Mix β, $MeSO_2NH_2$, t-BuOH:H_2O (1:1)	**I** (82), —	437
	AD-Mix α, $MeSO_2NH_2$, t-BuOH:H_2O (1:1) 0-5°, 24 h	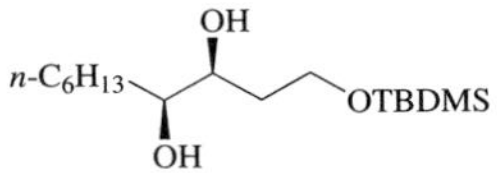 (98), 91	438
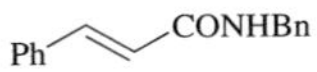	$(DHQ)_2PHAL$, $K_2OsO_2(OH)_4$, $K_3Fe(CN)_6$, K_2CO_3, $MeSO_2NH_2$, t-BuOH:H_2O (1:1), 0°, 18 h	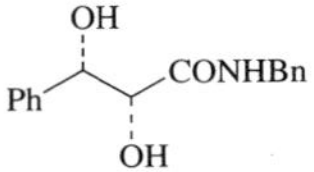 (88), 96	355
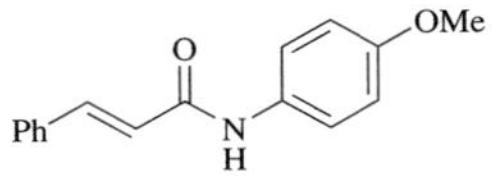	$(DHQ)_2PHAL$, $K_2OsO_2(OH)_4$, $K_3Fe(CN)_6$, K_2CO_3, $MeSO_2NH_2$, t-BuOH:H_2O (1:1), 0°, 24 h	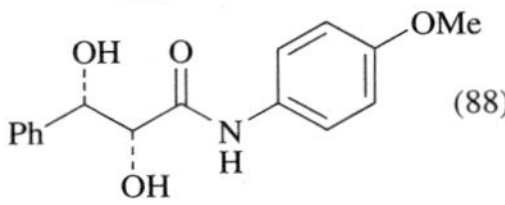 (88), 94	355

TABLE 3. REACTIONS OF TRANS 1,2-DISUBSTITUTED ALKENES (*Continued*)

Substrate	Conditions	Product(s) and Yield(s) (%), % ee	Refs.
C_{16} 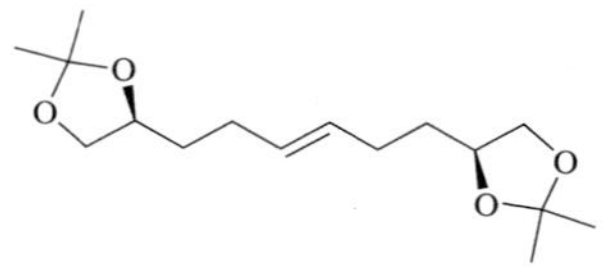	AD-Mix α, *t*-BuOH:H_2O (1:1), 0-20°, 12 h	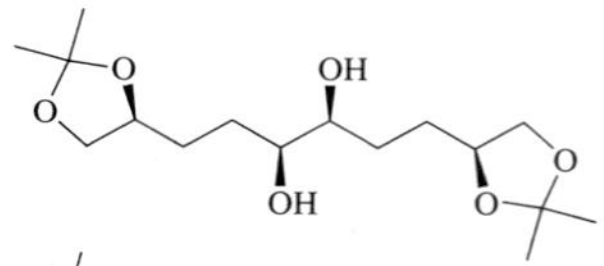 (84), 80 de	439
	AD-Mix β, *t*-BuOH:H_2O (1:1), 0-20°, 12 h	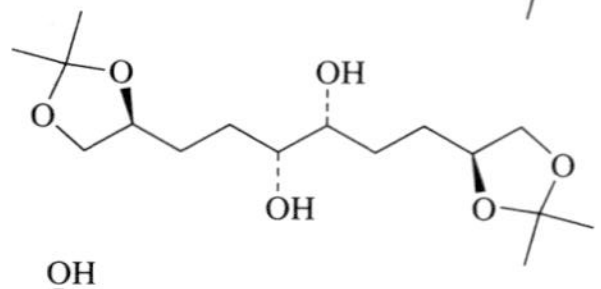 (98), 80 de	439
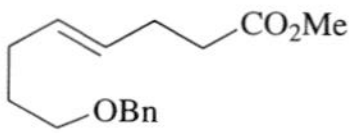	AD-Mix β, $MeSO_2NH_2$, *t*-BuOH:H_2O (1:1)	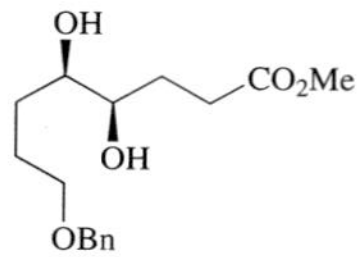 (100), —	440
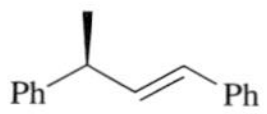	AD-Mix α, $MeSO_2NH_2$, *t*-BuOH:H_2O (1:1), 0°, 24 h	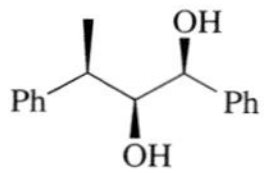 (85), 80 de	441
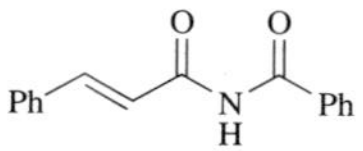	$(DHQ)_2PHAL$, OsO_4, $K_3Fe(CN)_6$, K_2CO_3, $MeSO_2NH_2$, *t*-BuOH:H_2O (1:1), 0°, 72 h	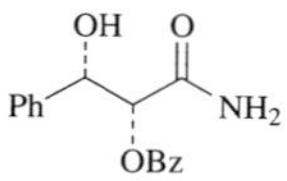 (42), —	355

C_{17}				
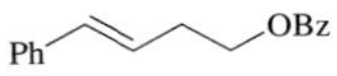	$(DHQD)_2PHAL$, $K_2OsO_2(OH)_4$, $K_3Fe(CN)_6$, K_2CO_3, $MeSO_2NH_2$, t-BuOH:H_2O (1:1), 0°	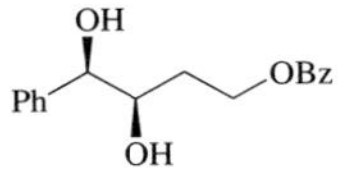	(—), 99	98
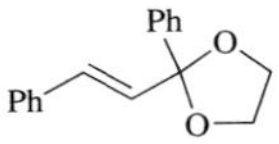	DHQD-*p*-chlorobenzoate, $K_2OsO_2(OH)_4$, NMO, acetone:H_2O (10:1), 0°, 21 h (slow addition of olefin)	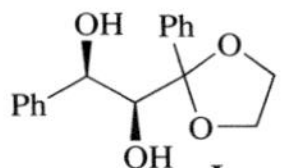 **I**	(80-95), 84[a]	112
	DHQD-*p*-chlorobenzoate, $K_2OsO_2(OH)_4$, NMO, TEAA, acetone:H_2O (10:1), 0°, 21 h (slow addition of olefin)	**I**	(80-95), 87	112
	DHQD-*p*-chlorobenzoate, $K_2OsO_2(OH)_4$, NMO, acetone:H_2O (10:1, 0.15 M), 0°	**I**	(—), 86	112
 E:Z = 92:8	DHQD-*p*-chlorobenzoate, $K_2OsO_2(OH)_4$, $K_3Fe(CN)_6$, K_2CO_3, $MeSO_2NH_2$, t-BuOH:H_2O (1:1), rt, 1-5 d	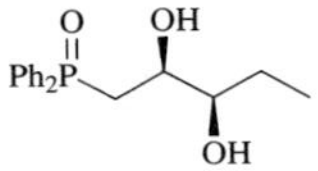	(95), 76	119
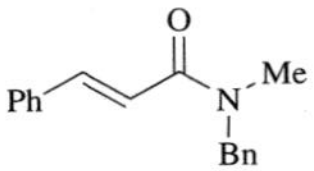	$(DHQD)_2PHAL$, $K_2OsO_2(OH)_4$, $K_3Fe(CN)_6$, K_2CO_3, $MeSO_2NH_2$, t-BuOH:H_2O (1:1), rt	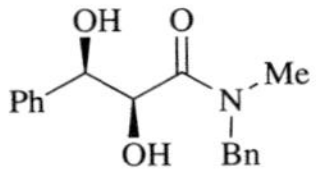	(95), 98	111
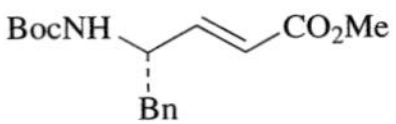	$(DHQ)_2PHAL$, $K_2OsO_2(OH)_4$, $K_3Fe(CN)_6$, K_2CO_3, t-BuOH:H_2O (1:1), rt, 19 h	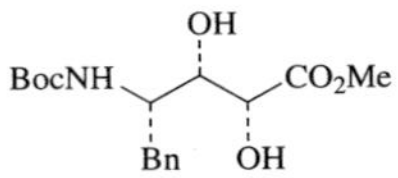	(93), 88 de	385
	$(DHQD)_2PHAL$, $K_2OsO_2(OH)_4$, $K_3Fe(CN)_6$, K_2CO_3, t-BuOH:H_2O (1:1), rt, 19 h	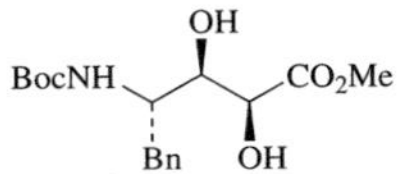	(84), 78 de	385

TABLE 3. REACTIONS OF TRANS 1,2-DISUBSTITUTED ALKENES (*Continued*)

Substrate	Conditions	Product(s) and Yield(s) (%), % ee	Refs.
C_{17}			
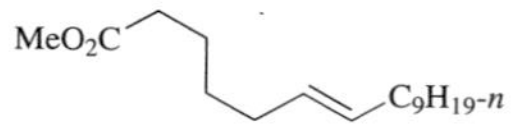	$(DHQ)_2PHAL$, $K_2OsO_2(OH)_4$, $K_2Fe(CN)_6$, K_2CO_3, *t*-BuOH:H_2O (1:1), 0°, 18 h	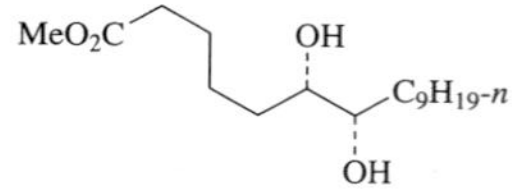 (97), 85	401
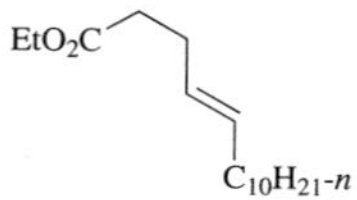	AD-Mix β, $MeSO_2NH_2$, *t*-BuOH:H_2O (1:1)	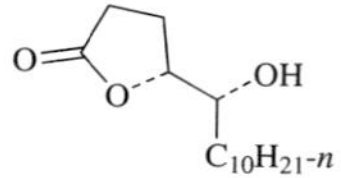 (89), —	442
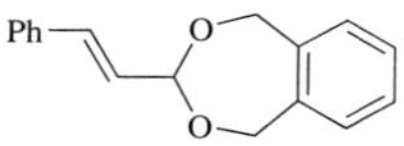	$(DHQD)_2PHAL$, $K_2OsO_2(OH)_4$, $K_3Fe(CN)_6$, K_2CO_3, *t*-BuOH:H_2O (1:1), rt, 48 h	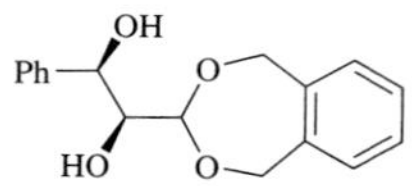 (91), >95	84
	$(DHQ)_2PHAL$, $K_2OsO_2(OH)_4$, $K_3Fe(CN)_6$, K_2CO_3, *t*-BuOH:H_2O (1:1), rt, 48 h	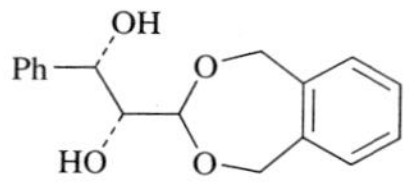 (100), 79	84
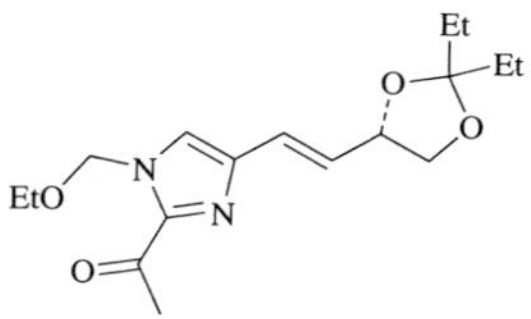	$(DHQD)_2PHAL$, $K_2OsO_2(OH)_4$, $K_3Fe(CN)_6$, K_2CO_3, $MeSO_2NH_2$, *t*-BuOH:H_2O (1:1), 0°, 3 d	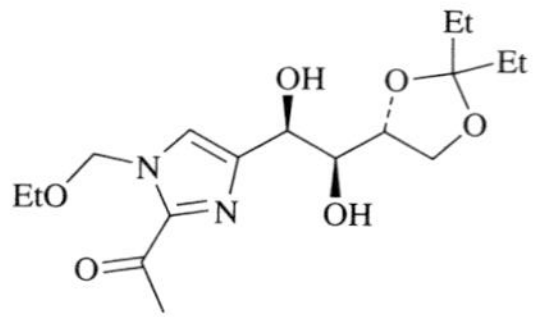 (80), 99 de	443
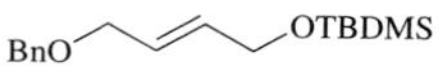	$(DHQD)_2PHAL$, $K_2OsO_2(OH)_4$, $K_3Fe(CN)_6$, K_2CO_3, $MeSO_2NH_2$, *t*-BuOH:H_2O (1:1)	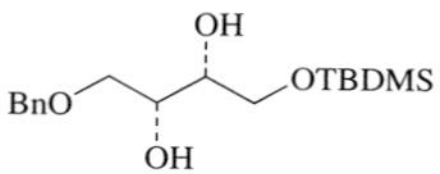 (89), 95	124
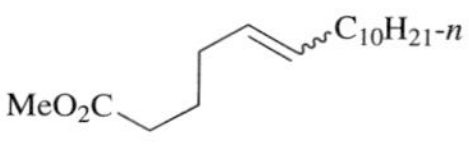	1. $(DHQ)_2PHAL$, OsO_4, K_2CO_3, $K_3Fe(CN)_6$, *t*-BuOH:H_2O (1:1) 2. Recrystallization	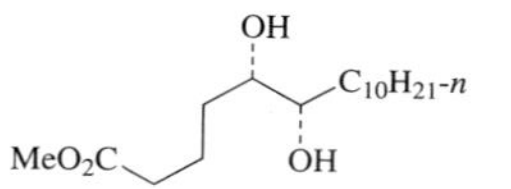 (62), 83	444

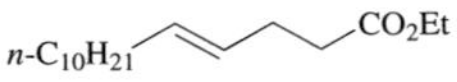	AD-Mix β, t-BuOH:H_2O (1:1), 0°	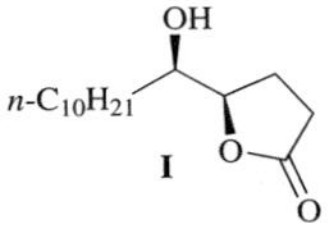 **I**	(94), 98	445
	AD-Mix β, $MeSO_2NH_2$, t-BuOH:H_2O (1:1), 0°, 24 h	**I** (89), 96		446
	AD-Mix α, $MeSO_2NH_2$, t-BuOH:H_2O (1:1), 0°, 24 h	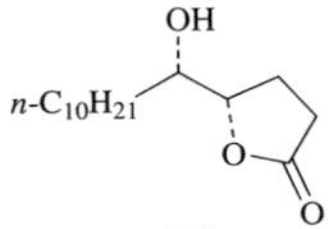	(86), 94	446
	AD-Mix β, $MeSO_2NH_2$, t-BuOH:H_2O (1:1), 0°, 24 h	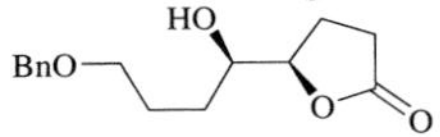	(75-93), 96	446, 447
	AD-Mix α, $MeSO_2NH_2$, t-BuOH:H_2O (1:1), 0°, 24 h	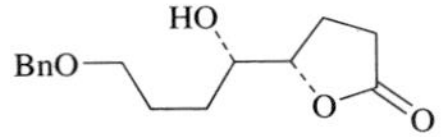	(72), 94	446
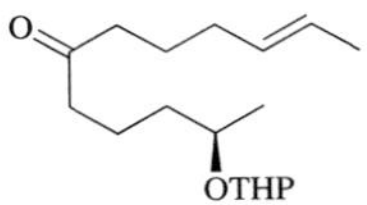	1. AD-Mix α, t-BuOH:H_2O (1:1), 0°, 6 days 2. H^+	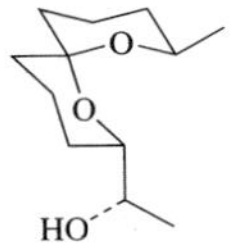	(33), —	268
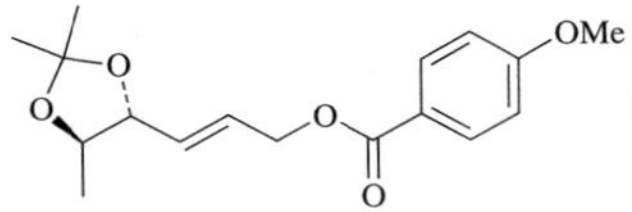	$(DHQD)_2PYDZ$, $K_2OsO_2(OH)_4$, $K_3Fe(CN)_6$, K_2CO_3, $MeSO_2NH_2$, t-BuOH:H_2O (1:1), 0°	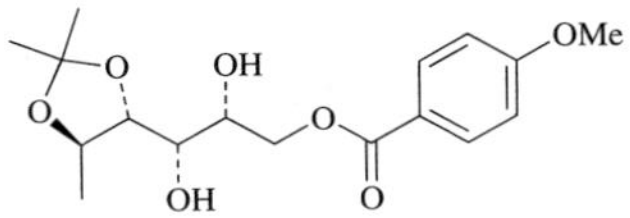	(86), 96 de	143
	$(DHQ)_2PHAL$, $K_2OsO_2(OH)_4$, $K_3Fe(CN)_6$, K_2CO_3, $MeSO_2NH_2$, t-BuOH:H_2O (1:1), 0°	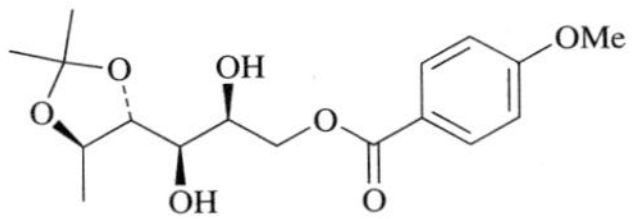	(86), 94 de	143

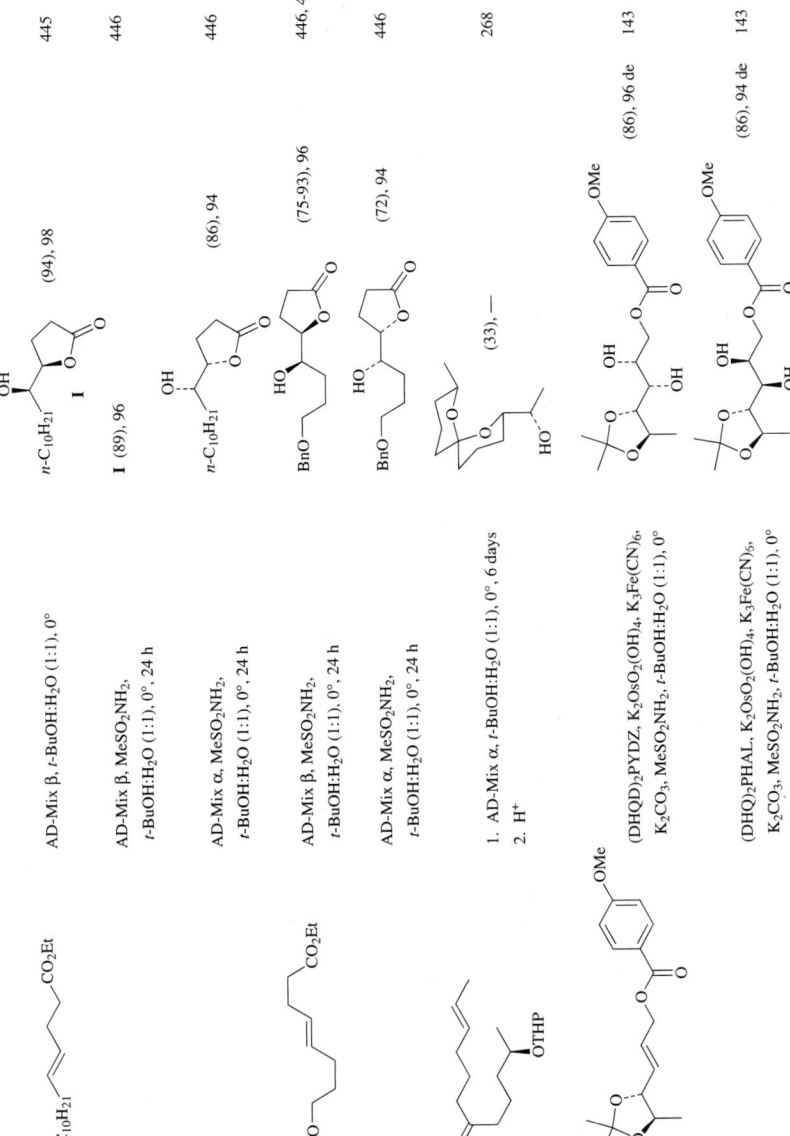

TABLE 3. REACTIONS OF TRANS 1,2-DISUBSTITUTED ALKENES (*Continued*)

Substrate	Conditions	Product(s) and Yield(s) (%), % ee	Refs.
C_{17} 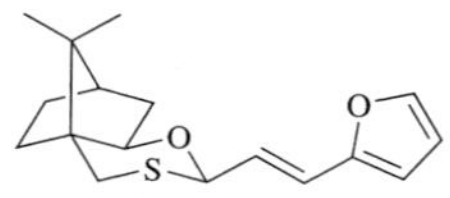	DHQD-OAc, OsO_4, toluene, –20°	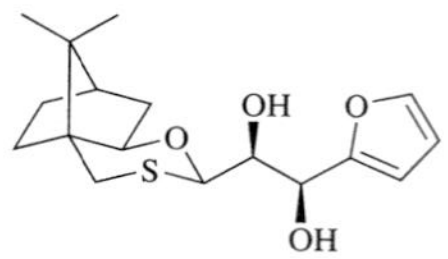 (66), 93	380
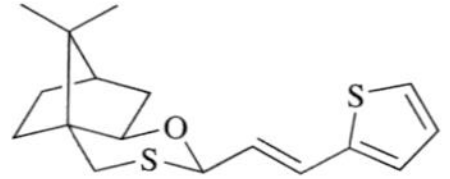	DHQD-OAc, OsO_4, toluene, –20°	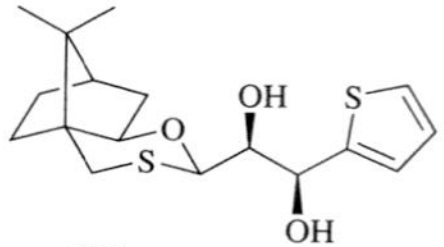 (68), 83	380
	$(DHQD)_2PHAL$, $K_2OsO_2(OH)_4$, $K_3Fe(CN)_6$, $MeSO_2NH_2$, K_2CO_3, *t*-BuOH:H_2O, 4°, 24 h	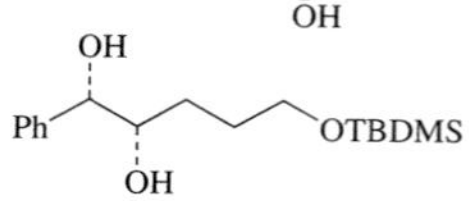 (94), 97	448
C_{18} 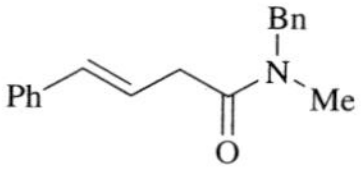	$(DHQD)_2PHAL$, $K_2OsO_2(OH)_4$, $K_3Fe(CN)_6$, K_2CO_3, $MeSO_2NH_2$, *t*-BuOH:H_2O (1:1), rt	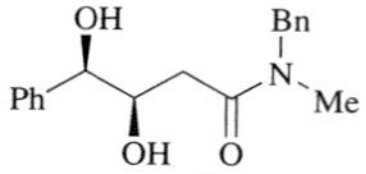 **I** (97), 98	111
	AD-Mix β, $MeSO_2NH_2$, *t*-BuOH:H_2O (1:1), 0°	**I** (97), 98	111
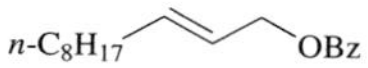	$(DHQD)_2PHAL$, $K_2OsO_2(OH)_4$, $K_3Fe(CN)_6$, K_2CO_3, $MeSO_2NH_2$, *t*-BuOH:H_2O (1:1), 0°	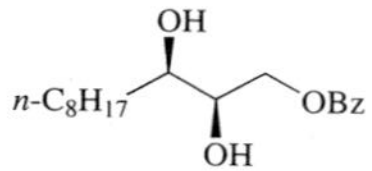 (—), 99	98
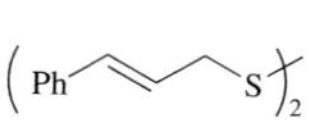	$(DHQD)_2PHAL$, $K_2OsO_2(OH)_4$, $K_3Fe(CN)_6$, K_2CO_3, *t*-BuOH:H_2O (1:1), 0°	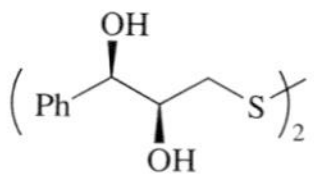 (80), 95 de	100

C$_{18}$

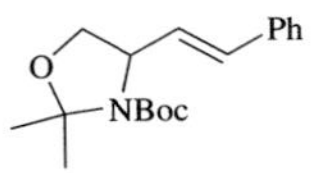	$(DHQD)_2PHAL$, $K_2OsO_2(OH)_4$, $K_3Fe(CN)_6$, K_2CO_3, $MeSO_2NH_2$, t-BuOH:H_2O (1:1), 0°-rt, 24 h	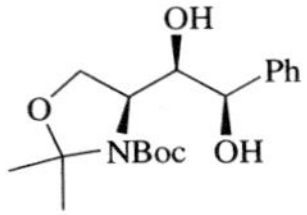 (60), 98 de	449
	$(DHQ)_2PHAL$, $K_2OsO_2(OH)_4$, $K_3Fe(CN)_6$, K_2CO_3, $MeSO_2NH_2$, t-BuOH:H_2O (1:1), 0°-rt, 24 h	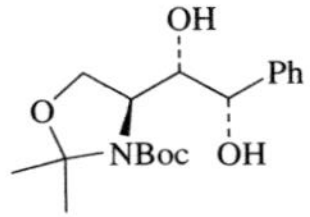 (70), 98 de	449
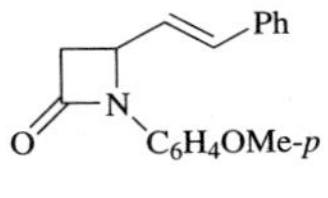	1. DHQD-p-chlorobenzoate, $K_2OsO_2(OH)_4$, $K_3Fe(CN)_6$, K_2CO_3, t-BuOH:H_2O (1:1), rt, 15 h 2. $(MeO)_2CMe_2$, TsOH	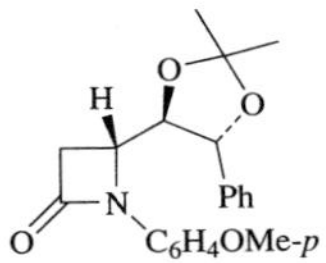 **I** 90% ee + 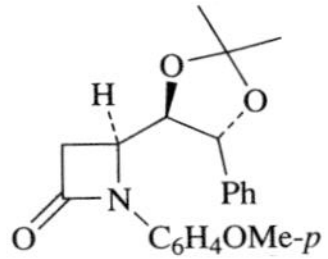 **II** 85% ee **I** + **II** = (70), **I**:**II** = 3:2	450
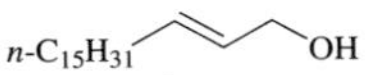	$(DHQD)_2PHAL$, OsO_4, $K_3Fe(CN)_6$, K_2CO_3, $MeSO_2NH_2$, t-BuOH:H_2O (1:1), 0°, 24 h	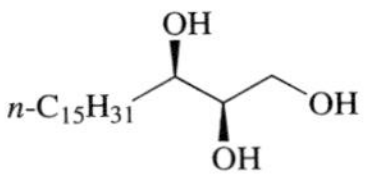 (91), —	451
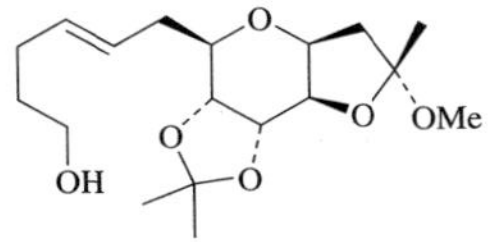	AD-Mix β, $MeSO_2NH_2$, t-BuOH:H_2O (1:1), 0°, 24 h	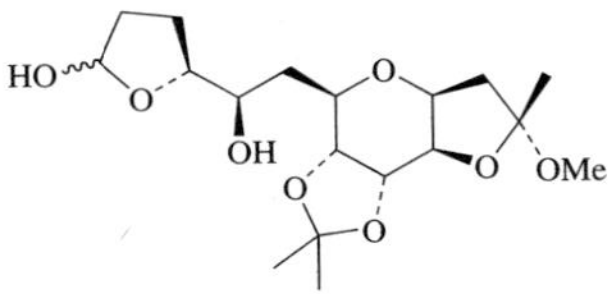 (96), >90 de	453

TABLE 3. REACTIONS OF TRANS 1,2-DISUBSTITUTED ALKENES (*Continued*)

Substrate	Conditions	Product(s) and Yield(s) (%), % ee	Refs.
C_{18} 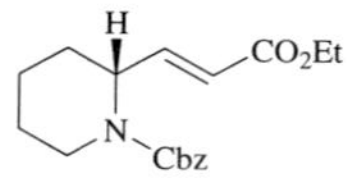	AD-Mix β, t-BuOH:H_2O (1:1), rt, 24 h	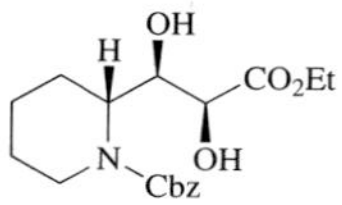 (—), 84 de	454
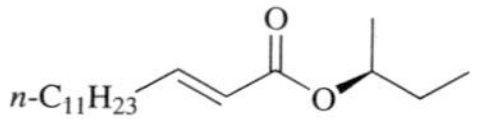	AD-Mix α, $MeSO_2NH_2$, t-BuOH:H_2O (1:1), 0°, 24 h	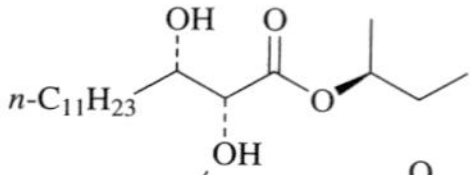 (74), —	455
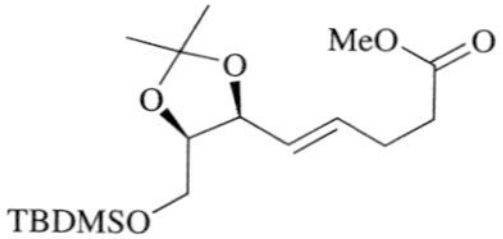	AD-Mix β, $MeSO_2NH_2$, t-BuOH:H_2O (1:1), rt, 18 h	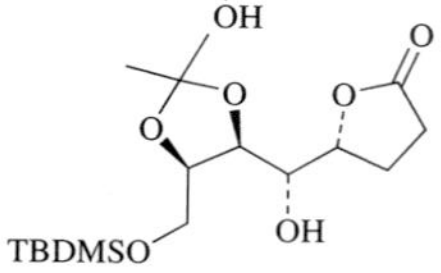 (79), 84 de	456
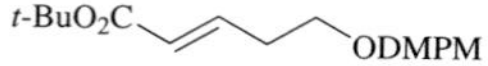	AD-Mix β, $MeSO_2NH_2$, t-BuOH:H_2O (1:1), 0°, 20 h	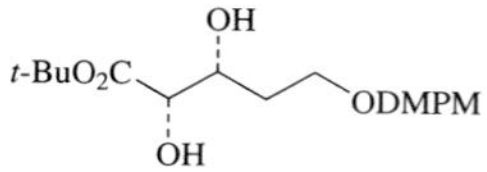 (99), 96	456a
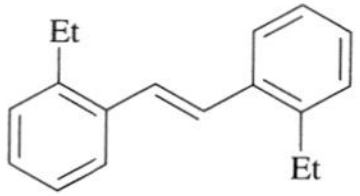	$(DHQD)_2PHAL$, $K_2OsO_2(OH)_4$, $K_3Fe(CN)_6$, K_2CO_3, $MeSO_2NH_2$, t-BuOH:H_2O (1:1), 20°, 4 days	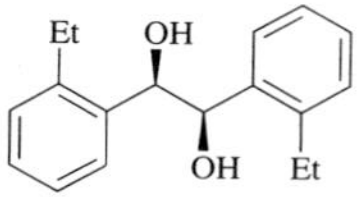 (91), —	456b
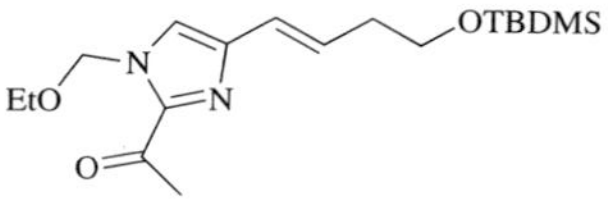	AD-Mix β, $MeSO_2NH_2$, t-BuOH:H_2O (1:1), 0°, 3 days	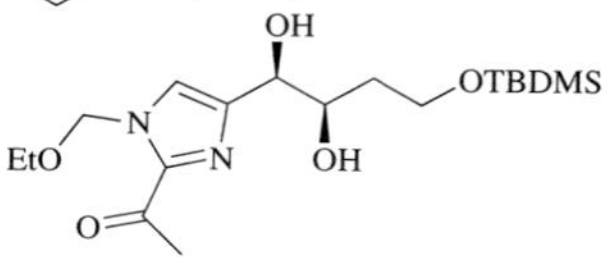 (65), 99	392
	AD-Mix α, $MeSO_2NH_2$, t-BuOH:H_2O (1:1), 0°, 3 days	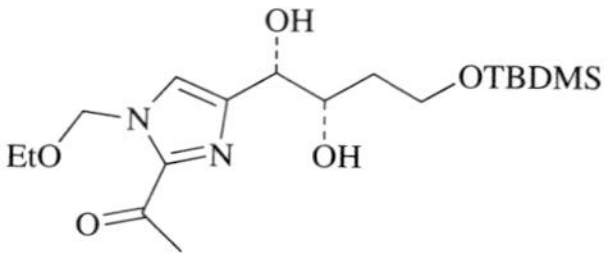 (56), 98	392

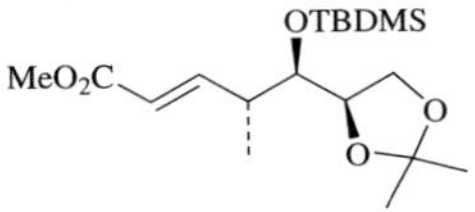	1. (DHQD)$_2$PHAL, OsO_4, $K_3Fe(CN)_6$, K_2CO_3, $MeSO_2NH_2$, *t*-BuOH:H_2O (1:1) 2. $(MeO)_2CMe_2$, acetone, TsOH	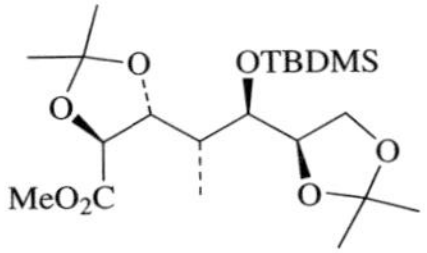	(66), 78 de	457
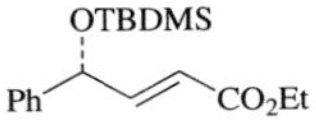	AD-Mix α, $MeSO_2NH_2$, *t*-BuOH:H_2O (1:1), rt, 36 h	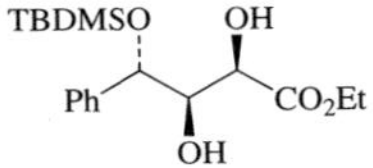	(97), >98de	458, 459
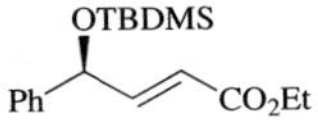	AD-Mix β, $MeSO_2NH_2$, *t*-BuOH:H_2O (1:1), rt, 14 h	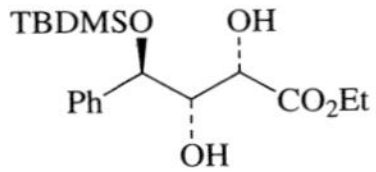	(97), >92 de	459
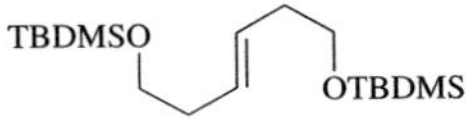	(DHQD)$_2$PHAL, $K_2OsO_2(OH)_4$, $K_3Fe(CN)_6$, K_2CO_3, $MeSO_2NH_2$, *t*-BuOH:H_2O (1:1) 0°, 19 h	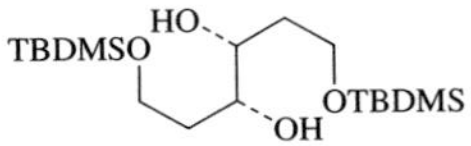	(93), 94	460
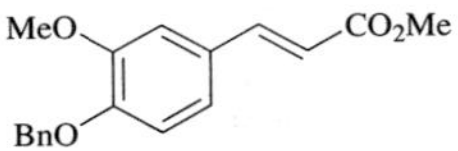	AD-Mix α, $MeSO_2NH_2$, *t*-BuOH:H_2O (1:1), 0°, 24 h	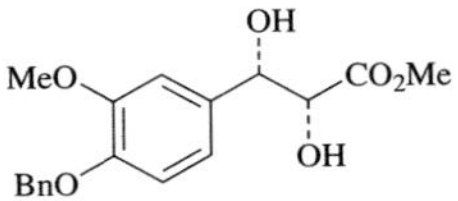	(68), 92	461
	AD-Mix β, $MeSO_2NH_2$, *t*-BuOH:H_2O (1:1), 0°, 24 h	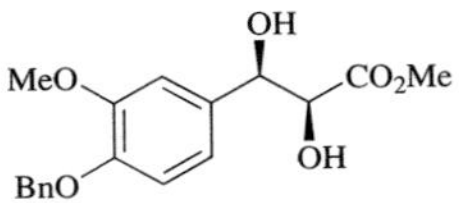	(—), 84	461
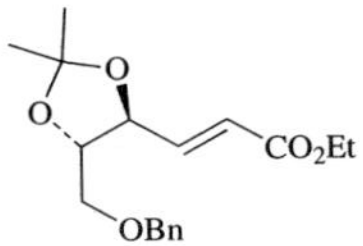	(DHQD)$_2$PHAL, $K_2OsO_2(OH)_4$, $K_3Fe(CN)_6$, K_2CO_3, $MeSO_2NH_2$, *t*-BuOH:H_2O (1:1) rt, 18 h	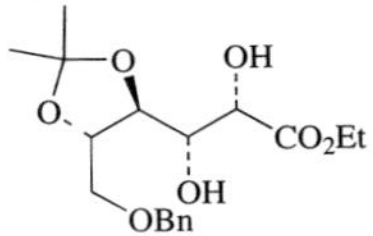	(74), 98 de	374

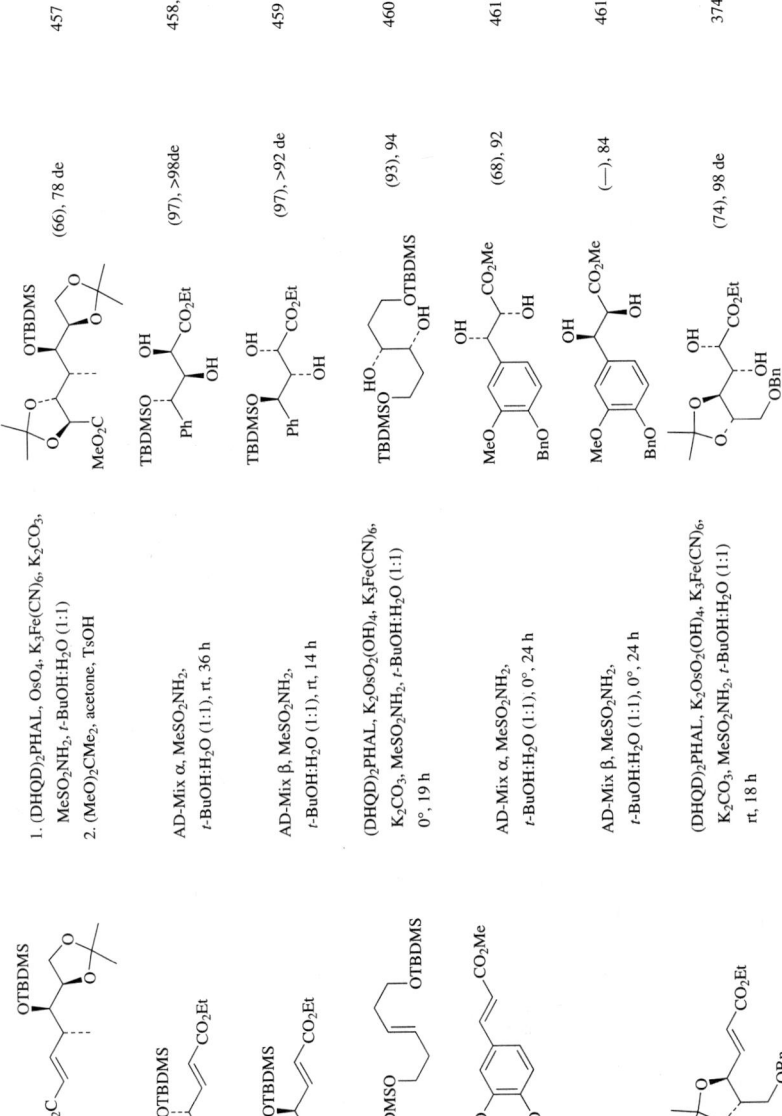

TABLE 3. REACTIONS OF TRANS 1,2-DISUBSTITUTED ALKENES (*Continued*)

Substrate	Conditions	Product(s) and Yield(s) (%), % ee	Refs.
C_{18}			
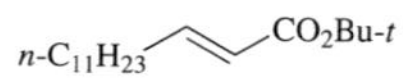	AD-Mix β, $MeSO_2NH_2$, t-BuOH:H_2O (1:1), 10°	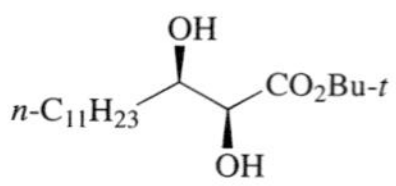 (99), >99	462
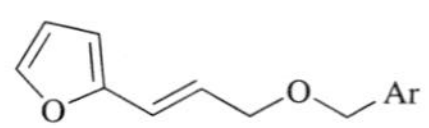 Ar = 2-naphthyl	AD-Mix β, $MeSO_2NH_2$, t-BuOH:H_2O (1:1), 0°, 46 h	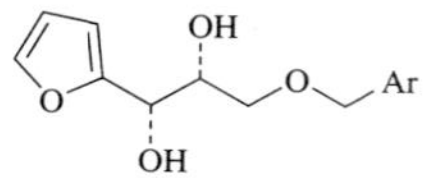 (90), >99	402
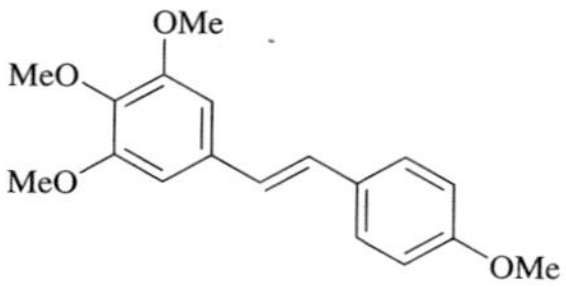	AD-Mix β, t-BuOH-H_2O-CH_2Cl_2	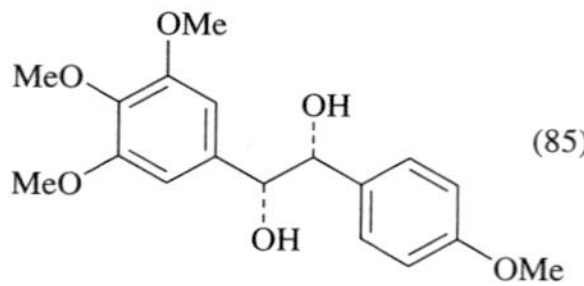 (85), >99	463
	AD-Mix α, t-BuOH-H_2O-CH_2Cl_2	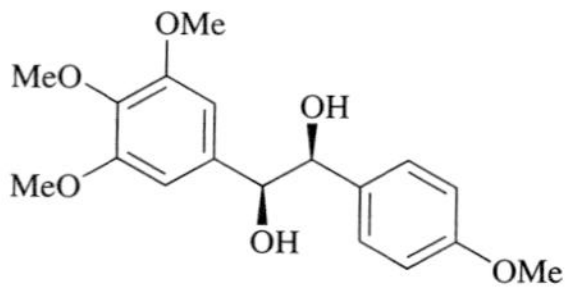 (81), >99	463
C_{19}			
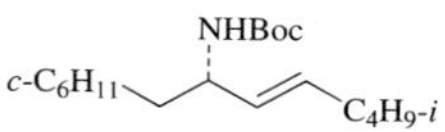	$(DHQD)_2PHAL$, $K_2OsO_2(OH)_4$, $K_3Fe(CN)_6$, K_2CO_3, $MeSO_2NH_2$, t-BuOH:H_2O (1:1), 0°, 6-24 h	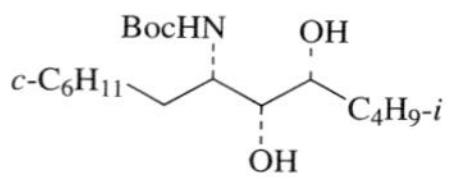 (—), 72	464
	$(DHQ)_2PHAL$, $K_2OsO_2(OH)_4$, $K_3Fe(CN)_6$, K_2CO_3, $MeSO_2NH_2$, t-BuOH:H_2O (1:1), 0°, 6-24 h	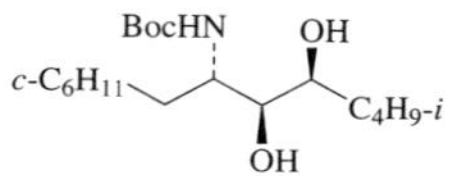 (—), 50	464

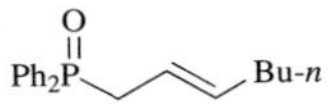	(DHQ)$_2$PHAL, $K_2OsO_2(OH)_4$, $K_3Fe(CN)_6$, K_2CO_3, $MeSO_2NH_2$, t-BuOH:H_2O (1:1), 0°, 1-3 d	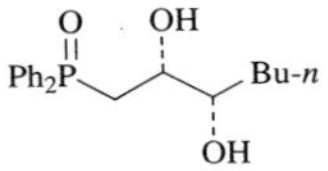 (61), 49	120, 119
	Ligand, K_2OsO_4, $K_3Fe(CN)_6$, K_2CO_3, $MeSO_2NH_2$, t-BuOH:H_2O (1:1), rt, 1-5 d	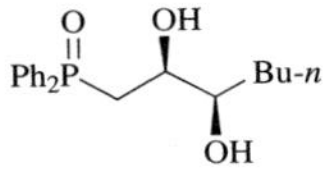	119, 120
	Ligand: DHQD-p-chlorobenzoate	(74), 76	
	Ligand: DHQD-PHN	(70), 75	
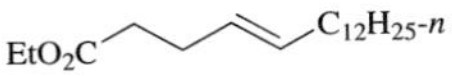	(DHQD)$_2$PHAL, $K_2OsO_2(OH)_4$, $K_3Fe(CN)_5$, K_2CO_3, $MeSO_2NH_2$, t-BuOH:H_2O (1:1), 0°, 24-36 h	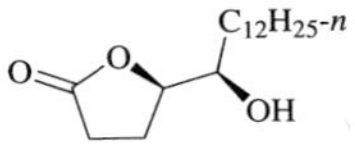 (84), 96	109
	(DHQ)$_2$PHAL, $K_2OsO_2(OH)_4$, $K_3Fe(CN)_6$, K_2CO_3, $MeSO_2NH_2$, t-BuOH:H_2O (1:1), 0°, 24-36 h	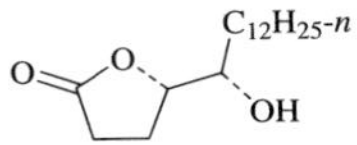 (82), 95	109
	(DHQD)$_2$PHAL, $K_2OsO_2(OH)_4$, $K_3Fe(CN)_6$, K_2CO_3, $MeSO_2NH_2$, t-BuOH:H_2O (1:1), 0°	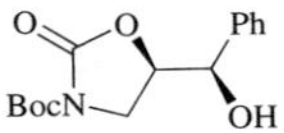 (78), 97	85
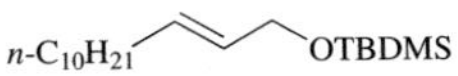	(DHQ)$_2$PHAL, $K_2OsO_2(OH)_4$, $K_3Fe(CN)_6$, K_2CO_3, $MeSO_2NH_2$, t-BuOH:H_2O (1:1), 0°	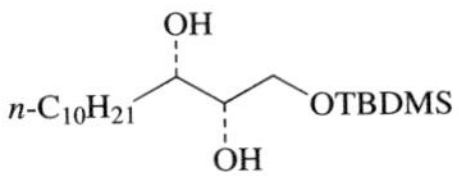 (92), 90	465
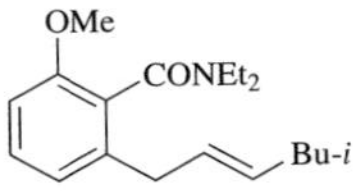	1. DHQ-PHN, K_2OsO_4, $K_3Fe(CN)_6$, K_2CO_3, $MeSO_2NH_2$, t-BuOH:H_2O (1:1), 0°, 18 h 2. NaOH (50% aq):EtOH (1:1), reflux, 36 h	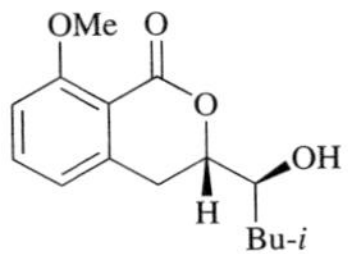 (70), 99	466

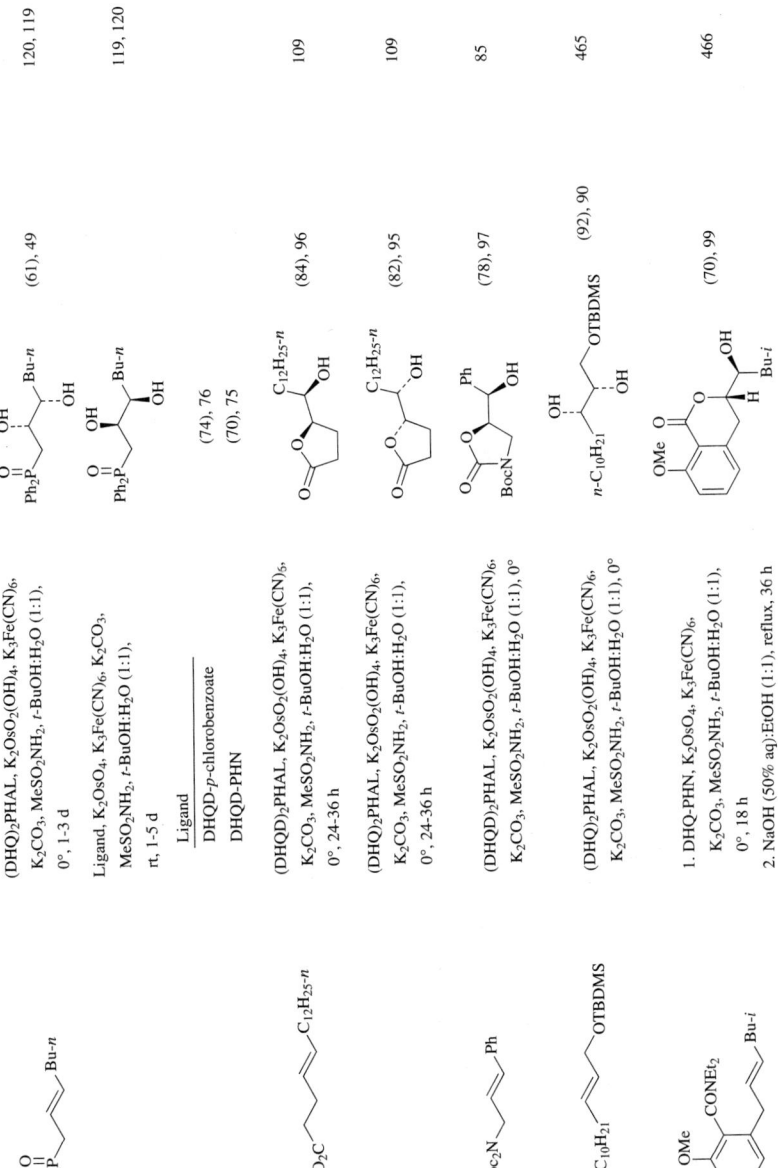

TABLE 3. REACTIONS OF TRANS 1,2-DISUBSTITUTED ALKENES (*Continued*)

Substrate	Conditions	Product(s) and Yield(s) (%), % ee	Refs.
C_{19} 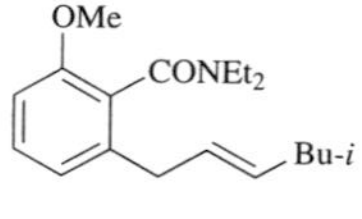	$(DHQ)_2PHAL$, $K_2OsO_2(OH)_4$, $K_3Fe(CN)_6$, K_2CO_3, *t*-BuOH:H_2O (1:1), 0°, 6-24 h	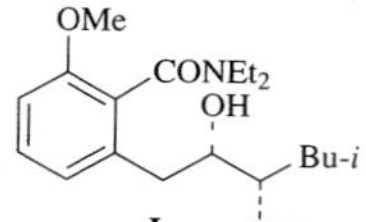 **I** (>60-71), 81	241
	DHQ-PHN, $K_2OsO_2(OH)_4$, $K_3Fe(CN)_6$, K_2CO_3, *t*-BuOH:H_2O (1:1), 0°, 18 h	**I** (>60-71), 86	241
	DHQD-PHN, $K_2OsO_2(OH)_4$, $K_3Fe(CN)_6$, K_2CO_3, *t*-BuOH:H_2O (1:1), 0°, 18 h	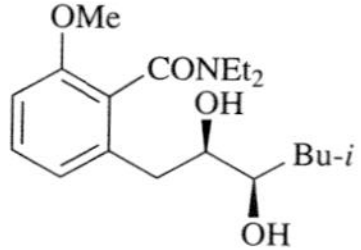 (>60-71), 90	241
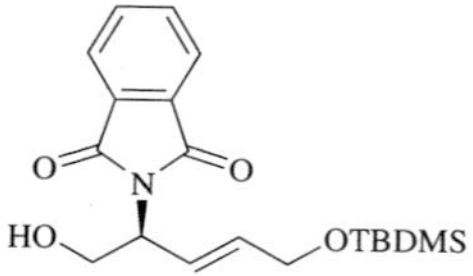	$(DHQ)_2PYR$, OsO_4, NMO, CH_2Cl_2-H_2O, rt	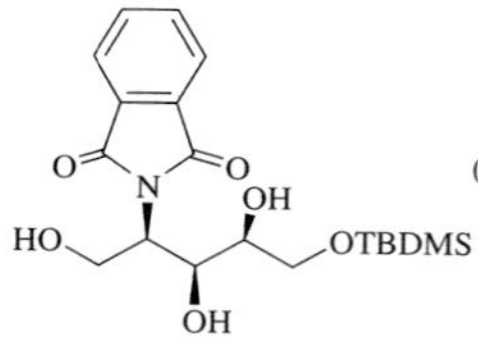 (86), 33 de	467
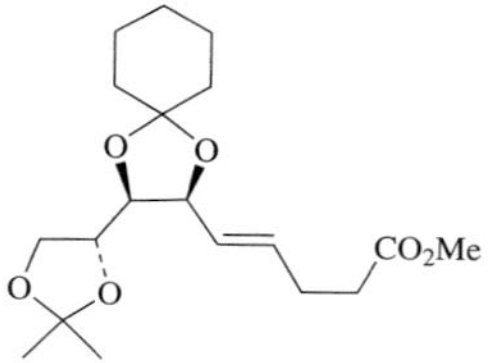	$(DHQD)_2PHAL$ K_2OsO_4, $K_3Fe(CN)_6$, K_2CO_3, $MeSO_2NH_2$, *t*-BuOH:H_2O (1:1), 0°, 36 h	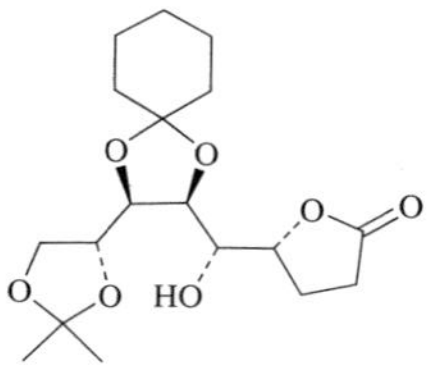 (72), 84 de	468

Substrate	Conditions	Product(s) and Yield(s) (%)	Refs.
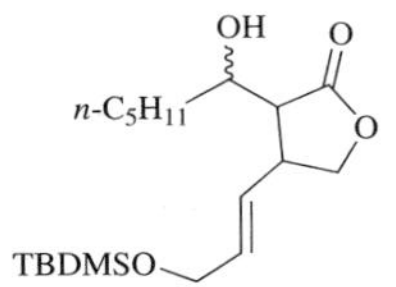	AD-Mix β, $MeSO_2NH_2$, t-BuOH:H_2O (1:1), 0°, 16 h	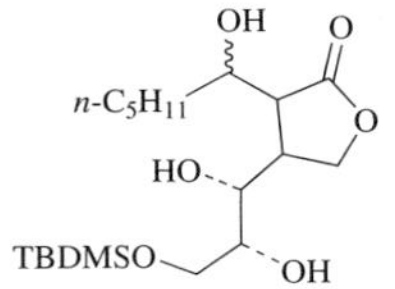 (38-57), 87-95	469
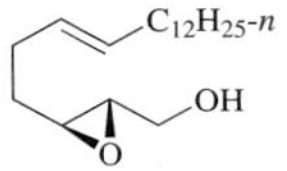	1. AD-Mix α, $MeSO_2NH_2$, t-BuOH:H_2O (1:1), 0°, 24 h 2. CSA, CH_2Cl_2	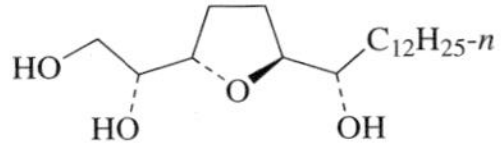 (92), 96 de	470
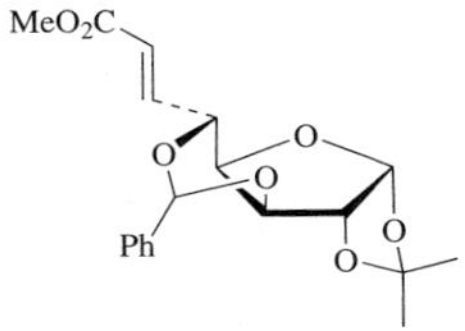	DHQD-p-chlorobenzoate, OsO_4, NMO acetone:H_2O (10:1), 0°, 12 h	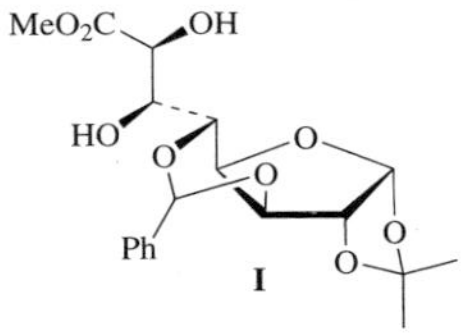 I (>90), 68 de	471
	DHQ-p-chlorobenzoate, OsO_4, NMO acetone:H_2O (10:1), 0°, 12 h	**I** (>90), 23 de	471
C_{19-27}			
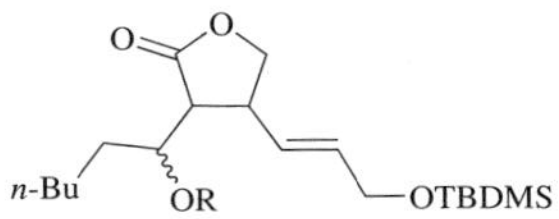 R: H, MPM	AD-Mix β, $MeSO_2NH_2$, t-BuOH:H_2O (1:1), 0°, 3 d	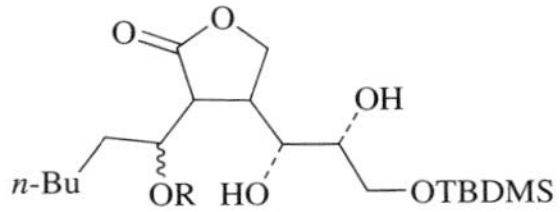 (38-57), 74-90 (42-53), 86-100	472
C_{19}			
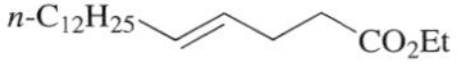	AD-Mix α, $MeSO_2NH_2$, t-BuOH:H_2O (1:1), 0°, 16 h	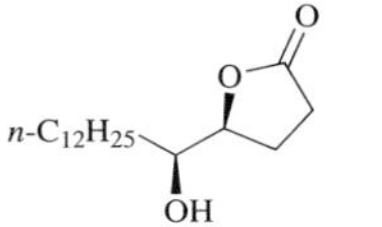 (92), —	473

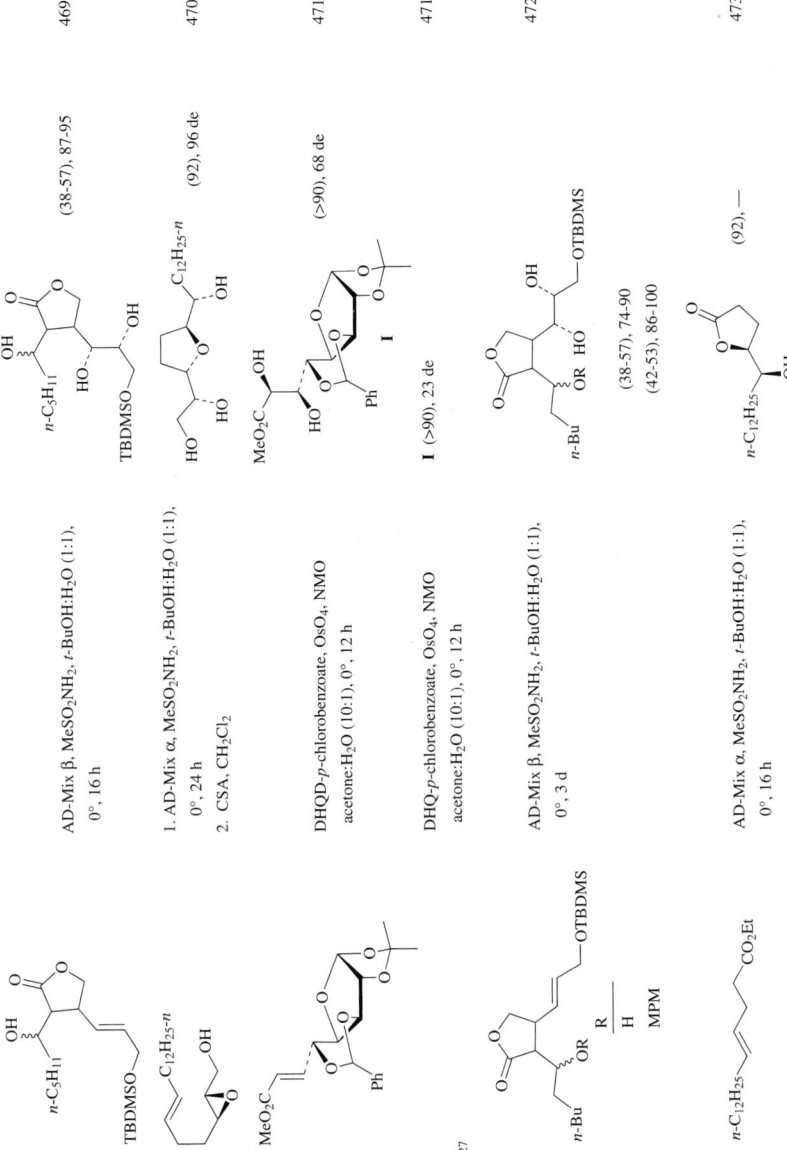

TABLE 3. REACTIONS OF TRANS 1,2-DISUBSTITUTED ALKENES (*Continued*)

Substrate	Conditions	Product(s) and Yield(s) (%), % ee	Refs.
C_{19} 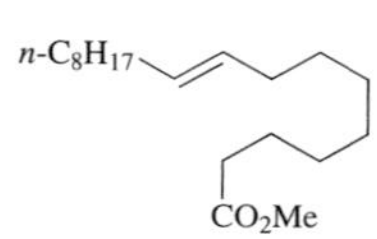	AD-Mix β, $MeSO_2NH_2$, t-BuOH:H_2O (1:1), 0°, 24 h	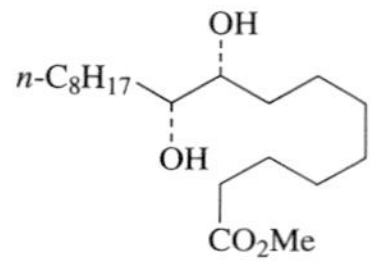 (95), 97	277
	AD-Mix α, $MeSO_2NH_2$, t-BuOH:H_2O (1:1), 0°, 24 h	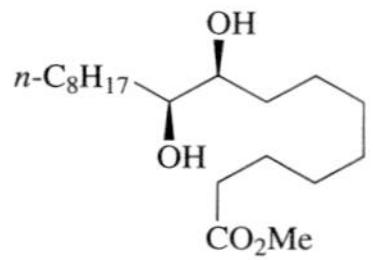 (97), 95	277
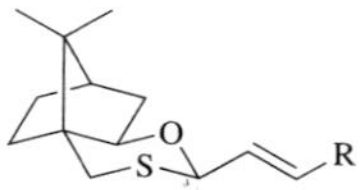 R: c-C_6H_{11}	DHQD-OAc, OsO_4, toluene, –20°	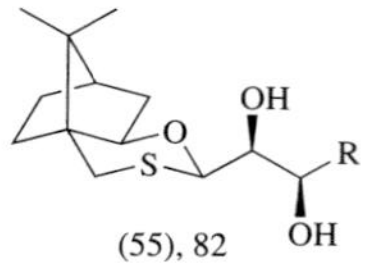 (55), 82	380
R: Ph		(63), 95	
R = Ph	DHQ-OAc, OsO_4, toluene, –20°	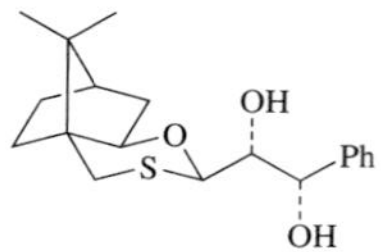 (90), 88	380
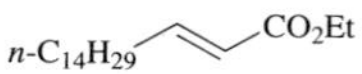	$(DHQ)_2$PHAL, OsO_4, $K_3Fe(CN)_6$, K_2CO_3, $MeSO_2NH_2$, t-BuOH:H_2O (1:1), 0°, 24 h	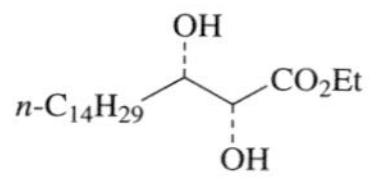 (94), 99	282
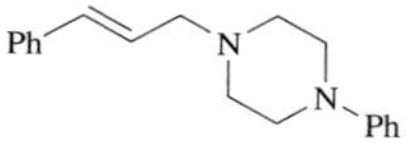	$(DHQD)_2$PHAL, $K_2OsO_2(OH)_4$, $K_3Fe(CN)_6$, K_2CO_3, $MeSO_2NH_2$, t-BuOH:H_2O (1:1), rt, 7 d	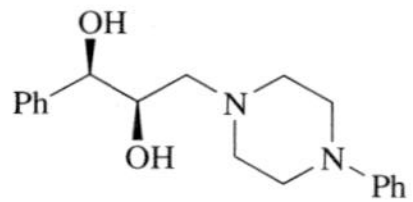 (70), 55	111

C_{20}

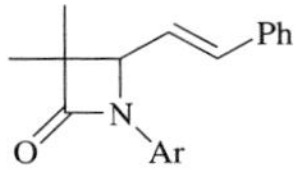 Ar = 4-MeOC$_6$H$_4$	1. DHQD-*p*-chlorobenzoate, $K_2OsO_2(OH)_4$, $K_3Fe(CN)_6$, K_2CO_3, *t*-BuOH:H_2O (1:1), rt, 15 h 2. $(MeO)_2CMe_2$, TsOH	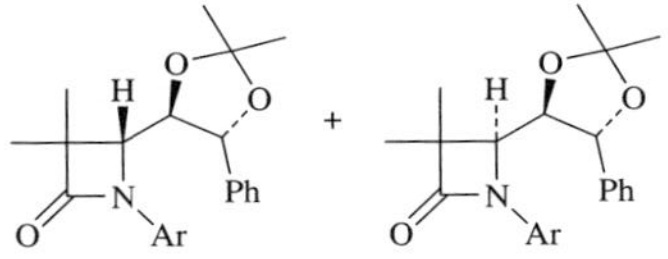 **I** 90% ee **II** 91% ee	**I + II** (61)
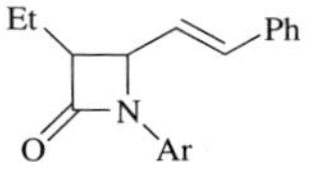 *trans*, Ar = 4-MeOC$_6$H$_4$	1. DHQD-*p*-chlorobenzoate, $K_2OsO_2(OH)_4$, $K_3Fe(CN)_6$, K_2CO_3, *t*-BuOH:H_2O (1:1), rt, 15 h 2. $(MeO)_2CMe_2$, TsOH	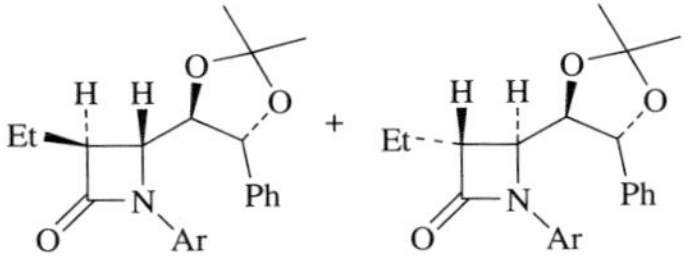 **I** 94% ee **II** 86% ee	**I + II** (73) 450
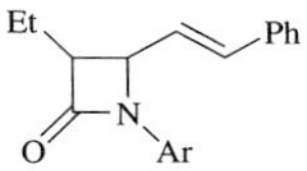 *cis*, Ar = 4-MeOC$_6$H$_4$	1. DHQD-*p*-chlorobenzoate, $K_2OsO_2(OH)_4$, $K_3Fe(CN)_6$, K_2CO_3, *t*-BuOH:H_2O (1:1), rt, 15 h 2. $(MeO)_2CMe_2$, TsOH	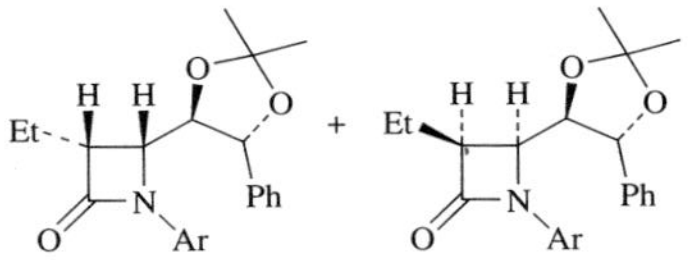 **I** 94% ee **II** 88% ee	**I + II** (63) 450
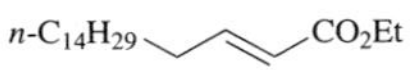 E:Z = 68:1	$(DHQD)_2PHAL$, $K_2OsO_2(OH)_4$, $K_3Fe(CN)_6$, K_2CO_3, $MeSO_2NH_2$, *t*-BuOH:H_2O (1:1), rt	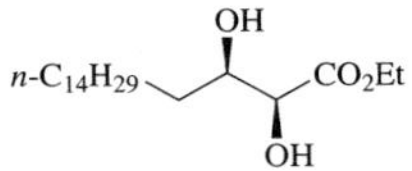 (95), 97	261

TABLE 3. REACTIONS OF TRANS 1,2-DISUBSTITUTED ALKENES (*Continued*)

Substrate	Conditions	Product(s) and Yield(s) (%), % ee		Refs.
C_{20}				
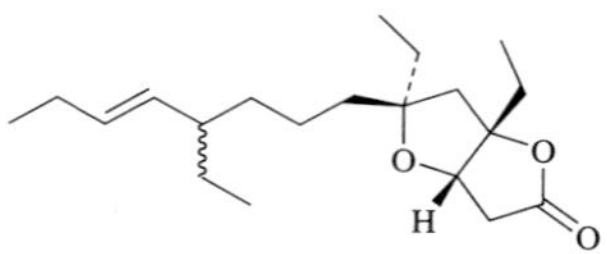	AD-Mix α, $MeSO_2NH_2$, t-BuOH:H_2O (1:1), 0°, 24 h	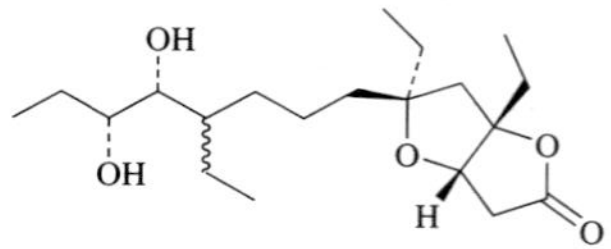	(95), 92 de	474
	AD-Mix β, $MeSO_2NH_2$, t-BuOH:H_2O (1:1), 0°, 24 h	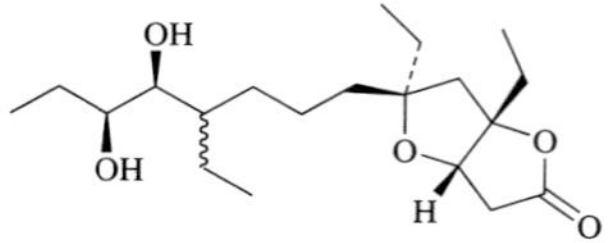	(100), 92 de	474
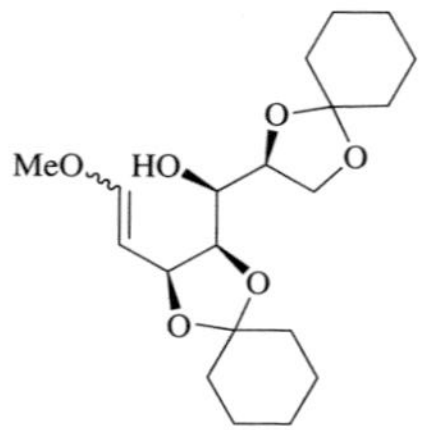	1. DHQD-p-chlorobenzoate, OsO_4, NMO, acetone-H_2O 2. Ac_2O, DMAP, pyridine	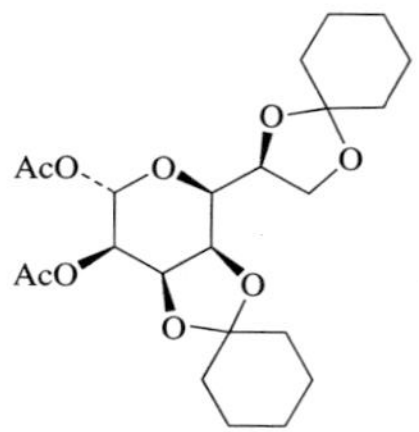	(75), 67	475
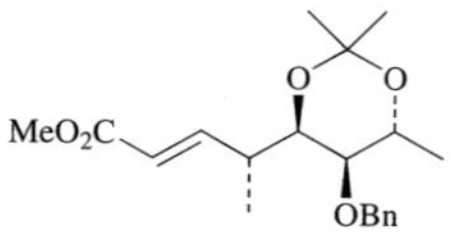	AD-Mix β, $MeSO_2NH_2$, t-BuOH:H_2O (1:1), 20°, 17 h	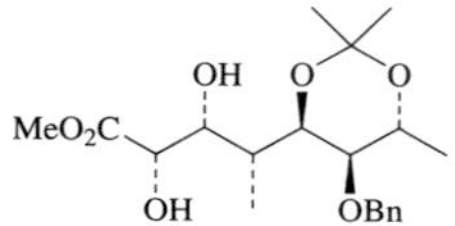	(98), >97 de	475a
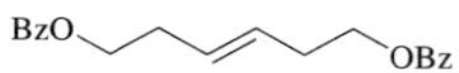	AD-Mix β, t-BuOH:H_2O (1:1), 0°	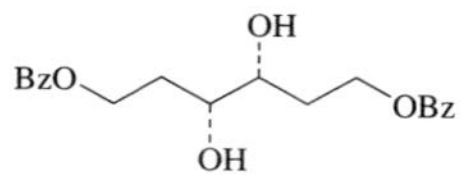	(84), >95	475b

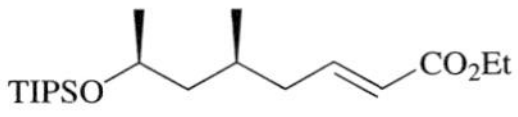	1. AD-Mix α, $NaHCO_3$, t-BuOH:H_2O (1:1), 5°, 24 h 2. 2,2-Dimethoxypropane, TsOH-H_2O	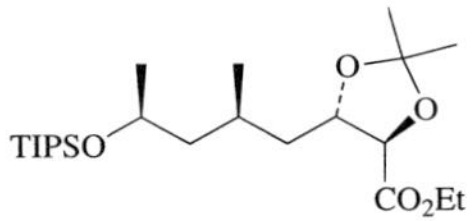	(86), — de	476
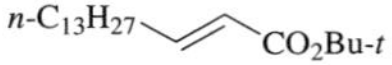	AD-Mix α, $MeSO_2NH_2$, t-BuOH:H_2O (1:1), 0°	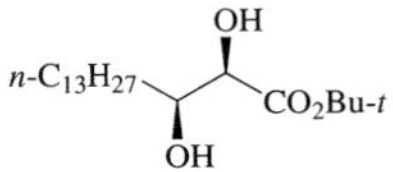	(83), >99	477, 478
	AD-Mix β, $MeSO_2NH_2$, t-BuOH:H_2O (1:1), 0°	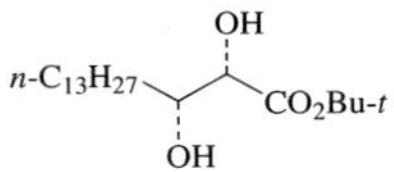	(99), >99	477
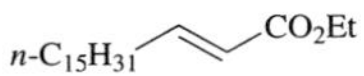	AD-Mix β, $MeSO_2NH_2$, t-BuOH:H_2O (1:1), 0°, 24 h	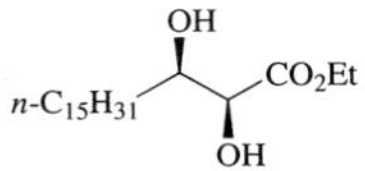	(85-95), 91-97	277, 261
	AD-Mix α, $MeSO_2NH_2$, t-BuOH:H_2O (1:1), 0°, 24 h	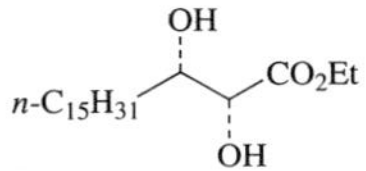	(85), 95	277
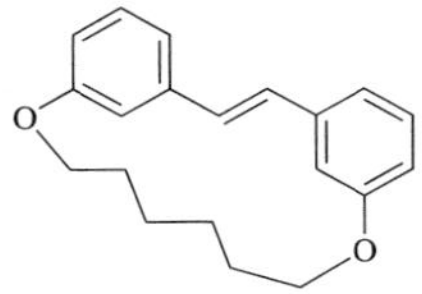	AD-Mix β, $MeSO_2NH_2$, t-BuOH:H_2O (1:1), 0°-rt, 30 h	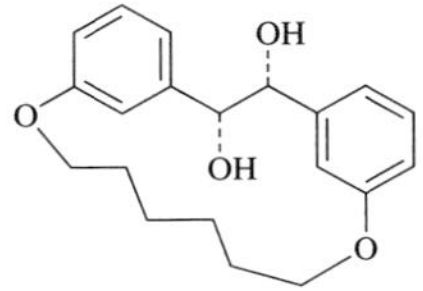	(75), 78	479
	$(DHQD)_2PHAL$, $K_2OsO_2(OH)_4$, $K_3Fe(CN)_6$, K_2CO_3, $MeSO_2NH_2$, t-BuOH:H_2O (1:1), 0°	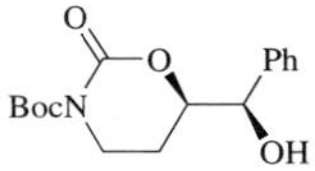	(73), 98	85

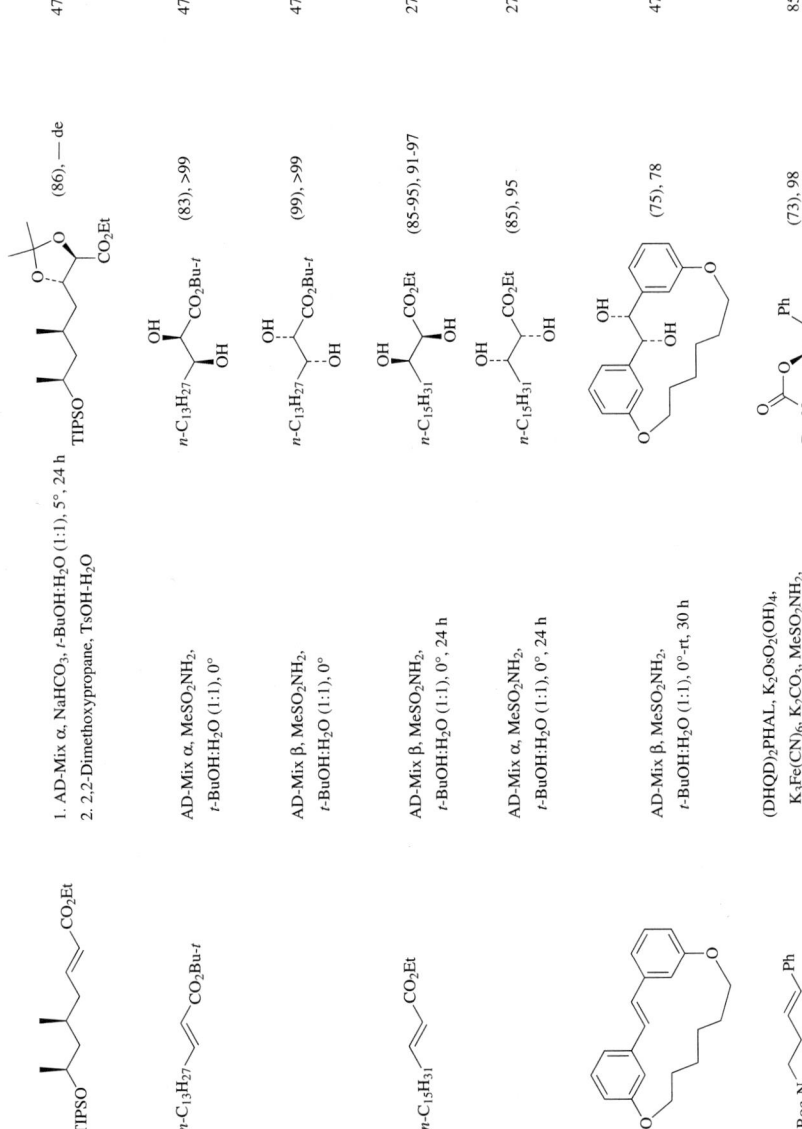

TABLE 3. REACTIONS OF TRANS 1,2-DISUBSTITUTED ALKENES (*Continued*)

Substrate	Conditions	Product(s) and Yield(s) (%), % ee	Refs.
C_{20}			
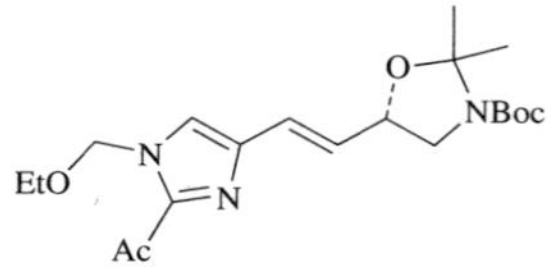	$(DHQD)_2PHAL$, $K_2OsO_2(OH)_4$, $K_3Fe(CN)_6$, K_2CO_3, $MeSO_2NH_2$, t-BuOH:H_2O (1:1), 0°, 4 d	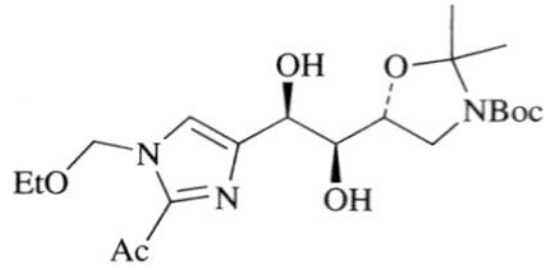 (81), >98 de	225
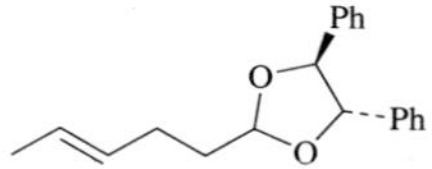	$(DHQ)_2PHAL$, $K_2OsO_2(OH)_4$, $K_3Fe(CN)_6$, K_2CO_3, $MeSO_2NH_2$, t-BuOH:H_2O (1:1), 0°, 6-24 h	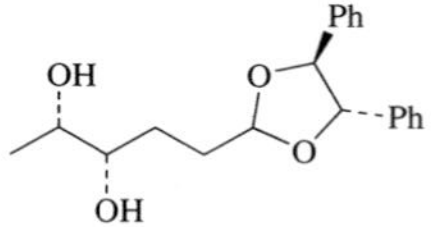 (95), 72 de	479a
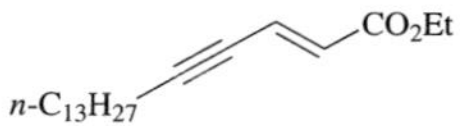	AD-Mix β, $MeSO_2NH_2$, t-BuOH:H_2O (1:1), rt	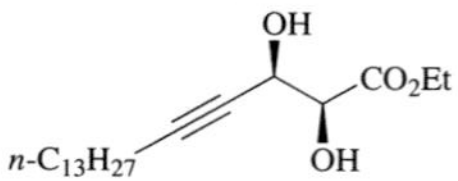 (90), 98	480
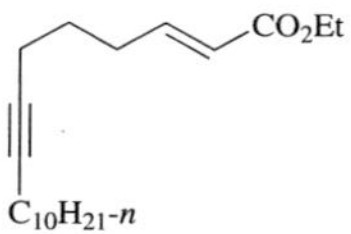	AD-Mix β, $MeSO_2NH_2$, t-BuOH:H_2O (1:1), rt	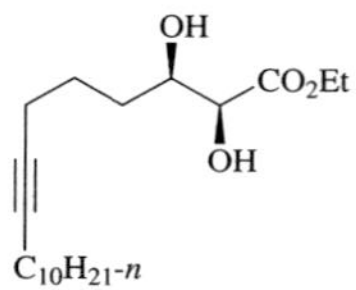 (93), 98	480
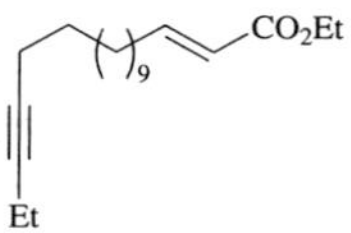	AD-Mix β, $MeSO_2NH_2$, t-BuOH:H_2O (1:1), rt	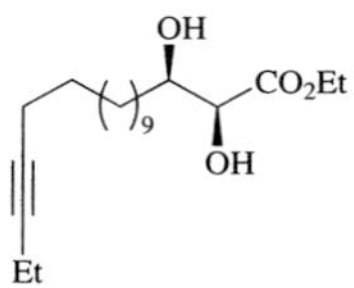 (95), 97	480

Reactant	Conditions	Product(s) and Yield(s) (%), % ee	Refs.
C_{21}			
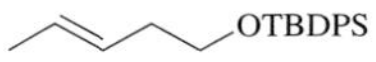	**35**, OsO_4, CH_2Cl_2, –90°, 2 h	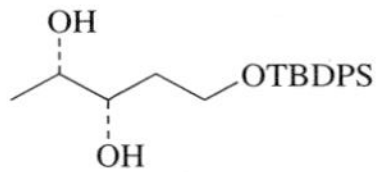 (87), 95	11
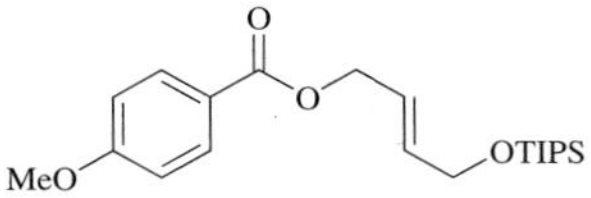	$(DHQD)_2PYDZ$, $K_2OsO_2(OH)_4$, $K_3Fe(CN)_6$, K_2CO_3, $MeSO_2NH_2$, t-BuOH:H_2O (1:1), 0°, 1.5-3.5 h	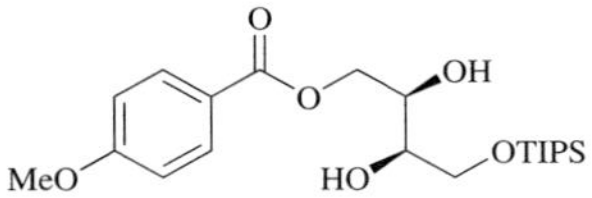 (96), >99	61
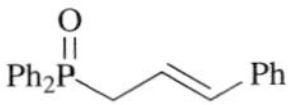	$(DHQ)_2PHAL$, $K_2OsO_2(OH)_4$, $K_3Fe(CN)_6$, K_2CO_3, $MeSO_2NH_2$, t-BuOH:H_2O (1:1), 0°, 1-5 d	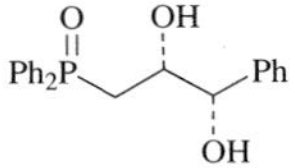 (66), 74	120, 119
	Ligand, $K_2OsO_2(OH)_4$, $K_3Fe(CN)_6$, K_2CO_3, $MeSO_2NH_2$, t-BuOH:H_2O (1:1), rt	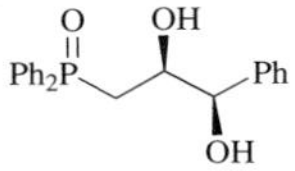	120, 119
	Ligand / Time		
	$(DHQD)_2PHAL$ / 6-24 h	(94), 85	
	DHQD-p-chlorobenzoate / 1-5 d	(79), 86	
	DHQD-PHN / 1-5 d	(79), 88	
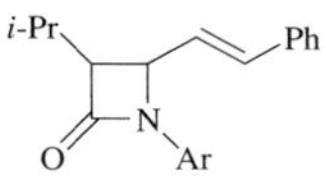 *trans*, Ar = 4-$MeOC_6H_4$	1. DHQD-p-chlorobenzoate, $K_2OsO_2(OH)_4$, $K_3Fe(CN)_6$, K_2CO_3, t-BuOH:H_2O (1:1), rt, 15 h 2. $(MeO)_2CMe_2$, TsOH	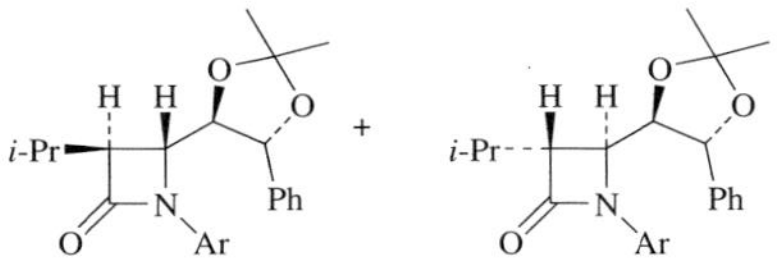 **I** 94% ee **I + II** (72) **II** 93% ee	450

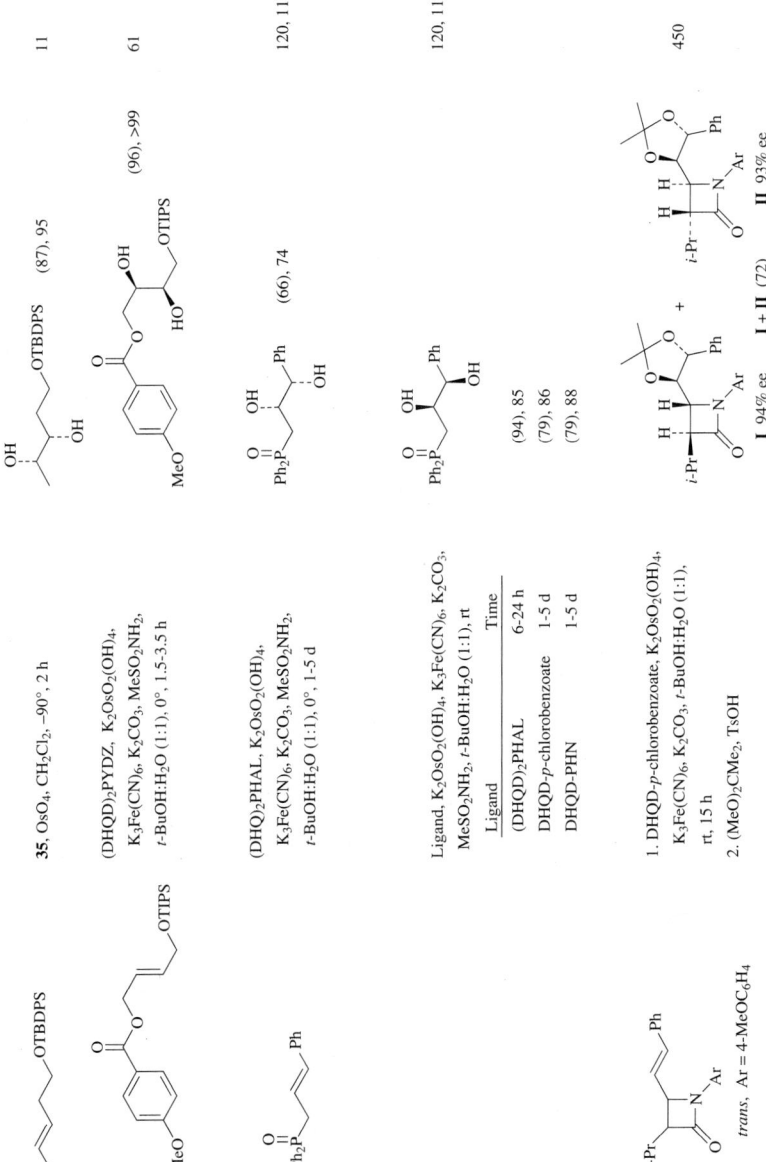

TABLE 3. REACTIONS OF TRANS 1,2-DISUBSTITUTED ALKENES (*Continued*)

Substrate	Conditions	Product(s) and Yield(s) (%), % ee	Refs.
C_{21}			
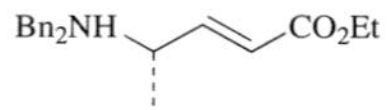	$(DHQ)_2PHAL$, $K_2OsO_2(OH)_4$, $K_3Fe(CN)_6$, K_2CO_3, t-BuOH:H_2O (1:1), rt, 4 d	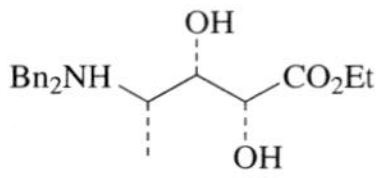 (79), 84 de	385
	$(DHQD)_2PHAL$, $K_2OsO_2(OH)_4$, $K_3Fe(CN)_6$, K_2CO_3, t-BuOH:H_2O (1:1), rt, 4 d	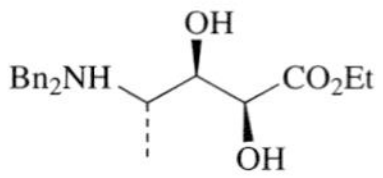 (68), 88 de	385
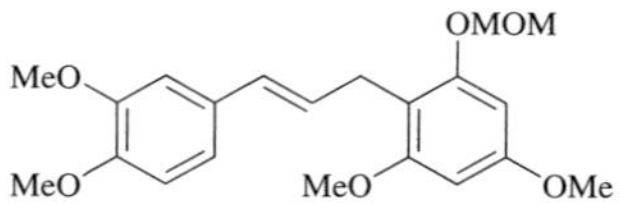	AD-Mix β, $MeSO_2NH_2$, t-BuOH:H_2O (1:1), 0°	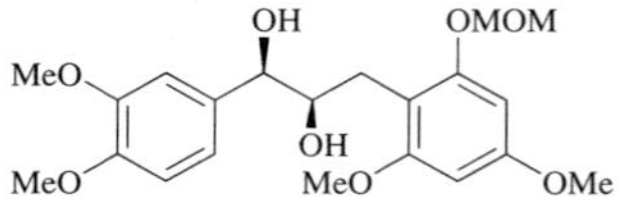 (83), >99	481
	AD-Mix α, $MeSO_2NH_2$, t-BuOH:H_2O (1:1), 0°	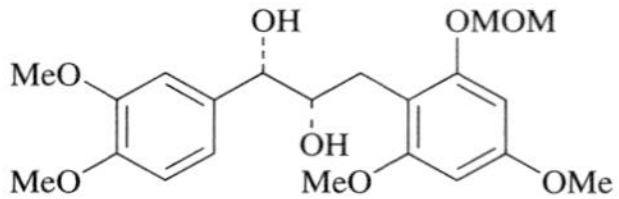 (85), >99	481
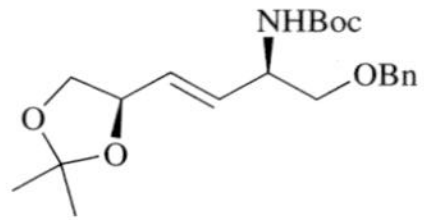	AD-Mix β, t-BuOH:H_2O (1:1), 0°, 48 h	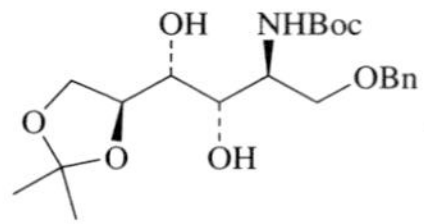 (40), 80 de	482
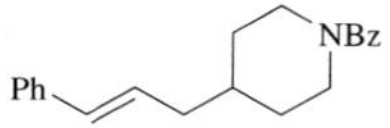	AD-Mix α, $MeSO_2NH_2$, t-BuOH:H_2O (1:1), 0°-rt, 24 h	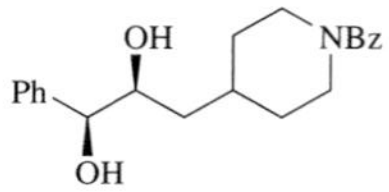 (72), >95	101
	AD-Mix β, $MeSO_2NH_2$, t-BuOH:H_2O (1:1), 0°-rt, 24 h	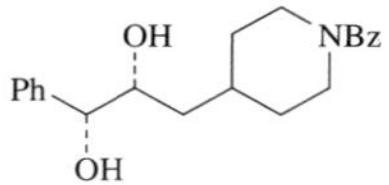 (72), >95	101

C_{22}				
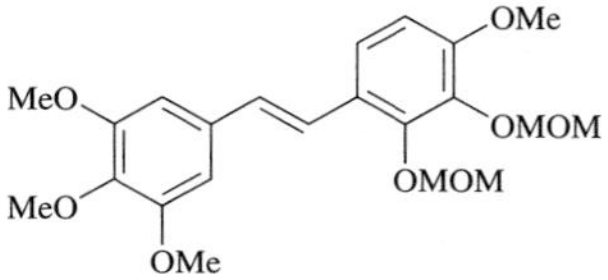	AD-Mix α, $MeSO_2NH_2$, *t*-BuOH:H_2O (1:1), 0°-rt, 32 h	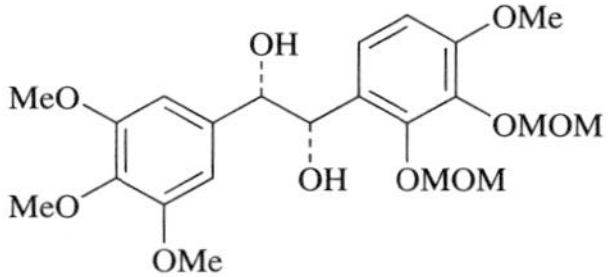	(82), —	483
	AD-Mix β, $MeSO_2NH_2$, *t*-BuOH:H_2O (1:1), 0°-rt, 32 h	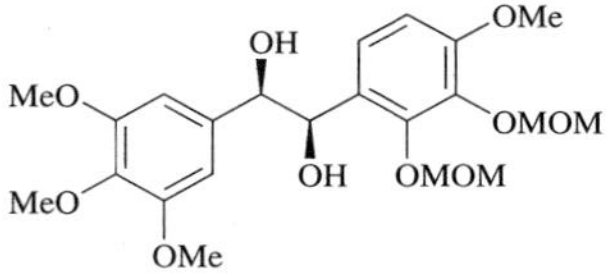	(89), —	483
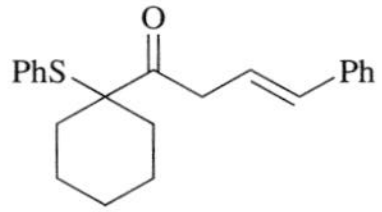	AD-Mix β, $MeSO_2NH_2$, *t*-BuOH:H_2O (1:1), rt	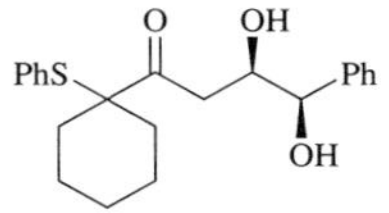	(92), >98	238
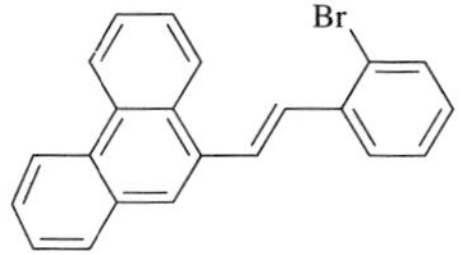	$(DHQD)_2PHAL$, $K_2OsO_2(OH)_4$, $K_3Fe(CN)_6$, K_2CO_3, $MeSO_2NH_2$, *t*-BuOH:H_2O (1:1), 0°, 6-24 h	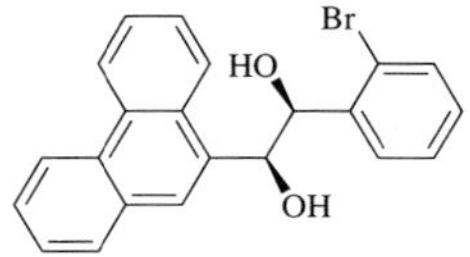	(—), >99	484
	$(DHQ)_2PHAL$, $K_2OsO_2(OH)_4$, $K_3Fe(CN)_6$, K_2CO_3, $MeSO_2NH_2$, *t*-BuOH:H_2O (1:1), 0°, 6-24 h	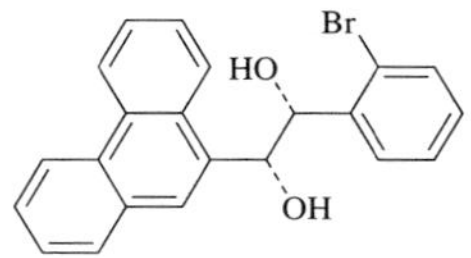	(—), >99	484

TABLE 3. REACTIONS OF TRANS 1,2-DISUBSTITUTED ALKENES (*Continued*)

Substrate	Conditions	Product(s) and Yield(s) (%), % ee	Refs.
C_{22} 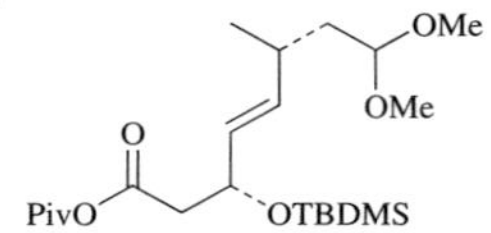	(DHQ)$_2$PHAL, $K_2OsO_2(OH)_4$, $K_3Fe(CN)_6$, K_2CO_3, $MeSO_2NH_2$, *t*-BuOH:H_2O (1:1), rt, 3 d	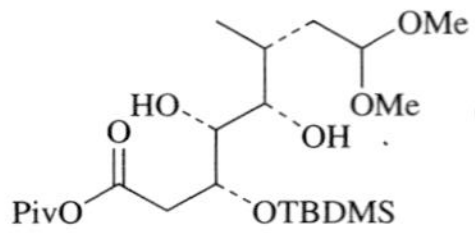 (70), 20 de	485
	(DHQD)$_2$PHAL, $K_2OsO_2(OH)_4$, $K_3Fe(CN)_6$, K_2CO_3, $MeSO_2NH_2$, *t*-BuOH:H_2O (1:1), rt, 3 d	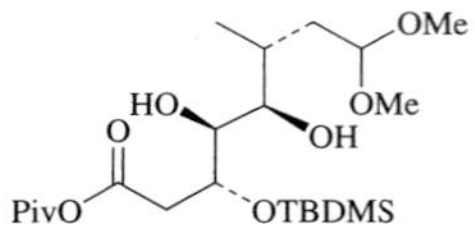 (79), 82 de	485
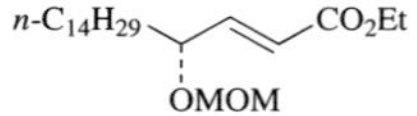 E:Z = 36:1	(DHQD)$_2$PHAL, $K_2OsO_2(OH)_4$, $K_3Fe(CN)_6$, K_2CO_3, $MeSO_2NH_2$, *t*-BuOH:H_2O (1:1), 0°	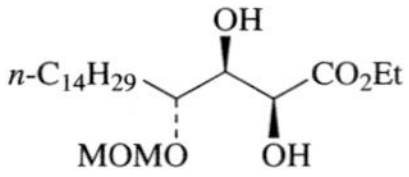 (92), 91 de	261
	AD-Mix β, *t*-BuOH:H_2O (1:1), 0°	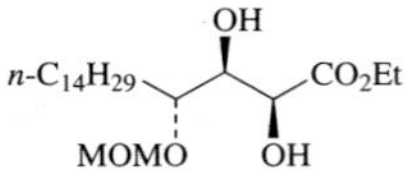 (92), 91 de	261
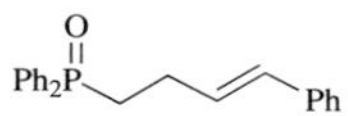	AD-Mix α, *t*-BuOH:H_2O (1:1), rt, 96 h	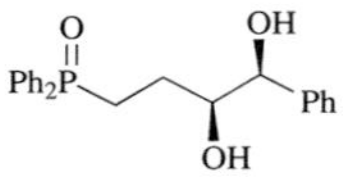 (97), >95	319
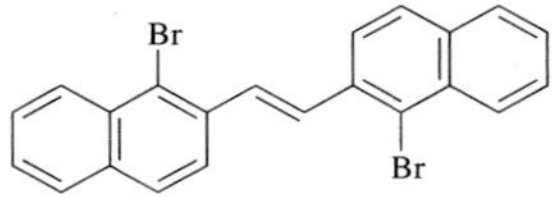	(DHQD)$_2$PHAL, $K_2OsO_2(OH)_4$, NMO, acetone-H_2O, rt, 1-2 d	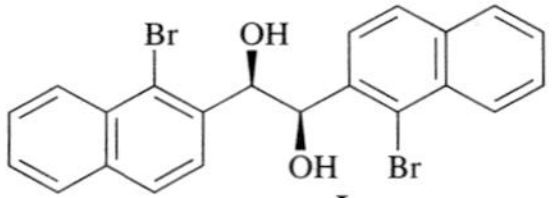 **I** (60-80), 95	486, 487
	DHQD-*p*-chlorobenzoate, $K_2OsO_2(OH)_4$, NMO, acetone-H_2O, rt, 1-2 d	**I** (60-80), 91	486

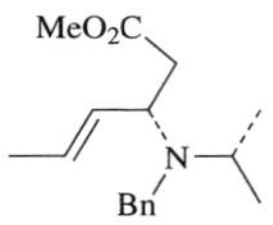	(DHQ)$_2$PHAL, OsO_4, $K_3Fe(CN)_6$, K_2CO_3, $MeSO_2NH_2$, *t*-BuOH:H_2O (1:1), 20°, 24 h	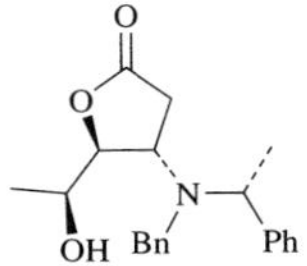	(—), 50 de	488
	(DHQD)$_2$PHAL, OsO_4, $K_3Fe(CN)_6$, K_2CO_3, $MeSO_2NH_2$, *t*-BuOH:H_2O (1:1), 20°	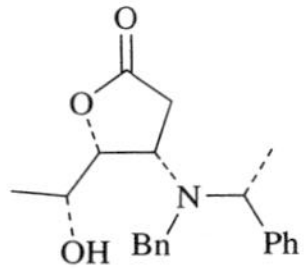	(—), 64 de	488
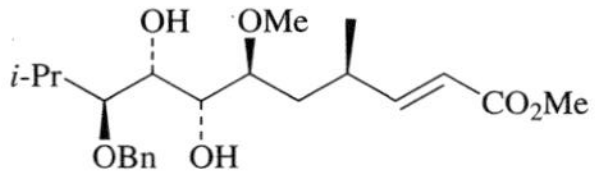	DHQD-OAc, OsO_4, toluene, rt	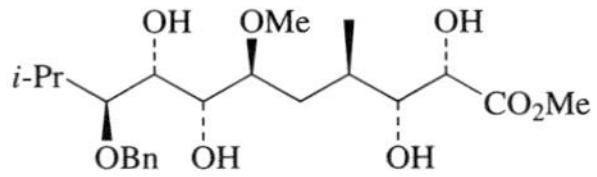	(60), >95	489
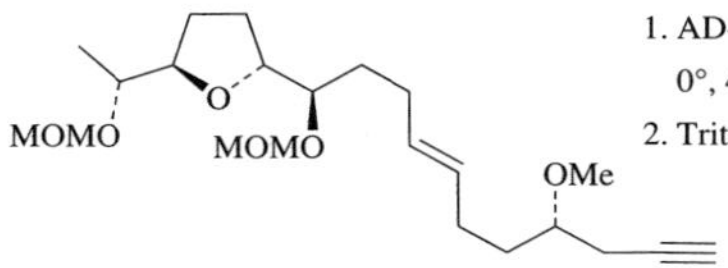	1. AD-mix α, (DHQ)$_2$PHAL, *t*-BuOH:H_2O (1:1), 0°, 48 h 2. Triton B, MeOH	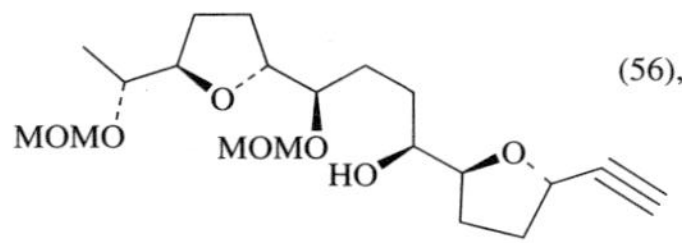	(56), 92 de	489a
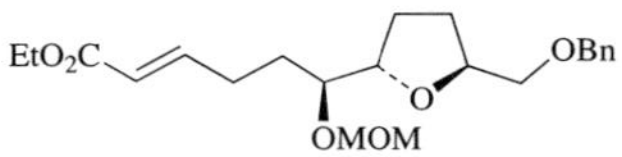	(DHQD)$_2$PHAL, $K_2OsO_2(OH)_4$, $K_3Fe(CN)_6$, K_2CO_3, *t*-BuOH:H_2O (1:1), 0°	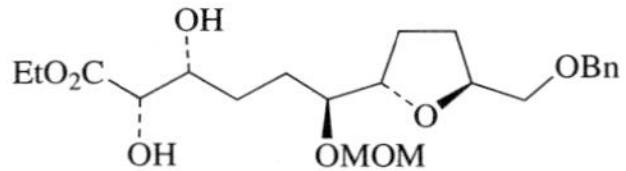	(97), — de	334
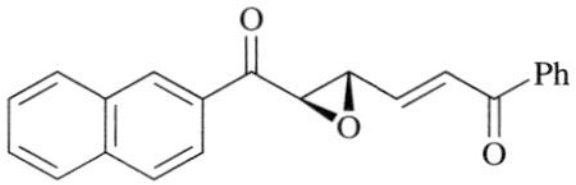	AD-Mix β, $MeSO_2NH_2$, *t*-BuOH:H_2O (1:1)	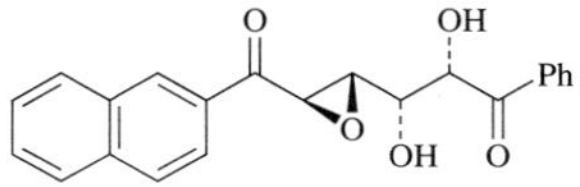	(95), — de	403

TABLE 3. REACTIONS OF TRANS 1,2-DISUBSTITUTED ALKENES (*Continued*)

Substrate	Conditions	Product(s) and Yield(s) (%), % ee	Refs.
C_{23}			
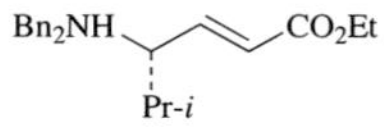	(DHQ)$_2$PHAL, $K_2OsO_2(OH)_4$, $K_3Fe(CN)_6$, K_2CO_3, t-BuOH:H_2O (1:1), rt, 10 d	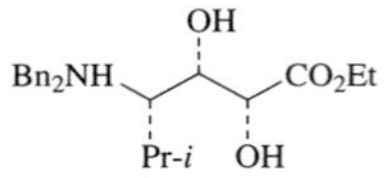 (7), 44 de	385
	(DHQD)$_2$PHAL, $K_2OsO_2(OH)_4$, $K_3Fe(CN)_6$, K_2CO_3, t-BuOH:H_2O (1:1), rt, 10 d	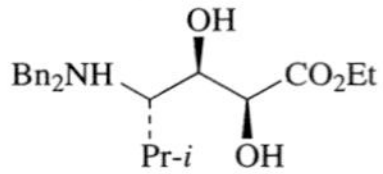 (21), 90 de	385
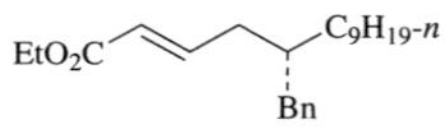	(DHQD)$_2$PHAL, $K_2OsO_2(OH)_4$, $K_3Fe(CN)_6$, K_2CO_3, $MeSO_2NH_2$, t-BuOH:H_2O (1:1), 4°, 18 h	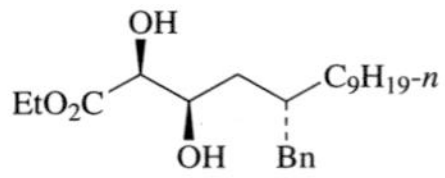 (100), 91 de	490
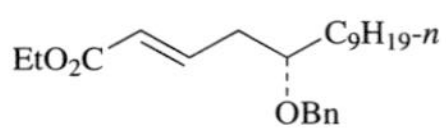	AD-Mix β, $MeSO_2NH_2$, t-BuOH:H_2O (1:1), 0°	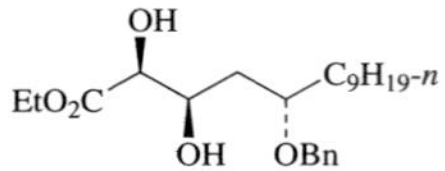 (100), 91 de	491
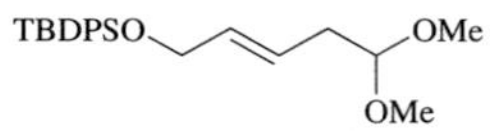	(DHQD)$_2$PHAL, $K_2OsO_2(OH)_4$, $K_3Fe(CN)_6$, K_2CO_3, $MeSO_2NH_2$, t-BuOH:H_2O (1:1)	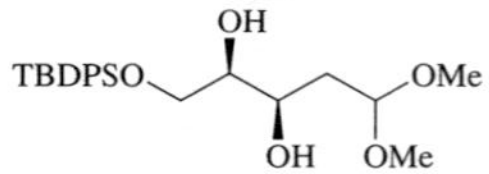 (81), 91	492
	(DHQ)$_2$PHAL, $K_2OsO_2(OH)_4$, $K_3Fe(CN)_6$, K_2CO_3, $MeSO_2NH_2$, t-BuOH:H_2O (1:1)	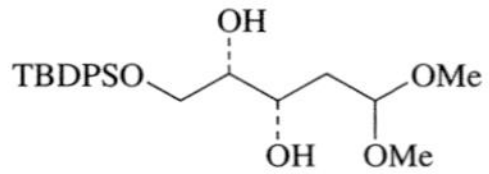 (81), 84-86	492
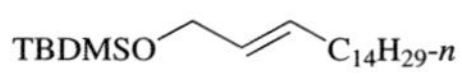	AD-Mix β, $MeSO_2NH_2$, t-BuOH:H_2O (1:1), 0°, 24 h	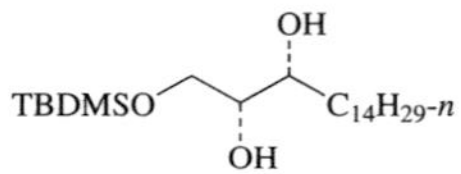 (99), 94	493, 306

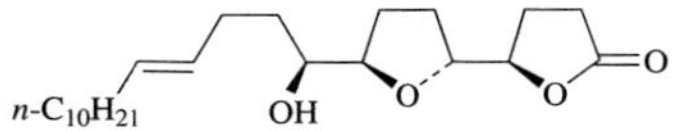	AD-Mix β, $MeSO_2NH_2$, t-BuOH:H_2O (1:1), 0°	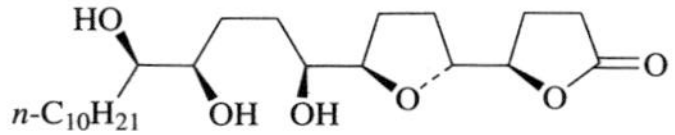	(92), >95	447
C_{24}				
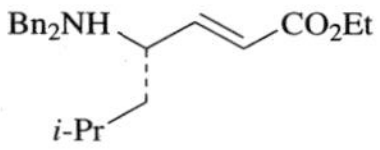	$(DHQ)_2PHAL$, $K_2OsO_2(OH)_4$, $K_3Fe(CN)_6$, K_2CO_3, t-BuOH:H_2O (1:1), rt, 11 d	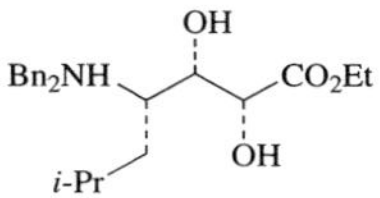	(15), 50 de	385
	$(DHQD)_2PHAL$, $K_2OsO_2(OH)_4$, $K_3Fe(CN)_6$, K_2CO_3, t-BuOH:H_2O (1:1), rt, 11 d	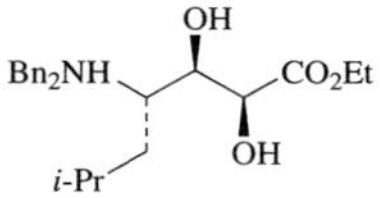	(25), 90 de	385
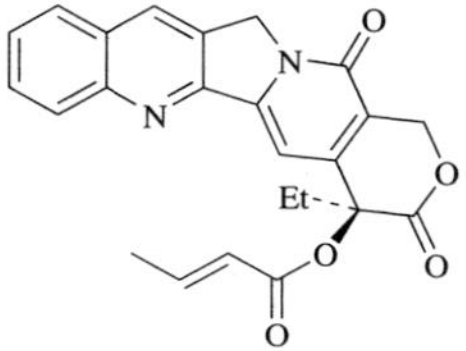	$(DHQD)_2PHAL$, $K_2OsO_2(OH)_4$, $K_3Fe(CN)_6$, K_2CO_3, $MeSO_2NH_2$, t-BuOH:H_2O (1:1), 0°	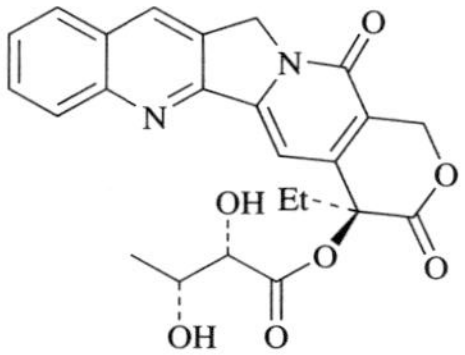	(67), 94	365
	$(DHQ)_2PHAL$, $K_2OsO_2(OH)_4$, NMO, acetone:H_2O (10:1), 0°	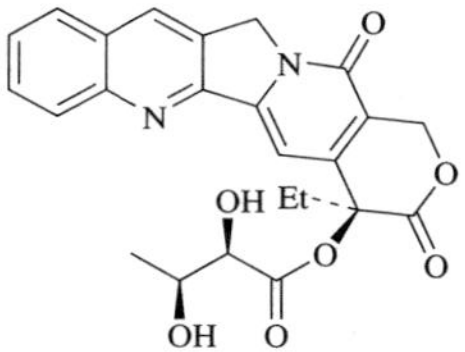	(68), 78	365

TABLE 3. REACTIONS OF TRANS 1,2-DISUBSTITUTED ALKENES (*Continued*)

Substrate	Conditions	Product(s) and Yield(s) (%), % ee		Refs.
C_{24} 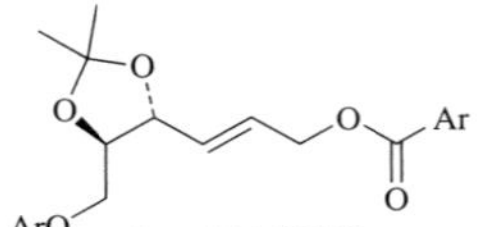 Ar = 4-$MeOC_6H_4$	$(DHQD)_2PYDZ$, $K_2OsO_2(OH)_4$, $K_3Fe(CN)_6$, K_2CO_3, $MeSO_2NH_2$, t-BuOH:H_2O (1:1), 0°	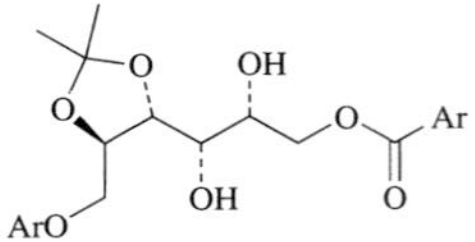	(90), 98 de	143
	$(DHQ)_2PHAL$, $K_2OsO_2(OH)_4$, $K_3Fe(CN)_6$, K_2CO_3, $MeSO_2NH_2$, t-BuOH:H_2O (1:1), 0°	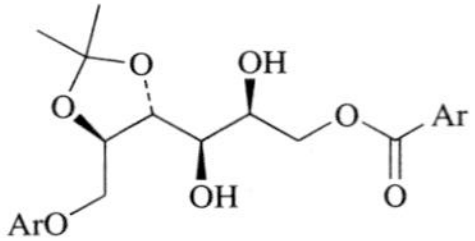	(93), >99 de	143
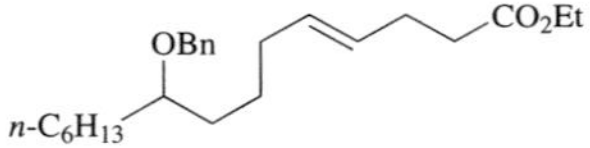	AD-Mix α, $MeSO_2NH_2$, t-BuOH:H_2O (1:1), 0°, 24 h	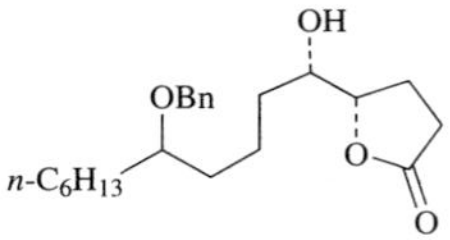	(52), — de	494
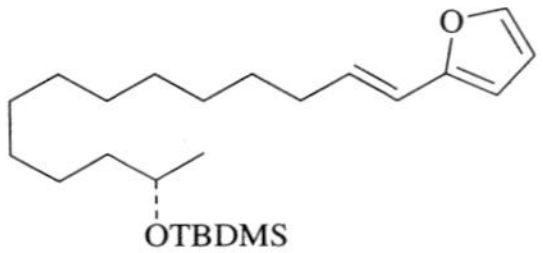	AD-Mix β, $MeSO_2NH_2$, t-BuOH:H_2O (1:1), 0°-rt, 15 h	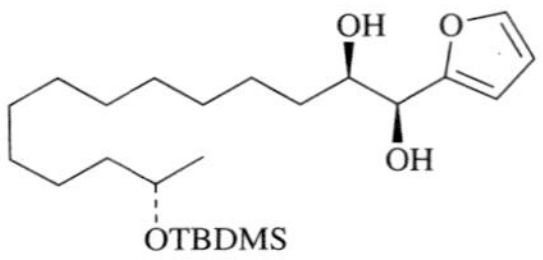	(91), 100 de	495, 496
	AD-Mix α, $MeSO_2NH_2$, t-BuOH:H_2O (1:1), 0°-rt	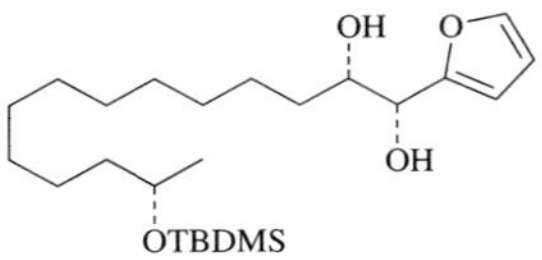	(85), >95 de	496

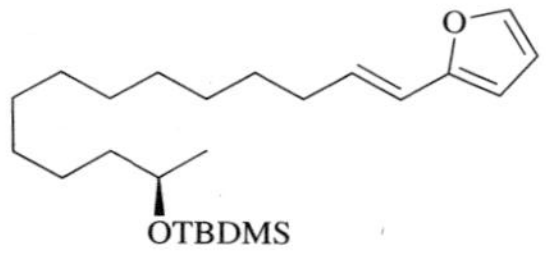	AD-Mix α, $MeSO_2NH_2$, *t*-BuOH:H_2O (1:1), 0°-rt, 15 h	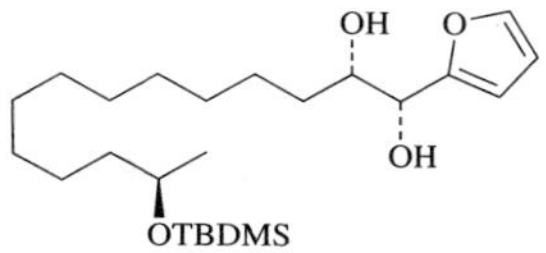 (91), 100 de	495
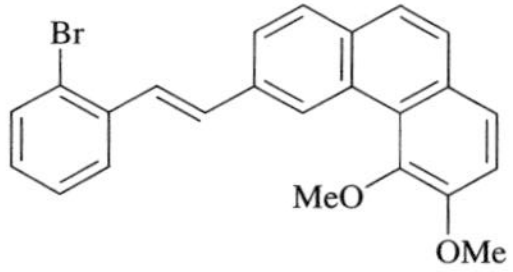	AD-Mix β, $MeSO_2NH_2$, *t*-BuOH:H_2O (1:1), rt, 24 h	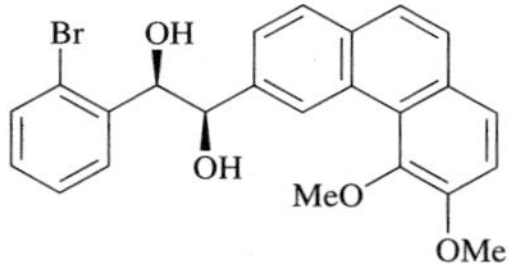 (93), —	497
C_{25}			
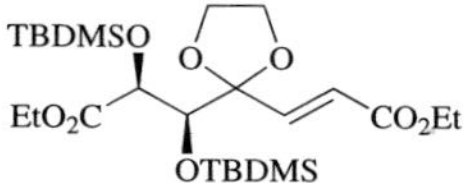	$(DHQD)_2PHAL$, OsO_4, NMO, *t*-BuOH:H_2O (10:1), 20°, 20 h	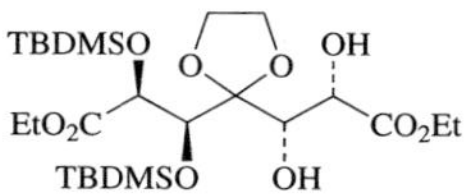 (93), 50 de	498
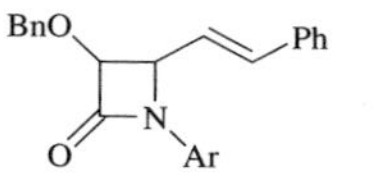 *cis* Ar = 4-$MeOC_6H_4$	1. DHQD-*p*-chlorobenzoate, $K_2OsO_2(OH)_4$, $K_3Fe(CN)_6$, K_2CO_3, THF:H_2O (1:1), rt, 15 h 2. $(MeO)_2CMe_2$, TsOH	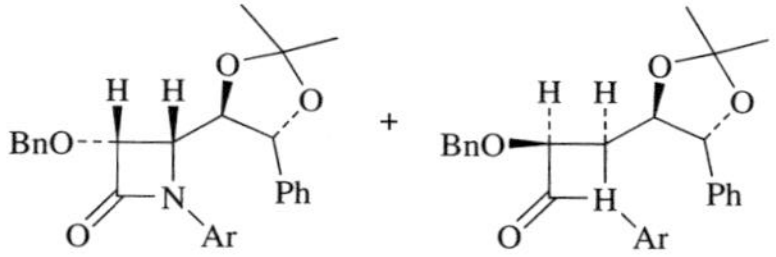 **I** 93% **I + II** (68) **II** 97%	450
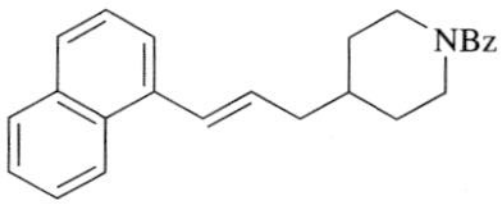	AD-Mix α, $MeSO_2NH_2$, *t*-BuOH:H_2O (1:1), 0°-rt, 24 h	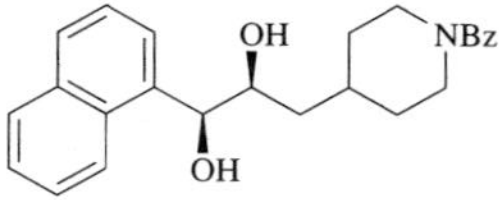 (93), >95	101

TABLE 3. REACTIONS OF TRANS 1,2-DISUBSTITUTED ALKENES (*Continued*)

Substrate	Conditions	Product(s) and Yield(s) (%), % ee	Refs.
C_{25} 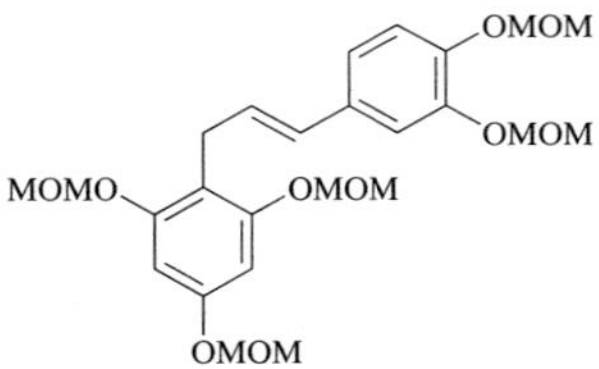	AD-Mix α, $MeSO_2NH_2$, t-BuOH:H_2O (1:1), 0°	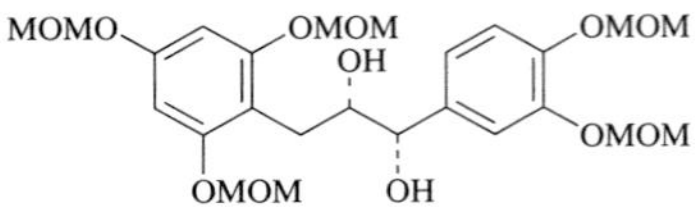 (75), 99	499
	AD-Mix β, $MeSO_2NH_2$, t-BuOH:H_2O (1:1), 0°	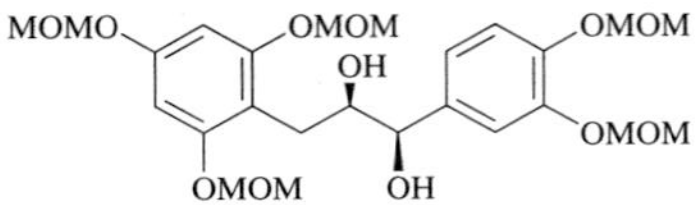 (79), 99	499
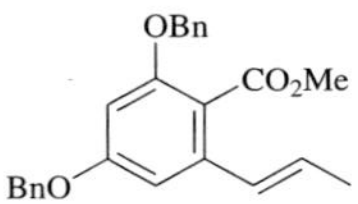	AD-Mix β, $MeSO_2NH_2$, t-BuOH:H_2O (1:1), 4°, 12 d	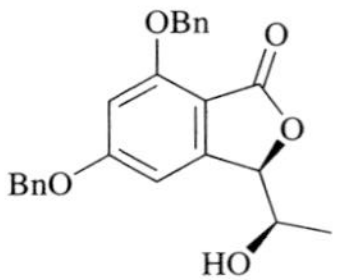 (75), 99	499a
	AD-Mix α, $MeSO_2NH_2$, t-BuOH:H_2O (1:1), 4°, 12 d	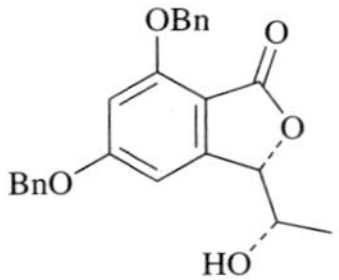 (79), 99	499a
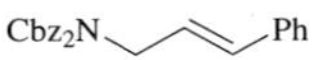	(DHQD)$_2$PHAL, $K_2OsO_2(OH)_4$, $K_3Fe(CN)_6$, K_2CO_3, $MeSO_2NH_2$, t-BuOH:H_2O (1:1), 0°	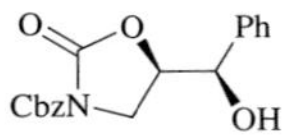 (75), 96	85

	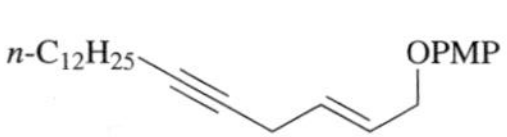	(DHQD)$_2$PHAL, $K_2OsO_2(OH)_4$, $K_3Fe(CN)_6$, K_2CO_3, $MeSO_2NH_2$, t-BuOH:H_2O (1:1), 0°, 48 h	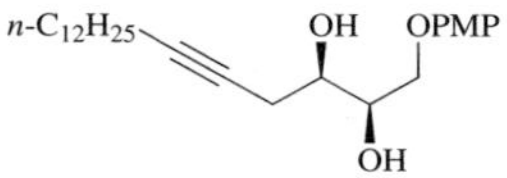	(96), 95	500
		(DHQ)$_2$PHAL, $K_2OsO_2(OH)_4$, $K_3Fe(CN)_6$, K_2CO_3, $MeSO_2NH_2$, t-BuOH:H_2O (1:1), 0°, 48 h	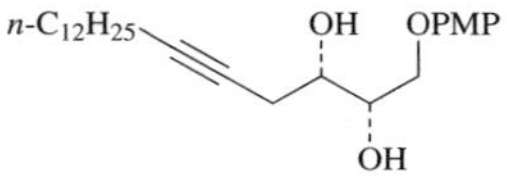	(—), 89	500
	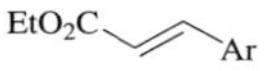	AD-Mix β, $MeSO_2NH_2$, t-BuOH:H_2O:toluene (1:2:1), rt, 72 h	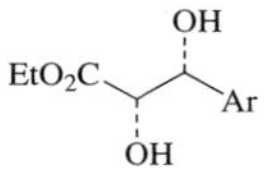		501
	Ar				
	3,5-$(BnO)_2C_6H_3$			(83), >99	
	3,4-$(BnO)_2C_6H_3$			(91), >99	
	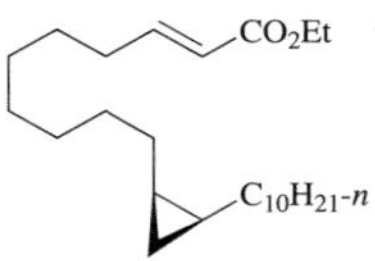	AD-Mix β, $MeSO_2NH_2$, t-BuOH:H_2O (1:1), 0°, 8 h	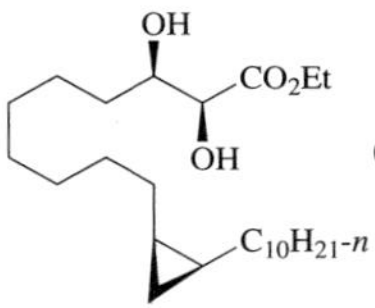	(96), — de	289
C_{26}	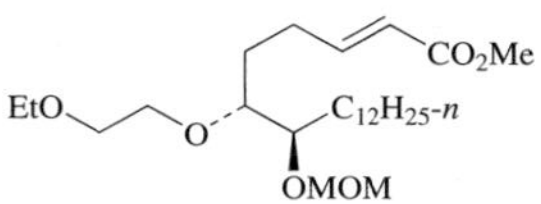	AD-Mix α, $MeSO_2NH_2$, t-BuOH:H_2O (1:1), 0°	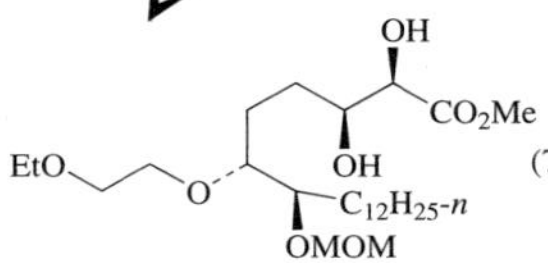	(72),100 de	502
	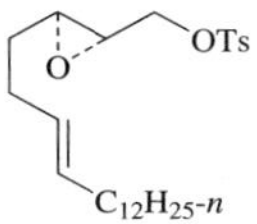	AD-Mix β, t-BuOH:H_2O (1:1)	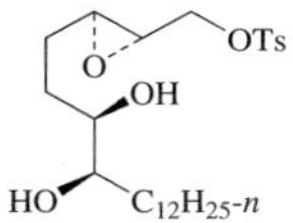	(97), —	503

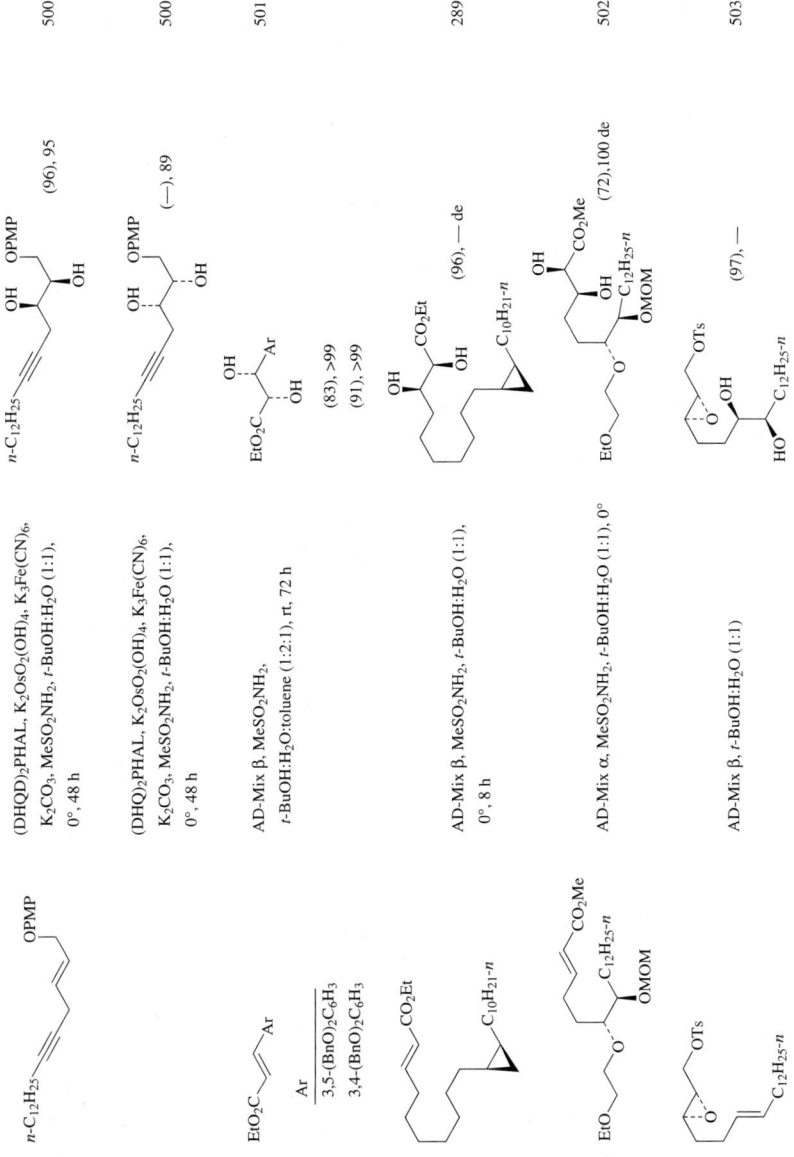

TABLE 3. REACTIONS OF TRANS 1,2-DISUBSTITUTED ALKENES (*Continued*)

Substrate	Conditions	Product(s) and Yield(s) (%), % ee	Refs.
C_{26}			
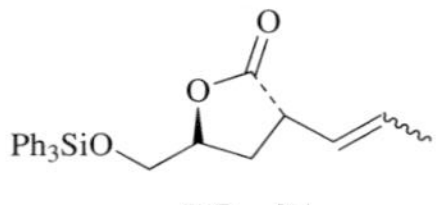 E:Z = 8:1	AD-Mix β, t-BuOH:H_2O (1:1), 0°	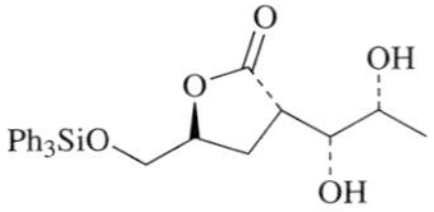 + (16), — 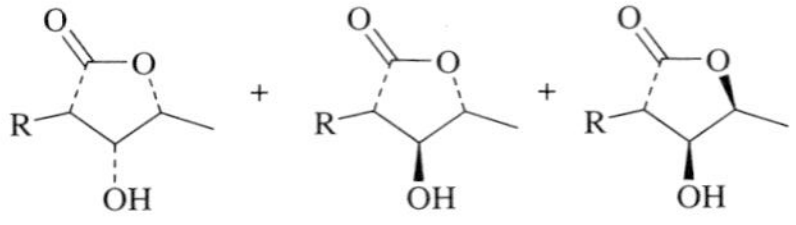 (54), — + (4), — + (3), — R = 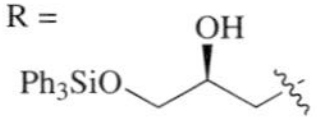	504
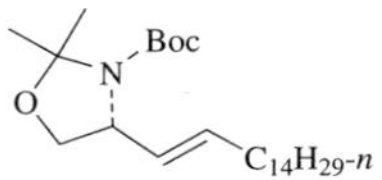	AD-Mix β, $MeSO_2NH_2$, t-BuOH:H_2O (1:1), 0°-rt, 2 d	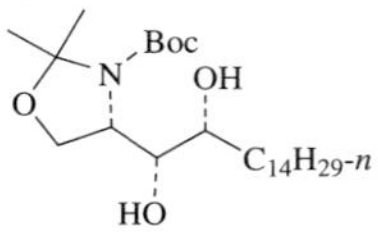 (85), 20 de	505
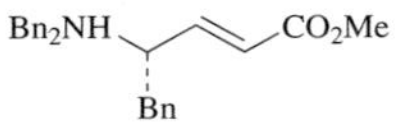	$(DHQ)_2PHAL$, $K_2OsO_2(OH)_4$, $K_3Fe(CN)_6$, K_2CO_3, t-BuOH:H_2O (1:1), rt, 10 d	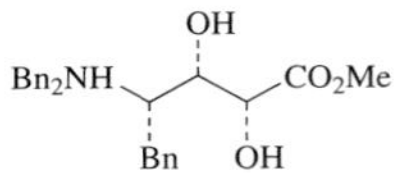 (74), 76 de	385
	$(DHQD)_2PHAL$, $K_2OsO_2(OH)_4$, $K_3Fe(CN)_6$, K_2CO_3, t-BuOH:H_2O (1:1), rt, 10 d	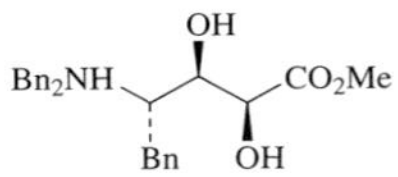 (49), 72 de	385
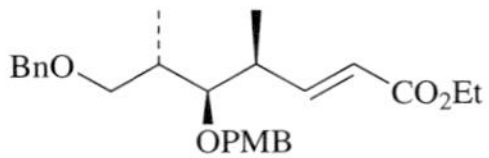	AD-Mix α, $MeSO_2NH_2$, t-BuOH:H_2O (1:1), 0°, 20 h	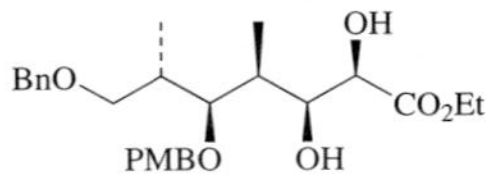 (85), 84 de	506

Reactant	Conditions	Product(s) and Yield(s) (%)	Refs.
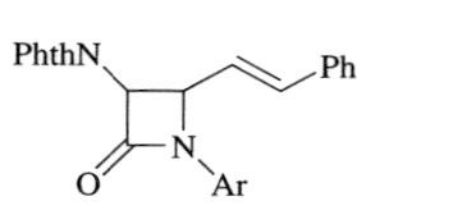 *cis* Ar = 4-$MeOC_6H_4$	1. DHQD-*p*-chlorobenzoate, $K_2OsO_2(OH)_4$, $K_3Fe(CN)_6$, K_2CO_3, *t*-BuOH:H_2O (1:1), rt, 15 h 2. $(MeO)_2CMe_2$, TsOH	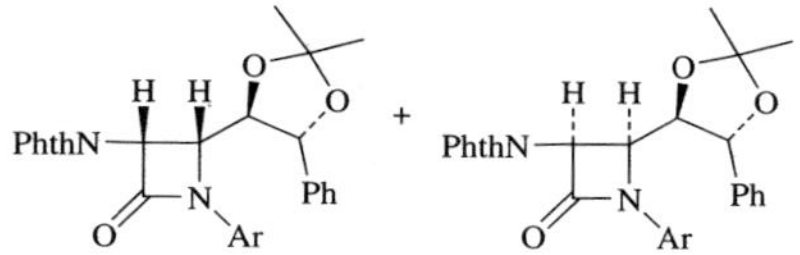 **I** 80% ee **I + II** (52) **II** 77% ee	450
C_{27}			
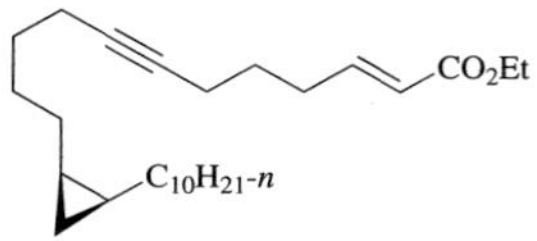	AD-Mix β, $MeSO_2NH_2$, *t*-BuOH:H_2O (1:1), 0°, 8 h	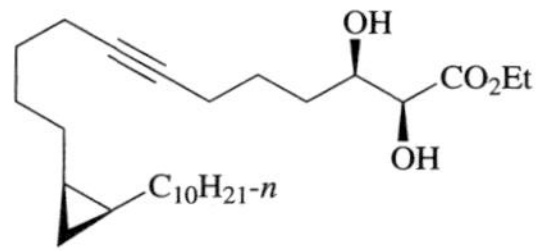 (97), — de	289
C_{28}			
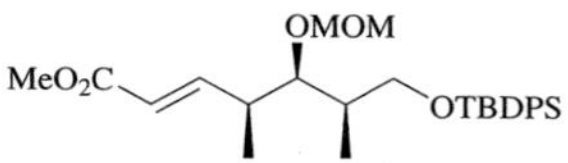	$(DHQ)_2PHAL$, $K_2OsO_2(OH)_4$, $K_3Fe(CN)_6$, K_2CO_3, $MeSO_2NH_2$, *t*-BuOH:H_2O (1:1), rt, 6 h	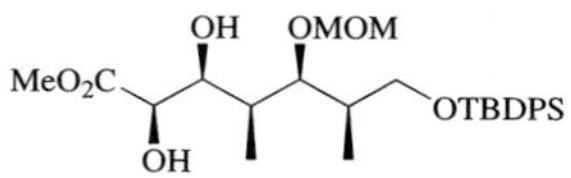 (89), 92 de	508
	$(DHQD)_2PHAL$, $K_2OsO_2(OH)_4$, $K_3Fe(CN)_6$, K_2CO_3, $MeSO_2NH_2$, *t*-BuOH:H_2O (1:1), rt, 6 h	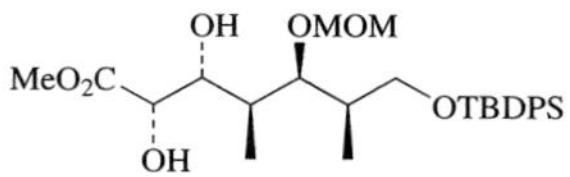 (37), 75 de	508

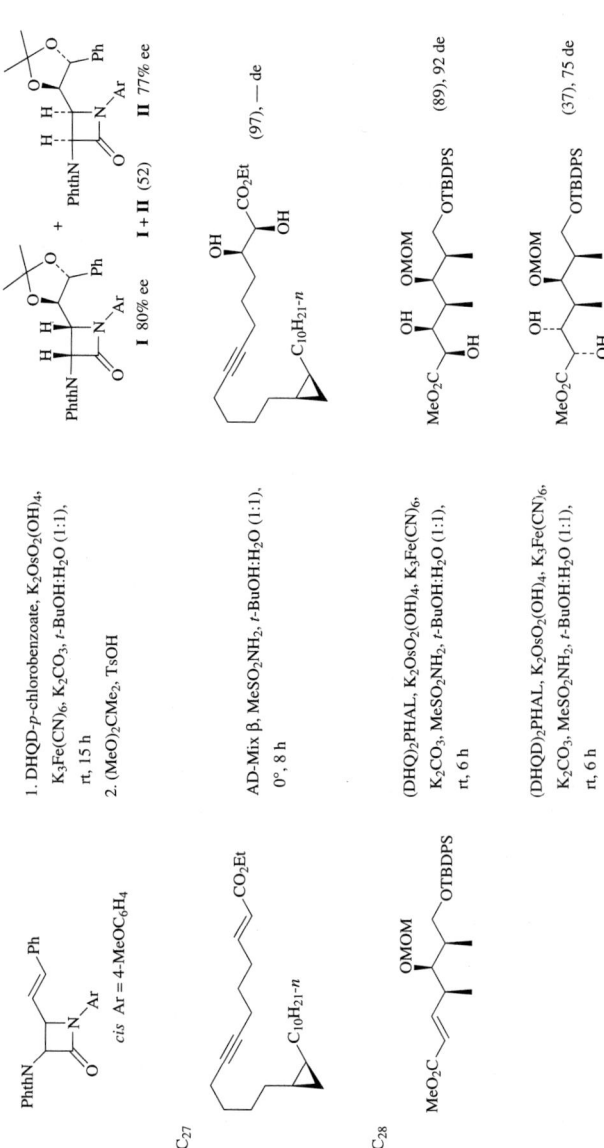

TABLE 3. REACTIONS OF TRANS 1,2-DISUBSTITUTED ALKENES (*Continued*)

Substrate	Conditions	Product(s) and Yield(s) (%), % ee	Refs.
C_{28} 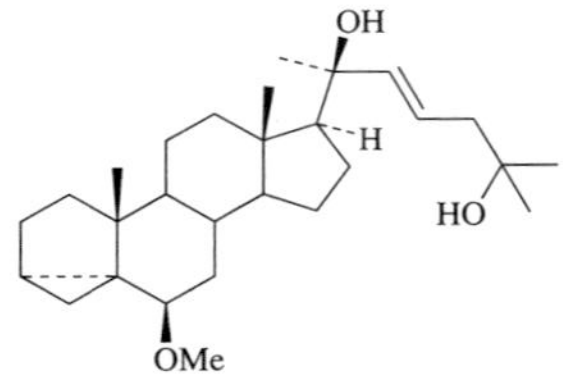	DHQD-*p*-chlorobenzoate, OsO_4	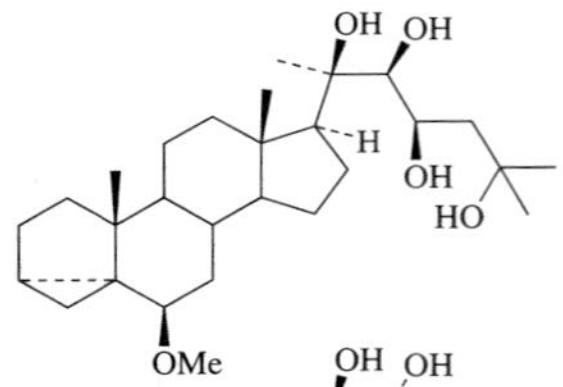 (—), 68 de	509
	DHQ-*p*-chlorobenzoate, OsO_4	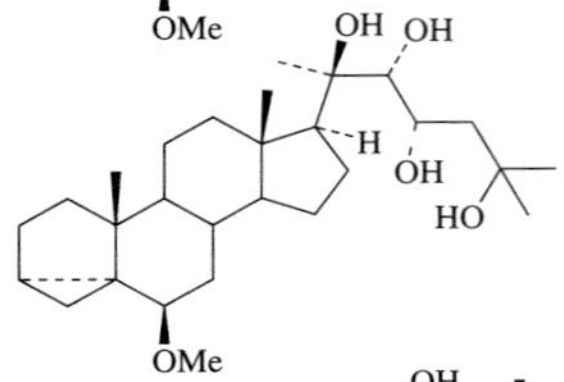 (—), 64 de	509
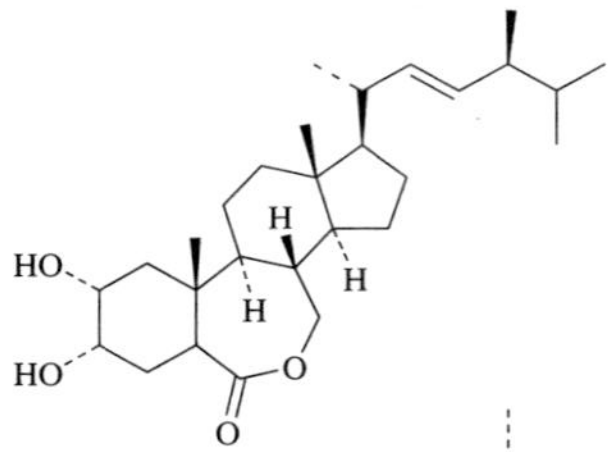	$(DHQD)_2PHAL$, $K_2OsO_2(OH)_4$, $K_3Fe(CN)_6$, K_2CO_3, $MeSO_2NH_2$, *t*-BuOH:H_2O (1:1), rt, 30 h	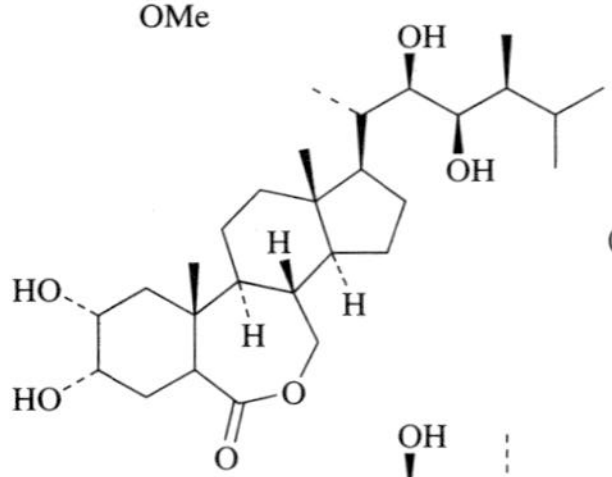 (70), 100 de	510
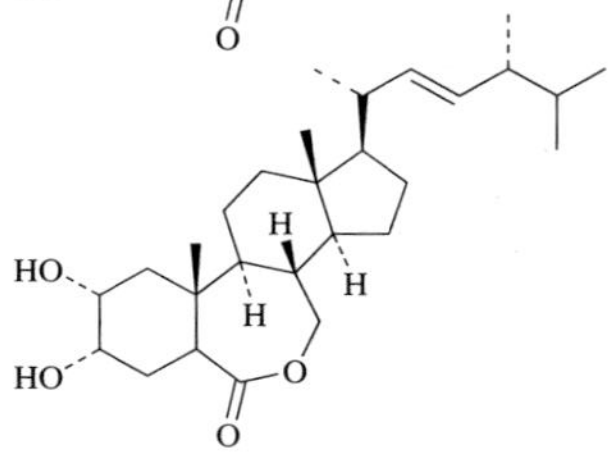	$(DHQD)_2PHAL$, $K_2OsO_2(OH)_4$, $K_3Fe(CN)_6$, K_2CO_3, $MeSO_2NH_2$, *t*-BuOH:H_2O (1:1), rt, 48 h	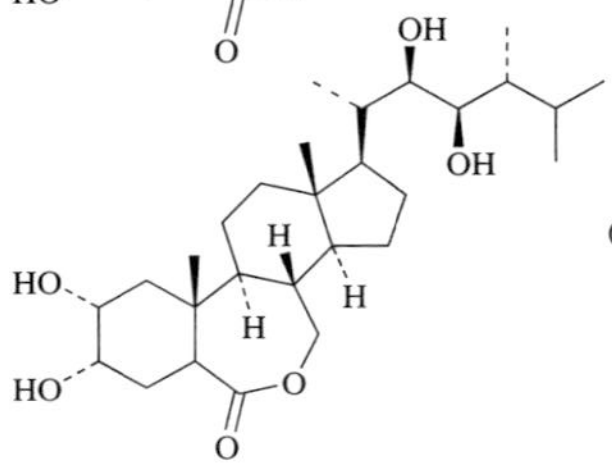 (67), 100 de	510

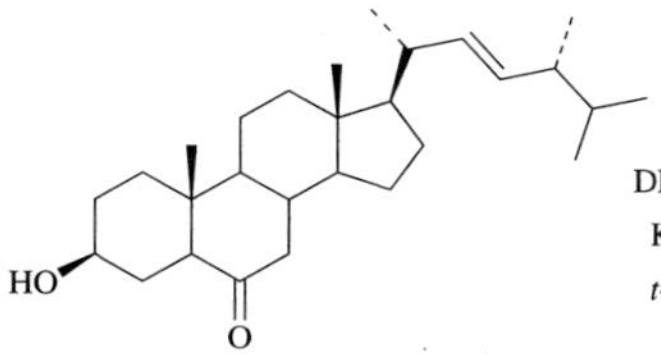	DHQD-*p*-chlorobenzoate, $K_2OsO_2(OH)_4$, $K_3Fe(CN)_6$, K_2CO_3, $MeSO_2NH_2$, *t*-BuOH:H_2O (1:1), rt, 5 d	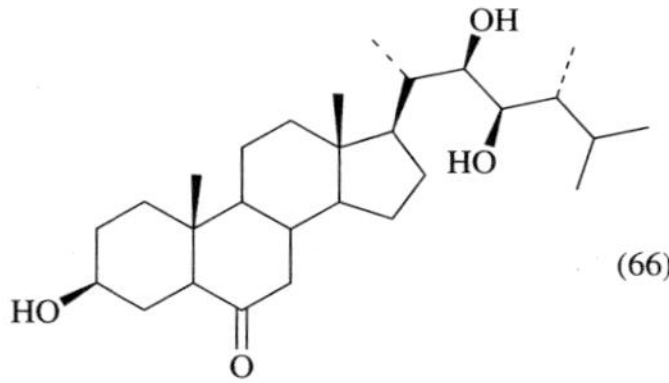 (66), — de	510a
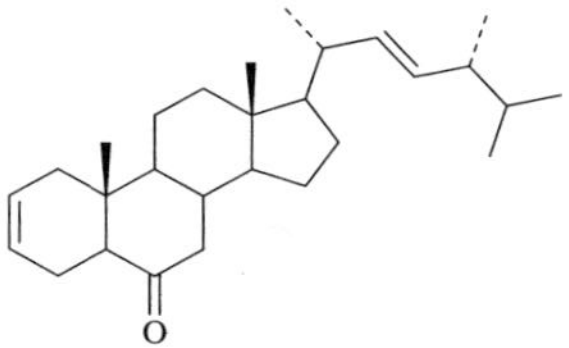	DHQD-*p*-chlorobenzoate, OsO_4, $K_3Fe(CN)_6$, *t*-BuOH:H_2O (1:1), rt, 3-5 d	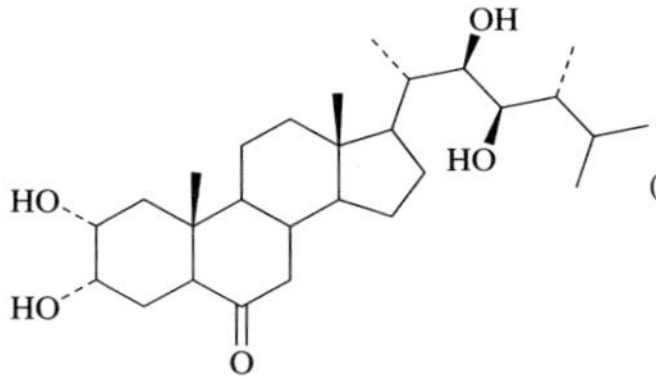 (90), 78 de	507
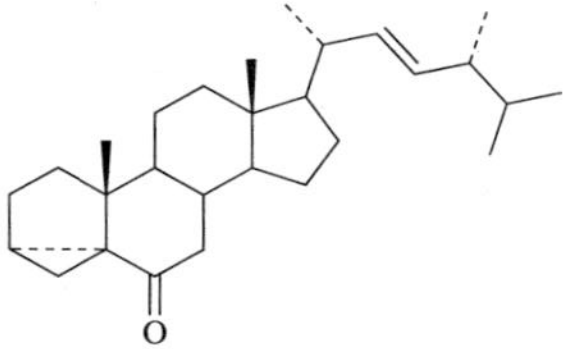	DHQD-*p*-chlorobenzoate, OsO_4, $K_3Fe(CN)_6$, *t*-BuOH:H_2O (1:1), rt, 3-5 d	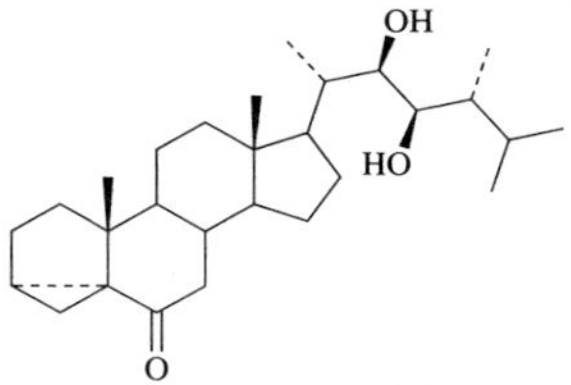 (94), 78 de	507
	DHQ-*p*-chlorobenzoate, OsO_4, $K_3Fe(CN)_6$, *t*-BuOH:H_2O (1:1), rt, 3-5 d	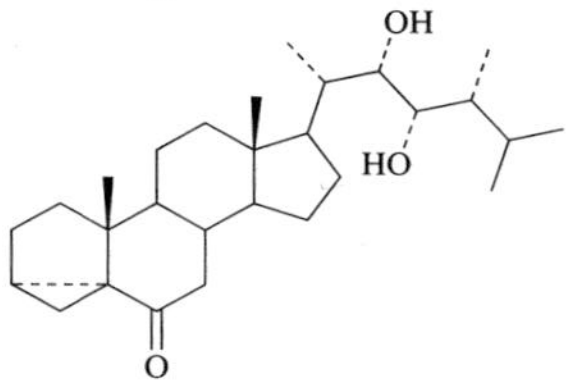 (84), 80 de	507

TABLE 3. REACTIONS OF TRANS 1,2-DISUBSTITUTED ALKENES (*Continued*)

Substrate	Conditions	Product(s) and Yield(s) (%), % ee	Refs.
C_{28}			
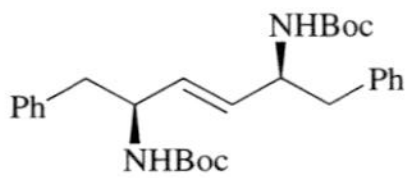	AD-Mix β, $MeSO_2NH_2$, *t*-BuOH:H_2O (1:1), 0°-rt	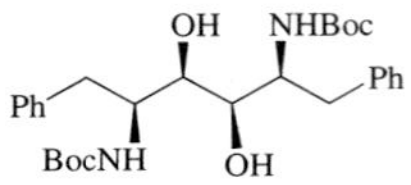 (—), 60 de	511
C_{29}			
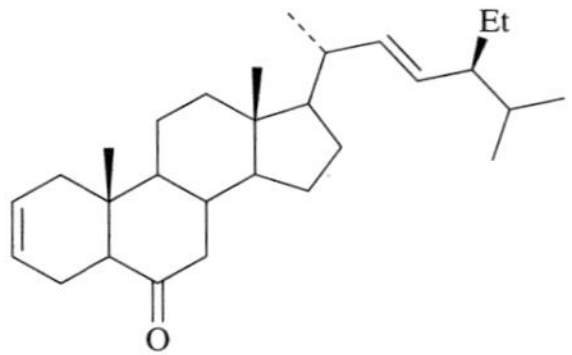	DHQD-*p*-chlorobenzoate, OsO_4, $K_3Fe(CN)_6$, K_2CO_3, *t*-BuOH:H_2O (1:1), rt, 3-5 d	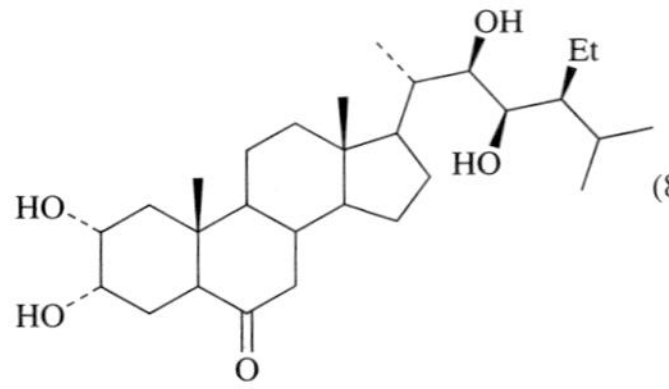 (89), 13 de	507
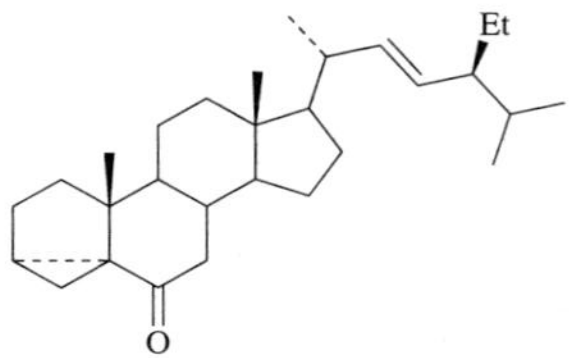	DHQD-*p*-chlorobenzoate, OsO_4, $K_3Fe(CN)_6$, *t*-BuOH:H_2O (1:1), rt, 3-5 d	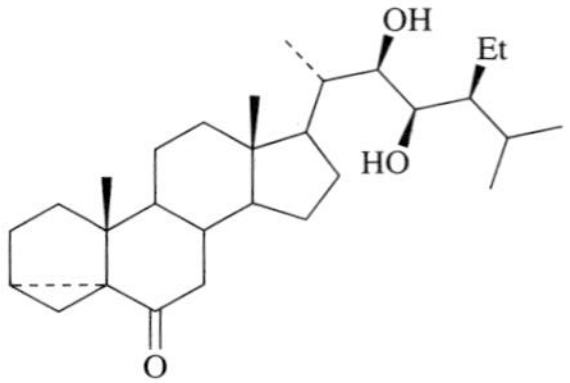 (93), 20 de	507
	DHQ-*p*-chlorobenzoate, OsO_4, $K_3Fe(CN)_6$, *t*-BuOH:H_2O (1:1), rt, 3-5 d	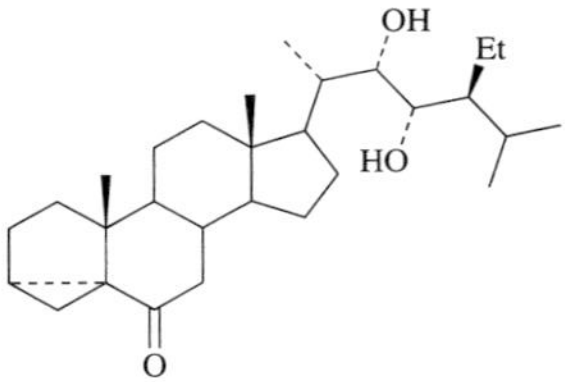 (83), >95 de	507

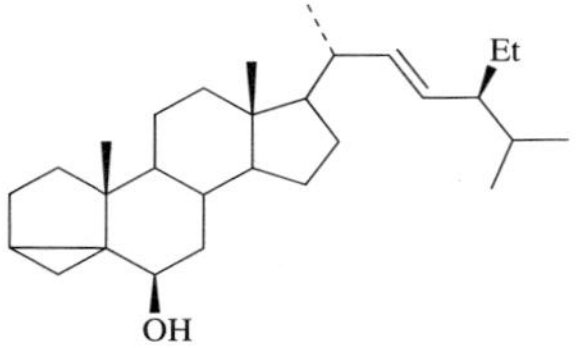	DHQD-*p*-chlorobenzoate, OsO_4, $K_3Fe(CN)_6$, K_2CO_3, $MeSO_2NH_2$, *t*-BuOH:H_2O (1:1), rt, 6 d	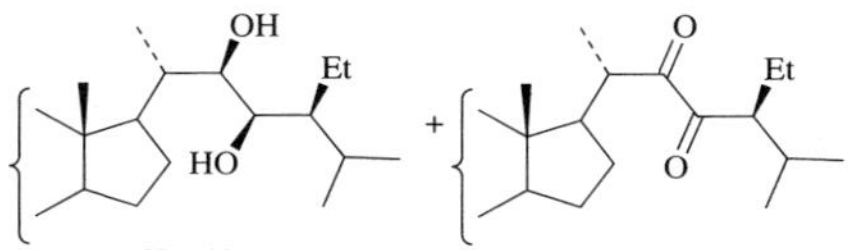 (57), 40 de + (25), — 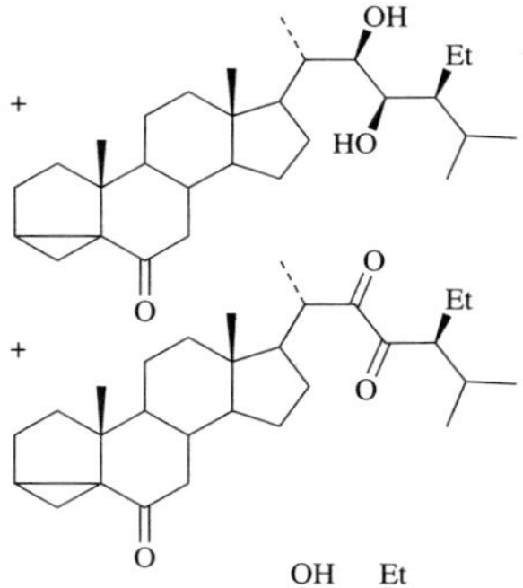 + (3), 33 de + (3), — de	512
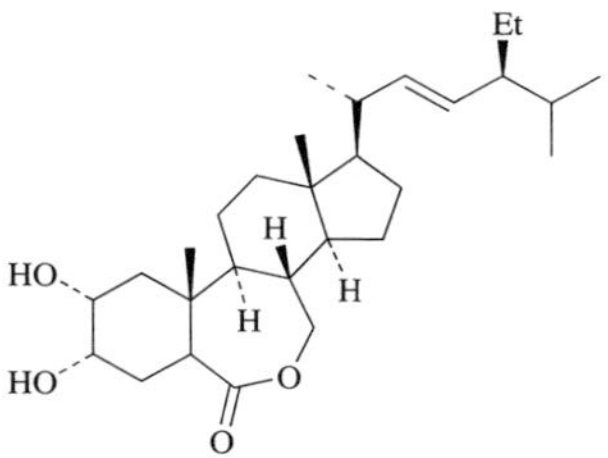	$(DHQD)_2PHAL$, $K_2OsO_2(OH)_4$, $K_3Fe(CN)_6$, K_2CO_3, $MeSO_2NH_2$, *t*-BuOH:H_2O (1:1), rt, 30 h	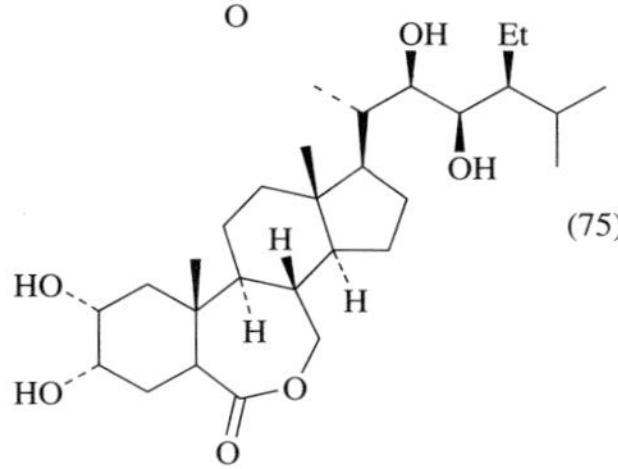 (75), 100 de	510
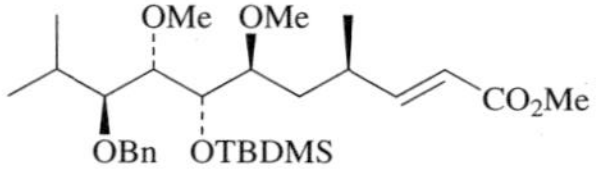	DHQD-OAc, OsO_4	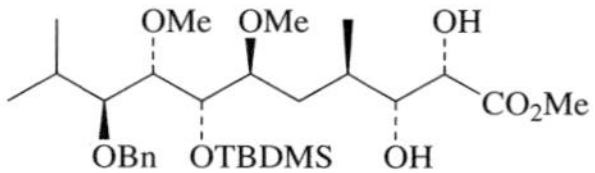 (—), —	513

TABLE 3. REACTIONS OF TRANS 1,2-DISUBSTITUTED ALKENES (*Continued*)

Substrate	Conditions	Product(s) and Yield(s) (%), % ee	Refs.
C_{29-31} 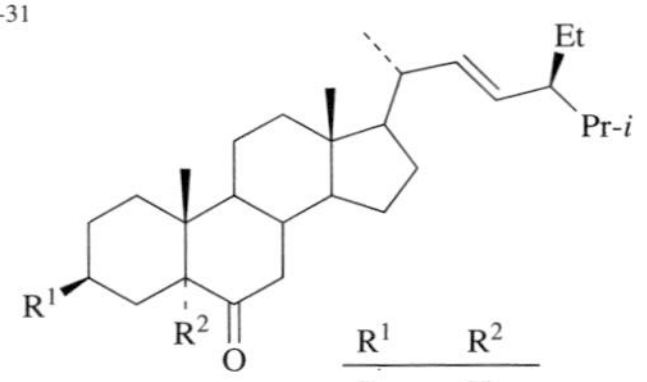	$(DHQD)_2PHAL$, $K_2OsO_2(OH)_4$, $K_3Fe(CN)_6$, K_2CO_3, $MeSO_2NH_2$, THF:*t*-BuOH:H_2O (1:2:2), rt, 9 d	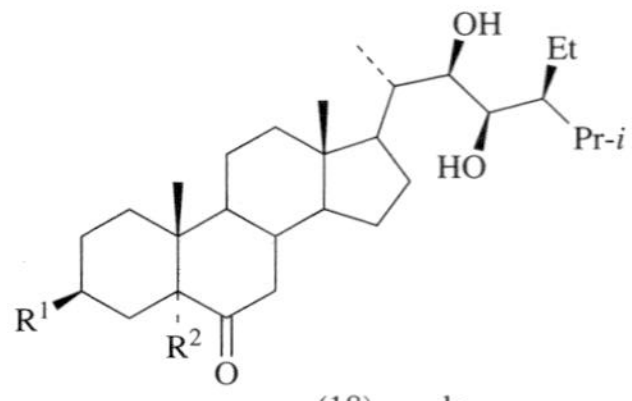	514
R^1 / R^2: Br / H		(18), — de	
R^1 / R^2: Br / OH		(19), — de	
R^1 / R^2: Ac / H		(22), 50 de	
R^1 / R^2: Ac / OH		(12), — de	
C_{29} 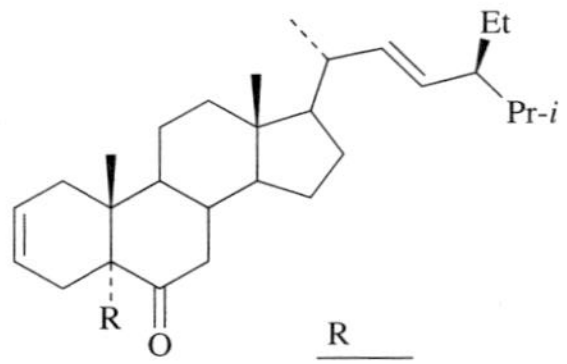	$(DHQD)_2PHAL$, $K_2OsO_2(OH)_4$, $K_3Fe(CN)_6$, K_2CO_3, $MeSO_2NH_2$, THF:*t*-BuOH:H_2O (1:2:2), rt, 9 d	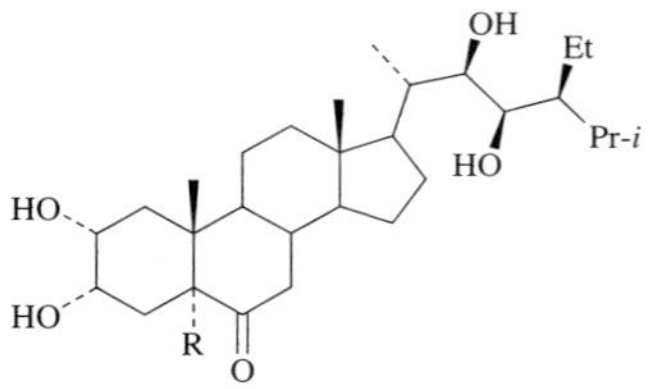	514
R: H		(34), — de	
R: OH		(12), — de	
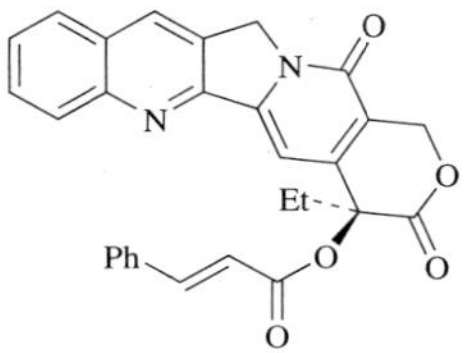	$(DHQD)_2PHAL$, $K_2OsO_2(OH)_4$, $K_3Fe(CN)_6$, K_2CO_3, $MeSO_2NH_2$, *t*-BuOH:H_2O (1:1), 0°	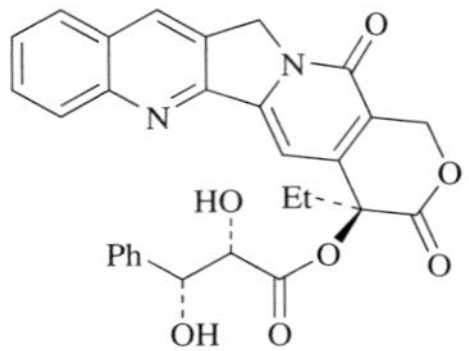 (60), 86 de	365

	(DHQ)$_2$PHAL, $K_2OsO_2(OH)_4$, NMO, acetone:H_2O (10:1), 0°	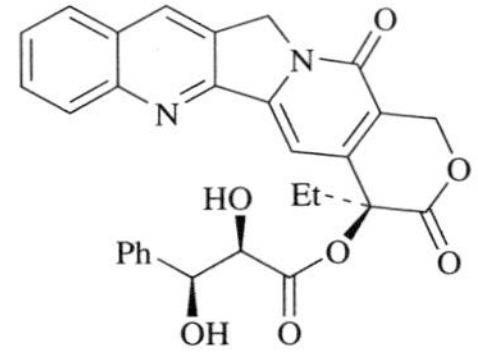 (57), 84 de	365
C_{30} 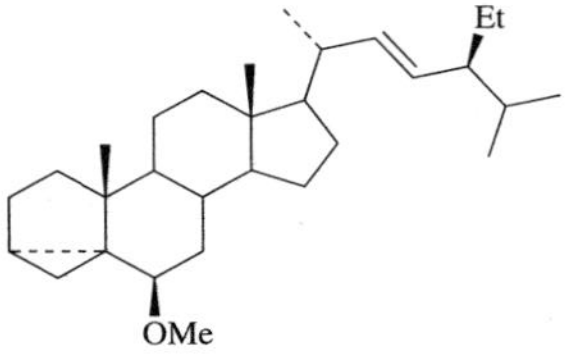	Ligand, OsO_4, NMO, THF:*t*-BuOH:H_2O (7.2:4:1.1), rt, 4-5 d Ligand DHQD-*p*-chlorobenzoate DHQD-PHN	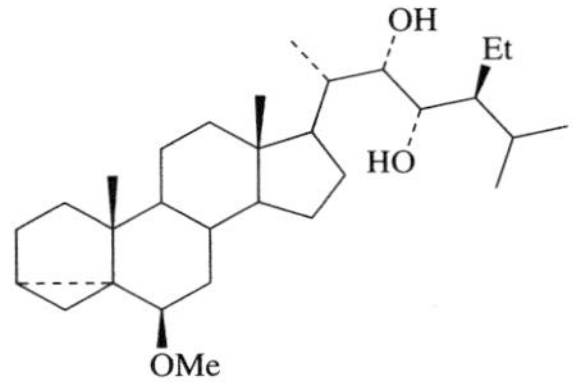 (—), 44 de (—), 20 de	515
	DHQD-PHN, OsO_4, NMO, THF:*t*-BuOH:H_2O (7.2:4:1.1), 0°-rt, 4-5 d (slow addition of olefin)	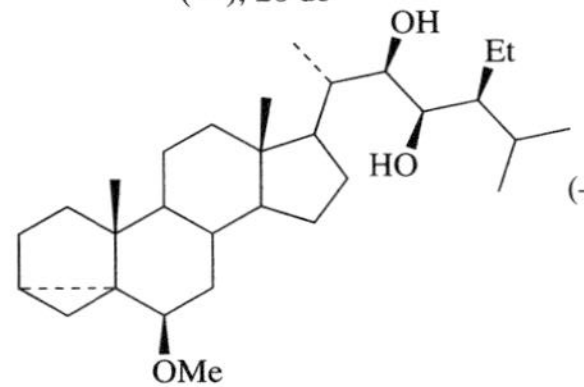 (—), 5-38 de	515
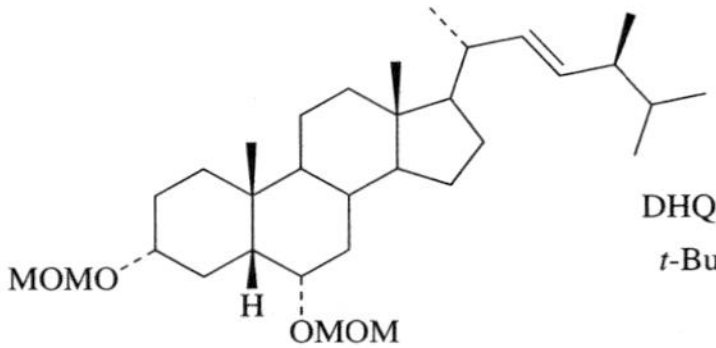	DHQD-*p*-chlorobenzoate, OsO_4, $K_3Fe(CN)_6$, *t*-BuOH:H_2O (1:1), rt, 3-5 d	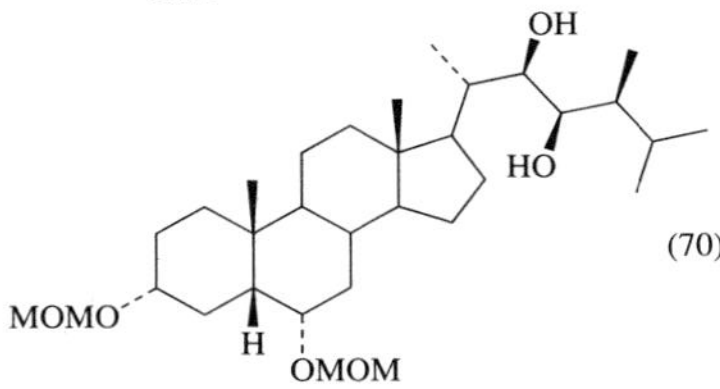 (70), 78 de	507

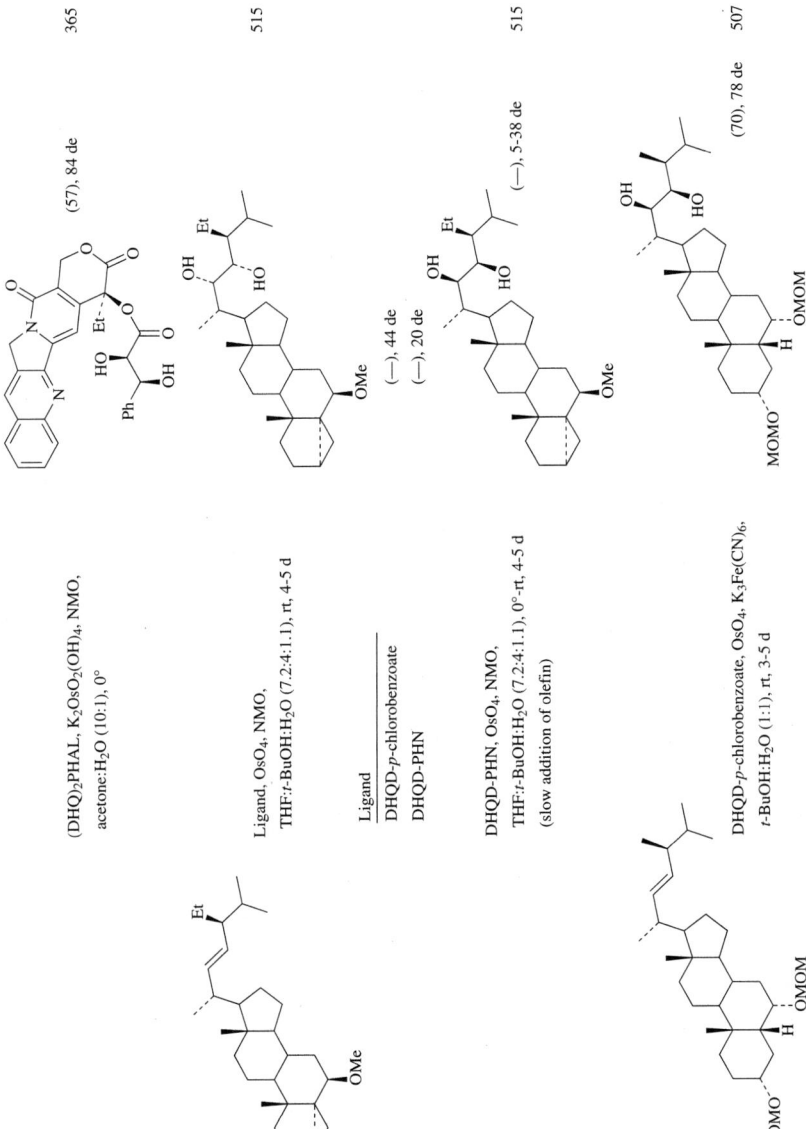

TABLE 3. REACTIONS OF TRANS 1,2-DISUBSTITUTED ALKENES (*Continued*)

Substrate	Conditions	Product(s) and Yield(s) (%), % ee	Refs.
C_{30} 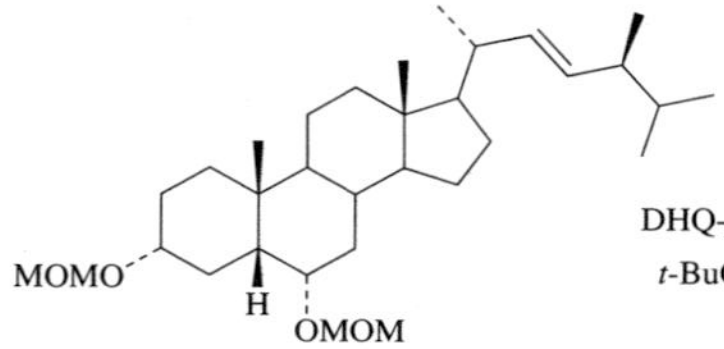	DHQ-*p*-chlorobenzoate, OsO_4, $K_3Fe(CN)_6$, *t*-BuOH:H_2O (1:1), rt, 3-5 d	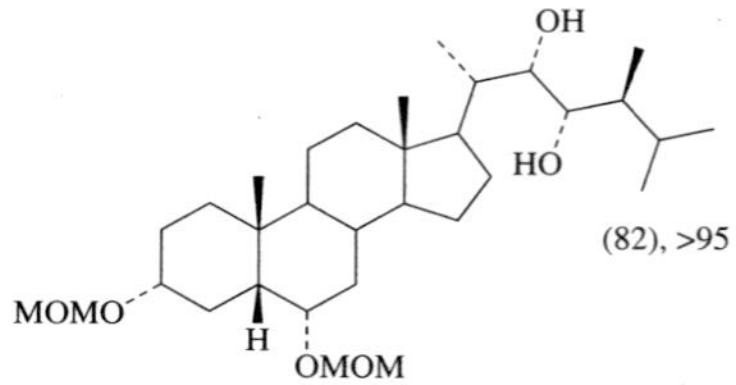 (82), >95 de	507
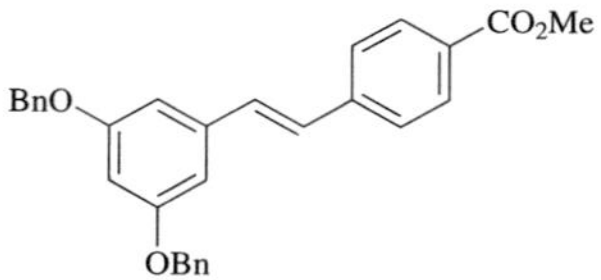	AD-Mix β, *t*-BuOH:H_2O (1:1), 0°	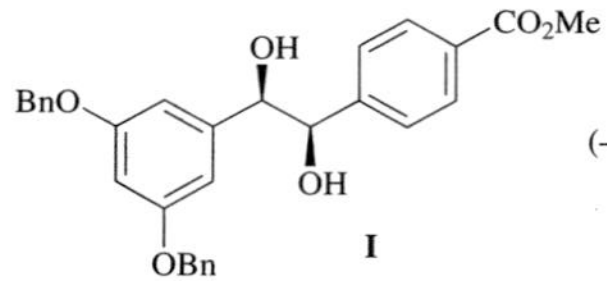 **I** (—), >97	516
	AD-Mix β, $MeSO_2NH_2$, *t*-BuOH:H_2O:toluene (1:2:1), 0°, 48 h	**I** (74), >97	501, 517
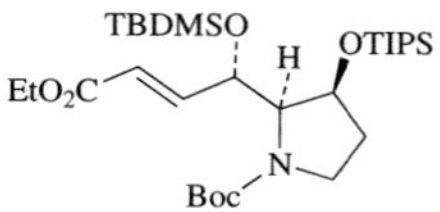	$(DHQD)_2PHAL$, OsO_4	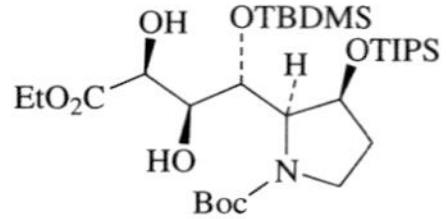 (88), 100 de	518
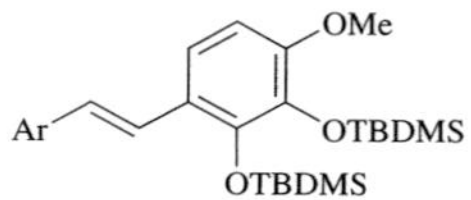 Ar = 3,4,5-$MeO_3C_6H_2$	AD-Mix β, $MeSO_2NH_2$, *t*-BuOH:H_2O (1:1), 0°-rt, 32 h	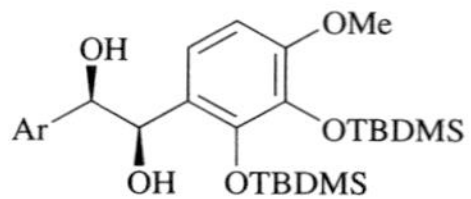 (75), —	519
	AD-Mix α, $MeSO_2NH_2$, *t*-BuOH:H_2O (1:1), 0°-rt, 32 h	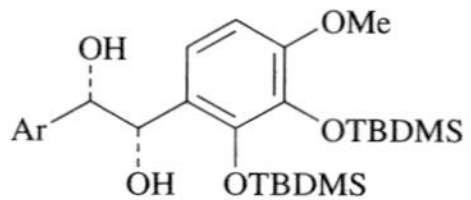 (74), —	519

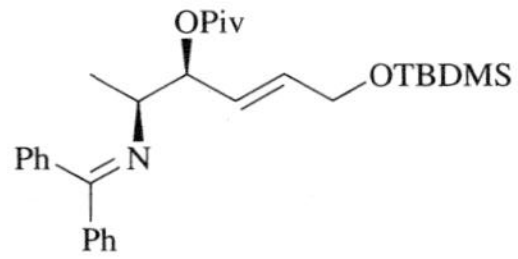	(DHQ)$_2$PHAL, $K_2OsO_2(OH)_4$, $K_2Fe(CN)_6$, K_2CO_3, $MeSO_2NH_2$, *t*-BuOH:H_2O (1:1), 0°, 12 h	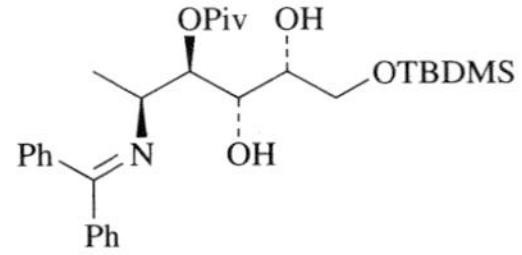 (60), 87	520
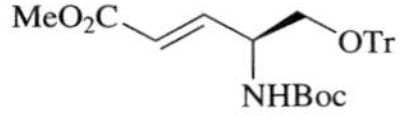	(DHQD)$_2$PHAL, $K_2OsO_2(OH)_4$, $K_2Fe(CN)_6$, K_2CO_3, $MeSO_2NH_2$, *t*-BuOH:H_2O (1:1), 0°, 12 h	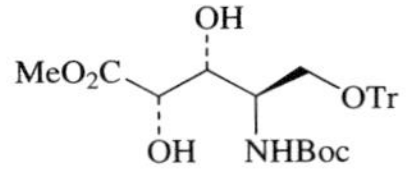 (84), >98 de	521
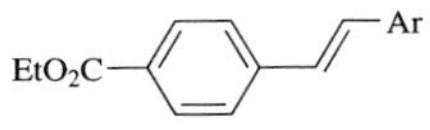 Ar = 3,4-$(BnO)_2C_6H_3$	AD-Mix β, $MeSO_2NH_2$, *t*-BuOH:H_2O:toluene (1:2:1), 0°, 48 h	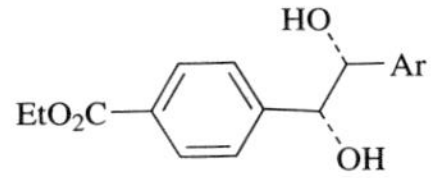 (90), >97	501
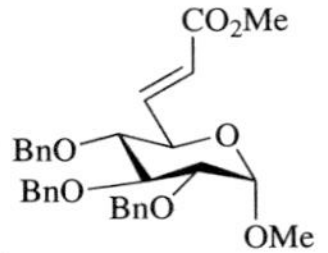	DHQD-*p*-chlorobenzoate, OsO_4, NMO, acetone:H_2O (10:1), 0°, 12 h	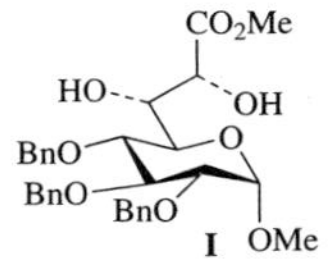 **I** (>90), 15 de	471
	DHQ-*p*-chlorobenzoate, OsO_4, NMO, acetone:H_2O (10:1), 0°, 12 h	**I** (>90), 91 de	471
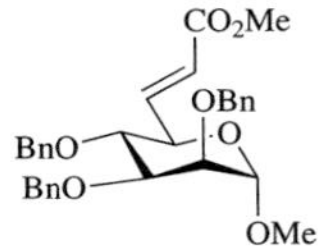	DHQD-*p*-chlorobenzoate, OsO_4, NMO, acetone:H_2O (10:1), 0°, 12 h	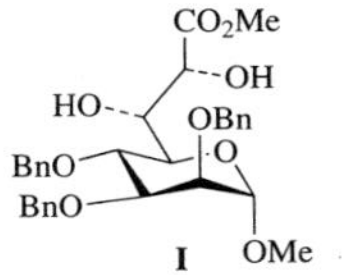 **I** (>90), 55 de	471
	DHQ-*p*-chlorobenzoate, OsO_4, NMO, acetone:H_2O (10:1), 0°, 12 h	**I** (>90), 88 de	471

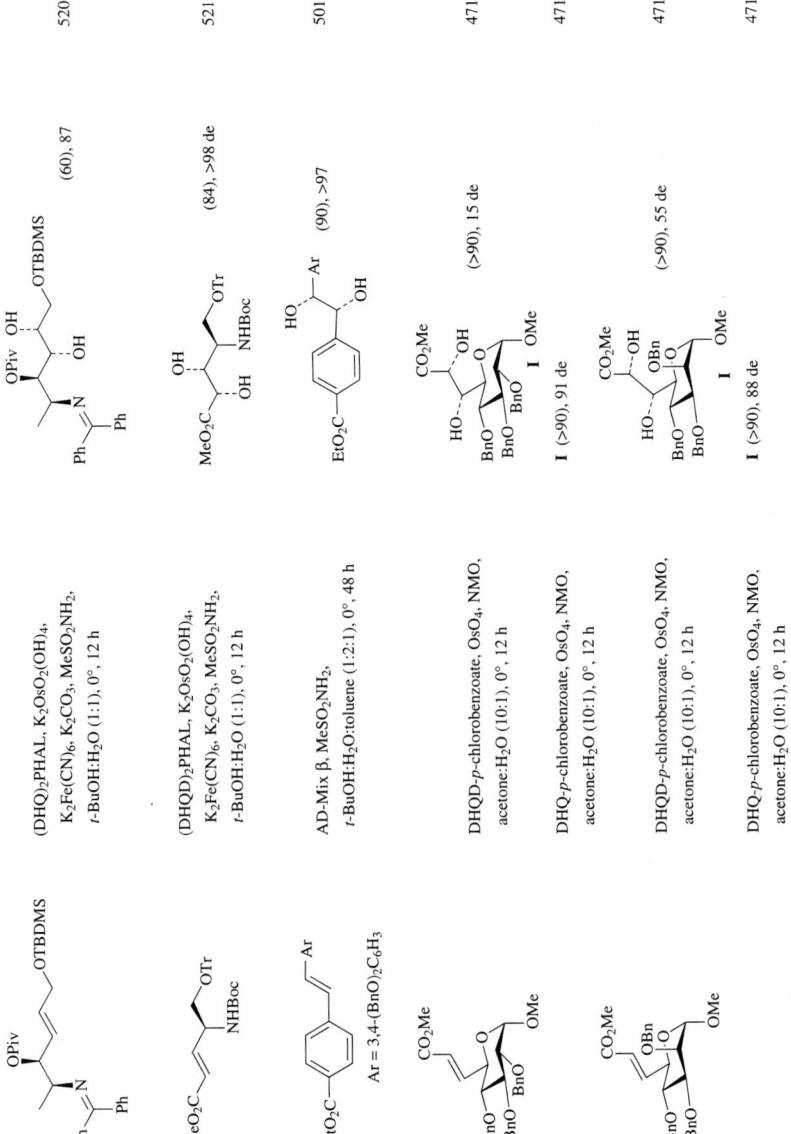

TABLE 3. REACTIONS OF TRANS 1,2-DISUBSTITUTED ALKENES (*Continued*)

Substrate	Conditions	Product(s) and Yield(s) (%), % ee	Refs.
C_{31}			
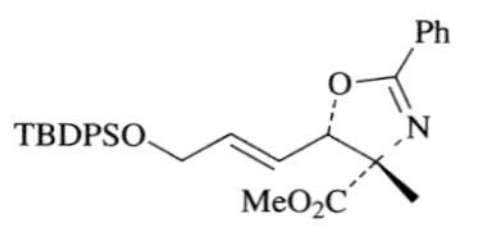	AD-Mix α, t-BuOH:H_2O (1:1)	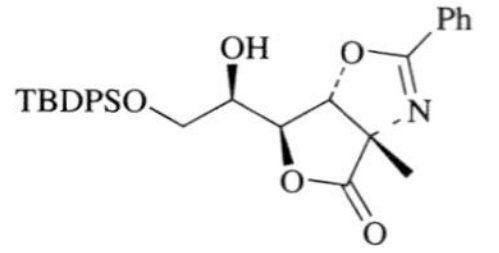 (100), 39 de	523
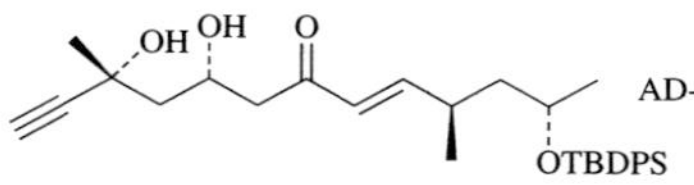	AD-Mix α, $MeSO_2NH_2$, t-BuOH:H_2O (1:1), 0°	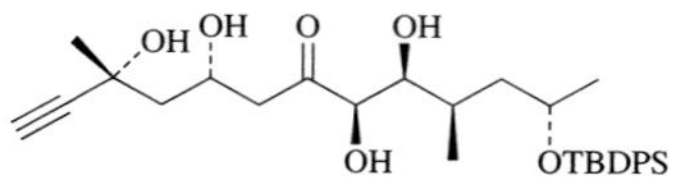 (57), 89	524
C_{32}			
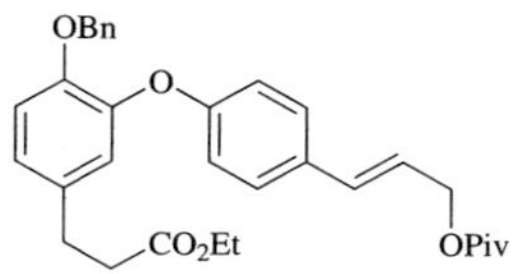	$(DHQD)_2PHAL$, $K_2OsO_2(OH)_4$, $K_3Fe(CN)_6$, K_2CO_3, $MeSO_2NH_2$, t-BuOH:H_2O (1:1) 0°, 12 h	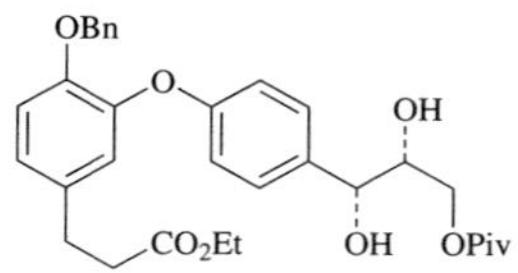 (87), 96	525, 526
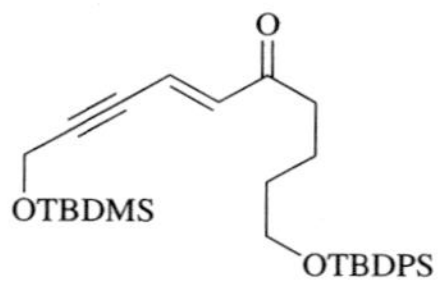	$(DHQ)_2PHAL$, $K_2OsO_2(OH)_4$, $K_3Fe(CN)_6$, K_2CO_3, $MeSO_2NH_2$, t-BuOH:H_2O (1:1)	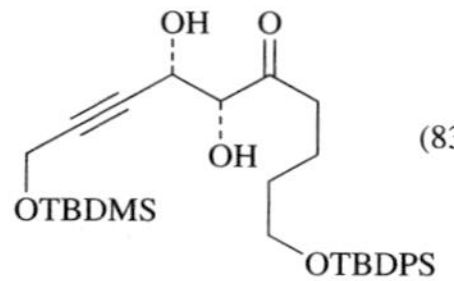 (83), 96	350

C_{33} 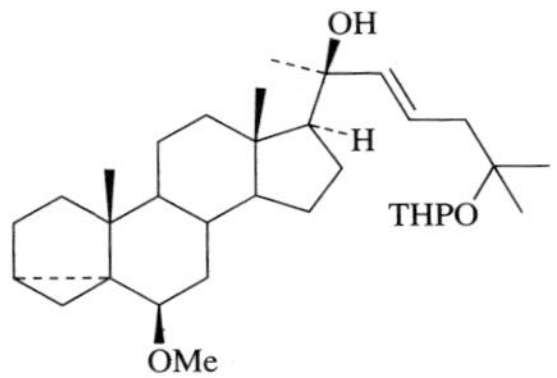	DHQ-*p*-chlorobenzoate, OsO_4	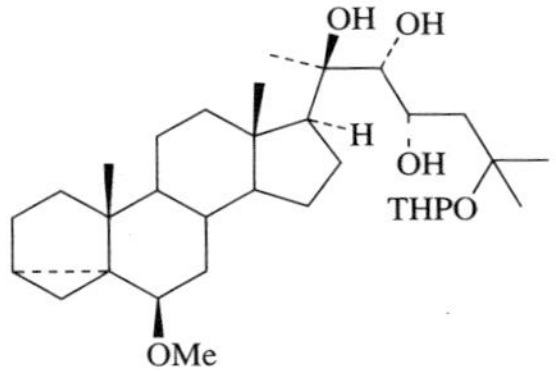	(—), 82 de	509
	DHQD-*p*-chlorobenzoate, OsO_4	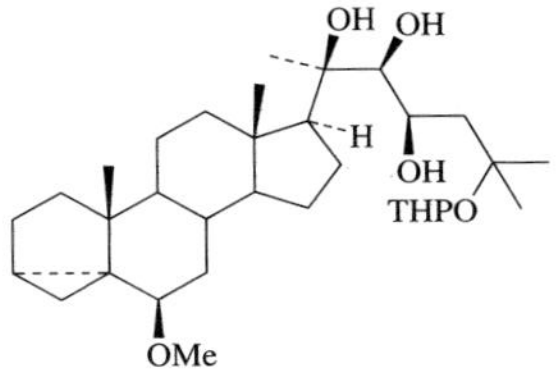	(—), 74 de	509
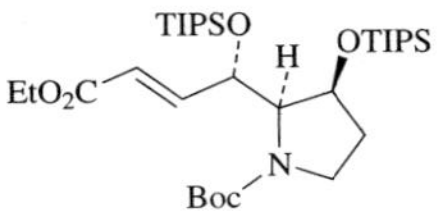	$(DHQD)_2PHAL$, OsO_4	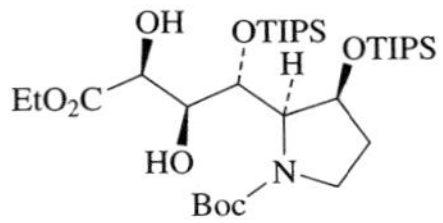	(88), 100 de	518

TABLE 3. REACTIONS OF TRANS 1,2-DISUBSTITUTED ALKENES (*Continued*)

Substrate	Conditions	Product(s) and Yield(s) (%), % ee	Refs.
C_{34-35}			
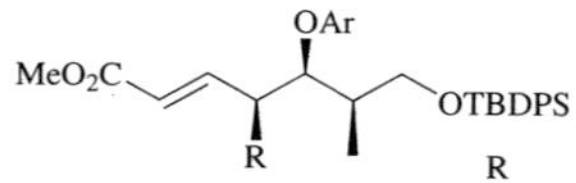 Ar = $CH_2C_6H_3(OMe)_2$-3,5 R: H, Me	(DHQ)$_2$PHAL, $K_2OsO_2(OH)_4$, $K_3Fe(CN)_6$, K_2CO_3, $MeSO_2NH_2$, *t*-BuOH:H_2O (1:1) rt, 6 h	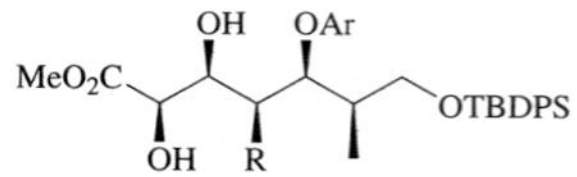 (83), 72 de (78), 98 de	527, 508
R: H, Me	(DHQD)$_2$PHAL, $K_2OsO_2(OH)_4$, $K_3Fe(CN)_6$, K_2CO_3, $MeSO_2NH_2$, *t*-BuOH:H_2O (1:1), rt, 6 h	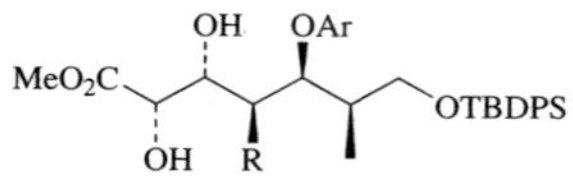 (93), 85 de (39), 78 de	527, 508
C_{34}			
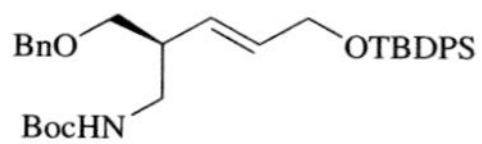	AD-Mix β, $MeSO_2NH_2$, *t*-BuOH:H_2O (1:1), 0°	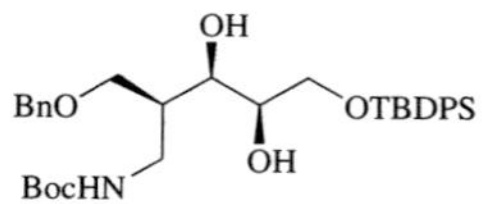	
	Temp 0°, Time —	(88), >97 de	528
	Temp 20°, Time 24 h	(88), >97 de	529

C_{35}			
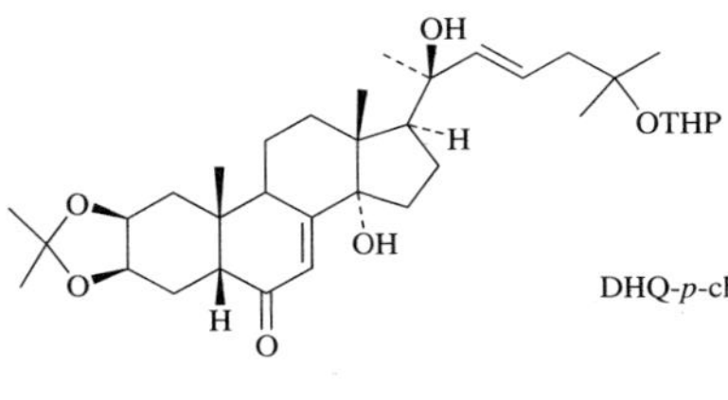	DHQ-*p*-chlorobenzoate, OsO_4	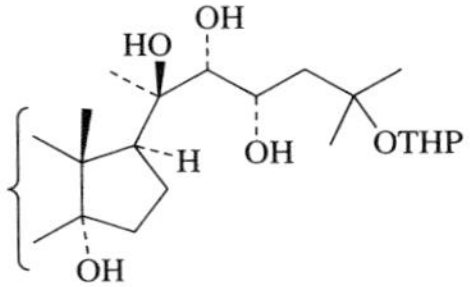 (65), 90 de	509
	DHQD-*p*-chlorobenzoate, OsO_4	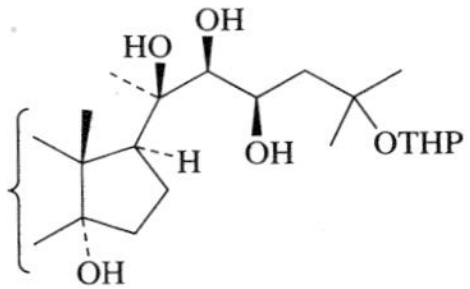 (54), 58 de	509
C_{36}			
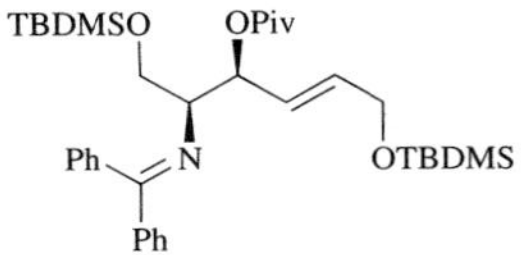	$(DHQD)_2PHAL$, $K_2OsO_2(OH)_4$, $K_2Fe(CN)_6$, K_2CO_3, $MeSO_2NH_2$, *t*-BuOH:H_2O (1:1), 0°, 12 h	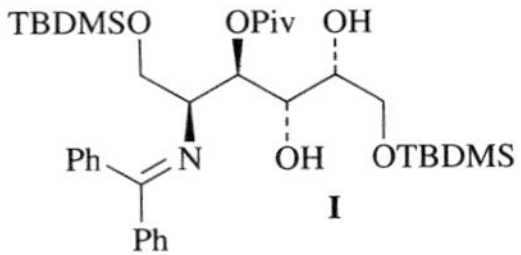 (60), 80	520
	$(DHQ)_2PHAL$, $K_2OsO_2(OH)_4$, $K_2Fe(CN)_6$, K_2CO_3, $MeSO_2NH_2$, *t*-BuOH:H_2O (1:1), 0°, 12 h	**I** (54-60), 83-90	520
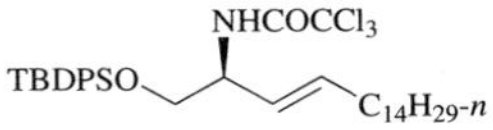	AD-Mix β, $MeSO_2NH_2$, *t*-BuOH:H_2O (1:1), 4 h	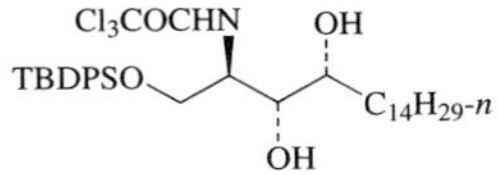 (80), 94	530

TABLE 3. REACTIONS OF TRANS 1,2-DISUBSTITUTED ALKENES (*Continued*)

Substrate	Conditions	Product(s) and Yield(s) (%), % ee	Refs.
C_{37} 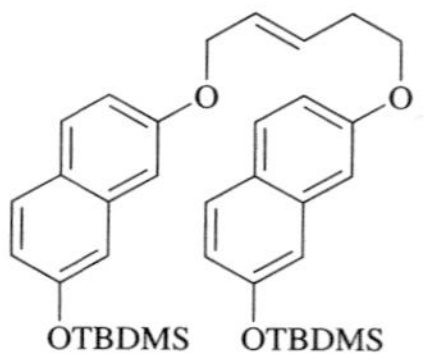	AD-Mix β, *t*-BuOH:H_2O (1:1), 0°, 5 h	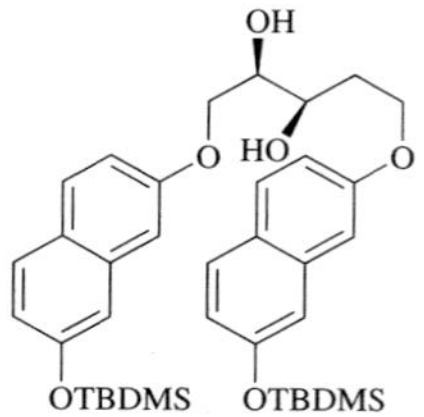 (80), >99	531
C_{38} 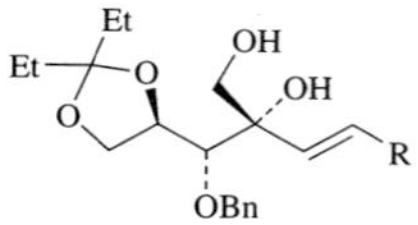	$(DHQD)_2$PHAL, OsO_4, NMO, acetone-H_2O	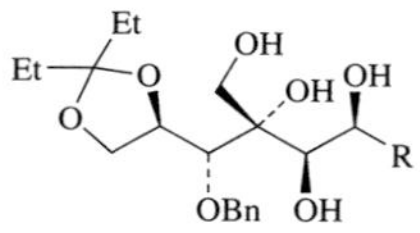 **I** (—), 26 de	532
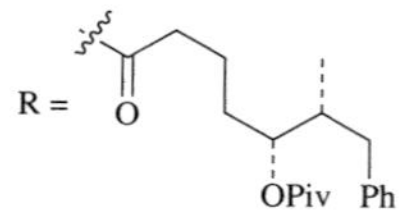	$(DHQ)_2$PHAL, OsO_4, NMO, acetone-H_2O	**I** (—), 26 de	532
C_{41} 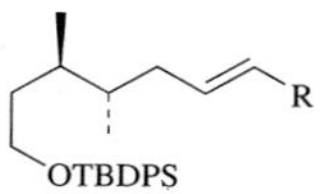	AD-Mix α, $MeSO_2NH_2$, *t*-BuOH:H_2O (1:1)	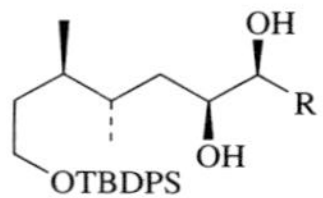 (53), —	533
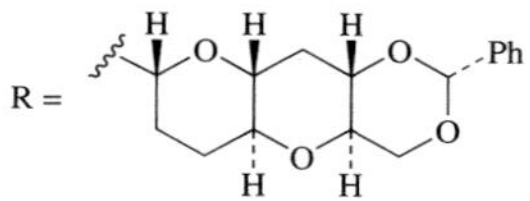	AD-Mix β, $MeSO_2NH_2$, *t*-BuOH:H_2O (1:1)	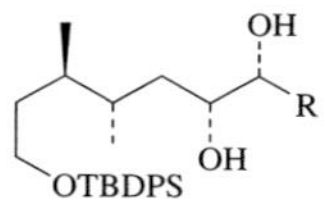 (—), >95 de	533

C_{45}	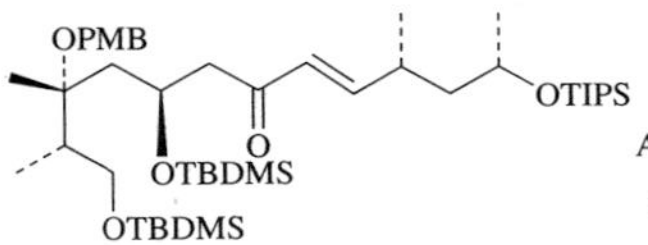	AD-Mix α, $NaHCO_3$, *t*-BuOH:H_2O (1:1), 5°, 36 h	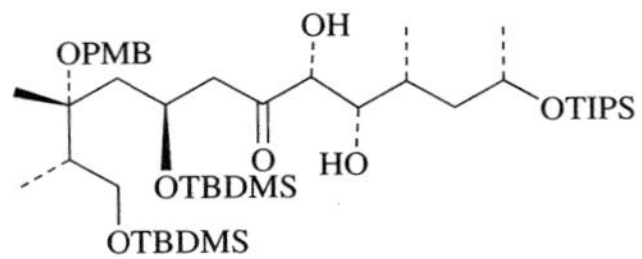	(—), 71 de	283
C_n	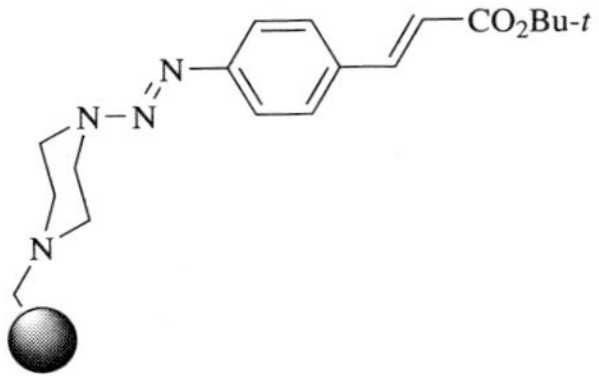	1. AD-Mix β, $MeSO_2NH_2$, acetone:H_2O (10:1), 72 h 2. THF, HCl (conc.), ultrasound, 50°, 5 min	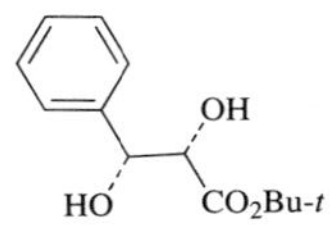	(53), >95	534

[a] The authors assigned product stereochemistry based on a mechanistic model.

[b] The yield given is that of the recrystallized product.

[c] This reaction gives the opposite enantiomer of that given in the previous entry, but the configurations for these products are not shown.

[d] The authors reported "good control of stereochemistry" but gave no % de values.

TABLE 4. REACTIONS OF CIS 1,2-DISUBSTITUTED ALKENES

Substrate	Conditions	Product(s) and Yield(s) (%) and Enatiomeric Excess	Refs.
*See Chart 1 at the beginning of the Tabular Survey for ligand structures that are indicated by **bold** numbers.*			
$C_{5\text{-}12}$ 	$(DHQD)_2PHAL$, $K_2OsO_2(OH)_4$, $K_3Fe(CN)_6$, t-BuOH:H_2O (1:1), 0°	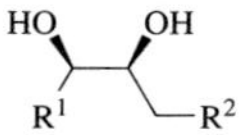	123
R^1 / R^2			
Et / OH		(—), 74	
Et / CH_2OH		(—), 54	
i-Pr / OH		(—), 64	
Et / CH_2OMe		(—), 0	
Ph / OH		(—), 71	
Ph / OMe		(—), 13	
OBn / OH		(—), 64	
OBn / OMe		(—), 23	
$C_{5\text{-}11}$ 	DHQD-IND, $K_2OsO_2(OH)_4$, $K_3Fe(CN)_6$, t-BuOH:H_2O (1:1), 0°	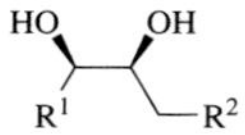	123
R^1 / R^2			
Et / OH		(—), 51	
Ph / OH		(—), 72	
OBn / OH		(—), 31	

C_{5-11}			
	(DHQ)$_2$PHAL, $K_2OsO_2(OH)_4$, $K_3Fe(CN)_6$, *t*-BuOH:H_2O (1:1), 0°	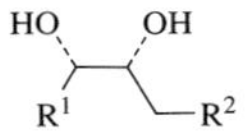	123
R^1 / R^2			
Et / OH		(—), 64	
Et / CH_2OH		(—), 45	
Ph / OH		(—), 73	
OBn / OH		(—), 57	
C_6			
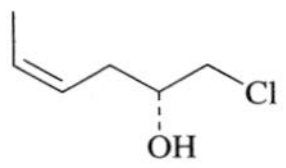	AD-Mix α, *t*-BuOH:H_2O (1:1)	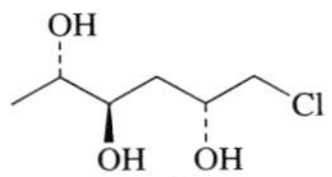 **I** (91), 39 de	342
	AD-Mix β, *t*-BuOH:H_2O (1:1)	**I** (85), 29 de	342
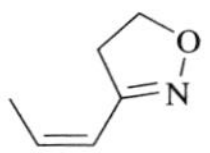	DHQD-*p*-chlorobenzoate, OsO_4, NMO, acetone-H_2O, 0°	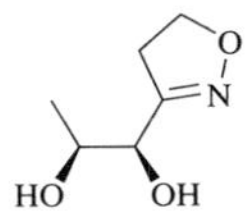 (—), 34	341

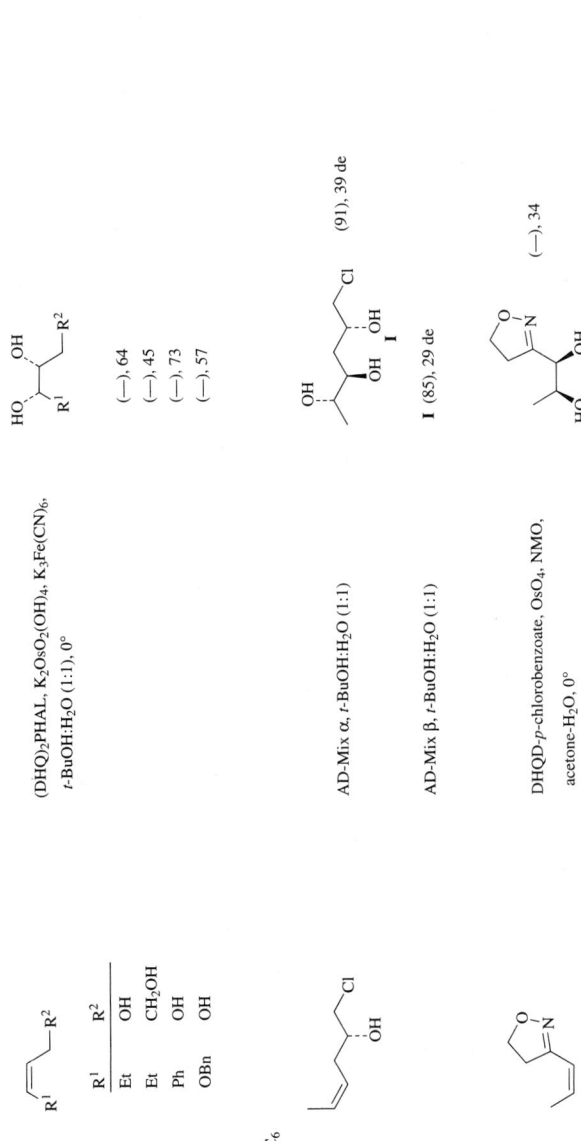

TABLE 4. REACTIONS OF CIS 1,2-DISUBSTITUTED ALKENES (*Continued*)

Substrate	Conditions	Product(s) and Yield(s) (%) and Enatiomeric Excess	Refs.
C_7			
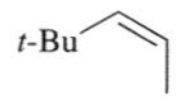	1. DHQD-OAc, OsO_4, toluene, rt, 12 h 2. LAH, ether	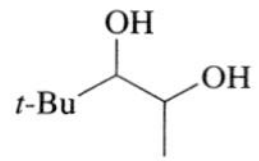 (78), <5	8
C_8			
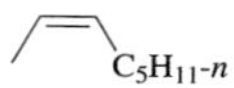	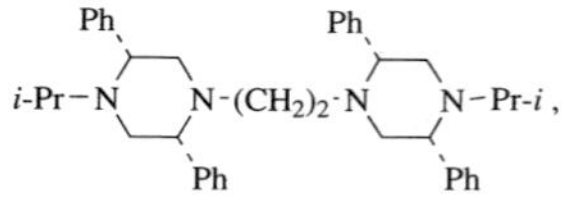 OsO_4, toluene, –78°	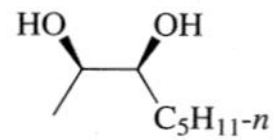 (81), 46	215
C_9			
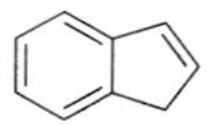	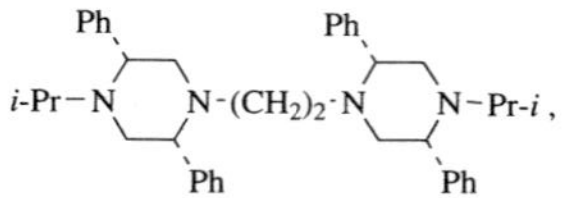 OsO_4, toluene, –78°	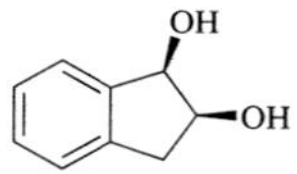 I (63), 14	215
	Ligand, $K_2OsO_2(OH)_4$, $K_3Fe(CN)_6$, K_2CO_3, t-BuOH:H_2O (1:1)	I	
	Ligand / Temp / Time	I	
	(DHQD)$_2$AQN / 0° / 12 h	(65-90), 63	86
	(DHQD)$_2$DPPHAL / — / —	(—), 53	86, 157
	(DHQD)$_2$PHAL / 0° / 4 h	(—), 42	223, 157, 147
	(DHQD)$_2$PYR / 0° / 4 h	(—), 35	157, 147
	(DHQD)$_2$DPP / 0° / 4 h	(—), 20	157
	DHQD-IND / 0° / 17-22	(—), 72	121

36, OsO_4, CH_2Cl_2, –78°	**I** (77), 35		96
33, OsO_4, toluene, –90°, 4 h	**I** (70), 80		9
(DHQ)$_2$PHAL, $K_2OsO_2(OH)_4$, $K_3Fe(CN)_6$, K_2CO_3, *t*-BuOH:H_2O (1:1), 0°, 4 h	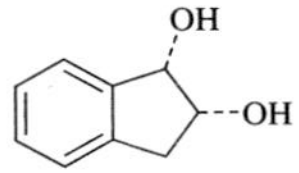	(—), 25-34	223
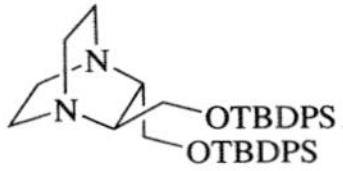 , OsO_4, $K_3Fe(CN)_6$, K_2CO_3, *t*-BuOH-H_2O (1:1), rt, 8-24 h	**I** (66), 12		193
1. 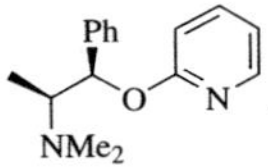 , OsO_4, toluene, –78°, 10 h 2. LAH-ether, –78° to rt, 8 h	**I** (94), 30		183
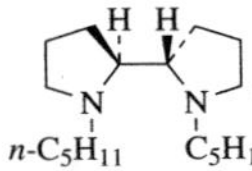 , OsO_4, CH_2Cl_2, –78°	**I** (77), 20		12
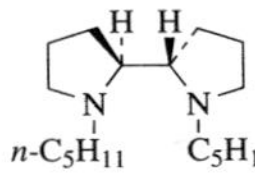 , OsO_4, toluene, –78°	**I** (85), 35		12
DHQ-IND, $K_2OsO_2(OH)_4$, $K_3Fe(CN)_6$, K_2CO_3, *t*-BuOH:H_2O (1:1), 0°, 17-22 h	**I** (—), 59		121

TABLE 4. REACTIONS OF CIS 1,2-DISUBSTITUTED ALKENES (*Continued*)

Substrate	Conditions	Product(s) and Yield(s) (%), % ee	Refs.
C_9			
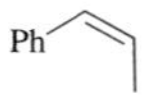	Ligand, $K_2OsO_2(OH)_4$, $K_3Fe(CN)_6$, K_2CO_3, t-BuOH:H_2O (1:1), 0°, 4 h	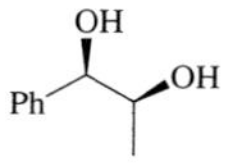 **I**	
	Ligand		
	$(DHQD)_2PHAL$	(—), 35	157
	$(DHQD)_2DPPHAL$	(—), 63	157
	$(DHQD)_2DPP$	(—), 68	157
	DHQD-IND	(—), 72	157, 86
			121
	1. DHQD-OAc, OsO_4, toluene, rt, 12 h 2. LAH, ether	**I** (85), 25.5	8
	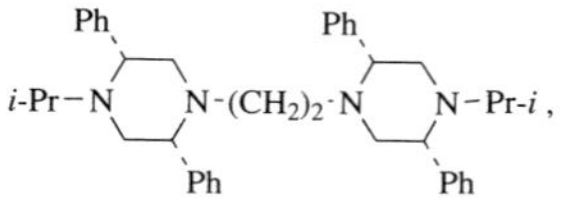 OsO_4, toluene, –78°	**I** (75), 81	215
	33, OsO_4, toluene, –90°, 2 h	**I** (80), 77	9
	1. DHQ-OAc, OsO_4, toluene, rt, 12 h 2. LAH, ether	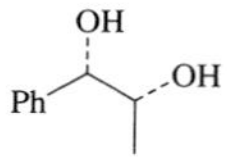 **I** (82), 26.8	8
	$(DHQ)_2PHAL$, $K_2OsO_2(OH)_4$, $K_3Fe(CN)_6$, K_2CO_3, t-BuOH:H_2O (1:1), 0°, 12 h	**I** (65-90), 45	86
	DHQ-IND, $K_2OsO_2(OH)_4$, $K_3Fe(CN)_6$, K_2CO_3, t-BuOH:H_2O (1:1), 0°, 12 h	**I** (—), 59	121

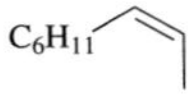	DHQD-IND, $K_2OsO_2(OH)_4$, $K_3Fe(CN)_6$, K_2CO_3, *t*-BuOH:H_2O (1:1), 0°, 17-22 h	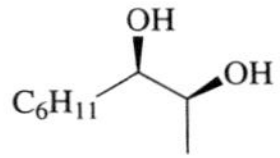	(—), 56	121
	DHQ-IND, $K_2OsO_2(OH)_4$, $K_3Fe(CN)_6$, K_2CO_3, *t*-BuOH:H_2O (1:1), 0°, 17-22 h	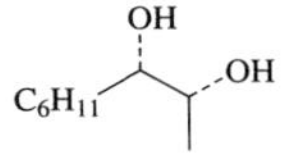	(—), 44	121
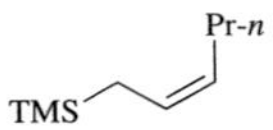	$(DHQD)_2PHAL$, $K_2OsO_2(OH)_4$, $K_3Fe(CN)_6$, K_2CO_3, $MeSO_2NH_2$, *t*-BuOH:H_2O (1:1), 0°, 24 h	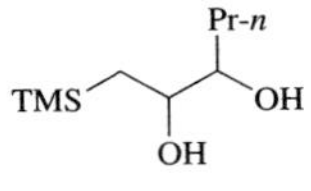	(83), 54[a]	113
	$(DHQ)_2PHAL$, $K_2OsO_2(OH)_4$, $K_3Fe(CN)_6$, K_2CO_3, $MeSO_2NH_2$, *t*-BuOH:H_2O (1:1), 0°, 24 h	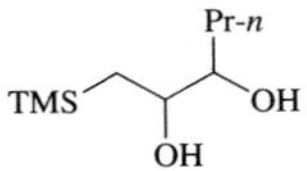	(81), 50[a]	113
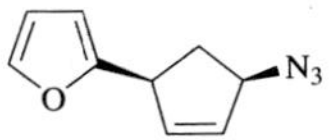	$(DHQ)_2PHAL$, $K_2OsO_2(OH)_4$, $K_3Fe(CN)_6$, K_2CO_3, $MeSO_2NH_2$, *t*-BuOH:H_2O (1:1), 0°, 24 h	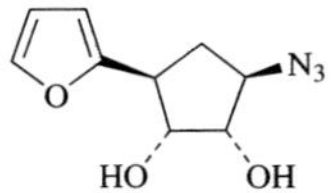	(—), 50	535
	$(DHQD)_2PHAL$, $K_2OsO_2(OH)_4$, $K_3Fe(CN)_6$, K_2CO_3, $MeSO_2NH_2$, *t*-BuOH:H_2O (1:1), 0°, 24 h	**I** (34), 71		535

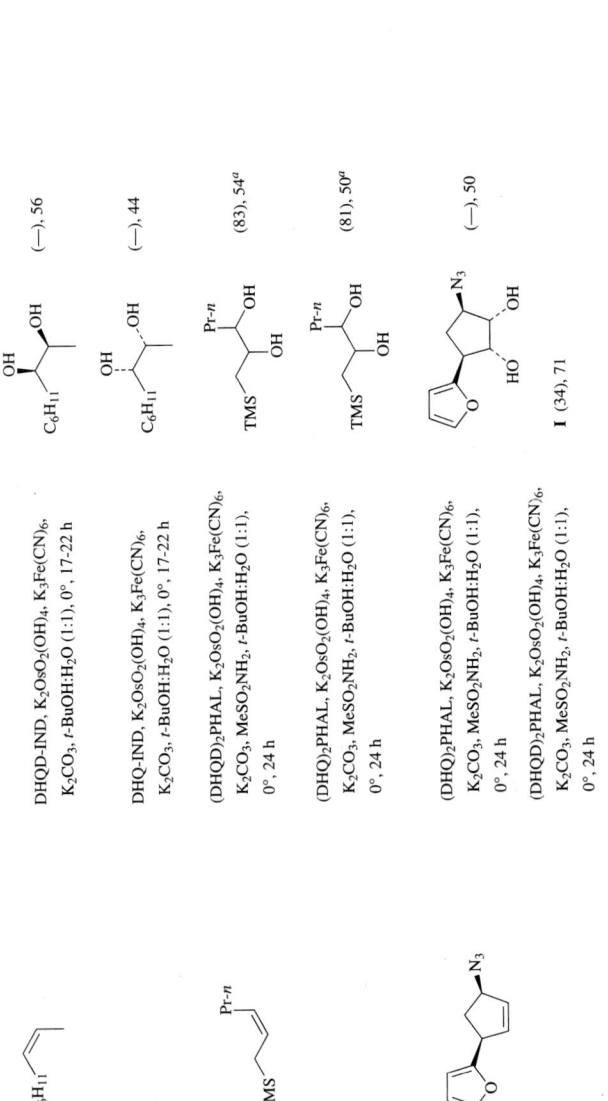

TABLE 4. REACTIONS OF CIS 1,2-DISUBSTITUTED ALKENES (*Continued*)

Substrate	Conditions	Product(s) and Yield(s) (%), % ee	Refs.
C_{9-11} 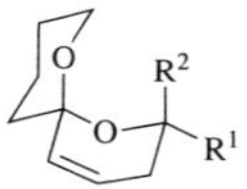	Ligand, $K_2OsO_2(OH)_4$, $K_3Fe(CN)_6$, K_2CO_3, $MeSO_2NH_2$, t-BuOH:H_2O (1:1), 0°, 257 h	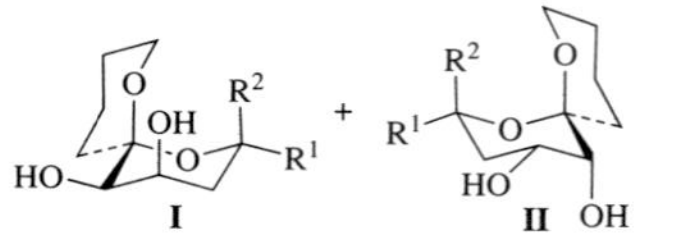	536

R^1	R^2	Ligand	I	II
H	H	$(DHQ)_2PHAL$	(2.8), 82	(44.9), 51.9
H	H	DHQ-IND	(3.2), 56	(40.7), 37
Me	H	$(DHQ)_2PHAL$	(2.4), 80.2	(36.1), 63.8
Me	H	DHQ-IND	(6.7), 71	(66.2), 34
Me	Me	$(DHQ)_2PHAL$	(2.5), 51.2	(47.8), 40.3
Me	Me	DHQ-IND	(5.6), 46	(65.2), 37

Substrate	Conditions	Product(s) and Yield(s) (%), % ee	Refs.
.	Ligand, $K_2OsO_2(OH)_4$, $K_3Fe(CN)_6$, K_2CO_3, $MeSO_2NH_2$, t-BuOH:H_2O (1:1), 0°, 257 h	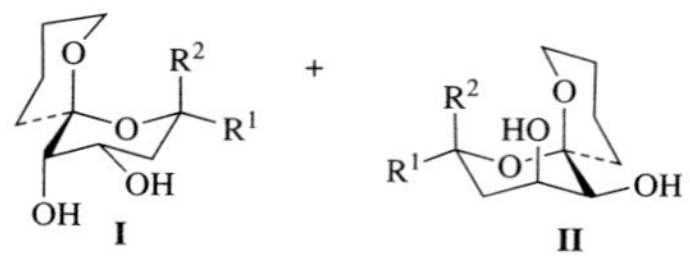	536

R^1	R^2	Ligand	I	II
H	H	DHQD-IND	(75.3), 25	(9.7), 78
H	H	$(DHQD)_2PHAL$	(84.6), 23.6	(15.4), 91.1
Me	H	$(DHQD)_2PHAL$	(41.7), 55.8	(4.6), 82.3
Me	H	DHQD-IND	(61.5), 33	(6.2), 49
Me	Me	$(DHQD)_2PHAL$	(31.9), 38.6	(1.6), 63.6
Me	Me	DHQD-IND	(78.5), 28	(9.6), 92

Reactant	Conditions	Product(s) and Yield(s) (%)	Refs.
C_{10} 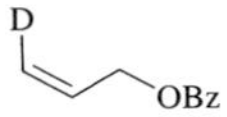	$(DHQD)_2PHAL$, $K_2OsO_2(OH)_4$, $K_3Fe(CN)_6$, K_2CO_3, t-BuOH:H_2O (1:1), 0°, 4 h	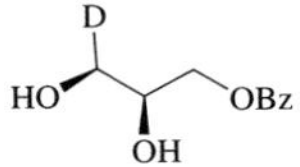 (94), 77	78
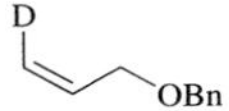	AD-Mix β, t-BuOH:H_2O (1:1), 0°, 12 h	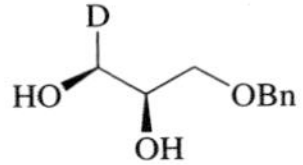 (91), 58	367
	AD-Mix α, t-BuOH:H_2O (1:1), 0°, 12 h	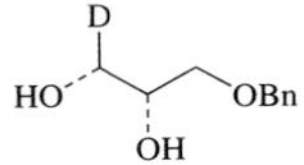 (51), 50	367
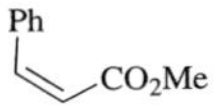	AD-Mix α, t-BuOH:H_2O (1:1)	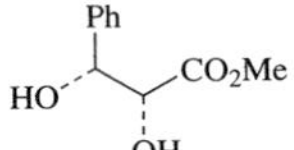 I (36), 7	537
	33, OsO_4, toluene, –90°, 5 h	**I** (60), 87	9
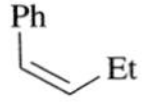	DHQD-IND, $K_2OsO_2(OH)_4$, $K_3Fe(CN)_6$, K_2CO_3, t-BuOH:H_2O (1:1), 0°, 17-22 h	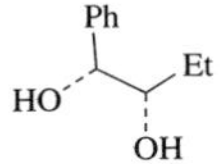 (—), 72	121
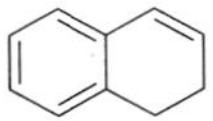	Ligand, $K_2OsO_2(OH)_4$, $K_3Fe(CN)_6$, K_2CO_3, t-BuOH:H_2O (1:1), 0°	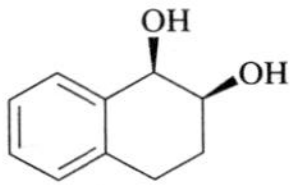 **I**	
	Ligand / **Time**	**I**	
	$(DHQD)_2PHAL$ / 4 h	(—), 15	157, 223
	$(DHQD)_2PYR$ / 4 h	(—), 7	157, 121
	$(DHQD)_2DPPHAL$ / 4 h	(—), 56	157
	$(DHQD)_2DPP$ / 4 h	(—), 40	157
	$(DHQD)_2IND$ / 17-22 h	(—), 16	157, 121
	33, OsO_4, toluene, –90°, 4 h	**I** (73), 70	9

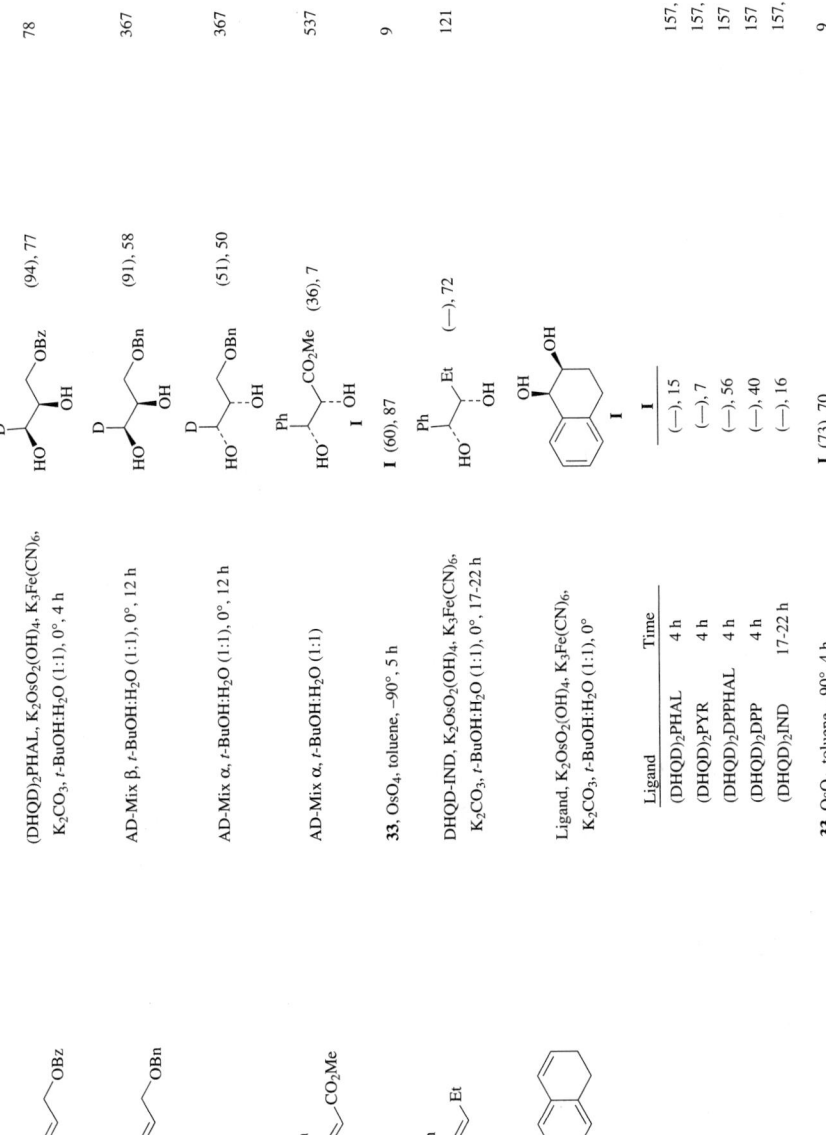

TABLE 4. REACTIONS OF CIS 1,2-DISUBSTITUTED ALKENES (*Continued*)

Substrate	Conditions	Product(s) and Yield(s) (%), % ee	Refs.
C_{11}			
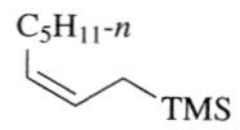	$(DHQD)_2PHAL$, $K_2OsO_2(OH)_4$, $K_3Fe(CN)_6$, K_2CO_3, $MeSO_2NH_2$, t-BuOH:H_2O (1:1), 0°, 24 h	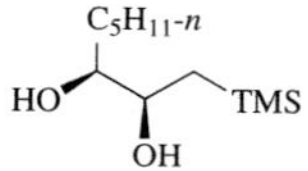 (82), 53	113
	$(DHQ)_2PHAL$, $K_2OsO_2(OH)_4$, $K_3Fe(CN)_6$, K_2CO_3, $MeSO_2NH_2$, t-BuOH:H_2O (1:1), 0°, 24 h	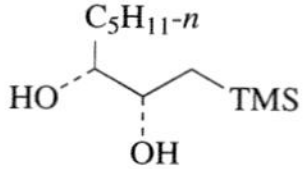 (84), 53	113
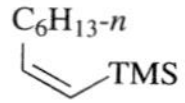	$(DHQD)_2PHAL$, $K_2OsO_2(OH)_4$, $K_3Fe(CN)_6$, K_2CO_3, $MeSO_2NH_2$, t-BuOH:H_2O (1:1), 0°, 26 h	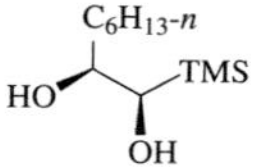 (81), 61	113
	$(DHQ)_2PHAL$, $K_2OsO_2(OH)_4$, $K_3Fe(CN)_6$, K_2CO_3, $MeSO_2NH_2$, t-BuOH:H_2O (1:1), 0°, 26 h	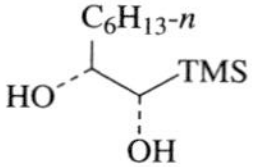 (78), 61	113
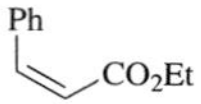	DHQD-IND, $K_2OsO_2(OH)_4$, K_2CO_3, $K_3Fe(CN)_6$, t-BuOH:H_2O (1:1)	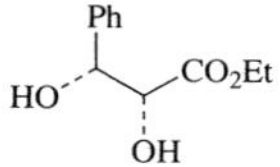 (—), 78	538, 121
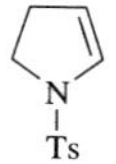	1. AD-Mix β, t-BuOH:H_2O (1:1) 2. TsOH, MeOH	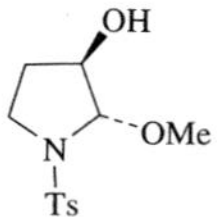 (48), 9, 69 de	539
	Ligand, $K_2OsO_2(OH)_4$, $K_3Fe(CN)_6$, K_2CO_3, t-BuOH:H_2O (1:1), 0°, 4 h	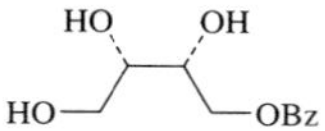	157
	Ligand		
	$(DHQD)_2PHAL$	(—), 64	
	$(DHQD)_2DPPHAL$	(—), 73	
	$(DHQD)_2DPP$	(—), 82	
	DHQD-IND	(—), 31	

C_{11-13}			
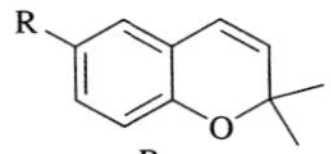	$(DHQD)_2PYR$, $K_2OsO_2(OH)_4$, $K_3Fe(CN)_6$, K_2CO_3, $MeSO_2NH_2$, t-BuOH:H_2O (1:1), 0°	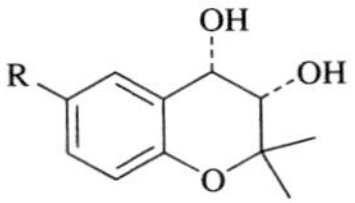	147
R			
Br		(—), 59	
CF_3		(—), 66	
CN		(—), 58	
Et		(—), 76	
	$(DHQ)_2PYR$, $K_2OsO_2(OH)_4$, $K_3Fe(CN)_6$, K_2CO_3, $MeSO_2NH_2$, t-BuOH:H_2O (1:1), 0°	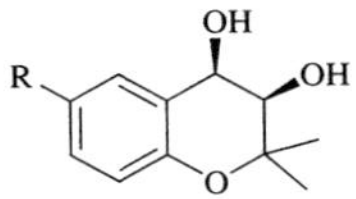	147
R			
Br		(—), 74	
CF_3		(—), 74	
CN		(—), 63	
Et		(—), 84	
C_{12}			
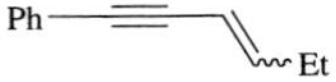 E:Z = 25:75	DHQD-PHN, $K_2OsO_2(OH)_4$, $K_3Fe(CN)_6$, K_2CO_3, t-BuOH:H_2O (1:1), 0°, 15-30 h	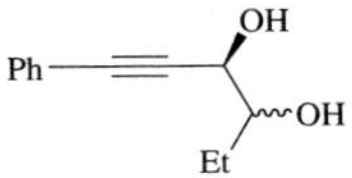 (94), 54	219
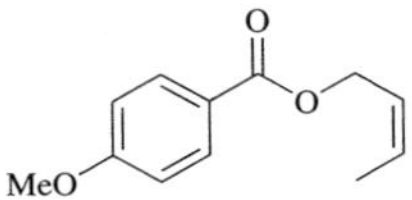	$(DHQD)_2PYDZ$, $K_2OsO_2(OH)_4$, $K_3Fe(CN)_6$, K_2CO_3, $MeSO_2NH_2$, t-BuOH:H_2O (1:1), 0°, 1.5-3.5 h	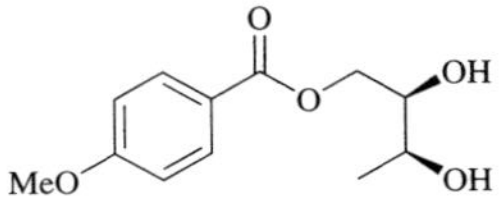 (99), 49	61
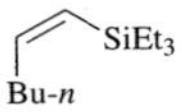	$(DHQ)_2PHAL$, $K_2OsO_2(OH)_4$, $K_3Fe(CN)_6$, K_2CO_3, $MeSO_2NH_2$, t-BuOH:H_2O (1:1), 0°, 96 h	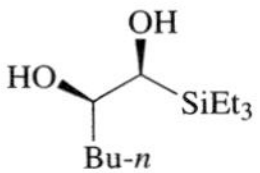 (90), 46	114

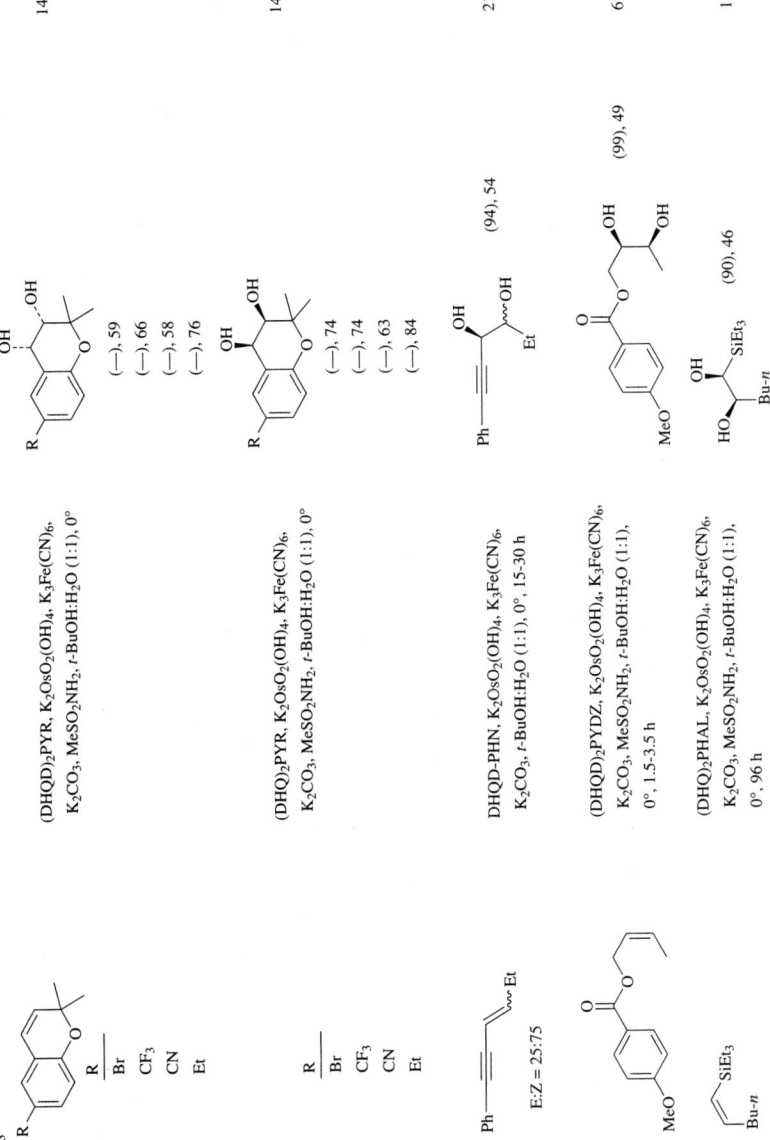

TABLE 4. REACTIONS OF CIS 1,2-DISUBSTITUTED ALKENES (*Continued*)

Substrate	Conditions	Product(s) and Yield(s) (%), % ee	Refs.
C_{12}			
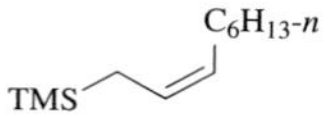	$(DHQD)_2PHAL$, $K_2OsO_2(OH)_4$, $K_3Fe(CN)_6$, K_2CO_3, $MeSO_2NH_2$, t-BuOH:H_2O (1:1), 0°, 24 h	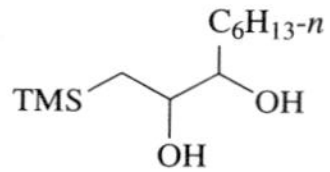 (81), 43[a]	113
	$(DHQ)_2PHAL$, $K_2OsO_2(OH)_4$, $K_3Fe(CN)_6$, K_2CO_3, $MeSO_2NH_2$, t-BuOH:H_2O (1:1), 0°, 24 h	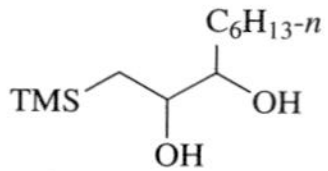 (79), 50[a]	113
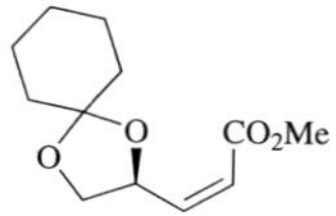	DHQD-OAc, OsO_4, toluene, –20°, 18 h	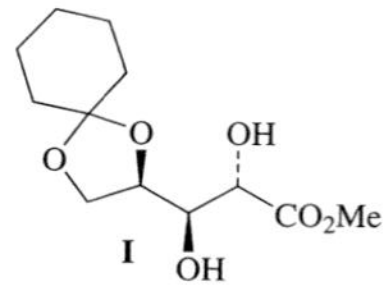 **I** (100), 20 de	384
	DHQD-*p*-chlorobenzoate, OsO_4, toluene, –20°, 18 h	**I** (84), 10 de	384
	DHQ-*p*-chlorobenzoate, OsO_4, toluene, –20°, 18 h	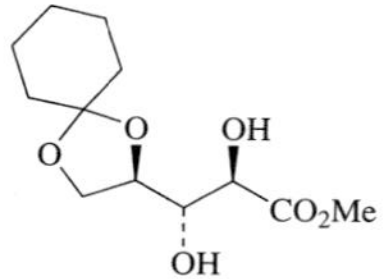 (91), 42 de	384
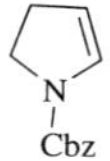	AD-Mix α, t-BuOH:H_2O (1:1)	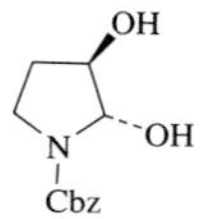 (90), 25, 46 de	540
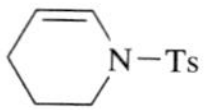	1. AD-Mix β, t-BuOH:H_2O (1:1) 2. TsOH, MeOH	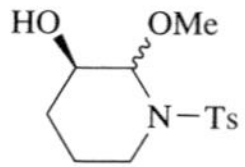 (83), 28, 0 de 1:1 cis:trans	539

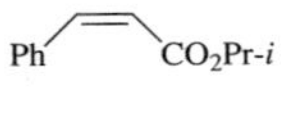	DHQD-IND, $K_2OsO_2(OH)_4$, K_2CO_3, $K_3Fe(CN)_6$, *t*-BuOH:H_2O (1:1), 17-22 h	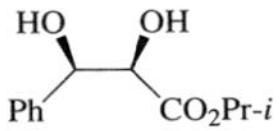	(—), 80	121
	DHQ-IND, $K_2OsO_2(OH)_4$, K_2CO_3, $K_3Fe(CN)_6$, *t*-BuOH:H_2O (1:1), 17-22 h	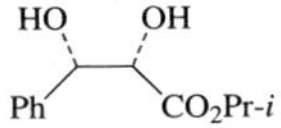	(—), 72	121
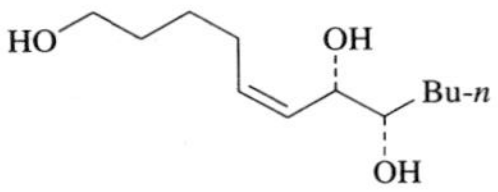	AD-Mix β, $MeSO_2NH_2$, *t*-BuOH:H_2O (1:1), rt, 24 h	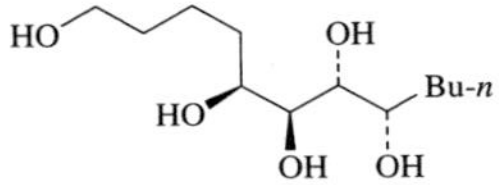	(78), 36 de	388
	AD-Mix α, $MeSO_2NH_2$, *t*-BuOH:H_2O (1:1), rt, 24 h	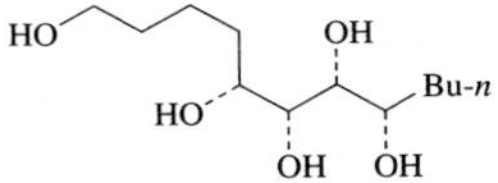	(78), 20 de	388
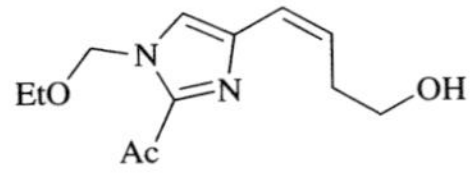	AD-Mix β, $MeSO_2NH_2$, *t*-BuOH:H_2O (1:1), 0°, 3 d	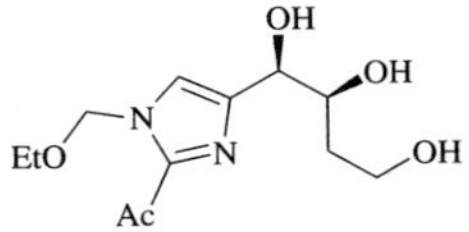	(75), 26	392
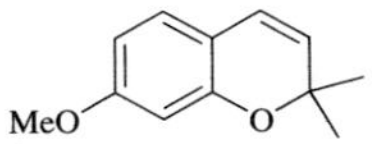	Ligand, $K_2OsO_2(OH)_4$, $K_3Fe(CN)_6$, K_2CO_3, $MeSO_2NH_2$, *t*-BuOH:H_2O (1:1), 0°	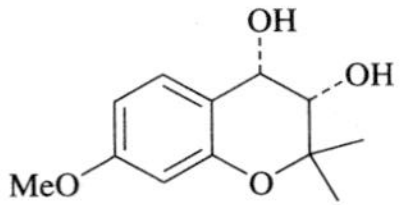		147
	Ligand			
	$(DHQD)_2PYR$	(—), 67		
	$(DHQD)_2PHAL$	(—), 21		
	DHQD-IND	(—), 32		
	$(DHQ)_2PYR$, $K_2OsO_2(OH)_4$, $K_3Fe(CN)_6$, K_2CO_3, $MeSO_2NH_2$, *t*-BuOH:H_2O (1:1), 0°	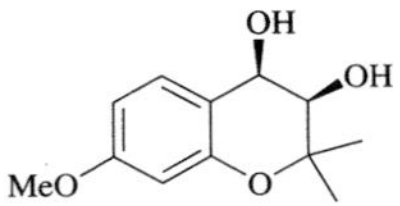	(—), 67	147

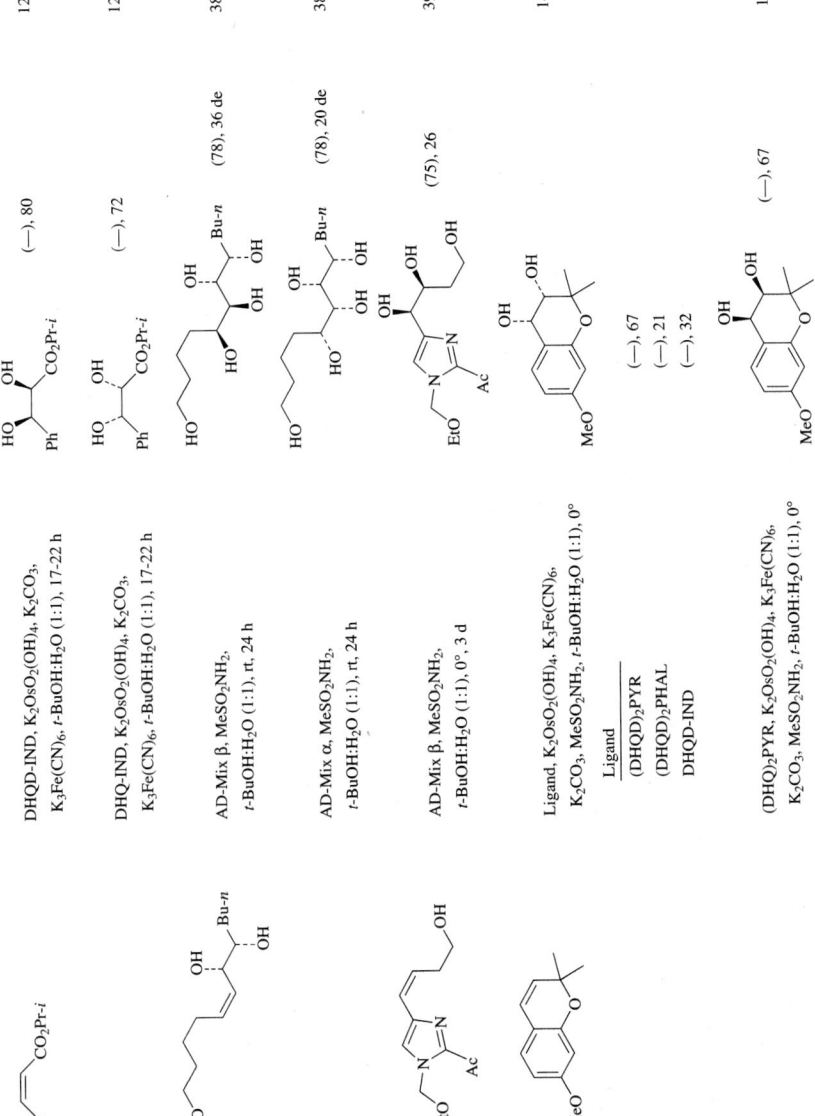

TABLE 4. REACTIONS OF CIS 1,2-DISUBSTITUTED ALKENES (*Continued*)

Substrate	Conditions	Product(s) and Yield(s) (%), % ee	Refs.
C_{13}			
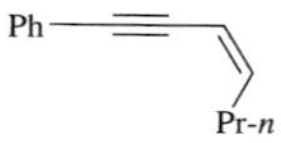	DHQD-PHN, $K_2OsO_2(OH)_4$, $K_3Fe(CN)_6$, K_2CO_3, *t*-BuOH:H_2O (1:1), 0°, 15-30 h	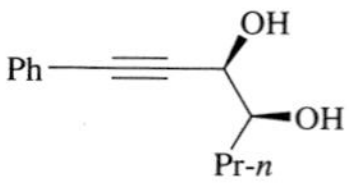 (61), 62	219
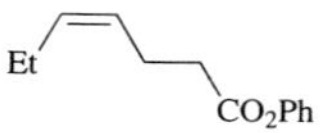	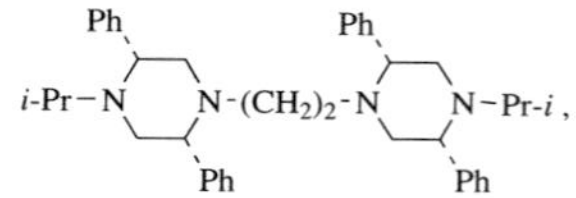 OsO_4, toluene, –78°	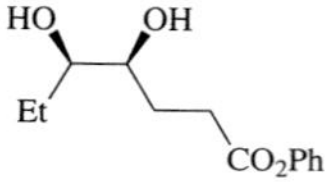 (83), 12	215
C_{14}			
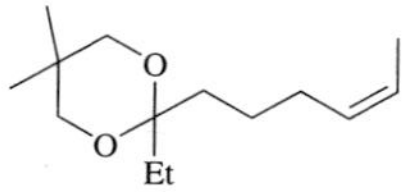	AD-Mix β, $MeSO_2NH_2$, *t*-BuOH:H_2O (1:1), 4°, 9 h	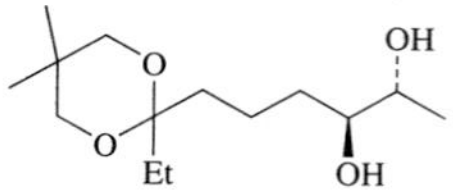 (94), 15	423
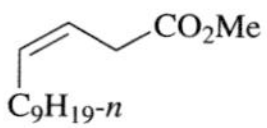	AD-Mix α, $MeSO_2NH_2$, *t*-BuOH:H_2O (1:1), 0°, 39 h	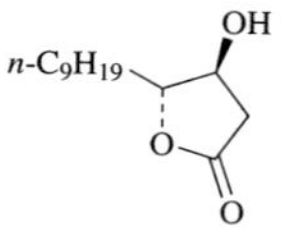 (84), low	424
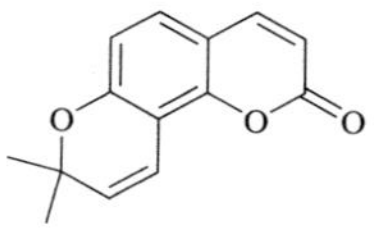	Ligand, $K_2OsO_2(OH)_4$, $K_3Fe(CN)_6$, K_2CO_3, *t*-BuOH:H_2O (1:1), 0°	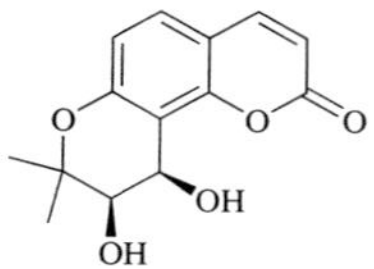	122

Ligand	Time	
DHQD-*p*-chlorobenzoate	4 d	(—), 30
DHQD-PHN	4 h	(—), 67
$(DHQD)_2PYR$	1 d	(—), 80

	Ligand, $K_2OsO_2(OH)_4$, $K_3Fe(CN)_6$, K_2CO_3, *t*-BuOH:H_2O (1:1), 0°	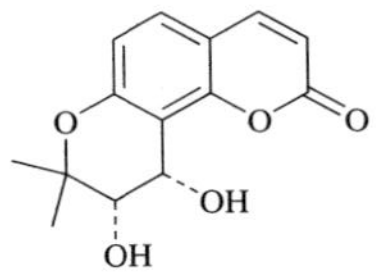		122
	Ligand — Time			
	DHQ-PHN — 4 d	(—), 59		
	DHQ-*p*-chlorobenzoate — 4 d	(—), 50		
	DHQ-MEQ — 2 d	(—), 50		
	$(DHQ)_2PYR$ — 2.5 d	(—), 86		
C_{15} 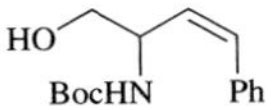	$(DHQD)_2PHAL$, $K_2OsO_2(OH)_4$, K_2CO_3, $K_3Fe(CN)_6$, *t*-BuOH:H_2O (1:1), 0°-rt, 24 h	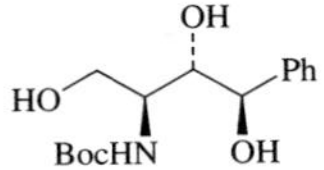	(55), 68 de	449
	$(DHQ)_2PHAL$, $K_2OsO_2(OH)_4$, K_2CO_3, $K_3Fe(CN)_6$, *t*-BuOH:H_2O (1:1), 0°-rt, 24 h	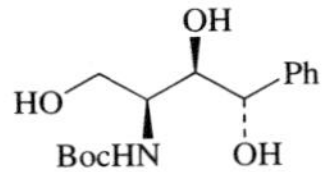	(55), 72 de	449
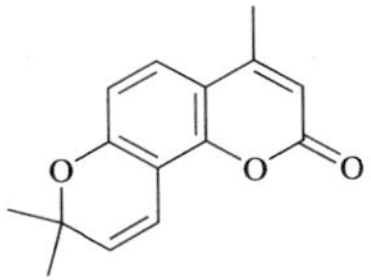	$(DHQ)_2PYR$, $K_2OsO_2(OH)_4$, $K_3Fe(CN)_6$, K_2CO_3, *t*-BuOH:H_2O (1:1), 0°	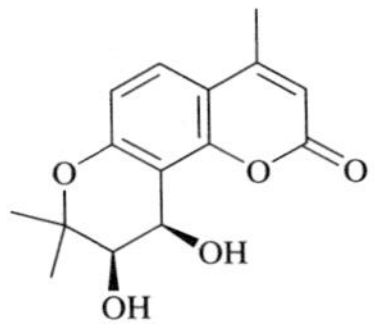	(—), 81-90	541
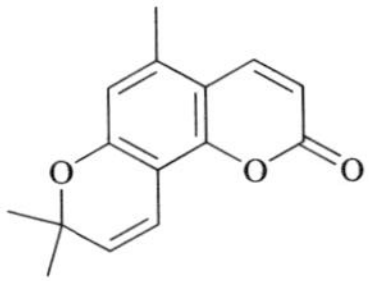	$(DHQ)_2PYR$, $K_2OsO_2(OH)_4$, $K_3Fe(CN)_6$, K_2CO_3, *t*-BuOH:H_2O (1:1), 0°	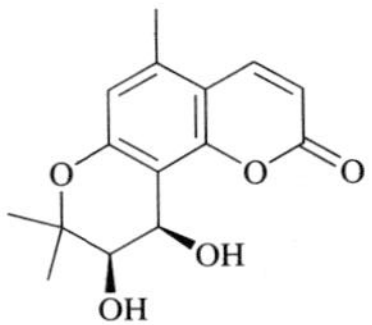	(—), 81-90	541

TABLE 4. REACTIONS OF CIS 1,2-DISUBSTITUTED ALKENES (*Continued*)

Substrate	Conditions	Product(s) and Yield(s) (%), % ee	Refs.
C_{15}			
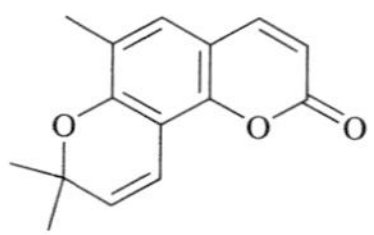	$(DHQ)_2PYR$, $K_2OsO_2(OH)_4$, $K_3Fe(CN)_6$, K_2CO_3, t-BuOH:H_2O (1:1), 0°	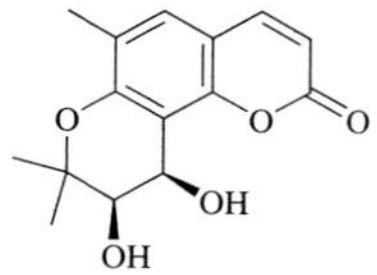 (—), 81-90	541
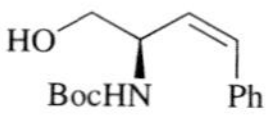	$(DHQD)_2PHAL$, $K_2OsO_2(OH)_4$, $K_3Fe(CN)_6$, K_2CO_3, $MeSO_2NH_2$, t-BuOH:H_2O (1:1), 0°-rt, 24 h	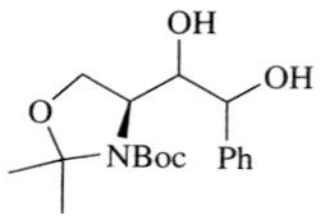 (4), 0 de	449
	$(DHQ)_2PHAL$, $K_2OsO_2(OH)_4$, $K_3Fe(CN)_6$, K_2CO_3, $MeSO_2NH_2$, t-BuOH:H_2O (1:1) 0°-rt, 24 h	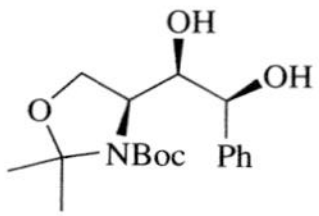 (13), 70 de	449
C_{17}			
 Z:E = 86:14	DHQD-p-chlorobenzoate, $K_2OsO_2(OH)_4$, $K_3Fe(CN)_6$, K_2CO_3, $MeSO_2NH_2$, t-BuOH:H_2O (1:1), rt, 1-5 d	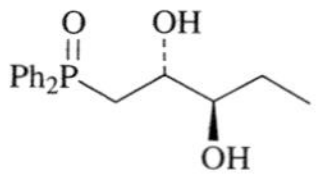 (55), 22	119
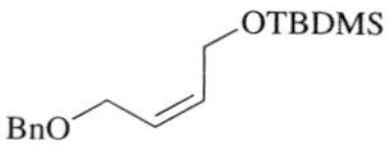	$(DHQD)_2PHAL$, $K_2OsO_2(OH)_4$, $K_3Fe(CN)_6$, K_2CO_3, $MeSO_2NH_2$, t-BuOH:H_2O (1:1)	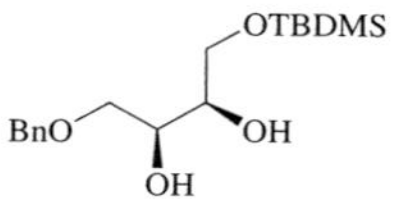 (94), 5	124

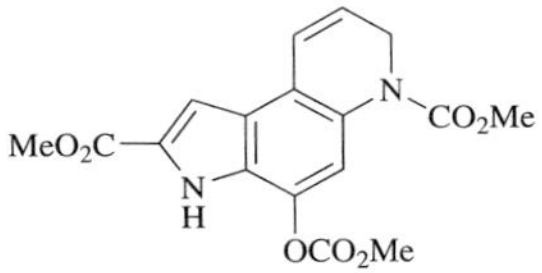	$(DHQ)_2PYR$, OsO_4, $K_3Fe(CN)_6$, K_2CO_3, $NaHCO_3$, $MeSO_2NH_2$, *t*-BuOH:H_2O (1:1), 0°-rt, 16 h	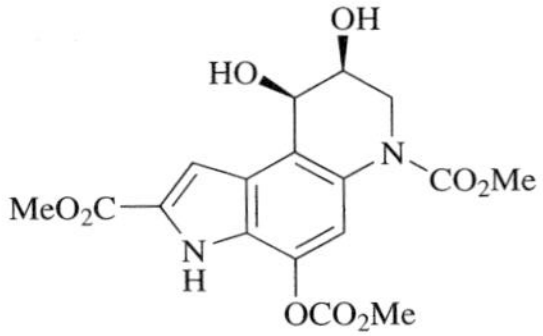	(69), 81	542
	$(DHQD)_2PYR$, OsO_4, $K_3Fe(CN)_6$, K_2CO_3, $NaHCO_3$, $MeSO_2NH_2$, *t*-BuOH:H_2O (1:1), 0°-rt, 23 h	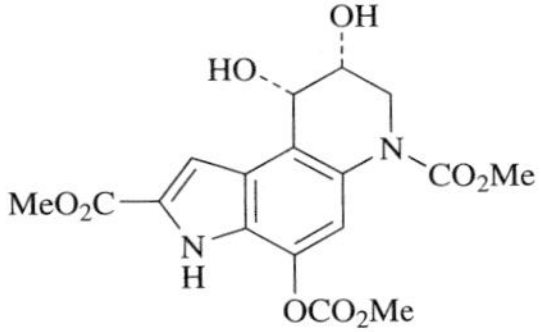	(75), 58	542
	1. $(DHQD)_2PYR$, OsO_4, $K_3Fe(CN)_6$, K_2CO_3, $NaHCO_3$, $MeSO_2NH_2$, *t*-BuOH:H_2O (1:1), 0°-rt, 22 h 2. Et_3SiH, BF_3•OEt_2, CH_2Cl_2, 0°-rt, 1 h	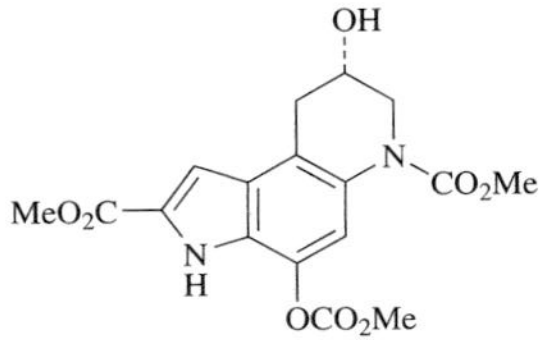	(76), 63	542
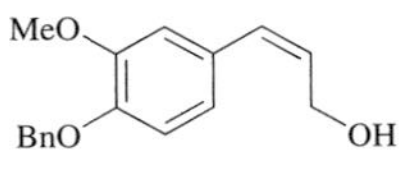	AD-Mix α, $MeSO_2NH_2$, *t*-BuOH:H_2O (1:1), 0°, 24 h	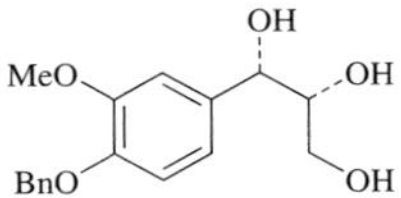	(66), 90	461
	AD-Mix β, $MeSO_2NH_2$, *t*-BuOH:H_2O (1:1), 0°, 24 h	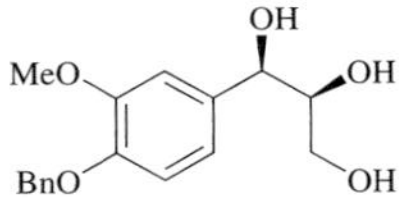	(—), 97	461

TABLE 4. REACTIONS OF CIS 1,2-DISUBSTITUTED ALKENES (*Continued*)

Substrate	Conditions	Product(s) and Yield(s) (%), % ee	Refs.
	(DHQ)$_2$PHAL, $K_2OsO_2(OH)_4$, $K_2Fe(CN)_6$, K_2CO_3, *t*-BuOH:H_2O (1:1), 0°, 18 h	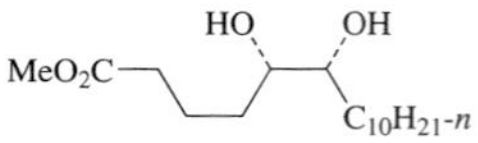 (50), 20	401
C_{18} 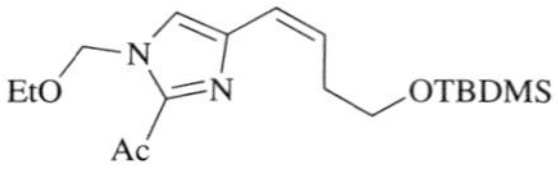	AD-Mix β, $MeSO_2NH_2$, *t*-BuOH:H_2O (1:1), 0°, 3 d	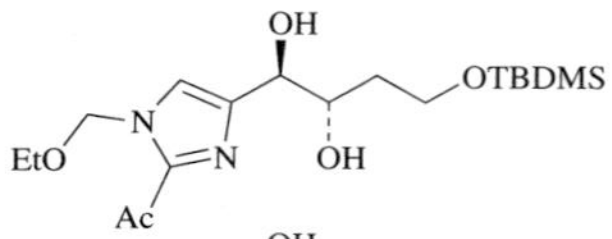 (73), 26	392
	AD-Mix α, $MeSO_2NH_2$, *t*-BuOH:H_2O (1:1), 0°, 3 d	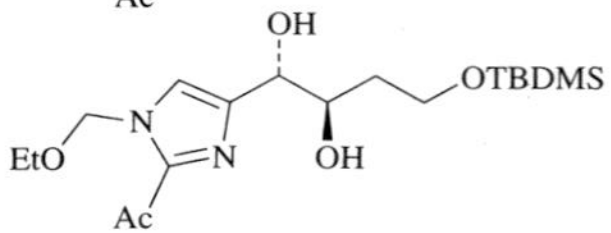 (58), 8	392
C_{19} 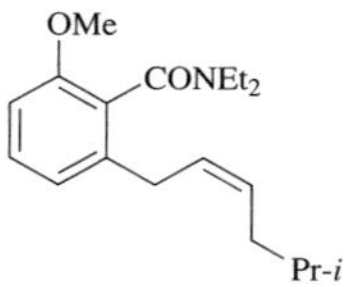	1. DHQ-PHN, $K_2OsO_2(OH)_4$, $K_3Fe(CN)_6$, K_2CO_3, $MeSO_2NH_2$, *t*-BuOH:H_2O (1:1), 0°, 18 h 2. NaOH (50% aq):EtOH (1:1), reflux, 36 h	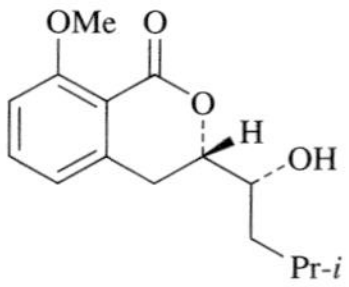 (63), 38	466
C_{21} 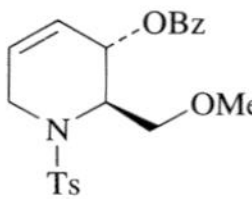	(DHQ)$_2$PHAL, OsO_4, $K_3Fe(CN)_6$, K_2CO_3, $MeSO_2NH_2$, *t*-BuOH:H_2O (1:1), rt, 2 d	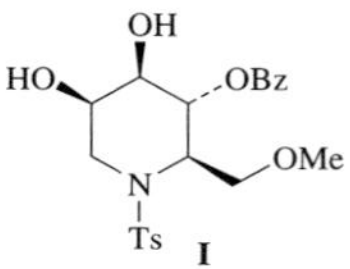 **I** (85), >95 de	543
	(DHQ)$_2$PHAL, OsO_4, $K_3Fe(CN)_6$, K_2CO_3, *t*-BuOH:H_2O (1:1), rt, 2 d	**I** (85), >90	544

C_{22} 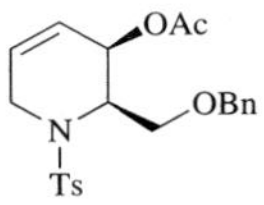	(DHQD)$_2$PHAL, $K_2OsO_2(OH)_4$, $K_3Fe(CN)_6$, K_2CO_3, *t*-BuOH:H_2O (1:1), 0°	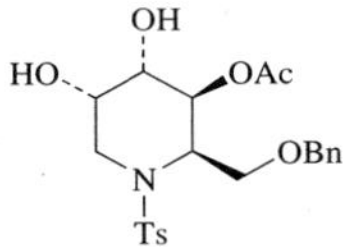 (94), 60 de	190
C_{23} 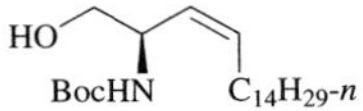	(DHQD)$_2$PHAL, $K_2OsO_2(OH)_4$, $K_3Fe(CN)_6$, K_2CO_3, $MeSO_2NH_2$, *t*-BuOH:H_2O (1:1), 0°-rt, 24 h	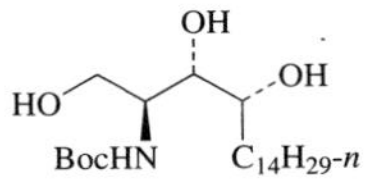 (89), 66 de	449
	(DHQ)$_2$PHAL, $K_2OsO_2(OH)_4$, $K_3Fe(CN)_6$, K_2CO_3, $MeSO_2NH_2$, *t*-BuOH:H_2O (1:1), 0°-rt, 24 h	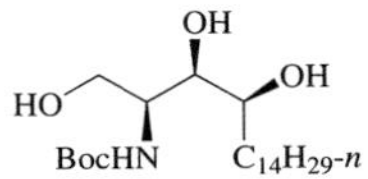 (86), 66 de	449
C_{25} 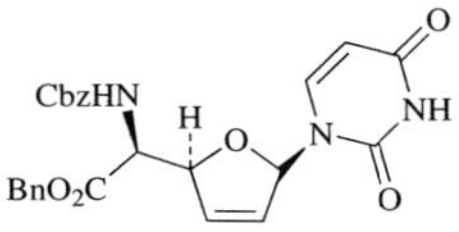	(DHQ)$_2$PHAL, OsO_4, NMO, THF	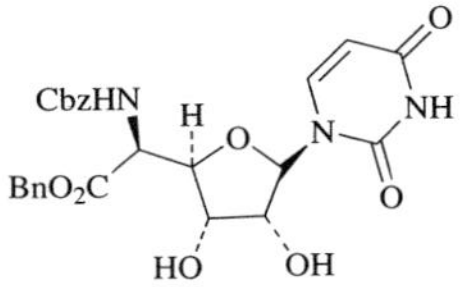 (30), — 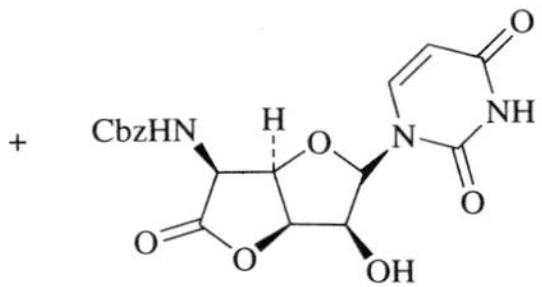 (57), —	545
C_{25-30} 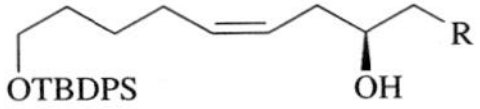 R: OH; N_3; OTHP	DHQD-IND, $K_2OsO_2(OH)_4$, $K_3Fe(CN)_6$, K_2CO_3, *t*-BuOH-H_2O	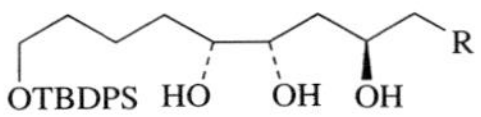 (100), 56 de; (89), 8 de; (85), 62 de	381

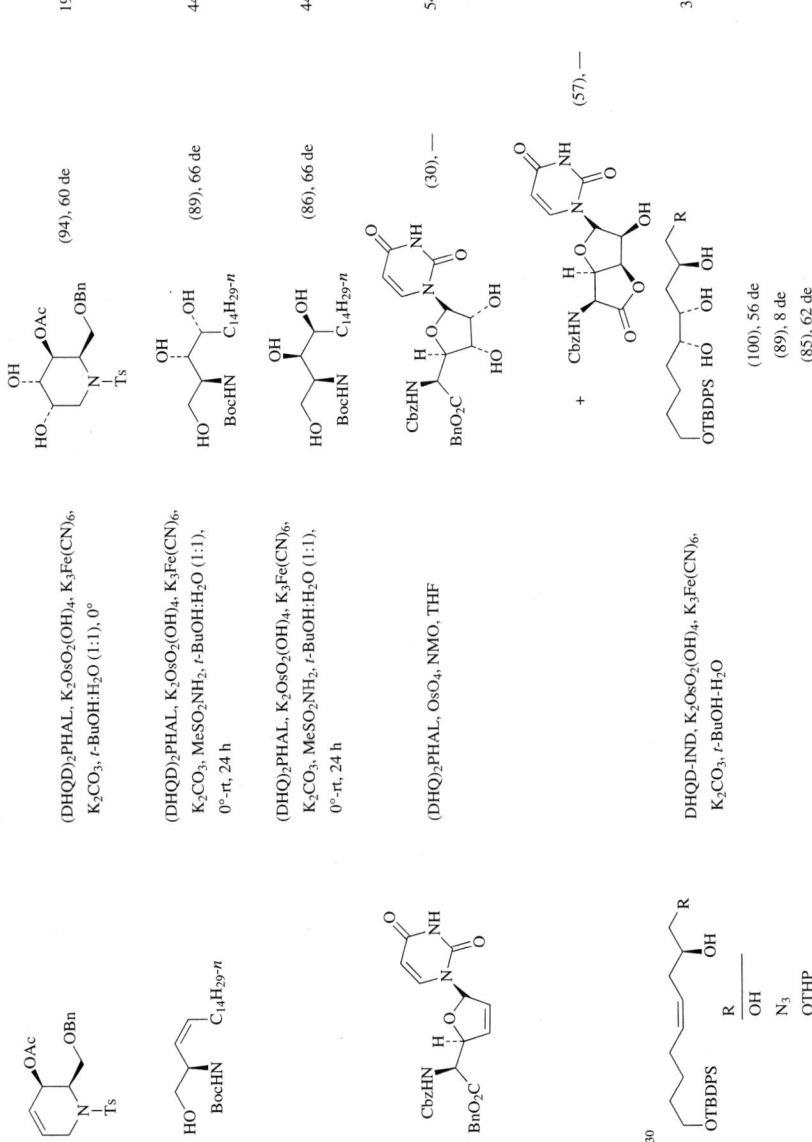

TABLE 4. REACTIONS OF CIS 1,2-DISUBSTITUTED ALKENES (*Continued*)

Substrate	Conditions	Product(s) and Yield(s) (%), % ee		Refs.
C_{26}				
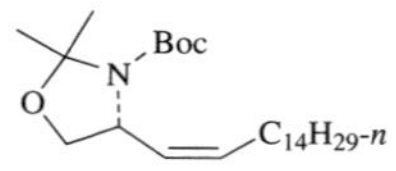	AD-Mix β, $MeSO_2NH_2$, t-BuOH:H_2O (1:1), 0°-rt, 2 d	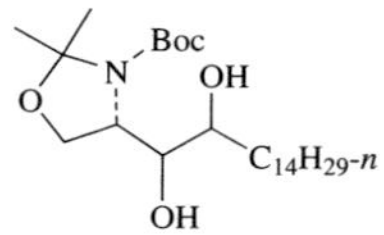	(71), 0 de	505
C_{28} 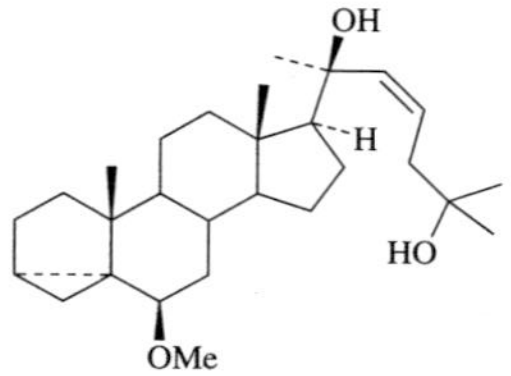	DHQ-p-chlorobenzoate, OsO_4	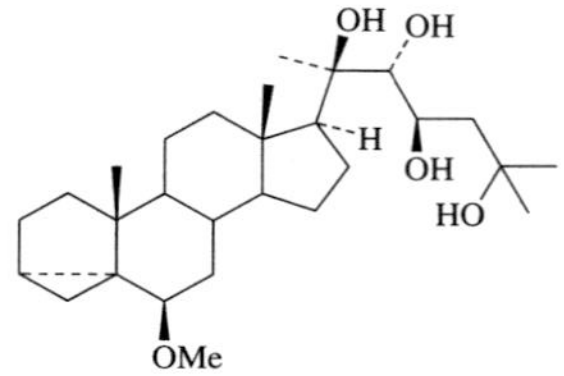	(—), 6 de	509
	DHQD-p-chlorobenzoate, OsO_4	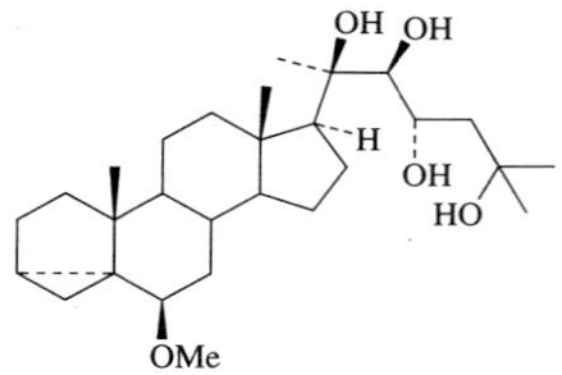	(—), 70 de	509
C_{30}				
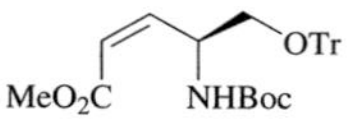	$(DHQD)_2PHAL$, $K_2OsO_2(OH)_4$, $K_3Fe(CN)_6$, K_2CO_3, $MeSO_2NH_2$, t-BuOH:H_2O (1:1), 0°, 12 h	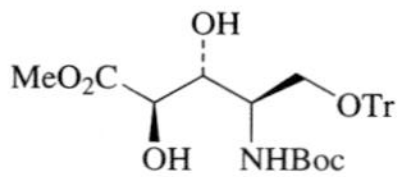	(—), 20 de	521

C_{31}	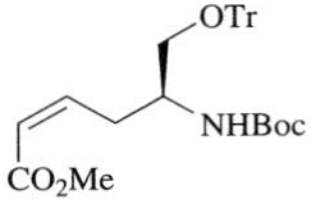	DHQD-PHN, $K_2OsO_2(OH)_4$, $K_3Fe(CN)_6$, K_2CO_3, $MeSO_2NH_2$, *t*-BuOH:H_2O (1:1), 0°, 24 h	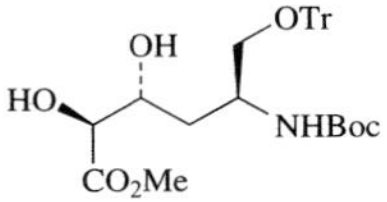	(66), 86 de	522
		DHQ-PHN, $K_2OsO_2(OH)_4$, $K_3Fe(CN)_6$, K_2CO_3, $MeSO_2NH_2$, *t*-BuOH:H_2O (1:1), 0°, 24 h	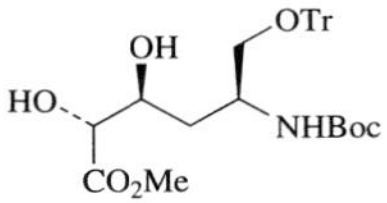	(62), 62 de	522
C_{33}	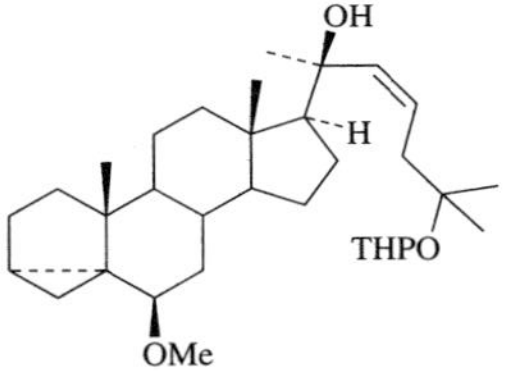	DHQ-*p*-chlorobenzoate, OsO_4	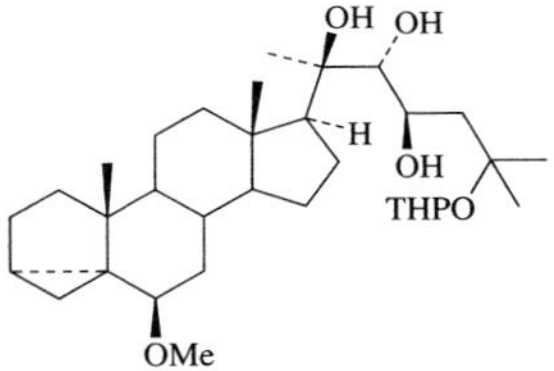	(—), 42 de	509
		DHQD-*p*-chlorobenzoate, OsO_4	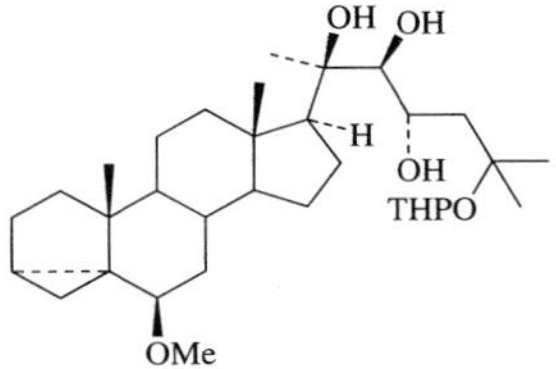	(—), 30 de	509
C_{36}	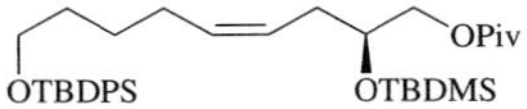	DHQD-IND, $K_2OsO_2(OH)_4$, $K_3Fe(CN)_6$, K_2CO_3, *t*-BuOH-H_2O	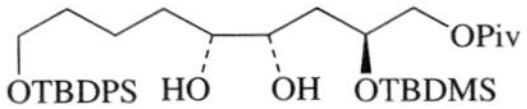	(73), 74 de	381

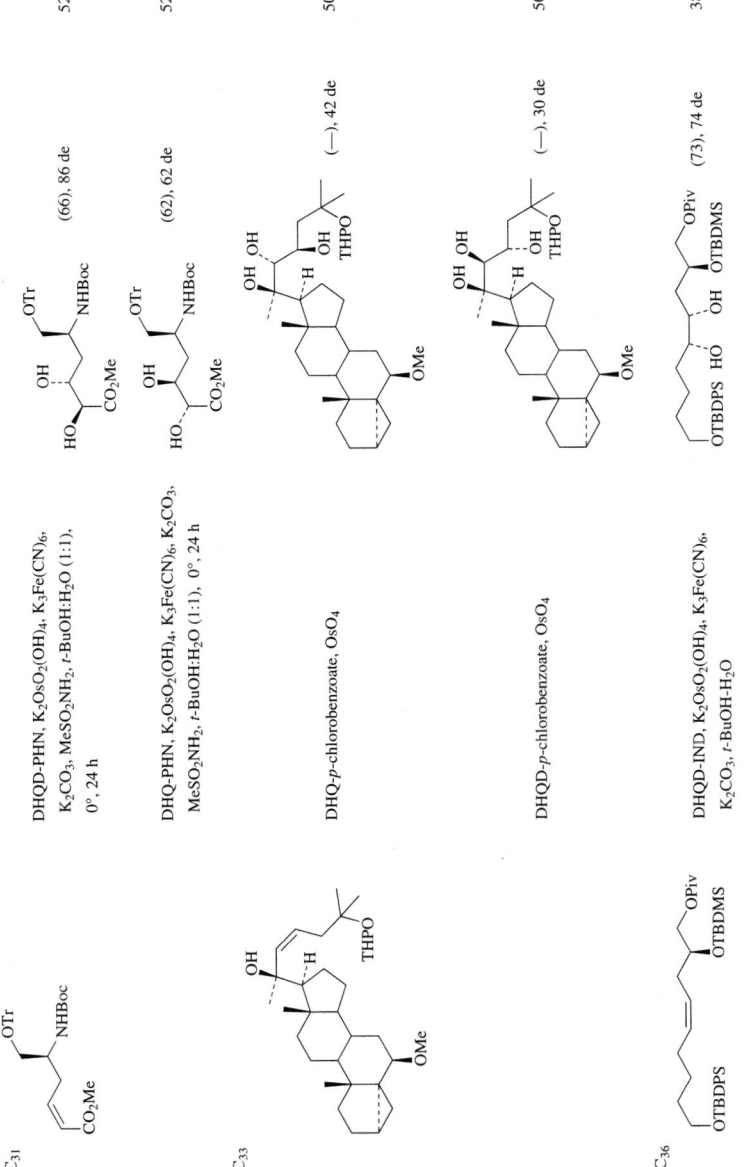

[a] The authors do not report the configuration of the product, but $(DHQD)_2PHAL$ is reported to give the enantiomer opposite to that given with $(DHQ)_2PHAL$.

TABLE 5. REACTIONS OF TRISUBSTITUTED ALKENES

Substrate	Conditions	Product(s) and Yield(s) (%), % ee		Refs.
*See Chart 1 at the beginning of the Tabular Survey for ligand structures that are indicated by **bold** numbers.*				
C_5				
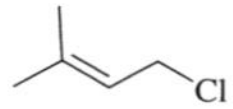	$(DHQD)_2PHAL$, $K_2OsO_2(OH)_4$, $K_3Fe(CN)_6$, K_2CO_3, $MeSO_2NH_2$, $NaHCO_3$, t-BuOH:H_2O (1:1), 0°	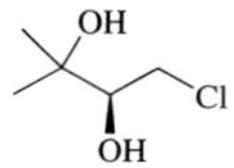	(50), 12	154
C_6				
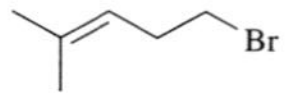	AD-Mix α, $MeSO_2NH_2$, t-BuOH:H_2O (1:1), 0°, 24 h	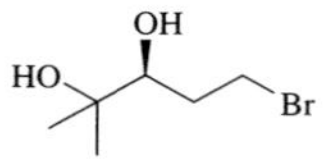	(55), >95	546
C_7				
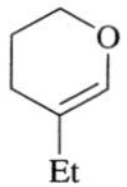	1. $(DHQD)_2PHAL$, $K_2OsO_2(OH)_4$, $K_3Fe(CN)_6$, K_2CO_3, $MeSO_2NH_2$, t-BuOH:H_2O (1:1), 0°, 24 h 2. I_2, $CaCO_3$, ether, rt, 32 h	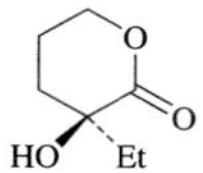	(70), 78	132
	1. $(DHQ)_2PHAL$, $K_2OsO_2(OH)_4$, $K_3Fe(CN)_6$, K_2CO_3, $MeSO_2NH_2$, t-BuOH:H_2O (1:1), 0°, 24 h 2. I_2, $CaCO_3$, ether, rt, 32 h	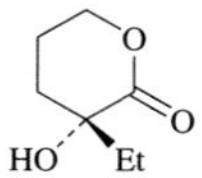	(69), 83	132
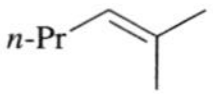	DHQD-p-chlorobenzoate, $K_2OsO_2(OH)_4$, NMO, $Me_4NOH{\bullet}5H_2O$, AcOH, acetone:H_2O (10:1), rt, 25 h (slow addition of olefin)	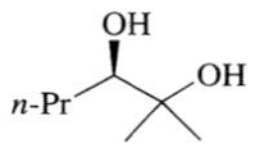	(80-95), 54[a]	112
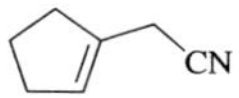	$(DHQD)_2PHAL$, $K_2OsO_2(OH)_4$, $K_3Fe(CN)_6$, K_2CO_3, $MeSO_2NH_2$, t-BuOH:H_2O (1:1), 0°, 8 h	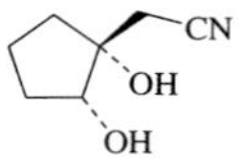	(97), 61	547

	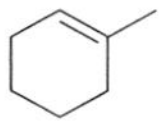	AD-Mix β, $MeSO_2NH_2$, *t*-BuOH:H_2O (1:1), 0°, 8-48 h	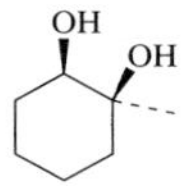 **I** (85), 50-52	548, 549
		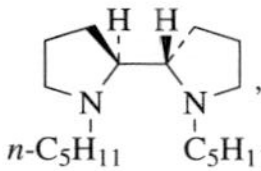, OsO_4, CH_2Cl_2, –78°	**I** (78), 12	12
		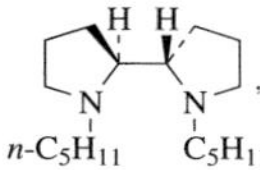, OsO_4, toluene, –78°	**I** (77), 31	12
		34, OsO_4, CH_2Cl_2, rt	**I** (71), 86	10
C_8	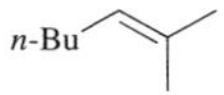	$(DHQD)_2PHAL$, $K_2OsO_2(OH)_4$, $K_3Fe(CN)_6$, K_2CO_3, $MeSO_2NH_2$, *t*-BuOH:H_2O (1:1), 0°, 6-24 h	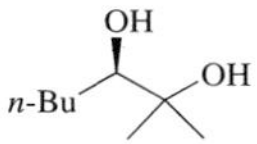 (80-98), 98	19, 200
		$(DHQ)_2PHAL$, $K_2OsO_2(OH)_4$, $K_3Fe(CN)_6$, K_2CO_3, $MeSO_2NH_2$, *t*-BuOH:H_2O (1:1), 0°, 6-24 h	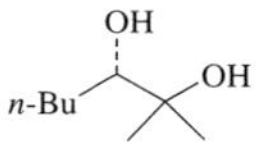 (80-98), 95	19, 200
	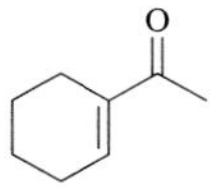	$(DHQD)_2PHAL$, $K_2OsO_2(OH)_4$, $K_3Fe(CN)_6$, $NaHCO_3$, K_2CO_3, $MeSO_2NH_2$, *t*-BuOH:H_2O (1:1), 0°, 24 h	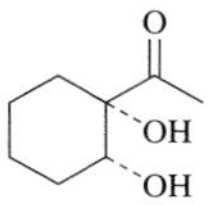 (73), 98	155
		DHQD-*p*-chlorobenzoate, $K_2OsO_2(OH)_4$, NMO, $Me_4NOH{\bullet}5H_2O$, AcOH, acetone:H_2O (10:1), rt, 25 h (slow addition of olefin)	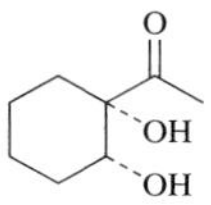 (80-95), 54[a]	112

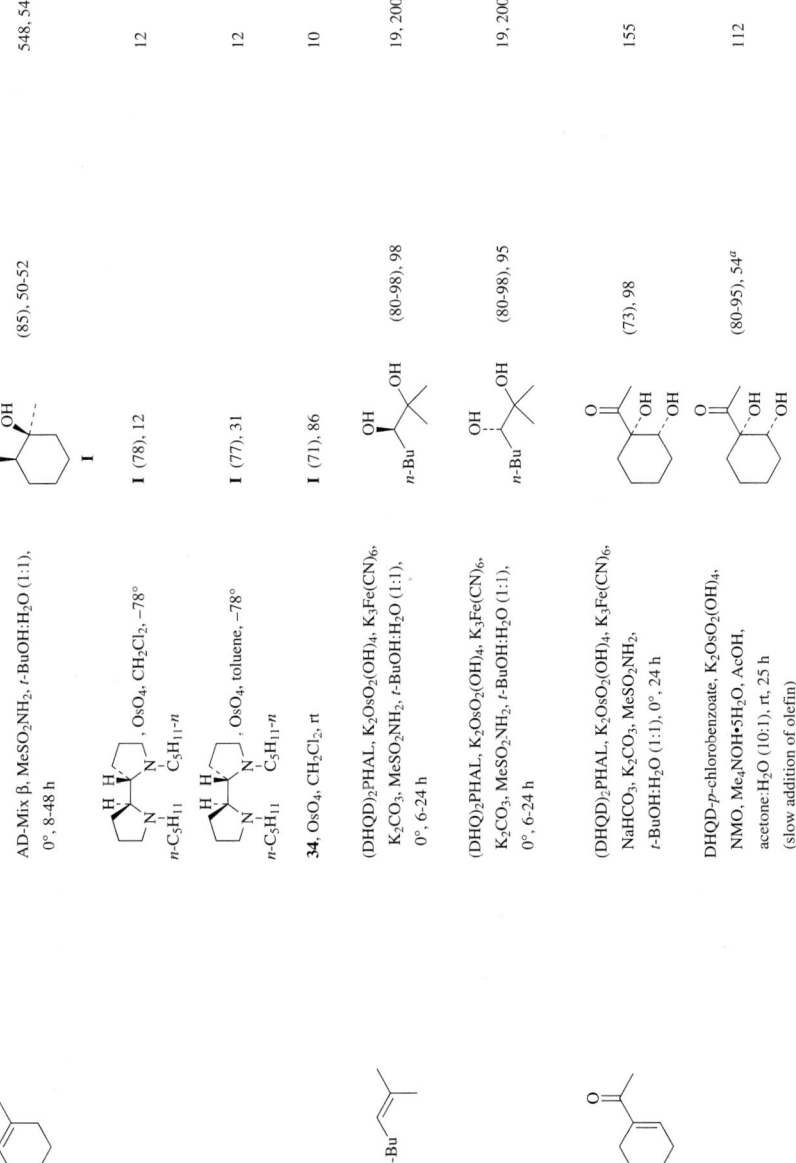

TABLE 5. REACTIONS OF TRISUBSTITUTED ALKENES (*Continued*)

Substrate	Conditions	Product(s) and Yield(s) (%), % ee		Refs.
C_8				
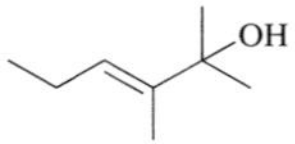	(DHQD)$_2$PHAL, $K_2OsO_2(OH)_4$, $K_3Fe(CN)_6$, K_2CO_3, $MeSO_2NH_2$, *t*-BuOH:H_2O (1:1), 0°, 6-36 h	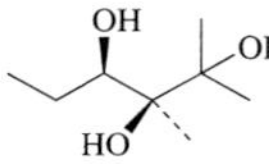	(75), 36	116
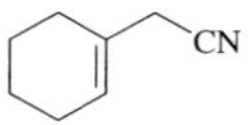	(DHQD)$_2$PHAL, $K_2OsO_2(OH)_4$, $K_3Fe(CN)_6$, K_2CO_3, $MeSO_2NH_2$, *t*-BuOH:H_2O (1:1), 0°, 8 h	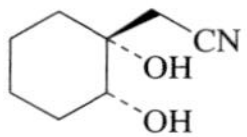	(94), 71	547
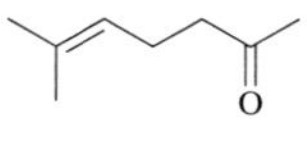	(DHQ)$_2$PHAL, OsO_4, $K_3Fe(CN)_6$, K_2CO_3, $MeSO_2NH_2$, *t*-BuOH:H_2O (1:1), 0°, 6 h	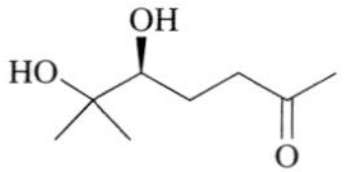	(72), 86	550
	(DHQD)$_2$PHAL, OsO_4, $K_3Fe(CN)_6$, K_2CO_3, $MeSO_2NH_2$, *t*-BuOH:H_2O (1:1), 0°, 6 h	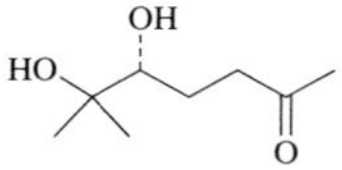	(77), 95	550
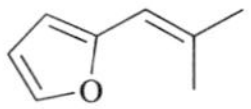	(DHQD)$_2$PHAL, $K_2OsO_2(OH)_4$, $K_3Fe(CN)_6$, K_2CO_3, *t*-BuOH:H_2O (1:1)	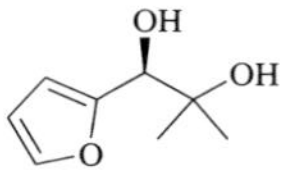	(—), >95	551
	(DHQ)$_2$PHAL, OsO_4, $K_3Fe(CN)_6$, K_2CO_3, *t*-BuOH:H_2O (1:1), rt, 2 d	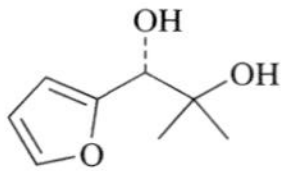	(—), >95	551
C_9				
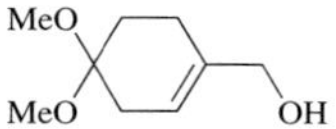	(DHQ)$_2$PHAL, $K_2OsO_2(OH)_4$, $K_3Fe(CN)_6$, K_2CO_3, $MeSO_2NH_2$, *t*-BuOH:H_2O (1:1), 0°	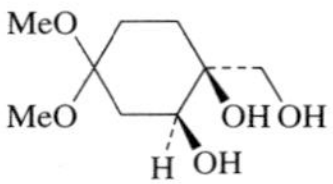	(—), 18	61, 133

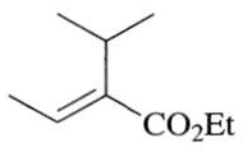	(DHQ)$_2$PHAL, $K_2OsO_2(OH)_4$, $K_3Fe(CN)_6$, K_2CO_3, $MeSO_2NH_2$, *t*-BuOH:H_2O (1:1), 0°, 42 h	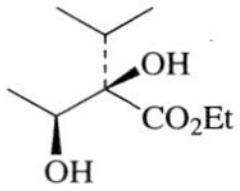	(81), 90	552
	(DHQD)$_2$PHAL, $K_2OsO_2(OH)_4$, $K_3Fe(CN)_6$, K_2CO_3, $MeSO_2NH_2$, *t*-BuOH:H_2O (1:1), 0°, 42 h	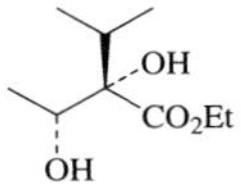	(83), 85	552
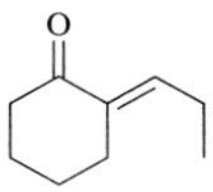	(DHQD)$_2$PHAL, $K_2OsO_2(OH)_4$, $K_2Fe(CN)_6$, K_2CO_3, *t*-BuOH:H_2O (1:1)	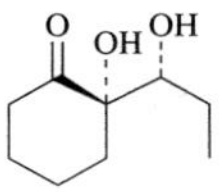	(>60), >90	553
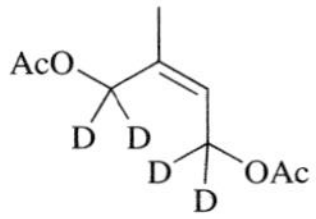	AD-Mix β, $MeSO_2NH_2$, *t*-BuOH:H_2O (1:1), 0°, 40 h	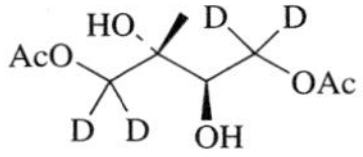	(96), 80	554
	AD-Mix α, $MeSO_2NH_2$, *t*-BuOH:H_2O (1:1), 0°, 40 h	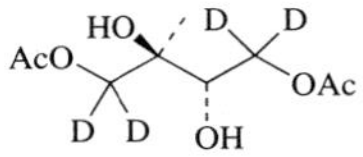	(96), 80	554
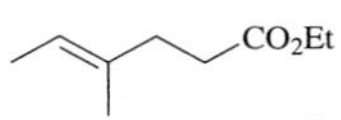	AD-Mix β, $MeSO_2NH_2$, *t*-BuOH:H_2O (1:1)	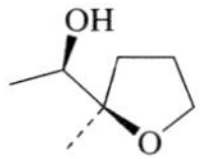	(—), 95	555

TABLE 5. REACTIONS OF TRISUBSTITUTED ALKENES (*Continued*)

Substrate	Conditions	Product(s) and Yield(s) (%), % ee	Refs.
C_9			
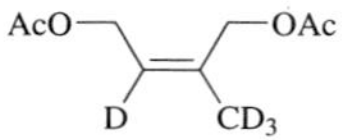	1. AD-Mix β, *t*-BuOH:H_2O (1:1), –15 to 4° 2. Ac_2O, DMAP, Et_3N, CH_2Cl_2	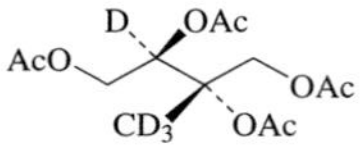 (96), 80	556
C_{10}			
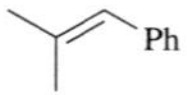	Ligand, $K_2OsO_2(OH)_4$, $K_3Fe(CN)_6$, K_2CO_3, *t*-BuOH:H_2O (1:1), rt, 24 h	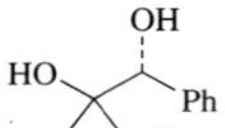 **I**	72
	Ligand	**I**	
	DHQD-PHN	(75-95), 84	
	DHQD-MEQ	(75-95), 81	
	DHQD-*p*-chlorobenzoate	(75-95), 74	
	$(DHQD)_2$PHAL, ABC MC OsO_4, NMO, acetone:H_2O:CH_3CN (1:1:1), rt, 24 h (slow addittion of olefin)	**I** (36), 85	170
	DHQD-*p*-chlorobenzoate, $K_2OsO_2(OH)_4$, NMO, acetone:H_2O (10:1), 0°, 13 h (slow addition of olefin),	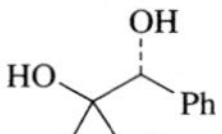 **I** (80-95), 53	112
	DHQD-*p*-chlorobenzoate, $K_2OsO_2(OH)_4$, NMO, toluene (0.15 M), 0°	**I** (—), 55	112
	Quinidine-*p*-chlorobenzoate, OsO_4, NMO, acetone-H_2O, rt	**I** (—), 10	557
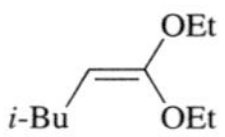	AD-Mix β, $MeSO_2NH_2$, *t*-BuOH:H_2O (1:1), 0°, 24 h	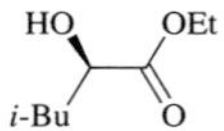 (27), 92	558
	AD-Mix α, $MeSO_2NH_2$, *t*-BuOH:H_2O (1:1), 0°, 24 h	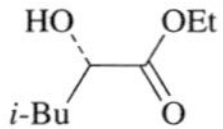 (49), 93	558

$C_{10\text{-}12}$ 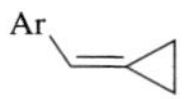	(DHQD)$_2$PHAL, $K_2OsO_2(OH)_4$, $K_3Fe(CN)_6$, K_2CO_3, $MeSO_2NH_2$, *t*-BuOH:H_2O (1:1)	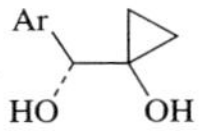		131
Ar				
2-FC_6H_4		(92), 98		
3-FC_6H_4		(51), 89		
4-FC_6H_4		(46), 86		
4-BrC_6H_4		(51), 86		
4-ClC_6H_4		(33), 87		
2-$MeOC_6H_4$		(50), 76		
4-$MeOC_6H_4$		(68), 84		
3,4-$(MeO)_2C_6H_4$		(76), 70		
	(DHQ)$_2$PHAL, $K_2OsO_2(OH)_4$, $K_3Fe(CN)_6$, K_2CO_3, $MeSO_2NH_2$, *t*-BuOH:H_2O (1:1)	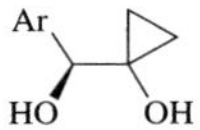		131
Ar				
2-FC_6H_4		(72), 85		
3-FC_6H_4		(42), 71		
4-FC_6H_4		(28), 68		
4-BrC_6H_4		(80), 63		
4-ClC_6H_4		(70), 47		
2-$MeOC_6H_4$		(55), 67		
4-$MeOC_6H_4$		(68), 63		
3,4-$(MeO)_2C_6H_4$		(90), 0		
C_{10} 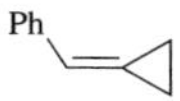	(DHQD)$_2$PHAL, $K_2OsO_2(OH)_4$, $K_3Fe(CN)_6$, K_2CO_3, $MeSO_2NH_2$, *t*-BuOH:H_2O (1:1), 0°, 6-24 h	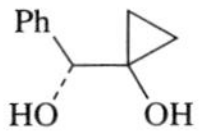	(91), 89	131, 130
	(DHQ)$_2$PHAL, $K_2OsO_2(OH)_4$, $K_3Fe(CN)_6$, K_2CO_3, $MeSO_2NH_2$, *t*-BuOH:H_2O (1:1)	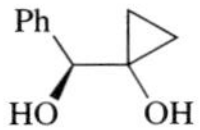	(55), 76	131
	1. AD-Mix β, $MeSO_2NH_2$, *t*-BuOH:H_2O (1:1), 0°, 7 d 2. $SOCl_2$, Et_3N, CH_2Cl_2	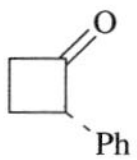	(63), 29	559

TABLE 5. REACTIONS OF TRISUBSTITUTED ALKENES (*Continued*)

Substrate	Conditions	Product(s) and Yield(s) (%), % ee	Refs.
C_{10}			
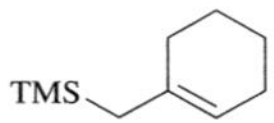	Ligand, $K_2OsO_2(OH)_4$, $K_3Fe(CN)_6$, K_2CO_3, $MeSO_2NH_2$, *t*-BuOH:H_2O (1:1), rt, 24 h	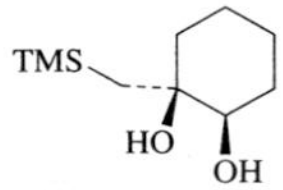	115
	Ligand		
	DHQD-*p*-chlorobenzoate	(—), 54	
	DHQD-PHN	(57), 35	
	$(DHQD)_2PHAL$	(55), 15	
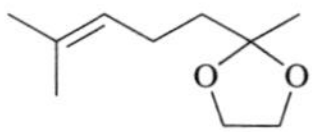	$(DHQ)_2PHAL$, OsO_4, $K_3Fe(CN)_6$, K_2CO_3, $MeSO_2NH_2$, *t*-BuOH:H_2O (1:1), 0°, 6 h	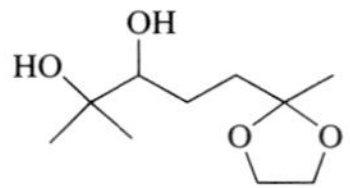 (94), —	550
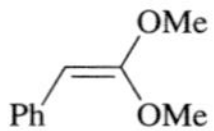	AD-Mix α, $MeSO_2NH_2$, *t*-BuOH:CH_3CN:H_2O (4:1:5), 0°, 6 h	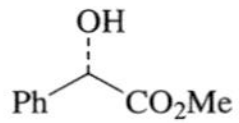 (90), 93	136
	AD-Mix β, $MeSO_2NH_2$, *t*-BuOH:CH_3CN:H_2O (4:1:5), 0°, 6 h	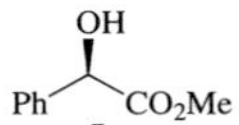 **I** (90), 96	136
	36, OsO_4, CH_2Cl_2, –100°	**I** (95), 66	96
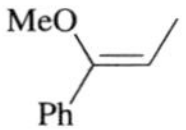	$(DHQD)_2PHAL$, $K_2OsO_2(OH)_4$, $K_3Fe(CN)_6$, K_2CO_3, $MeSO_2NH_2$, *t*-BuOH:H_2O (1:1)	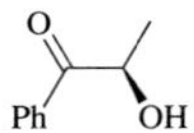 (—), 97	20
	$(DHQ)_2PHAL$, $K_2OsO_2(OH)_4$, $K_3Fe(CN)_6$, K_2CO_3, $MeSO_2NH_2$, *t*-BuOH:H_2O (1:1)	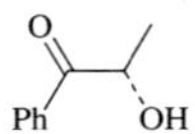 (—), 84	20

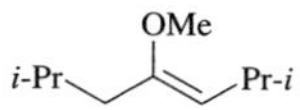	$(DHQD)_2PHAL$, $K_2OsO_2(OH)_4$, $K_3Fe(CN)_6$, K_2CO_3, $MeSO_2NH_2$, *t*-BuOH:H_2O (1:1)	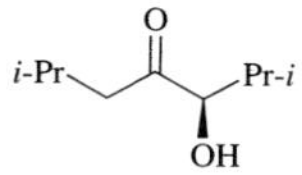	(—), 91	20
	$(DHQ)_2PHAL$, $K_2OsO_2(OH)_4$, $K_3Fe(CN)_6$, K_2CO_3, $MeSO_2NH_2$, *t*-BuOH:H_2O (1:1)	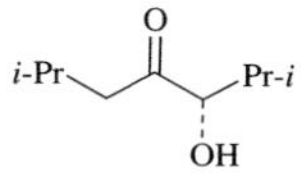	(—), 93	20
C_{11} 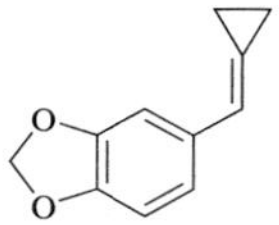	$(DHQD)_2PHAL$, $K_2OsO_2(OH)_4$, $K_3Fe(CN)_6$, K_2CO_3, $MeSO_2NH_2$, *t*-BuOH:H_2O (1:1)	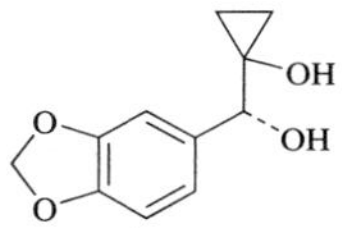	(66), 75	131
	$(DHQ)_2PHAL$, $K_2OsO_2(OH)_4$, $K_3Fe(CN)_6$, K_2CO_3, $MeSO_2NH_2$, *t*-BuOH:H_2O (1:1)	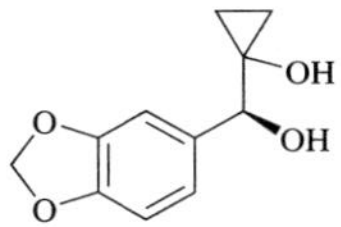	(67), 47	131
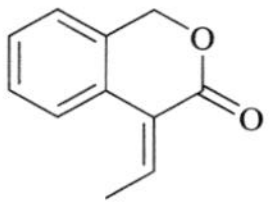	$(DHQD)_2PHAL$, $K_2OsO_2(OH)_4$, $K_3Fe(CN)_6$, K_2CO_3, $MeSO_2NH_2$, *t*-BuOH:H_2O (1:1), 0°, 24 h	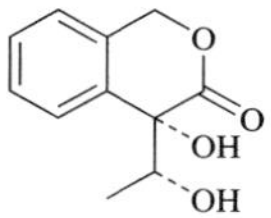	(78), 50	132
	$(DHQ)_2PHAL$, $K_2OsO_2(OH)_4$, $K_3Fe(CN)_6$, K_2CO_3, $MeSO_2NH_2$, *t*-BuOH:H_2O (1:1), 0°, 24 h	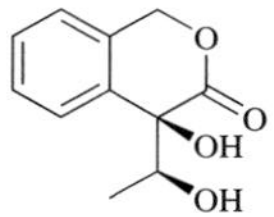	(70), 28	132

TABLE 5. REACTIONS OF TRISUBSTITUTED ALKENES (*Continued*)

Substrate	Conditions	Product(s) and Yield(s) (%), % ee	Refs.
C_{11} 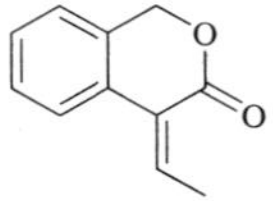	$(DHQD)_2PHAL$, $K_2OsO_2(OH)_4$, $K_3Fe(CN)_6$, K_2CO_3, $MeSO_2NH_2$, t-BuOH:H_2O (1:1), 0°, 24 h	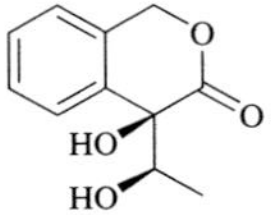 (68), 99	132
	$(DHQ)_2PHAL$, $K_2OsO_2(OH)_4$, $K_3Fe(CN)_6$, K_2CO_3, $MeSO_2NH_2$, t-BuOH:H_2O (1:1), 0°, 24 h	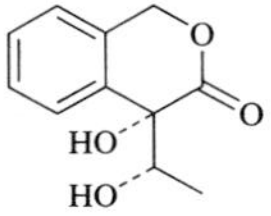 (73), 97	132
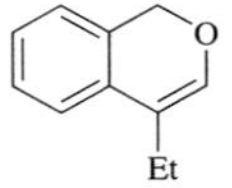	1. $(DHQD)_2PHAL$, $K_2OsO_2(OH)_4$, $K_3Fe(CN)_6$, K_2CO_3, $MeSO_2NH_2$, t-BuOH:H_2O (1:1), 0°, 24 h 2. I_2, $CaCO_3$, ether, rt, 32 h	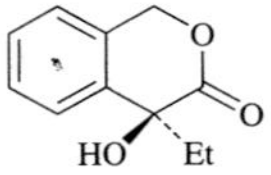 **I** (90), 74	132
	1. $(DHQD)_2PYR$, $K_2OsO_2(OH)_4$, $K_3Fe(CN)_6$, K_2CO_3, $MeSO_2NH_2$, t-BuOH:H_2O (1:1), 0°, 24 h 2. I_2, $CaCO_3$, ether, rt, 32 h	**I** (68), 91	132
	1. $(DHQ)_2PHAL$, $K_2OsO_2(OH)_4$, $K_3Fe(CN)_6$, K_2CO_3, $MeSO_2NH_2$, t-BuOH:H_2O (1:1), 0°, 24 h 2. I_2, $CaCO_3$, ether, rt, 32 h	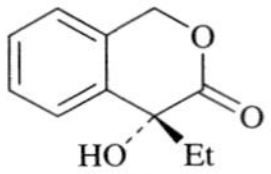 **I** (84), 63	132
	1. $(DHQ)_2PYR$, $K_2OsO_2(OH)_4$, $K_3Fe(CN)_6$, K_2CO_3, $MeSO_2NH_2$, t-BuOH:H_2O (1:1), 0°, 24 h 2. I_2, $CaCO_3$, ether, rt, 32 h	**I** (69), 78	132

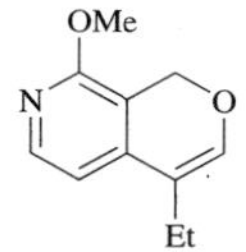	1. (DHQD)$_2$PYR, $K_2OsO_2(OH)_4$, $K_3Fe(CN)_6$, K_2CO_3, $MeSO_2NH_2$, t-BuOH:H_2O (1:1), 0°, 24 h 2. I_2, $CaCO_3$, ether, rt, 32 h	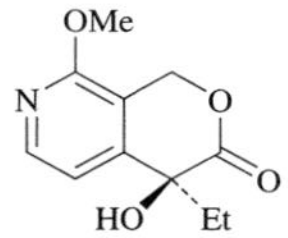	(74), 94	135, 134
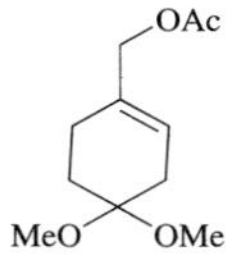	(DHQ)$_2$PHAL, $K_2OsO_2(OH)_4$, $K_3Fe(CN)_6$, K_2CO_3, $MeSO_2NH_2$, t-BuOH:H_2O (1:1), 0°	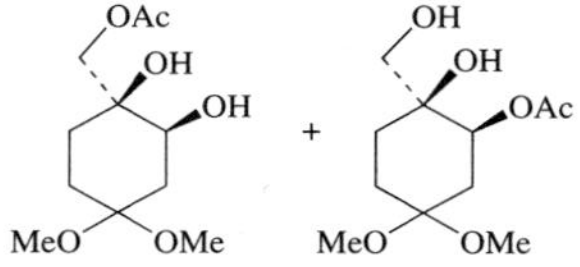 2.5 : 1	(—), 96	61
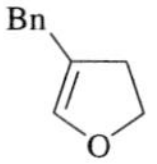	1. (DHQD)$_2$PHAL, $K_2OsO_2(OH)_4$, $K_3Fe(CN)_6$, K_2CO_3, $MeSO_2NH_2$, t-BuOH:H_2O (1:1), 0°, 24 h 2. I_2, $CaCO_3$, ether, rt, 32 h	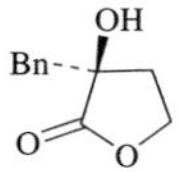	(87), 50	132
	1. (DHQ)$_2$PHAL, $K_2OsO_2(OH)_4$, $K_3Fe(CN)_6$, K_2CO_3, $MeSO_2NH_2$, t-BuOH:H_2O (1:1), 0°, 24 h 2. I_2, $CaCO_3$, ether, rt, 32 h	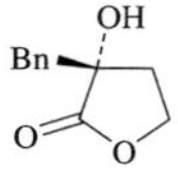	(87), 48	132
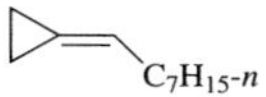	(DHQD)$_2$PHAL, $K_2OsO_2(OH)_4$, $K_3Fe(CN)_6$, K_2CO_3, $MeSO_2NH_2$, t-BuOH:H_2O (1:1), 0°, 6-24 h	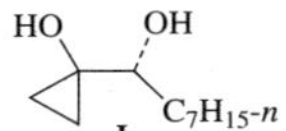 **I**	(85), 5	130
	AD-Mix β, $MeSO_2NH_2$, t-BuOH:H_2O (1:1), 0°, 12 h	**I**	(59), 58	560
	1. AD-Mix β, $MeSO_2NH_2$, t-BuOH:H_2O (1:1), 0°, 7 d 2. $SOCl_2$, Et_3N, CH_2Cl_2	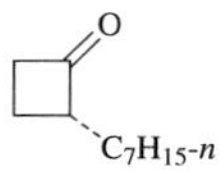	(77), 5	559
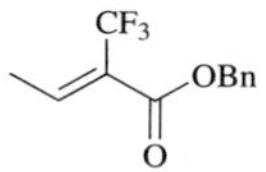	(DHQ)$_2$PHAL, OsO_4, $K_3Fe(CN)_6$, K_2CO_3, $MeSO_2NH_2$, t-BuOH:H_2O (1:1), 0°	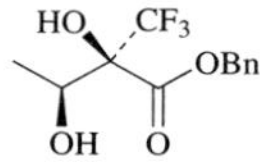	(92), 73	394

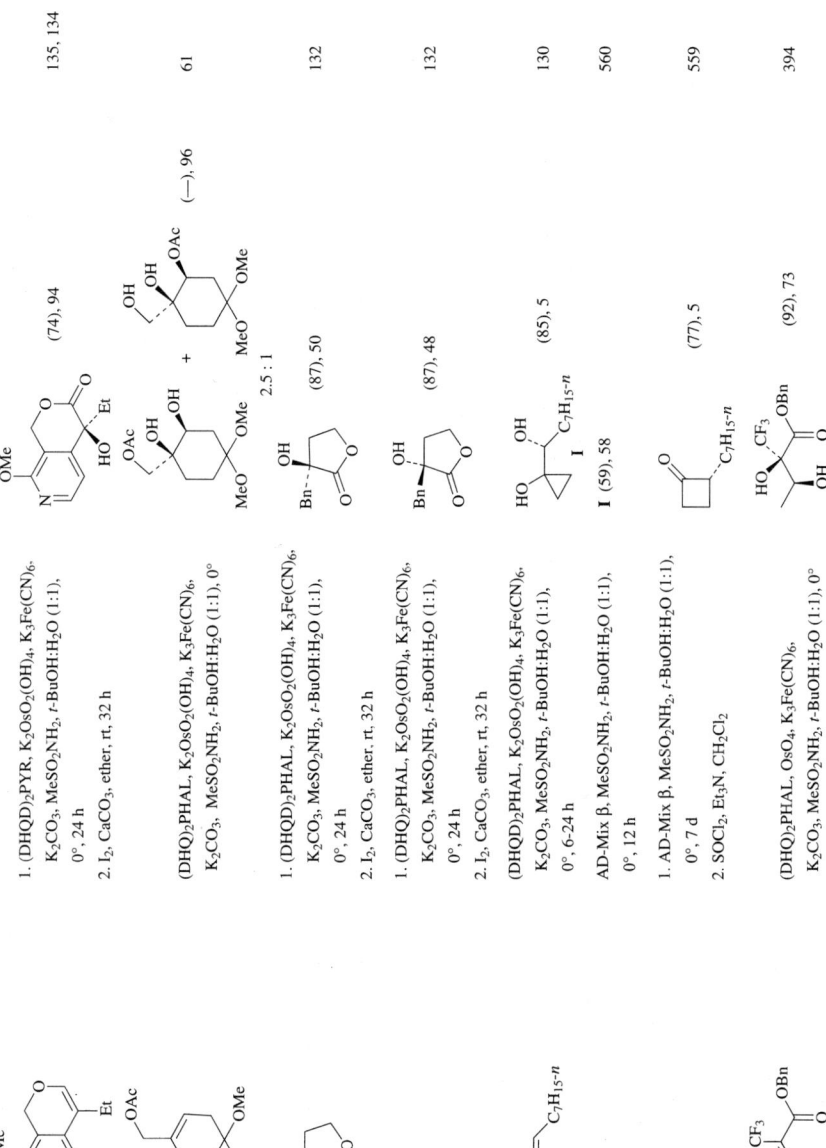

TABLE 5. REACTIONS OF TRISUBSTITUTED ALKENES (*Continued*)

Substrate	Conditions	Product(s) and Yield(s) (%), % ee	Refs.
C_{11}			
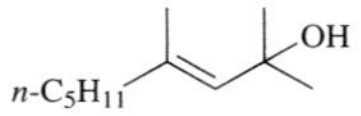	(DHQD)$_2$PHAL, $K_2OsO_2(OH)_4$, $K_3Fe(CN)_6$, K_2CO_3, $MeSO_2NH_2$, *t*-BuOH:H_2O (1:1), 0°, 6-36 h	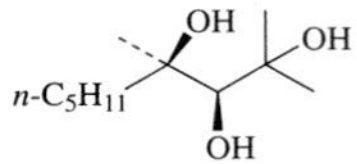 (91), 79	116
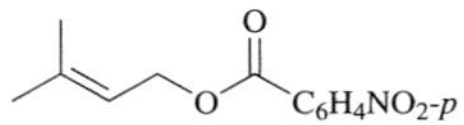	AD-Mix β, $MeSO_2NH_2$, *t*-BuOH:H_2O (1:1), 0°, 23 h	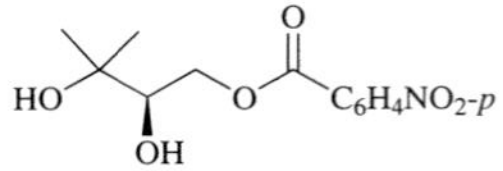 (59), 91	393
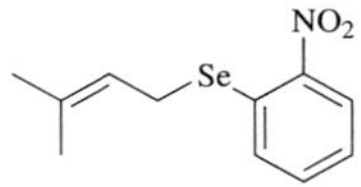	AD-Mix β, $MeSO_2NH_2$, *t*-BuOH:H_2O (1:1), 0°	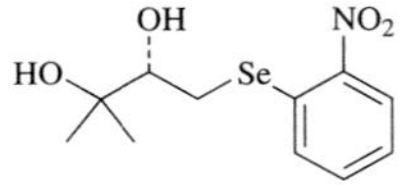 (50), —	561
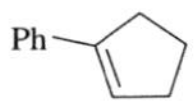	(DHQD)$_2$PHAL, $K_2OsO_2(OH)_4$, $K_3Fe(CN)_6$, K_2CO_3, $MeSO_2NH_2$, *t*-BuOH:H_2O (1:1), 0°, 6-24 h	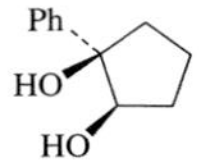 (—), 97	20
C_{12}			
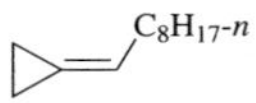	AD-Mix β, $MeSO_2NH_2$, *t*-BuOH:H_2O (1:1), 0°, 12 h	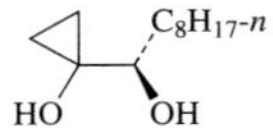 (86), 64	560
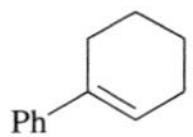	DHQD-PHN, $K_2OsO_2(OH)_4$, $K_3Fe(CN)_6$, K_2CO_3, *t*-BuOH:H_2O (1:1), rt, 24 h	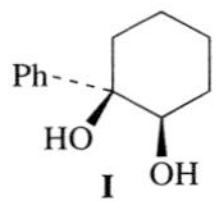 **I** (75-95), 93	72
	DHQD-MEQ, $K_2OsO_2(OH)_4$, $K_3Fe(CN)_6$, K_2CO_3, *t*-BuOH:H_2O (1:1), rt, 24 h	**I** (75-95), 92	72
	(DHQD)$_2$PHAL, $K_2OsO_2(OH)_4$, $K_3Fe(CN)_6$, K_2CO_3, $MeSO_2NH_2$, *t*-BuOH:H_2O (1:1), 0°, 6-24 h	**I** (80-98), 99	19, 20

DHQD-*p*-chlorobenzoate, $K_2OsO_2(OH)_4$, $K_3Fe(CN)_6$, K_2CO_3, *t*-BuOH:H_2O (1:1), rt, 24 h	**I** (75-95), 91	72
DHQD-*p*-chlorobenzoate, OsO_4, NMO, acetone:H_2O (10:1), 0°, 27 h (slow addition of olefin)	**I** (85-95), 78	69, 112
DHQD-*p*-chlorobenzoate, OsO_4, NMO, TEAA, acetone:H_2O (10:1), 0°, 17 h (slow addition of olefin)	**I** (85-95), 81	69, 112
DHQD-*p*-chlorobenzoate, $K_2OsO_2(OH)_4$, NMO, acetone:H_2O (0.15 M, 10:1), 0°	**I** (—), 79	69, 112
DHQD-*p*-chlorobenzoate, $K_2OsO_2(OH)_4$, NMO, TEAA, acetone-H_2O, 0°	**I** (—), 52	69
DHQD-*p*-chlorobenzoate, $K_2OsO_2(OH)_4$, NMO, acetone:H_2O (5:1), 0°, 7 d	**I** (—), 8	69
$(DHQD)_2PHAL$, $K_2OsO_2(OH)_4$, pH 10.4, *t*-BuOH-H_2O, 50°, 14-18 h	**I** (51), 86	178
33, OsO_4, toluene, –90°, 4 h	**I** (84), 90	9
45, OsO_4, $K_3Fe(CN)_6$, K_2CO_3, *t*-BuOH:H_2O (1:1), 0°, 24 h	**I** (85), 97	160
$(DHQD)_2PHAL$ (6%), OsO_4, H_2O_2, **54** (Chart 2), NMM, Et_4NOAc, *t*-BuOH:H_2O (3:1), rt, 20 h	**I** (50), 92	208
$(DHQ)_2PHAL$, $K_2OsO_2(OH)_4$, $K_3Fe(CN)_6$, K_2CO_3, $MeSO_2NH_2$, *t*-BuOH:H_2O (1:1), 0°, 6-24 h	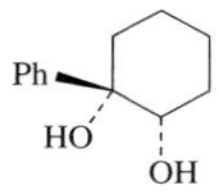 **I** (80-98), 97	19
DHQ-*p*-chlorobenzoate, OsO_4, NMO, acetone:H_2O (10:1), 0°, 27 h (slow addition of olefin)	**I** (85-95), 68-73	69
DHQ-*p*-chlorobenzoate, OsO_4, NMO, TEAA, acetone:H_2O (10:1), 0°, 27 h (slow addition of olefin)	**I** (85-95), 71-76	69

TABLE 5. REACTIONS OF TRISUBSTITUTED ALKENES (*Continued*)

Substrate	Conditions	Product(s) and Yield(s) (%), % ee	Refs.
C_{12} 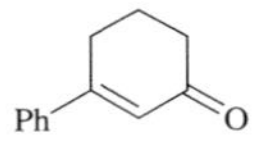	$(DHQD)_2PHAL$, $K_2OsO_2(OH)_4$, $K_3Fe(CN)_6$, $NaHCO_3$-K_2CO_3, $MeSO_2NH_2$, t-BuOH:H_2O (1:1), 0°, 24 h	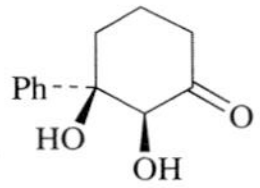 (79), 82	155
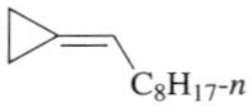	$(DHQD)_2PHAL$, $K_2OsO_2(OH)_4$, $K_3Fe(CN)_6$, K_2CO_3, $MeSO_2NH_2$, t-BuOH:H_2O (1:1), 0°, 4-24 h	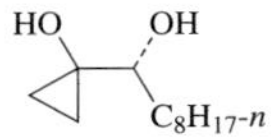 (86), 64	129, 130
	$(DHQD)_2PHAL$, $K_2OsO_2(OH)_4$, $K_3Fe(CN)_6$, K_2CO_3, $MeSO_2NH_2$, t-BuOH:H_2O (1:1), 0°, 4-5 h	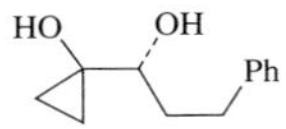 (82), 72	129
	$(DHQ)_2PHAL$, $K_2OsO_2(OH)_4$, $K_3Fe(CN)_6$, K_2CO_3, $MeSO_2NH_2$, t-BuOH:H_2O (1:1), 0°, 4-5 h	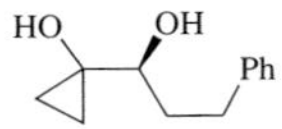 (66), 49	129
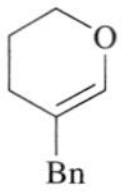	1. $(DHQD)_2PHAL$, $K_2OsO_2(OH)_4$, $K_3Fe(CN)_6$, K_2CO_3, $MeSO_2NH_2$, t-BuOH:H_2O (1:1), 0°, 24 h 2. I_2, $CaCO_3$, ether, rt, 32 h	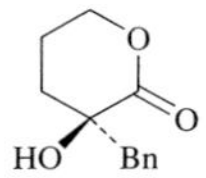 (81), 72	132
	1. $(DHQ)_2PHAL$, $K_2OsO_2(OH)_4$, $K_3Fe(CN)_6$, K_2CO_3, $MeSO_2NH_2$, t-BuOH:H_2O (1:1), 0°, 24 h 2. I_2, $CaCO_3$, ether, rt, 32 h	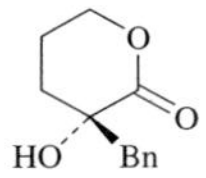 (81), 76	132
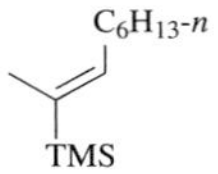	$(DHQD)_2PHAL$, $K_2OsO_2(OH)_4$, $K_3Fe(CN)_6$, K_2CO_3, $MeSO_2NH_2$, t-BuOH:H_2O (1:1), 20°, 168 h	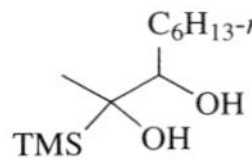 **I** (53), 85[b]	113
	$(DHQ)_2PHAL$, $K_2OsO_2(OH)_4$, $K_3Fe(CN)_6$, K_2CO_3, $MeSO_2NH_2$, t-BuOH:H_2O (1:1), 20°, 168 h	**I** (47), 80[b]	113

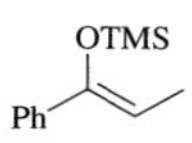	DHQD-*p*-chlorobenzoate, OsO_4, NMO, acetone:H_2O (10:1), 0°, 24 h (slow addition of olefin)	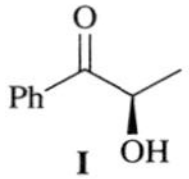 **I** (80-95), 72[a]	112
	DHQD-*p*-chlorobenzoate, OsO_4, NMO, TEAA, acetone:H_2O (10:1), 0°, 24 h (slow addition of olefin)	**I** (80-95), 78	112
	DHQD-*p*-chlorobenzoate, $K_2OsO_2(OH)_4$, NMO, acetone:H_2O (10:1), 0°	**I** (—), 85	112
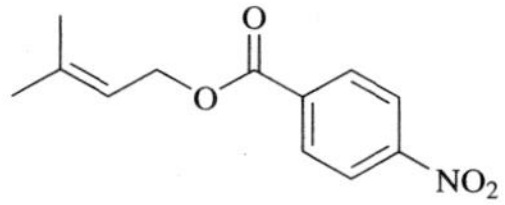	AD-Mix β, $MeSO_2NH_2$, *t*-BuOH:H_2O (1:1), 0°, 23 h	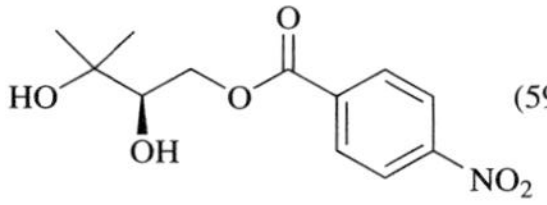 (59), 91	393
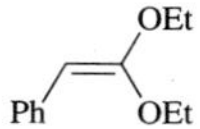	AD-Mix β, $MeSO_2NH_2$, *t*-BuOH:H_2O (1:1), 0°, 24 h	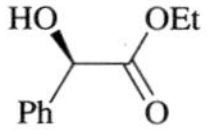 (68), 99	558
	AD-Mix α, $MeSO_2NH_2$, *t*-BuOH:H_2O (1:1), 0°, 24 h	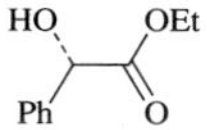 (76), 98	558
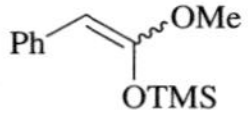	**36**, OsO_4, CH_2Cl_2, –78°	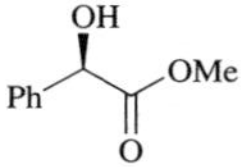 (45), 55	96
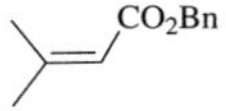	AD-Mix α, $NaHCO_3$, $MeSO_2NH_2$, *t*-BuOH:H_2O (1:1)	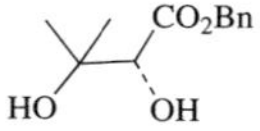 (80), >95	561a, 106
	AD-Mix β, $MeSO_2NH_2$, *t*-BuOH:H_2O (1:1)	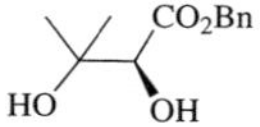 (86), 98	106
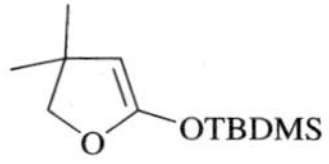	$(DHQD)_2PHAL$, $K_2OsO_2(OH)_4$, $K_3Fe(CN)_6$, K_2CO_3, *t*-BuOH:H_2O (1:1), rt, 18 h	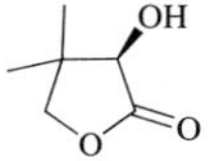 (60), 92	346

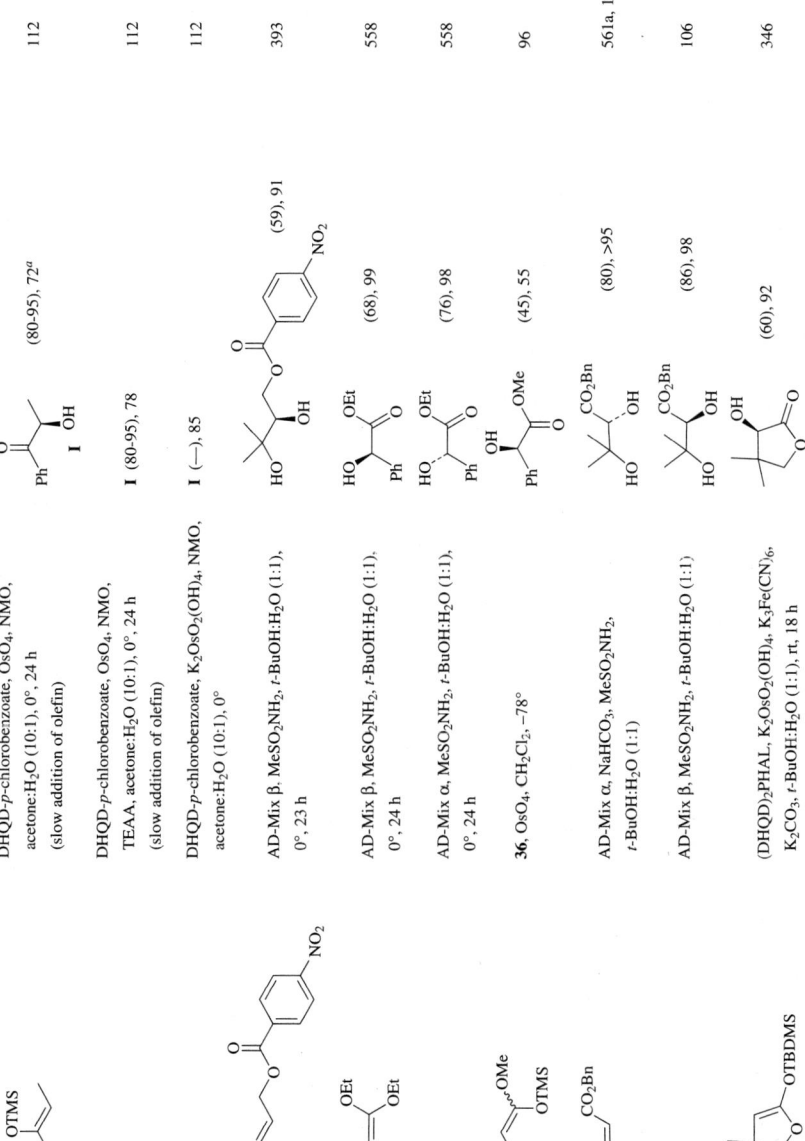

TABLE 5. REACTIONS OF TRISUBSTITUTED ALKENES (*Continued*)

Substrate	Conditions	Product(s) and Yield(s) (%), % ee		Refs.
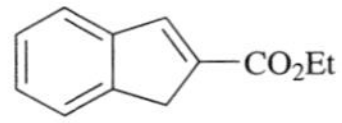	AD-Mix α, t-BuOH:H_2O (1:1), 0°	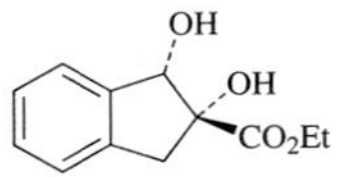	(65), 92	562
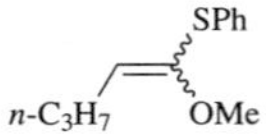	AD-Mix α, $MeSO_2NH_2$, t-BuOH:CH_3CN:H_2O (4:1:5), 0°, 6 h	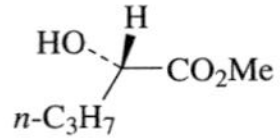	(7), 62	136
	AD-Mix β, $MeSO_2NH_2$, t-BuOH:CH_3CN:H_2O (4:1:5), 0°, 6 h	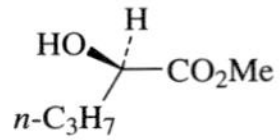	(7), 73	136
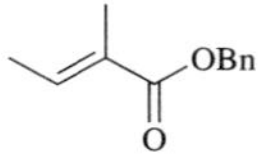	AD-Mix α, t-BuOH:H_2O (1:1), 4°, 2 d	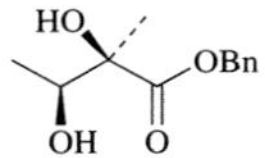	(91), >98	563
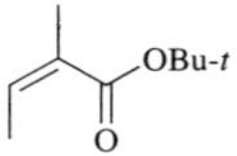	AD-Mix β, t-BuOH:H_2O (1:1), 4°, 2 d	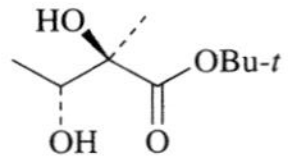	(85), 60	563
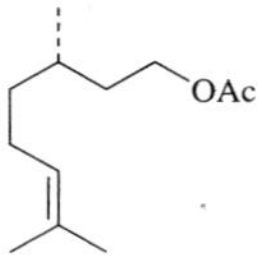	AD-Mix β, $MeSO_2NH_2$, t-BuOH:H_2O (1:1), 0°, 24 h	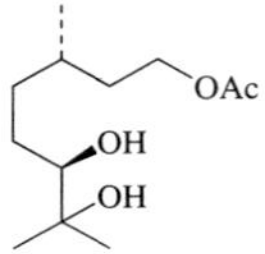	(97), >98, >98 de	564
	AD-Mix α, $MeSO_2NH_2$, t-BuOH:H_2O (1:1), 0°, 24 h	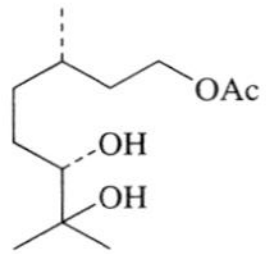	(95), >98, 96 de	564

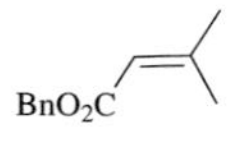	AD-Mix α, $MeSO_2NH_2$, $NaHCO_3$, t-BuOH:H_2O (1:1), 0°, 24 h	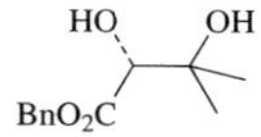	(80), >95	561b
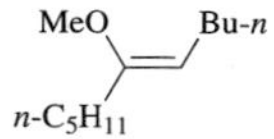	$(DHQD)_2PHAL$, $K_2OsO_2(OH)_4$, $K_3Fe(CN)_6$, K_2CO_3, $MeSO_2NH_2$, t-BuOH:H_2O (1:1)	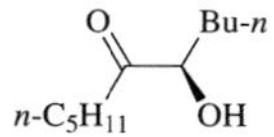	(—), 95	20
	$(DHQ)_2PHAL$, $K_2OsO_2(OH)_4$, $K_3Fe(CN)_6$, K_2CO_3, $MeSO_2NH_2$, t-BuOH:H_2O (1:1)	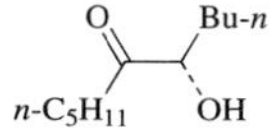	(—), 96	20
C_{13}				
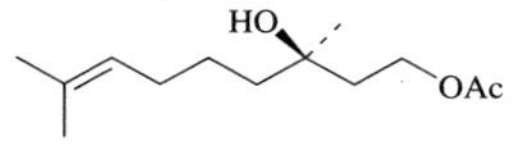	AD-Mix α, $MeSO_2NH_2$, $NaHCO_3$, t-BuOH:H_2O (1:1), 0°, 24 h	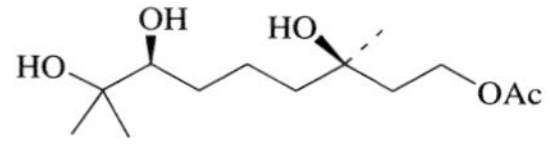	(80), 84 de	564
	AD-Mix β, $MeSO_2NH_2$, $NaHCO_3$, t-BuOH:H_2O (1:1), 0°, 24 h	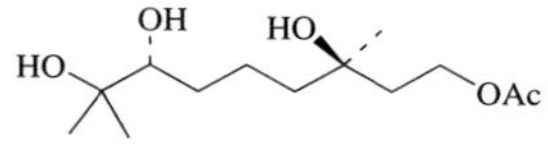	(77), 94 de	564
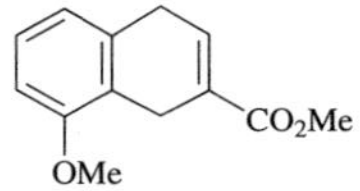	$(DHQ)_2PHAL$, $K_2OsO_2(OH)_4$, $K_3Fe(CN)_6$, K_2CO_3, $MeSO_2NH_2$, t-BuOH:H_2O (1:1), 4°, 5 h	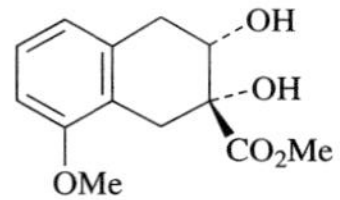	(85), 83	565
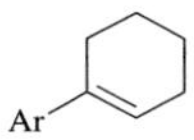	$(DHQD)_2PHAL$, $K_2OsO_2(OH)_4$, $K_3Fe(CN)_6$, K_2CO_3, $MeSO_2NH_2$, t-BuOH:H_2O (1:1), rt	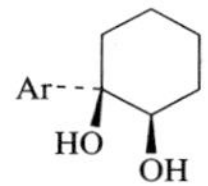		566

Ar	TIme	
4-MeC_6H_4	72 h	(84), 96
4-$MeOC_6H_4$	60 h	(95), 95
4-$CF_3C_6H_4$	96 h	(42), 91

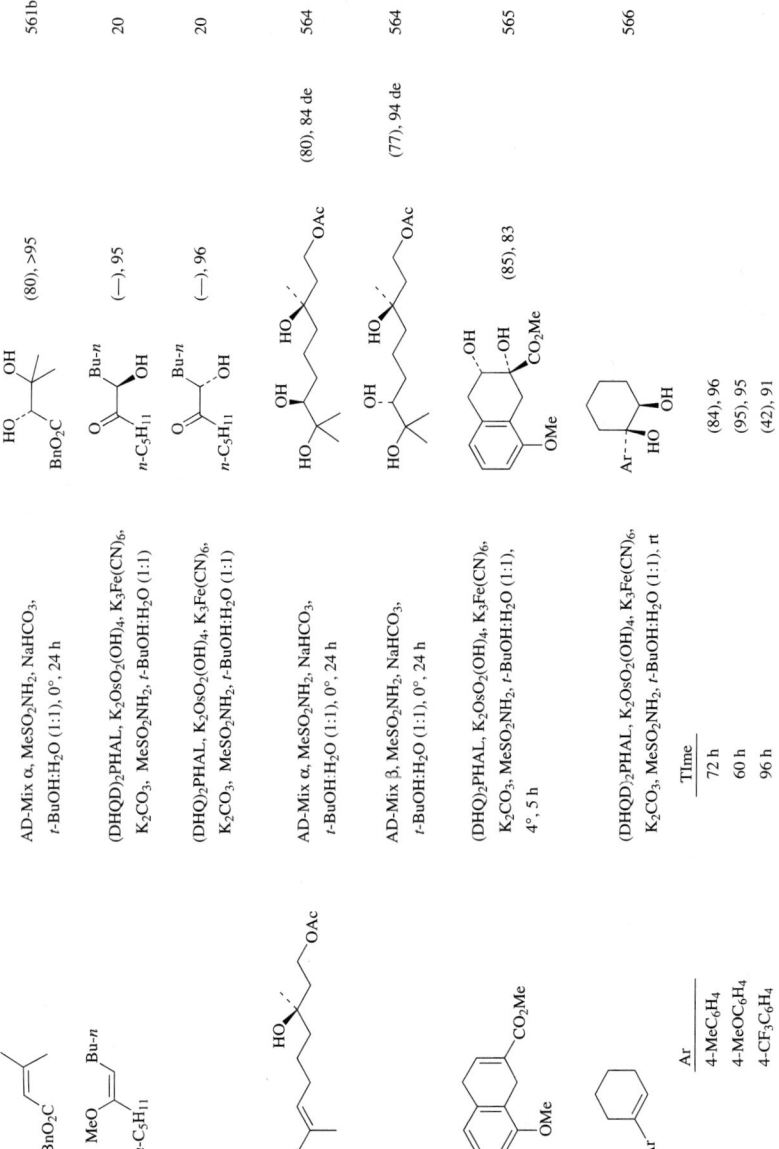

TABLE 5. REACTIONS OF TRISUBSTITUTED ALKENES (*Continued*)

Substrate	Conditions	Product(s) and Yield(s) (%), % ee	Refs.
C_{13} 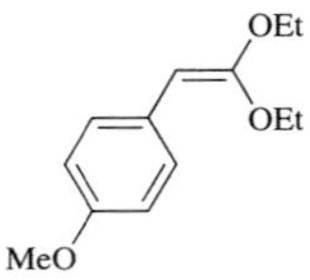	AD-Mix β, $MeSO_2NH_2$, t-BuOH:H_2O (1:1), 0°, 24 h	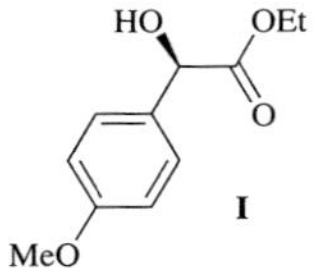 **I** (65), 88	558
	AD-Mix β, $MeSO_2NH_2$, t-BuOH:CH_3CN:H_2O (4:1:5), 0°, 6 h	**I** (71), 88	136
	AD-Mix α, $MeSO_2NH_2$, t-BuOH:H_2O (1:1), 0°, 24 h	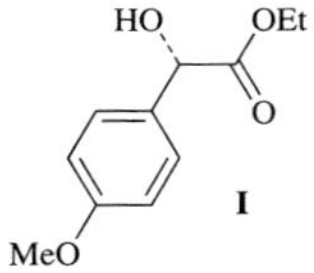 **I** (71), 86	558
	AD-Mix α, $MeSO_2NH_2$, t-BuOH:CH_3CN:H_2O (4:1:5), 0°, 6 h	**I** (71), 86	136
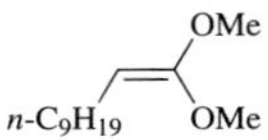	AD-Mix α, $MeSO_2NH_2$, t-BuOH:CH_3CN:H_2O (4:1:5), 0°, 6 h	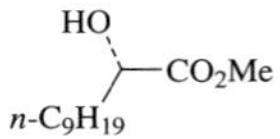 (70), 92	136
	AD-Mix β, $MeSO_2NH_2$, t-BuOH:CH_3CN:H_2O (4:1:5), 0°, 6 h	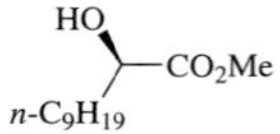 (70), 98	136
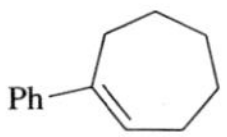	$(DHQD)_2PHAL$, $K_2OsO_2(OH)_4$, $K_3Fe(CN)_6$, K_2CO_3, $MeSO_2NH_2$, t-BuOH:H_2O (1:1), 0°, 6-24 h	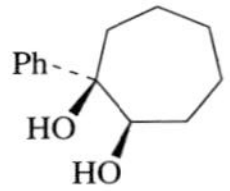 (—), 95	20
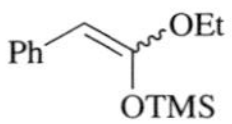	**36**, OsO_4, CH_2Cl_2, –78°	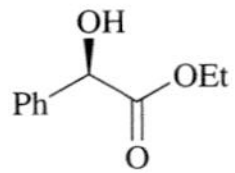 (47), 60	96

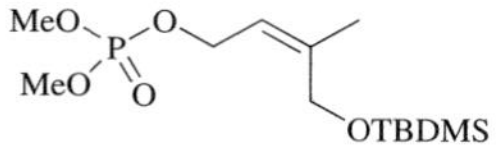	AD-Mix β, $NaHCO_3$, *t*-BuOH:H_2O (1:1)	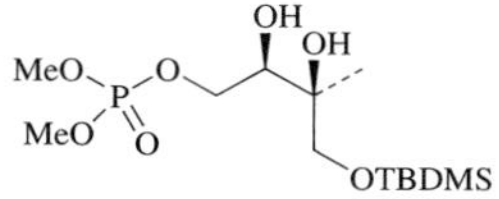	(60), 78	566a
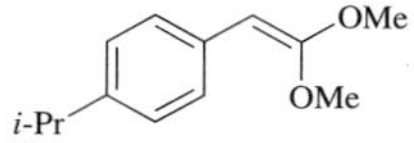	AD-Mix α, $MeSO_2NH_2$, *t*-BuOH:CH_3CN:H_2O (4:1:5), 0°, 6 h	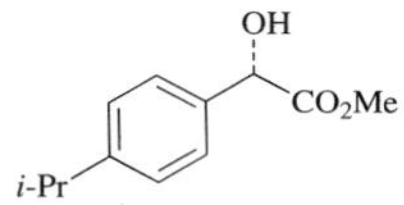	(93), 91	136
	AD-Mix β, $MeSO_2NH_2$, *t*-BuOH:CH_3CN:H_2O (4:1:5), 0°, 6 h	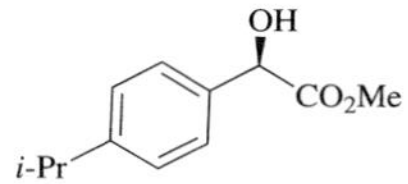	(93), 96	136
C_{14} 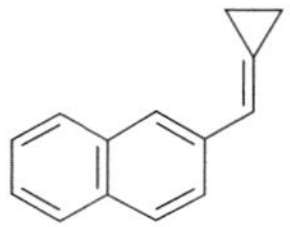	$(DHQD)_2PHAL$, $K_2OsO_2(OH)_4$, $K_3Fe(CN)_6$, K_2CO_3, $MeSO_2NH_2$, *t*-BuOH:H_2O (1:1)	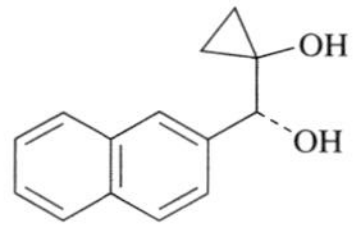	(64), 90	131
	$(DHQ)_2PHAL$, $K_2OsO_2(OH)_4$, $K_3Fe(CN)_6$, K_2CO_3, $MeSO_2NH_2$, *t*-BuOH:H_2O (1:1)	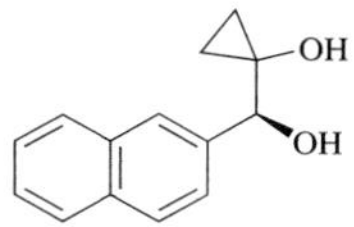	(68), 79	131
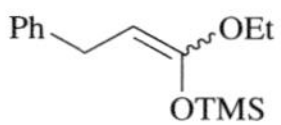	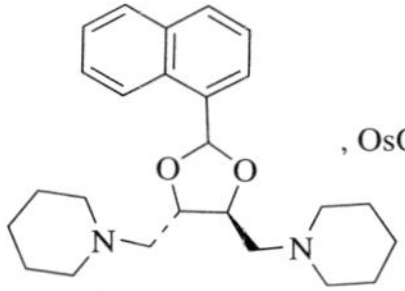 , OsO_4, CH_2Cl_2, –78°	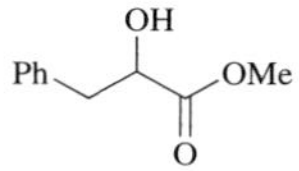	(42), 0	96
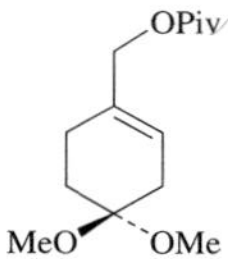	$(DHQ)_2PHAL$, $K_2OsO_2(OH)_4$, $K_3Fe(CN)_6$, K_2CO_3, $MeSO_2NH_2$, *t*-BuOH:H_2O (1:1), 0°	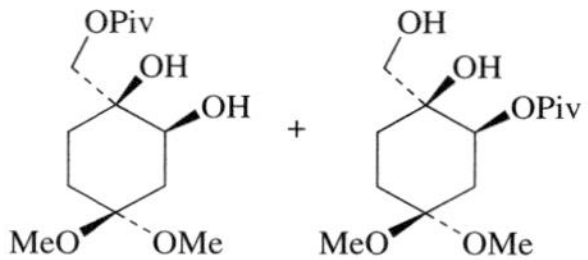 12 : 1	(—), 35	61

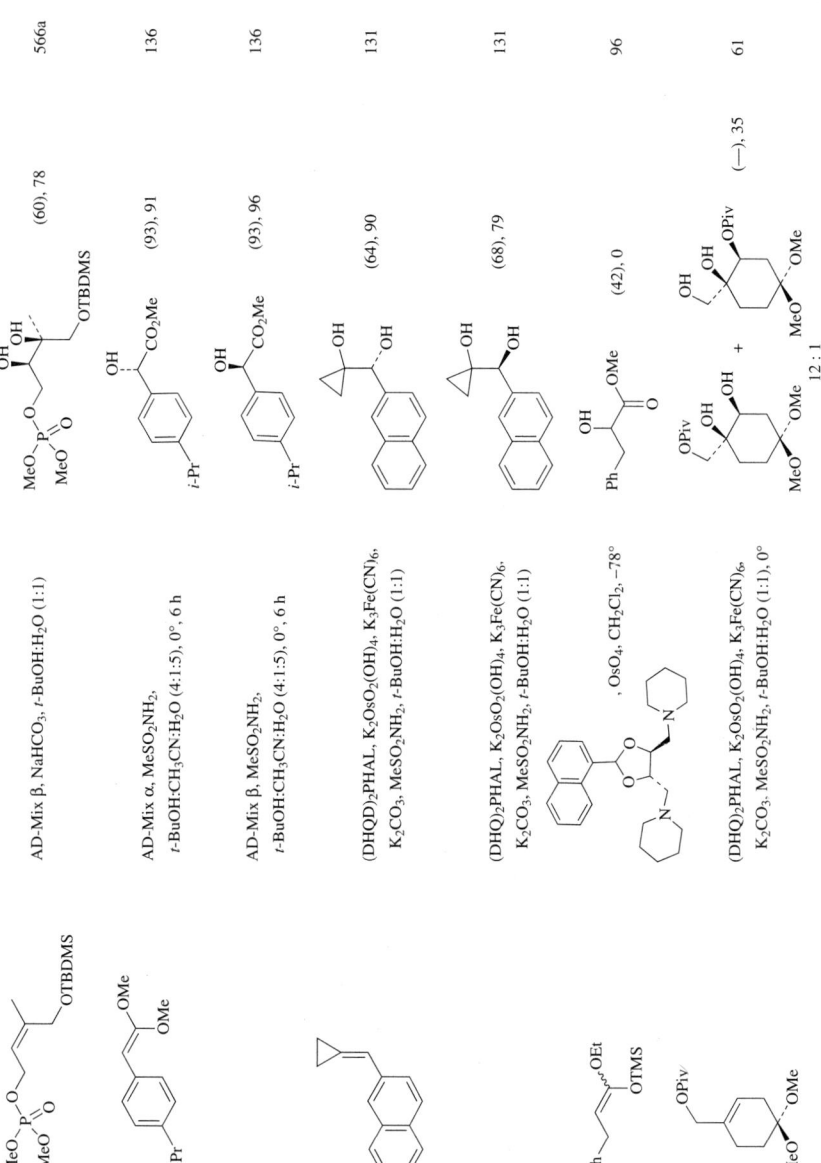

TABLE 5. REACTIONS OF TRISUBSTITUTED ALKENES (*Continued*)

Substrate	Conditions	Product(s) and Yield(s) (%), % ee	Refs.
C_{14}			
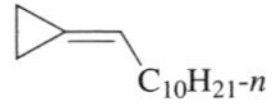	(DHQD)$_2$PHAL, $K_2OsO_2(OH)_4$, $K_3Fe(CN)_6$, K_2CO_3, $MeSO_2NH_2$, *t*-BuOH:H_2O (1:1), 0°, 4-5 h	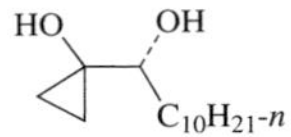 (81), 68	129
	(DHQ)$_2$PHAL, $K_2OsO_2(OH)_4$, $K_3Fe(CN)_6$, K_2CO_3, $MeSO_2NH_2$, *t*-BuOH:H_2O (1:1), 0°, 4-5 h	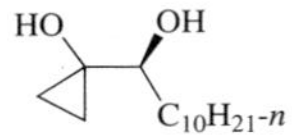 (85), 60	129
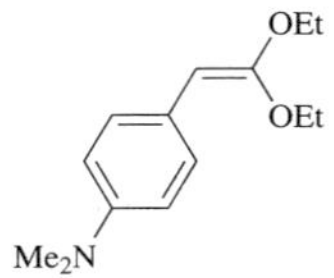	AD-Mix β, $Me_3SO_2NH_2$, *t*-BuOH:H_2O (1:1), 0°, 24 h	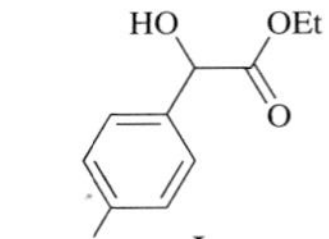 **I** (16), 0	558
	AD-Mix α, $MeSO_2NH_2$, *t*-BuOH:H_2O (1:1), 0°, 24 h	**I** (23), 0	558
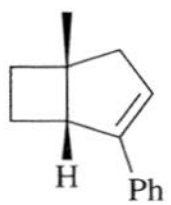	AD-Mix β, $MeSO_2NH_2$, *t*-BuOH:H_2O (1:1), 0°	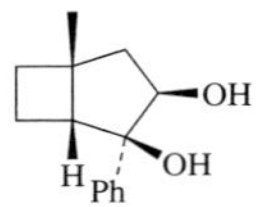 (—), 84 de	567
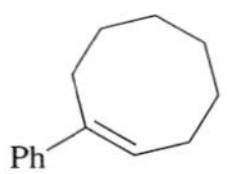	(DHQD)$_2$PHAL, $K_2OsO_2(OH)_4$, $K_3Fe(CN)_6$, K_2CO_3, $MeSO_2NH_2$, *t*-BuOH:H_2O (1:1), 0°, 6-24 h	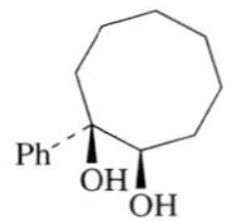 (—), 83	20

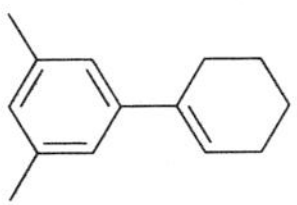	(DHQD)$_2$PHAL, $K_2OsO_2(OH)_4$, $K_3Fe(CN)_6$, K_2CO_3, $MeSO_2NH_2$, *t*-BuOH:H_2O (1:1), 0°, 6-24 h	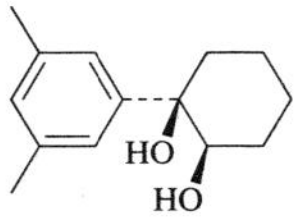	(—), 98	20
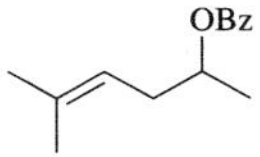	AD-Mix α, $MeSO_2NH_2$, *t*-BuOH:H_2O (1:1)	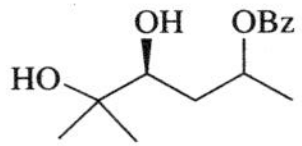	(95), 76, 0 de	438
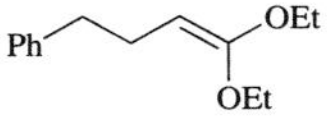	AD-Mix α, $MeSO_2NH_2$, *t*-BuOH:CH_3CN:H_2O (4:1:5), 0°, 6 h	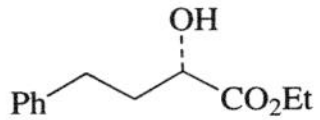	(34), 93	136
	AD-Mix β, $MeSO_2NH_2$, *t*-BuOH:CH_3CN:H_2O (4:1:5), 0°, 6 h	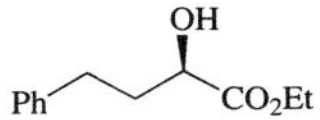	(34), 94	136
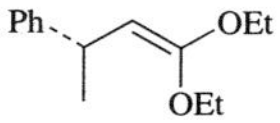	AD-Mix α, $MeSO_2NH_2$, *t*-BuOH:CH_3CN:H_2O (4:1:5), 0°, 6 h	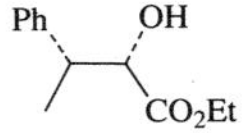	(88), 91 de	136
	AD-Mix β, $MeSO_2NH_2$, *t*-BuOH:CH_3CN:H_2O (4:1:5), 0°, 6 h	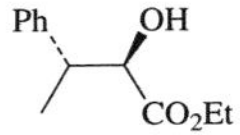	(88), 78 de	136
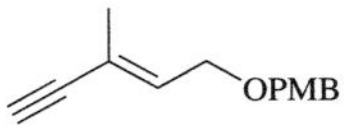	AD-Mix β	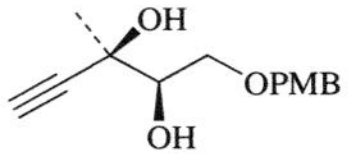	(76), 93	568

TABLE 5. REACTIONS OF TRISUBSTITUTED ALKENES (*Continued*)

Substrate	Conditions	Product(s) and Yield(s) (%), % ee	Refs.
C_{14} 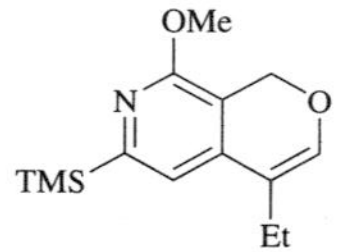	1. $(DHQD)_2PHAL$, OsO_4, $K_3Fe(CN)_6$, K_2CO_3, $MeSO_2NH_2$, t-BuOH:H_2O (1:1), 0°, 12 h 2. I_2, $CaCO_3$	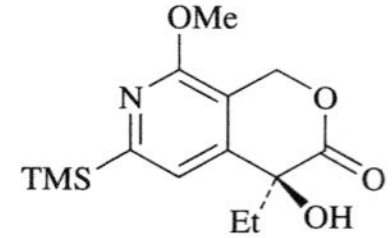 **I** (87), 32	569
	1. $(DHQD)_2PYR$, OsO_4, $K_3Fe(CN)_6$, K_2CO_3, $MeSO_2NH_2$, t-BuOH:H_2O (1:1), 0°, 12 h 2. I_2, $CaCO_3$	**I** (87), 94	569
	1. $(DHQ)_2PYR$, OsO_4, $K_3Fe(CN)_6$, K_2CO_3, $MeSO_2NH_2$, t-BuOH:H_2O (1:1), 0°, 12 h 2. I_2, $CaCO_3$	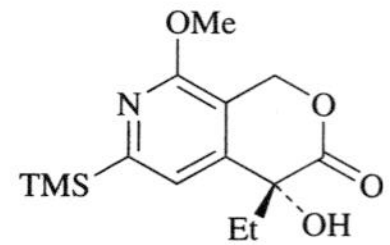 (87), 90	569
C_{15} 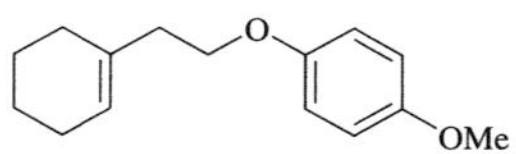	$(DHQD)_2PYDZ$, $K_2OsO_2(OH)_4$, $K_3Fe(CN)_6$, K_2CO_3, $MeSO_2NH_2$, t-BuOH:H_2O (1:1), 0°, 4-22 h	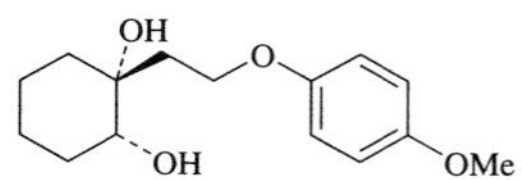 (99), 95	83, 61
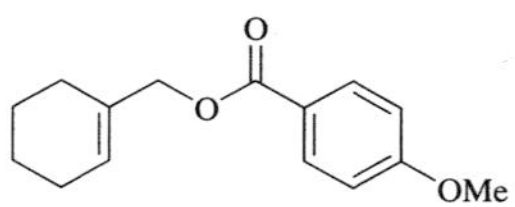	$(DHQD)_2PYDZ$, $K_2OsO_2(OH)_4$, $K_3Fe(CN)_6$, K_2CO_3, $MeSO_2NH_2$, t-BuOH:H_2O (1:1), 0°, 4-22 h	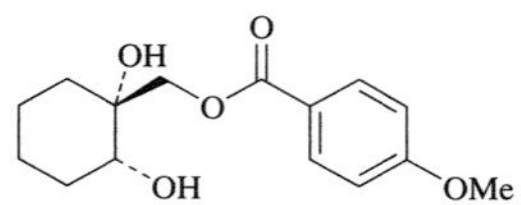 **I** (99), 98	63, 61, 133
	6, $K_2OsO_2(OH)_4$, $K_3Fe(CN)_6$, K_2CO_3, $MeSO_2NH_2$, t-BuOH:H_2O (1:1), 0°	**I** (99), 98	63
	17, $K_2OsO_2(OH)_4$, $K_3Fe(CN)_6$, K_2CO_3, $MeSO_2NH_2$, t-BuOH:H_2O (1:1), 0°	**I** (99), 95	63

Reactant	Conditions	Product(s) and Yield(s) (%)		Refs.
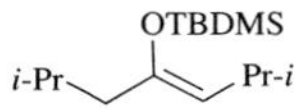	(DHQD)$_2$PHAL, $K_2OsO_2(OH)_4$, $K_3Fe(CN)_6$, K_2CO_3, $MeSO_2NH_2$, t-BuOH:H_2O (1:1)	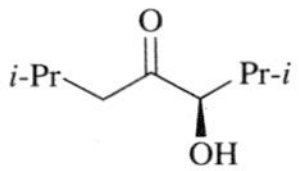	(—), 79	20
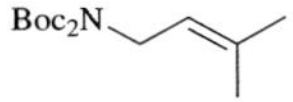	(DHQD)$_2$PHAL, $K_2OsO_2(OH)_4$, $K_3Fe(CN)_6$, $MeSO_2NH_2$, K_2CO_3, t-BuOH:H_2O (1:1), 0°	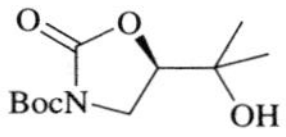	(44), 47	85
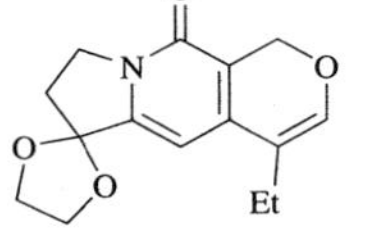	Ligand, $K_2OsO_2(OH)_4$, $K_3Fe(CN)_6$, K_2CO_3, $MeSO_2NH_2$, t-BuOH:H_2O (1:1), 0°	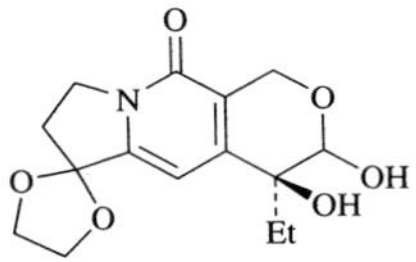		570

Ligand	Time	Yield
(DHQD)$_2$PHAL	96 h	(89), 82 de
(DHQD)$_2$PYR	18 h	(74), 84 de
DHQD-PHN	24 h	(54), 70 de
DHQD-MEQ	24 h	(51), 70 de
DHQD-p-chlorobenzoate	24 h	(29), 48 de

Reactant	Conditions	Product(s) and Yield(s) (%)		Refs.
	Ligand, $K_2OsO_2(OH)_4$, $K_3Fe(CN)_6$, K_2CO_3, $MeSO_2NH_2$, t-BuOH:H_2O (1:1), 0°	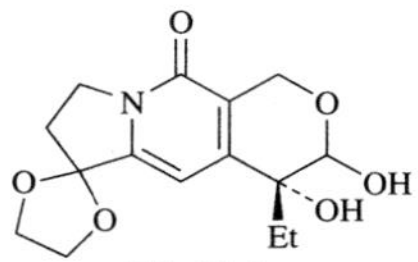		570

Ligand	Time	Yield
(DHQ)$_2$PHAL	5 d	(66), 78 de
(DHQ)$_2$PYR	18 h	(71), 94 de

Reactant	Conditions	Product(s) and Yield(s) (%)		Refs.
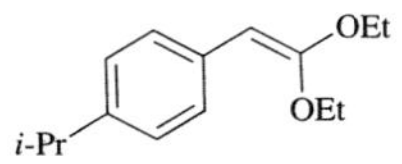	AD-Mix β, $MeSO_2NH_2$, t-BuOH:H_2O (1:1), 0°, 24 h	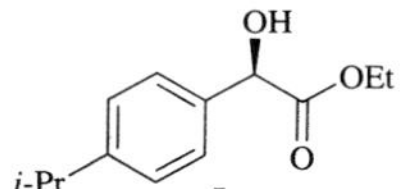 **I**	(63), 80	558
	AD-Mix β, $MeSO_2NH_2$, t-BuOH:CH_3CN:H_2O (4:1:5), 0°, 6 h	**I** (88), 80		136

TABLE 5. REACTIONS OF TRISUBSTITUTED ALKENES (*Continued*)

Substrate	Conditions	Product(s) and Yield(s) (%), % ee	Refs.
C_{15} 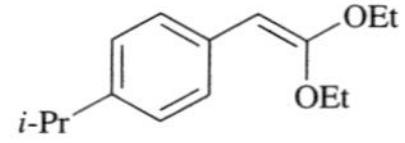	AD-Mix α, $Me_3SO_2NH_2$, t-BuOH:H_2O (1:1), 0°, 24 h	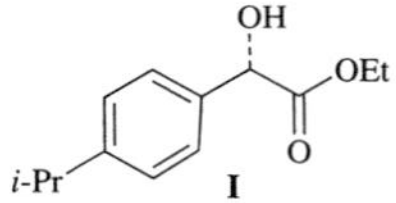 (88), 70	558
	AD-Mix α, $MeSO_2NH_2$, t-BuOH:CH_3CN:H_2O (4:1:5), 0°, 6 h	**I** (88), 70	136
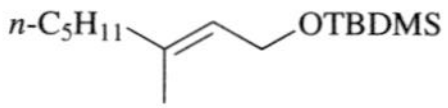	AD-Mix α, $MeSO_2NH_2$, t-BuOH:H_2O (1:1), 0°	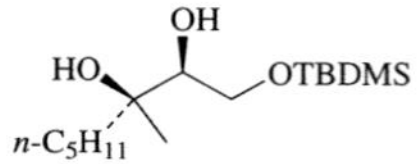 (97), 97	571
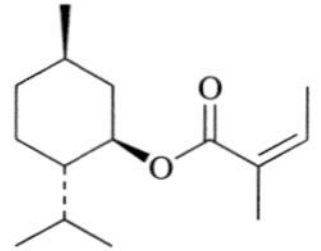	AD-Mix α, t-BuOH:H_2O (1:1)	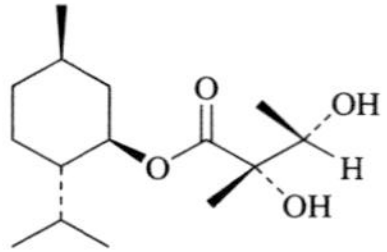 (91), 56 de	572, 573
	AD-Mix β, t-BuOH:H_2O (1:1)	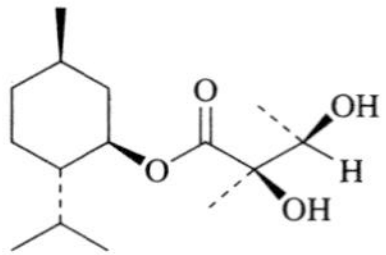 (92), 80 de	572, 573
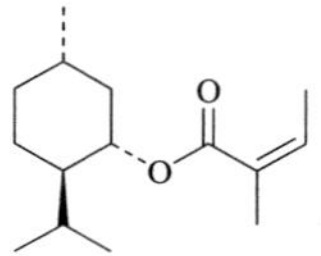	AD-Mix α, t-BuOH:H_2O (1:1)	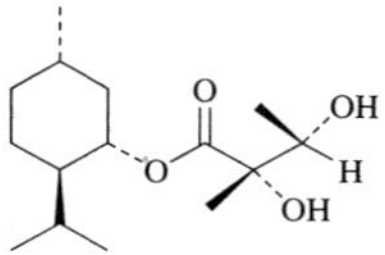 (88), 60 de	572, 573
	AD-Mix β, t-BuOH:H_2O (1:1)	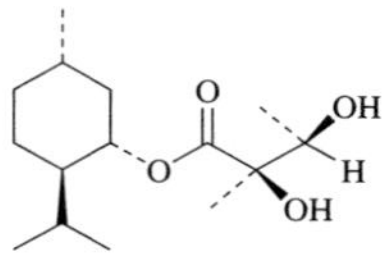 (90), 68 de	572, 573

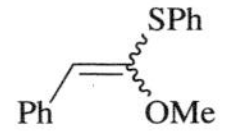	AD-Mix α, $MeSO_2NH_2$, t-BuOH:CH_3CN:H_2O (4:1:5), 0°, 6 h	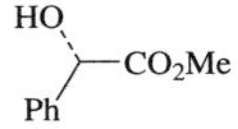	(37), 77	136
	AD-Mix β, $MeSO_2NH_2$, t-BuOH:CH_3CN:H_2O (4:1:5), 0°, 6 h	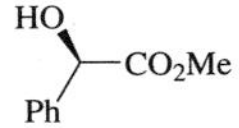	(37), 80	136
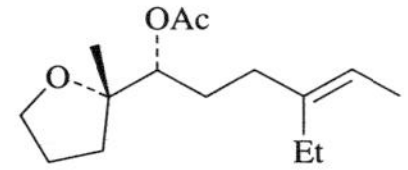	$(DHQD)_2PHAL$, $K_2OsO_2(OH)_4$, $K_3Fe(CN)_6$, $MeSO_2NH_2$, K_2CO_3, t-BuOH:H_2O, 0°, 36 h	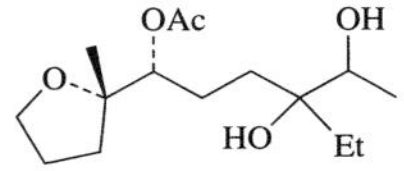	(73), 0 1:1 mixture	574
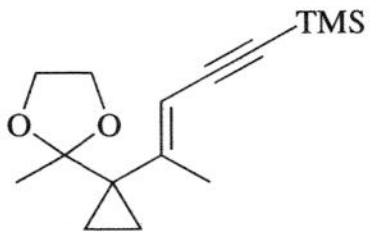	Ligand, $K_2OsO_2(OH)_4$, $K_3Fe(CN)_6$, K_2CO_3, $MeSO_2NH_2$, t-BuOH:H_2O (1:1), 48 h	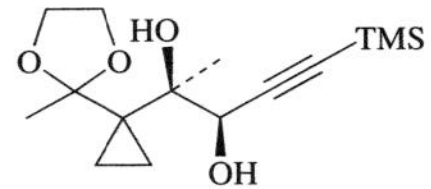		575
	Ligand			
	$(DHQD)_2PHAL$	(12-15), 83		
	$(DHQD)_2AQN$	(8), 51		
	DHQD-MEQ	(9), 82		
	$(DHQD)_2PYR$	(39), >95		
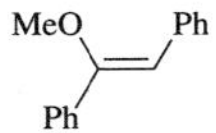	$(DHQD)_2PHAL$, $K_2OsO_2(OH)_4$, $K_3Fe(CN)_6$, K_2CO_3, $MeSO_2NH_2$, t-BuOH:H_2O (1:1)	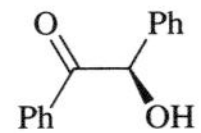	(—), 99	20
	$(DHQ)_2PHAL$, $K_2OsO_2(OH)_4$, $K_3Fe(CN)_6$, K_2CO_3, $MeSO_2NH_2$, t-BuOH:H_2O (1:1)	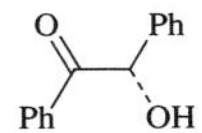	(—), 98	20
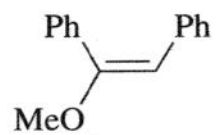	$(DHQD)_2PHAL$, $K_2OsO_2(OH)_4$, $K_3Fe(CN)_6$, K_2CO_3, $MeSO_2NH_2$, t-BuOH:H_2O (1:1)	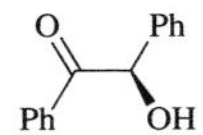	(—), 90	20

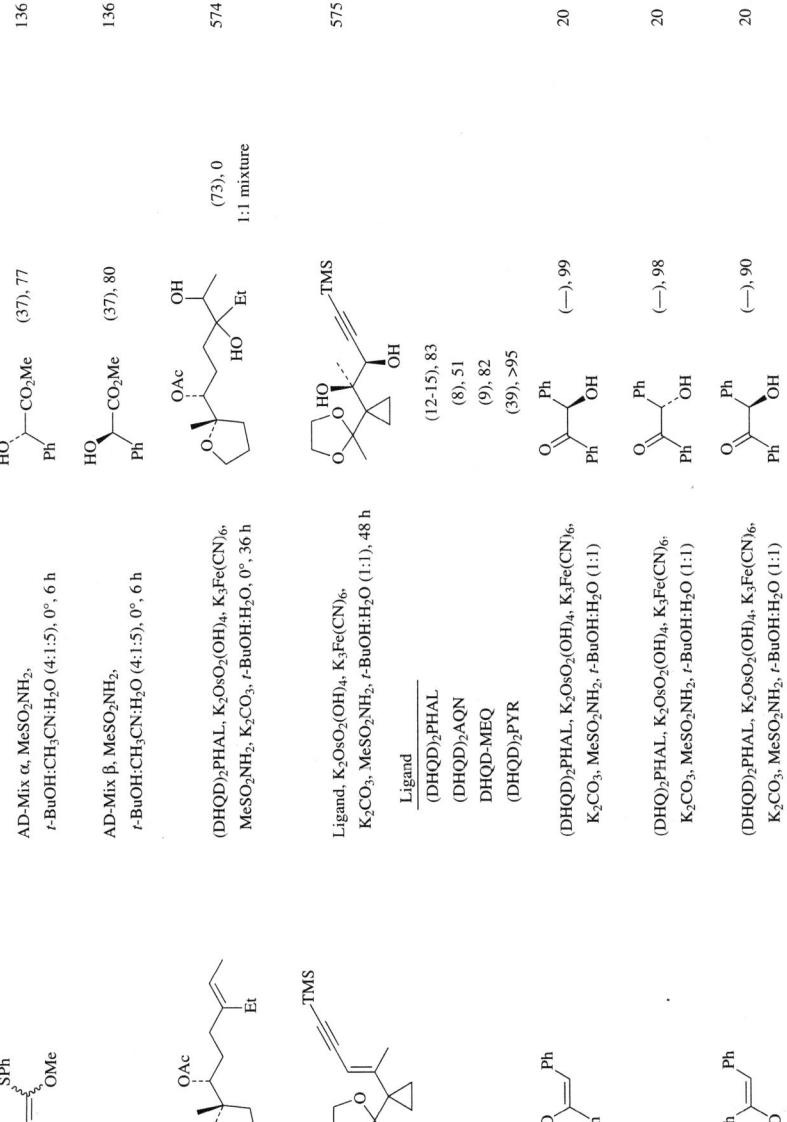

TABLE 5. REACTIONS OF TRISUBSTITUTED ALKENES (*Continued*)

Substrate	Conditions	Product(s) and Yield(s) (%), % ee	Refs.
C_{15} 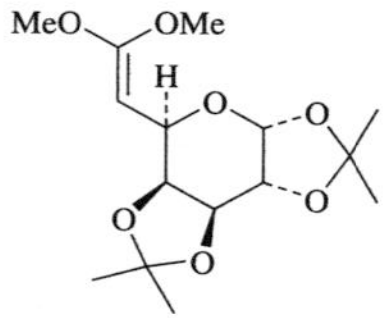	AD-Mix α, $MeSO_2NH_2$, t-BuOH:CH_3CN:H_2O (4:1:5), 0°, 6 h	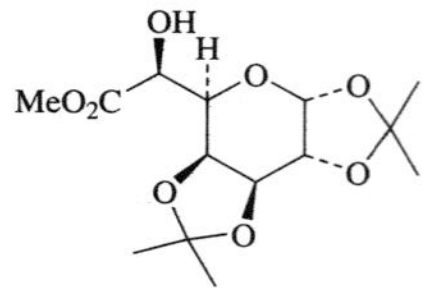 (81), >90 de	136
	AD-Mix β, $MeSO_2NH_2$, t-BuOH:CH_3CN:H_2O (4:1:5), 0°, 6 h	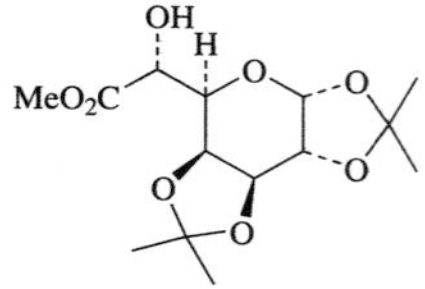 (81), >90 de	136
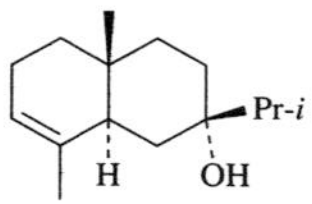	AD-Mix α, $MeSO_2NH_2$, t-BuOH:H_2O (1:1), 0°, 32 h	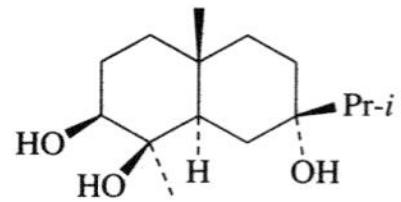 (73), >95 de	576
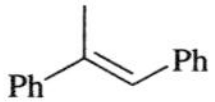	$(DHQD)_2PHAL$, $K_2OsO_2(OH)_4$, $K_3Fe(CN)_6$, K_2CO_3, t-BuOH:H_2O (1:1), rt, 21 h	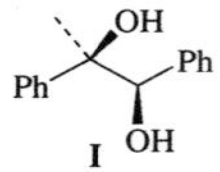 **I** (82), 99	71
	$(DHQD)_2PHAL$, $K_2OsO_2(OH)_4$, $K_3Fe(CN)_6$, pH 12, t-BuOH:H_2O (1:1), rt, 1.5 h	**I** (62), 99	71

Reactant	Conditions	Product(s) and Yield(s) (%)	Refs.
C_{16-17}			
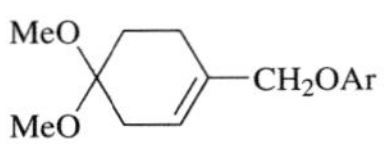	(DHQ)$_2$PHAL, $K_2OsO_2(OH)_4$, $K_3Fe(CN)_6$, K_2CO_3, $MeSO_2NH_2$, t-BuOH:H_2O (1:1), 0°	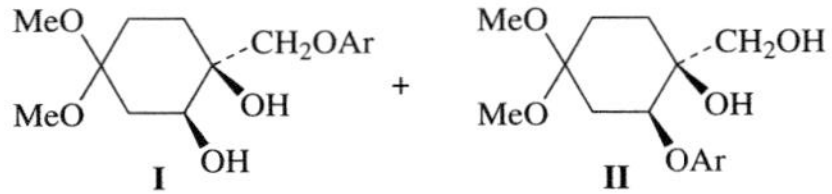 I + II	61

Ar	I	I+II	I:II
Bn	(—), 90	(—)	(—)
4-$MeOC_6H_4$	(—), 96	(—)	(—)
Bz	(—)	(—), 99	6.5:1
$COC_6H_4NO_2$-4	(—)	(—), 99	7:1
COC_6H_4Br-4	(—)	(—), 99	7:1
4-$MeOC_6H_4CO$	(—)	(93), 99	32:1

Reactant	Conditions	Product(s) and Yield(s) (%)	Refs.
C_{16}			
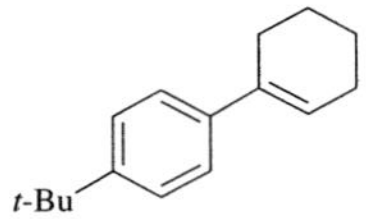	(DHQD)$_2$PHAL, $K_2OsO_2(OH)_4$, $K_3Fe(CN)_6$, K_2CO_3, $MeSO_2NH_2$, t-BuOH:H_2O (1:1), rt, 96 h	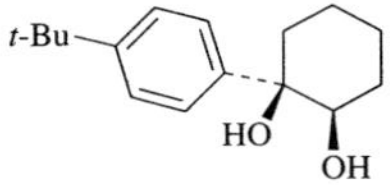 (49), 93	566
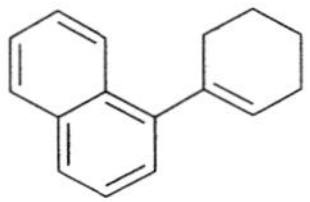	(DHQD)$_2$PHAL, $K_2OsO_2(OH)_4$, $K_3Fe(CN)_6$, K_2CO_3, $MeSO_2NH_2$, t-BuOH:H_2O (1:1), 0°-rt, 96 h	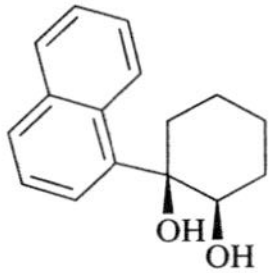 (~5), 86	566, 20
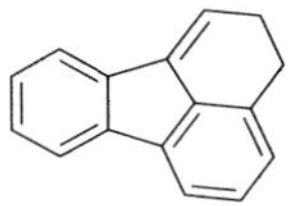	(DHQD)$_2$PYDZ, $K_2OsO_2(OH)_4$, $K_3Fe(CN)_6$, K_2CO_3, $MeSO_2NH_2$, t-BuOH:H_2O (1:1), 0°, 48 h	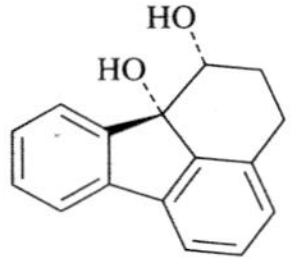 (91), 95	66
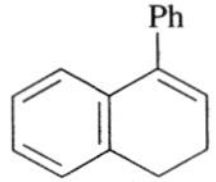	(DHQD)$_2$PYDZ, $K_2OsO_2(OH)_4$, $K_3Fe(CN)_6$, K_2CO_3, $MeSO_2NH_2$, t-BuOH:H_2O (1:1), 0°, 72 h	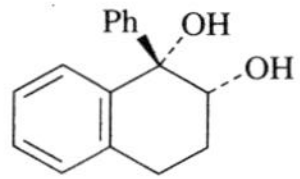 (79), 77	66

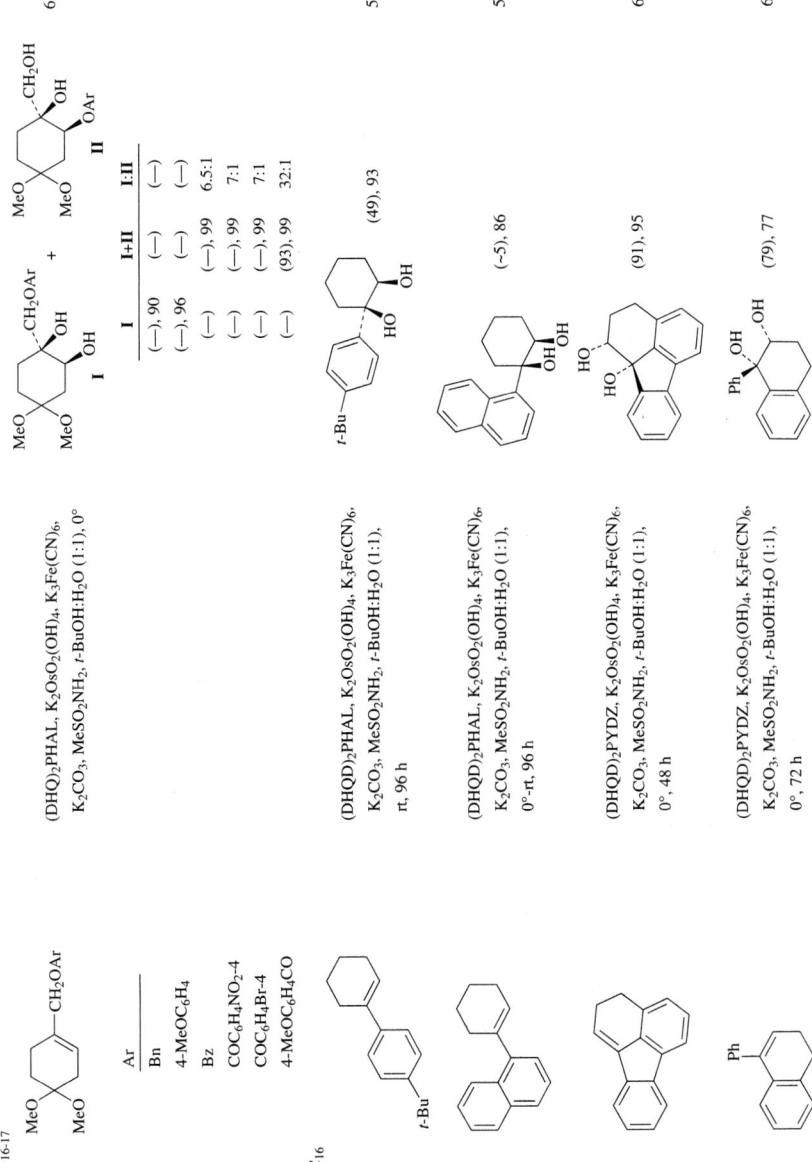

TABLE 5. REACTIONS OF TRISUBSTITUTED ALKENES (*Continued*)

Substrate	Conditions	Product(s) and Yield(s) (%), % ee	Refs.
C_{16} 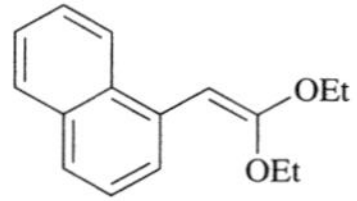	AD-Mix β, $MeSO_2NH_2$, *t*-BuOH:H_2O (1:1), 0°, 24 h	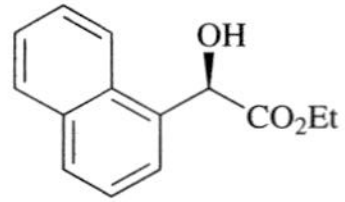 (57), 95	558
	AD-Mix α, $MeSO_2NH_2$, *t*-BuOH:H_2O (1:1), 0°, 24 h	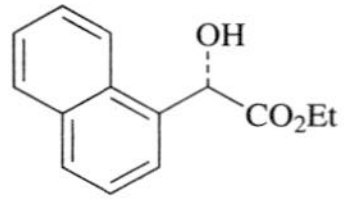 (68), 93	558
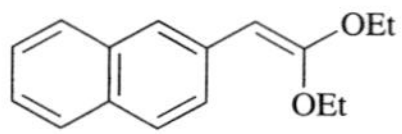	AD-Mix α, $MeSO_2NH_2$, *t*-BuOH:CH_3CN:H_2O (4:1:5), 0°, 6 h	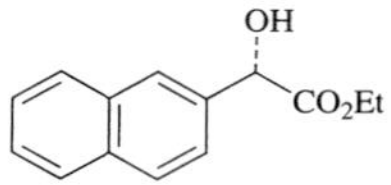 (68), 93	136
	AD-Mix β, $MeSO_2NH_2$, *t*-BuOH:CH_3CN:H_2O (4:1:5), 0°, 6 h	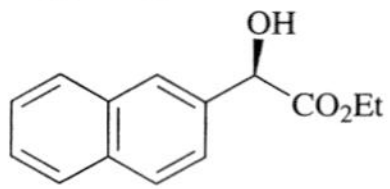 (68), 95	136
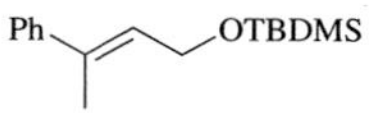	AD-Mix α, $MeSO_2NH_2$, *t*-BuOH:H_2O (1:1), 0°	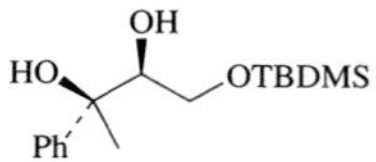 (99), >99	571
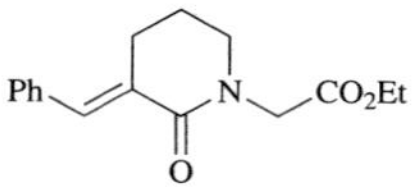	AD-Mix α, $MeSO_2NH_2$, *t*-BuOH:H_2O (1:1), rt	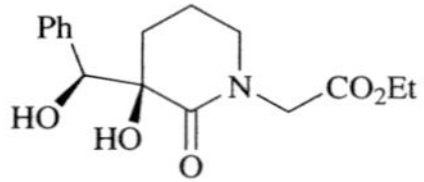 (69), >93	577
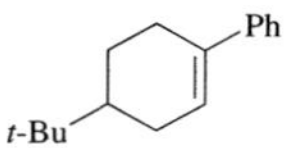	AD-Mix β, $MeSO_2NH_2$, *t*-BuOH:H_2O (1:1), 0°	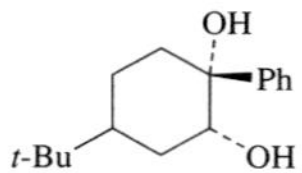 (84), >95 ee, 0 de	567

	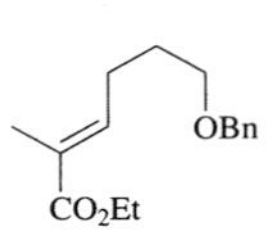	AD-Mix β, $MeSO_2NH_2$, t-BuOH:H_2O (1:1), rt, 12 h	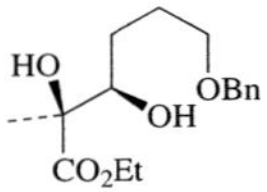	(96), >98	548
	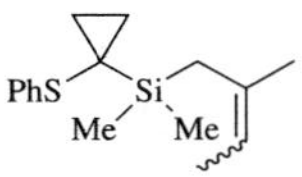	AD-Mix β, $MeSO_2NH_2$, t-BuOH:H_2O (1:1), rt, 48 h	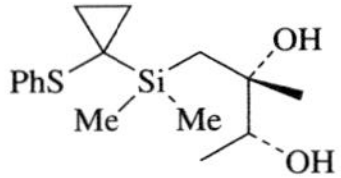	(96), 25	578
C_{17}		$(DHQ)_2PHAL$, $K_2OsO_2(OH)_4$, $K_3Fe(CN)_6$, K_2CO_3, $MeSO_2NH_2$, t-BuOH:H_2O (1:1), 0°	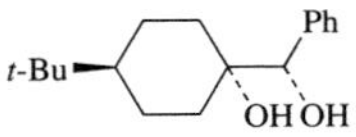	(78), 96 de	149
		$(DHQD)_2PHAL$, $K_2OsO_2(OH)_4$, $K_3Fe(CN)_6$, K_2CO_3, $MeSO_2NH_2$, t-BuOH:H_2O (1:1), 0°	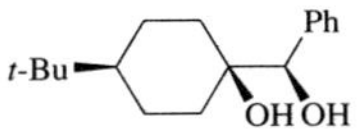	(67), 96 de	149
		$(DHQ)_2PHAL$, $K_2OsO_2(OH)_4$, $K_3Fe(CN)_6$, K_2CO_3, $MeSO_2NH_2$, t-BuOH:H_2O (1:1), 0°	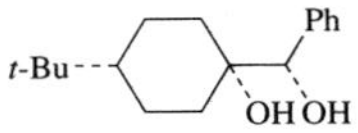	(60), 96 de	149
		$(DHQD)_2PHAL$, $K_2OsO_2(OH)_4$, $K_3Fe(CN)_6$, K_2CO_3, $MeSO_2NH_2$, t-BuOH:H_2O (1:1), 0°	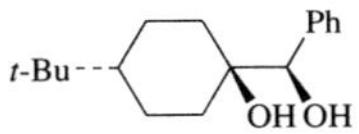	(67), 97 de	149
	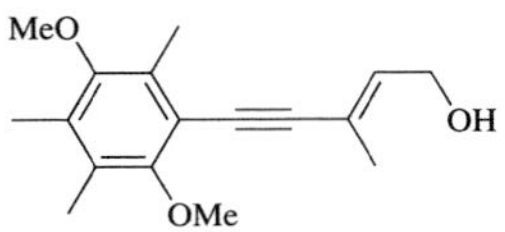	$(DHQ)_2PHAL$, $K_2OsO_2(OH)_4$, $K_3Fe(CN)_6$, K_2CO_3, t-BuOH:H_2O (1:1), 4°, 8-24 h	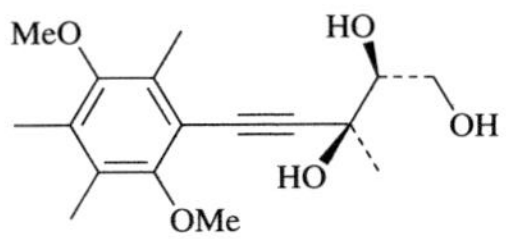	(—), >95	320

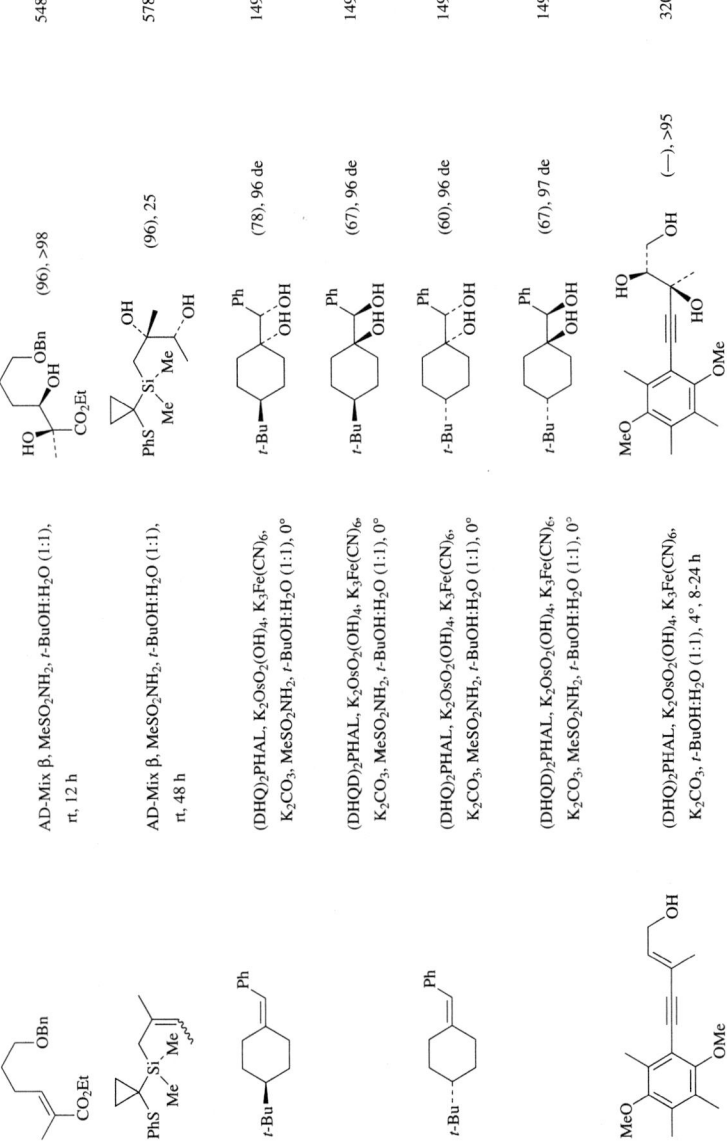

TABLE 5. REACTIONS OF TRISUBSTITUTED ALKENES (*Continued*)

Substrate	Conditions	Product(s) and Yield(s) (%), % ee	Refs.
C_{17}			
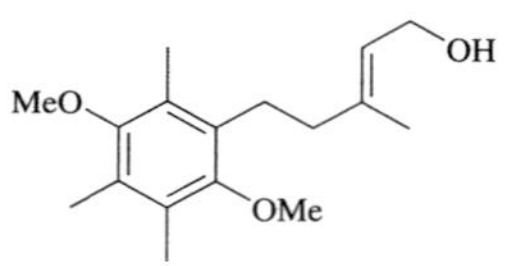	(DHQ)$_2$PHAL, $K_2OsO_2(OH)_4$, $K_3Fe(CN)_6$, K_2CO_3, *t*-BuOH:H_2O (1:1), 4°, 8-24 h	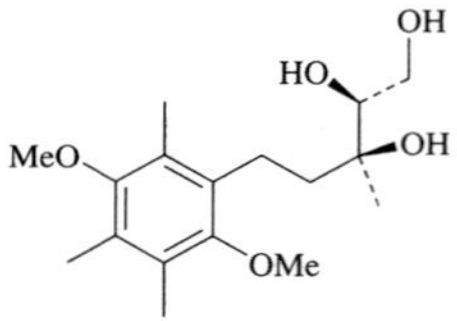 (92), 18	320
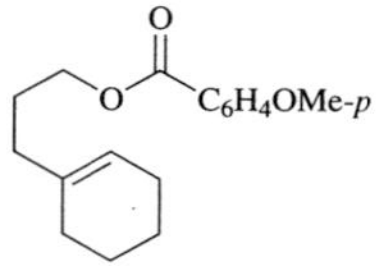	(DHQD)$_2$PYDZ, $K_2OsO_2(OH)_4$, $K_3Fe(CN)_6$, K_2CO_3, $MeSO_2NH_2$, *t*-BuOH:H_2O (1:1), 0°, 4-22 h	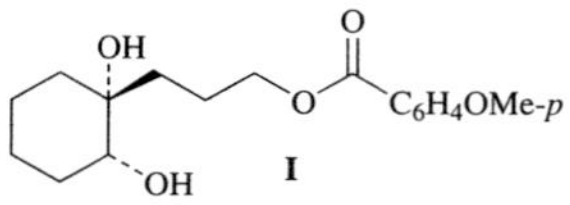 **I** (100), 89	82
	21, $K_2OsO_2(OH)_4$, $K_3Fe(CN)_6$, $MeSO_2NH_2$, K_2CO_3, *t*-BuOH:H_2O (1:1), 0°, 4-22 h	**I** (94), 91	82
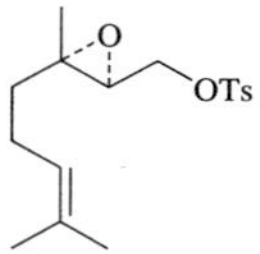	AD-Mix β, $MeSO_2NH_2$, *t*-BuOH:H_2O (1:1), rt, 72 h	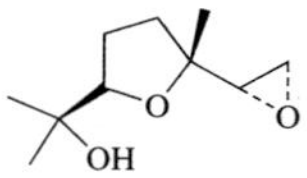 (84), 88 de	579
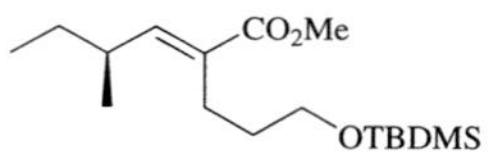	(DHQD)$_2$PHAL, $K_2OsO_2(OH)_4$, $K_2Fe(CN)_6$, K_2CO_3, $MeSO_2NH_2$, *t*-BuOH:H_2O (1:1), 0°, 6 d	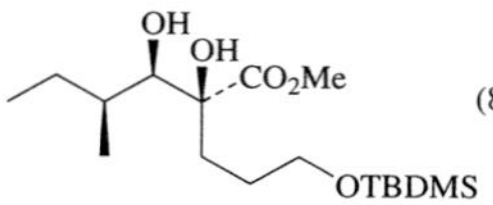 (89), 98	580
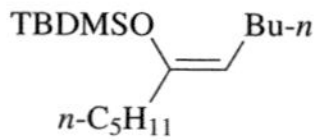	(DHQD)$_2$PHAL, $K_2OsO_2(OH)_4$, $K_3Fe(CN)_6$, K_2CO_3, $MeSO_2NH_2$, *t*-BuOH:H_2O (1:1)	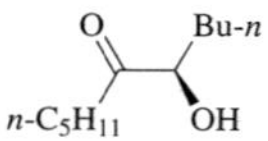 (—), 89	20
	(DHQ)$_2$PHAL, $K_2OsO_2(OH)_4$, $K_3Fe(CN)_6$, K_2CO_3, $MeSO_2NH_2$, *t*-BuOH:H_2O (1:1)	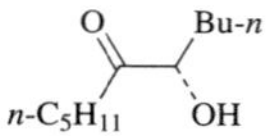 (—), 86	20

C_{18}					
	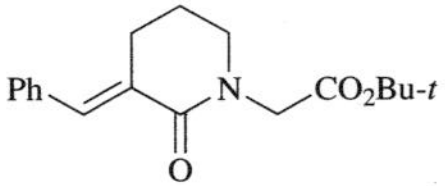	AD-Mix α, $MeSO_2NH_2$, t-BuOH:H_2O (1:1), rt	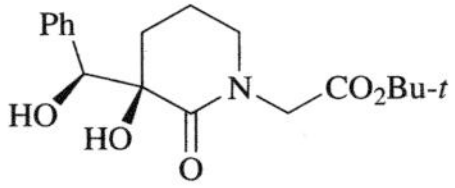	(46), >93	577
	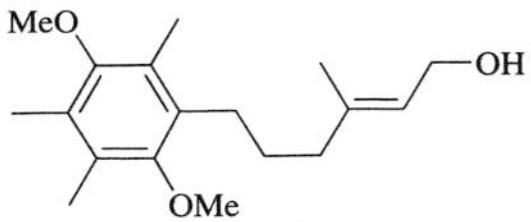	AD-Mix α, t-BuOH:H_2O (1:1), 4°, 48 h	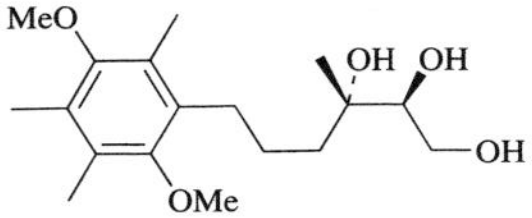	(92), 18	313
	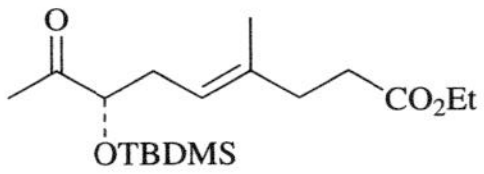	AD-Mix β, $MeSO_2NH_2$, t-BuOH:H_2O (1:1), rt, 24 h	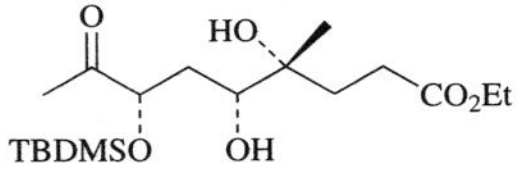	(>78), 50 de	578a
	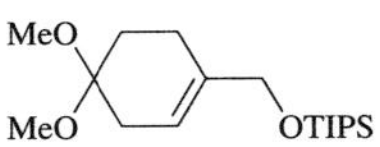	$(DHQ)_2PHAL$, $K_2OsO_2(OH)_4$, $K_3Fe(CN)_6$, K_2CO_3, $MeSO_2NH_2$, t-BuOH:H_2O (1:1), 0°	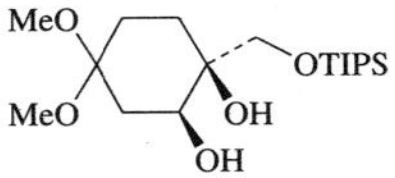	(—), 13	61, 133
C_{19}					
	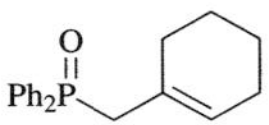	$(DHQ)_2PHAL$, $K_2OsO_2(OH)_4$, $K_3Fe(CN)_6$, K_2CO_3, $MeSO_2NH_2$, t-BuOH:H_2O (1:1), rt	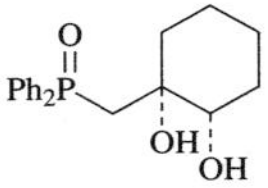	(73), 18	120, 119
		Ligand, $K_2OsO_2(OH)_4$, $K_3Fe(CN)_6$, K_2CO_3, $MeSO_2NH_2$, t-BuOH:H_2O (1:1), rt	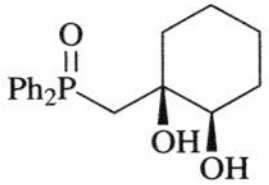		
		Ligand			
		$(DHQD)_2PHAL$		(62), 14	119
		$(DHQD)_2$-p-chlorobenzoate		(60), 62	119, 120
		$(DHQD)_2PHN$		(62), 38	119, 120

TABLE 5. REACTIONS OF TRISUBSTITUTED ALKENES (*Continued*)

Substrate	Conditions	Product(s) and Yield(s) (%), % ee	Refs.
C_{19}			
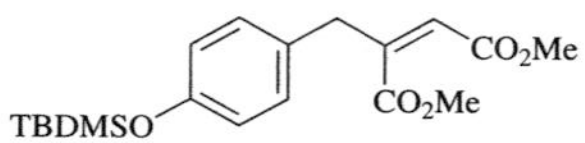	Ligand, $K_2OsO_2(OH)_4$, $K_3Fe(CN)_6$, K_2CO_3, $MeSO_2NH_2$, t-BuOH:H_2O (1:1), 4°	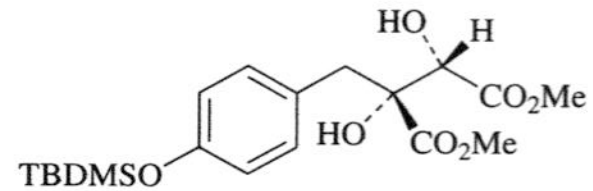	581
	Ligand / **Time**		
	$(DHQD)_2$PHAL / 14 h	(43), 27	
	$(DHQD)_2$PYR / 24 h	(92), 89	
	$(DHQD)_2$AQN / 24 h	(68), 93	
	(DHQD)-p-chlorobenzoate / 72 h	(61), 20	
	DHQD-IND / 72 h	(36), 74	
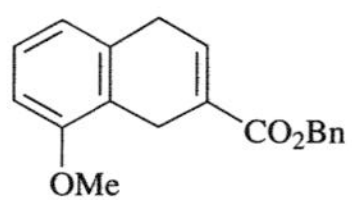	$(DHQ)_2$PHAL, $K_2OsO_2(OH)_2$, $K_3Fe(CN)_6$, K_2CO_3, $MeSO_2NH_2$, t-BuOH:H_2O:toluene (16.7:16.7:2), 4°, 5 h	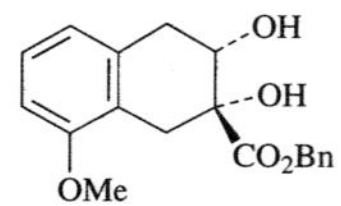 (74), 92	565
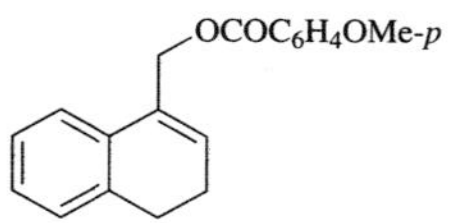	$(DHQD)_2$PYDZ, K_2OsO_4, $K_3Fe(CN)_6$, K_2CO_3, $MeSO_2NH_2$, t-BuOH:H_2O (1:1), 0°, 2.75 h	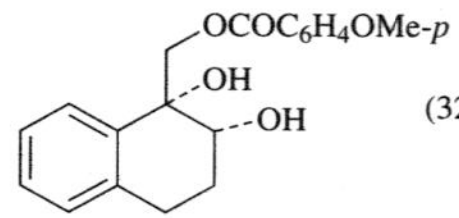 (32-90), 94-98	66
C_{20}			
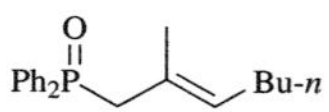	DHQD-p-chlorobenzoate, $K_2OsO_2(OH)_4$, $K_3Fe(CN)_6$, K_2CO_3, $MeSO_2NH_2$, t-BuOH:H_2O (1:1)	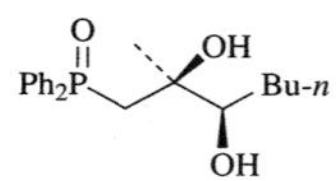 (103), 74	119

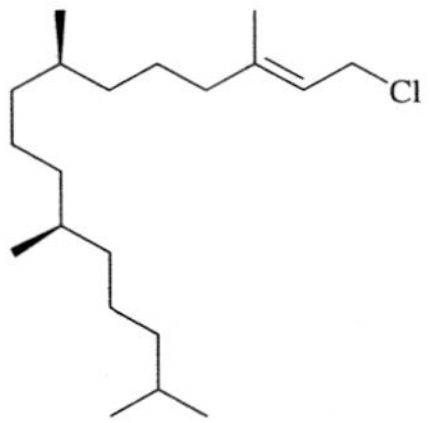	(DHQD)$_2$PHAL, $K_2OsO_2(OH)_4$, $K_3Fe(CN)_6$, K_2CO_3, $MeSO_2NH_2$, t-BuOH:H_2O (1:1), 0°, 32 h	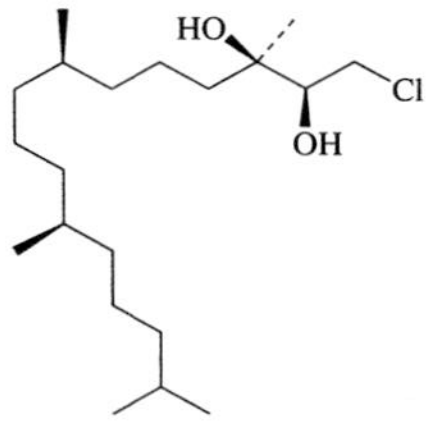 (78), >95 de	343
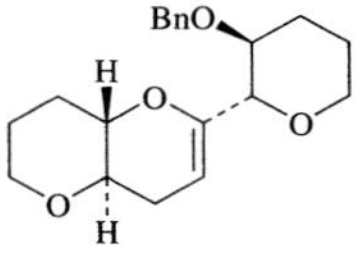	AD-Mix α, $MeSO_2NH_2$, t-BuOH:H_2O (1:1), 0°, 36 h	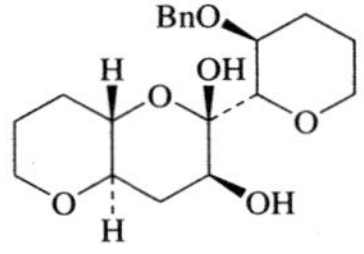 (98), 86 de	582
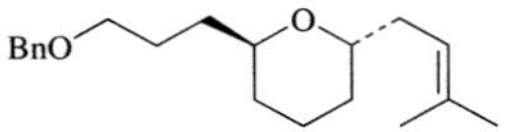	AD-Mix α, $MeSO_2NH_2$, t-BuOH:H_2O (1:1), 0°	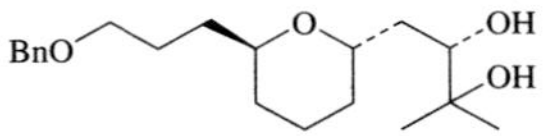 (70), 60 de	273
	AD-Mix β, $MeSO_2NH_2$, t-BuOH:H_2O (1:1), 0°	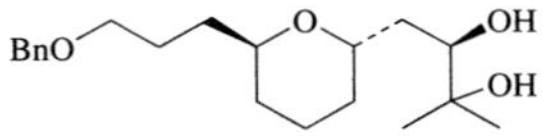 (77), 82 de	273
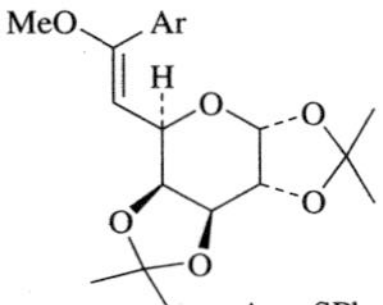 Ar = SPh	AD-Mix α, $MeSO_2NH_2$, t-BuOH:CH_3CN:H_2O (4:1:5), 0°, 6 h	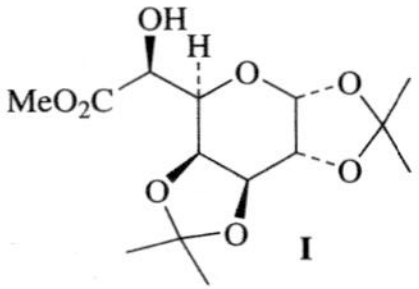 **I** (9), 33 de	136
Ar = S(O)Ph	AD-Mix β, $MeSO_2NH_2$, t-BuOH:CH_3CN:H_2O (4:1:5), 0°, 6 h	**I** (31), 0 de	136

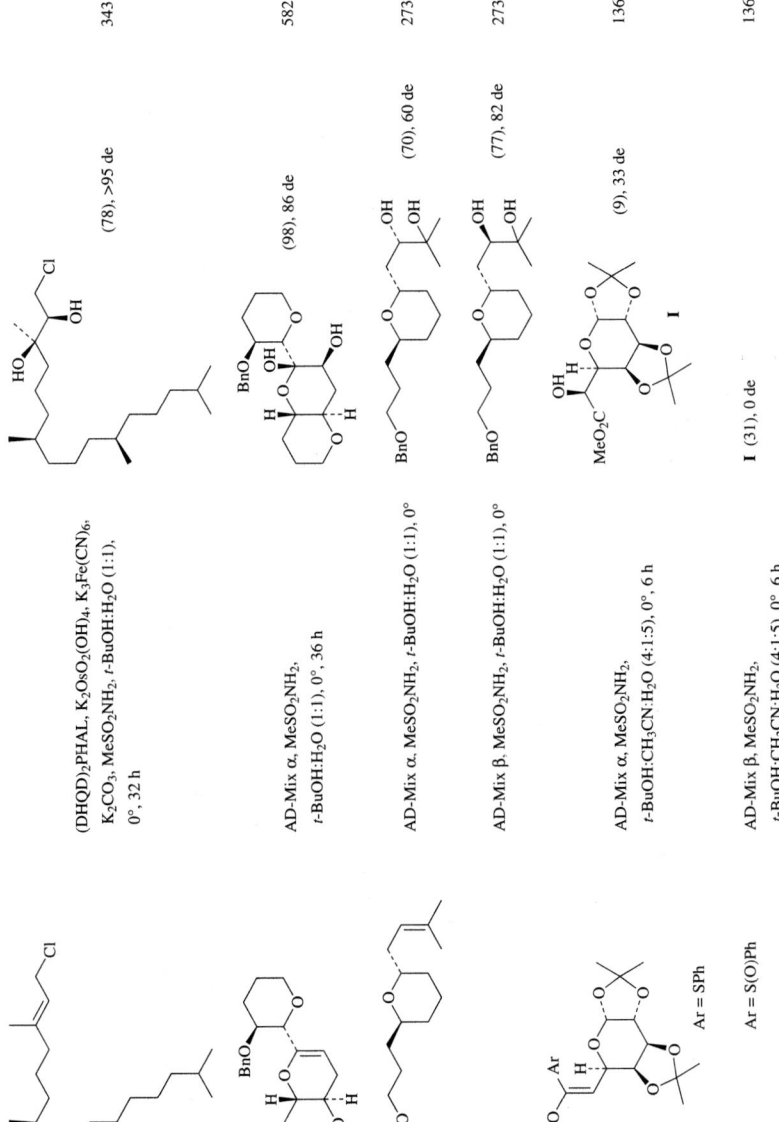

TABLE 5. REACTIONS OF TRISUBSTITUTED ALKENES (*Continued*)

Substrate	Conditions	Product(s) and Yield(s) (%), % ee	Refs.
C_{20}			
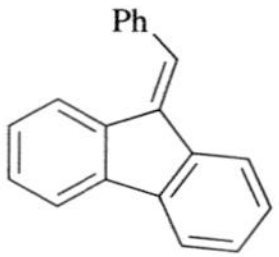	$(DHQD)_2PYDZ$, K_2OsO_4, $K_3Fe(CN)_6$, K_2CO_3, $MeSO_2NH_2$, *t*-BuOH:H_2O (1:1), 0°, 12 h	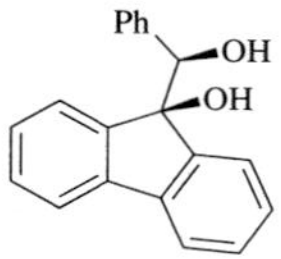 (61), 97	66
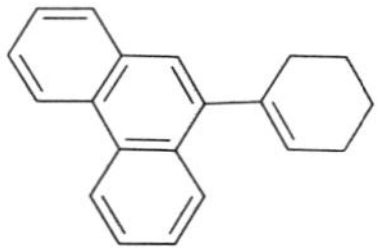	$(DHQD)_2PHAL$, $K_2OsO_2(OH)_4$, $K_3Fe(CN)_6$, K_2CO_3, $MeSO_2NH_2$, *t*-BuOH:H_2O (1:1), 0°, 6-24 h	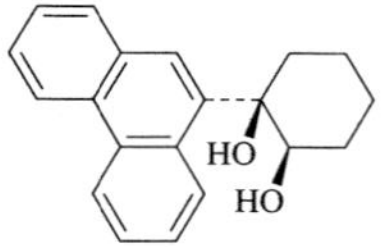 (—), 74	20
C_{21}			
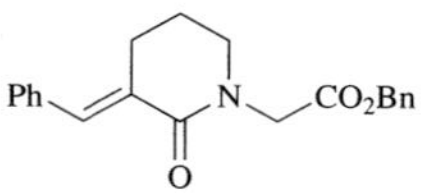	AD-Mix α, $MeSO_2NH_2$, *t*-BuOH:H_2O (1:1), rt	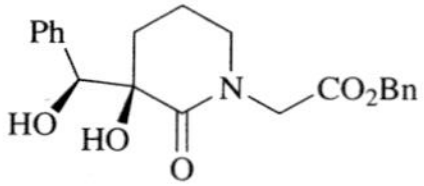 (61), >93	577
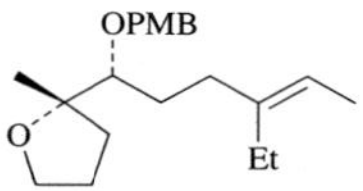	$(DHQD)_2PHAL$, $K_2OsO_2(OH)_4$, $K_3Fe(CN)_6$, $MeSO_2NH_2$, K_2CO_3, *t*-BuOH-H_2O, 0°, 120 h	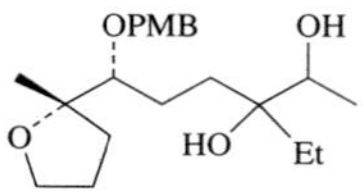 (92), 0 de	574
C_{22}			
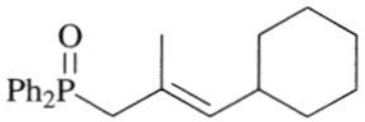	DHQD-*p*-chlorobenzoate, $K_2OsO_2(OH)_4$, $K_3Fe(CN)_6$, K_2CO_3, $MeSO_2NH_2$, *t*-BuOH:H_2O (1:1)	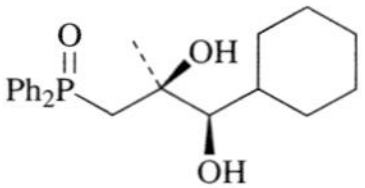 (96), 84	119

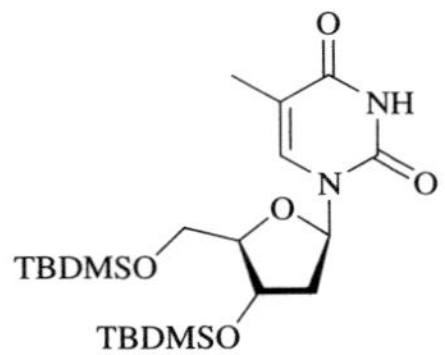	(DHQ)$_2$PHAL, $K_2OsO_2(OH)_4$, $K_3Fe(CN)_6$, K_2CO_3, $MeSO_2NH_2$, t-BuOH:H_2O (1:1), rt, 72 h	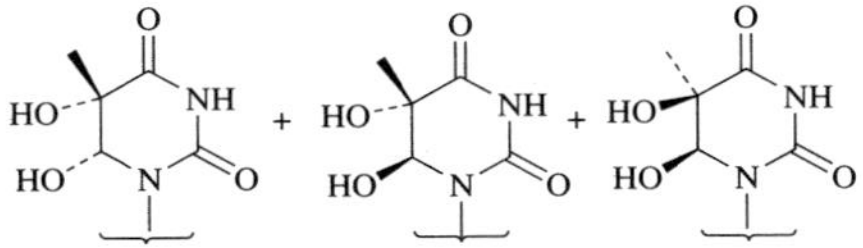 **I** + **II** + **III** (83) **I**:**II**:**III** = 13:4:1	583
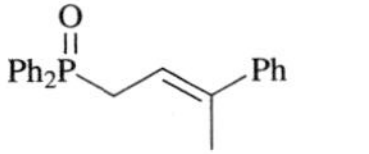	DHQD-p-chlorobenzoate, $K_2OsO_2(OH)_4$, $K_3Fe(CN)_6$, K_2CO_3, $MeSO_2NH_2$, t-BuOH:H_2O (1:1)	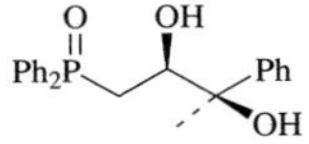 (64), 42	119
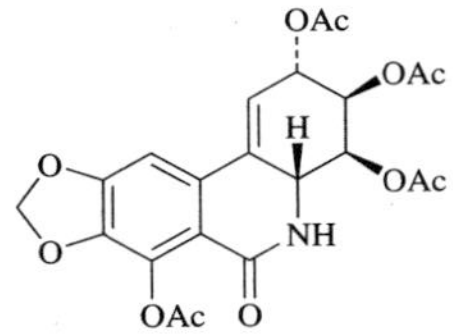	DHQ-p-chlorobenzoate, OsO_4, NMO, DMF:H_2O (9:1), 0°-rt	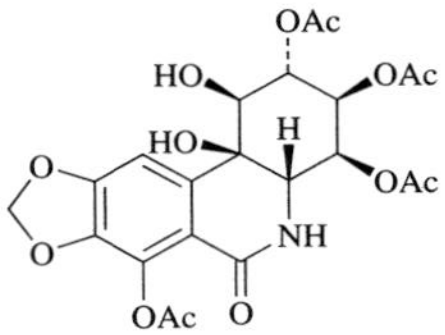 (72), >95 de	584
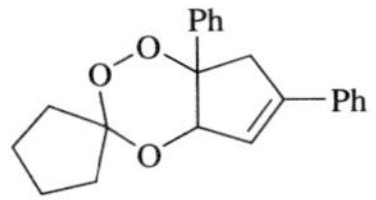	AD-Mix β, $MeSO_2NH_2$, t-BuOH:H_2O (1:1), 0°	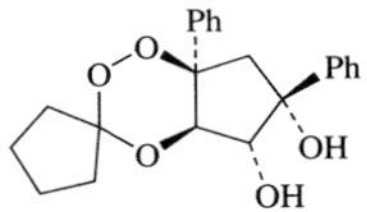 (—), 35 ee	585
	AD-Mix α, $MeSO_2NH_2$, t-BuOH:H_2O (1:1), 0°	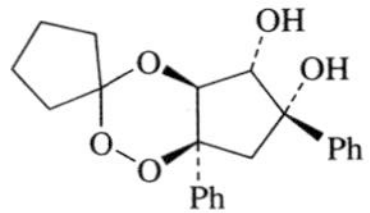 (—), 35 ee	585

TABLE 5. REACTIONS OF TRISUBSTITUTED ALKENES (*Continued*)

Substrate	Conditions	Product(s) and Yield(s) (%), % ee	Refs.
C_{24} 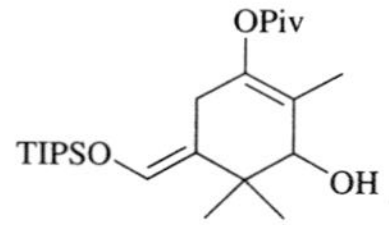	1. DHQ-PHN, $K_2OsO_2(OH)_4$, $K_3Fe(CN)_6$, K_2CO_3, $MeSO_2NH_2$, *t*-BuOH:H_2O (1:1), 0°, 11 h 2. $(MeNHCH_2)_2$, SiO_2, benzene, reflux, 0.5 h	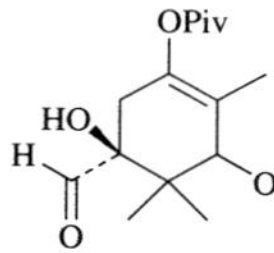 (96), 90	586
C_{25} 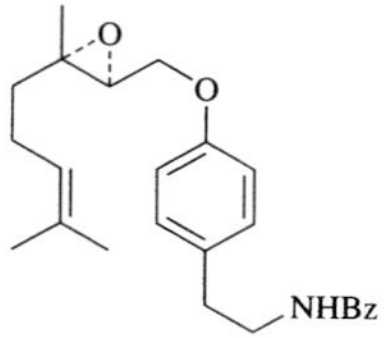	AD-Mix α, $MeSO_2NH_2$, *t*-BuOH:H_2O (1:1), rt, 72 h	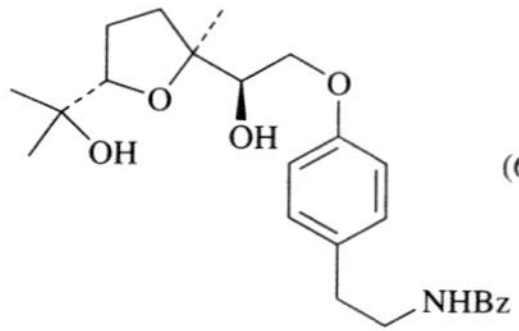 (65), 94 de	579
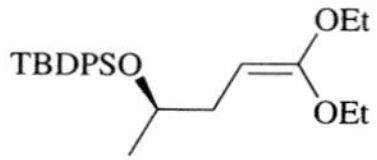	AD-Mix β, $MeSO_2NH_2$, *t*-BuOH:H_2O (1:1), 0°, 24 h	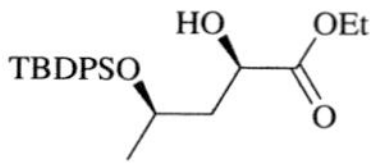 (32), >95 de	558
	AD-Mix α, $MeSO_2NH_2$, *t*-BuOH:H_2O (1:1), 0°, 24 h	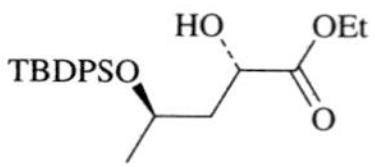 (94), >95 de	558
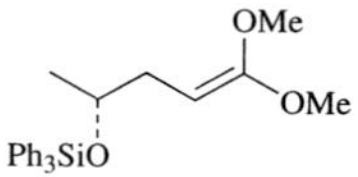	AD-Mix α, $MeSO_2NH_2$, *t*-BuOH:CH_3CN:H_2O (4:1:5), 0°, 6 h	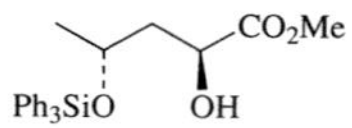 (94), >95 de	136
	AD-Mix β, $MeSO_2NH_2$, *t*-BuOH:CH_3CN:H_2O (4:1:5), 0°, 6 h	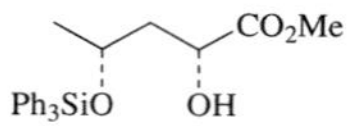 (94), >95 de	136

C_{26}

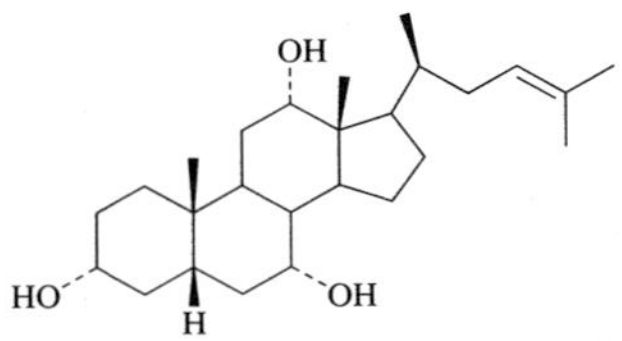

AD-Mix β, $MeSO_2NH_2$, t-BuOH:H_2O (1:1), 0°

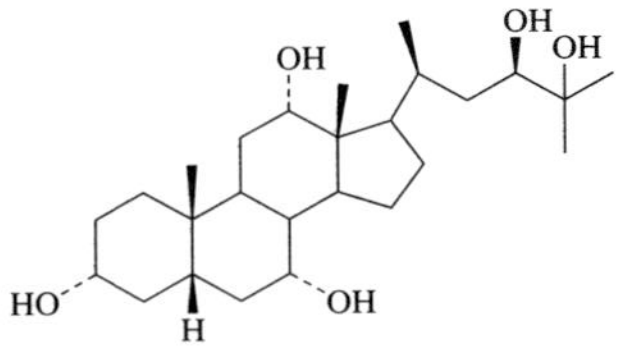

(80), 60 de 587

AD-Mix α, $MeSO_2NH_2$, t-BuOH:H_2O (1:1), 0°

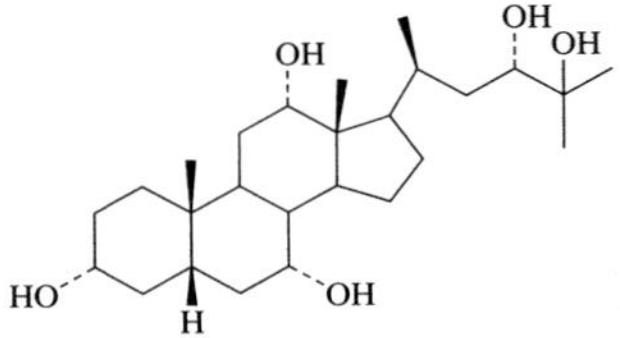

(55), 30 de 587

C_{27}

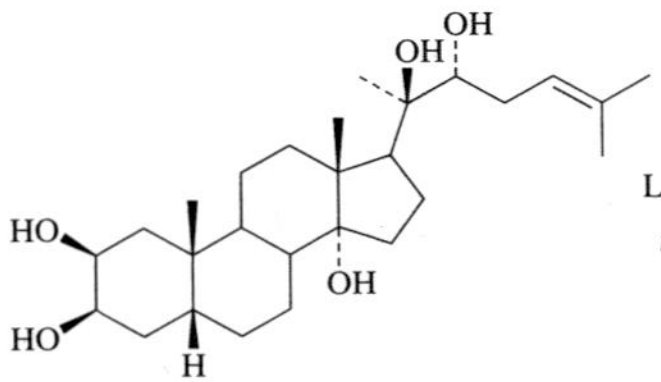

Ligand, OsO_4, t-BuOH:THF:H_2O (7:4:1), rt, 10 min

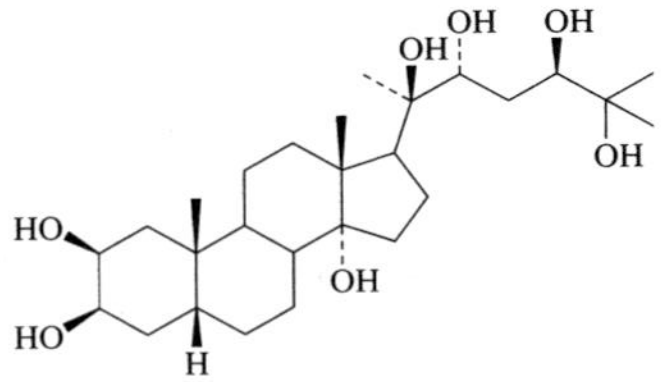

588

Ligand	
DHQD-MEQ	(89), 60
DHQD-PHN	(88), 60
$(DHQD)_2$PHAL	(84), 75
$(DHQD)_2$PYR	(82), 71
DHQ-MEQ	(80), 17
DHQ-PHN	(80), 17
$(DHQ)_2$PHAL	(76), 60
$(DHQ)_2$PYR	(76), 60

TABLE 5. REACTIONS OF TRISUBSTITUTED ALKENES (*Continued*)

Substrate	Conditions	Product(s) and Yield(s) (%), % ee		Refs.
C_{27} 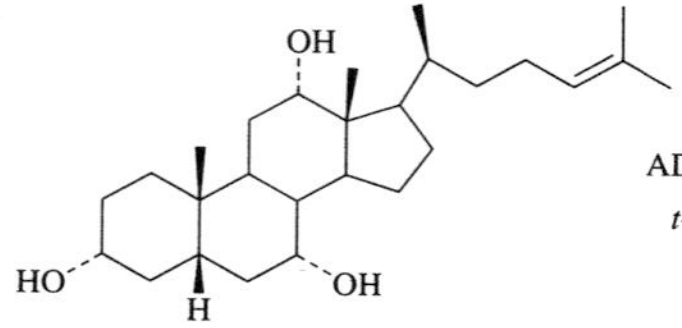	AD-Mix β, $MeSO_2NH_2$, *t*-BuOH:H_2O (1:1), 0°	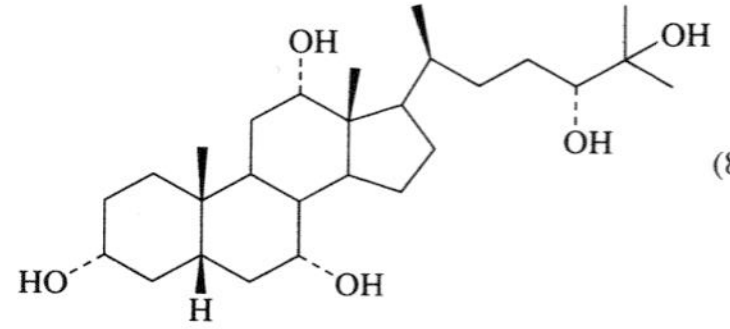	(89), —	587
	AD-Mix α, $MeSO_2NH_2$, *t*-BuOH:H_2O (1:1), 0°	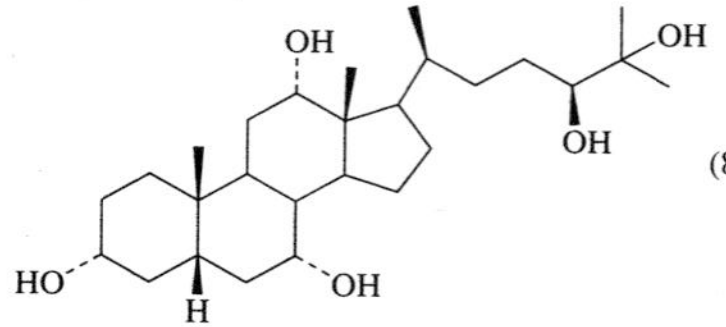	(89), —	587
C_{29} 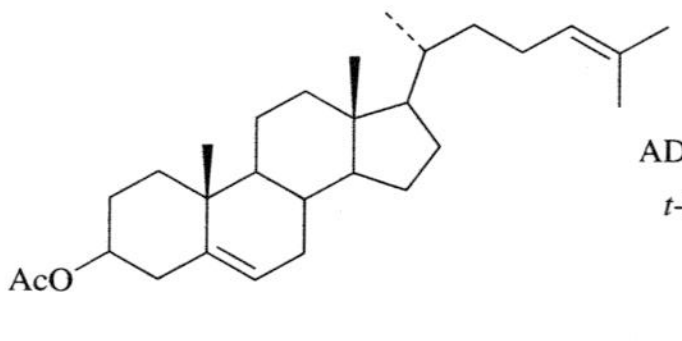	AD-Mix β, $MeSO_2NH_2$, *t*-BuOH:H_2O (1:1), rt, 4 d	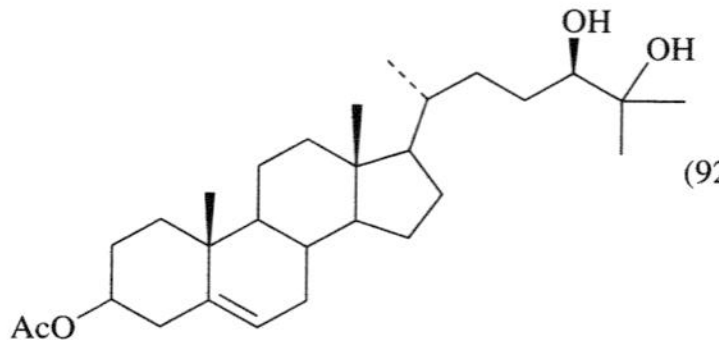	(92), >95 de	589
	AD-Mix α, $MeSO_2NH_2$, *t*-BuOH:H_2O (1:1), rt, 4 d	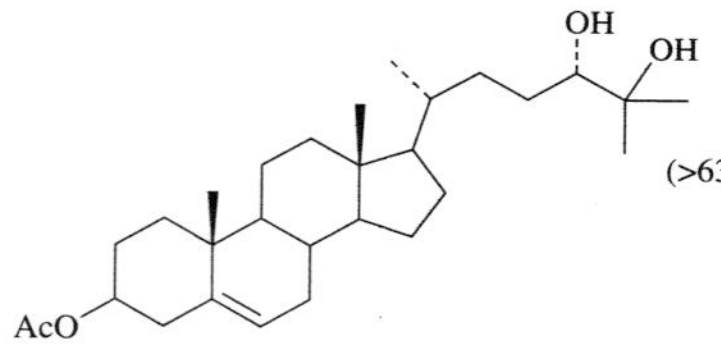	(>63), >95 de	589

C_{30} 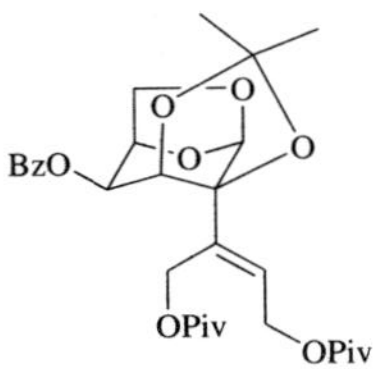	DHQD-*p*-chlorobenzoate, $K_2OsO_2(OH)_4$, $K_3Fe(CN)_6$, K_2CO_3, *t*-BuOH:H_2O (1:1), rt, 4 h	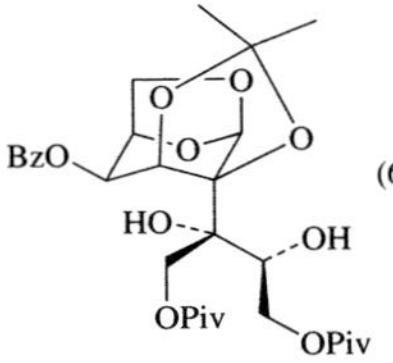 (62), 94 de	590
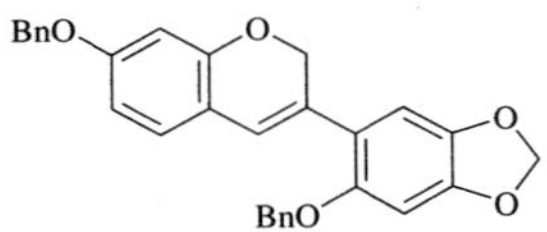	DHQ-*p*-chlorobenzoate, OsO_4, toluene, rt	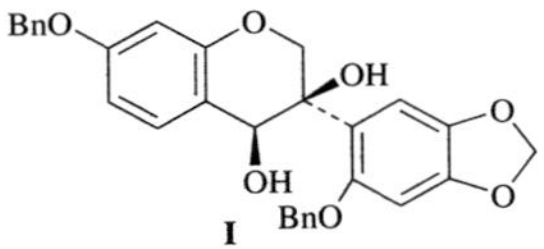 **I** (90), 80	591
	DHQ-*p*-chlorobenzoate, OsO_4, CH_2Cl_2, –78°	**I** (90), 94	591
C_{31} 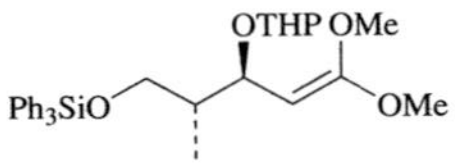	AD-Mix α, $MeSO_2NH_2$, *t*-BuOH:CH_3CN:H_2O (4:1:5), 0°, 6 h	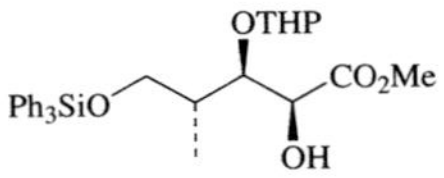 (65), >85 de	136
C_{33} 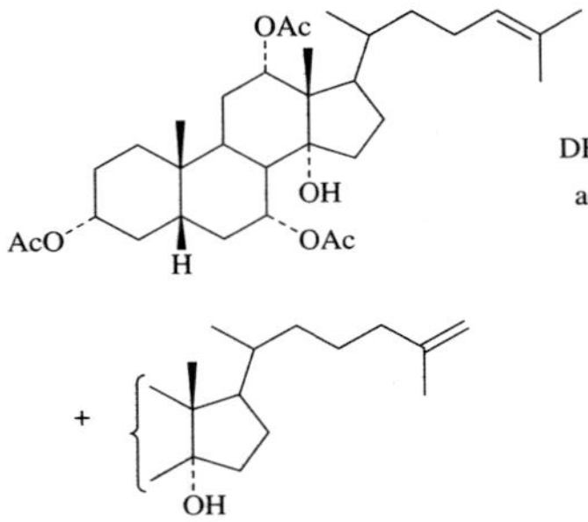	DHQD-*p*-chlorobenzoate, OsO_4, NMO acetone:H_2O (10:1), 0°, 14 h	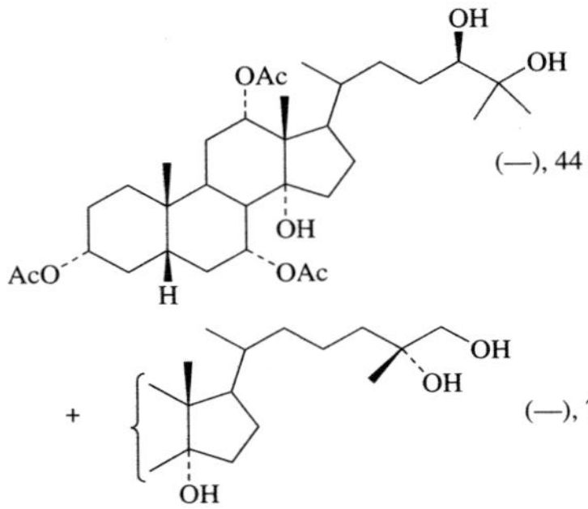 (—), 44 de; (—), 78 de	592

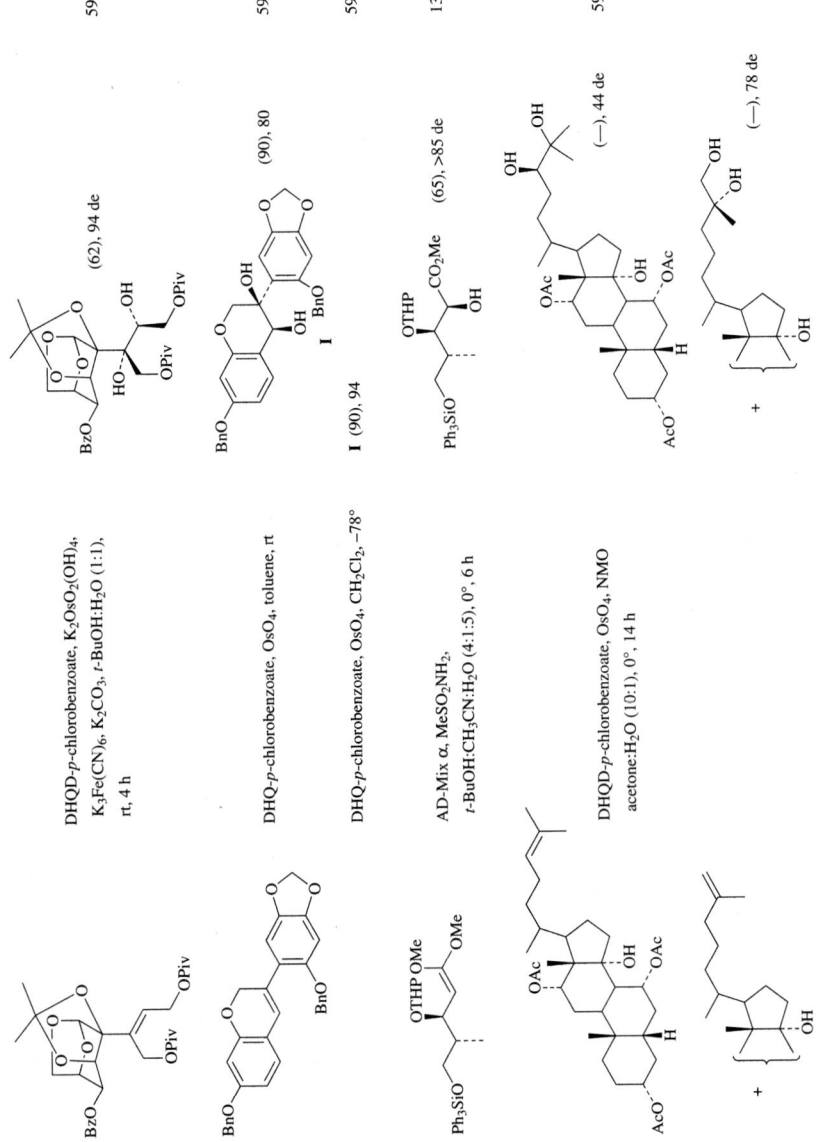

TABLE 5. REACTIONS OF TRISUBSTITUTED ALKENES (*Continued*)

Substrate	Conditions	Product(s) and Yield(s) (%), % ee	Refs.
C_{33} 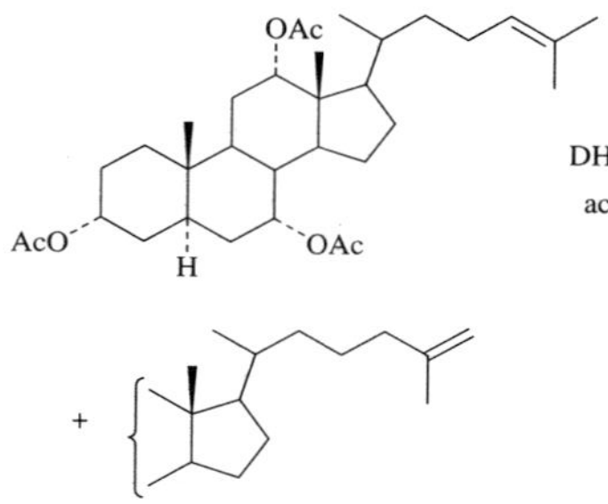	DHQ-*p*-chlorobenzoate, OsO_4, NMO acetone:H_2O (10:1), 0°, 14 h	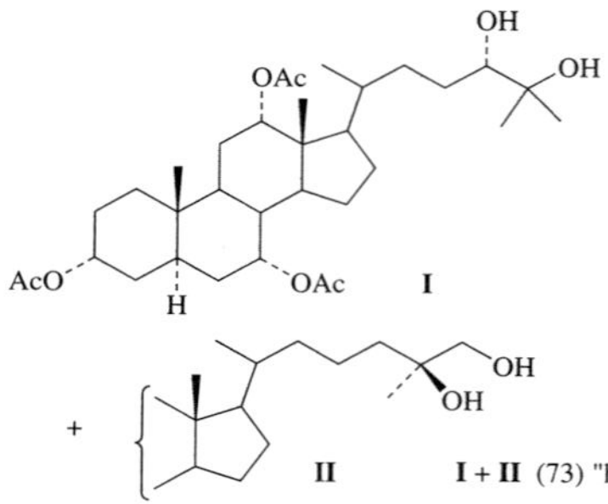 I, II; **I** + **II** (73) "high de"	592
C_{38} 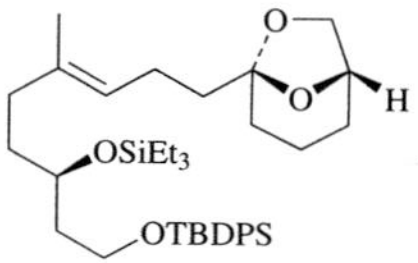	$(DHQ)_2PHAL$, OsO_4, $K_3Fe(CN)_6$, K_2CO_3, $MeSO_2NH_2$, *t*-BuOH:H_2O (1:1), 0°-rt, 5.5 h	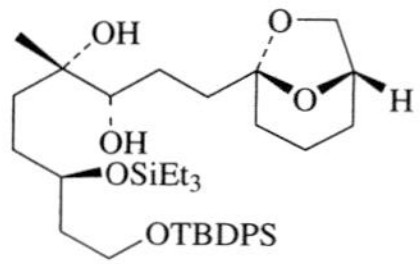 (100), 86 de	593
C_{49} 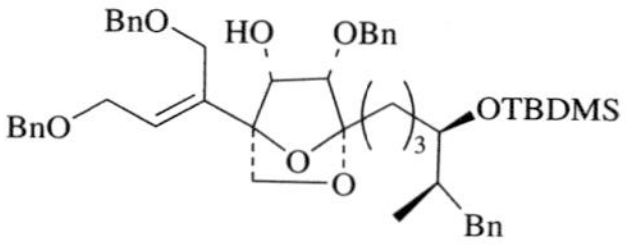	AD-Mix β, $MeSO_2NH_2$, *t*-BuOH:H_2O (1:1), 20°, 18 h	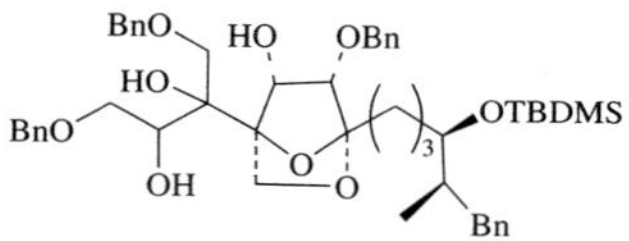 (77), 0 de	594

[a] The authors assigned the absolute configuration using a mechanistic model.

[b] The authors do not report the configuration of the product, but $(DHQD)_2PHAL$ is reported to give the enantiomer opposite to that given with $(DHQ)_2PHAL$.

TABLE 6. REACTIONS OF TETRASUBSTITUTED ALKENES

Substrate	Conditions	Product(s) and Yield(s) (%), % ee	Refs.
C_{10} 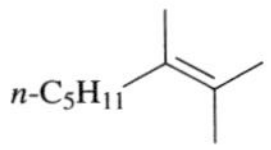	Ligand, $K_2OsO_2(OH)_4$, $K_3Fe(CN)_6$, K_2CO_3, $MeSO_2NH_2$, t-BuOH:H_2O (1:1), rt, 1-2 d	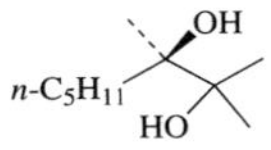	137
	Ligand		
	$(DHQD)_2PHAL$	(51-55), 20	
	$(DHQD)_2PYR$	(51-55), 22	
C_{11} 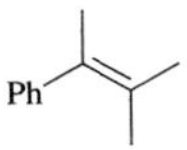	Ligand, $K_2OsO_2(OH)_4$, $K_3Fe(CN)_6$, K_2CO_3, $MeSO_2NH_2$, t-BuOH:H_2O (1:1), rt	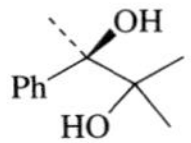 **I**	
	Ligand — Time	**I**	
	$(DHQD)_2PHAL$ — 1-2 d	(80-82), 39	137
	$(DHQD)_2PYR$ — 1-2 d	(80-82), 47	137
	$(DHQD)_2PHAL$ — 24 h	(28), 33	71
	Ligand, $K_2OsO_2(OH)_4$, $K_3Fe(CN)_6$, t-BuOH:H_2O (1:1), pH 12, 24 h	**I**	71
	Ligand — Temp	**I**	
	$(DHQD)_2PHAL$ — rt	(95), 23	
	$(DHQD)_2PHAL$ — 0°	(71), 52	
	$(DHQD)_2PYR$ — 0°	(73), 61	
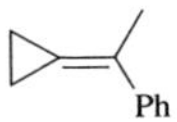	$(DHQD)_2PHAL$, $K_2OsO_2(OH)_4$, $K_3Fe(CN)_6$, K_2CO_3, $MeSO_2NH_2$, t-BuOH:H_2O (1:1), 0°, 6-24 h	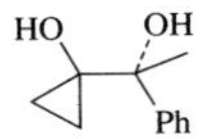 (57), 44	130
	1. AD-Mix β, $MeSO_2NH_2$, t-BuOH:H_2O (1:1), 0°, 7 d 2. $SOCl_2$, Et_3N, CH_2Cl_2	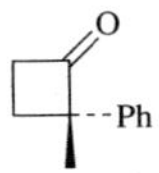 (48), 44	559

TABLE 6. REACTIONS OF TETRASUBSTITUTED ALKENES (*Continued*)

Substrate	Conditions	Product(s) and Yield(s) (%), % ee	Refs.
C_{12} 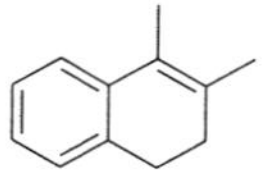	Ligand, $K_2OsO_2(OH)_4$, $K_3Fe(CN)_6$, K_2CO_3, $MeSO_2NH_2$, t-BuOH:H_2O (1:1), rt, 1-2 d	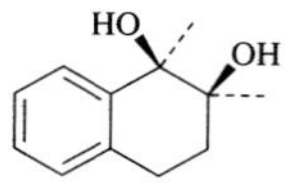	137
	Ligand		
	$(DHQD)_2PHAL$	(23-24), 59	
	$(DHQD)_2PYR$	(23-24), 56	
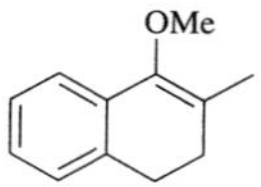	Ligand, $K_2OsO_2(OH)_4$, $K_3Fe(CN)_6$, K_2CO_3, $MeSO_2NH_2$, t-BuOH:H_2O (1:1), 0°, 1-2 d	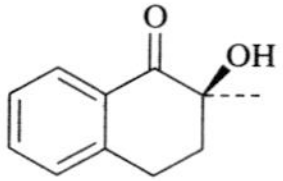	137
	Ligand		
	$(DHQD)_2PHAL$	(79-95), 64	
	$(DHQD)_2PYR$	(79-95), 41	
C_{16} 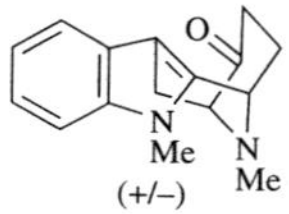	DHQ-*p*-chlorobenzoate, OsO_4, rt, 3 d	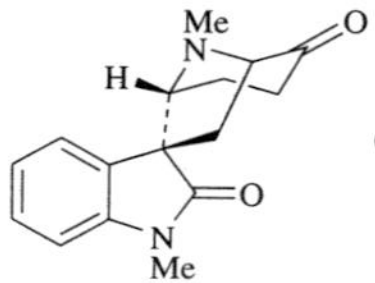 (66), 100 de	595

Reactant	Conditions	Product(s) and Yield(s) (%)	Refs.
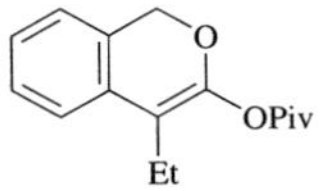	1. Ligand, $K_2OsO_2(OH)_4$, $K_3Fe(CN)_6$, K_2CO_3, $MeSO_2NH_2$, t-BuOH:H_2O (1:1), 0°, 24 h 2. I_2, $CaCO_3$, ether, rt, 32 h	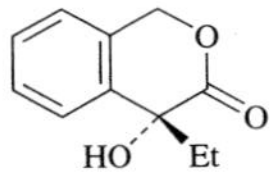	132
	Ligand: $(DHQD)_2PHAL$	(100), 78	
	Ligand: $(DHQD)_2PYR$	(73), 67	
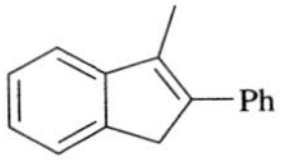	Ligand, $K_2OsO_2(OH)_4$, $K_3Fe(CN)_6$, K_2CO_3, $MeSO_2NH_2$, t-BuOH:H_2O (1:1), rt, 1-2 d	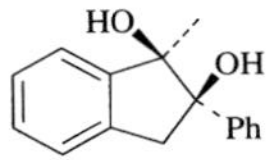	137
	Ligand: $(DHQD)_2PHAL$	(16-19), 75	
	Ligand: $(DHQD)_2PYR$	(16-19), 82	
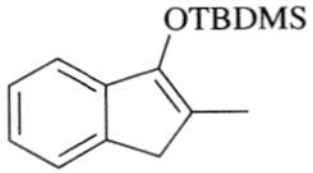	Ligand, $K_2OsO_2(OH)_4$, $K_3Fe(CN)_6$, K_2CO_3, $MeSO_2NH_2$, t-BuOH:H_2O (1:1), 0°, 12 h	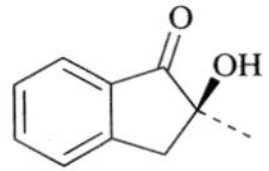	
	Ligand: $(DHQD)_2PHAL$; TIme: 12 h	(64-85), 85	137, 66
	Ligand: $(DHQD)_2PYR$; TIme: 1-2 d	(64-85), 59	137
	Ligand, $K_2OsO_2(OH)_4$, $K_3Fe(CN)_6$, K_2CO_3, $MeSO_2NH_2$, t-BuOH:H_2O (1:1), 0°, 1-2 d	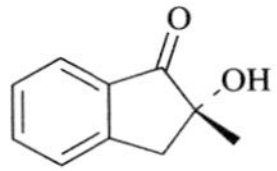	137
	Ligand: $(DHQ)_2PHAL$	(64-85), 81	
	Ligand: $(DHQ)_2PYR$	(64-85), 80	

TABLE 6. REACTIONS OF TETRASUBSTITUTED ALKENES (*Continued*)

Substrate	Conditions	Product(s) and Yield(s) (%) and Enatiomeric Excess	Refs.
C_{16}			
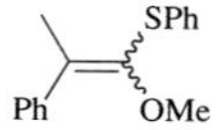	AD-Mix α, $MeSO_2NH_2$, t-BuOH:CH_3CN:H_2O (4:1:5), 0°, 6 h	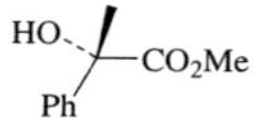 (18), 12	136
	AD-Mix β, $MeSO_2NH_2$, t-BuOH:CH_3CN:H_2O (4:1:5), 0°, 6 h	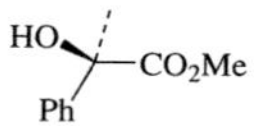 (18), 11	136
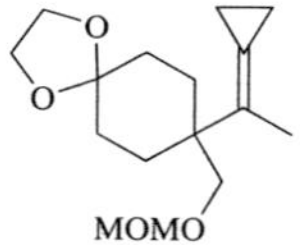	$(DHQ)_2PYR$, OsO_4, $K_3Fe(CN)_6$, $MeSO_2NH_2$, K_2CO_3, t-BuOH:H_2O (1:1), 0°, 8 h	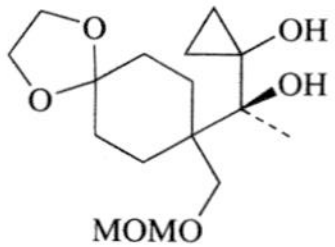 (79), >73	596
C_{17}			
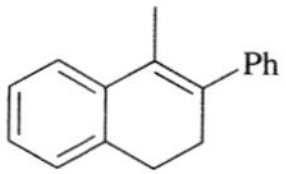	Ligand, $K_2OsO_2(OH)_4$, $K_3Fe(CN)_6$, K_2CO_3, $MeSO_2NH_2$, t-BuOH:H_2O (1:1), rt, 1-2 d	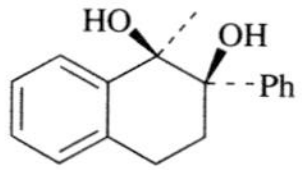	137
	Ligand		
	$(DHQD)_2PHAL$	(29-31), 83	
	$(DHQD)_2PYR$	(29-31), 85	
	Ligand, $K_2OsO_2(OH)_4$, $K_3Fe(CN)_6$, K_2CO_3, $MeSO_2NH_2$, t-BuOH:H_2O (1:1), rt, 1-2 d	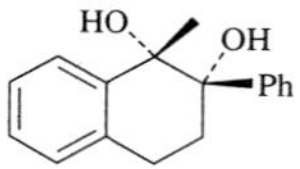	137
	Ligand		
	$(DHQ)_2PHAL$	(29-31), 85	
	$(DHQ)_2PYR$	(29-31), 89	

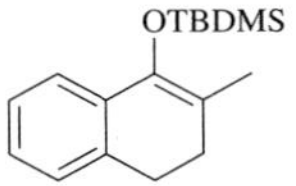	(DHQD)$_2$PHAL, $K_2OsO_2(OH)_4$, $K_3Fe(CN)_6$, K_2CO_3, $MeSO_2NH_2$, *t*-BuOH:H_2O (1:1), 0°, 1-2 d	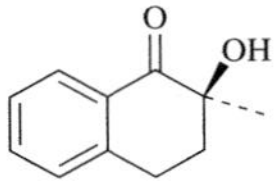	(89-92), 67	137
	Ligand, $K_2OsO_2(OH)_4$, $K_3Fe(CN)_6$, K_2CO_3, $MeSO_2NH_2$, *t*-BuOH:H_2O (1:1), 0°, 1-2 d	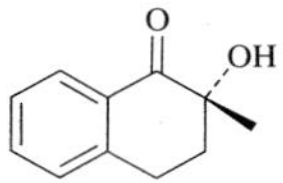		137
	Ligand			
	(DHQD)$_2$PYR	(89-92), 6		
	(DHQ)$_2$PHAL	(89-92), 65		
	(DHQ)$_2$PYR	(89-92), 37		
C_{17-20} 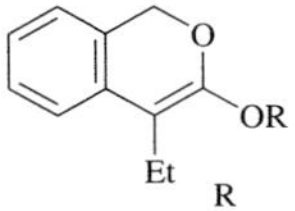	1. (DHQD)$_2$PHAL, $K_2OsO_2(OH)_4$, $K_3Fe(CN)_6$, K_2CO_3, $MeSO_2NH_2$, *t*-BuOH:H_2O (1:1), 0°, 24 h 2. I_2, $CaCO_3$, ether, rt., 32 h	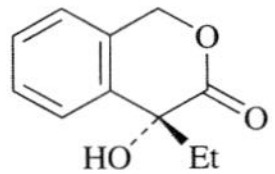		132
R				
TBDMS		(70), 40		
Bz		(82), 65		
TIPS		(67), 30		
C_{18} 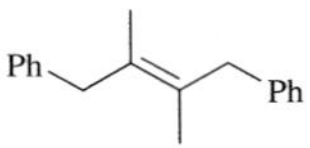	Ligand, $K_2OsO_2(OH)_4$, $K_3Fe(CN)_6$, K_2CO_3, $MeSO_2NH_2$, *t*-BuOH:H_2O (1:1), rt, 1-2 d	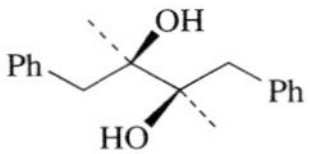		137
	Ligand			
	(DHQD)$_2$PHAL	(85-87), 29		
	(DHQD)$_2$PYR	(85-87), 31		
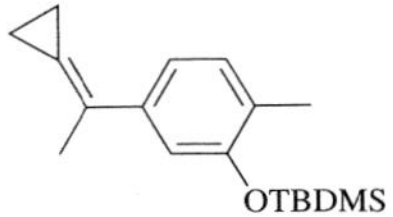	1. AD-Mix β, $MeSO_2NH_2$, *t*-BuOH:H_2O (1:1), 0°, 7 d 2. $SOCl_2$, Et_3N, CH_2Cl_2	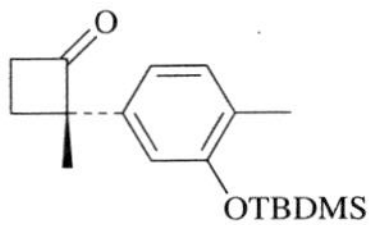	(50), 55	559

TABLE 6. REACTIONS OF TETRASUBSTITUTED ALKENES (*Continued*)

Substrate	Conditions	Product(s) and Yield(s) (%), % ee	Refs.
C_{19}			
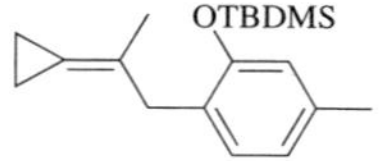	(DHQD)$_2$PHAL, $K_2OsO_2(OH)_4$, $K_3Fe(CN)_6$, K_2CO_3, $MeSO_2NH_2$, *t*-BuOH:H_2O (1:1), 0°, 6-24 h	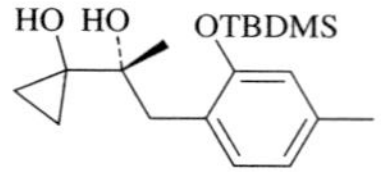 (52), 55	130
C_{21}			
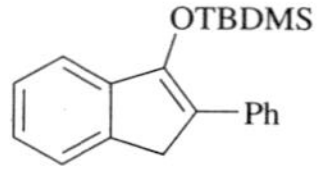	Ligand, $K_2OsO_2(OH)_4$, $K_3Fe(CN)_6$, K_2CO_3, $MeSO_2NH_2$, *t*-BuOH:H_2O (1:1), 0°	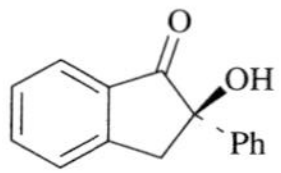	
	Ligand / Time		
	(DHQD)$_2$PYDZ / 12 h	(23-32), 89	66, 137
	(DHQD)$_2$PYR / 1-2 d	(23-32), 84	137
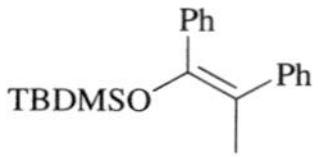	Ligand, $K_2OsO_2(OH)_4$, $K_3Fe(CN)_6$, K_2CO_3, $MeSO_2NH_2$, *t*-BuOH:H_2O (1:1), 0°, 1-2 d	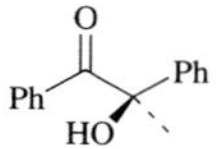	137
	Ligand		
	(DHQD)$_2$PHAL	(46-60), 53	
	(DHQD)$_2$PYR	(46-60), 60	
	Ligand, $K_2OsO_2(OH)_4$, $K_3Fe(CN)_6$, K_2CO_3, $MeSO_2NH_2$, *t*-BuOH:H_2O (1:1), 0°, 1-2 d	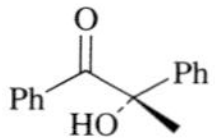	137
	Ligand		
	(DHQ)$_2$PHAL	(46-60), 63	
	(DHQ)$_2$PYR	(46-60), 33	

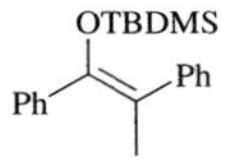	Ligand, $K_2OsO_2(OH)_4$, $K_3Fe(CN)_6$, K_2CO_3, $MeSO_2NH_2$, t-BuOH:H_2O (1:1), 0°, 1-2 d	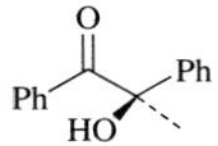	137
Ligand			
$(DHQD)_2PHAL$		(15-22), 75	
$(DHQD)_2PYR$		(15-22), 79	
	Ligand, $K_2OsO_2(OH)_4$, $K_3Fe(CN)_6$, K_2CO_3, $MeSO_2NH_2$, t-BuOH:H_2O (1:1), 0°, 1-2 d	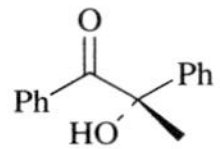	137
Ligand			
$(DHQ)_2PHAL$		(15-22), 81	
$(DHQ)_2PYR$		(15-22), 79	
C_{22} 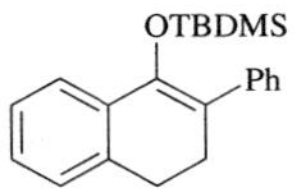	Ligand, $K_2OsO_2(OH)_4$, $K_3Fe(CN)_6$, K_2CO_3, $MeSO_2NH_2$, t-BuOH:H_2O (1:1), 0°, 1-2 d	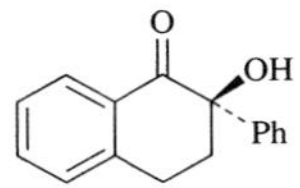	
Ligand			
$(DHQD)_2PHAL$		(94-98), 93	137, 66
$(DHQD)_2PYR$		(94-98), 95	137
	Ligand, $K_2OsO_2(OH)_4$, $K_3Fe(CN)_6$, K_2CO_3, $MeSO_2NH_2$, t-BuOH:H_2O (1:1), 0°, 1-2 d	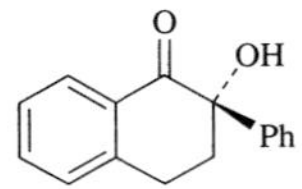	137
Ligand			
$(DHQ)_2PHAL$		(94-98), 95	
$(DHQ)_2PYR$		(94-98), 97	

TABLE 6. REACTIONS OF TETRASUBSTITUTED ALKENES (*Continued*)

Substrate	Conditions	Product(s) and Yield(s) (%), % ee	Refs.
C_{22}			
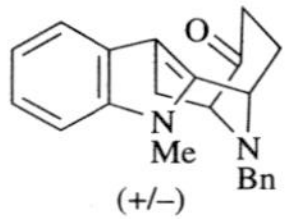 (+/–)	Ligand, OsO_4, rt, 3 d	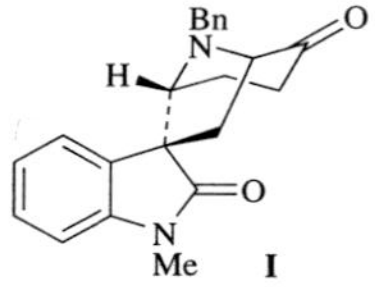 I	597, 595
	Ligand	**I**	
	DHQD-*p*-chlorobenzoate	(76), 4 de	
	DHQ-*p*-chlorobenzoate	(82), 52 de	
	$(DHQ)_2PHAL$	(76), 60 de	
	$(DHQD)_2PHAL$	(66), 6 de	
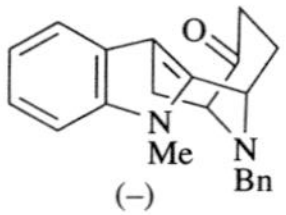 (–)	Ligand, OsO_4, rt, 3 d	I	597, 595
	Ligand	**I**	
	DHQ-*p*-chlorobenzoate	(91), 94 de	
	DHQD-*p*-chlorobenzoate	(77), 60 de	
	$(DHQ)_2PHAL$	(81), 50 de	
	$(DHQD)_2PHAL$	(82), 60 de	
C_{24}			
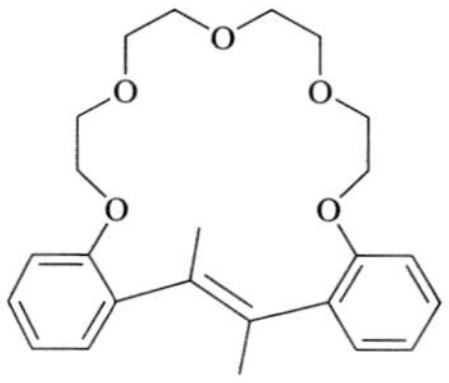	DHQD-*p*-chlorobenzoate, $K_2OsO_2(OH)_4$, $K_3Fe(CN)_6$, K_2CO_3:*t*-BuOH, $MeSO_2NH_2$, H_2O, rt, 24 h	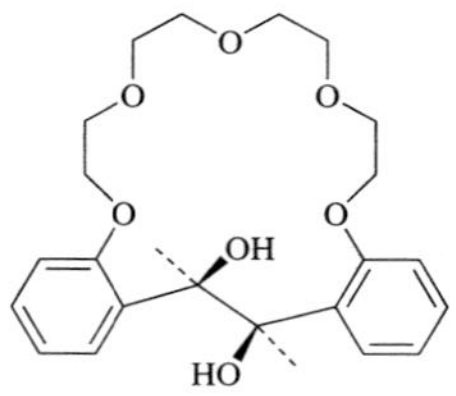 (62), >95	598

	DHQ-*p*-chlorobenzoate, $K_2OsO_2(OH)_4$, $K_3Fe(CN)_6$, K_2CO_3, *t*-BuOH-H_2O, rt, 24 h	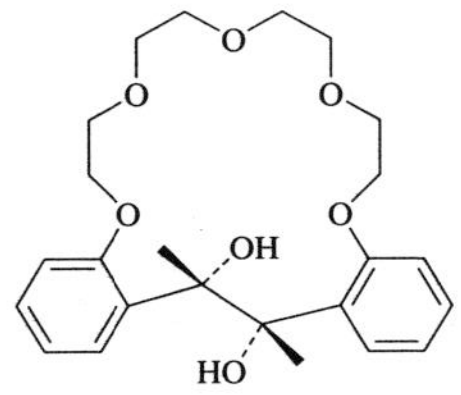	(56), >95	598

TABLE 7. REACTIONS OF CONJUGATED POLYALKENES

See Chart 1 at the beginning of the Tabular Survey for ligand structures that are indicated by ***bold*** *numbers.*

Substrate	Conditions	Product(s) and Yield(s) (%), % ee	Refs.
C_5 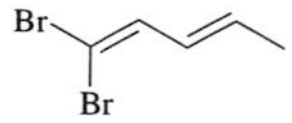	AD-Mix α, $MeSO_2NH_2$, t-BuOH:H_2O (1:1), 0°, 24 h	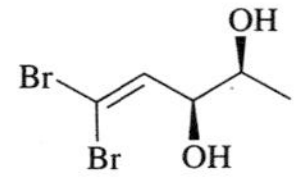 (68), 73	425
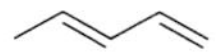	$(DHQD)_2PHAL$, $K_2OsO_2(OH)_4$, $K_3Fe(CN)_6$, K_2CO_3, $MeSO_2NH_2$, t-BuOH:H_2O (1:1), 0°	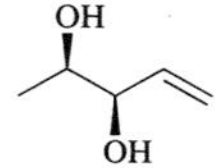 (36), 90 + 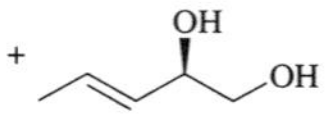 (12), 72	139
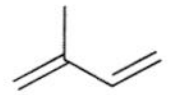	AD-Mix α, t-BuOH:H_2O (1:1), 0°, 24 h	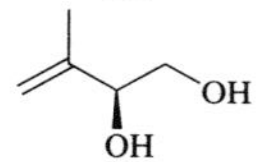 (60), — + 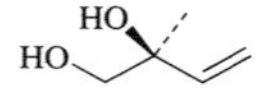 (40), 30	599
	Ligand, $K_2OsO_2(OH)_4$, $K_3Fe(CN)_6$, K_2CO_3, $MeSO_2NH_2$, t-BuOH:H_2O (1:1), 0°	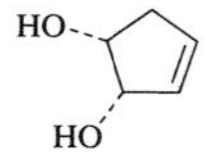	147
	Ligand		
	$(DHQD)_2PHAL$	(—), 38	
	$(DHQD)_2PYR$	(—), 7	
	DHQD-IND	(—), 29	
C_6 	AD-Mix β, $MeSO_2NH_2$, t-BuOH:H_2O (1:1), 0°, 60 h	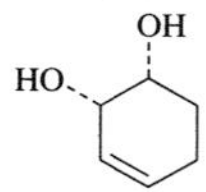 **I** (72), 33	148
	Ligand, $K_2OsO_2(OH)_4$, $K_3Fe(CN)_6$, K_2CO_3, $MeSO_2NH_2$, t-BuOH:H_2O (1:1), 0°	**I**	147
	Ligand	**I**	
	$(DHQD)_2PHAL$	(—), 37	
	$(DHQD)_2PYR$	(—), 23	
	DHQD-IND	(—), 24	
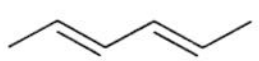	$(DHQD)_2PHAL$, $K_2OsO_2(OH)_4$, $K_3Fe(CN)_6$, K_2CO_3, $MeSO_2NH_2$, t-BuOH:H_2O (1:1), 0°	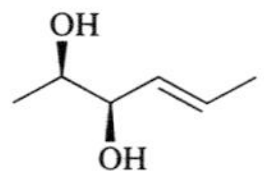 (78), 93	139

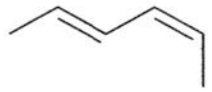	(DHQD)$_2$PHAL, $K_2OsO_2(OH)_4$, $K_3Fe(CN)_6$, K_2CO_3, $MeSO_2NH_2$, *t*-BuOH:H_2O (1:1), 0°	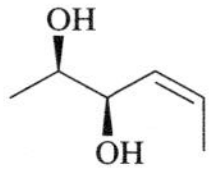 **I** + 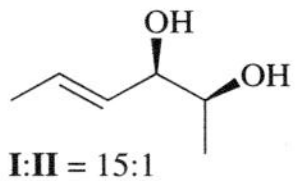 **II** **I** + **II** (88), 98; **I**:**II** = 15:1	139
	Ligand, $K_2OsO_2(OH)_4$, $K_3Fe(CN)_6$, K_2CO_3, $MeSO_2NH_2$, *t*-BuOH:H_2O (1:1), 0° Ligand (DHQD)$_2$PHAL (DHQD)$_2$PYR DHQD-IND	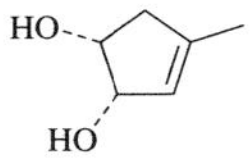 (—), 56 (—), 8 (—), 31	147
C$_7$ 	Ligand, $K_2OsO_2(OH)_4$, $K_3Fe(CN)_6$, K_2CO_3, $MeSO_2NH_2$, *t*-BuOH:H_2O (1:1), 0° Ligand (DHQD)$_2$PHAL (DHQD)$_2$PYR DHQD-IND	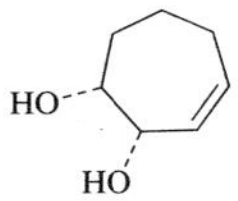 (—), 30 (—), 21 (—), 1	147
	Ligand, $K_2OsO_2(OH)_4$, $K_3Fe(CN)_6$, K_2CO_3, $MeSO_2NH_2$, *t*-BuOH:H_2O (1:1), 0° Ligand (DHQD)$_2$PHAL (DHQD)$_2$PYR DHQD-IND	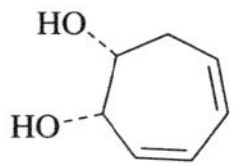 (—), 16 (—), 24 (—), 11	147
	Ligand, $K_2OsO_2(OH)_4$, $K_3Fe(CN)_6$, K_2CO_3, $MeSO_2NH_2$, *t*-BuOH:H_2O (1:1), 0° Ligand (DHQD)$_2$PHAL (DHQD)$_2$PYR DHQD-IND	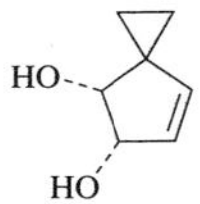 (—), 53 (—), 55 (—), 50	147

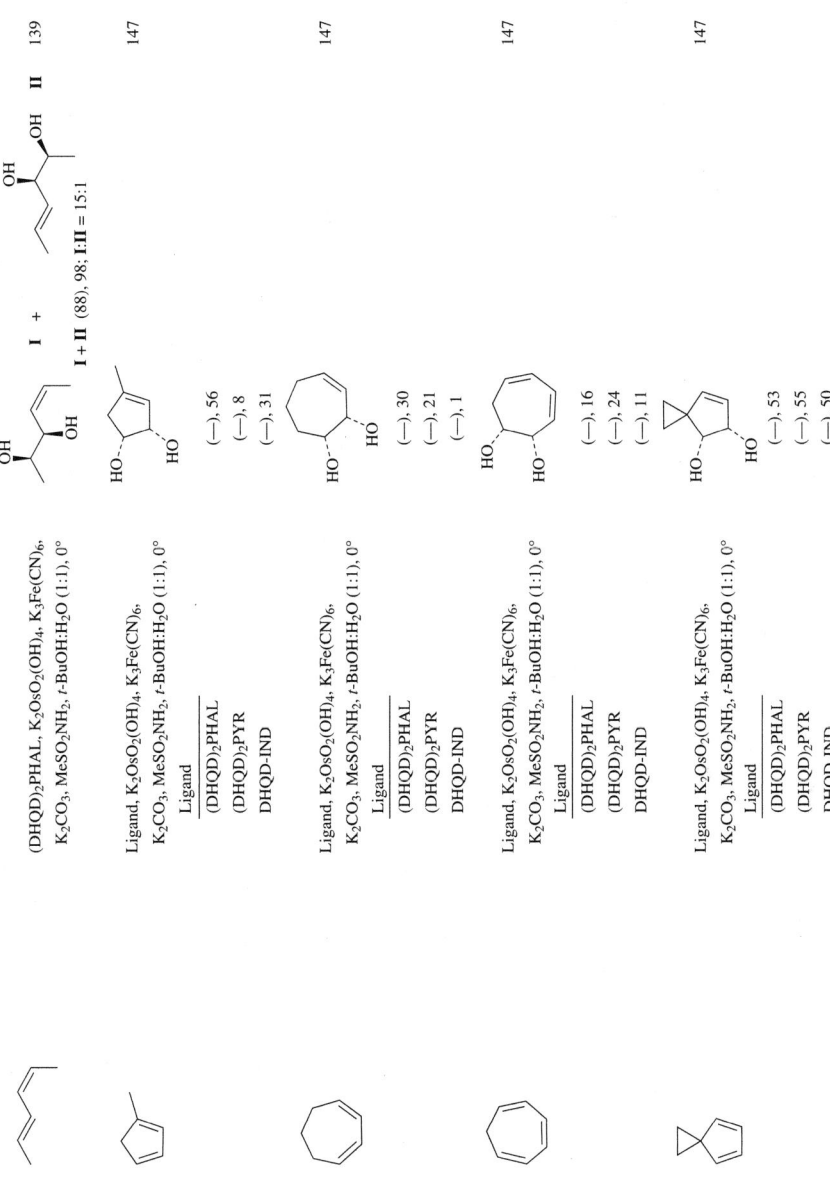

TABLE 7. REACTIONS OF CONJUGATED POLYALKENES (*Continued*)

Substrate	Conditions	Product(s) and Yield(s) (%), % ee	Refs.
C_7			
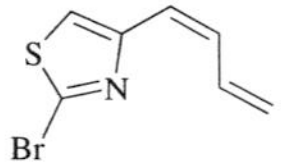	AD-mix α, $(DHQ)_2PHAL$, t-BuOH:H_2O (1:1), 0°, 18 h	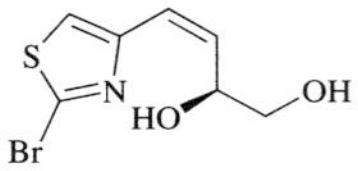 (89), 32	371
C_8			
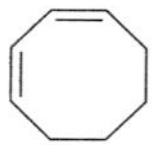	$(DHQD)_2PHAL$, $K_2OsO_2(OH)_4$, $K_3Fe(CN)_6$, K_2CO_3, $MeSO_2NH_2$, t-BuOH:H_2O (1:1), 0°	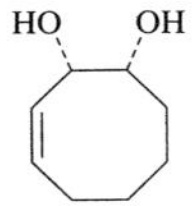 (—), 5	147
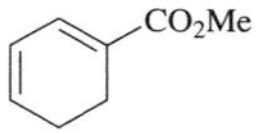	$(DHQD)_2PHAL$, $K_2OsO_2(OH)_4$, $K_3Fe(CN)_6$, K_2CO_3, $MeSO_2NH_2$, t-BuOH:H_2O (1:1), 0°	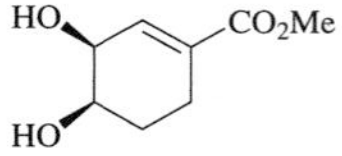 (—), 31	147
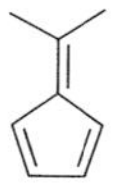	$(DHQD)_2PHAL$, $K_2OsO_2(OH)_4$, $K_3Fe(CN)_6$, K_2CO_3, t-BuOH:H_2O (1:1)	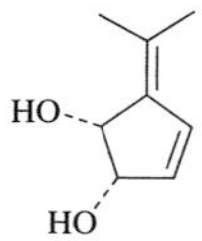 (—), 30	147
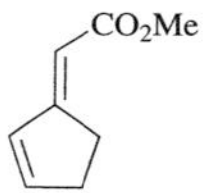	$(DHQD)_2PHAL$, $K_2OsO_2(OH)_4$, $K_3Fe(CN)_6$, K_2CO_3, t-BuOH:H_2O (1:1)	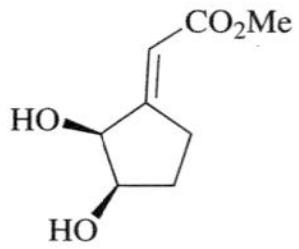 (—), 47	147
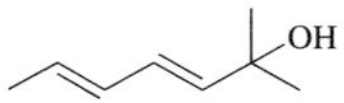	$(DHQD)_2PHAL$, $K_2OsO_2(OH)_4$, $K_3Fe(CN)_6$, K_2CO_3, $MeSO_2NH_2$, t-BuOH:H_2O (1:1), 0°, 6-36 h	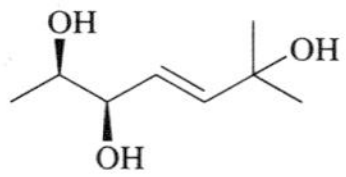 (74), 93	116

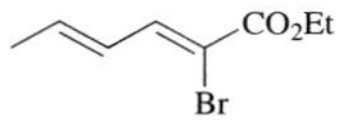	AD-Mix α, t-BuOH:H_2O (1:1), 0°	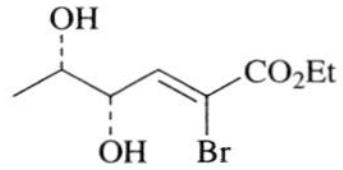 **I** (90), >99	600
	AD-Mix α, $MeSO_2NH_2$, t-BuOH:H_2O (1:1), rt, 16 h	**I** (73), 94	601
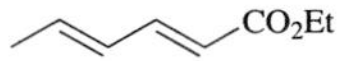	AD-Mix α, $MeSO_2NH_2$, t-BuOH:H_2O (1:1), 0°, 20 h	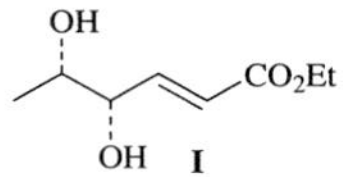 **I** (91), 80	602
	AD-Mix α, $MeSO_2NH_2$, t-BuOH:H_2O (1:1), 0°-rt	**I** (52), 80	603
	AD-Mix β, $MeSO_2NH_2$, t-BuOH:H_2O (1:1), 0°, 20 h	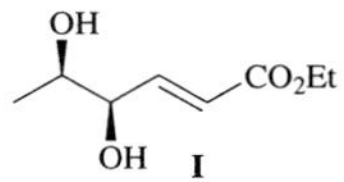 **I** (91), 92-96	602, 138
	$(DHQD)_2PHAL$, $K_2OsO_2(OH)_4$, $K_3Fe(CN)_6$, K_2CO_3, $MeSO_2NH_2$, t-BuOH:H_2O (1:1), 0°	**I** (93), 95	139
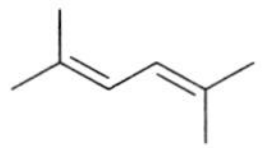	$(DHQD)_2PHAL$, $K_2OsO_2(OH)_4$, $K_3Fe(CN)_6$, K_2CO_3, $MeSO_2NH_2$, t-BuOH:H_2O (1:1), 0°	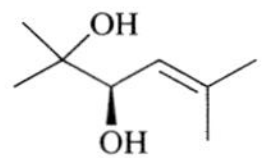 (94), 93	139
C_9			
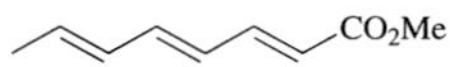	$(DHQD)_2PHAL$, $K_2OsO_2(OH)_4$, $K_3Fe(CN)_6$, K_2CO_3, $MeSO_2NH_2$, t-BuOH:H_2O (1:1), 0°	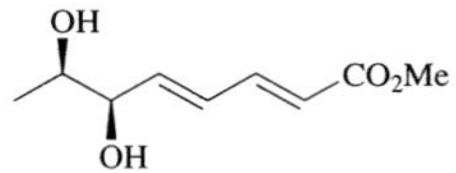 (93), 95	138
	Ligand, $K_2OsO_2(OH)_4$, $K_3Fe(CN)_6$, K_2CO_3, $MeSO_2NH_2$, t-BuOH:H_2O (1:1), 0°	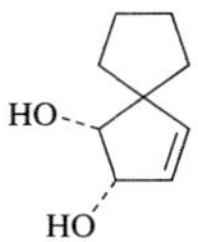	147
	Ligand		
	$(DHQD)_2PHAL$	(—), 70	
	$(DHQD)_2PYR$	(—), 70	
	DHQD-IND	(—), 73	

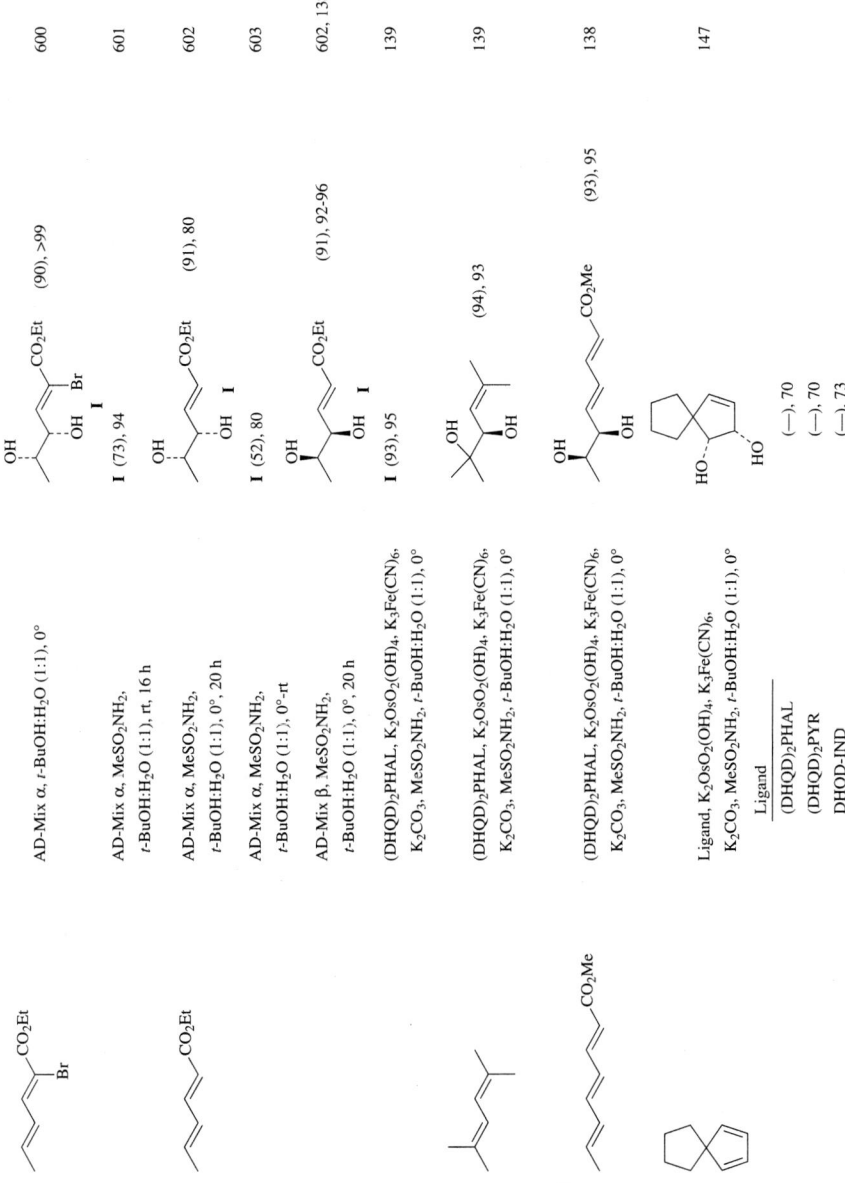

TABLE 7. REACTIONS OF CONJUGATED POLYALKENES (*Continued*)

Substrate	Conditions	Product(s) and Yield(s) (%), % ee	Refs.
C_9 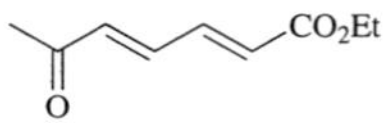	Ligand, $K_2OsO_2(OH)_4$, $K_3Fe(CN)_6$, K_2CO_3, $MeSO_2NH_2$, t-BuOH:H_2O (1:1), 0°	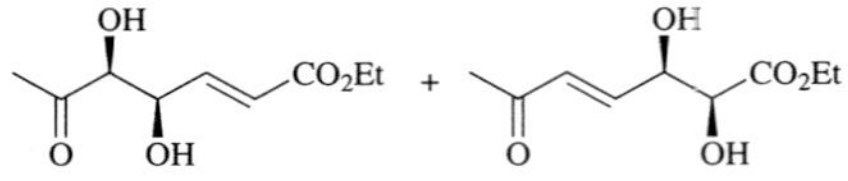 I + II	138
	Ligand	**I** / **II** / **I:II**	
	$(DHQD)_2PHAL$	(—), 97 / (—), 97 / 56:44	
	$(DHQD)_2PYR$	(—), 90 / (—), 89 / 83:17	
	Ligand, $K_2OsO_2(OH)_4$, $K_3Fe(CN)_6$, K_2CO_3, $MeSO_2NH_2$, t-BuOH:H_2O (1:1), 0°	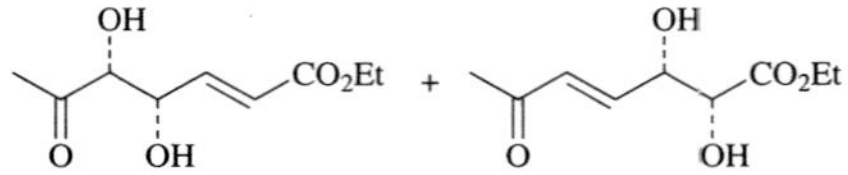 I + II	138
	Ligand	**I** / **II** / **I:II**	
	$(DHQ)_2PHAL$	(—), 97 / (—), 97 / 60:40	
	$(DHQ)_2PYR$	(—), 91 / (—), 90 / 86:14	
C_{10} 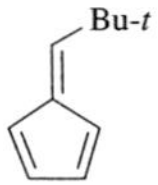	$(DHQD)_2PHAL$, $K_2OsO_2(OH)_4$, $K_3Fe(CN)_6$, K_2CO_3, t-BuOH:H_2O (1:1)	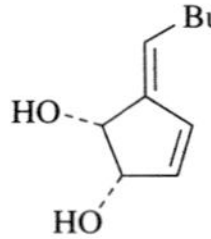 I (40), 83-84	604
	Ligand, $K_2OsO_2(OH)_4$, $K_3Fe(CN)_6$, K_2CO_3, $MeSO_2NH_2$, t-BuOH:H_2O (1:1), 0°	I	147
	Ligand	**I**	
	$(DHQD)_2PYR$	(—), 1	
	DHQD-IND	(—), 61	

C_{11}				
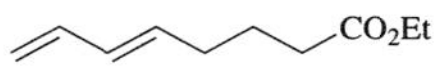	AD-Mix α, $MeSO_2NH_2$, t-BuOH:H_2O (1:1), 0°	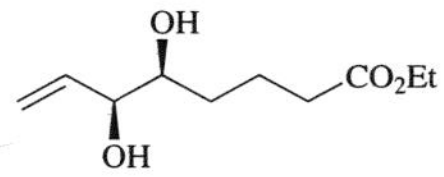 (50), 94		605
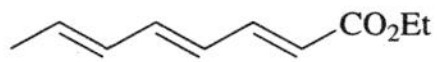	$(DHQD)_2PHAL$, $K_2OsO_2(OH)_4$, $K_3Fe(CN)_6$, K_2CO_3, $MeSO_2NH_2$, t-BuOH:H_2O (1:1), 0°	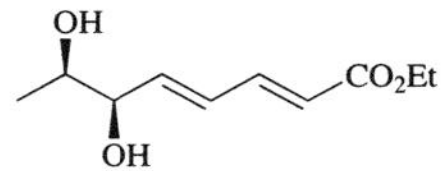 (93), 95		139
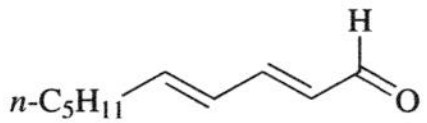	$(DHQD)_2PHAL$, $K_2OsO_2(OH)_4$, $K_3Fe(CN)_6$, K_2CO_3, $MeSO_2NH_2$, t-BuOH:H_2O (1:1), 0°	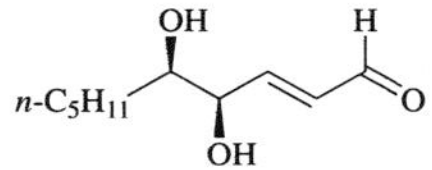 (50), 94		138
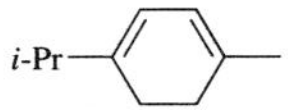	$(DHQD)_2PHAL$, $K_2OsO_2(OH)_4$, $K_3Fe(CN)_6$, K_2CO_3, $MeSO_2NH_2$, t-BuOH:H_2O (1:1), 0°	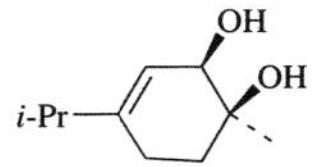 (69), 86 + 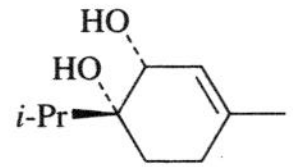 (9), —		138
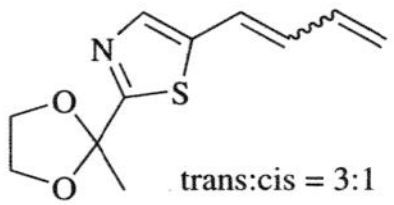 trans:cis = 3:1	$(DHQ)_2PHAL$, $K_2OsO_2(OH)_4$, $K_3Fe(CN)_6$, K_2CO_3, $MeSO_2NH_2$, t-BuOH:H_2O (1:1), 0°, 18 h	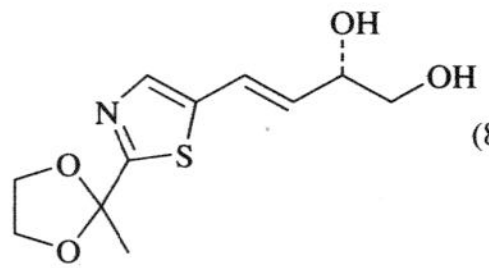 (85), 75 de		418
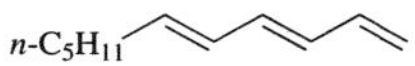	$(DHQD)_2PHAL$, $K_2OsO_2(OH)_4$, $K_3Fe(CN)_6$, K_2CO_3, $MeSO_2NH_2$, t-BuOH:H_2O (1:1), 0°	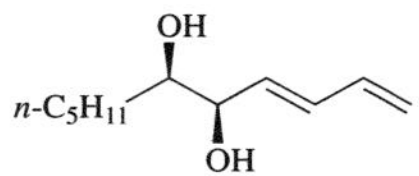 (52), — + 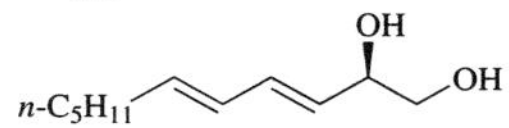 (8), —		138
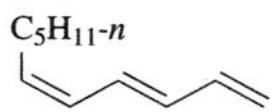	$(DHQD)_2PHAL$, $K_2OsO_2(OH)_4$, $K_3Fe(CN)_6$, K_2CO_3, $MeSO_2NH_2$, t-BuOH:H_2O (1:1), 0°	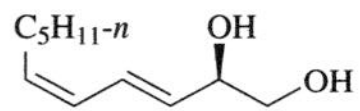 (48), 84		138

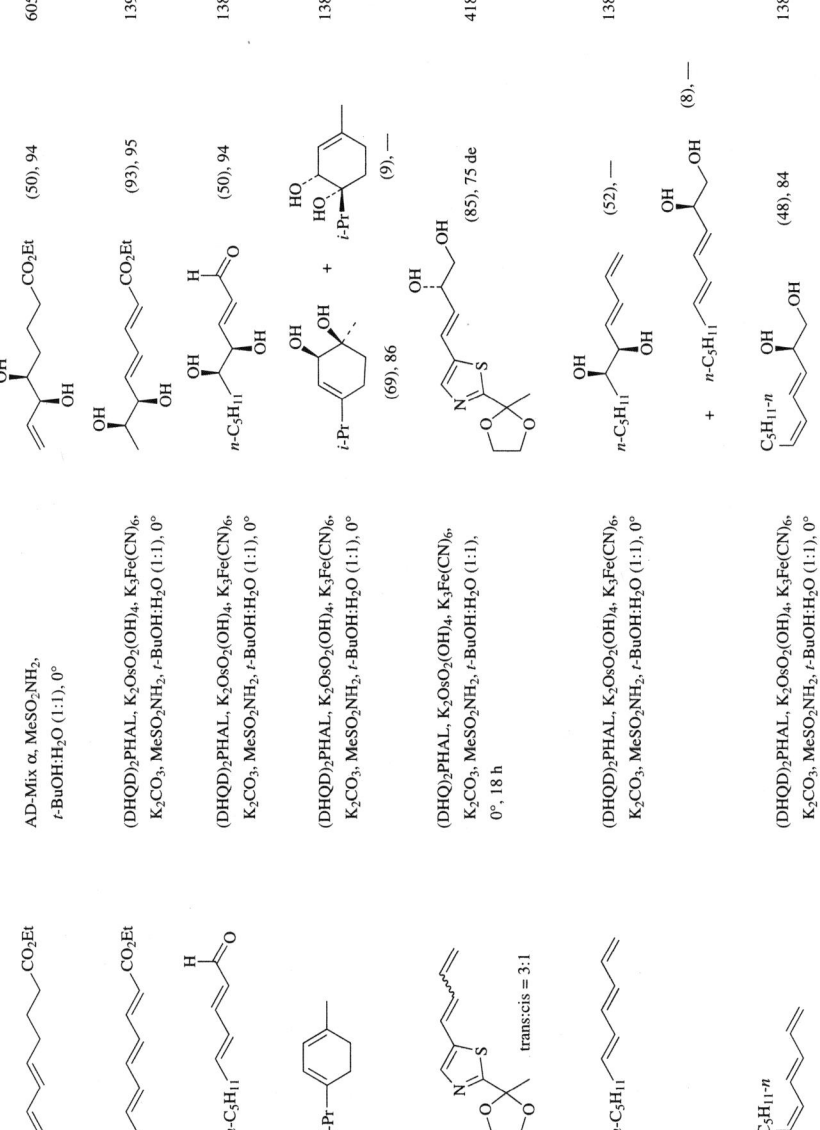

TABLE 7. REACTIONS OF CONJUGATED POLYALKENES (*Continued*)

Substrate	Conditions	Product(s) and Yield(s) (%), % ee	Refs.
C_{11}			
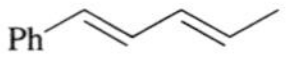	(DHQD)$_2$PHAL, $K_2OsO_2(OH)_4$, $K_3Fe(CN)_6$, K_2CO_3, $MeSO_2NH_2$, t-BuOH:H_2O (1:1), 0°	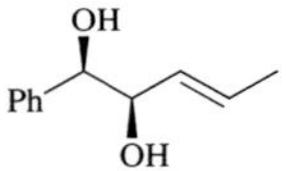 (16), — + 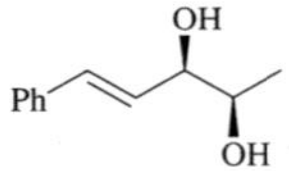 (66), 92	138
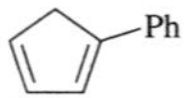	Ligand, $K_2OsO_2(OH)_4$, $K_3Fe(CN)_6$, K_2CO_3, $MeSO_2NH_2$, t-BuOH:H_2O (1:1), 0°	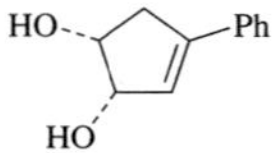	147
	Ligand		
	(DHQD)$_2$PHAL	(—), 99	
	(DHQD)$_2$PYR	(—), 97	
	DHQD-IND	(—), 95	
C_{12}			
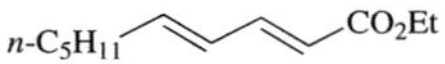	AD-Mix β, t-BuOH:H_2O (1:1), 0-4°, 24 h	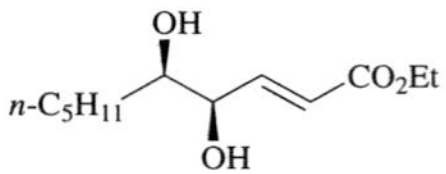 (88), 98	140
	AD-Mix α, t-BuOH:H_2O (1:1), 0-4°, 24 h	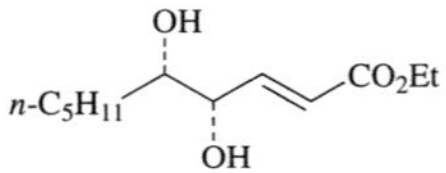 (88), 96	140
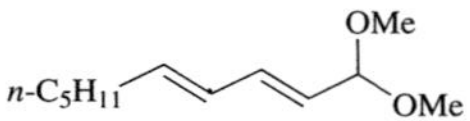	1. AD-Mix β, t-BuOH:H_2O (1:1), 0-4°, 24 h 2. H^+	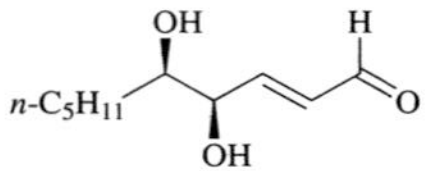 (52), 90	140

	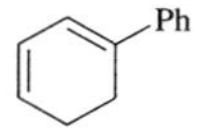	(DHQD)$_2$PHAL, $K_2OsO_2(OH)_4$, $K_3Fe(CN)_6$, K_2CO_3, $MeSO_2NH_2$, t-BuOH:H_2O (1:1), 0°	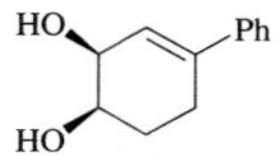 **I** (—), 91	147
		(DHQD)$_2$PYR, $K_2OsO_2(OH)_4$, $K_3Fe(CN)_6$, K_2CO_3, $MeSO_2NH_2$, t-BuOH:H_2O (1:1), 0°	**I** (—), 10	147
C_{13}	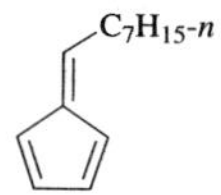	(DHQD)$_2$PHAL, $K_2OsO_2(OH)_4$, K_2CO_3, $K_3Fe(CN)_6$, t-BuOH:H_2O (1:1)	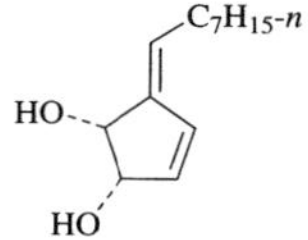 (36), 80-84	604, 147
	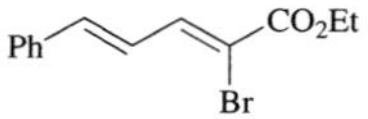	AD-Mix α, t-BuOH:H_2O (1:1), 0°	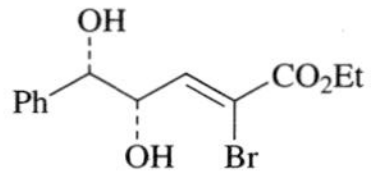 **I** (73), >94	600
		AD-Mix α, $MeSO_2NH_2$, t-BuOH:H_2O (1:1), rt, 16 h	**I** (90), 99	601
	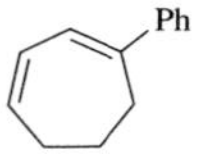	(DHQD)$_2$PHAL, $K_2OsO_2(OH)_4$, $K_3Fe(CN)_6$, K_2CO_3, $MeSO_2NH_2$, t-BuOH:H_2O (1:1), 0°	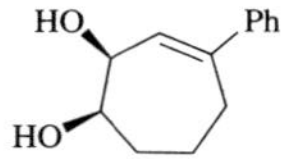 (—), 41	147
	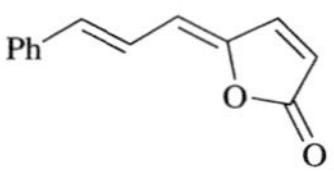	AD-Mix β, t-BuOH:H_2O (1:1), 0°, 4 h	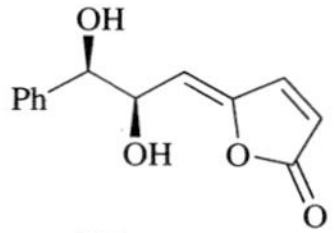 (81), 99	538
	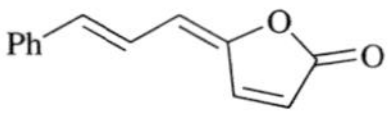	AD-Mix β, t-BuOH:H_2O (1:1), 0°, 4 h	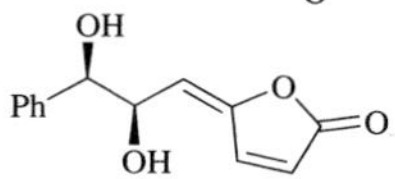 (85), 99	538
	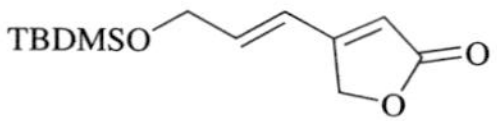	AD-Mix β, $MeSO_2NH_2$, t-BuOH:H_2O (1:1), 0°, 24 h	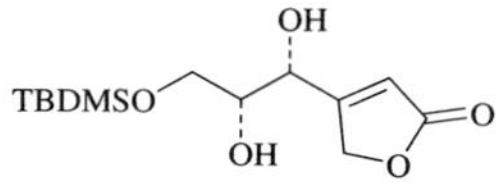 (85), >98	606

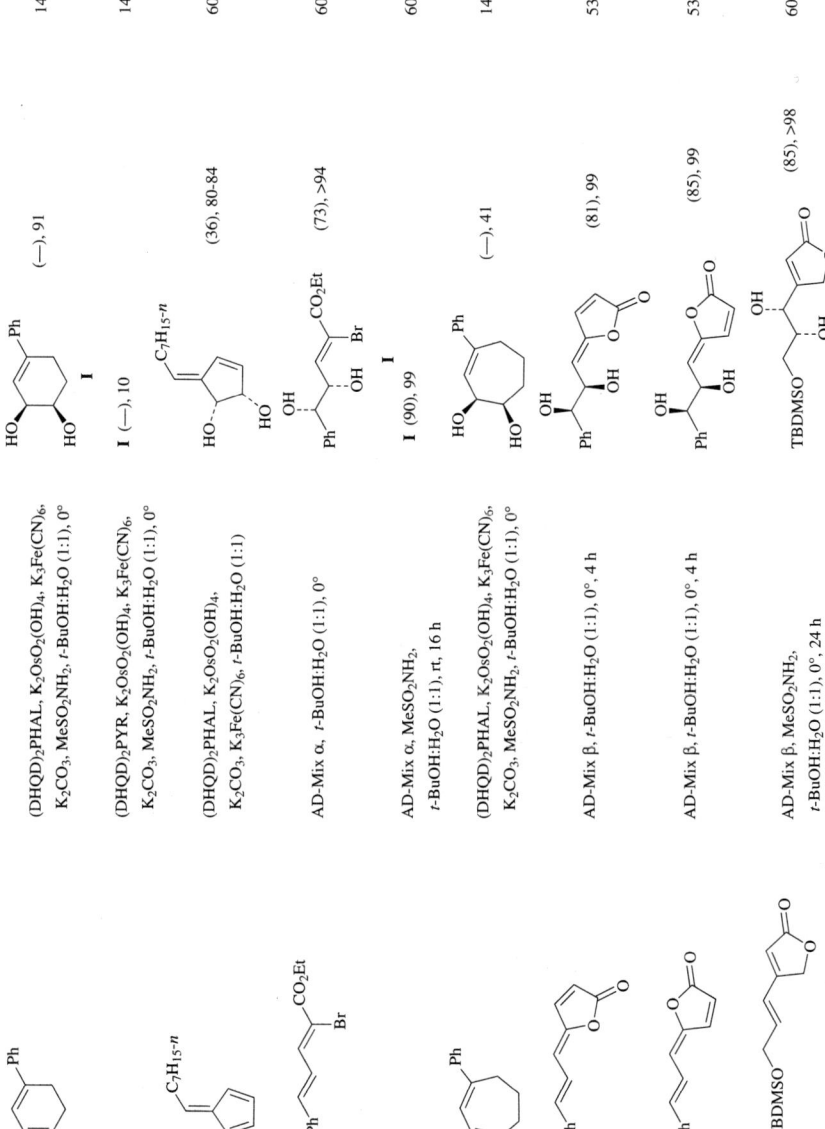

TABLE 7. REACTIONS OF CONJUGATED POLYALKENES (*Continued*)

Substrate	Conditions	Product(s) and Yield(s) (%), % ee	Refs.
C_{13} 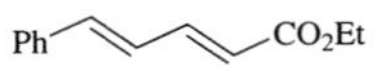	Ligand, $K_2OsO_2(OH)_4$, $K_3Fe(CN)_6$, K_2CO_3, $MeSO_2NH_2$, t-BuOH:H_2O (1:1), 0°	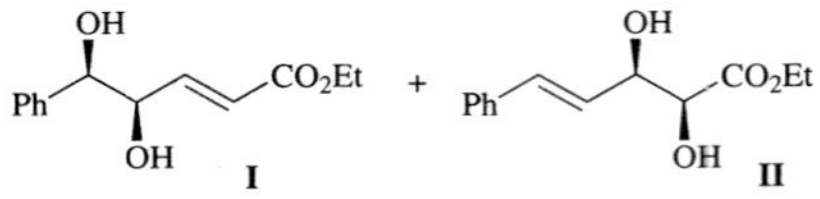	138
	Ligand $(DHQD)_2PHAL$ $(DHQD)_2PYR$	I / II (64), >99 / (12), — (51), >99 / (8), —	
	Ligand, $K_2OsO_2(OH)_4$, $K_3Fe(CN)_6$, K_2CO_3, $MeSO_2NH_2$, t-BuOH:H_2O (1:1), 0°	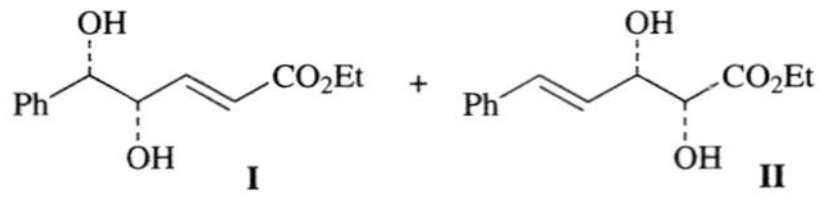	138
	Ligand $(DHQ)_2PHAL$ $(DHQ)_2PYR$	I / II (63), >97 / (10), — (57), >94 / (3), —	
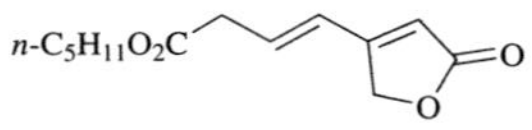	AD-Mix β, $MeSO_2NH_2$, t-BuOH:H_2O (1:1), 0°, 24 h	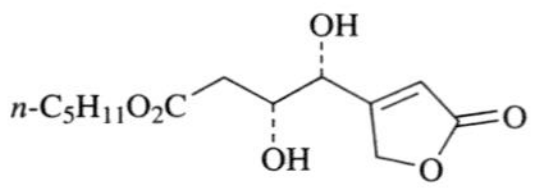 (72), >98	606
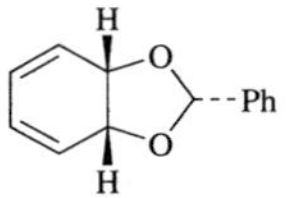	AD-Mix α, $MeSO_2NH_2$, t-BuOH:H_2O (1:1), 0°, 60 h	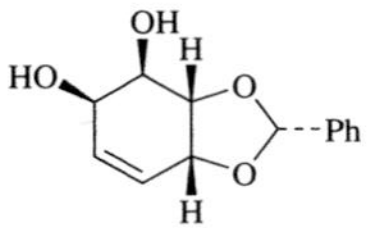 (85), 85	148, 147
	AD-Mix β, $MeSO_2NH_2$, t-BuOH:H_2O (1:1), 0°, 60 h	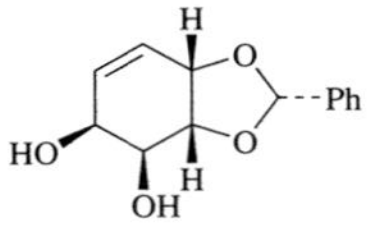 (81), 87	148, 147

Reactant	Conditions	Product(s) and Yield(s) (%), ee (%)	Refs.
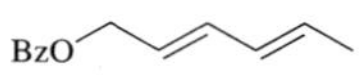	(DHQD)$_2$PHAL, $K_2OsO_2(OH)_4$, $K_3Fe(CN)_6$, K_2CO_3, $MeSO_2NH_2$, t-BuOH:H_2O (1:1), 0°	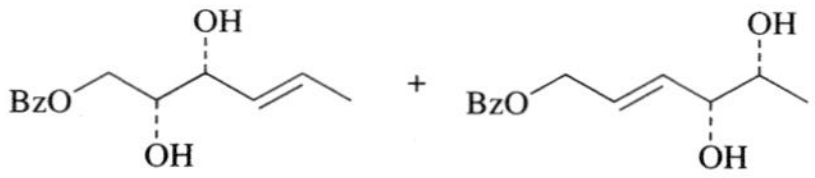 (6), — + (85), >98	138
C_{13-19}			
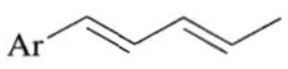	(DHQD)$_2$PHAL, $K_2OsO_2(OH)_4$, $K_3Fe(CN)_6$, K_2CO_3, $MeSO_2NH_2$, t-BuOH:H_2O (1:1), 0°	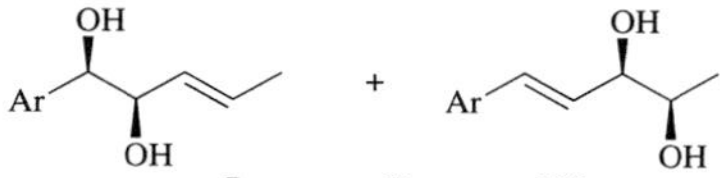	138

Ar	I	II	I:II
3,5-$Me_2C_6H_3$	(—), —	(—), 96	25:75
2-naphthyl	(—), 98	(—), —	80:20
3,5-(t-Bu)$_2C_6H_3$	(—), —	(—), 92	6:94

Reactant	Conditions	Product(s) and Yield(s) (%), ee (%)	Refs.
C_{13} 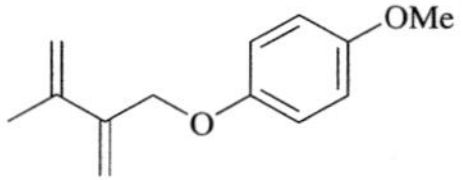	(DHQD)$_2$PYDZ, $K_2OsO_2(OH)_4$, $K_3Fe(CN)_6$, K_2CO_3, t-BuOH:H_2O (1:1), 0°	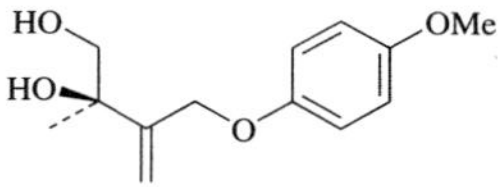 (70), 93	64
C_{14} 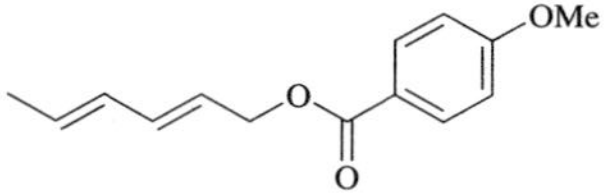	(DHQD)$_2$PYDZ, $K_2OsO_2(OH)_4$, $K_3Fe(CN)_6$, K_2CO_3, $MeSO_2NH_2$, t-BuOH:H_2O (1:1), 0°	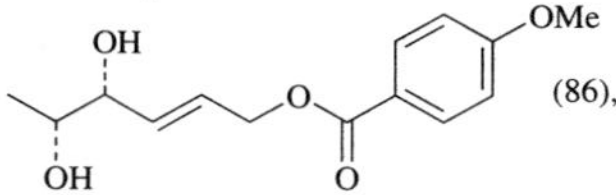 (86), 98	143
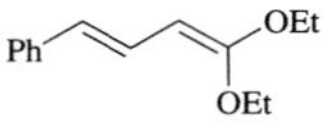	AD-Mix β, $MeSO_2NH_2$, t-BuOH:H_2O (1:1), 0°, 24 h	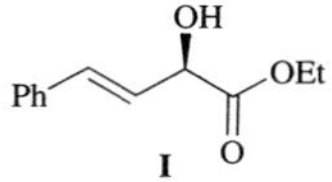 **I** (52), 94	558
	AD-Mix β, $MeSO_2NH_2$, t-BuOH:CH_3CN:H_2O (4:1:5), 0°, 6 h	**I** (73), 94	136
	AD-Mix α, $MeSO_2NH_2$, t-BuOH:H_2O (1:1), 0°, 24 h	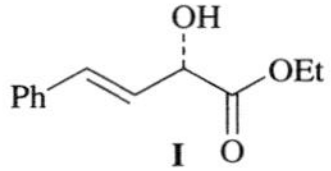 **I** (73), 96	558
	AD-Mix α, $MeSO_2NH_2$, t-BuOH:CH_3CN:H_2O (4:1:5), 0°, 6 h	**I** (73), 96	136

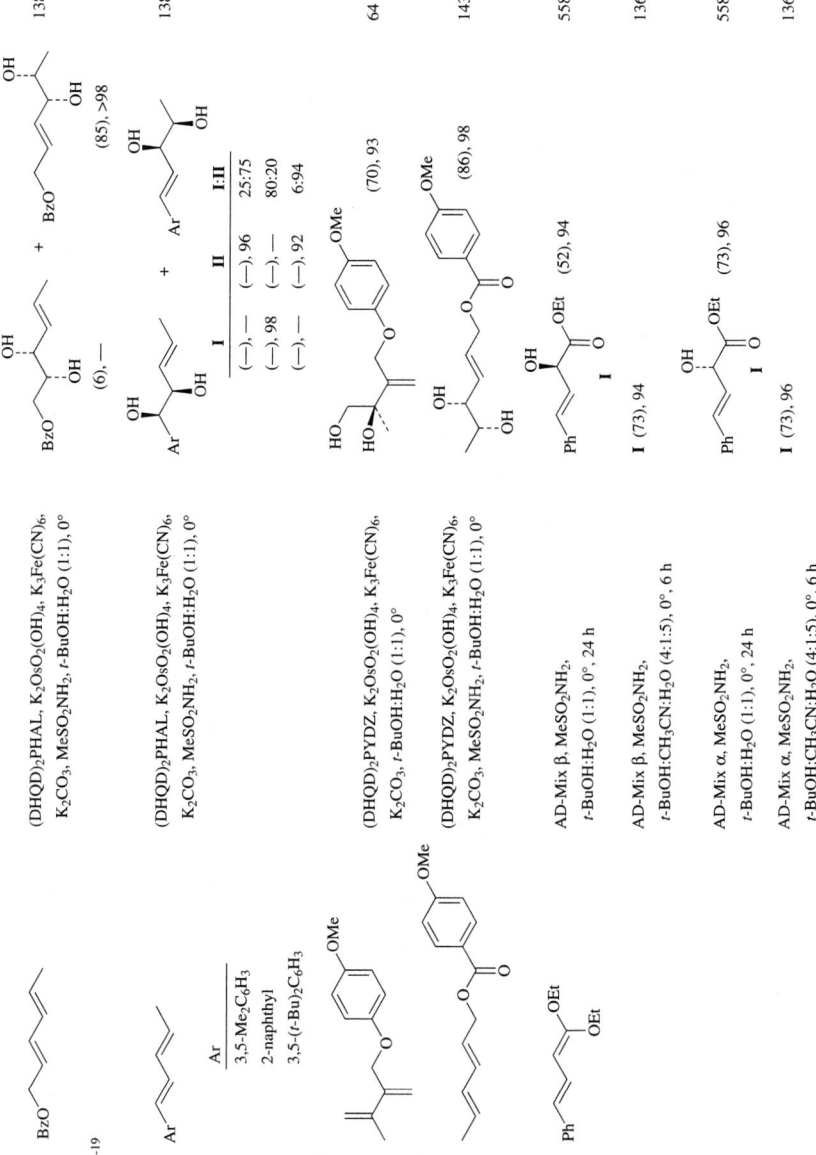

TABLE 7. REACTIONS OF CONJUGATED POLYALKENES (*Continued*)

Substrate	Conditions	Product(s) and Yield(s) (%), % ee	Refs.
C_{14}			
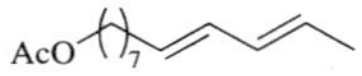	(DHQD)$_2$PHAL, $K_2OsO_2(OH)_4$, $K_3Fe(CN)_6$, K_2CO_3, $MeSO_2NH_2$, *t*-BuOH:H_2O (1:1), 0°	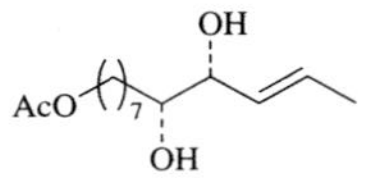 (55), 95 + 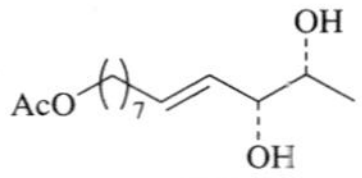 (27), 94	138
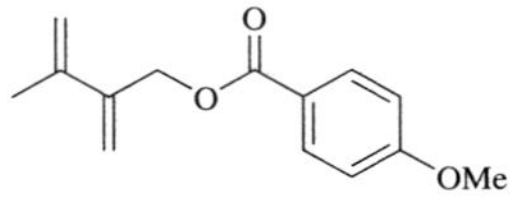	(DHQD)$_2$PYDZ, $K_2OsO_2(OH)_4$, $K_3Fe(CN)_6$, K_2CO_3, *t*-BuOH:H_2O (1:1), 0°	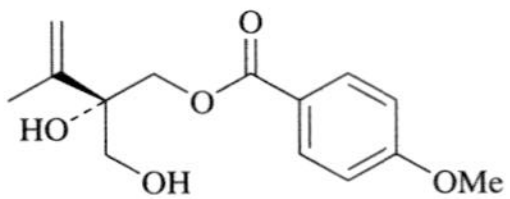 (80), 84	64
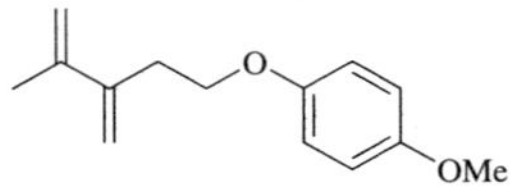	(DHQD)$_2$PYDZ, $K_2OsO_2(OH)_4$, $K_3Fe(CN)_6$, K_2CO_3, *t*-BuOH:H_2O (1:1), 0°	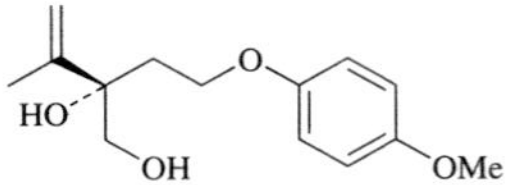 (70), 74	64
C_{15}			
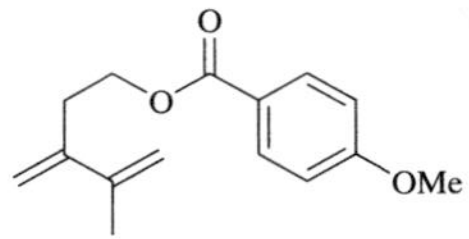	(DHQD)$_2$PYDZ, $K_2OsO_2(OH)_4$, $K_3Fe(CN)_6$, K_2CO_3, $MeSO_2NH_2$, *t*-BuOH:H_2O (1:1), 0°, 4-22 h	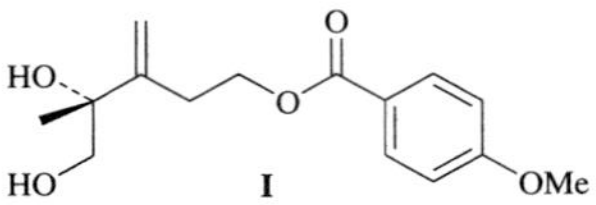 **I** (73), 89	82
	21, $K_2OsO_2(OH)_4$, $K_3Fe(CN)_6$, K_2CO_3, $MeSO_2NH_2$, *t*-BuOH:H_2O (1:1), 0°, 4-22 h	**I** (74), 97	64, 82
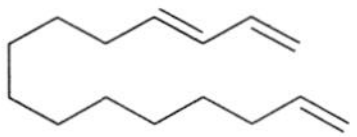	AD-Mix β, $MeSO_2NH_2$, *t*-BuOH:H_2O (1:1), 0°, 18 h	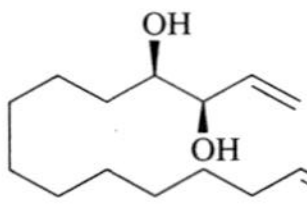 (52), >96 + 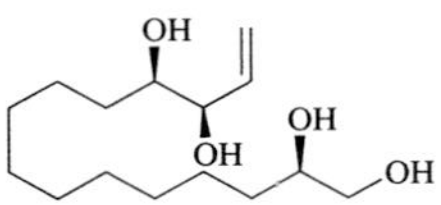 (36), 83	279

	AD-Mix β, $MeSO_2NH_2$, t-BuOH:H_2O (1:1), 0°, 8 h	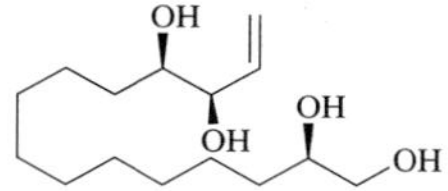 (89), 83	279
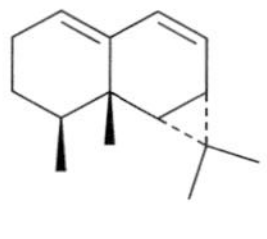	AD-Mix α, $MeSO_2NH_2$, t-BuOH:H_2O (1:1), 0°, 24 h	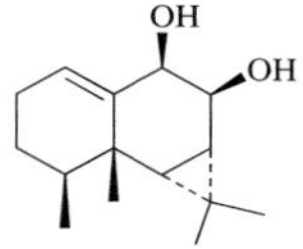 (85), >85 de	607
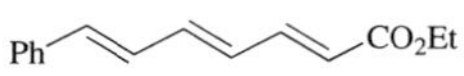	$(DHQD)_2$PHAL, $K_2OsO_2(OH)_4$, $K_3Fe(CN)_6$, K_2CO_3, $MeSO_2NH_2$, t-BuOH:H_2O (1:1), 0°	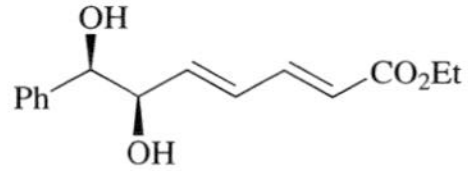 (77), >99	138
	$(DHQ)_2$PHAL, $K_2OsO_2(OH)_4$, $K_3Fe(CN)_6$, K_2CO_3, $MeSO_2NH_2$, t-BuOH:H_2O (1:1), 0°	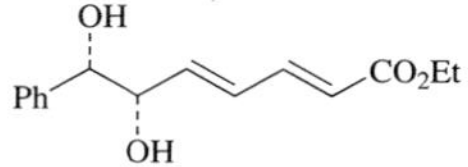 (75), >99	138
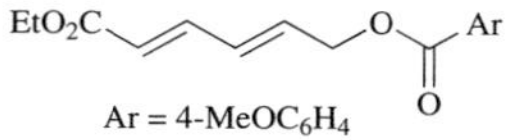 Ar = 4-$MeOC_6H_4$	$(DHQD)_2$PYDZ, $K_2OsO_2(OH)_4$, $K_3Fe(CN)_6$, K_2CO_3, $MeSO_2NH_2$, t-BuOH:H_2O (1:1), 0°	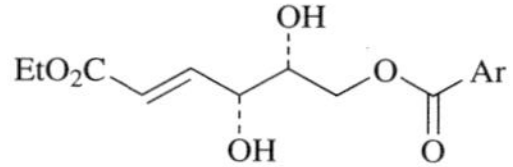 (82), 97	143
C_{16}			
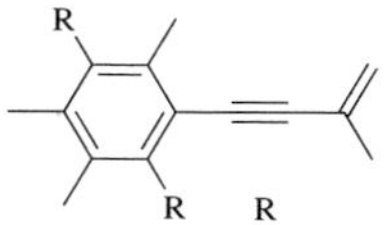 R: MeO; Me	$(DHQ)_2$PHAL, $K_2OsO_2(OH)_4$, $K_3Fe(CN)_6$, K_2CO_3, t-BuOH:H_2O (1:1), 4°, 18 h	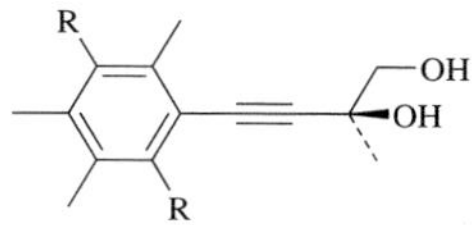 (93), 84; (95), 85	608
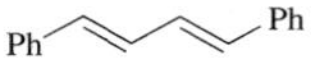	$(DHQD)_2$PHAL, $K_2OsO_2(OH)_4$, $K_3Fe(CN)_6$, K_2CO_3, $MeSO_2NH_2$, t-BuOH:H_2O (1:1), 0°	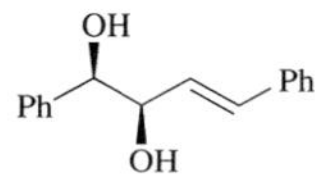 (85), >99	139

TABLE 7. REACTIONS OF CONJUGATED POLYALKENES (*Continued*)

Substrate	Conditions	Product(s) and Yield(s) (%), % ee	Refs.
C_{16}			
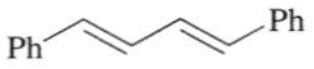	DHQD-*p*-chlorobenzoate, OsO_4, NMO, acetone:H_2O (5:1), 0°, 18 h	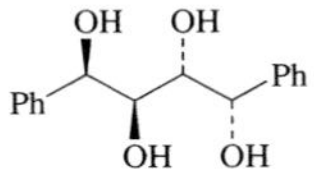 (—), 71 de	609
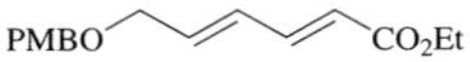	AD-Mix α, *t*-BuOH:H_2O (1:1), 0°	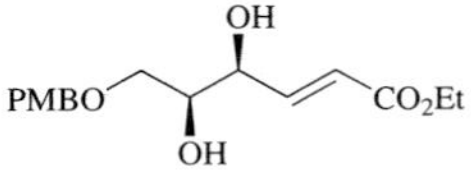 (62), 97	610
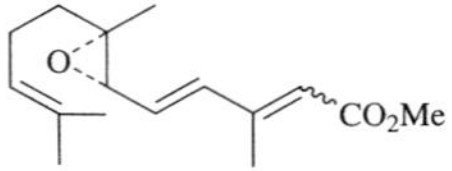	AD-Mix α, *t*-BuOH:H_2O (1:1)	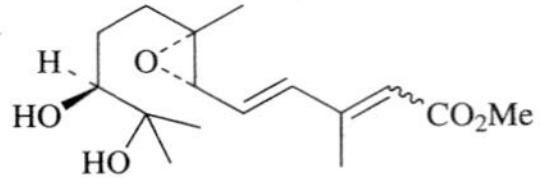 (82), 88 de	611
C_{17}			
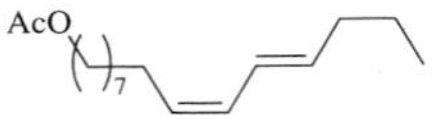	$(DHQD)_2PHAL$, $K_2OsO_2(OH)_4$, $K_3Fe(CN)_6$, K_2CO_3, $MeSO_2NH_2$, *t*-BuOH:H_2O (1:1), 0°	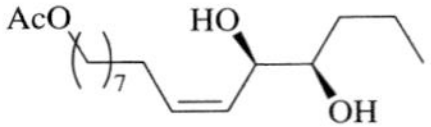 (82), 93	138
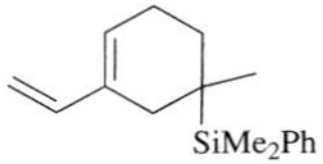	1. AD-Mix α, *t*-BuOH:H_2O (1:1), 0°, 12 h 2. $NaIO_4$, THF-H_2O, rt, 1.5 h	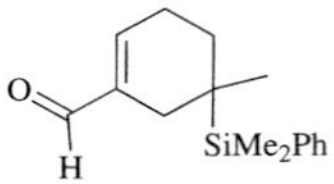 (81), —	612
	AD-Mix β, *t*-BuOH:H_2O (1:1), 0°, 12 h	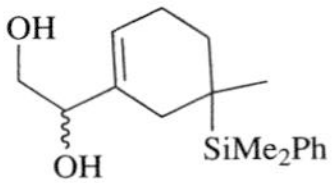 (92), —	613
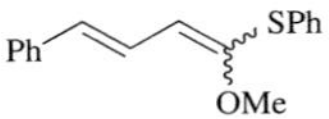	AD-Mix α, $MeSO_2NH_2$, *t*-BuOH:CH_3CN:H_2O (4:1:5), 0°, 6 h	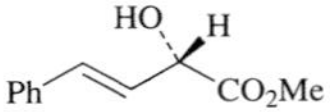 (30), 75	136
	AD-Mix β, $MeSO_2NH_2$, *t*-BuOH:CH_3CN:H_2O (4:1:5), 0°, 6 h	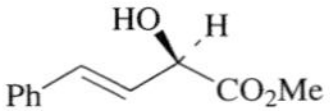 (30), 80	136

C_{22} 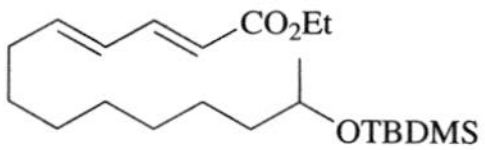	AD-Mix α, $MeSO_2NH_2$, t-BuOH:H_2O (1:1), 0°, 10 h	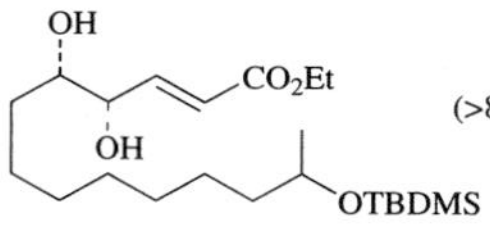	(>80), >89 de	614
C_{23} 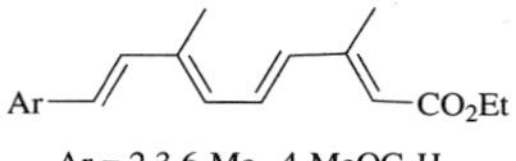 Ar = 2,3,6-Me_3-4-$MeOC_6H$	$(DHQD)_2PHAL$, $K_2OsO_2(OH)_4$, $K_3Fe(CN)_6$, K_2CO_3, $MeSO_2NH_2$, t-BuOH:H_2O (1:1), 0°	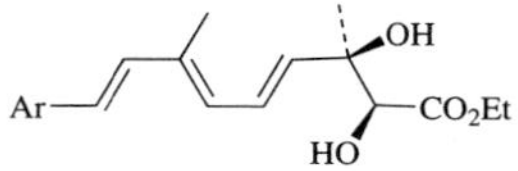	(42), 82	138
	$(DHQ)_2PHAL$, $K_2OsO_2(OH)_4$, $K_3Fe(CN)_6$, K_2CO_3, $MeSO_2NH_2$, t-BuOH:H_2O (1:1), 0°	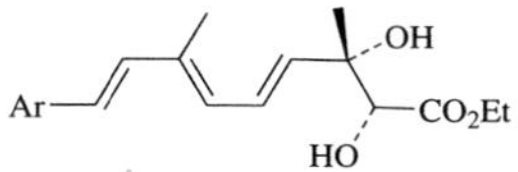	(28), 82	138
C_{24} 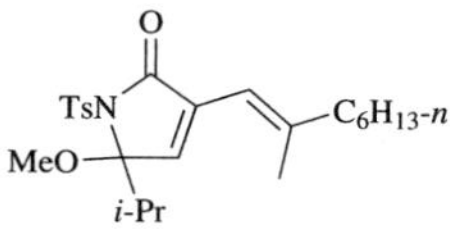	AD-Mix α, $MeSO_2NH_2$, t-BuOH:H_2O (1:1), 0°	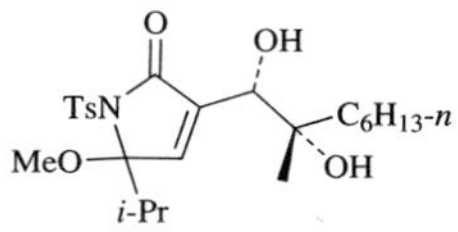	(77), 91-93, 0 de	615
C_{26-40} 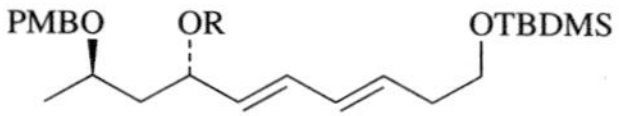	AD-Mix α, $MeSO_2NH_2$, t-BuOH:H_2O (1:1), 0°, 24 h	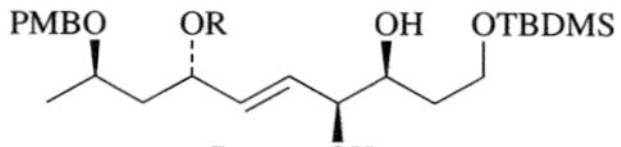 I + 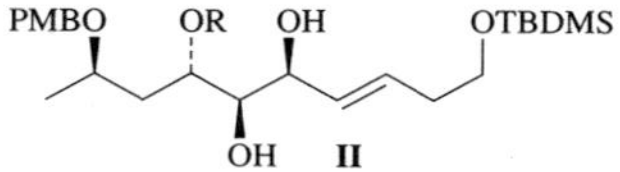 II		616

R	I	II
MOM	(78), 99 de	(5), —
Ac	(67), 90	(13), —
PMB	(43), 54 de	(15), —
TBDPS	(63), 88 de	(7), —

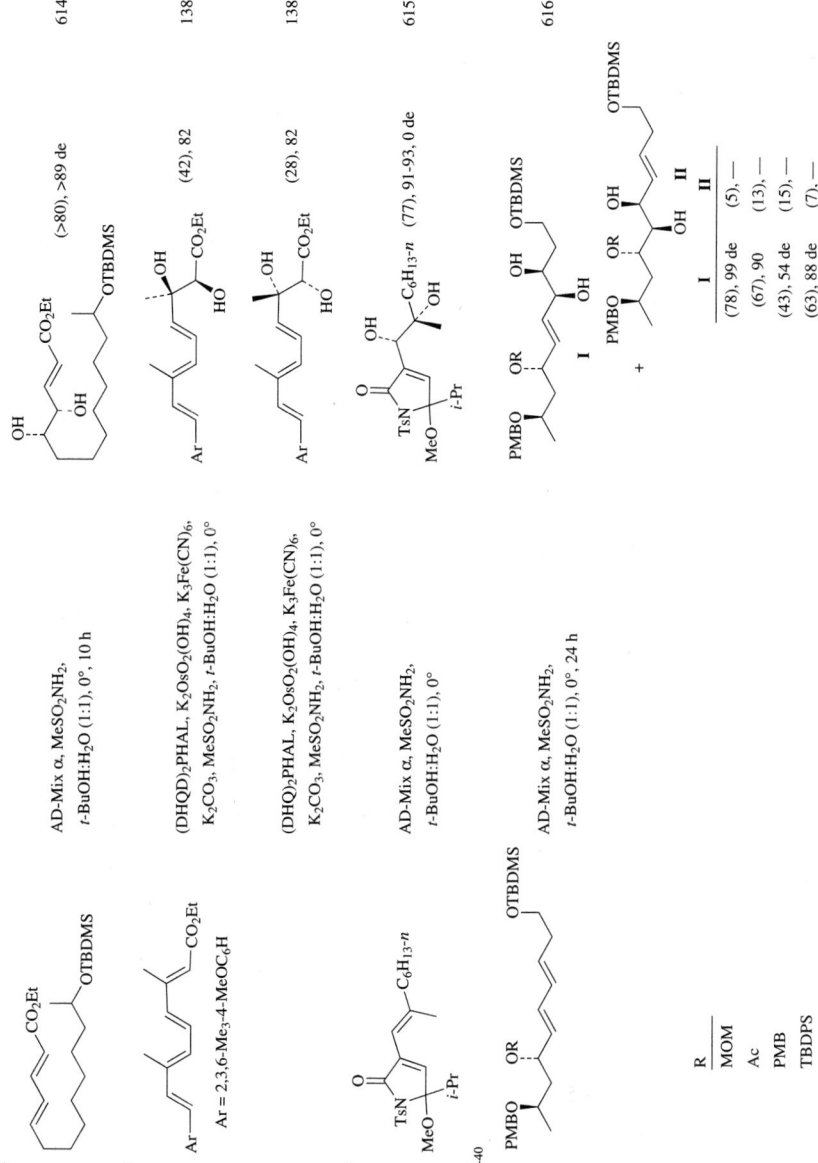

TABLE 7. REACTIONS OF CONJUGATED POLYALKENES (*Continued*)

	Substrate	Conditions	Product(s) and Yield(s) (%), % ee	Refs.
C_{29-37}	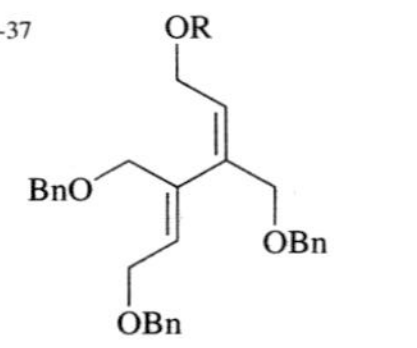	$(DHQD)_2PHAL$, OsO_4, NMO, acetone:H_2O (5:1), rt	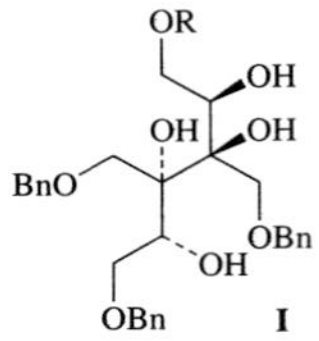 **I**	617
	R: H		**I**: (74), 24, 78 de	
	Piv		(37), 33 de	
	TBDMS		(37), 50 de	
	COC_6H_4OMe-*p*		(76), 29	
C_{29}	R = H	AD-Mix β, $MeSO_2NH_2$, $K_2S_2O_3$, *t*-BuOH:H_2O (1:1), 0°-rt, 4 d	**I** (45), 76, >80 de	617
C_{30}	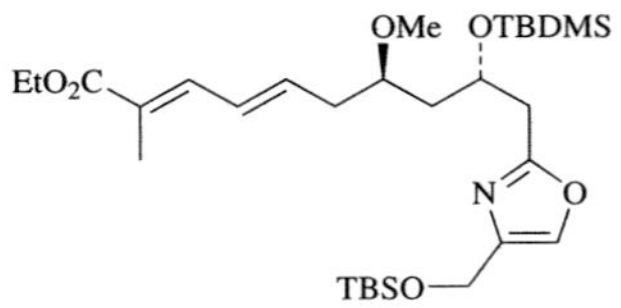	AD-Mix β, $MeSO_2NH_2$, *t*-BuOH:H_2O (2:3), rt, 3 d	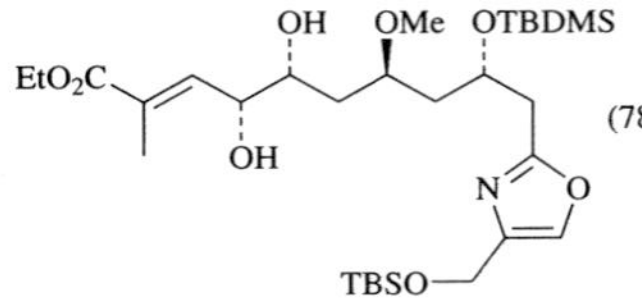 (78), 92 de	618
C_{31}	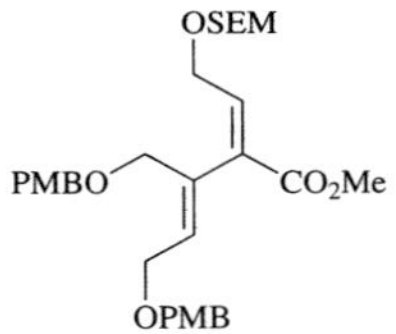	$(DHQD)_2PHAL$, $K_2OsO_2(OH)_4$, $K_3Fe(CN)_6$, K_2CO_3, $MeSO_2NH_2$, *t*-BuOH:H_2O (1:1), 0°, 36 h	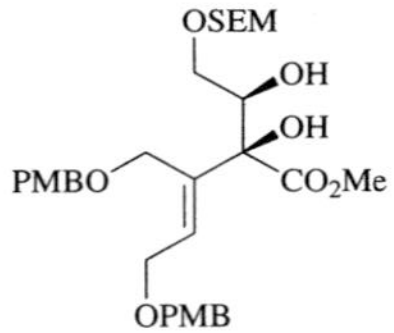 (20), —	619

C_{32}	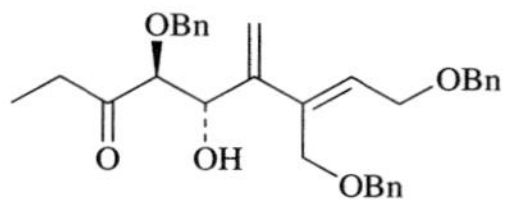	AD-Mix β, $MeSO_2NH_2$, *t*-BuOH:H_2O (1:1), 20°, 14 h	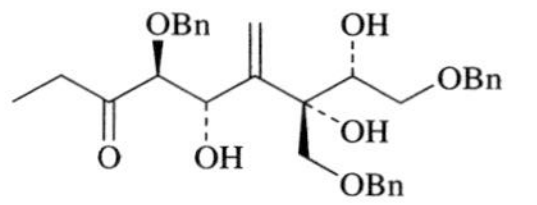 (61), —	620
C_{34}	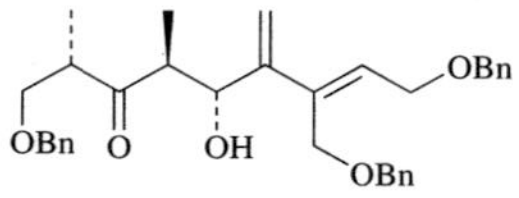	AD-Mix β, *t*-BuOH:H_2O (1:1), 20°, 14 h	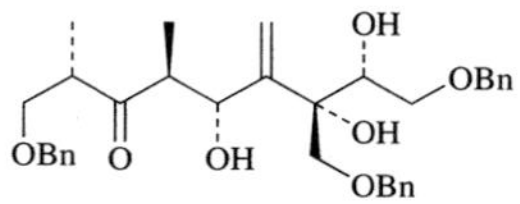 (63), >95 de	620
C_{49}	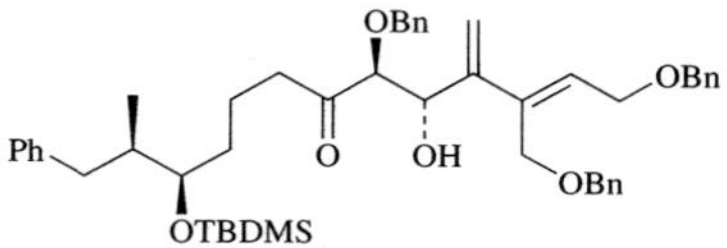	AD-Mix β, $MeSO_2NH_2$, *t*-BuOH:H_2O (1:1), 20°, 18 h	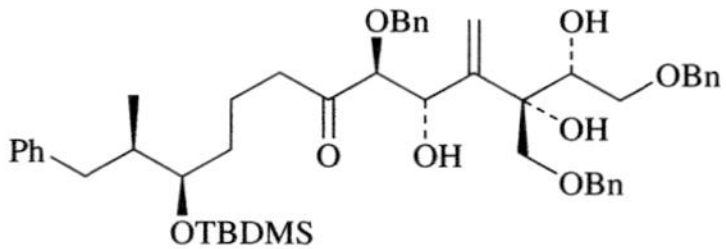 (54), >95 de	594

TABLE 7. REACTIONS OF CONJUGATED POLYALKENES (*Continued*)

Substrate	Conditions	Product(s) and Yield(s) (%), % ee	Refs.
C_{51} 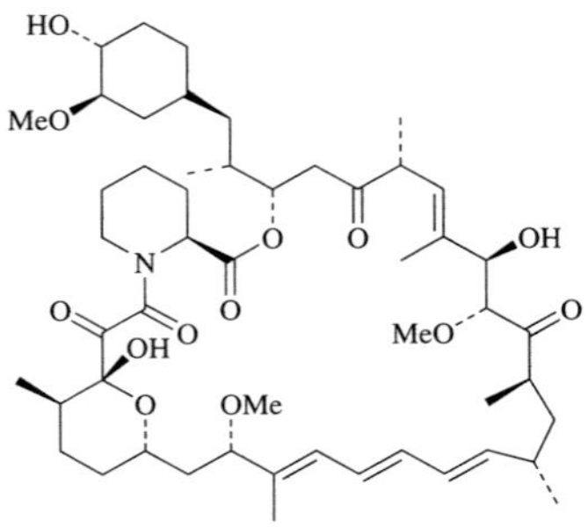	(DHQD)$_2$PHAL, $K_2OsO_2(OH)_2$, $K_3Fe(CN)_6$, K_2CO_3, $MeSO_2NH_2$, t-BuOH:H_2O (1:1), 0°, 12 h	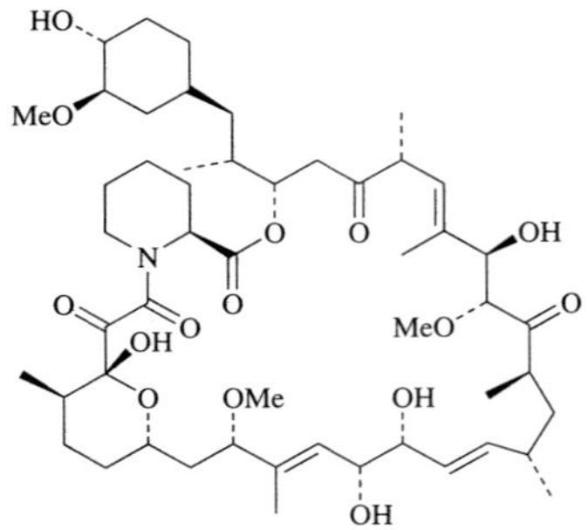 (39), >95 de + 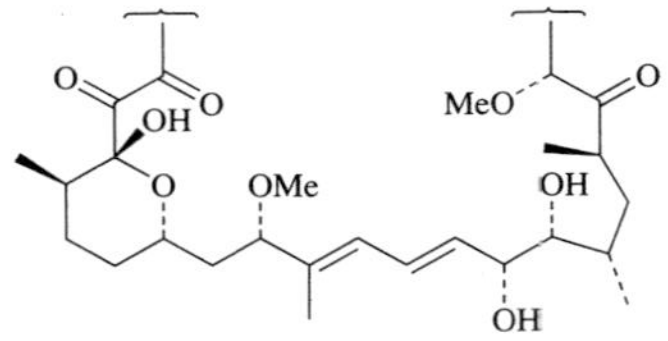 (23), >95 de	621

TABLE 8. REACTIONS OF UNCONJUGATED POLYALKENES

Substrate	Conditions	Product(s) and Yield(s) (%), % ee	Refs.
See Chart 1 at the beginning of the Tabular Survey for ligand structures that are indicated by ***bold*** *numbers.*			
C_6 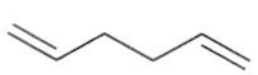	$(DHQD)_2PYR$, $K_2OsO_2(OH)_4$, $K_3Fe(CN)_6$, K_2CO_3, t-BuOH:H_2O (1:1), 0°, 24 h	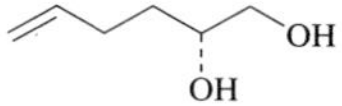 (64-65), 77-80	144, 622
	$(DHQ)_2PYR$, $K_2OsO_2(OH)_4$, $K_3Fe(CN)_6$, K_2CO_3, t-BuOH:H_2O (1:1), 0°, 24 h	 (42), 70	144
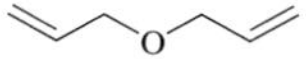	Ligand, $K_2OsO_2(OH)_4$, $K_3Fe(CN)_6$, K_2CO_3, t-BuOH:H_2O (1:1), 0°, 24 h	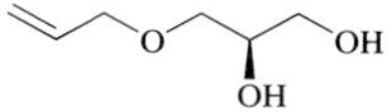	144
	Ligand		
	$(DHQ)_2PYR$	(—), 33	
	$(DHQ)_2AQN$	(76), 83	
	Ligand, $K_2OsO_2(OH)_4$, $K_3Fe(CN)_6$, K_2CO_3, t-BuOH:H_2O (1:1), 0°, 24 h	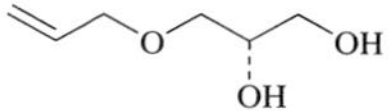	144
	Ligand		
	$(DHQD)_2PYR$	(—), 45	
	$(DHQD)_2AQN$	(96), 89	
	AD-Mix β, t-BuOH:H_2O (1:1), 0°, 24 h	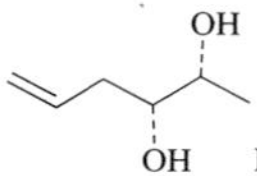 **I** (45-58), 96	623
	$(DHQD)_2PHAL$, $K_2OsO_2(OH)_4$, $K_3Fe(CN)_6$, K_2CO_3, $MeSO_2NH_2$, t-BuOH:H_2O (1:1), 0°	**I** + 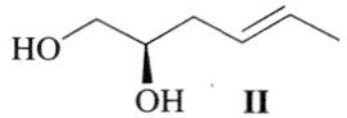 **II** **I** + **II** (56), 94; **I**:**II** = 13:1	139
C_7 	Ligand, $K_2OsO_2(OH)_4$, $K_3Fe(CN)_6$, K_2CO_3, t-BuOH:H_2O (1:1), 0°	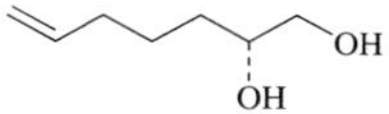	
	Ligand / Time		
	$(DHQD)_2PYR$ / 24 h	(81), 88	144
	$(DHQD)_2PHAL$ / 24 h	(—), 84	144
	$(DHQD)_2PYR$ / —	(81), 88	217

TABLE 8. REACTIONS OF UNCONJUGATED POLYALKENES (*Continued*)

Substrate	Conditions	Product(s) and Yield(s) (%), % ee	Refs.
C_7 	Ligand, $K_2OsO_2(OH)_4$, $K_3Fe(CN)_6$, K_2CO_3, *t*-BuOH:H_2O (1:1), 0°, 24 h	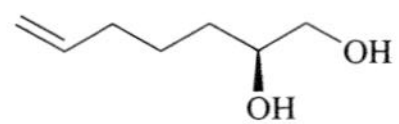	144
	Ligand		
	$(DHQ)_2PYR$	(81), 85	
	$(DHQ)_2PHAL$	(—), 83	
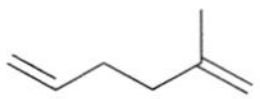	$(DHQD)_2PHAL$, $K_2OsO_2(OH)_4$, $K_3Fe(CN)_6$, K_2CO_3, $MeSO_2NH_2$, *t*-BuOH:H_2O (1:1), 0°	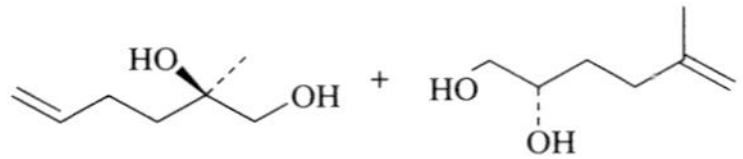 **I** + **II**; **I + II** (42), 74; **I**:**II** = 5:1	139
C_8 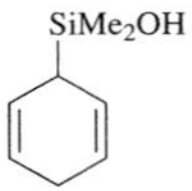	Ligand, $K_2OsO_2(OH)_4$, $K_3Fe(CN)_6$, K_2CO_3, *t*-BuOH:H_2O (1:1), 0°, 12 h	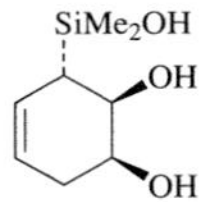	624
	Ligand		
	$(DHQD)_2PHAL$	(65-75), 44	
	DHQD-IND	(65-75), 40	
	$(DHQD)_2PYR$	(65-75), 52	
	$(DHQ)_2PYR$, $K_2OsO_2(OH)_4$, $K_3Fe(CN)_6$, K_2CO_3, *t*-BuOH:H_2O (1:1), 0°, 12 h	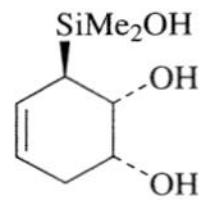 (65-75), 65	624
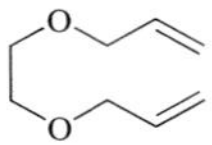	AD-Mix α, *t*-BuOH:H_2O (1:1)	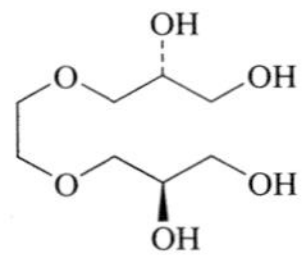 (49), 28	625

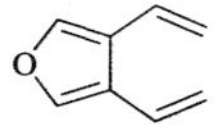	AD-Mix α, t-BuOH:H_2O (1:1), 0°-rt, 17 h	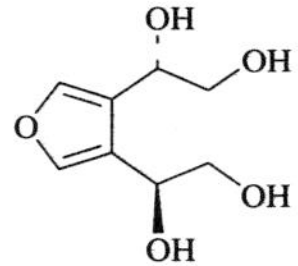	(51), >98, >80 de	321, 626
C_{8-9} 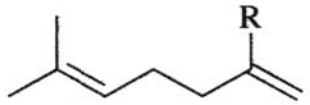	(DHQD)$_2$PHAL, $K_2OsO_2(OH)_4$, $K_3Fe(CN)_6$, K_2CO_3, $MeSO_2NH_2$, t-BuOH:H_2O (1:1), 0°	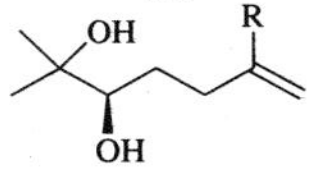	R: H (73), 98; Me (70), 98	139
C_9 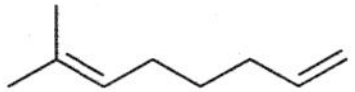	Ligand, $K_2OsO_2(OH)_4$, $K_3Fe(CN)_6$, K_2CO_3, $MeSO_2NH_2$, t-BuOH:H_2O (1:1), 0°	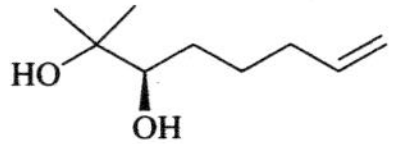		
	Ligand — Time			
	(DHQD)$_2$PHAL — 15 h	(74), 98		77, 66, 157, 627, 152
	(DHQD)$_2$PYDZ — 6-24 h	(—), 95		152, 66
	(DHQD)$_2$TP — 6-24 h	(—), 77		152
	(DHQD)$_2$PYR — 6-24 h	(—), 87		157
	(DHQD)$_2$DPP — 6-24 h	(—), 98		157
	(DHQD)$_2$DPPHAL — 6-24 h	(—), 99		157, 86
	(DHDQ)$_2$AQN — —	(—), 99		86
	Ligand, $K_2OsO_2(OH)_4$, $K_3Fe(CN)_6$, K_2CO_3, $MeSO_2NH_2$, t-BuOH:H_2O (1:1), 0°	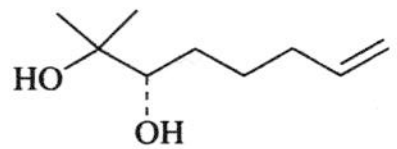		
	Ligand — Time			
	(DHQ)$_2$PHAL — 15 h	(69), 94-95		77, 627, 152, 157,
	(DHQ)$_2$PYDZ — 6-24 h	(—), 81		86
	(DHQ)$_2$TP — 6-24 h	(—), 65		152
	(DHQ)$_2$DPP — 6-24 h	(—), 94		152
	(DHQ)$_2$DPPHAL — 6-24 h	(—), 91		157
	(DHQ)$_2$AQN — —	(—), 96.5		86

TABLE 8. REACTIONS OF UNCONJUGATED POLYALKENES (*Continued*)

Substrate	Conditions	Product(s) and Yield(s) (%), % ee	Refs.
C_9			
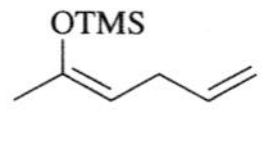	(DHQ)$_2$PHAL, $K_2OsO_2(OH)_4$, $K_3Fe(CN)_6$, K_2CO_3, t-BuOH:H_2O (1:1), 0°, 2 h	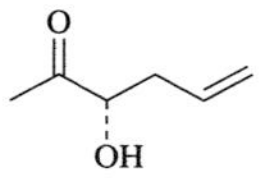 (87), 90	628
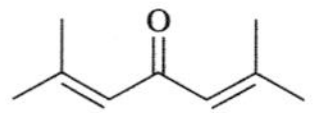	(DHQD)$_2$PHAL, $K_2OsO_2(OH)_4$, $K_3Fe(CN)_6$, $NaHCO_3$-K_2CO_3, $MeSO_2NH_2$, t-BuOH:H_2O (1:1), 0°, 24 h	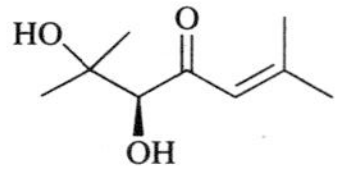 (59), 89	155
	(DHQD)$_2$PYR, $K_2OsO_2(OH)_4$, $K_3Fe(CN)_6$, K_2CO_3, t-BuOH:H_2O (1:1), 0°, 24 h	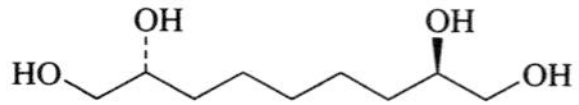 (45), >95	234
C_{10}			
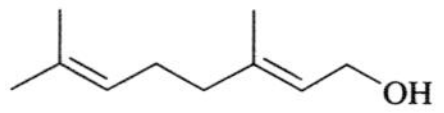	(DHQ)$_2$PHAL, $K_2OsO_2(OH)_4$, $K_3Fe(CN)_6$, K_2CO_3, $MeSO_2NH_2$, t-BuOH:H_2O (1:1), 0°, 24 h	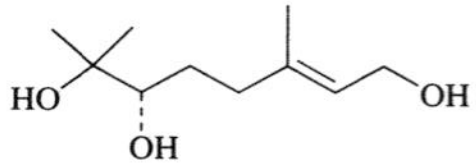 (52), >78[a]	141
	(DHQD)$_2$PHAL, $K_2OsO_2(OH)_4$, $K_3Fe(CN)_6$, K_2CO_3, $MeSO_2NH_2$, t-BuOH:H_2O (1:1), 0°, 12 h	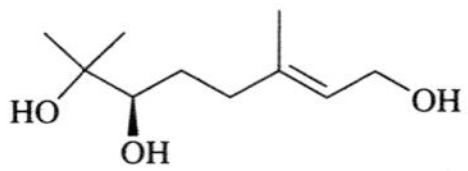 (89), 94	98
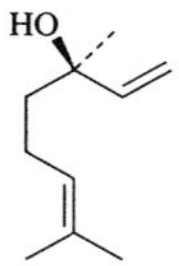	(DHQ)$_2$PHAL, $K_2OsO_2(OH)_4$, $K_3Fe(CN)_6$, K_2CO_3, $MeSO_2NH_2$, t-BuOH:H_2O (1:1), 0°, 24 h	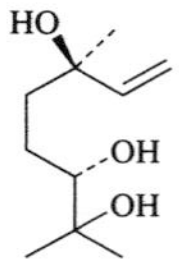 (80), 84 de	564

	(DHQD)$_2$PHAL, $K_2OsO_2(OH)_4$, $K_3Fe(CN)_6$, K_2CO_3, $MeSO_2NH_2$, t-BuOH:H_2O (1:1), 0°, 24 h	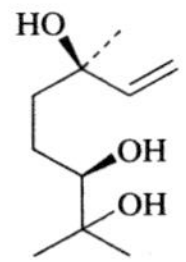	(77), 94 de	564
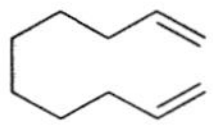	AD-Mix α, t-BuOH:H_2O (1:1)	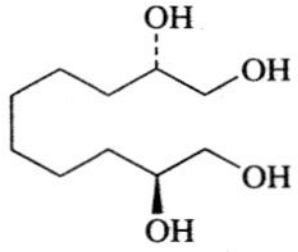	(95), 84	625
	(DHQD)$_2$PYR, $K_2OsO_2(OH)_4$, $K_3Fe(CN)_6$, K_2CO_3, $MeSO_2NH_2$, t-BuOH:H_2O (1:1), 0°	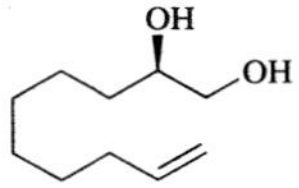	(75), 82	447
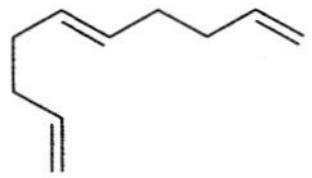	(DHQD)$_2$PHAL, $K_2OsO_2(OH)_4$, $K_3Fe(CN)_6$, K_2CO_3, $MeSO_2NH_2$, t-BuOH:H_2O (1:1), 0°	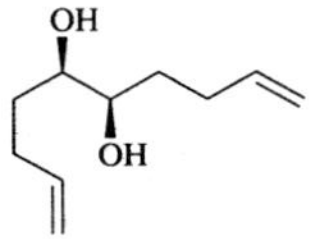	(51-59), ~95	629, 630
	(DHQ)$_2$PHAL, $K_2OsO_2(OH)_4$, $K_3Fe(CN)_6$, K_2CO_3, $MeSO_2NH_2$, t-BuOH:H_2O (1:1), 0°	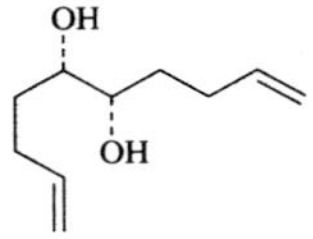	(57), 94	629, 630
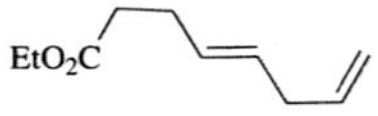	AD-Mix α, t-BuOH:H_2O (1:1)	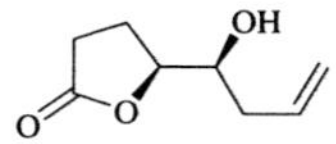	(59), 92	631
	AD-Mix β, t-BuOH:H_2O (1:1)	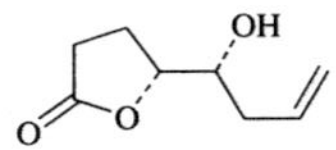	(64), 96	631

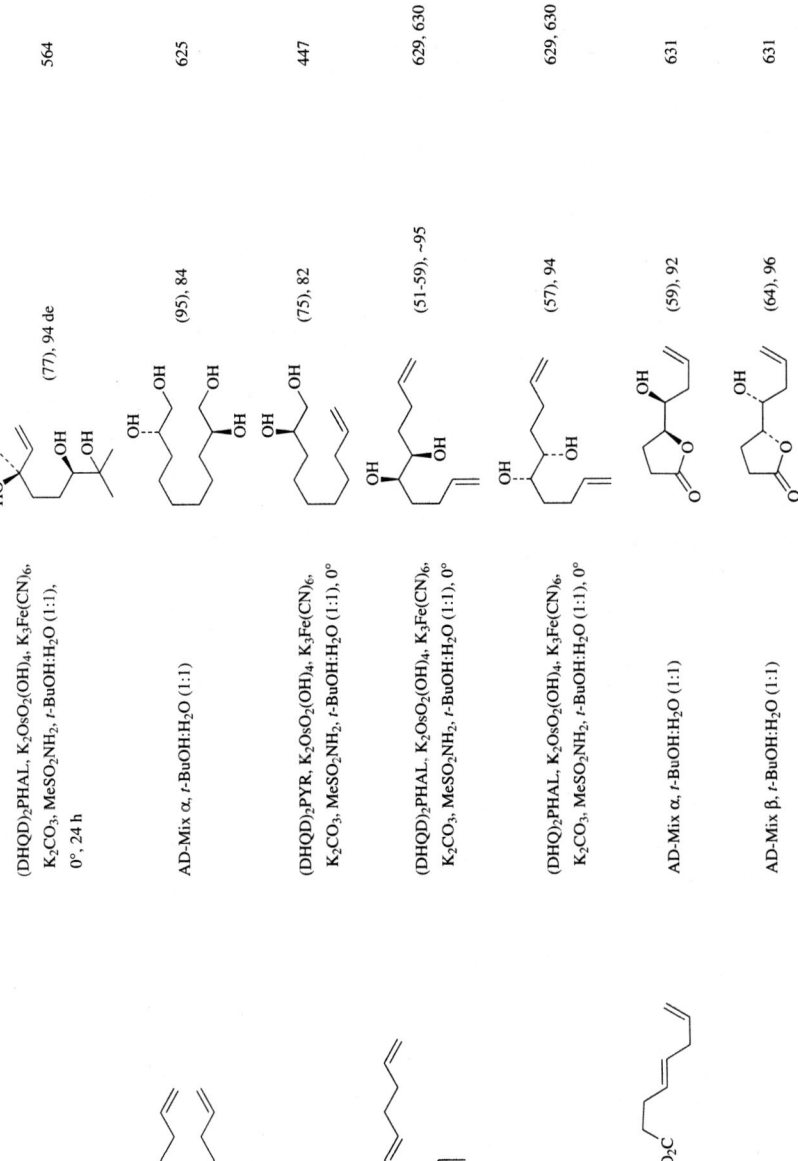

TABLE 8. REACTIONS OF UNCONJUGATED POLYALKENES (*Continued*)

Substrate	Conditions	Product(s) and Yield(s) (%), % ee	Refs.
C_{10}			
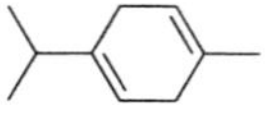	(DHQD)$_2$PHAL, $K_2OsO_2(OH)_4$, $K_3Fe(CN)_6$, K_2CO_3, $MeSO_2NH_2$, t-BuOH:H_2O (1:1), 0°	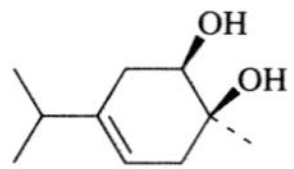 (84), 96	138
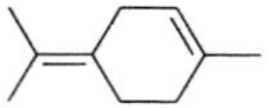	(DHQD)$_2$PHAL, $K_2OsO_2(OH)_4$, $K_3Fe(CN)_6$, K_2CO_3, $MeSO_2NH_2$, t-BuOH:H_2O (1:1), 0°	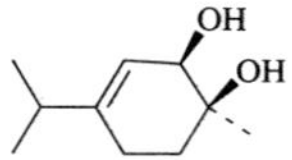 (67), 92 + 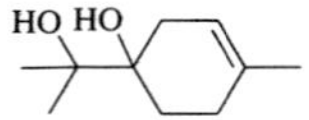 (13), —	138
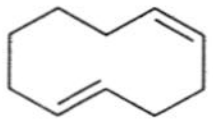	Ligand, $K_2OsO_2(OH)_4$, $K_3Fe(CN)_6$, K_2CO_3, $MeSO_2NH_2$, t-BuOH:H_2O (1:1), 0°	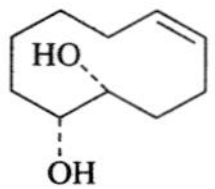	138
	Ligand		
	(DHQD)$_2$PHAL	(75-95), 51	
	(DHQD)$_2$PYR	(75-95), 94	
C_{11}			
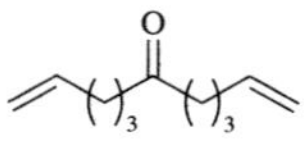	(DHQD)$_2$PHAL, $K_2OsO_2(OH)_4$, $K_3Fe(CN)_6$, K_2CO_3, $MeSO_2NH_2$, t-BuOH:H_2O (1:1), 4°, 18 h	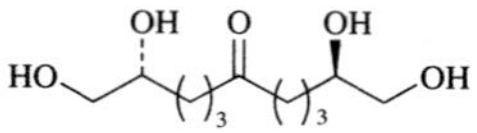 (—), >95	632
	(DHQ)$_2$PHAL, $K_2OsO_2(OH)_4$, $K_3Fe(CN)_6$, K_2CO_3, $MeSO_2NH_2$, t-BuOH:H_2O (1:1), 4°, 18 h	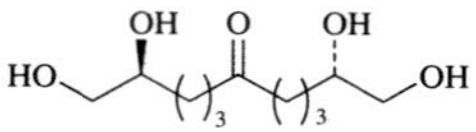 (—), >95	632

Substrate	Conditions	Product(s) and Yield(s) (%), ee	Refs.
C$_{12}$ 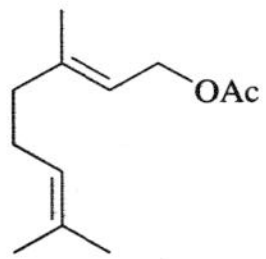	Ligand, $K_2OsO_2(OH)_4$, $K_3Fe(CN)_6$, K_2CO_3, $MeSO_2NH_2$, t-BuOH:H_2O (1:1)	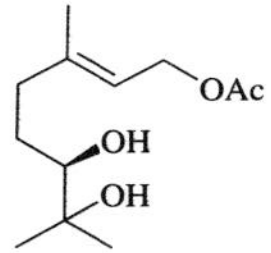	
	Ligand: $(DHQD)_2PYDZ$, Temp: 4°, Time: 12 h	(76), >95	632a
	Ligand: $(DHQD)_2PHAL$, Temp: 0°, Time: 24 h	(82), 92	141
	$(DHQ)_2PHAL$, $K_2OsO_2(OH)_4$, $K_3Fe(CN)_6$, K_2CO_3, $MeSO_2NH_2$, t-BuOH:H_2O (1:1), 0°, 24 h	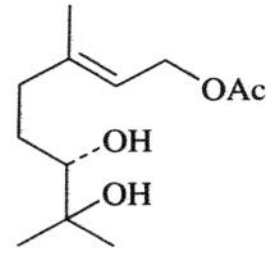 (78), 90	141
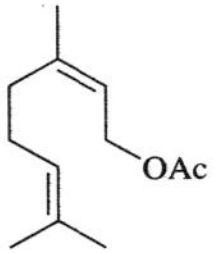	$(DHQ)_2PHAL$, $K_2OsO_2(OH)_4$, $K_3Fe(CN)_6$, K_2CO_3, $MeSO_2NH_2$, t-BuOH:H_2O (1:1), 0°, 24 h	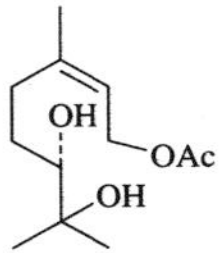 (92), 98 ee	141
	$(DHQD)_2PHAL$, $K_2OsO_2(OH)_4$, $K_3Fe(CN)_6$, K_2CO_3, $MeSO_2NH_2$, t-BuOH:H_2O (1:1), 0°, 24 h	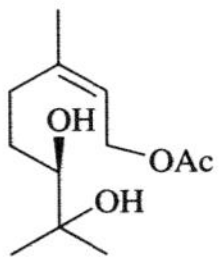 (94), 97 ee	141
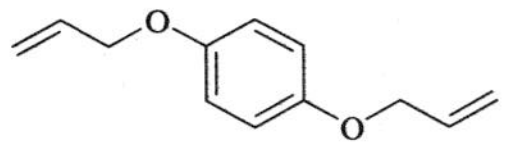	AD-Mix β, t-BuOH:H_2O (1:1), 0°, 15 h	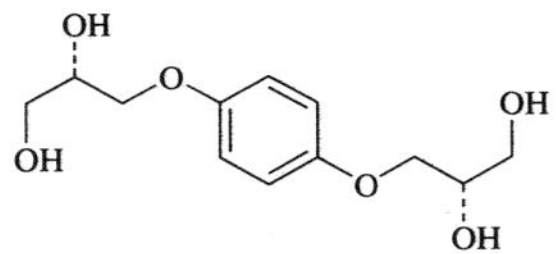 (80), 90 de	633

TABLE 8. REACTIONS OF UNCONJUGATED POLYALKENES (*Continued*)

Substrate	Conditions	Product(s) and Yield(s) (%), % ee	Refs.
C_{12} 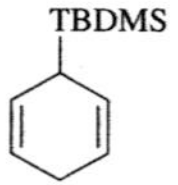	(DHQ)$_2$PYR, $K_2OsO_2(OH)_4$, $K_2Fe(CN)_6$, K_2CO_3, t-BuOH:H_2O (1:1), 0°, 12 h	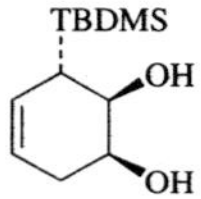 (>76), 71 , >98 de	634
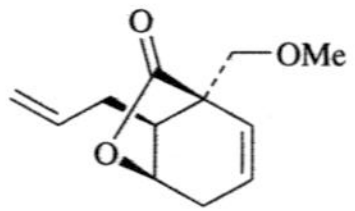	1. AD-Mix β, t-BuOH:H_2O (1:1) 2. $NaIO_4$	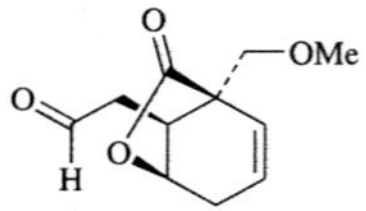 (56), —	635
	Ligand, $K_2OsO_2(OH)_4$, $K_3Fe(CN)_6$, K_2CO_3, $MeSO_2NH_2$, t-BuOH:H_2O (1:1), 0°	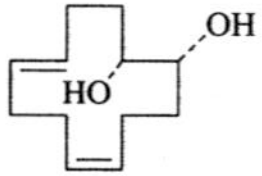	138
	Ligand		
	(DHQD)$_2$PHAL	(75-95), 69	
	(DHQD)$_2$PYR	(75-95), 88	
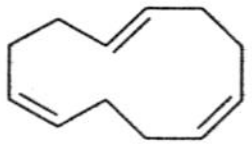	Ligand, $K_2OsO_2(OH)_4$, $K_3Fe(CN)_6$, K_2CO_3, $MeSO_2NH_2$, t-BuOH:H_2O (1:1), 0°	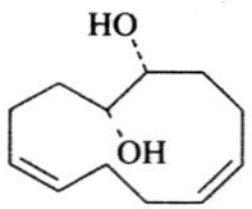	138
	Ligand		
	(DHQD)$_2$PHAL	(75-95), 22	
	(DHQD)$_2$PYR	(75-95), 89	
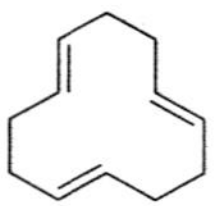	Ligand, $K_2OsO_2(OH)_4$, $K_3Fe(CN)_6$, K_2CO_3, $MeSO_2NH_2$, t-BuOH:H_2O (1:1), 0°; stop at 10% conversion	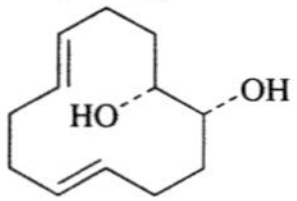	138
	Ligand		
	(DHQD)$_2$PHAL	(10), 65	
	(DHQD)$_2$PYR	(10), 95	

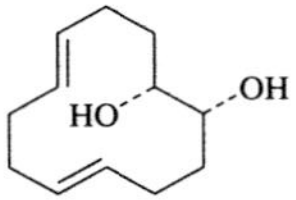	(DHQ)$_2$PYR, $K_2OsO_2(OH)_4$, $K_3Fe(CN)_6$, K_2CO_3, $MeSO_2NH_2$, t-BuOH:H_2O (1:1), 0°	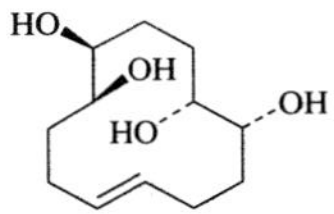	(72), 99.4 de	138
	(DHQD)$_2$PYR, $K_2OsO_2(OH)_4$, $K_3Fe(CN)_6$, K_2CO_3, $MeSO_2NH_2$, t-BuOH:H_2O (1:1), 0°	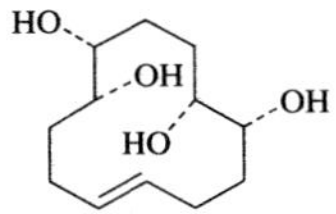	(65), 50 de	138
	Ligand, $K_2OsO_2(OH)_4$, $K_3Fe(CN)_6$ (6 eq), K_2CO_3 (6 eq), $MeSO_2NH_2$ (3 eq), t-BuOH:H_2O (1:1), 0°	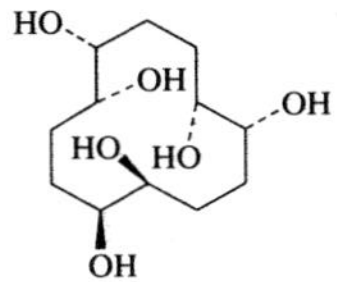		138
	Ligand			
	(DHQ)$_2$PYR	(93), 99.2 de		
	(DHQD)$_2$PYR	(89), 28 de		
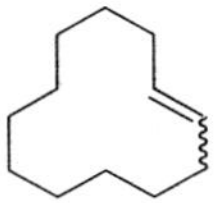	Ligand, $K_2OsO_2(OH)_4$, $K_3Fe(CN)_6$, K_2CO_3, $MeSO_2NH_2$, t-BuOH:H_2O (1:1), 0°	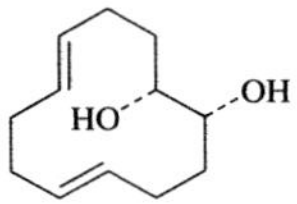		138
	Ligand			
	(DHQD)$_2$PHAL	(75-95), 60		
	(DHQD)$_2$PYR	(75-95), 97		

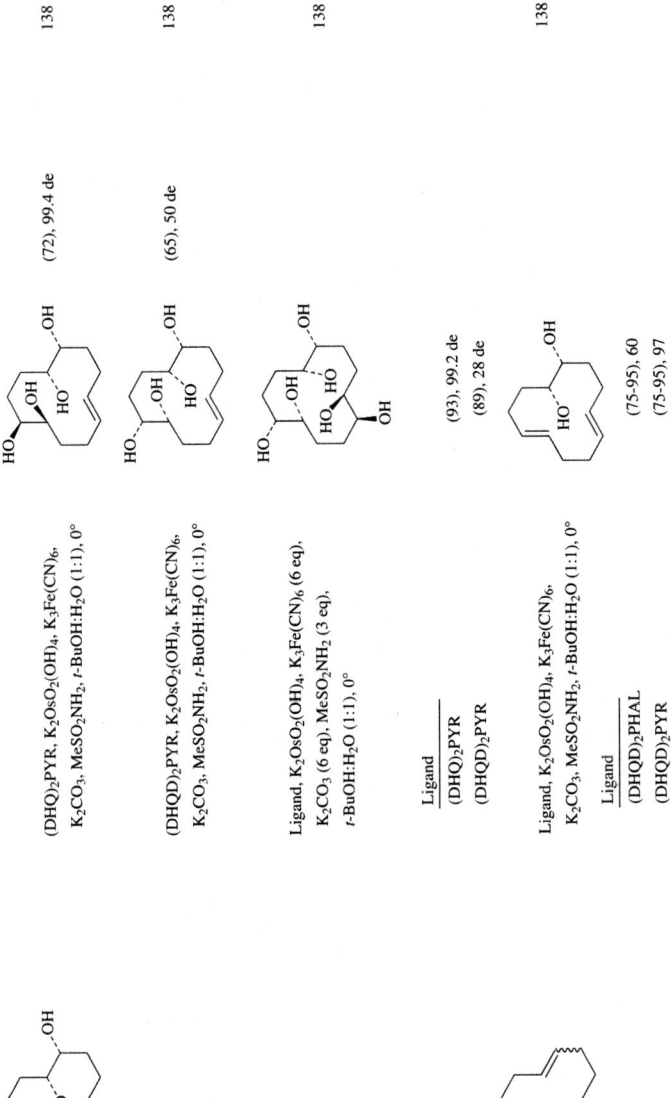

TABLE 8. REACTIONS OF UNCONJUGATED POLYALKENES (*Continued*)

Substrate	Conditions	Product(s) and Yield(s) (%), % ee	Refs.
C_{13-15}			
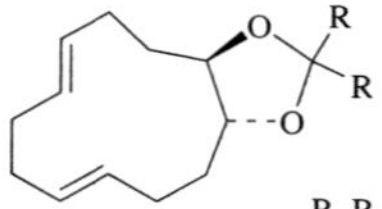 R, R: Me, Me; =O	(DHQ)$_2$PYR, $K_2OsO_2(OH)_4$, $K_2Fe(CN)_6$, K_2CO_3, $MeSO_2NH_2$, t-BuOH:H_2O (1:1), 0°	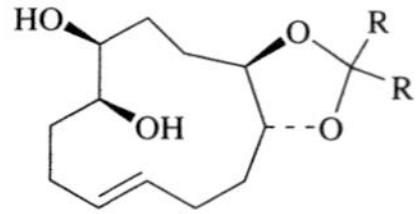 (85), 98 de; (85), 98 de	138
R, R: Me, Me; =O	(DHQD)$_2$PYR, $K_2OsO_2(OH)_4$, $K_2Fe(CN)_6$, K_2CO_3, $MeSO_2NH_2$, t-BuOH:H_2O (1:1), 0°	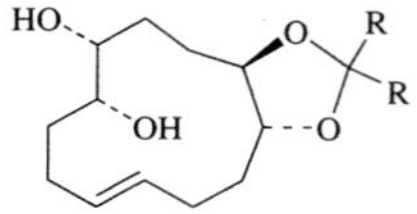 (81), 78 de; (80), 91 de	138
C_{13}			
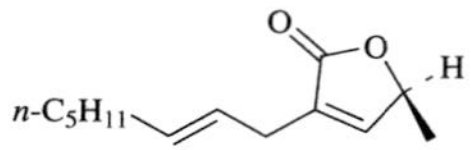	(DHQ)$_2$PHAL, $K_2OsO_2(OH)_4$, $K_2Fe(CN)_6$, K_2CO_3, $MeSO_2NH_2$, t-BuOH:H_2O (1:1)	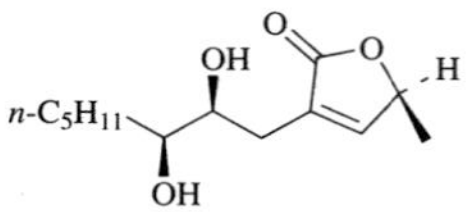 (93), >95 de	636
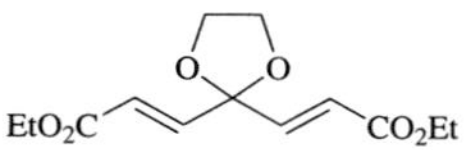	(DHQD)$_2$PHAL, OsO_4, NMO, t-BuOH:H_2O (9:1), 20°, 5 h	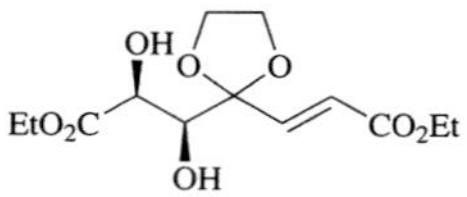 (82), 78	498
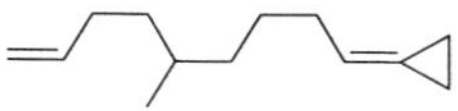	AD-Mix β, $MeSO_2NH_2$, t-BuOH:H_2O (1:1), 20°, 1 h	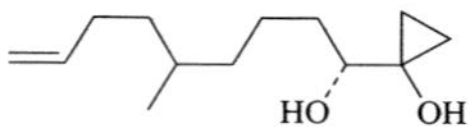 (92), —	637

C_{14} 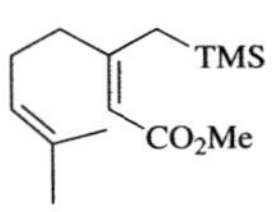	AD-Mix β, $MeSO_2NH_2$, t-BuOH:H_2O (9:1), 0°, 6 h	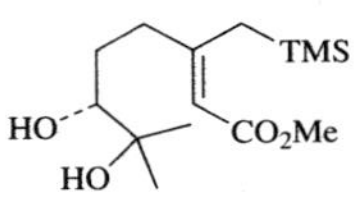 (76), —	638
C_{15} 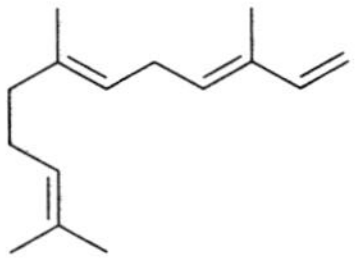	(DHQ)$_2$PHAL, K_2OsO_4, $K_3Fe(CN)_6$, K_2CO_3, $MeSO_2NH_2$, t-BuOH:H_2O (1:1), 0°, 6 h	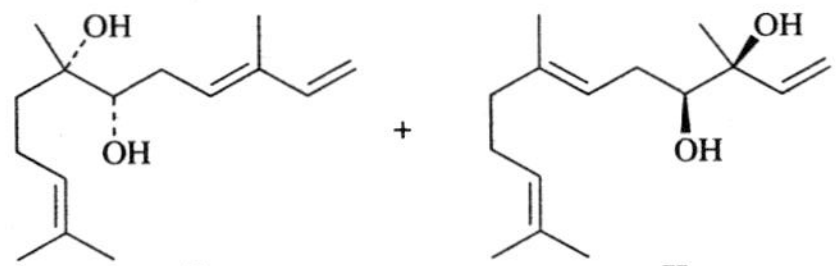 I + II **I+II** (28), >95; **I:II** = 20:1 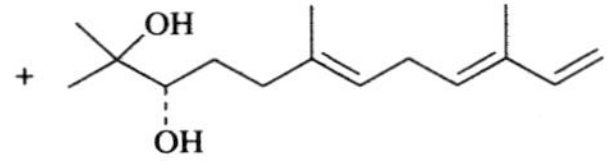 (11), 95 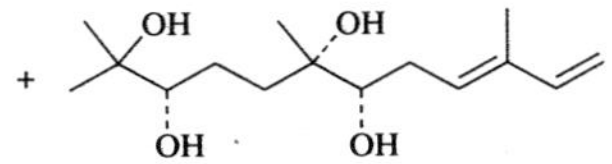 (7), >95	639
	(DHQD)$_2$PHAL, K_2OsO_4, $K_3Fe(CN)_6$, K_2CO_3, $MeSO_2NH_2$, t-BuOH:H_2O (1:1), 0°, 6 h	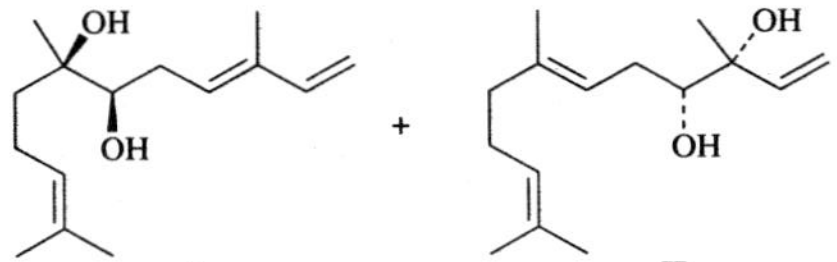 I + II **I+II** (18), >95; **I:II** = 20:1 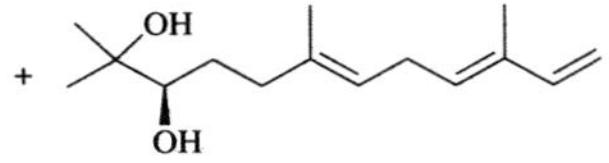 (15), >95 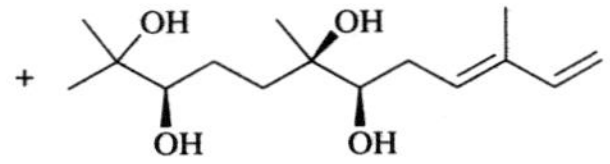 (14), >95	639

TABLE 8. REACTIONS OF UNCONJUGATED POLYALKENES (*Continued*)

Substrate	Conditions	Product(s) and Yield(s) (%), % ee	Refs.
C_{15}			
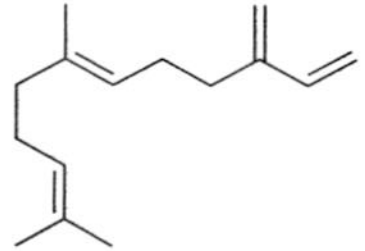	(DHQ)$_2$PHAL, K_2OsO_4, $K_3Fe(CN)_6$, K_2CO_3, $MeSO_2NH_2$, t-BuOH:H_2O (1:1), 0°, 6 h	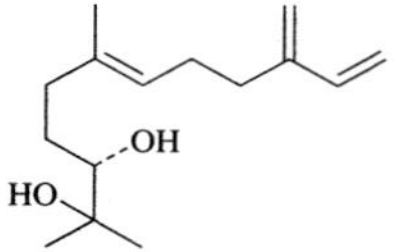 (18), —[a] + 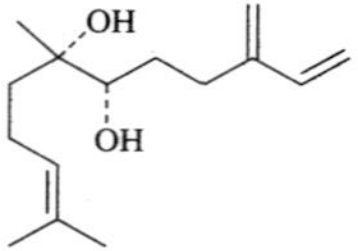 (9), —[a]	639
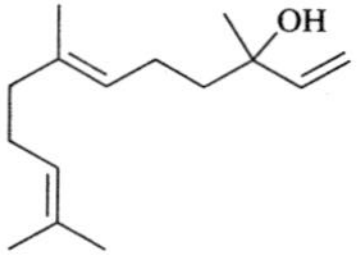	(DHQ)$_2$PHAL, K_2OsO_4, $K_3Fe(CN)_6$, K_2CO_3, $MeSO_2NH_2$, t-BuOH:H_2O (1:1), 0°, 50 h	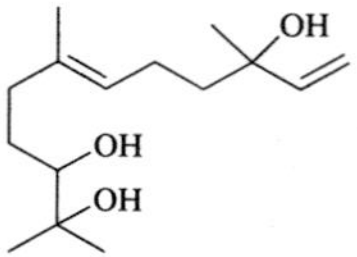 **I** + 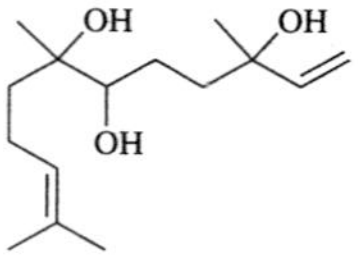 **II** **I+II** (60), —; **I**:**II** = 1.2:1 + 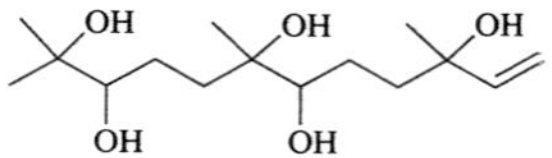 (3), —	564
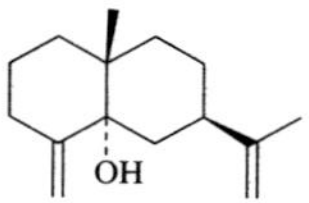	AD-Mix β, t-BuOH:H_2O (1:1), 0°, 24 h	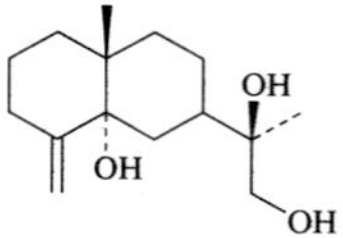 (90), 25 de	316
	AD-Mix α, t-BuOH:H_2O (1:1), 0°, 24 h	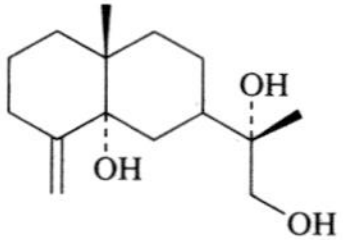 (85), 25 de	316

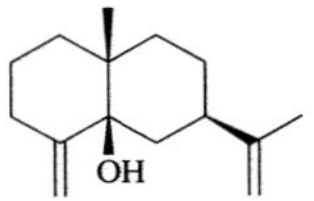	AD-Mix β, t-BuOH:H_2O (1:1), 0°, 24 h	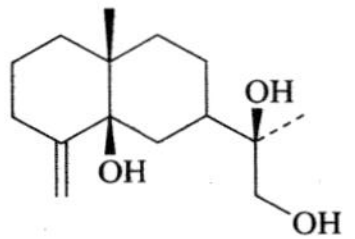 (92), 23 de	316
	AD-Mix α, t-BuOH:H_2O (1:1), 0°, 24 h	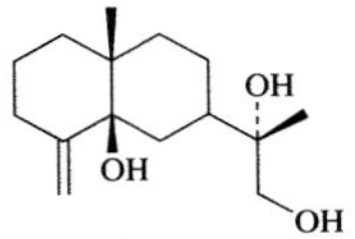 (95), 29 de	316
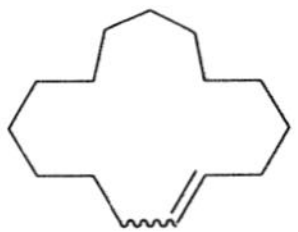	Ligand, K_2OsO_4, $K_3Fe(CN)_6$, K_2CO_3, $MeSO_2NH_2$, t-BuOH:H_2O (1:1), 0°	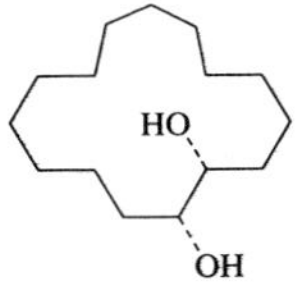	138
	Ligand: $(DHQD)_2PHAL$	(75-95), 65	
	Ligand: $(DHQD)_2PYR$	(75-95), 95	
C_{15-16} 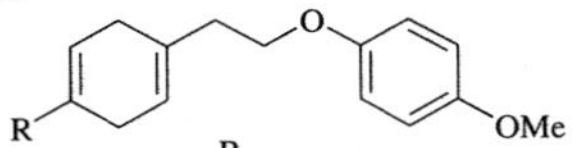	$(DHQD)_2PYDZ$, K_2OsO_4, $K_3Fe(CN)_6$, K_2CO_3, $MeSO_2NH_2$, t-BuOH:H_2O (1:1), 0°, 4 h	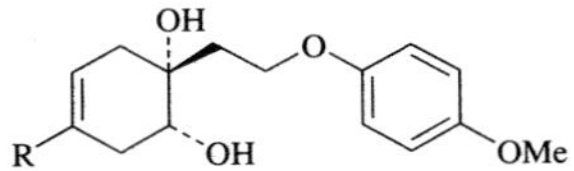	66
R = H		(74), 95	
R = Me		(62), 95	
C_{15} 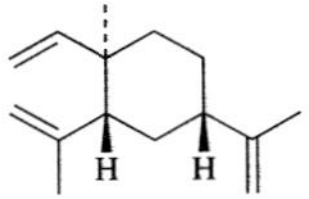	$(DHQD)_2PHAL$, $K_2OsO_2(OH)_4$, $K_3Fe(CN)_6$, K_2CO_3, t-BuOH:H_2O (1:1), 0°, 11 h	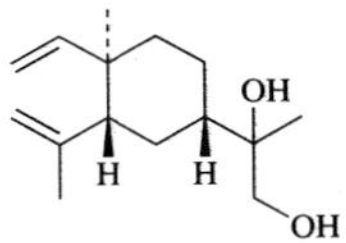 (76), 50	640

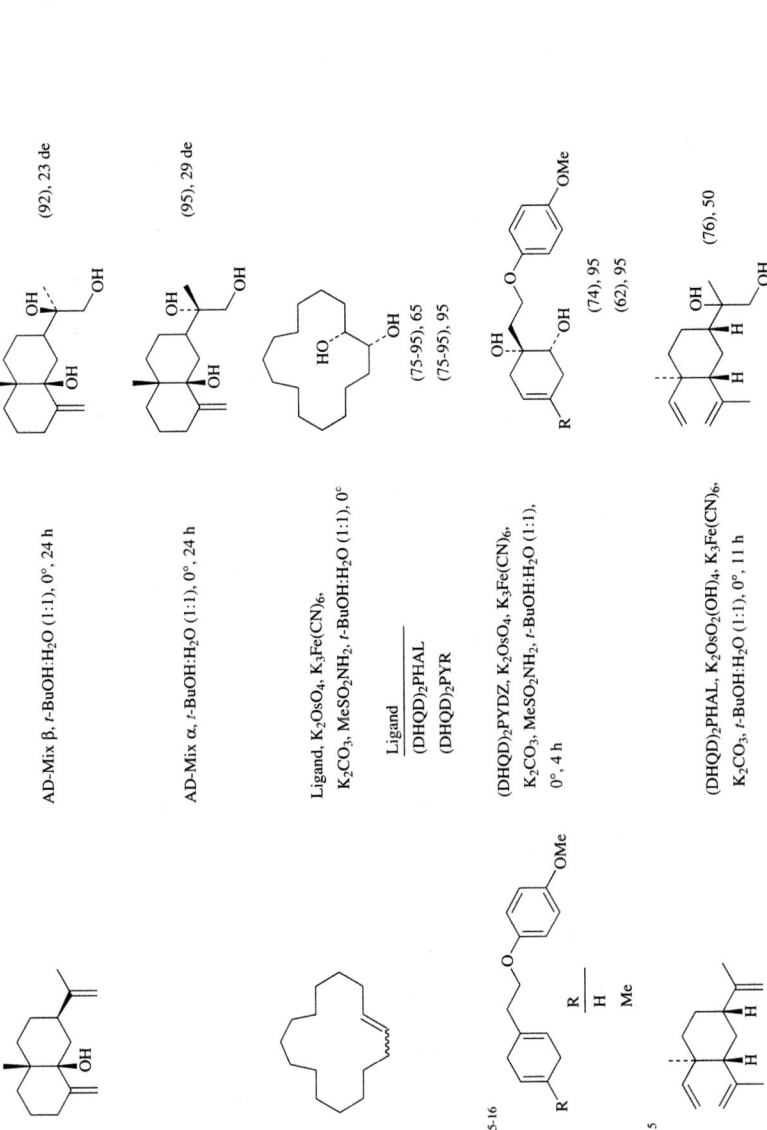

Substrate	Conditions	Product(s) and Yield(s) (%), % ee	Refs.
C_{15} 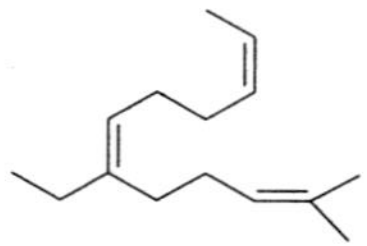	AD-Mix β, $MeSO_2NH_2$, *t*-BuOH:H_2O (1:1), 0°, 7.5 h	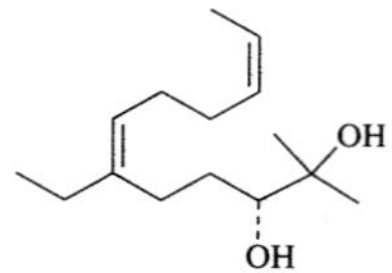 (44), 89	641, 642
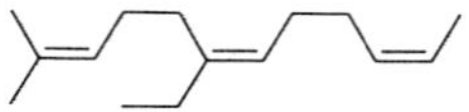	AD-Mix β, $MeSO_2NH_2$, *t*-BuOH:H_2O (1:1)	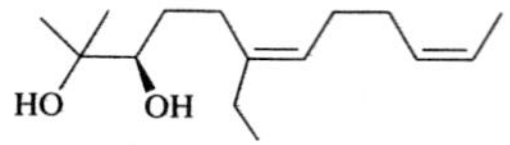 (—), —	643
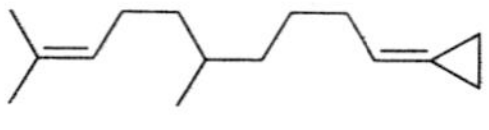	AD-Mix β, $MeSO_2NH_2$, *t*-BuOH:H_2O (1:1), 20°, 1 h	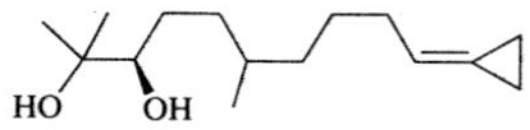 (60), —	637
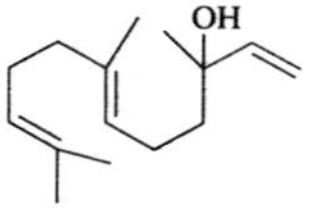	AD-Mix α, $MeSO_2NH_2$, *t*-BuOH:H_2O (1:1), 0°, 24 h	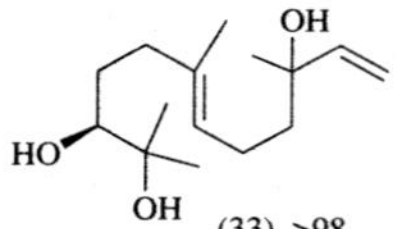 (33), >98 + 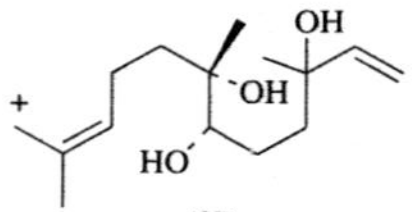 (27), — + 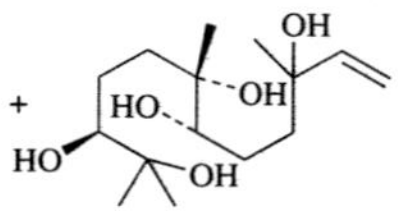 (3), —	564
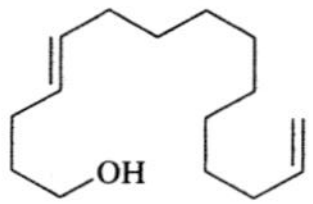	$(DHQD)_2PHAL$, OsO_4, $K_3Fe(CN)_6$, K_2CO_3, $MeSO_2NH_2$, *t*-BuOH:H_2O (1:1), 0°	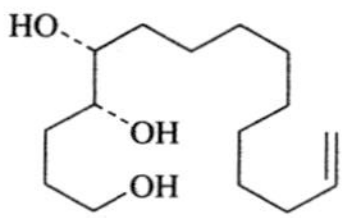 (71), 94	644
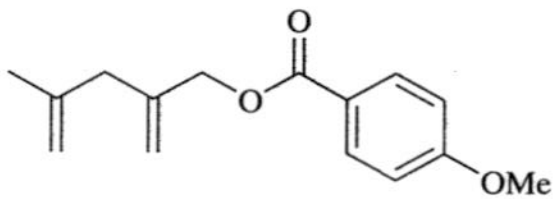	$(DHQD)_2PYDZ$, K_2OsO_4, $K_3Fe(CN)_6$, K_2CO_3, *t*-BuOH:H_2O (1:1), 0°	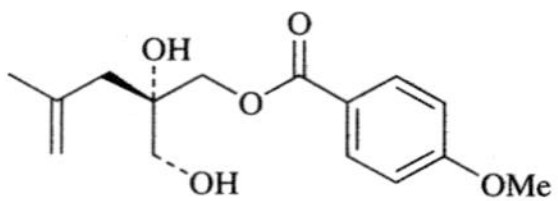 (72), 93	64

C_{16}

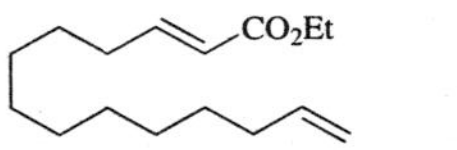	AD-Mix α, $MeSO_2NH_2$, *t*-BuOH:H_2O (1:1), 0°, 24 h	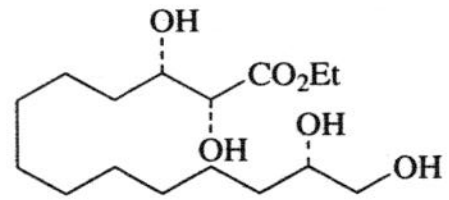 (61), —	234
	AD-Mix β (2 eq), $MeSO_2NH_2$, *t*-BuOH:H_2O (1:1), 0°, 8 h	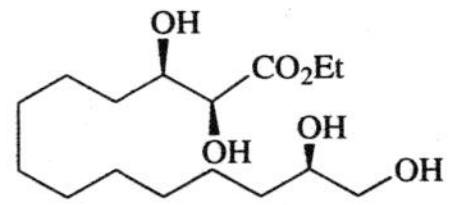 (>87), 83	279
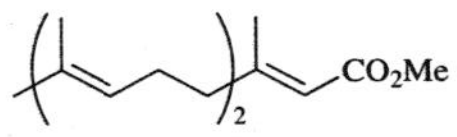	$(DHQD)_2PHAL$, $K_2OsO_2(OH)_4$, $K_3Fe(CN)_6$, K_2CO_3, $MeSO_2NH_2$, *t*-BuOH:H_2O (1:1), 0°, 24 h	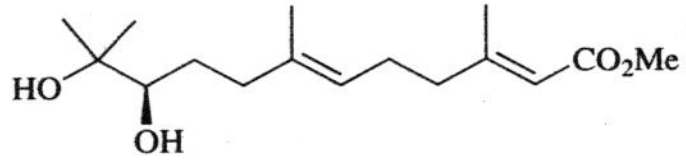 (32), 98 + 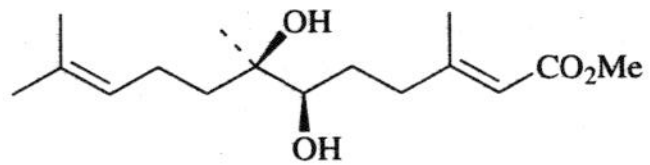 (2), — + 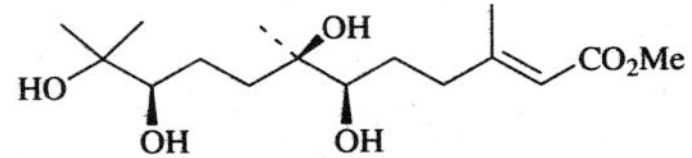 (37), >95 de	645
	$(DHQ)_2PHAL$, $K_2OsO_2(OH)_4$, $K_3Fe(CN)_6$, K_2CO_3, $MeSO_2NH_2$, *t*-BuOH:H_2O (1:1), 0°, 24 h	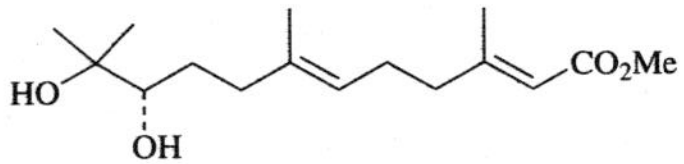 (28), 92 + 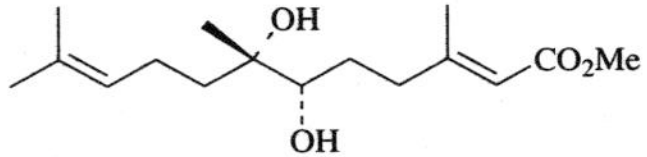 (3), — + 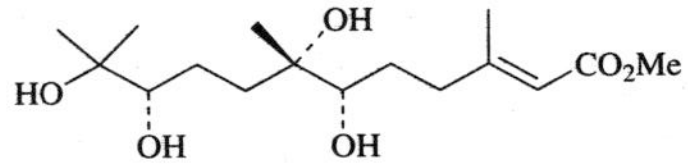 (29), >95 de	645

TABLE 8. REACTIONS OF UNCONJUGATED POLYALKENES (*Continued*)

Substrate	Conditions	Product(s) and Yield(s) (%), % ee	Refs.
C_{16}			
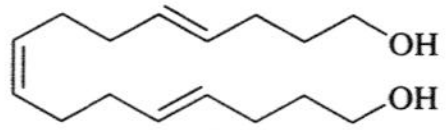	AD-Mix β, $MeSO_2NH_2$, t-BuOH:H_2O (1:1), 0°	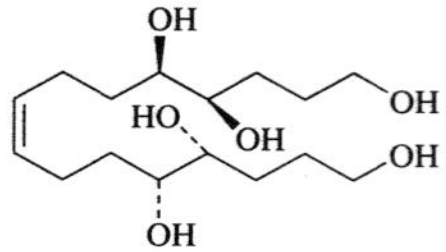 (62), —, >90 de	646
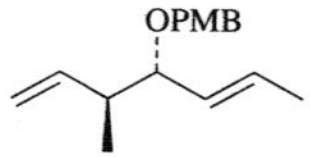	AD-Mix α, t-BuOH:H_2O (1:1), rt, 19 h	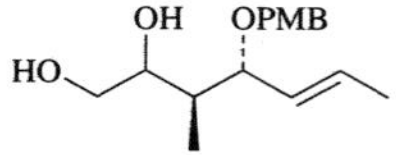 (50), —	647
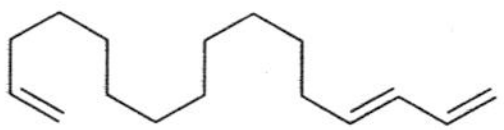	AD-Mix β, $MeSO_2NH_2$, t-BuOH:H_2O (1:1), 0°, 8 h	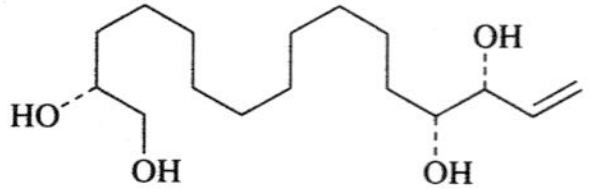 (89), 80	145
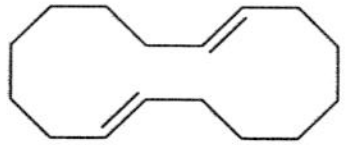	Ligand, $K_2OsO_2(OH)_4$, $K_3Fe(CN)_6$, K_2CO_3, $MeSO_2NH_2$, t-BuOH:H_2O (1:1), 0°	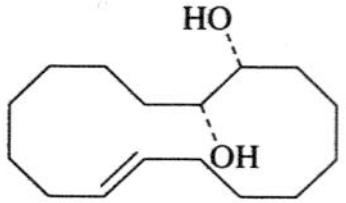	138
	Ligand		
	$(DHQD)_2PHAL$	(75-95), 50	
	$(DHQD)_2PYR$	(75-95), 92	
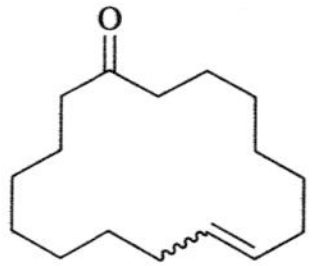	Ligand, $K_2OsO_2(OH)_4$, $K_3Fe(CN)_6$, K_2CO_3, $MeSO_2NH_2$, t-BuOH:H_2O (1:1), 0°	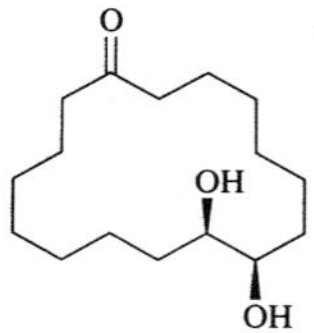	138
	Ligand		
	$(DHQD)_2PHAL$	(75-95), 58	
	$(DHQD)_2PYR$	(75-95), 95	

C_{17}	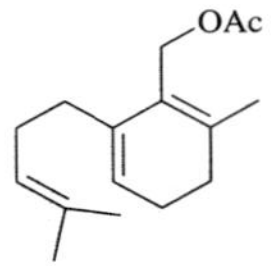	(DHQD)$_2$PHAL, $K_2OsO_2(OH)_4$, $K_3Fe(CN)_6$, K_2CO_3, $MeSO_2NH_2$, *t*-BuOH:H_2O (1:1), 0°, 24 h	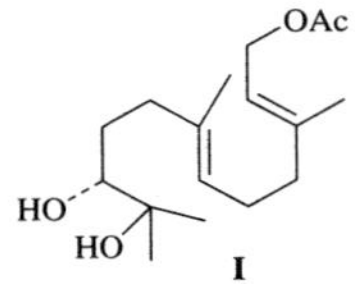 **I**	(34), —	141
		AD-Mix β, (DHQD)$_2$PHAL, $K_2OsO_2(OH)_4$, $MeSO_2NH_2$, *t*-BuOH:H_2O, 0°, 7 h	**I** (34), —		648
		20, $K_2OsO_2(OH)_4$, $K_3Fe(CN)_6$, K_2CO_3, $MeSO_2NH_2$, *t*-BuOH:H_2O, rt, 45 min	**I** (80), 96		62
	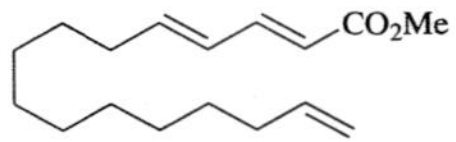	AD-Mix α, $MeSO_2NH_2$, *t*-BuOH:H_2O (1:1), 0°, 24 h	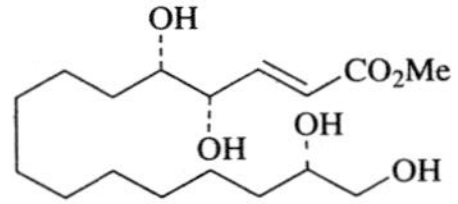	(53), —	234
C_{18}	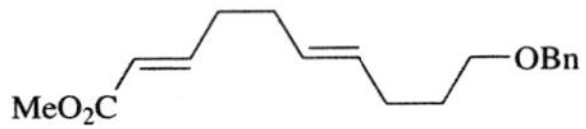	(DHQD)$_2$PHAL, $K_2OsO_2(OH)_4$, $K_3Fe(CN)_6$, K_2CO_3, *t*-BuOH:H_2O (1:1), 0°, 3 d	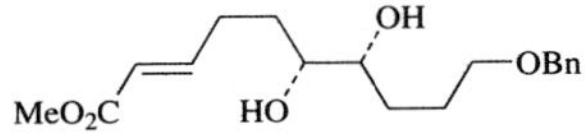	(90), >95	649
C_{19}	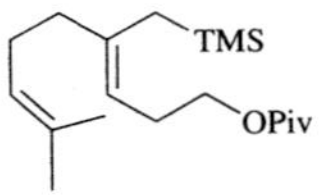	AD-Mix α, $MeSO_2NH_2$, *t*-BuOH:H_2O (1:1), 0°, 6-8 h	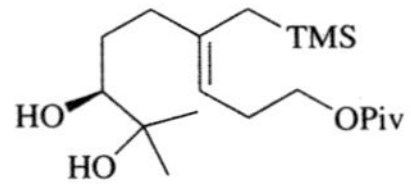	(66), 92	650
	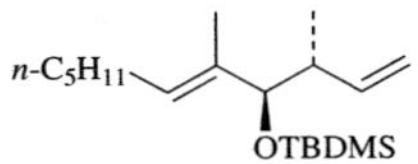	AD-Mix β, *t*-BuOH:H_2O (1:1), 0°	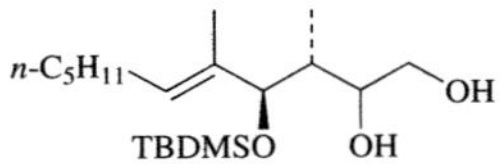	(70-76), — de	651, 652

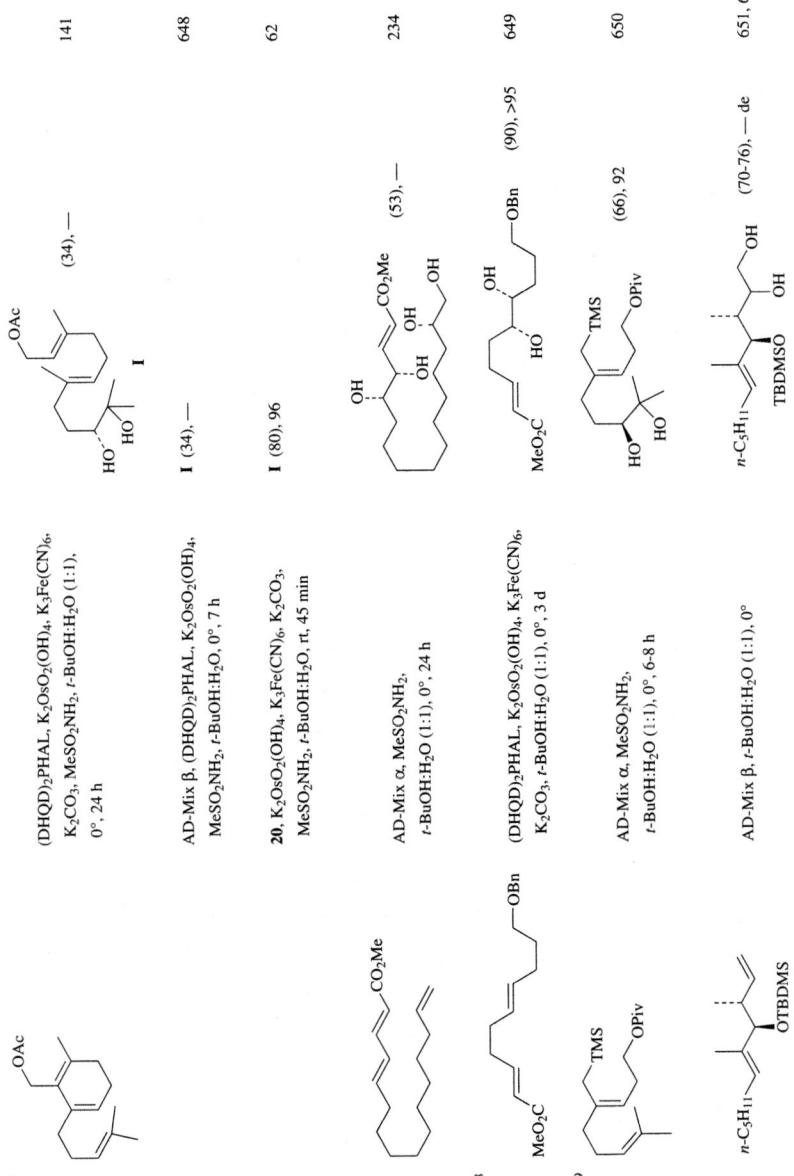

Substrate	Conditions	Product(s) and Yield(s) (%), % ee	Refs.
C_{20}			
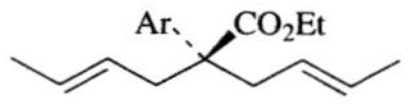 Ar = 3,4-$MeO_2C_6H_3$	AD-Mix β (2 eq), $MeSO_2NH_2$, t-BuOH:H_2O (1:1), 0°	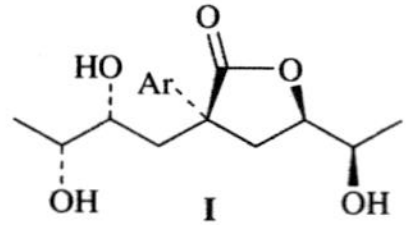 I (89), 35	653
	AD-Mix β (1 eq), $MeSO_2NH_2$, t-BuOH:H_2O (1:1), 0°	I (54), — + 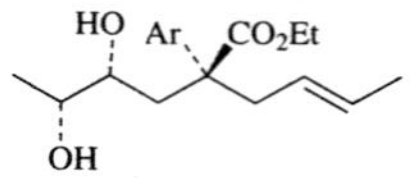 (30), 50	653
C_{21}			
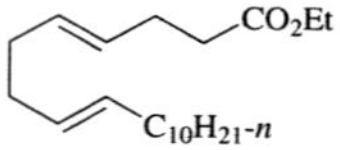	1. AD-Mix β, t-BuOH:H_2O (1:1) 2. DMP, TsOH, acetone 3. Recrystallization	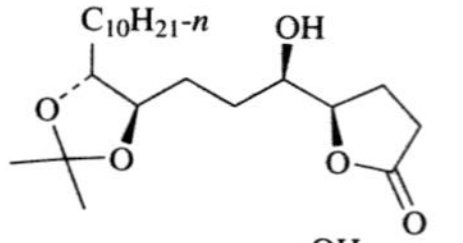 (72), >95	199
	1. AD-Mix β, $MeSO_2NH_2$, t-BuOH:H_2O (1:1), 0°, 16 h 2. KOH 3. HCl	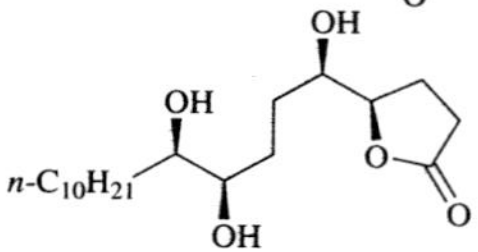 (65), 100	335
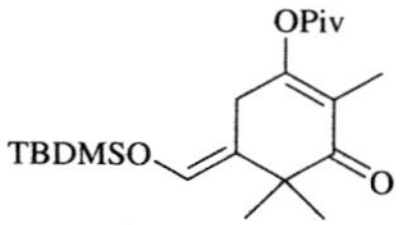	DHQD-PHN, $K_2OsO_2(OH)_4$, $K_3Fe(CN)_6$, K_2CO_3, t-BuOH:H_2O (1:1)	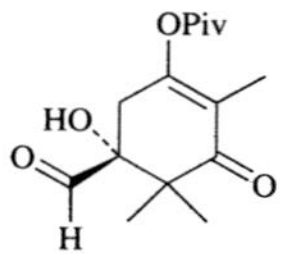 (78), 78	654
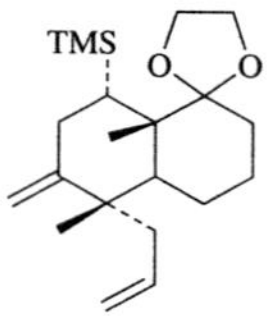	$(DHQD)_2PYDZ$, $K_2OsO_2(OH)_4$, $K_3Fe(CN)_6$, K_2CO_3, t-BuOH:H_2O (1:1), 0°, 4 h	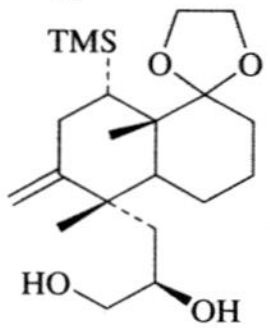 (97), 45	655

C_{22} 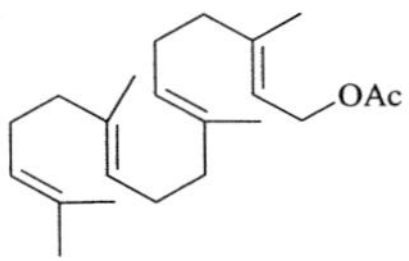	**20**, $K_2OsO_2(OH)_4$, $K_3Fe(CN)_6$, K_2CO_3, $MeSO_2NH_2$, *t*-BuOH:H_2O (1:1), 0°	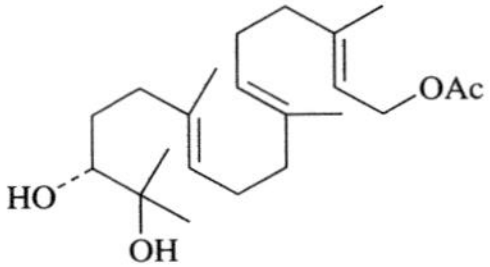	(54), 95	62
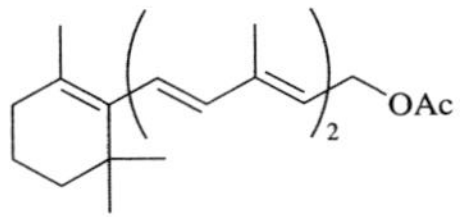	$(DHQ)_2PHAL$, $K_2OsO_2(OH)_4$, $K_3Fe(CN)_6$, K_2CO_3, $MeSO_2NH_2$, *t*-BuOH:H_2O (1:1), 0°, 24 h	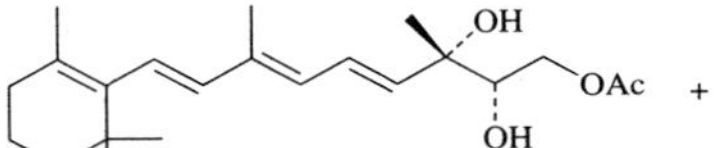 +	(90), >95	656
		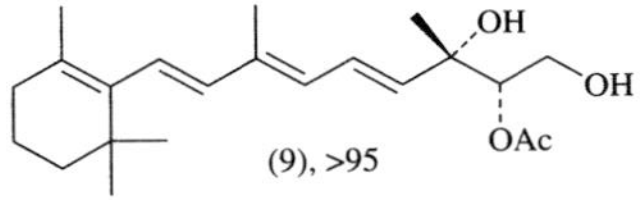	(9), >95	
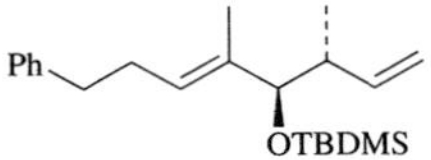	AD-Mix β, *t*-BuOH:H_2O (1:1), rt, 30 h	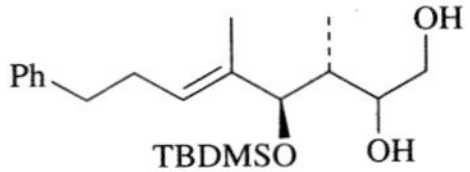	(70-86), —	657, 651, 652
	AD-Mix α, *t*-BuOH:H_2O (1:1), rt, 30 h	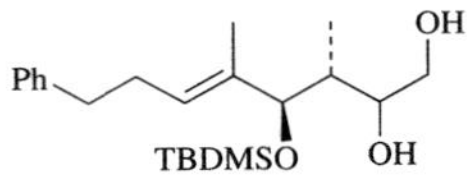	(35-45), —	657
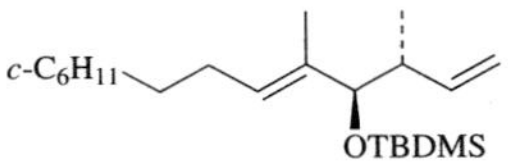	AD-Mix β, *t*-BuOH:H_2O (1:1), 0°	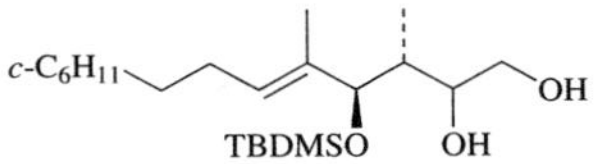	(70-86), — de	651, 652

TABLE 8. REACTIONS OF UNCONJUGATED POLYALKENES (*Continued*)

Substrate	Conditions	Product(s) and Yield(s) (%), % ee	Refs.
C_{22} 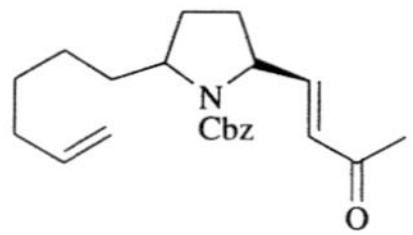	AD-Mix β, substituting $(DHQD)_2PYR$	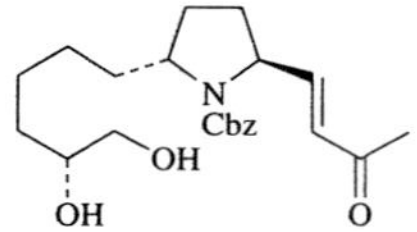 (92),—	622
	AD-Mix α, substituting $(DHQ)_2PYR$	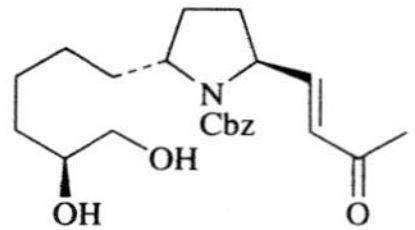 (>33),—	622
$C_{22\text{-}24}$ 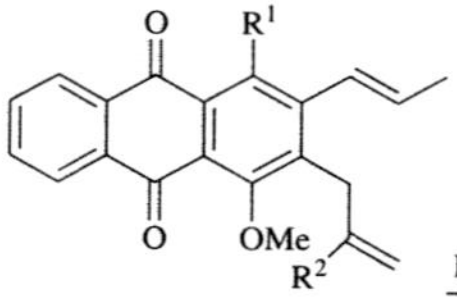	AD-Mix α, $MeSO_2NH_2$, t-BuOH:H_2O (1:1), rt	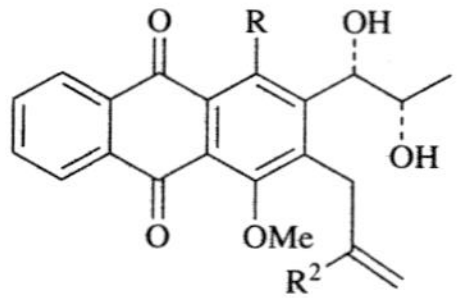	329
R^1 = H, R^2 = Me	Time 24 h	(41), —	
R^1 = OMe, R^2 = Me	Time 24 h	(77), —	
R^1 = OMe, R^2 = Et	Time 17 h	(60), —	
C_{23} 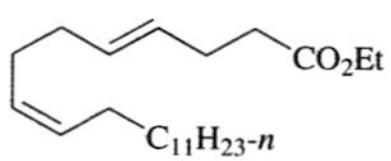	AD-Mix β, t-BuOH:H_2O (1:1)	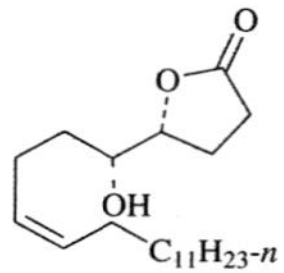 (52), —	658
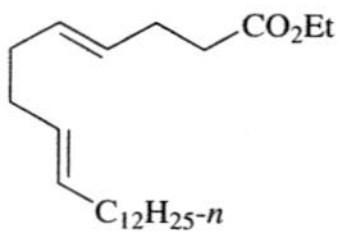	AD-Mix β, $MeSO_2NH_2$, t-BuOH:H_2O (1:1), 0°, 16 h	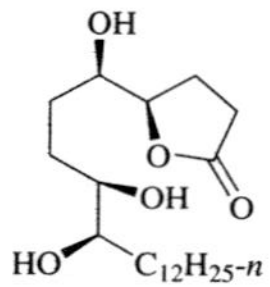 (66), >96	659

C_{24}	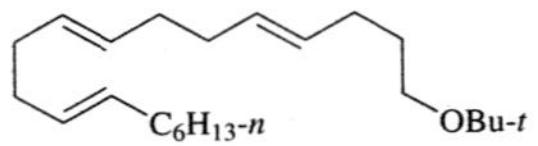	AD-Mix β, $MeSO_2NH_2$, t-BuOH:H_2O (1:1), 0°	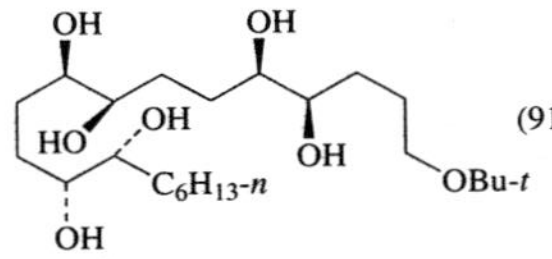	(91), 100 de	660
	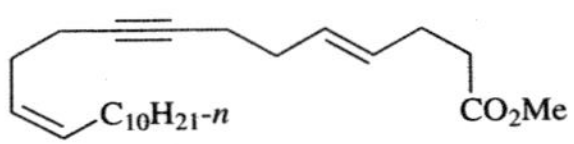	1. AD-Mix β, $MeSO_2NH_2$, t-BuOH:H_2O (1:1), 0°, 16 h 2. KOH (aq) 3. HCl (3N) 4. TsOH, CH_2Cl_2	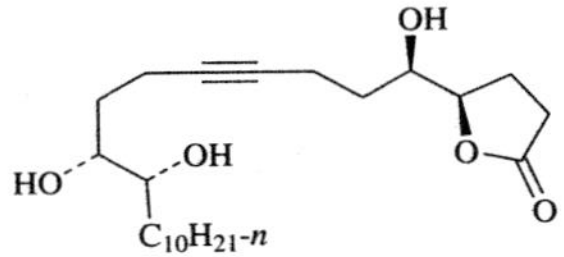	(80), —	661
	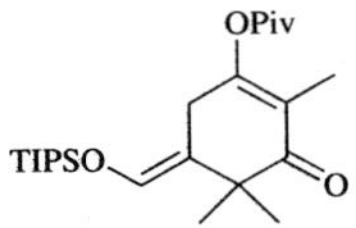	DHQD-PHN, $K_2OsO_2(OH)_4$, $K_3Fe(CN)_6$, K_2CO_3, t-BuOH:H_2O (1:1)	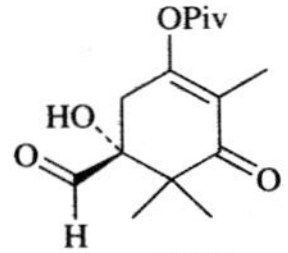	(88), 94	654
		DHQ-PHN, $K_2OsO_2(OH)_4$, $K_3Fe(CN)_6$, K_2CO_3, t-BuOH:H_2O (1:1)	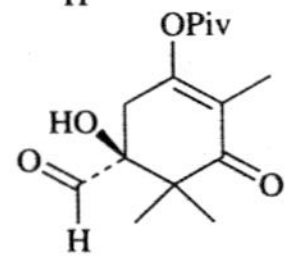	(96-98), 90	654, 662
	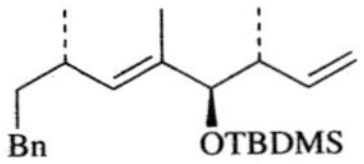	AD-Mix α, t-BuOH:H_2O (1:1)	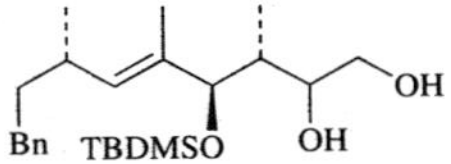	(87), 0 de	663
		AD-Mix α, t-BuOH:H_2O (1:1), rt, 30 h	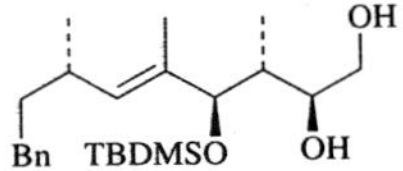	(87), —	657
	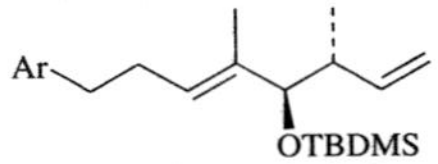 Ar = 3,4-$(MeO)_2C_6H_3$	AD-Mix β, t-BuOH:H_2O (1:1), 0°	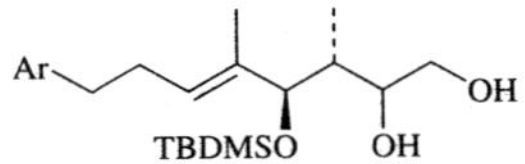	(70-86), — de	651, 652

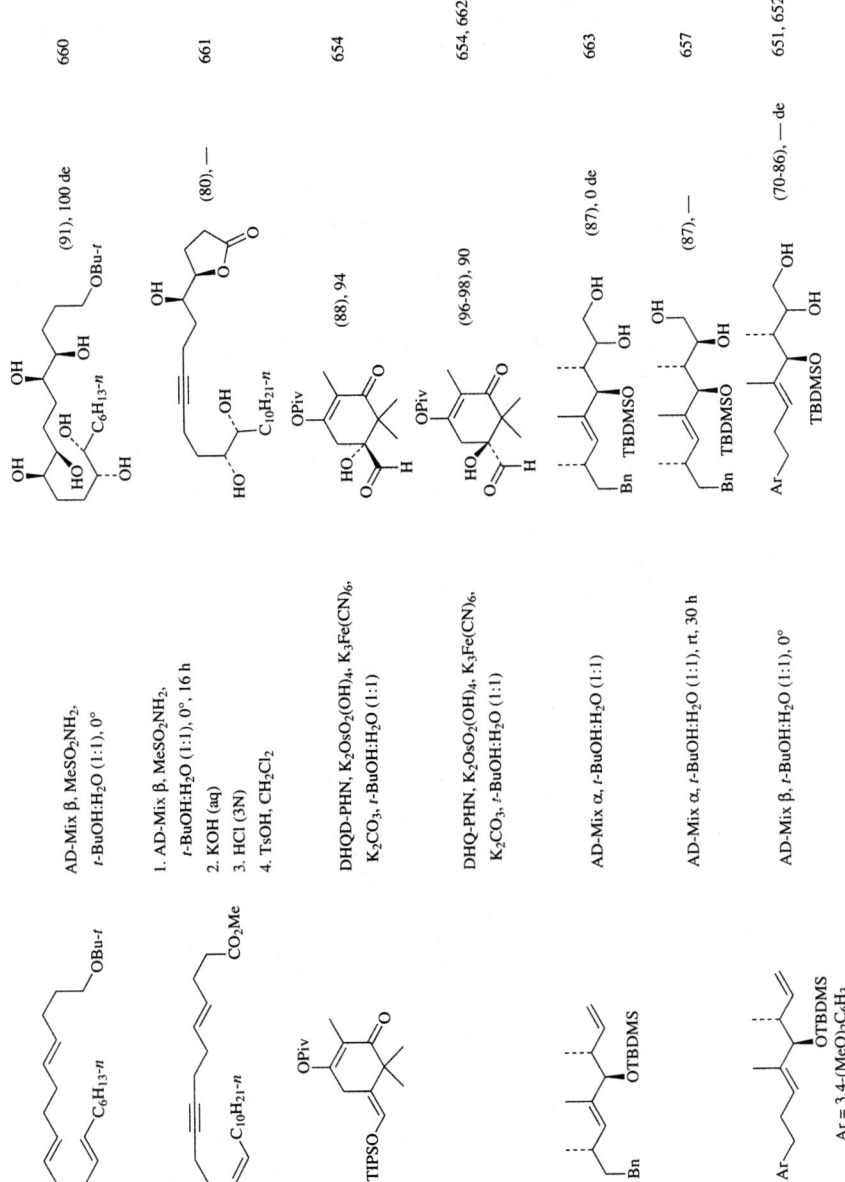

TABLE 8. REACTIONS OF UNCONJUGATED POLYALKENES (*Continued*)

Substrate	Conditions	Product(s) and Yield(s) (%), % ee	Refs.
C_{25}			
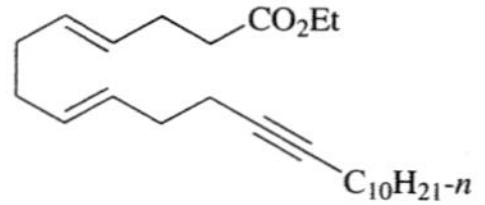	1. AD-Mix β, $MeSO_2NH_2$, *t*-BuOH:H_2O (1:1), 0°, 16 h 2. KOH (aq) 3. HCl (3N) 4. TsOH, CH_2Cl_2	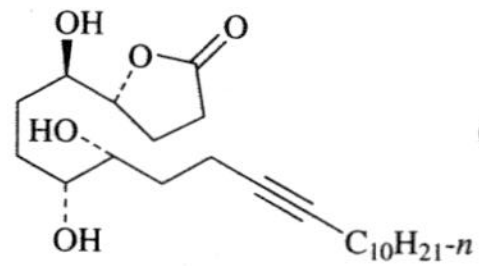 (72-82), >99	661, 447
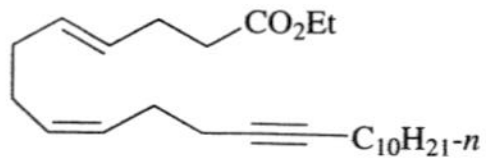	1. AD-Mix β, $MeSO_2NH_2$, *t*-BuOH:H_2O (1:1), 0°, 16 h 2. KOH (aq) 3. HCl (3N) 4. TsOH, CH_2Cl_2	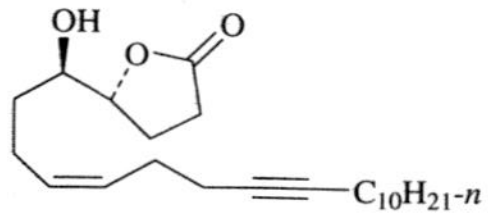 (57), >99	661, 447
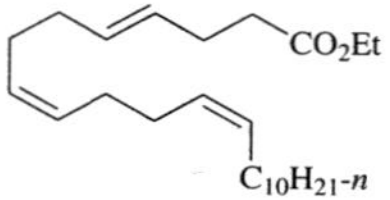	AD-Mix β, *t*-BuOH:H_2O (1:1)	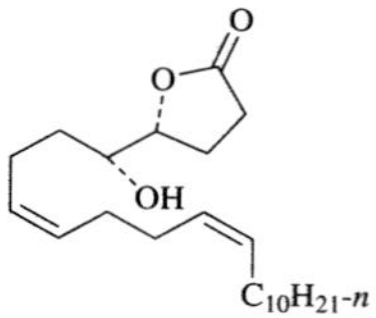 (52), —	658
C_{26}			
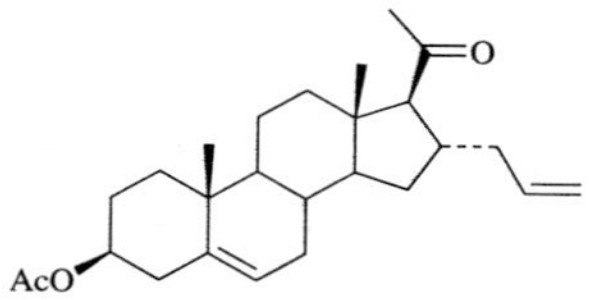	AD-Mix β, *t*-BuOH:H_2O (1:1), rt, 48 h	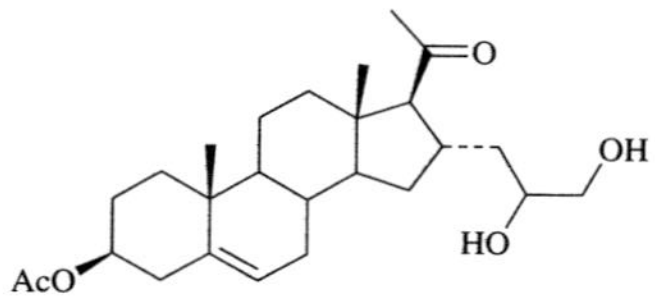 (70), — de	664
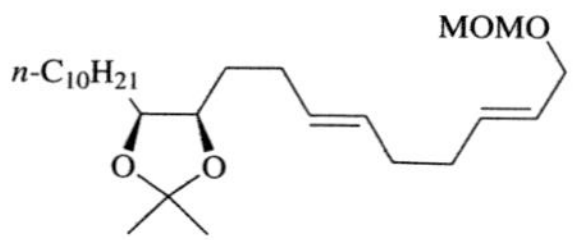	AD-Mix α, $MeSO_2NH_2$, *t*-BuOH:H_2O (1:1), 0°, 16 h	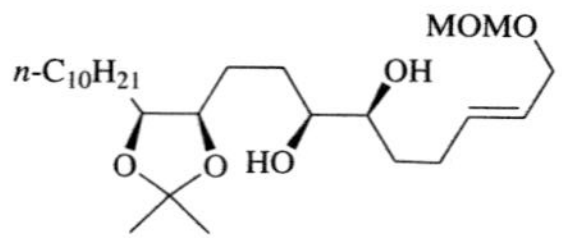 (63), >90 de	442

	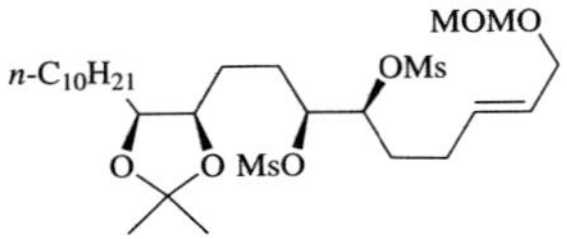	AD-Mix β, $MeSO_2NH_2$, t-BuOH:H_2O (1:1), 0°, 16 h	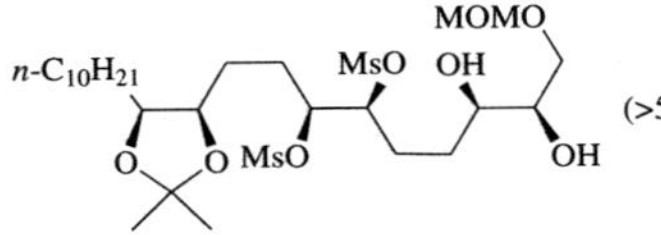 (>50), >90 de	442
	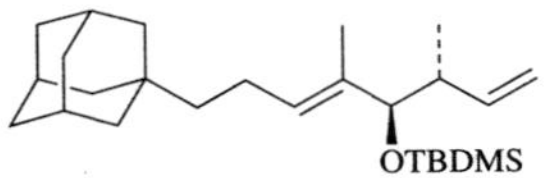	AD-Mix β, t-BuOH:H_2O (1:1), 0°	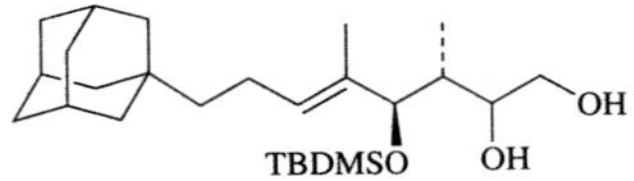 (70-86), — de	651, 652
	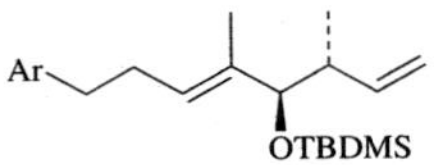 Ar = 2-naphthyl	AD-Mix β, t-BuOH:H_2O (1:1), 0°	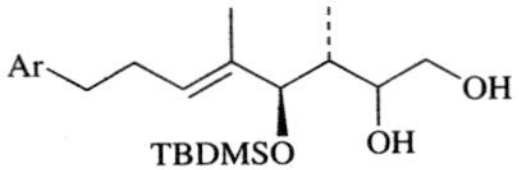 (70-86), — de	651, 652
C_{27}	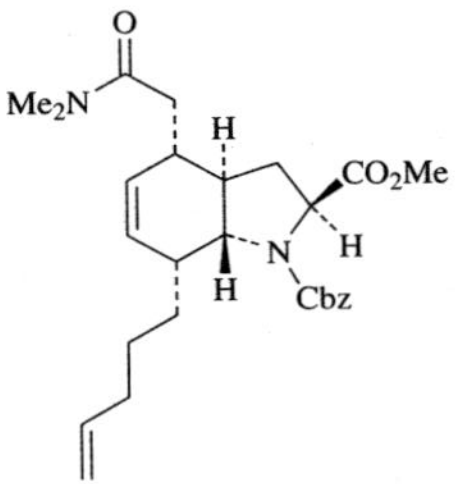	AD-Mix β, $NaIO_4$, t-BuOH:H_2O (1:1)	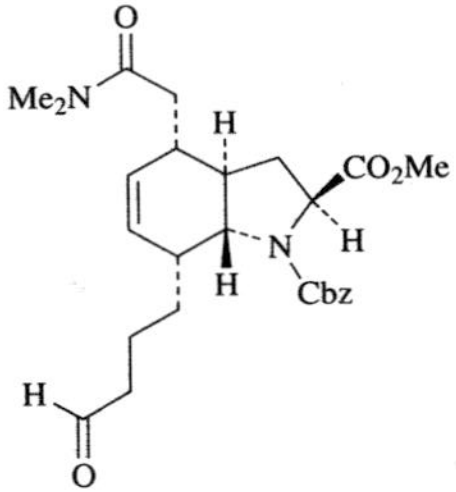 (82), —	665
	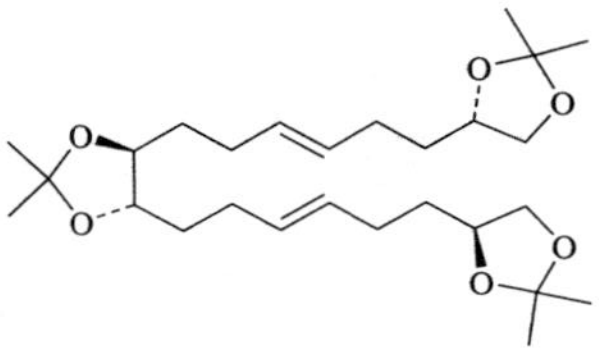	AD-Mix β, t-BuOH:H_2O (1:1), 0-20°, 12 h	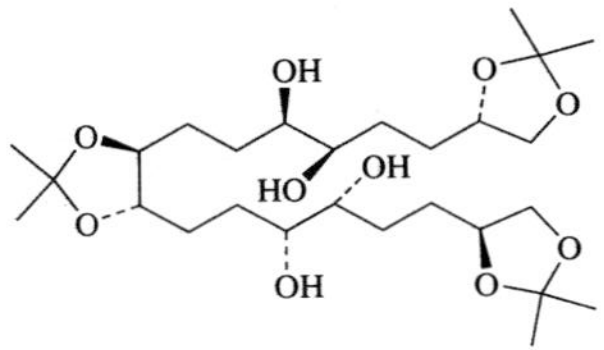 (98), 62 de	439

TABLE 8. REACTIONS OF UNCONJUGATED POLYALKENES (*Continued*)

Substrate	Conditions	Product(s) and Yield(s) (%), % ee	Refs.
C_{27} 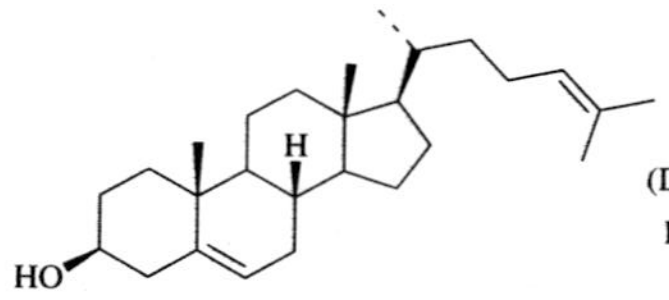	$(DHQD)_2PYDZ$, OsO_4, $K_3Fe(CN)_6$, K_2CO_3, *t*-BuOH:H_2O (1:1), 0°, 52 h	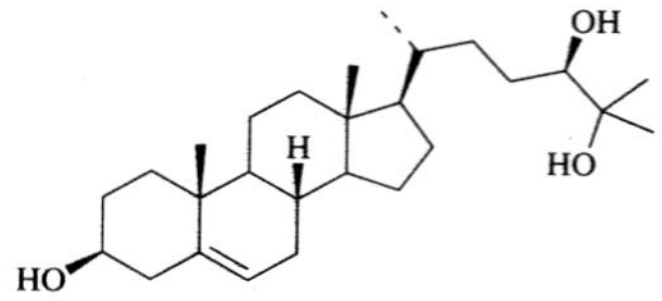 (82), 92	666
	$(DHQ)_2PHAL$, OsO_4, $K_3Fe(CN)_6$, K_2CO_3, *t*-BuOH:H_2O (1:1), 0°, 52 h	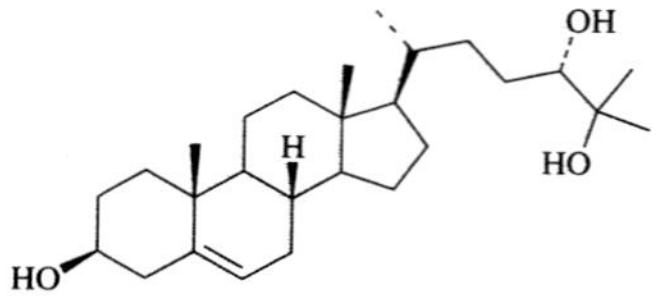 (82), 92	666
C_{28} 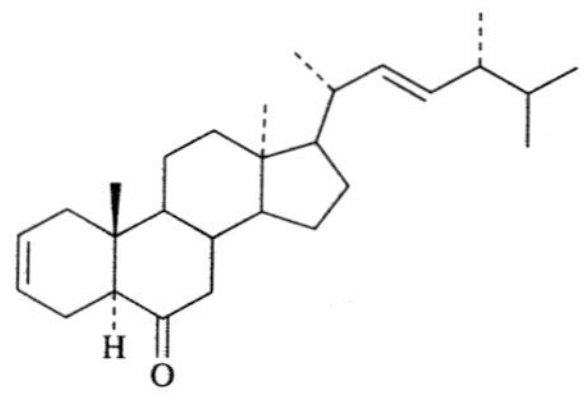	Ligand, $K_2OsO_2(OH)_4$, $K_3Fe(CN)_6$, K_2CO_3, $MeSO_2NH_2$, *t*-BuOH:H_2O (1:1), rt, 6 d Ligand DHQD-*p*-chlorobenzoate $(DHQD)_2PHAL$	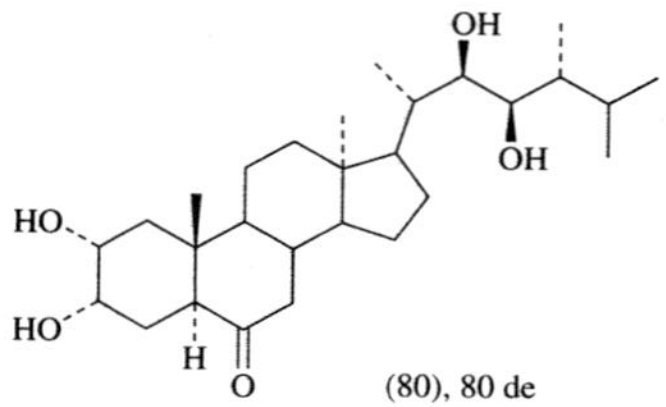 (80), 80 de (80), 82 de	667

Substrate	Conditions	Product	Yield (%), selectivity	Ref.
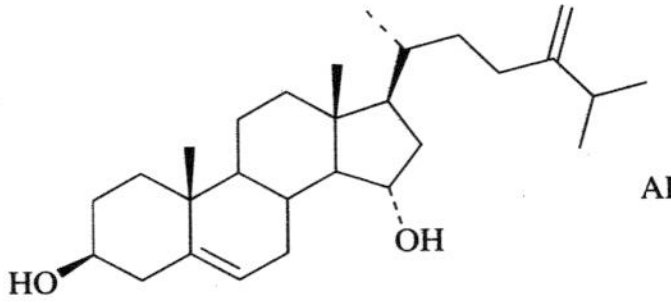	AD-Mix α, *t*-BuOH:H_2O (1:1), rt, 72 h	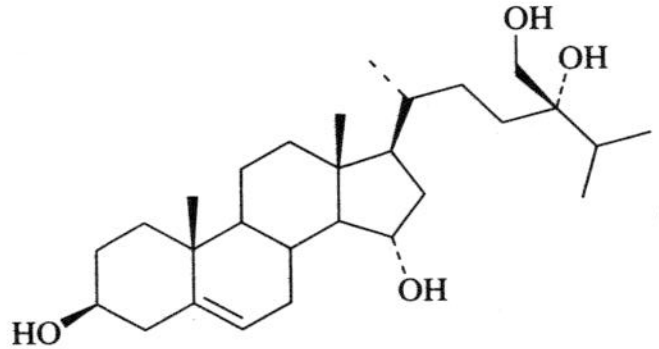	(—), 40 de	668
	AD-Mix β, *t*-BuOH:H_2O (1:1), rt, 72 h	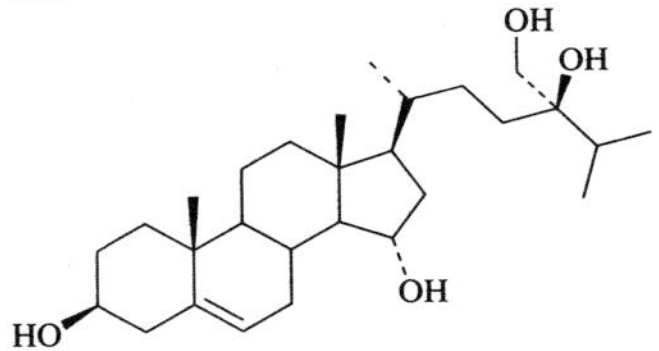	(—), 40 de	668
C_{29} 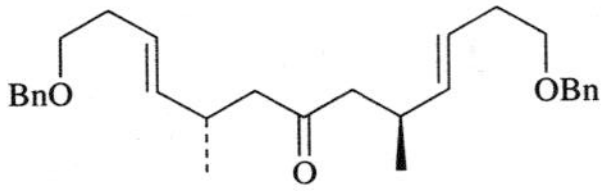	AD-Mix α, *t*-BuOH:H_2O (1:1), 0°	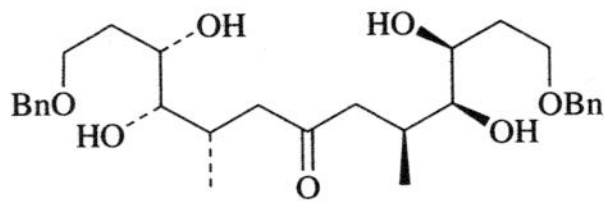	(>85), >95 de	669
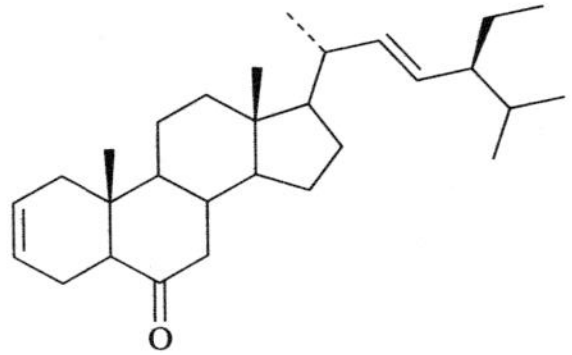	DHQD-PHN (1 eq), OsO_4, NMO, THF:*t*-BuOH:H_2O (7.2:4:1.1), rt, 5 d	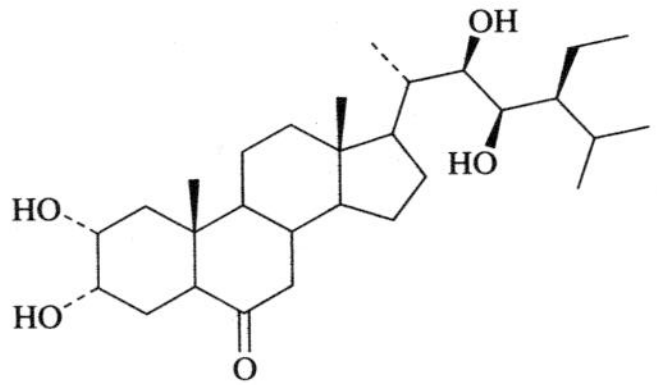	(—), 44 de	515

TABLE 8. REACTIONS OF UNCONJUGATED POLYALKENES (*Continued*)

Substrate	Conditions	Product(s) and Yield(s) (%), % ee	Refs.
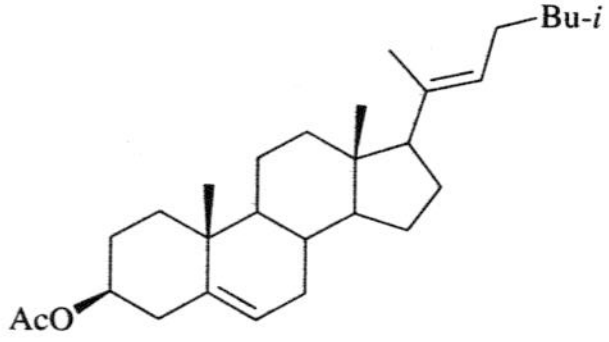	AD-Mix α, $MeSO_2NH_2$, *t*-BuOH:CH_3CN:H_2O (1:1:1), 4°, 24 h	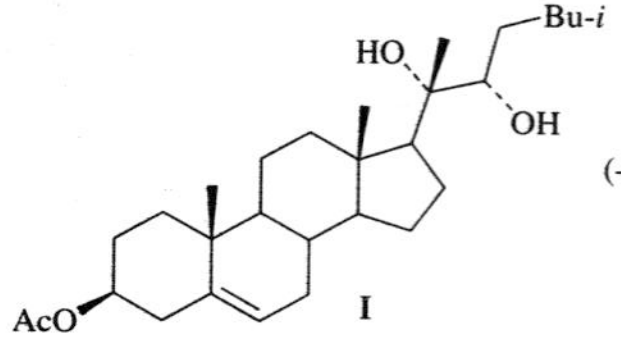 **I** (—), 98 de	670
	AD-Mix β, $MeSO_2NH_2$, *t*-BuOH:CH_3CN:H_2O (1:1:1), 4°, 24 h	**I** (67), 0 de	670
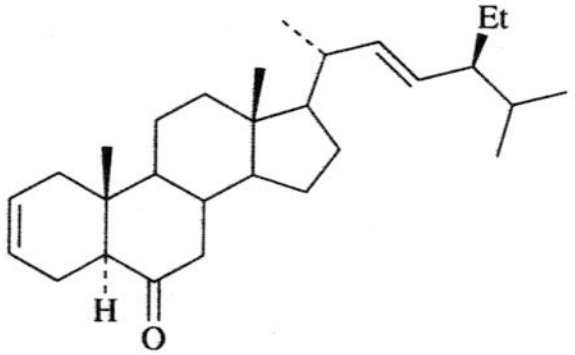	DHQD-*p*-chlorobenzoate, OsO_4, $K_3Fe(CN)_6$, K_2CO_3, $MeSO_2NH_2$, *t*-BuOH:H_2O (1:1), rt, 6 d	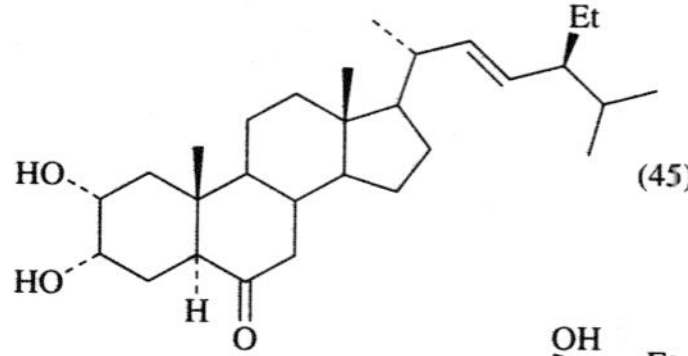 (45), >95 de + 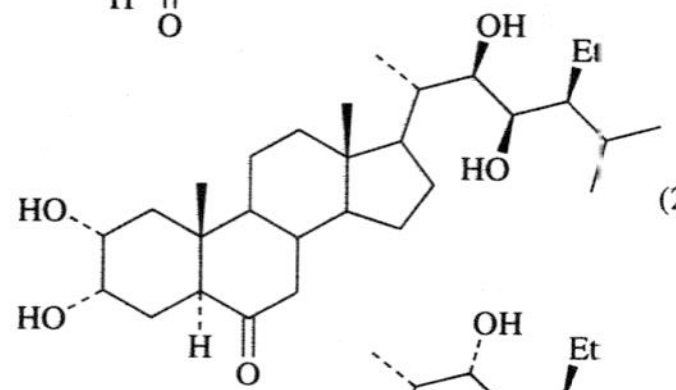 (27.5), >95 de + 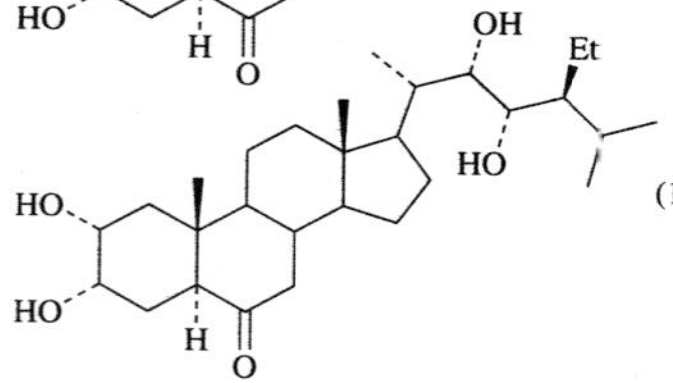 (19), >95 de	671

C_{29-37}	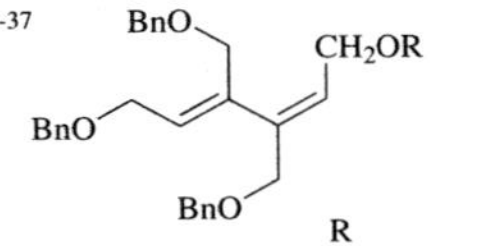		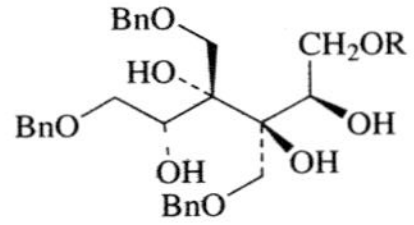		672
	R				
	H	AD-Mix β, $MeSO_2NH_2$, t-BuOH:H_2O (1:1), 0°-rt, 4 d	(45), 76, >80 de		
	Piv	$(DHQD)_2PHAL$, OsO_4, NMO, acetone:H_2O (5:1), rt	(37), —. 33 de		
	TBDMS	"	(59), —, 50 de		
	COC_6H_4OMe-*p*	"	(76), 26		
C_{30}	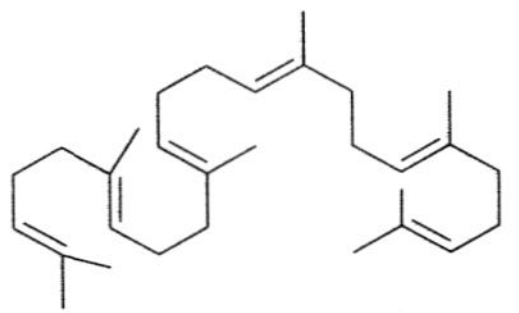	**20**, $K_2OsO_2(OH)_4$, $K_3Fe(CN)_6$, K_2CO_3, $MeSO_2NH_2$, t-BuOH:H_2O:methylcyclohexane (3.3:3.3:1), 0°	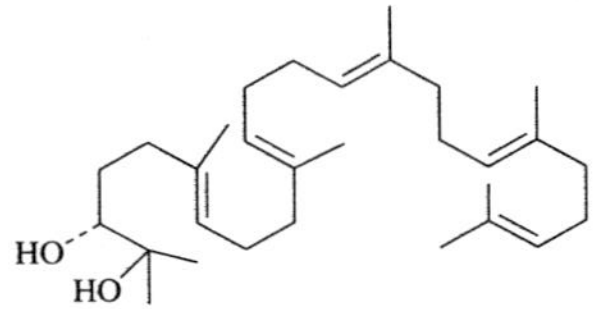	(32), 90	62
C_{32}	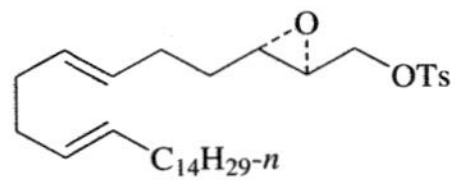	AD-Mix β, $MeSO_2NH_2$, t-BuOH:H_2O (1:1), 0°, 24 h	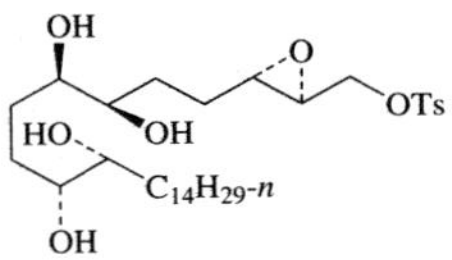	(91), >90 de	673
C_{33}	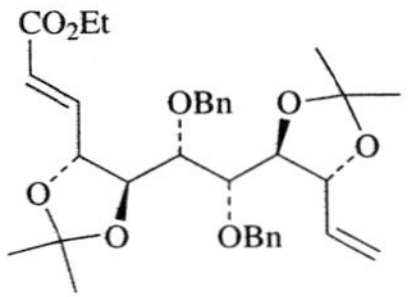	DHQ-*p*-chlorobenzoate, OsO_4, NMO, acetone-H_2O	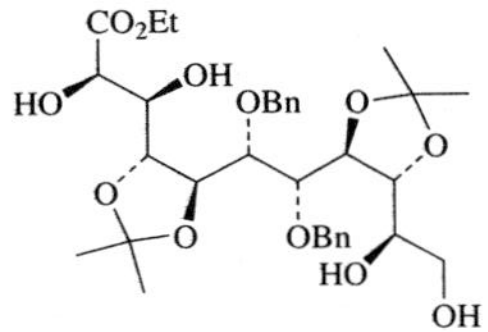	(88), 50 de	674, 675

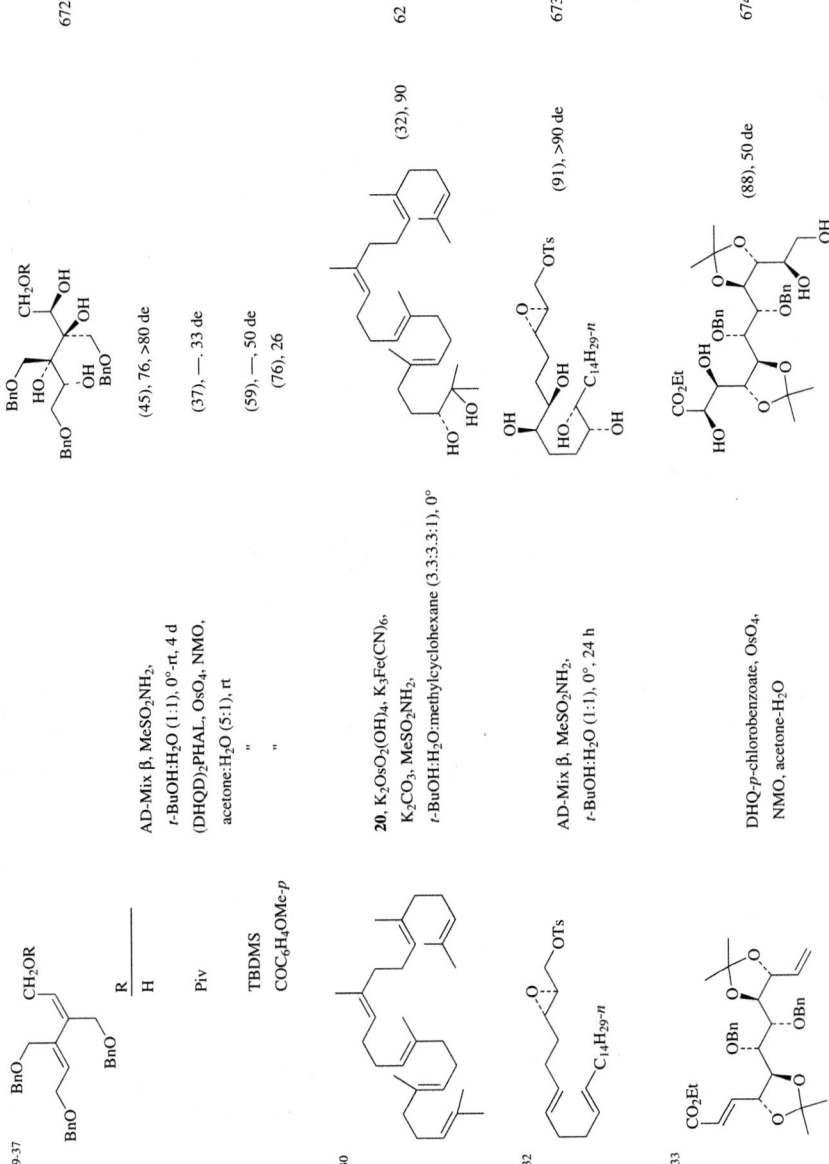

TABLE 8. REACTIONS OF UNCONJUGATED POLYALKENES (*Continued*)

Substrate	Conditions	Product(s) and Yield(s) (%), % ee	Refs.
C_{34} 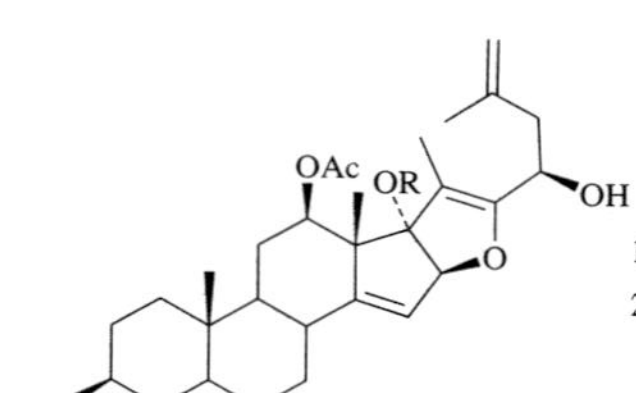 R= TMS	1. **35**, OsO_4, –95°, 1 h 2. $NaHSO_3$, THF-H_2O, reflux, 11 h	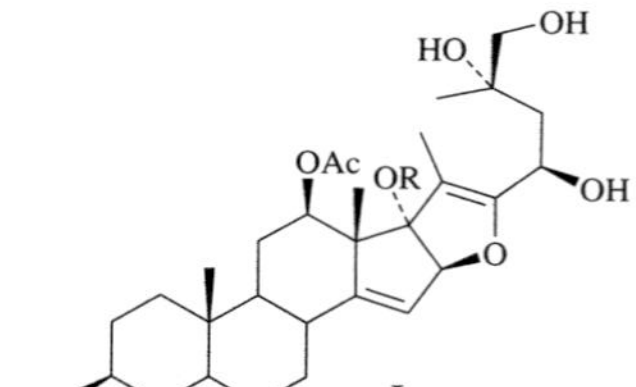 **I** (95), 33 de	676, 677
	AD-Mix α, *t*-BuOH:H_2O (1:1), rt, 24 h	**I** (25), 33 de	676, 677
	AD-Mix β, *t*-BuOH:H_2O (1:1), rt, 24 h	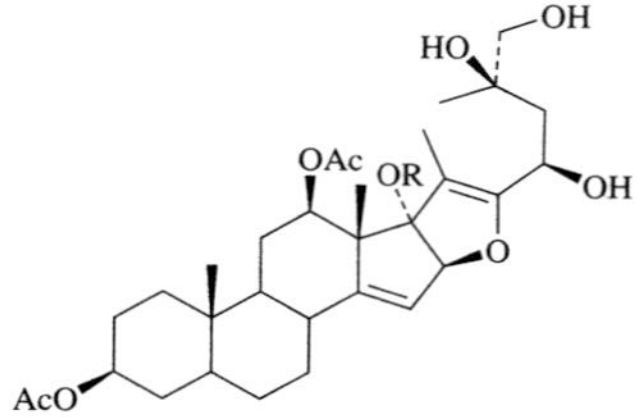 (25), 60 de	676, 677
C_{35} 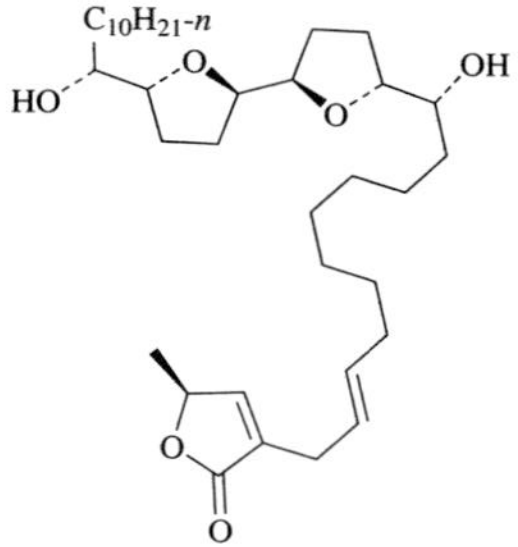	$(DHQ)_2PHAL$, OsO_4, $K_3Fe(CN)_6$, K_2CO_3, $MeSO_2NH_2$, *t*-BuOH:H_2O (1:1), 0°	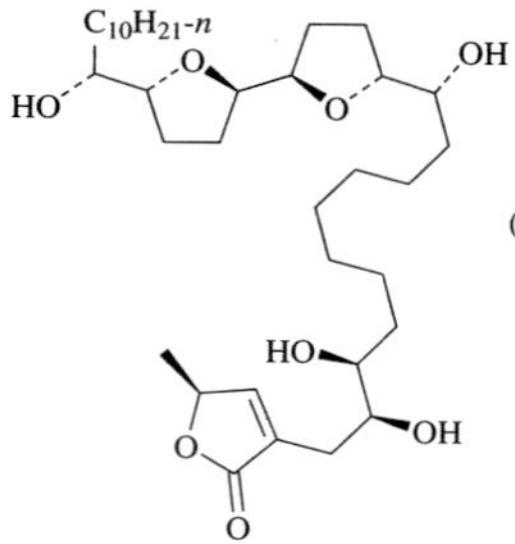 (78), >98	644

Substrate	Conditions	Product(s) and Yield(s) (%)	Refs.
C_{36} 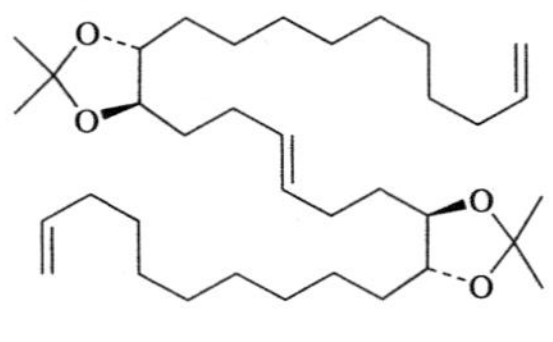	$(DHQ)_2PHAL$, OsO_4, $K_3Fe(CN)_6$, K_2CO_3, $MeSO_2NH_2$, t-BuOH:H_2O (1:1), 0°	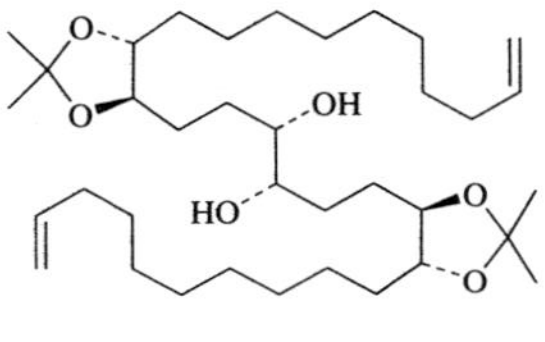 (82), >98 de	644
C_{38} 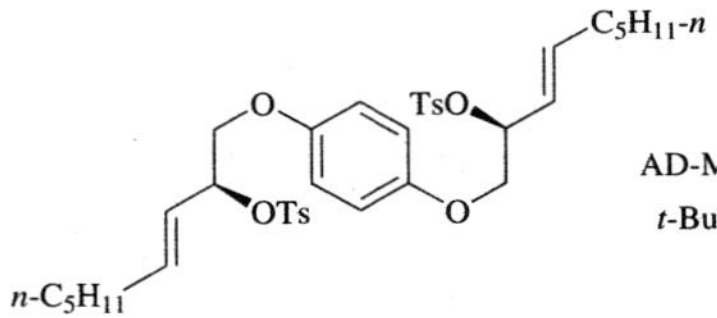	AD-Mix β, $NaHCO_3$, $MeSO_2NH_2$, t-BuOH:H_2O (1:1), rt, 24 h	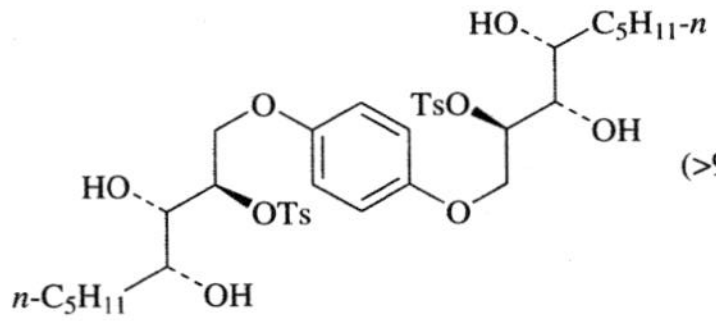 (>90), 88 de	633
C_{42} 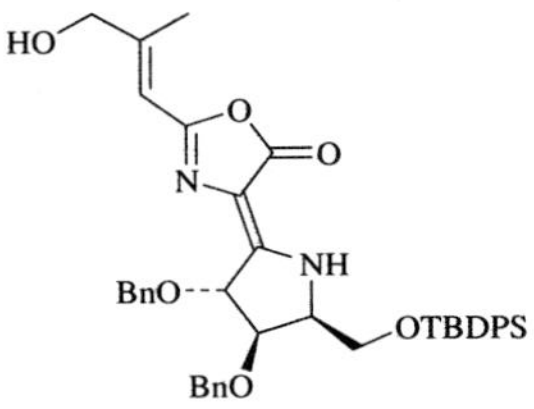	1. $(DHQ)_2PHAL$, OsO_4, CH_2Cl_2, −15° 2. H_2S	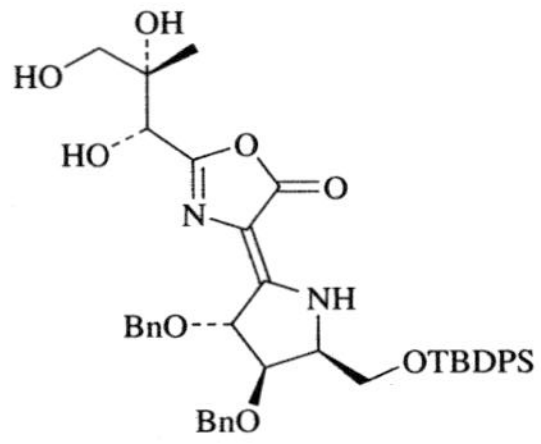 (59), >95 de	678

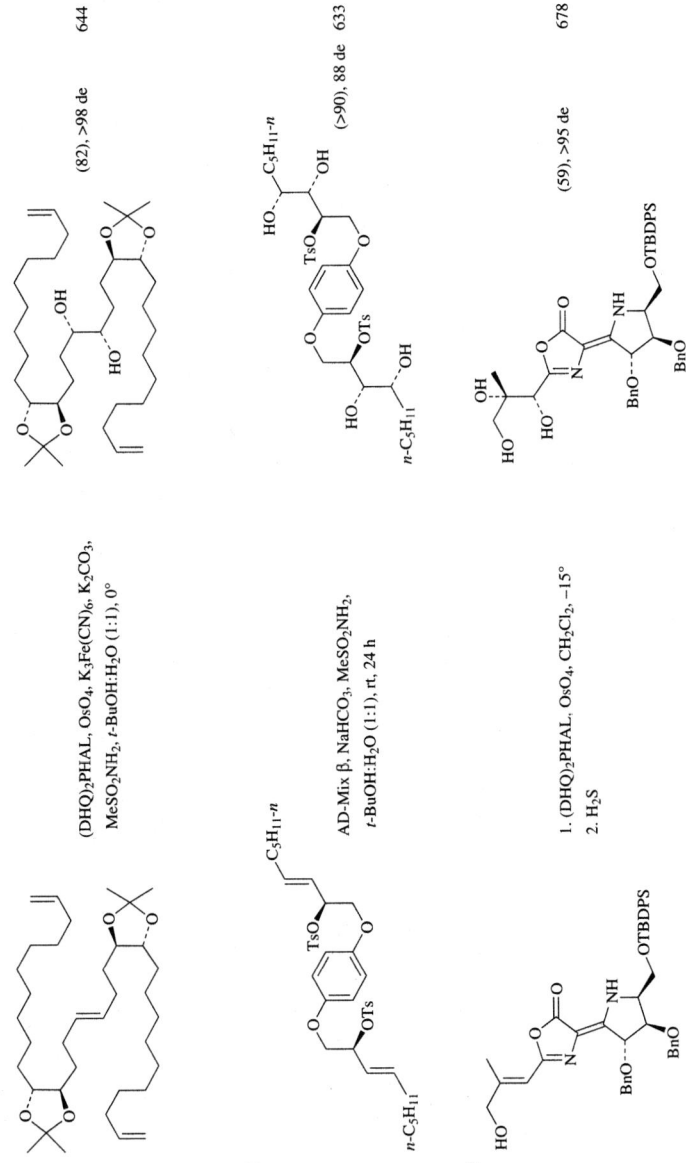

TABLE 8. REACTIONS OF UNCONJUGATED POLYALKENES (*Continued*)

Substrate	Conditions	Product(s) and Yield(s) (%), % ee	Refs.
C_{43} 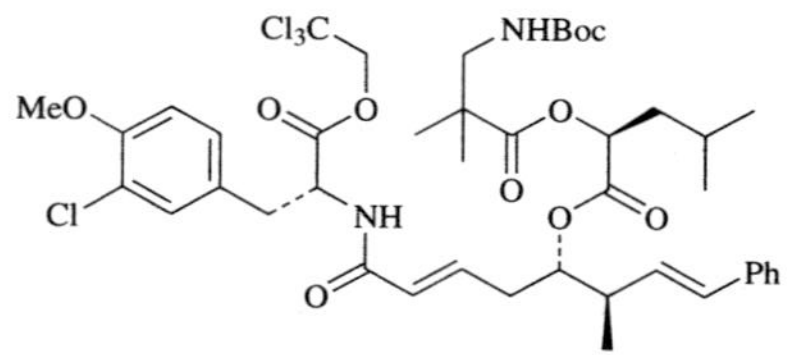			
	(DHQD)$_2$PHAL, $K_2OsO_2(OH)_4$, $MeSO_2NH_2$, $K_3Fe(CN)_6$, K_2CO_3, t-BuOH:H_2O (1:1), rt, 21 h	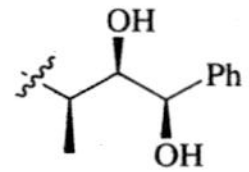 (61), 93 de	679
	(DHQD)$_2$PHAL, $K_2OsO_2(OH)_4$, $MeSO_2NH_2$, $K_3Fe(CN)_6$, K_2CO_3, $NaHCO_3$, t-BuOH:H_2O (1:1), rt, 21 h	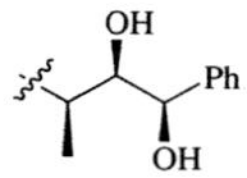 (60), 93 de	679
C_{47} 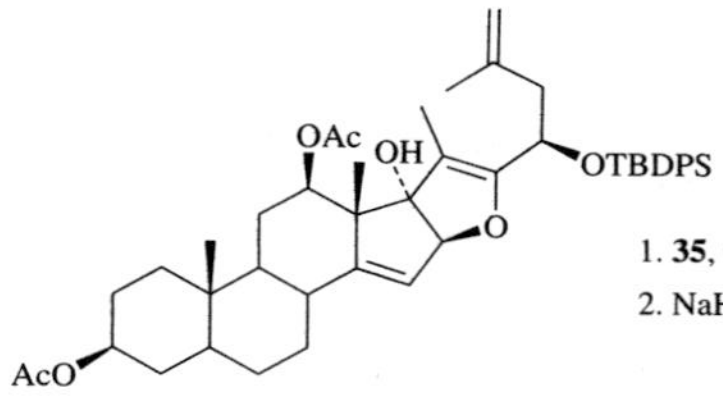	1. **35**, OsO_4, –100°, 30 min 2. $NaHSO_3$, THF-H_2O, reflux, 11 h	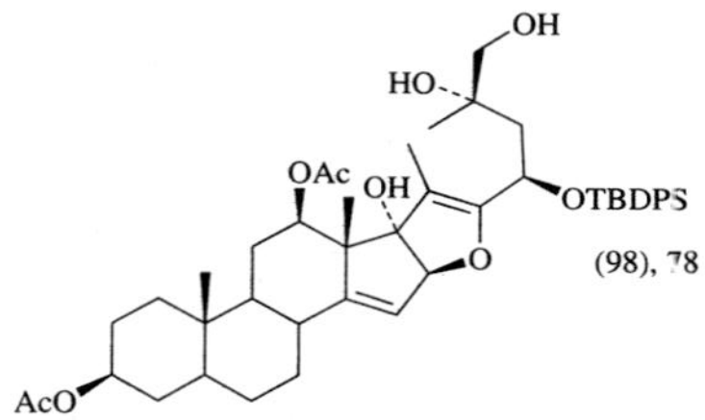 (98), 78 de	676, 677

C_{48} 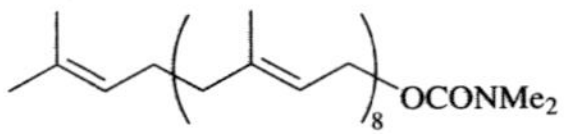	**20**, $K_2OsO_2(OH)_4$, $K_3Fe(CN)_6$, K_2CO_3, $MeSO_2NH_2$, *t*-BuOH:H_2O:methylcyclohexane (3.3:3.3:1)	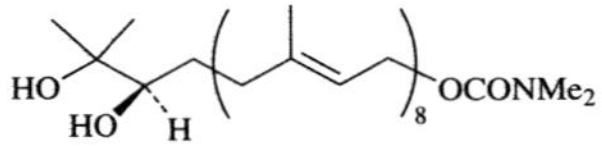	(23), 95	62
C_{50} 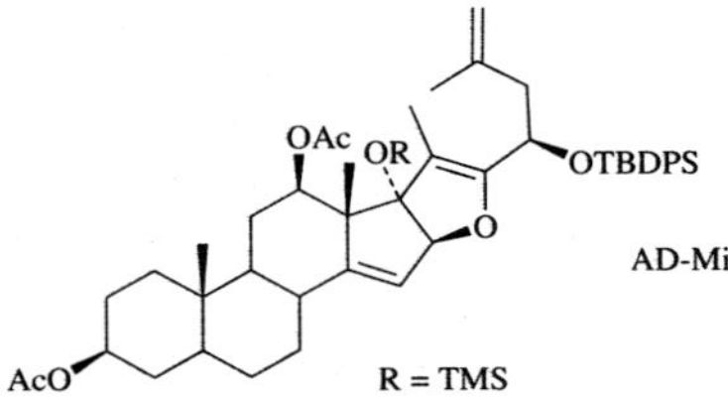 R = TMS	AD-Mix α, *t*-BuOH:H_2O (1:1), rt, 24 h	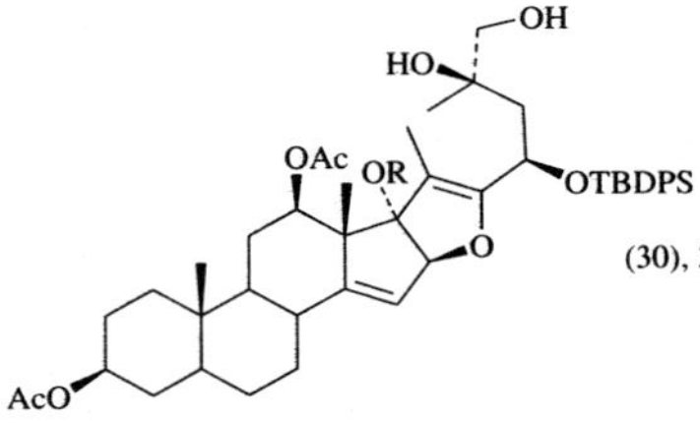	(30), 33 de	676, 677
	1. **35**, OsO_4, –95°, 1 h 2. $NaHSO_3$, THF-H_2O, reflux, 11 h	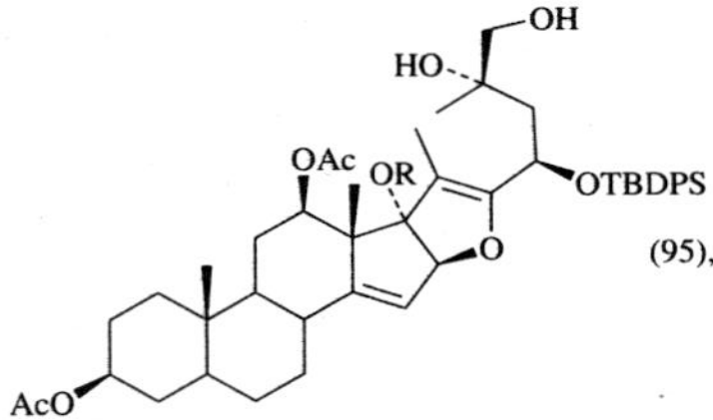	(95), 60 de	676, 677

[a] The authors assigned the absolute configuration using a mechanistic model.

TABLE 9. KINETIC RESOLUTIONS

Substrate	Conditions	Product(s) and (K_{rel})	Refs.
See Chart 1 at the beginning of the Tabular Survey for ligand structures that are indicated by ***bold*** *numbers.*			
C_9			
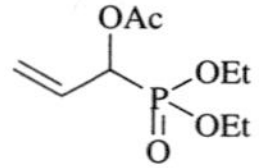	AD-Mix β, $MeSO_2NH_2$, t-BuOH:H_2O (1:1), 0°	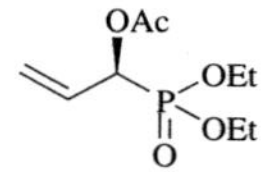 (5.1)	150
C_{10}			
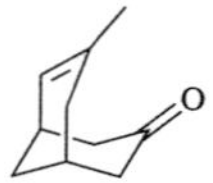	AD-Mix α, $MeSO_2NH_2$, t-BuOH:H_2O (1:1), rt, 72 h	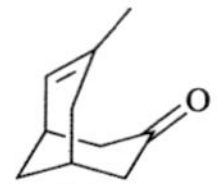 (–) (<5)	680
C_{12}			
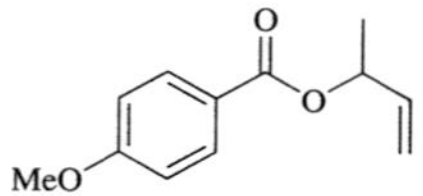	**17**, $K_2OsO_2(OH)_4$, $K_3Fe(CN)_6$, K_2CO_3, t-BuOH:H_2O (1:1), 0°, 2 h	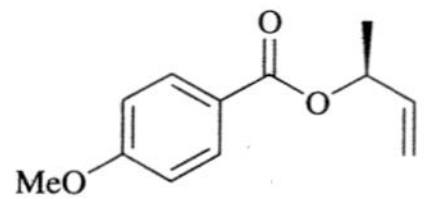 (20)	63
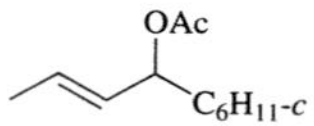	$(DHQD)_2TP$, $K_2OsO_2(OH)_4$, $K_3Fe(CN)_6$, K_2CO_3, t-BuOH:H_2O (1:1), 20°, 48 h	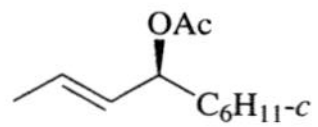 (12)	151
	$(DHQ)_2TP$, $K_2OsO_2(OH)_4$, $K_3Fe(CN)_6$, K_2CO_3, t-BuOH:H_2O (1:1), 20°, 48 h	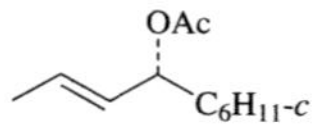 (6.7)	151
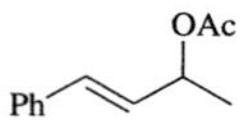	$(DHQD)_2TP$, $K_2OsO_2(OH)_4$, $K_3Fe(CN)_6$, K_2CO_3, t-BuOH:H_2O (1:1), 20°, 48 h	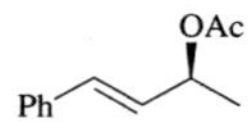 (8.8)	151
C_{13}			
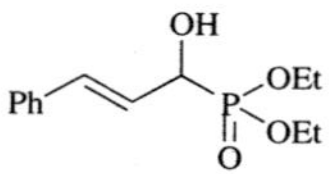	AD-Mix β, $MeSO_2NH_2$, t-BuOH:H_2O (1:1), 0°	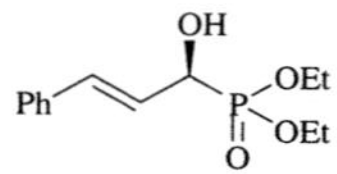 (1)	150

C_{14}			
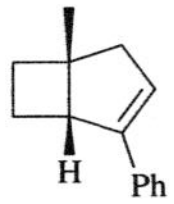	AD-Mix β, $MeSO_2NH_2$, t-BuOH:H_2O (1:1), 0°	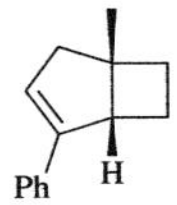 (24)	567
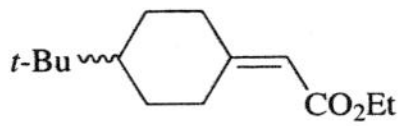	(DHQD)$_2$PHAL, $K_2OsO_2(OH)_4$, $K_3Fe(CN)_6$, K_2CO_3, t-BuOH:H_2O (1:1), 0°, 2 h	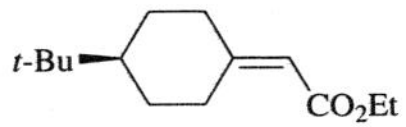 (32)	149
	(DHQ)$_2$PHAL, $K_2OsO_2(OH)_4$, $K_3Fe(CN)_6$, K_2CO_3, t-BuOH:H_2O (1:1), 0°, 2 h	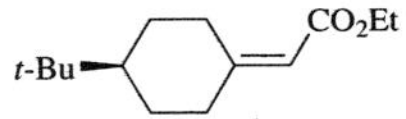 (26.5)	149
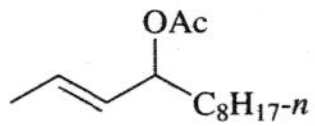	(DHQ)$_2$TP, $K_2OsO_2(OH)_4$, $K_3Fe(CN)_6$, K_2CO_3, t-BuOH:H_2O (1:1), 20°, 48 h	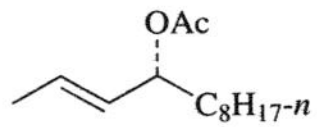 (9.7)	151
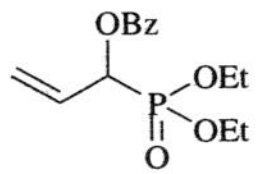	AD-Mix β, $MeSO_2NH_2$, t-BuOH:H_2O (1:1), 0°	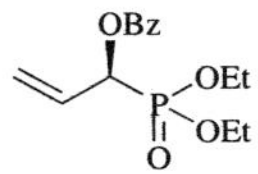 (1.6)	150
C_{15}			
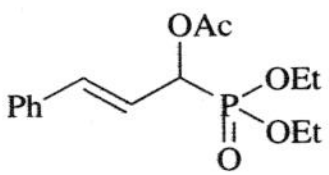	AD-Mix β, $MeSO_2NH_2$, t-BuOH:H_2O (1:1), 0°	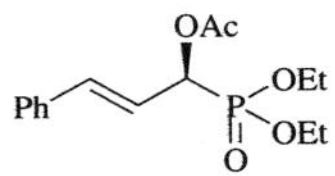 (15.6)	150
	AD-Mix α, $MeSO_2NH_2$, t-BuOH:H_2O (1:1), 0°	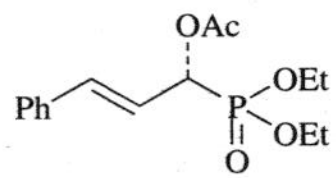 (4.9)	150

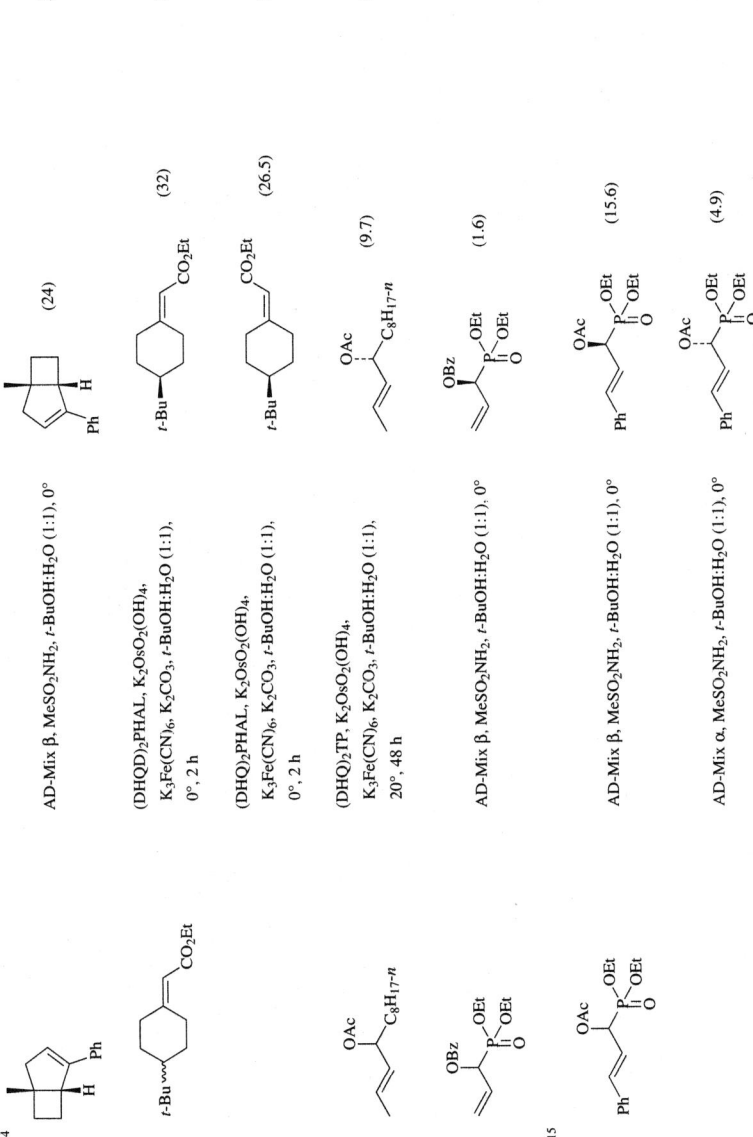

TABLE 9. KINETIC RESOLUTIONS (*Continued*)

Substrate	Conditions	Product(s) and (K_{rel})	Refs.
C_{16}			
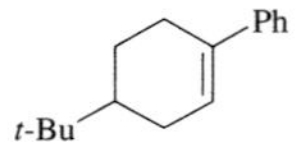	AD-Mix β, $MeSO_2NH_2$, *t*-BuOH:H_2O (1:1), 0°	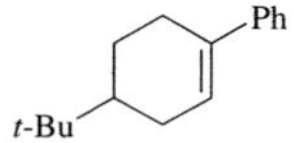 (2)	567
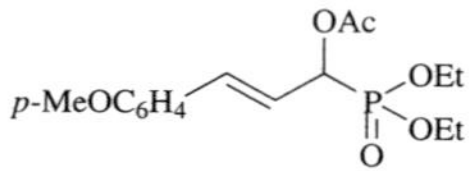	AD-Mix β, $MeSO_2NH_2$, *t*-BuOH:H_2O (1:1), 0°	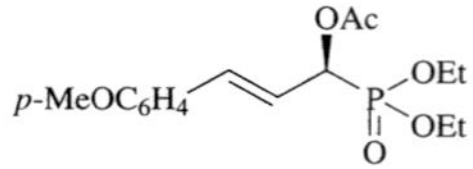 (10.7)	150
C_{17}			
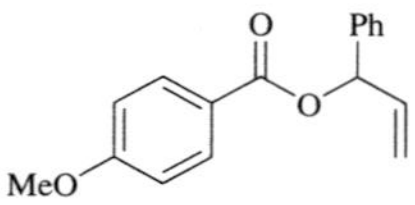	**17**, $K_2OsO_2(OH)_4$, $K_3Fe(CN)_6$, K_2CO_3, *t*-BuOH:H_2O (1:1), 0°, 2 h	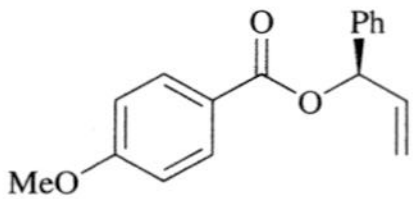 (79)	63
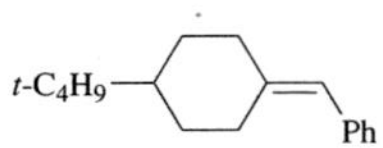	$(DHQD)_2PHAL$, $K_2OsO_2(OH)_4$, $K_3Fe(CN)_6$, K_2CO_3, *t*-BuOH:H_2O (1:1), 0°, 2 h	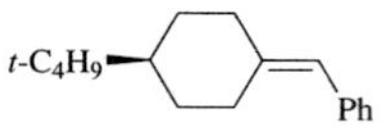 (9.7)	149
	$(DHQ)_2PHAL$, $K_2OsO_2(OH)_4$, $K_3Fe(CN)_6$, K_2CO_3, *t*-BuOH:H_2O (1:1), 0°, 2 h	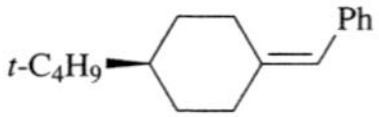 (5.0)	149
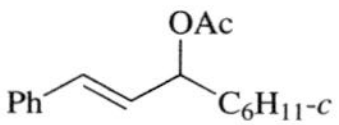	$(DHQD)_2TP$, $K_2OsO_2(OH)_4$, $K_3Fe(CN)_6$, K_2CO_3, *t*-BuOH:H_2O (1:1), 20°, 48 h	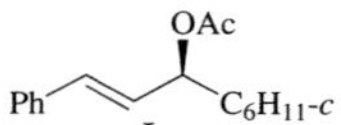 **I** (25)	151, 152
	$(DHQD)_2PHAL$, $K_2OsO_2(OH)_4$, $K_3Fe(CN)_6$, K_2CO_3, *t*-BuOH:H_2O (1:1)	**I** (2.3)	152
	DHQD-*p*-chlorobenzoate, $K_2OsO_2(OH)_4$, $K_3Fe(CN)_6$, K_2CO_3, *t*-BuOH:H_2O (1:1)	**I** (8.0)	152

C_{17-26}			
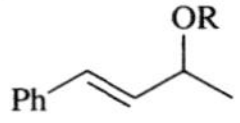 R: Bn; TBDPS	(DHQD)$_2$TP, $K_2OsO_2(OH)_4$, $K_3Fe(CN)_6$, K_2CO_3, *t*-BuOH:H_2O (1:1), 20°, 48 h	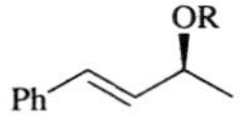 (4.5) (4.7)	151
C_{17}			
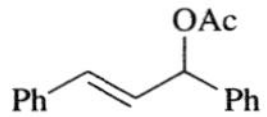	(DHQD)$_2$TP, $K_2OsO_2(OH)_4$, $K_3Fe(CN)_6$, K_2CO_3, *t*-BuOH:H_2O (1:1), 20°, 48 h	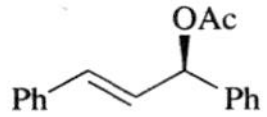 (4.9)	151
C_{20}			
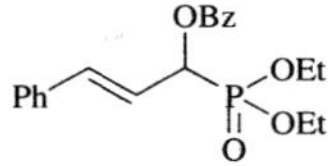	AD-Mix β, $MeSO_2NH_2$, *t*-BuOH:H_2O (1:1), 0°	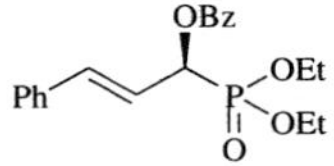 (6.6)	150
	AD-Mix α, $MeSO_2NH_2$, *t*-BuOH:H_2O (1:1), 0°	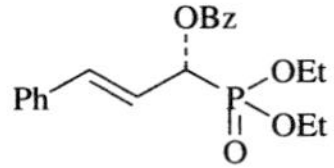 (4.6)	150
C_{22}			
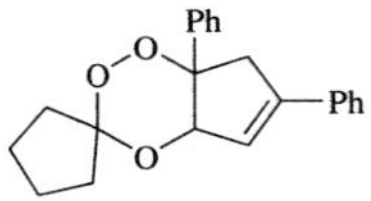	AD-Mix β, $MeSO_2NH_2$, *t*-BuOH:H_2O (1:1), 0°	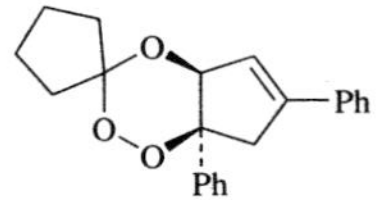 (71)[a], 95% ee	585
	AD-Mix α, $MeSO_2NH_2$, *t*-BuOH:H_2O (1:1), 0°	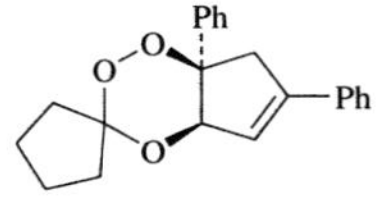 (71)[a], 41% ee	585
C_{76}			
(+/–) C_{76} Buckyball	DHQD-PHN (5 eq), OsO_4 (0.9 eq)	(–) C_{76} (33)[a], 28% ee	681
	DHQ-PHN (5 eq), OsO_4 (0.9 eq)	(+) C_{76} (33)[a], 28% ee	681

[a] This number is the % conversion.

TABLE 10. SUPPLEMENTAL TABLE ENTRIES: 2001-2004

A. REACTIONS OF TERMINAL ALKENES

Substrate	Conditions	Product(s) and Yield(s) (%), % ee	Refs.
*See Chart 2 at the beginning of the Tabular Survey for ligand and additive structures that are indicated by **bold** numbers.*			
C_3 	Wool-OsO_4, $K_3Fe(CN)_6$, K_2CO_3, t-BuOH:H_2O (1:1), rt, 24 h	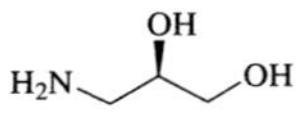 (80), 83.7	682
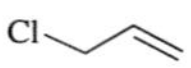	Wool-OsO_4, $K_3Fe(CN)_6$, K_2CO_3, t-BuOH:H_2O (1:1), rt, 24 h	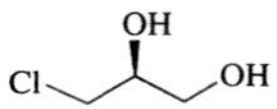 (68), 57.2	682
C_4 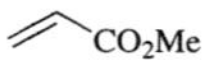	1. LDH-PdOsW, Et_3N, PhI, 70°, 16 h 2. $(DHQD)_2$PHAL, NMM, H_2O_2, t-BuOH:H_2O, rt, 12 h	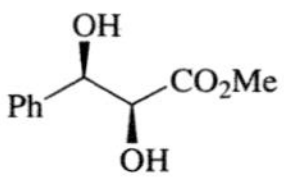 (93), 99	683
	1. LDH-PdOsW, Et_3N, PhBr, 70°, 16 h, 2. $(DHQ)_2$PHAL, NMM, H_2O_2, t-BuOH:H_2O (5:1), rt, 12 h	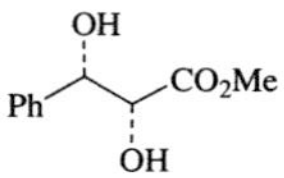 (90), 99	683
C_6 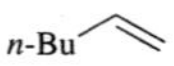	Ligand, $K_2OsO_2(OH)_4$, K_2CO_3 oxidant, solvent, rt, 24 h	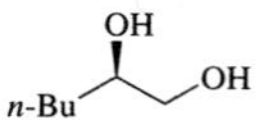 I	684

Ligand	Oxidant	Solvent	I
$(DHQD)_2$PHAL	$K_3Fe(CN)_6$	**89**:H_2O (1:2)	(71), 90
$(DHQD)_2$PYR	$K_3Fe(CN)_6$	**89**:H_2O (1:2)	(96), 90
$(DHQD)_2$PHAL	NMO	**89**:H_2O (1:2)	(92), 82
$(DHQD)_2$PHAL	$K_3Fe(CN)_6$	**94**:H_2O (1:2)	(77), 80
$(DHQD)_2$PYR	$K_3Fe(CN)_6$	**94**:H_2O (1:2)	(94), 89
$(DHQD)_2$PHAL	$K_3Fe(CN)_6$	**89**:H_2O:t-BuOH (1:1:2)	(88), 89
$(DHQD)_2$PYR	$K_3Fe(CN)_6$	**89**:H_2O:t-BuOH (1:1:2)	(96), 91
$(DHQD)_2$PHAL	NMO	**89**:H_2O:t-BuOH (1:1:2)	(85), 94
$(DHQD)_2$PHAL	$K_3Fe(CN)_6$	**94**:H_2O:t-BuOH (1:1:2)	(89), 97
$(DHQD)_2$PYR	$K_3Fe(CN)_6$	**94**:H_2O:t-BuOH (1:1:2)	(95), 91

C8

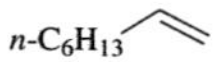

(DHQD)$_2$PHAL, $K_2OsO_2(OH)_4$, $NaClO_2$, NaOH, t-BuOH:H_2O (1:1), 0°, 5 h — **I** (80), 78 — 685

(DHQD)$_2$AQN, Resin-OsO_4, $K_3Fe(CN)_6$, K_2CO_3, t-BuOH:H_2O (1:1), 0°, 15 h

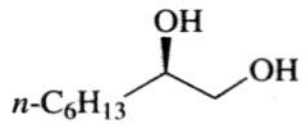

(85), 84 — 686

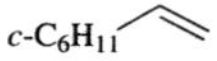

(DHQD)$_2$AQN, Resin-OsO_4, $K_3Fe(CN)_6$, K_2CO_3, t-BuOH:H_2O (1:1), 0°, 15 h

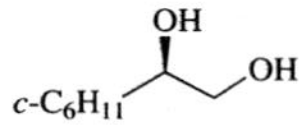

(80), 70 — 686

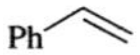

Ligand, $K_2OsO_2(OH)_4$, oxidant, solvent, 24 h — 687

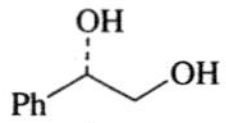

I

Ligand	Oxidant	Solvent	Temp	I
74	NMO	t-BuOH:H_2O (1:1)	rt	(92), 95
74	$K_3Fe(CN)_6$, K_2CO_3	t-BuOH:H_2O (1:1)	rt	(88), 96
74	O_2	t-BuOH:pH 10.4 buffer (1:2)	50°	(42), 88

76, OsO_4, NMO, solvent, 0°, 10 h — **I** — 688

Solvent	I
$Me_2CO:H_2O$:PEG (10:1:1)	(88), 80
$Me_2CO:H_2O$:**89** (10:1:1)	(85), 81

51, OsO_4, **89**, NMO, $Me_2CO:H_2O$ (10:1), rt, 20 h — **I** (89), 72 — 689

(DHQ)$_2$PHAL, PEM-MC-OsO_4, $K_3Fe(CN)_6$, K_2CO_3, $Me_2CO:H_2O$ (1:1), 30°, 5 h — **I** (80), 82 — 690

TABLE 10. SUPPLEMENTAL TABLE ENTRIES: 2001-2004 (*Continued*)

A. REACTIONS OF TERMINAL ALKENES (*Continued*)

Substrate	Conditions	Product(s) and Yield(s) (%), % ee	Refs.
C_8 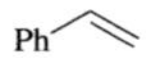	$(DHQ)_2PHAL$, [Os(VIII)], $K_3Fe(CN)_6$, K_2CO_3, $MeSO_2NH_2$, t-BuOH:H_2O (1:1), rt	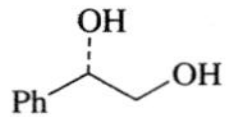 **I**	691

Ligand %	[Os(VIII)]	Time	**I**
1	XAD-4-OsO_4 (1%)	0.5 h	(92), 95
0.5	XAD-4-OsO_4 (0.2%)	2 h	(91), 93
0.25	XAD-4-OsO_4 (0.1%)	5 h	(91), 93
1	XAD-7-OsO_4 (1%)	1.5 h	(97), 95
0.5	XAD-7-OsO_4 (0.2%)	3 h	(98), 93
0.25	XAD-7-OsO_4 (0.1%)	6 h	(96), 92

Ligand, OsO_4,$K_3Fe(CN)_6$, K_2CO_3 t-BuOH:H_2O (1:1) — **I**

Ligand	Additive	Temp	Time	**I**	Refs.
53	-	0°	24 h	(91), 99	692
61	-	0°	24 h	(100), 72	693
62	-	0°	24 h	(100), 87	693
63	-	0°	24 h	(100), 96	693
74	K_2CO_3	rt	22 h	(87), 78	694
75	-	0°	24 h	(94), 97	695
88	K_2CO_3	5°	-	(91), 30	696

Conditions	Product(s) and Yield(s) (%), % ee	Refs.
83, OsO_4, $K_2CO_3K_3Fe(CN)_6$, t-BuOH:H_2O (1:1), 0°, 18 h	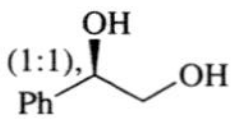 **I** (73), 85	697

Ligand, $K_2OsO_2(OH)_4$, $K_2CO_3K_3Fe(CN)_6$, solvent, rt, 24 h → I | 698

Ligand	Solvent	I
$(DHQD)_2$PHAL	*t*-BuOH:H_2O (1:2)	(88), 97
$(DHQD)_2$PHAL	**89**:H_2O (1:2)	(87), 62
$(DHQD)_2$PHAL	**89**:H_2O:*t*-BuOH (1:1:2)	(86), 94
$(DHQD)_2$PYR	*t*-BuOH:H_2O (1:2)	(91), 93
$(DHQD)_2$PYR	**89**:H_2O (1:2)	(86), 75
$(DHQD)_2$PYR	**89**:H_2O:*t*-BuOH (1:1:2)	(90), 89

Ligand, $K_2OsO_2(OH)_4$, $K_3Fe(CN)_6$, K_2CO_3, solvent, rt, 24 h → I | 684

Ligand	Solvent	I
$(DHQD)_2$PHAL	**89**:H_2O (1:1)	(84), 71
$(DHQD)_2$PHAL	**89**:H_2O (1:2)	(86), 62
$(DHQD)_2$PYR	**89**:H_2O (1:5)	(86), 75
$(DHQD)_2$PHAL	**89**:H_2O (1:2)	(85), 67
$(DHQD)_2$PHAL	**90**:H_2O (1:2)	(45), 71
$(DHQD)_2$PHAL	**91**:H_2O (1:2)	(67), 77
$(DHQD)_2$PHAL	**94**:H_2O (1:2)	(81), 78
$(DHQD)_2$PYR	**94**:H_2O (1:2)	(83), 87
$(DHQD)_2$PHAL	**95**:H_2O (1:2)	(64), 79
$(DHQD)_2$PYR	**95**:H_2O (1:2)	(82), 88
$(DHQD)_2$PHAL	**96**:H_2O (1:2)	(72), 82
$(DHQD)_2$PHAL	**98**:H_2O (1:2)	(70), 74
$(DHQD)_2$PHAL	**99**:H_2O (1:2)	(61), 74
$(DHQD)_2$PHAL	**100**:H_2O (1:2)	(35), 84
$(DHQD)_2$PHAL	**89**:H_2O:*t*-BuOH (1:1:2)	(86), 94
$(DHQD)_2$PHAL	**89**:H_2O:*t*-BuOH (1:2:4)	(88), 92
$(DHQD)_2$PYR	**89**:H_2O:*t*-BuOH (1:1:2)	(90), 90
$(DHQD)_2$PYR	**89**:H_2O:*t*-BuOH (1:2:4)	(92), 93
$(DHQD)_2$PHAL	**89**:H_2O:EtOH (1:1:2)	(68), 82

TABLE 10. SUPPLEMENTAL TABLE ENTRIES: 2001-2004 (*Continued*)

A. REACTIONS OF TERMINAL ALKENES (*Continued*)

Substrate	Conditions	Product(s) and Yield(s) (%), % ee	Refs.
C_8 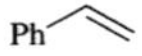	Ligand, $K_2OsO_2(OH)_4$, $K_3Fe(CN)_6$, K_2CO_3, solvent, rt, 24 h	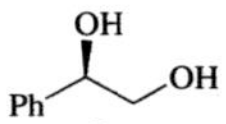 **I**	684

Ligand	Solvent	I
$(DHQD)_2$PHAL	**91**:H_2O:*t*-BuOH (1:1:2)	(69), 99
$(DHQD)_2$PHAL	**93**:H_2O:*t*-BuOH (1:2:4)	(68), 82
$(DHQD)_2$PHAL	**94**:H_2O:*t*-BuOH (1:1:2)	(85), 92
$(DHQD)_2$PYR	**94**:H_2O:*t*-BuOH (1:1:2)	(87), 96
$(DHQD)_2$PHAL	**95**:H_2O:*t*-BuOH (1:1:2)	(71), 94
$(DHQD)_2$PYR	**95**:H_2O:*t*-BuOH (1:1:2)	(83), 88
$(DHQD)_2$PHAL	**97**:H_2O:*t*-BuOH (1:1:2)	(71), 96
$(DHQD)_2$PHAL	**97**:H_2O:*t*-BuOH (1:2:4)	(66), 93
$(DHQD)_2$PYR	**97**:H_2O:*t*-BuOH (1:1:2)	(79), 96
$(DHQD)_2$PHAL	**98**:H_2O:*t*-BuOH (1:2:4)	(73), 76
$(DHQD)_2$PHAL	**99**:H_2O:*t*-BuOH (1:2:4)	(67), 77
$(DHQD)_2$PHAL	**100**:H_2O:*t*-BuOH (1:2:4)	(53), 86
$(DHQD)_2$PHAL	**101**:H_2O:*t*-BuOH (1:2:4)	(23), 89
$(DHQD)_2$PHAL	H_2O:*t*-BuOH (1:1)	(90), 97
$(DHQD)_2$PHAL	H_2O:*t*-BuOH (1:2)	(88), 97
$(DHQD)_2$PHAL	H_2O:*t*-BuOH (1:5)	(85), 91
$(DHQD)_2$PYR	H_2O:*t*-BuOH (1:1)	(93), 95
$(DHQD)_2$PYR	H_2O:*t*-BuOH (1:2)	(91), 93

Substrate	Conditions	Product(s) and Yield(s) (%), % ee	Refs.
	Ligand, $K_2OsO_2(OH)_4$, NMO, solvent, rt, 24 h	**I**	684

Ligand	Solvent	I
$(DHQD)_2$PHAL	**89**:H_2O (1:1)	(84), 75
$(DHQD)_2$PHAL	**89**:H_2O (1:2)	(56), 88

$(DHQD)_2PYR$	**89**:H_2O (1:1)	(52), 85	
$(DHQD)_2PYR$	**89**:H_2O (1:2)	(85), 89	
$(DHQD)_2PHAL$	**94**:H_2O (1:2)	(86), 94	
$(DHQD)_2PYR$	**94**:H_2O (1:2)	(39), 76	
$(DHQD)_2PHAL$	**89**:H_2O:*t*-BuOH (1:1:2)	(89), 85	
$(DHQD)_2PYR$	**89**:H_2O:*t*-BuOH (1:1:2)	(85), 94	
$(DHQD)_2PHAL$	**94**:H_2O:*t*-BuOH (1:1:2)	(64), 67	
$(DHQD)_2PYR$	**94**:H_2O:*t*-BuOH (1:1:2)	(61), 74	
$(DHQD)_2PHAL$	H_2O:*t*-BuOH (1:1)	(96), 98	
$(DHQD)_2PHAL$	H_2O:*t*-BuOH (1:2)	(88), 91	
$(DHQD)_2PYR$	H_2O:*t*-BuOH (1:1)	(87), 64	

77, $K_2OsO_2(OH)_4$, oxidant(s), K_2CO_3, *t*-BuOH:H_2O (1:1), 0° — **I** — 699

Oxidant(s)	Time	I
$K_3Fe(CN)_6$	8 h	(94), 93
$K_3Fe(CN)_6$, $NaClO_2$	16 h	(97), 90

$(DHQD)_2PHAL$, $K_2OsO_2(OH)_4$, $NaClO_2$, NaOH, *t*-BuOH:H_2O (1:1), 0°, 2.5 h — **I** (73), 96 — 685

Ligand, $K_2OsO_2(OH)_4$, $K_3Fe(CN)_6$, K_2CO_3, *t*-BuOH:H_2O (1:1) — **I**

Ligand	Temp	Time	I	Refs.
64	0°	24 h	(87), 82	700
65	rt	-	(73), 92	701
66	rt	-	(85), 97	701

TABLE 10. SUPPLEMENTAL TABLE ENTRIES: 2001-2004 (*Continued*)

A. REACTIONS OF TERMINAL ALKENES (*Continued*)

Substrate	Conditions	Product(s) and Yield(s) (%), % ee	Refs.
C_8			
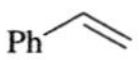	$(DHQD)_2PHAL$, PEM-MC OsO_4, $K_3Fe(CN)_6$, K_2CO_3, surfactant, H_2O, 30°, 24 h	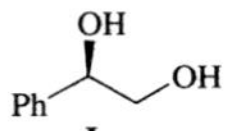 **I**	702
	Surfactant (conc. or amt.)	**I**	
	n-$C_{12}H_{25}NMe_2^+CH_2CH_2SO_3^-$ (10 mol%)	(82), 68	
	Span® #20 (10 mol%)	(77), 54	
	Tween® #20 (10 mol%)	(100), 77	
	Triton®-WR 1339 (2 mg/mL)	(91), 76	
	Triton®-CF 10 (2 mg/mL)	(84), 76	
	Triton® X-100 (10 mol%)	(84), 75	
	Triton® X-405 (10 mol%)	(68), 75	
	Triton® X-405 (3 mol%)	(76), 74	
	$(DHQD)_2PHAL$, Os EnCat, K_2CO_3, $K_3Fe(CN)_6$, $MeSO_2NH_2$, THF:H_2O (1:1), rt, 48 h	**I** (90), 95	703
	$(DHQD)_2PHAL$, PEM-MC-OsO_4, $K_3Fe(CN)_6$, K_2CO_3, Me_2CO:H_2O(1:1), 30°, 5 h	**I** (85), 78	690

(DHQD)$_2$PHAL, $K_2OsO_2(OH)_4$, Na_2WO_4, NMM, H_2O_2, Et_4NOAc, rt, 20 h	**I** (90), 95	683
(DHQD)$_2$PHAL, LDH-OsW, NMM, H_2O_2, Et_4NOAc, rt, 20 h	**I** (92), 96	683
(DHQD)$_2$PHAL, LDH-OsO_4, NMO, *t*-BuOH:H_2O (1:1), rt	**I** (94), 95	704
(DHQD)$_2$PHAL, Resin-OsO_4, NMO, *t*-BuOH:H_2O (3:1), rt, 12-24 h	**I**	705

Oxidant	Solvent	I
NMO	*t*-BuOH:H_2O (3:1)	(92), 95
$K_3Fe(CN)_6$	*t*-BuOH:H_2O (1:1)	(89), 97
O_2	*t*-BuOH:pH 10.4 buffer (1:2.5)	(50), 89

52, OsO_4, titanium silicalite, H_2O_2, NMM, t-BuOH:H_2O (3:1), rt, 12 h	**I** (55), 93	706
Ligand, OsO_4, NMO, *t*-BuOH:H_2O (1:1), rt	**I** (—)	707

Ligand	% ee
54	25
55	7
56	29
57	42
58	5
59	25

(DHQD)$_2$PHAL, OsO_4, $MeReO_3$, H_2O_2, *t*-BuOH:H_2O (3:1), 0°, 16 h	**I** (90), 95	708

TABLE 10. SUPPLEMENTAL TABLE ENTRIES: 2001-2004 (*Continued*)

A. REACTIONS OF TERMINAL ALKENES (*Continued*)

Substrate	Conditions	Product(s) and Yield(s) (%), % ee	Refs.
C_8 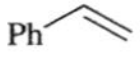	$(DHQD)_2PHAL$, $K_2OsO_2(OH)_4$, $PhSe(O)CH_2Ph$, t-BuOH:H_2O(1:1), rt, 12 h	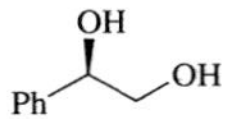 **I** (93), 97	709
	$(DHQD)_2PHAL$, OsO_4, **60**, H_2O_2, Et_4NOAc, t-BuOH:H_2O (3:1), 0°, 16 h	**I** (75), 95	710
	$(DHQD)_2PHAL$-*N*-oxide, OsO_4, Et_4NOAc, t-BuOH:H_2O (3:1), 0°	**I** (71), 98	710
	1. LDH-PdOsW, Et_3N, PhI, 70°, 16 h 2. $(DHQD)_2PHAL$, NMM, H_2O_2, t-BuOH:H_2O, rt, 12 h	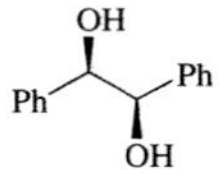 (85), 99	683
C_9 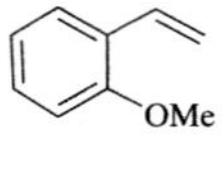	AD-Mix α, t-BuOH:H_2O (1:1), 0°, 3 h	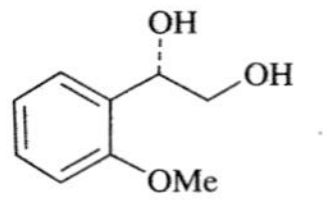 (93), 94	711
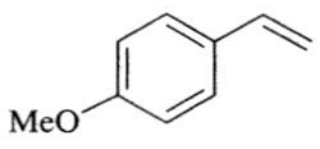	$(DHQD)_2PHAL$, PEM-MC OsO_4, $K_3Fe(CN)_6$, K_2CO_3, Triton® X-405, H_2O, 30°, 24 h	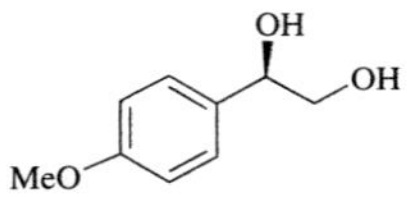 (68), 89	702
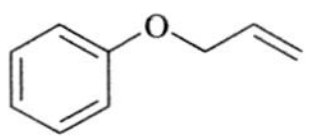	$(DHQD)_2PHAL$, PEM-MC OsO_4, $K_3Fe(CN)_6$, K_2CO_3 , Triton® X-405, H_2O, 30°, 24 h	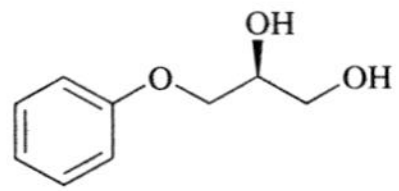 (79), 55	702

	$(DHQD)_2$PHAL, Os EnCat, K_2CO_3, $K_3Fe(CN)_6$, $MeSO_2NH_2$, THF:H_2O (1:1), rt, 48 h	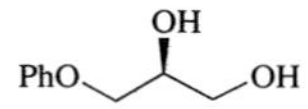	(83), 51	703
	$(DHQD)_2$PHAL, $K_2OsO_2(OH)_4$, PhSe(O)CH_2Ph, *t*-BuOH:H_2O (1:1), rt, 20 h	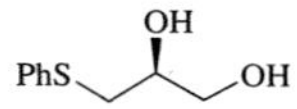	(80), 59	709
C_{10}				
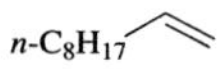	$(DHQD)_2$AQN, Resin-OsO_4, $K_3Fe(CN)_6$, K_2CO_3, *t*-BuOH:H_2O (1:1), 0°, 15 h	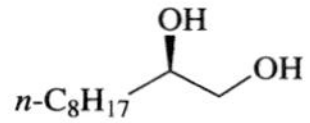 **I**	(80), 74	686
	$(DHQD)_2$PHAL, $K_2OsO_2(OH)_4$, PhSe(O)CH_2Ph, *t*-BuOH:H_2O (1:1), rt, 20 h	**I**	(69), 70	709
	88, OsO_4, $K_3Fe(CN)_6$, K_2CO_3, *t*-BuOH:H_2O (1:1), 5°	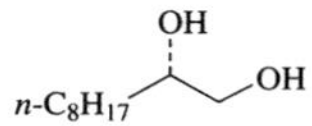	(88), 38	696
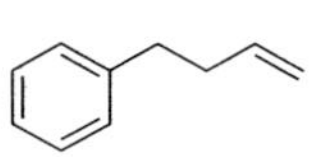	$(DHQD)_2$PHAL, PEM-MC OsO_4, $K_3Fe(CN)_6$, K_2CO_3, Triton® X-405, H_2O, 30°, 24 h	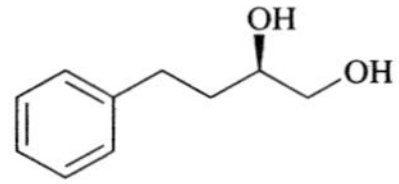	(76), 66	702
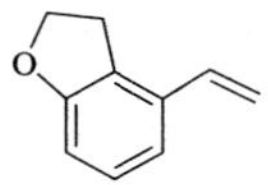	AD-Mix α, $MeSO_2NH_2$, *i*-PrOH:H_2O (1:1), 0°, 24 h	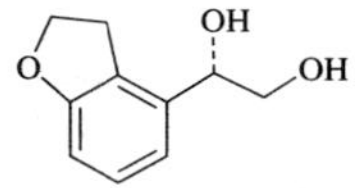	(90), >99	712
C_{12}				
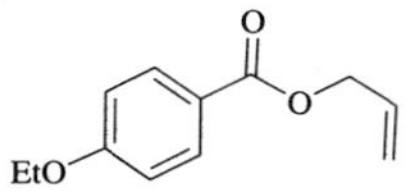	**88**, OsO_4, $K_3Fe(CN)_6$, K_2CO_3, *t*-BuOH:H_2O (1:1), 5°	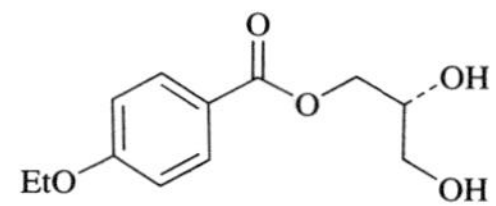	(85), 5	696

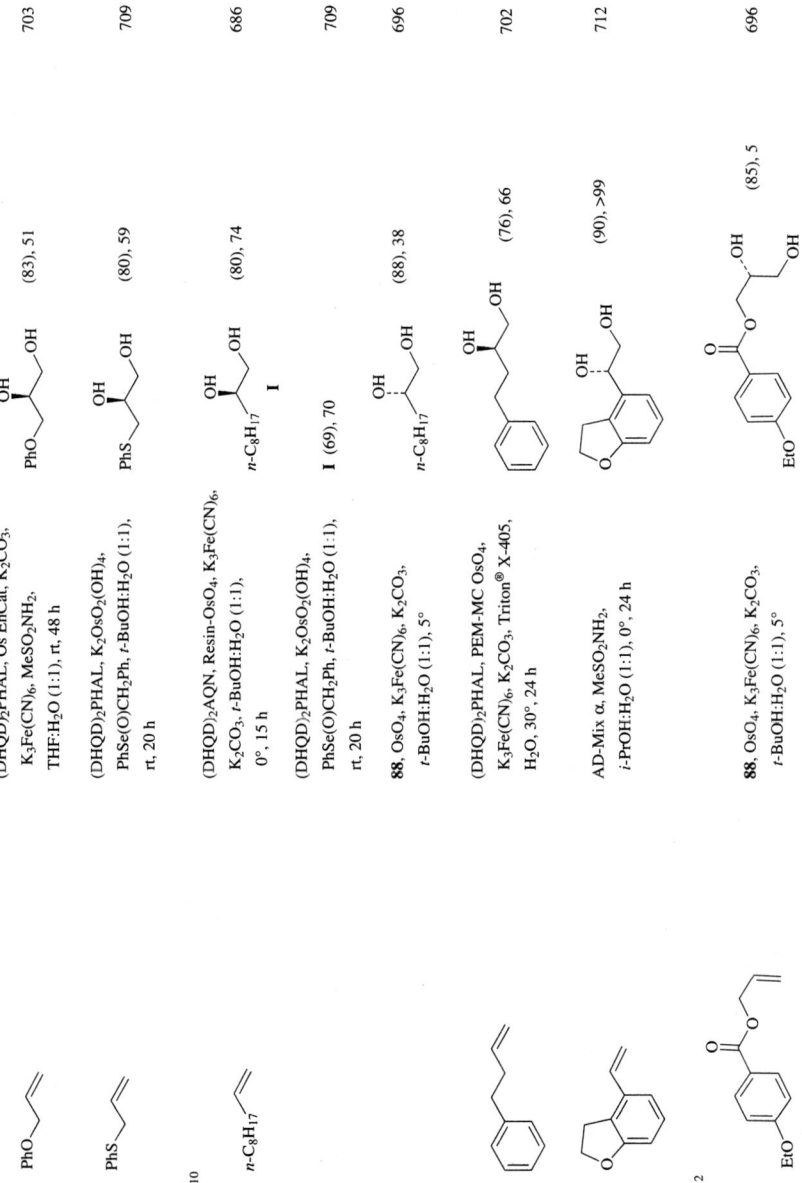

TABLE 10. SUPPLEMENTAL TABLE ENTRIES: 2001-2004 (*Continued*)

A. REACTIONS OF TERMINAL ALKENES (*Continued*)

Substrate	Conditions	Product(s) and Yield(s) (%), % ee	Refs.
C_{13} 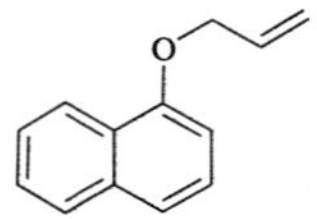	Ligand, OsO_4, $K_3Fe(CN)_6$, *t*-BuOH:H_2O (1:1), 0°, 24 h	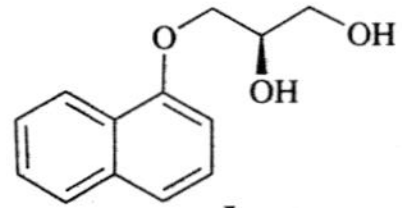 I	
	Ligand: **53**	I: (93), 89	692
	Ligand: **75**	I: (87), 84	695
	64, $K_2OsO_2(OH)_4$, $K_3Fe(CN)_6$, K_2CO_3, *t*-BuOH:H_2O (1:1), 0°, 24 h	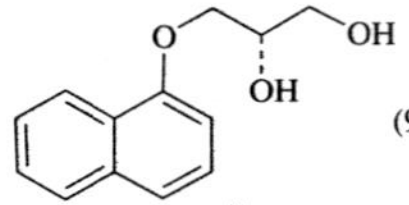 I (91), 90	700
	$(DHQD)_2PHAL$, LDH-OsO_4, Et_4NOAc, NMO, *t*-BuOH:H_2O (1:1), rt	**I** (94), 77	704
C_{14} 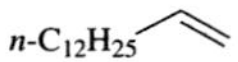	$(DHQD)_2PHAL$, OsO_4, NMO, PEG-400, rt, 2 h	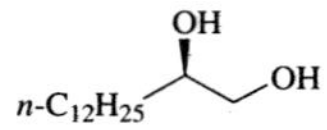 (96), 44	713
C_{16} 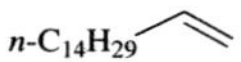	$(DHQD)_2PHAL$, $K_2OsO_2(OH)_4$, $NaClO_2$, NaOH, *t*-BuOH:H_2O (1:1), 0°, 6 h	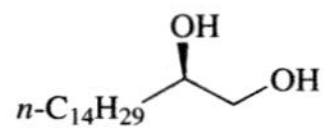 (77), 87	685
C_{17} 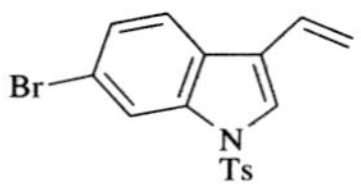	AD-Mix α, *t*-BuOH:H_2O (1:1), 0°, 10 h	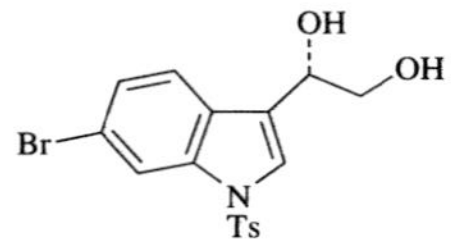 (92), 99	714

C_{17} 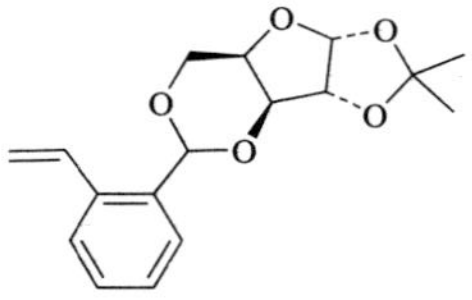	AD-Mix α, *t*-BuOH-H_2O	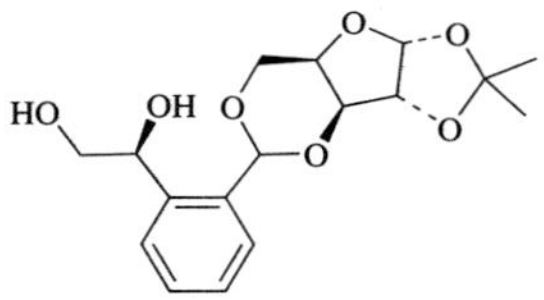	(—), 99.2	715
	AD-Mix β, *t*-BuOH-H_2O	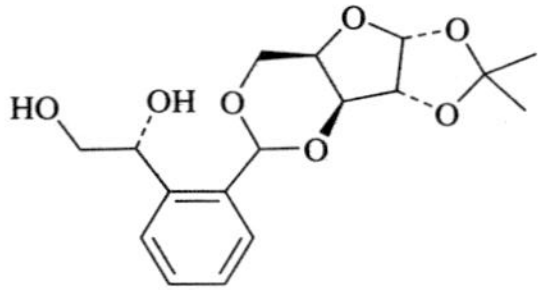	(—), 97.4	715
C_{17-23} 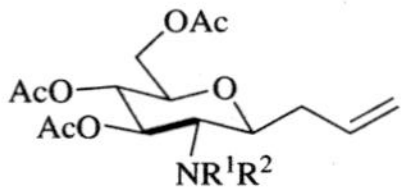	1. AD-Mix β, *t*-BuOH:H_2O (1:1), 0°, 10 h 2. $Me_2C(OMe)_2$, *p*-TsOH	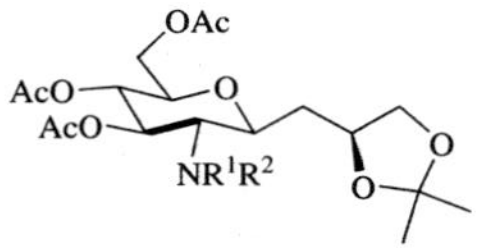 R^1R^2 — H, $COCH_3$: (89), 44 (de); *o*-$C_6H_4(CO)_2$: (73), 26 (de)		716
C_{18} 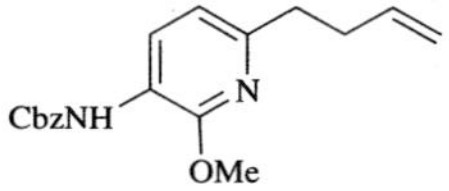	Ligand, $K_2OsO_4(OH)_2$, $K_3Fe(CN)_6$, K_2CO_3, *t*-BuOH:H_2O (1:1), 0°, 20 h Ligand: $(DHQD)_2PHAL$; $(DHQD)_2PYR$; $(DHQD)_2AQN$	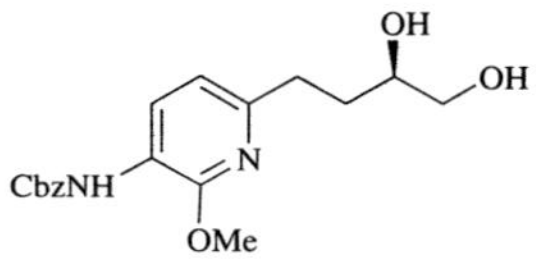 (96), 55.5; (91), 71.1; (98), 83.4		717

TABLE 10. SUPPLEMENTAL TABLE ENTRIES: 2001-2004 (*Continued*)

A. REACTIONS OF TERMINAL ALKENES (*Continued*)

Substrate	Conditions	Product(s) and Yield(s) (%), % ee		Refs.
C_{18}				
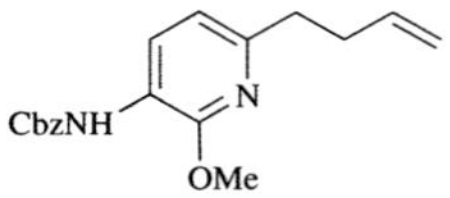	$(DHQ)_2PHAL$, $K_2OsO_4(OH)_2$, $K_3Fe(CN)_6$, K_2CO_3, t-BuOH:H_2O (1:1), 0°, 20 h	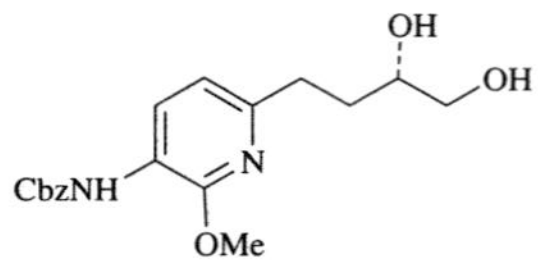	(96), 78	717
C_{21}				
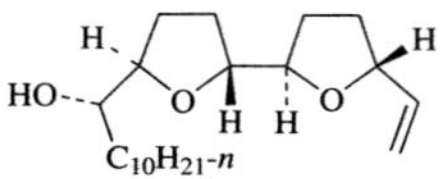	$(DHQ)_2AQN$, $K_2OsO_2(OH)_4$, $K_3Fe(CN)_6$, K_2CO_3, t-BuOH-H_2O	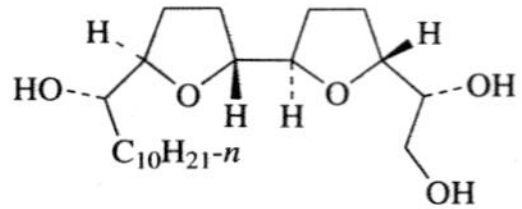	(86), 100	759

B. REACTIONS OF 1,1-DISUBSTITUTED ALKENES

Substrate	Conditions	Product(s) and Yield(s) (%), % ee	Refs.
C_5 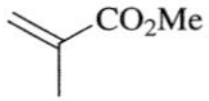	1. LDH-PdOsW, Et_3N, PhI, 70°, 16 h 2. $(DHQD)_2PHAL$, NMM, H_2O_2, *t*-BuOH:H_2O, rt, 12 h	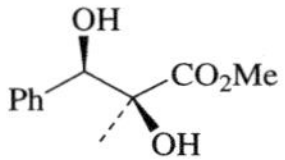 (92), 98	683
C_8 	$(DHQD)_2AQN$, Resin-OsO_4, K_2CO_3, $K_3Fe(CN)_6$, *t*-BuOH:H_2O (1:1), 0°, 15 h	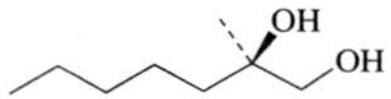 (85), 82	686
C_9 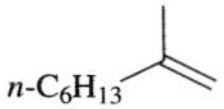	Ligand, OsO_4, $K_3Fe(CN)_6$, K_2CO_3, $MeSO_2NH_2$, *t*-BuOH:H_2O (1:1), 0°, 16 h	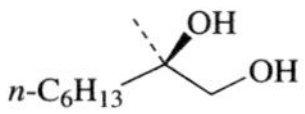 **I**	718
	Ligand: $(DHQD)_2PHAL$	**I**: (88), 78	
	85	(86), 75	
	86	(82), 69	
	87	(82), 5	
	$(DHQD)_2PHAL$, $K_2OsO_2(OH)_4$, $PhSe(O)CH_2Ph$, *t*-BuOH:H_2O (1:1), rt, 2.5 h	**I** (82), 50	709
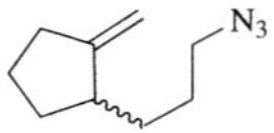	$(DHQD)_2PHAL$, $K_2OsO_2(OH)_4$, $K_3Fe(CN)_6$, K_2CO_3, $MeSO_2NH_2$, *t*-BuOH:H_2O (1:1), 6°, 3.5 h	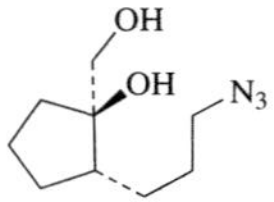 (54), >77 + 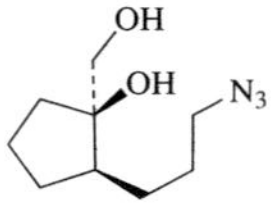 (41), —	719
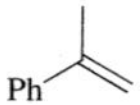	**51**, OsO_4, **89**, NMO, Me_2CO:H_2O (10:1), rt, 20 h	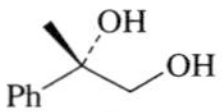 **I** (98), 63	689

TABLE 10. SUPPLEMENTAL TABLE ENTRIES: 2001-2004 (*Continued*)

B. REACTIONS OF 1,1-DISUBSTITUTED ALKENES (*Continued*)

Substrate	Conditions	Product(s) and Yield(s) (%), % ee	Refs.
C_9 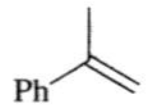	**76**, OsO_4, NMO, $Me_2CO:H_2O$:cosolvent (10:1:1), 0°, 10 h	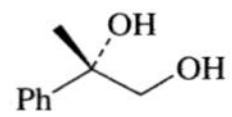 **I**	688
	Cosolvent: PEG	(89), 78	
	Cosolvent: **89**	(89), 66	
	74, $K_2OsO_2(OH)_4$, Additive cooxidant, solvent, 24 h	**I**	687
	Cooxidant: NMO; Additive: —; Solvent: t-BuOH:H_2O (1:1); Temp: rt	(94), 91	
	Cooxidant: $K_3Fe(CN)_6$; Additive: K_2CO_3; Solvent: t-BuOH:H_2O (1:1); Temp: rt	(90), 94	
	Cooxidant: O_2; Additive: —; Solvent: t-BuOH:pH 10.4 buffer (1:2); Temp: 50	(99), 79	
	Ligand, OsO_4, $K_3Fe(CN)_6$, t-BuOH:H_2O (1:1), 0°, 24 h	**I**	
	Ligand: **53**	(82), 91	692
	Ligand: **61**	(100), 57	693
	Ligand: **62**	(100), 75	693
	Ligand: **63**	(100), 90	693
	Ligand: **74**	(82), 91	695
	Ligand, OsO_4, $K_3Fe(CN)_6$, K_2CO_3, t-BuOH:H_2O (1:1), 0°, 14 h	**I**	720
	Ligand: **69**	(85), 64	
	Ligand: **70**	(—), 64	

(DHQ)$_2$PHAL (1%), [Os(VIII)], $K_3Fe(CN)_6$, K_2CO_3, $MeSO_2NH_2$, t-BuOH:H_2O (1:1), rt, 1.5 h — **I** — 691

[Os(VIII)]	Time	I
XAD-4-OsO_4	1.5 h	(96), 89
XAD-7-OsO_4	3.5 h	(97), 88

Ligand, $K_2OsO_2(OH)_4$, oxidant, additive, solvent, rt, 24 h — 684

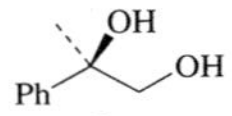

I

Ligand	Oxidant	Additive	Solvent	I
(DHQD)$_2$PHAL	$K_3Fe(CN)_6$	K_2CO_3	**89**:H_2O (1:2)	(80), 66
(DHQD)$_2$PYR	$K_3Fe(CN)_6$	K_2CO_3	**89**:H_2O (1:2)	(70), 71
(DHQD)$_2$PHAL	NMO	—	**89**:H_2O (1:2)	(53), 81
(DHQD)$_2$PHAL	$K_3Fe(CN)_6$	K_2CO_3	**93**:H_2O (1:2)	(77), 67
(DHQD)$_2$PYR	$K_3Fe(CN)_6$	K_2CO_3	**93**:H_2O (1:2)	(76), 76
(DHQD)$_2$PHAL	$K_3Fe(CN)_6$	K_2CO_3	**89**:H_2O:t-BuOH (1:1:2)	(85), 84
(DHQD)$_2$PYR	$K_3Fe(CN)_6$	K_2CO_3	**89**:H_2O:t-BuOH (1:1:2)	(97), 80
(DHQD)$_2$PHAL	NMO	—	**89**:H_2O:t-BuOH (1:1:2)	(63), 90
(DHQD)$_2$PHAL	$K_3Fe(CN)_6$	K_2CO_3	**94**:H_2O:t-BuOH (1:1:2)	(65), 86
(DHQD)$_2$PYR	$K_3Fe(CN)_6$	K_2CO_3	**94**:H_2O:t-BuOH (1:1:2)	(84), 80

52, OsO_4, titanium silicalite, H_2O_2, NMM, t-BuOH:H_2O (3:1), rt, 12 h — **I** (90), 91 — 706

(DHQD)$_2$PHAL, LDH-OsO_4, NMO, t-BuOH:H_2O (1:1), rt — **I** (89), 90 — 704

TABLE 10. SUPPLEMENTAL TABLE ENTRIES: 2001-2004 (*Continued*)

B. REACTIONS OF 1,1-DISUBSTITUTED ALKENES (*Continued*)

Substrate	Conditions	Product(s) and Yield(s) (%), % ee	Refs.
C_9 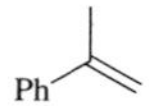	$(DHQD)_2PHAL$, [Os(VIII)], Additive oxidant, solvent, rt, 12-24 h	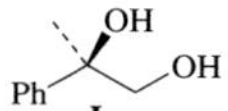 **I**	705
	[Os(VIII)] / Oxidant / Additive / Solvent	**I**	
	Resin-OsO_4 / NMO / — / t-BuOH:H_2O (3:1)	(93), 91	
	Resin-OsO_4 / $K_3Fe(CN)_6$ / K_2CO_3 / t-BuOH:H_2O (1:1)	(92), 93	
	Resin-OsO_4 / O_2 / — / t-BuOH:pH 10.4 buffer (1:2.5)	(99), 84	
	Silica-OsO_4 / NMO / — / t-BuOH:H_2O (3:1)	(90), 90	
	Silica-OsO_4 / $K_3Fe(CN)_6$ / K_2CO_3 / t-BuOH:H_2O (1:1)	(88), 93	
	Silica-OsO_4 / O_2 / — / t-BuOH:pH 10.4 buffer (1:2.5)	(97), 82	
	$(DHQD)_2PHAL$, $K_2OsO_2(OH)_4$, Na_2WO_4, NMM, H_2O_2, Et_4NOAc, t-BuOH:H_2O (3:1), rt, 20 h	**I** (93), 92	683
	$(DHQD)_2PHAL$, LDH-OsW, NMM, H_2O_2, Et_4NOAc, t-BuOH:H_2O, rt, 20 h	**I** (95), 93	683
	Ligand, OsO_4, $K_3Fe(CN)_6$, t-BuOH:H_2O (1:1), 0°	**I**	697
	Ligand / Time	**I**	
	83 / 1 h	(91), 92	
	84 / 2 h	(93), 90	
	Ligand, OsO_4, NMO, Et_4NOAc t-BuOH:H_2O (1:1), 0°, 2 h	**I**	697
	Ligand	**I**	
	83	(89), 64	
	84	(86), 76	
	$(DHQD)_2PHAL$, PEM-MC OsO_4, $K_3Fe(CN)_6$, K_2CO_3, Triton® X-405, H_2O, 30°, 24 h	**I** (85), 80	702

(DHQD)$_2$PHAL, OsO_4, $MeReO_3$, H_2O_2, *t*-BuOH:H_2O (3:1), 0°, 16 h	**I** (85), 64	708
(DHQD)$_2$PHAL, OsO_4, **60**, H_2O_2, Et_4NOAc, *t*-BuOH:H_2O (3:1), 0°, 16 h	**I** (81), 90	710
(DHQD)$_2$PHAL, $K_2OsO_2(OH)_4$, $NaClO_2$, NaOH, *t*-BuOH:H_2O (1:1), 0°, 1 h	**I** (73), 93	685
(DHQD)$_2$PHAL, $K_2OsO_2(OH)_4$, PhSe(O)CH_2Ph, *t*-BuOH:H_2O (1:1), rt, 24 h	**I** (96), 97	709
(DHQD)$_2$PHAL, PEM-MC-OsO_4, $K_3Fe(CN)_6$, K_2CO_3, Me_2CO:H_2O (1:1), 30°, 5 h	**I** (85), 76	690
64, $K_2OsO_2(OH)_4$, $K_3Fe(CN)_6$, K_2CO_3, *t*-BuOH:H_2O (1:1), 0°, 24 h	**I** (89), 79	700
Ligand, OsO_4, $K_3Fe(CN)_6$, K_2CO_3, *t*-BuOH:H_2O (1:1), 0°, 14 h	**I**	720
Ligand	**I**	
71	(—), 64	
72	(—), 74	
1. LDH-PdOsW, Et_3N, PhI, 70°, 16 h 2. (DHQD)$_2$PHAL, NMM, H_2O_2, *t*-BuOH:H_2O (1:1), rt, 12 h	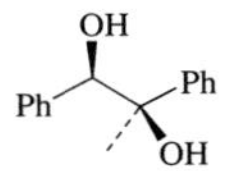 (90), 47	683

TABLE 10. SUPPLEMENTAL TABLE ENTRIES: 2001-2004 (*Continued*)

B. REACTIONS OF 1,1-DISUBSTITUTED ALKENES (*Continued*)

Substrate	Conditions	Product(s) and Yield(s) (%), % ee	Refs.
C_{14} 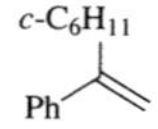	$(DHQ)_2PHAL$, OsO_4, $K_3Fe(CN)_6$, K_2CO_3, t-BuOH:H_2O (1:1), 0°, 18 h	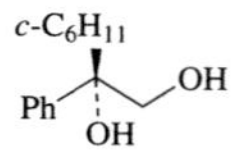 (70), 92	721
C_{30-36} 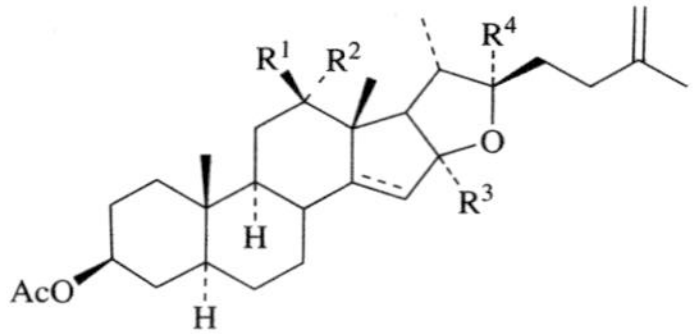	Ligand, OsO_4, $K_3Fe(CN)_6$, K_2CO_3, t-BuOH:H_2O (1:1), 0°	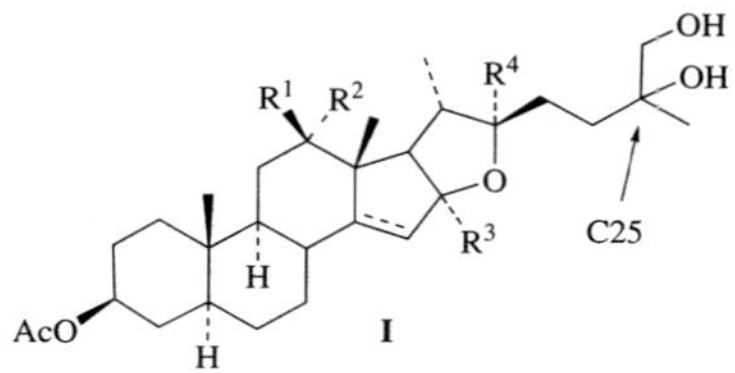 I	722

R^1	R^2	R^3	R^4	C14-C15	Ligand	I	25*S*:25*R*
H	OBz	H	H	Δ^{14}	$(DHQ)_2PHAL$	(96)	5.9:1
H	OBz	H	H	Δ^{14}	$(DHQ)_2PYR$	(98)	1.2:1
OBz	H	H	H	Δ^{14}	$(DHQ)_2PHAL$	(96)	5.4:1
OBz	H	OH	H	14α-H	$(DHQ)_2PHAL$	(95)	1:1
=O		H	OMe	14α-H	$(DHQ)_2PHAL$	(94)	1.5:1

C. REACTIONS OF TRANS 1,2-DISUBSTITUTED ALKENES

Substrate	Conditions	Product(s) and Yield(s) (%), % ee	Refs.
C_4 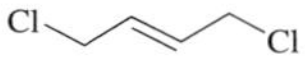	$(DHQD)_2PHAL$, $LDH\text{-}OsO_4$, Et_4NOAc, NMO, $t\text{-}BuOH:H_2O$ (1:1), rt	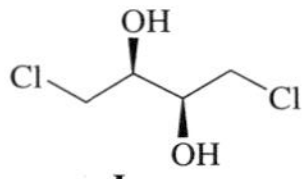 **I** (90), 82	704
	$(DHQD)_2AQN$, $Resin\text{-}OsO_4$, $K_3Fe(CN)_6$, $MeSO_2NH_2$, K_2CO_3, $t\text{-}BuOH:H_2O$ (1:1), 0°, 15 h	**I** (90), 75	686
C_5 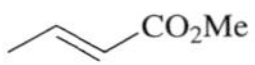	**77**, $K_2OsO_2(OH)_4$, $K_3Fe(CN)_6$, K_2CO_3, $t\text{-}BuOH:H_2O$ (1:1), 0°, 18 h	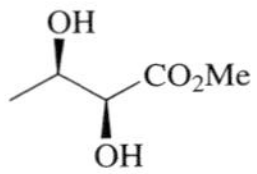 (67), 84	699
C_6 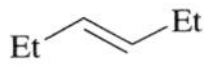	**88**, OsO_4, $K_3Fe(CN)_6$, K_2CO_3, $t\text{-}BuOH:H_2O$ (1:1), 5°	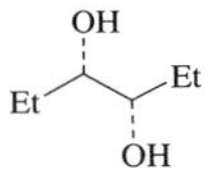 (87), 53	696
C_8 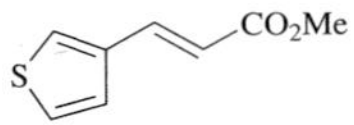	$(DHQ)_2PHAL$, $K_2OsO_2(OH)_4$, $K_3Fe(CN)_6$, K_2CO_3, $t\text{-}BuOH:H_2O$ (1:1), rt, 18 h	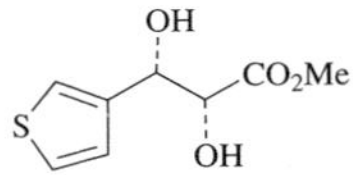 (59), 99	723
$C_{8\text{-}16}$ 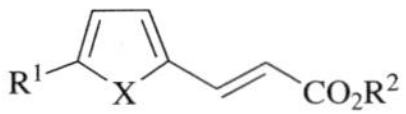	$(DHQ)_2PHAL$, $K_2OsO_2(OH)_4$, $K_3Fe(CN)_6$, K_2CO_3, $t\text{-}BuOH:H_2O$ (1:1)	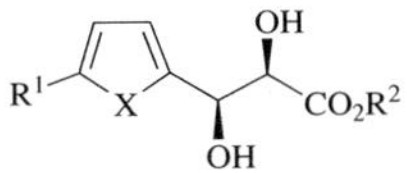	

R^1	X	R^2	Temp	Time	Yield (%), % ee	Refs.
H	S	Me	0°	6 h	(49), 99	723
NO_2	S	Et	rt	48 h	(60), 99	723
H	O	Et	rt	12 h	(75), 99.2	724
Me	O	Et	rt	12 h	(70), 90.1	724
H	NH	Et	rt	12 h	(15), —	724
H	NBn	Et	rt	12 h	(89), 99.0	724
H	NTs	Et	rt	12 h	(91), 99.3	724

TABLE 10. SUPPLEMENTAL TABLE ENTRIES: 2001-2004 (*Continued*)

C. REACTIONS OF TRANS 1,2-DISUBSTITUTED ALKENES (*Continued*)

Substrate	Conditions	Product(s) and Yield(s) (%), % ee	Refs.
C_{9-16}			
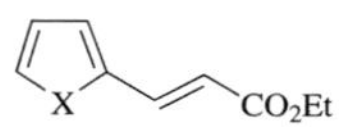	(DHQD)$_2$PHAL, $K_2OsO_2(OH)_4$, $K_3Fe(CN)_6$, K_2CO_3, t-BuOH:H_2O (1:1), rt	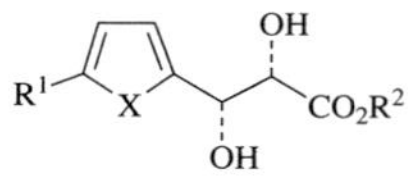	724
X	Time		
O	12 h	(78), 99.9	
NH	12 h	(13), —	
NBn	24 h	(93), 98.8	
NTs	24 h	(95), 98.9	
C_9			
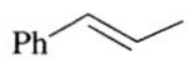	**76**, OsO_4, NMO, solvent, additive, 0°, 10 h	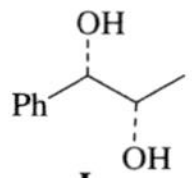 **I**	688
	Solvent / Additive	**I**	
	Me_2CO:H_2O:PEG (10:1:1) / —	(89), 96	
	Me_2CO:H_2O:**89** (10:1:1) / —	(82), 92	
	Me_2CO:H_2O:**89** (10:1:1) / Et_4NOAc	(80), 95	
	(DHQ)$_2$PHAL, OsO_4, **89**, NMO, Me_2CO:H_2O (10:1), rt, 20 h	**I** (92), 90	689
	Ligand, OsO_4, $K_3Fe(CN)_6$, K_2CO_3, t-BuOH:H_2O (1:1)	**I**	

Ligand	Temp	Time	I	
53	0°	15 h	(90), 98	692
61	0°	24 h	(100), 90	693
62	0°	24 h	(100), 95	693
63	0°	24 h	(100), 98	693
69	0°	14 h	(78), 93	720
70	0°	14 h	(—), 94	720
73	0°	14 h	(—), 76	720
74	rt	24 h	(91), 96	694

$(DHQ)_2PHAL$, [Os(VIII)], $K_3Fe(CN)_6$, K_2CO_3, $MeSO_2NH_2$, *t*-BuOH:H_2O (1:1), rt — **I** — 691

[Os(VIII)]	Time	I
XAD-4-OsO4	1.5 h	(97), 95
XAD-7-OsO4	3.5 h	(98), 94

$(DHQD)_2PHAL$, Resin-OsO_4, oxidant, Additive solvent, *t*-BuOH:H_2O (3:1), rt, 12-24 h — 705

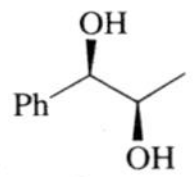

I

Oxidant	Solvent	Additive	I
NMO	*t*-BuOH:H_2O (3:1)	—	(95), 96
$K_3Fe(CN)_6$	*t*-BuOH:H_2O (1:1)	K_2CO_3, $MeSO_2NH_2$	(92), 98
O_2	*t*-BuOH:pH 10.4 buffer (1:2.5)	—	(47), 86

$(DHQD)_2PHAL$, LDH-OsO_4, NMO, *t*-BuOH:H_2O (1:1), rt — **I** (97), 97 — 704

$(DHQD)_2PHAL$, $K_2OsO_2(OH)_4$, Na_2WO_4, NMM, H_2O_2, Et_4NOAc, — **I** (92), 94 — 683

TABLE 10. SUPPLEMENTAL TABLE ENTRIES: 2001-2004 (*Continued*)

C. REACTIONS OF TRANS 1-2-DISUBSTITUTED ALKENES (*Continued*)

Substrate	Conditions	Product(s) and Yield(s) (%), % ee	Refs.
C_9 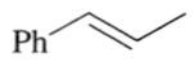	(DHQD)$_2$PHAL, LDH-OsW, NMM, H_2O_2, Et_4NOAc, rt, 20 h	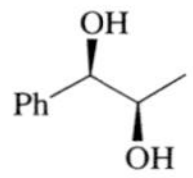 **I** (93), 96	683
	Ligand,OsO_4, $K_3Fe(CN)_6$, K_2CO_3, $MeSO_2NH_2$, *t*-BuOH:H_2O (1:1), 0°	**I**	697
	Ligand / Time: **83** / 18 h; **84** / 14 h	**I**: (92), 92; (93), 92	
	Ligand, OsO_4, NMO, Et_4NOAc, Me_2CO:H_2O (10:1), 0°	**I**	697
	Ligand / Time: **83** / 2.3 h; **84** / 2.5 h	**I**: (92), 84; (91), 89	
	(DHQD)$_2$PHAL, OsO_4, $MeReO_3$, H_2O_2, *t*-BuOH:H_2O (3:1), 0°, 16 h	**I** (87), 90	708
	(DHQD)$_2$PHAL, OsO_4, **60**, H_2O_2, Et_4NOAc, *t*-BuOH:H_2O (3:1), 0°, 16 h	**I** (61), 99	710
	(DHQD)$_2$PHAL, $K_2OsO_2(OH)_4$, PhSe(O)CH_2Ph, *t*-BuOH:H_2O (1:1), rt, 20 h	**I** (92), 99	709

	$(DHQD)_2PHAL$, PEM-MC OsO_4, $K_3Fe(CN)_6$, K_2CO_3, Triton® X-405, H_2O, 30°, 24 h	**I** (76), 92	702
	77, $K_2OsO_2(OH)_4$, $K_3Fe(CN)_6$, K_2CO_3, t-BuOH:H_2O (1:1), 0°, 18 h	**I** (94), 97	699
	$(DDHQ)_2PHAL$, PEM-MC-OsO_4, $K_3Fe(CN)_6$, K_2CO_3, Me_2CO:H_2O (1:1), 30°, 5 h	**I** (86), 94	690
	$(DHQD)_2PHAL$, Os EnCat, K_2CO_3, $K_3Fe(CN)_6$, $MeSO_2NH_2$, THF:H_2O, (1:1), rt, 48 h	**I** (98), 94	703
	Ligand, $K_2OsO_2(OH)_4$, $K_3Fe(CN)_6$, K_2CO_3, t-BuOH:H_2O (1:1), rt	**I**	701

Ligand	I
65	(98), 98
66	(85), 96

	Ligand, OsO_4, $K_3Fe(CN)_6$, K_2CO_3, $MeSO_2NH_2$, t-BuOH:H_2O (1:1), 0°, 14 h	**I**	720

Ligand	I
71	(—), 86
72	(—), 92

C_{10}

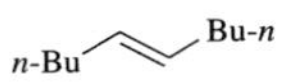

74, $K_2OsO_2(OH)_4$, oxidant, additive, 24 h

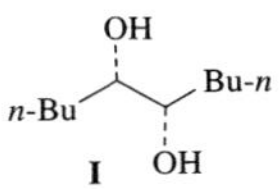

I

687

Oxidant	Additive	Temp	I
NMO	Et_4NOAc	rt	(96), 99
$K_3Fe(CN)_6$	K_2CO_3, $MeSO_2NH_2$	rt	(95), 99
O_2	—	50°	(15), 99

Substrate	Conditions	Product(s) and Yield(s) (%), % ee	Refs.
C_{10} 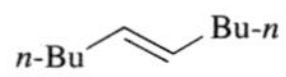	Ligand, OsO_4, $K_3Fe(CN)_6$, K_2CO_3, Additive t-BuOH:H_2O (1:1), 0°, 24 h	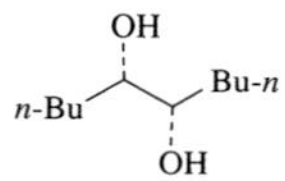	

Ligand	Additive	Yield (%), % ee	Refs.
53	$MeSO_2NH_2$	(85), 95	692
62	—	(100), 75	693
63	—	(100), 87	693

Ligand, $K_2OsO_2(OH)_4$, oxidant, additive, rt, 24 h

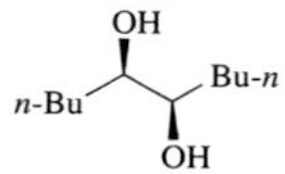 **I**

684

Ligand	Oxidant	Additive	Solvent	I
$(DHQD)_2$PHAL	$K_3Fe(CN)_6$	K_2CO_3	**89**:H_2O (1:2)	(69), 87
$(DHQD)_2$PYR	$K_3Fe(CN)_6$	K_2CO_3	**89**:H_2O (1:2)	(52), 63
$(DHQD)_2$PHAL	NMO	—	**89**:H_2O (1:2)	(73), 60
$(DHQD)_2$PHAL	$K_3Fe(CN)_6$	K_2CO_3	**93**:H_2O (1:2)	(66), 66
$(DHQD)_2$PYR	$K_3Fe(CN)_6$	K_2CO_3	**93**:H_2O (1:2)	(65), 60
$(DHQD)_2$PHAL	$K_3Fe(CN)_6$	K_2CO_3	**89**:H_2O:t-BuOH (1:1:2)	(96), 92
$(DHQD)_2$PYR	$K_3Fe(CN)_6$	K_2CO_3	**89**:H_2O:t-BuOH (1:1:2)	(92), 96
$(DHQD)_2$PHAL	NMO	—	**89**:H_2O:t-BuOH (1:1:2)	(81), 78
$(DHQD)_2$PHAL	$K_3Fe(CN)_6$	K_2CO_3	**94**:H_2O:t-BuOH (1:1:2)	(81), 79
$(DHQD)_2$PYR	$K_3Fe(CN)_6$	K_2CO_3	**94**:H_2O:t-BuOH (1:1:2)	(74), 74

$(DHDQ)_2$PHAL, Resin-OsO_4, NMO, oxidant, additive, rt, 12-24 h — **I** — 705

Oxidant	Additive	Solvent	I
NMO	Et_4NOAc	t-BuOH:H_2O (3:1)	(94), 67
$K_3Fe(CN)_6$	K_2CO_3, $MeSO_2NH_2$	t-BuOH:H_2O (3:1)	(92), 95
O_2	—	t-BuOH:pH 10.4 buffer (1:2.5)	(91), 88

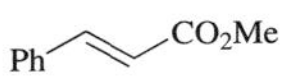

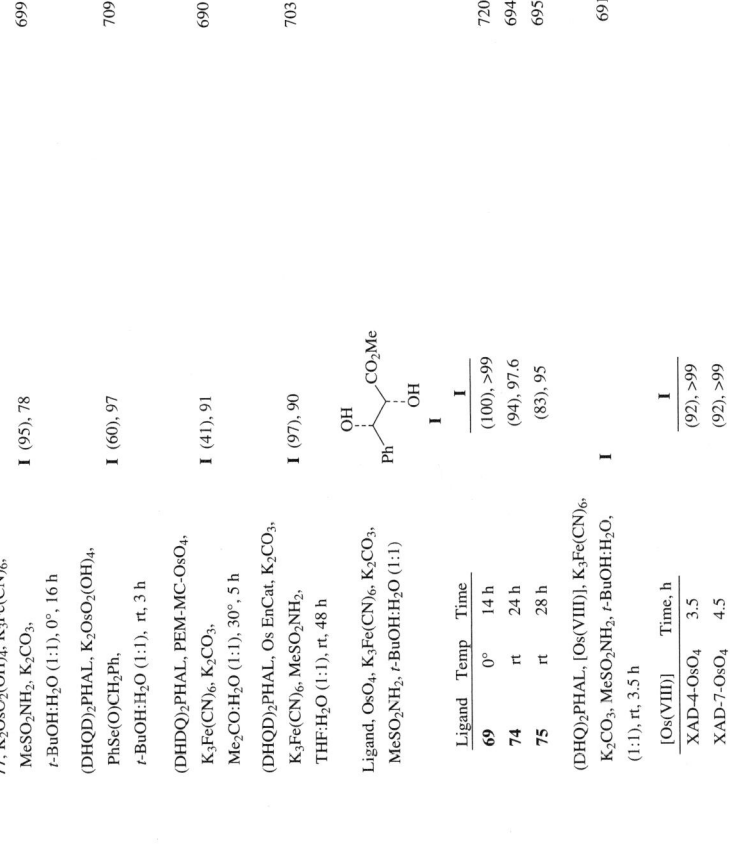

Conditions	Product(s) and Yield(s) (%), % ee	Refs.
77, $K_2OsO_2(OH)_4$, $K_3Fe(CN)_6$, $MeSO_2NH_2$, K_2CO_3, t-BuOH:H_2O (1:1), 0°, 16 h	**I** (95), 78	699
(DHQD)$_2$PHAL, $K_2OsO_2(OH)_4$, PhSe(O)CH_2Ph, t-BuOH:H_2O (1:1), rt, 3 h	**I** (60), 97	709
(DHDQ)$_2$PHAL, PEM-MC-OsO_4, $K_3Fe(CN)_6$, K_2CO_3, Me_2CO:H_2O (1:1), 30°, 5 h	**I** (41), 91	690
(DHQD)$_2$PHAL, Os EnCat, K_2CO_3, $K_3Fe(CN)_6$, $MeSO_2NH_2$, THF:H_2O (1:1), rt, 48 h	**I** (97), 90	703
Ligand, OsO_4, $K_3Fe(CN)_6$, K_2CO_3, $MeSO_2NH_2$, t-BuOH:H_2O (1:1)	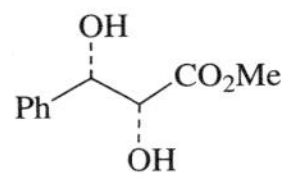 **I**	

Ligand	Temp	Time	I	Refs.
69	0°	14 h	(100), >99	720
74	rt	24 h	(94), 97.6	694
75	rt	28 h	(83), 95	695

Conditions	Product(s) and Yield(s) (%), % ee	Refs.
(DHQ)$_2$PHAL, [Os(VIII)], $K_3Fe(CN)_6$, K_2CO_3, $MeSO_2NH_2$, t-BuOH:H_2O, (1:1), rt, 3.5 h	**I**	691

[Os(VIII)]	Time, h	I
XAD-4-OsO_4	3.5	(92), >99
XAD-7-OsO_4	4.5	(92), >99

TABLE 10. SUPPLEMENTAL TABLE ENTRIES: 2001-2004 (*Continued*)

C. REACTIONS OF TRANS 1,2-DISUBSTITUTED ALKENES (*Continued*)

Substrate	Conditions	Product(s) and Yield(s) (%), % ee	Refs.
C_{10} 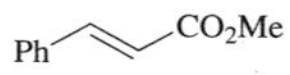	Ligand, OsO_4, NMO, solvent, rt	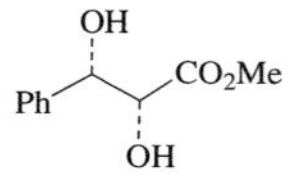 **I**	

Ligand	Solvent	Time	**I**	Refs.
76	$Me_2CO:H_2O$:PEG (10:1:1)	10 h	(78), 96	688
76	$Me_2CO:H_2O$:**83** (10:1:1)	10 h	(67), 94	688
$(DHQ)_2PHAL$	$Me_2CO:H_2O$ (10:1)	20 h	(96), 95	689
51	$Me_2CO:H_2O$ (10:1)	20 h	(96), 94	689

$(DHQ)_2PHAL$, OsO_4, NMO, pressure, $Me_2CO:H_2O$ (10:1), 10° — **I** — 725

OsO_4 (mol%)	Pressure (bar)	Time	TON	**I**
0.05	1	48 h	680	(34), 96.2
0.05	20	48 h	1440	(72), 96.1
0.01	1	24 h	1300	(13), 94.1
0.01	1	168 h	1900	(19), 94.1
0.01	10,000	24 h	2600	(26), 85.5
0.01	10,000	168 h	4400	(44), 85.0

Ligand, $K_2OsO_2(OH)_4$, $K_3Fe(CN)_6$, K_2CO_3, additive, *t*-BuOH:H_2O (1:1), rt

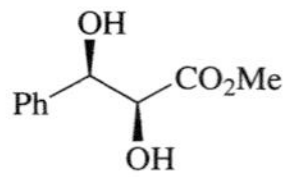

I

Ligand	Additive	Time	**I**	Refs.
64	$MeSO_2NH_2$	30 h	(81), 96	700
65	—	—	(67), 98	690
66	—	—	(72), 94	690

Conditions	Product(s) and Yield(s) (%), % ee	Refs.
(DHQD)$_2$PHAL, $K_2OsO_2(OH)_4$, $NaClO_2$, NaOH, *t*-BuOH:H_2O (1:1), rt, 2 h	**I** (72), >99.5	685
77, $K_2OsO_2(OH)_4$, $K_3Fe(CN)_6$, $MeSO_2NH_2$, $NaClO_2$, K_2CO_3, *t*-BuOH:H_2O (1:1), 0°, 24 h	**I** (75), 90	699
(DHQD)$_2$PHAL, Os EnCat, K_2CO_3, $K_3Fe(CN)_6$, $MeSO_2NH_2$, THF:H_2O (1:1), rt, 48 h	**I** (94), 99	703
52, OsO_4, titanium silicalite, H_2O_2, NMM, t-BuOH:H_2O (3:1), rt, 12 h	**I** (94), 98	706
(DHQD)$_2$PHAL, [Os(VIII)], Additive, oxidant, solvent, rt, 12-24 h	**I**	

[Os(VIII)]	Additive	Oxidant	Solvent	**I**	Refs.
Resin-OsO_4	—	NMO	*t*-BuOH:H_2O (1:1)	(94), 98	705
Resin-OsO_4	K_2CO_3, $MeSO_2NH_2$	$K_3Fe(CN)_6$	*t*-BuOH:H_2O (3:1)	(92), 99	705
LDH-OsO_4	—	NMO	*t*-BuOH:H_2O (1:1)	(96), 99	704

Ligand, OsO_4, NMO, *t*-BuOH:H_2O (1:1) pH 5, rt — **I** — 707

Ligand	**I**
57	(—), 48
59	(—), 51

(DHQD)$_2$PHAL, [Os(VIII)], additive, NMM, H_2O_2, Et_4NOAc, *t*-BuOH:H_2O (3:1), rt, 20 h — **I** — 683

[Os(VIII)]	Additive	**I**
$K_2OsO_2(OH)_4$	Na_2WO_4	(91), 99
LDH-OsW	—	(92), 99

(DHQD)$_2$PHAL, OsO_4, $MeReO_3$, H_2O_2, *t*-BuOH:H_2O (3:1), 0°, 16 h	**I** (87), 98	708

TABLE 10. SUPPLEMENTAL TABLE ENTRIES: 2001-2004 (*Continued*)

C. REACTIONS OF TRANS 1,2-DISUBSTITUTED ALKENES (*Continued*)

	Substrate	Conditions			Refs.
C_{10}	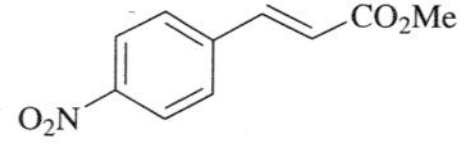	Ligand, OsO_4, NMO, t-BuOH:H_2O (1:1), pH 5, rt	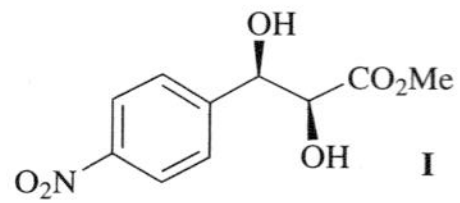 **I**		707
		Ligand	**I**		
		57	(—), 40		
		59	(—), 40		
C_{11}	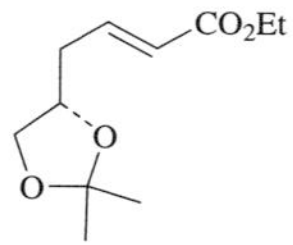	$(DHQD)_2PHAL$, OsO_4, $K_3Fe(CN)_6$, K_2CO_3, $MeSO_2NH_2$, t-BuOH:H_2O (1:1), 0°, 24 h	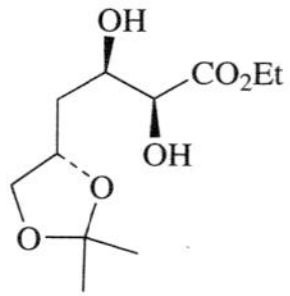	(84), 94	726
	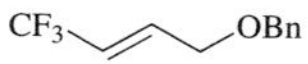	AD-Mix β, $MeSO_2NH_2$, t-BuOH:H_2O (1:1), 0°, 4 d	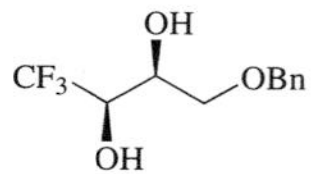	(95), 93	727
		AD-Mix α, $MeSO_2NH_2$, t-BuOH:H_2O (1:1), 0°, 4 d	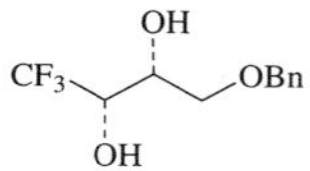	(—), 93	727
	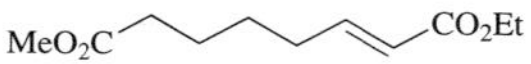	$(DHQD)_2PHAL$, OsO_4, $K_3Fe(CN)_6$, K_2CO_3, t-BuOH:H_2O (1:1), rt	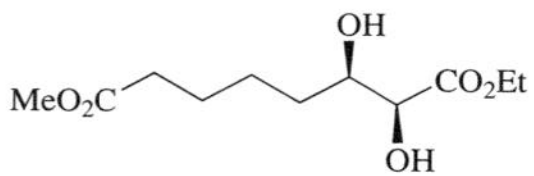	(95), 96	728

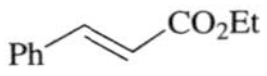

76, OsO_4, NMO, additive, solvent, rt, 10 h

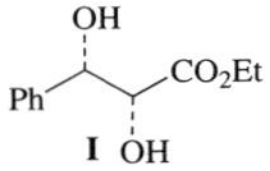

I

Solvent	Additive	I
$Me_2CO:H_2O$:PEG (10:1:1)	—	(82), 95
$Me_2CO:H_2O$:**89** (10:1:1)	—	(73), 94
$Me_2CO:H_2O$:**89** (10:1:1)	Et_4NOAc	(74), 91

(688)

Ligand, OsO_4, $K_3Fe(CN)_6$, K_2CO_3, Additive t-BuOH:H_2O (1:1) — I

Ligand	Temp	Time	Additive	I	Refs.
62	0°	24 h	—	(72), 98	693
63	0°	24 h	—	(72), 98	693
75	rt	28 h	$MeSO_2NH_2$	(78), 92	695

88, OsO_4, $K_3Fe(CN)_6$, K_2CO_3, t-BuOH:H_2O (1:1), 5° — I (90), 12 — 696

Ligand, OsO_4, NMO, t-BuOH:H_2O (1:1), pH 5, rt — 707

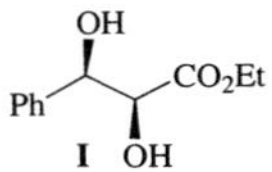

I

Ligand	I
57	(—), 50
59	(—), 48

(DHQD)$_2$PHAL, OsO_4, NMO, PEG-400, rt, 2 h — I (92), 91 — 713

Ligand, OsO_4, $K_3Fe(CN)_6$, t-BuOH:H_2O (1:1), 0° — I — 697

Ligand	Time	I
83	24 h	(91), 96
84	12.5 h	(90), 91

TABLE 10. SUPPLEMENTAL TABLE ENTRIES: 2001-2004 (*Continued*)

C. REACTIONS OF TRANS 1,2-DISUBSTITUTED ALKENES (*Continued*)

Substrate	Conditions	Product(s) and Yield(s) (%), % ee	Refs.
C_{11}			
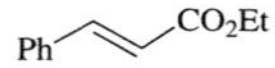	(DHQD)$_2$PHAL, PEM-MC-OsO$_4$, $K_3Fe(CN)_6$, K_2CO_3, Me_2CO:H_2O (1:1), 30°, 5 h	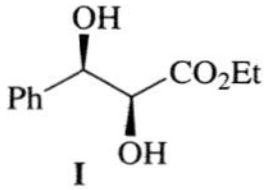 **I** (51), >99	690
	64, $K_2OsO_2(OH)_4$, $K_3Fe(CN)_6$, $MeSO_2NH_2$, K_2CO_3, *t*-BuOH:H_2O (1:1), rt, 30 h	**I** (83), 93	700
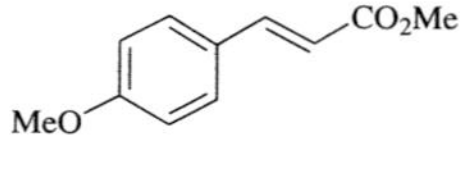	**51**, OsO_4, NMO, **89**, Me_2CO:H_2O (10:1), rt, 20 h	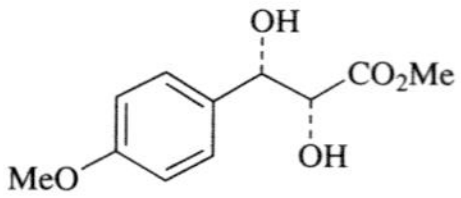 (93), 96	689
C_{12}			
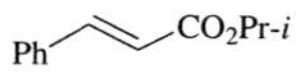	Ligand, OsO_4, NMO, *t*-BuOH:H_2O (1:1), pH 5, rt Ligand: **57** Ligand: **59**	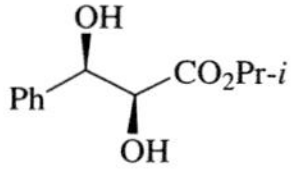 **I** **I**: (—), 53 **I**: (—), 44	707
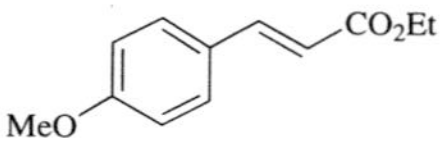	**52**, OsO_4, titanium silicalite, H_2O_2, NMM, t-BuOH:H_2O (3:1), rt, 12 h	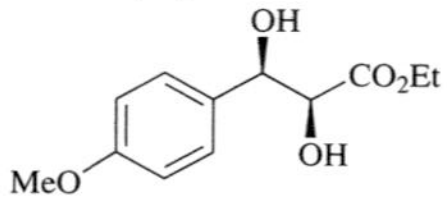 **I** (91), 99	706
	(DHQD)$_2$PHAL, LDH-OsO$_4$, NMO, *t*-BuOH:H_2O (1:1), rt	**I** (93), 99	704

(DHQD)$_2$PHAL, [Os(VIII)], additive, NMM, H_2O_2, Et_4NOAc, t-BuOH:H_2O (3:1), rt, 20 h — **I** — 683

[Os(VIII)]	Additive	**I**
$K_2OsO_2(OH)_4$	Na_2WO_4	(82), 99
LDH-OsW	—	(89), 99

(DHQD)$_2$PHAL, OsO_4, NMO, PEG-400, rt, 2 h — **I** (94), 96 — 713

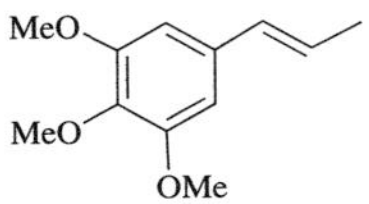

AD-Mix α, $MeSO_2NH_2$, t-BuOH:H_2O (1:1), 0°, 16 h

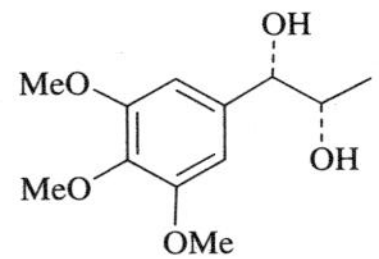

(83), 96 — 703

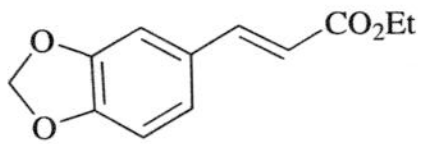

(DHQ)$_2$PHAL, OsO_4, $K_3Fe(CN)_6$, K_2CO_3, t-BuOH:H_2O (1:1), 0°

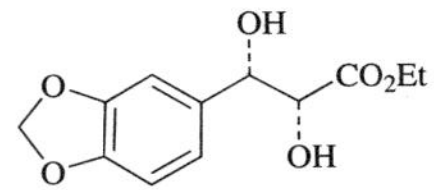

(85), >95 — 729

C_{12-13}

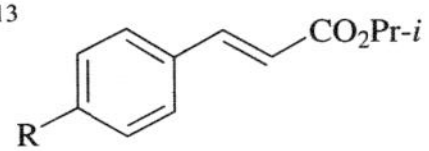

(DHQ)$_2$PHAL, OsO_4, NMO, pressure, Me_2CO:H_2O (10:1), 10°

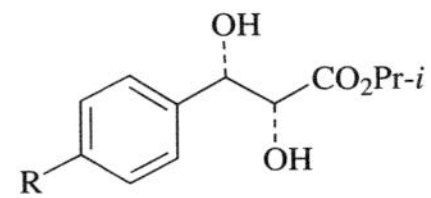

725

R	OsO_4 (mol%)	Presuure (bar)	Time	TON	
NO_2	0.05	1	48 h	1460	(67), 95.5
NO_2	0.05	20	48 h	1640	(69), 96.0
NO_2	0.01	1	168 h	1200	(12), 95.8
NO_2	0.01	10,000	168 h	7800	(75), 81.2
MeO	0.05	1	48 h	1460	(73), 97.0
MeO	0.05	20	48 h	1640	(82), 97.3
MeO	0.01	1	168 h	1200	(12), 96.1
MeO	0.01	10,000	168 h	7800	(78), 91.1

TABLE 10. SUPPLEMENTAL TABLE ENTRIES: 2001-2004 *(Continued)*

C. REACTIONS OF TRANS 1,2-DISUBSTITUTED ALKENES *(Continued)*

Substrate	Conditions	Product(s) and Yield(s) (%), % ee	Refs.
C_{13}			
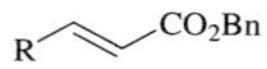	AD-Mix α, $MeSO_2NH_2$	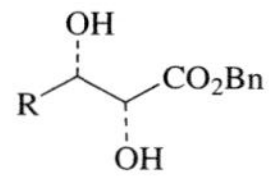 R: *n*-Pr (85), >94; *i*-Pr (88), >94	730
C_{14}			
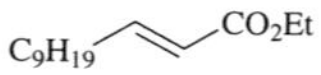	$(DHQ)_2PHAL$, OsO_4, $K_3Fe(CN)_6$, K_2CO_3, $MeSO_2NH_2$, *t*-BuOH:H_2O (1:1), 0°, 24 h	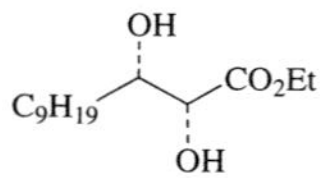 (94), >99	731
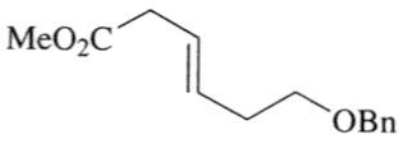	AD-Mix β, $MeSO_2NH_2$, *t*-BuOH:H_2O (1:1), 0°, 20 h	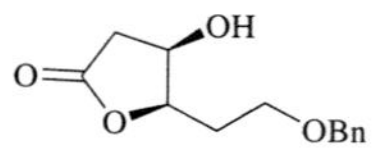 (90), 93	732
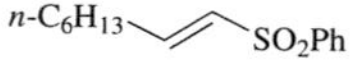	1. AD-Mix α, $MeSO_2NH_2$, *t*-BuOH:H_2O (1:1), rt, 24 h 2. $(EtO)_2P(O)CH_2CO_2Me$, NaH, THF, rt	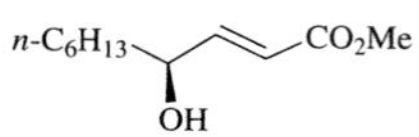 (52), 90	733
	1. AD-Mix α, $MeSO_2NH_2$, *t*-BuOH:H_2O (1:1), rt, 24 h 2. $(CF_3CH_2O)_2P(O)CH_2CO_2Me$, NaH, THF, –78° to rt	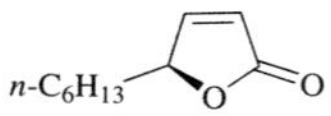 (26), 88	733
	1. AD-Mix β, $MeSO_2NH_2$, *t*-BuOH:H_2O (1:1), rt, 24 h 2. $(EtO)_2P(O)CH_2CO_2Me$, NaH, THF, rt	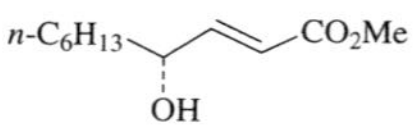 (49), 90	733
	1. AD-Mix β, $MeSO_2NH_2$, *t*-BuOH:H_2O (1:1), rt, 24 h 2. $(CF_3CH_2O)_2P(O)CH_2CO_2Me$, NaH, THF, –78° to rt	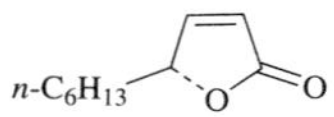 (23), 95	733

C14

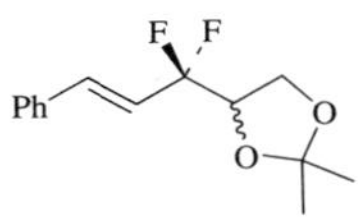

1. AD-Mix β, $MeSO_2NH_2$
2. Me_2CO, p-TsOH, $CuSO_4$

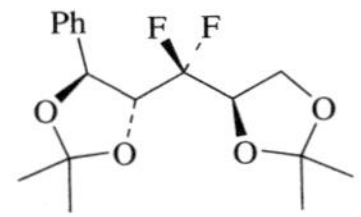

(43), 95 — 734

1. AD-Mix α, $MeSO_2NH_2$
2. Me_2CO, p-TsOH, $CuSO_4$

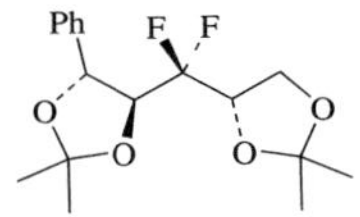

(46), 99.5 — 734

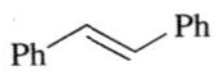

Ligand, OsO_4, $K_3Fe(CN)_6$, K_2CO_3, Additive t-BuOH:H_2O (1:1), 0°, 24 h

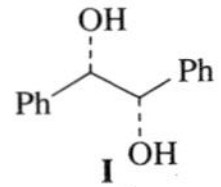

I

Ligand	Additive	I	Ref.
53	$MeSO_2NH_2$	(91), 99	692
61	—	(100), 97	693
62	—	(100), 99	693
63	—	(100), >99.5	693
75	$MeSO_2NH_2$	(85), 99	695

Ligand, OsO_4, $K_3Fe(CN)_6$, K_2CO_3, $MeSO_2NH_2$, t-BuOH:H_2O (1:1) — I

Ligand	Temp	Time	I	Ref.
69	0°	14 h	(75), >99	720
74	rt	24 h	(93), 99	694
88	5°	—	(95), 72	696

$(DHQ)_2PHAL$, [Os(VIII)], $K_3Fe(CN)_6$, K_2CO_3, $MeSO_2NH_2$, t-BuOH:H_2O (1:1), rt, 8 h — I — 691

[Os(VIII)]	I
XAD-4-OsO_4	(94), >99
XAD-7-OsO_4	(94), >99

TABLE 10. SUPPLEMENTAL TABLE ENTRIES: 2001-2004 (*Continued*)

C. REACTIONS OF TRANS 1,2-DISUBSTITUTED ALKENES (*Continued*)

Substrate	Conditions	Product(s) and Yield(s) (%), % ee	Refs.
C_{14}			
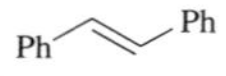	**74**, $K_2OsO_2(OH)_4$, oxidant, 24 h	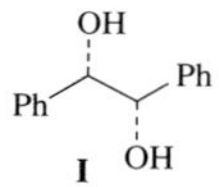 **I**	687
	Oxidant / Temp		
	NMO / rt	(96), 99	
	$K_3Fe(CN)_6$ / rt	(95), 99	
	O_2 / 50°	(15), 99	
	Ligand, OsO_4, NMO, additive Me_2CO:H_2O (10:1), rt, 20 h	**I**	689
	Ligand / Additive	**I**	
	$(DHQ)_2PHAL$ / **89**	(94), 97	
	51 / **89**	(95), 97	
	51 / **92**	(92), 94	
	76, OsO_4, NMO, solvent, additive, 0°, 10 h	**I**	688
	Solvent / Additive	**I**	
	Me_2CO:H_2O:PEG (10:1:1) / —	(88), 95	
	Me_2CO:H_2O:**89** (10:1:1) / —	(86), 95	
	Me_2CO:H_2O:**89** (10:1:1) / Et_4NOAc	(88), >99	
	64, $K_2OsO_2(OH)_4$, $K_3Fe(CN)_6$, $MeSO_2NH_2$, K_2CO_3, t-BuOH:H_2O (1:1), 0°, 18 h	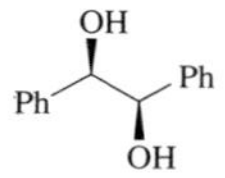 **I** (93), 98	700

Ligand, $K_2OsO_2(OH)_4$, $K_3Fe(CN)_6$, K_2CO_3, *t*-BuOH:H_2O (1:1)	**I**	735

Ligand (mol%)	[Os] mol%	Time	Temp	I
68 (1)	0.4	24 h	rt	(80), 20
67 (1)	0.4	24 h	rt	(72), 42
67 (3)	1.2	24 h	rt	(89), 61
67 (5)	2.0	24 h	rt	(89), 59
67 (3)	1.2	12 h	0°	(79), 85
68 (3)	1.2	12 h	0°	(59), 52

$(DHQD)_2PHAL$, PEM-MC-OsO_4, $K_3Fe(CN)_6$, K_2CO_3, Me_2CO:H_2O (1:1), 30°, 9 h	**I** (66), >99	690
65, $K_2OsO_2(OH)_4$, $K_3Fe(CN)_6$, K_2CO_3, *t*-BuOH:H_2O (1:1), rt	**I** (97), >99	701
Ligand, OsO_4, $K_3Fe(CN)_6$, K_2CO_3, $MeSO_2NH_2$, *t*-BuOH:H_2O (1:1), 0°, 24 h	**I**	718

Ligand	I
$(DHQD)_2PHAL$	(89), 95
85	(86), 97
86	(83), 97
87	(82), 94

$(DHQD)_2PHAL$, Os EnCat, K_2CO_3, $K_3Fe(CN)_6$, $MeSO_2NH_2$, THF:H_2O (1:1), rt, 48 h	**I** (88), >99	703
77, $K_2OsO_2(OH)_4$, $K_3Fe(CN)_6$, cooxidant, $MeSO_2NH_2$, K_2CO_3, *t*-BuOH:H_2O (1:1), 0°, 20 h	**I**	699

Cooxidant	I
—	(>99), 99.5
$NaClO_2$	(73), 99

TABLE 10. SUPPLEMENTAL TABLE ENTRIES: 2001-2004 (*Continued*)

C. REACTIONS OF TRANS 1,2-DISUBSTITUTED ALKENES (*Continued*)

Substrate	Conditions	Product(s) and Yield(s) (%), % ee	Refs.
C_{14} 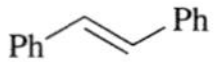	(DHQD)$_2$PHAL, [Os(VIII)], Additive oxidant, solvent, rt, 12-24 h	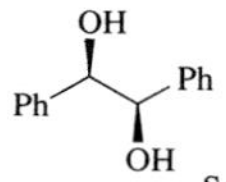 **I**	705

[Os(VIII)]	Oxidant	Additive	Solvent	I
Resin-OsO_4	NMO	—	t-BuOH:H_2O (5:1)	(92), 99
Resin-OsO_4	$K_3Fe(CN)_6$	K_2CO_3, $MeSO_2NH_2$	t-BuOH:H_2O (2:1)	(95), 99
Resin-OsO_4	O_2	—	t-BuOH:pH 10.4 buffer (1:1)	(20), 92
SiO_2-OsO_4	NMO	—	t-BuOH:H_2O (5:1)	(89), 99
SiO_2-OsO_4	$K_3Fe(CN)_6$	K_2CO_3, $MeSO_2NH_2$	t-BuOH:H_2O (2:1)	(91), 99
SiO_2-OsO_4	O_2	—	t-BuOH:pH 10.4 buffer (1:1)	(17), 91

Substrate	Conditions	Product(s) and Yield(s) (%), % ee	Refs.
	Ligand, $K_2OsO_2(OH)_4$, oxidant, additive, solvent, rt, 24 h	**I**	684

Ligand	Oxidant	Additive	Solvent System	I
(DHQD)$_2$PHAL	$K_3Fe(CN)_6$	K_2CO_3	**89**:H_2O (1:2)	(87), 98
(DHQD)$_2$PYR	$K_3Fe(CN)_6$	K_2CO_3	**89**:H_2O (1:2)	(81), 96
(DHQD)$_2$PHAL	NMO	—	**89**:H_2O (1:2)	(56), 91
(DHQD)$_2$PHAL	$K_3Fe(CN)_6$	K_2CO_3	**94**:H_2O (1:2)	(97), 96
(DHQD)$_2$PYR	$K_3Fe(CN)_6$	K_2CO_3	**94**:H_2O (1:2)	(73), 84
(DHQD)$_2$PHAL	$K_3Fe(CN)_6$	K_2CO_3	**89**:H_2O:t-BuOH (1:1:2)	(92), 99
(DHQD)$_2$PYR	$K_3Fe(CN)_6$	K_2CO_3	**89**:H_2O:t-BuOH (1:1:2)	(79), 77
(DHQD)$_2$PHAL	NMO	—	**89**:H_2O:t-BuOH (1:1:2)	(65), 99
(DHQD)$_2$PHAL	$K_3Fe(CN)_6$	K_2CO_3	**94**:H_2O:t-BuOH (1:1:2)	(81), 97
(DHQD)$_2$PYR	$K_3Fe(CN)_6$	K_2CO_3	**94**:H_2O:t-BuOH (1:1:2)	(89), 96

Substrate	Conditions	Product(s) and Yield(s) (%), % ee	Refs.
	(DHQD)$_2$PHAL, $K_2OsO_2(OH)_4$, $NaClO_2$, NaOH, t-BuOH:H_2O (1:1), 0°, 3 h	**I** (63), >99.5	685
	52, OsO_4, titanium silicalite, H_2O_2, NMM, t-BuOH:H_2O (3:1), rt, 12 h	**I** (75), 99	706
	(DHQD)$_2$PHAL, LDH-OsO_4, NMO, t-BuOH:H_2O (1:1), rt	**I** (97), 97	704

(DHQD)$_2$PHAL,[Os(VIII)], additive, Na_2WO_4, NMM, H_2O_2, Et_4NOAc, *t*-BuOH:H_2O, rt, 20 h — **I** — 683

[Os(VIII)]	Additive	**I**
$K_2OsO_2(OH)_4$	—	(85), 99
LDH-OsW	Na_2WO_4	(93), 99

(DHQD)$_2$PHAL, OsO_4, **60**, H_2O_2, Et_4NOAc, Me_2CO:H_2O (4:1), 0°, 16 h — **I** (89), 90 — 710

(DHQD)$_2$PHAL, OsO_4, NMO, PEG-400, rt, 2 h — **I** (95), 94 — 713

Ligand, OsO_4, oxidant, additive, solvent, 0° — **I** — 697

Ligand (mol%)	OsO_4 (mol%)	Oxidant	Additive	Solvent	Time	**I**
78 (25)	1	NMO	Et_4NOAc	Me_2CO:H_2O (10:1)	2.5 h	(85), 83
79 (25)	1	NMO	Et_4NOAc	Me_2CO:H_2O (10:1)	2.0 h	(77), 79
80 (25)	1	NMO	Et_4NOAc	Me_2CO:H_2O (10:1)	2.0 h	(72), 79
81 (25)	1	NMO	Et_4NOAc	Me_2CO:H_2O (10:1)	2.0 h	(83), 82
82 (25)	1	NMO	Et_4NOAc	Me_2CO:H_2O (10:1)	3.0 h	(86), 79
78 (25)	0.1	NMO	Et_4NOAc	Me_2CO:H_2O (10:1)	46 h	(79), 69
78 (25)	0.5	NMO	Et_4NOAc	Me_2CO:H_2O (10:1)	3.5 h	(85), 82
79 (25)	0.1	NMO	Et_4NOAc	Me_2CO:H_2O (10:1)	29 h	(72), 76
79 (25)	0.5	NMO	Et_4NOAc	Me_2CO:H_2O (10:1)	3.0 h	(73), 85
83 (25)	1	NMO	Et_4NOAc	Me_2CO:H_2O (10:1)	3.0 h	(87), 93
83 (25)	1	$K_3Fe(CN)_6$	K_2CO_3	*t*-BuOH:H_2O (1:1)	24 h	(94), 99
83 (10)	1	$K_3Fe(CN)_6$	K_2CO_3	*t*-BuOH:H_2O (1:1)	24 h	(92), 95
83 (5)	1	$K_3Fe(CN)_6$	K_2CO_3	*t*-BuOH:H_2O (1:1)	24 h	(98), 93
83 (5)	1	NMO	Et_4NOAc	Me_2CO:H_2O (10:1)	3.0 h	(85), 91
84 (25)	1	NMO	Et_4NOAc	Me_2CO:H_2O (10:1)	2.5 h	(65), 96
84 (10)	1	NMO	Et_4NOAc	Me_2CO:H_2O (10:1)	3.0 h	(71), 91
84 (5)	1	NMO	Et_4NOAc	Me_2CO:H_2O (10:1)	3.0 h	(34), 92
84 (5)	1	$K_3Fe(CN)_6$	K_2CO_3	*t*-BuOH:H_2O (1:1)	21 h	(38), 83

TABLE 10. SUPPLEMENTAL TABLE ENTRIES: 2001-2004 (*Continued*)
C. REACTIONS OF TRANS 1,2-DISUBSTITUTED ALKENES (*Continued*)

Substrate	Conditions	Product(s) and Yield(s) (%), % ee	Refs.
C_{14}			
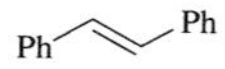	(DHQD)$_2$PHAL, OsO_4, $MeReO_3$, H_2O_2, Me_2CO:H_2O (4.4:1), 0°, 16 h	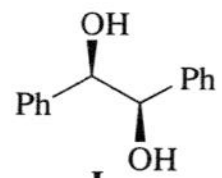 **I** (85), 97	708
	(DHQD)$_2$PHAL, $K_2OsO_2(OH)_4$, PhSe(O)CH_2Ph, *t*-BuOH:H_2O (1:1), rt, 72 h	**I** (84), 99	709
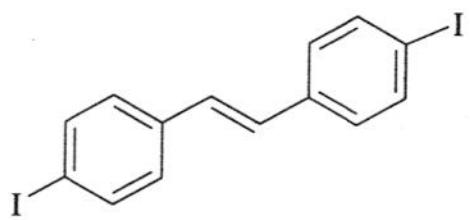	(DHQD)$_2$PHAL, $K_2OsO_2(OH)_4$, $K_3Fe(CN)_6$, K_2CO_3, $MeSO_2NH_2$, THF:H_2O (1:1), rt, 64 h	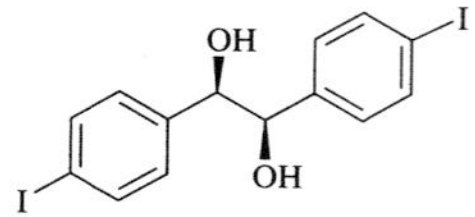 (80), 98.7	736
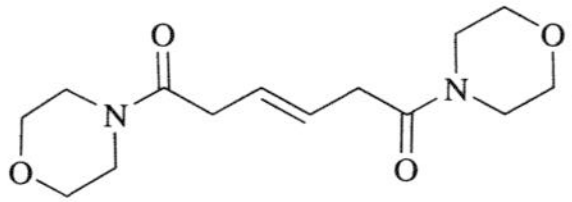	(DHQD)$_2$PHAL, $K_2OsO_2(OH)_4$, $K_3Fe(CN)_6$, K_2CO_3, $MeSO_2NH_2$, *t*-BuOH:H_2O (1:1), 0°, 48 h	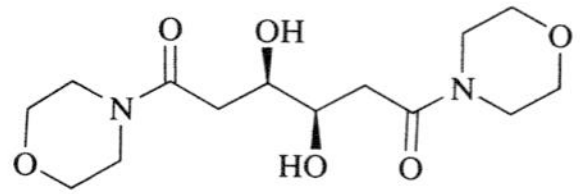 (67), —	737
C_{15}			
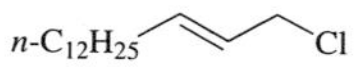	AD-Mix α, $MeSO_2NH_2$, $NaHCO_3$, *t*-BuOH:H_2O (1:1), 0°, 8 h	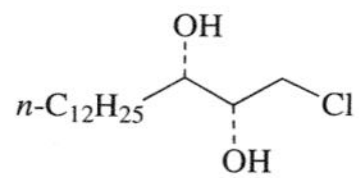 (88), 93	738

C_{15} 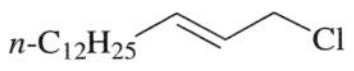	AD-Mix α, $MeSO_2NH_2$, $NaHCO_3$, t-BuOH:H_2O (1:1), 0°, 8 h	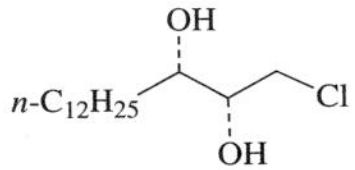	(88), 93	738
C_{16} 	$(DHQ)_2PHAL$, OsO_4, $K_3Fe(CN)_6$, K_2CO_3, $MeSO_2NH_2$, t-BuOH:H_2O (1:1), rt	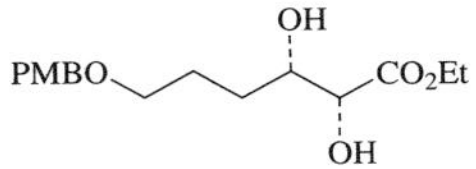	(85), 97	739
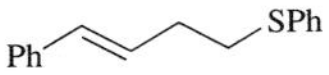	AD-mix α, $MeSO_2NH_2$, t-BuOH:H_2O (1:1), 0°, 48 h	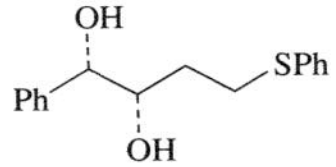	(—), 71	740
	AD-mix β, $MeSO_2NH_2$, t-BuOH:H_2O (1:1), 0°, 48 h	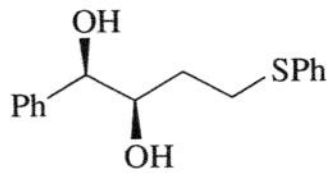	(—), 76	740
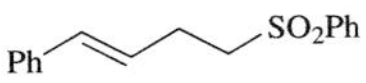	AD-mix α, $MeSO_2NH_2$, t-BuOH:H_2O (1:1), 0°, 48 h	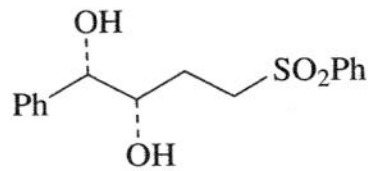	(—), 40	740
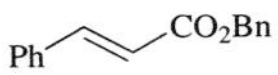	AD-Mix α, $MeSO_2NH_2$, t-BuOH:H_2O (1:1), rt, 24 h	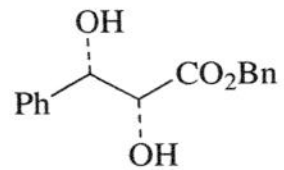	(90), >95	741, 730
C_{17} 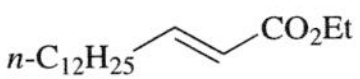	AD-Mix β, $MeSO_2NH_2$, t-BuOH:H_2O (1:1), rt	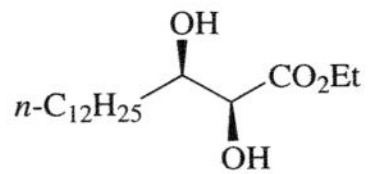	(90), 96	738
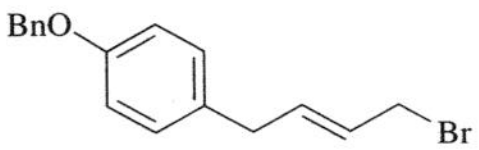	AD-Mix α, $NaHCO_3$, $MeSO_2NH_2$, t-BuOH:H_2O (1:1), 0°, 18 h	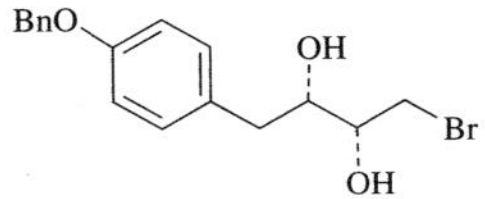	(73), 99	742

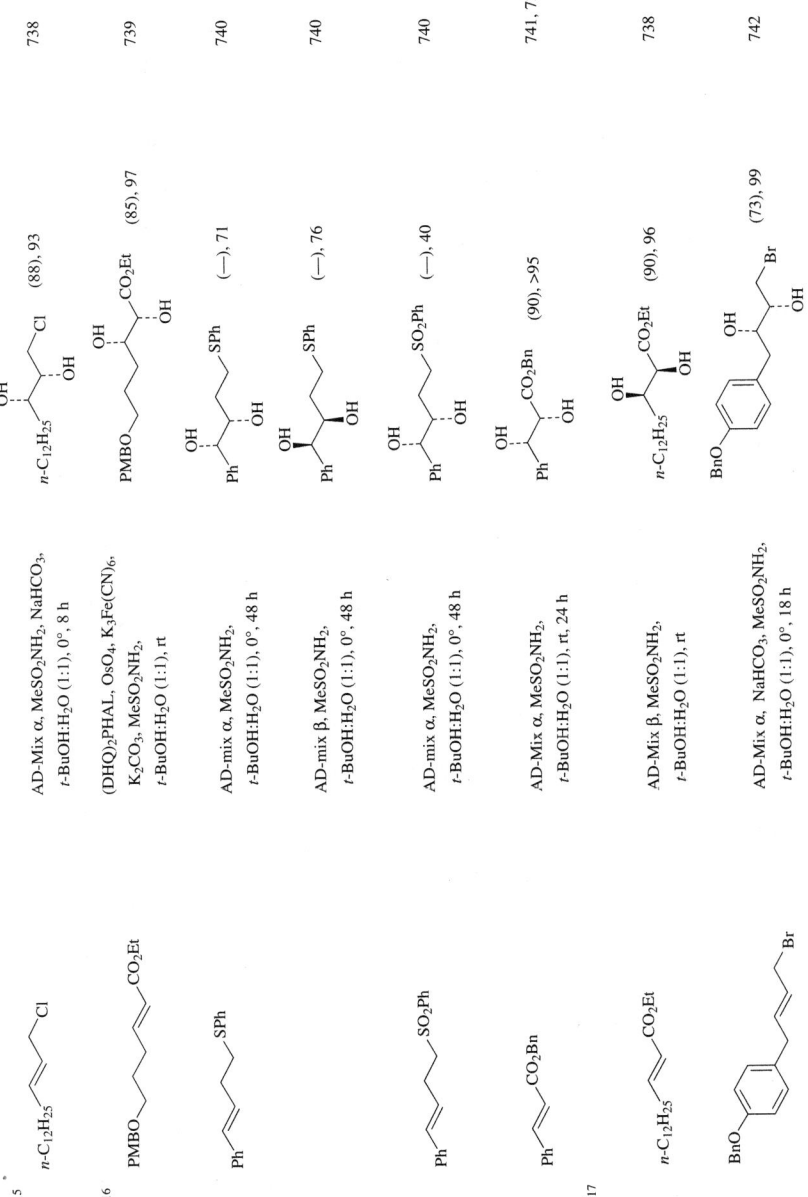

TABLE 10. SUPPLEMENTAL TABLE ENTRIES: 2001-2004 (*Continued*)

C. REACTIONS OF TRANS 1,2-DISUBSTITUTED ALKENES (*Continued*)

Substrate	Conditions	Product(s) and Yield(s) (%), % ee	Refs.
C_{17} 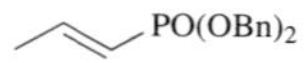	$(DHQ)_2PHAL$, $K_2OsO_2(OH)_4$, $K_3Fe(CN)_6$, K_2CO_3, t-BuOH:H_2O (1:1), rt, 8 d	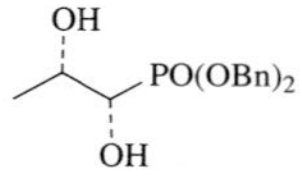	743
	Ligand % 1, [Os] % 0.2	(<30), 62	
	Ligand % 3, [Os] % 0.6	(95), 78	
C_{18} 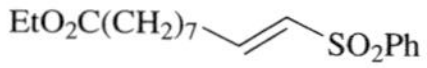	1. AD-Mix α, $MeSO_2NH_2$, t-BuOH:H_2O (1:1), rt, 24 h 2. $(CF_3CH_2O)_2P(O)CH_2CO_2Me$, NaH, THF, –78° to rt	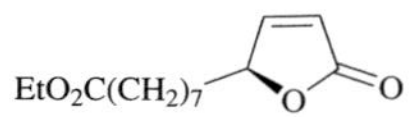 (22), 95	733
	1. AD-Mix β, $MeSO_2NH_2$, t-BuOH:H_2O (1:1), rt, 24 h 2. $(CF_3CH_2O)_2P(O)CH_2CO_2Me$, NaH, THF, –78° to rt	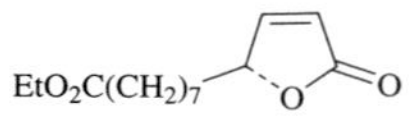 (28), 96	733
C_{19} 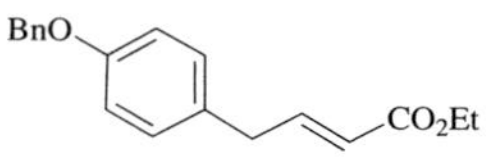	AD-Mix α, $NaHCO_3$, $MeSO_2NH_2$, t-BuOH:H_2O (1:1), 0°, 18 h	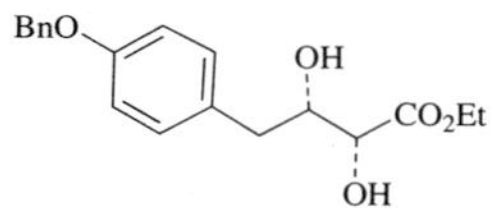 (91), 99	742
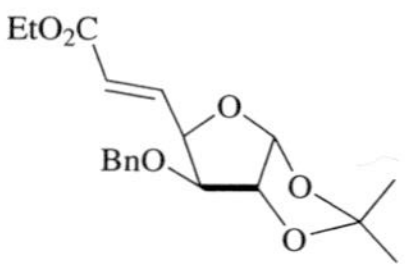	$(DHQ)_2PHAL$, $K_2OsO_2(OH)_4$, $K_3Fe(CN)_6$, t-BuOH:H_2O (1:1), rt, 24 h	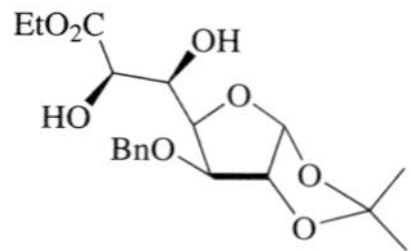 (—), 94 de	744
	$(DHQD)_2PHAL$, $K_2OsO_2(OH)_4$, $K_3Fe(CN)_6$, t-BuOH:H_2O (1:1), rt, 24 h	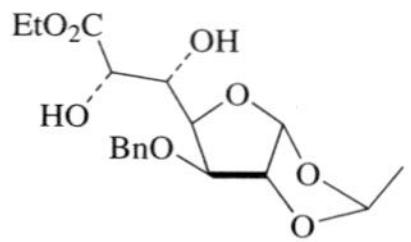 (—), 36 de	744

Substrate	Conditions	Product(s) and Yield(s) (%), % ee	Refs.
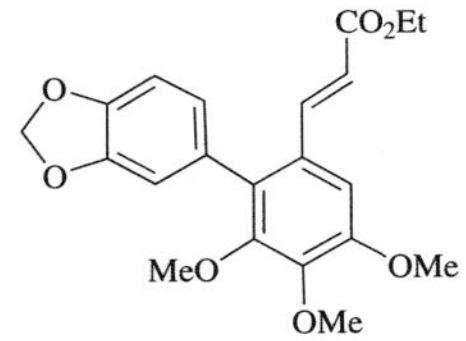	$(DHQD)_2PHAL$, $K_2OsO_2(OH)_4$, K_2CO_3, $MeSO_2NH_2$, t-BuOH:H_2O (1:1), 0°, 18 h	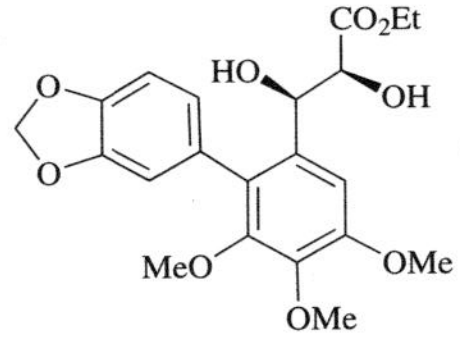 (80), >95	745
C_{23} 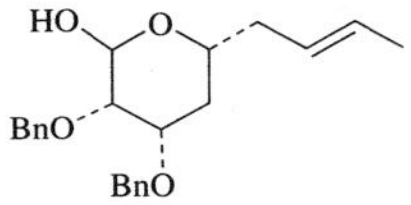	Ligand, OsO_4, $K_3Fe(CN)_6$, K_2CO_3, $MeSO_2NH_2$, t-BuOH:H_2O (1:1), 24 h	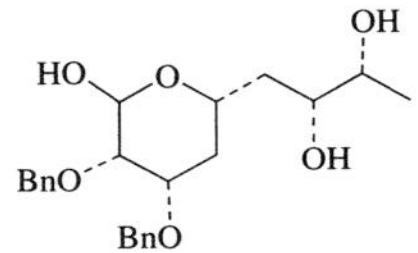	718

Ligand	Temp	
$(DHQD)_2PHAL$	20°	(59), 69 de
$(DHQD)_2PHAL$	0°	(75), 76 de
$(DHQD)_2PYR$	20°	(63), 60 de
DHQD-*p*-chlorobenzoate	20°	(60), 52 de
85	0°	(47), >80 de
86	0°	(37), 76 de
87	0°	(27), 60 de

Substrate	Conditions	Product(s) and Yield(s) (%), % ee	Refs.
C_{24} 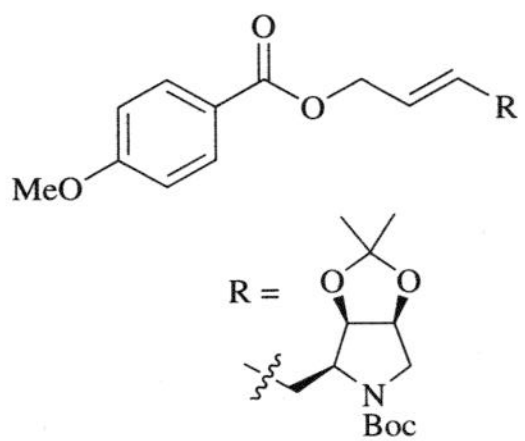	AD-Mix α, $MeSO_2NH_2$, t-BuOH:H_2O (1:1), 0°, 24 h	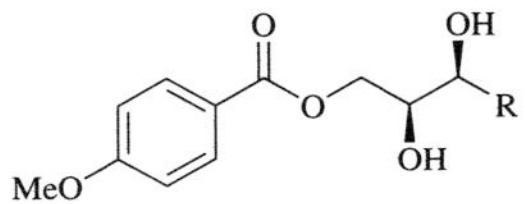 (72), 97	746
	AD-Mix β, $MeSO_2NH_2$, t-BuOH:H_2O (1:1), 0°, 48 h	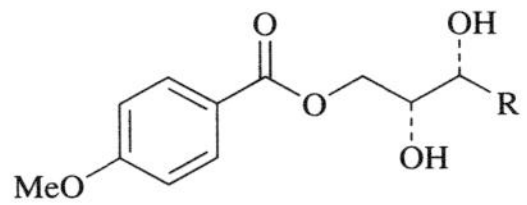 (60), 91	746

Substrate	Conditions	Product(s) and Yield(s) (%), % ee	Refs.
C_{26}-C_{33} 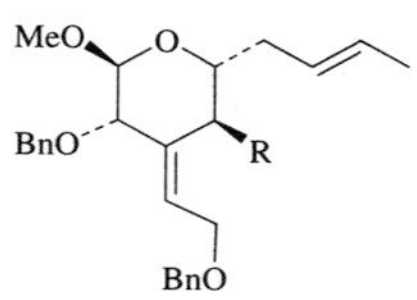	Ligand, OsO_4, $K_3Fe(CN)_6$, K_2CO_3, $MeSO_2NH_2$, t-BuOH:H_2O (1:1), 0°	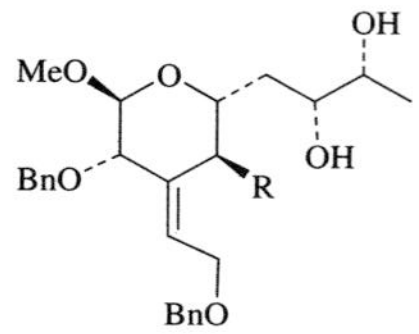	718

R	Ligand (mol%)	OsO_4 (eq)	Time	
H	$(DHQD)_2PHAL$ (1)	0.09	15 h	(64), 57 de
H	DHQD-*p*-chlorobenzoate (10)	0.1	14 h	(80), 46 de
H	$(DHQD)_2PHAL$ (10)	0.1	20 h	(79), 68 de
H	$(DHQD)_2AQN$ (10)	0.1	24 h	(77), 78 de
H	$(DHQD)_2PYR$ (10)	0.1	14 h	(97), 78 de
H	$(DHQD)_2PYR$ (50)	0.01	14 h	(99), 84 de
OH	$(DHQD)_2PYR$ (10)	0.01	16 h	(99), 77 de
OH	$(DHQD)_2AQN$ (10)	0.01	16 h	(99), 78 de
OBn	$(DHQD)_2PHAL$ (1)	0.09	11 h	(75), 50 de
OBn	$(DHQD)_2PYR$ (10)	0.01	6 h	(88), 64 de
OBn	$(DHQD)_2AQN$ (10)	0.01	7 h	(90), 88 de

Substrate	Conditions	Product(s) and Yield(s) (%), % ee	Refs.
C_{33}-C_{42} 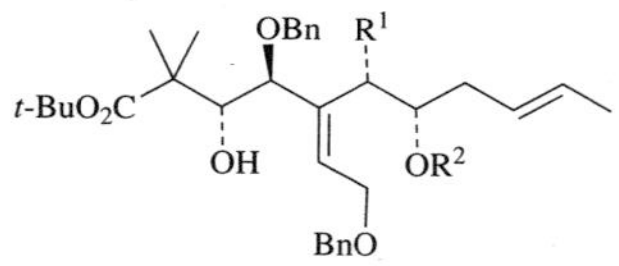	Ligand (10 mol%), OsO_4, $K_3Fe(CN)_6$, $MeSO_2NH_2$, t-BuOH:H_2O (1:1), 0°	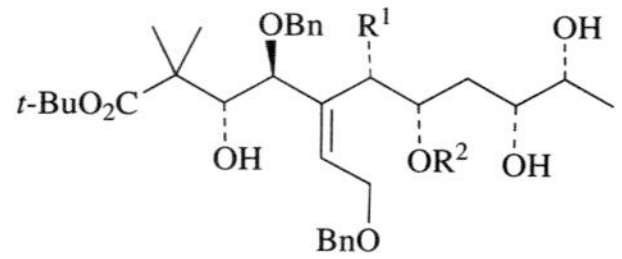	718

R^1	R^2	Ligand	OsO_4 (eq)	Time	
H	H	$(DHQD)_2PHAL$	0.1	16 h	(85), 46 de
H	H	DHQD-*p*-chlorobenzoate	0.1	15 h	(87), 44 de
H	H	$(DHQD)_2AQN$	0.1	14 h	(87), 49 de
H	H	$(DHQD)_2PYR$	0.01	16 h	(97), 66 de
OBn	H	$(DHQD)_2AQN$	0.01	10 h	(91), 82 de
OBn	H	$(DHQD)_2PYR$	0.01	7 h	(95), 55 de
H	TBDMS	$(DHQD)_2PYR$	0.1	14 h	(81), 23 de
H	TIPS	$(DHQD)_2PYR$	0.1	15 h	(99), 13 de
H	TIPS	DHQD-*p*-chlorobenzoate	0.1	14 h	(96), 39 de

C_{37}

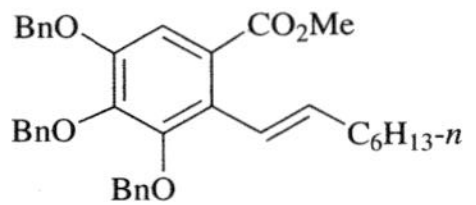

AD-Mix β, $MeSO_2NH_2$, *t*-BuOH:H_2O (1:1), 0°, 48 h

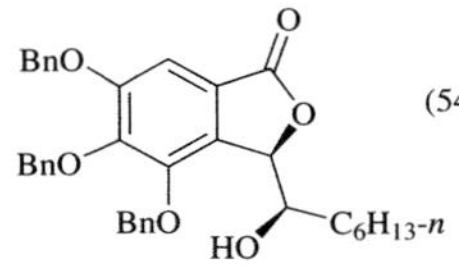

(54), 45 — 747

AD-Mix α, $MeSO_2NH_2$, *t*-BuOH:H_2O (1:1), 0°, 48 h

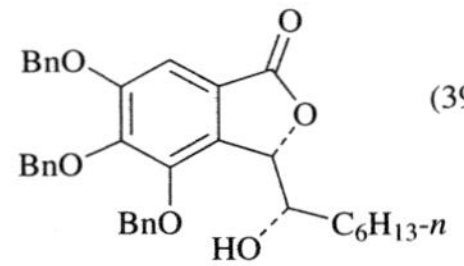

(39), 36 — 747

C_{43}

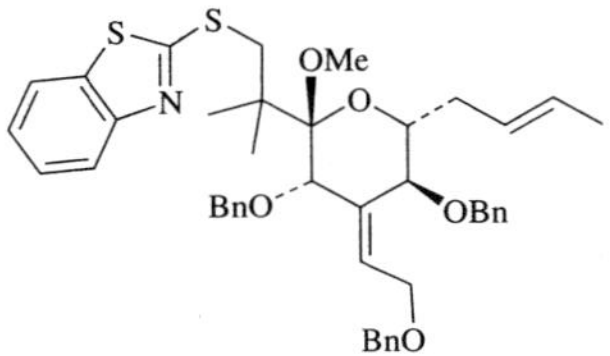

Ligand, OsO_4, $K_3Fe(CN)_6$, K_2CO_3, $MeSO_2NH_2$, *t*-BuOH:THF:H_2O (2:1:3), 0°

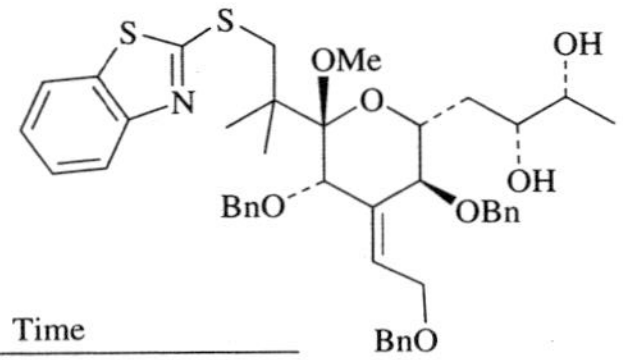

718

Ligand (mol%)	OsO_4 (eq)	Time	
$(DHQD)_2PHAL$ (10)	0.1	46 h	(98), 35 de
DHQD-*p*-chlorobenzoate (10)	0.1	37 h	(97), 52 de
$(DHQD)_2PYR$ (10)	0.1	25 h	(89), 51 de
$(DHQD)_2AQN$ (10)	0.1	26 h	(90), 75 de

TABLE 10. SUPPLEMENTAL TABLE ENTRIES: 2001-2004 (*Continued*)

C. REACTIONS OF TRANS 1,2-DISUBSTITUTED ALKENES (*Continued*)

Substrate	Conditions	Product(s) and Yield(s) (%), % ee	Refs.
C_{45} 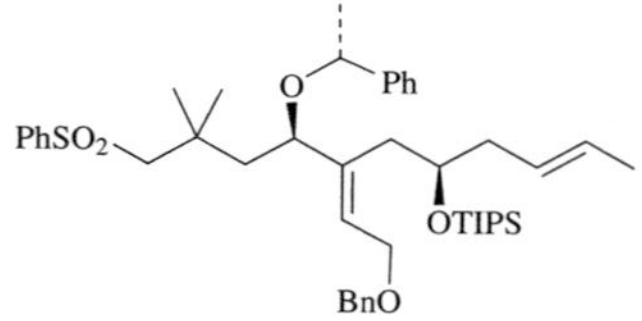	Ligand (10 mol%), OsO_4, K_2CO_3, $K_3Fe(CN)_6$, $MeSO_2NH_2$, *t*-BuOH:H_2O (1:1)	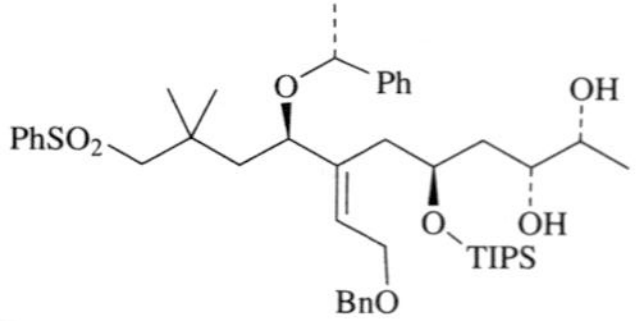	718
	Ligand / Temp / Time: (DHQD)$_2$PHAL, 0°, 24 h	(92), 31 de	
	(DHQD)$_2$PYR, 0°, 6 h	(83), 17 de	
	DHQD-*p*-chlorobenzoate, 20°, 6 h	(74), 37 de	
	DHQD-*p*-chlorobenzoate, 0°, 60 h	(97), 46 de	
C_{56} 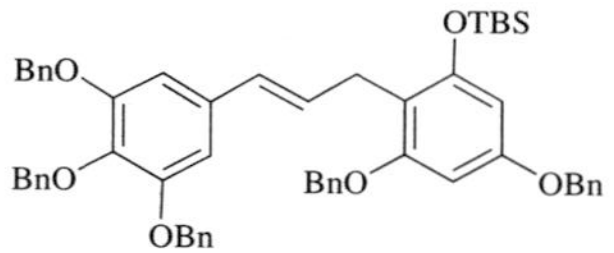	AD-Mix α, $MeSO_2NH_2$, *t*-BuOH-H_2O-CH_2Cl_2, 0°	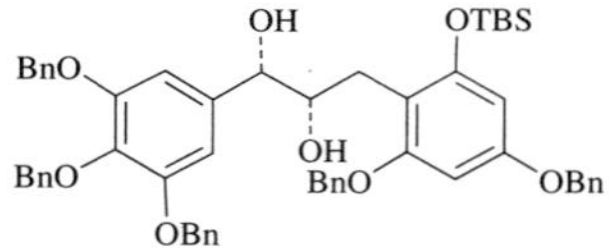 (—), >95	748
	AD-Mix β, $MeSO_2NH_2$, *t*-BuOH-H_2O-CH_2Cl_2, 0°	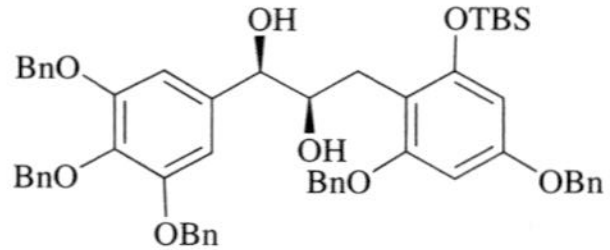 (—), >95	748

TABLE 10. SUPPLEMENTAL TABLE ENTRIES: 2001-2004 *(Continued)*

D. REACTIONS OF CIS-1,2-DISUBSTITUTED ALKENES

Substrate	Conditions	Product(s) and Yield(s) (%), % ee	Refs.
C_5 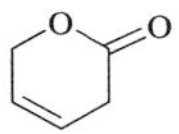	$(DHQD)_2PHAL$, $K_2OsO_2(OH)_4$, $K_3Fe(CN)_6$, *t*-BuOH:pH 10.1 phosphate buffer (1:1), rt	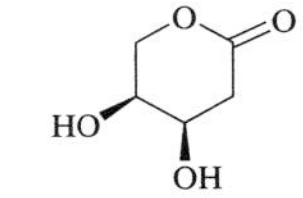 (80), 95	749
	$(DHQ)_2PHAL$, $K_2OsO_2(OH)_4$, $K_3Fe(CN)_6$, *t*-BuOH:pH 10.1 phosphate buffer (1:1), rt	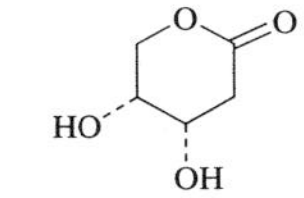 (78), 91	749
C_6 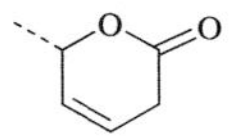	$(DHQD)_2PHAL$, $K_2OsO_2(OH)_4$, $K_3Fe(CN)_6$, *t*-BuOH:pH 10.1 phosphate buffer (1:1), rt	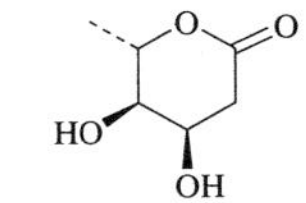 (88), 97	749
	$(DHQ)_2PHAL$, $K_2OsO_2(OH)_4$, $K_3Fe(CN)_6$, *t*-BuOH:pH 10.1 phosphate buffer (1:1), rt	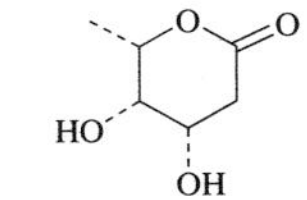 (25), 33	749
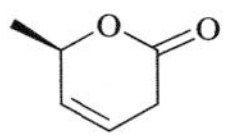	$(DHQD)_2PHAL$, $K_2OsO_2(OH)_4$, $K_3Fe(CN)_6$, *t*-BuOH:pH 10.1 phosphate buffer (1:1), rt	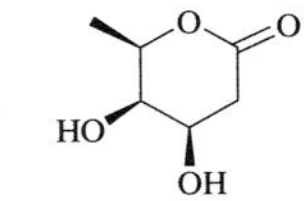 (20), 33	749
	$(DHQ)_2PHAL$, $K_2OsO_2(OH)_4$, $K_3Fe(CN)_6$, *t*-BuOH:pH 10.1 phosphate buffer (1:1), rt	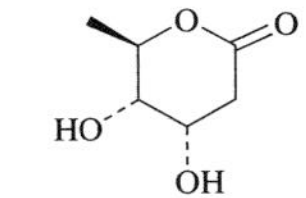 (88), 98	749
C_7 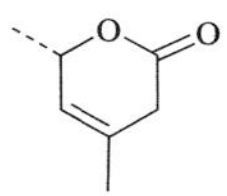	$(DHQD)_2PHAL$, $K_2OsO_2(OH)_4$, $K_3Fe(CN)_6$, *t*-BuOH:pH 10.1 phosphate buffer (1:1), rt	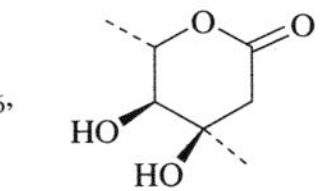 (83), 97	749

TABLE 10. SUPPLEMENTAL TABLE ENTRIES: 2001-2004 (*Continued*)

D. REACTIONS OF CIS-1,2-DISUBSTITUTED ALKENES (*Continued*)

Substrate	Conditions	Product(s) and Yield(s) (%), % ee	Refs.
C_7 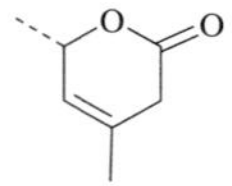	$(DHQ)_2PHAL$, $K_2OsO_2(OH)_4$, $K_3Fe(CN)_6$, *t*-BuOH:pH 10.1 phosphate buffer (1:1), rt	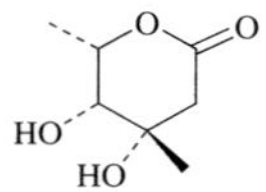 (35), 40	749
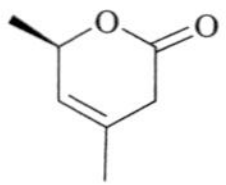	$(DHQD)_2PHAL$, $K_2OsO_2(OH)_4$, $K_3Fe(CN)_6$, *t*-BuOH:pH 10.1 phosphate buffer (1:1), rt	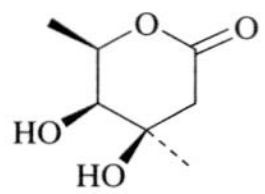 (40), 32	749
	$(DHQ)_2PHAL$, $K_2OsO_2(OH)_4$, $K_3Fe(CN)_6$, *t*-BuOH:pH 10.1 phosphate buffer (1:1), rt	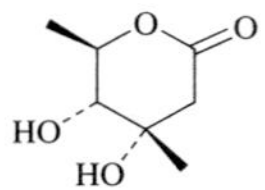 (92), 98	749
C_9 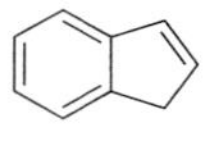	**83**, OsO_4, $K_3Fe(CN)_6$, *t*-BuOH:H_2O (1:1), 0°, 18 h	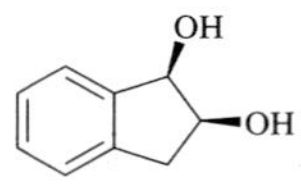 **I** (70), 36	697
	$(DHQD)_2PHAL$, $K_2OsO_2(OH)_4$, $NaClO_2$, NaOH, *t*-BuOH:H_2O (1:1), 0°, 2 h	**I** (65), 41	685
	Ligand, OsO_4, $K_3Fe(CN)_6$, K_2CO_3, *t*-BuOH:H_2O (1:1), 0°, 14 h Ligand **71** **72**	**I** **I** (—), 29 (—), 22	720

Substrate	Conditions	Product(s) and Yield(s) (%)	Refs.
	Ligand, OsO_4, $K_3Fe(CN)_6$, K_2CO_3, t-BuOH:H_2O (1:1), 0°, 14 h	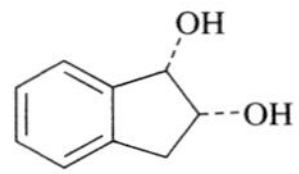	720
	Ligand **69**	(90), 26	
	Ligand **70**	(—), 24	
	Ligand **73**	(—), 2	
C_9 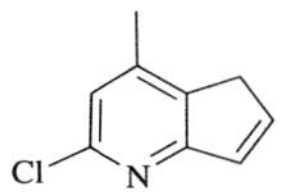	1. AD-Mix β, t-BuOH:H_2O (1:1), 0°, 12 h 2. 3-pentanone, p-TsOH, C_6H_6	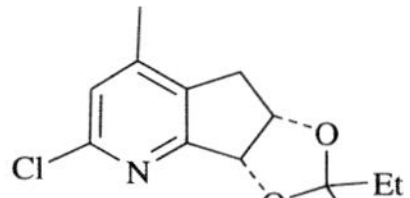 **I** (57), 90	750
	1. $(DHQD)_2PHAL$ (5 mol%), $K_2OsO_2(OH)_4$ (1 mol%), 0°, $K_3Fe(CN)_6$, K_2CO_3, t-BuOH:H_2O(1:1), 2 h 2. 3-pentanone, p-TsOH, C_6H_6	**I** (81), 90	750
	1. AD-Mix α, t-BuOH:H_2O (1:1), 0°, 12 h 2. 3-pentanone, p-TsOH, C_6H_6	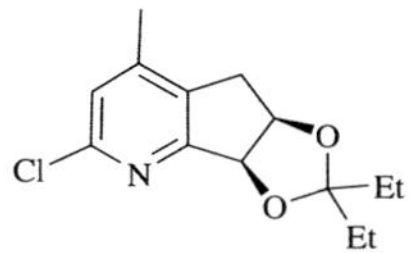 (48), 82	750
C_{10} 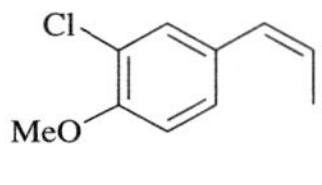	AD-Mix β, $MeSO_2NH_2$, t-BuOH:H_2O (1:1), 0°, 24 h	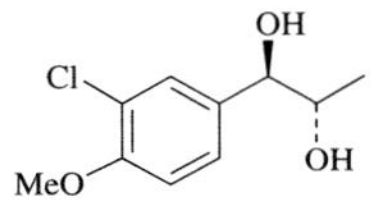 (89), 45	751

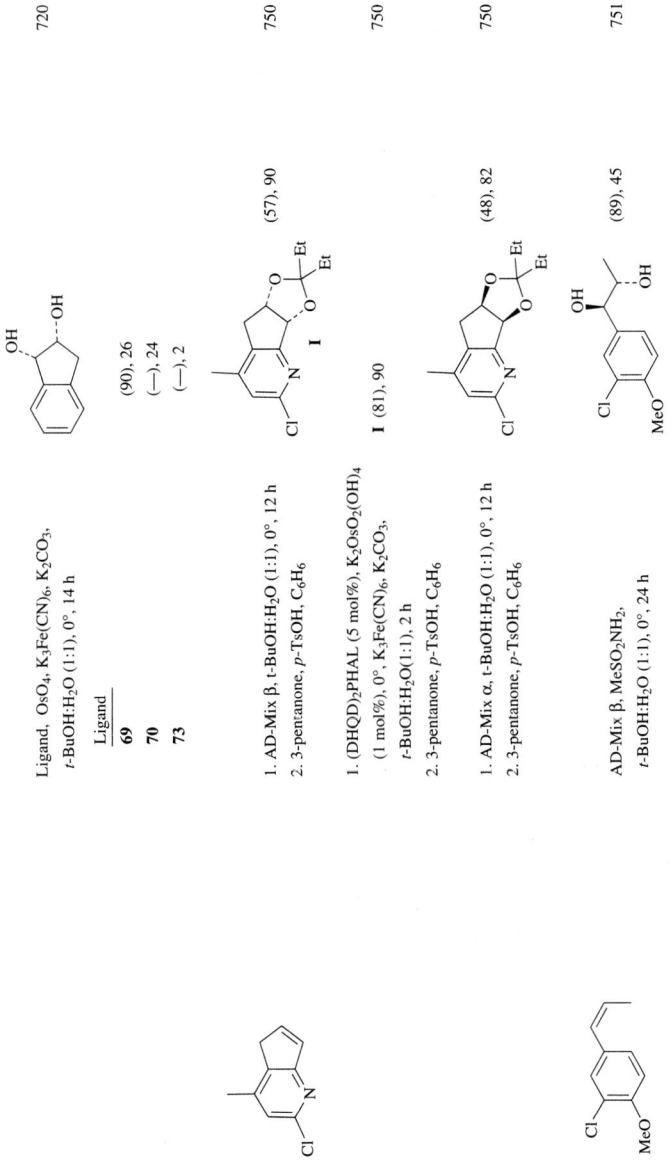

TABLE 10. SUPPLEMENTAL TABLE ENTRIES: 2001-2004 (*Continued*)

E. REACTIONS OF TRISUBSTITUTED ALKENES

Substrate	Conditions	Product(s) and Yield(s) (%), % ee	Refs.
C_7 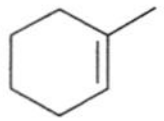	Ligand, $K_2OsO_2(OH)_4$, oxidant, additive, solvent, rt, 24 h	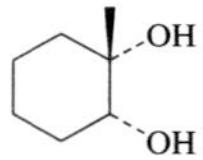 **I**	684

Ligand	Oxidant	Additive	Solvent	I
$(DHQD)_2PHAL$	$K_3Fe(CN)_6$	K_2CO_3	**89**:H_2O (1:2)	(47), 92
$(DHQD)_2PYR$	$K_3Fe(CN)_6$	K_2CO_3	**89**:H_2O (1:2)	(66), 86
$(DHQD)_2PHAL$	NMO	—	**89**:H_2O (1:2)	(92), 68
$(DHQD)_2PHAL$	$K_3Fe(CN)_6$	K_2CO_3	**94**:H_2O (1:2)	(56), 92
$(DHQD)_2PYR$	$K_3Fe(CN)_6$	K_2CO_3	**94**:H_2O (1:2)	(67), 86
$(DHQD)_2PHAL$	$K_3Fe(CN)_6$	K_2CO_3	**94**:H_2O:*t*-BuOH (1:1:2)	(53), 87
$(DHQD)_2PYR$	$K_3Fe(CN)_6$	K_2CO_3	**89**:H_2O:*t*-BuOH (1:1:2)	(57), 83
$(DHQD)_2PHAL$	NMO	—	**89**:H_2O:*t*-BuOH (1:1:2)	(95), 88
$(DHQD)_2PHAL$	$K_3Fe(CN)_6$	K_2CO_3	**94**:H_2O:*t*-BuOH (1:1:2)	(87), 86
$(DHQD)_2PYR$	$K_3Fe(CN)_6$	K_2CO_3	**94**:H_2O:*t*-BuOH (1:1:2)	(89), 69

Conditions	Product(s) and Yield(s) (%), % ee	Refs.
$(DHQD)_2PHAL$, $K_2OsO_2(OH)_4$, $NaClO_2$, NaOH, solvent, H_2O (1:1), 0°	**I**	685

Solvent	Time	I
t-BuOH	3.5 h	(75), 52
MeCOEt	2.5 h	(63), 75

Conditions	Product(s) and Yield(s) (%), % ee	Refs.
Ligand, OsO_4, $K_3Fe(CN)_6$, K_2CO_3, $MeSO_2NH_2$, *t*-BuOH:H_2O (1:1), 0°, 16 h	**I**	718

Ligand	I
$(DHQD)_2PHAL$	(87), 51
85	(83), 33

Substrate	Conditions	Product(s) and Yield(s) (%), ee	Refs.
C_{7-11} 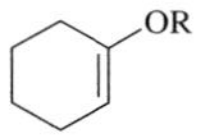	$(DHQD)_2PHAL$, $K_2OsO_2(OH)_4$, $K_3Fe(CN)_6$, K_2CO_3, $MeSO_2NH_2$, $MeCN:H_2O$ (1:1), 0°, 18 h	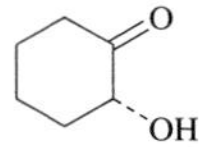 R: Me (47), 79; n-C_5H_{11} (31), 92	752
	$(DHQ)_2PHAL$, $K_2OsO_2(OH)_4$, $K_3Fe(CN)_6$, K_2CO_3, $MeSO_2NH_2$, $MeCN:H_2O$ (1:1), 0°, 18 h	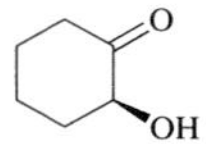 R: Me (48), 76; n-C_5H_{11} (49), 90	752
C_8 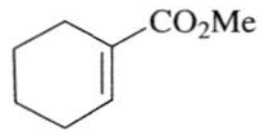	**77**, $K_2OsO_2(OH)_4$, $K_3Fe(CN)_6$, $MeSO_2NH_2$, K_2CO_3, t-BuOH:H_2O (1:1), 0°, 16 h	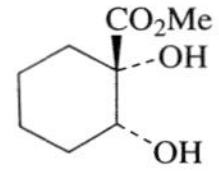 (72), 89	699
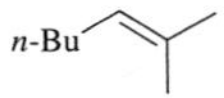	$(DHQD)_2PHAL$, $K_2OsO_2(OH)_4$, $PhSe(O)CH_2Ph$, t-BuOH:H_2O (1:1), rt, 20 h	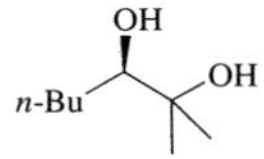 (85), 99	709
C_{8-12} 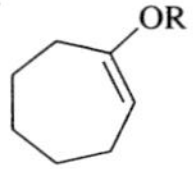	$(DHQD)_2PHAL$, $K_2OsO_2(OH)_4$, $K_3Fe(CN)_6$, K_2CO_3, $MeSO_2NH_2$, $MeCN:H_2O$ (1:1), 0°, 18 h	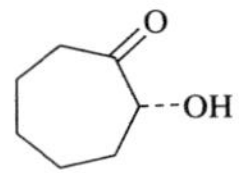 R: Me (78), 87; n-C_5H_{11} (50), 80	752
	$(DHQ)_2PHAL$, $K_2OsO_2(OH)_4$, $K_3Fe(CN)_6$, K_2CO_3, $MeSO_2NH_2$, $MeCN:H_2O$ (1:1), 0°, 18 h	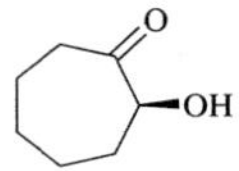 R: Me (49), 83; n-C_5H_{11} (47), 92	752
C_{10} 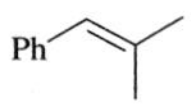	$(DHQD)_2PHAL$, Os EnCat, K_2CO_3, $K_3Fe(CN)_6$, $MeSO_2NH_2$, $THF:H_2O$ (1:1), rt, 48 h	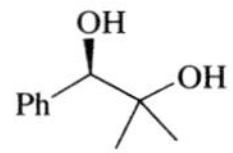 **I** (89), 87	703
	$(DHQD)_2PHAL$, resin-OsO_4, oxidant, additive, t-BuOH:H_2O (3:1), rt, 12-24 h	**I**	705

Oxidant	Additive	Solvent	
NMO	Et_4NOAc	t-BuOH:H_2O (3:1)	(95), 92
$K_3Fe(CN)_6$	K_2CO_3, $MeSO_2NH_2$	t-BuOH:H_2O (1:1)	(90), 92
O_2	—	t-BuOH:pH 10.4 buffer (1:2.5)	(53), 87

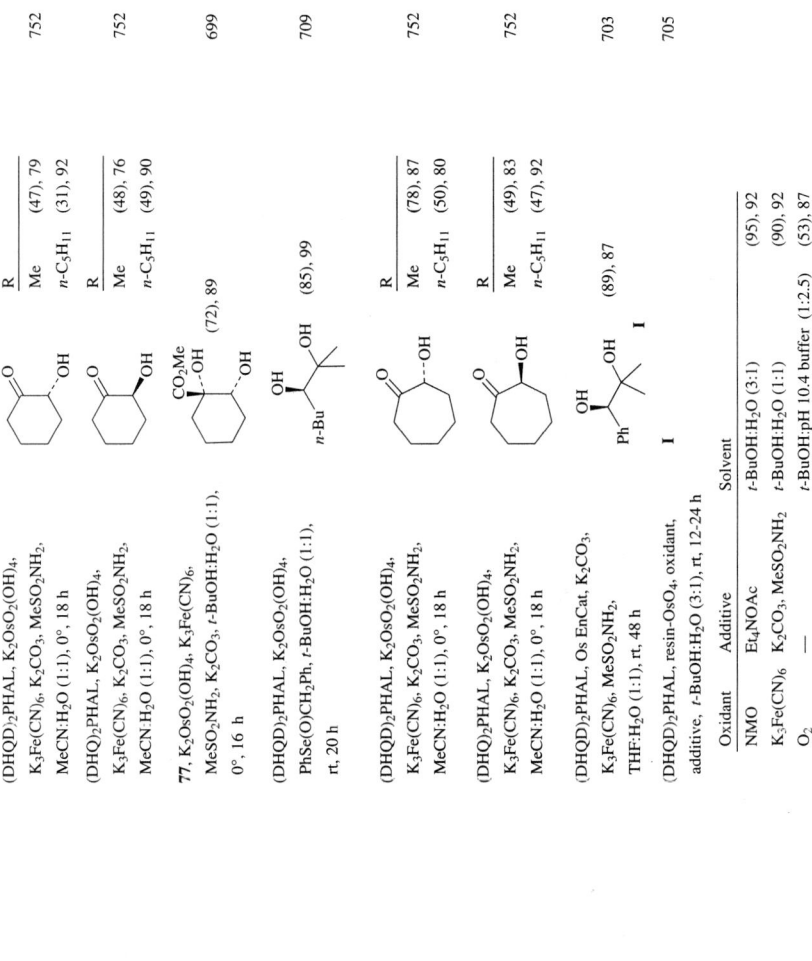

TABLE 10. SUPPLEMENTAL TABLE ENTRIES: 2001-2004 (*Continued*)

E. REACTIONS OF TRISUBSTITUTED ALKENES (*Continued*)

Substrate	Conditions	Product(s) and Yield(s) (%), % ee	Refs.
C_{10} 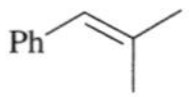	$(DHQD)_2PHAL$, LDH-OsO_4, Et_4NOAc, NMO, *t*-BuOH:H_2O (1:1), rt	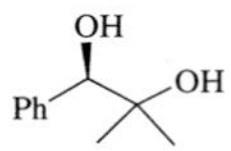 **I** (92), 91	704
	$(DHQD)_2PHAL$, $K_2OsO_2(OH)_4$, Na_2WO_4, NMM, H_2O_2, Et_4NOAc, rt, 20 h	**I** (87), 92	683
	$(DHQD)_2PHAL$, LDH-OsW, NMM, H_2O_2, Et_4NOAc, rt, 20 h	**I** (90), 93	683
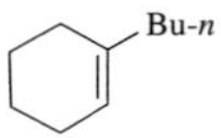	**77**, $K_2OsO_2(OH)_4$, $K_3Fe(CN)_6$, $MeSO_2NH_2$, K_2CO_3, *t*-BuOH:H_2O (1:1), 0°, 16 h	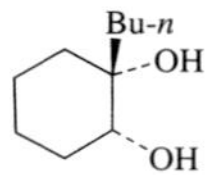 (96), 79	699
C_{10-14} 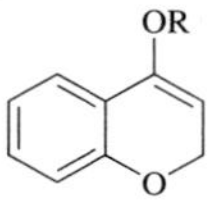	$(DHQ)_2PHAL$, $K_2OsO_2(OH)_4$, $K_3Fe(CN)_6$, K_2CO_3, $MeSO_2NH_2$, MeCN:H_2O (1:1), 0°, 18 h	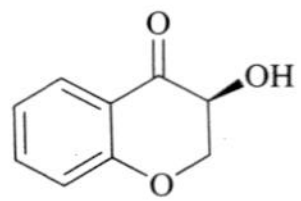 R: Me (69), 66; *i*-Pr (70), 12; *n*-Bu (73), 80; *n*-C_5H_{11} (77), 92	752
C_{10-19}	$(DHQD)_2PHAL$, $K_2OsO_2(OH)_4$, $K_3Fe(CN)_6$, K_2CO_3, $MeSO_2NH_2$, MeCN:H_2O (1:1), 0°, 18 h	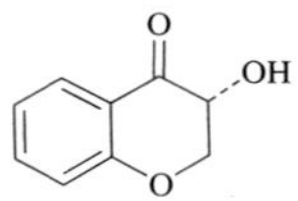	752

C_{11-17}

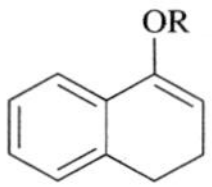

(DHQD)$_2$PHAL, $K_2OsO_2(OH)_4$, $K_3Fe(CN)_6$, K_2CO_3, $MeSO_2NH_2$, MeCN:H_2O (1:1), 0°, 18 h

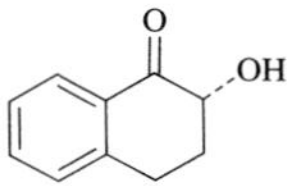

R	
Me	(89), 93
n-Pr	(79), 94
i-Pr	(85), 36
i-Bu	(95), 94
n-C_5H_{11}	(82), 97
Ph	(70), 38

752

(DHQ)$_2$PHAL, $K_2OsO_2(OH)_4$, $K_3Fe(CN)_6$, K_2CO_3, $MeSO_2NH_2$, MeCN:H_2O (1:1), 0°, 18 h

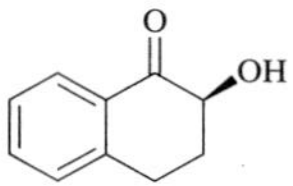

R	
Me	(88), 92
i-Pr	(68), 26
n-C_5H_{11}	(53), 96

752

C_{11}

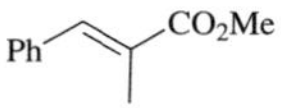

77, $K_2OsO_2(OH)_4$, $K_3Fe(CN)_6$, $MeSO_2NH_2$, K_2CO_3, t-BuOH:H_2O (1:1), 0°, 40 h

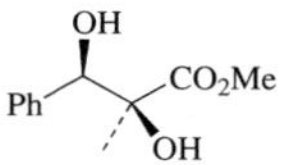

(86), 85

699

TABLE 10. SUPPLEMENTAL TABLE ENTRIES: 2001-2004 (*Continued*)

E. REACTIONS OF TRISUBSTITUTED ALKENES (*Continued*)

Substrate	Conditions	Product(s) and Yield(s) (%), % ee	Refs.
C_{12} 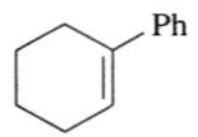	**74**, $K_2OsO_2(OH)_4$, cooxidant, additive, 24 h	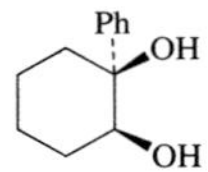 **I**	687
	Cooxidant / Additive / Temp: NMO / Et_4NOAc / rt	(92), 88	
	$K_3Fe(CN)_6$ / K_2CO_3, $MeSO_2NH_2$ / rt	(90), 94	
	O_2 / — / 50°	(78), 83	
	Ligand, OsO_4, $K_3Fe(CN)_6$, *t*-BuOH:H_2O (1:1), 0°, 24 h	**I**	693
	Ligand **61**	**I** (100), 57	
	Ligand **62**	(100), 82	
	Ligand **63**	(100), 94	
	$(DHQ)_2PHAL$, XAD-4-OsO_4, $K_3Fe(CN)_6$, K_2CO_3, $MeSO_2NH_2$, *t*-BuOH:H_2O (1:1), rt	**I**	691
	[Os(VIII)] / Time: XAD-4-OsO_4 / 4.5 h	**I** (93), 97	
	XAD-7-OsO_4 / 6 h	(88), 93	
	$(DHQD)_2PHAL$ (0.25%), $K_2OsO_2(OH)_4$ (0.05%), $K_3Fe(CN)_6$, K_2CO_3, $MeSO_2NH_2$, *t*-BuOH:H_2O (1:1.5), rt, 2 d	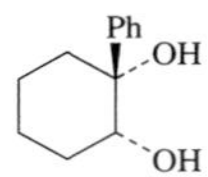 **I** (99), 99.4	753

Ligand, $K_2OsO_2(OH)_4$, $K_3Fe(CN)_6$, solvent, rt, 24 h | **I** | 698

Ligand	Solvent	**I**
$(DHQD)_2$PHAL	*t*-BuOH:H_2O (1:2)	(91), 97
$(DHQD)_2$PHAL	**89**:H_2O (1:2)	(47), 92
$(DHQD)_2$PHAL	**89**:H_2O:*t*-BuOH (1:1:2)	(53), 87
$(DHQD)_2$PYR	*t*-BuOH:H_2O (1:2)	(63), 90
$(DHQD)_2$PYR	**89**:H_2O (1:2)	(66), 86
$(DHQD)_2$PYR	**89**:H_2O:*t*-BuOH (1:1:2)	(57), 83

$(DHQD)_2$PHAL, resin-OsO_4, oxidant, additive, solvent, rt, 12-24 h | **I** | 705

Oxidant	Additive	Solvent	**I**
NMO	Et_4NOAc	*t*-BuOH:H_2O (3:1)	(90), 91
$K_3Fe(CN)_6$	K_2CO_3	*t*-BuOH:H_2O (1:1)	(88), 99
O_2	—	*t*-BuOH:pH 10.4 buffer (1:2.5)	(80), 90

$(DHQD)_2$PHAL, OsO_4, **60**, H_2O_2, Et_4NOAc, *t*-BuOH:H_2O (3:1), 0°, 16 h | **I** (58), 70 | 710

52, OsO_4, titanium silicalite, H_2O_2, NMM, *t*-BuOH:H_2O (3:1), rt, 20 h | **I** (72), 89 | 706

$(DHQD)_2$PHAL, [Os(VIII)], additive, NMM, H_2O_2, Et_4NOAc, *t*-BuOH:H_2O (1:1), rt, 20 h | **I** | 683

[Os(VIII)]	Additive	**I**
$K_2OsO_2(OH)_4$	Na_2WO_4	(91), 90
LDH-OsW	—	(89), 91

TABLE 10. SUPPLEMENTAL TABLE ENTRIES: 2001-2004 (*Continued*)

E. REACTIONS OF TRISUBSTITUTED ALKENES (*Continued*)

Substrate	Conditions	Product(s) and Yield(s) (%), % ee	Refs.
C_{12} 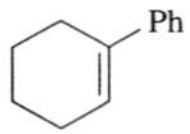	$(DHQD)_2PHAL$, OsO_4, $MeReO_3$, H_2O_2, *t*-BuOH:H_2O (3:1), 0°, 20 h	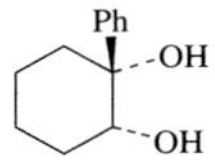 **I** (68), 77	708
	$(DHQD)_2PHAL$, $K_2OsO_2(OH)_4$, PhSe(O)CH_2Ph, *t*-BuOH:H_2O (1:1), rt, 96 h	**I** (95), 99	709
	$(DHQD)_2PHAL$, PEM-MC-OsO_4, $K_3Fe(CN)_6$, K_2CO_3, Me_2CO:H_2O (1:1), 30°, 9 h	**I** (85), 95	690
	$(DHQD)_2PHAL$, Os EnCat, K_2CO_3, $K_3Fe(CN)_6$, $MeSO_2NH_2$, THF:H_2O (1:1), rt, 48 h	**I** (91), 97	703
C_{12-16} 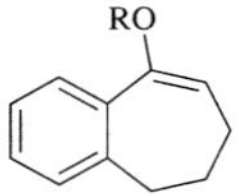	$(DHQD)_2PHAL$, $K_2OsO_2(OH)_4$, $K_3Fe(CN)_6$, K_2CO_3, $MeSO_2NH_2$, MeCN:H_2O (1:1), 0°, 18 h	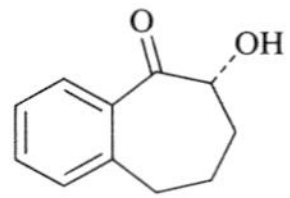 R: Me (87), 93; *n*-Pr (99), 93; *i*-Pr (89), 36; *n*-C_5H_{11} (25), 92	752
	$(DHQ)_2PHAL$, $K_2OsO_2(OH)_4$, $K_3Fe(CN)_6$, K_2CO_3, $MeSO_2NH_2$, MeCN:H_2O (1:1), 0°, 18 h	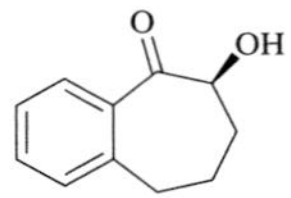 R: Me (80), 74; *i*-Pr (68), 26; *n*-C_5H_{11} (53), 94	752

Substrate	Conditions	Product(s) and Yield(s) (%), % ee	Refs.
C_{15} 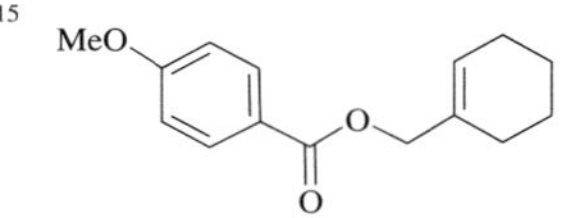	**77**, $K_2OsO_2(OH)_4$, oxidants, $MeSO_2NH_2$, K_2CO_3, t-BuOH:H_2O (1:1), 0°	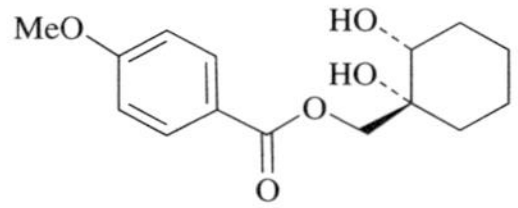	699
	Oxidants — Time		
	$K_3Fe(CN)_6$ — 24 h	(95), 97	
	$K_3Fe(CN)_6$, $NaClO_2$ — 16 h	(85), 94	
C_{17} 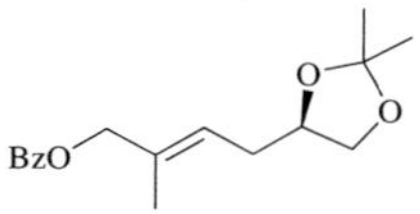	Ligand, $K_2OsO_2(OH)_4$, $K_3Fe(CN)_6$, $MeSO_2NH_2$, K_2CO_3, t-BuOH-H_2O, 0°	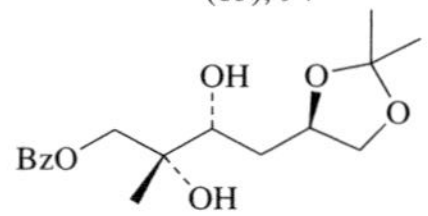	754
	Ligand		
	DHQD	(80), 72 de	
	$(DHQD)_2PHAL$	(92), 95 de	
	$(DHQ)_2PHAL$, $K_2OsO_2(OH)_4$, $K_3Fe(CN)_6$, $MeSO_2NH_2$, K_2CO_3, t-BuOH-H_2O, 0°	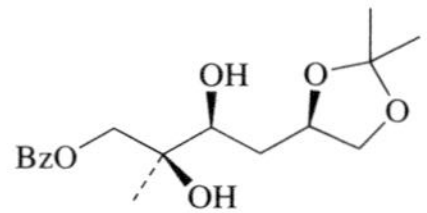 (90), 83 de	754
C_{19} 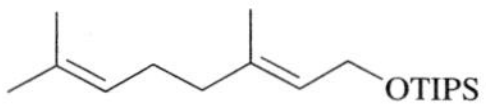	**77**, $K_2OsO_2(OH)_4$, $K_3Fe(CN)_6$, $MeSO_2NH_2$, K_2CO_3, t-BuOH:H_2O (1:1), 0°, 20 h	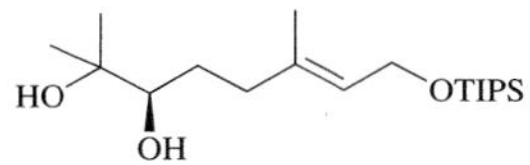 (75), 81	699
C_{31} 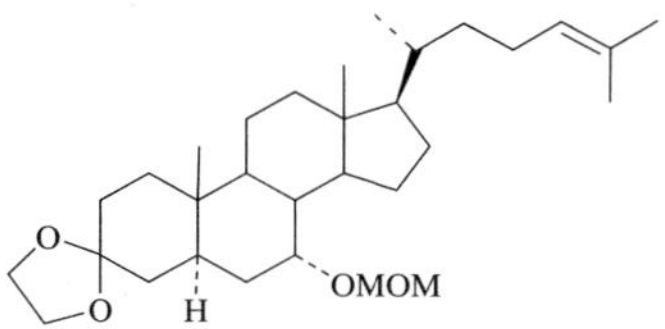	$(DHQD)_2PHAL$, $K_2OsO_2(OH)_4$, $K_3Fe(CN)_6$, K_2CO_3, $MeSO_2NH_2$, t-BuOH:t-BuOMe:H_2O (5:6:5), rt	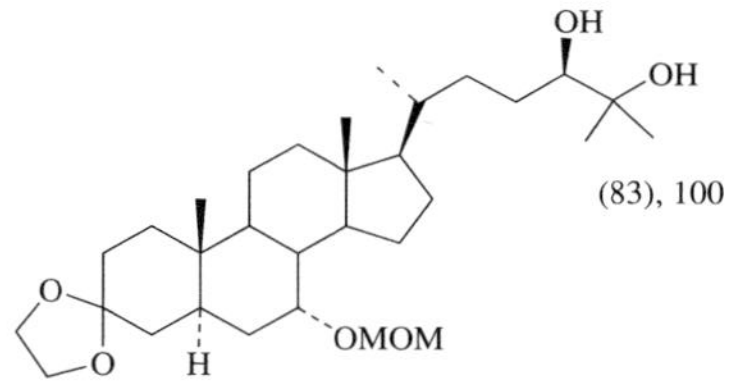 (83), 100	755

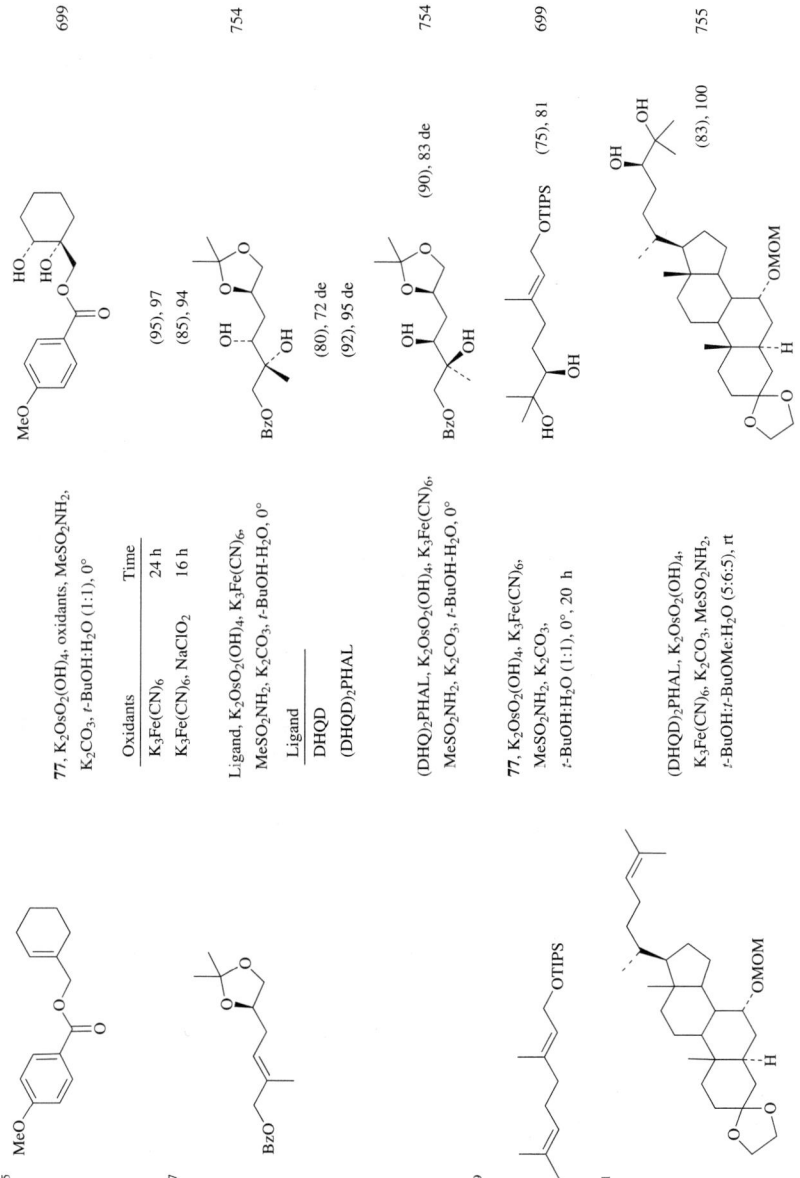

TABLE 10. SUPPLEMENTAL TABLE ENTRIES: 2001-2004 (*Continued*)

F. REACTIONS OF CONJUGATED POLYALKENES

Substrate	Conditions	Product(s) and Yield(s) (%), % ee	Refs.
C_8 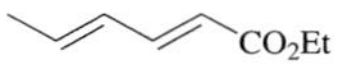	AD-Mix α, $MeSO_2NH_2$, t-BuOH:H_2O (1:1)	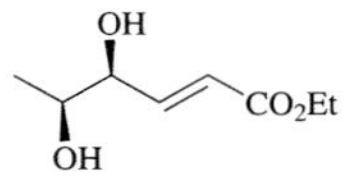 (85), 80	756
	AD-Mix β, $MeSO_2NH_2$, t-BuOH:H_2O (1:1)	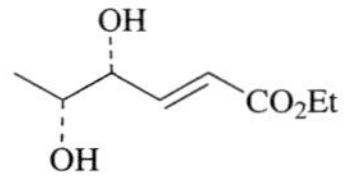 (85), >90	756
C_{18} 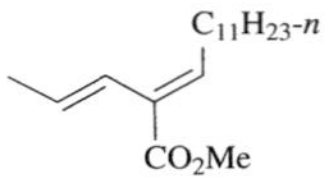	AD-Mix α, $MeSO_2NH_2$, t-BuOH:H_2O (1:1), 0°, 6 d	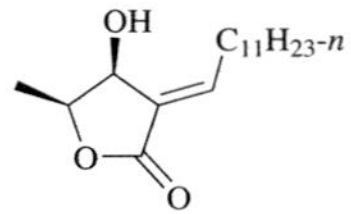 (28), —	757
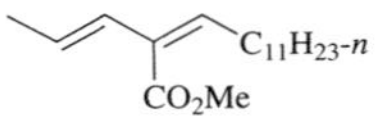	AD-Mix α, $MeSO_2NH_2$, t-BuOH:H_2O (1:1), 0°, 6 d	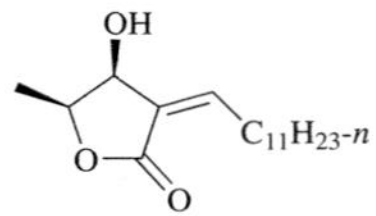 (65), —	757
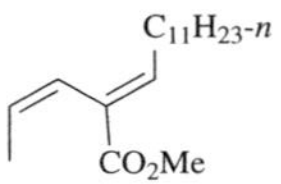	AD-Mix α, $MeSO_2NH_2$, t-BuOH:H_2O (1:1), 5°, 7 d	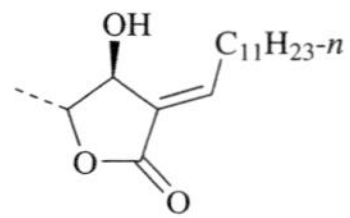 (22[a]), 28	757
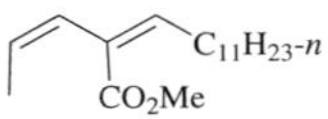	AD-Mix α, $MeSO_2NH_2$, t-BuOH:H_2O (1:1), 5°, 7 d	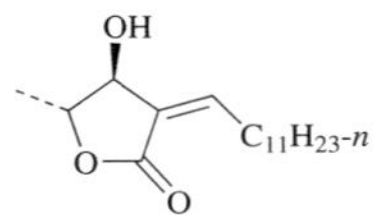 (28[b]), 16	757

C_{21}				
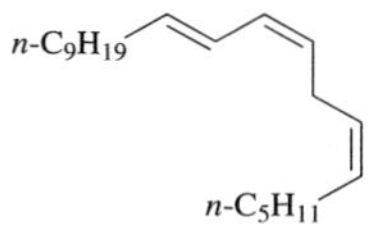	$(DHQ)_2PHAL$, OsO_4, $K_3Fe(CN)_6$, K_2CO_3, $MeSO_2NH_2$, t-BuOH:H_2O (1:1), 0°, 10 h	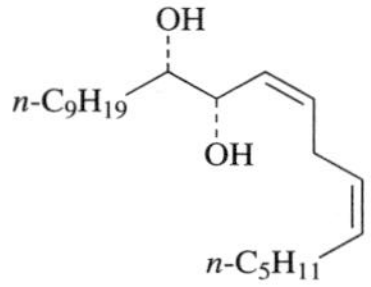	(78), >97	731
	$(DHQD)_2PHAL$, OsO_4, $K_3Fe(CN)_6$, K_2CO_3, $MeSO_2NH_2$, t-BuOH:H_2O (1:1), 0°, 10 h	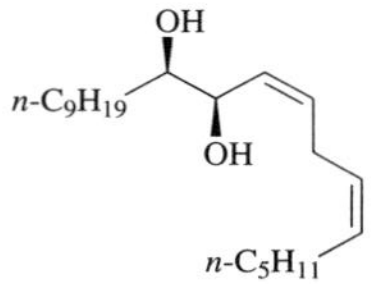	(79), >97	731
C_{22}				
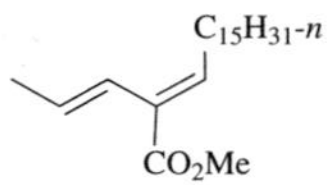	AD-Mix α, $MeSO_2NH_2$, t-BuOH:H_2O (1:1), 0°, 6 d	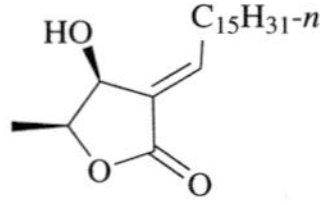	(21[c]), 78	757
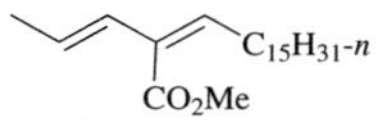	AD-Mix α, $MeSO_2NH_2$, t-BuOH:H_2O (1:1), 0°, 6 d	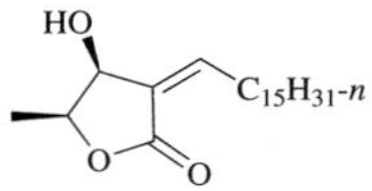	(29), 66	757

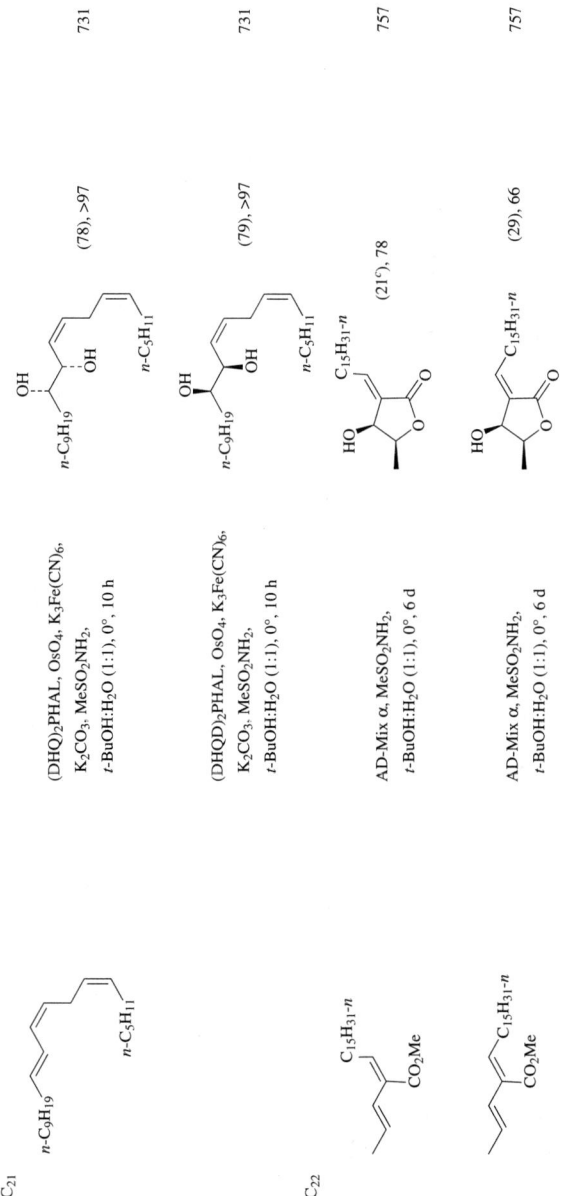

[a] The yield is 47% based on reacted starting material.
[b] The yield is 43% based on reacted starting material.
[c] The yield is 45% based on reacted starting material.

TABLE 10. SUPPLEMENTAL TABLE ENTRIES: 2001-2004 (*Continued*)

G. REACTIONS OF UNCONJUGATED POLYALKENES

Substrate	Conditions	Product(s) and Yield(s) (%), % ee	Refs.
*For Table 10G, see Chart 1 at the beginning of the Tabular Survey for ligand structures that are indicated by **bold** numbers.*			
C_8 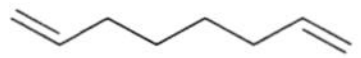	AD-Mix β, *t*-BuOH:H_2O (2:1), 0°, 18 h	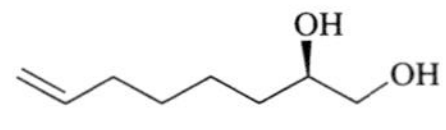 (34), —	758
C_{12} 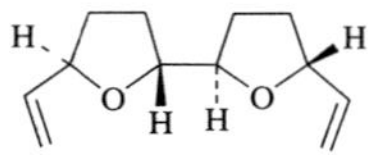	$(DHQD)_2AQN$, $K_2OsO_2(OH)_4$, $K_3Fe(CN)_6$, K_2CO_3, *t*-BuOH-H_2O	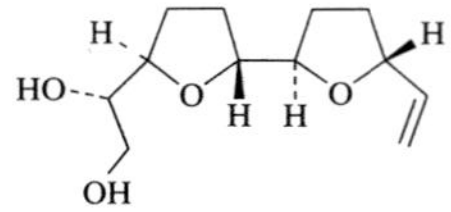 (70), 100	759
C_{14} 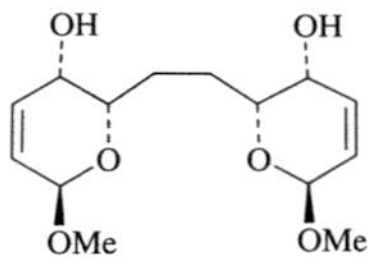	1. **35**, OsO_4, CH_2Cl_2, –20°, 6 d 2. Ac_2O, pyridine	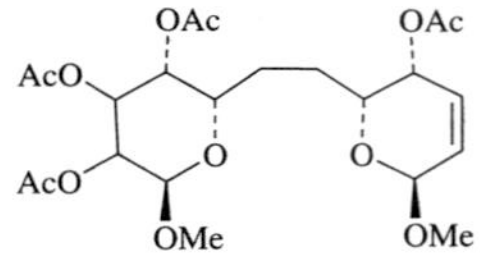 (87), 93	760
C_{17} 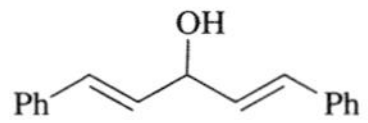	Ligand, $K_2OsO_2(OH)_4$, $K_3Fe(CN)_6$, K_2CO_3, $MeSO_2NH_2$, *t*-BuOH:H_2O:toluene (1:1:0.2), 20 d	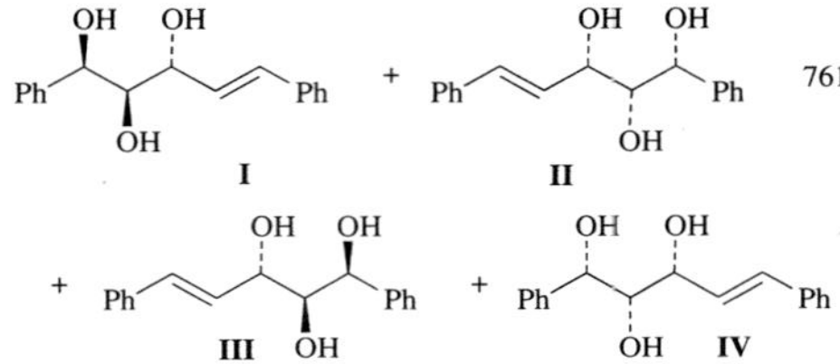 I + II + III + IV	761

Ligand	I	II	III	IV
DHQD-*p*-chlorobenzoate	(64), 80.1	(24), 67.2	(7), —	(5), —
$(DHQD)_2PHAL$	(63), 93.3	(33), 93.3	(2), —	(1), —
$(DHQ)_2PHAL$	(5), —	(2), —	(55), 82.7	(38), 90.0
$(DHQD)_2PYR$	(32), 34.9	(42), 58.3	(15), —	(11), —
$(DHQ)_2PYR$	(13), —	(9), —	(37), 47.4	(41), 64.9

C_{18}

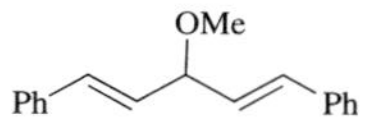

Ligand, $K_2OsO_2(OH)_4$, $K_3Fe(CN)_6$, K_2CO_3, $MeSO_2NH_2$, *t*-BuOH:H_2O:toluene (1:1:0.2), 20 d

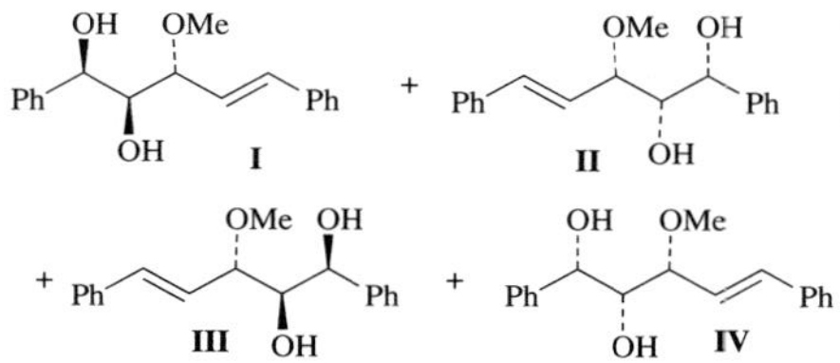

761

Ligand	I	II	III	IV
DHQD-*p*-chlorobenzoate	(67), 89.8	(27), 81.5	(4), —	(3), —
DHQ-*p*-chlorobenzoate	(6), —	(3), —	(68), 93.9	(24), 77.8
$(DHQD)_2PHAL$	(66), 98.3	(33), 97.2	(1), —	(0)
$(DHQ)_2PHAL$	(0)	(0)	(69), 98.9	(30), 98.5
$(DHQD)_2PYR$	(39), 48.0	(36), 54.5	(14), —	(11), —
$(DHQ)_2PYR$	(9), —	(7), —	(45), 65.8	(39), 69.2

C_{23}

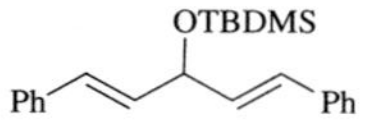

Ligand, $K_2OsO_2(OH)_4$, $K_3Fe(CN)_6$, K_2CO_3, $MeSO_2NH_2$, *t*-BuOH:H_2O:toluene (1:1:0.2), 20 d

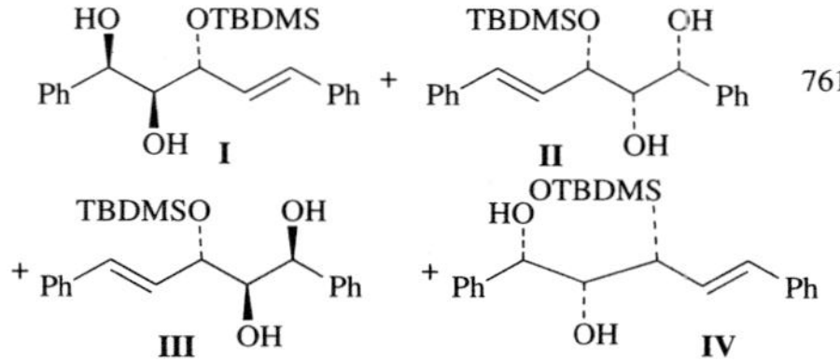

761

Ligand	I	II	III	IV
DHQD-*p*-chlorobenzoate	(76), 93.1	(19), 80.4	(3), —	(2), —
DHQ-*p*-chlorobenzoate	(6), —	(1), —	(70), 83.4	(22), 88.8
$(DHQD)_2PHAL$	(75), 99.5	(24), 97.3	(0)	(0)
$(DHQ)_2PHAL$	(8), —	(0)	(46), 71.7	(46), 98.5
$(DHQD)_2AQN$	(65), 84.1	(27), 84.8	(6), —	(2), —
$(DHQ)_2AQN$	(16), —	(2), —	(56), 56.1	(26), 84.8
$(DHQD)_2PYR$	(49), 36.7	(16), 13.0	(23), —	(12), —
$(DHQ)_2PYR$	(14), —	(12), —	(59), 61.9	(15), 13.0

TABLE 10. SUPPLEMENTAL TABLE ENTRIES: 2001-2004 (*Continued*)

G. REACTIONS OF UNCONJUGATED POLYALKENES (*Continued*)

Substrate	Conditions	Product(s) and Yield(s) (%), % ee	Refs.
C_{24} 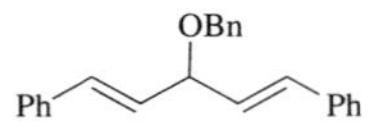	Ligand, $K_2OsO_2(OH)_4$, $K_3Fe(CN)_6$, K_2CO_3, $MeSO_2NH_2$, *t*-BuOH:H_2O:toluene (1:1:0.2), 20 d	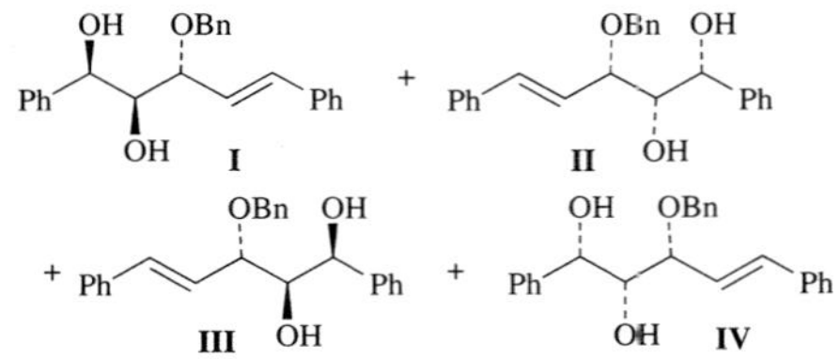	761

Ligand	I	II	III	IV
DHQD-*p*-chlorobenzoate	(66), 85.7	(26), 80.8	(5), —	(3), —
$(DHQD)_2PHAL$	(67), 92.9	(30), 97.1	(2), —	(0)
$(DHQD)_2PYR$	(42), 37.5	(26), 33.3	(19), —	(13), —

Substrate	Conditions	Product(s) and Yield(s) (%), % ee	Refs.
C_{26} 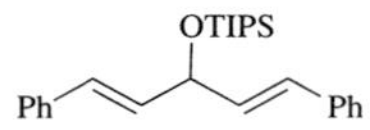	Ligand, $K_2OsO_2(OH)_4$, $K_3Fe(CN)_6$, K_2CO_3, $MeSO_2NH_2$, *t*-BuOH:H_2O:toluene (1:1:0.2), 20 d	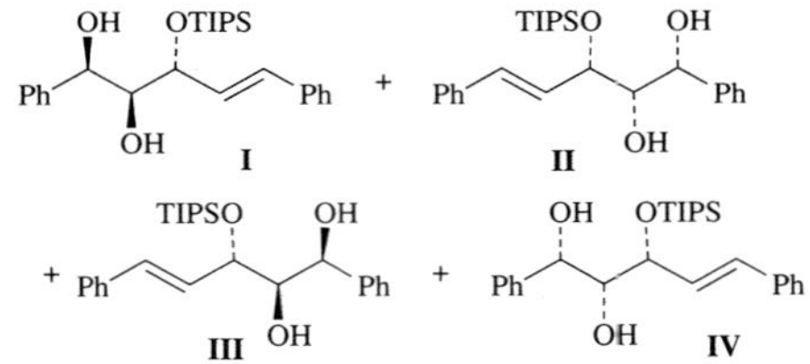	761

Ligand	I	II	III	IV
DHQD-*p*-chlorobenzoate	(73), 94.8	(22), 77.6	(2), —	(3), —
DHQ-*p*-chlorobenzoate	(4), —	(4), —	(67), 88.8	(24), 71.8
$(DHQD)_2PHAL$	(83), 99.0	(16), 94.8	(0)	(0)
$(DHQ)_2PHAL$	(4), —	(1), —	(61), 87.9	(34), 94.7
$(DHQD)_2PYR$	(41), 29.4	(19), 0	(22), —	(19), —
$(DHQ)_2PYR$	(20), —	(17), —	(47), 40.0	(17), 0

C_{27} 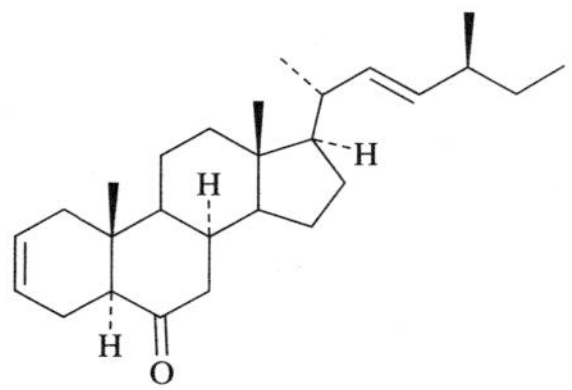	$(DHQD)_2PHAL$, $K_2OsO_2(OH)_4$, K_2CO_3, $MeSO_2NH_2$, K_2CO_3, *t*-BuOH:H_2O (1:1), rt, 3 d	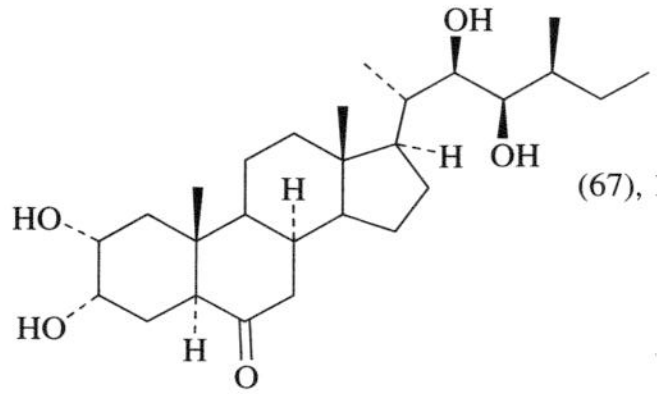	(67), 100 de	762
C_{28} 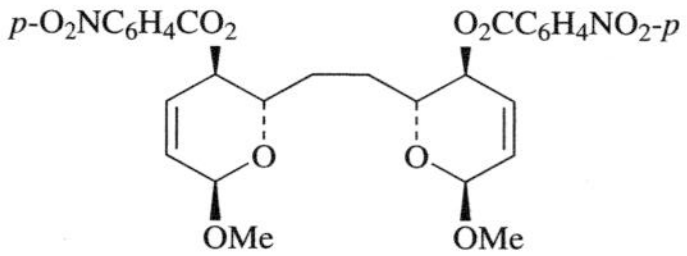	**117**, $OsCl_3$, $K_3Fe(CN)_6$, $MeSO_2NH_2$, *t*-BuOH-H_2O	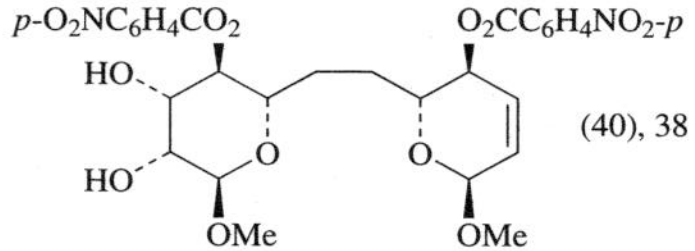	(40), 38	760
	35, OsO_4, CH_2Cl_2, –20°, 2 d	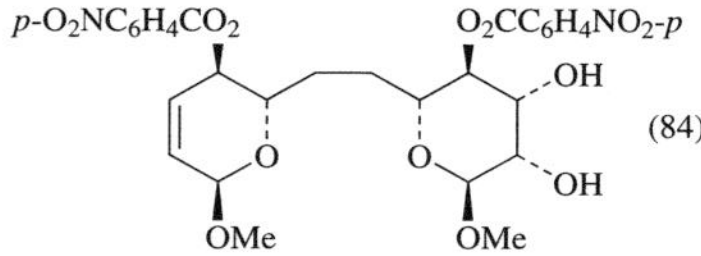	(84), 60	760
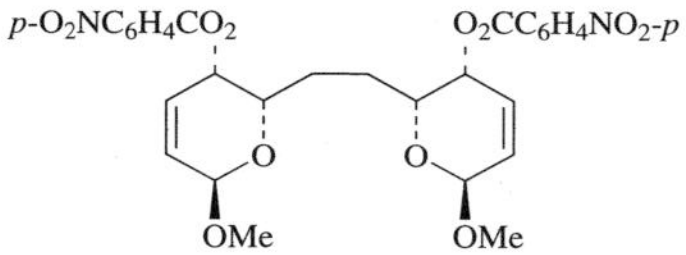	**35**, OsO_4, CH_2Cl_2, –20°, 6 d	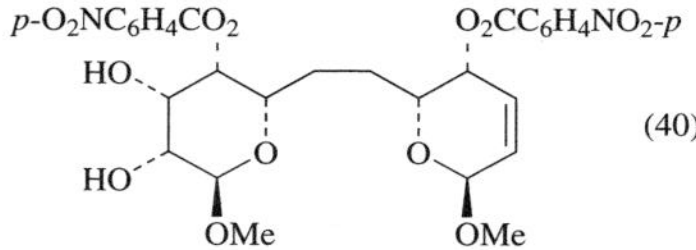	(40), 38	760

TABLE 10. SUPPLEMENTAL TABLE ENTRIES: 2001-2004 (*Continued*)

G. REACTIONS OF UNCONJUGATED POLYALKENES (*Continued*)

Substrate	Conditions	Product(s) and Yield(s) (%), % ee	Refs.
C_{36} 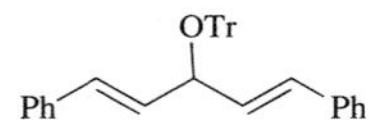	Ligand, $K_2OsO_2(OH)_4$, $K_3Fe(CN)_6$, K_2CO_3, $MeSO_2NH_2$, t-BuOH:H_2O:toluene (1:1:0.2), 20 d	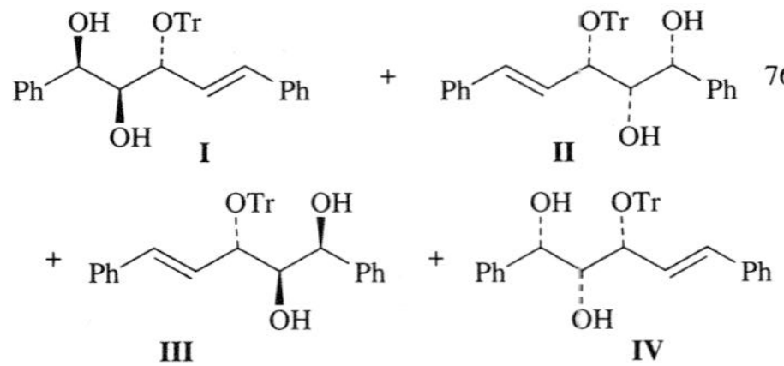 I + II + III + IV	761

Ligand	I	II	III	IV
DHQD-*p*-chlorobenzoate	(47), 85.5	(40), 65.2	(4), —	(9), —
DHQ-*p*-chlorobenzoate	(5), —	(11), —	(56), 82.5	(28), 45.2
$(DHQD)_2PHAL$	(63), 91.9	(28), 62.8	(3), —	(6), —
$(DHQ)_2PHAL$	(2), —	(5), —	(73), 94.7	(20), 55.2
$(DHQD)_2PYR$	(27), 31.0	(30), 0	(14), —	(30), —
$(DHQ)_2PYR$	(17), —	(27), —	(27), 23.1	(30), 5.9

TABLE 10. SUPPLEMENTAL TABLE ENTRIES: 2001-2004 (*Continued*)

H. REACTIONS OF ALLENES

Substrate	Conditions	Product(s) and Yield(s) (%), % ee	Refs.
C_{9-10} 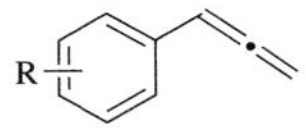	AD-Mix β	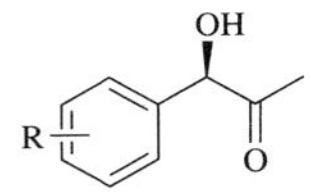 **I** R — **I** H — (45), 88 4-Cl — (58), 82 2-Me — (52), 88 4-Me — (49), 89 4-MeO — (79), 92	763
C_{13} 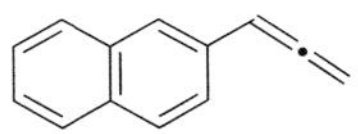	AD-Mix β	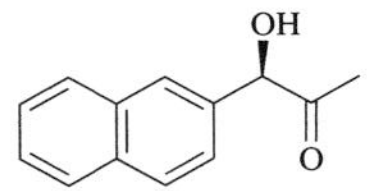 (63), 92	763

TABLE 10. SUPPLEMENTAL TABLE ENTRIES: 2001-2004 (*Continued*)

I. KINETIC RESOLUTIONS

Substrate	Conditions	Product(s) and (K_{rel})	Refs.
C_{21-31} 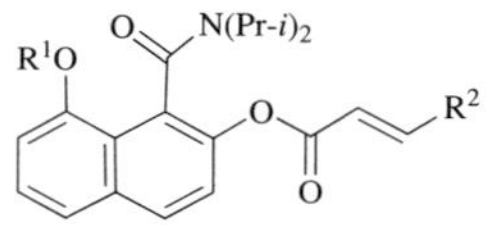	Ligand, $K_2OsO_2(OH)_4$, $K_3Fe(CN)_6$, K_2CO_3, *t*-BuOH:H_2O (1:1), 0°	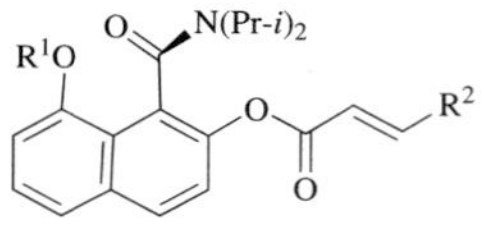	764

Ligand	R^1	R^2	
$(DHQD)_2PHAL$	Me	Me	(11.2)
$(DHQD)_2PHAL$	TBDMS	Me	(1.0)
$(DHQD)_2PHAL$	*n*-Bu	Me	(2.5)
$(DHQD)_2PHAL$	Me	Ph	(1.1)
$(DHQD)_2PHAL$	TBDMS	Ph	(1.1)
$(DHQ)_2PHAL$	Me	Me	(4.1)
$(DHQ)_2PHAL$	TBDMS	Me	(1.2)
$(DHQ)_2PHAL$	*n*-Bu	Me	(1.8)
$(DHQ)_2PHAL$	Me	Ph	(1.1)
$(DHQ)_2PHAL$	TBDMS	Ph	(1.1)

REFERENCES

[1] Schröder, M. *Chem. Rev.* **1980**, *80*, 187.
[2] Gao, Y.; Sharpless, K. B. *J. Am. Chem. Soc.* **1988**, *110*, 7538.
[3] Lohray, B. B. *Synthesis* **1992**, 1035.
[4] Lohray, B. B.; Bhushan, V. *Adv. Heterocycl. Chem.* **1997**, *68*, 89.
[5] Byun, H.-S.; He, L.; Bittman, R. *Tetrahedron* **2000**, *56*, 7051.
[6] Subbaraman, L. R.; Subbaraman, J.; Behrman, E. J. *Inorg. Chem.* **1972**, *11*, 2621.
[7] Clark, R. L.; Behrman, E. J. *Inorg. Chem.* **1975**, *14*, 1425.
[8] Hentges, S. G.; Sharpless, K. B. *J. Am. Chem. Soc.* **1980**, *102*, 4263.
[9] Hanessian, S.; Meffre, P.; Girard, M.; Beaudoin, S.; Sanceau, J. Y.; Bennani, Y. *J. Org. Chem.* **1993**, *58*, 1991.
[10] Tokles, M.; Snyder, J. K. *Tetrahedron Lett.* **1986**, *27*, 3951.
[11] Corey, E. J.; Jardine, P. D.; Virgil, S.; Yuen, P. W.; Connell, R. D. *J. Am. Chem. Soc.* **1989**, *111*, 9243.
[12] Hirama, M.; Oishi, T.; Ito, S. *J. Chem. Soc., Chem. Commun.* **1989**, 665.
[13] Oishi, T.; Hirama, M. *J. Org. Chem.* **1989**, *54*, 5834.
[14] Tomioka, K.; Nakajima, M.; Koga, K. *J. Am. Chem. Soc.* **1987**, *109*, 6213.
[15] Tomioka, K.; Nakajima, M.; Iitaka, Y.; Koga, K. *Tetrahedron Lett.* **1988**, *29*, 573.
[16] Tomioka, K.; Nakajima, M.; Koga, K. *Tetrahedron Lett.* **1990**, *31*, 1741.
[17] Nakajima, M.; Tomioka, K.; Iitaka, Y.; Koga, K. *Tetrahedron* **1993**, *49*, 10793.
[18] Jacobsen, E. N.; Marko, I.; Mungall, W. S.; Schroder, G.; Sharpless, K. B. *J. Am. Chem. Soc.* **1988**, *110*, 1968.
[19] Sharpless, K. B.; Amberg, W.; Bennani, Y. L.; Crispino, G. A.; Hartung, J.; Jeong, K. S.; Kwong, H. L.; Morikawa, K.; Wang, Z. M.; Xu, D. Q.; Zhang, X. L. *J. Org. Chem.* **1992**, *57*, 2768.
[20] Kolb, H. C.; Vannieuwenhze, M. S.; Sharpless, K. B. *Chem. Rev.* **1994**, *94*, 2483.
[21] Becker, H.; Sharpless, K. B. In *Asymmetric Oxidation Reactions*; Katsuki, T., Ed.; Oxford University Press: Oxford, UK, 2001, p. 81.
[22] Johnson, R. A.; Sharpless, K. B. In *Catalytic Asymmetric Synthesis*; Second ed.; Ojima, I., Ed.; Wiley-VCH: New York, NY, 2000, p. 357.
[23] Johnson, R.; Sharpless, K. B. In *Catalytic Asymmetric Synthesis*; Ojima, I., Ed.; VCH: New York, NY, 1993, p. 227.
[24] Bolm, C.; Hildebrand, J. P.; Muniz, K. In *Catalytic Asymmetric Synthesis*; Second ed.; Ojima, I., Ed.; Wiley-VCH: New York, NY, 2000, p. 399.
[25] Beller, M.; Sharpless, K. B. In *Applied Homogeneous Catalysis with Organometallic Compounds*; Cornils, B., Hermann, W. A., Eds.; VCH: Weinheim, Germany, 1996; Vol. 2, p. 1009.
[26] Waldmann, H. In *Organic Synthesis Highlights II*; Waldmann, H., Ed.; VCH: Weinheim, Germany, 1995, p. 9.
[27] Lohray, B. B. *Tetrahedron: Asymmetry* **1992**, *3*, 1317.
[28] Marko, I. E.; Svendsen, J. E. *Comprehensive Asymmetric Catalysis* **1999**, *2*, 713.
[29] Kolb, H. C.; Sharpless, K. B. *Transition Metals in Organic Synthesis* **1998**, *2*, 219.
[30] Cha, J. K.; Kim, N. S. *Chem. Rev.* **1995**, *95*, 1761.
[31] Criegee, R. *Liebigs Ann. Chem.* **1936**, *522*, 75.
[32] Criegee, R. *Angew. Chem.* **1937**, *50*, 153.
[33] Criegee, R.; Marchand, B.; Wannowius, H. *Liebigs Ann. Chem.* **1942**, *550*, 99.
[34] Cartwright, B. A.; P., G. W.; Schröder, M.; Skapski, A. C. *J. Chem. Soc., Chem. Commun.* **1978**, 853.
[35] Griffith, W. P.; Skapski, A. C.; Woode, K. A.; Wright, M. J. *Inorg. Chim. Acta* **1978**, *31*, L413.
[36] Prange, T.; Pascard, C. *Acta Cryst.* **1977**, *B33*, 621.
[37] Conn, J. F.; Kim, J. J.; Suddath, F. L.; Blattman, P.; Rich, A. *J. Am. Chem. Soc.* **1974**, *96*, 7152.
[38] Pearlstein, R. M.; Blackburn, B. K.; Davis, W. M.; Sharpless, K. B. *Angew. Chem., Int. Ed. Engl.* **1990**, *29*, 639.
[39] Sharpless, K. B.; Teranishi, A. Y.; Backvall, J.-E. *J. Am. Chem. Soc.* **1977**, *99*, 3120.
[40] Jorgensen, K. A. *Tetrahedron Lett.* **1990**, *31*, 6417.
[41] Jorgensen, K. A. *Tetrahedron: Asymmetry* **1991**, *2*, 515.
[42] Pidun, U.; Boehme, C.; Frenking, G. *Angew. Chem., Int. Ed. Engl.* **1997**, *35*, 2817.

[43] Dapprich, S.; Ujaque, G.; Maseras, F.; Lledos, A.; Musaev, D. G.; Morokuma, K. *J. Am. Chem. Soc.* **1996**, *118*, 11660.

[44] Torrent, M.; Deng, L. Q.; Duran, M.; Sola, M.; Ziegler, T. *Organometallics* **1997**, *16*, 13.

[45] Corey, E. J.; Sarshar, S.; Azimioara, M. D.; Newbold, R. C.; Noe, M. C. *J. Am. Chem. Soc.* **1996**, *118*, 7851.

[46] Jacobsen, E. N.; Marko, I.; France, M. B.; Svendsen, J. S.; Sharpless, K. B. *J. Am. Chem. Soc.* **1989**, *111*, 737.

[47] Kolb, H. C.; Andersson, P. G.; Bennani, Y. L.; Crispino, G. A.; Jeong, K. S.; Kwong, H. L.; Sharpless, K. B. *J. Am. Chem. Soc.* **1993**, *115*, 12226.

[48] Norrby, P. O.; Kolb, H. C.; Sharpless, K. B. *J. Am. Chem. Soc.* **1994**, *116*, 8470.

[49] Corey, E. J.; Noe, M. C. *J. Am. Chem. Soc.* **1996**, *118*, 319.

[50] Andersson, P. G.; Sharpless, K. B. *J. Am. Chem. Soc.* **1993**, *115*, 7047.

[51] Gobel, T.; Sharpless, K. B. *Angew. Chem., Int. Ed. Engl.* **1993**, *32*, 1329.

[52] Norrby, P. O.; Gable, K. P. *J. Chem. Soc., Perkin Trans. 2* **1996**, 171.

[53] Corey, E. J.; Noe, M. C.; Grogan, M. J. *Tetrahedron Lett.* **1996**, *37*, 4899.

[54] DelMonte, A. J.; Haller, J.; Houk, K. N.; Sharpless, K. B.; Singleton, D. A.; Strassner, T.; Thomas, A. A. *J. Am. Chem. Soc.* **1997**, *119*, 9907.

[55] Dijkstra, G. D. H.; Kellogg, R. M.; Wynberg, H.; Svendsen, J. S.; Marko, I.; Sharpless, K. B. *J. Am. Chem. Soc.* **1989**, *111*, 8069.

[56] Svendsen, J. S.; Marko, I.; Jacobsen, E. N.; Rao, C. P.; Bott, S.; Sharpless, K. B. *J. Org. Chem.* **1989**, *54*, 2263.

[57] Amberg, W.; Bennani, Y. L.; Chadha, R. K.; Crispino, G. A.; Davis, W. D.; Hartung, J.; Jeong, K. S.; Ogino, Y.; Shibata, T.; Sharpless, K. B. *J. Org. Chem.* **1993**, *58*, 844.

[58] Corey, E. J.; Noe, M. C. *J. Am. Chem. Soc.* **1993**, *115*, 12579.

[59] Corey, E. J.; Noe, M. C.; Sarshar, S. *Tetrahedron Lett.* **1994**, *35*, 2861.

[60] Becker, H.; Ho, P. T.; Kolb, H. C.; Loren, S.; Norrby, P. O.; Sharpless, K. B. *Tetrahedron Lett.* **1994**, *35*, 7315.

[61] Corey, E. J.; Guzman-Perez, A.; Noe, M. C. *J. Am. Chem. Soc.* **1995**, *117*, 10805.

[62] Corey, E. J.; Noe, M. C.; Lin, S. Z. *Tetrahedron Lett.* **1995**, *36*, 8741.

[63] Corey, E. J.; Noe, M. C.; Guzman-Perez, A. *J. Am. Chem. Soc.* **1995**, *117*, 10817.

[64] Noe, M. C.; Corey, E. J. *Tetrahedron Lett.* **1996**, *37*, 1739.

[65] Norrby, P. O.; Becker, H.; Sharpless, K. B. *J. Am. Chem. Soc.* **1996**, *118*, 35.

[66] Corey, E. J.; Noe, M. C. *J. Am. Chem. Soc.* **1996**, *118*, 11038.

[67] Ujaque, G.; Maseras, F.; Lledos, A. *J. Org. Chem.* **1997**, *62*, 7892.

[68] Ujaque, G.; Maseras, F.; Lledos, A. *J. Am. Chem. Soc.* **1999**, *121*, 1317.

[69] Wai, J. S. M.; Marko, I.; Svendsen, J. S.; Finn, M. G.; Jacobsen, E. N.; Sharpless, K. B. *J. Am. Chem. Soc.* **1989**, *111*, 1123.

[70] Kwong, H. L.; Sorato, C.; Ogino, Y.; Hou, C.; Sharpless, K. B. *Tetrahedron Lett.* **1990**, *31*, 2999.

[71] Mehltretter, G. M.; Döbler, C.; Sundermeier, U.; Beller, M. *Tetrahedron Lett.* **2000**, *41*, 8083.

[72] Sharpless, K. B.; Amberg, W.; Beller, M.; Chen, H.; Hartung, J.; Kawanami, Y.; Lubben, D.; Manoury, E.; Ogino, Y.; Shibata, T.; Ukita, T. *J. Org. Chem.* **1991**, *56*, 4585.

[73] Corey, E. J.; Noe, M. C.; Grogan, M. J. *Tetrahedron Lett.* **1994**, *35*, 6427.

[74] Taniguchi, T.; Nakamura, K.; Ogasawara, K. *Synlett* **1996**, 971.

[75] Harris, J. M.; Keranen, M. D.; O'Doherty, G. A. *J. Org. Chem.* **1999**, *64*, 2982.

[76] Philippo, C. M. G.; Mougenot, P.; Braun, A.; Defosse, G.; Auboussier, S.; Bovy, P. R. *Synthesis* **2000**, 127.

[77] Crispino, G. A.; Jeong, K. S.; Kolb, H. C.; Wang, Z. M.; Xu, D. Q.; Sharpless, K. B. *J. Org. Chem.* **1993**, *58*, 3785.

[78] Nieschalk, J.; O'Hagan, D. *Tetrahedron: Asymmetry* **1997**, *8*, 2325.

[79] Wang, Z. M.; Zhang, X. L.; Sharpless, K. B. *Tetrahedron Lett.* **1993**, *34*, 2267.

[80] Vilcheze, C.; Bittman, R. *J. Lipid Res.* **1994**, *35*, 734.

[81] Kawashima, E.; Naito, Y.; Ishido, Y. *Tetrahedron Lett.* **2000**, *41*, 3903.

[82] Corey, E. J.; Noe, M. C.; Ting, A. Y. *Tetrahedron Lett.* **1996**, *37*, 1735.

[83] Corey, E. J.; Guzman-Perez, A.; Noe, M. C. *Tetrahedron Lett.* **1995**, *36*, 3481.

[84] Henderson, I.; Sharpless, K. B.; Wong, C. H. *J. Am. Chem. Soc.* **1994**, *116*, 558.
[85] Walsh, P. J.; Bennani, Y. L.; Sharpless, K. B. *Tetrahedron Lett.* **1993**, *34*, 5545.
[86] Becker, H.; Sharpless, K. B. *Angew. Chem., Int. Ed. Engl.* **1996**, *35*, 448.
[87] Wang, Z. M.; Sharpless, K. B. *Synlett* **1993**, 603.
[88] Vanhessche, K. P. M.; Sharpless, K. B. *J. Org. Chem.* **1996**, *61*, 7978.
[89] Bennani, Y. L.; Vanhessche, K. P. M.; Sharpless, K. B. *Tetrahedron: Asymmetry* **1994**, *5*, 1473.
[90] Crawley, G. C.; Briggs, M. T. *J. Med. Chem.* **1995**, *38*, 3951.
[91] Ishibashi, H.; Maeki, M.; Yagi, J.; Ohba, M.; Kanai, T. *Tetrahedron* **1999**, *55*, 6075.
[92] King, S. B.; Sharpless, K. B. *Tetrahedron Lett.* **1994**, *35*, 5611.
[93] Krysan, D. J. *Tetrahedron Lett.* **1996**, *37*, 1375.
[94] Hale, K. J.; Manaviazar, S.; Peak, S. A. *Tetrahedron Lett.* **1994**, *35*, 425.
[95] Tietze, L. F.; Gorlitzer, J. *Synthesis* **1998**, 873.
[96] Yamada, T.; Narasaka, K. *Chem. Lett.* **1986**, 131.
[97] Nymann, K.; Jensen, L.; Svendsen, J. S. *Acta Chem. Scand.* **1996**, *50*, 832.
[98] Xu, D. Q.; Park, C. Y.; Sharpless, K. B. *Tetrahedron Lett.* **1994**, *35*, 2495.
[99] Hale, K. J.; Bhatia, G. S.; Peak, S. A.; Manaviazar, S. *Tetrahedron Lett.* **1993**, *34*, 5343.
[100] Walsh, P. J.; Ho, P. T.; King, S. B.; Sharpless, K. B. *Tetrahedron Lett.* **1994**, *35*, 5129.
[101] Lygo, B.; Crosby, J.; Lowdon, T.; Wainwright, P. G. *Tetrahedron* **1999**, *55*, 2795.
[102] Woltering, T. J.; Weitzschmidt, G.; Wong, C. H. *Tetrahedron Lett.* **1996**, *37*, 9033.
[103] Boger, D. L.; Schule, G. *J. Org. Chem.* **1998**, *63*, 6421.
[104] Harada, N.; Hashiyama, T.; Ozaki, K.; Yamaguchi, T.; Ando, A.; Tsujihara, K. *Heterocycles* **1997**, *44*, 305.
[105] Hughes, P. F.; Smith, S. H.; Olson, J. T. *J. Org. Chem.* **1994**, *59*, 5799.
[106] Shao, H.; Goodman, M. *J. Org. Chem.* **1996**, *61*, 2582.
[107] Wang, Z. M.; Kolb, H. C.; Sharpless, K. B. *J. Org. Chem.* **1994**, *59*, 5104.
[108] Harcken, C.; Bruckner, R. *Angew. Chem., Int. Ed. Engl.* **1997**, *36*, 2750.
[109] Wang, Z. M.; Zhang, X. L.; Sharpless, K. B.; Sinha, S. C.; Sinhabagchi, A.; Keinan, E. *Tetrahedron Lett.* **1992**, *33*, 6407.
[110] Maier, M. E.; Reuter, S. *Synlett* **1995**, 1029.
[111] Bennani, Y. L.; Sharpless, K. B. *Tetrahedron Lett.* **1993**, *34*, 2079.
[112] Lohray, B. B.; Kalantar, T. H.; Kim, B. M.; Park, C. Y.; Shibata, T.; Wai, J. S. M.; Sharpless, K. B. *Tetrahedron Lett.* **1989**, *30*, 2041.
[113] Bassindale, A. R.; Taylor, P. G.; Xu, Y. L. *J. Chem. Soc., Perkin Trans. 1* **1994**, 1061.
[114] Soderquist, J. A.; Rane, A. M.; Lopez, C. J. *Tetrahedron Lett.* **1993**, *34*, 1893.
[115] Okamoto, S.; Tani, K.; Sato, F.; Sharpless, K. B.; Zargarian, D. *Tetrahedron Lett.* **1993**, *34*, 2509.
[116] Wang, Z. M.; Sharpless, K. B. *Tetrahedron Lett.* **1993**, *34*, 8225.
[117] Jary, W. G.; Baumgartner, J. *Tetrahedron: Asymmetry* **1998**, *9*, 2081.
[118] Lohray, B. B.; Maji, D. K.; Nandanan, E. *Indian J. Chem., Sect. B: Org. Chem. Incl. Med. Chem.* **1995**, *34B*, 1023.
[119] Nelson, A.; Warren, S. *J. Chem. Soc., Perkin Trans. 1* **1997**, 2645.
[120] Nelson, A.; Obrien, P.; Warren, S. *Tetrahedron Lett.* **1995**, *36*, 2685.
[121] Wang, L.; Sharpless, K. B. *J. Am. Chem. Soc.* **1992**, *114*, 7568.
[122] Xie, L.; Crimmins, M. T.; Lee, K. H. *Tetrahedron Lett.* **1995**, *36*, 4529.
[123] Vannieuwenhze, M. S.; Sharpless, K. B. *Tetrahedron Lett.* **1994**, *35*, 843.
[124] Ko, S. Y.; Malik, M. *Tetrahedron Lett.* **1993**, *34*, 4675.
[125] Ko, S. Y.; Malik, M.; Dickinson, A. F. *J. Org. Chem.* **1994**, *59*, 2570.
[126] Hashiyama, T.; Morikawa, K.; Sharpless, K. B. *J. Org. Chem.* **1992**, *57*, 5067.
[127] Wohl, R. A. *Synthesis* **1974**, 38.
[128] Ireland, R. E.; Mueller, R. H.; Willard, A. K. *J. Am. Chem. Soc.* **1976**, *98*, 2868.
[129] Krief, A.; Ronvaux, A.; Tuch, A. *Bull. Soc. Chim. Belg.* **1997**, *106*, 699.
[130] Nemoto, H.; Miyata, J.; Hakamata, H.; Fukumoto, K. *Tetrahedron Lett.* **1995**, *36*, 1055.
[131] Diffendal, J. M.; Filan, J.; Spoors, P. G. *Tetrahedron Lett.* **1999**, *40*, 6137.
[132] Curran, D. P.; Ko, S. B. *J. Org. Chem.* **1994**, *59*, 6139.
[133] Corey, E. J.; Guzman-Perez, A.; Noe, M. C. *J. Am. Chem. Soc.* **1994**, *116*, 12109.

[134] Fang, F. G.; Xie, S.; Lowery, M. W. *J. Org. Chem.* **1994**, *59*, 6142.
[135] Fang, F. G.; Bankston, D. D.; Huie, E. M.; Johnson, M. R.; Kang, M. C.; LeHoullier, C. S.; Lewis, G. C.; Lovelace, T. C.; Lowery, M. W.; McDougald, D. L.; Meerholz, C. A.; Partridge, J. J.; Sharp, M. J.; Xie, S. P. *Tetrahedron* **1997**, *53*, 10953.
[136] Monenschein, H.; Drager, G.; Jung, A.; Kirschning, A. *Chem. Eur. J.* **1999**, *5*, 2270.
[137] Morikawa, K.; Park, J.; Andersson, P. G.; Hashiyama, T.; Sharpless, K. B. *J. Am. Chem. Soc.* **1993**, *115*, 8463.
[138] Becker, H.; Soler, M. A.; Sharpless, K. B. *Tetrahedron* **1995**, *51*, 1345.
[139] Xu, D. Q.; Crispino, G. A.; Sharpless, K. B. *J. Am. Chem. Soc.* **1992**, *114*, 7570.
[140] Allevi, P.; Tarocco, G.; Longo, A.; Anastasia, M.; Cajone, F. *Tetrahedron: Asymmetry* **1997**, *8*, 1315.
[141] Vidari, G.; Dapiaggi, A.; Zanoni, G.; Garlaschelli, L. *Tetrahedron Lett.* **1993**, *34*, 6485.
[142] Corey, E. J.; Luo, G. L.; Lin, L. S. Z. *J. Am. Chem. Soc.* **1997**, *119*, 9927.
[143] Guzman-Perez, A.; Corey, E. J. *Tetrahedron Lett.* **1997**, *38*, 5941.
[144] Takahata, H.; Takahashi, S.; Kouno, S.; Momose, T. *J. Org. Chem.* **1998**, *63*, 2224.
[145] Sinha, S. C.; Keinan, E. *J. Org. Chem.* **1994**, *59*, 949.
[146] Hoye, T. R.; Mayer, M. J.; Vos, T. J.; Ye, Z. X. *J. Org. Chem.* **1998**, *63*, 8554.
[147] Wang, Z. M.; Kakiuchi, K.; Sharpless, K. B. *J. Org. Chem.* **1994**, *59*, 6895.
[148] Takano, S.; Yoshimitsu, T.; Ogasawara, K. *J. Org. Chem.* **1994**, *59*, 54.
[149] Vannieuwenhze, M. S.; Sharpless, K. B. *J. Am. Chem. Soc.* **1993**, *115*, 7864.
[150] Yokomatsu, T.; Yamagishi, T.; Sada, T.; Suemune, K.; Shibuya, S. *Tetrahedron* **1998**, *54*, 781.
[151] Lohray, B. B.; Bhushan, V. *Tetrahedron Lett.* **1993**, *34*, 3911.
[152] Crispino, G. A.; Makita, A.; Wang, Z. M.; Sharpless, K. B. *Tetrahedron Lett.* **1994**, *35*, 543.
[153] Krief, A.; Colaux, C.; Dumont, W. *Tetrahedron Lett.* **1997**, *38*, 3315.
[154] Vanhessche, K. P. M.; Wang, Z. M.; Sharpless, K. B. *Tetrahedron Lett.* **1994**, *35*, 3469.
[155] Walsh, P. J.; Sharpless, K. B. *Synlett* **1993**, 605.
[156] Corey, E. J.; Noe, M. C.; Sarshar, S. *J. Am. Chem. Soc.* **1993**, *115*, 3828.
[157] Becker, H.; King, S. B.; Taniguchi, M.; Vanhessche, K. P. M.; Sharpless, K. B. *J. Org. Chem.* **1995**, *60*, 3940.
[158] Bolm, C.; Gerlach, A. *Eur. J. Org. Chem.* **1998**, 21.
[159] Song, C. E.; Roh, E. J.; Lee, S. G.; Kim, I. O. *Tetrahedron: Asymmetry* **1995**, *6*, 2687.
[160] Salvadori, P.; Pini, D.; Petri, A. *J. Am. Chem. Soc.* **1997**, *119*, 6929.
[161] Pini, D.; Petri, A.; Salvadori, P. *Tetrahedron* **1994**, *50*, 11321.
[162] Petri, A.; Pini, D.; Rapaccini, S.; Salvadori, P. *Chirality* **1999**, *11*, 745.
[163] Petri, A.; Pini, D.; Salvadori, P. *Tetrahedron Lett.* **1995**, *36*, 1549.
[164] Nandanan, E.; Sudalai, A.; Ravindranathan, T. *Tetrahedron Lett.* **1997**, *38*, 2577.
[165] Song, C. E.; Yang, J. W.; Ha, H. J. *Tetrahedron: Asymmetry* **1997**, *8*, 841.
[166] Bolm, C.; Maischak, A.; Gerlach, A. *Chem. Commun.* **1997**, 2353.
[167] Han, H.; Janda, K. D. *Angew. Chem., Int. Ed. Engl.* **1997**, *36*, 1731.
[168] Han, H. S.; Janda, K. D. *Tetrahedron Lett.* **1997**, *38*, 1527.
[169] Bolm, C.; Gerlach, A. *Angew. Chem., Int. Ed. Engl.* **1997**, *36*, 741.
[170] Kobayashi, S.; Endo, M.; Nagayama, S. *J. Am. Chem. Soc.* **1999**, *121*, 11229.
[171] Ahrgren, L.; Sutin, L. *Org. Process Res. Dev.* **1997**, *1*, 425.
[172] Wang, Z. M.; Sharpless, K. B. *J. Org. Chem.* **1994**, *59*, 8302.
[173] Bergstad, K.; Piet, J. J. N.; Backvall, J. E. *J. Org. Chem.* **1999**, *64*, 2545.
[174] Bergstad, K.; Jonsson, S. Y.; Backvall, J.-E. *J. Am. Chem. Soc.* **1999**, *121*, 10424.
[175] Torii, S.; Liu, P.; Bhuvaneswari, N.; Amatore, C.; Jutand, A. *J. Org. Chem.* **1996**, *61*, 3055.
[176] Amundsen, A. R.; Balko, E. N. *J. Appl. Electrochem.* **1992**, *22*, 810.
[177] Torii, S.; Liu, P.; Tanaka, H. *Chem. Lett.* **1995**, 319.
[178] Döbler, C.; Mehltretter, G.; Beller, M. *Angew. Chem., Int. Ed. Eng.* **1999**, *38*, 3026.
[179] Hill, J. H. M.; Ehrlich, J. H. *J. Org. Chem.* **1971**, *36*, 3248.
[180] Beller, M.; Döbler, C.; Mehltretter, G. M.; Sundermeier, U. *J. Am. Chem. Soc.* **2000**, *122*, 10289.
[181] Kolb, H. C.; Bennani, Y. L.; Sharpless, K. B. *Tetrahedron: Asymmetry* **1993**, *4*, 133.
[182] Vanhessche, K. P. M.; Sharpless, K. B. *Chem.-Eur. J.* **1997**, *3*, 517.
[183] Tomioka, K.; Shinmi, Y.; Shiina, K.; Nakajima, M.; Koga, K. *Chem. Pharm. Bull.* **1990**, *38*, 2133.

[184] Dinsmore, A.; Garner, C. D.; Joule, J. A. *Tetrahedron* **1998**, *54*, 3291.
[185] Vanmaarseveen, J. H.; Oberye, E. H. H.; Bolster, M. B.; Scheeren, H. W.; Kruse, C. G. *Recl. Trav. Chim. Pays-Bas* **1995**, *114*, 27.
[186] Ni, C. Y.; Matile, S. *Chem. Commun.* **1998**, 755.
[187] Takahata, H.; Ihara, K.; Kubota, M.; Momose, T. *Heterocycles* **1997**, *46*, 349.
[188] Nelson, A.; Warren, S. *J. Chem. Soc., Perkin Trans. 1* **1999**, 1963.
[189] Takahata, H.; Kubota, M.; Takahashi, S.; Momose, T. *Tetrahedron: Asymmetry* **1996**, *7*, 3047.
[190] Liao, L. X.; Wang, Z. M.; Zhang, H. X.; Zhou, W. S. *Tetrahedron: Asymmetry* **1999**, *10*, 3649.
[191] Harris, J. M.; Keranen, M. D.; Nguyen, H.; Young, V. G.; O'Doherty, G. A. *Carbohydr. Res.* **2000**, *328*, 17.
[192] Pini, D.; Petri, A.; Nardi, A.; Rosini, C.; Salvadori, P. *Tetrahedron Lett.* **1991**, *32*, 5175.
[193] Oishi, T.; Hirama, M. *Tetrahedron Lett.* **1992**, *33*, 639.
[194] Pais, G. C. G.; Fernandes, R. A.; Kumar, P. *Tetrahedron* **1999**, *55*, 13445.
[195] Smith, A. B III.; Chen, S. S. Y.; Nelson, F. C.; Reichert, J. M.; Salvatore, B. A. *J. Am. Chem. Soc.* **1995**, *117*, 12013.
[196] Smith, A. B. III; Chen, S. S. Y.; Nelson, F. C.; Reichert, J. M.; Salvatore, B. A. *J. Am. Chem. Soc.* **1997**, *119*, 10935.
[197] Simons, K. E.; Wang, G. Z.; Heinz, T.; Giger, T.; Mallat, T.; Pfaltz, A.; Baiker, A. *Tetrahedron: Asymmetry* **1995**, *6*, 505.
[198] Turpin, J. A.; Weigel, L. O. *Tetrahedron Lett.* **1992**, *33*, 6563.
[199] Hoye, T. R.; Tan, L. S. *Tetrahedron Lett.* **1995**, *36*, 1981.
[200] Nakajima, Y. *J. Synth. Org. Chem. Jpn.* **1998**, *56*, 694.
[201] Zhang, K. S.; Sun, J. T.; He, B. L. *Chemical J. of Chinese Univ.* **1999**, *20*, 900.
[202] Pini, D.; Petri, A.; Salvadori, P. *Tetrahedron: Asymmetry* **1993**, *4*, 2351.
[203] Imada, Y.; Saito, T.; Kawakami, T.; Murahashi, S. I. *Tetrahedron Lett.* **1992**, *33*, 5081.
[204] Dvorak, C. A.; Rawal, V. H. *Tetrahedron Lett.* **1998**, *39*, 2925.
[205] Han, H. S.; Janda, K. D. *J. Am. Chem. Soc.* **1996**, *118*, 7632.
[206] Petri, A.; Pini, D.; Rapaccini, S.; Salvadori, P. *Chirality* **1995**, *7*, 580.
[207] Lohray, B. B.; Thomas, A.; Chittari, P.; Ahuja, J. R.; Dhal, P. K. *Tetrahedron Lett.* **1992**, *33*, 5453.
[208] Jonsson, S.; Färnegårdh, K.; Bäckvall, J. *J. Am. Chem. Soc.* **2001**, *123*, 1365.
[209] Barlow, T.; Dipple, A. *Chem. Res. Toxicol.* **1998**, *11*, 44.
[210] Kawasaki, K.; Katsuki, T. *Tetrahedron* **1997**, *53*, 6337.
[211] Hattori, K.; Nagano, M.; Kato, T.; Nakanishi, I.; Imai, K.; Kinoshita, T.; Sakane, K. *Bioorg. Med. Chem. Lett.* **1995**, *5*, 2821.
[212] Difabio, R.; Pietra, C.; Thomas, R. J.; Ziviani, L. *Bioorg. Med. Chem. Lett.* **1995**, *5*, 551.
[213] Rao, A. V. R.; Gurjar, M. K.; Lakshmipathi, P.; Reddy, M. M.; Nagarajan, M.; Pal, S.; Sarma, B.; Tripathy, N. K. *Tetrahedron Lett.* **1997**, *38*, 7433.
[214] Chakraborty, T. K.; Suresh, V. R. *Chem. Lett.* **1997**, 565.
[215] Fuji, K.; Tanaka, K.; Miyamoto, H. *Tetrahedron Lett.* **1992**, *33*, 4021.
[216] Jones, G. B.; Guzel, M. *Tetrahedron: Asymmetry* **1998**, *9*, 2023.
[217] Takahata, H.; Yotsui, Y.; Momose, T. *Tetrahedron* **1998**, *54*, 13505.
[218] Ogren, M.; Langstrom, B. *Acta Chem. Scand.* **1998**, *52*, 1137.
[219] Jeong, K. S.; Sjo, P.; Sharpless, K. B. *Tetrahedron Lett.* **1992**, *33*, 3833.
[220] Ogino, Y.; Chen, H.; Manoury, E.; Shibata, T.; Beller, M.; Lubben, D.; Sharpless, K. B. *Tetrahedron Lett.* **1991**, *32*, 5761.
[221] Arrington, M. P.; Bennani, Y. L.; Gobel, T.; Walsh, P.; Zhao, S. H.; Sharpless, K. B. *Tetrahedron Lett.* **1993**, *34*, 7375.
[222] Miao, G. B.; Rossiter, B. E. *J. Org. Chem.* **1995**, *60*, 8424.
[223] Spivey, A. C.; Hanson, R.; Scorah, N.; Thorpe, S. J. *J. Chem. Educ.* **1999**, *76*, 655.
[224] Byun, H. S.; Kumar, E. R.; Bittman, R. *J. Org. Chem.* **1994**, *59*, 2630.
[225] Marinoalbernas, J. R.; Bittman, R.; Peters, A.; Mayhew, E. *J. Med. Chem.* **1996**, *39*, 3241.
[226] Fensholdt, J.; Wengel, J. *Acta Chem. Scand.* **1996**, *50*, 1157.
[227] Mohan, H. R.; Rao, A. S. *Indian J. Chem., Sect. B:* **1998**, *37B*, 78.
[228] Ramacciotti, A.; Fiaschi, R.; Napolitano, E. *Tetrahedron: Asymmetry* **1996**, *7*, 1101.

[229] Lohray, B. B.; Reddy, A. S.; Bhushan, V. *Tetrahedron: Asymmetry* **1996**, *7*, 2411.
[230] Oi, R.; Sharpless, K. B. *Tetrahedron Lett.* **1992**, *33*, 2095.
[231] Lindner, N.; Kolbel, M.; Sauer, C.; Diele, S.; Jokiranta, J.; Tschierske, C. *J. Phys. Chem. B* **1998**, *102*, 5261.
[232] Riedl, R.; Tappe, R.; Berkessel, A. *J. Am. Chem. Soc.* **1998**, *120*, 8994.
[233] Ewing, D. F.; Len, C.; Mackenzie, G.; Ronco, G.; Villa, P. *Tetrahedron: Asymmetry* **2000**, *11*, 4995.
[234] Sinha, S. C.; Sinhabagchi, A.; Keinan, E. *J. Org. Chem.* **1993**, *58*, 7789.
[235] Wang, G. Z.; Heinz, T.; Pfaltz, A.; Minder, B.; Mallat, T.; Baiker, A. *J. Chem. Soc., Chem. Commun.* **1994**, 2047.
[236] Liu, D. G.; Lin, G. Q. *Tetrahedron Lett.* **1999**, *40*, 337.
[237] Philippo, C.; Fett, E.; Bovy, P.; Barras, M.; Angel, I.; Georges, G.; Ochsenbein, P. *Eur. J. Med. Chem.* **1997**, *32*, 881.
[238] House, D.; Kerr, F.; Warren, S. *Chem. Commun.* **2000**, 1783.
[239] Evina, C. M.; Guillerm, G. *Tetrahedron Lett.* **1996**, *37*, 163.
[240] Muraoka, O.; Zheng, B. Z.; Fujiwara, N.; Tanabe, G. *J. Chem. Soc., Perkin Trans. 1* **1996**, 405.
[241] Salvadori, P.; Superchi, S.; Minutolo, F. *J. Org. Chem.* **1996**, *61*, 4190.
[242] Boger, D. L.; Hikota, M.; Lewis, B. M. *J. Org. Chem.* **1997**, *62*, 1748.
[243] Hodgkinson, T. J.; Shipman, M. *Synthesis* **1998**, 1141.
[244] Phukan, P.; Sudalai, A. *J. Chem. Soc., Perkin Trans. 1* **1999**, 3015.
[245] Chakraborty, T. K.; Thippeswamy, D. *Synlett* **1999**, 150.
[246] Berger, D.; Overman, L. E.; Renhowe, P. A. *J. Am. Chem. Soc.* **1997**, *119*, 2446.
[247] Moitessier, N.; Chretien, F.; Chapleur, Y. *Tetrahedron: Asymmetry* **1997**, *8*, 2889.
[248] Moitessier, N.; Maigret, B.; Chretien, F.; Chapleur, Y. *Eur. J. Org. Chem* **2000**, 995.
[249] Liao, L. X.; Zhou, W. S. *Tetrahedron Lett.* **1996**, *37*, 6371.
[250] Takahata, H.; Inose, K.; Araya, N.; Momose, T. *Heterocycles* **1994**, *38*, 1961.
[251] Takahata, H.; Kubota, M.; Momose, T. *Tetrahedron: Asymmetry* **1997**, *8*, 2801.
[252] Swindell, C. S.; Fan, W. M.; Klimko, P. G. *Tetrahedron Lett.* **1994**, *35*, 4959.
[253] Branalt, J.; Kvarnstrom, I.; Classon, B.; Samuelsson, B.; Nillroth, U.; Danielson, U. H.; Karlen, A.; Hallberg, A. *Tetrahedron Lett.* **1997**, *38*, 3483.
[254] Swindell, C. S.; Fan, W. M. *J. Org. Chem.* **1996**, *61*, 1109.
[255] House, D.; Kerr, F.; Warren, S. *Chem. Commun.* **2000**, 1779.
[256] Nakata, T.; Fukui, H.; Nakagawa, T.; Matsukura, H. *Heterocycles* **1996**, *42*, 159.
[257] Couladouros, E. A.; Moutsos, V. I. *Tetrahedron Lett.* **1999**, *40*, 7027.
[258] Liao, L. X.; Zhou, W. S. *Tetrahedron* **1998**, *54*, 12571.
[259] Takahata, H.; Kubota, M.; Momose, T. *Tetrahedron Lett.* **1997**, *38*, 3451.
[260] Takahata, H.; Kubota, M.; Ikota, N. *J. Org. Chem.* **1999**, *64*, 8594.
[261] He, L.; Byun, H.; Bittman, R. *J. Org. Chem.* **2000**, *65*, 7618.
[261a] Li, S. R.; Pang, J. H.; Wilson, W. K.; Schroepfer, G. J. *Tetrahedron: Asymmetry* **1999**, 10, 1697.
[262] Marron, T. G.; Roush, W. R. *Tetrahedron Lett.* **1995**, *36*, 1581.
[263] Peci, L. M.; Stick, R. V.; Tilbrook, D. M. G.; Winslade, M. L. *Aust. J. Chem.* **1997**, *50*, 1105.
[264] Gurjar, M. K.; Mainkar, A. S. *Tetrahedron: Asymmetry* **1992**, *3*, 21.
[265] Fairweather, J. K.; Stick, R. V.; Tilbrook, D. M. G. *Aust. J. Chem.* **1998**, *51*, 471.
[266] Gurjar, M. K.; Mainkar, A. S.; Syamala, M. *Tetrahedron: Asymmetry* **1993**, *4*, 2343.
[267] Hubieki, M. P.; Gandour, R. D.; Ashendel, C. L. *J. Org. Chem.* **1996**, *61*, 9379.
[268] Jacobs, M. F.; Suthers, B. D.; Hubener, A.; Kitching, W. *J. Chem. Soc., Perkin Trans. 1* **1995**, 901.
[269] Takahata, H.; Kubota, M.; Ihara, K.; Okamoto, N.; Momose, T.; Azer, N.; Eldefrawi, A. T.; Eldefrawi, M. E. *Tetrahedron: Asymmetry* **1998**, *9*, 3289.
[270] Takahata, H.; Okamoto, N. *Bioorg. Med. Chem. Lett.* **2000**, *10*, 1799.
[271] Woodard, S. S.; Finn, M. G.; Sharpless, K. B. *J. Am. Chem. Soc.* **1991**, *113*, 106.
[272] Bittman, R.; Kasireddy, C. R.; Mattjus, P.; Slotte, J. P. *Biochemistry* **1994**, *33*, 11776.
[273] McDonald, F. E.; Vadapally, P. *Tetrahedron Lett.* **1999**, *40*, 2235.
[274] Oka, T.; Murai, A. *Chem. Lett.* **1994**, 1611.
[275] Bando, T.; Shishido, K. *Heterocycles* **1997**, *46*, 111.

[276] Jirousek, M. R.; Cheung, A. W. H.; Babine, R. E.; Sass, P. M.; Schow, S. R.; Wick, M. M. *Tetrahedron Lett.* **1993**, *34*, 3671.
[277] Plate, M.; Overs, M.; Schafer, H. J. *Synthesis* **1998**, 1255.
[278] Zheng, T.; Flippen-Anderson, J.; Yu, P.; Wang, T.; Mirghani, R.; Cook, J. M. *J. Org. Chem.* **2001**, *66*, 1509.
[279] Sinha, S. C.; Keinan, E. *J. Org. Chem.* **1997**, *62*, 377.
[280] Anderson, J. C.; McDermott, B. P.; Griffin, E. J. *Tetrahedron* **2000**, *56*, 8747.
[281] Lin, G. Q.; Shi, Z. C. *Tetrahedron* **1996**, *52*, 2187.
[282] Fernandes, R.; Kumar, P. *Tetrahedron Lett.* **2000**, *41*, 10309.
[283] Eng, H. M.; Myles, D. C. *Tetrahedron Lett.* **1999**, *40*, 2279.
[284] Schroder, D.; Schwarz, H. *Angew. Chem., Int. Ed. Engl.* **1995**, *34*, 1973.
[285] Boger, D. L.; Borzilleri, R. M.; Nukui, S. *J. Org. Chem.* **1996**, *61*, 3561.
[286] Ishiyama, H.; Ishibashi, M.; Kobayashi, J. *Chem. Pharm. Bull.* **1996**, *44*, 1819.
[287] Chakraborty, T. K.; Suresh, V. R. *Tetrahedron Lett.* **1998**, *39*, 7775.
[288] Rychnovsky, S. D.; Yang, G.; Hu, Y. Q.; Khire, U. R. *J. Org. Chem.* **1997**, *62*, 3022.
[289] Nicolaou, K. C.; Li, J.; Zenke, G. *Helv. Chim. Acta* **2000**, *83*, 1977.
[290] Lawrence, A. J.; Pavey, J. B. J.; Cosstick, R.; Oneil, I. A. *J. Org. Chem.* **1996**, *61*, 9213.
[291] Rao, A. V. R.; Gurjar, M. K.; Reddy, A. B.; Khare, V. B. *Tetrahedron Lett.* **1993**, *34*, 1657.
[292] Hossain, N.; Rozenski, J.; Declercq, E.; Herdewijn, P. *Tetrahedron* **1996**, *52*, 13655.
[293] Hossain, N.; Rozenski, J.; DeClercq, E.; Herdewijn, P. *J. Org. Chem.* **1997**, *62*, 2442.
[294] Ren, T.; Liu, D. *Bioorg. Med. Chem. Lett.* **1999**, *9*, 1247.
[295] Staroske, T.; Hennig, L.; Wetzel, P.; Hofmann, H. J.; Muller, D.; Hausler, T.; Sheldrick, W. S.; Zillikens, S.; Gretzer, B.; Pusch, H.; Glitsch, H. G. *Tetrahedron* **1996**, *52*, 12723.
[296] Aoyagi, Y.; Williams, R. M. *Tetrahedron* **1998**, *54*, 13045.
[297] Boger, D. L.; McKie, J. A.; Nishi, T.; Ogiku, T. *J. Am. Chem. Soc.* **1997**, *119*, 311.
[298] Denmark, S. E.; Martinborough, E. A. *J. Am. Chem. Soc.* **1999**, *121*, 3046.
[298a] Boger, D. L.; McKie, J. A.; Nishi, T.; Ogiku, T. *J. Am. Chem. Soc.* **1996**, *118*, 2301.
[299] Oka, T.; Murai, A. *Tetrahedron* **1998**, *54*, 1.
[300] Lin, G. Q.; Liu, D. G. *Heterocycles* **1998**, *47*, 337.
[301] Rao, A. V. R.; Gurjar, M. K.; Kaiwar, V.; Khare, V. B. *Tetrahedron Lett.* **1993**, *34*, 1661.
[302] Sutherlin, D. P.; Armstrong, R. W. *J. Am. Chem. Soc.* **1996**, *118*, 9802.
[303] Sutherlin, D. P.; Armstrong, R. W. *J. Org. Chem.* **1997**, *62*, 5267.
[304] Moussou, P.; Archelas, A.; Furstoss, R. *Tetrahedron* **1998**, *54*, 1563.
[305] Aladro, F. J.; Guerra, F. M.; Moreno-Dorado, F. J.; Bustamante, J. M.; Jorge, Z. D.; Massanet, G. M. *Tetrahedron Lett.* **2000**, *41*, 3209.
[306] Konno, H.; Makabe, H.; Tanaka, A.; Oritani, T. *Biosci., Biotechnol., Biochem.* **1995**, *59*, 2355.
[307] Wirth, T. *Angew. Chem., Int. Ed. Eng.* **2000**, *39*, 334.
[308] Santiago, B.; Soderquist, J. A. *J. Org. Chem.* **1992**, *57*, 5844.
[309] Blundell, P.; Ganguly, A. K.; Girijavallabhan, V. M. *Synlett* **1994**, 263.
[310] Tovar-Miranda, R.; Cortes-Garcia, R.; Santos-Sanchez, N. F.; Joseph-Nathan, P. *J. Nat. Prod.* **1998**, *61*, 1216.
[311] Wang, X.; Zak, M.; Maddess, M.; O'Shea, P.; Tillyer, R.; Grabowski, E. J. J.; Reider, P. *Tetrahedron Lett.* **2000**, *41*, 4865.
[312] Mechelke, M. F.; Meyers, A. I. *Tetrahedron Lett.* **2000**, *41*, 9377.
[313] Tietze, L. F.; Gorlitzer, J.; Schuffenhauer, A.; Hubner, M. *Eur. J. Org. Chem.* **1999**, 1075.
[314] Gardiner, J. M.; Bruce, S. E. *Tetrahedron Lett.* **1998**, *39*, 1029.
[315] Zhou, G.; Gao, X.; Zhang, Z.; Li, Y. *J. Chem. Res. (S)* **2000**, 174.
[316] Zhou, G.; Gao, X.; Zhang, Z.; Li, W. Z.; Li, Y. *Tetrahedron: Asymmetry* **2000**, *11*, 1819.
[317] Obrien, P.; Warren, S. *Tetrahedron Lett.* **1995**, *36*, 2681.
[318] Obrien, P.; Warren, S. *J. Chem. Soc., Perkin Trans. 1* **1996**, 2129.
[319] Eames, J.; Mitchell, H. J.; Nelson, A.; Obrien, P.; Warren, S.; Wyatt, P. *Tetrahedron Lett.* **1995**, *36*, 1719.
[320] Tietze, L. F.; Gorlitzer, J. *Synlett* **1997**, 1049.
[321] Nishi, T.; Ishibashi, K.; Nakajima, R.; Iio, Y.; Fukazawa, T. *Tetrahedron: Asymmetry* **1998**, *9*, 3251.

[322] Nishi, T.; Fukazawa, T.; Ishibashi, K.; Nakajima, K.; Sugioka, Y.; Iio, Y.; Kurata, H.; Itoh, K.; Mukaiyama, O.; Satoh, Y.; Yamaguchi, T. *Bioorg. Med. Chem. Lett.* **1999**, *9*, 875.
[323] Tsuruoka, A.; Kaku, Y. Y.; Kakinuma, H.; Tsukada, I.; Yanagisawa, M.; Naito, T. *Chem. Pharm. Bull.* **1997**, *45*, 1169.
[324] Jimenez, O.; Bosch, M. P.; Guerrero, A. *J. Org. Chem.* **1997**, *62*, 3496.
[325] Takikawa, H.; Sano, S.; Mori, K. *Liebigs Ann./Recl.* **1997**, 2495.
[326] Allen, P. A.; Brimble, M. A.; Prabaharan, H. *Synlett* **1999**, 295.
[327] Maezaki, N.; Gijsen, H. J. M.; Sun, L. Q.; Paquette, L. A. *J. Org. Chem.* **1996**, *61*, 6685.
[328] Beauregard, D. A.; Cambie, R. C.; Dansted, P. C.; Rutledge, P. S.; Woodgate, P. D. *Aust. J. Chem.* **1995**, *48*, 669.
[329] Cambie, R. C.; Clark, R. B.; Rustenhoven, J. J.; Rutledge, P. S. *Aust. J. Chem.* **1999**, *52*, 781.
[330] Rizzo, C. J.; Smith, A. B. III *J. Chem. Soc., Perkin Trans. 1* **1991**, 969.
[331] Micalizio, G. C.; Roush, W. R. *Tetrahedron Lett.* **1999**, *40*, 3351.
[332] Banwell, M.; DeSavi, C.; Hockless, D.; Watson, K. *Chem. Commun.* **1998**, 645.
[333] Banwell, M.; DeSavi, C.; Watson, K. *J. Chem. Soc., Perkin Trans. 1* **1998**, 2251.
[334] Wang, Z. M.; Tian, S. K.; Shi, M. *Tetrahedron: Asymmetry* **1999**, *10*, 667.
[334a] Wang, Z. M.; Tian, S. K.; Shi, M. *Tetrahedron Lett.* **1999**, *40*, 977.
[335] Sinha, A.; Sinha, S. C.; Keinan, E. *J. Org. Chem.* **1999**, *64*, 2381.
[336] Richardson, T. I.; Rychnovsky, S. D. *J. Org. Chem.* **1996**, *61*, 4219.
[337] Shibata, T.; Gilheany, D. G.; Blackburn, B. K.; Sharpless, K. B. *Tetrahedron Lett.* **1990**, *31*, 3817.
[338] Harcken, C.; Bruckner, R.; Rank, E. *Chem. Eur. J.* **1998**, *4*, 2342.
[339] Berkenbusch, T.; Bruckner, R. *Tetrahedron* **1998**, *54*, 11461.
[340] Wade, P. A.; D'Ambrosia, S. G.; Rao, J. A.; ShahPatel, S.; Cole, D. T.; Murray, J. K.; Carroll, P. J. *J. Org. Chem.* **1997**, *62*, 3671.
[341] Wade, P. A.; Rao, J. A.; Bereznak, J. F.; Yuan, C. K. *Tetrahedron Lett.* **1989**, *30*, 5969.
[342] Ohshiba, Y.; Yoshimitsu, T.; Ogasawara, K. *Chem. Pharm. Bull.* **1995**, *43*, 1067.
[343] Takano, S.; Yoshimitsu, T.; Ogasawara, K. *Synlett* **1994**, 119.
[344] Yokomatsu, T.; Yoshida, Y.; Suemune, K.; Yamagishi, T.; Shibuya, S. *Tetrahedron: Asymmetry* **1995**, *6*, 365.
[345] Yokomatsu, T.; Yamagishi, T.; Suemune, K.; Yoshida, Y.; Shibuya, S. *Tetrahedron* **1998**, *54*, 767.
[346] Upadhya, T. T.; Gurunath, S.; Sudalai, A. *Tetrahedron: Asymmetry* **1999**, *10*, 2899.
[347] Andersson, P. G.; Schink, H. E.; Osterlund, K. *J. Org. Chem.* **1998**, *63*, 8067.
[348] Wang, Z. M.; Shen, M. *Tetrahedron: Asymmetry* **1997**, *8*, 3393.
[349] Hale, K. J.; Manaviazar, S.; Delisser, V. M. *Tetrahedron* **1994**, *50*, 9181.
[350] Zhang, Z. B.; Wang, Z. M.; Wang, Y. X.; Liu, H. Q.; Lei, G. X.; Shi, M. *Tetrahedron: Asymmetry* **1999**, *10*, 837.
[351] Daccolti, L.; Detomaso, A.; Fusco, C.; Rosa, A.; Curci, R. *J. Org. Chem.* **1993**, *58*, 3600.
[352] Haddad, M.; Larcheveque, M. *Tetrahedron Lett.* **1996**, *37*, 4525.
[353] Soderquist, J. A.; Rane, A. M. *Tetrahedron Lett.* **1993**, *34*, 5031.
[354] Miyazaki, Y.; Hotta, H.; Sato, F. *Tetrahedron Lett.* **1994**, *35*, 4389.
[355] Song, C. E.; Lee, S. W.; Roh, E. J.; Lee, S. G.; Lee, W. K. *Tetrahedron: Asymmetry* **1998**, *9*, 983.
[356] Tanner, D.; Birgersson, C.; Gogoll, A.; Luthman, K. *Tetrahedron* **1994**, *50*, 9797.
[357] Deng, J. G.; Hamada, Y.; Shioiri, T. *J. Am. Chem. Soc.* **1995**, *117*, 7824.
[358] Andersson, P. G.; Guijarro, D.; Tanner, D. *J. Org. Chem.* **1997**, *62*, 7364.
[359] Lu, X.; Xu, Z.; Yang, G. *Org. Process Res. Dev.* **2000**, *4*, 575.
[360] Denis, J. N.; Correa, A.; Greene, A. E. *J. Org. Chem.* **1990**, *55*, 1957.
[361] Hu, Z. Y.; Erhardt, P. W. *Org. Process Res. Dev.* **1997**, *1*, 387.
[362] Palomo, C.; Oiarbide, M.; Landa, A. *J. Org. Chem.* **2000**, *65*, 41.
[363] Song, C. E.; Yang, J. W.; Ha, H. J.; Lee, S. G. *Tetrahedron: Asymmetry* **1996**, *7*, 645.
[364] Wade, P. A.; Cole, D. T.; Dambrosio, S. G. *Tetrahedron Lett.* **1994**, *35*, 53.
[365] Tagami, K.; Takagi, S.; Sano, S.; Shiro, M.; Nagao, Y. *Heterocycles* **1997**, *45*, 1663.
[366] Kim, N. S.; Choi, J. R.; Cha, J. K. *J. Org. Chem.* **1993**, *58*, 7096.
[367] Yagi, K.; Oikawa, H.; Ichihara, A. *Biosci., Biotechnol., Biochem.* **1997**, *61*, 1038.
[368] Garcia, C.; Soler, M. A.; Martin, V. S. *Tetrahedron Lett.* **2000**, *41*, 4127.

[369] Rao, A. V. R.; Rao, S. P.; Bhanu, M. N. *J. Chem. Soc., Chem. Commun.* **1992**, 859.
[370] Kazmaier, U.; Schneider, C. *Synlett* **1996**, 975.
[371] Ung, A. T.; Pyne, S. G. *Tetrahedron: Asymmetry* **1998**, *9*, 1395.
[372] Bennani, Y. L.; Sharpless, K. B. *Tetrahedron Lett.* **1993**, *34*, 2083.
[373] Morikawa, K.; Sharpless, K. B. *Tetrahedron Lett.* **1993**, *34*, 5575.
[374] Schneider, C.; Kazmaier, U. *Synthesis-Stuttgart* **1998**, 1314.
[375] Nandanan, E.; Phukan, P.; Sudalai, A. *Indian J. Chem.,* **1999**, *38B*, 283.
[376] Yokoyama, Y.; Mori, K. *Liebigs Ann./Recl.* **1997**, 845.
[377] Boger, D. L.; Patane, M. A.; Zhou, J. C. *J. Am. Chem. Soc.* **1994**, *116*, 8544.
[378] Watson, K. G.; Fung, Y. M.; Gredley, M.; Bird, G. J.; Jackson, W. R.; Gountzos, H.; Matthews, B. R. *J. Chem. Soc., Chem. Commun.* **1990**, 1018.
[379] Harada, N.; Ozaki, K.; Yamaguchi, T.; Arakawa, H.; Ando, A.; Oda, K.; Nakanishi, N.; Ohashi, M.; Hashiyama, T.; Tsujihara, K. *Heterocycles* **1997**, *46*, 241.
[380] Annunziata, R.; Cinquini, M.; Cozzi, F.; Raimondi, L.; Stefanelli, S. *Tetrahedron Lett.* **1987**, *28*, 3139.
[381] Oikawa, H.; Kagawa, T.; Kobayashi, T.; Ichihara, A. *Tetrahedron Lett.* **1996**, *37*, 6169.
[382] Takikawa, H.; Shimbo, K.; Mori, K. *Liebigs Ann./Recl.* **1997**, 821.
[383] Watanabe, T.; Uemura, M. *Chem. Commun.* **1998**, 871.
[384] Annunziata, R.; Cinquini, M.; Cozzi, F.; Raimondi, L. *Tetrahedron* **1988**, *44*, 6897.
[385] Reetz, M. T.; Strack, T. J.; Mutulis, F.; Goddard, R. *Tetrahedron Lett.* **1996**, *37*, 9293.
[386] Maier, M. E.; Hermann, C. *Tetrahedron* **2000**, *56*, 557.
[387] Bollbuck, B.; Kraft, P.; Tochtermann, W. *Tetrahedron* **1996**, *52*, 4581.
[388] Honda, T.; Horiuchi, S.; Mizutani, H.; Kanai, K. *J. Org. Chem.* **1996**, *61*, 4944.
[389] Sakamoto, Y.; Shiraishi, A.; Seonhee, J.; Nakata, T. *Tetrahedron Lett.* **1999**, *40*, 4203.
[390] Matsuura, F.; Hamada, Y.; Shioiri, T. *Tetrahedron* **1994**, *50*, 11303.
[391] Gurjar, M. K.; Kumar, V. S.; Rao, B. V. *Tetrahedron* **1999**, *55*, 12563.
[392] Cliff, M. D.; Pyne, S. G. *J. Org. Chem.* **1995**, *60*, 2378.
[393] Traber, B.; Pfander, H. *Helv. Chim. Acta* **1996**, *79*, 499.
[394] Wipf, P.; Henninger, T. C.; Geib, S. J. *J. Org. Chem.* **1998**, *63*, 6088.
[394a] Yoshimitsu, T.; Song, J. J.; Wang, G. Q.; Masamune, S. *J. Org. Chem.* **1997**, *62*, 8978.
[395] Cooper, A. J.; Salomon, R. G. *Tetrahedron Lett.* **1990**, *31*, 3813.
[396] Pilli, R. A.; Victor, M. M. *Tetrahedron Lett.* **1998**, *39*, 4421.
[397] Filippov, D.; Timmers, C. M.; Roerdink, A. R.; vanderMarel, G. A.; vanBoom, J. H. *Tetrahedron Lett.* **1998**, *39*, 4891.
[398] Uchida, K.; Watanabe, H.; Kitahara, T. *Tetrahedron* **1998**, *54*, 8975.
[399] Blakemore, P. R.; Kocienski, P. J.; Morley, A.; Muir, K. *J. Chem. Soc., Perkin Trans. 1* **1999**, 955.
[400] Pearson, W. H.; Lian, B. W. *Angew. Chem., Int. Ed. Eng.* **1998**, *37*, 1724.
[401] Lohray, B. B.; Rao, B. S.; Baskaran, S.; Venkateswarlu, S.; Bhushan, V. *Indian J. Chem.,* **1998**, *37B*, 209.
[402] Takeuchi, M.; Taniguchi, T.; Ogasawara, K. *Synthesis-Stuttgart* **1999**, 341.
[403] Allen, J. V.; Cappi, M. W.; Kary, P. D.; Roberts, S. M.; Williamson, N. M.; Wu, L. E. *J. Chem. Soc., Perkin Trans. 1* **1997**, 3297.
[404] Mizuno, M.; Kanai, M.; Iida, A.; Tomioka, K. *Tetrahedron* **1997**, *53*, 10699.
[405] Reichardt, C.; Blum, A.; Harms, K.; Schafer, G. *Liebigs Ann./Recl.* **1997**, 707.
[406] Haubenstock, H.; Subasinghe, K. *Chirality* **1992**, *4*, 300.
[407] Kim, B. M.; Sharpless, K. B. *Tetrahedron Lett.* **1990**, *31*, 3003.
[408] Superchi, S.; Donnoli, M. I.; Proni, G.; Spada, G. P.; Rosini, C. *J. Org. Chem.* **1999**, *64*, 4762.
[409] MoyeSherman, D.; Jin, S.; Ham, I.; Lim, D. Y.; Scholtz, J. M.; Burgess, K. *J. Am. Chem. Soc.* **1998**, *120*, 9435.
[410] Tanner, D.; Johansson, F.; Harden, A.; Andersson, P. G. *Tetrahedron* **1998**, *54*, 15731.
[411] Tanner, D.; Harden, A.; Johansson, F.; Wyatt, P.; Andersson, P. G. *Acta Chem. Scand.* **1996**, *50*, 361.
[412] Zhang, S. Y.; Sun, X. L.; Li, X. Y. *Chemical J. of Chinese Univ.* **1998**, *19*, 1277.
[413] Zhang, S. Y.; Girard, C.; Kagan, H. B. *Tetrahedron: Asymmetry* **1995**, *6*, 2637.
[414] Terfort, A.; Brunner, H. *J. Chem. Soc., Perkin Trans. 1* **1996**, 1467.
[415] Kelly, T. R.; Li, Q.; Bhushan, V. *Tetrahedron Lett.* **1990**, *31*, 161.

[416] Broady, S. D.; Rexhausen, J. E.; Thomas, E. J. *J. Chem. Soc., Perkin Trans. 1* **1999**, 1083.
[417] Wallace, T. W.; Wardell, I.; Li, K. D.; Leeming, P.; Redhouse, A. D.; Challand, S. R. *J. Chem. Soc., Perkin Trans. 1* **1995**, 2293.
[418] Ung, A. T.; Pyne, S. G.; Skelton, B. W.; White, A. H. *Tetrahedron* **1996**, *52*, 14069.
[419] Ghosh, A. K.; Shin, D.; Mathivanan, P. *Chem. Commun.* **1999**, 1025.
[420] Budzinska, A.; Wojciech, S. *Tetrahedron Lett.* **2001**, *42*, 105.
[421] Lear, M. J.; Hirama, M. *Tetrahedron Lett.* **1999**, *40*, 4897.
[422] Lygo, B.; Crosby, J.; Lowdon, T. R.; Wainwright, P. G. *Tetrahedron Lett.* **1997**, *38*, 2343.
[423] Mori, K. J.; Takikawa, H.; Nishimura, Y.; Horikiri, H. *Liebigs Ann./Recl.* **1997**, 327.
[424] Berkenbusch, T.; Bruckner, R. *Tetrahedron* **1998**, *54*, 11471.
[425] Taber, D. F.; Yu, H.; Incarvito, C. D.; Rheingold, A. L. *J. Am. Chem. Soc.* **1998**, *120*, 13285.
[426] Davis, F. A.; Kasu, P. V. N.; Sundarababu, G.; Qi, H. Y. *J. Org. Chem.* **1997**, *62*, 7546.
[427] Ko, S. Y.; Lerpiniere, J. *Tetrahedron Lett.* **1995**, *36*, 2101.
[428] Iwasawa, Y.; Shibata, J.; Nonoshita, K.; Arai, S.; Masaki, H.; Tomimoto, K. *Tetrahedron* **1996**, *52*, 13881.
[429] Iwasawa, Y.; Nonoshita, K.; Tomimoto, K. *Tetrahedron Lett.* **1995**, *36*, 7459.
[430] Ohmori, K.; Nishiyama, S.; Yamamura, S. *Tetrahedron Lett.* **1995**, *36*, 6519.
[431] Chakraborty, T. K.; Thippeswamy, D.; Suresh, V. R.; Jayaprakash, S. *Chem. Lett.* **1997**, 563.
[432] Lemaire-Audoire, S.; Vogel, P. *Tetrahedron Lett.* **1998**, *39*, 1345.
[433] Lemaire-Audoire, S.; Vogel, P. *J. Org. Chem.* **2000**, *65*, 3346.
[434] Marshall, J. A.; Beaudoin, S. *J. Org. Chem.* **1994**, *59*, 6614.
[435] Marshall, J. A.; Jiang, H. J. *J. Org. Chem.* **1999**, *64*, 971.
[436] Li, P.; Yang, J.; Zhao, K. *J. Org. Chem.* **1999**, *64*, 2259.
[437] Andersson, P. G.; Johansson, F.; Tanner, D. *Tetrahedron* **1998**, *54*, 11549.
[438] Mori, K.; Abe, K. *Liebigs Ann. Chem.* **1995**, 943.
[439] Wagner, H.; Koert, U. *Angew. Chem., Int. Ed. Engl.* **1994**, *33*, 1873.
[440] Marshall, J. A.; Hinkle, K. W. *Tetrahedron Lett.* **1998**, *39*, 1303.
[441] Schwink, L.; Knochel, P. *Chem. Eur. J.* **1998**, *4*, 950.
[442] Yazbak, A.; Sinha, S. C.; Keinan, E. *J. Org. Chem.* **1998**, *63*, 5863.
[443] Matthew, D. C.; Pyne, S. G. *Tetrahedron Lett.* **1995**, *36*, 5969.
[444] Lohray, B. B.; Venkateswarlu, S. *Tetrahedron: Asymmetry* **1997**, *8*, 633.
[445] Sinhabagchi, A.; Sinha, S. C.; Keinan, E. *Tetrahedron: Asymmetry* **1995**, *6*, 2889.
[446] Sinha, S. C.; Sinha, A.; Yazbak, A.; Keinan, E. *J. Org. Chem.* **1996**, *61*, 7640.
[447] Avedissian, H.; Sinha, S. C.; Yazbak, A.; Sinha, A.; Neogi, P.; Sinha, S. C.; Keinan, E. *J. Org. Chem.* **2000**, *65*, 6035.
[448] Trost, B. M.; Vidal, B.; Thommen, M. *Chem. Eur. J.* **1999**, *5*, 1055.
[449] Imashiro, R.; Sakurai, O.; Yamashita, T.; Horikawa, H. *Tetrahedron* **1998**, *54*, 10657.
[450] Annunziata, R.; Benaglia, M.; Cinquini, M.; Cozzi, F.; Ponzini, F. *Bioorg. Med. Chem. Lett.* **1993**, *3*, 2397.
[451] Fernandes, R. A.; Kumar, P. *Eur. J. Org. Chem.* **2000**, 3447.
[452] Reference deleted.
[453] Nakamura, H.; Fujimaki, K.; Murai, A. *Tetrahedron Lett.* **1996**, *37*, 3153.
[454] Gurjar, M. K.; Ghosh, L.; Syamala, M.; Jayasree, V. *Tetrahedron Lett.* **1994**, *35*, 8871.
[455] Bewley, C. A.; He, H. Y.; Williams, D. H.; Faulkner, D. J. *J. Am. Chem. Soc.* **1996**, *118*, 4314.
[456] Pearson, W. H.; Hembre, E. J. *J. Org. Chem.* **1996**, *61*, 7217.
[456a] Oikawa, M.; Ueno, T.; Oikawa, H.; Ichihara, A. *J. Org. Chem.* **1995**, *60*, 5048.
[456b] Hayes, R.; Li, K. D.; Leeming, P.; Wallace, T. W.; Williams, R. C. *Tetrahedron* **1999**, *55*, 12907.
[457] Hermitage, S. A.; Roberts, S. M. *Tetrahedron Lett.* **1998**, *39*, 3563.
[458] Surivet, J. P.; Vatele, J. M. *Tetrahedron Lett.* **1998**, *39*, 9681.
[459] Surivet, J. P.; Vatele, J. M. *Tetrahedron* **1999**, *55*, 13011.
[460] Lee, S. G.; Lee, S. H.; Song, C. E.; Chung, B. Y. *Tetrahedron: Asymmetry* **1999**, *10*, 1795.
[461] DellaGreca, M.; Fiorentino, A.; Monaco, P.; Previtera, L. *Synth. Commun.* **1998**, *28*, 3693.
[462] Oikawa, M.; Kusumoto, S. *Tetrahedron: Asymmetry* **1995**, *6*, 961.
[463] Shirai, R.; Okabe, T.; Iwasaki, S. *Heterocycles* **1997**, *46*, 145.

[464] Krysan, D. J.; Rockway, T. W.; Haight, A. R. *Tetrahedron: Asymmetry* **1994**, *5*, 625.
[465] Ko, S. Y. *Tetrahedron Lett.* **1994**, *35*, 3601.
[466] Superchi, S.; Minutolo, F.; Pini, D.; Salvadori, P. *J. Org. Chem.* **1996**, *61*, 3183.
[467] Trost, B. M.; Krueger, A. C.; Bunt, R. C.; Zambrano, J. *J. Am. Chem. Soc.* **1996**, *118*, 6520.
[468] Hembre, E. J.; Pearson, W. H. *Tetrahedron* **1997**, *53*, 11021.
[469] Ishihara, J.; Sugimoto, T.; Murai, A. *Synlett* **1996**, 335.
[470] Makabe, H.; Tanimoto, H.; Tanaka, A.; Oritani, T. *Heterocycles* **1996**, *43*, 2229.
[471] Brimacombe, J. S.; McDonald, G.; Rahman, M. A. *Carbohydr. Res.* **1990**, *205*, 422.
[472] Ishihara, J.; Sugimoto, T.; Murai, A. *Tetrahedron* **1997**, *53*, 16029.
[473] Sinha, S. C.; Keinan, E. *J. Org. Chem.* **1999**, *64*, 7067.
[474] Patil, A. D.; Freyer, A. J.; Bean, M. F.; Carte, B. K.; Westley, J. W.; Johnson, R. K.; Lahouratate, P. *Tetrahedron* **1996**, *52*, 377.
[475] Stamos, D. P.; Kishi, Y. *Tetrahedron Lett.* **1996**, *37*, 8643.
[475a] Paterson, I.; Keown, L. E. *Tetrahedron Lett.* **1997**, *38*, 5727.
[475b] Lipshutz, B. H.; Shin, Y. J. *Tetrahedron Lett.* **1998**, *39*, 7017.
[476] Eng, H. M.; Myles, D. C. *Tetrahedron Lett.* **1999**, *40*, 2275.
[477] Toshima, H.; Maru, K.; Jiao, Y.; Yoshihara, T.; Ichihara, A. *Tetrahedron Lett.* **1999**, *40*, 935.
[478] Toshima, H.; Maru, K.; Saito, M.; Ichihara, A. *Tetrahedron* **1999**, *55*, 5793.
[479] Dyker, G.; Korning, J.; Stirner, W. *Eur. J. Org. Chem.* **1998**, 149.
[479a] Sobti, A.; Sulikowski, G. A. *Tetrahedron Lett.* **1995**, *36*, 4193.
[480] He, L.; Byun, H.-S.; Bittman, R. *J. Org. Chem.* **2000**, *65*, 7627.
[481] vanRensburg, H.; vanHeerden, P. S.; Bezuidenhoudt, B. C. B.; Ferreira, D. *Tetrahedron Lett.* **1997**, *38*, 3089.
[482] Ermolenko, L.; Sasaki, N. A.; Potier, P. *Tetrahedron Lett.* **1999**, *40*, 5187.
[483] Petit, G. R.; Lippert III, J. W.; Boyd, M. R.; Verdier-Pinard, P.; Hamel, E. *Anti-Cancer Drug Design* **2000**, *15*, 361.
[484] Tang, X. Q.; Harvey, R. G. *Tetrahedron Lett.* **1995**, *36*, 6037.
[485] Oddon, G.; Uguen, D.; DeCian, A.; Fischer, J. *Tetrahedron Lett.* **1998**, *39*, 1149.
[486] Rawal, V. H.; Florjancic, A. S.; Singh, S. P. *Tetrahedron Lett.* **1994**, *35*, 8985.
[487] Rosini, C.; Superchi, S.; Donnoli, M. I. *Enantiomer* **1999**, *4*, 3.
[488] Davies, S. G.; Smyth, G. D. *Tetrahedron: Asymmetry* **1996**, *7*, 1273.
[489] Somers, P. K.; Wandless, T. J.; Schreiber, S. L. *J. Am. Chem. Soc.* **1991**, *113*, 8045.
[489a] Makabe, H.; Tanaka, A.; Oritani, T. *Tetrahedron* **1998**, *54*, 6329.
[490] Toshima, H.; Sato, H.; Ichihara, A. *Tetrahedron* **1999**, *55*, 2581.
[491] Toshima, H.; Watanabe, A.; Sato, H.; Ichihara, A. *Tetrahedron Lett.* **1998**, *39*, 9223.
[492] Jung, M. E.; Gardiner, J. M. *Tetrahedron Lett.* **1994**, *35*, 6755.
[493] Konno, H.; Makabe, H.; Tanaka, A.; Oritani, T. *Tetrahedron* **1996**, *52*, 9399.
[494] Tsai, S. H.; Hsieh, P. C.; Wei, L. L.; Chiu, H. F.; Wu, Y. C.; Wu, M. J. *Tetrahedron Lett.* **1999**, *40*, 1975.
[495] Kobayashi, Y.; Nakano, M.; Kumar, G. B.; Kishihara, K. *J. Org. Chem.* **1998**, *63*, 7505.
[496] Kobayashi, Y.; Nakano, M.; Okui, H. *Tetrahedron Lett.* **1997**, *38*, 8883.
[497] Zhang, J. T.; Dai, W.; Harvey, R. G. *J. Org. Chem.* **1998**, *63*, 8125.
[498] LemaireAudoire, S.; Vogel, P. *Tetrahedron: Asymmetry* **1999**, *10*, 1283.
[499] Nel Reinier, J. J.; vanRensburg, H.; vanHeerden, P. S.; Ferreira, D. *J. Chem. Res. (S)* **1999**, 606.
[499a] Uchida, K.; Watanabe, H.; Usui, T.; Osada, H.; Kitahara, T. *Heterocycles* **1998**, *48*, 2049.
[500] He, L. L.; Byun, H. S.; Smit, J.; Wilschut, J.; Bittman, R. *J. Am. Chem. Soc.* **1999**, *121*, 3897.
[501] McElhanon, J. R.; Wu, M. J.; Escobar, M.; Chaudhry, U.; Hu, C. L.; McGrath, D. V. *J. Org. Chem.* **1997**, *62*, 908.
[502] Konno, H.; Makabe, H.; Tanaka, A.; Oritani, T. *Tetrahedron Lett.* **1996**, *37*, 5393.
[503] Jan, S. T.; Li, K. Q.; Vig, S.; Rudolph, A.; Uckun, F. M. *Tetrahedron Lett.* **1999**, *40*, 193.
[504] Warmerdam, E.; Tranoy, I.; Renoux, B.; Gesson, J. P. *Tetrahedron Lett.* **1998**, *39*, 8077.
[505] Shirota, O.; Nakanishi, K.; Berova, N. *Tetrahedron* **1999**, *55*, 13643.
[506] Gurjar, M. K.; Mainkar, A. S.; Srinivas, P. *Tetrahedron Lett.* **1995**, *36*, 5967.
[507] Sun, L.; Zhou, W.; Pan, X. *Tetrahedron: Asymmetry* **1991**, *2*, 973.

[508] Matsushima, T.; Mori, M.; Zheng, B. Z.; Maeda, H.; Nakajima, N.; Uenishi, J.; Yonemitsu, O. *Chem. Pharm. Bull.* **1999**, *47*, 308.
[509] Tsubuki, M.; Takada, H.; Katoh, T.; Miki, S.; Honda, T. *Tetrahedron* **1996**, *52*, 14515.
[510] Marino, J. P.; Dedios, A.; Anna, L. J.; Delapradilla, R. F. *J. Org. Chem.* **1996**, *61*, 109.
[510a] Voigt, B.; Schmidt, J.; Adam, G. *Tetrahedron* **1996**, *52*, 1997.
[511] Rao, A. V. R.; Gurjar, M. K.; Pal, S.; Pariza, R. J.; Chorghade, M. S. *Tetrahedron Lett.* **1995**, *36*, 2505.
[512] Hellrung, B.; Voigt, B.; Schmidt, J.; Adam, G. *Steroids* **1997**, *62*, 415.
[513] Bierer, B. E.; Somers, P. K.; Wandless, T. J.; Burakoff, S. J.; Schreiber, S. L. *Science* **1990**, *250*, 556.
[514] Ramirez, J. A.; Temme Centurion, O. M.; Gros, E. G.; Galagovsky, L. R. *Steroids* **2000**, *65*, 329.
[515] Brosa, C.; Peracaula, R.; Puig, R.; Ventura, M. *Tetrahedron Lett.* **1992**, *33*, 7057.
[516] McGrath, D. V.; Wu, M. J.; Chaudhry, U. *Tetrahedron Lett.* **1996**, *37*, 6077.
[517] McElhanon, J. R.; Wu, M. J.; Escobar, M.; McGrath, D. V. *Macromolecules* **1996**, *29*, 8979.
[518] Kim, N. S.; Kang, C. H.; Cha, J. K. *Tetrahedron Lett.* **1994**, *35*, 3489.
[519] Pettit, G. R.; Lippert III, J. W.; Herald, D. L.; Hamel, E.; Pettit, R. K. *J. Nat. Prod.* **2000**, *63*, 969.
[520] Polt, R.; Sames, D.; Chruma, J. *J. Org. Chem.* **1999**, *64*, 6147.
[521] Ikota, N.; Hirano, J.; Gamage, R.; Nakagawa, H.; HamaInaba, H. *Heterocycles* **1997**, *46*, 637.
[522] Ikota, N. *Heterocycles* **1995**, *41*, 983.
[523] Trost, B. M.; Lee, C. B. *J. Am. Chem. Soc.* **1998**, *120*, 6818.
[524] Ohi, K.; Nishiyama, S. *Synlett* **1999**, 573.
[525] Couladouros, E. A.; Soufli, I. C. *Tetrahedron Lett.* **1995**, *36*, 9369.
[526] Couladouros, E. A.; Soufli, I. C.; Moutsos, V. I.; Chadha, R. K. *Chem. Eur. J.* **1998**, *4*, 33.
[527] Matsushima, T.; Mori, M.; Nakajima, N.; Maeda, H.; Uenishi, J.; Yonemitsu, O. *Chem. Pharm. Bull.* **1998**, *46*, 1335.
[528] Kazmaier, U.; Schneider, C. *Tetrahedron Lett.* **1998**, *39*, 817.
[529] Schneider, C.; Kazmaier, U. *Eur. J. Org. Chem.* **1998**, 1155.
[530] Martin, C.; Prunck, W.; Bortolussi, M.; Bloch, R. *Tetrahedron: Asymmetry* **2000**, *11*, 1585.
[531] Lipshutz, B. H.; James, B.; Vance, S.; Carrico, I. *Tetrahedron Lett.* **1997**, *38*, 753.
[532] Carreira, E. M.; Dubois, J. *J. Am. Chem. Soc.* **1995**, *117*, 8106.
[533] Zheng, W. J.; Demattei, J. A.; Wu, J. P.; Duan, J. J. W.; Cook, L. R.; Oinuma, H.; Kishi, Y. *J. Am. Chem. Soc.* **1996**, *118*, 7946.
[534] Brase, S.; Enders, D.; Kobberling, J.; Avemaria, F. *Angew. Chem., Int. Ed. Eng.* **1998**, *37*, 3413.
[535] Tokoro, Y.; Kobayashi, Y. *Chem. Commun.* **1999**, 807.
[536] Brimble, M. A.; Johnston, A. D. *Tetrahedron: Asymmetry* **1997**, *8*, 1661.
[537] Koskinen, A. M. P.; Karvinen, E. K.; Siirila, J. P. *J. Chem. Soc., Chem. Commun.* **1994**, 21.
[538] Xu, D. Q.; Sharpless, K. B. *Tetrahedron Lett.* **1994**, *35*, 4685.
[539] Sunose, M.; Anderson, K. M.; Orpen, A. G.; Gallagher, T.; Macdonald, S. J. F. *Tetrahedron Lett.* **1998**, *39*, 8885.
[540] Sugisaki, C. H.; Carroll, P. J.; Correia, C. R. D. *Tetrahedron Lett.* **1998**, *39*, 3413.
[541] Xie, L.; Takeuchi, Y.; Cosentino, L. M.; Lee, K. H. *Bioorg. Med. Chem. Lett.* **1998**, *8*, 2151.
[542] Muratake, H.; Tonegawa, M.; Natsume, M. *Chem. Pharm. Bull.* **1998**, *46*, 400.
[543] Xu, Y. M.; Zhou, W. S. *Tetrahedron Lett.* **1996**, *37*, 1461.
[544] Xu, Y. M.; Zhou, W. S. *J. Chem. Soc., Perkin Trans. 1* **1997**, 741.
[545] Aggarwal, V. K.; Monteiro, N. *J. Chem. Soc., Perkin Trans. 1* **1997**, 2531.
[546] Vidari, G.; Lanfranchi, G.; Sartori, P.; Serra, S. *Tetrahedron: Asymmetry* **1995**, *6*, 2977.
[547] Devaux, J. M.; Gore, J.; Vatele, J. M. *Tetrahedron: Asymmetry* **1998**, *9*, 1619.
[548] Mulzer, J.; Karig, G.; Pojarliev, P. *Tetrahedron Lett.* **2000**, *41*, 7635.
[549] Nelson, D. W.; Gypser, A.; Ho, P. T.; Kolb, H. C.; Kondo, T.; Kwong, H. L.; McGrath, D. V.; Rubin, A. E.; Norrby, P. O.; Gable, K. P.; Sharpless, K. B. *J. Am. Chem. Soc.* **1997**, *119*, 1840.
[550] Brimble, M. A.; Rowan, D. D.; Spicer, J. A. *Synthesis* **1995**, 1263.
[551] Tinaowooldridge, L. V.; Hsiang, B. C. H.; Latifi, T. N.; Ferrendelli, J. A.; Covey, D. F. *Bioorg. Med. Chem. Lett.* **1995**, *5*, 265.
[552] Nambu, M.; White, J. D. *Chem. Commun.* **1996**, 1619.
[553] Cho, S. Y.; Lee, J. C.; Cha, J. K. *J. Org. Chem,* **1999**, *64*, 3394.
[554] Duvold, T.; Cali, P.; Bravo, J. M.; Rohmer, M. *Tetrahedron Lett.* **1997**, *38*, 6181.

[555] Kobayashi, Y.; Nakayama, Y.; Yoshida, S. *Tetrahedron Lett.* **2000**, *41*, 1465.
[556] Charon, L.; Hoeffler, J. F.; Pale-Grosdemange, C.; Rohmer, M. *Tetrahedron Lett.* **1999**, *40*, 8369.
[557] Rebiere, F.; Kagan, H. B. *Tetrahedron Lett.* **1989**, *30*, 3659.
[558] Kirschning, A.; Drager, G.; Jung, A. *Angew. Chem., Int. Ed. Engl.* **1997**, *36*, 253.
[559] Nemoto, H.; Miyata, J.; Hakamata, H.; Nagamochi, M.; Fukumoto, K. *Tetrahedron* **1995**, *51*, 5511.
[560] Krief, A.; Ronvaux, A.; Tuch, A. *Tetrahedron* **1998**, *54*, 6903.
[561] Shipman, M.; Thorpe, H. R.; Clemens, I. R. *Tetrahedron* **1999**, *55*, 10845.
[561a] Bryant, H. J.; Dardonville, C. Y.; Hodgkinson, T. J.; Shipman, M.; Slawin, A. M. Z. *Synlett* **1996**, 973.
[561b] Bryant, II. J.; Dardonvillc, C. Y.; Hodgkinson, T. J.; Hursthouse, M. B.; Malik, K. M. A.; Shipman, M. *J. Chem. Soc., Perkin Trans. 1* **1998**, 1249.
[562] Barboni, L.; Lambertucci, C.; Ballini, R.; Appendino, G.; Bombardelli, E. *Tetrahedron Lett.* **1998**, *39*, 7177.
[563] Shao, H.; Rueter, J. K.; Goodman, M. *J. Org. Chem.* **1998**, *63*, 5240.
[564] Vidari, G.; Giori, A.; Dapiaggi, A.; Lanfranchi, G. *Tetrahedron Lett.* **1993**, *34*, 6925.
[565] Mander, L. N.; Morris, J. C. *J. Org. Chem.* **1997**, *62*, 7497.
[566] McIntosh, J. M.; Kiser, E. J. *Synlett* **1997**, 1283.
[566a] Koppisch, A. T.; Blagg, B. S. J.; Poulter, C. D. *Org. Lett.* **2000**, *2*, 215.
[567] Christie, H. S.; Hamon, D. P. G.; Tuck, K. L. *Chem. Commun.* **1999**, 1989.
[568] Nakatani, K.; Okamoto, A.; Saito, I. *Angew. Chem., Int. Ed. Engl.* **1998**, *36*, 2794.
[569] Josien, H.; Ko, S. B.; Bom, D.; Curran, D. P. *Chem. Eur. J.* **1998**, *4*, 67.
[570] Jew, S. S.; Ok, K. D.; Kim, H. J.; Kim, M. G.; Kim, J. M.; Hah, J. M.; Cho, Y. S. *Tetrahedron: Asymmetry* **1995**, *6*, 1245.
[571] Matsushita, M.; Maeda, H.; Kodama, M. *Tetrahedron Lett.* **1998**, *39*, 3749.
[572] Torres-Valencia, J. M.; Cerda-Garcia-Rojas, C. M.; Joseph-Nathan, P. *Tetrahedron: Asymmetry* **1998**, *9*, 757.
[573] Cravotto, G.; Giovenzana, G. B.; Pagliarin, R.; Palmisano, G.; Sisti, M. *Tetrahedron: Asymmetry* **1998**, *9*, 745.
[574] Brimble, M. A.; Prabaharan, H. *Tetrahedron* **1998**, *54*, 2113.
[575] Brummond, K. M.; Lu, J.; Petersen, J. *J. Am. Chem. Soc.* **2000**, *122*, 4915.
[576] Toyota, M.; Saito, T.; Asakawa, Y. *Phytochemistry* **1999**, *51*, 913.
[577] Minami, N. K.; Reiner, J. E.; Semple, J. E. *Bioorg. Med. Chem. Lett.* **1999**, *9*, 2625.
[578] Martin, T.; Martin, V. S. *Tetrahedron Lett.* **2000**, *41*, 2503.
[578a] Martin, H. J.; Drescher, M.; Mulzer, J. *Angew. Chem., Int. Ed. Eng.* **2000**, *39*, 581.
[579] Taber, D. F.; Bhamidipati, R. S.; Thomas, M. L. *J. Org. Chem.* **1994**, *59*, 3442.
[580] Trost, B. M.; Probst, G. D.; Schoop, A. *J. Am. Chem. Soc.* **1998**, *120*, 9228.
[581] Toshima, H.; Saito, M.; Yoshihara, T. *Biosci., Biotechnol., Biochem.* **1999**, *63*, 964.
[582] Nicolaou, K. C.; Postema, M. H. D.; Yue, E. W.; Nadin, A. *J. Am. Chem. Soc.* **1996**, *118*, 10335.
[583] Barvian, M. R.; Greenberg, M. M. *J. Org. Chem.* **1993**, *58*, 6151.
[584] Pettit, G. R.; Melody, N.; Osullivan, M.; Thompson, M. A.; Herald, D. L.; Coates, B. *J. Chem. Soc., Chem. Commun.* **1994**, 2725.
[585] Jefford, C. W.; Misra, D.; Dishington, A. P.; Timari, G.; Rossier, J. C.; Bernardinelli, G. *Tetrahedron Lett.* **1994**, *35*, 6275.
[586] Kusama, H.; Hara, R.; Kawahara, S.; Nishimori, T.; Kashima, H.; Nakamura, N.; Morihira, K.; Kuwajima, I. *J. Am. Chem. Soc.* **2000**, *12*, 3811.
[587] Ertel, N. H.; Dayal, B.; Rao, K.; Salen, G. *Lipids* **1999**, *34*, 395.
[588] Yingyongnarongkul, B. E.; Suksamrarn, A. *Tetrahedron* **1998**, *54*, 2795.
[589] Tomkinson, N. C. O.; Willson, T. M.; Russel, J. S.; Spencer, T. A. *J. Org. Chem.* **1998**, *63*, 9919.
[590] Abdelrahman, H.; Adams, J. P.; Boyes, A. L.; Kelly, M. J.; Lamont, R. B.; Mansfield, D. J.; Procopiou, P. A.; Roberts, S. M.; Slee, D. H.; Watson, N. S. *J. Chem. Soc., Perkin Trans. 1* **1994**, 1259.
[591] Pinard, E.; Gaudry, M.; Henot, F.; Thellend, A. *Tetrahedron Lett.* **1998**, *39*, 2739.
[592] Dayal, B.; Salen, G.; Padia, J.; Shefer, S.; Tint, G. S.; Williams, T. H.; Toome, V.; Sasso, G. *Chem. Phys. Lipids* **1992**, *61*, 271.
[593] Noda, T.; Ishiwata, A.; Uemura, S.; Sakamoto, S.; Hirama, M. *Synlett* **1998**, 298.
[594] Paterson, I.; Fessner, K.; Finlay, M. R. V. *Tetrahedron Lett.* **1997**, *38*, 4301.

[595] Peterson, A. C.; Cook, J. M. *J. Org. Chem.* **1995**, *60*, 120.
[596] Miyata, J.; Nemoto, H.; Ihara, M. *J. Org. Chem.* **2000**, *65*, 504.
[597] Peterson, A. C.; Cook, J. M. *Tetrahedron Lett.* **1994**, *35*, 2651.
[598] Merz, A.; Karl, A.; Futterer, T.; Stacherdinger, N.; Schneider, O.; Lex, J.; Luboch, E.; Biernat, J. F. *Liebigs Ann. Chem.* **1994**, 1199.
[599] Chiappe, C.; DeRubertis, A.; Tinagli, V.; Amato, G.; Gervasi, P. G. *Chem. Res. Toxicol.* **2000**, *13*, 831.
[600] Braun, N. A.; Burkle, U.; Klein, I.; Spitzner, D. *Tetrahedron Lett.* **1997**, *38*, 7057.
[601] Braun, N. A.; Burkle, U.; Feth, M. P.; Klein, I.; Spitzner, D. *Eur. J. Org. Chem.* **1998**, 1569.
[602] Xu, Z. M.; Johannes, C. W.; Houri, A. F.; La, D. S.; Cogan, D. A.; Hofilena, G. E.; Hoveyda, A. H. *J. Am. Chem. Soc.* **1997**, *119*, 10302.
[603] Xu, Z. M.; Johannes, C. W.; La, D. S.; Hofilena, G. E.; Hoveyda, A. H. *Tetrahedron* **1997**, *53*, 16377.
[604] Armstrong, A.; Hayter, B. R. *Tetrahedron: Asymmetry* **1997**, *8*, 1677.
[605] Nokami, J.; Furukawa, A.; Okuda, Y.; Hazato, A.; Kurozumi, S. *Tetrahedron Lett.* **1998**, *39*, 1005.
[606] Honda, T.; Mizutani, H.; Kanai, K. *J. Org. Chem.* **1996**, *61*, 9374.
[607] Toyota, M.; Konoshima, M.; Yoshinori, A. *Phytochemistry* **1999**, *52*, 105.
[608] Tietze, L. F.; Gorlitzer, J. *Synlett* **1996**, 1041.
[609] Park, C. Y.; Kim, B. M.; Sharpless, K. B. *Tetrahedron Lett.* **1991**, *32*, 1003.
[610] Lindstrom, U. M.; Somfai, P. *Tetrahedron Lett.* **1998**, *39*, 7173.
[611] Rickards, R. W.; Thomas, R. D. *Tetrahedron Lett.* **1993**, *34*, 8369.
[612] Matsuo, G.; Miki, Y.; Nakata, M.; Matsumura, S.; Toshima, K. *Chem. Commun.* **1996**, 225.
[613] Matsuo, G.; Miki, Y.; Nakata, M.; Matsumura, S.; Toshima, K. *J. Org. Chem.* **1999**, *64*, 7101.
[614] Nagarajan, M. *Tetrahedron Lett.* **1999**, *40*, 1207.
[615] Iwasawa, N.; Maeyama, K. *J. Org. Chem.* **1997**, *62*, 1918.
[616] Andrus, M. B.; Shih, T. L. *J. Org. Chem.* **1996**, *61*, 8780.
[617] Armstrong, A.; Barsanti, P. A. *Synlett* **1995**, 903.
[618] Wolbers, P.; Hoffmann, H. M. R. *Synthesis* **1999**, 797.
[619] Nicolaou, K. C.; Nadin, A.; Leresche, J. E.; Lagreca, S.; Tsuri, T.; Yue, E. W.; Yang, Z. *Angew. Chem., Int. Ed. Engl.* **1994**, *33*, 2187.
[620] Paterson, I.; Fessner, K.; Finlay, M. R. V.; Jacobs, M. F. *Tetrahedron Lett.* **1996**, *37*, 8803.
[621] Sedrani, R.; Thai, B.; France, J.; Cottens, S. *J. Org. Chem.* **1998**, *63*, 10069.
[622] Takahata, H.; Takahashi, S.; Azer, N.; Eldefrawi, A. T.; Eldefrawi, M. E. *Bioorg. Med. Chem. Lett.* **2000**, *10*, 1293.
[623] Hale, K. J.; Lennon, J. A.; Manaviazar, S.; Javaid, M. H.; Hobbs, C. J. *Tetrahedron Lett.* **1995**, *36*, 1359.
[624] Angelaud, R.; Landais, Y. *J. Org. Chem.* **1996**, *61*, 5202.
[625] Hoye, T. R.; Tan, L. S. *Synlett* **1996**, 615.
[626] Freeman-Cook, K. D.; Halcomb, R. L. *J. Org. Chem.* **2000**, *65*, 6153.
[627] Crispino, G. A.; Sharpless, K. B. *Synlett* **1993**, 47.
[628] Quitschalle, M.; Kalesse, M. *Tetrahedron Lett.* **1999**, *40*, 7765.
[629] Tian, S. K.; Wang, Z. M.; Jiang, J. K.; Shi, M. *Tetrahedron: Asymmetry* **1999**, *10*, 2551.
[630] Wang, Z.-M.; Tian, S.-K.; Shi, M. *Eur. J. Org. Chem.* **2000**, 349.
[631] Fan, X. D.; Flentke, G. R.; Rich, D. H. *J. Am. Chem. Soc.* **1998**, *120*, 8893.
[632] Sauret, S.; Cuer, A.; Gourcy, J. G.; Jeminet, G. *Tetrahedron: Asymmetry* **1995**, *6*, 1995.
[632a] Corey, E. J.; Noe, M. C.; Shieh, W. C. *Tetrahedron Lett.* **1993**, *34*, 5995.
[633] Wang, Z. M.; Shen, M. *J. Org. Chem.* **1998**, *63*, 1414.
[634] Angelaud, R.; Landais, Y. *Tetrahedron Lett.* **1997**, *38*, 8841.
[635] Hart, D. J.; Ellis, D. A. *Heterocycles* **1998**, *49*, 117.
[636] Trost, B. M.; Calkins, T. L. *Tetrahedron Lett.* **1995**, *36*, 6021.
[637] Krief, A.; Ronvaux, A. *Synlett* **1998**, 491.
[638] Stork, G.; Doi, T.; Liu, L. B. *Tetrahedron Lett.* **1997**, *38*, 7471.
[639] Brimble, M. A.; Rowan, D. D.; Spicer, J. A. *Tetrahedron Lett.* **1994**, *35*, 9445.
[640] Corey, E. J.; Roberts, B. E.; Dixon, B. R. *J. Am. Chem. Soc.* **1995**, *117*, 193.
[641] McDonald, F. E.; Schultz, C. C. *Tetrahedron* **1997**, *53*, 16435.
[642] Sanders, W. J.; Manning, D. D.; Koeller, K. M.; Kiessling, L. L. *Tetrahedron* **1997**, *53*, 16391.
[643] McDonald, F. E.; Towne, T. B.; Schultz, C. C. *Pure Appl. Chem.* **1998**, *70*, 355.

644 Trost, B. M.; Calkins, T. L.; Bochet, C. G. *Angew. Chem., Int. Ed. Engl.* **1997**, *36*, 2632.
645 Crispino, G. A.; Sharpless, K. B. *Synthesis* **1993**, 777.
646 Ruan, Z. M.; Mootoo, D. R. *Tetrahedron Lett.* **1999**, *40*, 49.
647 Jyojima, T.; Katohno, M.; Miyamoto, N.; Nakata, M.; Matsumura, S.; Toshima, K. *Tetrahedron Lett.* **1998**, *39*, 6003.
648 Madden, B. A.; Prestwich, G. D. *Bioorg. Med. Chem. Lett.* **1997**, *7*, 309.
649 Zhang, H. P.; Seepersaud, M.; Seepersaud, S.; Mootoo, D. R. *J. Org. Chem.* **1998**, *63*, 2049.
650 Vidari, G.; Lanfranchi, G.; Masciaga, F.; Moriggi, J. D. *Tetrahedron: Asymmetry* **1996**, *7*, 3009.
651 Andrus, M. B.; Turner, T. M.; Asgari, D.; Li, W. K. *J. Org. Chem.* **1999**, *64*, 2978.
652 Andrus, M. B.; Turner, T. M.; Sauna, Z. E.; Ambudkar, S. V. *J. Org. Chem.* **2000**, *65*, 4973.
653 Yoshimitsu, T.; Ogasawara, K. *Heterocycles* **1996**, *42*, 135.
654 Nakamura, T.; Waizumi, N.; Horiguchi, Y.; Kuwajima, I. *Tetrahedron Lett.* **1994**, *35*, 7813.
655 Corey, E. J.; Roberts, B. E. *J. Am. Chem. Soc.* **1997**, *119*, 12425.
656 Corey, E. J.; Noe, M. C.; Guzman-Perez, A. *Tetrahedron Lett.* **1995**, *36*, 4171.
657 Andrus, M. B.; Lepore, S. D.; Turner, T. M. *J. Am. Chem. Soc.* **1997**, *119*, 12159.
658 Sinha, S. C.; Sinhabagchi, A.; Keinan, E. *J. Am. Chem. Soc.* **1995**, *117*, 1447.
659 Sinha, S. C.; Keinan, E. *J. Am. Chem. Soc.* **1993**, *115*, 4891.
660 Beauchamp, T. J.; Powers, J. P.; Rychnovsky, S. D. *J. Am. Chem. Soc.* **1995**, *117*, 12873.
661 Sinha, S. C.; Sinhabagchi, A.; Yazbak, A.; Keinan, E. *Tetrahedron Lett.* **1995**, *36*, 9257.
662 Kusama, H.; Hara, R.; Kawahara, S.; Nishimori, T.; Kashima, H.; Nakamura, N.; Morihara, K.; Kuwajima, I. *J. Am. Chem. Soc.* **2000**, *122*, 3811.
663 Andrus, M. B.; Lepore, S. D. *J. Am. Chem. Soc.* **1997**, *119*, 2327.
664 Katoch, R.; Korde, S. S.; Deodhar, K. D.; Trivedi, G. K. *Tetrahedron* **1999**, *55*, 1741.
665 Wipf, P.; Kim, Y.; Goldstein, D. M. *J. Am. Chem. Soc.* **1995**, *117*, 11106.
666 Corey, E. J.; Grogan, M. J. *Tetrahedron Lett.* **1998**, *39*, 9351.
667 McMorris, T. C.; Patil, P. A. *J. Org. Chem.* **1993**, *58*, 2338.
668 Iorizzi, M.; Bryan, P.; McClintock, J.; Minale, L.; Palagiano, E.; Maurelli, S.; Riccio, R.; Zollo, F. *J. Nat. Prod. (Lloydia)* **1995**, *58*, 653.
669 Burke, S. D.; Austad, B. C.; Hart, A. C. *J. Org. Chem.* **1998**, *63*, 6770.
670 Ruan, B. F.; Wilson, W. K.; Schroepfer, G. J. *Steroids* **1999**, *64*, 385.
671 McMorris, T. C.; Patil, P. A.; Chavez, R. G.; Baker, M. E.; Clouse, S. D. *Phytochemistry* **1994**, *36*, 585.
672 Armstrong, A.; Barsanti, P. A.; Jones, L. H.; Ahmed, G. *J. Org. Chem.* **2000**, *65*, 7020.
673 Wang, T. L.; Hu, X. E.; Cassady, J. M. *Tetrahedron Lett.* **1995**, *36*, 9301.
674 Ikemoto, N.; Schreiber, S. L. *J. Am. Chem. Soc.* **1990**, *112*, 9657.
675 Ikemoto, N.; Schreiber, S. L. *J. Am. Chem. Soc.* **1992**, *114*, 2524.
676 Kim, S. K.; Sutton, S. C.; Fuchs, P. L. *Tetrahedron Lett.* **1995**, *36*, 2427.
677 Kim, S.; Sutton, S. C.; Guo, C. X.; LaCour, T. G.; Fuchs, P. L. *J. Am. Chem. Soc.* **1999**, *121*, 2056.
678 Hashimoto, M.; Terashima, S. *Heterocycles* **1998**, *47*, 59.
679 Liang, J.; Moher, E. D.; Moore, R. E.; Hoard, D. W. *J. Org. Chem.* **2000**, *65*, 3143.
680 Camps, P.; Contreras, J.; Font-Bardia, M.; Morral, J.; Munoz-Torrero, D.; Solans, X. *Tetrahedron: Asymmetry* **1998**, *9*, 835.
681 Hawkins, J. M.; Meyer, A. *Science* **1993**, *260*, 1918.
682 Miao, J. H.; Yang, J. H.; Chen, L. Y.; Tu, B. X.; Huang, M. Y.; Jiang, Y. Y. *Polym. Adv. Technol.* **2004**, *15*, 221.
683 Choudary, B. M.; Chowdari, N. S.; Madhi, S.; Kantam, M. L. *J. Org. Chem.* **2003**, *68*, 1736.
684 Branco, L. C.; Afonso, C. A. M. *J. Org. Chem.* **2004**, *69*, 4381.
685 Junttila, M. H.; Hormi, O. E. O. *J. Org. Chem.* **2004**, *69*, 4816.
686 Choudary, B. M.; Jyothi, K.; Madhi, S.; Kantam, M. *Adv. Synth. Catal.* **2003**, *345*, 1190.
687 Choudary, B. M.; Chowdari, N. S.; Jyothi, K.; Kantam, M. L. *Catalysis Lett.* **2002**, *82*, 99.
688 Jiang, R.; Kuang, Y.; Sun, X.; Zhang, S. *Tetrahedron: Asymmetry* **2004**, *15*, 743.
689 Song, C. E.; Jung, D.; Roh, E. J.; Lee, S.; Chi, D. Y. *Chem. Commun.* **2002**, 3038.
690 Kobayashi, S.; Ishida, T.; Akiyama, R. *Org. Lett.* **2001**, *3*, 2649.
691 Yang, J. W.; Han, H.; Roh, E. J.; Lee, S.; Song, C. E. *Org. Lett.* **2002**, *4*, 4685.
692 Kuang, Y. Q.; Zhang, S. Y.; Jiang, R.; Wei, L. L. *Tetrahedron Lett.* **2002**, *43*, 3669.

[693] Lee, H. M.; Kim, S. W.; Hyeon, T.; Kim, B. M. *Tetrahedron: Asymmetry* **2001**, *12*, 1537.
[694] Kim, S. H.; Jin, M. J. *Stud. Surf. Sci. Catal.* **2003**, *146*, 677.
[695] Kuang, Y. Q.; Zhang, S. Y.; Wei, L. L. *Synth. Commun.* **2003**, *33*, 3545.
[696] Lohray, B. B.; Singh, S. K.; Bhushan, V. *Indian J. Chem.* **2002**, *41B*, 1226.
[697] DeClue, M. S.; Siegel, J. S. *Org. Biomol. Chem.* **2004**, *2*, 2287.
[698] Branco, L. C.; Afonso, C. A. M. *Chem. Commun.* **2002**, 3036.
[699] Huang, J.; Corey, E. *J. Org. Lett.* **2003**, *5*, 3455.
[700] Kuang, Y. Q.; Zhang, S. Y.; Wei, L. L. *Tetrahedron Lett.* **2001**, *42*, 5925.
[701] Motorina, I.; Crudden, C. M. *Org. Lett.* **2001**, *3*, 2325.
[702] Ishida, T.; Akiyama, R.; Kobayashi, S. *Adv. Synth. Catal.* **2003**, *345*, 576.
[703] Lee, A. I.; Ley, S. V. *Org. Biomol. Chem.* **2003**, *1*, 3957S.
[704] Choudary, B. M.; Chowdari, N. S.; Kantam, M. L.; Raghavan, K. V. *J. Am. Chem. Soc.* **2001**, *123*, 9220.
[705] Choudary, B. M.; Chowdari, N. S.; Jyothi, K.; Kantam, M. L. *J. Am. Chem. Soc.* **2002**, *124*, 5341.
[706] Choudary, B. M.; Chowdari, N. S.; Jyothi, K.; Madhi, S.; Kantam, M. L. *Adv. Synth. Catal.* **2002**, *344*, 503.
[707] Andersson, M. A.; Epple, R.; Fokin, V. V.; Sharpless, K. B. *Angew. Chem., Int. Edit. Engl.* **2002**, *41*, 472.
[708] Jonsson, S. Y.; Adolfsson, H.; Bäeckvall, J. E. *Chem. Eur. J.* **2003**, *9*, 2783.
[709] Krief, A.; Castillo-Colaux, C. *Synlett* **2001**, 501.
[710] Jonsson, S. Y.; Adolfsson, H.; Bäeckvall, J. E. *Org. Lett.* **2001**, *3*, 3463.
[711] Wipf, P.; Hopkins, C. R. *J. Org. Chem.* **2001**, *66*, 3133.
[712] Prasad, J. S.; Vu, T.; Totleben, M. J.; Crispino, G. A.; Kacsur, D. J.; Swaminathan, S.; Thornton, J. E.; Fritz, A.; Singh, A. K. *Org. Process Res. Dev. J.* **2003**, *7*, 821.
[713] Chandrasekhar, S.; Narsihmulu, C.; Sultana, S. S.; Reddy, N. R. *Chem. Commun.* **2003**, 1716.
[714] Jiang, B.; Yang, C. G.; Wang, J. *J. Org. Chem.* **2001**, *66*, 4865.
[715] Pilard, S.; Riboul, D.; Glacon, V.; Moitessier, N.; Chapleur, Y.; Postel, D.; Len, C. *Tetrahedron: Asymmetry* **2002**, *13*, 529.
[716] Machinami, T.; Itaba, Y.; Kayama, A.; Fujimoto, T.; Suami, T. *Carbohydr. Res.* **2002**, *337*, 1917.
[717] Dragovich, P. S.; Zhou, R.; Prins, T. *J. Org. Chem.* **2002**, *67*, 741.
[718] Seidel, M. C.; Smits, R.; Stark, C. B. W.; Frackenpohl, J.; Gaertzen, O.; Hoffmann, H. M. R. *Synthesis* **2004**, 1391.
[719] Reddy, P. G.; Baskaran, S. *J. Org. Chem.* **2004**, *69*, 3093.
[720] McNamara, C. A.; King, F.; Bradley, M. *Tetrahedron Lett.* **2004**, *45*, 8527.
[721] Gupta, P.; Fernandes, R. A.; Kumar, P. *Tetrahedron Lett.* **2003**, *44*, 4231.
[722] Lee, S.; LaCour, T. G.; Lantrip, D.; Fuchs, P. L. *Org. Lett.* **2002**, *4*, 313.
[723] Bonini, C.; D'Auria, M.; Fedeli, P. *Tetrahedron Lett.* **2002**, *43*, 3813.
[724] Feng, Z. X.; Zhou, W. S. *Tetrahedron Lett.* **2003**, *44*, 493.
[725] Song, C. E.; Oh, C. R.; Roh, E. J.; Choi, J. H. *Tetrahedron: Asymmetry* **2001**, *12*, 1533.
[726] Fernandes, R. A.; Kumar, P. *Eur. J. Org. Chem.* **2002**, 2921.
[727] Jiang, Z. X.; Qin, Y. Y.; Qing, F. L. *J. Org. Chem.* **2003**, *68*, 7544.
[728] Upadhya, T. T.; Nikalje, M. D.; Sudalai, A. *Tetrahedron Lett.* **2001**, *42*, 4891.
[729] Sayyed, I. A.; Sudalai, A. *Tetrahedron: Asymmetry* **2004**, *15*, 3111.
[730] Xiong, C.; Wang, W.; Hruby, V. R. *J. Org. Chem.* **2002**, *67*, 3514.
[731] Fernandes, R. A.; Kumar, P. *Tetrahedron* **2002**, *58*, 6685.
[732] García, C.; Martín, T.; Martín, V. S. *J. Org. Chem.* **2001**, *66*, 1420.
[733] Evans, P.; Leffray, M. *Tetrahedron* **2003**, *59*, 7973.
[734] Audouard, C.; Barsukov, I.; Fawcett, J.; Griffith, G. A.; Percy, J. M.; Pintat, S.; Smith, C. A. *Chem. Commun.* **2004**, 1526.
[735] Periasamy, M.; Ramanathan, C. R.; Kumar, N.; Thirumalaikumar, M. *J. Chem. Res. (S)* **2001**, 512.
[736] Bieging, A.; Liao, L. X.; McGrath, D. V. *Chirality* **2002**, *14*, 258.
[737] Hodgson, R.; Nelson, A. *Org. Biomol. Chem.* **2004**, *2*, 373.
[738] Chun, J.; Byun, H. S.; Bittman, R. *J. Org. Chem.* **2003**, *68*, 348.
[739] Bodas, M. S.; Kumar, P. *Tetrahedron Lett.* **2004**, *45*, 8461.

[740] Skarżewski, J.; Wojaczyńska, E.; Turowska-Tyrk, I. *Tetrahedron: Asymmetry* **2002**, *13*, 369.
[741] Xiong, C.; Wang, W.; Cai, C.; Hruby, V. J. *J. Org. Chem.* **2002**, *67*, 1399.
[742] Fernandes, R. A.; Bodas, M. S.; Kumar, P. *Tetrahedron* **2002**, *58*, 1223.
[743] Kobayashi, Y.; William, A. D.; Tokoro, Y. *J. Org. Chem.* **2001**, *66*, 7903.
[744] Dhavale, D.; Markad, S. D.; Karanjule, N. S.; Prakasha-Reddy, J. *J. Org. Chem.* **2004**, *69*, 4760.
[745] Gurjar, M. K.; Cherian, J.; Ramana, C. V. *Org. Lett.* **2004**, *6*, 317.
[746] Carmona, A. T.; Fuentes, J.; Robina, I.; Garcia, E. R.; Demange, R.; Vogel, P.; Winters, A. L. *J. Org. Chem.* **2003**, *68*, 3874.
[747] Ohzeki, T.; Mori, K. *Biosci., Biotechnol., Biochem.*. **2003**, *67*, 2584.
[748] Li, L.; Chan, T. H. *Org. Lett.* **2001**, *3*, 739.
[749] Andreana, P. R.; McLellan, J. S.; Chen, Y.; Wang, P. G. *Org. Lett.* **2002**, *4*, 3875.
[750] Lyle, M. P. A.; Wilson, P. D. *Org. Lett.* **2004**, *6*, 855.
[751] Kousaka, T.; Mori, K. *Biosci., Biotechnol., Biochem.* **2002**, *66*, 697.
[752] Marcune, B. F.; Karady, S.; Reider, P. J.; Miller, R. A.; Biba, M.; DiMichele, L.; Reamer, R. A. *J. Org. Chem.* **2003**, *68*, 8088.
[753] Gonzalez, J.; Aurigemma, C.; Truesdale, L. *Org. Synth.* **2002**, *79*, 93.
[754] Miyashita, K.; Ikejiri, M.; Kawasaki, H.; Maemura, S.; Imanishi, T. *J. Am. Chem. Soc.* **2003**, *125*, 8238.
[755] Zhou, X. D.; Cai, F.; Zhou, W. S. *Tetrahedron Lett.* **2001**, *42*, 2537.
[756] Garaas, S. D.; Hunter, T. J.; O'Doherty, G. A. *J. Org. Chem.* **2002**, *67*, 2682.
[757] Harcken, C.; Brückner, R. *Tetrahedron Lett.* **2001**, *42*, 3967.
[758] Han, H.; Sinha, M. K.; D'Souza, L. J.; Keinan, E.; Sinha, S. C. *Chem. Eur. J.* **2004**, *10*, 2149.
[759] Burke, S. D.; Jiang, L. *Org. Lett.* **2001**, *3*, 1953.
[760] Hodgson, R.; Majid, T.; Nelson, A. *J. Chem. Soc., Perkin Trans. 1* **2002**, 1631.
[761] Bayer, A.; Svendsen, J. S. *Eur. J. Org. Chem.* **2001**, 1769.
[762] Watanabe, T.; Noguchi, T.; Yokota, T.; Shibata, K.; Koshino, H.; Seto, H.; Kim, S.; Takatsuto, S. *Phytochemistry* **2001**, *58*, 343.
[763] Fleming, S. A.; Carroll, S. M.; Hirschi, J.; Liu, R.; Pace, J. L.; Redd, J. T. *Tetrahedron Lett.* **2004**, *45*, 3341.
[764] Dai, W. M.; Zhang, Y.; Zhang, Y. *Tetrahedron: Asymmetry* **2004**, *15*, 525.

CUMULATIVE CHAPTER TITLES BY VOLUME

Volume 1 (1942)

1. **The Reformatsky Reaction**: Ralph L. Shriner
2. **The Arndt-Eistert Reaction**: W. E. Bachmann and W. S. Struve
3. **Chloromethylation of Aromatic Compounds**: Reynold C. Fuson and C. H. McKeever
4. **The Amination of Heterocyclic Bases by Alkali Amides**: Marlin T. Leffler
5. **The Bucherer Reaction**: Nathan L. Drake
6. **The Elbs Reaction**: Louis F. Fieser
7. **The Clemmensen Reduction**: Elmore L. Martin
8. **The Perkin Reaction and Related Reactions**: John R. Johnson
9. **The Acetoacetic Ester Condensation and Certain Related Reactions**: Charles R. Hauser and Boyd E. Hudson, Jr.
10. **The Mannich Reaction**: F. F. Blicke
11. **The Fries Reaction**: A. H. Blatt
12. **The Jacobson Reaction**: Lee Irvin Smith

Volume 2 (1944)

1. **The Claisen Rearrangement**: D. Stanley Tarbell
2. **The Preparation of Aliphatic Fluorine Compounds**: Albert L. Henne
3. **The Cannizzaro Reaction**: T. A. Geissman
4. **The Formation of Cyclic Ketones by Intramolecular Acylation**: William S. Johnson
5. **Reduction with Aluminum Alkoxides (The Meerwein-Ponndorf-Verley Reduction)**: A. L. Wilds

6. **The Preparation of Unsymmetrical Biaryls by the Diazo Reaction and the Nitrosoacetylamine Reaction**: Werner E. Bachmann and Roger A. Hoffman

7. **Replacement of the Aromatic Primary Amino Group by Hydrogen**: Nathan Kornblum

8. **Periodic Acid Oxidation**: Ernest L. Jackson

9. **The Resolution of Alcohols**: A. W. Ingersoll

10. **The Preparation of Aromatic Arsonic and Arsinic Acids by the Bart, Béchamp, and Rosenmund Reactions**: Cliff S. Hamilton and Jack F. Morgan

Volume 3 (1946)

1. **The Alkylation of Aromatic Compounds by the Friedel-Crafts Method**: Charles C. Price

2. **The Willgerodt Reaction**: Marvin Carmack and M. A. Spielman

3. **Preparation of Ketenes and Ketene Dimers**: W. E. Hanford and John C. Sauer

4. **Direct Sulfonation of Aromatic Hydrocarbons and Their Halogen Derivatives**: C. M. Suter and Arthur W. Weston

5. **Azlactones**: H. E. Carter

6. **Substitution and Addition Reactions of Thiocyanogen**: John L. Wood

7. **The Hofmann Reaction**: Everett L. Wallis and John F. Lane

8. **The Schmidt Reaction**: Hans Wolff

9. **The Curtius Reaction**: Peter A. S. Smith

Volume 4 (1948)

1. **The Diels-Alder Reaction with Maleic Anhydride**: Milton C. Kloetzel

2. **The Diels-Alder Reaction: Ethylenic and Acetylenic Dienophiles**: H. L. Holmes

3. **The Preparation of Amines by Reductive Alkylation**: William S. Emerson

4. **The Acyloins**: S. M. McElvain

5. **The Synthesis of Benzoins**: Walter S. Ide and Johannes S. Buck

6. **Synthesis of Benzoquinones by Oxidation**: James Cason

7. **The Rosenmund Reduction of Acid Chlorides to Aldehydes**: Erich Mosettig and Ralph Mozingo

8. **The Wolff-Kishner Reduction**: David Todd

Volume 5 (1949)

1. **The Synthesis of Acetylenes**: Thomas L. Jacobs
2. **Cyanoethylation**: Herman L. Bruson
3. **The Diels-Alder Reaction: Quinones and Other Cyclenones**: Lewis L. Butz and Anton W. Rytina
4. **Preparation of Aromatic Fluorine Compounds from Diazonium Fluoborates: The Schiemann Reaction**: Arthur Roe
5. **The Friedel and Crafts Reaction with Aliphatic Dibasic Acid Anhydrides**: Ernst Berliner
6. **The Gattermann-Koch Reaction**: Nathan N. Crounse
7. **The Leuckart Reaction**: Maurice L. Moore
8. **Selenium Dioxide Oxidation**: Norman Rabjohn
9. **The Hoesch Synthesis**: Paul E. Spoerri and Adrien S. DuBois
10. **The Darzens Glycidic Ester Condensation**: Melvin S. Newman and Barney J. Magerlein

Volume 6 (1951)

1. **The Stobbe Condensation**: William S. Johnson and Guido H. Daub
2. **The Preparation of 3,4-Dihydroisoquinolines and Related Compounds by the Bischler-Napieralski Reaction**: Wilson M. Whaley and Tutucorin R. Govindachari
3. **The Pictet-Spengler Synthesis of Tetrahydroisoquinolines and Related Compounds**: Wilson M. Whaley and Tutucorin R. Govindachari
4. **The Synthesis of Isoquinolines by the Pomeranz-Fritsch Reaction**: Walter J. Gensler
5. **The Oppenauer Oxidation**: Carl Djerassi
6. **The Synthesis of Phosphonic and Phosphinic Acids**: Gennady M. Kosolapoff
7. **The Halogen-Metal Interconversion Reaction with Organolithium Compounds**: Reuben G. Jones and Henry Gilman
8. **The Preparation of Thiazoles**: Richard H. Wiley, D. C. England, and Lyell C. Behr
9. **The Preparation of Thiophenes and Tetrahydrothiophenes**: Donald E. Wolf and Karl Folkers
10. **Reductions by Lithium Aluminum Hydride**: Weldon G. Brown

Volume 7 (1953)

1. **The Pechmann Reaction**: Suresh Sethna and Ragini Phadke

2. **The Skraup Synthesis of Quinolines**: R. H. F. Manske and Marshall Kulka

3. **Carbon-Carbon Alkylations with Amines and Ammonium Salts**: James H. Brewster and Ernest L. Eliel

4. **The von Braun Cyanogen Bromide Reaction**: Howard A. Hageman

5. **Hydrogenolysis of Benzyl Groups Attached to Oxygen, Nitrogen, or Sulfur**: Walter H. Hartung and Robert Simonoff

6. **The Nitrosation of Aliphatic Carbon Atoms**: Oscar Touster

7. **Epoxidation and Hydroxylation of Ethylenic Compounds with Organic Peracids**: Daniel Swern

Volume 8 (1954)

1. **Catalytic Hydrogenation of Esters to Alcohols**: Homer Adkins

2. **The Synthesis of Ketones from Acid Halides and Organometallic Compounds of Magnesium, Zinc, and Cadmium**: David A. Shirley

3. **The Acylation of Ketones to Form β-Diketones or β-Keto Aldehydes**: Charles R. Hauser, Frederic W. Swamer, and Joe T. Adams

4. **The Sommelet Reaction**: S. J. Angyal

5. **The Synthesis of Aldehydes from Carboxylic Acids**: Erich Mosettig

6. **The Metalation Reaction with Organolithium Compounds**: Henry Gilman and John W. Morton, Jr.

7. **β-Lactones**: Harold E. Zaugg

8. **The Reaction of Diazomethane and Its Derivatives with Aldehydes and Ketones**: C. David Gutsche

Volume 9 (1957)

1. **The Cleavage of Non-enolizable Ketones with Sodium Amide**: K. E. Hamlin and Arthur W. Weston

2. **The Gattermann Synthesis of Aldehydes**: William E. Truce

3. **The Baeyer-Villiger Oxidation of Aldehydes and Ketones**: C. H. Hassall

4. **The Alkylation of Esters and Nitriles**: Arthur C. Cope, H. L. Holmes, and Herbert O. House

5. **The Reaction of Halogens with Silver Salts of Carboxylic Acids**: C. V. Wilson

6. **The Synthesis of β-Lactams**: John C. Sheehan and Elias J. Corey

7. **The Pschorr Synthesis and Related Diazonium Ring Closure Reactions**: DeLos F. DeTar

Volume 10 (1959)

1. **The Coupling of Diazonium Salts with Aliphatic Carbon Atoms**: Stanley J. Parmerter

2. **The Japp-Klingemann Reaction**: Robert R. Phillips

3. **The Michael Reaction**: Ernst D. Bergmann, David Ginsburg, and Raphael Pappo

Volume 11 (1960)

1. **The Beckmann Rearrangement**: L. Guy Donaruma and Walter Z. Heldt

2. **The Demjanov and Tiffeneau-Demjanov Ring Expansions**: Peter A. S. Smith and Donald R. Baer

3. **Arylation of Unsaturated Compounds by Diazonium Salts**: Christian S. Rondestvedt, Jr.

4. **The Favorskii Rearrangement of Haloketones**: Andrew S. Kende

5. **Olefins from Amines: The Hofmann Elimination Reaction and Amine Oxide Pyrolysis**: Arthur C. Cope and Elmer R. Trumbull

Volume 12 (1962)

1. **Cyclobutane Derivatives from Thermal Cycloaddition Reactions**: John D. Roberts and Clay M. Sharts

2. **The Preparation of Olefins by the Pyrolysis of Xanthates. The Chugaev Reaction**: Harold R. Nace

3. **The Synthesis of Aliphatic and Alicyclic Nitro Compounds**: Nathan Kornblum

4. **Synthesis of Peptides with Mixed Anhydrides**: Noel F. Albertson

5. **Desulfurization with Raney Nickel**: George R. Pettit and Eugene E. van Tamelen

Volume 13 (1963)

1. **Hydration of Olefins, Dienes, and Acetylenes via Hydroboration**: George Zweifel and Herbert C. Brown

2. **Halocyclopropanes from Halocarbenes**: William E. Parham and Edward E. Schweizer

3. **Free Radical Addition to Olefins to Form Carbon-Carbon Bonds**: Cheves Walling and Earl S. Huyser

4. **Formation of Carbon-Heteroatom Bonds by Free Radical Chain Additions to Carbon-Carbon Multiple Bonds**: F. W. Stacey and J. F. Harris, Jr.

Volume 14 (1965)

1. **The Chapman Rearrangement**: J. W. Schulenberg and S. Archer

2. **α-Amidoalkylations at Carbon**: Harold E. Zaugg and William B. Martin

3. **The Wittig Reaction**: Adalbert Maercker

Volume 15 (1967)

1. **The Dieckmann Condensation**: John P. Schaefer and Jordan J. Bloomfield

2. **The Knoevenagel Condensation**: G. Jones

Volume 16 (1968)

1. **The Aldol Condensation**: Arnold T. Nielsen and William J. Houlihan

Volume 17 (1969)

1. **The Synthesis of Substituted Ferrocenes and Other π-Cyclopentadienyl-Transition Metal Compounds**: Donald E. Bublitz and Kenneth L. Rinehart, Jr.

2. **The γ-Alkylation and γ-Arylation of Dianions of β-Dicarbonyl Compounds**: Thomas M. Harris and Constance M. Harris

3. **The Ritter Reaction**: L. I. Krimen and Donald J. Cota

Volume 18 (1970)

1. **Preparation of Ketones from the Reaction of Organolithium Reagents with Carboxylic Acids**: Margaret J. Jorgenson

2. **The Smiles and Related Rearrangements of Aromatic Systems**: W. E. Truce, Eunice M. Kreider, and William W. Brand

3. **The Reactions of Diazoacetic Esters with Alkenes, Alkynes, Heterocyclic, and Aromatic Compounds**: Vinod Dave and E. W. Warnhoff

4. **The Base-Promoted Rearrangements of Quaternary Ammonium Salts**: Stanley H. Pine

Volume 19 (1972)

1. **Conjugate Addition Reactions of Organocopper Reagents**: Gary H. Posner
2. **Formation of Carbon-Carbon Bonds via π-Allylnickel Compounds**: Martin F. Semmelhack
3. **The Thiele-Winter Acetoxylation of Quinones**: J. F. W. McOmie and J. M. Blatchly
4. **Oxidative Decarboxylation of Acids by Lead Tetraacetate**: Roger A. Sheldon and Jay K. Kochi

Volume 20 (1973)

1. **Cyclopropanes from Unsaturated Compounds, Methylene Iodide, and Zinc-Copper Couple**: H. E. Simmons, T. L. Cairns, Susan A. Vladuchick, and Connie M. Hoiness
2. **Sensitized Photooxygenation of Olefins**: R. W. Denny and A. Nickon
3. **The Synthesis of 5-Hydroxyindoles by the Nenitzescu Reaction**: George R. Allen, Jr.
4. **The Zinin Reaction of Nitroarenes**: H. K. Porter

Volume 21 (1974)

1. **Fluorination with Sulfur Tetrafluoride**: G. A. Boswell, Jr., W. C. Ripka, R. M. Scribner, and C. W. Tullock
2. **Modern Methods to Prepare Monofluoroaliphatic Compounds**: Clay M. Sharts and William A. Sheppard

Volume 22 (1975)

1. **The Claisen and Cope Rearrangements**: Sara Jane Rhoads and N. Rebecca Raulins
2. **Substitution Reactions Using Organocopper Reagents**: Gary H. Posner
3. **Clemmensen Reduction of Ketones in Anhydrous Organic Solvents**: E. Vedejs
4. **The Reformatsky Reaction**: Michael W. Rathke

Volume 23 (1976)

1. **Reduction and Related Reactions of α,β-Unsaturated Compounds with Metals in Liquid Ammonia**: Drury Caine
2. **The Acyloin Condensation**: Jordan J. Bloomfield, Dennis C. Owsley, and Janice M. Nelke
3. **Alkenes from Tosylhydrazones**: Robert H. Shapiro

Volume 24 (1976)

1. **Homogeneous Hydrogenation Catalysts in Organic Solvents**: Arthur J. Birch and David H. Williamson

2. **Ester Cleavages via S_N2-Type Dealkylation**: John E. McMurry

3. **Arylation of Unsaturated Compounds by Diazonium Salts (The Meerwein Arylation Reaction)**: Christian S. Rondestvedt, Jr.

4. **Selenium Dioxide Oxidation**: Norman Rabjohn

Volume 25 (1977)

1. **The Ramberg-Bäcklund Rearrangement**: Leo A. Paquette

2. **Synthetic Applications of Phosphoryl-Stabilized Anions**: William S. Wadsworth, Jr.

3. **Hydrocyanation of Conjugated Carbonyl Compounds**: Wataru Nagata and Mitsuru Yoshioka

Volume 26 (1979)

1. **Heteroatom-Facilitated Lithiations**: Heinz W. Gschwend and Herman R. Rodriguez

2. **Intramolecular Reactions of Diazocarbonyl Compounds**: Steven D. Burke and Paul A. Grieco

Volume 27 (1982)

1. **Allylic and Benzylic Carbanions Substituted by Heteroatoms**: Jean-François Biellmann and Jean-Bernard Ducep

2. **Palladium-Catalyzed Vinylation of Organic Halides**: Richard F. Heck

Volume 28 (1982)

1. **The Reimer-Tiemann Reaction**: Hans Wynberg and Egbert W. Meijer

2. **The Friedländer Synthesis of Quinolines**: Chia-Chung Cheng and Shou-Jen Yan

3. **The Directed Aldol Reaction**: Teruaki Mukaiyama

Volume 29 (1983)

1. **Replacement of Alcoholic Hydroxy Groups by Halogens and Other Nucleophiles via Oxyphosphonium Intermediates**: Bertrand R. Castro

2. **Reductive Dehalogenation of Polyhalo Ketones with Low-Valent Metals and Related Reducing Agents**: Ryoji Noyori and Yoshihiro Hayakawa

3. **Base-Promoted Isomerizations of Epoxides**: Jack K. Crandall and Marcel Apparu

Volume 30 (1984)

1. **Photocyclization of Stilbenes and Related Molecules**: Frank B. Mallory and Clelia W. Mallory

2. **Olefin Synthesis via Deoxygenation of Vicinal Diols**: Eric Block

Volume 31 (1984)

1. **Addition and Substitution Reactions of Nitrile-Stabilized Carbanions**: Siméon Arseniyadis, Keith S. Kyler, and David S. Watt

Volume 32 (1984)

1. **The Intramolecular Diels-Alder Reaction**: Engelbert Ciganek

2. **Synthesis Using Alkyne-Derived Alkenyl- and Alkynylaluminum Compounds**: George Zweifel and Joseph A. Miller

Volume 33 (1985)

1. **Formation of Carbon-Carbon and Carbon-Heteroatom Bonds via Organoboranes and Organoborates**: Ei-Ichi Negishi and Michael J. Idacavage

2. **The Vinylcyclopropane-Cyclopentene Rearrangement**: Tomáš Hudlický, Toni M. Kutchan, and Saiyid M. Naqvi

Volume 34 (1985)

1. **Reductions by Metal Alkoxyaluminum Hydrides**: Jaroslav Málek

2. **Fluorination by Sulfur Tetrafluoride**: Chia-Lin J. Wang

Volume 35 (1988)

1. **The Beckmann Reactions: Rearrangements, Elimination-Additions, Fragmentations, and Rearrangement-Cyclizations**: Robert E. Gawley

2. **The Persulfate Oxidation of Phenols and Arylamines (The Elbs and the Boyland-Sims Oxidations)**: E. J. Behrman

3. **Fluorination with Diethylaminosulfur Trifluoride and Related Aminofluorosulfuranes**: Miloš Hudlický

Volume 36 (1988)

1. **The [3 + 2] Nitrone-Olefin Cycloaddition Reaction**: Pat N. Confalone and Edward M. Huie

2. **Phosphorus Addition at sp^2 Carbon**: Robert Engel

3. **Reduction by Metal Alkoxyaluminum Hydrides. Part II. Carboxylic Acids and Derivatives, Nitrogen Compounds, and Sulfur Compounds**: Jaroslav Málek

Volume 37 (1989)

1. **Chiral Synthons by Ester Hydrolysis Catalyzed by Pig Liver Esterase**: Masaji Ohno and Masami Otsuka

2. **The Electrophilic Substitution of Allylsilanes and Vinylsilanes**: Ian Fleming, Jacques Dunoguès, and Roger Smithers

Volume 38 (1990)

1. **The Peterson Olefination Reaction**: David J. Ager

2. **Tandem Vicinal Difunctionalization: β-Addition to α,β-Unsaturated Carbonyl Substrates Followed by α-Functionalization**: Marc J. Chapdelaine and Martin Hulce

3. **The Nef Reaction**: Harold W. Pinnick

Volume 39 (1990)

1. **Lithioalkenes from Arenesulfonylhydrazones**: A. Richard Chamberlin and Steven H. Bloom

2. **The Polonovski Reaction**: David Grierson

3. **Oxidation of Alcohols to Carbonyl Compounds via Alkoxysulfonium Ylides: The Moffatt, Swern, and Related Oxidations**: Thomas T. Tidwell

Volume 40 (1991)

1. **The Pauson-Khand Cycloaddition Reaction for Synthesis of Cyclopentenones**: Neil E. Schore

2. **Reduction with Diimide**: Daniel J. Pasto and Richard T. Taylor

3. **The Pummerer Reaction of Sulfinyl Compounds**: Ottorino DeLucchi, Umberto Miotti, and Giorgio Modena

4. **The Catalyzed Nucleophilic Addition of Aldehydes to Electrophilic Double Bonds**: Hermann Stetter and Heinrich Kuhlmann

Volume 41 (1992)

1. **Divinylcyclopropane-Cycloheptadiene Rearrangement**: Tomáš Hudlický, Rulin Fan, Josephine W. Reed, and Kumar G. Gadamasetti

2. **Organocopper Reagents: Substitution, Conjugate Addition, Carbo/Metallocupration, and Other Reactions**: Bruce H. Lipshutz and Saumitra Sengupta

Volume 42 (1992)

1. **The Birch Reduction of Aromatic Compounds**: Peter W. Rabideau and Zbigniew Marcinow

2. **The Mitsunobu Reaction**: David L. Hughes

Volume 43 (1993)

1. **Carbonyl Methylenation and Alkylidenation Using Titanium-Based Reagents**: Stanley H. Pine

2. **Anion-Assisted Sigmatropic Rearrangements**: Stephen R. Wilson

3. **The Baeyer-Villiger Oxidation of Ketones and Aldehydes**: Grant R. Krow

Volume 44 (1993)

1. **Preparation of α,β-Unsaturated Carbonyl Compounds and Nitriles by Selenoxide Elimination**: Hans J. Reich and Susan Wollowitz

2. **Enone Olefin [2 + 2] Photochemical Cyclizations**: Michael T. Crimmins and Tracy L. Reinhold

Volume 45 (1994)

1. **The Nazarov Cyclization**: Karl L. Habermas, Scott E. Denmark, and Todd K. Jones

2. **Ketene Cycloadditions**: John A. Hyatt and Peter W. Raynolds

Volume 46 (1994)

1. **Tin(II) Enolates in the Aldol, Michael, and Related Reactions**: Teruaki Mukaiyama and Shū Kobayashi

2. **The [2,3]-Wittig Reaction**: Takeshi Nakai and Koichi Mikami

3. **Reductions with Samarium(II) Iodide**: Gary A. Molander

Volume 47 (1995)

1. **Lateral Lithiation Reactions Promoted by Heteroatomic Substituents**: Robin D. Clark and Alam Jahangir

2. **The Intramolecular Michael Reaction**: R. Daniel Little, Mohammad R. Masjedizadeh, Olof Wallquist (in part), and Jim I. McLoughlin (in part)

Volume 48 (1996)

1. **Asymmetric Epoxidation of Allylic Alcohols: The Katsuki–Sharpless Epoxidation Reaction**: Tsutomu Katsuki and Victor S. Martin

2. **Radical Cyclization Reactions**: B. Giese, B. Kopping, T. Göbel, J. Dickhaut, G. Thoma, K. J. Kulicke, and F. Trach

Volume 49 (1997)

1. **The Vilsmeier Reaction of Fully Conjugated Carbocycles and Heterocycles**: Gurnos Jones and Stephen P. Stanforth

2. **[6 + 4] Cycloaddition Reactions**: James H. Rigby

3. **Carbon–Carbon Bond-Forming Reactions Promoted by Trivalent Manganese**: Gagik G. Melikyan

Volume 50 (1997)

1. **The Stille Reaction**: Vittorio Farina, Venkat Krishnamurthy, and William J. Scott

Volume 51 (1997)

1. **Asymmetric Aldol Reactions Using Boron Enolates**: Cameron J. Cowden and Ian Paterson

2. **The Catalyzed α-Hydroxylation and α-Aminoalkylation of Activated Olefins (The Morita–Baylis–Hillman Reaction)**: Engelbert Ciganek

3. **[4 + 3] Cycloaddition Reactions**: James H. Rigby and F. Christopher Pigge

Volume 52 (1998)

1. **The Retro–Diels–Alder Reaction. Part I. C—C Dienophiles**: Bruce Rickborn

2. **Enantioselective Reduction of Ketones**: Shinichi Itsuno

Volume 53 (1998)

1. **The Oxidation of Alcohols by Modified Oxochromium(VI)-Amine Reagents**: Frederick A. Luzzio

2. **The Retro–Diels–Alder Reaction. Part II. Dienophiles with One or More Heteroatoms**: Bruce Rickborn

Volume 54 (1999)

1. **Aromatic Substitution by the $S_{RN}1$ Reaction**: Roberto Rossi, Adriana B. Pierini, and Ana N. Santiago

2. **Oxidation of Carbonyl Compounds with Organohypervalent Iodine Reagents**: Robert M. Moriarty and Om Prakash

Volume 55 (1999)

1. **Synthesis of Nucleosides**: Helmut Vorbrüggen and Carmen Ruh-Pohlenz

Volume 56 (2000)

1. **The Hydroformylation Reaction**: Iwao Ojima, Chung-Ying Tsai, Maria Tzamarioudaki, and Dominique Bonafoux

2. **The Vilsmeier Reaction. 2. Reactions with Compounds Other Than Fully Conjugated Carbocycles and Heterocycles**: Gurnos Jones and Stephen P. Stanforth

Volume 57 (2001)

1. **Intermolecular Metal-Catalyzed Carbenoid Cyclopropanations:** Huw M. L. Davies and Evan G. Antoulinakis

2. **Oxidation of Phenolic Compounds with Organohypervalent Iodine Reagents:** Robert M. Moriarty and Om Prakash

3. **Synthetic Uses of Tosylmethyl Isocyanide (TosMIC):** Daan van Leusen and Albert M. van Leusen

Volume 58 (2001)

1. **Simmons-Smith Cyclopropanation Reaction:** André B. Charette and André Beauchemin

2. **Preparation and Applications of Functionalized Organozine Compounds:** Paul Knochel, Nicolas Millot, Alain L. Rodriguez, and Charles E. Tucker

Volume 59 (2002)

1. **Reductive Aminations of Carbonyl Compounds with Borohydride and Borane Reducing Agents:** Ellen W. Baxter and Allen B. Reitz

Volume 60 (2002)

1. **Epoxide Migration (Payne Rearrangement) and Related Reactions:** Robert M. Hanson

2. **The Intramolecular Heck Reaction:** J. T. Link

Volume 61 (2002)

1. **[3 + 2] Cycloaddition of Trimethylenemethane and its Synthetic Equivalents:** Shigeru Yamago and Eiichi Nakamura

2. **Dioxirane Epoxidation of Alkenes:** Waldemar Adam, Chantu R. Saha-Möller, and Cong-Gui Zhao

Volume 62 (2003)

1. **The α-Hydroxylation of Enolates and Silyl Enol Ethers**: Bang-Chi Chen, Ping Zhou, Franklin A. Davis, and Engelbert Ciganek

2. **The Ramberg-Bäcklund Reaction**: Richard J. K. Taylor and Guy Casy

3. **The α-Hydroxy Ketone (α-Ketol) and Related Rearrangements**: Leo A. Paquette and John E. Hofferberth

4. **Transformation of Glycals Into 2,3-Unsaturated Glycosyl Derivatives**: Robert J. Ferrier and Oleg A. Zubkov

Volume 63 (2004)

1. **The Biginelli Dihydropyrimidine Synthesis:** C. Oliver Kappe and Alexander Stadler

2. **Microbial Arene Oxidations:** Roy A. Johnson

3. **Cu, Ni, and Pd Mediated Homocoupling Reactions in Biaryl Syntheses: The Ullmann Reaction:** Todd D. Nelson and R. David Crouch

Volume 64 (2004)

1. **Additions of Allyl, Allenyl, and Propargylstannanes to Aldehydes and Imines:** Benjamin W. Gung

2. **Glycosylation with Sulfoxides and Sulfinates as Donors or Promoters:** David Crich and Linda B. L. Lim

3. **Addition of Organochromium Reagents to Carbonyl Compounds:** Kazuhiko Takai

Volume 65 (2005)

1. **The Passerini Reaction:** Luca Banfi and Renata Riva

2. **Diels-Alder Reactions of Imino Dienophiles:** Geoffrey R. Heintzelman, Ivona R. Meigh, Yogesh R. Mahajan, and Steven M. Weinreb

AUTHOR INDEX, VOLUMES 1–66

Volume number only is designated in this index.

Adam, Waldemar, 61
Adams, Joe T., 8
Adkins, Homer, 8
Ager, David J., 38
Albertson, Noel F., 12
Allen, George R., Jr., 20
Angyal, S. J., 8
Antoulinkis, Evan G., 57
Apparu, Marcel, 29
Archer, S., 14
Arseniyadis, Siméon, 31

Bachmann, W. E., 1, 2
Baer, Donald R., 11
Banfi, Luca, 65
Baxter, Ellen W., 59
Beauchemin, André, 58
Behr, Lyell C., 6
Behrman, E. J., 35
Bergmann, Ernst D., 10
Berliner, Ernst, 5
Biellmann, Jean-François, 27
Birch, Arthur J., 24
Blatchly, J. M., 19
Blatt, A. H., 1
Blicke, F. F., 1
Block, Eric, 30
Bloom, Steven H., 39
Bloomfield, Jordan J., 15, 23
Bonafoux, Dominique, 56
Boswell, G. A., Jr., 21
Brand, William W., 18
Brewster, James H., 7
Brown, Herbert C., 13
Brown, Weldon G., 6
Bruson, Herman Alexander, 5
Bublitz, Donald E., 17
Buck, Johannes S., 4
Burke, Steven D., 26
Butz, Lewis W., 5

Caine, Drury, 23
Cairns, Theodore L., 20
Carmack, Marvin, 3
Carpenter, Nancy E., 66
Carter, H. E., 3
Cason, James, 4
Castro, Bertrand R., 29
Casy, Guy, 62
Chamberlin, A. Richard, 39
Chapdelaine, Marc J., 38
Charette, André B., 58
Chen, Bang-Chi, 62
Cheng, Chia-Chung, 28
Ciganek, Engelbert, 32, 51, 62
Clark, Robin D., 47
Confalone, Pat N., 36
Cope, Arthur C., 9, 11
Corcy, Elias J., 9
Cota, Donald J., 17
Cowden, Cameron J., 51
Crandall, Jack K., 29
Crich, David, 64
Crimmins, Michael T., 44
Crouch, R. David, 63
Crounse, Nathan N., 5

Daub, Guido H., 6
Dave, Vinod, 18
Davies, Huw M. L., 57
Davis, Franklin A., 62
Denmark, Scott E., 45
Denny, R. W., 20
DeLucchi, Ottorino, 40
DeTar, DeLos F., 9
Dickhaut, J., 48
Djerassi, Carl, 6
Donaruma, L. Guy, 11
Drake, Nathan L., 1
DuBois, Adrien S., 5
Ducep, Jean-Bernard, 27
Dunoguès, Jacques, 37

Eliel, Ernest L., 7
Emerson, William S., 4
Engel, Robert, 36

England, D. C., 6

Fan, Rulin, 41
Farina, Vittorio, 50
Ferrier, Robert J., 62
Fieser, Louis F., 1
Fleming, Ian, 37
Folkers, Karl, 6
Fuson, Reynold C., 1

Gadamasetti, Kumar G., 41
Gawley, Robert E., 35
Geissman, T. A., 2
Gensler, Walter J., 6
Giese, B., 48
Gilman, Henry, 6, 8
Ginsburg, David, 10
Göbel, T., 48
Govindachari, Tuticorin R., 6
Grieco, Paul A., 26
Grierson, David, 39
Gschwend, Heinz W., 26
Gung, Benjamin W., 64
Gutsche, C. David, 8

Habermas, Karl L., 45
Hageman, Howard A., 7
Hamilton, Cliff S., 2
Hamlin, K. E., 9
Hanford, W. E., 3
Hanson, Robert M., 60
Harris, Constance M., 17
Harris, J. F., Jr., 13
Harris, Thomas M., 17
Hartung, Walter H., 7
Hassall, C. H., 9
Hauser, Charles R., 1, 8
Hayakawa, Yoshihiro, 29
Heck, Richard F., 27
Heldt, Walter Z., 11
Heintzelman, Geoffrey R., 65
Henne, Albert L., 2
Hofferberth, John E., 62
Hoffman, Roger A., 2
Hoiness, Connie M., 20
Holmes, H. L., 4, 9
Houlihan, William J., 16
House, Herbert O., 9
Hudlický, Miloš, 35
Hudlický, Tomáš, 33, 41
Hudson, Boyd E., Jr., 1
Hughes, David L., 42
Huie, E. M., 36
Hulce, Martin, 38

Huyser, Earl S., 13
Hyatt, John A., 45

Idacavage, Michael J., 33
Ide, Walter S., 4
Ingersoll, A. W., 2
Itsuno, Shinichi, 52

Jackson, Ernest L., 2
Jacobs, Thomas L., 5
Jahangir, Alam, 47
Johnson, John R., 1
Johnson, Roy A., 63
Johnson, William S., 2, 6
Jones, Gurnos, 15, 49, 56
Jones, Reuben G., 6
Jones, Todd K., 45
Jorgenson, Margaret J., 18

Kappe, C. Oliver, 63
Katsuki, Tsutomu, 48
Kende, Andrew S., 11
Kloetzel, Milton C., 4
Knochel, Paul, 58
Kobayashi, Shū, 46
Kochi, Jay K., 19
Kopping, B., 48
Kornblum, Nathan, 2, 12
Kosolapoff, Gennady M., 6
Kreider, Eunice M., 18
Krimen, L. I., 17
Krishnamurthy, Venkat, 50
Krow, Grant R., 43
Kuhlmann, Heinrich, 40
Kulicke, K. J., 48
Kulka, Marshall, 7
Kutchan, Toni M., 33
Kyler, Keith S., 31

Lane, John F., 3
Leffler, Marlin T., 1
Letavic, Michael A., 66
Lim, Linda B. L., 64
Link, J. T., 60
Little, R. Daniel, 47
Lipshutz, Bruce H., 41
Luzzio, Frederick A., 53

McCombie, Stuart W., 66
McElvain, S. M., 4
McKeever, C. H., 1
McLoughlin, Jim I., 47
McMurry, John E., 24
McOmie, J. F. W., 19

Maercker, Adalbert, 14
Magerlein, Barney J., 5
Mahajan, Yogesh R., 65
Málek, Jaroslav, 34, 36
Mallory, Clelia W., 30
Mallory, Frank B., 30
Manske, Richard H. F., 7
Marcinow, Zbigniew, 42
Martin, Elmore L., 1
Martin, Victor S., 48
Martin, William B., 14
Masjedizadeh, Mohammad R., 47
Meigh, Ivona R., 65
Meijer, Egbert W., 28
Melikyan, Gagik G., 49
Mikami, Koichi, 46
Miller, Joseph A., 32
Millot, Nicolas, 58
Miotti, Umberto, 40
Modena, Giorgio, 40
Molander, Gary, 46
Moore, Maurice L., 5
Morgan, Jack F., 2
Moriarty, Robert M., 54, 57
Morton, John W., Jr., 8
Mosettig, Erich, 4, 8
Mozingo, Ralph, 4
Mukaiyama, Teruaki, 28, 46

Nace, Harold R., 12
Nagata, Wataru, 25
Nakai, Takeshi, 46
Nakamura, Eiichi, 61
Naqvi, Saiyid M., 33
Negishi, Ei-Ichi, 33
Nelke, Janice M., 23
Nelson, Todd D., 63
Newman, Melvin S., 5
Nickon, A., 20
Nielsen, Arnold T., 16
Noe, Mark C., 66
Noyori, Ryoji, 29

Ohno, Masaji, 37
Ojima, Iwao, 56
Otsuka, Masami, 37
Overman, Larry E., 66
Owsley, Dennis C., 23

Pappo, Raphael, 10
Paquette, Leo A., 25, 62
Parham, William E., 13
Parmerter, Stanley M., 10
Pasto, Daniel J., 40
Paterson, Ian, 51
Pettit, George R., 12
Phadke, Ragini, 7
Phillips, Robert R., 10
Pierini, Adriana B., 54
Pigge, F. Christopher, 51
Pine, Stanley H., 18, 43
Pinnick, Harold W., 38
Porter, H. K., 20
Posner, Gary H., 19, 22
Prakash, Om, 54, 57
Price, Charles C., 3

Rabideau, Peter W., 42
Rabjohn, Norman, 5, 24
Rathke, Michael W., 22
Raulins, N. Rebecca, 22
Raynolds, Peter W., 45
Reed, Josephine W., 41
Reich, Hans J., 44
Reinhold, Tracy L., 44
Reitz, Allen B., 59
Rhoads, Sara Jane, 22
Rickborn, Bruce, 52, 53
Rigby, James H., 49, 51
Rinehart, Kenneth L., Jr., 17
Ripka, W. C., 21
Riva, Renata, 65
Roberts, John D., 12
Rodriguez, Alain L., 58
Rodriguez, Herman R., 26
Roe, Arthur, 5
Rondestvedt, Christian S., Jr., 11, 24
Rossi, Roberto A., 54
Ruh-Pohlenz, Carmen, 55
Rytina, Anton W., 5

Saha-Möller, Chantu R., 61
Santiago, Ana N., 54
Sauer, John C., 3
Schaefer, John P., 15
Schore, Neil E., 40
Schulenberg, J. W., 14
Schweizer, Edward E., 13
Scott, William J., 50
Scribner, R. M., 21
Semmelhack, Martin F., 19
Sengupta, Saumitra, 41
Sethna, Suresh, 7
Shapiro, Robert H., 23
Sharts, Clay M., 12, 21
Sheehan, John C., 9
Sheldon, Roger A., 19
Sheppard, W. A., 21

Shirley, David A., 8
Shriner, Ralph L., 1
Simmons, Howard E., 20
Simonoff, Robert, 7
Smith, Lee Irvin, 1
Smith, Peter A. S., 3, 11
Smithers, Roger, 37
Snow, Sheri L., 66
Spielman, M. A., 3
Spoerri, Paul E., 5
Stacey, F. W., 13
Stadler, Alexander, 63
Stanforth, Stephen P., 49, 56
Stetter, Hermann, 40
Struve, W. S., 1
Suter, C. M., 3
Swamer, Frederic W., 8
Swern, Daniel, 7

Takai, Kazuhiko, 64
Tarbell, D. Stanley, 2
Taylor, Richard J. K., 62
Taylor, Richard T., 40
Thoma, G., 48
Tidwell, Thomas T., 39
Todd, David, 4
Touster, Oscar, 7
Trach, F., 48
Truce, William E., 9, 18
Trumbull, Elmer R., 11
Tsai, Chung-Ying, 56
Tucker, Charles, E., 58
Tullock, C. W., 21
Tzamarioudaki, Maria, 56

van Leusen, Albert M., 57
van Leusen, Daan, 57
van Tamelen, Eugene E., 12
Vedejs, E., 22
Vladuchick, Susan A., 20
Vorbrüggen, Helmut, 55

Wadsworth, William S., Jr., 25
Walling, Cheves, 13
Wallis, Everett S., 3
Wallquist, Olof, 47
Wang, Chia-Lin L., 34
Warnhoff, E. W., 18
Watt, David S., 31
Weinreb, Steven M., 65
Weston, Arthur W., 3, 9
Whaley, Wilson M., 6
Wilds, A. L., 2
Wiley, Richard H., 6
Williamson, David H., 24
Wilson, C. V., 9
Wilson, Stephen R., 43
Wolf, Donald E., 6
Wolff, Hans, 3
Wollowitz, Susan, 44
Wood, John L., 3
Wynberg, Hans, 28

Yamago, Shigeru, 61
Yan, Shou-Jen, 28
Yoshioka, Mitsuru, 25

Zaugg, Harold E., 8, 14
Zhao, Cong-Gui, 61
Zhou, Ping, 62
Zubkov, Oleg A., 62
Zweifel, George, 13, 32

CHAPTER AND TOPIC INDEX, VOLUMES 1–66

Many chapters contain brief discussions of reactions and comparisons of alternative synthetic methods related to the reaction that is the subject of the chapter. These related reactions and alternative methods are not usually listed in this index. In this index, the volume number is in **boldface**, the chapter number is in ordinary type.

Acetoacetic ester condensation, **1**, 9
Acetylenes, synthesis of, **5**, 1; **23**, 3; **32**, 2
Acid halides:
  reactions with esters, **1**, 9
  reactions with organometallic compounds, **8**, 2
α-Acylamino acid mixed anhydrides, **12**, 4
α-Acylamino acids, azlactonization of, **3**, 5
Acylation:
  of esters with acid chlorides, **1**, 9
  intramolecular, to form cyclic ketones, **2**, 4; **23**, 2
  of ketones to form diketones, **8**, 3
Acyl fluorides, synthesis of, **21**, 1; **34**, 2; **35**, 3
Acyl hypohalites, reactions of, **9**, 5
Acyloins, **4**, 4; **15**, 1; **23**, 2
Alcohols:
  conversion to fluorides, **21**, 1, 2; **34**, 2; **35**, 3
  conversion to olefins, **12**, 2
  oxidation of, **6**, 5; **39**, 3; **53**, 1
  replacement of hydroxy group by nucleophiles, **29**, 1; **42**, 2
  resolution of, **2**, 9
Alcohols, synthesis:
  by allylstannane addition to aldehydes, **64**, 1
  by base-promoted isomerization of epoxides, **29**, 3
  by hydroboration, **13**, 1
  by hydroxylation of ethylenic compounds, **7**, 7
  by organochromium reagents to carbonyl compounds, **64**, 3
  from organoboranes, **33**, 1
  by reduction, **6**, 10; **8**, 1
Aldehydes, additions of allyl, allenyl, propargyl stannanes, **64**, 1
Aldehydes, catalyzed addition to double bonds, **40**, 4
Aldehydes, synthesis of, **4**, 7; **5**, 10; **8**, 4, 5; **9**, 2; **33**, 1
Aldol condensation, **16**
  directed, **28**, 3
  with boron enolates, **51**, 1
Aliphatic fluorides, **2**, 2; **21**, 1, 2; **34**, 2; **35**, 3
Alkanes, by reduction of alkyl halides with organochromium reagents, **64**, 3
Alkenes:
  arylation of, **11**, 3; **24**, 3; **27**, 2
  asymmetric dihydroxylation, **66**, 2
  cyclopropanes from, **20**, 1
  cyclization in intramolecular Heck reactions **60**, 2
  from carbonyl compounds with organochromium reagents, **64**, 3
  dioxirane epoxidation of, **61**, 2
  epoxidation and hydroxylation of, **7**, 7
  free-radical additions to, **13**, 3, 4
  hydroboration of, **13**, 1
  hydrogenation with homogeneous catalysts, **24**, 1
  reactions with diazoacetic esters, **18**, 3
  reactions with nitrones, **36**, 1
  reduction by alkoxyaluminum hydrides, **34**, 1
Alkenes, synthesis:
  from amines, **11**, 5
  from aryl and vinyl halides, **27**, 2
  by Bamford–Stevens reaction, **23**, 3
  by Claisen and Cope rearrangements, **22**, 1
  by dehydrocyanation of nitriles, **31**
  by deoxygenation of vicinal diols, **30**, 2
  from α-halosulfones, **25**, 1; **62**, 2
  by palladium-catalyzed vinylation, **27**, 2
  from phosphoryl-stabilized anions, **25**, 2
  by pyrolysis of xanthates, **12**, 2
  from silicon-stabilized anions, **38**, 1
  from tosylhydrazones, **23**, 3; **39**, 1
  by Wittig reaction, **14**, 3
Alkene reduction by diimide, **40**, 2
Alkenyl- and alkynylaluminum reagents, **32**, 2
Alkenyllithiums, formation of **39**, 1
Alkoxyaluminum hydride reductions, **34**, 1; **36**, 3
Alkoxyphosphonium cations, nucleophilic displacements on, **29**, 1
Alkylation:
  of allylic and benzylic carbanions, **27**, 1
  with amines and ammonium salts, **7**, 3

of aromatic compounds, **3**, 1
of esters and nitriles, **9**, 4
γ-, of dianions of β-dicarbonyl compounds, **17**, 2
of metallic acetylides, **5**, 1
of nitrile-stabilized carbanions, **31**
with organopalladium complexes, **27**, 2
Alkylidenation by titanium-based reagents, **43**, 1
Alkylidenesuccinic acids, synthesis and reactions of, **6**, 1
Alkylidene triphenylphosphoranes, synthesis and reactions of, **14**, 3
Allenylsilanes, electrophilic substitution reactions of, **37**, 2
Allylic alcohols, synthesis:
from epoxides, **29**, 3
by Wittig rearrangement, **46**, 2
Allylic and benzylic carbanions, heteroatom-substituted, **27**, 1
Allylic hydroperoxides, in photooxygenations, **20**, 2
Allylic rearrangements, transformation of glycols into 2,3-unsaturated glycosyl derivatives, **62**, 4
Allylic rearrangements, trihaloacetimidate, **66**, 1
π-Allylnickel complexes, **19**, 2
Allylphenols, synthesis by Claisen rearrangement, **2**, 1; **22**, 1
Allylsilanes, electrophilic substitution reactions of, **37**, 2
Aluminum alkoxides:
in Meerwein–Ponndorf–Verley reduction, **2**, 5
in Oppenauer oxidation, **6**, 5
Amide formation by oxime rearrangement, **35**, 1
α-Amidoalkylations at carbon, **14**, 2
Amination:
of heterocyclic bases by alkali amides, **1**, 4
of hydroxy compounds by Bucherer reaction, **1**, 5
Amine oxides:
Polonovski reaction of, **39**, 2
pyrolysis of, **11**, 5
Amines:
from allylstannane addition to imines, **64**, 1
synthesis from organoboranes, **33**, 1
synthesis by reductive alkylation, **4**, 3; **5**, 7
synthesis by Zinin reaction, **20**, 4
reactions with cyanogen bromide, **7**, 4
α-Aminoalkylation of activated olefins, **51**, 2
Aminophenols from anilines, **35**, 2
Anhydrides of aliphatic dibasic acids, Friedel–Crafts reaction with, **5**, 5
Anion-assisted sigmatropic rearrangements, **43**, 2
Anthracene homologs, synthesis of, **1**, 6
Anti-Markownikoff hydration of alkenes, **13**, 1
π-Arenechromium tricarbonyls, reaction with nitrile-stabilized carbanions, **31**
Arndt–Eistert reaction, **1**, 2
Aromatic aldehydes, synthesis of, **5**, 6; **28**, 1
Aromatic compounds, chloromethylation of, **1**, 3
Aromatic fluorides, synthesis of, **5**, 4
Aromatic hydrocarbons, synthesis of, **1**, 6; **30**, 1
Aromatic substitution by the $S_{RN}1$ reaction, **54**, 1
Arsinic acids, **2**, 10
Arsonic acids, **2**, 10
Arylacetic acids, synthesis of, **1**, 2; **22**, 4
β-Arylacrylic acids, synthesis of, **1**, 8
Arylamines, synthesis and reactions of, **1**, 5
Arylation:
by aryl halides, **27**, 2
by diazonium salts, **11**, 3; **24**, 3
γ-, of dianions of β-dicarbonyl compounds, **17**, 2
of nitrile-stabilized carbanions, **31**
of alkenes, **11**, 3; **24**, 3; **27**, 2
Arylglyoxals, condensation with aromatic hydrocarbons, **4**, 5
Arylsulfonic acids, synthesis of, **3**, 4
Aryl halides, homocoupling of, **63**, 3
Aryl thiocyanates, **3**, 6
Asymmetric aldol reactions using boron enolates, **51**, 1
Asymmetric cyclopropanation, **57**, 1
Asymmetric dihydroxylation, **66**, 2
Asymmetric epoxidation, **48**, 1; **61**, 2
Atom transfer preparation of radicals, **48**, 2
Aza-Payne rearrangements, **60**, 1
Azaphenanthrenes, synthesis by photocyclization, **30**, 1
Azides, synthesis and rearrangement of, **3**, 9
Azlactones, **3**, 5

Baeyer–Villiger reaction, **9**, 3; **43**, 3
Bamford–Stevens reaction, **23**, 3
Barbier reaction, **58**, 2
Bart reaction, **2**, 10
Barton fragmentation reaction, **48**, 2
Béchamp reaction, **2**, 10
Beckmann rearrangement, **11**, 1; **35**, 1
Benzils, reduction of, **4**, 5
Benzoin condensation, **4**, 5
Benzoquinones:
acetoxylation of, **19**, 3
in Nenitzescu reaction, **20**, 3
synthesis of, **4**, 6

Benzylic carbanions, **27**, 1
Biaryls, synthesis of, **2**, 6; **63**, 3
Bicyclobutanes, from cyclopropenes, **18**, 3
Biginelli dihydropyrimidine synthesis, **63**, 1
Birch reaction, **23**, 1; **42**, 1
Bischler–Napieralski reaction, **6**, 2
Bis(chloromethyl) ether, **1**, 3; **19**, *warning*
Borane reduction, chiral, **52**, 2
Borohydride reduction, chiral, **52**, 2
in reductive aminations, **59**, 1
Boron enolates, **51**, 1
Boyland–Sims oxidation, **35**, 2
Bucherer reaction, **1**, 5

Cannizzaro reaction, **2**, 3
Carbenes, **13**, 2; **26**, 2; **28**, 1
Carbenoid cyclopropanation reactions, **57**, 1; **58**, 1
Carbohydrates, deoxy, synthesis of, **30**, 2
Carbo/metallocupration, **41**, 2
Carbon–carbon bond formation:
by acetoacetic ester condensation, **1**, 9
by acyloin condensation, **23**, 2
by aldol condensation, **16**; **28**, 3; **46**, 1
by alkylation with amines and ammonium salts, **7**, 3
by γ-alkylation and arylation, **17**, 2
by allylic and benzylic carbanions, **27**, 1
by amidoalkylation, **14**, 2
by Cannizzaro reaction, **2**, 3
by Claisen rearrangement, **2**, 1; **22**, 1
by Cope rearrangement, **22**, 1
by cyclopropanation reaction, **13**, 2; **20**, 1
by Darzens condensation, **5**, 10
by diazonium salt coupling, **10**, 1; **11**, 3; **24**, 3
by Dieckmann condensation, **15**, 1
by Diels–Alder reaction, **4**, 1, 2; **5**, 3; **32**, 1
by free-radical additions to alkenes, **13**, 3
by Friedel–Crafts reaction, **3**, 1; **5**, 5
by Knoevenagel condensation, **15**, 2
by Mannich reaction, **1**, 10; **7**, 3
by Michael addition, **10**, 3
by nitrile-stabilized carbanions, **31**
by organoboranes and organoborates, **33**, 1
by organocopper reagents, **19**, 1; **38**, 2; **41**, 2
by organopalladium complexes, **27**, 2
by organozinc reagents, **20**, 1
by rearrangement of α-halosulfones, **25**, 1; **62**, 2
by Reformatsky reaction, **1**, 1; **28**, 3
by trivalent manganese, **49**, 3
by Vilsmeier reaction, **49**, 1; **56**, 2
by vinylcyclopropane-cyclopentene rearrangement, **33**, 2
Carbon–halogen bond formation, by replacement of hydroxy groups, **29**, 1
Carbon–heteroatom bond formation:
by free-radical chain additions to carbon–carbon multiple bonds, **13**, 4
by organoboranes and organoborates, **33**, 1
Carbon–nitrogen bond formation, by reductive amination, **59**, 1
Carbon–phosphorus bond formation, **36**, 2
Carbonyl compounds, addition of organochromium reagents, **64**, 3
Carbonyl compounds, α,β-unsaturated:
formation by selenoxide elimination, **44**, 1
vicinal difunctionalization of, **38**, 2
Carbonyl compounds, from nitro compounds, **38**, 3
in the Passerini reaction, **65**, 1
oxidation with hypervalent iodine reagents, **54**, 2
reductive amination of, **59**, 1
Carbonylation as part of intramolecular Heck reaction, **60**, 2
Carboxylic acid derivatives, conversion to fluorides, **21**, 1, 2; **34**, 2; **35**, 3
Carboxylic acids:
reaction with organolithium reagents, **18**, 1
synthesis from organoboranes, **33**, 1
Chapman rearrangement, **14**, 1; **18**, 2
Chloromethylation of aromatic compounds, **2**, 3; **9**, *warning*
Cholanthrenes, synthesis of, **1**, 6
Chromium reagents, **64**, 3
Chugaev reaction, **12**, 2
Claisen condensation, **1**, 8
Claisen rearrangement, **2**, 1; **22**, 1
Cleavage:
of benzyl–oxygen, benzyl–nitrogen, and benzyl–sulfur bonds, **7**, 5
of carbon–carbon bonds by periodic acid, **2**, 8
of esters via S_N2-type dealkylation, **24**, 2
of non-enolizable ketones with sodium amide, **9**, 1
in sensitized photooxidation, **20**, 2
Clemmensen reduction, **1**, 7; **22**, 3
Collins reagent, **53**, 1
Condensation:
acetoacetic ester, **1**, 9
acyloin, **4**, 4; **23**, 2
aldol, **16**
benzoin, **4**, 5
Biginelli, **63**, 1
Claisen, **1**, 8
Darzens, **5**, 10; **31**

Dieckmann, **1**, 9; **6**, 9; **15**, 1
directed aldol, **28**, 3
Knoevenagel, **1**, 8; **15**, 2
Stobbe, **6**, 1
Thorpe–Ziegler, **15**, 1; **31**
Conjugate addition:
of hydrogen cyanide, **25**, 3
of organocopper reagents, **19**, 1; **41**, 2
Cope rearrangement, **22**, 1; **41**, 1; **43**, 2
Copper–Grignard complexes, conjugate additions of, **19**, 1; **41**, 2
Corey–Winter reaction, **30**, 2
Coumarins, synthesis of, **7**, 1; **20**, 3
Coupling reaction of organostannanes, **50**, 1
Cuprate reagents, **19**, 1; **38**, 2; **41**, 2
Curtius rearrangement, **3**, 7, 9
Cyanoborohydride, in reductive aminations, **59**, 1
Cyanoethylation, **5**, 2
Cyanogen bromide, reactions with tertiary amines, **7**, 4
Cyclic ketones, formation by intramolecular acylation, **2**, 4; **23**, 2
Cyclization:
of alkyl dihalides, **19**, 2
of aryl-substituted aliphatic acids, acid chlorides, and anhydrides, **2**, 4; **23**, 2
of α-carbonyl carbenes and carbenoids, **26**, 2
cycloheptenones from α-bromoketones, **29**, 2
of diesters and dinitriles, **15**, 1
Fischer indole, **10**, 2
intramolecular by acylation, **2**, 4
intramolecular by acyloin condensation, **4**, 4
intramolecular by Diels–Alder reaction, **32**, 1
intramolecular by Heck reaction, **60**, 2
intramolecular by Michael reaction, **47**, 2
Nazarov, **45**, 1
by radical reactions, **48**, 2
of stilbenes, **30**, 1
tandem cyclization by Heck reaction, **60**, 2
Cycloaddition reactions:
of cyclenones and quinones, **5**, 3
cyclobutanes, synthesis of, **12**, 1; **44**, 2
Diels–Alder, acetylenes and alkenes, **4**, 2
Diels–Alder, imino dienophiles, **65**, 2
Diels–Alder, intramolecular, **32**, 1
Diels–Alder, maleic anhydride, **4**, 1
[4 + 3], **51**, 3
of enones, **44**, 2
of ketenes, **45**, 2
of nitrones and alkenes, **36**, 1
Pauson–Khand, **40**, 1
photochemical, **44**, 2
retro Diels–Alder reaction, **52**, 1; **53**, 2
[6 + 4], **49**, 2
[3 + 2], **61**, 1
Cyclobutanes, synthesis:
from nitrile-stabilized carbanions, **31**
by thermal cycloaddition reactions, **12**, 1
Cycloheptadienes, from:
divinylcyclopropanes, **41**, 1
polyhalo ketones, **29**, 2
π-Cyclopentadienyl transition metal carbonyls, **17**, 1
Cyclopentenones:
annulation, **45**, 1
synthesis, **40**, 1; **45**, 1
Cyclopropane carboxylates, from diazoacetic esters, **18**, 3
Cyclopropanes:
from α-diazocarbonyl compounds, **26**, 2
from metal-catalyzed decomposition of diazo compounds, **57**, 1
from nitrile-stabilized carbanions, **31**
from tosylhydrazones, **23**, 3
from unsaturated compounds, methylene iodide, and zinc-copper couple, **20**, 1; **58**, 1; **58**, 2
Cyclopropenes, synthesis of, **18**, 3

Darzens glycidic ester condensation, **5**, 10; **31**
DAST, **34**, 2; **35**, 3
Deamination of aromatic primary amines, **2**, 7
Debenzylation, **7**, 5; **18**, 4
Decarboxylation of acids, **9**, 5; **19**, 4
Dehalogenation of α-haloacyl halides, **3**, 3
Dehydrogenation:
in synthesis of ketenes, **3**, 3
in synthesis of acetylenes, **5**, 1
Demjanov reaction, **11**, 2
Deoxygenation of vicinal diols, **30**, 2
Desoxybenzoins, conversion to benzoins, **4**, 5
Dess-Martin oxidation, **53**, 1
Desulfurization:
of α-(alkylthio)nitriles, **31**
in alkene synthesis, **30**, 2
with Raney nickel, **12**, 5
Diazo compounds, carbenoids derived from, **57**, 1
Diazoacetic esters, reactions with alkenes, alkynes, heterocyclic and aromatic compounds, **18**, 3; **26**, 2
α-Diazocarbonyl compounds, insertion and addition reactions, **26**, 2
Diazomethane:
in Arndt–Eistert reaction, **1**, 2
reactions with aldehydes and ketones, **8**, 8
Diazonium fluoroborates, synthesis and decomposition, **5**, 4

Diazonium salts:
coupling with aliphatic compounds, **10**, 1, 2
in deamination of aromatic primary amines, **2**, 7
in Meerwein arylation reaction, **11**, 3; **24**, 3
in ring closure reactions, **9**, 7
in synthesis of biaryls and aryl quinones, **2**, 6
Dieckmann condensation, **1**, 9; **15**, 1
for synthesis of tetrahydrothiophenes, **6**, 9
Diels–Alder reaction:
intramolecular, **32**, 1
retro–Diels–Alder reaction, **52**, 1; **53**, 2
with alkynyl and alkenyl dienophiles, **4**, 2
with cyclenones and quinones, **5**, 3
with imines, **65**, 2
with maleic anhydride, **4**, 1
Dihydrodiols, **63**, 2
Dihydropyrimidine synthesis, **63**, 1
Dihydroxylation of alkenes, asymmetric, **66**, 2
Diimide, **40**, 2
Diketones:
pyrolysis of diaryl, **1**, 6
reduction by acid in organic solvents, **22**, 3
synthesis by acylation of ketones, **8**, 3
synthesis by alkylation of β-diketone anions, **17**, 2
Dimethyl sulfide, in oxidation reactions, **39**, 3
Dimethyl sulfoxide, in oxidation reactions, **39**, 3
Diols:
deoxygenation of, **30**, 2
oxidation of, **2**, 8
Dioxetanes, **20**, 2
Dioxiranes, **61**, 2
Dioxygenases, **63**, 2
Divinyl-aziridines, -cyclopropanes, -oxiranes, and -thiiranes, rearrangements of, **41**, 1
Doebner reaction, **1**, 8

Eastwood reaction, **30**, 2
Elbs reaction, **1**, 6; **35**, 2
Enamines, reaction with quinones, **20**, 3
Ene reaction, in photosensitized oxygenation, **20**, 2
Enolates:
α-Hydroxylation of, **62**, 1
in directed aldol reactions, **28**, 3; **46**, 1; **51**, 1
Enone cycloadditions, **44**, 2
Enzymatic reduction, **52**, 2
Enzymatic resolution, **37**, 1
Epoxidation:
of alkenes, **61**, 2
of allylic alcohols, **48**, 1
with organic peracids, **7**, 7
Epoxide isomerizations, **29**, 3
Epoxide:
formation, **61**, 2
migration, **60**, 1
Esters:
acylation with acid chlorides, **1**, 9
alkylation of, **9**, 4
alkylidenation of, **43**, 1
cleavage via S_N2-type dealkylation, **24**, 2
dimerization, **23**, 2
glycidic, synthesis of, **5**, 10
hydrolysis, catalyzed by pig liver esterase, **37**, 1
β-hydroxy, synthesis of, **1**, 1; **22**, 4
β-keto, synthesis of, **15**, 1
reaction with organolithium reagents, **18**, 1
reduction of, **8**, 1
synthesis from diazoacetic esters, **18**, 3
synthesis by Mitsunobu reaction, **42**, 2
Ethers, synthesis by Mitsunobu reaction, **42**, 2
Exhaustive methylation, Hofmann, **11**, 5

Favorskii rearrangement, **11**, 4
Ferrocenes, **17**, 1
Fischer indole cyclization, **10**, 2
Fluorination of aliphatic compounds, **2**, 2; **21**, 1, 2; **34**, 2; **35**, 3
Fluorination by DAST, **35**, 3
Fluorination by sulfur tetrafluoride, **21**, 1; **34**, 2
Formylation:
by hydroformulation, **56**, 1
of alkylphenols, **28**, 1
of aromatic hydrocarbons, **5**, 6
of aromatic compounds, **49**, 1
of nonaromatic compounds, **56**, 2
Free radical additions:
to alkenes and alkynes to form carbon–heteroatom bonds, **13**, 4
to alkenes to form carbon-carbon bonds, **13**, 3
Friedel-Crafts catalysts, in nucleoside synthesis, **55**, 1
Friedel–Crafts reaction, **2**, 4; **3**, 1; **5**, 5; **18**, 1
Friedländer synthesis of quinolines, **28**, 2
Fries reaction, **1**, 11

Gattermann aldehyde synthesis, **9**, 2
Gattermann–Koch reaction, **5**, 6
Germanes, addition to alkenes and alkynes, **13**, 4
Glycals, transformation in glycosyl derivatives, **62**, 4
Glycosides, synthesis of, **64**, 2
Glycosylation, with sulfoxides and sulfinates, **64**, 2
Glycidic esters, synthesis and reactions of, **5**, 10

Gomberg–Bachmann reaction, **2**, 6; **9**, 7
Grundmann synthesis of aldehydes, **8**, 5

Halides, displacement reactions of, **22**, 2; **27**, 2
Halide-metal exchange, **58**, 2
Halides, synthesis:
  from alcohols, **34**, 2
  by chloromethylation, **1**, 3
  from organoboranes, **33**, 1
  from primary and secondary alcohols, **29**, 1
Haller–Bauer reaction, **9**, 1
Halocarbenes, synthesis and reactions of, **13**, 2
Halocyclopropanes, reactions of, **13**, 2
Halogen-metal interconversion reactions, **6**, 7
α-Haloketones, rearrangement of, **11**, 4
α-Halosulfones, synthesis and reactions of, **25**, 1; **62**, 2
Heck reaction, intramolecular, **60**, 2
Helicenes, synthesis by photocyclization, **30**, 1
Heterocyclic aromatic systems, lithiation of, **26**, 1
Heterocyclic bases, amination of, **1**, 4
  in nucleosides, **55**, 1
Heterodienophiles, **53**, 2
Hilbert-Johnson method, **55**, 1
Hoesch reaction, **5**, 9
Hofmann elimination reaction, **11**, 5; **18**, 4
Hofmann reaction of amides, **3**, 7, 9
Homocouplings mediated by Cu, Ni, and Pd, **63**, 3
Homogeneous hydrogenation catalysts, **24**, 1
Hunsdiecker reaction, **9**, 5; **19**, 4
Hydration of alkenes, dienes, and alkynes, **13**, 1
Hydrazoic acid, reactions and generation of, **3**, 8
Hydroboration, **13**, 1
Hydrocyanation of conjugated carbonyl compounds, **25**, 3
Hydroformylation, **56**, 1
Hydrogenation catalysts, homogeneous, **24**, 1
Hydrogenation of esters, with copper chromite and Raney nickel, **8**, 1
Hydrohalogenation, **13**, 4
Hydroxyaldehydes, aromatic, **28**, 1
α-Hydroxyalkylation of activated olefins, **51**, 2
α-Hydroxyketones:
  rearrangement, **62**, 3
  synthesis of, **23**, 2
Hydroxylation:
  of enolates, **62**, 1
  of ethylenic compounds with organic peracids, **7**, 7
Hypervalent iodine reagents, **54**, 2; **57**, 2

Imidates, rearrangement of, **14**, 1
Imines, additions of allyl, allenyl, propargyl stannanes, **64**, 1
  as dienophiles, **65**, 2
Iminium ions, **39**, 2; **65**, 2
Imino Diels-Alder reactions, **65**, 2
Indoles, by Nenitzescu reaction, **20**, 3
  via reaction with TosMIC, **57**, 3
Isocyanides, in the Passerini reaction, **65**, 1
  sulfonylmethyl, reactions of, **57**, 3
Isoquinolines, synthesis of, **6**, 2, 3, 4; **20**, 3

Jacobsen reaction, **1**, 12
Japp–Klingemann reaction, **10**, 2

Katsuki–Sharpless epoxidation, **48**, 1
Ketene cycloadditions, **45**, 2
Ketenes and ketene dimers, synthesis of, **3**, 3; **45**, 2
α-Ketol rearrangement, **62**, 3
Ketones:
  acylation of, **8**, 3
  alkylidenation of, **43**, 1
  Baeyer–Villiger oxidation of, **9**, 3; **43**, 3
  cleavage of non-enolizable, **9**, 1
  comparison of synthetic methods, **18**, 1
  conversion to amides, **3**, 8; **11**, 1
  conversion to fluorides, **34**, 2; **35**, 3
  cyclic, synthesis of, **2**, 4; **23**, 2
  cyclization of divinyl ketones, **45**, 1
  synthesis from acid chlorides and organometallic compounds, **8**, 2; **18**, 1
  synthesis from organoboranes, **33**, 1
  synthesis from α,β-unsaturated carbonyl compounds and metals in liquid ammonia, **23**, 1
  reaction with diazomethane, **8**, 8
  reduction to aliphatic compounds, **4**, 8
  reduction by alkoxyaluminum hydrides, **34**, 1
  reduction in anhydrous organic solvents, **22**, 3
  synthesis from organolithium reagents and carboxylic acids, **18**, 1
  synthesis by oxidation of alcohols, **6**, 5; **39**, 3
Kindler modification of Willgerodt reaction, **3**, 2
Knoevenagel condensation, **1**, 8; **15**, 2; **57**, 3
Koch–Haaf reaction, **17**, 3
Kornblum oxidation, **39**, 3
Kostaneki synthesis of chromanes, flavones, and isoflavones, **8**, 3

β-Lactams, synthesis of, **9**, 6; **26**, 2
β-Lactones, synthesis and reactions of, **8**, 7
Leuckart reaction, **5**, 7
Lithiation:
  of allylic and benzylic systems, **27**, 1
  by halogen-metal exchange, **6**, 7
  heteroatom facilitated, **26**, 1; **47**, 1
  of heterocyclic and olefinic compounds, **26**, 1

Lithioorganocuprates, **19**, 1; **22**, 2; **41**, 2
Lithium aluminum hydride reductions, **6**, 2
chirally modified, **52**, 2
Lossen rearrangement, **3**, 7, 9

Mannich reaction, **1**, 10; **7**, 3
Meerwein arylation reaction, **11**, 3; **24**, 3
Meerwein–Ponndorf–Verley reduction, **2**, 5
Mercury hydride method to prepare radicals, **48**, 2
Metalations with organolithium compounds, **8**, 6; **26**, 1; **27**, 1
Methylenation of carbonyl groups, **43**, 1
Methylenecyclopropane, in cycloaddition reactions, **61**, 1
Methylene-transfer reactions, **18**, 3; **20**, 1; **58**, 1
Michael reaction, **10**, 3; **15**, 1, 2; **19**, 1; **20**, 3; **46**, 1; **47**, 2
Microbiological oxygenations, **63**, 2
Mitsunobu reaction, **42**, 2
Moffatt oxidation, **39**, 3; **53**, 1
Morita–Baylis–Hillman reaction, **51**, 2

Nazarov cyclization, **45**, 1
Nef reaction, **38**, 3
Nenitzescu reaction, **20**, 3
Nitriles:
formation from oximes, **35**, 2
synthesis from organoboranes, **33**, 1
α,β-unsaturated:
by elimination of selenoxides, **44**, 1
Nitrile-stabilized carbanions:
alkylation and arylation of, **31**
Nitroamines, **20**, 4
Nitro compounds, conversion to carbonyl compounds, **38**, 3
Nitro compounds, synthesis of, **12**, 3
Nitrone-olefin cycloadditions, **36**, 1
Nitrosation, **2**, 6; **7**, 6
Nucleosides, synthesis of, **55**, 1

Olefins, hydroformylation of, **56**, 1
Oligomerization of 1,3-dienes, **19**, 2
Oppenauer oxidation, **6**, 5
Organoboranes:
formation of carbon–carbon and carbon–heteroatom bonds from, **33**, 1
isomerization and oxidation of, **13**, 1
reaction with anions of α-chloronitriles, **31**, 1
Organochromium reagents, addition to carbonyl compounds, **64**, 3
Organohypervalent iodine reagents, **54**, 2; **57**, 2
Organometallic compounds:
of aluminum, **25**, 3
of chromium, **64**, 3
of copper, **19**, 1; **22**, 2; **38**, 2; **41**, 2
of lithium, **6**, 7; **8**, 6; **18**, 1; **27**, 1
of magnesium, zinc, and cadmium, **8**, 2;
of palladium, **27**, 2
of tin, **50**, 1; **64**, 1
of zinc, **1**, 1; **20**, 1; **22**, 4; **58**, 2
Osmium Tetroxide asymmetric dihydroxylation, **66**, 2
Overman rearrangement of allylic imidates, **66**, 1
Oxidation:
by dioxiranes, **61**, 2
of alcohols and polyhydroxy compounds, **6**, 5; **39**, 3; **53**, 1
of aldehydes and ketones, Baeyer–Villiger reaction, **9**, 3; **43**, 3
of amines, phenols, aminophenols, diamines, hydroquinones, and halophenols, **4**, 6; **35**, 2
of enolates and silyl enol ethers, **62**, 1
of α-glycols, α-amino alcohols, and polyhydroxy compounds by periodic acid, **2**, 8
with hypervalent iodine reagents, **54**, 2
of organoboranes, **13**, 1
of phenolic compounds, **57**, 2
with peracids, **7**, 7
by photooxygenation, **20**, 2
with selenium dioxide, **5**, 8; **24**, 4
Oxidative decarboxylation, **19**, 4
Oximes, formation by nitrosation, **7**, 6
Oxochromium(VI)-amine complexes, **53**, 1
Oxo process, **56**, 1
Oxygenation of arenes by dioxygenases, **63**, 2

Palladium-catalyzed vinylic substitution, **27**, 2
Palladium-catalyzed coupling of organostannane, **50**, 1
Palladium intermediates in Heck reactions, **60**, 2
Passerini reaction, **65**, 1
Pauson–Khand reaction to prepare cyclopentenones, **40**, 1
Payne rearrangement, **60**, 1
Pechmann reaction, **7**, 1
Peptides, synthesis of, **3**, 5; **12**, 4
Peracids, epoxidation and hydroxylation with, **7**, 7
in Baeyer–Villiger oxidation, **9**, 3; **43**, 3
Periodic acid oxidation, **2**, 8
Perkin reaction, **1**, 8
Persulfate oxidation, **35**, 2
Peterson olefination, **38**, 1
Phenanthrenes, synthesis by photocyclization, **30**, 1
Phenols, dihydric from phenols, **35**, 2
oxidation of **57**, 2
Phosphinic acids, synthesis of, **6**, 6
Phosphonic acids, synthesis of, **6**, 6

Phosphonium salts:
halide synthesis, use in, **29**, 1
synthesis and reactions of, **14**, 3
Phosphorus compounds, addition to carbonyl group, **6**, 6; **14**, 3; **25**, 2; **36**, 2
addition reactions at imine carbon, **36**, 2
Phosphoryl-stabilized anions, **25**, 2
Photochemical cycloadditions, **44**, 2
Photocyclization of stilbenes, **30**, 1
Photooxygenation of olefins, **20**, 2
Photosensitizers, **20**, 2
Pictet–Spengler reaction, **6**, 3
Pig liver esterase, **37**, 1
Polonovski reaction, **39**, 2
Polyalkylbenzenes, in Jacobsen reaction, **1**, 12
Polycyclic aromatic compounds, synthesis by photocyclization of stilbenes, **30**, 1
Polyhalo ketones, reductive dehalogenation of, **29**, 2
Pomeranz–Fritsch reaction, **6**, 4
Prévost reaction, **9**, 5
Pschorr synthesis, **2**, 6; **9**, 7
Pummerer reaction, **40**, 3
Pyrazolines, intermediates in diazoacetic ester reactions, **18**, 3
Pyridinium chlorochromate, **53**, 1
Pyrolysis:
of amine oxides, phosphates, and acyl derivatives, **11**, 5
of ketones and diketones, **1**, 6
for synthesis of ketenes, **3**, 3
of xanthates, **12**, 2

Quaternary ammonium salts, rearrangements of, **18**, 4
Quinolines, synthesis of:
by Friedländer synthesis, **28**, 2
by Skraup synthesis, **7**, 2
Quinones:
acetoxylation of, **19**, 3
diene additions to, **5**, 3
synthesis of, **4**, 6
in synthesis of 5-hydroxyindoles, **20**, 3

Radical formation and cyclization, **48**, 2
Ramberg–Bäcklund rearrangement, **25**, 1; **62**, 2
Rearrangements:
allylic trihaloacetamidate, **66**, 1
anion-assisted sigmatropic, **43**, 2
Beckmann, **11**, 1; **35**, 1
Chapman, **14**, 1; **18**, 2
Claisen, **2**, 1; **22**, 1
Cope, **22**, 1; **41**, 1, **43**, 2
Curtius, **3**, 7, 9
divinylcyclopropane, **41**, 1
Favorskii, **11**, 4
Lossen, **3**, 7, 9
Ramberg–Bäcklund, **25**, 1; **62**, 2
Smiles, **18**, 2
Sommelet–Hauser, **18**, 4
Stevens, **18**, 4
[2,3] Wittig, **46**, 2
vinylcyclopropane-cyclopentene, **33**, 2
Reduction:
of acid chlorides to aldehydes, **4**, 7; **8**, 5
of aromatic compounds, **42**, 1
of benzils, **4**, 5
of ketones, enantioselective, **52**, 2
Clemmensen, **1**, 7; **22**, 3
desulfurization, **12**, 5
with diimide, **40**, 2
by dissolving metal, **42**, 1
by homogeneous hydrogenation catalysts, **24**, 1
by hydrogenation of esters with copper chromite and Raney nickel, **8**, 1
hydrogenolysis of benzyl groups, **7**, 5
by lithium aluminum hydride, **6**, 10
by Meerwein–Ponndorf–Verley reaction, **2**, 5
chiral, **52**, 2
by metal alkoxyaluminum hydrides, **34**, 1; **36**, 3
of mono- and polynitroarenes, **20**, 4
of olefins by diimide, **40**, 2
of α,β-unsaturated carbonyl compounds, **23**, 1
by samarium(II) iodide, **46**, 3
by Wolff–Kishner reaction, **4**, 8
Reductive alkylation, synthesis of amines, **4**, 3; **5**, 7
Reductive amination of carbonyl compounds, **59**, 1
Reductive cyanation, **57**, 3
Reductive desulfurization of thiol esters, **8**, 5
Reformatsky reaction, **1**, 1; **22**, 4
Reimer–Tiemann reaction, **13**, 2; **28**, 1
Resolution of alcohols, **2**, 9
Retro–Diels–Alder reactions, **52**, 1; **53**, 2
Ritter reaction, **17**, 3
Rosenmund reaction for synthesis of arsonic acids, **2**, 10
Rosenmund reduction, **4**, 7

Samarium(II) iodide, **46**, 3
Sandmeyer reaction, **2**, 7
Schiemann reaction, **5**, 4
Schmidt reaction, **3**, 8, 9
Selenium dioxide oxidation, **5**, 8; **24**, 4
Seleno–Pummerer reaction, **40**, 3
Selenoxide elimination, **44**, 1
Shapiro reaction, **23**, 3; **39**, 1

Silanes:
addition to olefins and acetylenes, **13**, 4
electrophilic substitution reactions, **37**, 2
Sila–Pummerer reaction, **40**, 3
Silyl carbanions, **38**, 1
Silyl enol ether, α-hydroxylation, **62**, 1
Simmons–Smith reaction, **20**, 1; **58**, 1
Simonini reaction, **9**, 5
Singlet oxygen, **20**, 2
Skraup synthesis, **7**, 2; **28**, 2
Smiles rearrangement, **18**, 2
Sommelet–Hauser rearrangement, **18**, 4
Sommelet reaction, **8**, 4
$S_{RN}1$ reactions of aromatic systems, **54**, 1
Stevens rearrangement, **18**, 4
Stetter reaction of aldehydes with olefins, **40**, 4
Stilbenes, photocyclization of, **30**, 1
Stille reaction, **50**, 1
Stobbe condensation, **6**, 1
Substitution reactions using organocopper reagents, **22**, 2; **41**, 2
Sugars, synthesis by glycosylation with sulfoxides and sulfinates, **64**, 2
Sulfide reduction of nitroarenes, **20**, 4
Sulfonation of aromatic hydrocarbons and aryl halides, **3**, 4
Swern oxidation, **39**, 3; **53**, 1

Tetrahydroisoquinolines, synthesis of, **6**, 3
Tetrahydrothiophenes, synthesis of, **6**, 9
Thia-Payne rearrangement, **60**, 1
Thiazoles, synthesis of, **6**, 8
Thiele–Winter acetoxylation of quinones, **19**, 3
Thiocarbonates, synthesis of, **17**, **3**
Thiocyanation of aromatic amines, phenols, and polynuclear hydrocarbons, **3**, 6
Thiophenes, synthesis of, **6**, 9
Thorpe–Ziegler condensation, **15**, 1; **31**
Tiemann reaction, **3**, 9
Tiffeneau–Demjanov reaction, **11**, 2
Tin(II) enolates, **46**, 1
Tin hydride method to prepare radicals, **48**, 2
Tipson–Cohen reaction, **30**, 2
Tosylhydrazones, **23**, 3; **39**, 1
Tosylmethyl isocyanide (TosMIC), **57**, 3
Transmetallation reactions, **58**, 2
Trihaloacetimidate, allylic rearrangements, **66**, 1
Trimethylenemethane, [3 + 2] cycloaddition of, **61**, 1

Ullmann reaction:
homocoupling mediated by Cu, Ni, and Pd, **63**, 3
in synthesis of diphenylamines, **14**, 1
in synthesis of unsymmetrical biaryls, **2**, 6
Unsaturated compounds, synthesis with alkenyl- and alkynylaluminum reagents, **32**, 2

Vilsmeier reaction, **49**, 1; **56**, 2
Vinylcyclopropanes, rearrangement to cyclopentenes, **33**, 2
Vinyllithiums, from sulfonylhydrazones, **39**, 1
Vinylsilanes, electrophilic substitution reactions of, **37**, 2
Vinyl substitution, catalyzed by palladium complexes, **27**, 2
von Braun cyanogen bromide reaction, **7**, 4
Vorbrüggen reaction, **55**, 1

Willgerodt reaction, **3**, 2
Wittig reaction, **14**, 3; **31**
[2,3]-Wittig rearrangement, **46**, 2
Wolff–Kishner reaction, **4**, 8

Xanthates, synthesis and pyrolysis of, **12**, 2

Ylides:
in Stevens rearrangement, **18**, 4
in Wittig reaction, structure and properties, **14**, 3

Zinc–copper couple, **20**, 1; **58**, 1, 2
Zinin reduction of nitroarenes, **20**, 4